HANDBOOK OF APPLIED MATHEMATICS

HANDBOOK OF APPLIED MATHEMATICS

FOURTH EDITION

Edited by

EDWARD E. GRAZDA, B.E.E.
Editor, Electronic Design

MORRIS BRENNER, B.E.E., M.Ad.E., Assoc. Editor
Licensed Professional Engineer
Burndy Engineering Co., Inc., Norwalk, Conn.

WILLIAM R. MINRATH, A.B., Ch.E.
Vice President, D. Van Nostrand Co., Inc.

BASED ON THE ORIGINAL WORK BY
MARTIN E. JANSSON
HERBERT D. HARPER PETER L. AGNEW

D. VAN NOSTRAND COMPANY, INC.
PRINCETON, NEW JERSEY
TORONTO LONDON
NEW YORK

D. VAN NOSTRAND COMPANY, INC.
120 Alexander St., Princeton, New Jersey (*Principal office*)
24 West 40 Street, New York, New York

D. VAN NOSTRAND COMPANY, LTD.
358, Kensington High Street, London, W.14, England

D. VAN NOSTRAND COMPANY (Canada), LTD.
25 Hollinger Road, Toronto, Canada

Published simultaneously in Canada by
D. VAN NOSTRAND COMPANY (Canada), LTD.

Library of Congress Catalogue Card No. 55–5634

First Edition, December 1933
Three Reprintings

Second Edition, February 1936
Twenty Reprintings

Third Edition, February 1955
Reprinted April 1956, February 1958,
May 1961

Fourth Edition, September 1966

PRINTED IN THE UNITED STATES OF AMERICA

Preface to Fourth Edition

This book has been prepared to demonstrate how readily mathematics lends itself to the solution of practical problems. While it does not illustrate every type of problem, it seeks to develop logical reasoning which, if properly cultivated, will enable the reader to analyze his own problems and arrive at their solution by the most direct method. It is also a reference book with a wealth of specific information on many subjects. Whether the book is used for reference or as a text for self-instruction, the reader is urged to read the Introduction with some care for it contains the key to the handling of mathematical problems.

Every effort has been made to retain the working usefulness of the original book in this fourth edition. At the same time a vast amount of new material has been added to meet the present day needs of workers in many fields. The new developments include plate glass in home construction, new methods of building-block construction, new insulating materials, complete information on transistors, television in black-and-white and color, and stereophonic music systems.

Most of the other chapters have been extensively revised and brought up to date. New tables and detailed estimating data have been added on materials such as plastic tile, plywood, rubber-base paints, glass brick, copper tubing, modern heating systems, and electrical wiring. Several topics in the section on Business Mathematics including insurance, foreign exchange, small loans, stock and bond transactions, and loans were revised in accordance with present day rates and practices. A feature which should greatly assist the user of this handbook is a separate index of all the tables.

Many manufacturers, technical societies, trade associations, and the University of Illinois Small Homes Council have provided a great deal of valuable reference material. Special mention is made of Mr. William M. Milazzo of the Triangle Conduit and Cable Co., the American Plywood Assn., the Flat Glass Jobbers Assn., Mr. J. F. McCullough of the Union Carbide Corp., the Cast Iron Soil Pipe Inst., the Libbey-Owens-Ford Glass Co., and Mr. Fred Apostolos of the Herbert L. Jamison Co.

W. R. M.

Princeton, New Jersey
September, 1966

Contents

I

Introduction

HOW TO USE THIS BOOK

Mathematics underlies virtually all phases of the complex age we live in. From the intricate calculations involved in designing an atomic power plant to the everyday additions and subtractions performed on the job, in a small business, or in figuring out a home budget, mathematics plays a vital role. A working knowledge of this important science can be a valuable aid in job advancement, increasing business profits, and in simplifying "do-it-yourself" tasks around the home.

The purpose of this book is to promote a better understanding of this rewarding subject and to show how it is actually applied in solving practical problems in many fields. The first six chapters are devoted to a review of the operations of arithmetic, algebra, geometry, trigonometry, and differential calculus. Detailed descriptions of the principles of these branches of mathematics have been left to more specialized books, many of which are cited as references. These chapters lay the foundation for a sound understanding of the remainder of the book. The reader who has not been a constant user of mathematics will do well to read these chapters carefully, since they contain the key to the applications that follow.

The remainder of the book is divided into sections each covering some special trade or art, such as carpentry, electricity, machine-shop work, etc. Each of these sections has been made to cover the subject with a logical development from the elementary to the more complex. If then, for example, an electrical worker wishes to study the applications of mathematics to his entire field of work, he can do no better than to start at the beginning of the electrical section and follow it through to its conclusion. On the other hand, the man who is interested in the solution of a specific problem will find the index the best guide to the section dealing with this or a similar problem. Liberal use of the index is recommended because some subjects are covered in widely separated parts; for example, roofing occurs in both the carpentry and sheet metal sections.

Any man who operates an automobile or a machine of any kind, knows that it is to his advantage to "know what it can do" when operating under various conditions. Similarly, this book will increase in reference value to the man who knows what information it contains and where it is located. No man can carry a great amount of statistical information in his head, and, more important than knowing facts, is knowing where to find facts quickly.

The first step in solving a problem by any method is to picture the problem in its entirety and determine in what terms the final result is desired. This is particularly true in mathematics. The final result must be kept constantly in mind or energy may be needlessly expended in arriving at unnecessary partial results.

Having determined what is wanted in the way of a solution, the next step is to examine the data from which the problem is to be solved. Perhaps this is not sufficiently complete. Then it must be supplemented by information contained in this book or obtained from some other source, or the solution must proceed based on assumptions.

Simple problems may be solved most conveniently by setting them up as one expression. As an illustration, consider the problem of finding the distance which a train will travel in $2\frac{1}{4}$ hours when running at an average speed of 36 miles per hour. If we let S represent the speed, t the time, and D the distance, then

$$D = S \times t$$

The problem is now completely set up and all that is required to find the answer is to substitute the correct values and perform the indicated operations. "Correct values" implies proper units. We may then substitute as follows:

$$D = S \times t = 36 \times 2\tfrac{1}{4} = 81 \text{ miles.}$$

This example illustrates another very important principle in the use of mathematics. That is, if a problem is set up as above, how will one know in what units the answer will result? This is simple, because the unit designations may be cancelled, raised to powers or have their roots extracted in a manner similar to the operations performed on numbers. Thus, if we are finding the length of a surface whose area is 136 square inches and whose breadth is 8 inches, we may write, Length $= \frac{136}{8} = 17$. To find the units of the answer we may set up the units as we did the numerical problem. Thus we have, Length $= \frac{\text{in.}^2}{\text{in.}}$ or $\frac{\text{in.} \times \cancel{\text{in.}}}{\cancel{\text{in.}}}$. Cancelling "in." in both the numerator and the denominator, the answer is in inches.

In these considerations the word "per" has the same significance as the bar or line of a decimal. In fact, "miles per hour" may be written, "miles/hour." Then, in the previous illustration, when $D = S \times t$, we may write $D = \frac{\text{miles}}{\text{hour}} \times \text{hour}$. The hours cancel and the answer is in miles.

More involved problems and particularly those requiring the addition of many parts, can best be solved by attacking them step by step. Thus, in estimating the quantities of material required for a building construction job, it is necessary to compute the separate quantities required for the various parts of the building and then find the sum of these quantities for the final result.

Once a problem has been set up and the steps and operations determined, the processes of multiplication and division and the finding of powers, roots and reciprocals of numbers may proceed by any one of several methods. They may be performed by arithmetic, by algebra, by the use of tables, by logarithms, by the logarithmic slide rule, or by a computing machine.

This book does not attempt to dictate which method should be used, but generally shows the problem set up for arithmetical solution with the understanding that the reader will select the method which he can handle most readily and which is most suitable for his particular problem. Arithmetical solution is the longest process, except for simple calculations, and the practical man will do well to acquaint himself with other shorter methods and the types of problems to which they are most applicable.

Logarithms may be used most effectively when a problem calls for the multiplication or division or the handling of roots and powers of several factors. Thus, the operations indicated by $\dfrac{(25.136)^2 \times 728 \times 1728 \times 0.005679}{33{,}485 \times 36}$ may be performed logarithmically with greater ease than by any other method. Logarithms are also particularly adapted to the extraction of roots. Thus, solving $\sqrt[5]{\dfrac{838.75}{0.658}}$ is a very simple matter with this method. If, however, the addition and subtraction of a number of terms is interposed in an expression also involving multiplication and division, the use of logarithms may *not* be a time-saver. For example, the operations indicated by

$$\frac{0.125 \times 367 + 36.25 \times 450.3 + 0.825 \times 380}{750 \times 45.38}$$

are a border case, since the finding of the anti-logarithms to perform the additions may consume more than the time saved by performing the multiplications by logarithms. The use of logarithms is recommended whenever the multiplication or division of trigonometric functions is involved. Thus, in solving

$$252.67 \times \cos 67^\circ\, 36',$$

the logarithm of the cosine of $67^\circ\, 36'$ may be found from the tables with no more effort than would be required in finding the cosine itself.

The ordinary slide rule is a convenient instrument for multiplying and dividing when accuracy to greater than three significant figures is not a matter of great concern. Calculations may be made very rapidly with a slide rule and it is of great value in making rough estimates and checking results. It is not to be assumed from these remarks that a slide rule is a crude instrument and inherently inaccurate. This is not true, but it is rather a case of the inability of the human eye to evaluate the relative lengths of short distances with greater accuracy which limits its usefulness.

Computing machines have come into considerable favor in many offices and where one is available it will pay a man to learn how to perform the various operations on it. One of their particular merits is that addition and subtraction may be performed on them as well as multiplication, division, raising to powers and extracting square root. When computing machines are used to compute quantities which must be checked by another person within very narrow limits of discrepancy, as when computing certain land measurements, it is necessary to record all of the figures which appear on the machine, no matter if this results in nine or ten decimal places; and also to decide on a convention as to whether the nearest even or the nearest odd number should be recorded when the last figure of an eliminated decimal is 5.

As with the use of many other tools, the application of a liberal amount of common sense is necessary with the use of mathematics. Thus, it would be foolish to compute to the nearest cubic foot the quantity of sand required for a job, when sand is sold by five-ton truckloads. Also it would be wasted effort to measure farm land worth $50.00 an acre with the same care as would be used in measuring city property worth thousands of dollars an acre.

The illustrations of the last paragraph indicate that there is an economic reason for the use of mathematics. Such is the case. Correct application of mathematics leads to accuracy, accuracy results in less waste and fewer rejections, hence a higher return for work done.

There is another field of applied mathematics which is not governed by economic considerations. That is the calculation of the strength, proportions, or security of machinery or structures on which the safety of life and property depend. Here not only must the most care be exercised, but computations must be checked by responsible persons and should then be preserved in legible form. In the event of disaster, a court of inquiry to fix responsibility will ask, "Was accepted practice followed; and was due diligence exercised in arriving at results?" Accurate and well-preserved computations may be a big aid in establishing affirmative answers to these questions.

One of the points in dealing with figures at which common sense comes into greatest play is in the evaluation of the true accuracy of figures. "Figures do not lie," is a common expression but not always a true one.

Let us illustrate with an example. Suppose a piece of lumber is measured with a carpenter's rule and is found to be $3\frac{3}{8}$ inches wide and $1\frac{5}{8}$ inches thick. Changing these figures to decimals, as is common in performing computations, they become 3.375 inches and 1.625 inches, respectively. Now, if we want to obtain the cross-sectional area of this piece of lumber, we multiply the breadth of the board by the thickness and obtain $3.375 \times 1.625 = 5.484375$ square inches. Many of the figures of this decimal have no significance and the retention of the right number of figures requires the exercise of judgment and a knowledge of the accuracy with which the measurements were made and the purpose for which the figures are to be used. In the illustration we were told that the wood was measured with a carpenter's rule and presumably only to the nearest $\frac{1}{16}$ inch. Then, it will be entirely accurate to state that the cross-sectional area is $5\frac{1}{2}$ square inches.

To be precise the preceding problem would be written $3.375 \times 1.625 = 5.484375$, the number of square inches. However, in practical problems correct mathematical notation is usually disregarded for the sake of brevity. Throughout this work the answers are given in units which, theoretically, would not result from operations with abstract numbers.

The units of a dimension often indicate the degree of accuracy. Thus, if we are told without further qualification that a man is 5 feet 8 inches or 68 inches tall we know that he is between $67\frac{1}{2}$ inches and $68\frac{1}{2}$ inches tall. In other words, his height has been measured or estimated to the nearest one-half inch and we have no right to assume a more exact measurement. However, if we are told that he is $68\frac{3}{16}$ inches tall we know that his height has been measured to the nearest $\frac{1}{32}$ inch and that the actual height is between $68\frac{5}{32}$ and $68\frac{7}{32}$ inches.

When dimensions are stated in decimals, the decimal is an index of its accuracy. Thus, if we are told that a bolt is 0.318 inch in diameter we can feel reasonably sure that the measurement is correct to the nearest half of a thousandth of an inch or to 0.0005 inch. This would indicate that the measurement had been made with a micrometer caliper. However, if the diameter is given as 0.325 inch, the 5 in the last place raises a question as to whether the measurement was actually made to thousandths or to half-hundredths or to quarter-tenths. As a matter of fact, vernier calipers would give such a measurement since they are usually graduated to spaces 0.025 inch long.

It is equally important that the final results of a problem be expressed in rational practical units. Thus, quantities of lumber should be given in board feet or thousand board feet, cement in barrels, sand and gravel in cubic yards, etc. This does not imply that it is not perfectly proper to deal with fractional quantities during the course of the solution of a problem. This is particularly true when arriving at *unit quantities.* For instance, we may state that the quantities of materials required for one cubic yard of concrete are: 0.61 bbl. cement, 7.32 cu. ft. sand, 10.93 cu. ft. stone. These are unit quantities, but after being multiplied by the number of cubic yards of concrete to be made, the quantity of cement should be given to the next nearest whole barrel and the quantities of aggregates to the next nearest whole cubic yards.

These brief remarks indicate that clear logical thinking is a necessary adjunct to the use of mathematics. It may be added that nothing stimulates such mental procedure more than does mathematics and hence its use will result in many indirect benefits.

II

Arithmetic

Definitions.—Arithmetic is the science and application of numbers. Numbers are said to be *concrete* when they apply to *things, objects,* or *quantities* (examples, 12 bolts, 8 bricks, and 25 watts) and *abstract* when they do not so apply (examples, 12, 8, and 25).

The four *fundamental operations* of mathematics are *addition, subtraction, multiplication* and *division*; all necessary in performing *calculations.* A *proposition* is a statement set forth either with or without *demonstration.* It may be (1) an *axiom,* or self-evident truth, without demonstration; (2) a *theorem,* or truth by demonstration; (3) a *problem,* or question for solution; (4) an *hypothesis,* or tentative or preliminary proposition.

Signs and Symbols Used in Arithmetic.—Mathematical operations are largely indicated by signs and symbols. Thus $+$ placed between two numbers means that they are to be added and $\times$ between two numbers means that they are to be multiplied by each other.

The common mathematical symbols of arithmetic together with illustrations of their use are as follows:

$=$ Equals, sign of equality, is equal to, as 100 cents $=$ 1 dollar

$+$ Plus, sign of addition, as $3 + 4 = 7$; positive, as $+\frac{1}{2} = +0.5$

$-$ Minus, sign of subtraction, as $4 - 1 = 3$; negative, as $-\frac{1}{2} = -0.5$; contraction, as $\frac{1}{6} = 0.17-$

$\pm$ Plus or minus, as $\sqrt{4} = \pm 2$

$\times$ Times, multiplication sign, multiplied by, as $3 \times 2 = 6$

$\div$ Divided by, division sign, as $8 \div 2 = 4$; also $\frac{8}{2} = 4$; and $8/2 = 4$

$\therefore$ Therefore, hence, as if $2 + 2 = 4 \therefore 4 - 2 = 2$

$\because$ Because

. Decimal point

: Is to, sign of division, in ratio as $3 : 6$

:: Formerly used in proportion for the equality sign as $2 : 3 :: 4 : 6$ (read " 2 is to 3 as 4 is to 6 "), which means $2 : 3 = 4 : 6$ or $\frac{2}{3} = \frac{4}{6}$

$>$ Is greater than, as $4 > 3$; reads " 4 is greater than 3 "

$<$ Is less than, as $3 < 4$; reads " 3 is less than 4 "

$\cong$ Congruent sign, coincides with

∞ Infinity, as $\frac{3}{0} = \infty$

Symbol		These symbols denote that quantities covered or enclosed must be "taken together"	
|	Bar		$3 \times \overline{4 + 3} = 3 \times 7 = 21$
—	Vinculum		$3 \times (4 + 3) = 3 \times 7 = 21$
( )	Parentheses		$2[3 \times (4 + 3)] = 2[3 \times 7] = 42$
[ ]	Brackets		$4\{2[5(7 + 3) + 8] + 6$ $= 4\{2[50 + 8] + 6\}$
{ }	Braces		$= 4\{116 + 6\} = 488$

$\sqrt{}$ Radical sign or square root, as $\sqrt{9} = 3$

$\sqrt[3]{}$ Cube root

$\sqrt[n]{}$ *n*th root

a^2 A squared or second power of a, as $a \times a$

a^3 A cubed or third power of a, as $a \times a \times a$

a^n *n*th power of a

$\frac{1}{n}$ Reciprocal value of n

π Pi $= 3.1416$ (more accurately 3.14159265359) $= \dfrac{\text{circumference}}{\text{diameter}}$

Notation and Numeration.—Notation is a system of representing numbers by symbols while numeration is a system of naming or reading numbers.

There are two methods of notation in use, (1) the Roman and (2) the Arabic. The Roman has little use, the Arabic being the notation commonly used.

Roman notation is a method of notation by letters,

I	V	X	L	C	D	M
1	5	10	50	100	500	1,000

Repeating a letter repeats its value, i.e., I = 1, II = 2, III = 3.

Placing a letter of less value before one of greater value diminishes the value of the greater by the lesser, i.e.,

IX = 9, XC = 90

Placing the lesser after the greater increases the value of the greater by that of the lesser, i.e.,

VIII = 8, XIV = 14, LXX = 70

Placing a vinculum or horizontal line over a letter increases its value one thousand times, i.e.:

$$\overline{\mathrm{V}} = 5000, \qquad \overline{\mathrm{X}} = 10{,}000, \qquad \overline{\mathrm{M}} = 1{,}000{,}000$$

Arabic method of notation uses ten characters or figures, i.e.:

1	2	3	4	5	6	7	8	9	0
one	two	three	four	five	six	seven	eight	nine	zero

Numeration.—In the Arabic method of reading numbers, the value of numbers increases from left to right in a ten-fold ratio. The successive figures from right to left or from left to right are called orders of units, the value of any order being ten times the value of one of the order next to its right, and one-tenth the value of one of the order next to its left.

Billions Period			Millions Period			Thousands Period			Units Period		
Hundreds of Billions	Tens of Billions	Billions	Hundreds of Millions	Tens of Millions	Millions	Hundreds of Thousands	Tens of Thousands	Thousands	Hundreds	Tens	Units
	4	9	5	8	6	7	5	0	1	3	2

To read an integral number expressed in figures, begin at the right and separate the figures by commas into periods of three figures each. Then begin at the left and read each period as if it stood alone, adding the name of each period except the name of the period of units.

ILLUSTRATION: Read the number 49,586,750,132.

Forty-nine billion, five hundred eighty-six million, seven hundred fifty thousand, one hundred thirty-two. Note: The names beyond billions are in order: trillions, quadrillions, quintillions, sextillions, etc.

Addition.—Addition is the process of finding the sum of two or more numbers. To add several numbers, place the numbers in a vertical column with units under units, tens under tens, hundreds under hundreds, etc. Then add the figures in the right-hand column (column of units) and place the sum under this column. If there be more than one figure in this sum write down only the right-hand one and "carry" the others to the next column to the left. Repeat until each column has been added.

```
 438
1273
  46
 391
----
2148 Ans.
121  carried
```

The accuracy of the addition may be checked by writing the sums of the columns as shown below and adding

Sum of column (1)	18
Sum of column (2)	23
Sum of column (3)	9
Sum of column (4)	1
Sum =	2148

Subtraction.—Subtraction is the process of finding the difference of two numbers by taking one number from another. Example: $15 - 7 = 8$. The *minuend* is the number from which the other is to be taken (15 is the minuend in the example). The *subtrahend* is the number which is to be taken from the minuend (7 is the subtrahend in the example). The *remainder* is the number which remains after the subtrahend has been taken from the minuend (8 is the remainder in the example).

In order to subtract two figures write the subtrahend under the minuend so that the units of one are under the units of the other, tens under tens, etc. Take the figure in the subtrahend from the corresponding figure in the minuend and write the remainder directly underneath as follows:

Minuend:	56387
Subtrahend:	−12265
Remainder:	44122

If, however, the figure in the subtrahend is larger than the figure directly above it, it is necessary to borrow one unit from the next figure to the left. This is illustrated in the following operation:

		4		8	
Minuend	4	$\not{5}$	$^{1}3$	$\not{9}$	$^{1}6$
Subtrahend	−2	4	7	5	8
Remainder	2	0	6	3	8

A subtraction may be checked by adding the subtrahend to

the remainder. This sum should always equal the minuend. The following example illustrates this operation:

Minuend	6356	
Subtrahend	−1728	
Remainder	4628	
Subtrahend	+1728	
Minuend	6356	(Check)

Multiplication.—Multiplication is the process of taking or increasing one number a certain number of times and the result is called the *product*. The number which is multiplied or taken a certain number of times is called the *multiplicand* and the number by which it is multiplied is the *multiplier*. The multiplicand and the multiplier are known as *factors*.

In performing multiplication, the multiplier is written below the multiplicand, the units of one under the units of the other, tens under tens, etc. Each figure of the multiplicand, beginning at the right, is multiplied by each figure of the multiplier and the right-hand figure of each partial product is placed in turn directly under the figure used as a multiplier. Partial products are placed on different lines. The sum of the partial products will equal the required product.

Illustration:

```
  1653 multiplicand
   247 multiplier
------
 11571
 6612
3306
------
408291 product
```

Division.—Division is the process of finding how many times one number is contained in another. The number to be divided is called the *dividend* and the number by which it is divided, the *divisor*. The result of the operation, or the number of times the divisor is contained in the dividend, is known as the *quotient.*

When the divisor contains but one figure, the method commonly used is known as *short division.* In performing this, place the divisor to the left of the dividend, separated by a line, and draw a line under the dividend. Divide the first or the first two figures of the dividend, as is necessary, by the divisor and place the quotient under the line. If the divisor does not go a whole number of times, the remainder is prefixed to the next figure in the dividend and the process is repeated.

Illustration: divide 21372 by 6

Solution:
```
6)21372
   3562 quotient
```

When the divisor contains two or more figures, the method used is known as *long division.* This is performed as follows: Place the divisor at the left of the dividend, separated by a line, and place the quotient either above or to the right of the dividend. Divide the first group of figures which gives a number larger than the divisor by the divisor, place the first figure of the quotient above the dividend, multiply this figure by the divisor and place this product below the figures divided into and subtract. The remainder prefixed to the next figure brought down from the dividend forms the new trial dividend. Repeat until all figures of the dividend are brought down.

Illustration: divide 2841020 by 364

```
        7805   quotient
364)2841020
    2548
    ----
     2930
     2912
     ----
       1820
       1820
       ----
```

It is very common in both short and long division that the

divisor will not go into the last trial dividend a whole number of times. It is then necessary to express the remainder as a fraction.

EXAMPLE: Divide 327 by 18

$$\begin{array}{r} 18\frac{1}{6} \text{ quotient} \\ 18\overline{)327} \\ 18 \\ \hline 147 \\ 144 \\ \hline \end{array}$$

SOLUTION:

$$\text{Remainder} = \frac{3}{18} = \frac{1}{6}$$

Fractions.—A fraction is a part of any object or unit. It consists of three essential elements, a number called a *denominator* which denotes the number of equal parts into which the object or unit is divided, a horizontal line above the denominator, called the *fraction line,* and a number above the line known as the *numerator* which denotes how many of the equal parts are to be taken. Thus in the fraction $\frac{3}{4}$, 3 is the numerator and 4 the denominator. This type of fraction is usually called a common fraction. To read a common fraction, read the numerator and then the denominator.

ILLUSTRATION: $\frac{1}{2}$, $\frac{3}{4}$, $\frac{7}{11}$ are read, one-half, three-fourths or three-quarters, seven-elevenths.

A *proper fraction* is one whose numerator is less than the denominator, as $\frac{2}{3}$.

An *improper fraction* is one whose numerator is greater than the denominator, as $\frac{5}{2}$.

A *mixed number* consists of a whole number and a fraction written together, as $2\frac{1}{2}$.

Reduction of Fractions.—A fraction may be reduced to its lowest form (without changing its value) by dividing both the numerator and the denominator by their *greatest common divisor*

(G.C.D.). Thus the G.C.D. of $\frac{12}{30}$ is 6 and if the numerator and denominator are both divided by this number, the fraction becomes, $\dfrac{12}{30} = \dfrac{12 \div 6}{30 \div 6} = \dfrac{2}{5}$.

A mixed number may be reduced to an improper fraction by multiplying the whole number by the denominator and adding the numerator to form a new numerator. Thus,

$$4\tfrac{1}{2} = \frac{(2 \times 4) + 1}{2} = \frac{9}{2}.$$

To change an improper fraction to a mixed number, divide the numerator by the denominator. The quotient is the whole number and the remainder is the new numerator. Thus, $\frac{177}{32} = 5\frac{17}{32}$.

Addition and Subtraction of Fractions.—To add fractions, the *least common multiple* (L.C.M.) of all denominators must first be determined to find a common denominator. The L.C.M. is found by multiplying the product of the prime factors (numbers divisible only by themselves and one) of the largest denominator by the product of the prime factors which occur in the other denominators but not in the largest. Thus the prime factors of the denominators of the fractions $\dfrac{3}{4}$, $\dfrac{5}{6}$, and $\dfrac{7}{12}$, are $\dfrac{3}{2 \times 2}$, $\dfrac{5}{3 \times 2}$, and $\dfrac{7}{2 \times 2 \times 3}$. In this case $2 \times 2 \times 3$ contains all of the factors in the required number of times, so 12 is the least common denominator of these fractions.

The next step is to expand both terms of each fraction proportionately so that their denominators will be equal. Thus, $\dfrac{3 \times 3}{4 \times 3} = \dfrac{9}{12}$, $\dfrac{5 \times 2}{6 \times 2} = \dfrac{10}{12}$, and $\dfrac{7 \times 1}{12 \times 1} = \dfrac{7}{12}$. Then all of the expanded numerators may be placed over the common denominator and the numerators added, thus,

$$\frac{9 + 10 + 7}{12} = \frac{26}{12} = 2\frac{2}{12} = 2\frac{1}{6}.$$

When mixed numbers are added they may be changed to

improper fractions and then the same procedure as above followed. Thus,

$$2\frac{1}{2} + 3\frac{3}{4} = \frac{5}{2} + \frac{15}{4} = \frac{5 \times 2}{2 \times 2} + \frac{15}{4} = \frac{10}{4} + \frac{15}{4} = \frac{10 + 15}{4} = \frac{25}{4} = 6\frac{1}{4}.$$

Fractions are subtracted by reducing to the smallest common denominator as for addition and then finding the difference of the new numerators. Thus, $\frac{15}{16} - \frac{3}{8} = \frac{15}{16} - \frac{6}{16} = \frac{9}{16}$; and $6\frac{1}{4} - 3\frac{7}{16} = \frac{25}{4} - \frac{55}{16} = \frac{100}{16} - \frac{55}{16} = \frac{45}{16} = 2\frac{13}{16}$.

Multiplication and Division of Fractions.—To multiply a fraction by a whole number, multiply the numerator by the whole number. The product will be the new numerator over the old denominator. Thus, $5 \times \frac{3}{4} = \frac{5 \times 3}{4} = \frac{15}{4} = 3\frac{3}{4}$.

To divide a fraction by a whole number, multiply the denominator of the fraction by the whole number. The quotient will be the old numerator over the new denominator. Thus,

$$\frac{1}{2} \div 5 = \frac{1}{2 \times 5} = \frac{1}{10}; \frac{7}{8} \div 3 = \frac{7}{8 \times 3} = \frac{7}{24}.$$

To multiply one fraction by another fraction, place the product of the numerators over the product of the denominators and reduce to required form. Thus,

$$\frac{5}{8} \times \frac{2}{3} = \frac{5 \times 2}{8 \times 3} = \frac{10}{24} = \frac{5}{12}; \frac{3}{16} \times \frac{1}{2} = \frac{3 \times 1}{16 \times 2} = \frac{3}{32}.$$

When mixed numbers are to be multiplied by fractions, it is advisable to change the mixed numbers to improper fractions before using the above procedure. Thus,

$$5\frac{3}{8} \times \frac{3}{4} = \frac{40 + 3}{8} \times \frac{3}{4} = \frac{43 \times 3}{8 \times 4} = \frac{129}{32} = 4\frac{1}{32}.$$

To divide a whole number or a fraction by a fraction, invert the divisor and multiply. Thus, $\frac{1}{8} \div \frac{1}{2} = \frac{1}{8} \times \frac{2}{1} = \frac{2}{8} = \frac{1}{4}$; $\frac{5}{9} \div \frac{2}{3} = \frac{5}{9} \times \frac{3}{2} = \frac{15}{18} = \frac{5}{6}$; $12\frac{1}{4} \div \frac{1}{6} = \frac{49}{4} \div \frac{1}{6} = \frac{49}{4} \times \frac{6}{1} = \frac{294}{4} = 73\frac{1}{2}$.

Cancellation.—In practical operations where the multiplication of various kinds of numbers, including fractions is expressed the process may often be shortened by cancellation. This consists of taking out common factors above and below the fraction line before multiplying. As an example, take the expression $\frac{10 \times 4 \times 12}{25 \times 3 \times 8}$. This would require several operations to simplify without cancellation. However, it will be noted that 5 is a common factor in the 10 and the 25, and that 4 can be factored out of the 4 and the 8 and the 3 can be factored out of the 12. The operation is performed by striking out the numbers and writing above or below the remaining portion as follows:

$$\frac{\overset{2}{\not{10}} \times \not{4} \times \overset{4}{\not{12}}}{\underset{5}{\not{25}} \times \not{3} \times \underset{2}{\not{8}}} = \frac{4}{5}.$$

Cancellation can be made as long as factors remain which will cancel each other, but there is, however, a medium point where cancellation may sometimes cease for simplicity of operation.

Decimal Fractions.—A fraction which has for its denominator the number 10, 100, 1000, etc., may be expressed by writing only one number and using a period or a decimal point to indicate whether the fraction is tenths, hundredths, etc. Thus, .1 is $\frac{1}{10}$, .01 is $\frac{1}{100}$, .17 is $\frac{17}{100}$, .125 is $\frac{125}{1000}$, etc. These are called decimal fractions or simply decimals. When written alone, 0 is usually placed to the left of the decimal point, 0.125.

To read a decimal expressed in figures, read the decimal as if a whole number, and add the fractional name of the lowest place. For example, 6.18 is read 6 and 18 hundredths; 6.0018, 6 and 18 ten-thousandths.

Changing Common Fractions into Decimals.—Since the fraction line indicates division it is easy to see that a fraction can be reduced to a decimal simply by performing the indicated operation and dividing the numerator by the denominator and writing the quotient in decimal form. Thus, $\frac{3}{8} = \frac{8)3.000}{0.375} = 0.375.$

In this example the quotient came out exactly and the decimal is the exact equivalent of the fraction. Some decimals will not come out exactly and the division should then be carried out only as far as the nature of the work requires. Decimals are seldom carried out to more than five places. When the value of a decimal correct to the nearest tenth, hundredth, thousandth, etc., is required, 1 is added to the last required figure if the next figure is

TABLE 1

Decimal Equivalents of Common Fractions

Fraction				Decimal	Fraction				Decimal
			1/64	0.015625				33/64	0.515625
		1/32		.03125			17/32		.53125
			3/64	.046875				35/64	.546875
	1/16			.0625		9/16			.5625
			5/64	.078125				37/64	.578125
		3/32		.09375			19/32		.59375
			7/64	.109375				39/64	.609375
1/8				.125	5/8				.625
			9/64	.140625				41/64	.640625
		5/32		.15625			21/32		.65625
			11/64	.171875				43/64	.671875
	3/16			.1875		11/16			.6875
			13/64	.203125				45/64	.703125
		7/32		.21875			23/32		.71875
			15/64	.234375				47/64	.734375
1/4				.25	3/4				.75
			17/64	.265625				49/64	.765625
		9/32		.28125			25/32		.78125
			19/64	.296875				51/64	.796875
	5/16			.3125		13/16			.8125
			21/64	.328125				53/64	.828125
		11/32		.34375			27/32		.84375
			23/64	.359375				55/64	.859375
3/8				.375	7/8				.875
			25/64	.390625				57/64	.890625
		13/32		.40625			29/32		.90625
			27/64	.421875				59/64	.921875
	7/16			.4375		15/16			.9375
			29/64	.453125				61/64	.953125
		15/32		.46875			31/32		.96875
			31/64	.484375				63/64	.984375
1/2				.5	1				1.

five or more. Thus, 0.375 correct to the nearest tenth is 0.4; correct to the nearest hundredth is 0.38.

Addition and Subtraction of Decimals.—In the addition and subtraction of decimals, the numbers are written one above the other in such a manner that the decimal points are always directly in a vertical column. The operations are then performed in the ordinary manner, care being taken that the decimal point in the sum or the remainder is also directly in line with those above.

Example of addition:
```
  2.0625
315.25
  0.0375
--------
317.3500
```

Zeros to the right of the last significant figure in a decimal may be stricken out when they have no significance without changing the value of the number.

Example of subtraction:
```
24.325
 5.7036
-------
18.6214
```

Multiplication of Decimals.—In multiplication of decimals, the points are not required to fall under each other and the fractions are placed so that the right-hand figures of the multiplier and multiplicand are in the same column as when dealing with whole numbers. The multiplication is then performed as with whole numbers and the product has as many decimal places as the multiplicand and the multiplier combined. That is, if the multiplicand has three figures to the right of the decimal point and the multiplier has two figures to the right of the decimal point, then the product will have $3 + 2 = 5$ figures to the right of the decimal point.

Examples:
```
    8.475                1.26
     2.25              0.0012
---------              ------
    42375                 252
   16950                 126
  16950                ------
---------          0.001512 product
19.06875 product
```

Decimals, or any other number, may be multiplied by 10 by simply moving the decimal point one place to the right; by 100 by moving the decimal point two places to the right, etc. Examples: $10 \times 46.75 = 467.5$; $1000 \times 0.0627 = 62.7$.

Division of Decimals.—To divide decimals, multiply or divide the divisor and the dividend by some power of 10 (10, 100, 1000, etc.) so as to make the divisor a whole number. Mark the new decimal point in the dividend by a caret (∧) and proceed as with the division of whole numbers, placing the decimal point of the quotient above or below the caret depending on whether long or short division is used. The quotient will then have as many decimal places as the new dividend.

Examples: Divide 43.28 by 400.

```
4ØØ)∧43.28
     0.1082  (Ans.)
```

Divide 43.28 by 0.004.

```
0.ØØ4)43.280∧
      10820.  (Ans.)
```

Divide 1728.5 by 1.356 to the nearest thousandth.

```
               1274.705  (Ans.)
1x356)1728x500∧000∧
      1356
       3725
       2712
       10130
        9492
         6380
         5424
          9560
          9492
            6800
            6780
```

Changing Decimals to Common Fractions.—*Exact Decimals,* that is, a decimal whose denominator is contained in the numerator without a remainder. For the numerator of the fraction, use the significant figures of the decimal, the denominator being 1 with as many ciphers as there are decimal places in the decimal; reduce to lowest terms.

Examples: $0.75 = \frac{75}{100} = \frac{3}{4}$; $0.375 = \frac{375}{1000} = \frac{3}{8}$.

Table 1 will be found convenient for finding the equivalent fraction to many decimals.

Repeating Decimals.—A common fraction can be expressed exactly by a decimal if the denominator contains no other factors than 2 or 5; otherwise it cannot. For example, when the fraction $\frac{3}{11}$ is expressed as a decimal the quotient obtained by dividing 3 by 11 is 0.27272727, etc., however far it is carried.

A decimal that contains a constantly recurring figure or series of figures is called a repeating decimal. In the case given above, 0.27272727, etc. is a repeating decimal, the series of figures constantly recurring being 27. In writing a repeating decimal dots are usually placed over the first and last figures of the repetend, i.e., the figure or series of figures that constantly recurs. Thus, 0.272727 .... would be written $0.\dot{2}\dot{7}$ and 0.333 ... $0.\dot{3}$.

To Reduce a Repeating Decimal to a Common Fraction.—Treat the *non-repeating* and the *first repeating* groups as a whole number; subtract from this the non-repeating group treated as a whole number; the difference will be the numerator of the fraction. The denominator will be composed of as many 9's as there are repeating figures in the group, followed by as many 0's as there are non-repeating figures. Reduce to lowest terms.

Example: Reduce $0.\dot{3}$ to a fraction

$$\begin{array}{r} 3 \\ -0 \\ \hline 3 \end{array}$$

$$\frac{\text{numer.}}{\text{denom.}} \quad \frac{3}{9} = \frac{1}{3} \text{ (Ans.)}$$

Example: Reduce $0.\dot{2}\dot{7}$ to a fraction

$$\begin{array}{r} 27 \\ -00 \\ \hline 27 \end{array}$$

$$\frac{\text{numer.}}{\text{denom.}} \quad \frac{27}{99} = \frac{3}{11} \text{ (Ans.)}$$

Example: Reduce 0.79054054

$$\begin{array}{ll} & -79 \\ \hline \text{numer.} & 78975 \\ \hline \text{denom.} & 99900 \end{array} = \frac{117}{148} \text{ (Ans.)}$$

Compound or Denominate Numbers.—A quantity expressed in units of two or more denominations is called a compound quantity or a compound denominate number. Thus, $4\frac{1}{2}$ feet is a simple quantity; but its equivalent 4 feet 6 inches is a compound quantity.

The process of changing the denomination in which a quantity is expressed, without changing the value of the quantity, is called reduction.

Reduction Descending.—To reduce a compound number to a lower denomination, multiply the number by as many units of the lower denomination as makes one of the higher.

Examples: Reduce $4\frac{1}{2}$ feet to inches: $4\frac{1}{2} \times 12 = 54$ inches.

Reduce $3\frac{1}{4}$ pecks to quarts: $3\frac{1}{4} \times 8 = 26$ quarts.

When the given number is expressed in more than one denomination, proceed in steps from the highest denomination to the next lower, and so on to the lowest, adding in the units of each denomination as the operation proceeds.

Example: Reduce 10 gallons, 1 quart, 1 pint, to pints.

$10 \times 4 = 40, + 1 = 41, 41 \times 2 = 82, + 1 = 83$ pints. (Ans.)

Reduction Ascending.—To express a number of a lower denomination in terms of a higher, divide the number by the number of units of the lower denomination contained in one of the next higher; the quotient is in the higher denomination, and the remainder, if any, is in the lower.

Example: Reduce 227 pints to higher units.

$227 \div 2 = 113$ qts., $+1$ pt., $113 \div 4 = 28$ gal. $+ 1$ qt.
28 gal. 1 qt. 1 pt. (Ans.)

To express the results in decimals of the higher denomination, divide the given number by the number of units of the given denomination contained in one of the required denomination, carrying the result to as many places as required.

Example: Reduce 1 inch to feet. Give result in ten-thousandths. $1 \div 12 = 0.0833$ ft. (Ans.)

Addition of Compound Quantities.—

Example: Add 12 feet $4\frac{1}{4}$ inches, 6 feet $8\frac{5}{8}$ inches, and 15 feet $3\frac{1}{2}$ inches.

ft.	in.	8
12	$4\frac{1}{4}$	2
6	$8\frac{5}{8}$	5
15	$3\frac{1}{2}$	4
33	15	$\frac{11}{8} = 1\frac{3}{8}$
	$+ 1\frac{3}{8}$	
	$16\frac{3}{8}$	$= 1$ ft. $4\frac{3}{8}$ in.

33 ft. + 1 ft. $4\frac{3}{8}$ in. = 34 ft. $4\frac{3}{8}$ in. (Ans.)

Subtraction of Compound Quantities.—

Example: Subtract 4 yds. 1 ft. 3 in. from 6 yd. 7 ft. 1 in.

yd.	ft.	in.	
	6		
6	7	12 + 1	(1 ft. or 12 inches is borrowed from 7 ft.)
4	1	3	
2	5	10	

Therefore, the required difference is 2 yds. 5 ft. 10 in. (Ans.)

Multiplication of Compound Quantities.—

Example: Multiply 3 ft. $4\frac{5}{16}$ in. by 8.

$$\begin{array}{ll} 3' & 4\frac{5}{16}'' \\ \times 8 & \\ \hline 24' & 32 \\ & +\ 2\frac{1}{2} \\ \hline & 34\frac{1}{2}'' = 2'\ 10\frac{1}{2}'' \end{array} \qquad \frac{5}{16} \times 8 = \frac{40}{16} = 2\frac{1}{2}''$$

$$24' + \quad 2'\ 10\frac{1}{2}'' = 26'\ 10\frac{1}{2}''$$

Therefore, the product is 26 ft. $10\frac{1}{2}$ in. (Ans.)

Division of Compound Quantities.—

Example: Divide 122 bu. 2 pk. 7 qt. 1 pt. by 5.

$$\begin{array}{r|rrrr} & \text{bu.} & \text{pk.} & \text{qt.} & \text{pt.} \\ 5 & 122 & 2 & 7 & 1 \\ \hline & 24 & 2 & 1 & 1 \end{array}$$

Therefore, the quotient is 24 bu. 2 pk. 1 qt. 1 pt. (Ans.)

Example: Divide 12 ft. 4 in. by 5 to the nearest $\frac{1}{16}$ in.

$$12'\ 4'' \times 12 = 148''$$

$$148'' \div 5 = \frac{148}{5} = 29\frac{3''}{5} = 2'\ 5\frac{3''}{5}$$

$$\frac{3}{5} \times 16 = \frac{48}{5} \quad \text{or} \quad \frac{9\frac{3}{5}}{16} = \frac{10}{16} = \frac{5}{8}$$

Therefore, the quotient is 2 ft. $5\frac{5}{8}$ in. (Ans.)

Powers.—When a number is multiplied by itself once it is said to be *squared* and the product is called the square of the number. Thus, in $3 \times 3 = 9$, the 9 is the square of 3. The same number has been used twice as a factor. The operation of squaring a number is usually indicated by a small number called an *exponent,* thus $3^2 = 9$. A number multiplied by itself once is said to have been raised to the second power.

Similarly, a number may be multiplied by itself twice. It is then used three times as a factor and is said to have been *cubed* or raised to the third power and the operation is indicated thus: $3^3 = 27$, which means $3 \times 3 \times 3 = 27$.

A number may be raised to any power, the power being indicated by the proper exponent. Thus, 4^6 is four to the sixth power or $4 \times 4 \times 4 \times 4 \times 4 \times 4$, 3^{10} is three to the tenth, etc.

Roots.—A number may be divided into several equal factors. Thus, 36 is the product of 6×6. Each of the equal factors of a number is called a *root* of the number. If a number is divided into two equal factors, the root is said to be the *square root*; if three equal factors, the *cube-root*; if four equal factors, the fourth root, etc.

A root is indicated by the symbol $\sqrt{\ }$ called the *radical sign* and the degree of the root is indicated by a small number called the *root index* thus $\sqrt[3]{\ }$. When the radical sign has no index number, the square root is meant, which could also be indicated by writing $\sqrt[2]{\ }$. Thus, $\sqrt{25} = 5$ or $\sqrt[2]{25} = 5$. In all other cases an index number must be used, as $\sqrt[3]{27} = 3$ and $\sqrt[4]{16} = 2$.

The values of roots may be determined by arithmetical computation, by the use of logarithms, or by reference to tables containing values already computed. Square roots and cube roots are those most commonly needed and for most practical purposes, the average man will find that the tables of these values fill his needs. Such a table will be found on pages 23 to 36 and the values may be read directly. The computation of square root is described in the next paragraph, and the finding of roots by logarithms (the most convenient method for higher roots) is dealt with on page 62.

Square Root.—The square root of a number is extracted as follows:

Point off the number into periods of two figures each, beginning with the units; if there are decimals, begin at the decimal point, separating the whole number to the left and the decimal to the right into such periods, supplying as many ciphers in groups of two as may be desired in the decimal.

Find the greatest number whose square is less than the first left-hand period and place this to the right of the given number as the first figure of the root. Subtract its square from the first left-hand period and to the remainder annex the second period for a dividend.

Place before this as a partial divisor, double the root figure

just found. Find how many times the dividend, exclusive of its right-hand figure, contains the divisor, and place the quotient as the second figure of the root, and also at the right of the partial divisor.

Multiply the divisor thus completed, by the second root figure and subtract the product from the dividend. To this remainder annex the next period for a new dividend, and double the two root figures for a new partial divisor. Proceed as before until all the periods have been brought down.

Example: Extract the square root of 5386.3928 to 3 decimal places.

```
          53'86.39'28'00 (73.392   (Ans.)
          49
  143   ) 486
          429
  1463  ) 5739
          4389
  14669 ) 135028
          132021
  146782 ) 300700
           293564
             7136
```

Extracting Square Root of a Fraction.—The square root of a fraction is the square root of its numerator over the square root of its denominator. Thus, $\sqrt{\frac{9}{16}} = \frac{\sqrt{9}}{\sqrt{16}} = \frac{3}{4}$. When neither the numerator nor the denominator is a perfect square a convenient short cut is to multiply both by a common number to convert one or the other to a perfect square. Thus,

$$\sqrt{\frac{2}{3}} = \sqrt{\frac{2 \times 3}{3 \times 3}} = \sqrt{\frac{6}{9}} = \frac{\sqrt{6}}{\sqrt{9}} = \frac{\sqrt{6}}{3} = \frac{2.449}{3}.$$

Since the square root of a fraction often results in decimals, it is

TABLE 2

SQUARES, CUBES, SQUARE ROOTS, CUBE ROOTS, OF NUMBERS 1 TO 1600.

No.	Square	Cube.	Sq. Rt.	Cu. Rt.	No.	Square	Cube.	Sq. Rt.	Cu. Rt.
0	0	0	0.0000000	0.0000000	65	42 25	274 625	8.0622577	4.0207256
1	1	1	1.0000000	1.0000000	6	43 56	287 496	.1240384	.0412401
2	4	8	.4142136	.2599210	7	44 89	300 763	.1853528	.0615480
3	9	27	.7320508	.4422496	8	46 24	314 432	.2462113	.0816551
4	16	64	2.0000000	5874011	9	47 61	328 509	.3066239	.1015661
5	25	125	2.2360680	1.7099759	70	49 00	343 000	8.3666003	4.1212853
6	36	216	.4494897	.8171206	1	50 41	357 911	.4261498	.1408178
7	49	343	.6457513	.9129312	2	51 84	373 248	.4852814	.1601676
8	64	512	.8284271	2.0000000	3	53 29	389 017	.5440037	.1793392
9	81	729	3.0000000	.0800837	4	54 76	405 224	.6023253	.1983364
10	1 00	1 000	3.1622777	2.1544347	75	56 25	421 875	8.6602540	4.2171633
11	1 21	1 331	.3166248	.2239801	6	57 76	438 976	.7177979	.235823[illegible]
12	1 44	1 728	.4641016	.2894286	7	59 29	456 533	.7749644	.2543210
13	1 69	2 197	.6055513	.3513347	8	60 84	474 552	.8317609	.2726586
14	1 96	2 744	.7416574	.4101422	9	62 41	493 039	.8881944	2908404
15	2 25	3 375	3.8729833	2.4662121	80	64 00	512 000	8.9442719	4.3088695
16	2 56	4 096	4.0000000	.5198421	1	65 61	531 441	9.0000000	.3267487
17	2 89	4 913	.1231056	.5712816	2	67 24	551 368	.0553851	.3444815
18	3 24	5 832	.2426407	.6207414	3	68 89	571 787	.1104336	.3620707
19	3 61	6 859	.3588989	.6684016	4	70 56	592 704	.1651514	.3795191
20	4 00	8 000	4.4721360	2.7144177	85	72 25	614 125	9.2195445	4.3968296
1	4 41	9 261	.5825757	2.7589243	6	73 96	636 056	.2736185	.4140049
2	4 84	10 648	.6904158	.8020393	7	75 69	658 503	.3273791	.4310476
3	5 29	12 167	7958315	.8438670	8	77 44	681 472	.3808315	.4479602
4	5 76	13 824	.8989795	.8844991	9	79 21	704 969	.4339811	.4647451
25	6 25	15 625	5.0000000	2.9240177	90	81 00	729 000	9.4868330	4.4814047
6	6 76	17 576	.0990195	.9624960	1	82 81	753 571	.5393920	.4979414
7	7 29	19 683	.1961524	3.0000000	2	84 64	778 688	.5916630	.5143574
8	7 84	21 952	.2915026	.0365889	3	86 49	804 357	.6436508	.5306549
9	8 41	24 389	.3851648	.0723168	4	88 36	830 584	.6953597	.5468359
30	9 00	27 000	5.4772256	3.1072325	95	90 25	857 375	9.7467943	4.5629026
1	9 61	29 791	.5677644	.1413806	6	92 16	884 736	.7979590	.5788570
2	10 24	32 768	.6568542	.1748021	7	94 09	912 673	.8488578	.5947009
3	10 89	35 937	.7445626	.2075343	8	96 04	941 192	.8994949	.6104363
4	11 56	39 304	.8309519	.2396118	9	98 01	970 299	.9498744	.6260650
35	12 25	42 875	5.9160798	3.2710663	100	1 00 00	1 000 000	10.0000000	4.6415888
6	12 96	46 656	6.0000000	.3019272	1	1 02 01	1 030 301	.0498756	.6570095
7	13 69	50 653	.0827625	.3322218	2	1 04 04	1 061 208	.0995049	.6723287
8	14 44	54 872	.1644140	.3619754	3	1 06 09	1 092 927	.1488916	.6875482
9	15 21	59 319	2449980	.3912114	4	1 08 16	1 124 864	.1980390	.7026694
40	16 00	64 000	6.3245553	3.4199519	105	1 10 25	1 157 625	10.2469508	4.7176940
1	16 81	68 921	.4031242	.4482172	6	1 12 36	1 191 016	.2956301	.7326235
2	17 64	74 088	.4807407	.4760266	7	1 14 49	1 225 043	.3440804	.7474594
3	18 49	79 507	.5574385	.5033981	8	1 16 64	1 259 712	.3923048	.7622032
4	19 36	85 184	.6332496	.5303483	9	1 18 81	1 295 029	.4403065	.7768562
45	20 25	91 125	6.7082039	3.5568933	110	1 21 00	1 331 000	10.4880885	4.7914199
6	21 16	97 336	.7823300	.5830479	11	1 23 21	1 367 631	.5356538	.8058955
7	22 09	103 823	.8556546	.6088261	12	1 25 44	1 404 928	.5830052	.8202845
8	23 04	110 592	.9282032	.6342411	13	1 27 69	1 442 897	.6301458	.8345881
9	24 01	117 649	7.0000000	.6593057	14	1 29 96	1 481 544	.6770783	.8488076
50	25 00	125 000	7.0710678	3.6840314	115	1 32 25	1 520 875	10.7238053	4.8629442
1	26 01	132 651	.1414284	.7084298	16	1 34 56	1 560 896	7703296	.8769990
2	27 04	140 608	.2111026	.7325111	17	1 36 89	1 601 613	8166538	.8909732
3	28 09	148 877	.2801099	.7562858	18	1 39 24	1 643 032	.8627805	.9048681
4	29 16	157 464	.3484692	.7797631	19	1 41 61	1 685 159	.9087121	.9186847
55	30 25	166 375	7.4161985	3.8029525	120	1 44 00	1 728 000	10.9544512	4.9324242
6	31 36	175 616	.4833148	.8258624	1	1 46 41	1 771 561	11.0000000	9460874
7	32 49	185 193	.5498344	.8485011	2	1 48 84	1 815 848	.0453610	.9596757
8	33 64	195 112	.6157731	.8708766	3	1 51 29	1 860 867	.0905365	.9731898
9	34 81	205 379	.6811457	.8929965	4	1 53 76	1 906 624	.1355287	.9866310
60	36 00	216 000	7.7459667	3.9148676	125	1 56 25	1 953 125	11.1803399	5.0000000
1	37 21	226 981	.8102497	.9364972	6	1 58 76	2 000 376	.2249722	.0132979
2	38 44	238 328	.8740079	.9578915	7	1 61 29	2 048 383	.2694277	.0265257
3	39 69	250 047	9372539	.9790571	8	1 63 84	2 097 152	3137085	.0396842
4	40 96	262 144	8.0000000	4.0000000	9	1 66 41	2 146 689	.3578167	.0527748
65	42 25	274 625	8.0622577	4.0207256	130	1 69 00	2 197 000	11.4017543	5.0657970

2.—SQUARES, CUBES, SQUARE ROOTS, CUBE ROOTS, OF NUMBERS 1 TO 1600—Continued.

No.	Square	Cube.	Sq. Rt.	Cu. Rt.	No.	Square	Cube.	Sq. Rt.	Cu. Rt.
130	1 69 00	2 197 000	11.4017543	5.0657970	195	3 80 25	7 414 875	13.9642400	5.7988900
1	1 71 61	2 248 091	.4455231	.0787531	6	3 84 16	7 529 536	14.0000000	.8087857
2	1 74 24	2 299 968	.4891253	.0916434	7	3 88 09	7 645 373	.0356688	.8186479
3	1 76 89	2 352 637	.5325626	.1044687	8	3 92 04	7 762 392	.0712473	.8284767
4	1 79 56	2 406 104	.5758369	.1172299	9	3 96 01	7 880 599	.1067360	.8382725
135	1 82 25	2 460 375	11.6189500	5.1299278	200	4 00 00	8 000 000	14.1421356	5.8480355
6	1 84 96	2 515 456	.6619038	.1425632	1	4 04 01	8 120 601	.1774469	.8577660
7	1 87 69	2 571 353	.7046999	.1551367	2	4 08 04	8 242 408	.2126704	.8674643
8	1 90 44	2 628 072	.7473401	.1676493	3	4 12 09	8 365 427	.2478068	.8771307
9	1 93 21	2 685 619	.7898261	.1801015	4	4 16 16	8 489 664	.2828569	.8867653
140	1 96 00	2 744 000	11.8321596	5.1924941	205	4 20 25	8 615 125	14.3178211	5.8963685
1	1 98 81	2 803 221	.8743422	.2048279	6	4 24 36	8 741 816	.3527001	.9059406
2	2 01 64	2 863 288	.9163753	.2171034	7	4 28 49	8 869 743	.3874946	.9154817
3	2 04 49	2 924 207	.9582607	.2293215	8	4 32 64	8 998 912	.4222051	.9249921
4	2 07 36	2 985 984	12.0000000	.2414828	9	4 36 81	9 129 329	.4568323	.9344721
145	2 10 25	3 048 625	12.0415946	5.2535879	210	4 41 00	9 261 000	14.4913767	5.9439220
6	2 13 16	3 112 136	.0830460	.2656374	11	4 45 21	9 393 931	.5258390	.9533418
7	2 16 09	3 176 523	.1243557	.2776321	12	4 49 44	9 528 128	.5602198	.9627320
8	2 19 04	3 241 792	.1655251	.2895725	13	4 53 69	9 663 597	.5945195	.9720926
9	2 22 01	3 307 949	.2065556	.3014592	14	4 57 96	9 800 344	.6287388	.9814240
150	2 25 00	3 375 000	12.2474487	5.3132928	215	4 62 25	9 938 375	14.6628783	5.9907264
1	2 28 01	3 442 951	.2882057	.3250740	16	4 66 56	10 077 696	.6969385	6.0000000
2	2 31 04	3 511 808	.3288280	.3368033	17	4 70 89	10 218 313	.7309199	.0092450
3	2 34 09	3 581 577	.3693169	.3484812	18	4 75 24	10 360 232	.7648231	.0184617
4	2 37 16	3 652 264	.4096736	.3601084	19	4 79 61	10 503 459	.7986486	.0276502
155	2 40 25	3 723 875	12.4498996	5.3716854	220	4 84 00	10 648 000	14.8323970	6.0368107
6	2 43 36	3 796 416	.4899960	.3832126	1	4 88 41	10 793 861	.8660687	.0459435
7	2 46 49	3 869 893	.5299641	.3946907	2	4 92 84	10 941 048	.8996644	.0550489
8	2 49 64	3 944 312	.5698051	.4061202	3	4 97 29	11 089 567	.9331845	.0641270
9	2 52 81	4 019 679	.6095202	.4175015	4	5 01 76	11 239 424	.9666295	.0731779
160	2 56 00	4 096 000	12.6491106	5.4288352	225	5 06 25	11 390 625	15.0000000	6.0822020
1	2 59 21	4 173 281	.6885775	.4401218	6	5 10 76	11 543 176	.0332964	.0911994
2	2 62 44	4 251 528	.7279221	.4513618	7	5 15 29	11 697 083	.0665192	.1001702
3	2 65 69	4 330 747	.7671453	.4625556	8	5 19 84	11 852 352	.0996689	.1091147
4	2 68 96	4 410 944	.8062485	.4737037	9	5 24 41	12 008 989	.1327460	.1180332
165	2 72 25	4 492 125	12.8452326	5.4848066	230	5 29 00	12 167 000	15.1657509	6.1269257
6	2 75 56	4 574 296	.8840987	.4958647	1	5 33 61	12 326 391	.1986842	.1357924
7	2 78 89	4 657 463	.9228480	.5068784	2	5 38 24	12 487 168	.2315462	.1446337
8	2 82 24	4 741 632	.9614814	.5178484	3	5 42 89	12 649 337	.2643375	.1534495
9	2 85 61	4 826 809	13.0000000	.5287748	4	5 47 56	12 812 904	.2970585	.1622401
170	2 89 00	4 913 000	13.0384048	5.5396583	235	5 52 25	12 977 875	15.3297097	6.1710058
1	2 92 41	5 000 211	.0766968	.5504991	6	5 56 96	13 144 256	.3622915	.1797466
2	2 95 84	5 088 448	.1148770	.5612978	7	5 61 69	13 312 053	.3948043	.1884628
3	2 99 29	5 177 717	.1529464	.5720546	8	5 66 44	13 481 272	.4272486	.1971544
4	3 02 76	5 268 024	.1909060	.5827702	9	5 71 21	13 651 919	.4596248	.2058218
175	3 06 25	5 359 375	13.2287566	5.5934447	240	5 76 00	13 824 000	15.4919334	6.2144650
6	3 09 76	5 451 776	.2664992	.6040787	1	5 80 81	13 997 521	.5241747	.2230843
7	3 13 29	5 545 233	.3041347	.6146724	2	5 85 64	14 172 488	.5563492	.2316797
8	3 16 84	5 639 752	.3416641	.6252263	3	5 90 49	14 348 907	.5884573	.2402515
9	3 20 41	5 735 339	.3790882	.6357408	4	5 95 36	14 526 784	.6204994	.2487998
180	3 24 00	5 832 000	13.4164079	5.6462162	245	6 00 25	14 706 125	15.6524758	6.2573248
1	3 27 61	5 929 741	.4536240	.6566528	6	6 05 16	14 886 936	.6843871	.2658266
2	3 31 24	6 028 568	.4907376	.6670511	7	6 10 09	15 069 223	.7162336	.2743054
3	3 34 89	6 128 487	.5277493	.6774114	8	6 15 04	15 252 992	.7480157	.2827613
4	3 38 56	6 229 504	.5646600	.6877340	9	6 20 01	15 438 249	.7797338	.2911946
185	3 42 25	6 331 625	13.6014705	5.6980192	250	6 25 00	15 625 000	15.8113883	6.2996053
6	3 45 96	6 434 856	.6381817	.7082675	1	6 30 01	15 813 251	.8429795	.3079935
7	3 49 69	6 539 203	.6747943	.7184791	2	6 35 04	16 003 008	.8745079	.3163596
8	3 53 44	6 644 672	.7113092	.7286543	3	6 40 09	16 194 277	.9059737	.3247035
9	3 57 21	6 751 269	.7477271	.7387936	4	6 45 16	16 387 064	.9373775	.3330256
190	3 61 00	6 859 000	13.7840488	5.7488971	255	6 50 25	16 581 375	15.9687194	6.3413257
1	3 64 81	6 967 871	.8202750	.7589652	6	6 55 36	16 777 216	16.0000000	.3496042
2	3 68 64	7 077 888	.8564065	.7689982	7	6 60 49	16 974 593	.0312195	.3578611
3	3 72 49	7 189 057	.8924440	.7789966	8	6 65 64	17 173 512	.0623784	.3660968
4	3 76 36	7 301 384	.9283883	.7889604	9	6 70 81	17 373 979	.0934769	.3743111
195	3 80 25	7 414 875	13.9642400	5.7988900	260	6 76 00	17 576 000	16.1245155	6.3825043

2.—Squares, Cubes, Square Roots, Cube Roots, of Numbers 1 to 1600—Continued.

No.	Square	Cube.	Sq. Rt.	Cu. Rt.	No.	Square	Cube.	Sq. Rt.	Cu. Rt.
260	6 76 00	17 576 000	16.1245155	6.3825043	325	10 56 25	34 328 125	18.0277564	6.8753443
1	6 81 21	17 779 581	.1554944	.3906765	6	10 62 76	34 645 976	.0554701	.8823888
2	6 86 44	17 984 728	.1864141	.3988279	7	10 69 29	34 965 783	.0831413	.8894188
3	6 91 69	18 191 447	.2172747	.4069585	8	10 75 84	35 287 552	.1107703	.8964345
4	6 96 96	18 399 744	.2480768	.4150687	9	10 82 41	35 611 289	.1383571	.9034359
265	7 02 25	18 609 625	16.2788206	6.4231583	330	10 89 00	35 937 000	18.1659021	6.9104232
6	7 07 56	18 821 096	.3095064	.4312276	1	10 95 61	36 264 691	.1934054	.9173964
7	7 12 89	19 034 163	.3401346	.4392767	2	11 02 24	36 594 368	.2208672	.9243556
8	7 18 24	19 248 832	.3707055	.4473057	3	11 08 89	36 926 037	.2482876	.9313008
9	7 23 61	19 465 109	.4012195	.4553148	4	11 15 56	37 259 704	.2756669	.9382321
270	7 29 00	19 683 000	16.4316767	6.4633041	335	11 22 25	37 595 375	18.3030052	6.9451496
1	7 34 41	19 902 511	.4620776	.4712736	6	11 28 96	37 933 056	.3303028	.9520533
2	7 39 84	20 123 648	.4924225	.4792236	7	11 35 69	38 272 753	.3575598	.9589434
3	7 45 29	20 346 417	.5227116	.4871541	8	11 42 44	38 614 472	.3847763	.9658198
4	7 50 76	20 570 824	.5529454	.4950653	9	11 49 21	38 958 219	.4119526	.9726826
275	7 56 25	20 796 875	16.5831240	6.5029572	340	11 56 00	39 304 000	18 4390889	6.9795321
6	7 61 76	21 024 576	.6132477	.5108300	1	11 62 81	39 651 821	.4661853	.9863681
7	7 67 29	21 253 933	.6433170	.5186839	2	11 69 64	40 001 688	.4932420	.9931906
8	7 72 84	21 484 952	.6733320	.5265189	3	11 76 49	40 353 607	.5202592	7.0000000
9	7 78 41	21 717 639	.7032931	.5343351	4	11 83 36	40 707 584	.5472370	.0067962
280	7 84 00	21 952 000	16.7332005	6.5421326	345	11 90 25	41 063 625	18.5741756	7.0135791
1	7 89 61	22 188 041	.7630546	.5499116	6	11 97 16	41 421 736	.6010752	.0203490
2	7 95 24	22 425 768	.7928556	.5576722	7	12 04 09	41 781 923	.6279360	.0271058
3	8 00 89	22 665 187	.8226038	.5654144	8	12 11 04	42 144 192	.6547581	.0338497
4	8 06 56	22 906 304	.8522995	.5731385	9	12 18 01	42 508 549	.6815417	.0405806
285	8 12 25	23 149 125	16.8819430	6.5808443	350	12 25 00	42 875 000	18.7082869	7.0472987
6	8 17 96	23 393 656	.9115345	.5885323	1	12 32 01	43 243 551	.7349940	.0540041
7	8 23 69	23 639 903	.9410743	.5962023	2	12 39 04	43 614 208	.7616630	.0606967
8	8 29 44	23 887 872	.9705627	.6038545	3	12 46 09	43 986 977	.7882942	.0673767
9	8 35 21	24 137 569	17.0000000	.6114890	4	12 53 16	44 361 864	.8148877	.0740440
290	8 41 00	24 389 000	17.0293864	6.6191060	355	12 60 25	44 738 875	18.8414437	7.0806988
1	8 46 81	24 642 171	.0587221	.6267054	6	12 67 36	45 118 016	.8679623	.0873411
2	8 52 64	24 897 088	.0880075	.6342874	7	12 74 49	45 499 293	.8944436	.0939709
3	8 58 49	25 153 757	.1172428	.6418522	8	12 81 64	45 882 712	.9208879	.1005885
4	8 64 36	25 412 184	.1464282	.6493998	9	12 88 81	46 268 279	.9472953	.1071937
295	8 70 25	25 672 375	17.1755640	6.6569302	360	12 96 00	46 656 000	18.9736660	7.1137866
6	8 76 16	25 934 336	2046505	.6644437	1	13 03 21	47 045 881	19,0000000	.1203674
7	8 82 09	26 198 073	.2336879	.6719403	2	13 10 44	47 437 928	.0262976	.1269360
8	8 88 04	26 463 592	.2626765	.6794200	3	13 17 69	47 832 147	.0525589	.1334925
9	8 94 01	26 730 899	.2916165	.6868831	4	13 24 96	48 228 544	.0787840	.1400370
300	9 00 00	27 000 000	17.3205081	6.6943295	365	13 32 25	48 627 125	19.1049732	7.0465695
1	9 06 01	27 270 901	.3493516	.7017593	6	13 39 56	49 027 896	.1311265	.1530901
2	9 12 04	27 543 608	.3781472	.7091729	7	13 46 89	49 430 863	.1572441	.1595988
3	9 18 09	27 818 127	.4068952	.7165700	8	13 54 24	49 836 032	.1833261	.1660957
4	9 24 16	28 094 464	.4355958	.7239508	9	13 61 61	50 243 409	.2093727	.1725809
305	9 30 25	28 372 625	17.4642492	6.7313155	370	13 69 00	50 653 000	19.2353841	7.1790544
6	9 36 36	28 652 616	.4928557	.7386641	1	13 76 41	51 064 811	.2613603	.1855162
7	9 42 49	28 934 443	.5214155	.7459967	2	13 83 84	51 478 848	.2873015	.1919663
8	9 48 64	29 218 112	.5499288	.7533134	3	13 91 29	51 895 117	.3132079	.1984050
9	9 54 81	29 503 629	.5783958	.7606143	4	13 98 76	52 313 624	.3390796	.2048322
310	9 61 00	29 791 000	17.6068169	6.7678995	375	14 06 25	52 734 375	19.3649167	7.2112479
11	9 67 21	30 080 231	.6351921	.7751690	6	14 13 76	53 157 376	.3907194	.2176522
12	9 73 44	30 371 328	.6635217	.7824229	7	14 21 29	53 582 633	.4164878	.2240450
13	9 79 69	30 664 297	.6918060	.7896613	8	14 28 84	54 010 152	.4422221	.2304268
14	9 85 96	30 959 144	.7200451	.7968844	9	14 36 41	54 439 939	.4679223	.2367972
315	9 92 25	31 255 875	17.7482393	6.8040921	380	14 44 00	54 872 000	19.4935887	7.2431565
16	9 98 56	31 554 496	.7763888	.8112847	1	14 51 61	55 306 341	.5192213	.2495045
17	10 04 89	31 855 013	.8044938	.8184620	2	14 59 24	55 742 968	5448203	.2558415
18	10 11 24	32 157 432	.8325545	.8256242	3	14 66 89	56 181 887	.5703858	.2621675
19	10 17 61	32 461 759	.8605711	.8327714	4	14 74 56	56 623 104	.5959179	.2684824
320	10 24 00	32 768 000	17.8885438	6.8399037	385	14 82 25	57 066 625	19 6214169	7.2747864
1	10 30 41	33 076 161	.9184729	.8470213	6	14 89 96	57 512 456	.6468827	.2810794
2	10 36 84	33 386 248	.9443584	.8541240	7	14 97 69	57 960 603	.6723156	.2873617
3	10 43 29	33 658 267	.9722008	.8612120	8	15 05 44	58 411 072	.6977156	.2936330
4	10 49 76	34 012 224	18.0000000	.8682855	9	15 13 21	58 863 869	.7230829	.2998936
325	10 56 25	34 328 125	18.0277564	6.8753443	390	15 21 00	59 319 000	19.7484177	7.3061436

2.—Squares, Cubes, Square Roots, Cube Roots, of Numbers 1 to 1600—Continued.

No.	Square	Cube.	Sq. Rt.	Cu. Rt.	No.	Square	Cube.	Sq. Rt.	Cu. Rt.
390	15 21 00	59 319 000	19.7484177	7.3061436	455	20 70 25	94 196 375	21.3307290	7.6913717
1	15 28 81	59 776 471	.7737199	.3123828	6	20 79 36	94 818 816	.3541565	.6970023
2	15 36 64	60 236 288	.7989899	.3186114	7	20 88 49	95 443 993	.3775583	.7026246
3	15 44 49	60 698 457	.8242276	.3248295	8	20 97 64	96 071 912	.4009346	.7082388
4	15 52 36	61 162 984	.8494332	.3310369	9	21 06 81	96 702 579	.4242853	.7138448
395	15 60 25	61 629 875	19.8746069	7.3372339	460	21 16 00	97 336 000	21.4476106	7.7194426
6	15 68 16	82 099 136	.8997487	.3434205	1	21 25 21	97 972 181	.4709106	.7250325
7	15 76 09	62 570 773	.9248588	.3495966	2	21 34 44	98 611 128	.4941853	.7306141
8	15 84 04	63 044 792	.9499373	.3557624	3	21 43 69	99 252 847	.5174348	.7361877
9	15 92 01	63 521 199	.9749844	.3619178	4	21 52 96	99 897 344	.5406592	.7417532
400	16 00 00	64 000 000	20.0000000	7.3680630	465	21 62 25	100 544 625	21.5638587	7.7473109
1	16 08 01	64 481 201	.0249844	.3741979	6	21 71 56	101 194 696	.5870331	.7528606
2	16 16 04	64 964 808	.0499377	.3803227	7	21 80 89	101 847 563	.6101828	.7584023
3	16 24 09	65 450 827	.0748599	.3864373	8	21 90 24	102 503 232	.6333077	.7639361
4	16 32 16	65 939 264	.0997512	.3925418	9	21 99 61	103 161 709	.6564078	.7694620
405	16 40 25	66 430 125	20.1246118	7.3986363	470	22 09 00	103 823 000	21.6794834	7.7749801
6	16 48 36	66 923 416	.1494417	.4047206	1	22 18 41	104 487 111	.7025344	.7804904
7	16 56 49	67 419 143	.1742410	.4107950	2	22 27 84	105 154 048	.7255610	.7859928
8	16 64 64	67 917 312	.1990099	.4168595	3	22 37 29	105 823 817	.7485632	.7914875
9	16 72 81	68 417 929	.2237484	.4229142	4	22 46 76	106 496 424	.7715411	.7969745
410	16 81 00	68 921 000	20.2484567	7.4289589	475	22 56 25	107 171 875	21.7944947	7.8024538
11	16 89 21	69 426 531	.2731349	.4349938	6	22 65 76	107 850 176	.8174242	.8079254
12	16 97 44	69 934 528	.2977831	.4410189	7	22 75 29	108 531 333	.8403297	.8133892
13	17 05 69	70 444 997	.3224014	.4470342	8	22 84 84	109 215 352	.8632111	.8188456
14	17 13 96	70 957 944	.3469899	.4530399	9	22 94 41	109 902 239	.8860686	.8242942
415	17 22 25	71 473 375	20.3715488	7.4590359	480	23 04 00	110 592 000	21.9089023	7.8297353
16	17 30 56	71 991 296	.3960781	.4650223	1	23 13 61	111 284 641	.9317122	.8351688
17	17 38 89	92 511 713	.4205779	.4709991	2	23 23 24	111 980 168	.9544984	.8405949
18	17 47 24	73 034 632	.4450483	.4769664	3	23 32 89	112 678 587	.9772610	.8460134
19	17 55 61	73 560 059	.4694895	.4829242	4	23 42 56	113 379 904	22.0000000	.8514244
420	17 64 00	74 088 000	20.4939015	7.4888724	485	23 52 25	114 084 125	22.0227155	7.8568281
1	17 72 41	74 618 461	.5182845	.4948113	6	23 61 96	114 791 256	.0454077	.8622242
2	17 80 84	75 151 448	.5426386	.5007406	7	23 71 69	115 501 303	.0680765	.8676130
3	17 89 29	75 686 967	.5669638	.5066607	8	23 81 44	116 214 272	.0907220	.8729944
4	17 97 76	76 225 024	.5912603	.5125715	9	23 91 21	116 930 169	.1133444	.8783684
425	18 06 25	76 765 625	20.6155281	7.5184730	490	24 01 00	117 649 000	22.1359436	7.8837352
6	18 14 76	77 308 776	.6397674	.5243652	1	24 10 81	118 370 771	.1585198	.8890946
7	18 23 29	77 854 483	.6639783	.5302482	2	24 20 64	119 095 488	.1810730	.8944468
8	18 31 84	78 402 752	.6881609	.5361221	3	24 30 49	119 823 157	.2036033	.8997917
9	18 40 41	78 953 589	.7123152	.5419867	4	24 40 36	120 553 784	.2261108	.9051294
430	18 49 00	79 507 000	20.7364414	7.5478423	495	24 50 25	121 287 375	22.2485955	7.9104599
1	18 57 61	80 062 991	.7605395	.5536888	6	24 60 16	122 023 936	.2710575	.9157832
2	18 66 24	80 621 568	.7846097	.5595263	7	24 70 09	122 763 473	.2934968	.9210994
3	18 74 89	81 182 737	.8086520	.5653548	8	24 80 04	123 505 992	.3159136	.9264085
4	18 83 56	81 746 504	.8326667	.5711743	9	24 90 01	124 251 499	.3383079	.9317104
435	18 92 25	82 312 875	20.8566536	7.5769849	500	25 00 00	125 000 000	22.3606798	7.9370053
6	19 00 96	82 881 856	.8806130	.5827865	1	25 10 01	125 751 501	.3830293	.9422931
7	19 09 69	83 453 453	.9045450	.5885793	2	25 20 04	126 506 008	.4053565	.9475739
8	19 18 44	84 027 672	.9284495	.5943633	3	25 30 09	127 263 527	.4276615	.9528477
9	19 27 21	84 604 519	.9523268	.6001385	4	25 40 16	128 024 064	.4499443	.9581144
440	19 36 00	85 184 000	20.9761770	7.6059049	505	25 50 25	128 787 625	22.4722051	7.9633743
1	19 44 81	85 766 121	21.0000000	.6116626	6	25 60 36	129 554 216	.4944438	.9686271
2	19 53 64	86 350 888	.0237960	.6174116	7	25 70 49	130 323 843	.5166605	.9738731
3	19 62 49	86 938 307	.0475652	.6231519	8	25 80 64	131 096 512	.5388553	.9791122
4	19 71 36	87 528 384	.0713075	.6288837	9	25 90 81	131 872 229	.5610283	.9843444
445	19 80 25	88 121 125	21.0950231	7.6346067	510	26 01 00	132 651 000	22.5831796	7.7895697
6	19 89 16	88 716 536	.1187121	.6403213	11	26 11 21	133 432 831	.6053091	.9947883
7	19 98 09	89 314 623	.1423745	.6460272	12	26 21 44	134 217 728	.6274170	8.0000000
8	20 07 04	89 915 392	.1660105	.6517247	13	26 31 69	135 005 697	.6495033	.0052049
9	20 16 01	90 518 849	.1896201	.6574138	14	26 41 96	135 796 744	.6715681	.0104032
450	20 25 00	91 125 000	21.2132034	7.6630943	515	26 52 25	136 590 875	22.6936114	8.0155946
1	20 34 01	91 733 851	.2367606	.6687665	16	26 62 56	137 388 096	.7156334	.0207794
2	20 43 04	92 345 408	.2602916	.6744303	17	26 72 89	138 188 413	.7376340	.0259574
3	20 52 09	92 959 677	.2837967	.6800857	18	26 83 24	138 991 832	.7596134	.0311287
4	20 61 16	93 576 664	.3072758	.6857328	19	26 93 61	139 798 359	.7815715	.0362935
455	20 70 25	94 196 375	21.3307290	.6913717	520	27 04 00	140 608 000	22.8035085	8.0414515

2.—Squares, Cubes, Square Roots, Cube Roots, of Numbers 1 to 1600—Continued.

No.	Square	Cube.	Sq. Rt.	Cu. Rt.	No.	Square	Cube.	Sq. Rt.	Cu. Rt.
520	27 04 00	140 608 000	22.8035085	8.0414515	585	34 22 25	200 201 625	24.1867732	8.3634466
1	27 14 41	141 420 761	.8254244	.0466030	6	34 33 96	201 230 056	.2074369	.3682095
2	27 24 84	142 236 648	.8473193	.0517479	7	34 45 69	202 262 003	2280829	.3729668
3	27 35 29	143 055 667	.8691933	.0568862	8	34 57 44	203 297 472	.2487113	.3777188
4	27 45 76	143 877 824	.8910463	.0620180	9	34 69 21	204 336 469	.2693222	.3824653
525	27 56 25	144 703 125	22.9128785	8.0671432	590	34 81 00	205 379 000	24.2899156	8.3872065
6	27 66 76	145 531 576	.9346899	.0722620	1	34 92 81	206 425 071	.3104916	.3919423
7	27 77 29	146 363 183	.9564806	.0773743	2	35 04 64	207 474 688	.3310501	.3966729
8	27 87 84	147 197 952	.9782506	.0824800	3	35 16 49	208 527 857	.3515913	.4013981
9	27 98 41	148 035 889	23.0000000	.0875794	4	35 28 36	209 584 584	.3721152	.4061180
530	28 09 00	148 877 000	23.0217289	8.0926723	595	35 40 25	210 644 875	24.3926218	8.4108326
1	28 19 61	149 721 291	.0434372	0977589	6	35 52 16	211 708 736	.4131112	.4155419
2	28 30 24	150 568 768	.0651252	.1028390	7	35 64 09	212 776 173	.4335834	.4202460
3	28 40 89	151 419 437	.0867928	.1079128	8	35 76 04	213 847 192	.4540385	.4249448
4	28 51 56	152 273 304	.1084400	.1129803	9	35 88 01	214 921 799	.4744765	.4296383
535.	28 62 25	153 130 375	23.1300670	8 1180414	600	36 00 00	216 000 000	24.4948974	8.4343267
6	28 72 96	153 990 656	.1516738	.1230962	1	36 12 01	217 081 801	.5153013	.4390098
7	28 83 69	154 854 153	.1732605	.1281447	2	36 24 04	218 167 208	.5356883	.4436877
8	28 94 44	155 720 872	.1948270	.1331870	3	36 36 09	219 256 227	.5560583	.4483605
9	29 05 21	156 590 819	.2163735	.1382230	4	36 48 16	220 348 864	.5764115	.4530281
540	29 16 00	157 464 000	23.2379001	8.1432529	605	36 60 25	221 445 125	24.5967478	8.4576906
1	29 26 81	158 340 421	.2594067	.1482765	6	36 72 36	222 545 016	.6170673	.4623479
2	29 37 64	159 220 088	.2808935	.1532939	7	36 84 49	223 648 543	.6373700	.4670000
3	29 48 49	160 103 007	.3023604	.1583051	8	36 96 64	224 755 712	.6576560	.4716471
4	29 59 36	160 989 184	.3238076	.1633102	9	37 08 81	225 866 529	.6779254	.4762892
545	29 70 25	161 878 625	23.3452351	8.1683092	610	37 21 00	226 981 000	24.6981781	8.4809261
6	29 81 16	162 771 336	.3666429	.1733020	11	37 33 21	228 099 131	.7184142	.4855579
7	29 92 09	163 667 323	.3880311	.1782888	12	37 45 44	229 220 928	.7386338	.4901848
8	30 03 04	164 566 592	.4093998	.1832695	13	37 57 69	230 346 397	.7588368	.4948065
9	30 14 01	165 469 149	.4307490	.1882441	14	37 69 96	231 475 544	.7790234	.4994233
550	30 25 00	166 375 000	23.4520788	8.1932127	615	37 82 25	232 608 375	24.7991935	8.5040350
1	30 36 01	167 284 151	.4733892	.1981753	16	37 94 56	233 744 896	.8193473	.5086417
2	30 47 04	168 196 608	.4946802	.2031319	17	38 06 89	234 885 113	.8394847	.5132435
3	30 58 09	169 112 377	.5159520	.2080825	18	38 19 24	236 029 032	.8596058	.5178403
4	30 69 16	170 031 464	.5372046	.2130271	19	38 31 61	237 176 659	.8797106	.5224321
555	30 80 25	170 953 875	23.5584380	8.2179657	620	38 44 00	238 328 000	24.8997992	8.5270189
6	30 91 36	171 879 616	.5796522	.2228985	1	38 56 41	239 483 061	.9198716	.5316009
7	31 02 49	172 808 693	.6008474	.2278254	2	38 68 84	240 641 848	.9399278	.5361780
8	31 13 64	173 741 112	.6220236	.2327463	3	38 81 29	241 804 367	.9599679	5407501
9	31 24 81	174 676 879	.6431808	.2376614	4	38 93 76	242 970 624	.9799920	.5453173
560	31 36 00	175 616 000	23.6643191	8.2425706	625	39 06 25	244 140 625	25.0000000	8.5498797
1	31 47 21	176 558 481	.6854386	.2474740	6	39 18 76	245 314 376	.0199920	.5544372
2	31 58 44	177 504 328	.7065392	.2523715	7	39 31 29	246 491 883	.0399681	.5589899
3	31 69 69	178 453 547	.7276210	.2572633	8	39 43 84	247 673 152	.0599282	.5635377
4	31 80 96	179 406 144	.7486842	..2621492	9	39 56 41	248 858 189	.0798724	.5686807
565	31 92 25	180 362 125	23.7697286	8.2670294	630	39 69 00	250 047 000	25.0998008	8.5726189
6	32 03 56	181 321 496	.7907545	.2719039	1	39 81 61	251 239 591	.1197134	.5771523
7	32 14 89	182 284 263	.8117618	.2767726	2	39 94 24	252 435 968	.1396102	.5816809
8	32 26 24	183 250 432	.8327506	.2816355	3	40 06 89	253 636 137	.1594913	.5862047
9	32 37 61	184 220 009	.8537209	.2864928	4	40 19 56	254 840 104	.1793566	.5907238
570	32 49 00	185 193 000	23.8746728	8.2913444	635	40 32 25	256 047 875	25.1992063	8.5952380
1	32 60 41	186 169 411	.8956063	.2961903	6	40 44 96	257 259 456	.2190404	.5997476
2	32 71 84	187 149 248	.9165215	.3010304	7	40 57 69	258 474 853	.2388589	.6042525
3	32 83 29	188 132 517	.9374184	.3058651	8	40 70 44	259 694 072	.2586619	.6087526
4	32 94 76	189 119 224	.9582971	.3106941	9	40 83 21	260 917 119	.2784493	.6132480
575	33 06 25	190 109 375	23.9791576	8.3155175	640	40 96 00	262 144 000	25.2982213	8.6177388
6	33 17 76	191 102 976	24.0000000	.3203353	1	41 08 81	263 374 721	.3179778	.6222248
7	33 29 29	192 100 033	.0208243	.3251475	2	41 21 64	264 609 288	.3377189	.6267063
8	33 40 84	193 100 552	.0416306	.3299542	3	41 34 49	265 847 707	.3574447	.6311830
9	33 52 41	194 104 539	.0624188	.3347553	4	41 47 36	267 089 984	.3771551	.6356551
580	33 64 00	195 112 000	24.0831891	8.3395509	645	41 60 25	268 336 125	25.3968502	8.6401226
1	33 75 61	196 122 941	.1039416	.3443410	6	41 73 16	269 586 136	.4165301	.6445855
2	33 87 24	197 137 368	.1246762	.3491256	7	41 86 09	270 840 023	.4361947	.6490437
3	33 98 89	198 155 287	.1453929	.3539047	8	41 99 04	272 097 792	.4558441	.6534974
4	34 10 56	199 176 704	.1660919	.3586784	9	42 12 01	273 359 449	.4754784	.6579465
585	34 22 25	200 201 625	24.1867732	8.3634466	650	42 25 00	274 625 000	25.4950976	8.6623[illegible]

2.—Squares, Cubes, Square Roots, Cube Roots, of Numbers 1 to 1600—Continued.

No.	Square	Cube.	Sq. Rt.	Cu. Rt.	No.	Square	Cube.	Sq. Rt.	Cu. Rt.
650	42 25 00	274 625 000	25.4950976	8.6623911	715	51 12 25	365 525 875	26.7394839	8.9420140
1	42 38 01	275 894 451	.5147016	.6668310	16	51 26 56	367 061 696	.7581763	.9461809
2	42 51 04	277 167 808	.5342907	.6712665	17	51 40 89	368 601 813	.7768557	.9503438
3	42 64 09	278 445 077	.5538647	.6756974	18	51 55 24	370 146 232	.7955220	.9545029
4	42 77 16	279 726 264	.5734237	.6801237	19	51 69 61	371 694 959	.8141754	.9586581
655	42 90 25	281 011 375	25.5929678	8.6845456	720	51 84 00	373 248 000	26.8328157	8.9628095
6	43 03 36	282 300 416	.6124969	.6889630	1	51 98 41	374 805 361	.8514432	.9669570
7	43 16 49	283 593 393	.6320112	.6933759	2	52 12 84	376 367 048	.8700577	.9711007
8	43 29 64	284 890 312	.6515107	.6977843	3	52 27 29	377 933 067	.8886593	.9752406
9	43 42 81	286 191 179	.6709953	.7021882	4	52 41 76	379 503 424	.9072481	.9793766
660	43 56 00	287 496 000	25.6904652	8.7065877	725	52 56 25	381 078 125	26.9258240	8.9835089
1	43 69 21	288 804 781	.7099203	.7109827	6	52 70 76	382 657 176	.9443872	.9876373
2	43 82 44	290 117 528	.7293607	.7153734	7	52 85 29	384 240 583	.9629375	.9917620
3	43 95 69	291 434 247	.7487864	.7197596	8	52 99 84	385 828 352	.9814751	.9958829
4	44 08 96	292 754 944	.7681975	.7241414	9	53 14 41	387 420 489	27.0000000	9.0000000
665	44 22 25	294 079 625	25.7875939	8.7285187	730	53 29 00	389 017 000	27.0185122	9.0041134
6	44 35 56	295 408 296	.8069758	.7328918	1	53 43 61	390 617 891	.0370117	.0082229
7	44 48 89	296 740 963	.8263431	.7372604	2	53 58 24	392 223 168	.0554985	.0123288
8	44 62 24	298 077 632	.8456960	.7416246	3	53 72 89	393 832 837	.0739727	.0164309
9	44 75 61	299 418 309	.8650343	.7459846	4	53 87 56	395 446 904	.0924344	.0205293
670	44 89 00	300 763 000	25.8843582	8.7503401	735	54 02 25	397 065 375	27.1108834	9.0246239
1	45 02 41	302 111 711	.9036677	.7546913	6	54 16 96	398 688 256	.1293199	.0287149
2	45 15 84	303 464 448	.9229628	.7590383	7	54 31 69	400 315 553	.1477439	.0328021
3	45 29 29	304 821 217	.9422435	.7633809	8	54 46 44	401 947 272	.1661554	.0368857
4	45 42 76	306 182 024	.9615100	.7677192	9	54 61 21	403 583 419	.1845544	.0409655
675	45 56 25	307 546 875	25.9807621	8.7720532	740	54 76 00	405 224 000	27.2029410	9.0450417
6	45 69 76	308 915 776	26.0000000	.7763830	1	54 90 81	406 869 021	.2213152	.0491142
7	45 83 29	310 288 733	.0192237	.7807084	2	55 05 64	408 518 488	.2396769	.0531831
8	45 96 84	311 665 752	.0384331	.7850296	3	55 20 49	410 172 407	.2580263	.0572482
9	46 10 41	313 046 839	.0576284	.7893466	4	55 35 36	411 830 784	.2763634	.0613098
680	46 24 00	314 432 000	26,0768096	8.7936593	745	55 50 25	413 493 625	27.2946881	9.0653677
1	46 37 61	315 821 241	.0959767	.7979679	6	55 65 16	415 160 936	.3130006	.0694220
2	46 51 24	317 214 568	.1151297	.8022721	7	55 80 09	416 832 723	.3313007	.0734726
3	46 64 89	318 611 987	.1342687	.8065722	8	55 95 04	418 508 992	.3495887	.0775197
4	46 78 56	320 013 504	.1533937	.8108681	9	56 10 01	420 189 749	.3678644	.0815631
685	46 92 25	321 419 125	26.1725047	8.8151598	750	56 25 00	421 875 000	27.3861279	9.0856030
6	47 05 96	322 828 856	.1916017	.8194474	1	56 40 01	423 564 751	.4043792	.0896392
7	47 19 69	324 242 703	.2106848	.8237307	2	56 55 04	425 259 008	.4226184	.0936719
8	47 33 44	325 660 672	.2297541	.8280099	3	56 70 09	426 957 777	.4408455	.0977010
9	47 47 21	327 082 769	.2488095	.8322850	4	56 85 16	428 661 064	.4590604	.1017265
690	47 61 00	328 509 000	26.2678511	8.8365559	755	57 00 25	430 368 875	27.4772633	9.1057485
1	47 74 81	329 939 371	.2868789	.8408227	6	57 15 36	432 081 216	.4954542	.1097669
2	47 88 64	331 373 888	.3058929	.8450854	7	57 30 49	433 798 093	.5136330	.1137818
3	48 02 49	332 812 557	.3248932	.8493440	8	57 45 64	435 519 512	.5317998	.1177931
4	48 16 36	334 255 384	.3438797	.8535985	9	57 60 81	437 245 479	.5499546	.1218010
695	48 30 25	335 702 375	26.3628527	8.8578489	760	57 76 00	438 976 000	27.5680975	9.1258053
6	48 44 16	337 153 536	.3818119	.8620952	1	57 91 21	440 711 081	.5862284	.1298061
7	48 58 09	338 608 873	.4007576	.8663375	2	58 06 44	442 450 728	.6043475	.1338034
8	48 72 04	340 068 392	.4196896	.8705757	3	58 21 69	444 194 947	.6224546	.1377971
9	48 86 01	341 532 099	.4386081	.8748099	4	58 36 96	445 943 744	.6405499	.1417874
700	49 00 00	343 000 000	26.4575131	8.8790400	765	58 52 25	447 697 125	27.6586334	9.1457742
1	49 14 01	344 472 101	.4764046	.8832661	6	58 67 56	449 455 096	.6767050	.1497576
2	49 28 04	345 948 408	.4952826	.8874882	7	58 82 89	451 217 663	.6947648	.1537375
3	49 42 09	347 428 927	.5141472	.8917063	8	58 98 24	452 984 832	.7128129	.1577139
4	49 56 16	348 913 664	.5329983	.8959204	9	59 13 61	454 756 609	.7308492	.1616869
705	49 70 25	350 402 625	26.5518361	8.9001304	770	59 29 00	456 533 000	27.7488739	9.1656565
6	49 84 36	351 895 816	.5706605	.9043366	1	59 44 41	458 314 011	.7668868	.1696225
7	49 98 49	353 393 243	.5894716	.9085387	2	59 59 84	460 099 648	.7848880	.1735852
8	50 12 64	354 894 912	.6082694	.9127369	3	59 75 29	461 889 917	.8028775	.1775445
9	50 26 81	356 400 829	6270539	.9169311	4	59 90 76	463 684 824	.8208555	.1815003
710	50 41 00	357 911 000	26.6458252	8.9211214	775	60 06 25	465 484 375	27.8388218	9.1854527
11	50 55 21	359 425 431	.6645833	.9253078	6	60 21 76	467 288 576	.8567766	.1894018
12	50 69 44	360 944 128	.6833281	.9294902	7	60 37 29	469 097 433	.8747197	.1933474
13	50 83 69	362 467 097	.7020598	.9336687	8	60 52 84	470 910 952	.8926514	.1972897
14	50 97 96	363 994 344	.7207784	.9378433	9	60 68 41	472 729 139	.9105715	.2012286
715	51 12 25	365 525 875	26.7394839	8.9420140	780	60 84 00	474 552 000	27.9284801	9.2051641

2.—SQUARES, CUBES, SQUARE ROOTS, CUBE ROOTS, OF NUMBERS 1 TO 1600—Continued.

No.	Square	Cube.	Sq. Rt.	Cu. Rt.	No.	Square	Cube.	Sq. Rt.	Cu. Rt.
780	60 84 00	474 552 000	27.9284801	9.2051641	845	71 40 25	603 351 125	29.0688837	9.4540719
1	60 99 61	476 379 541	.9463772	.2090962	6	71 57 16	605 495 736	.0860791	.4577999
2	61 15 24	478 211 768	.9642629	.2130250	7	71 74 09	607 645 423	.1032644	.4615249
3	61 30 89	480 048 687	.9821372	.2169505	8	71 91 04	609 800 192	.1204396	.4652470
4	61 46 56	481 890 304	28.0000000	.2208726	9	72 08 01	611 960 049	.1376046	.4689661
785	61 62 25	483 736 625	28.0178515	9.2247914	850	72 25 00	614 125 000	29.1547595	9.4726824
6	61 77 96	485 587 656	.0356915	.2287068	1	72 42 01	616 295 051	.1719043	.4763957
7	61 93 69	487 443 403	.0535203	.2326189	2	72 59 04	618 470 208	.1890390	.4801061
8	62 09 44	489 303 872	.0713377	.2365277	3	72 76 09	620 650 477	.2061637	.4838136
9	62 25 21	491 169 069	.0891438	.2404333	4	72 93 16	622 835 864	.2232784	.4875182
790	62 41 00	493 039 000	28.1069386	9.2443355	855	73 10 25	625 026 375	29.2403830	9.4912200
1	62 56 81	494 913 671	.1247222	.2482344	6	73 27 36	627 222 016	.2574777	.4949188
2	62 72 64	496 793 088	.1424946	.2521300	7	73 44 49	629 422 793	.2745623	.4986147
3	62 88 49	498 677 257	.1602557	.2560224	8	73 61 64	631 628 712	.2916370	.5023078
4	63 04 36	500 566 184	.1780056	.2599114	9	73 78 81	633 839 779	.3087018	.5059980
795	63 20 25	502 459 875	28.1957444	9.2637973	860	73 96 00	636 056 000	29.3257566	9.5096854
6	63 36 16	504 358 336	.2134720	.2676798	1	74 13 21	638 277 381	.3428015	.5133699
7	63 52 09	506 261 573	.2311884	.2715592	2	74 30 44	640 503 928	.3598365	.5170515
8	63 68 04	508 169 592	.2488938	.2754352	3	74 47 69	642 735 647	.3768616	.5207303
9	63 84 01	510 082 399	.2665881	.2793081	4	74 64 96	644 972 544	.3938769	.5244063
800	64 00 00	512 000 000	28.2842712	9.2831777	865	74 82 25	647 214 625	29.4108823	9.5280794
1	64 16 01	513 922 401	.3019434	.2870440	6	74 99 56	649 461 896	.4278779	.5317497
2	64 32 04	515 849 608	.3196045	.2909072	7	75 16 89	651 714 363	.4448637	.5354172
3	64 48 09	517 781 627	.3372546	.2947671	8	75 34 24	653 972 032	.4618397	.5390818
4	64 64 16	519 718 464	.3548938	.2986239	9	75 51 61	656 234 909	.4788059	.5427437
805	64 80 25	521 660 125	28.3725219	9.3024775	870	75 69 00	658 503 000	29.4957624	9.5464027
6	64 96 36	523 606 616	.3901391	.3063278	1	75 86 41	660 776 311	.5127091	.5500589
7	65 12 49	525 557 943	.4077454	.3101750	2	76 03 84	663 054 848	.5296461	.5537123
8	65 28 64	527 514 112	.4253408	.3140190	3	76 21 29	665 338 617	.5465734	.5573630
9	65 44 81	529 475 129	.4429253	.3178599	4	76 38 76	667 627 624	.5634910	.5610108
810	65 61 00	531 441 000	28.4604989	9.3216975	875	76 56 25	669 921 875	29.5803989	9.5646559
11	65 77 21	533 411 731	.4780617	.3255320	6	76 73 76	672 221 376	.5972972	.5682982
12	65 93 44	535 387 328	.4956137	.3293634	7	76 91 29	674 526 133	.6141858	.5719377
13	66 09 69	537 367 797	.5131549	.3331916	8	77 08 84	676 836 152	.6310648	.5755745
14	66 25 96	539 353 144	.5306852	.3370167	9	77 26 41	679 151 439	.6479342	.5792085
815	66 42 25	541 343 375	28.5482048	9.3408386	880	77 44 00	681 472 000	29.6647939	9.5828397
16	66 58 56	543 338 496	.5657137	.3446575	1	77 61 61	683 797 841	.6816442	.5864682
17	66 74 89	545 338 513	.5832119	.3484731	2	77 79 24	686 128 968	.6984848	.5900939
18	66 91 24	547 343 432	.6006993	.3522857	3	77 96 89	688 465 387	.7153159	.5937169
19	67 07 61	549 353 259	.6181760	.3560952	4	78 14 56	690 807 104	.7321375	.5973373
820	67 24 00	551 368 000	28.6356421	9.3599016	885	78 32 25	693 154 125	29.7489496	9.6009548
1	67 40 41	553 387 661	.6530976	.3637049	6	78 49 96	695 506 456	.7657521	.6045696
2	67 56 84	555 412 248	.6705424	.3675051	7	78 67 69	697 864 103	.7825452	.6081817
3	67 73 29	557 441 767	.6879766	.3713022	8	78 85 44	700 227 072	.7993289	.6117911
4	67 89 76	559 476 224	.7054002	.3750963	9	79 03 21	702 595 369	.8161030	.6153977
825	68 06 25	561 515 625	28.7228132	9.3788873	890	79 21 00	704 969 000	29.8328678	9.6190017
6	68 22 76	563 559 976	.7402157	.3826752	1	79 38 81	707 347 971	.8496231	.6226030
7	68 39 29	565 609 283	.7576077	.3864600	2	79 56 64	709 732 288	.8663690	.6262016
8	68 55 84	567 663 552	.7749891	.3902419	3	79 74 49	712 121 957	.8831056	.6297975
9	68 72 41	569 722 789	.7923601	.3940206	4	79 92 36	714 516 984	.8998328	.6333907
830	68 89 00	571 787 000	28.8097206	9.3977964	895	80 10 25	716 917 375	29.9165506	9.6369812
1	69 05 61	573 856 191	.8270706	.4015691	6	80 28 16	719 323 136	.9332591	.6405690
2	69 22 24	575 930 368	.8444102	.4053387	7	80 46 09	721 734 273	.9499583	.6441542
3	69 38 89	578 009 537	.8617394	.4091054	8	80 64 04	724 150 792	.9666481	.6477367
4	69 55 56	580 093 704	.8790582	.4128690	9	80 82 01	726 572 699	.9833287	.6513166
835	69 72 25	582 182 875	28.8963666	9 4166297	900	81 00 00	729 000 000	30.0000000	9.6548938
6	69 88 96	584 277 056	.9136646	.4203873	1	81 18 01	731 432 701	.0166620	.6584684
7	70 05 69	586 376 253	.9309523	.4241420	2	81 36 04	733 870 808	.0333148	.6620403
8	70 22 44	588 480 472	.9482297	.4278936	3	81 54 09	736 314 327	.0499584	.6656096
9	70 39 21	590 589 719	.9654967	.4316423	4	81 72 16	738 763 264	.0665928	.6691762
840	70 56 00	592 704 000	28.9827535	9.4353880	905	81 90 25	741 217 625	30.0832179	9.6727403
1	70 72 81	594 823 321	29.0000000	.4391307	6	82 08 36	743 677 416	.0998339	.6763017
2	70 89 64	596 947 688	.0172363	.4428704	7	82 26 49	746 142 643	.1164407	.6798604
3	71 06 49	599 077 107	.0344623	.4466072	8	82 44 64	748 613 312	.1330383	.6834166
4	71 23 36	601 211 584	.0516781	.4503410	9	82 62 81	751 089 429	.1496269	.6869701
845	71 40 25	603 351 125	29.0688837	9.4540719	910	82 81 00	753 571 000	30.1662063	9.6905211

2.—Squares, Cubes, Square Roots, Cube Roots, of Numbers 1 to 1600—Continued.

No.	Square	Cube.	Sq. Rt.	Cu. Rt.	No.	Square.	Cube.	Sq. Rt.	Cu. Rt.
910	82 81 00	753 571 000	30.1662063	9.6905211	975	95 06 25	926 859 375	31.2249900	9.9159624
11	82 99 21	756 058 031	.1827765	.6940694	6	95 25 76	929 714 176	.2409987	.9193513
12	83 17 44	758 550 528	.1993377	.6976151	7	95 45 29	932 574 833	.2569992	.9227379
13	83 35 69	761 048 497	.2158899	.7011583	8	95 64 84	935 441 352	.2729915	.9261222
14	83 53 96	763 551 944	.2324329	.7046989	9	95 84 41	938 313 739	.2889757	.9295042
915	83 72 25	766 060 875	30.2489669	9.7082369	980	96 04 00	941 192 000	31.3049517	9.9328839
16	83 90 56	768 575 296	.2654919	.7117723	1	96 23 61	944 076 141	.3209195	.9362613
17	84 08 89	771 095 213	.2820079	.7153051	2	96 43 24	946 966 168	.3368792	.9396363
18	84 27 24	773 620 632	.2985148	.7188354	3	96 62 89	949 862 087	.3528308	.9430092
19	84 45 61	776 151 559	.3150128	.7223631	4	96 82 56	952 763 904	.3687743	.9463797
920	84 64 00	778 688 000	30.3315018	9.7258883	985	97 02 25	955 671 625	31.3847097	9.9497479
1	84 82 41	781 229 961	.3479818	.7294109	6	97 21 96	958 585 256	.4006369	.9531138
2	85 00 84	783 777 448	.3644529	.7329309	7	97 41 69	961 504 803	.4165561	.9564775
3	85 19 29	786 330 467	.3809151	.7364484	8	97 61 44	964 430 272	.4324673	.9598389
4	85 37 76	788 889 024	.3973683	.7399634	9	97 81 21	967 361 669	.4483704	.9631981
925	85 56 25	791 453 125	30.4138127	9.7434758	990	98 01 00	970 299 000	31.4642654	9.9665549
6	85 74 76	794 022 776	.4302481	.7469857	1	98 20 81	973 242 271	.4801525	.9699095
7	85 93 29	796 597 983	.4466747	.7504930	2	98 40 64	976 191 488	.4960315	.9732619
8	86 11 84	799 178 752	.4630924	.7539979	3	98 60 49	979 146 657	.5119025	.9766120
9	86 30 41	801 765 089	.4795013	.7575002	4	98 80 36	982 107 784	.5277655	.9799599
930	86 49 00	804 357 000	30.4959014	9.7610001	995	99 00 25	985 074 875	31.5436206	9.9833055
1	86 67 61	806 954 491	.5122926	.7644974	6	99 20 16	988 047 936	.5594677	.9866488
2	86 86 24	809 557 568	.5286750	.7679922	7	99 40 09	991 026 973	.5753068	.9899900
3	87 04 89	812 166 237	.5450487	.7714845	8	99 60 04	994 011 992	.5911380	.9933289
4	87 23 56	814 780 504	.5614136	.7749743	9	99 80 01	997 002 999	.6069613	.9966656
935	87 42 25	817 400 375	30.5777697	9.7784616	1000	1 00 00 00	1 000 000 000	31.6227766	10.0000000
6	87 60 96	820 025 856	.5941171	.7819466	1	1 00 20 01	1 003 003 001	.6385840	.0033322
7	87 79 69	822 656 953	.6104557	.7854288	2	1 00 40 04	1 006 012 008	.6543836	.0066622
8	87 98 44	825 293 672	.6267857	.7889087	3	1 00 60 09	1 009 027 027	.6701752	.0099899
9	88 17 21	827 936 019	.6431069	.7923861	4	1 00 80 16	1 012 048 064	.6859590	.0133155
940	88 36 00	830 584 000	30.6594194	9.7958611	1005	1 01 00 25	1 015 075 125	31.7017349	10.0166389
1	88 54 81	833 237 621	.6757233	.7993336	6	1 01 20 36	1 018 108 216	.7175030	.0199601
2	88 73 64	835 896 888	.6920185	.8028036	7	1 01 40 49	1 021 147 343	.7332633	.0232791
3	88 92 49	838 561 807	.7083051	.8062711	8	1 01 60 64	1 024 192 512	.7490157	.0265958
4	89 11 36	841 232 384	.7245830	.8097362	9	1 01 80 81	1 027 243 729	.7647603	.0299104
945	89 30 25	843 908 625	30.7408523	9.8131989	1010	1 02 01 00	1 030 301 000	31.7804972	10.0332228
6	89 49 16	846 590 536	.7571130	.8166591	11	1 02 21 21	1 033 364 331	.7962262	.0365330
7	89 68 09	849 278 123	.7733651	.8201169	12	1 02 41 44	1 036 433 728	.8119474	.0398410
8	89 87 04	851 971 392	.7896086	.8235723	13	1 02 61 69	1 039 509 197	.8276609	.0431469
9	90 06 01	854 670 349	.8058436	.8270252	14	1 02 81 96	1 042 590 744	.8433666	.0464506
950	90 25 00	857 375 000	30.8220700	9.8304757	1015	1 03 02 25	1 045 678 375	31.8590646	10.0497521
1	90 44 01	860 085 351	.8382879	.8339238	16	1 03 22 56	1 048 772 096	.8747549	.0530514
2	90 63 04	862 801 408	.8544972	.8373695	17	1 03 42 89	1 051 871 913	.8904374	.0563485
3	90 82 09	865 523 177	.8706981	.8408127	18	1 03 63 24	1 054 977 832	.9061123	.0596435
4	91 01 16	868 250 664	.8868904	.8442536	19	1 03 83 61	1 058 089 859	.9217794	.0629364
955	91 20 25	870 983 875	30.9030743	9.8476920	1020	1 04 04 00	1 061 208 000	31.9374388	10.0662271
6	91 39 36	873 722 816	.9192497	.8511280	21	1 04 24 41	1 064 332 261	.9530906	.0695156
7	91 58 49	876 467 493	.9354166	.8545617	22	1 04 44 84	1 067 462 648	.9687347	.0728020
8	91 77 64	879 217 912	.9515751	.8579929	23	1 04 65 29	1 070 599 167	.9843712	.0760863
9	91 96 81	881 974 079	.9677251	.8614218	24	1 04 85 76	1 073 741 824	32.0000000	.0793684
960	92 16 00	884 736 000	30.9838668	9.8648483	1025	1 05 06 25	1 076 890 625	32.0156212	10.0826484
1	92 35 21	887 503 681	31.0000000	.8682724	26	1 05 26 76	1 080 045 576	.0312348	.0859262
2	92 54 44	890 277 128	.0161248	.8716941	27	1 05 47 29	1 083 206 683	.0468407	.0892019
3	92 73 69	893 056 347	.0322413	.8751135	28	1 05 67 84	1 086 373 952	.0624391	.0924755
4	92 92 96	895 841 344	.0483494	.8785305	29	1 05 88 41	1 089 547 389	.0780298	.0957469
965	93 12 25	898 632 125	31.0644491	9.8819451	1030	1 06 09 00	1 092 727 000	32.0936131	10.0990163
6	93 31 56	901 428 696	.0805405	.8853574	31	1 06 29 61	1 095 912 791	.1091887	.1022835
7	93 50 89	904 231 063	.0966236	.8887673	32	1 06 50 24	1 099 104 768	.1247568	.1055487
8	93 70 24	907 039 232	.1126984	.8921749	33	1 06 70 89	1 102 302 937	.1403173	.1088117
9	93 89 61	909 853 209	.1287648	.8955801	34	1 06 91 56	1 105 507 304	.1558704	.1120726
970	94 09 00	912 673 000	31.1448230	9.8989830	1035	1 07 12 25	1 108 717 875	32.1714159	10.1153314
1	94 28 41	915 498 611	.1608729	.9023835	36	1 07 32 96	1 111 934 656	.1869539	.1185882
2	94 47 84	918 330 048	.1769145	.9057817	37	1 07 53 69	1 115 157 653	.2024844	.1218428
3	94 67 29	921 167 317	.1929479	.9091776	38	1 07 74 44	1 118 386 872	.2180074	.1250953
4	94 86 76	924 010 424	.2089731	.9125712	39	1 07 95 21	1 121 622 319	.2335229	.1283457
975	95 06 25	926 859 375	31.2249900	9.9159624	1040	1 08 16 00	1 124 864 000	32.2490310	10.1315941

2.—Squares, Cubes, Square Roots, Cube Roots, of Numbers 1 to 1600—Continued.

No.	Square	Cube.	Sq. Rt.	Cu. Rt.	No.	Square	Cube.	Sq. Rt.	Cu. Rt.
1040	1 08 16 00	1 124 864 000	32.2490310	10.1315941	1105	1 22 10 25	1 349 232 625	33.2415403	10.3384181
41	1 08 36 81	1 128 111 921	.2645316	.1348403	6	1 22 32 36	1 352 899 016	.2565783	.3415358
42	1 08 57 64	1 131 366 088	.2800248	.1380845	7	1 22 54 49	1 356 572 043	.2716095	.3446517
43	1 08 78 49	1 134 626 507	.2955105	.1413266	8	1 22 76 64	1 360 251 712	.2866339	.3477657
44	1 08 99 36	1 137 893 184	.3109888	.1445667	9	1 22 98 81	1 363 938 029	.3016516	.3508778
1045	1 09 20 25	1 141 166 125	32.3264598	10.1478047	1110	1 23 21 00	1 367 631 000	33.3166625	10.3539880
46	1 09 41 16	1 144 445 336	.3419233	.1510406	11	1 23 43 21	1 371 330 631	.3316666	.3570964
47	1 09 62 09	1 147 730 823	.3573794	.1542744	12	1 23 65 44	1 375 036 928	.3466640	.3602029
48	1 09 83 04	1 151 022 592	.3728281	.1575062	13	1 23 87 69	1 378 749 897	.3616546	.3633076
49	1 10 04 01	1 154 320 649	.3882695	.1607359	14	1 24 09 96	1 382 469 544	.3766385	.3664103
1050	1 10 25 00	1 157 625 000	32.4037035	10.1639636	1115	1 24 32 25	1 386 195 875	33.3916157	10.3695113
51	1 10 46 01	1 160 935 651	.4191301	.1671893	16	1 24 54 56	1 389 928 896	.4065862	.3726103
52	1 10 67 04	1 164 252 608	.4345495	.1704129	17	1 24 76 89	1 393 668 613	.4215499	.3757076
53	1 10 88 09	1 167 575 877	.4499615	.1736344	18	1 24 99 24	1 397 415 032	.4365070	.3788030
54	1 11 09 16	1 170 905 464	.4653662	.1768539	19	1 25 21 61	1 401 168 159	.4514573	.3818965
1055	1 11 30 25	1 174 241 375	32.4807635	10.1800714	1120	1 25 44 00	1 404 928 000	33.4664011	10.3849882
56	1 11 51 36	1 117 583 616	.4961536	.1832868	21	1 25 66 41	1 408 694 561	.4813381	.3880781
57	1 11 72 49	1 180 932 193	.5115364	.1865002	22	1 25 88 84	1 412 467 848	.4962684	.3911661
58	1 11 93 64	1 184 287 112	.5269119	.1897116	23	1 26 11 29	1 416 247 867	.5111921	.3942523
59	1 12 14 81	1 187 648 379	.5422802	.1929209	24	1 26 33 76	1 420 034 624	.5261092	.3973366
1060	1 12 36 00	1 191 016 000	32.5576412	10.1961283	1125	1 26 56 25	1 423 828 125	33.5410196	10.4004192
61	1 12 57 21	1 194 389 981	.5729949	.1993336	26	1 26 78 76	1 427 628 376	.5559234	.4034999
62	1 12 78 44	1 197 770 328	.5883415	.2025369	27	1 27 01 29	1 431 435 383	.5708206	.4065787
63	1 12 99 69	1 201 157 047	.6036807	.2057382	28	1 27 23 84	1 435 249 152	.5857112	.4096557
64	1 13 20 96	1 204 550 144	.6190129	.2089375	29	1 27 46 41	1 439 069 689	.6005952	.4127310
1065	1 13 42 25	1 207 949 625	32.6343377	10.2121347	1130	1 27 69 00	1 442 897 000	33.6154726	10.4158044
66	1 13 63 56	1 211 355 496	.6496554	.2153300	31	1 27 91 61	1 446 731 091	.6303434	.4188760
67	1 13 84 89	1 214 767 763	.6649659	.2185233	32	1 28 14 24	1 450 571 968	.6452077	.4219458
68	1 14 06 24	1 218 186 432	.6802693	.2217146	33	1 28 36 89	1 454 419 637	.6600653	.4250138
69	1 14 27 61	1 221 611 509	.6955654	.2249039	34	1 28 59 56	1 458 274 104	.6749165	.4280800
1070	1 14 49 00	1 225 043 000	32.7108544	10.2280912	1135	1 28 82 25	1 462 135 375	33.6897610	10.4311443
71	1 14 70 41	1 228 480 911	.7261363	.2312766	36	1 29 04 96	1 466 003 456	.7045991	.4342069
72	1 14 91 84	1 231 925 248	.7414111	.2344599	37	1 29 27 69	1 469 878 353	.7194306	.4372677
73	1 15 13 29	1 235 376 017	.7566787	.2376413	38	1 29 50 44	1 473 760 072	.7342556	.4403267
74	1 15 34 76	1 238 833 224	.7719392	.2408207	39	1 29 73 21	1 477 648 619	.7490741	.4433839
1075	1 15 56 25	1 242 296 875	32.7871926	10.2439981	1140	1 29 96 00	1 481 544 000	33.7638860	10.4464393
76	1 15 77 76	1 245 766 976	.8024389	.2471735	41	1 30 18 81	1 485 446 221	.7786915	.4494929
77	1 15 99 29	1 249 243 533	.8176782	.2503470	42	1 30 41 64	1 489 355 288	7934905	4525448
78	1 16 20 84	1 252 726 552	.8329103	.2535186	43	1 30 64 49	1 493 271 207	.8082830	.4555948
79	1 16 42 41	1 256 216 039	.8481354	.2566881	44	1 30 87 36	1 497 193 984	.8230691	.4586431
1080	1 16 64 00	1 259 712 000	32.8633535	10.2598557	1145	1 31 10 25	1 501 123 625	33.8378486	10.4616896
81	1 16 85 61	1 263 214 441	.8785644	.2630213	46	1 31 33 16	1 505 060 136	.8526218	.4647343
82	1 17 07 24	1 266 723 368	.8937684	.2661850	47	1 31 56 09	1 509 003 523	.8673884	.4677773
83	1 17 28 89	1 270 238 787	.9089653	.2693467	48	1 31 79 04	1 512 953 792	.8821487	.4708185
84	1 17 50 56	1 273 760 704	.9241553	.2725065	49	1 32 02 01	1 516 910 949	.8969025	.4738579
1085	1 17 72 25	1 277 289 125	32.9393382	10.2756644	1150	1 32 25 00	1 520 875 000	33.9116499	10.4768955
86	1 17 93 96	1 280 824 056	.9545141	.2788203	51	1 32 48 01	1 524 845 951	.9263909	.4799314
87	1 18 15 69	1 284 365 503	.9696830	.2819743	52	1 32 71 04	1 528 823 808	.9411255	.4829656
88	1 18 37 44	1 287 913 472	.9848450	.2851264	53	1 32 94 09	1 532 808 577	.9558537	.4859980
89	1 18 59 21	1 291 467 969	33.0000000	.2882765	54	1 33 17 16	1 536 800 264	.9705755	.4890286
1090	1 18 81 00	1 295 029 000	33.0151480	10.2914247	1155	1 33 40 25	1 540 798 875	33.9852910	10.4920575
91	1 19 02 81	1 298 596 571	.0302891	.2945709	56	1 33 63 36	1 544 804 416	34.0000000	.4950847
92	1 19 24 64	1 302 170 688	.0454233	.2977153	57	1 33 86 49	1 548 816 893	.0147027	.4981101
93	1 19 46 49	1 305 751 357	.0605505	.3008577	58	1 34 09 64	1 552 836 312	.0293990	.5011337
94	1 19 68 36	1 309 338 584	.0756708	.3039982	59	1 34 32 81	1 556 862 679	.0440890	.5041556
1095	1 19 90 25	1 312 932 375	33.0907842	10.3071368	1160	1 34 56 00	1 560 896 000	34.0587727	10.5071757
96	1 20 12 16	1 316 532 736	.1058907	.3102735	61	1 34 79 21	1 564 936 281	.0734501	.5101942
97	1 20 34 09	1 320 139 673	.1209903	.3134083	62	1 35 02 44	1 568 983 528	.0881211	.5132109
98	1 20 56 04	1 323 753 192	.1360830	.3165411	63	1 35 25 69	1 573 037 747	.1027858	.5162259
99	1 20 78 01	1 327 373 299	.1511689	.3196721	64	1 35 48 96	1 577 098 944	.1174442	.5192391
1100	1 21 00 00	1 331 000 000	33.1662479	10.3228012	1165	1 35 72 25	1 581 167 125	34.1320963	10.5222506
1	1 21 42 01	1 334 633 301	.1813200	.3259284	66	1 35 95 56	1 585 242 296	.1467422	.5252604
2	1 21 44 04	1 338 273 208	.1963853	.3290537	67	1 36 18 89	1 589 324 463	.1613817	.5282685
3	1 21 66 09	1 341 919 727	.2114438	.3321770	68	1 36 42 24	1 593 413 632	.1760150	.5312749
4	1 21 88 16	1 345 572 864	.2264955	.3352985	69	1 36 65 61	1 597 509 809	.1906420	.5342795
1105	1 22 10 25	1 349 232 625	33.2415403	10.3384181	1170	1 36 89 00	1 601 613 000	34.2052627	10.5372825

2.—Squares, Cubes, Square Roots, Cube Roots, of Numbers 1 to 1600—Continued.

No.	Square	Cube.	Sq. Rt.	Cu. Rt.	No.	Square	Cube.	Sq. Rt.	Cu. Rt.
1170	1 36 89 00	1 601 613 000	34.2052627	10.5372825	1235	1 52 52 25	1 883 652 875	35.1425668	10.7289112
71	1 37 12 41	1 605 723 211	.2198773	.5402837	36	1 52 76 96	1 888 232 256	.1567917	.7318062
72	1 37 35 84	1 609 840 448	.2344855	.5432832	37	1 53 01 69	1 892 819 053	.1710108	.7346997
73	1 37 59 29	1 613 964 717	.2490875	.5462810	38	1 53 26 44	1 897 413 272	.1852242	.7375916
74	1 37 82 76	1 618 096 024	.2636834	.5492771	39	1 53 51 21	1 902 014 919	.1994318	.7404819
1175	1 38 06 25	1 622 234 375	34.2782730	10.5522715	1240	1 53 76 00	1 906 624 000	35.2136337	10.7433707
76	1 38 29 76	1 626 379 776	.2928564	.5552642	41	1 54 00 81	1 911 240 521	.2278299	.7462579
77	1 38 53 29	1 630 532 233	.3074336	.5582552	42	1 54 25 64	1 915 864 488	.2420204	.7491436
78	1 38 76 84	1 634 691 752	.3220046	.5612445	43	1 54 50 49	1 920 495 907	.2562051	.7520277
79	1 39 00 41	1 638 858 339	.3365694	.5642322	44	1 54 75 36	1 925 134 784	.2703842	.7549103
1180	1 39 24 00	1 643 032 000	34.3511281	10.5672181	1245	1 55 00 25	1 929 781 125	35.2845575	10.7577913
81	1 39 47 61	1 647 212 741	.3656805	.5702024	46	1 55 25 16	1 934 434 936	.2987252	.7606708
82	1 39 71 24	1 651 400 568	.3802268	.5731849	47	1 55 50 09	1 939 096 223	.3128872	.7635488
83	1 39 94 89	1 655 595 487	.3947670	.5761658	48	1 55 75 04	1 943 764 992	.3270435	.7664252
84	1 40 18 56	1 659 797 504	.4093011	.5791449	49	1 56 00 01	1 948 441 249	.3411941	.7693001
1185	1 40 42 25	1 664 006 625	34.4238289	10.5821225	1250	1 56 25 00	1 953 125 000	35.3553391	10.7721735
86	1 40 65 96	1 668 222 856	.4383507	.5850983	51	1 56 50 01	1 957 816 251	.3694784	.7750453
87	1 40 89 69	1 672 446 203	.4528663	.5880725	52	1 56 75 04	1 962 515 008	.3836120	.7779156
88	1 41 13 44	1 676 676 672	.4673759	.5910450	53	1 57 00 09	1 967 221 277	.3977400	.7807843
89	1 41 37 21	1 680 914 269	.4818793	.5940158	54	1 57 25 16	1 971 935 064	.4118624	.7836516
1190	1 41 61 00	1 685 159 000	34.4963766	10.5969850	1255	1 57 50 25	1 976 656 375	35.4259792	10.7865173
91	1 41 84 81	1 689 410 871	.5108678	.5999525	56	1 57 75 36	1 981 385 216	.4400903	.7893815
92	1 42 08 64	1 693 669 888	.5253530	.6029184	57	1 58 00 49	1 986 121 593	.4541958	.7922441
93	1 42 32 49	1 697 936 057	.5398321	.6058826	58	1 58 25 64	1 990 865 512	.4682957	.7951053
94	1 42 56 36	1 702 209 384	.5543051	.6088451	59	1 58 50 81	1 995 616 979	.4823900	.7979649
1195	1 42 80 25	1 706 489 875	34.5687720	10.6118060	1260	1 58 76 00	2 000 376 000	35.4964787	10.8008230
96	1 43 04 16	1 710 777 536	.5832329	.6147652	61	1 59 01 21	2 005 142 581	.5105618	.8036797
97	1 43 28 09	1 715 072 373	.5976879	.6177228	62	1 59 26 44	2 009 916 728	.5246393	.8065348
98	1 43 52 04	1 719 374 392	.6121366	.6206788	63	1 59 51 69	2 014 698 447	.5387113	.8093884
99	1 43 76 01	1 723 683 599	.6265794	.6236331	64	1 59 76 96	2 019 487 744	.5527777	.8122404
1200	1 44 00 00	1 728 000 000	34.6410162	10.6265857	1265	1 60 02 25	2 024 284 625	35.5668385	10.8150909
1	1 44 24 01	1 732 323 601	.6554469	.6295367	66	1 60 27 56	2 029 089 096	.5808937	.8179400
2	1 44 48 04	1 736 654 408	.6698716	.6324860	67	1 60 52 89	2 033 901 163	.5949434	.8207876
3	1 44 72 09	1 740 992 427	.6842904	.6354338	68	1 60 78 24	2 038 720 832	.6089876	.8236336
4	1 44 96 16	1 745 337 664	.6987031	.6383799	69	1 61 03 61	2 043 548 109	.6230262	.8264782
1205	1 45 20 25	1 749 690 125	34.7131099	10.6413244	1270	1 61 29 00	2 048 383 000	35.6370593	10.8293213
6	1 45 44 36	1 754 049 816	.7275107	.6442672	71	1 61 54 41	2 053 225 511	.6510869	.8321629
7	1 45 68 49	1 758 416 743	.7419055	.6472085	72	1 61 79 84	2 058 075 648	.6651090	.8350030
8	1 45 92 64	1 762 790 912	.7562944	.6501480	73	1 62 05 29	2 062 933 417	.6791255	.8378416
9	1 46 16 81	1 767 172 329	.7706773	.6530860	74	1 62 30 76	2 067 798 824	.6931366	.8406788
1210	1 46 41 00	1 771 561 000	34.7850543	10.6560223	1275	1 62 56 25	2 072 671 875	35.7071421	10.8435144
11	1 46 65 21	1 775 956 931	.7994253	.6589570	76	1 62 81 76	2 077 552 576	.7211422	.8463485
12	1 46 89 44	1 780 360 128	.8137904	.6618902	77	1 63 07 29	2 082 440 933	.7351367	.8491812
13	1 47 13 69	1 784 770 597	.8281495	.6648217	78	1 63 32 84	2 087 336 952	.7491258	.8520125
14	1 47 37 96	1 789 188 344	.8425028	.6677516	79	1 63 58 41	2 092 240 639	.7631095	.8548422
1215	1 47 62 25	1 793 613 375	34.8568501	10.6706799	1280	1 63 84 00	2 097 152 000	35.7770876	10.8576704
16	1 47 86 56	1 798 045 696	.8711915	.6736066	81	1 64 09 61	2 102 071 041	.7910603	.8604972
17	1 48 10 89	1 802 485 313	.8855271	.6765317	82	1 64 35 24	2 106 997 768	.8050276	.8633225
18	1 48 35 24	1 806 932 232	.8998567	.6794552	83	1 64 60 89	2 111 932 187	.8189894	.8661464
19	1 48 59 61	1 811 386 459	.9141805	.6823771	84	1 64 86 56	2 116 874 304	.8329457	.8689687
1220	1 48 84 00	1 815 848 000	34.9284984	10.6852973	1285	1 65 12 25	2 121 824 125	35.8468966	10.8717897
21	1 49 08 41	1 820 316 861	.9428104	.6882160	86	1 65 37 96	2 126 781 656	.8608421	.8746091
22	1 49 32 84	1 824 793 048	.9571166	.6911331	87	1 65 63 69	2 131 746 903	.8747822	.8774271
23	1 49 57 29	1 829 276 567	.9714169	.6940486	88	1 65 89 44	2 136 719 872	.8887169	.8802436
24	1 49 81 76	1 833 767 424	.9857114	.6969625	89	1 66 15 21	2 141 700 569	.9026461	.8830587
1225	1 50 06 25	1 838 265 625	35.0000000	10.6998748	1290	1 66 41 00	2 146 689 000	35.9165699	10.8858723
26	1 50 30 76	1 842 771 176	.0142828	.7027855	91	1 66 66 81	2 151 685 171	.9304884	.8886845
27	1 50 55 29	1 847 284 083	.0285598	.7056947	92	1 66 92 64	2 156 689 088	.9444015	.8914952
28	1 50 79 84	1 851 804 352	.0428309	.7086023	93	1 67 18 49	2 161 700 757	.9583092	.8943044
29	1 51 04 41	1 856 331 989	.0570963	.7115083	94	1 67 44 36	2 166 720 184	.9722115	.8971123
1230	1 51 29 00	1 860 867 000	35.0713558	10.7144127	1295	1 67 70 25	2 171 747 375	35.9861084	10.8999186
31	1 51 53 61	1 865 409 391	.0856096	.7173155	96	1 67 96 16	2 176 782 336	36.0000000	.9027235
32	1 51 78 24	1 869 959 168	.0998575	.7202168	97	1 68 22 09	2 181 825 073	.0138862	.9055269
33	1 52 02 89	1 874 516 337	.1140997	.7231165	98	1 68 48 04	2 186 875 592	.0277671	.9083290
34	1 52 27 56	1 879 080 904	.1283361	.7260146	99	1 68 74 01	2 191 933 899	.0416426	.9111296
1235	1 52 52 25	1 883 652 875	35.1425668	10.7289112	1300	1 69 00 00	2 197 000 000	36.0555128	10.9139287

2.—SQUARES, CUBES, SQUARE ROOTS, CUBE ROOTS, OF NUMBERS 1 TO 1600—Continued.

No.	Square	Cube.	Sq. Rt.	Cu. Rt.	No.	Square	Cube.	Sq. Rt.	Cu. Rt.
1300	1 69 00 00	2 197 000 000	36.0555128	10.9139287	1365	1 86 32 25	2 543 302 125	36.9459064	11.0928775
1	1 69 26 01	2 202 073 901	.0693776	.9167265	66	1 86 59 56	2 548 895 896	.9594372	.0955857
2	1 69 52 04	2 207 155 608	.0832371	.9195228	67	1 86 86 89	2 554 497 863	.9729631	.0982926
3	1 69 78 09	2 212 245 127	.0970913	.9223177	68	1 87 14 24	2 560 108 032	.9864840	.1009982
4	1 70 04 16	2 217 342 464	.1109402	.9251111	69	1 87 41 61	2 565 726 409	37.0000000	.1037025
1305	1 70 30 25	2 222 447 625	36.1247837	10.9279031	1370	1 87 69 00	2 571 353 000	37.0135110	11.1064054
6	1 70 56 36	2 227 560 616	.1386220	.9306937	71	1 87 96 41	2 576 987 811	.0270172	.1091070
7	1 70 82 49	2 232 681 443	.1524550	.9334829	72	1 88 23 84	2 582 630 848	.0405184	.1118073
8	1 71 08 64	2 237 810 112	.1662826	.9362706	73	1 88 51 29	2 588 282 117	.0540146	.1145064
9	1 71 34 81	2 242 946 629	.1801050	.9390569	74	1 88 78 76	2 593 941 624	.0675060	.1172041
1310	1 71 61 00	2 248 091 000	36.1939221	10.9418418	1375	1 89 06 25	2 599 609 375	37.0809924	11.1199004
11	1 71 87 21	2 253 243 231	.2077340	.9446253	76	1 89 33 76	2 605 285 376	.0944740	.1225955
12	1 72 13 44	2 258 403 328	.2215406	.9474074	77	1 89 61 29	2 610 969 633	.1079506	.1252893
13	1 72 39 69	2 263 571 297	.2353419	.9501880	78	1 89 88 84	2 616 662 152	.1214224	.1279817
14	1 72 65 96	2 268 747 144	.2491379	.9529673	79	1 90 16 41	2 622 362 939	.1348893	.1306729
1315	1 72 92 25	2 273 930 875	36.2629287	10.9557451	1380	1 90 44 00	2 628 072 000	37.1483512	11.1333628
16	1 73 18 56	2 279 122 496	.2767143	.9585215	81	1 90 71 61	2 633 789 341	.1618084	.1360514
17	1 73 44 89	2 284 322 013	.2904946	.9612965	82	1 90 99 24	2 639 514 968	.1752606	.1387386
18	1 73 71 24	2 289 529 432	.3042697	.9640701	83	1 91 26 89	2 645 248 887	.1887079	.1414246
19	1 73 97 61	2 294 744 759	.3180396	.9668423	84	1 91 54 56	2 650 991 104	.2021505	.1441093
1320	1 74 24 00	2 299 968 000	36.3318042	10.9696131	1385	1 91 82 25	2 656 741 625	37.2155881	11.1467926
21	1 74 50 41	2 305 199 161	.3455637	.9723825	86	1 92 09 96	2 662 500 456	.2290209	.1494747
22	1 74 76 84	2 310 438 248	.3593179	.9751505	87	1 92 37 69	2 668 267 603	.2424489	.1521555
23	1 75 03 29	2 315 685 267	.3730670	.9779171	88	1 92 65 44	2 674 043 072	.2558720	.1548350
24	1 75 29 76	2 320 940 224	.3868108	.9806823	89	1 92 93 21	2 679 826 869	.2692903	.1575133
1325	1 75 56 25	2 326 203 125	36.4005494	10.9834462	1390	1 93 21 00	2 685 619 000	37.2827037	11.1601903
26	1 75 82 76	2 331 473 976	.4142829	.9862086	91	1 93 48 81	2 691 419 471	.2961124	.1628659
27	1 76 09 29	2 336 752 783	.4280112	.9889696	92	1 93 76 64	2 697 228 288	.3095162	.1655403
28	1 76 35 84	2 342 039 552	.4417343	.9917293	93	1 94 04 49	2 703 045 457	.3229152	.1682134
29	1 76 62 41	2 347 334 289	.4554523	.9944876	94	1 94 32 36	2 708 870 984	.3363094	.1708852
1330	1 76 89 00	2 352 637 000	36.4691650	10.9972445	1395	1 94 60 25	2 714 704 875	37.3496988	11.1735558
31	1 77 15 61	2 357 947 691	.4828727	11.0000000	96	1 94 88 16	2 720 547 136	.3630834	.1762250
32	1 77 42 24	2 363 266 368	.4965752	.0027541	97	1 95 16 09	2 726 397 773	.3764632	.1788930
33	1 77 68 89	2 368 593 037	.5102725	.0055069	98	1 95 44 04	2 732 256 792	.3898382	.1815598
34	1 77 95 56	2 373 927 704	.5239647	.0082583	99	1 95 72 01	2 738 124 199	.4032084	.1842252
1335	1 78 22 25	2 379 270 375	36.5376518	11.0110082	1400	1 96 00 00	2 744 000 000	37.4165738	11.1868894
36	1 78 48 96	2 384 621 056	.5513338	.0137569	1	1 96 28 01	2 749 884 201	.4299345	.1895523
37	1 78 75 69	2 389 979 753	.5650106	.0165041	2	1 96 56 04	2 755 776 808	.4432904	.1922139
38	1 79 02 44	2 395 346 472	.5786823	.0192500	3	1 96 84 09	2 761 677 827	.4566416	.1948743
39	1 79 29 21	2 400 721 219	.5923489	.0219945	4	1 97 12 16	2 767 587 264	.4699880	.1975334
1340	1 79 56 00	2 406 104 000	36.6060104	11.0247377	1405	1 97 40 25	2 773 505 125	37.4833296	11.2001913
41	1 79 82 81	2 411 494 821	.6196668	.0274795	6	1 97 68 36	2 779 431 416	.4966665	.2028479
42	1 80 09 64	2 416 893 688	.6333181	.0302199	7	1 97 96 49	2 785 366 143	.5099987	.2055032
43	1 80 36 49	2 422 300 607	.6469644	.0329590	8	1 98 24 64	2 791 309 312	.5233261	.2081573
44	1 80 63 36	2 427 715 584	.6606056	.0356967	9	1 98 52 81	2 797 260 929	.5366487	.2108101
1345	1 80 90 25	2 433 138 625	36.6742416	11.0384330	1410	1 98 81 00	2 803 221 000	37.5499667	11.2134617
46	1 81 17 16	2 438 569 736	.6878726	.0411680	11	1 99 09 21	2 809 189 531	.5632799	.2161120
47	1 81 44 09	2 444 008 923	.7014986	.0439017	12	1 99 37 44	2 815 166 528	.5765885	.2187611
48	1 81 71 04	2 449 456 192	.7151195	.0466339	13	1 99 65 69	2 821 151 997	.5898922	.2214089
49	1 81 98 01	2 454 911 549	.7287353	.0493649	14	1 99 93 96	2 827 145 944	.6031913	.2240554
1350	1 82 25 00	2 460 375 000	36.7423461	11.0520945	1415	2 00 22 25	2 833 148 375	37.6164857	11.2267007
51	1 82 52 01	2 465 846 551	.7559519	.0548227	16	2 00 50 56	2 839 159 296	.6297754	.2293448
52	1 82 79 04	2 471 326 208	.7695526	.0575497	17	2 00 78 89	2 845 178 713	.6430604	.2319876
53	1 83 06 09	2 476 813 977	.7831483	.0602752	18	2 01 07 24	2 851 206 632	.6563407	.2346292
54	1 83 33 16	2 482 309 864	.7967390	.0629994	19	2 01 35 61	2 857 243 059	.6696164	.2372696
1355	1 83 60 25	2 487 813 875	36.8103246	11.0657222	1420	2 01 64 00	2 863 288 000	37.6828874	11.2399087
56	1 82 87 36	2 493 326 016	.8239053	.0684437	21	2 01 92 41	2 869 341 461	.6961536	.2425465
57	1 84 14 49	2 498 846 293	.8374809	.0711639	22	2 02 20 84	2 875 403 448	.7094153	.2451831
58	1 84 41 64	2 504 374 712	.8510515	.0738828	23	2 02 49 29	2 881 473 967	.7226722	.2478185
59	1 84 68 81	2 509 911 279	.8646172	.0766003	24	2 02 77 76	2 887 553 024	.7359245	.2504527
1360	1 84 96 00	2 515 456 000	36.8781778	11.0793165	1425	2 03 06 25	2 893 640 625	37.7491722	11.2530856
61	1 85 23 21	2 521 008 881	.8917335	.0820314	26	2 03 34 76	2 899 736 776	.7624152	.2557173
62	1 85 50 44	2 526 569 928	.9052842	.0847449	27	2 03 63 29	2 905 841 483	.7756535	.2583478
63	1 85 77 69	2 532 139 147	.9188299	.0874571	28	2 03 91 84	2 911 954 752	.7888873	.2609770
64	1 86 04 96	2 537 716 544	.9323706	.0901679	29	2 04 20 41	2 918 076 589	.8021163	.2636050
1365	1 86 32 25	2 543 302 125	36.9459064	11.0928775	1430	2 04 49 00	2 924 207 000	37.8153408	11.2662318

2. —Squares, Cubes, Square Roots, Cube Roots, of Numbers 1 to 1600—Continued.

No.	Square	Cube.	Sq. Rt.	Cu. Rt.	No.	Square	Cube.	Sq. Rt.	Cu. Rt.
1430	2 04 49 00	2 924 207 000	37.8153408	11.2662318	1495	2 23 50 25	3 341 362 375	38.6652299	11.4344092
31	2 04 77 61	2 930 345 991	.8285606	.2688573	96	2 23 80 16	3 348 071 936	.6781593	.4369581
32	2 05 06 24	2 936 493 568	.8417759	.2714816	97	2 24 10 09	3 354 790 473	.6910843	.4395059
33	2 05 34 89	2 942 649 737	.8549864	.2741047	98	2 24 40 04	3 361 517 992	.7040050	.4420525
34	2 05 63 56	2 948 814 504	.8681924	.2767266	99	2 24 70 01	3 368 254 499	.7169214	.4445980
1435	2 05 92 25	2 954 987 875	37.8813938	11.2793472	1500	2 25 00 00	3 375 000 000	38.7298335	11.4471424
36	2 06 20 96	2 961 169 856	.8945906	.2819666	1	2 25 30 01	3 381 754 501	.7427412	.4496857
37	2 06 49 69	2 967 360 453	.9077828	.2845849	2	2 25 60 04	3 388 518 008	.7556447	.4522278
38	2 06 78 44	2 973 559 672	.9209704	.2872019	3	2 25 90 09	3 395 290 527	.7685439	.4547688
39	2 07 07 21	2 979 767 519	.9341535	.2898177	4	2 26 20 16	3 402 072 064	.7814389	.4573087
1440	2 07 36 00	2 985 984 000	37.9473319	11.2924323	1505	2 26 50 25	3 408 862 625	38.7943294	11.4598474
41	2 07 64 81	2 992 209 121	.9605058	.2950457	6	2 26 80 36	3 415 662 216	.8072158	.4623850
42	2 07 93 64	2 998 442 888	.9736751	.2976579	7	2 27 10 49	3 422 470 843	.8200978	.4649215
43	2 08 22 49	3 004 685 307	.9868398	.3002688	8	2 27 40 64	3 429 288 512	.8329757	.4674568
44	2 08 51 36	3 010 936 384	38.0000000	.3028786	9	2 27 70 81	3 436 115 229	.8458491	.4699911
1445	2 08 80 25	3 017 196 125	38.0131556	11.3054871	1510	2 28 01 00	3 442 951 000	38.8587184	11.4725242
46	2 09 09 16	3 023 464 536	.0263067	.3080945	11	2 28 31 21	3 449 795 831	.8715834	.4750562
47	2 09 38 09	3 029 741 623	.0394532	.3107006	12	2 28 61 44	3 456 649 728	.8844442	.4775871
48	2 09 67 04	3 036 027 392	.0525952	.3133056	13	2 28 91 69	3 463 512 697	.8973006	.4801169
49	2 09 96 01	3 042 321 849	.0657326	.3159094	14	2 29 21 96	3 470 384 744	.9101529	.4826455
1450	2 10 25 00	3 048 625 000	38.0788655	11.3185119	1515	2 29 52 25	3 477 265 875	38.9230009	11.4851731
51	2 10 54 01	3 054 936 851	.0919939	.3211132	16	2 29 82 56	3 484 156 096	.9358447	.4876995
52	2 10 83 04	3 061 257 408	.1051178	.3237134	17	2 30 12 89	3 491 055 413	.9486841	.4902249
53	2 11 12 09	3 067 586 677	.1182371	.3263124	18	2 30 43 24	3 497 963 832	.9615194	.4927491
54	2 11 41 16	3 073 924 664	.1313519	.3289102	19	2 30 73 61	3 504 881 359	.9743505	.4952722
1455	2 11 70 25	3 080 271 375	38.1444622	11.3315067	1520	2 31 04 00	3 511 808 000	38.9871774	11.4977942
56	2 11 99 36	3 086 626 816	.1575681	.3341022	21	2 31 34 41	3 518 743 761	39.0000000	.5003151
57	2 12 28 49	3 092 990 993	.1706693	.3366964	22	2 31 64 84	3 525 688 648	.0128184	.5028348
58	2 12 57 64	3 099 363 912	.1837662	.3392894	23	2 31 95 29	3 532 642 667	.0256326	.5053535
59	2 12 86 81	3 105 745 579	.1968585	.3418813	24	2 32 25 76	3 539 605 824	.0384426	.5078711
1460	2 13 16 00	3 112 136 000	38.2099463	11.3444719	1525	2 32 56 25	3 546 578 125	39.0512483	11.5103876
61	2 13 45 21	3 118 535 181	.2230297	.3470614	26	2 32 86 76	3 553 559 576	.0640499	.5129030
62	2 13 74 44	3 124 943 128	.2361085	.3496497	27	2 33 17 29	3 560 550 183	.0768473	.5154173
63	2 14 03 69	3 131 359 847	.2491829	.3522368	28	2 33 47 84	3 567 549 952	.0896406	.5179305
64	2 14 32 96	3 137 785 344	.2622529	.3548227	29	2 33 78 41	3 574 558 889	.1024296	.5204425
1465	2 14 62 25	3 144 219 625	38.2753184	11.3574075	1530	2 34 09 00	3 581 577 000	39.1152144	11.5229535
66	2 14 91 56	3 150 662 696	.2883794	.3599911	31	2 34 39 61	3 588 604 291	.1279951	.5254634
67	2 15 20 89	3 157 114 563	.3014360	.3625735	32	2 34 70 24	3 595 640 768	.1407716	.5279722
68	2 15 50 24	3 163 575 232	.3144881	.3651547	33	2 35 00 89	3 602 686 437	.1535439	.5304799
69	2 15 79 61	3 170 044 709	.3275358	.3677347	34	2 35 31 56	3 609 741 304	.1663120	.5329865
1470	2 16 09 00	3 176 523 000	38.3405790	11.3703136	1535	2 35 62 25	3 616 805 375	39.1790760	11.5354920
71	2 16 38 41	3 183 010 111	.3536178	.3728914	36	2 35 92 96	3 623 878 656	.1918359	.5379965
72	2 16 67 84	3 189 506 048	.3666522	.3754679	37	2 36 23 69	3 630 961 153	.2045915	.5404998
73	2 16 97 29	3 196 010 817	.3796821	.3780433	38	2 36 54 44	3 638 052 872	.2173431	.5430021
74	2 17 26 76	3 202 524 424	.3927076	.3806175	39	2 36 85 21	3 645 153 819	.2300905	.5455033
1475	2 17 56 25	3 209 046 875	38.4057287	11.3831906	1540	2 37 16 00	3 652 264 000	39.2428337	11.5480034
76	2 17 85 76	3 215 578 176	.4187454	.3857625	41	2 37 46 81	3 659 383 421	.2555728	.5505025
77	2 18 15 29	3 222 118 333	.4317577	.3883332	42	2 37 77 64	3 666 512 088	.2683078	.5530004
78	2 18 44 84	3 228 667 352	.4447656	.3909028	43	2 38 08 49	3 673 650 007	.2810387	.5554973
79	2 18 74 41	3 235 225 239	.4577691	.3934712	44	2 38 39 36	3 680 797 184	.2937654	.5579931
1480	2 19 04 00	3 241 792 000	38.4707681	11.3960384	1545	2 38 70 25	3 687 953 625	39.3064880	11.5604878
81	2 19 33 61	3 248 367 641	.4837627	.3986045	46	2 39 01 16	3 695 119 336	.3192065	.5629815
82	2 19 63 24	3 254 952 168	.4967530	.4011695	47	2 39 32 09	3 702 294 323	.3319208	.5654740
83	2 19 92 89	3 261 545 587	.5097390	.4037332	48	2 39 63 04	3 709 478 592	.3446311	.5679655
84	2 20 22 56	3 268 147 904	.5227206	.4062959	49	2 39 94 01	3 716 672 149	.3573373	.5704559
1485	2 20 52 25	3 274 759 125	38.5356977	11.4088574	1550	2 40 25 00	3 723 875 000	39.3700394	11.5729453
86	2 20 81 96	3 281 379 256	.5486705	.4114177	51	2 40 56 01	3 731 087 151	.3827373	.5754336
87	2 21 11 69	3 288 008 303	.5616389	.4139769	52	2 40 87 04	3 738 308 608	.3954312	.5779208
88	2 21 41 44	3 294 646 272	.5746030	.4165349	53	2 41 18 09	3 745 539 377	.4081210	.5804069
89	2 21 71 21	3 301 293 169	.5875627	.4190918	54	2 41 49 16	3 752 779 464	.4208067	.5828919
1490	2 22 01 00	3 307 949 000	38.6005181	11.4216476	1555	2 41 80 25	3 760 028 875	39.4334883	11.5853759
91	2 22 30 81	3 314 613 771	.6134691	.4242022	56	2 42 11 36	3 767 287 616	.4461658	.5878588
92	2 22 60 64	3 321 287 488	.6264158	.4267556	57	2 42 42 49	3 774 555 693	.4588393	.5903407
93	2 22 90 49	3 327 970 157	.6393582	.4293079	58	2 42 73 64	3 781 833 112	.4715087	.5928215
94	2 23 20 36	3 334 661 784	.6522962	.4318591	59	2 43 04 81	3 789 119 879	.4841740	.5953013
1495	2 23 50 25	3 341 362 375	38.6652299	11.4344092	1560	2 43 36 06	3 796 416 000	39.4968353	11.5977799

2.—Squares, Cubes, Square Roots, Cube Roots, of Numbers 1 to 1600—Concluded.

No.	Square	Cube.	Sq. Rt	Cu. Rt	No.	Square	Cube.	Sq. Rt.	Cu. Rt.
1560	2 43 36 00	3 796 416 000	39.4968353	11.5977799	1580	2 49 64 00	3 944 312 000	39.7492138	11.6471329
61	2 43 67 21	3 803 721 481	.5094925	.6002576	81	2 49 95 61	3 951 805 941	.7617907	.6495895
62	2 43 98 44	3 811 036 328	.5221457	.6027342	82	2 50 27 24	3 959 309 368	.7743636	.6520452
63	2 44 29 69	3 818 360 547	.5347948	.6052097	83	2 50 58 89	3 966 822 287	.7869325	.6544998
64	2 44 60 96	3 825 694 144	.5474399	.6076841	84	2 50 90 56	3 974 344 704	.7994975	.6569534
1565	2 44 92 25	3 833 037 125	39.5600809	11.6101575	1585	2 51 22 25	3 981 876 625	39.8120585	11.6594059
66	2 45 23 56	3 840 389 496	.5727179	.6126299	86	2 51 53 96	3 989 418 056	.8246155	.6618574
67	2 45 54 89	3 847 751 263	.5853508	.6151012	87	2 51 85 69	3 996 969 003	.8371686	.6643079
68	2 45 86 24	3 855 122 432	.5979797	.6175715	88	2 52 17 44	4 004 529 472	.8497177	.6667574
69	2 46 17 61	3 862 503 009	.6106046	.6200407	89	2 52 49 21	4 012 099 469	.8622628	.6692058
1570	2 46 49 00	3 869 893 000	39.6232255	11.6225088	1590	2 52 81 00	4 019 679 000	39.8748040	11.6716532
71	2 46 80 41	3 877 292 411	.6358424	.6249759	91	2 53 12 81	4 027 268 071	.8873413	.6740996
72	2 47 11 84	3 884 701 248	.6484552	.6274420	92	2 53 44 64	4 034 866 688	.8998747	.6765449
73	2 47 43 29	3 892 119 517	.6610640	.6299070	93	2 53 76 49	4 042 474 857	.9124041	.6789892
74	2 47 74 76	3 899 547 224	.6736688	.6323710	94	2 54 08 36	4 050 092 584	.9249295	.6814325
1575	2 48 06 25	3 906 984 375	39.6862696	11.6348339	1595	2 54 40 25	4 057 719 875	39.9374511	11.6838748
76	2 48 37 76	3 914 430 976	.6988665	.6372957	96	2 54 72 16	4 065 356 736	.9499687	.6863161
77	2 48 69 29	3 921 887 033	.7114593	.6397566	97	2 55 04 09	4 073 003 173	.9624824	.6887563
78	2 49 00 84	3 929 352 552	.7240481	.6422164	98	2 55 36 04	4 080 659 192	.9749922	.6911955
79	2 49 32 41	3 936 827 539	.7366329	.6446751	99	2 55 68 01	4 088 324 799	.9874980	.6936337
1580	2 49 64 00	3 944 312 000	39.7492138	11.6471329	1600	2 56 00 00	4 096 000 000	40.0000000	11.6960709

2a.—Squares of Numbers 1600 to 1810.

No.	Square.	No.	Square.	No.	Square.	No.	Square.	No.	Square.	No.	Square.
1600	2560000	1635	2673225	1670	2788900	1705	2907025	1740	3027600	1775	3150625
01	2563201	36	2676496	71	2792241	06	2910436	41	3031081	76	3154176
02	2566404	37	2679769	72	2795584	07	2913849	42	3034564	77	3157729
03	2569609	38	2683044	73	2798929	08	2917264	43	3038049	78	3161284
04	2572816	39	2686321	74	2802276	09	2920681	44	3041536	79	3164841
1605	2576025	1640	2689600	1675	2805625	1710	2924100	1745	3045025	1780	3168400
06	2579236	41	2692881	76	2808976	11	2927521	46	3048516	81	3171961
07	2582449	42	2696164	77	2812329	12	2930944	47	3052009	82	3175524
08	2585664	43	2699449	78	2815684	13	2934369	48	3055504	83	3179089
09	2588881	44	2702736	79	2819041	14	2937796	49	3059001	84	3182656
1610	2592100	1645	2706025	1680	2822400	1715	2941225	1750	3062500	1785	3186225
11	2595321	46	2709316	81	2825761	16	2944656	51	3066001	86	3189796
12	2598544	47	2712609	82	2829124	17	2948089	52	3069504	87	3193369
13	2601769	48	2715904	83	2832489	18	2951524	53	3073009	88	3196944
14	2604996	49	2719201	84	2835856	19	2954961	54	3076516	89	3200521
1615	2608285	1650	2722500	1685	2839225	1720	2958400	1755	3080025	1790	3204100
16	2611456	51	2725801	86	2842596	21	2961841	56	3083536	91	3207681
17	2614689	52	2729104	87	2845969	22	2965284	57	3087049	92	3211264
18	2617924	53	2732409	88	2849344	23	2968729	58	3090564	93	3214849
19	2621161	54	2735716	89	2852721	24	2972176	59	3094081	94	3218436
1620	2624400	1655	2739025	1690	2856100	1725	2975625	1760	3097600	1795	3222025
21	2627641	56	2742336	91	2859481	26	2979076	61	3101121	96	3225616
22	2630884	57	2745649	92	2862864	27	2982529	62	3104644	97	3229209
23	2634129	58	2748964	93	2866249	28	2985984	63	3108169	98	3232804
24	2637376	59	2752281	94	2869636	29	2989441	64	3111696	99	3236401
1625	2640625	1660	2755600	1695	2873025	1730	2992900	1765	3115225	1800	3240000
26	2643876	61	2758921	96	2876416	31	2996361	66	3118756	01	3243601
27	2647129	62	2762244	97	2879809	32	2999824	67	3122289	02	3247204
28	2650384	63	2765569	98	2883204	33	3003289	68	3125824	03	3250809
29	2653641	64	2768896	99	2886601	34	3006756	69	3129361	04	3254416
1630	2656900	1665	2772225	1700	2890000	1735	3010225	1770	3132900	1805	3258025
31	2660161	66	2775556	01	2893401	36	3013696	71	3136441	06	3261636
32	2663424	67	2778889	02	2896804	37	3017169	72	3139984	07	3265249
33	2666689	68	2782224	03	2900209	38	3020644	73	3143529	08	3268864
34	2669956	69	2785561	04	2903616	39	3024121	74	3147076	09	3272481
1635	2673225	1670	2788900	1705	2907025	1740	3027600	1775	3150625	1810	3276100

2b.—Square Roots and Cube Roots of Numbers 1600 to 1860.

No.	Sq. Rt.	Cu. Rt.	No.	Sq. Rt.	Cu. Rt.	No.	Sq. Rt.	Cu. Rt.	No.	Sq. Rt.	Cu. Rt.
1600	40.0000	11.6961	1665	40.8044	11.8524	1730	41.5933	12.0046	1795	42.3674	12.1531
1	.0125	.6985	66	.8167	.8547	31	.6053	.0069	96	.3792	.1554
2	.0250	.7009	67	.8289	.8571	32	.6173	.0093	97	.3910	.1576
3	.0375	.7034	68	.8412	.8595	33	.6293	.0116	98	.4028	.1599
4	.0500	.7058	69	.8534	.8618	34	.6413	.0139	99	.4146	.1622
1605	40.0625	11.7082	1670	40.8656	11.8642	1735	41.6533	12.0162	1800	42.4264	12.1644
6	.0749	.7107	71	.8779	.8666	36	.6653	.0185	1	.4382	.1667
7	.0874	.7131	72	.8901	.8689	37	.6773	.0208	2	.4500	.1689
8	.0999	.7155	73	.9023	.8713	38	.6893	.0231	3	.4617	.1712
9	.1123	.7180	74	.9145	.8737	39	.7013	.0254	4	.4735	.1734
1610	40.1248	11.7204	1675	40.9268	11.8760	1740	41.7133	12.0277	1805	42.4853	12.1757
11	.1373	.7228	76	.9390	.8784	41	.7253	.0300	6	.4971	.1779
12	.1497	.7252	77	.9512	.8808	42	.7373	.0323	7	.5088	.1802
13	.1622	.7277	78	.9634	.8831	43	.7493	.0346	8	.5206	.1824
14	.1746	.7301	79	.9756	.8855	44	.7612	.0369	9	.5323	.1846
1615	40.1871	11.7325	1680	40.9878	11.8878	1745	41.7732	12.0392	1810	42.5441	12.1869
16	.1995	.7350	81	41.0000	.8902	46	.7852	.0415	11	.5558	.1891
17	.2119	.7373	82	.0122	.8926	47	.7971	.0438	12	.5676	.1914
18	.2244	.7398	83	.0244	.8949	48	.8091	.0461	13	.5793	.1936
19	.2368	.7422	84	.0366	.8973	49	.8210	.0484	14	.5911	.1959
1620	40.2492	11.7446	1685	41.0488	11.8996	1750	41.8330	12.0507	1815	42.6028	12.1981
21	.2616	.7470	86	.0609	.9020	51	.8450	.0530	16	.6146	.2003
22	.2741	.7494	87	.0731	.9043	52	.8569	.0553	17	.6263	.2026
23	.2865	.7518	88	.0853	.9067	53	.8688	.0576	18	.6380	.2048
24	.2989	.7543	89	.0974	.9090	54	.8808	.0599	19	.6497	.2071
1625	40.3113	11.7567	1690	41.1096	11.9114	1755	41.8927	12.0622	1820	42.6615	12.2093
26	.3237	.7591	91	.1218	.9137	56	.9047	.0645	21	.6732	.2115
27	.3361	.7615	92	.1339	.9161	57	.9166	.0668	22	.6849	.2138
28	.3485	.7639	93	.1461	.9184	58	.9285	.0690	23	.6966	.2160
29	.3609	.7663	94	.1582	.9208	59	.9404	.0713	24	.7083	.2182
1630	40.3733	11.7687	1695	41.1704	11.9231	1760	41.9524	12.0736	1825	42.7200	12.2205
31	.3856	.7711	96	.1825	.9255	61	.9643	.0759	26	.7317	.2227
32	.3980	.7735	97	.1947	.9278	62	.9762	.0782	27	.7434	.2249
33	.4104	.7759	98	.2068	.9301	63	.9881	.0805	28	.7551	.2272
34	.4228	.7783	99	.2189	.9325	64	42.0000	.0828	29	.7668	.2294
1635	40.4351	11.7807	1700	41.2311	11.9348	1765	42.0119	12.0850	1830	42.7785	12.2316
36	.4475	.7831	1	.2432	.9372	66	.0238	.0873	31	.7902	.2338
37	.4599	.7855	2	.2553	.9395	67	.0357	.0896	32	.8019	.2361
38	.4722	.7879	3	.2674	9418	68	.0476	.0919	33	.8135	.2383
39	.4846	.7903	4	.2795	.9442	69	.0595	.0942	34	.8252	.2405
1640	40.4969	11.7927	1705	41.2916	11.9465	1770	42.0714	12.0964	1835	42.8369	12.2427
41	.5093	.7951	6	.3038	.9489	71	.0833	.0987	36	.8486	.2450
42	.5216	.7975	7	.3159	.9512	72	.0951	.1010	37	.8602	.2472
43	.5339	.7999	8	.3280	.9535	73	.1070	.1033	38	.8719	.2494
44	.5463	.8023	9	.3401	.9559	74	.1189	.1056	39	.8836	.2516
1645	40.5586	11.8047	1710	41.3521	11.9582	1775	42.1307	12.1078	1840	42.8952	12.2539
46	.5709	.8071	11	.3642	.9605	76	.1426	.1101	41	.9069	.2561
47	.5832	.8095	12	.3763	.9628	77	.1545	.1124	42	.9185	.2583
48	.5956	.8119	13	.3884	.9652	78	.1663	.1146	43	.9302	.2605
49	.6079	.8143	14	.4005	.9675	79	.1782	.1169	44	.9418	.2627
1650	40.6202	11.8167	1715	41.4126	11.9698	1780	42.1900	12.1192	1845	42.9535	12.2649
51	.6325	.8190	16	.4246	.9722	81	.2019	.1215	46	.9651	.2672
52	.6448	.8214	17	.4367	.9745	82	.2137	.1237	47	.9767	.2694
53	.6571	.8238	18	.4488	.9768	83	.2256	.1260	48	.9884	.2716
54	.6694	.8262	19	.4608	.9791	84	.2374	.1283	49	43.0000	.2738
1655	40.6817	11.8286	1720	41.4729	11.9815	1785	42.2493	12.1305	1850	43.0116	12.2760
56	.6940	.8310	21	.4849	.9838	86	.2611	.1328	51	.0232	.2782
57	.7063	.8333	22	.4970	.9861	87	.2729	.1350	52	.0349	.2804
58	.7185	.8357	23	.5090	.9884	88	.2847	.1373	53	.0465	.2826
59	.7308	.8381	24	.5211	.9907	89	.2966	.1396	54	.0581	.2849
1660	40.7431.	11.8405	1725	41.5331	11.9931	1790	42.3084	12.1418	1855	43.0697	12.2871
61	.7554	.8429	26	.5452	.9954	91	.3202	.1441	56	.0813	.2893
62	.7676	.8452	27	.5572	.9977	92	.3320	.1464	57	.0929	.2915
63	.7799	.8476	28	.5692	12.0000	93	.3438	.1486	58	.1045	.2937
64	7922	.8500	29	.5812	.0023	94	.3556	.1509	59	.1161	.2959
1665	40.8044	11.8524	1730	41.5933	12.0046	1795	42.3674	12.1531	1860	43.1277	12.2981

2c.—Squares of Mixed Numbers from $\frac{1}{64}$ to 12, by 64ths

I. Squares of Mixed Numbers from $\frac{1}{64}$ to 6.

	0	1	2	3	4	5
1/64	0.00024	1.03149	4.06274	9.09399	16.12524	25.15649
1/32	0.00098	1.06348	4.12598	9.18848	16.25098	25.31348
3/64	0.00220	1.09595	4.18970	9.28345	16.37720	25.47095
1/16	0.00391	1.12891	4.25391	9.37891	16.50391	25.62891
5/64	0.00610	1.16235	4.31860	9.47485	16.63110	25.78735
3/32	0.00879	1.19629	4.38379	9.57129	16.75879	25.94629
7/64	0.01196	1.23071	4.44946	9.66821	16.88696	26.10571
1/8	0.01562	1.26562	4.51562	9.76562	17.01562	26.26562
9/64	0.01978	1.30103	4.58228	9.86353	17.14478	26.42603
5/32	0.02441	1.33691	4.64941	9.96191	17.27441	26.58691
11/64	0.02954	1.37329	4.71704	10.06079	17.40454	26.74829
3/16	0.03516	1.41016	4.78516	10.16016	17.53516	26.91016
13/64	0.04126	1.44751	4.85376	10.26001	17.66626	27.07251
7/32	0.04785	1.48535	4.92285	10.36035	17.79785	27.23535
15/64	0.05493	1.52368	4.99243	10.46118	17.92993	27.39868
1/4	0.06250	1.56250	5.06250	10.56250	18.06250	27.56250
17/64	0.07056	1.60181	5.13306	10.66431	18.19556	27.72681
9/32	0.07910	1.64160	5.20410	10.76660	18.32910	27.89160
19/64	0.08813	1.68188	5.27563	10.86938	18.46313	28.05688
5/16	0.09766	1.72266	5.34766	10.97266	18.59766	28.22266
21/64	0.10767	1.76392	5.42017	11.07642	18.73267	28.38892
11/32	0.11816	1.80566	5.49316	11.18066	18.86816	28.55566
23/64	0.12915	1.84790	5.56663	11.28540	19.00415	28.72290
3/8	0.14062	1.89062	5.64062	11.39062	19.14062	28.89062
25/64	0.15259	1.93384	5.71509	11.49634	19.27759	29.05884
13/32	0.16504	1.97754	5.79004	11.60254	19.41504	29.22754
27/64	0.17798	2.02173	5.86548	11.70923	19.55298	29.39673
7/16	0.19141	2.06641	5.94141	11.81641	19.69141	29.56641
29/64	0.20532	2.11157	6.01782	11.92407	19.83032	29.73657
15/32	0.21973	2.15723	6.09473	12.03223	19.96973	29.90723
31/64	0.23462	2.20337	6.17212	12.14087	20.10962	30.07837
1/2	0.25000	2.25000	6.25000	12.25000	20.25000	30.25000
33/64	0.26587	2.29712	6.32837	12.35962	20.39087	30.42212
17/32	0.28223	2.34473	6.40723	12.46973	20.53223	30.59473
35/64	0.29907	2.39282	6.48657	12.58032	20.67407	30.76782
9/16	0.31641	2.44141	6.56641	12.69141	20.81641	30.94141
37/64	0.33423	2.49048	6.64673	12.80298	20.95923	31.11548
19/32	0.35254	2.54004	6.72754	12.91504	21.10254	31.29004
39/64	0.37134	2.59009	6.80884	13.02759	21.24634	31.46509
5/8	0.39062	2.64062	6.89062	13.14062	21.39062	31.64062
41/64	0.41040	2.69165	6.97290	13.25415	21.53540	31.81665
21/32	0.43066	2.74316	7.05566	13.36816	21.68066	31.99316

2c.—SQUARES OF MIXED NUMBERS FROM $\frac{1}{64}$ TO 6—*Continued.*

	0	1	2	3	4	5
43/64	0.45142	2.79517	7.13892	13.48267	21.82642	32.17017
11/16	0.47266	2.84766	7.22266	13.59766	21.97266	32.34766
45/64	0.49438	2.90063	7.30688	13.71313	22.11938	32.52563
23/32	0.51660	2.95410	7.39160	13.82910	22.26660	32.70410
47/64	0.53931	3.00806	7.47681	13.94556	22.41431	32.88306
3/4	0.56250	3.06250	7.56250	14.06250	22.56250	33.06250
49/64	0.58618	3.11743	7.64868	14.17993	22.71118	33.24243
25/32	0.61035	3.17285	7.73535	14.29785	22.86035	33.42285
51/64	0.63501	3.22876	7.82251	14.41626	23.01001	33.60376
13/16	0.66016	3.28516	7.91016	14.53516	23.16016	33.78516
53/64	0.68579	3.34204	7.99829	14.65454	23.31079	33.96704
27/32	0.71191	3.39941	8.08691	14.77441	23.46191	34.14941
55/64	0.73853	3.45728	8.17603	14.89478	23.61363	34.33228
7/8	0.76562	3.51562	8.26562	15.01562	23.76562	34.51562
57/64	0.79321	3.57446	8.35571	15.13696	23.91821	34.69946
29/32	0.82129	3.63379	8.44629	15.25879	24.07129	34.88379
59/64	0.84985	3.69360	8.53735	15.38110	24.22485	35.06860
15/16	0.87891	3.75391	8.62891	15.50391	24.37891	35.25391
61/64	0.90845	3.81470	8.72095	15.62720	24.53345	35.43970
31/32	0.93848	3.87598	8.81348	15.75098	24.68848	35.62598
63/64	0.96899	3.93774	8.90649	15.87524	24.84399	35.81274

2d.—II. SQUARES OF MIXED NUMBERS FROM $6\frac{1}{64}$ TO 12

	6	7	8	9	10	11
1/64	36.18774	49.21899	64.25024	81.28149	100.31274	121.34399
1/32	36.37598	49.43848	64.50098	81.56348	100.62598	121.68848
3/64	36.56470	49.65845	64.75220	81.84595	100.93970	122.03345
1/16	36.75391	49.87891	65.00391	82.12891	101.25391	122.37891
5/64	36.94360	50.09985	65.25610	82.41235	101.56860	122.72485
3/32	37.13379	50.32129	65.50879	82.69629	101.88379	123.07129
7/64	37.32446	50.54321	65.76196	82.98071	102.19946	123.41821
1/8	37.51562	50.76562	66.01562	83.26562	102.51562	123.76562
9/64	37.70728	50.98853	66.26978	83.55103	102.83228	124.11353
5/32	37.89941	51.21191	66.52441	83.83691	103.14941	124.46191
11/64	38.09204	51.43579	66.77954	84.12329	103.46704	124.81079
3/16	38.28516	51.66016	67.03516	84.41016	103.78516	125.16016
13/64	38.47876	51.88501	67.29126	84.69751	104.10376	125.51001
7/32	38.67285	52.11035	67.54785	84.98535	104.42285	125.86035
15/64	38.86743	52.33618	67.80493	85.27368	104.74243	126.21110
1/4	39.06250	52.56250	68.06250	85.56250	105.06250	126.56250

2*d*.—Squares of Mixed Numbers from $6\frac{1}{64}$ to 12—*Continued*

	6	7	8	9	10	11
17/64	39.25806	52.78931	68.32056	85.85181	105.38306	126.91431
9/32	39.45410	53.01660	68.57910	86.14160	105.70410	127.26660
19/64	39.65063	53.24438	68.83813	86.43188	106.02563	127.61938
5/16	39.84766	53.47266	69.09766	86.72266	106.34766	127.97266
21/64	40.04517	53.70142	69.35767	87.01392	106.67017	128.32642
11/32	40.24316	53.93066	69.61816	87.30566	106.99316	128.68066
23/64	40.44165	54.16040	69.87915	87.59790	107.31665	129.03540
3/8	40.64062	54.39062	70.14062	87.89062	107.64062	129.39062
25/64	40.84009	54.62134	70.40259	88.18384	107.96509	129.74634
13/32	41.04004	54.85254	70.66504	88.47754	108.29004	130.10254
27/64	41.24048	55.08423	70.92798	88.77173	108.61548	130.45923
7/16	41.44141	55.31641	71.19141	89.06641	108.94141	130.81641
29/64	41.64282	55.54907	71.45532	89.36157	109.26782	131.17407
15/32	41.84473	55.78223	71.71973	89.65723	109.59473	131.53223
31/64	42.04712	56.01587	71.98462	89.95337	109.92212	131.89087
1/2	42.25000	56.25000	72.25000	90.25000	110.25000	132.25000
33/64	42.45337	56.48462	72.51587	90.54712	110.57837	132.60962
17/32	42.65723	56.71973	72.78223	90.84473	110.90723	132.96973
35/64	42.86157	56.95532	73.04907	91.14282	111.23657	133.33032
9/16	43.06641	57.19141	73.31641	91.44141	111.56641	133.69141
37/64	43.27173	57.42798	73.58423	91.74048	111.89673	134.05298
19/32	43.47754	57.66504	73.85254	92.04004	112.22754	134.41504
39/64	43.68384	57.90259	74.12134	92.34009	112.55884	134.77759
5/8	43.89062	58.14062	74.39062	92.64062	112.89062	135.14062
41/64	44.09790	58.37915	74.66040	92.94165	113.22290	135.50415
21/32	44.30566	58.61816	74.93066	93.24316	113.55566	135.86816
43/64	44.51392	58.85767	75.20142	93.54517	113.88892	136.23267
11/16	44.72266	59.09766	75.47266	93.84766	114.22266	136.59766
45/64	44.93188	59.33813	75.74438	94.15063	114.55688	136.96313
23/32	45.14160	59.57910	76.01660	94.45410	114.89160	137.32910
47/64	45.35181	59.82056	76.28931	94.75806	115.22681	137.69556
3/4	45.56250	60.06250	76.56250	95.06250	115.56250	138.06250
49/64	45.77368	60.30493	76.83618	95.36743	115.89868	138.42993
25/32	45.98535	60.54785	77.11035	95.67285	116.23535	138.79785
51/64	46.19751	60.79126	77.38501	95.97876	116.57251	139.16626
13/16	46.41016	61.03516	77.66016	96.28516	116.91016	139.53516
53/64	46.62329	61.27954	77.93579	96.59204	117.24829	139.90454
27/32	46.83691	61.52441	78.21191	96.89941	117.58691	140.27441
55/64	47.05103	61.76978	78.48853	97.20728	117.92603	140.64478
7/8	47.26562	62.01562	78.76562	97.51562	118.26562	141.01562
57/64	47.48071	62.26196	79.04321	97.82446	118.60571	141.38696
29/32	47.69629	62.50879	79.32129	98.13379	118.94629	141.75879
59/64	47.91235	62.75610	79.59985	98.44360	119.28735	142.13110
15/16	48.12891	63.00391	79.87891	98.75391	119.62891	142.50391
61/64	48.34595	63.25220	80.15845	99.06470	119.97095	142.87720
31/32	48.56348	63.50098	80.43848	99.37598	120.31348	143.25098
63/64	48.78149	63.75024	80.71899	99.68774	120.65649	143.62524

frequently most convenient to change the fraction to a decimal first and then extract the square root.

Ratio and Proportion.—The *ratio* of two numbers is the relation which the value of the first bears to the value of the second and this relation is indicated by the sign (:). Thus, 3 : 4 is the ratio of 3 to 4. Ratio is equivalent to the fraction obtained by dividing the first number by the second. Thus, $\frac{3}{4}$ also expresses the ratio of 3 to 4.

An expression consisting of two equal ratios is called a *proportion.* It is written, 3 : 4 = 9 : 12, and read, "3 is to 4 as 9 is to 12." The first and last, or the "end," numbers are called the extremes and the second and third, or the middle, numbers are called the *means.* Since a ratio may also be expressed as a fraction, then a proportion may also be set upon, $\frac{1}{8} = \frac{9}{12}$.

Illustration: If the diameter of a gear is 13.53 inches and the circumference is 42.5 inches, find the ratio of the diameter to the circumference.

$$\frac{13.53}{42.5} = 0.3183$$

Therefore, the ratio of the diameter to the circumference is 0.3183. The above value is the same as that obtained by dividing 1 by 3.1416; that is, in any circle the ratio of the diameter to the circumference is $1 \div \pi$. Thus, it is evident that ratio is always the quotient obtained by dividing the first number by the second.

Proportion is one of the most useful tools in mathematical calculation. It is the *key* to many of its operations. Indeed, practically all mathematical problems may be expressed in proportion.

Rules of Proportion.—Proportion derives its great usefulness from the fundamental rule which states that *the product of the means equals the product of the extremes.* Thus, in the proportion 3 : 4 = 9 : 12, according to the rule, 4×9 (the product of the means) $= 3 \times 12$ (the product of the extremes) = 36. Then, when three terms of a proportion are known, the fourth can be found. For example, if it takes twenty days to build five lathes, how long will it take to build fifteen lathes at the same rate?

$$x : 20 = 15 : 5$$

whence
$$x = \frac{20 \times \overset{3}{\cancel{15}}}{\cancel{5}} = 60 \text{ days (Ans.)}$$

Where one extreme and both means are known, to find the other extreme, divide the product of the means by the known extreme.

Where both extremes and one mean are known, to find the other mean, divide the product of the extremes by the known mean.

For the purpose of illustrating these rules, replace the figures in a proportion by the letters A, B, C, D, and write $A : B = C : D$; then

$$A \times D = B \times C, \frac{A}{B} = \frac{C}{D}, \quad A = \frac{B \times C}{D},$$

$$D = \frac{B \times C}{A}, \quad B = \frac{A \times D}{C}, \quad C = \frac{A \times D}{B}.$$

Triangles may be used advantageously in illustrating ratio and proportion. Thus, let us say, if a train travels 260 miles in 8 hours, how far will it travel in 5 hours? Draw a triangle letting the base represent the distance (260 mi.) and a leg the time 8 hours. Then draw another leg parallel to the first and of a length in proportion to the first as 5 is to 8. Then the distance x represents the distance which the train will travel in 5 hours because from similar triangles, and

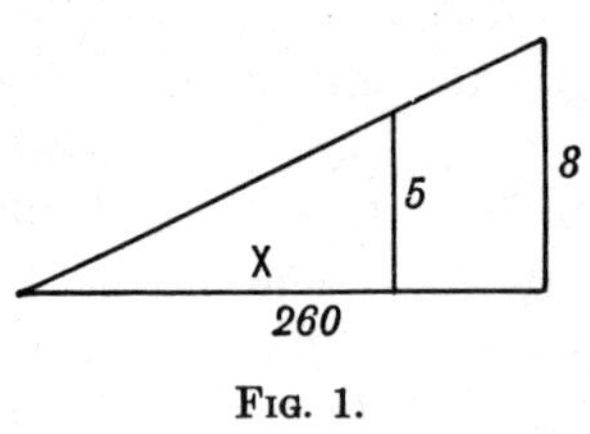

Fig. 1.

$$x : 260 = 5 : 8$$

whence
$$x = \frac{5 \times 260}{8} = 162.5 \text{ miles (Ans.)}$$

Inverse Proportion.—In the preceding problems the ratio of the elements of one figure was equal to the ratio of the corresponding elements of the other figure, that is, directly proportional. When the ratio is equal to the inverse of that ratio the elements are said to be inversely proportional.

The speed of pulleys connected by belts are inversely proportional to their diameters, i.e., the smaller pulley rotates faster than the larger pulley.

ILLUSTRATION: A 24-inch pulley fixed to a line shaft which makes 400 revolutions per minute (R.P.M.) is belted to a 6-inch pulley. Find the number of R.P.M. of the smaller pulley.

R.P.M. of Driven Pulley		R.P.M. of Driving Pulley		Diameter of Driving Pulley		Diameter of Driven Pulley
x	:	400	=	24	:	6

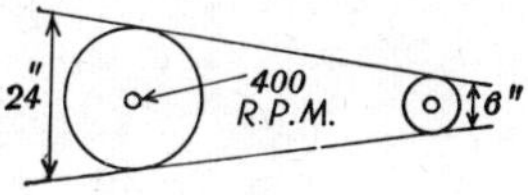

FIG. 2.

$$\text{whence } x = \frac{400 \times \overset{4}{\cancel{24}}}{\cancel{6}} = 1600 \text{ R.P.M. (Ans.)}$$

Likewise, the speeds of gears running together are inversely proportional to their number of teeth.

ILLUSTRATION: A driving gear with 48 teeth meshes with a driven gear with 16 teeth. If the driving gear makes 100 R.P.M. find the number of R.P.M. made by the driven gear.

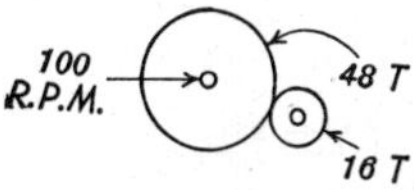

FIG. 3.

R.P.M. of Driven Gear		R.P.M. of Driving Gear		No. of Teeth of Driving Gear		No. of Teeth on Driven Gear
x	:	100	=	48	:	16

$$\text{whence} \qquad x = \frac{100 \times \overset{3}{\cancel{48}}}{\cancel{16}} = 300 \text{ R.P.M. (Ans.)}$$

Pulley Train.—A pulley train is a series of pulleys connected by belting, the power coming from one of the pulleys.

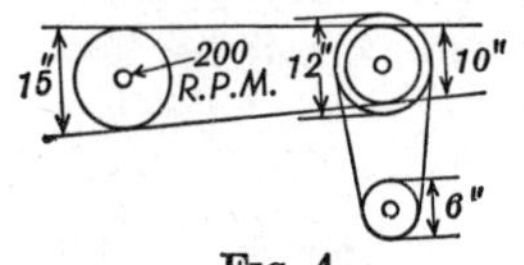

Fig. 4.

Illustration: In the sketch at the left, find the R.P.M. of the 6-inch pulley.

R.P.M. of Last Driven Pulley	:	R.P.M. of First Driving Pulley	=	Product of Diameters of All Driving Pulleys	:	Product of Diameters of All Driven Pulleys
x	:	200	=	(15×12)	:	(10×6)

$$\text{whence} \qquad x = \frac{\overset{20}{\cancel{200}} \times 15 \times \overset{2}{\cancel{12}}}{\cancel{10} \times \cancel{6}} = 600 \text{ R.P.M.} \quad \text{(Ans.)}$$

Gear Train.—A gear train is a series of gears running together.

Illustration: In the sketch at the right find the R.P.M. of the 36 T. gear.

75 R.P.M.
72 T
64 T
24 T
36 T

Fig. 5.

R.P.M. of Last Driven Gear	:	R.P.M. of First Driving Gear	=	Product of Number of Teeth of Driving Gears	:	Product of Number of Teeth of Driven Gears
x	:	75	=	(72×64)	:	(24×36)

$$\text{whence} \qquad x - \frac{\overset{25}{\cancel{75}} \times \overset{2}{\cancel{72}} \times \overset{8}{\cancel{64}}}{\cancel{24} \times \underset{3}{\cancel{36}}} = 400 \text{ R.P.M.} \quad \text{(Ans.)}$$

Inverse proportion can be used to solve other types of problems. For instance in manufacturing plants the time per week is in an inverse proportion to the number of men employed; the shorter the time, the more men.

ILLUSTRATION: A factory employing 300 men completes a given number of vacuum cleaners weekly, the number of working hours being 40 per week. How many men would be required for the same production if the working hours were reduced to 30 per week?

$$x : 300 = 40 : 30$$

whence

$$x = \frac{\overset{10}{\cancel{300}} \times 40}{\cancel{30}} = 400$$

therefore, 400 men would be needed for the same production.

Compound Proportion.—A compound proportion is a proportion which has one of its ratios a compound ratio, that is, a ratio expressed by a fraction that is the product of fractions representing given ratios. Thus, the ratios 3 : 4 and 5 : 7 are represented by the fractions $\frac{3}{4}$ and $\frac{5}{7}$; and the ratio 15 : 28 which is represented by $\frac{15}{28}$, the product of $\frac{3}{4}$ and $\frac{5}{7}$, is said to be compounded of the ratios 3 : 4 and 5 : 7.

Problems in compound proportion are solved by the cause and effect method which is based on the following principle. Like causes produce like effects; and the ratio between any two causes equals the ratio between the effects produced.

ILLUSTRATION: If a mechanic who machines 70 pieces in an 8-hour day is paid \$1.47 per hour, find how much a man ought to be paid who machines 80 similar pieces in a 7-hour day if paid in the same proportion.

Make up a table with four columns headed "First Cause," "First Effect," "Second Cause," "Second Effect," and place under each the respective factors given in the problem. In the example above, the table would be as follows:

First Cause	First Effect	Second Cause	Second Effect
1 man 8 hours 147 cents	70 pieces	1 man 7 hours x cents	80 pieces

whence $$(1 \times 8 \times 147) : 70 = (1 \times 7 \times x) : 80$$

and $$x = \frac{1 \times 8 \times \overset{\overset{3}{\cancel{21}}}{\cancel{147}} \times 8\cancel{0}}{7\cancel{0} \times 1 \times \cancel{7}} = 192 \text{ cents} = \$1.92$$

Therefore, the second operator should receive $1.92 an hour.

Reciprocals.—The use of reciprocals facilitates computations in long division particularly when many different dividends are to be divided by the same divisor.

ILLUSTRATION: 7246 ÷ 1572.

From the table on page 58 find the reciprocal of 1572. *Ans.*: 0.000636132.

TABLE 3

3.—Reciprocals, 1 to 200

No.	Reciprocal	No.	Reciprocal	No.	Reciprocal	No.	Reciprocal
1	1.0000000	51	0.0196078	101	0.0099010	151	0.0066225
2	0.5000000	52	0.0192308	102	0.0098039	152	0.0065789
3	0.3333333	53	0.0188679	103	0.0097087	153	0.0065359
4	0.2500000	54	0.0185185	104	0.0096154	154	0.0064935
5	0.2000000	55	0.0181818	105	0.0095238	155	0.0064516
6	0.1666667	56	0.0178571	106	0.0094340	156	0.0064103
7	0.1428571	57	0.0175439	107	0.0093458	157	0.0063694
8	0.1250000	58	0.0172414	108	0.0092593	158	0.0063291
9	0.1111111	59	0.0169492	109	0.0091743	159	0.0062893
10	0.1000000	60	0.0166667	110	0.0090909	160	0.0062500
11	0.0909091	61	0.0163934	111	0.0090090	161	0.0062112
12	0.0833333	62	0.0161290	112	0.0089286	162	0.0061728
13	0.0769231	63	0.0158730	113	0.0088496	163	0.0061350
14	0.0714286	64	0.0156250	114	0.0087719	164	0.0060976
15	0.0666667	65	0.0153846	115	0.0086957	165	0.0060606
16	0.0625000	66	0.0151515	116	0.0086207	166	0.0060241
17	0.0588235	67	0.0149254	117	0.0085470	167	0.0059880
18	0.0555556	68	0.0147059	118	0.0084746	168	0.0059524
19	0.0526316	69	0.0144928	119	0.0084034	169	0.0059172
20	0.0500000	70	0.0142857	120	0.0083333	170	0.0058823
21	0.0476190	71	0.0140845	121	0.0082645	171	0.0058480
22	0.0454545	72	0.0138889	122	0.0081967	172	0.0058140
23	0.0434783	73	0.0136986	123	0.0081301	173	0.0057803
24	0.0416667	74	0.0135135	124	0.0080645	174	0.0057471
25	0.0400000	75	0.0133333	125	0.0080000	175	0.0057143
26	0.0384615	76	0.0131579	126	0.0079365	176	0.0056818
27	0.0370370	77	0.0129870	127	0.0078740	177	0.0056497
28	0.0357143	78	0.0128205	128	0.0078125	178	0.0056180
29	0.0344828	79	0.0126582	129	0.0077519	179	0.0055866
30	0.0333333	80	0.0125000	130	0.0076923	180	0.0055556
31	0.0322581	81	0.0123457	131	0.0076336	181	0.0055249
32	0.0312500	82	0.0121951	132	0.0075758	182	0.0054945
33	0.0303030	83	0.0120482	133	0.0075188	183	0.0054645
34	0.0294118	84	0.0119048	134	0.0074627	184	0.0054348
35	0.0285714	85	0.0117647	135	0.0074074	185	0.0054054
36	0.0277778	86	0.0116279	136	0.0073529	186	0.0053763
37	0.0270270	87	0.0114943	137	0.0072993	187	0.0053476
38	0.0263158	88	0.0113636	138	0.0072464	188	0.0053191
39	0.0256410	89	0.0112360	139	0.0071942	189	0.0052910
40	0.0250000	90	0.0111111	140	0.0071429	190	0.0052632
41	0.0243902	91	0.0109890	141	0.0070922	191	0.0052356
42	0.0238095	92	0.0108696	142	0.0070423	192	0.0052083
43	0.0232558	93	0.0107527	143	0.0069930	193	0.0051813
44	0.0227273	94	0.0106383	144	0.0069444	194	0.0051546
45	0.0222222	95	0.0105263	145	0.0068966	195	0.0051282
46	0.0217391	96	0.0104167	146	0.0068493	196	0.0051020
47	0.0212766	97	0.0103093	147	0.0068027	197	0.0050761
48	0.0208333	98	0.0102041	148	0.0067568	198	0.0050505
49	0.0204082	99	0.0101010	149	0.0067114	199	0.0050251
50	0.0200000	100	0.0100000	150	0.0066667	200	0.0050000

3.—Reciprocals, 201 to 400

No.	Reciprocal	No.	Reciprocal	No.	Reciprocal	No.	Reciprocal
201	0.0049751	251	0.0039841	301	0.0033223	351	0.0028490
202	0.0049505	252	0.0039683	302	0.0033113	352	0.0028409
203	0.0049261	253	0.0039526	303	0.0033003	353	0.0028329
204	0.0049020	254	0.0039370	304	0.0032895	354	0.0028249
205	0.0048780	255	0.0039216	305	0.0032787	355	0.0028169
206	0.0048544	256	0.0039063	306	0.0032680	356	0.0028090
207	0.0048309	257	0.0038911	307	0.0032573	357	0.0028011
208	0.0048077	258	0.0038760	308	0.0032468	358	0.0027933
209	0.0047847	259	0.0038610	309	0.0032362	359	0.0027855
210	0.0047619	260	0.0038462	310	0.0032258	360	0.0027778
211	0.0047393	261	0.0038314	311	0.0032154	361	0.0027701
212	0.0047170	262	0.0038168	312	0.0032051	362	0.0027624
213	0.0046948	263	0.0038023	313	0.0031949	363	0.0027548
214	0.0046729	264	0.0037879	314	0.0031847	364	0.0027473
215	0.0046512	265	0.0037736	315	0.0031746	365	0.0027397
216	0.0046296	266	0.0037594	316	0.0031646	366	0.0027322
217	0.0046083	267	0.0037453	317	0.0031546	367	0.0027248
218	0.0045872	268	0.0037313	318	0.0031447	368	0.0027174
219	0.0045662	269	0.0037175	319	0.0031348	369	0.0027100
220	0.0045455	270	0.0037037	320	0.0031250	370	0.0027027
221	0.0045249	271	0.0036900	321	0.0031153	371	0.0026954
222	0.0045045	272	0.0036765	322	0.0031056	372	0.0026882
223	0.0044843	273	0.0036630	323	0.0030960	373	0.0026810
224	0.0044643	274	0.0036496	324	0.0030864	374	0.0026738
225	0.0044444	275	0.0036364	325	0.0030769	375	0.0026667
226	0.0044248	276	0.0036232	326	0.0030675	376	0.0026596
227	0.0044053	277	0.0036101	327	0.0030581	377	0.0026525
228	0.0043860	278	0.0035971	328	0.0030488	378	0.0026455
229	0.0043668	279	0.0035842	329	0.0030395	379	0.0026385
230	0.0043478	280	0.0035714	330	0.0030303	380	0.0026316
231	0.0043290	281	0.0035587	331	0.0030211	381	0.0026247
232	0.0043103	282	0.0035461	332	0.0030120	382	0.0026178
233	0.0042918	283	0.0035336	333	0.0030030	383	0.0026110
234	0.0042735	284	0.0035211	334	0.0029940	384	0.0026042
235	0.0042553	285	0.0035088	335	0.0029851	385	0.0025974
236	0.0042373	286	0.0034965	336	0.0029762	386	0.0025907
237	0.0042194	287	0.0034843	337	0.0029674	387	0.0025840
238	0.0042017	288	0.0034722	338	0.0029586	388	0.0025773
239	0.0041841	289	0.0034602	339	0.0029499	389	0.0025707
240	0.0041667	290	0.0034483	340	0.0029412	390	0.0025641
241	0.0041494	291	0.0034364	341	0.0029326	391	0.0025575
242	0.0041322	292	0.0034247	342	0.0029240	392	0.0025510
243	0.0041152	293	0.0034130	343	0.0029155	393	0.0025445
244	0.0040984	294	0.0034014	344	0.0029070	394	0.0025381
245	0.0040816	295	0.0033898	345	0.0028986	395	0.0025316
246	0.0040650	296	0.0033784	346	0.0028902	396	0.0025253
247	0.0040486	297	0.0033670	347	0.0028818	397	0.0025189
248	0.0040323	298	0.0033557	348	0.0028736	398	0.0025126
249	0.0040161	299	0.0033445	349	0.0028653	399	0.0025063
250	0.0040000	300	0.0033333	350	0.0028571	400	0.0025000

3.—Reciprocals, 401 to 600

No.	Reciprocal	No.	Reciprocal	No.	Reciprocal	No.	Reciprocal
401	0.0024938	451	0.0022173	501	0.0019960	551	0.0018149
402	0.0024876	452	0.0022124	502	0.0019920	552	0.0018116
403	0.0024814	453	0.0022075	503	0.0019881	553	0.0018083
404	0.0024752	454	0.0022026	504	0.0019841	554	0.0018051
405	0.0024691	455	0.0021978	505	0.0019802	555	0.0018018
406	0.0024631	456	0.0021930	506	0.0019763	556	0.0017986
407	0.0024570	457	0.0021882	507	0.0019724	557	0.0017953
408	0.0024510	458	0.0021834	508	0.0019685	558	0.0017921
409	0.0024450	459	0.0021786	509	0.0019646	559	0.0017889
410	0.0024390	460	0.0021739	510	0.0019608	560	0.0017857
411	0.0024331	461	0.0021692	511	0.0019569	561	0.0017825
412	0.0024272	462	0.0021645	512	0.0019531	562	0.0017794
413	0.0024213	463	0.0021598	513	0.0019493	563	0.0017762
414	0.0024155	464	0.0021552	514	0.0019455	564	0.0017731
415	0.0024096	465	0.0021505	515	0.0019417	565	0.0017699
416	0.0024038	466	0.0021459	516	0.0019380	566	0.0017668
417	0.0023981	467	0.0021413	517	0.0019342	567	0.0017637
418	0.0023923	468	0.0021368	518	0.0019305	568	0.0017606
419	0.0023866	469	0.0021322	519	0.0019268	569	0.0017575
420	0.0023810	470	0.0021277	520	0.0019231	570	0.0017544
421	0.0023753	471	0.0021231	521	0.0019194	571	0.0017513
422	0.0023697	472	0.0021186	522	0.0019157	572	0.0017483
423	0.0023641	473	0.0021142	523	0.0019120	573	0.0017452
424	0.0023585	474	0.0021097	524	0.0019084	574	0.0017422
425	0.0023529	475	0.0021053	525	0.0019048	575	0.0017391
426	0.0023474	476	0.0021008	526	0.0019011	576	0.0017361
427	0.0023419	477	0.0020964	527	0.0018975	577	0.0017331
428	0.0023364	478	0.0020921	528	0.0018939	578	0.0017301
429	0.0023310	479	0.0020877	529	0.0018904	579	0.0017271
430	0.0023256	480	0.0020833	530	0.0018868	580	0.0017241
431	0.0023202	481	0.0020790	531	0.0018832	581	0.0017212
432	0.0023148	482	0.0020747	532	0.0018797	582	0.0017182
433	0.0023095	483	0.0020704	533	0.0018762	583	0.0017153
434	0.0023041	484	0.0020661	534	0.0018727	584	0.0017123
435	0.0022989	485	0.0020619	535	0.0018692	585	0.0017094
436	0.0022936	486	0.0020576	536	0.0018657	586	0.0017065
437	0.0022883	487	0.0020534	537	0.0018622	587	0.0017036
438	0.0022831	488	0.0020492	538	0.0018587	588	0.0017007
439	0.0022779	489	0.0020450	539	0.0018553	589	0.0016978
440	0.0022727	490	0.0020408	540	0.0018519	590	0.0016949
441	0.0022676	491	0.0020367	541	0.0018484	591	0.0016920
442	0.0022624	492	0.0020325	542	0.0018450	592	0.0016892
443	0.0022573	493	0.0020284	543	0.0018416	593	0.0016863
444	0.0022523	494	0.0020243	544	0.0018382	594	0.0016835
445	0.0022472	495	0.0020202	545	0.0018349	595	0.0016807
446	0.0022422	496	0.0020161	546	0.0018315	596	0.0016779
447	0.0022371	497	0.0020121	547	0.0018282	597	0.0016750
448	0.0022321	498	0.0020080	548	0.0018248	598	0.0016722
449	0.0022272	499	0.0020040	549	0.0018215	599	0.0016694
450	0.0022222	500	0.0020000	550	0.0018182	600	0.0016667

3.—Reciprocals, 601 to 800

No.	Reciprocal	No.	Reciprocal	No.	Reciprocal	No.	Reciprocal
601	0.0016639	651	0.0015361	701	0.0014265	751	0.0013316
602	0.0016611	652	0.0015337	702	0.0014245	752	0.0013298
603	0.0016584	653	0.0015314	703	0.0014225	753	0.0013280
604	0.0016556	654	0.0015291	704	0.0014205	754	0.0013263
605	0.0016529	655	0.0015267	705	0.0014184	755	0.0013245
606	0.0016502	656	0.0015244	706	0.0014164	756	0.0013228
607	0.0016474	657	0.0015221	707	0.0014144	757	0.0013210
608	0.0016447	658	0.0015198	708	0.0014124	758	0.0013193
609	0.0016420	659	0.0015175	709	0.0014104	759	0.0013175
610	0.0016393	660	0.0015152	710	0.0014085	760	0.0013158
611	0.0016367	661	0.0015129	711	0.0014065	761	0.0013141
612	0.0016340	662	0.0015106	712	0.0014045	762	0.0013123
613	0.0016313	663	0.0015083	713	0.0014025	763	0.0013106
614	0.0016287	664	0.0015060	714	0.0014006	764	0.0013089
615	0.0016260	665	0.0015038	715	0.0013986	765	0.0013072
616	0.0016234	666	0.0015015	716	0.0013966	766	0.0013055
617	0.0016207	667	0.0014993	717	0.0013947	767	0.0013038
618	0.0016181	668	0.0014970	718	0.0013928	768	0.0013021
619	0.0016155	669	0.0014948	719	0.0013908	769	0.0013004
620	0.0016129	670	0.0014925	720	0.0013889	770	0.0012987
621	0.0016103	671	0.0014903	721	0.0013870	771	0.0012970
622	0.0016077	672	0.0014881	722	0.0013850	772	0.0012953
623	0.0016051	673	0.0014859	723	0.0013831	773	0.0012937
624	0.0016026	674	0.0014837	724	0.0013812	774	0.0012920
625	0.0016000	675	0.0014815	725	0.0013793	775	0.0012903
626	0.0015974	676	0.0014793	726	0.0013774	776	0.0012887
627	0.0015949	677	0.0014771	727	0.0013755	777	0.0012870
628	0.0015924	678	0.0014749	728	0.0013736	778	0.0012853
629	0.0015898	679	0.0014728	729	0.0013717	779	0.0012837
630	0.0015873	680	0.0014706	730	0.0013699	780	0.0012821
631	0.0015848	681	0.0014684	731	0.0013680	781	0.0012804
632	0.0015823	682	0.0014663	732	0.0013661	782	0.0012788
633	0.0015798	683	0.0014641	733	0.0013643	783	0.0012771
634	0.0015773	684	0.0014620	734	0.0013624	784	0.0012755
635	0.0015748	685	0.0014599	735	0.0013605	785	0.0012739
636	0.0015723	686	0.0014577	736	0.0013587	786	0.0012723
637	0.0015699	687	0.0014556	737	0.0013569	787	0.0012706
638	0.0015674	688	0.0014535	738	0.0013550	788	0.0012690
639	0.0015649	689	0.0014514	739	0.0013532	789	0.0012674
640	0.0015625	690	0.0014493	740	0.0013514	790	0.0012658
641	0.0015601	691	0.0014472	741	0.0013495	791	0.0012642
642	0.0015576	692	0.0014451	742	0.0013477	792	0.0012626
643	0.0015552	693	0.0014430	743	0.0013459	793	0.0012610
644	0.0015528	694	0.0014409	744	0.0013441	794	0.0012594
645	0.0015504	695	0.0014388	745	0.0013423	795	0.0012579
646	0.0015480	696	0.0014368	746	0.0013405	796	0.0012563
647	0.0015456	697	0.0014347	747	0.0013387	797	0.0012547
648	0.0015432	698	0.0014327	748	0.0013369	798	0.0012531
649	0.0015408	699	0.0014306	749	0.0013351	799	0.0012516
650	0.0015385	700	0.0014286	750	0.0013333	800	0.0012500

3.—Reciprocals, 801 to 1000

No.	Reciprocal	No.	Reciprocal	No.	Reciprocal	No.	Reciprocal
801	0.0012484	851	0.0011751	901	0.0011099	951	0.0010515
802	0.0012469	852	0.0011737	902	0.0011086	952	0.0010504
803	0.0012453	853	0.0011723	903	0.0011074	953	0.0010493
804	0.0012438	854	0.0011710	904	0.0011062	954	0.0010482
805	0.0012422	855	0.0011696	905	0.0011050	955	0.0010471
806	0.0012407	856	0.0011682	906	0.0011038	956	0.0010460
807	0.0012392	857	0.0011669	907	0.0011025	957	0.0010449
808	0.0012376	858	0.0011655	908	0.0011013	958	0.0010438
809	0.0012361	859	0.0011641	909	0.0011001	959	0.0010428
810	0.0012346	860	0.0011628	910	0.0010989	960	0.0010417
811	0.0012330	861	0.0011614	911	0.0010977	961	0.0010406
812	0.0012315	862	0.0011601	912	0.0010965	962	0.0010395
813	0.0012300	863	0.0011587	913	0.0010953	963	0.0010384
814	0.0012285	864	0.0011574	914	0.0010941	964	0.0010373
815	0.0012270	865	0.0011561	915	0.0010929	965	0.0010363
816	0.0012255	866	0.0011547	916	0.0010917	966	0.0010352
817	0.0012240	867	0.0011534	917	0.0010905	967	0.0010341
818	0.0012225	868	0.0011521	918	0.0010893	968	0.0010331
819	0.0012210	869	0.0011507	919	0.0010881	969	0.0010320
820	0.0012195	870	0.0011494	920	0.0010870	970	0.0010309
821	0.0012180	871	0.0011481	921	0.0010858	971	0.0010299
822	0.0012165	872	0.0011468	922	0.0010846	972	0.0010288
823	0.0012151	873	0.0011455	923	0.0010834	973	0.0010277
824	0.0012136	874	0.0011442	924	0.0010823	974	0.0010267
825	0.0012121	875	0.0011429	925	0.0010811	975	0.0010256
826	0.0012107	876	0.0011416	926	0.0010799	976	0.0010246
827	0.0012092	877	0.0011403	927	0.0010787	977	0.0010235
828	0.0012077	878	0.0011390	928	0.0010776	978	0.0010225
829	0.0012063	879	0.0011377	929	0.0010764	979	0.0010215
830	0.0012048	880	0.0011364	930	0.0010753	980	0.0010204
831	0.0012034	881	0.0011351	931	0.0010741	981	0.0010194
832	0.0012019	882	0.0011338	932	0.0010730	982	0.0010183
833	0.0012005	883	0.0011325	933	0.0010718	983	0.0010173
834	0.0011990	884	0.0011312	934	0.0010707	984	0.0010163
835	0.0011976	885	0.0011299	935	0.0010695	985	0.0010152
836	0.0011962	886	0.0011287	936	0.0010684	986	0.0010142
837	0.0011947	887	0.0011274	937	0.0010672	987	0.0010132
838	0.0011933	888	0.0011261	938	0.0010661	988	0.0010121
839	0.0011919	889	0.0011249	939	0.0010650	989	0.0010111
840	0.0011905	890	0.0011236	940	0.0010638	990	0.0010101
841	0.0011891	891	0.0011223	941	0.0010627	991	0.0010091
842	0.0011876	892	0.0011211	942	0.0010616	992	0.0010081
843	0.0011862	893	0.0011198	943	0.0010604	993	0.0010070
844	0.0011848	894	0.0011186	944	0.0010593	994	0.0010060
845	0.0011834	895	0.0011173	945	0.0010582	995	0.0010050
846	0.0011820	896	0.0011161	946	0.0010571	996	0.0010040
847	0.0011806	897	0.0011148	947	0.0010560	997	0.0010030
848	0.0011792	898	0.0011136	948	0.0010549	998	0.0010020
849	0.0011779	899	0.0011123	949	0.0010537	999	0.0010010
850	0.0011765	900	0.0011111	950	0.0010526	1000	0.0010000

3.—Reciprocals, 1001 to 1200

No.	Reciprocal	No.	Reciprocal	No.	Reciprocal	No.	Reciprocal
1001	0.0009990	1051	0.0009515	1101	0.0009083	1151	0.0008688
1002	0.0009980	1052	0.0009506	1102	0.0009074	1152	0.0008681
1003	0.0009970	1053	0.0009497	1103	0.0009066	1153	0.0008673
1004	0.0009960	1054	0.0009488	1104	0.0009058	1154	0.0008666
1005	0.0009950	1055	0.0009479	1105	0.0009050	1155	0.0008658
1006	0.0009940	1056	0.0009470	1106	0.0009042	1156	0.0008651
1007	0.0009930	1057	0.0009461	1107	0.0009033	1157	0.0008643
1008	0.0009921	1058	0.0009452	1108	0.0009025	1158	0.0008636
1009	0.0009911	1059	0.0009443	1109	0.0009017	1159	0.0008628
1010	0.0009901	1060	0.0009434	1110	0.0009009	1160	0.0008621
1011	0.0009891	1061	0.0009425	1111	0.0009001	1161	0.0008613
1012	0.0009881	1062	0.0009416	1112	0.0008993	1162	0.0008606
1013	0.0009872	1063	0.0009407	1113	0.0008985	1163	0.0008598
1014	0.0009862	1064	0.0009398	1114	0.0008977	1164	0.0008591
1015	0.0009852	1065	0.0009390	1115	0.0008969	1165	0.0008584
1016	0.0009843	1066	0.0009381	1116	0.0008961	1166	0.0008576
1017	0.0009833	1067	0.0009372	1117	0.0008953	1167	0.0008569
1018	0.0009823	1068	0.0009363	1118	0.0008945	1168	0.0008562
1019	0.0009814	1069	0.0009355	1119	0.0008937	1169	0.0008554
1020	0.0009804	1070	0.0009346	1120	0.0008929	1170	0.0008547
1021	0.0009794	1071	0.0009337	1121	0.0008921	1171	0.0008540
1022	0.0009785	1072	0.0009328	1122	0.0008913	1172	0.0008532
1023	0.0009775	1073	0.0009320	1123	0.0008905	1173	0.0008525
1024	0.0009766	1074	0.0009311	1124	0.0008897	1174	0.0008518
1025	0.0009756	1075	0.0009302	1125	0.0008889	1175	0.0008511
1026	0.0009747	1076	0.0009294	1126	0.0008881	1176	0.0008503
1027	0.0009737	1077	0.0009285	1127	0.0008873	1177	0.0008496
1028	0.0009728	1078	0.0009276	1128	0.0008865	1178	0.0008489
1029	0.0009718	1079	0.0009268	1129	0.0008857	1179	0.0008482
1030	0.0009709	1080	0.0009259	1130	0.0008850	1180	0.0008475
1031	0.0009699	1081	0.0009251	1131	0.0008842	1181	0.0008467
1032	0.0009690	1082	0.0009242	1132	0.0008834	1182	0.0008460
1033	0.0009681	1083	0.0009234	1133	0.0008826	1183	0.0008453
1034	0.0009671	1084	0.0009225	1134	0.0008818	1184	0.0008446
1035	0.0009662	1085	0.0009217	1135	0.0008811	1185	0.0008439
1036	0.0009653	1086	0.0009208	1136	0.0008803	1186	0.0008432
1037	0.0009643	1087	0.0009200	1137	0.0008795	1187	0.0008425
1038	0.0009634	1088	0.0009191	1138	0.0008787	1188	0.0008418
1039	0.0009625	1089	0.0009183	1139	0.0008780	1189	0.0008410
1040	0.0009615	1090	0.0009174	1140	0.0008772	1190	0.0008403
1041	0.0009606	1091	0.0009166	1141	0.0008764	1191	0.0008396
1042	0.0009597	1092	0.0009158	1142	0.0008757	1192	0.0008389
1043	0.0009588	1093	0.0009149	1143	0.0008749	1193	0.0008382
1044	0.0009579	1094	0.0009141	1144	0.0008741	1194	0.0008375
1045	0.0009569	1095	0.0009132	1145	0.0008734	1195	0.0008368
1046	0.0009560	1096	0.0009124	1146	0.0008726	1196	0.0008361
1047	0.0009551	1097	0.0009116	1147	0.0008718	1197	0.0008354
1048	0.0009542	1098	0.0009107	1148	0.0008711	1198	0.0008347
1049	0.0009533	1099	0.0009099	1149	0.0008703	1199	0.0008340
1050	0.0009524	1100	0.0009091	1150	0.0008696	1200	0.0008333

3.—Reciprocals, 1201 to 1400

No.	Reciprocal	No.	Reciprocal	No.	Reciprocal	No.	Reciprocal
1201	0.0008326	1251	0.0007994	1301	0.0007686	1351	0.0007402
1202	0.0008319	1252	0.0007987	1302	0.0007680	1352	0.0007396
1203	0.0008313	1253	0.0007981	1303	0.0007675	1353	0.0007391
1204	0.0008306	1254	0.0007974	1304	0.0007669	1354	0.0007386
1205	0.0008299	1255	0.0007968	1305	0.0007663	1355	0.0007380
1206	0.0008292	1256	0.0007962	1306	0.0007657	1356	0.0007375
1207	0.0008285	1257	0.0007955	1307	0.0007651	1357	0.0007369
1208	0.0008278	1258	0.0007949	1308	0.0007645	1358	0.0007364
1209	0.0008271	1259	0.0007943	1309	0.0007639	1359	0.0007358
1210	0.0008264	1260	0.0007937	1310	0.0007634	1360	0.0007353
1211	0.0008258	1261	0.0007930	1311	0.0007628	1361	0.0007348
1212	0.0008251	1262	0.0007924	1312	0.0007622	1362	0.0007342
1213	0.0008244	1263	0.0007918	1313	0.0007616	1363	0.0007337
1214	0.0008237	1264	0.0007911	1314	0.0007610	1364	0.0007331
1215	0.0008230	1265	0.0007905	1315	0.0007605	1365	0.0007326
1216	0.0008224	1266	0.0007899	1316	0.0007599	1366	0.0007321
1217	0.0008217	1267	0.0007893	1317	0.0007593	1367	0.0007315
1218	0.0008210	1268	0.0007886	1318	0.0007587	1368	0.0007310
1219	0.0008203	1269	0.0007880	1319	0.0007582	1369	0.0007305
1220	0.0008197	1270	0.0007874	1320	0.0007576	1370	0.0007299
1221	0.0008190	1271	0.0007868	1321	0.0007570	1371	0.0007294
1222	0.0008183	1272	0.0007862	1322	0.0007564	1372	0.0007289
1223	0.0008177	1273	0.0007855	1323	0.0007559	1373	0.0007283
1224	0.0008170	1274	0.0007849	1324	0.0007553	1374	0.0007278
1225	0.0008163	1275	0.0007843	1325	0.0007547	1375	0.0007273
1226	0.0008157	1276	0.0007837	1326	0.0007541	1376	0.0007267
1227	0.0008150	1277	0.0007831	1327	0.0007536	1377	0.0007262
1228	0.0008143	1278	0.0007825	1328	0.0007530	1378	0.0007257
1229	0.0008137	1279	0.0007819	1329	0.0007524	1379	0.0007252
1230	0.0008130	1280	0.0007813	1330	0.0007519	1380	0.0007246
1231	0.0008123	1281	0.0007806	1331	0.0007513	1381	0.0007241
1232	0.0008117	1282	0.0007800	1332	0.0007508	1382	0.0007236
1233	0.0008110	1283	0.0007794	1333	0.0007502	1383	0.0007231
1234	0.0008104	1284	0.0007788	1334	0.0007496	1384	0.0007225
1235	0.0008097	1285	0.0007782	1335	0.0007491	1385	0.0007220
1236	0.0008091	1286	0.0007776	1336	0.0007485	1386	0.0007215
1237	0.0008084	1287	0.0007770	1337	0.0007479	1387	0.0007210
1238	0.0008078	1288	0.0007764	1338	0.0007474	1388	0.0007205
1239	0.0008071	1289	0.0007758	1339	0.0007468	1389	0.0007199
1240	0.0008065	1290	0.0007752	1340	0.0007463	1390	0.0007194
1241	0.0008058	1291	0.0007746	1341	0.0007457	1391	0.0007189
1242	0.0008052	1292	0.0007740	1342	0.0007452	1392	0.0007184
1243	0.0008045	1293	0.0007734	1343	0.0007446	1393	0.0007179
1244	0.0008039	1294	0.0007728	1344	0.0007440	1394	0.0007174
1245	0.0008032	1295	0.0007722	1345	0.0007435	1395	0.0007168
1246	0.0008026	1296	0.0007716	1346	0.0007429	1396	0.0007163
1247	0.0008019	1297	0.0007710	1347	0.0007424	1397	0.0007158
1248	0.0008013	1298	0.0007704	1348	0.0007418	1398	0.0007153
1249	0.0008006	1299	0.0007698	1349	0.0007413	1399	0.0007148
1250	0.0008000	1300	0.0007692	1350	0.0007407	1400	0.0007143

3.—Reciprocals, 1401 to 1600

No.	Reciprocal	No.	Reciprocal	No.	Reciprocal	No.	Reciprocal
1401	0.0007138	1451	0.0006892	1501	0.0006662	1551	0.0006447
1402	0.0007133	1452	0.0006887	1502	0.0006658	1552	0.0006443
1403	0.0007128	1453	0.0006882	1503	0.0006653	1553	0.0006439
1404	0.0007123	1454	0.0006878	1504	0.0006649	1554	0.0006435
1405	0.0007117	1455	0.0006873	1505	0.0006645	1555	0.0006431
1406	0.0007112	1456	0.0006868	1506	0.0006640	1556	0.0006427
1407	0.0007107	1457	0.0006863	1507	0.0006636	1557	0.0006423
1408	0.0007102	1458	0.0006859	1508	0.0006631	1558	0.0006418
1409	0.0007097	1459	0.0006854	1509	0.0006627	1559	0.0006414
1410	0.0007092	1460	0.0006849	1510	0.0006623	1560	0.0006410
1411	0.0007087	1461	0.0006845	1511	0.0006618	1561	0.0006406
1412	0.0007082	1462	0.0006840	1512	0.0006614	1562	0.0006402
1413	0.0007077	1463	0.0006835	1513	0.0006609	1563	0.0006398
1414	0.0007072	1464	0.0006831	1514	0.0006605	1564	0.0006394
1415	0.0007067	1465	0.0006826	1515	0.0006601	1565	0.0006390
1416	0.0007062	1466	0.0006821	1516	0.0006596	1566	0.0006386
1417	0.0007057	1467	0.0006817	1517	0.0006592	1567	0.0006382
1418	0.0007052	1468	0.0006812	1518	0.0006588	1568	0.0006378
1419	0.0007047	1469	0.0006807	1519	0.0006583	1569	0.0006373
1420	0.0007042	1470	0.0006803	1520	0.0006579	1570	0.0006369
1421	0.0007037	1471	0.0006798	1521	0.0006575	1571	0.0006365
1422	0.0007032	1472	0.0006793	1522	0.0006570	1572	0.0006361
1423	0.0007027	1473	0.0006789	1523	0.0006566	1573	0.0006357
1424	0.0007022	1474	0.0006784	1524	0.0006562	1574	0.0006353
1425	0.0007018	1475	0.0006780	1525	0.0006557	1575	0.0006349
1426	0.0007013	1476	0.0006775	1526	0.0006553	1576	0.0006345
1427	0.0007008	1477	0.0006770	1527	0.0006549	1577	0.0006341
1428	0.0007003	1478	0.0006766	1528	0.0006545	1578	0.0006337
1429	0.0006998	1479	0.0006761	1529	0.0006540	1579	0.0006333
1430	0.0006993	1480	0.0006757	1530	0.0006536	1580	0.0006329
1431	0.0006988	1481	0.0006752	1531	0.0006532	1581	0.0006325
1432	0.0006983	1482	0.0006748	1532	0.0006527	1582	0.0006321
1433	0.0006978	1483	0.0006743	1533	0.0006523	1583	0.0006317
1434	0.0006974	1484	0.0006739	1534	0.0006519	1584	0.0006313
1435	0.0006969	1485	0.0006734	1535	0.0006515	1585	0.0006309
1436	0.0006964	1486	0.0006729	1536	0.0006510	1586	0.0006305
1437	0.0006959	1487	0.0006725	1537	0.0006506	1587	0.0006301
1438	0.0006954	1488	0.0006720	1538	0.0006502	1588	0.0006297
1439	0.0006949	1489	0.0006716	1539	0.0006498	1589	0.0006293
1440	0.0006944	1490	0.0006711	1540	0.0006494	1590	0.0006289
1441	0.0006940	1491	0.0006707	1541	0.0006489	1591	0.0006285
1442	0.0006935	1492	0.0006702	1542	0.0006485	1592	0.0006281
1443	0.0006930	1493	0.0006698	1543	0.0006481	1593	0.0006277
1444	0.0006925	1494	0.0006693	1544	0.0006477	1594	0.0006274
1445	0.0006920	1495	0.0006689	1545	0.0006472	1595	0.0006270
1446	0.0006916	1496	0.0006684	1546	0.0006468	1596	0.0006266
1447	0.0006911	1497	0.0006680	1547	0.0006464	1597	0.0006262
1448	0.0006906	1498	0.0006676	1548	0.0006460	1598	0.0006258
1449	0.0006901	1499	0.0006671	1549	0.0006456	1599	0.0006254
1450	0.0006897	1500	0.0006667	1550	0.0006452	1600	0.0006250

3.—RECIPROCALS, 1601 TO 1800

No.	Reciprocal	No.	Reciprocal	No.	Reciprocal	No.	Reciprocal
1601	0.0006246	1651	0.0006057	1701	0.0005879	1751	0.0005711
1602	0.0006242	1652	0.0006053	1702	0.0005875	1752	0.0005708
1603	0.0006238	1653	0.0006050	1703	0.0005872	1753	0.0005705
1604	0.0006234	1654	0.0006046	1704	0.0005869	1754	0.0005701
1605	0.0006231	1655	0.0006042	1705	0.0005865	1755	0.0005698
1606	0.0006227	1656	0.0006039	1706	0.0005862	1756	0.0005695
1607	0.0006223	1657	0.0006035	1707	0.0005858	1757	0.0005692
1608	0.0006219	1658	0.0006031	1708	0.0005855	1758	0.0005688
1609	0.0006215	1659	0.0006028	1709	0.0005851	1759	0.0005685
1610	0.0006211	1660	0.0006024	1710	0.0005848	1760	0.0005682
1611	0.0006207	1661	0.0006020	1711	0.0005845	1761	0.0005679
1612	0.0006203	1662	0.0006017	1712	0.0005841	1762	0.0005675
1613	0.0006200	1663	0.0006013	1713	0.0005838	1763	0.0005672
1614	0.0006196	1664	0.0006010	1714	0.0005834	1764	0.0005669
1615	0.0006192	1665	0.0006006	1715	0.0005831	1765	0.0005666
1616	0.0006188	1666	0.0006002	1716	0.0005828	1766	0.0005663
1617	0.0006184	1667	0.0005999	1717	0.0005824	1767	0.0005659
1618	0.0006180	1668	0.0005995	1718	0.0005821	1768	0.0005656
1619	0.0006177	1669	0.0005992	1719	0.0005817	1769	0.0005653
1620	0.0006173	1670	0.0005988	1720	0.0005814	1770	0.0005650
1621	0.0006169	1671	0.0005984	1721	0.0005811	1771	0.0005647
1622	0.0006165	1672	0.0005981	1722	0.0005807	1772	0.0005643
1623	0.0006161	1673	0.0005977	1723	0.0005804	1773	0.0005640
1624	0.0006158	1674	0.0005974	1724	0.0005800	1774	0.0005637
1625	0.0006154	1675	0.0005970	1725	0.0005797	1775	0.0005634
1626	0.0006150	1676	0.0005967	1726	0.0005794	1776	0.0005631
1627	0.0006146	1677	0.0005963	1727	0.0005790	1777	0.0005627
1628	0.0006143	1678	0.0005959	1728	0.0005787	1778	0.0005624
1629	0.0006139	1679	0.0005956	1729	0.0005784	1779	0.0005621
1630	0.0006135	1680	0.0005952	1730	0.0005780	1780	0.0005618
1631	0.0006131	1681	0.0005949	1731	0.0005777	1781	0.0005615
1632	0.0006127	1682	0.0005945	1732	0.0005774	1782	0.0005612
1633	0.0006124	1683	0.0005942	1733	0.0005770	1783	0.0005609
1634	0.0006120	1684	0.0005938	1734	0.0005767	1784	0.0005605
1635	0.0006116	1685	0.0005935	1735	0.0005764	1785	0.0005602
1636	0.0006112	1686	0.0005931	1736	0.0005760	1786	0.0005599
1637	0.0006109	1687	0.0005928	1737	0.0005757	1787	0.0005596
1638	0.0006105	1688	0.0005924	1738	0.0005754	1788	0.0005593
1639	0.0006101	1689	0.0005921	1739	0.0005750	1789	0.0005590
1640	0.0006098	1690	0.0005917	1740	0.0005747	1790	0.0005587
1641	0.0006094	1691	0.0005914	1741	0.0005744	1791	0.0005583
1642	0.0006090	1692	0.0005910	1742	0.0005741	1792	0.0005580
1643	0.0006086	1693	0.0005907	1743	0.0005737	1793	0.0005577
1644	0.0006083	1694	0.0005903	1744	0.0005734	1794	0.0005574
1645	0.0006079	1695	0.0005900	1745	0.0005731	1795	0.0005571
1646	0.0006075	1696	0.0005896	1746	0.0005727	1796	0.0005568
1647	0.0006072	1697	0.0005893	1747	0.0005724	1797	0.0005565
1648	0.0006068	1698	0.0005889	1748	0.0005721	1798	0.0005562
1649	0.0006064	1699	0.0005886	1749	0.0005718	1799	0.0005559
1650	0.0006061	1700	0.0005882	1750	0.0005714	1800	0.0005556

3.—Reciprocals, 1801 to 2000

No.	Reciprocal	No.	Reciprocal	No.	Reciprocal	No.	Reciprocal
1801	0.0005552	1851	0.0005402	1901	0.0005260	1951	0 0005126
1802	0.0005549	1852	0.0005400	1902	0.0005258	1952	0.0005123
1803	0.0005546	1853	0.0005397	1903	0.0005255	1953	0.0005120
1804	0.0005543	1854	0.0005394	1904	0.0005252	1954	0.0005118
1805	0.0005540	1855	0.0005391	1905	0.0005249	1955	0.0005115
1806	0.0005537	1856	0.0005388	1906	0.0005247	1956	0.0005112
1807	0.0005534	1857	0.0005385	1907	0.0005244	1957	0.0005110
1808	0.0005531	1858	0.0005382	1908	0.0005241	1958	0.0005107
1809	0.0005528	1859	0.0005379	1909	0.0005238	1959	0.0005105
1810	0.0005525	1860	0.0005376	1910	0.0005236	1960	0.0005102
1811	0.0005522	1861	0.0005373	1911	0.0005233	1961	0.0005099
1812	0.0005519	1862	0.0005371	1912	0.0005230	1962	0.0005097
1813	0.0005516	1863	0.0005368	1913	0.0005227	1963	0.0005094
1814	0.0005513	1864	0.0005365	1914	0.0005225	1964	0.0005092
1815	0.0005510	1865	0.0005362	1915	0.0005222	1965	0.0005089
1816	0.0005507	1866	0.0005359	1916	0.0005219	1966	0.0005086
1817	0.0005504	1867	0.0005356	1917	0.0005216	1967	0.0005084
1818	0.0005501	1868	0.0005353	1918	0.0005214	1968	0.0005081
1819	0.0005498	1869	0.0005350	1919	0.0005211	1969	0.0005079
1820	0.0005495	1870	0.0005348	1920	0.0005208	1970	0.0005076
1821	0.0005491	1871	0.0005345	1921	0.0005206	1971	0.0005074
1822	0.0005488	1872	0.0005342	1922	0.0005203	1972	0.0005071
1823	0.0005485	1873	0.0005339	1923	0.0005200	1973	0.0005068
1824	0.0005482	1874	0.0005336	1924	0.0005198	1974	0.0005066
1825	0 0005479	1875	0.0005333	1925	0.0005195	1975	0.0005063
1826	0.0005476	1876	0.0005330	1926	0.0005192	1976	0.0005061
1827	0.0005473	1877	0.0005328	1927	0.0005189	1977	0.0005058
1828	0.0005470	1878	0.0005325	1928	0.0005187	1978	0.0005056
1829	0.0005467	1879	0.0005322	1929	0.0005184	1979	0.0005053
1830	0.0005464	1880	0.0005319	1930	0.0005181	1980	0.0005051
1831	0.0005461	1881	0.0005316	1931	0.0005179	1981	0.0005048
1832	0.0005459	1882	0.0005313	1932	0.0005176	1982	0.0005045
1833	0.0005456	1883	0.0005311	1933	0.0005173	1983	0.0005043
1834	0.0005453	1884	0.0005308	1934	0.0005171	1984	0.0005040
1835	0.0005450	1885	0.0005305	1935	0.0005168	1985	0.0005038
1836	0.0005447	1886	0.0005302	1936	0.0005165	1986	0.0005035
1837	0.0005444	1887	0.0005299	1937	0.0005163	1987	0.0005033
1838	0.0005441	1888	0.0005297	1938	0.0005160	1988	0.0005030
1839	0.0005438	1889	0.0005294	1939	0.0005157	1989	0.0005028
1840	0.0005435	1890	0.0005291	1940	0.0005155	1990	0.0005025
1841	0.0005432	1891	0.0005288	1941	0.0005152	1991	0.0005023
1842	0.0005429	1892	0.0005285	1942	0.0005149	1992	0.0005020
1843	0.0005426	1893	0.0005283	1943	0.0005147	1993	0.0005018
1844	0.0005423	1894	0.0005280	1944	0.0005144	1994	0.0005015
1845	0.0005420	1895	0.0005277	1945	0.0005141	1995	0.0005013
1846	0.0005417	1896	0.0005274	1946	0.0005139	1996	0.0005010
1847	0.0005414	1897	0.0005271	1947	0.0005136	1997	0.0005008
1848	0.0005411	1898	0.0005269	1948	0.0005133	1998	0.0005005
1849	0.0005408	1899	0.0005266	1949	0.0005131	1999	0.0005003
1850	0.0005405	1900	0.0005263	1950	0.0005128	2000	0.0005000

Then arrange a small table of its multiples up to nine times and use this as a multiplication table.

$$0.000636132 \times 1 = 0.000636132$$
$$0.000636132 \times 2 = 0.001272264$$
$$0.000636132 \times 3 = 0.001908396$$
$$0.000636132 \times 4 = 0.002544528$$
$$0.000636132 \times 5 = 0.003180660$$
$$0.000636132 \times 6 = 0.003816792$$
$$0.000636132 \times 7 = 0.004452924$$
$$0.000636132 \times 8 = 0.005088956$$
$$0.000636132 \times 9 = 0.005696188$$

Dividend 7246

Take from above table		
	6..........	.003816792
	4..........	0.02544528
	2..........	00.1272264
	7..........	004.452924
		4.609412472

Correct quotient by direct division to hundred thousandths 4.60941.

Percentage.—*Percent* means *hundredths* and rate percent means any given number of hundredths. Thus, 5 per cent, or 5%, means .05 or $\frac{5}{100}$, in which 5 is the rate. It may also be expressed in true ratio, 5 : 100, meaning 5 *parts* of the 100, both terms being of the same denomination. The percents commonly used may be written in fractional form as follows:

$6\frac{1}{4}\% = \frac{1}{16}$	$12\frac{1}{2}\% = \frac{1}{8}$	$25\ \% = \frac{1}{4}$	$62\frac{1}{2}\% = \frac{5}{8}$
$6\frac{2}{3}\% = \frac{1}{15}$	$14\frac{2}{7}\% = \frac{1}{7}$	$33\frac{1}{3}\% = \frac{1}{3}$	$66\frac{2}{3}\% = \frac{2}{3}$
$8\frac{1}{3}\% = \frac{1}{12}$	$16\frac{2}{3}\% = \frac{1}{6}$	$50\ \% = \frac{1}{2}$	$83\frac{1}{3}\% = \frac{5}{6}$
$10\ \% = \frac{1}{10}$	$20\ \% = \frac{1}{5}$	$37\frac{1}{2}\% = \frac{3}{8}$	$100\ \% = 1$

Percentage covers the operations of finding the part of a given number at a given rate percent, as 4 percent of 650, $650 \times .04 = 26$; of finding what percent one number is of another; as, what percent of 560 is 32?

$$32 \div 560 = .057 = 5.7 \text{ percent;}$$

of ascertaining a number when an amount is given, which is a given percent of that number; as, 112 is 24 percent of what number?

$$112 \div .24 = 467.$$

Logarithms of Numbers.—This section will not attempt to describe in detail the principles upon which logarithms are founded but will confine itself to a brief exposition of the *use* of logarithms.

The *logarithm* of any given number is the exponent of the power to which another fixed number, called the *base*, must be raised in order to produce the given number. A system of logarithms may be founded on any base. Two systems are in use, namely, *common logarithms* and *Naperian* or *natural logarithms.* Common logarithms are on the base 10. In other words, the logarithm of a number indicates the power to which 10 must be raised to produce the given number. In this system

$$
\begin{array}{ll}
10^0 = 1 & \log\ 1 = 0 \\
10^1 = 10 & \log\ 10 = 1 \\
10^2 = 100 & \log\ 100 = 2 \\
10^3 = 1000, \text{etc.} & \log\ 1000 = 3, \text{etc.}
\end{array}
$$

This system is in general use for all practical purposes. When logarithms are mentioned without further qualification, common logarithms are meant.

Natural or Naperian logarithms are founded on a base $e = 2.7182818+$. It is used in pure mathematical discussion and in steam and electrical engineering.

Common Logarithms.—The logarithm of a number is composed of the *characteristic,* or integral portion to the left of the decimal point, and the *mantissa* or decimal fraction. The mantissa is all that appears in any table of logarithms and the degree of accuracy is dependent upon the number of decimal places used in the mantissa. Table 4, following, to five decimal places will be found compact and convenient, where the result to five significant figures is sufficiently accurate. Where greater accuracy is required, *Vega's* tables to seven decimal places are recommended.

In the logarithm of any number, the mantissa is independent of the position of the decimal point, while on the contrary the characteristic is dependent only on the position of the first significant figure of the number with relation to the decimal point. Thus in the following examples:

(a) log 3456.2 = 3.53859
(b) log 345.62 = 2.53859
(c) log 34.562 = 1.53589
(d) log 3.4562 = 0.53859
(e) log .34562 = $\bar{1}$.53859 = 9.53859 − 10
(f) log .034562 = $\bar{2}$.53859 = 8.53859 − 10

The use of the positive characteristic is generally preferred, omitting the (−10), in ordinary cases.

it will be seen that the characteristic is equal, *algebraically*, to the number of places minus one, which the first significant figure of the number occupies to the *left* of the decimal point. In (a) the characteristic is 3; in (b), 2; in (d) 0; in (e), −1; and in (f), −2. Some mathematicians prefer the use of the negative characteristic, but most of them employ the "positive," by algebraically adding 10 to the integer and placing −10 to the right of the mantissa or omitting the latter (−10) altogether. For example, log .040217 = 8.60441, the −10 being understood and the value of the characteristic being, of course, −2. In the case of finding the root of (or dividing) a pure decimal, however, the −10 must be employed.

To Find the Logarithm of a Number.—Example: Find the log of 357.46. Solution: The characteristic is 3 − 1 = 2. The mantissa for the first four figures, 3574, is read directly from Table 4 and is .55315. To this, however, must be added $\frac{6}{10}$ (the next figure of the number is 6) of the difference between .55315 and the log of 3575, or .55328. This difference is 13 and in the proportional parts (P.P.) column under 13 and opposite 6 will be found the value 8, which, added to .55315 in the last place, gives .55323. Hence, the log of 357.46 is 2.55323 (Ans.).

To Find the Anti-logarithm (number corresponding to a logarithm).—Example: What is the number whose logarithm is 1.73821? Solution: This is the reverse of finding the logarithm of a number. Neglecting, for the present, the characteristic, the next lower mantissa to .73821 is .73815 and the number corresponding is 5472. The difference between .73815 and the next higher mantissa in Table 4, .73823, is 8, and the proportional difference $\frac{.73821 - .73815}{.73823 - .73815} = \frac{6}{8}$ calls for .8 to be added to the fourth figure, i.e., 8 to the fifth place of the number, disregarding the decimal point, is 54728. The characteristic, 1, calls for two places to the left of the decimal point, hence the antilog of 1.73821 is 54.728 (Ans.).

Multiplication with Logarithms.—*To multiply two or more numbers, add the logarithms of the numbers and the sum is the logarithm of the product.*

Example: Multiply 25.316 by 42.18

$$\begin{aligned} \text{Solution:}\quad \log 25.316 &= 1.40339 \\ \log 42.18 &= \underline{1.62511} \\ \text{Sum} &= 3.02850 \end{aligned}$$

$$\text{Product} = \text{antilog } 3.02851 = 1067.9 \text{ (Ans.).}$$

Division with Logarithms.—*To divide one number by another, subtract the logarithm of the divisor from the logarithm of the dividend; the difference is the logarithm of the quotient.*

Example: Divide 458.62 by 86.25

$$\begin{aligned} \text{Solution:}\quad \log 458.62 &= 2.66145 \\ \log\ \ 86.25 &= \underline{1.93576} \\ \text{Difference} &= 0.72569 \end{aligned}$$

$$\text{Quotient} = \text{antilog } 0.72569 = 5.3173 \text{ (Ans.).}$$

Raising to Powers with Logarithms.—*To raise a number to a certain power, multiply the logarithm of the number by the exponent of the power; the product is the logarithm of the number raised to the required power.*

Example: What is the value of 4.53^5?

Solution:

$$\begin{aligned} \log 4.53 &= 0.65610 \\ \text{Exponent of power} &= \underline{5} \\ \text{Product} &= 3.28050 \end{aligned}$$

Number raised to the 5th power = antilog 3.28050 = 1907.65 (Ans.).

To Extract the Root of a Number.—*To extract the root of a number, divide the logarithm of the number by the index of the root; the quotient is the logarithm of the root.*

Example: What is $\sqrt[5]{356.07}$?

Solution: log 356.07 = 2.55153

$$5\overline{)2.55153}$$
$$.51031$$

Root = antilog .51031 = 3.2382 (Ans.).

Example: What is $\sqrt{.2516}$?

Solution: log .2516 = 9.40071 − 10

$$2\overline{)9.40071 - 10}$$
$$4.70035 - 5 = \bar{1}.70035$$

Root = antilog $\bar{1}.70035$ = .50159 (Ans.).

TABLE 4

Logarithms

No.	Common Logarithms of Numbers.														Naperian.		
	L. 0	1	2	3	4	5	6	7	8	9	P. P.				No.	Log.	Dif.
100	00 000	043	087	130	173	217	260	303	346	389		**44**	**43**	**42**	1.00	.00000	995
1	432	475	518	561	604	647	689	732	775	817	1	4	4	4	1.01	.00995	985
2	860	903	945	988	030	072	115	157	199	242	2	9	9	8	1.02	.01980	976
3	01 284	326	368	410	452	494	536	578	620	662	3	13	13	13	1.03	.02956	966
4	703	745	787	828	870	912	953	995	036	078	4	18	17	17	1.04	.03922	957
105	02 119	160	202	243	284	325	366	407	449	490	5	22	22	21	1.05	.04879	948
6	531	572	612	653	694	735	776	816	857	898	6	26	26	25	1.06	.05827	939
7	938	979	019	060	100	141	181	222	262	302	7	31	30	29	1.07	.06766	930
8	03 342	383	423	463	503	543	583	623	663	703	8	35	34	34	1.08	.07696	922
9	743	782	822	862	902	941	981	021	060	100	9	40	39	38	1.09	.08618	913
110	04 139	179	218	258	297	336	376	415	454	493		**41**	**40**	**39**	1.10	.09531	905
1	532	571	610	650	689	727	766	805	844	883	1	4	4	4	1.11	.10436	897
2	922	961	999	038	077	115	154	192	231	269	2	8	8	8	1.12	.11333	889
3	05 308	346	385	423	461	500	538	576	614	652	3	12	12	12	1.13	.12222	881
4	690	729	767	805	843	881	918	956	994	032	4	16	16	16	1.14	.13103	873
115	06 070	108	145	183	221	258	296	333	371	408	5	21	20	20	1.15	.13976	866
6	446	483	521	558	595	633	670	707	744	781	6	25	24	23	1.16	.14842	858
7	819	856	893	930	967	004	041	078	115	151	7	29	28	27	1.17	.15700	851
8	07 188	225	262	298	335	372	408	445	482	518	8	33	32	31	1.18	.16551	844
9	555	591	628	664	700	737	773	809	846	882	9	37	36	35	1.19	.17395	837
120	918	954	990	027	063	099	135	171	207	243		**38**	**37**	**36**	1.20	.18232	830
1	08 279	314	350	386	422	458	493	529	565	600	1	4	4	4	1.21	.19062	823
2	636	672	707	743	778	814	849	884	920	955	2	8	7	7	1.22	.19885	816
3	991	026	061	096	132	167	202	237	272	307	3	11	11	11	1.23	.20701	810
4	09 342	377	412	447	482	517	552	587	621	656	4	15	15	14	1.24	.21511	803
125	691	726	760	795	830	864	899	934	968	003	5	19	19	18	1.25	.22314	797
6	10 037	072	106	140	175	209	243	278	312	346	6	23	23	22	1.26	.23111	791
7	380	415	449	483	517	551	585	619	653	687	7	27	26	25	1.27	.23902	784
8	721	755	789	823	857	890	924	958	992	025	8	30	30	29	1.28	.24686	778
9	11 059	093	126	160	193	227	261	294	327	361	9	34	33	32	1.29	.25464	772
130	394	428	461	494	528	561	594	628	661	694		**35**	**34**	**33**	1.30	.26236	767
1	727	760	793	826	860	893	926	959	992	024	1	4	3	3	1.31	.27003	760
2	12 057	090	123	156	189	222	254	287	320	352	2	7	7	7	1.32	.27763	755
3	385	418	450	483	516	548	581	613	646	678	3	11	10	10	1.33	.28518	749
4	710	743	775	808	840	872	905	937	969	001	4	14	14	13	1.34	.29267	743
135	13 033	066	098	130	162	194	226	258	290	322	5	18	17	17	1.35	.30010	738
6	354	386	418	450	481	513	545	577	609	640	6	21	20	20	1.36	.30748	733
7	672	704	735	767	799	830	862	893	925	956	7	25	24	23	1.37	.31481	727
8	988	019	051	082	114	145	176	208	239	270	8	28	27	26	1.38	.32208	722
9	14 301	333	364	395	426	457	489	520	551	582	9	32	31	30	1.39	.32930	717
140	613	644	675	706	737	768	799	829	860	891		**32**	**31**	**30**	1.40	.33647	712
1	922	953	983	014	045	076	106	137	168	198	1	3	3	3	1.41	.34359	707
2	15 229	259	290	320	351	381	412	442	473	503	2	6	6	6	1.42	.35066	701
3	534	564	594	625	655	685	715	746	776	806	3	10	9	9	1.43	.35767	697
4	836	866	897	927	957	987	017	047	077	107	4	13	12	12	1.44	.36464	692
145	16 137	167	197	227	256	286	316	346	376	406	5	16	16	15	1.45	.37156	688
6	435	465	495	524	554	584	613	643	673	702	6	19	19	18	1.46	.37844	682
7	732	761	791	820	850	879	909	938	967	997	7	22	22	21	1.47	.38526	678
8	17 026	056	085	114	143	173	202	231	260	289	8	26	25	24	1.48	.39204	674
9	319	348	377	406	435	464	493	522	551	580	9	29	28	27	1.49	.39878	669
150	609	638	667	696	725	754	782	811	840	869					1.50	.40547	

4.—Logarithms—*Continued*

No.	Common Logarithms of Numbers.													Naperian.		
	L. 0	1	2	3	4	5	6	7	8	9	P. P.			No.	Log.	Dif.
150	17 609	638	667	696	725	754	782	811	840	869		**29**	**28**	1.50	.40547	664
1	898	926	955	984	0̄13	0̄41	0̄70	0̄99	1̄27	1̄56	1	3	3	1.51	.41211	660
2	18 184	213	241	270	298	327	355	384	412	441	2	6	6	1.52	.41871	656
3	469	498	526	554	583	611	639	667	696	724	3	9	8	1.53	.42527	651
4	752	780	808	837	865	893	921	949	977	0̄05	4	12	11	1.54	.43178	
																647
155	19 033	061	089	117	145	173	201	229	257	285	5	15	14	1.55	.43825	644
6	312	340	368	396	424	451	479	507	535	562	6	17	17	1.56	.44469	639
7	590	618	645	673	700	728	756	783	811	838	7	20	20	1.57	.45108	634
8	866	893	921	948	976	0̄03	0̄30	0̄58	0̄85	1̄12	8	23	22	1.58	.45742	631
9	20 140	167	194	222	249	276	303	330	358	385	9	26	25	1.59	.46373	
																627
160	412	439	466	493	520	548	575	602	629	656		**27**	**26**	1.60	.47000	623
1	683	710	737	763	790	817	844	871	898	925	1	3	3	1.61	.47623	620
2	952	978	0̄05	0̄32	0̄59	0̄85	1̄12	1̄39	1̄65	1̄92	2	5	5	1.62	.48243	615
3	21 219	245	272	299	325	352	378	405	431	458	3	8	8	1.63	.48858	612
4	484	511	537	564	590	617	643	669	696	722	4	11	10	1.64	.49470	
																608
165	748	775	801	827	854	880	906	932	958	985	5	14	13	1.65	.50078	604
6	22 011	037	063	089	115	141	167	194	220	246	6	16	16	1.66	.50682	600
7	272	298	324	350	376	401	427	453	479	505	7	19	18	1.67	.51282	597
8	531	557	583	608	634	660	686	712	737	763	8	22	21	1.68	.51879	594
9	789	814	840	866	891	917	943	968	994	0̄19	9	24	23	1.69	.52473	
																590
170	23 045	070	096	121	147	172	198	223	249	274		**25**	**24**	1.70	.53063	586
1	300	325	350	376	401	426	452	477	502	528	1	3	2	1.71	.53649	583
2	553	578	603	629	654	679	704	729	754	779	2	5	5	1.72	.54232	580
3	805	830	855	880	905	930	955	980	0̄05	0̄30	3	8	7	1.73	.54812	577
4	24 055	080	105	130	155	180	204	229	254	279	4	10	10	1.74	.55389	
																573
175	304	329	353	378	403	428	452	477	502	527	5	13	12	1.75	.55962	569
6	551	576	601	625	650	674	699	724	748	773	6	15	14	1.76	.56531	567
7	797	822	846	871	895	920	944	969	993	0̄18	7	18	17	1.77	.57098	563
8	25 042	066	091	115	139	164	188	212	237	261	8	20	19	1.78	.57661	561
9	285	310	334	358	382	406	431	455	479	503	9	23	22	1.79	.58222	
																557
180	527	551	575	600	624	648	672	696	720	744		**24**	**23**	1.80	.58779	554
1	768	792	816	840	864	888	912	935	959	983	1	2	2	1.81	.59333	551
2	26 007	031	055	079	102	126	150	174	198	221	2	5	5	1.82	.59884	548
3	245	269	293	316	340	364	387	411	435	458	3	7	7	1.83	.60432	545
4	482	505	529	553	576	600	623	647	670	694	4	10	9	1.84	.60977	
																542
185	717	741	764	788	811	834	858	881	905	928	5	12	12	1.85	.61519	539
6	951	975	998	0̄21	0̄45	0̄68	0̄91	1̄14	1̄38	1̄61	6	14	14	1.86	.62058	536
7	27 184	207	231	254	277	300	323	346	370	393	7	17	16	1.87	.62594	533
8	416	439	462	485	508	531	554	577	600	623	8	19	18	1.88	.63127	531
9	646	669	692	715	738	761	784	807	830	852	9	22	21	1.89	.63658	
																527
190	875	898	921	944	967	989	0̄12	0̄35	0̄58	0̄81		**22**	**21**	1.90	.64185	525
1	28 103	126	149	171	194	217	240	262	285	307	1	2	2	1.91	.64710	523
2	330	353	375	398	421	443	466	488	511	533	2	4	4	1.92	.65233	519
3	556	578	601	623	646	668	691	713	735	758	3	7	6	1.93	.65752	517
4	780	803	825	847	870	892	914	937	959	981	4	9	8	1.94	.66269	
																514
195	29 003	026	048	070	092	115	137	159	181	203	5	11	11	1.95	.66783	511
6	226	248	270	292	314	336	358	380	403	425	6	13	13	1.96	.67294	509
7	447	469	491	513	535	557	579	601	623	645	7	15	15	1.97	.67803	507
8	667	688	710	732	754	776	798	820	842	863	8	18	17	1.98	.68310	503
9	885	907	929	951	973	994	0̄16	0̄38	0̄60	0̄81	9	20	19	1.99	.68813	
																502
200	30 103	125	146	168	190	211	233	255	276	298				2.00	.69315	

4.—Logarithms—*Continued*

No.	Common Logarithms of Numbers.												Naperian.		
	L. O	1	2	3	4	5	6	7	8	9	P.	P.	No.	Log.	Dif.
200	30 103	125	146	168	190	211	233	255	276	298		**22**	2.00	.69315	498
1	320	341	363	384	406	428	449	471	492	514	1	2	2.01	.69813	497
2	535	557	578	600	621	643	664	685	707	728	2	4	2.02	.70310	494
3	750	771	792	814	835	856	878	899	920	942	3	7	2.03	.70804	491
4	963	984	$\overline{0}06$	$\overline{0}27$	$\overline{0}48$	$\overline{0}69$	$\overline{0}91$	$\overline{1}12$	$\overline{1}33$	$\overline{1}54$	4	9	2.04	.71295	489
205	31 175	197	218	239	260	281	302	323	345	366	5	11	2.05	.71784	487
6	387	408	429	450	471	492	513	534	555	576	6	13	2.06	.72271	484
7	597	618	639	660	681	702	723	744	765	785	7	15	2.07	.72755	482
8	806	827	848	869	890	911	931	952	973	994	8	18	2.08	.73237	479
9	32 015	035	056	077	098	118	139	160	181	201	9	20	2.09	.73716	478
210	222	243	263	284	305	325	346	366	387	408		**21**	2.10	.74194	475
1	428	449	469	490	510	531	552	572	593	613	1	2	2.11	.74669	473
2	634	654	675	695	715	736	756	777	797	818	2	4	2.12	.75142	470
3	838	858	879	899	919	940	960	980	$\overline{0}01$	$\overline{0}21$	3	6	2.13	.75612	469
4	33 041	062	082	102	122	143	163	183	203	224	4	8	2.14	.76081	466
215	244	264	284	304	325	345	365	385	405	425	5	11	2.15	.76547	464
6	445	465	486	506	526	546	566	586	606	626	6	13	2.16	.77011	462
7	646	666	686	706	726	746	766	786	806	826	7	15	2.17	.77473	459
8	846	866	885	905	925	945	965	985	$\overline{0}05$	$\overline{0}25$	8	17	2.18	.77932	458
9	34 044	064	084	104	124	143	163	183	203	223	9	19	2.19	.78390	456
220	242	262	282	301	321	341	361	380	400	420		**20**	2.20	.78846	453
1	439	459	479	498	518	537	557	577	596	616	1	2	2.21	.79299	452
2	635	655	674	694	713	733	753	772	792	811	2	4	2.22	.79751	449
3	830	850	869	889	908	928	947	967	986	$\overline{0}05$	3	6	2.23	.80200	448
4	35 025	044	064	083	102	122	141	160	180	199	4	8	2.24	.80648	445
225	218	238	257	276	295	315	334	353	372	392	5	10	2.25	.81093	443
6	411	430	449	468	488	507	526	545	564	583	6	12	2.26	.81536	442
7	603	622	641	660	679	698	717	736	755	774	7	14	2.27	.81978	440
8	793	813	832	851	870	889	908	927	946	965	8	16	2.28	.82418	437
9	984	$\overline{0}03$	$\overline{0}21$	$\overline{0}40$	$\overline{0}59$	$\overline{0}78$	$\overline{0}97$	$\overline{1}16$	$\overline{1}35$	$\overline{1}54$	9	18	2.29	.82855	436
230	36 173	192	211	229	248	267	286	305	324	342		**19**	2.30	.83291	434
1	361	380	399	418	436	455	474	493	511	530	1	2	2.31	.83725	432
2	549	568	586	605	624	642	661	680	698	717	2	4	2.32	.84157	430
3	736	754	773	791	810	829	847	866	884	903	3	6	2.33	.84587	428
4	922	940	959	977	996	$\overline{0}14$	$\overline{0}33$	$\overline{0}51$	$\overline{0}70$	$\overline{0}88$	4	8	2.34	.85015	427
235	37 107	125	144	162	181	199	218	236	254	273	5	10	2.35	.85442	424
6	291	310	328	346	365	383	401	420	438	457	6	11	2.36	.85866	423
7	475	493	511	530	548	566	585	603	621	639	7	13	2.37	.86289	421
8	658	676	694	712	731	749	767	785	803	822	8	15	2.38	.86710	419
9	840	858	876	894	912	931	949	967	985	$\overline{0}03$	9	17	2.39	.87129	418
240	38 021	039	057	075	093	112	130	148	166	184		**18**	2.40	.87547	416
1	202	220	238	256	274	292	310	328	346	364	1	2	2.41	.87963	414
2	382	399	417	435	453	471	489	507	525	543	2	4	2.42	.88377	412
3	561	578	596	614	632	650	668	686	703	721	3	5	2.43	.88789	411
4	739	757	775	792	810	828	846	863	881	899	4	7	2.44	.89200	409
245	917	934	952	970	987	$\overline{0}05$	$\overline{0}23$	$\overline{0}41$	$\overline{0}58$	$\overline{0}76$	5	9	2.45	.89609	407
6	39 094	111	129	146	164	182	199	217	235	252	6	11	2.46	.90016	406
7	270	287	305	322	340	358	375	393	410	428	7	13	2.47	.90422	404
8	445	463	480	498	515	533	550	568	585	602	8	14	2.48	.90826	402
9	620	637	655	672	690	707	724	742	759	777	9	16	2.49	.91228	401
250	794	811	829	846	863	881	898	915	933	950			2.50	.91629	

4.—Logarithms—*Continued*

No.	Common Logarithms of Numbers.													Naperian.		
	L.	0	1	2	3	4	5	6	7	8	9	P. P.		No.	Log.	Dif.
250	39	794	811	829	846	863	881	898	915	933	950		**18**	2.50	.91629	399
1		967	985	**0̄02**	0̄19	0̄37	0̄54	0̄71	0̄88	1̄06	1̄23	1	2	2.51	.92028	398
2	40	140	157	175	192	209	226	243	261	278	295	2	4	2.52	.92426	396
3		312	329	346	364	381	398	415	432	449	466	3	5	2.53	.92822	394
4		483	500	518	535	552	569	586	603	620	637	4	7	2.54	.93216	393
255		654	671	688	705	722	739	756	773	790	807	5	9	2.55	.93609	392
6		824	841	858	875	892	909	926	943	960	976	6	11	2.56	.94001	390
7		993	**0̄10**	0̄27	0̄44	0̄61	0̄78	0̄95	1̄11	1̄28	1̄45	7	13	2.57	.94391	388
8	41	162	179	196	212	229	246	263	280	296	313	8	14	2.58	.94779	387
9		330	347	363	380	397	414	430	447	464	481	9	16	2.59	.95166	385
260		497	514	531	547	564	581	597	614	631	647		**17**	2.60	.95551	384
1		664	681	697	714	731	747	764	780	797	814	1	2	2.61	.95935	382
2		830	847	863	880	896	913	929	946	963	979	2	3	2.62	.96317	381
3		996	**0̄12**	0̄29	0̄45	0̄62	0̄78	0̄95	1̄11	1̄27	1̄44	3	5	2.63	.96698	380
4	42	160	177	193	210	226	243	259	275	292	308	4	7	2.64	.97078	378
265		325	341	357	374	390	406	423	439	455	472	5	9	2.65	.97456	377
6		488	504	521	537	553	570	586	602	619	635	6	10	2.66	.97833	375
7		651	667	684	700	716	732	749	765	781	797	7	12	2.67	.98208	374
8		813	830	846	862	878	894	911	927	943	959	8	14	2.68	.98582	372
9		975	991	**0̄08**	0̄24	0̄40	0̄56	0̄72	0̄88	1̄04	1̄20	9	15	2.69	.98954	371
270	43	136	152	169	185	201	217	233	249	265	281		**16**	2.70	.99325	370
1		297	313	329	345	361	377	393	409	425	441	1	2	2.71	.99695	368
2		457	473	489	505	521	537	553	569	584	600	2	3	2.72	1.00063	367
3		616	632	648	664	680	696	712	727	743	759	3	5	2.73	1.00430	366
4		775	791	807	823	838	854	870	886	902	917	4	6	2.74	1.00796	364
275		933	949	965	981	996	**0̄12**	0̄28	0̄44	0̄59	0̄75	5	8	2.75	1.01160	363
6	44	091	107	122	138	154	170	185	201	217	232	6	10	2.76	1.01523	362
7		248	264	279	295	311	326	342	358	373	389	7	11	2.77	1.01885	360
8		404	420	436	451	467	483	498	514	529	545	8	13	2.78	1.02245	359
9		560	576	592	607	623	638	654	669	685	700	9	14	2.79	1.02604	358
280		716	731	747	762	778	793	809	824	840	855		**15**	2.80	1.02962	356
1		871	886	902	917	932	948	963	979	994	**0̄10**	1	2	2.81	1.03318	356
2	45	025	040	056	071	086	102	117	133	148	163	2	3	2.82	1.03674	354
3		179	194	209	225	240	255	271	286	301	317	3	5	2.83	1.04028	352
4		332	347	362	378	393	408	423	439	454	469	4	6	2.84	1.04380	352
285		484	500	515	530	545	561	576	591	606	621	5	8	2.85	1.04732	350
6		637	652	667	682	697	712	728	743	758	773	6	9	2.86	1.05082	349
7		788	803	818	834	849	864	879	894	909	924	7	11	2.87	1.05431	348
8		939	954	969	984	**0̄00**	0̄15	0̄30	0̄45	0̄60	0̄75	8	12	2.88	1.05779	347
9	46	090	105	120	135	150	165	180	195	210	225	9	14	2.89	1.06126	345
290		240	255	270	285	300	315	330	345	359	374		**14**	2.90	1.06471	344
1		389	404	419	434	449	464	479	494	509	523	1	1	2.91	1.06815	343
2		538	553	568	583	598	613	627	642	657	672	2	3	2.92	1.07158	342
3		687	702	716	731	746	761	776	790	805	820	3	4	2.93	1.07500	341
4		835	850	864	879	894	909	923	938	953	967	4	6	2.94	1.07841	340
295		982	997	**0̄12**	0̄26	0̄41	0̄56	0̄70	0̄85	1̄00	1̄14	5	7	2.95	1.08181	338
6	47	129	144	159	173	188	202	217	232	246	261	6	8	2.96	1.08519	337
7		276	290	305	319	334	349	363	378	392	407	7	10	2.97	1.08856	336
8		422	436	451	465	480	494	509	524	538	553	8	11	2.98	1.09192	335
9		567	582	596	611	625	640	654	669	683	698	9	13	2.99	1.09527	334
300		712	727	741	756	770	784	799	813	828	842			3.00	1.09861	

4.—Logarithms—*Continued*

No.	L. 0	1	2	3	4	5	6	7	8	9	P.	P.	No.	Log.	Dif.
	Common Logarithms of Numbers.												Naperian.		
300	47 712	727	741	756	770	784	799	813	828	842		**15**	3.00	1.09861	333
1	857	871	885	900	914	929	943	958	972	986	1	2	3.01	1.10194	332
2	48 001	015	029	044	058	073	087	101	116	130	2	3	3.02	1.10526	330
3	144	159	173	187	202	216	230	244	259	273	3	5	3.03	1.10856	330
4	287	302	316	330	344	359	373	387	401	416	4	6	3.04	1.11186	328
305	430	444	458	473	487	501	515	530	544	558	5	8	3.05	1.11514	327
6	572	586	601	615	629	643	657	671	686	700	6	9	3.06	1.11841	327
7	714	728	742	756	770	785	799	813	827	841	7	11	3.07	1.12168	325
8	855	869	883	897	911	926	940	954	968	982	8	12	3.08	1.12493	324
9	996	$\bar{0}$10	$\bar{0}$24	$\bar{0}$38	$\bar{0}$52	$\bar{0}$66	$\bar{0}$80	$\bar{0}$94	$\bar{1}$08	$\bar{1}$22	9	14	3.09	1.12817	323
310	49 136	150	164	178	192	206	220	234	248	262		**14**	3.10	1.13140	322
1	276	290	304	318	332	346	360	374	388	402	1	1	3.11	1.13462	321
2	415	429	443	457	471	485	499	513	527	541	2	3	3.12	1.13783	320
3	554	568	582	596	610	624	638	651	665	679	3	4	3.13	1.14103	319
4	693	707	721	734	748	762	776	790	803	817	4	6	3.14	1.14422	318
315	831	845	859	872	886	900	914	927	941	955	5	7	3.15	1.14740	317
6	969	982	996	$\bar{0}$10	$\bar{0}$24	$\bar{0}$37	$\bar{0}$51	$\bar{0}$65	$\bar{0}$79	$\bar{0}$92	6	8	3.16	1.15057	316
7	50 106	120	133	147	161	174	188	202	215	229	7	10	3.17	1.15373	315
8	243	256	270	284	297	311	325	338	352	365	8	11	3.18	1.15688	314
9	379	393	406	420	433	447	461	474	488	501	9	13	3.19	1.16002	313
320	515	529	542	556	569	583	596	610	623	637		**13**	3.20	1.16315	312
1	651	664	678	691	705	718	732	745	759	772	1	1	3.21	1.16627	311
2	786	799	813	826	840	853	866	880	893	907	2	3	3.22	1.16938	310
3	920	934	947	961	974	987	$\bar{0}$01	$\bar{0}$14	$\bar{0}$28	$\bar{0}$41	3	4	3.23	1.17248	309
4	51 055	068	081	095	108	121	135	148	162	175	4	5	3.24	1.17557	308
325	188	202	215	228	242	255	268	282	295	308	5	7	3.25	1.17865	308
6	322	335	348	362	375	388	402	415	428	441	6	8	3.26	1.18173	306
7	455	468	481	495	508	521	534	548	561	574	7	9	3.27	1.18479	305
8	587	601	614	627	640	654	667	680	693	706	8	10	3.28	1.18784	305
9	720	733	746	759	772	786	799	812	825	838	9	12	3.29	1.19089	303
330	851	865	878	891	904	917	930	943	957	970		**13**	3.30	1.19392	303
1	983	996	$\bar{0}$09	$\bar{0}$22	$\bar{0}$35	$\bar{0}$48	$\bar{0}$61	$\bar{0}$75	$\bar{0}$88	$\bar{1}$01	1	1	3.31	1.19695	301
2	52 114	127	140	153	166	179	192	205	218	231	2	3	3.32	1.19996	301
3	244	257	270	284	297	310	323	336	349	362	3	4	3.33	1.20297	300
4	375	388	401	414	427	440	453	466	479	492	4	5	3.34	1.20597	299
335	504	517	530	543	556	569	582	595	608	621	5	7	3.35	1.20896	298
6	634	647	660	673	686	699	711	724	737	750	6	8	3.36	1.21194	297
7	763	776	789	802	815	827	840	853	866	879	7	9	3.37	1.21491	297
8	892	905	917	930	943	956	969	982	994	$\bar{0}$07	8	10	3.38	1.21788	295
9	53 020	033	046	058	071	084	097	110	122	135	9	12	3.39	1.22083	295
340	148	161	173	186	199	212	224	237	250	263		**12**	3.40	1.22378	293
1	275	288	301	314	326	339	352	364	377	390	1	1	3.41	1.22671	293
2	403	415	428	441	453	466	479	491	504	517	2	2	3.42	1.22964	292
3	529	542	555	567	580	593	605	618	631	643	3	4	3.43	1.23256	291
4	656	668	681	694	706	719	732	744	757	769	4	5	3.44	1.23547	290
345	782	794	807	820	832	845	857	870	882	895	5	6	3.45	1.23837	290
6	908	920	933	945	958	970	983	995	$\bar{0}$08	$\bar{0}$20	6	7	3.46	1.24127	288
7	54 033	045	058	070	083	095	108	120	133	145	7	8	3.47	1.24415	288
8	158	170	183	195	208	220	233	245	258	270	8	10	3.48	1.24703	287
9	283	295	307	320	332	345	357	370	382	394	9	11	3.49	1.24990	286
350	407	419	432	444	456	469	481	494	506	518			3.50	1.25276	

4.—Logarithms—*Continued*

No.	Common Logarithms of Numbers.												Naperian.		
	L. 0	1	2	3	4	5	6	7	8	9	P.	P.	No.	Log.	Dif.
350	**54** 407	419	432	444	456	469	481	494	506	518		**13**	3.50	1.25276	286
1	531	543	555	568	580	593	605	617	630	642	1	1	3.51	1.25562	284
2	654	667	679	691	704	716	728	741	753	765	2	3	3.52	1.25846	284
3	777	790	802	814	827	839	851	864	876	888	3	4	3.53	1.26130	283
4	900	913	925	937	949	962	974	986	998	0̄11	4	5	3.54	1.26413	282
355	**55** 023	035	047	060	072	084	096	108	121	133	5	7	3.55	1.26695	281
6	145	157	169	182	194	206	218	230	242	255	6	8	3.56	1.26976	281
7	267	279	291	303	315	328	340	352	364	376	7	9	3.57	1.27257	279
8	388	400	413	425	437	449	461	473	485	497	8	10	3.58	1.27536	279
9	509	522	534	546	558	570	582	594	606	618	9	12	3.59	1.27815	278
360	630	642	654	666	678	691	703	715	727	739			3.60	1.28093	278
1	751	763	775	787	799	811	823	835	847	859	1		3.61	1.28371	276
2	871	883	895	907	919	931	943	955	967	979	2		3.62	1.28647	276
3	991	0̄03	0̄15	0̄27	0̄38	0̄50	0̄62	0̄74	0̄86	0̄98	3		3.63	1.28923	275
4	**56** 110	122	134	146	158	170	182	194	205	217	4		3.64	1.29198	275
365	229	241	253	265	277	289	301	312	324	336	5		3.65	1.29473	273
6	348	360	372	384	396	407	419	431	443	455	6		3.66	1.29746	273
7	467	478	490	502	514	526	538	549	561	573	7		3.67	1.30019	272
8	585	597	608	620	632	644	656	667	679	691	8		3.68	1.30291	272
9	703	714	726	738	750	761	773	785	797	808	9		3.69	1.30563	270
370	820	832	844	855	867	879	891	902	914	926		**12**	3.70	1.30833	270
1	937	949	961	972	984	996	0̄08	0̄19	0̄31	0̄43	1	1	3.71	1.31103	269
2	**57** 054	066	078	089	101	113	124	136	148	159	2	2	3.72	1.31372	269
3	171	183	194	206	217	229	241	252	264	276	3	4	3.73	1.31641	268
4	287	299	310	322	334	345	357	368	380	392	4	5	3.74	1.31909	267
375	403	415	426	438	449	461	473	484	496	507	5	6	3.75	1.32176	266
6	519	530	542	553	565	576	588	600	611	623	6	7	3.76	1.32442	265
7	634	646	657	669	680	692	703	715	726	738	7	8	3.77	1.32707	265
8	749	761	772	784	795	807	818	830	841	852	8	10	3.78	1.32972	265
9	864	875	887	898	910	921	933	944	955	967	9	11	3.79	1.33237	263
380	978	990	0̄01	0̄13	0̄24	0̄35	0̄47	0̄58	0̄70	0̄81			3.80	1.33500	263
1	**58** 092	104	115	127	138	149	161	172	184	195	1		3.81	1.33763	262
2	206	218	229	240	252	263	274	286	297	309	2		3.82	1.34025	261
3	320	331	343	354	365	377	388	399	410	422	3		3.83	1.34286	261
4	433	444	456	467	478	490	501	512	524	535	4		3.84	1.34547	260
385	546	557	569	580	591	602	614	625	636	647	5		3.85	1.34807	260
6	659	670	681	692	704	715	726	737	749	760	6		3.86	1.35067	258
7	771	782	794	805	816	827	838	850	861	872	7		3.87	1.35325	259
8	883	894	906	917	928	939	950	961	973	984	8		3.88	1.35584	257
9	995	0̄06	0̄17	0̄28	0̄40	0̄51	0̄62	0̄73	0̄84	0̄95	9		3.89	1.35841	257
390	**59** 106	118	129	140	151	162	173	184	195	207		**11**	3.90	1.36098	256
1	218	229	240	251	262	273	284	295	306	318	1	1	3.91	1.36354	255
2	329	340	351	362	373	384	395	406	417	428	2	2	3.92	1.36609	255
3	439	450	461	472	483	494	506	517	528	539	3	3	3.93	1.36864	254
4	550	561	572	583	594	605	616	627	638	649	4	4	3.94	1.37118	254
395	660	671	682	693	704	715	726	737	748	759	5	6	3.95	1.37372	252
6	770	780	791	802	813	824	835	846	857	868	6	7	3.96	1.37624	253
7	879	890	901	912	923	934	945	956	966	977	7	8	3.97	1.37877	251
8	988	999	0̄10	0̄21	0̄32	0̄43	0̄54	0̄65	0̄76	0̄86	8	9	3.98	1.38128	251
9	**60** 097	108	119	130	141	152	163	173	184	195	9	10	3.99	1.38379	250
400	206	217	228	239	249	260	271	282	293	304			4.00	1.38629	

4.—Logarithms—*Continued*

No.	Common Logarithms of Numbers.													Naperian.		
	L.	0	1	2	3	4	5	6	7	8	9	P. P.		No.	Log.	Dif.
400	60	206	217	228	239	249	260	271	282	293	304		11	4.00	1.38629	250
1		314	325	336	347	358	369	379	390	401	412	1	1	4.01	1.38879	249
2		423	433	444	455	466	477	487	498	509	520	2	2	4.02	1.39128	249
3		531	541	552	563	574	584	595	606	617	627	3	3	4.03	1.39377	247
4		638	649	660	670	681	692	703	713	724	735	4	4	4.04	1.39624	248
405		746	756	767	778	788	799	810	821	831	842	5	6	4.05	1.39872	246
6		853	863	874	885	895	906	917	927	938	949	6	7	4.06	1.40118	246
7		959	970	981	991	0̄02	0̄13	0̄23	0̄34	0̄45	0̄55	7	8	4.07	1.40364	246
8	61	066	077	087	098	109	119	130	140	151	162	8	9	4.08	1.40610	244
9		172	183	194	204	215	225	236	247	257	268	9	10	4.09	1.40854	245
410		278	289	300	310	321	331	342	352	363	374			4.10	1.41099	243
1		384	395	405	416	426	437	448	458	469	479	1		4.11	1.41342	243
2		490	500	511	521	532	542	553	563	574	584	2		4.12	1.41585	243
3		595	606	616	627	637	648	658	669	679	690	3		4.13	1.41828	242
4		700	711	721	731	742	752	763	773	784	794	4		4.14	1.42070	241
415		805	815	826	836	847	857	868	878	888	899	5		4.15	1.42311	241
6		909	920	930	941	951	962	972	982	993	0̄03	6		4.16	1.42552	240
7	62	014	024	034	045	055	066	076	086	097	107	7		4.17	1.42792	239
8		118	128	138	149	159	170	180	190	201	211	8		4.18	1.43031	239
9		221	232	242	252	263	273	284	294	304	315	9		4.19	1.43270	238
420		325	335	346	356	366	377	387	397	408	418		10	4.20	1.43508	238
1		428	439	449	459	469	480	490	500	511	521	1	1	4.21	1.43746	238
2		531	542	552	562	572	583	593	603	613	624	2	2	4.22	1.43984	236
3		634	644	655	665	675	685	696	706	716	726	3	3	4.23	1.44220	236
4		737	747	757	767	778	788	798	808	818	829	4	4	4.24	1.44456	236
425		839	849	859	870	880	890	900	910	921	931	5	5	4.25	1.44692	235
6		941	951	961	972	982	992	0̄02	0̄12	0̄22	0̄33	6	6	4.26	1.44927	234
7	63	043	053	063	073	083	094	104	114	124	134	7	7	4.27	1.45161	234
8		144	155	165	175	185	195	205	215	225	236	8	8	4.28	1.45395	234
9		246	256	266	276	286	296	306	317	327	337	9	9	4.29	1.45629	232
430		347	357	367	377	387	397	407	417	428	438			4.30	1.45861	233
1		448	458	468	478	488	498	508	518	528	538	1		4.31	1.46094	232
2		548	558	568	579	589	599	609	619	629	639	2		4.32	1.46326	231
3		649	659	669	679	689	699	709	719	729	739	3		4.33	1.46557	230
4		749	759	769	779	789	799	809	819	829	839	4		4.34	1.46787	231
435		849	859	869	879	889	899	909	919	929	939	5		4.35	1.47018	229
6		949	959	969	979	988	998	0̄08	0̄18	0̄28	0̄38	6		4.36	1.47247	229
7	64	048	058	068	078	088	098	108	118	128	137	7		4.37	1.47476	229
8		147	157	167	177	187	197	207	217	227	237	8		4.38	1.47705	228
9		246	256	266	276	286	296	306	316	326	335	9		4.39	1.47933	227
440		345	355	365	375	385	395	404	414	424	434		9	4.40	1.48160	227
1		444	454	464	473	483	493	503	513	523	532	1	1	4.41	1.48387	227
2		542	552	562	572	582	591	601	611	621	631	2	2	4.42	1.48614	226
3		640	650	660	670	680	689	699	709	719	729	3	3	4.43	1.48840	225
4		738	748	758	768	777	787	797	807	816	826	4	4	4.44	1.49065	225
445		836	846	856	865	875	885	895	904	914	924	5	4.5	4.45	1.49290	225
6		933	943	953	963	972	982	992	0̄02	0̄11	0̄21	6	5.4	4.46	1.49515	224
7	65	031	040	050	060	070	079	089	099	108	118	7	6	4.47	1.49739	223
8		128	137	147	157	167	176	186	196	205	215	8	7	4.48	1.49962	223
9		225	234	244	254	263	273	283	292	302	312	9	8	4.49	1.50185	223
450		321	331	341	350	360	369	379	389	398	408			4.50	1.50408	

4.—Logarithms—*Continued*

No.	Common Logarithms of Numbers.												Naperian.		
	L.	0	1	2	3	4	5	6	7	8	9	P.	P.	No.	Log.	Dif.
450	65	321	331	341	350	360	369	379	389	398	408		10	4.50	1.50408	222
1		418	427	437	447	456	466	475	485	495	504	1	1	4.51	1.50630	221
2		514	523	533	543	552	562	571	581	591	600	2	2	4.52	1.50851	221
3		610	619	629	639	648	658	667	677	686	696	3	3	4.53	1.51072	221
4		706	715	725	734	744	753	763	772	782	792	4	4	4.54	1.51293	220
455		801	811	820	830	839	849	858	868	877	887	5	5	4.55	1.51513	219
6		896	906	916	925	935	944	954	963	973	982	6	6	4.56	1.51732	219
7		992	$\overline{0}01$	$\overline{0}11$	$\overline{0}20$	$\overline{0}30$	$\overline{0}39$	$\overline{0}49$	$\overline{0}58$	$\overline{0}68$	$\overline{0}77$	7	7	4.57	1.51951	219
8	66	087	096	106	115	124	134	143	153	162	172	8	8	4.58	1.52170	218
9		181	191	200	210	219	229	238	247	257	266	9	9	4.59	1.52388	218
460		276	285	295	304	314	323	332	342	351	361			4.60	1.52606	217
1		370	380	389	398	408	417	427	436	445	455	1		4.61	1.52823	216
2		464	474	483	492	502	511	521	530	539	549	2		4.62	1.53039	217
3		558	567	577	586	596	605	614	624	633	642	3		4.63	1.53256	215
4		652	661	671	680	689	699	708	717	727	736	4		4.64	1.53471	216
465		745	755	764	773	783	792	801	811	820	829	5		4.65	1.53687	215
6		839	848	857	867	876	885	894	904	913	922	6		4.66	1.53902	214
7		932	941	950	960	969	978	987	997	$\overline{0}06$	$\overline{0}15$	7		4.67	1.54116	214
8	67	025	034	043	052	062	071	080	089	099	108	8		4.68	1.54330	213
9		117	127	136	145	154	164	173	182	191	201	9		4.69	1.54543	213
470		210	219	228	237	247	256	265	274	284	293		9	4.70	1.54756	213
1		302	311	321	330	339	348	357	367	376	385	1	1	4.71	1.54969	212
2		394	403	413	422	431	440	449	459	468	477	2	2	4.72	1.55181	212
3		486	495	504	514	523	532	541	550	560	569	3	3	4.73	1.55393	211
4		578	587	596	605	614	624	633	642	651	660	4	4	4.74	1.55604	210
475		669	679	688	697	706	715	724	733	742	752	5	4.5	4.75	1.55814	211
6		761	770	779	788	797	806	815	825	834	843	6	5.4	4.76	1.56025	210
7		852	861	870	879	888	897	906	916	925	934	7	6	4.77	1.56235	209
8		943	952	961	970	979	988	997	$\overline{0}06$	$\overline{0}15$	$\overline{0}24$	8	7	4.78	1.56444	209
9	68	034	043	052	061	070	079	088	097	106	115	9	8	4.79	1.56653	209
480		124	133	142	151	160	169	178	187	196	205			4.80	1.56862	208
1		215	224	233	242	251	260	269	278	287	296	1		4.81	1.57070	207
2		305	314	323	332	341	350	359	368	377	386	2		4.82	1.57277	208
3		395	404	413	422	431	440	449	458	467	476	3		4.83	1.57485	206
4		485	494	502	511	520	529	538	547	556	565	4		4.84	1.57691	207
485		574	583	592	601	610	619	628	637	646	655	5		4.85	1.57898	206
6		664	673	681	690	699	708	717	726	735	744	6		4.86	1.58104	205
7		753	762	771	780	789	797	806	815	824	833	7		4.87	1.58309	206
8		842	851	860	869	878	886	895	904	913	922	8		4.88	1.58515	204
9		931	940	949	958	966	975	984	993	$\overline{0}02$	$\overline{0}11$	9		4.89	1.58719	205
490	69	020	028	037	046	055	064	073	082	090	099		8	4.90	1.58924	203
1		108	117	126	135	144	152	161	170	179	188	1	1	4.91	1.59127	204
2		197	205	214	223	232	241	249	258	267	276	2	1.6	4.92	1.59331	203
3		285	294	302	311	320	329	338	346	355	364	3	2.4	4.93	1.59534	203
4		373	381	390	399	408	417	425	434	443	452	4	3	4.94	1.59737	202
495		461	469	478	487	496	504	513	522	531	539	5	4	4.95	1.59939	202
6		548	557	566	574	583	592	601	609	618	627	6	5	4.96	1.60141	201
7		636	644	653	662	671	679	688	697	705	714	7	5.6	4.97	1.60342	201
8		723	732	740	749	758	767	775	784	793	801	8	6.4	4.98	1.60543	201
9		810	819	827	836	845	854	862	871	880	888	9	7	4.99	1.60744	200
500		897	906	914	923	932	940	949	958	966	975			5.00	1.60944	

4.—Logarithms—*Continued*

No.	Common Logarithms of Numbers.												Naperian.		
	L. O	1	2	3	4	5	6	7	8	9	P.	P.	No.	Log.	Dif.
500	69 897	906	914	923	932	940	949	958	966	975		9	5.00	1.60944	200
1	984	992	$\overline{0}01$	$\overline{0}10$	$\overline{0}18$	$\overline{0}27$	$\overline{0}36$	$\overline{0}44$	$\overline{0}53$	$\overline{0}62$	1	1	5.01	1.61144	199
2	70 070	079	088	096	105	114	122	131	140	148	2	2	5.02	1.61343	199
3	157	165	174	183	191	200	209	217	226	234	3	3	5.03	1.61542	199
4	243	252	260	269	278	286	295	303	312	321	4	4	5.04	1.61741	
															198
505	329	338	346	355	364	372	381	389	398	406	5	4.5	5.05	1.61939	198
6	415	424	432	441	449	458	467	475	484	492	6	5.4	5.06	1.62137	197
7	501	509	518	526	535	544	552	561	569	578	7	6	5.07	1.62334	197
8	586	595	603	612	621	629	638	646	655	663	8	7	5.08	1.62531	197
9	672	680	689	697	706	714	723	731	740	749	9	8	5.09	1.62728	
															196
510	757	766	774	783	791	800	808	817	825	834			5.10	1.62924	196
1	842	851	859	868	876	885	893	902	910	919	1		5.11	1.63120	195
2	927	935	944	952	961	969	978	986	995	$\overline{0}03$	2		5.12	1.63315	196
3	71 012	020	029	037	046	054	063	071	079	088	3		5.13	1.63511	194
4	096	105	113	122	130	139	147	155	164	172	4		5.14	1.63705	
															195
515	181	189	198	206	214	223	231	240	248	257	5		5.15	1.63900	194
6	265	273	282	290	299	307	315	324	332	341	6		5.16	1.64094	193
7	349	357	366	374	383	391	399	408	416	425	7		5.17	1.64287	194
8	433	441	450	458	466	475	483	492	500	508	8		5.18	1.64481	192
9	517	525	533	542	550	559	567	575	584	592	9		5.19	1.64673	
															193
520	600	609	617	625	634	642	650	659	667	675		8	5.20	1.64866	192
1	684	692	700	709	717	725	734	742	750	759	1	1	5.21	1.65058	192
2	767	775	784	792	800	809	817	825	834	842	2	1.6	5.22	1.65250	191
3	850	858	867	875	883	892	900	908	917	925	3	2.4	5.23	1.65441	191
4	933	941	950	958	966	975	983	991	999	$\overline{0}08$	4	3	5.24	1.65632	
															191
525	72 016	024	032	041	049	057	066	074	082	090	5	4	5.25	1.65823	190
6	099	107	115	123	132	140	148	156	165	173	6	5	5.26	1.66013	190
7	181	189	198	206	214	222	230	239	247	255	7	5.6	5.27	1.66203	190
8	263	272	280	288	296	304	313	321	329	337	8	6.4	5.28	1.66393	189
9	346	354	362	370	378	387	395	403	411	419	9	7	5.29	1.66582	
															189
530	428	436	444	452	460	469	477	485	493	501			5.30	1.66771	188
1	509	518	526	534	542	550	558	567	575	583	1		5.31	1.66959	188
2	591	599	607	616	624	632	640	648	656	665	2		5.32	1.67147	188
3	673	681	689	697	705	713	722	730	738	746	3		5.33	1.67335	188
4	754	762	770	779	787	795	803	811	819	827	4		5.34	1.67523	
															187
535	835	843	852	860	868	876	884	892	900	908	5		5.35	1.67710	186
6	916	925	933	941	949	957	965	973	981	989	6		5.36	1.67896	187
7	997	$\overline{0}06$	$\overline{0}14$	$\overline{0}22$	$\overline{0}30$	$\overline{0}38$	$\overline{0}46$	$\overline{0}54$	$\overline{0}62$	$\overline{0}70$	7		5.37	1.68083	186
8	73 078	086	094	102	111	119	127	135	143	151	8		5.38	1.68269	186
9	159	167	175	183	191	199	207	215	223	231	9		5.39	1.68455	
															185
540	239	247	255	263	272	280	288	296	304	312		7	5.40	1.68640	185
1	320	328	336	344	352	360	368	376	384	392	1	0.7	5.41	1.68825	185
2	400	408	416	424	432	440	448	456	464	472	2	1.4	5.42	1.69010	184
3	480	488	496	504	512	520	528	536	544	552	3	2	5.43	1.69194	184
4	560	568	576	584	592	600	608	616	624	632	4	3	5.44	1.69378	
															184
545	640	648	656	664	672	679	687	695	703	711	5	3.5	5.45	1.69562	183
6	719	727	735	743	751	759	767	775	783	791	6	4.2	5.46	1.69745	183
7	799	807	815	823	830	838	846	854	862	870	7	5	5.47	1.69928	183
8	878	886	894	902	910	918	926	933	941	949	8	5.6	5.48	1.70111	182
9	957	965	973	981	989	997	005	$\overline{0}13$	$\overline{0}20$	$\overline{0}28$	9	6.3	5.49	1.70293	
															182
550	74 036	044	052	060	068	076	084	092	099	107			5.50	1.70475	

4.—Logarithms—*Continued*

No.	Common Logarithms of Numbers. L. 0	1	2	3	4	5	6	7	8	9	P. P.		Naperian. No.	Log.	Dif.
550	74 036	044	052	060	068	076	084	092	099	107			5.50	1.70475	181
1	115	123	131	139	147	155	162	170	178	186	1		5.51	1.70656	182
2	194	202	210	218	225	233	241	249	257	265	2		5.52	1.70838	181
3	273	280	288	296	304	312	320	327	335	343	3		5.53	1.71019	180
4	351	359	367	374	382	390	398	406	414	421	4		5.54	1.71199	181
555	429	437	445	453	461	468	476	484	492	500	5		5.55	1.71380	180
6	507	515	523	531	539	547	554	562	570	578	6		5.56	1.71560	180
7	586	593	601	609	617	624	632	640	648	656	7		5.57	1.71740	179
8	663	671	679	687	695	702	710	718	726	733	8		5.58	1.71919	179
9	741	749	757	764	772	780	788	796	803	811	9		5.59	1.72098	179
560	819	827	834	842	850	858	865	873	881	889		8	5.60	1.72277	178
1	896	904	912	920	927	935	943	950	958	966	1	1	5.61	1.72455	178
2	974	981	989	997	0̄05	0̄12	0̄20	0̄28	0̄35	0̄43	2	1.6	5.62	1.72633	178
3	75 051	059	066	074	082	089	097	105	113	120	3	2.4	5.63	1.72811	177
4	128	136	143	151	159	166	174	182	189	197	4	3	5.64	1.72988	178
565	205	213	220	228	236	243	251	259	266	274	5	4	5.65	1.73166	176
6	282	289	297	305	312	320	328	335	343	351	6	5	5.66	1.73342	177
7	358	366	374	381	389	397	404	412	420	427	7	5.6	5.67	1.73519	176
8	435	442	450	458	465	473	481	488	496	504	8	6.4	5.68	1.73695	176
9	511	519	526	534	542	549	557	565	572	580	9	7	5.69	1.73871	176
570	587	595	603	610	618	626	633	641	648	656			5.70	1.74047	175
1	664	671	679	686	694	702	709	717	724	732	1		5.71	1.74222	175
2	740	747	755	762	770	778	785	793	800	808	2		5.72	1.74397	175
3	815	823	831	838	846	853	861	868	876	884	3		5.73	1.74572	174
4	891	899	906	914	921	929	937	944	952	959	4		5.74	1.74746	174
575	967	974	982	989	997	0̄05	0̄12	0̄20	0̄27	0̄35	5		5.75	1.74920	174
6	76 042	050	057	065	072	080	087	095	103	110	6		5.76	1.75094	173
7	118	125	133	140	148	155	163	170	178	185	7		5.77	1.75267	173
8	193	200	208	215	223	230	238	245	253	260	8		5.78	1.75440	173
9	268	275	283	290	298	305	313	320	328	335	9		5.79	1.75613	173
580	343	350	358	365	373	380	388	395	403	410		7	5.80	1.75786	172
1	418	425	433	440	448	455	462	470	477	485	1	0.7	5.81	1.75958	172
2	492	500	507	515	522	530	537	545	552	559	2	1.4	5.82	1.76130	172
3	567	574	582	589	597	604	612	619	626	634	3	2	5.83	1.76302	171
4	641	649	656	664	671	678	686	693	701	708	4	3	5.84	1.76473	171
585	716	723	730	738	745	753	760	768	775	782	5	3.5	5.85	1.76644	171
6	790	797	805	812	819	827	834	842	849	856	6	4.2	5.86	1.76815	170
7	864	871	879	886	893	901	908	916	923	930	7	5	5.87	1.76985	171
8	938	945	953	960	967	975	982	989	997	0̄04	8	5.6	5.88	1.77156	170
9	77 012	019	026	034	041	048	056	063	070	078	9	6.3	5.89	1.77326	169
590	085	093	100	107	115	122	129	137	144	151			5.90	1.77495	170
1	159	166	173	181	188	195	203	210	217	225	1		5.91	1.77665	169
2	232	240	247	254	262	269	276	283	291	298	2		5.92	1.77834	168
3	305	313	320	327	335	342	349	357	364	371	3		5.93	1.78002	169
4	379	386	393	401	408	415	422	430	437	444	4		5.94	1.78171	168
595	452	459	466	474	481	488	495	503	510	517	5		5.95	1.78339	168
6	525	532	539	546	554	561	568	576	583	590	6		5.96	1.78507	168
7	597	605	612	619	627	634	641	648	656	663	7		5.97	1.78675	167
8	670	677	685	692	699	706	714	721	728	735	8		5.98	1.78842	167
9	743	750	757	764	772	779	786	793	801	808	9		5.99	1.79009	167
600	815	822	830	837	844	851	859	866	873	880			6.00	1.79176	

4.—Logarithms—*Continued*

No.	Common Logarithms of Numbers.												Naperian.		
	L. 0	1	2	3	4	5	6	7	8	9	P.	P.	No.	Log.	Dif.
600	77 815	822	830	837	844	851	859	866	873	880		8	6.00	1.79176	166
1	887	895	902	909	916	924	931	938	945	952	1	1	6.01	1.79342	167
2	960	967	974	981	988	996	0̄03	0̄10	0̄17	0̄25	2	1.6	6.02	1.79509	166
3	78 032	039	046	053	061	068	075	082	089	097	3	2.4	6.03	1.79675	165
4	104	111	118	125	132	140	147	154	161	168	4	3	6.04	1.79840	166
605	176	183	190	197	204	211	219	226	233	240	5	4	6.05	1.80006	165
6	247	254	262	269	276	283	290	297	305	312	6	5	6.06	1.80171	165
7	319	326	333	340	347	355	362	369	376	383	7	5.6	6.07	1.80336	164
8	390	398	405	412	419	426	433	440	447	455	8	6.4	6.08	1.80500	165
9	462	469	476	483	490	497	504	512	519	526	9	7	6.09	1.80665	164
610	533	540	547	554	561	569	576	583	590	597			6.10	1.80829	164
1	604	611	618	625	633	640	647	654	661	668	1		6.11	1.80993	163
2	675	682	689	696	704	711	718	725	732	739	2		6.12	1.81156	163
3	746	753	760	767	774	781	789	796	802	810	3		6.13	1.81319	163
4	817	824	831	838	845	852	859	866	873	880	4		6.14	1.81482	163
615	888	895	902	909	916	923	930	937	944	951	5		6.15	1.81645	163
6	858	965	972	979	986	993	0̄00	0̄07	0̄14	0̄21	6		6.16	1.81808	162
7	79 029	036	043	050	057	064	071	078	085	092	7		6.17	1.81970	162
8	099	106	113	120	127	134	141	148	155	162	8		6.18	1.82132	162
9	169	176	183	190	197	204	211	218	225	232	9		6.19	1.82294	161
620	239	246	253	260	267	274	281	288	295	302		7	6.20	1.82455	161
1	309	316	323	330	337	344	351	358	365	372	1	0.7	6.21	1.82616	161
2	379	386	393	400	407	414	421	428	435	442	2	1.4	6.22	1.82777	161
3	449	456	463	470	477	484	491	498	505	511	3	2	6.23	1.82938	160
4	518	525	532	539	546	553	560	567	574	581	4	3	6.24	1.83098	160
625	588	595	602	609	616	623	630	637	644	650	5	3.5	6.25	1.83258	160
6	657	664	671	678	685	692	699	706	713	720	6	4.2	6.26	1.83418	160
7	727	734	741	748	754	761	768	775	782	789	7	5	6.27	1.83578	159
8	796	803	810	817	824	831	837	844	851	858	8	5.6	6.28	1.83737	159
9	865	872	879	886	893	900	906	913	920	927	9	6.3	6.29	1.83896	159
630	934	941	948	955	962	969	975	982	989	996			6.30	1.84055	159
1	80 003	010	017	024	030	037	044	051	058	065	1		6.31	1.84214	158
2	072	079	085	092	099	106	113	120	127	134	2		6.32	1.84372	158
3	140	147	154	161	168	175	182	188	195	202	3		6.33	1.84530	158
4	209	216	223	229	236	243	250	257	264	271	4		6.34	1.84688	157
635	277	284	291	298	305	312	318	325	332	339	5		6.35	1.84845	158
6	346	353	359	366	373	380	387	393	400	407	6		6.36	1.85003	157
7	414	421	428	434	441	448	455	462	468	475	7		6.37	1.85160	157
8	482	489	496	502	509	516	523	530	536	543	8		6.38	1.85317	156
9	500	557	564	570	577	584	591	598	604	611	9		6.39	1.85473	157
640	618	625	632	638	645	652	659	665	672	679		6	6.40	1.85630	156
1	686	693	699	706	713	720	726	733	740	747	1	0.6	6.41	1.85786	156
2	754	760	767	774	781	787	794	801	808	814	2	1.2	6.42	1.85942	155
3	821	828	835	841	848	855	862	868	875	882	3	1.8	6.43	1.86097	156
4	889	895	902	909	916	922	929	936	943	949	4	2.4	6.44	1.86253	155
645	956	963	969	976	983	990	996	0̄03	0̄10	0̄17	5	3	6.45	1.86408	155
6	81 023	030	037	043	050	057	064	070	077	084	6	3.6	6.46	1.86563	155
7	090	097	104	111	117	124	131	137	144	151	7	4.2	6.47	1.86718	154
8	158	164	171	178	184	191	198	024	211	218	8	4.8	6.48	1.86872	154
9	224	231	238	245	251	258	265	271	278	285	9	5.4	6.49	1.87026	154
650	291	298	305	311	318	325	331	338	345	351			6.50	1.87180	

4.—LOGARITHMS—*Continued*

No.	Common Logarithms of Numbers.													Naperian.		
	L.	0	1	2	3	4	5	6	7	8	9	P. P.		No.	Log.	Dif.
650	81	291	298	305	311	318	325	331	338	345	351			6.50	1.87180	154
1		358	365	371	378	385	391	398	405	411	418	1		6.51	1.87334	153
2		425	431	438	445	451	458	465	471	478	485	2		6.52	1.87487	154
3		491	498	505	511	518	525	531	538	544	551	3		6.53	1.87641	153
4		558	564	571	578	584	591	598	604	611	617	4		6.54	1.87794	153
655		624	631	637	644	651	657	664	671	677	684	5		6.55	1.87947	152
6		690	697	704	710	717	723	730	737	743	750	6		6.56	1.88099	152
7		757	763	770	776	783	790	796	803	809	816	7		6.57	1.88251	152
8		823	829	836	842	849	856	862	869	875	882	8		6.58	1.88403	152
9		889	895	902	908	915	921	928	935	941	948	9		6.59	1.88555	152
660		954	961	968	974	981	987	994	**000**	$\bar{0}07$	$\bar{0}14$		**7**	6.60	1.88707	151
1	82	020	027	033	040	046	053	060	066	073	079	1	0.7	6.61	1.88858	152
2		086	092	099	105	112	119	125	132	138	145	2	1.4	6.62	1.89010	150
3		151	158	164	171	178	184	191	197	204	210	3	2	6.63	1.89160	151
4		217	223	230	236	243	249	256	263	269	276	4	3	6.64	1.89311	151
665		282	289	295	302	308	315	321	328	334	341	5	3.5	6.65	1.89462	150
6		347	354	360	367	373	380	387	393	400	406	6	4.2	6.66	1.89612	150
7		413	419	426	432	439	445	452	458	465	471	7	5	6.67	1.89762	150
8		478	484	491	497	504	510	517	523	530	536	8	5.6	6.68	1.89912	149
9		543	549	556	562	569	575	582	588	595	601	9	6.3	6.69	1.90061	150
670		607	614	620	627	633	640	646	653	659	666			6.70	1.90211	149
1		672	679	685	692	698	705	711	718	724	730	1		6.71	1.90360	149
2		737	743	750	756	763	769	776	782	789	795	2		6.72	1.90509	149
3		802	808	814	821	827	834	840	847	853	860	3		6.73	1.90658	148
4		866	872	879	885	892	898	905	911	918	924	4		6.74	1.90806	148
675		930	937	943	950	956	963	969	975	982	988	5		6.75	1.90954	148
6		995	$\bar{0}$**01**	008	$\bar{0}14$	$\bar{0}20$	$\bar{0}27$	$\bar{0}33$	$\bar{0}40$	$\bar{0}46$	$\bar{0}52$	6		6.76	1.91102	148
7	83	059	065	072	078	085	091	097	104	110	117	7		6.77	1.91250	148
8		123	129	136	142	149	155	161	168	174	181	8		6.78	1.91398	147
9		187	193	200	206	213	219	225	232	238	245	9		6.79	1.91545	147
680		251	257	264	270	276	283	289	296	302	308		**6**	6.80	1.91692	147
1		315	321	327	334	340	347	353	359	366	372	1	0.6	6.81	1.91839	147
2		378	385	391	398	404	410	417	423	429	436	2	1.2	6.82	1.91986	146
3		442	448	455	461	467	474	480	487	493	499	3	1.8	6.83	1.92132	147
4		506	512	518	525	531	537	544	550	556	563	4	2.4	6.84	1.92279	146
685		569	575	582	588	594	601	607	613	620	626	5	3	6.85	1.92425	146
6		632	639	645	651	658	664	670	677	683	689	6	3.6	6.86	1.92571	145
7		696	702	708	715	721	727	734	740	746	753	7	4.2	6.87	1.92716	146
8		759	765	771	778	784	790	797	803	809	816	8	4.8	6.88	1.92862	145
9		822	828	835	841	847	853	860	866	872	879	9	5.4	6.89	1.93007	145
690		885	891	897	904	910	916	923	929	935	942			6.90	1.93152	145
1		948	954	960	967	973	979	985	992	998	$\bar{0}$**04**	1		6.91	1.93297	145
2	84	011	017	023	029	036	042	048	055	061	067	2		6.92	1.93442	144
3		073	080	086	092	098	105	111	117	123	130	3		6.93	1.93586	144
4		136	142	148	155	161	167	173	180	186	192	4		6.94	1.93730	144
695		198	205	211	217	223	230	236	242	248	255	5		6.95	1.93874	144
6		261	267	273	280	286	292	298	305	311	317	6		6.96	1.94018	144
7		323	330	336	342	348	354	361	367	373	379	7		6.97	1.94162	143
8		386	392	398	404	410	417	423	429	435	442	8		6.98	1.94305	143
9		448	454	460	466	473	479	485	491	497	504	9		6.99	1.94448	143
700		510	516	522	528	535	541	547	553	559	566			7.00	1.94591	

4.—LOGARITHMS—*Continued*

No.	Common Logarithms of Numbers.												Naperian.		
	L. 0	1	2	3	4	5	6	7	8	9	P. P.		No.	Log.	Dif.
700	84 510	516	522	528	535	541	547	553	559	566		7	7.00	1.94591	
1	572	578	584	590	597	603	609	615	621	628	1	0.7	7.01	1.94734	143
2	634	640	646	652	658	665	671	677	683	689	2	1.4	7.02	1.94876	142
3	696	702	708	714	720	726	733	739	745	751	3	2	7.03	1.95019	143
4	757	763	770	776	782	788	794	800	807	813	4	3	7.04	1.95161	142
															142
705	819	825	831	837	844	850	856	862	868	874	5	3.5	7.05	1.95303	
6	880	887	893	899	905	911	917	924	930	936	6	4.2	7.06	1.95444	141
7	942	948	954	960	967	973	979	985	991	997	7	5	7.07	1.95586	142
8	85 003	009	016	022	028	034	040	046	052	058	8	5.6	7.08	1.95727	141
9	065	071	077	083	089	095	101	107	114	120	9	6.3	7.09	1.95869	142
															140
710	126	132	138	144	150	156	163	169	175	181			7.10	1.96009	
1	187	193	199	205	211	217	224	230	236	242	1		7.11	1.96150	141
2	248	254	260	266	272	278	285	291	297	303	2		7.12	1.96291	141
3	309	315	321	327	333	339	345	352	358	364	3		7.13	1.96431	140
4	370	376	382	388	394	400	406	412	418	425	4		7.14	1.96571	140
															140
715	431	437	443	449	455	461	467	473	479	485	5		7.15	1.96711	
6	491	497	503	509	516	522	528	534	540	546	6		7.16	1.96851	140
7	552	558	564	570	576	582	588	594	600	606	7		7.17	1.96991	140
8	612	618	625	631	637	643	649	655	661	667	8		7.18	1.97130	139
9	673	679	685	691	697	703	709	715	721	727	9		7.19	1.97269	139
															139
720	733	739	745	751	757	763	769	775	781	788		6	7.20	1.97408	
1	794	800	806	812	818	824	830	836	842	848	1	0.6	7.21	1.97547	139
2	854	860	866	872	878	884	890	896	902	908	2	1.2	7.22	1.97685	138
3	914	920	926	932	938	944	950	956	962	968	3	1.8	7.23	1.97824	139
4	974	980	986	992	998	$\overline{004}$	$\overline{010}$	$\overline{016}$	$\overline{022}$	$\overline{028}$	4	2.4	7.24	1.97962	138
															138
725	86 034	040	046	052	058	064	070	076	082	088	5	3	7.25	1.98100	
6	094	100	106	112	118	124	130	136	141	147	6	3.6	7.26	1.98238	138
7	153	159	165	171	177	183	189	195	201	207	7	4.2	7.27	1.98376	138
8	213	219	225	231	237	243	249	255	261	267	8	4.8	7.28	1.98513	137
9	273	279	285	291	297	303	308	314	320	326	9	5.4	7.29	1.98650	137
															137
730	332	338	344	350	356	362	368	374	380	386			7.30	1.98787	
1	392	398	404	410	415	421	427	433	439	445	1		7.31	1.98924	137
2	451	457	463	469	475	481	487	493	499	504	2		7.32	1.99061	137
3	510	516	522	528	534	540	546	552	558	564	3		7.33	1.99198	137
4	570	576	581	587	593	599	605	611	617	623	4		7.34	1.99334	136
															136
735	629	635	641	646	652	658	664	670	676	682	5		7.35	1.99470	
6	688	694	700	705	711	717	723	729	735	741	6		7.36	1.99606	136
7	747	753	759	764	770	776	782	788	794	800	7		7.37	1.99742	136
8	806	812	817	823	829	835	841	847	853	859	8		7.38	1.99877	135
9	864	870	876	882	888	894	900	906	911	917	9		7.39	2.00013	136
															135
740	923	929	935	941	947	953	958	964	970	976		5	7.40	2.00148	
1	982	988	994	999	$\overline{005}$	$\overline{011}$	$\overline{017}$	$\overline{023}$	$\overline{029}$	$\overline{035}$	1	0.5	7.41	2.00283	135
2	87 040	046	052	058	064	070	075	081	087	093	2	1	7.42	2.00418	135
3	099	105	111	116	122	128	134	140	146	151	3	1.5	7.43	2.00553	135
4	157	163	169	175	181	186	192	198	204	210	4	2	7.44	2.00687	134
															134
745	216	221	227	233	239	245	251	256	262	268	5	2.5	7.45	2.00821	
6	274	280	286	291	297	303	309	315	320	326	6	3	7.46	2.00956	135
7	332	338	344	349	355	361	367	373	379	384	7	3.5	7.47	2.01089	133
8	390	396	402	408	413	419	425	431	437	442	8	4	7 48	2.01223	134
9	448	454	460	466	471	477	483	489	495	500	9	4.5	7.49	2.01357	134
															133
750	506	512	518	523	529	535	541	547	552	558			7.50	2.01490	

4.—Logarithms—*Continued*

No.	Common Logarithms of Numbers.													Naperian.		
	L.	0	1	2	3	4	5	6	7	8	9	P.	P.	No.	Log.	Dif.
750	87	506	512	518	523	529	535	541	547	552	558			7.50	2.01490	134
1		564	570	576	581	587	593	599	604	610	616	1		7.51	2.01624	133
2		622	628	633	639	645	651	656	662	668	674	2		7.52	2.01757	133
3		679	685	691	697	703	708	714	720	726	731	3		7.53	2.01890	132
4		737	743	749	754	760	766	772	777	783	789	4		7.54	2.02022	133
755		795	800	806	812	818	823	829	835	841	846	5		7.55	2.02155	132
6		852	858	864	869	875	881	887	892	898	904	6		7.56	2.02287	132
7		910	915	921	927	933	938	944	950	955	961	7		7.57	2.02419	132
8		967	973	978	984	990	996	$\bar{0}01$	$\bar{0}07$	$\bar{0}13$	$\bar{0}18$	8		7.58	2.02551	132
9	88	024	030	036	041	047	053	058	064	070	076	9		7.59	2.02683	132
760		081	087	093	098	104	110	116	121	127	133		6	7.60	2.02815	131
1		138	144	150	156	161	167	173	178	184	190	1	0.6	7.61	2.02946	132
2		195	201	207	213	218	224	230	235	241	247	2	1.2	7.62	2.03078	131
3		252	258	264	270	275	281	287	292	298	304	3	1.8	7.63	2.03209	131
4		309	315	321	326	332	338	343	349	355	360	4	2.4	7.64	2.03340	131
765		366	372	377	383	389	395	400	406	412	417	5	3	7.65	2.03471	130
6		423	429	434	440	446	451	457	463	468	474	6	3.6	7.66	2.03601	131
7		480	485	491	497	502	508	513	519	525	530	7	4.2	7.67	2.03732	130
8		536	542	547	553	559	564	570	576	581	587	8	4.8	7.68	2.03862	130
9		593	598	604	610	615	621	627	632	638	643	9	5.4	7.69	2.03992	130
770		649	655	660	666	672	677	683	689	694	700			7.70	2.04122	130
1		705	711	717	722	728	734	739	745	750	756	1		7.71	2.04252	129
2		762	767	773	779	784	790	795	801	807	812	2		7.72	2.04381	130
3		818	824	829	835	840	846	852	857	863	868	3		7.73	2.04511	129
4		874	880	885	891	897	902	908	913	919	925	4		7.74	2.04640	129
775		930	936	941	947	953	958	964	969	975	981	5		7.75	2.04769	129
6		986	992	997	$\bar{0}03$	$\bar{0}09$	$\bar{0}14$	$\bar{0}20$	$\bar{0}25$	$\bar{0}31$	$\bar{0}37$	6		7.76	2.04898	129
7	89	042	048	053	059	064	070	076	081	087	092	7		7.77	2.05027	129
8		098	104	109	115	120	126	131	137	143	148	8		7.78	2.05156	128
9		154	159	165	170	176	182	187	193	198	204	9		7.79	2.05284	128
780		209	215	221	226	232	237	243	248	254	260		5	7.80	2.05412	128
1		265	271	276	282	287	293	298	304	310	315	1	0.5	7.81	2.05540	128
2		321	326	332	337	343	348	354	360	365	371	2	1	7.82	2.05668	128
3		376	382	387	393	398	404	409	415	421	426	3	1.5	7.83	2.05796	128
4		432	437	443	448	454	459	465	470	476	481	4	2	7.84	2.05924	127
785		487	492	498	504	509	515	520	526	531	537	5	2.5	7.85	2.06051	128
6		542	548	553	559	564	570	575	581	586	592	6	3	7.86	2.06179	127
7		597	603	609	614	620	625	631	636	642	647	7	3.5	7.87	2.06306	127
8		653	658	664	669	675	680	686	691	697	702	8	4	7.88	2.06433	127
9		708	713	719	724	730	735	741	746	752	757	9	4.5	7.89	2,06560	126
790		763	768	774	779	785	790	796	801	807	812			7.90	2.06686	127
1		818	823	829	834	840	845	851	856	862	867	1		7.91	2.06813	126
2		873	878	883	889	894	900	905	911	916	922	2		7.92	2.06939	126
3		927	933	938	944	949	955	960	966	971	977	3		7.93	2.07065	126
4		982	988	993	998	$\bar{0}04$	$\bar{0}09$	$\bar{0}15$	$\bar{0}20$	$\bar{0}26$	$\bar{0}31$	4		7.94	2.07191	126
795	90	037	042	048	053	059	064	069	075	080	086	5		7.95	2.07317	126
6		091	097	102	108	113	119	124	129	135	140	6		7.96	2.07443	125
7		146	151	157	162	168	173	179	184	189	195	7		7.97	2.07568	126
8		200	206	211	217	222	227	233	238	244	249	8		7.98	2.07694	125
9		255	260	266	271	276	282	287	293	298	304	9		7.99	2.07819	125
800		309	314	320	325	331	336	342	347	352	358			8.00	2.07944	

4.—Logarithms—*Continued*

No.	Common Logarithms of Numbers.													Naperian.		
	L.	0	1	2	3	4	5	6	7	8	9	P.	P.	No.	Log.	Dif.
800	90	309	314	320	325	331	336	342	347	352	358			8.00	2.07944	
1		363	369	374	380	385	390	396	401	407	412	1		8.01	2.08069	125
2		417	423	428	434	439	445	450	455	461	466	2		8.02	2.08194	125
3		472	477	482	488	493	499	504	509	515	520	3		8.03	2.08318	124
4		526	531	536	542	547	553	558	563	569	574	4		8.04	2.08443	125
																124
805		580	585	590	596	601	607	612	617	623	628	5		8.05	2.08567	
6		634	639	644	650	655	660	666	671	677	682	6		8.06	2.08691	124
7		687	693	698	703	709	714	720	725	730	736	7		8.07	2.08815	124
8		741	747	752	757	763	768	773	779	784	789	8		8.08	2.08939	124
9		795	800	806	811	816	822	827	832	838	843	9		8.09	2.09063	124
																123
810		849	854	859	865	870	875	881	886	891	897		6	8.10	2.09186	
1		902	907	913	918	924	929	934	940	945	950	1	0.6	8.11	2.09310	124
2		956	961	966	972	977	982	988	993	998	004	2	1.2	8.12	2.09433	123
3	91	009	014	020	025	030	036	041	046	052	057	3	1.8	8.13	2.09556	123
4		062	068	073	078	084	089	094	100	105	110	4	2.4	8.14	2.09679	123
																123
815		116	121	126	132	137	142	148	153	158	164	5	3	8.15	2.09802	
6		169	174	180	185	190	196	201	206	212	217	6	3.6	8.16	2.09924	122
7		222	228	233	238	243	249	254	259	265	270	7	4.2	8.17	2.10047	123
8		275	281	286	291	297	302	307	312	318	323	8	4.8	8.18	2.10169	122
9		328	334	339	344	350	355	360	365	371	376	9	5.4	8.19	2.10291	122
																122
820		381	387	392	397	403	408	413	418	424	429			8.20	2.10413	
1		434	440	445	450	455	461	466	471	477	482	1		8.21	2.10535	122
2		487	492	498	503	508	514	519	524	529	535	2		8.22	2.10657	122
3		540	545	551	556	561	566	572	577	582	587	3		8.23	2.10779	122
4		593	598	603	609	614	619	624	630	635	640	4		8.24	2.10900	121
																121
825		645	651	656	661	666	672	677	682	687	693	5		8.25	2.11021	
6		698	703	709	714	719	724	730	735	740	745	6		8.26	2.11142	121
7		751	756	761	766	772	777	782	787	793	798	7		8.27	2.11263	121
8		803	808	814	819	824	829	834	840	845	850	8		8.28	2.11384	121
9		855	861	866	871	876	882	887	892	897	903	9		8.29	2.11505	121
																121
830		908	913	918	924	929	934	939	944	950	955		5	8.30	2.11626	
1		960	965	971	976	981	986	991	997	002	007	1	0.5	8.31	2.11746	120
2	92	012	018	023	028	033	038	044	049	054	059	2	1	8.32	2.11866	120
3		065	070	075	080	085	091	096	101	106	111	3	1.5	8.33	2.11986	120
4		117	122	127	132	137	143	148	153	158	163	4	2	8.34	2.12106	120
																120
835		169	174	179	184	189	195	200	205	210	215	5	2.5	8.35	2.12226	
6		221	226	231	236	241	247	252	257	262	267	6	3	8.36	2.12346	120
7		273	278	283	288	293	298	304	309	314	319	7	3.5	8.37	2.12465	119
8		324	330	335	340	345	350	355	361	366	371	8	4	8.38	2.12585	120
9		376	381	387	392	397	402	407	412	418	423	9	4.5	8.39	2.12704	119
																119
840		428	433	438	443	449	454	459	464	469	474			8.40	2.12823	
1		480	485	490	495	500	505	511	516	521	526	1		8.41	2.12942	119
2		531	536	542	547	552	557	562	567	572	578	2		8.42	2.13061	119
3		583	588	593	598	603	609	614	619	624	629	3		8.43	2.13180	119
4		634	639	645	650	655	660	665	670	675	681	4		8.44	2.13298	118
																119
845		686	691	696	701	706	711	716	722	727	732	5		8.45	2.13417	
6		737	742	747	752	758	763	768	773	778	783	6		8.46	2.13535	118
7		788	793	799	804	809	814	819	824	829	834	7		8.47	2.13653	118
8		840	845	850	855	860	865	870	875	881	886	8		8.48	2.13771	118
9		891	896	901	906	911	916	921	927	932	937	9		8.49	2.13889	118
																118
850		942	947	952	957	962	967	973	978	983	988			8.50	2.14007	

4.—Logarithms—*Continued*

No.	Common Logarithms of Numbers.												Naperian.		
	L. 0	1	2	3	4	5	6	7	8	9	P.	P.	No.	Log.	Dif.
850	92 942	947	952	957	962	967	973	978	983	988		6	8.50	2.14007	117
1	993	998	$\bar{0}$03	$\bar{0}$08	$\bar{0}$13	$\bar{0}$18	$\bar{0}$24	$\bar{0}$29	$\bar{0}$34	$\bar{0}$39	1	0.6	8.51	2.14124	118
2	93 044	049	054	059	064	069	075	080	085	090	2	1.2	8.52	2.14242	117
3	095	100	105	110	115	120	125	131	136	141	3	1.8	8.53	2.14359	117
4	146	151	156	161	166	171	176	181	186	192	4	2.4	8.54	2.14476	117
855	197	202	207	212	217	222	227	232	237	242	5	3	8.55	2.14593	117
6	247	252	258	263	268	273	278	283	288	293	6	3.6	8.56	2.14710	117
7	298	303	308	313	318	323	328	334	339	344	7	4.2	8.57	2.14827	116
8	349	354	359	364	369	374	379	384	389	394	8	4.8	8.58	2.14943	117
9	399	404	409	414	420	425	430	435	440	445	9	5.4	8.59	2.15060	116
860	450	455	460	465	470	475	480	485	490	495			8.60	2.15176	116
1	500	505	510	515	520	526	531	536	541	546	1		8.61	2.15292	117
2	551	556	561	566	571	576	581	586	591	596	2		8.62	2.15409	115
3	601	606	611	616	621	626	631	636	641	646	3		8.63	2.15524	116
4	651	656	661	666	671	676	682	687	692	697	4		8.64	2.15640	116
865	702	707	712	717	722	727	732	737	742	747	5		8.65	2.15756	115
6	752	757	762	767	772	777	782	787	792	797	6		8.66	2.15871	116
7	802	807	812	817	822	827	832	837	842	847	7		8.67	2.15987	115
8	852	857	862	867	872	877	882	887	892	897	8		8.68	2.16102	115
9	902	907	912	917	922	927	932	937	942	947	9		8.69	2.16217	115
870	952	957	962	967	972	977	982	987	992	997		5	8.70	2.16332	115
1	94 002	007	012	017	022	027	032	037	042	047	1	0.5	8.71	2.16447	115
2	052	057	062	067	072	077	082	086	091	096	2	1	8.72	2.16562	115
3	101	106	111	116	121	126	131	136	141	146	3	1.5	8.73	2.16677	114
4	151	156	161	166	171	176	181	186	191	196	4	2	8.74	2.16791	114
875	201	206	211	216	221	226	231	236	240	245	5	2.5	8.75	2.16905	115
6	250	255	260	265	270	275	280	285	290	295	6	3	8.76	2.17020	114
7	300	305	310	315	320	325	330	335	340	345	7	3 5	8.77	2.17134	114
8	349	354	359	364	369	374	379	384	389	394	8	4	8.78	2.17248	113
9	399	404	409	414	419	424	429	433	438	443	9	4 5	8.79	2.17361	114
880	448	453	458	463	468	473	478	483	488	493			8.80	2.17475	114
1	498	503	507	512	517	522	527	532	537	542	1		8.81	2.17589	113
2	547	552	557	562	567	571	576	581	586	591	2		8.82	2.17702	114
3	596	601	606	611	616	621	626	630	635	640	3		8.83	2.17816	113
4	645	650	655	660	665	670	675	680	685	689	4		8.84	2.17929	113
885	694	699	704	709	714	719	724	729	734	738	5		8.85	2.18042	113
6	743	748	753	758	763	768	773	778	783	787	6		8.86	2.18155	112
7	792	797	802	807	812	817	822	827	832	836	7		8.87	2.18267	113
8	841	846	851	856	861	866	871	876	880	885	8		8.88	2.18380	113
9	890	895	900	905	910	915	919	924	929	934	9		8.89	2.18493	112
890	939	944	949	954	959	963	968	973	978	983		4	8.90	2.18605	112
1	988	993	998	$\bar{0}$02	$\bar{0}$07	$\bar{0}$12	$\bar{0}$17	$\bar{0}$22	$\bar{0}$27	$\bar{0}$32	1	0.4	8.91	2.18717	113
2	95 036	041	046	051	056	061	066	071	075	080	2	0.8	8.92	2.18830	112
3	085	090	095	100	105	109	114	119	124	129	3	1.2	8.93	2.18942	112
4	134	139	143	148	153	158	163	168	173	177	4	1.6	8.94	2.19054	111
895	182	187	192	197	202	207	211	216	221	226	5	2	8.95	2.19165	112
6	231	236	240	245	250	255	260	265	270	274	6	2.4	8.96	2.19277	112
7	279	284	289	294	299	303	308	313	318	323	7	2.8	8.97	2.19389	111
8	328	332	337	342	347	352	357	361	366	371	8	3.2	8.98	2.19500	111
9	376	381	386	390	395	400	405	410	415	419	9	3.6	8.99	2.19611	111
900	424	429	434	439	444	448	453	458	463	468			9.00	2.19722	

4.—Logarithms—*Continued*

No.	Common Logarithms of Numbers.												Naperian.		
	L. 0	1	2	3	4	5	6	7	8	9	P. P.		No.	Log.	Dif.
900	95 424	429	434	439	444	448	453	458	463	468			9.00	2.19722	112
1	472	477	482	487	492	497	501	506	511	516	1		9.01	2.19834	110
2	521	525	530	535	540	545	550	554	559	564	2		9.02	2.19944	111
3	569	574	578	583	588	593	598	602	607	612	3		9.03	2.20055	111
4	617	622	626	631	636	641	646	650	655	660	4		9.04	2.20166	110
905	665	670	674	679	684	689	694	698	703	708	5		9.05	2.20276	111
6	713	718	722	727	732	737	742	746	751	756	6		9.06	2.20387	110
7	761	766	770	775	780	785	789	794	799	804	7		9.07	2.20497	110
8	809	813	818	823	828	832	837	842	847	852	8		9.08	2.20607	110
9	856	861	866	871	875	880	885	890	895	899	9		9.09	2.20717	110
910	904	909	914	918	923	928	933	938	942	947		5	9.10	2.20827	110
1	952	957	961	966	971	976	980	985	990	995	1	0.5	9.11	2.20937	110
2	999	0̄04	0̄09	0̄14	0̄19	0̄23	0̄28	0̄33	0̄38	0̄42	2	1	9.12	2.21047	110
3	96 047	052	057	061	066	071	076	080	085	090	3	1.5	9.13	2.21157	109
4	095	099	104	109	114	118	123	128	133	137	4	2	9.14	2.21266	109
915	142	147	152	156	161	166	171	175	180	185	5	2.5	9.15	2.21375	110
6	190	194	199	204	209	213	218	223	227	232	6	3	9.16	2.21485	109
7	237	242	246	251	256	261	265	270	275	280	7	3.5	9.17	2.21594	109
8	284	289	294	298	303	308	313	317	322	327	8	4	9.18	2.21703	109
9	332	336	341	346	350	355	360	365	369	374	9	4.5	9.19	2.21812	108
920	379	384	388	393	398	402	407	412	417	421			9.20	2.21920	109
1	426	431	435	440	445	450	454	459	464	468	1		9.21	2.22029	109
2	473	478	483	487	492	497	501	506	511	515	2		9.22	2.22138	108
3	520	525	530	534	539	544	548	553	558	562	3		9.23	2.22246	108
4	567	572	577	581	586	591	595	600	605	609	4		9.24	2.22354	108
925	614	619	624	628	633	638	642	647	652	656	5		9.25	2.22462	108
6	661	666	670	675	680	685	689	694	699	703	6		9.26	2.22570	108
7	708	713	717	722	727	731	736	741	745	750	7		9.27	2.22678	108
8	755	759	764	769	774	778	783	788	792	797	8		9.28	2.22786	108
9	802	806	811	816	820	825	830	834	839	844	9		9.29	2.22894	107
930	848	853	858	862	867	872	876	881	886	890		4	9.30	2.23001	108
1	895	900	904	909	914	918	923	928	932	937	1	0.4	9.31	2.23109	107
2	942	946	951	956	960	965	970	974	979	984	2	0.8	9.32	2.23216	108
3	988	993	997	0̄02	0̄07	0̄11	0̄16	0̄21	0̄25	0̄30	3	1.2	9.33	2.23324	107
4	97 035	039	044	049	053	058	063	067	072	077	4	1.6	9.34	2.23431	107
935	081	086	090	095	100	104	109	114	118	123	5	2	9.35	2.23538	107
6	128	132	137	142	146	151	155	160	165	169	6	2.4	9.36	2.23645	106
7	174	179	183	188	192	197	202	206	211	216	7	2.8	9.37	2.23751	107
8	220	225	230	234	239	243	248	253	257	262	8	3.2	9.38	2.23858	107
9	267	271	276	280	285	290	294	299	304	308	9	3.6	9.39	2.23965	106
940	313	317	322	327	331	336	340	345	350	354			9.40	2.24071	106
1	359	364	368	373	377	382	387	391	396	400	1		9.41	2.24177	107
2	405	410	414	419	424	428	433	437	442	447	2		9.42	2.24284	106
3	451	456	460	465	470	474	479	483	488	493	3		9.43	2.24390	106
4	497	502	506	511	516	520	525	529	534	539	4		9.44	2.24496	105
945	543	548	552	557	562	566	571	575	580	585	5		9.45	2.24601	106
6	589	594	598	603	607	612	617	621	626	630	6		9.46	2.24707	106
7	635	640	644	649	653	658	663	667	672	676	7		9.47	2.24813	105
8	681	685	690	695	699	704	708	713	717	722	8		9.48	2.24918	106
9	727	731	736	740	745	749	754	759	763	768	9		9.49	2.25024	105
950	772	777	782	786	791	795	800	804	809	813			9.50	2.25129	

4.—Logarithms—*Concluded*

No.	Common Logarithms of Numbers.												Naperian.		
	L. 0	1	2	3	4	5	6	7	8	9	P.	P.	No.	Log.	Dif.
950	97 772	777	782	786	791	795	800	804	809	813			9.50	2.25129	105
1	818	823	827	832	836	841	845	850	855	859	1		9.51	2.25234	105
2	864	868	873	877	882	886	891	896	900	905	2		9.52	2.25339	105
3	909	914	918	923	928	932	937	941	946	950	3		9.53	2.25444	105
4	955	959	964	968	973	978	982	987	991	996	4		9.54	2.25549	105
955	98 000	005	009	014	019	023	028	032	037	041	5		9.55	2.25654	105
6	046	050	055	059	064	068	073	078	082	087	6		9.56	2.25759	104
7	091	096	100	105	109	114	118	123	127	132	7		9.57	2.25863	105
8	137	141	146	150	155	159	164	168	173	177	8		9.58	2.25968	104
9	182	186	191	195	200	204	209	214	218	223	9		9.59	2.26072	104
960	227	232	236	241	245	250	254	259	263	268		5	9.60	2.26176	104
1	272	277	281	286	290	295	299	304	308	313	1	0.5	9.61	2.26280	104
2	318	322	327	331	336	340	345	349	354	358	2	1	9.62	2.26384	104
3	363	367	372	376	381	385	390	394	399	403	3	1.5	9.63	2.26488	104
4	408	412	417	421	426	430	435	439	444	448	4	2	9.64	2.26592	104
965	453	457	462	466	471	475	480	484	489	493	5	2.5	9.65	2.26696	103
6	498	502	507	511	516	520	525	529	534	538	6	3	9.66	2.26799	104
7	543	547	552	556	561	565	570	574	579	583	7	3.5	9.67	2.26903	103
8	588	592	597	601	605	610	614	619	623	628	8	4	9.68	2.27006	103
9	632	637	641	646	650	655	659	664	668	673	9	4.5	9.69	2.27109	104
970	677	682	686	691	695	700	704	709	713	717			9.70	2.27213	103
1	722	726	731	735	740	744	749	753	758	762	1		9.71	2.27316	103
2	767	771	776	780	784	789	793	798	802	807	2		9.72	2.27419	102
3	811	816	820	825	829	834	838	843	847	851	3		9.73	2.27521	103
4	856	860	865	869	874	878	883	887	892	896	4		9.74	2.27624	103
975	900	905	909	914	918	923	927	932	936	941	5		9.75	2.27727	102
6	945	949	954	958	963	967	972	976	981	985	6		9.76	2.27829	103
7	989	994	998	$\bar{0}$03	$\bar{0}$07	$\bar{0}$12	$\bar{0}$16	$\bar{0}$21	$\bar{0}$25	$\bar{0}$29	7		9.77	2.27932	102
8	99 034	038	043	047	052	056	061	065	069	074	8		9.78	2.28034	102
9	078	083	087	092	096	100	105	109	114	118	9		9.79	2.28136	102
980	123	127	131	136	140	145	149	154	158	162		4	9.80	2.28238	102
1	167	171	176	180	185	189	193	198	202	207	1	0.4	9.81	2.28340	102
2	211	216	220	224	229	233	238	242	247	251	2	0.8	9.82	2.28442	102
3	255	260	264	269	273	277	282	286	291	295	3	1.2	9.83	2.28544	102
4	300	304	308	313	317	322	326	330	335	339	4	1.6	9.84	2.28646	101
985	344	348	352	357	361	366	370	374	379	383	5	2	9.85	2.28747	102
6	388	392	396	401	405	410	414	419	423	427	6	2.4	9.86	2.28849	101
7	432	436	441	445	449	454	458	463	467	471	7	2.8	9.87	2.28950	101
8	476	480	484	489	493	498	502	506	511	515	8	3.2	9.88	2.29051	101
9	520	524	528	533	537	542	546	550	555	559	9	3.6	9.89	2.29152	101
990	564	568	572	577	581	585	590	594	599	603			9.90	2.29253	101
1	607	612	616	621	625	629	634	638	642	647	1		9.91	2.29354	101
2	651	656	660	664	669	673	677	682	686	691	2		9.92	2.29455	101
3	695	699	704	708	712	717	721	726	730	734	3		9.93	2.29556	101
4	739	743	747	752	756	760	765	769	774	778	4		9.94	2.29657	100
995	782	787	791	795	800	804	808	813	817	822	5		9.95	2.29757	101
6	826	830	835	839	843	848	852	856	861	865	6		9.96	2.29858	100
7	870	874	878	883	887	891	896	900	904	909	7		9.97	2.29958	100
8	913	917	922	926	930	935	939	944	948	952	8		9.98	2.30058	100
9	957	961	965	970	974	978	983	987	991	996	9		9.99	2.30158	
1000	000 000	043	087	130	174	217	260	304	347	391			10.00	2.302585	

SLIDE RULE

The Slide Rule.—The fact that we are able to accomplish multiplication with logarithms by the *addition* of two numbers suggests the possibility of a simple mechanical instrument to carry this out. Such an instrument is the *slide rule.*

Let us consider first an ordinary ruler. If we place our finger on the two-inch mark and then *add* three inches to these two inches, our finger moves to the five-inch mark. Then, if we have a ruler which, instead of being graduated in inches, is graduated by the *logarithms* of numbers, it is evident that if we add to a distance representing the logarithm of 2 a distance representing the logarithm of 3, the sum of these logarithms is the logarithm of 6, and by the addition of these logarithms we have performed *multiplication.* This is the principle of the slide rule and the operation just described is illustrated in Fig. 6.

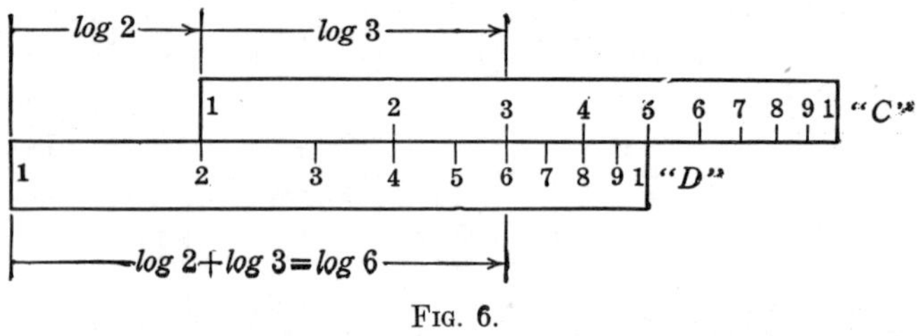

FIG. 6.

The addition of one distance to another is readily accomplished by having the slide rule made in two parts, one of which is called the *stock* and remains stationary, while the other is the *slide* which is free to move parallel with the stock. (See Fig. 7.)

Both the stock and the slide have graduations, but while the distances between graduations are proportioned according to logarithms, the *numbers* on the scales represent antilogs. Thus the answers to the operations are read directly without the need of conversion.

There are many forms of slide rules. With regard to shape, they may be straight, circular or cylindrical. With regard to size they may be short or long. Then there are many types of special graduations. The most common slide rule is the 10-inch straight rule of the Mannheim type illustrated in Fig. 8. It will be noted that this has four scales, two on the stock and two on the slide. These we shall refer to as the A, B, C, and D scales reading downward, and they are usually so stamped on the rules.

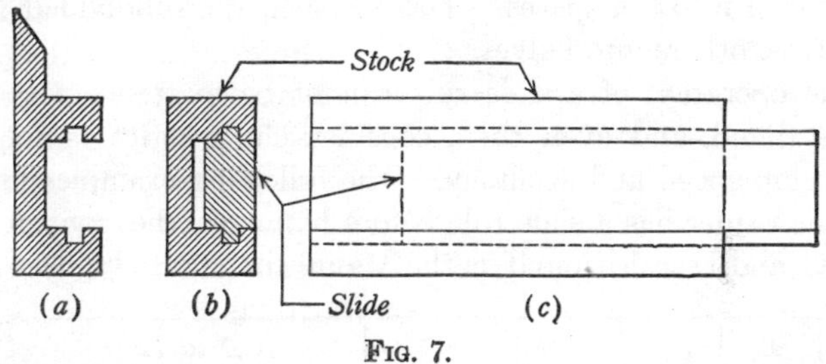

Fig. 7.

Straight slide rules other than the Mannheim usually have scales corresponding to these and differ only in that they have in addition a number of other scales. At the extreme end of each scale is the figure "1." At the left of the scale, this point is called the *left index*; at the right is the *right index*. A glass with a hair-line stretching across the scales is attached to the rule in such a

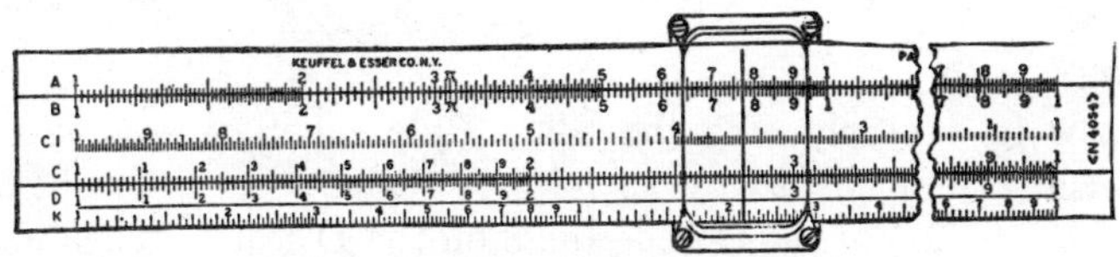

Fig. 8.—Mannheim Slide Rule.

manner that it may be moved along the scale to any position. This is called the indicator or runner and is a great aid in setting and reading values.

It will be noted that on the two lower scales on the slide rule in Fig. 8 the numbers begin at the left with 1, 1, 2, 3, 4, etc. These numbers represent 10, 11, 12, 13, 14, respectively, or 1.0, 1.1, 1.2, 1.3, 1.4, or 100, 110, 120, 130, 140. The extra "1" before the small numbers is omitted to save space. The space between each of these numbers is divided into ten spaces for the next significant figure. To the right of this first series of figures which terminate with "9" or "19" by the above representation, the numbers continue 2, 3, 4, 5, etc. If the preceding "9" is taken to represent "19" (not "1.9" or "190") then these numbers represent 20, 30, 40, 50, etc. The spaces between these numbers are divided into ten spaces. These are again subdivided, some into fifths, others into halves.

The operation of a slide rule cannot be mastered without a rule at hand, and even then, considerable practice is required to develop speed and accuracy. The following examples assume that the reader has a slide rule before him with the conventional A, B, C, and D scales found on the Mannheim and Polyphase rules.

C	*Set 1*	*R to 12*
D	*To 28*	*Read 336*

(a)

C	*Set 1*	*R to Multiplier*
D	*To Multiplicand*	*Read Product*

(b)

FIG. 9.

How to Multiply with the Slide Rule.—Let us assume that we wish to multiply 28 by 12. Move the slide to the right and set **1** (the index) of the C scale to 28 on the D scale. Then move the runner to 12 on the C scale and read the product on the D scale at this point. It is 336. These operations can be set up in the form of a diagram as shown in Fig. 9*a*. From this we can derive a general form for *all* multiplication as shown in Fig. 9*b*. Expressed in words we may say multiplication is carried out as follows:

(1) To *Multiplicand* on D set C index.
(2) To *Multiplier* on C set runner indicator.
(3) At indicator on D read *Product*.

Let us take another example. Multiply 52 by 25. Proceeding as in the previous example we find that by moving the slide to the right, the multiplier (25) falls beyond the end of the D scale. It is necessary then in this and all similar cases that the slide be moved to the *left* and the right index of the C scale set on the multiplicand. The answer on the D scale is then 1300.

C	Set 1	R to 25
D	to 52	Read 1300 (Ans.)

Fig. 10.

By using the runner, *R*, it is possible to perform continued multiplication without having to read the intermediate products.

Illustration: Multiply 12 × 8 × 18.

C	Set 1	R to 8	1 to R	Under 18
D	to 12			Read 1728 (Ans.)

Fig. 11.

With a ten-inch slide rule most numbers can be read directly to only two significant figures and the third figure must be estimated. Thus, if we multiply 854 by 537 we find in setting the index to the multiplicand that there is no line which represents 854. There is one which represents 850 and another 855 and it is necessary to set the index as closely as possible to a point which is estimated by eye to be $\frac{4}{5}$ of the distance from the smaller to the larger of these numbers. Similarly, the position to place the runner to represent 537 must be estimated by eye for the last figure. The product reads 459,000 (the last significant figure being estimated). With practice and careful operation the last place can be determined with remarkable accuracy.

C	Set Divisor	Under 1 Right or Left
D	Over Dividend	Read Quotient

C	25	Under 1
D	1300	Read 52 (Ans.)

Fig. 12.

Division with the Slide Rule.—Division is, of course, the reverse of multiplication and it is to be expected that it is carried out on the slide rule by performing the operations for multiplication in the reverse order. This is the case. For example, let us divide 1300 by 25

To dividend (1300) on the D scale set divisor (25) on the C scale. Under 1 on the C scale read the quotient (52).

Another example, divide 1648 by 536. To 1648 on the D scale, set 536 on the C scale. Under the index on the C scale read the quotient. This appears to be 3.075.

Calculations involving continued multiplication and division can be performed on the slide rule without having to read the intermediate results.

Illustration: Find the value of

$$\frac{150 \times 72 \times 10}{8 \times 6}$$

There are two methods:

C	8	R to 72	6 to R	Under 10
D	150			Read 2250 (Ans.)

C	Set 1	R to 72	1 to R	R to 10	8 to R	R to 1	6 to R	Under 1
D	150							Read 2250 (Ans.)

Fig. 13.

Locating Decimal Point in Slide Rule Multiplication and Division.—The preceding examples have illustrated the manipulation of the slide rule in arriving at products and quotients of numbers without any mention of how the decimal point is located in the result. We shall state briefly the rules governing this and then illustrate by a few examples. First, a definition is necessary In the rules we shall use the word *characteristic* of a number, which is not to be confused with the characteristic of a logarithm. The characteristic of a number is the number of digits before the decimal point, the characteristic of a decimal fraction is the number of ciphers immediately after the decimal point and is negative.

Rule I: *When the slide projects to the right in multiplication the characteristic of the product is one less than the sum of the characteristics of the factors.*

Thus, in the first example of multiplication we found the product of 28 and 12, the slide projecting to the right. The characteristic of each of these numbers is 2, the sum is 4, one less than the sum is 3, which is the characteristic of the product. Thus the product has three figures to the left of the decimal point (336).

In another example, 23×0.415, the characteristics are 2 and 0, respectively, and the sum less 1 is $2 + 0 - 1 = 1$. Thus the product (9.55) has one digit to the left of the decimal point.

In still another example, 0.0328×0.0024, the characteristics are -1 and -2, respectively. The sum less 1 is $-1 - 2 - 1 = -4$. Then the product is 0.0000787 with four ciphers following the decimal point.

These examples illustrate the cases which are apt to occur.

RULE II. *When the slide projects to the left in multiplication the characteristic of the product equals the sum of the characteristics of the factors.*

This rule requires no illustration in view of the foregoing.

RULE III: *When the slide projects to the right in division the characteristic of the quotient equals the characteristic of the dividend minus that of the divisor, plus 1.*

As an illustration, divide 6850 by 37.2. The characteristic of the dividend is 4 and that of the divisor 2. Then, according to the rule, $4 - 2 + 1 = 3$ and the quotient has three digits to the left of the decimal point—in this case 184.1.

As an illustration involving decimals, take the division of 47 by 0.024. The characteristic of the dividend is 2 and that of the divisor -1. Then $2 - (-1) + 1 = +4$ as the characteristic of the quotient. The quotient is then 1957.

The division of one decimal by another is illustrated by the following $0.0074 \div 0.026$. The characteristic of the dividend is -2 and that of the divisor -1. Then, $-2 - (-1) + 1 = 0$ and thus there are no digits to the left of the decimal point and no ciphers to the right, the quotient being 0.2847.

RULE IV: *When the slide projects to the left in division the characteristic of the quotient equals the characteristic of the dividend minus that of the divisor.*

The four rules may be combined into the following chart for ready reference:

Characteristic of result	Slide LEFT	Slide RIGHT
Multiplication	Sum of Characteristics of 2 Factors	Sum −1
Division	Characteristic of Dividend − that of divisor	Difference + 1

Squares, Cubes, and Roots.—The square of a number can, of course, be computed with a slide rule by multiplying the number by itself with the C and D scales. Likewise, the cube may be determined by multiplying the square so found by the original number. However, by the use of the D and A scales the square of any number on the D scale can be found by simply moving the runner to that number and reading the square on the A scale at the cross-line on the runner. Thus 4 on the A scale is directly opposite 2 on the D scale, 9 opposite 3, etc.

The following examples indicate how the slide rule can be used for evaluating such expressions as x^2y and $\sqrt{\frac{a}{b}}$.

ILLUSTRATION: Find the value of $6^2 \times 5$.

A		Read 180 (Ans.)
B	Set 1 (right)	Over 5
C		
D	Over 6	

ILLUSTRATION: Find the value of $\sqrt{\frac{3}{4}}$.

A	Under 3	
B	Set 4	Under 1 (right)
C		
D		Read 0.866 (Ans.)

The cube of a number is found by setting the runner on the number on the D scale, then setting either the left or right index of the B scale on the cross-line of the runner and reading the cube on the A scale opposite the original number on the B scale. Thus, the process consists of finding the square of the number and then performing a multiplication on the A and B scales of the square with the original number.

It will be noted that A and B scales have indexes not only at the left and right ends but also one in the middle. If the left index is taken as 1, the middle index is 10 and the right index is 100, or if the left index is taken as 100, the middle index is 1000 and the right index is 10,000. The left index may never be taken as 10 or 1000 because the square roots of these numbers are 3.16 and 31.6, respectively, and this occurs on the D scale only at the middle.

Thus it becomes apparent that whenever the square root of a number is to be found with the A and D scales it is very important to decide whether this number should be selected on the left- or the right-hand portions of the A scale. This is determined by first pointing off the digits of the number whose square root is to be found into groups of two's, beginning at the decimal point, and moving to the right. For example, 25,346 pointed off is 2,53,46. The square root will have as many digits to the left of the decimal point as there are groups. Decimal fractions are pointed off to the right from the decimal point thusly: 0.02758 becomes .02,75,8. The last group may have either one or two digits. Then if we call the left half of the A scale A1 and the right half A2, we may write the

> RULE: *If the last group contains one figure, use A1 for finding the square root, if it contains two use A2. In either case the characteristic of the square root read on D equals the number of groups in the given number.*

As an illustration, take the first number cited above, 2,53,46. This has one figure in the last group, so it is located on the left-hand (A1) scale and the runner placed on it. The square read on the D scale appears to be 159.1 and since there are three groups of figures in the original number, the root has three digits to the left of the decimal point. For decimal fractions we have the following.

> RULE: *After pairing off the digits to the right of the decimal point, at first disregard the groups immediately following the point which contain only ciphers. If in the first group containing other figures the first figure is a cipher, use A1. If the first figure is not a cipher, use A2. In the root there is one cipher immediately after the decimal point for each group consisting wholly of ciphers in the given number.*

As an example find the square root of the number 0.000625. Pointing off, .00,06,25. The first group containing significant figures (06) has first a cipher so the A1 scale is used. The square root is 0.025. Since the first group in the original number has only ciphers, the first digit of the root is a cipher.

By reversing the above rules we obtain a rule for locating the decimal point when computing squares.

> RULE: *If the square is on A1 the characteristic of the square is 1 less than 2 times that of the number, if on A2 it is twice that of the number. This applies to both positive and negative characteristics.*

References—The reader who is interested in a more comprehensive treatment of arithmetic is referred to the book ARITHMETIC FOR THE PRACTICAL MAN, by Mr. J. E. Thompson, published by the D. Van Nostrand Company. The same author has also written an excellent book entitled A MANUAL OF THE SLIDE RULE, which is also published by the D. Van Nostrand Company.

III

Algebra

Algebraic Symbols.—Algebra is the shorthand of mathematics. Letters and symbols take the place of cumbersome numbers, and many of the ordinary operations of arithmetic take a simpler and more compact form. In addition, algebra can be used to advantage in some problems where arithmetical solution would be extremely involved. In arithmetic the Greek letter π is used to designate the number 3.14159+ and multiplication, division, etc., is performed with it. Similarly, the letters, a, b, c, . . . can be used to represent certain quantities. The first letters of the alphabet are usually used to represent known quantities and the last letters, x, y, z, to represent unknown quantities.

The number of times that a single algebraic quantity is to be taken is indicated by a number before the letter. This number is called the *coefficient*. Thus, in $3b$, the 3 is the coefficient and the expression equivalent to $b + b + b$.

Signs of Algebra.—Whereas in arithmetic it is common to deal only with positive numbers, both positive and negative numbers are used in algebra and it thus becomes necessary to employ symbols to indicate the sign of the quantity. Thus, $+a$, $+b$, etc., denotes that the quantity is positive and $-a$, $-b$, etc., denotes that the quantity is negative. When no sign precedes a number or quantity it is understood to be positive. Powers and roots are indicated as in arithmetic.

Parentheses.—When a number of quantities are enclosed in parentheses with a positive sign before, the parentheses may be removed without altering the expression. Thus, $+(a + b)$ becomes $+a + b$. However, if the sign before is negative, the sign of each quantity must be changed when the parentheses are removed. Thus, $-(a + b)$ becomes $-a - b$, and $-(a - b)$ becomes $-a + b$.

Addition of Algebraic Quantities.—A number of like algebraic terms of like sign may be added by arranging in a column and adding together the coefficients, the sum having the same sign as the parts. Thus,

$$\begin{array}{r} +\ 7b \\ +\ 3b \\ +\ 5b \\ \hline +15b \end{array} \qquad \text{and} \qquad \begin{array}{r} -12c \\ -\ 4c \\ -\ 6c \\ \hline -22c \end{array}$$

If some of the quantities are unlike in sign, proceed as before, but regard the negative coefficients as being subtracted from the positive. Thus,

$$\begin{array}{r} +9a \\ +3a \\ -4a \\ \hline +8a \end{array} \qquad \text{and} \qquad \begin{array}{r} +\ 5b \\ +\ 4b \\ -12b \\ \hline -\ 3b \end{array}$$

When compound quantities (that is, quantities containing more than one term, as $2a - 4b$) are to be added, like terms must be placed in the same column and then added as above. For example, if $5a + 14b + 10c$, $2b - 6c$, $3a - 9c + 3x$, and $-12b - 11c - x$ are to be added, the procedure is as follows:

$$\begin{array}{rrrr} 5a & +\ 14b & +\ 10c & \\ & 2b & -\ \ 6c & \\ 3a & & -\ \ 9c & +\ 3x \\ & -\ 12b & -\ 11c & -\ \ x \\ \hline 8a & +\ \ 4b & -\ 16c & +\ 2x \end{array} \text{ (Ans.)}$$

Subtraction of Algebraic Quantities.—To subtract algebraic quantities, change the sign of the number to be subtracted and then combine the two numbers as in addition.

Example: Subtract $6x$ from $15x$.

$$\begin{array}{r} 15x \\ 6x \\ \hline 9x \end{array}$$

changing the sign of $6x$ makes it $-6x$. Adding $15x$ and $-6x$ gives $9x$.

Example: Subtract $6x$ from $-15x$.

$$\begin{array}{r} -15x \\ 6x \\ \hline -21x \end{array}$$

changing the sign of $6x$ makes it $-6x$. Adding $-15x$ and $-6x$ gives $-21x$.

Example: From $7x - 3y$ take $5x + 12y$.

$$\begin{array}{r} 7x - 3y \\ 5x + 12y \\ \hline 2x - 15y \end{array}$$

write like terms under each other and proceed with each pair of like terms as explained above.

Multiplication of Simple Quantities.—The parts of an algebraic expression separated by plus and minus signs are called *terms*. An expression consisting of one term is known as *monomial*, one of two terms, a *binomial*, one of three terms, a *trinomial*, and one of many terms a *polynomial*.

If two quantities to be multiplied have like signs, the sign of the product is plus; if they have unlike signs, that of the product is minus. Thus, $+a$ multiplied by $+b$ is $+ab$ (the multiplication sign ($\times$) is usually omitted between letters of a term in algebra), $-a$ multiplied by $-b$ is $+ab$, but $-a$ multiplied by $+b$ is $-ab$.

When multiplying monomial expressions, multiply the coefficients together and prefix the product by the proper sign as outlined above. Examples:

Multiply $-a$ $-b$. Product equals $+ab$.
Multiply $+4b$ by $-c$. Product equals $-4bc$.
Multiply $+6b$ by $+3c$. Product equals $+18bc$.
Multiply $-4ax$ by $+5ab$. Product equals $-20aabx = -20a^2bx$

Multiplication of Compound Quantities.—To multiply one polynomial by another, it is necessary to multiply each term of the multiplicand by all of the terms of the multiplier one after the other as by the former rule. The products are then collected into one sum for the required product.

Example: Multiply $3x - 2y$ by $x + 4y$.

Solution:

$$\begin{array}{llll} 3x & - 2y & & \\ x & + 4y & & \\ \hline 3x^2 & - 2xy & & \\ & + 12xy & - 8y^2 & \\ \hline 3x^2 & + 10xy & - 8y^2 & \text{(Ans.)} \end{array}$$

Example: Multiply $x - y + z$ by $x + y - z$.

Solution:

$$\begin{array}{llllll} x & - y + z & & & & \\ x & + y - z & & & & \\ \hline x^2 & - xy + xz & & & & \\ & + xy & & - y^2 & + yz & \\ & & - xz & & + yz & - z^2 \\ \hline x^2 & & & - y^2 & + 2yz & - z^2 \text{ (Ans.)} \end{array}$$

Division of Monomials.—One monomial is divided by another by simply writing the dividend over the divisor as a fraction and cancelling out common factors as in arithmetic. Thus,

$$12ax \div 6a = \frac{\overset{2}{\cancel{12}}ax}{\cancel{6}a} = 2x, \quad \text{and} \quad \frac{\overset{3}{\cancel{9}\cancel{6}}}{\cancel{36}x} = \frac{3}{x}$$

Since $x^2 = x \times x$ and $y^3 = y \times y \times y$, powers may be factored and the common factors cancelled. Then,

$$\frac{x^2y^4}{xy^2} \text{ may be written } \frac{\cancel{x} \times x \times \cancel{y} \times \cancel{y} \times y \times y}{\cancel{x} \times \cancel{y} \times \cancel{y}} = xy^2$$

It is evident from this example that the same result can be arrived at by subtracting the exponent of the smaller number from the exponent of the larger. Thus,

$$\frac{x^2y^4}{xy^2} = x^{(2-1)}y^{(4-2)} = xy^2$$

This is the method actually used in dividing monomials higher than the first power.

$$\text{Examples:} \quad \frac{4a^2b^5}{a^3b^2x^2} = \frac{4b^3}{ax^2}; \quad \frac{a^2}{a^4} = \frac{1}{a^2}; \quad \frac{3ab^2x}{ab^2x} = 3$$

Division of Polynomials.—A polynomial may be divided by a monomial by dividing each term of the polynomial by the monomial. Thus, $2a^2x^3 + 3ax^2 + 5x$ divided by ax may be written $\frac{2a^2x^3}{ax} + \frac{3ax^2}{ax} + \frac{5x}{ax}$. Cancelling out like terms, the quotient becomes $2ax^2 + 3x + \frac{5}{a}$.

To divide a polynomial by a polynomial, arrange both the dividend and divisor according to the ascending or descending powers of some letter and keep this arrangement throughout the operation. Divide the first term of the dividend by the first term of the divisor, and write the result as the first term of the quotient.

Multiply all the terms of the divisor by the first term of the quotient and subtract the product from the dividend. If there is a remainder, consider it as a new dividend and proceed as before.

Example: Divide $2x^3 + 4x^2y - xy - 2y^2$ by $x + 2y$

Solution: These expressions are already arranged according to descending powers of x. Then,

$$\begin{array}{lllll|l}
x + 2y) & 2x^3 & + 4x^2y & - xy & - 2y^2 & \underline{2x^2 - y \text{ (Ans.)}} \\
 & \underline{-+2x^3} & \underline{-+4x^2y} & & & \\
 & & & - xy & - 2y^2 & \\
 & & & \underline{+- xy} & \underline{+- 2y^2} &
\end{array}$$

Multiply $x + 2y$ by $2x^2$ and obtain $+2x^3 + 4x^2y$, which is to be subtracted from $2x^3 + 4x^2y$ in the dividend. Changing the signs of $2x^3 + 4x^2y$ so that this term becomes $-2x^3 - 4x^2y$ proceed as in addition. Then multiply $x + 2y$ by $-y$ and obtain $-xy - 2y^2$ which is to be subtracted from $-xy - 2y^2$. Changing the signs so that this term becomes $+xy + 2y^2$ proceed as in addition.

If the division is not exact and there is a remainder after the last operation has been performed, write the divisor beneath it to form a fraction which is the last term of the quotient.

Example: Divide $4x^2y - 3xy + 6y^2$ by $x^2 - y$.

Solution :

$$\begin{array}{r|l} & 4y \\ \hline x^2 - y & 4x^2y - 3xy + 6y^2 \\ & 4x^2y \qquad\quad - 4y^2 \\ \hline & -3xy + 10y^2 \end{array}$$

$$= 4y + \frac{-3xy + 10y^2}{x^2 - y} \quad \text{(Ans.)}$$

Factoring.—When a number is the product of two other numbers, the component parts are known as *factors*. Thus, in the expression $3a^2$, 3, a, and a, are the factors. Separating a number into its factors is called *factoring.*

Factoring is useful in solving equations, as will be discussed later, and also in simplifying complicated expressions. The operation of removing a monomial factor consists of scrutinizing each term of an expression with a view to determining common factors and then dividing each term by the common factor and placing it before the parentheses which contain the several quotients.

Example: Factor $12a^3x^2 + 33a^2x^2 - 18ax^3 + 9ax$.

Solution: Inspection reveals that a factor common to each term is $3ax$. Then, dividing each term by $3ax$, the expression becomes, $3ax(4a^2x + 11ax - 6x^2 + 3)$.

It is often the case that no single factor can be found common to all the terms of an expression. Then the terms must be examined and compared with a view to grouping them and removing factors common to the group. Thus, in the expression $3x^2 + 9bx + 24xy + 4ax + 12ab + 32ay$, there is no factor common to all terms, but a further examination shows that the first three terms have the common factor $3x$ and the last three terms the common factor $4a$. Removing these factors from the respective terms, the expression becomes,

$$3x(x + 3b + 8y) + 4a(x + 3b + 8y)$$

which may then be consolidated to,

$$(3x + 4a)(x + 3b + 8y)$$

Certain trinomials which are the product of two binomials lend themselves to ready recognition and factoring. Examples of such trinomials are, $(x + 5)(x + 2) = x^2 + 7x + 10$; $(x - 3)(x + 6) = x^2 + 3x - 18$; $(x + y)(x + y) = x^2 + 2xy + y^2$; and $(x - y)(x - y) = x^2 - 2xy + y^2$.

The first of these trinomials, $x^2 + 7x + 10$, could be written $x^2 + 5x + 2x + 10$ and the first two and the last two groups factored as, $x(x + 5) + 2(x + 5) = (x + 2)(x + 5)$. Further examination of this example leads to the observation that the coefficient of the middle term of the trinomial is the sum of the last terms $(2 + 5 = 7)$ of the factors, and the last term of the trinomial is the product of these last terms $(2 \times 5 = 10)$. This is the key to the factoring of factorable expressions of this type. Thus:

$$x^2 + 2x - 8 = (x + 4)(x - 2)$$
$$x^2 + x - 20 = (x + 5)(x - 4)$$
$$x^2 + 3xy + 2y^2 = (x + y)(x + 2y)$$

A ready recognition of a few other special forms is also valuable. These are,

$$x^2 + 2xy + y^2 = (x + y)(x + y) = (x + y)^2$$
$$x^2 - 2xy + y^2 = (x - y)(x - y) = (x - y)^2$$
$$x^2 - y^2 = (x + y)(x - y)$$

Powers and Exponents.—When a quantity is multiplied by itself several times, the resulting product is called a *power* and the quantity itself is called the *root.* Thus, in $ax \times ax \times ax \times ax = a^4x^4$, ax is the root and a^4x^4 is the power. A small number called the *exponent* is used to indicate how many times a number has been multiplied by itself.

The sign of the product of two positive numbers is plus $(+a \times + a = + a^2)$ and the sign of the product of two negative numbers is also plus $(-a \times - a = + a^2)$, but the product of a positive and a negative number is minus $(+a \times - a = - a^2)$. If, then, we raise a negative number to an odd power, for example to the third, as in $-a \times - a \times - a$ it is evident that the first product of $-a \times - a$ results in a positive number and then when this is multiplied again by $-a$ the product becomes negative. Hence, we derive the rule that the sign of an even power of a negative number is positive and the sign of an odd power of a negative number is negative. Examples: $(-a)^2 = + a^2$; $(-a)^3 = - a^3$; $(-a)^4 = + a^4$; $(-a)^5 = - a^5$, etc. The sign of any power of a positive number is, of course, plus.

The product of two or more powers of any quantity is the quantity with an exponent equal to the sum of the exponents of the powers. Examples: $x^2 \times x^3 = x^5$; $x^2y \times xy = x^3y^2$; $4xy \times (-3xz) = - 12x^2yz$.

In a similar manner, the quotient of two powers is the difference of their exponents. Thus, $x^5 \div x^3 = x^{5-3} = x^2$, and $6x^4 \div 2x^3 = \dfrac{6x^4}{2x^3} = 3x$. Then it is apparent that if the exponent of the divisor is greater than the exponent of the dividend, the exponent of the quotient becomes a negative number. Thus, $x^2 \div x^3 = x^{2-3} = x^{-1}$, or $\dfrac{x^2}{x^3} = \dfrac{1}{x} = x^{-1}$. In other words, if a power appears in the denominator with a positive exponent it may be shifted to the numerator by changing the sign of the exponent, as $\dfrac{2ab}{x^3} = 2abx^{-3}$. The law holds equally true for the reverse operation

If we divide one power by an equal power we have this interesting situation $x^3 \div x^3 = x^{3-3} = x^0$. But $\dfrac{x^3}{x^3} = 1$. Then $x^0 = 1$ and the general rule may be stated, that any quantity raised to the zero power is equal to 1.

When a quantity with an exponent is raised to a power, the exponent of the resulting quantity is the product of the exponent of the original quantity and the exponent of the power to which it was raised. This can be well understood from the following illustrations:

$$(x^2)^3 = x^2 \times x^2 \times x^2 = x^6; \quad (y^5)^2 = y^5 \times y^5 = y^{10}.$$

The square of the sum of two quantities is the sum of their squares plus twice their product. Thus,

$$(x + y)^2 = x^2 + y^2 + 2xy; \quad (3x + 4y)^2 = 9x^2 + 16y^2 + 24xy.$$

The square of the difference of two quantities is the sum of their squares minus twice their product. Thus,

$$(x - y)^2 = x^2 + y^2 - 2xy; \quad (2x - 5y)^2 = 4x^2 + 25y^2 - 20xy.$$

The square of a trinomial is equal to the sum of the squares of each term plus twice the product of each term by each of the other terms. Examples:

$$(x + y + z)^2 = x^2 + y^2 + z^2 + 2xy + 2xz + 2yz$$
$$(x - y - z)^2 = x^2 + y^2 + z^2 - 2xy - 2xz + 2yz$$

Roots.—The opposite operation to finding the power of an expression is called finding or extracting a root. The symbol used is the radical sign the same as in arithmetic, $\sqrt{\ }$, with a small number called the root index, $\sqrt[3]{\ }$, to indicate the number of times the root is contained as a factor in the power. When no index number is shown in the hook of the radical sign, the square root is intended.

The root of a product is equal to the product of the roots of the factors. Thus, $\sqrt{144} = \sqrt{9 \times 16} = \sqrt{9} \times \sqrt{16} = 3 \times 4 = 12$, $\sqrt{xy} = \sqrt{x} \times \sqrt{y}$, and $\sqrt{a^2b} = \sqrt{a^2} \times \sqrt{b} = a\sqrt{b}$. However, the root of the *sum* of several terms is *not* the sum of the roots of the individual terms. Thus, $\sqrt{x+y}$ is not $\sqrt{x}+\sqrt{y}$. A polynomial expression under a radical sign must be treated as a whole unless it can be simplified.

In the preceding section it was shown that when a quantity with an exponent is raised to a power the exponent of the resulting quantity is the product of the exponent of the original quantity and the exponent to which it was raised, as $(a^3)^6 = a^{18}$. Then, if we give a quantity a fractional exponent, for example $\frac{1}{2}$, and square the quantity we get this interesting result: $(x^{\frac{1}{2}})^2 = x^{\frac{2}{2}} = x$. But $(\sqrt{x})^2$ also equals x: Then $\sqrt{x} = x^{\frac{1}{2}}$ and the exponent $\frac{1}{2}$ is another way of indicating square root. Similarly, it can be shown that $x^{\frac{1}{3}} = \sqrt[3]{x}$, $x^{\frac{1}{4}} = \sqrt[4]{x}$, etc.

If we multiply, for example, $x^{\frac{1}{3}}$ by $x^{\frac{1}{3}}$ we obtain $x^{\frac{1}{3}} \times x^{\frac{1}{3}} = (x^{\frac{1}{3}})^2 = x^{\frac{2}{3}}$. Expressed in words this is, " the cube root of the square of x " and can be written $\sqrt[3]{x^2}$. Other fractional exponents can be similarly expressed, as $a^{\frac{3}{2}} = \sqrt{a^3}$, $b^{\frac{3}{4}} = \sqrt[4]{b^3}$.

In the preceding section it was shown that while the square of a positive number is positive, the square of a negative number is also positive. Then, if we are confronted with a positive power, as 25, it is impossible to tell whether its square root is positive or negative. Therefore, when the square root of a number has been found, it is necessary to precede it by a plus or minus sign. Thus, $\sqrt{25} = \pm 5$, and $\sqrt{x^2} = \pm x$. It was also found that the odd power of a negative number was negative. Then the odd root of a negative number is negative, as $\sqrt[3]{-8} = -2$, $\sqrt[5]{-243} = -3$. The odd root of a positive number is always positive, but the even root of a positive number may be either negative or positive.

The even root of a negative number cannot be determined and is said to be an *imaginary* number. Thus, the square root of -25 does not exist. Such expressions do, however, sometimes occur and then for the sake of simplicity may be treated as follows: $\sqrt{-25} = \sqrt{25 \times (-1)} = \sqrt{25} \times \sqrt{-1} = 5\sqrt{-1} = 5i$. The letter i is a symbol used to designate $\sqrt{-1}$.

Simple Equations.—If one algebraic expression is equal in value to another, the two, if written with an equality sign between them, constitute an algebraic *equation*, as $a + b = c + d$.

Both sides of an equation may be changed equally by addition, subtraction, multiplication, or division without disturbing the equality. To illustrate, if

$$a + b = c + d$$

then $$a + b + x = c + d + x,$$

$$a + b - x = c + d - x,$$

$$x(a + b) = x(c + d)$$

and $$\frac{a + b}{x} = \frac{c + d}{x}$$

Thus, if we have the equation, $x + 3y = 10$, and want to know the value of x, it is only necessary to subtract $3y$ from both sides of the equation. Then

$$x + 3y - 3y = 10 - 3y$$

$$x = 10 - 3y$$

From this it is apparent that any term of an equation may be changed from one side to the other provided its sign is moved. This is called transposition.

Solution of Simple Equations.—When the value of an unknown symbol in an equation is determined, the equation is said to be solved. Equations containing only one unknown quantity may be solved as follows: Transpose all the terms containing the unknown quantity to the left side of the equation, and all the other terms to the right side. Combine like terms, and divide both sides of the equation by the coefficient of the unknown quantity.

ILLUSTRATIONS:

$$9x - 18 = 12 - 6 + 3x$$

$$9x - 3x = 12 - 6 + 18 \quad \text{(transposing)}$$

$$6x = 24 \quad \text{(collecting terms)}$$

$$\frac{6x}{6} = \frac{24}{6} = \quad \text{(dividing by coefficient)}$$

$$x = 4$$

$$
\begin{aligned}
3y + 4 &= 8y + 36 \\
3y - 8y &= 36 - 4 && \text{(transposing)} \\
-5y &= 32 && \text{(collecting terms)} \\
-\frac{5y}{5} &= 32 && \text{(dividing by coefficient)} \\
y &= -6\frac{2}{5} && \text{(changing signs of both sides)}
\end{aligned}
$$

$$
\begin{aligned}
3\tfrac{1}{2}z - 14 &= 8 + 3z \\
3\tfrac{1}{2}z - 3z &= 8 + 14 \\
\tfrac{1}{2}z &= 22 \\
\frac{\tfrac{1}{2}z}{\tfrac{1}{2}} &= \frac{22}{\tfrac{1}{2}} \\
z &= \frac{22}{\tfrac{1}{2}} = 22 \times \frac{2}{1} = 44
\end{aligned}
$$

Solution of Simultaneous Simple Equations.—If an equation contains two unknown quantities, an indefinite number of pairs of values for them may be found, which will satisfy the equation. For example, in the equation, $x + y = 12$, when x is 4, y is 8; when x is 9, y is 3; when x is 16, y is -4; etc. However, if a second equation containing the same unknowns is given, a single pair of values may be found which will satisfy both equations. Equations solved for common values of their unknowns are called simultaneous equations.

The process of solving two simultaneous equations of two unknowns is to eliminate temporarily one of the unknowns by combining the two equations into one equation containing the other unknown only. One method of doing this is *elimination by addition or subtraction.* This proceeds as follows: Multiply the equations by such a number as will make the coefficients of one of the unknown quantities equal in both. Add or subtract the two equations according to whether the unknown quantities of equal coefficients have unlike or like signs. Solve the resulting equation of the remaining unknown in the regular manner and substitute the value found in one of the original equations to determine the value of the second unknown.

ILLUSTRATION: Find the values of x and y in the simultaneous equations

$$3x - 2y = 30$$
$$4x + 4y = 20$$

Multiply 1st by 4 $\quad 12x - 8y = 120$

Multiply 2nd by 3 $\quad 12x + 12y = 60$

$$-20y = 60$$
$$y = -3$$

Substituting value of y in first equation

$$3x + 6 = 30$$
$$3x = 24$$
$$x = 8$$

Substituting the values found, $x = 8$, $y = -3$ in the other original equation to check results,

$$4 \times 8 + 4(-3) = 20$$
$$32 - 12 = 20$$
$$20 = 20$$

Another method is *elimination by comparison.* From each equation obtain the value of one of the unknown quantities in terms of the other. Form an equation from these equal values of the same unknown quantity and reduce and solve in the regular manner and substitute the value found in one of the original equations to determine the value of the second unknown.

ILLUSTRATION: Find the values of x and y in the simultaneous equations

$$2x + 3y = 7 \quad (1)$$
$$4x - 5y = 3 \quad (2)$$

From (1) $\quad x = \frac{7 - 3y}{2}$

From (2) $\quad x = \frac{3 + 5y}{4}$

Equating these, $\frac{7 - 3y}{2} = \frac{3 + 5y}{4}$

Multiplying by 4, $14 - 6y = 3 + 5y$

$$11y = 11$$
$$y = 1$$

Substituting in one of the original equations:

$$2x + (3 \times 1) = 7$$
$$2x = 4$$
$$x = 2$$

The answer is, $x = 2$, $y = 1$, and may be checked by substituting these values in the two original equations.

A third method is *elimination by substitution*. From one of the original equations obtain the value of one of the unknown quantities in terms of the other. Substitute this value of this unknown quantity for it in the other equation and reduce the resulting equations.

ILLUSTRATION: Find the values of x and y in the simultaneous equations

$$4x - 6y = 28 \quad (1)$$
$$2x - 8y = 24 \quad (2)$$

From (1) $x = \frac{28 + 6y}{4}$

Substituting this value in (2)

$$2 \times \frac{28 + 6y}{4} - 8y = 24$$
$$14 + 3y - 8y = 24$$
$$-5y = 10$$
$$y = -2$$

Substituting this value in (1)

$$4x + 12 = 28$$
$$4x = 16$$
$$x = 4$$

The answer is, $x = 4$, $y = -2$.

The solution of equations containing three unknowns requires three simultaneous equations. Essentially the same methods may be applied as for the solution of two simultaneous equations. One of the unknown quantities must be eliminated between two pairs of the equations, then a second between the two resulting equations.

Quadratic Equations.—Equations containing the square or the second power of the unknown quantity but no higher power are called *quadratic equations.* A *pure quadratic* contains only the square; an *affected* or *complete quadratic* contains both the square and the first power. The equation $25x^2 + 18 = 3x^2 - 8$ is a pure quadratic; $50x^2 - 5x = 125$ is a complete or affected quadratic.

Solution of Pure Quadratic Equations.—To solve a pure quadratic collect the unknown quantities on the left side and the known quantities on the right side; divide by the coefficient of the unknown quantity and extract the square root of each side of the resulting equation. Examples:

$$\begin{aligned} \text{Solve}\quad 6x^2 - 2x^2 &= 64 \\ 4x^2 &= 64 \quad \text{(Combining terms)} \\ x^2 &= 16 \quad \text{(Dividing by coefficient)} \\ x &= \pm 4 \quad \text{(Extracting square root)} \end{aligned}$$

$$\begin{aligned} \text{Solve}\quad 5x^2 - 55 &= 0 \\ 5x^2 &= 55 \\ x^2 &= 11 \\ x &= \pm \sqrt{11} \end{aligned}$$

The root which is indicated, but can only be found approximately, is called a *surd.*

$$\begin{aligned} \text{Solve}\quad 8x^2 + 64 &= 0 \\ 8x^2 &= -64 \\ x^2 &= -8 \\ x &= \sqrt{-8} \end{aligned}$$

The square root of a negative number cannot be found even approximately and the root which is indicated is called *imaginary.*

Solution of Affected or Complete Quadratics.—Several methods of solution are applicable to complete quadratics. We shall consider first equations which may be solved by *factoring.* All of the terms are first transposed to the left-hand side leaving zero on the right and we obtain an equation of this type.

$$x^2 + 8x + 15 = 0$$

By the process previously described, the middle term may be separated into the sum of two terms. We then have

$$x^2 + 3x + 5x + 15 = 0$$

then grouping $\qquad (x^2 + 3x) + (5x + 15) = 0$

and factoring $\qquad x(x + 3) + 5(x + 3) = 0$

$$(x + 5)(x + 3) = 0$$

Any number multiplied by zero is equal to zero. Then in order for the product of these two factors to equal zero, either $(x + 5)$ or $(x + 3)$ or both must equal zero.

If $x + 5 = 0$, then $x = -5$

If $x + 3 = 0$, then $x = -3$

If we substitute $x = -5$ into the original equation we obtain

$$(-5)^2 + 8(-5) + 15 = 0$$

$$25 - 40 + 15 = 0$$

Similarly, if we substitute $x = -3$,

$$(-3)^2 + 8(-3) + 15 = 0$$

$$9 - 24 + 15 = 0$$

Thus, there are *two* solutions to the equation since either $x = -5$ or $x = -3$ satisfy it.

All complete quadratics may be solved by the method of *completing the square.* First transpose all of the terms containing the unknown to the left-hand side of the equation and the known quantities to the right-hand side. Arrange the unknown quantities in the order of their exponents and change signs, if necessary, so that the term containing the square will be positive. Divide all terms by the coefficient of the square of the unknown quantity. Complete the square by adding to both sides of the equation the square of half the coefficient of the first power of the unknown. The left-hand side will then be a perfect square. Extract the square root of both sides of the equation and solve the resulting simple equation. Examples:

Solve $2x^2 + 4x - 70 = 0$.

$$2x^2 + 4x = 70 \qquad \text{(Transposition)}$$
$$x^2 + 2x = 35 \qquad \text{(Dividing by coefficient of } x^2\text{)}$$
$$x^2 + 2x + 1 = 35 + 1 \qquad \text{(Adding square of } \tfrac{1}{2} \text{ coefficient of } x\text{)}$$
$$(x + 1)^2 = 36$$
$$x + 1 = \pm 6 \qquad \text{(Extracting square root)}$$
$$x = -1 \pm 6$$
$$\left.\begin{aligned} x &= -7 \\ \text{or } x &= +5 \end{aligned}\right\} \text{(Ans.)}$$

Here again we find that the equation has two solutions. Both solutions may be correct. Moreover, in some practical problems one answer may be correct and the other inconsistent with the conditions of the problem.

Example: A park which is in the form of a right triangle has one side twenty-five feet longer than the other. If the area ($\frac{1}{2}$[base × height]) is 625 square feet, find the length of the sides.

Let x = shorter side

$$x + 25 = \text{longer side}$$
$$\frac{x(x + 25)}{2} = 625$$
$$x^2 + 25x = 1250$$
$$x^2 + 25x - 1250 = 0$$
$$(x + 50)(x - 25) = 0$$

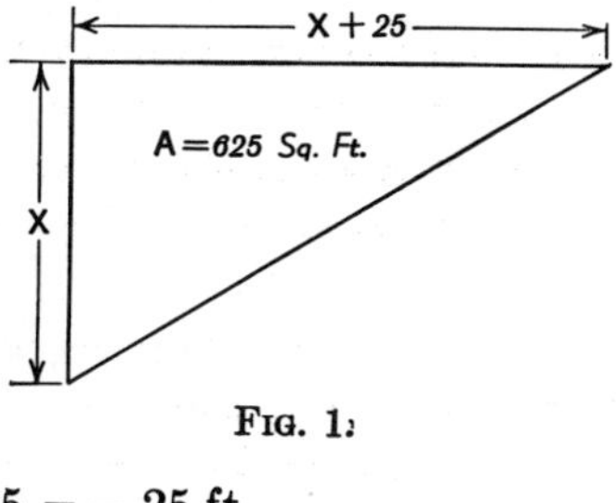

FIG. 1.

$$x = -50 \text{ ft.}, \; x + 25 = -25 \text{ ft.}$$
$$x = 25 \text{ ft.}, \; x + 25 = 50 \text{ ft.}$$

The -50 and -25 do not satisfy the conditions of the problem and therefore should be neglected.

A third method of solution is by the use of the *quadratic formula.* The terms of a complete quadratic equation when collected on one side of the equality sign constitute a trinomial consisting of one term with the unknown to the second power, one term with the unknown to the first power, and the third term of known quantities. This may be written in the general form

$$ax^2 + bx + c = 0$$

The coefficients a and b and the term c may be numerical or literal numbers, positive or negative, monomials or polynomials. The roots of this equation by the quadratic formula are

$$x = \frac{-b + \sqrt{b^2 - 4ac}}{2a}$$

$$x = \frac{-b - \sqrt{b^2 - 4ac}}{2a}$$

Examples:

Solve $2x^2 + 3x + 1 = 0$.

$$x = \frac{-3 + \sqrt{(3)^2 - 4 \times 2 \times 1}}{2 \times 2}$$

$$= \frac{-3 + \sqrt{9 - 8}}{4} = \frac{-3 + 1}{4} = -\tfrac{1}{2}$$

$$x = \frac{-3 - \sqrt{(3)^2 - 4 \times 2 \times 1}}{2 \times 2}$$

$$= \frac{-3 - \sqrt{9 - 8}}{4} = \frac{-3 - 1}{4} = -1$$

The roots of the equation are, $x = -\frac{1}{2}$, $x = -1$, both real and rational numbers.

Solve $3x^2 + 5x - 4 = 0.$

$$x = \frac{-5 + \sqrt{25 + 48}}{6} = \frac{-5 + \sqrt{73}}{6}$$

$$= \frac{-5 + 8.544+}{6} = \frac{3.544}{6} = .590+$$

$$x = \frac{-5- \sqrt{25 - 48}}{6} = \frac{-5 - \sqrt{73}}{6}$$

$$= \frac{-5-8.544+}{6} = \frac{-13.544}{6} = - 2.257+$$

In this example the roots are real, but since $(b^2 - 4ac)$ is not a perfect square, they are not rational, that is, they terminate in never-ending decimals.

Solve $-4x^2 + 4x - 8 = 0.$

$$-x^2 + x - 2 = 0$$

$$x^2 - x + 2 = 0$$

$$x = \frac{1 + \sqrt{1 - 8}}{2} = \frac{1 + \sqrt{-7}}{2}$$

$$x = \frac{1 - \sqrt{1 - 8}}{2} = \frac{1 - \sqrt{-7}}{2}$$

In this example $(b^2 - 4ac)$ is less than zero (negative) and since the square root of a negative number is an imaginary, the roots of the equation are imaginary.

Reference.—ALGEBRA FOR THE PRACTICAL MAN, by Mr. J. E. Thompson (D. Van Nostrand Company), covers the subjects dealt with above, as well as many others, with a simplicity particularly suited for home study.

IV

Geometry

Geometry is the science which treats of the properties of lines, angles, surfaces, and solids. It is based on a number of theorems and constructions for which formal proofs have been developed. These proofs are of little concern to the practical man. Hence, this section will present the most important definitions and conclusions without proofs, and then pass on to mensuration or the measurement of lines, areas and volumes, which is of great practical value to everyone, and then to geometrical construction which is very useful to the man in the shop and at the drafting table.

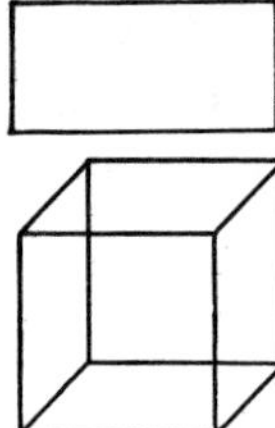

Fig. 1.

Definitions.—A *point* indicates position but has no magnitude, nor dimensions; neither length, breadth, nor thickness.

A *line* has length but no breadth or thickness. It may be straight, curved, or mixed. A straight line is the shortest distance between two points. A curve continually changes its direction between its extreme points. When a line is mentioned simply, it means a straight line.

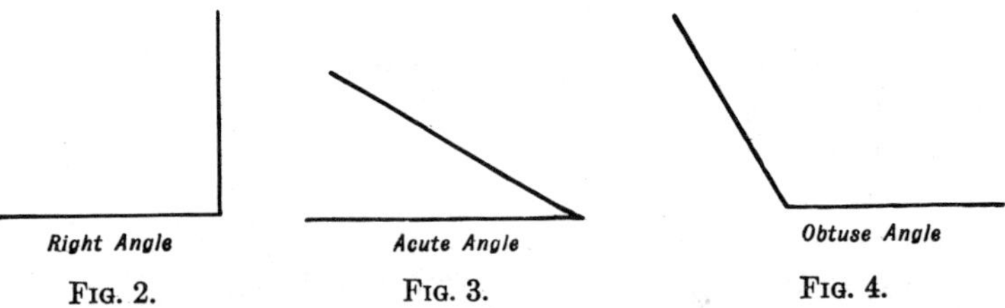

Fig. 2. Fig. 3. Fig. 4.

A *surface* has length and breadth but no thickness. It may be either plane or curved.

A *solid* or body is a figure of three *dimensions*, namely, length, breadth, and depth or thickness.

An *angle* is formed by the intersection of two lines. The point of intersection is called the vertex.

A *right* angle is formed when one of the lines is perpendicular to or makes an angle of 90 degrees with the other line. An *acute* angle is less than a right angle. An *obtuse* angle is greater than a right angle. Acute and obtuse angles are also said to be *oblique*.

A *plane* is that with which a straight line may every way coincide, or, if the line touches the plane at two points, it will touch it at every point.

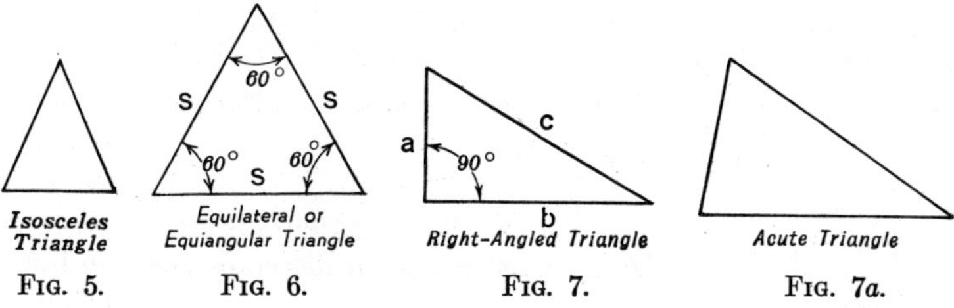

Isosceles Triangle — Fig. 5. *Equilateral or Equiangular Triangle* — Fig. 6. *Right-Angled Triangle* — Fig. 7. *Acute Triangle* — Fig. 7*a*.

Plane figures are bounded either by straight lines or curves. Plane figures that are bounded by straight lines have names according to their number of sides or of their angles, for they have as many sides as angles, the least number being three.

A plane figure bounded by three sides is called a *triangle*.

An *equilateral triangle* has three sides "S" equal. Its three angles are also equal and each has a value of 60 degrees.

Angle A = Angle B

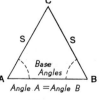

Fig. 8.

An *isosceles triangle* has two equal sides, called its legs. The angles between each leg of the isosceles triangle and the third side are called the base angles and are equal.

A *scalene triangle* has no sides equal.

A *right-angled triangle* has two sides perpendicular to each other making the angle between them a right angle or 90 degrees. The side opposite the right angle is called the *hypotenuse*, the other two sides are called the *legs*. The square of the length of the hypotenuse is equal to the sum of the squares of the lengths of the legs, or in Fig. 7, $c^2 = a^2 + b^2$.

All triangles other than right-angled triangles are *oblique-angled* and are *obtuse-angled* if they have one obtuse angle and *acute-angled* if all three angles are acute.

A figure of four sides and angles is called a *quadrangle* or *quadrilateral.*

A *parallelogram* is a quadrilateral which has both of its pairs of opposite sides parallel, and it takes the following particular names: rectangle, square, rhomboid, and rhombus.

Fig. 9.

A *rectangle* is a parallelogram, having right angles.

Fig. 10.

A *square* is an equilateral rectangle, having its length and breadth equal.

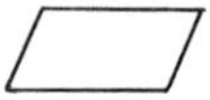

Fig. 11.

A *rhomboid* is an oblique-angled parallelogram.

Fig. 12.

A *rhombus* is an equilateral rhomboid, having all its sides equal but its angles oblique.

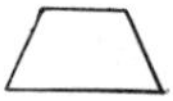

Fig. 13.

A *trapezoid* is a quadrilateral which has only one pair of opposite sides parallel.

Fig. 14.

A *trapezium* is a quadrilateral which has no opposite sides parallel.

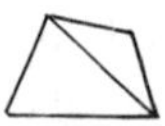

Fig. 15.

A *diagonal* is a line joining any two opposite angles of a quadrilateral.

Plane figures having more than four sides are, in general, called *polygons* and they receive their names according to their number of sides or angles. Thus, a *pentagon* is a polygon of five sides; a *hexagon* of six sides; a *heptagon*, seven; an *octagon*, eight; a *nonagon*, nine; a *decagon*, ten, etc. A *regular polygon* has all its sides equal and all its angles equal.

A *circle* is a plane figure bounded by a curved line called the *circumference* or periphery which is everywhere equidistant from a certain point within called its *center* (point *c* in Fig. 16).

The *radius* of a circle is a line drawn from the center to the circumference (*cf* in Fig. 16).

The *diameter* of a circle is a line drawn through the center and terminating at the circumference on both sides (*ecd* in Fig. 16). It is equal to twice the radius.

An *arc* of a circle is any part of the circumference (as *ab* or *bd* in Fig. 16).

A *chord* is a straight line joining the extremities of an arc (*ab* in Fig. 16).

A *segment* is any part of a circle bounded by an arc and its chord (as shaded area between *a* and *b*, Fig. 16).

A *sector* is any part of a circle bounded by an arc and two radii drawn to its extremities (as shaded area between *cd*, *cf*, and *fd*, Fig. 16).

A *semicircle* is half the circle, or a segment cut off by a diameter. The half circumference is sometimes called the semicircumference.

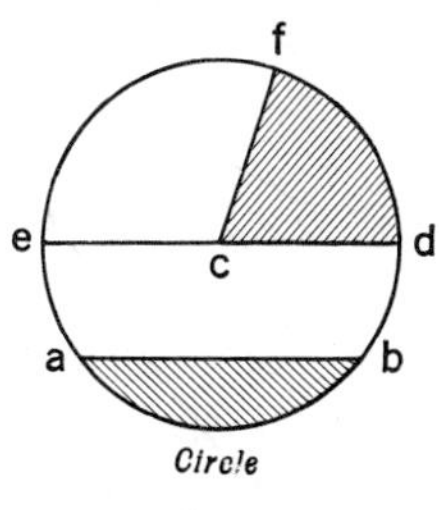

FIG. 16.

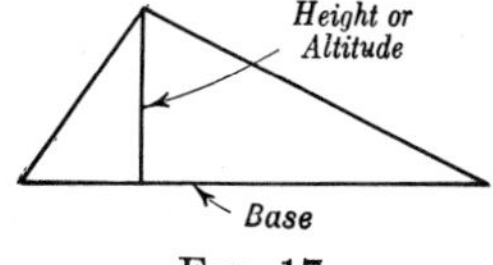

FIG. 17.

The *height* or *altitude* of a figure is a perpendicular let fall from an angle or its vertex, to the opposite side, called the *base*.

Geometrical Propositions.—A great many of the practical

problems in this book are based upon the following geometrical propositions:

If a triangle is equilateral, it is equiangular, and vice versa.

If a straight line from the vertex of an isosceles triangle bisects the base it bisects the vertical angle and is perpendicular to the base.

The sum of the three angles in a triangle always equals 180 degrees.

If two triangles are mutually equiangular, they are similar and their corresponding sides are proportional.

In every triangle, that angle is greater which is opposite a longer side. In every triangle, that side is greater which is opposite a greater angle.

In every triangle, the sum of the lengths of two sides is always greater than the length of the third side.

In a right triangle the square on the hypotenuse is equal to the sum of the squares on the other two sides.

The areas of triangles having equal base and equal height are equal.

If a triangle is inscribed in a semicircle, it is right-angled.

In a quadrilateral, the sum of the interior angles equals four right angles or 360 degrees.

In a parallelogram, the opposite sides are equal; the opposite angles are equal; it is bisected by its diagonal and its diagonals bisect each other.

The areas of two parallelograms which have equal base and height are equal.

If the diameter of a circle is at right angles to a chord, then it bisects or divides the chord into two equal parts. If two chords intersect each other in a circle, the rectangle of the segments of the one equals the rectangle of the segments of the other.

If an angle is formed by a tangent of any chord, it is measured by one-half of the arc intercepted by the chord; that is, it is equal to half the angle at the center subtended by the chord.

If two circles are tangent to each other, then the straight line which passes through the centers of the two circles must also pass through the point of tangency.

The length of circular arcs of the same circle are proportional to the corresponding angles at the center.

The circumference of two circles are proportional to their radii.

The areas of two circles are proportional to the squares of their radii.

Mensuration.—This subject deals with the finding of lengths, areas, and volumes, of lines, surfaces, and solids, respectively. We need a few more definitions of solids before proceeding.

A *prism* is a solid of which the sides are parallelograms and the ends equal, similar, and parallel plane figures. The figure of the ends gives the name to the prism; if the ends are triangular, the prism is triangular, etc. If the sides and ends of a prism be all equal squares, the prism is called a *cube*; and if the base or ends be parallelograms, the prism is called a *parallelepiped*. The *cylinder* is a round prism having circular ends. A *right prism* has its axis perpendicular to the base.

The *pyramid* has any plane figure for its base, and its sides triangles of which all the vertices meet in a point at the top called the *vertex* of the pyramid. A *right pyramid* has its axis perpendicular to the base.

A *cone* is a solid figure having a circle for its base and terminated in a vertex.

A *sphere* or globe is a solid bounded by one continued curved surface, every point of which is equally distant from a point within the sphere called the *center*.

The *axis* of a solid is a straight line drawn through the solid, from the middle of one end to the middle of the opposite.

The *height* of a solid is a line drawn from the vertex perpendicular to the base or the plane on which the base rests.

The *segment* of a solid is a part cut off by a plane, parallel to the base; and the *frustum* is the part remaining after the segment is cut off.

Properties of the Circle.—The *circumference* of a circle is divided into 360 equal parts, called *degrees*; each degree into 60 *minutes*, each minute into 60 *seconds*. Hence a semicircle contains 180 degrees, and a quarter of a circle, or a *quadrant*, 90 degrees.

The ratio of the length of the circumference of a circle to its diameter is a constant and has the value, 3.14159265+. For nearly all practical computations, this number is shortened to 3.1416. This ratio is called *pi* and is represented by the Greek letter π. If we let D represent the diameter of a circle and r the radius, then we may write

$$\text{circumference} = \pi \times D = 3.1416D$$

or,

$$\text{circumference} = \pi \times 2r = 2 \times 3.1416r$$

ILLUSTRATION: What is the circumference of a circle whose radius is 6 inches?

$$\text{circumference} = \pi \times 2r = 2 \times 6 \times 3.1416 = 37.7 \text{ in. (Ans.)}$$

The *area* of a circle is equal to $\frac{1}{4}\pi D^2$ or πr^2.

ILLUSTRATION: What is the area of a circle whose diameter is 5 inches?

$$\text{area} = \tfrac{1}{4}\pi D^2 = \tfrac{1}{4} \times 3.1416 \times 25 = 19.6 \text{ sq.in.}$$

ILLUSTRATION: What is the area of a circle whose radius is $\frac{1}{8}$ inch?

$$\text{area} = \pi \times r^2 = 3.1416 \times \tfrac{1}{64} = 0.049 \text{ sq.in. (Ans.)}$$

To find the *area of a sector* when (I) the length of the arc is known, and (II) when the angle of the sector is known:

CASE I. Multiply the length of the arc by $\frac{1}{2}$ the radius. Then, when A = area, l = length of arc, and r = radius,

$$A = \frac{rl}{2}.$$

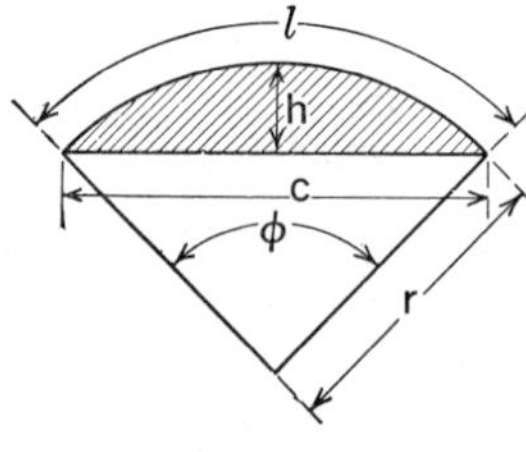

FIG. 18.

ILLUSTRATION: The length of arc of a sector is 40 feet on a circle whose diameter is 300 feet. What is the area of the sector?

$$A = \frac{rl}{2} = \frac{150 \times 40}{2} = 3000 \text{ sq.ft. (Ans.)}$$

Case II. The area of a sector of a circle is to the area of the whole circle as the number of degrees in the arc of the sector is to 360 degrees. Then if ϕ = angle of sector, and area of circle = πr^2,

$$\frac{A}{\pi r^2} = \frac{\phi}{360}, \quad A = \frac{\phi}{360} \pi r^2$$

Illustration: What is the area of a 60-degree sector of a circle whose diameter is 12 inches?

$$A = \frac{\phi}{360} \pi r^2 = \frac{6\not{0}}{36\not{0}} \times 3.1416 \times \not{6} \times \not{6} =$$

$$6 \times 3.1416 = 18.85 \text{ sq. in. (Ans.)}$$

The *area of a segment* of a circle in terms of its height, h, length of arc, l, length of chord, c, and radius of circle, r, is

$$A = \tfrac{1}{2}[r(l - c) + hc]$$

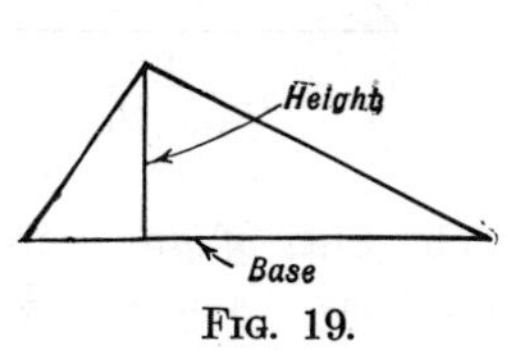

Fig. 19.

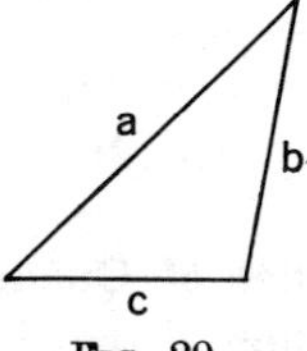

Fig. 20.

Properties of Triangles.—The *area of any triangle* is one-half the product of the base and the height

$$\text{Area} = \tfrac{1}{2}(\text{base} \times \text{height})$$

Illustration: What is the area of a triangular lot whose base is 40 feet and whose height is 48 feet?

$$A = \tfrac{1}{2}(b \times h) = \tfrac{1}{2}(40 \times 48) = 960 \text{ sq.ft. (Ans.)}$$

The *area of a right triangle* is one-half of the product of the two legs.

The *area of any triangle* whose three sides are known can be found by subtracting from one-half the sum of the three sides each side severally, then extracting the square root of the product of the three remainders and the half-sum of the sides. Thus when

$$s = \tfrac{1}{2}(a + b + c)$$
$$\text{Area} = \sqrt{s(s - a)(s - b)(s - c)}$$

ILLUSTRATION: What is the area of a triangle whose sides are 5, 7, and 8 inches long?

$$s = \frac{a + b + c}{2} = \frac{5 + 7 + 8}{2} = 10$$

$$A = \sqrt{10(10 - 5)(10 - 7)(10 - 8)}$$
$$= \sqrt{10 \times 5 \times 3 \times 2} = \sqrt{300} = 17.32 \text{ sq.in. (Ans.)}$$

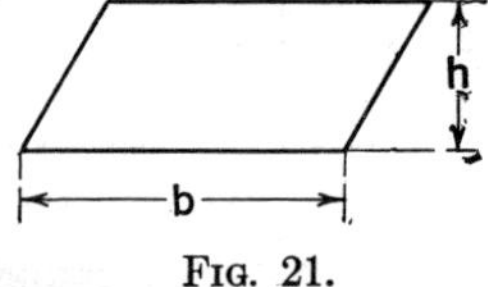

FIG. 21.

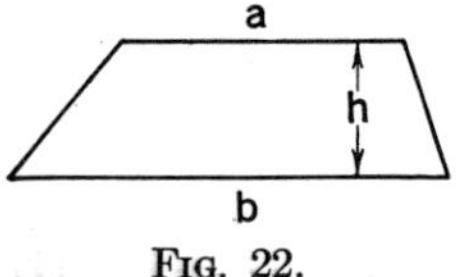

FIG. 22.

Properties of Quadrilaterals.—The *area of any parallelogram* is the product of the altitude and the base. $A = b \times h$.

ILLUSTRATION: What is the area of a rhomboid whose base is 8 inches and whose height is $3\frac{1}{2}$ inches?

$$A = b \times h = 8 \times 3\tfrac{1}{2} = 28 \text{ sq.in. (Ans.)}$$

ILLUSTRATION: What is the area of a square whose side is $4\frac{1}{4}$ inches?

$$A = b \times h = 4\tfrac{1}{4} \times 4\tfrac{1}{4} = 18.0625 \text{ sq.in. (Ans.)}$$

The *area of a trapezoid* is the product of one-half the sum of the two parallel sides and the height. $A = \frac{1}{2}(a + b) \times h$.

The *area of a trapezium* can only be found by drawing the trapezium to scale and then drawing a diagonal the length of which is measured by the same scale and then solving for the separate areas of the two resulting triangles by

$$A = \sqrt{S(S - a)(S - b)(S - c)}$$

Areas of Regular Polygons.—The areas of regular polygons may readily be calculated with the use of Table 1. The area is equal to the product of the square of the length of one side and the corresponding factor in the third column of the table.

TABLE 1

No. of Sides	Name of Polygon	Factor (F)
3	Triangle	0.4330127
4	Tetragon	1.0000000
5	Pentagon	1.7204774
6	Hexagon	2.5980762
7	Heptagon	3.6339124
8	Octagon	4.8284271
9	Nonagon	6.1818242
10	Decagon	7.6942088
11	Undecagon	9.3656405
12	Dodecagon	11.1961524

ILLUSTRATION: What is the area of a regular octagon the ength of whose side is 6 inches?

$$A = s^2 \times F = 6 \times 6 \times 4.828 = 173.81 \text{ sq. in. (Ans.)}$$

Properties of Prisms and Cylinders.—The *volume of any prism* or *cylinder* is the product of the area of the base and the altitude.

The *volume of a circular cylinder* is then, $V = \pi r^2 h$, when h is the altitude and r the radius of the base.

ILLUSTRATION: What is the volume of an oil drum 20 inches in diameter and 30 inches high?

$$V = \pi r^2 h = \pi 10^2 \times 30 = 3000 \times 3.1418 = 9{,}425 \text{ cu. in.} \quad \text{(Ans.)}$$

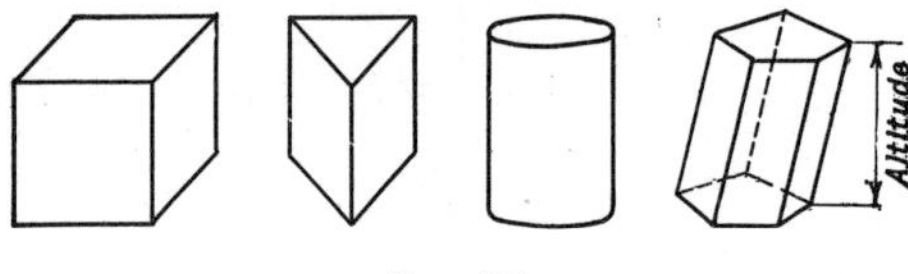

FIG. 23.

ILLUSTRATION: What is the volume of a prism whose height is 12 inches and whose base is a right triangle with legs 5 inches and 8 inches long?

Area of base $= \frac{1}{2} \times 5 \times 8 = 20$ sq. in. Volume $= A \times h = 20 \times 12 = 240$ cu. in. (Ans.)

The *surface area* of a right prism or cylinder is the product of the height and the perimeter of a base plus the area of the two bases. The surface area of a cylinder is then, $A = 2\pi rh + 2\pi r^2 = 2\pi r(h + r)$ or $\pi Dh + \frac{1}{2}\pi D^2 = \pi D\left(h + \frac{D}{2}\right)$.

ILLUSTRATION: What is the surface area of pole 12 inches in diameter and 9 feet long?

$$A = \pi D\left(h + \frac{D}{2}\right) = \pi \times 1 \times (9 + \tfrac{1}{2})$$

$$= 9.5 \times 3.1416 = 29.8 \text{ sq. ft.} \quad \text{(Ans.)}$$

Illustration: What is the surface area of a hexagonal bar 1 inch on the side and 8 inches long?

Area of end $= S^2 \times F = 1^2 \times 2.598 = 2.6$ sq. in. Area of 2 ends $= 5.2$ sq in.

Perimeter $= 6 \times 1 = 6$ in. Area of sides $= 6 \times 8 = 48$ sq in. Total area $= 48 + 5.2 = 53.2$ sq. in. (Ans.)

Properties of the Sphere.—The volume of a sphere is $\frac{4}{3}\pi r^3$ or $\frac{1}{6}\pi D^3$.

Illustration: What are the cubical contents of a spherical balloon 50 feet in diameter?

$$V = \frac{1}{6}\pi D^3 = \frac{125{,}000}{6} \times 3.1416$$
$$= 65{,}450 \text{ cu. ft. (Ans.)}$$

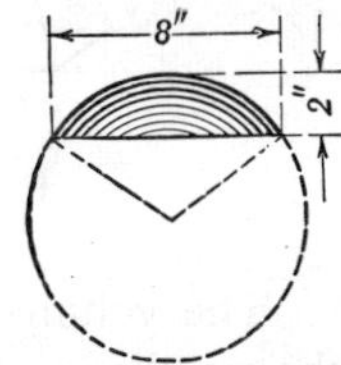

Segment of a Sphere

Fig. 24.

The surface of a sphere is πD^2 or $4\pi r^2$.

Illustration: What is the area of a spherical water tank 22 feet in diameter?

$$A = \pi D^2 = 3.1416 \times 22 \times 22 = 1521 \text{ sq. ft. (Ans.)}$$

The volume of a segment of a sphere is three times the square of the radius of the base plus the square of the height, this sum multiplied by the height and by 0.5236. If r is the radius of the base and h is the height, then volume $= 0.5236h(3r^2 + h^2)$.

Illustration: What is the volume of the segment shown in Fig. 24?

Here, $r = 4$ in., $h = 2$ in. Then,

$$V = 0.5236h(3r^2 + h^2) = 0.5236 \times 2(3 \times 16 + 4)$$
$$= 54.45 \text{ cu. in. (Ans.)}$$

Properties of Pyramids and Frustums of Pyramids.—The volume of any pyramid is one-third the product of the area of the base and the altitude. $V = \frac{1}{3}Ah$.

Illustration: What is the volume of a pyramid whose base is a square, 8 feet on a side, and whose altitude is 4 feet?

$$V = \frac{1}{3} \times Ah = \frac{1}{3} \times 8 \times 8 \times 4 = 85.33 \text{ cu. ft. (Ans.)}$$

The slanted surface of a regular pyramid is one-half the product of the perimeter of the base and the slant height of a side (not the slant height of an edge).

The total surface area of a pyramid is the sum of the slanted surface and the area of the base.

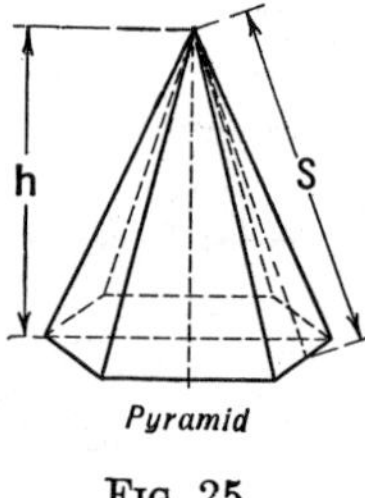

FIG. 25.

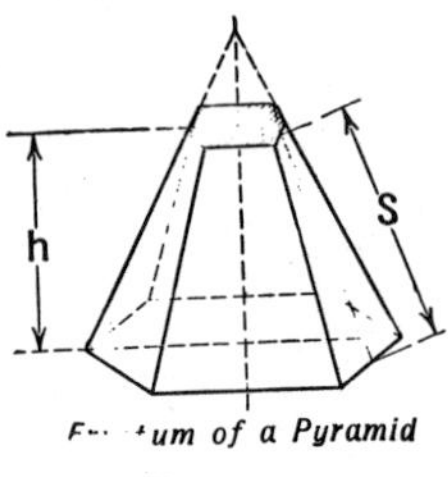

FIG. 26.

The volume of a frustum of a pyramid when a is the area of the small end, A the area of the large end, and h the perpendicular distance between the ends is, $V = \frac{h}{3}(a + A + \sqrt{Aa})$.

The area of the slanted surface of a frustum of a pyramid is the sum of the perimeter of the small end and the perimeter of the large end multiplied by the slant height and divided by two.

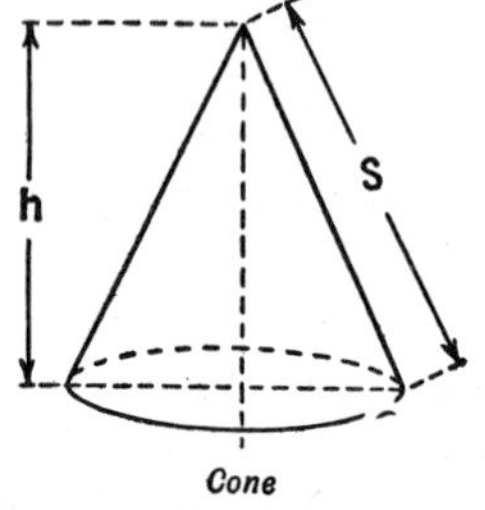

FIG. 27.

Properties of Cones and Frustums of Cones.—The volume of a cone is one-third the product of the area of the base and the altitude. Then, $V = \frac{1}{3}\pi r^2 h$ or $\frac{1}{12}\pi D^2 h$.

ILLUSTRATION: What is the volume of a conical pile of coal 30 feet in diameter and 14 feet high?

$$V = \tfrac{1}{12}\pi D^2 h = \tfrac{1}{12} \times 3.1416 \times 30^2 \times 14 = 3299 \text{ cu.ft. (Ans.)}$$

The area of the curved surface of a cone is one-half the product of the circumference and the slant height. If S = slant height, then, $A = \frac{1}{2}\pi DS$.

The volume of a frustum of a cone when R is the radius of the large end, r the radius of the small end, and h the perpendicular distance between the ends is, $V = (R^2 + r^2 + Rr)\pi\frac{h}{3}$.

The area of the curved surface of the frustum of a cone when R, r, and h have the same significance as above, is,

$$\text{Curved area} = (R + r)\pi\sqrt{(R - r)^2 + h^2}$$

Conic Sections.—A cone has already been defined as a solid figure having a circle for its base and terminated in a vertex. Conic sections are the figures made by a plane cutting a cone. Depending on the different positions of the cutting plane, there arise five different figures or sections, namely, a triangle, a circle,

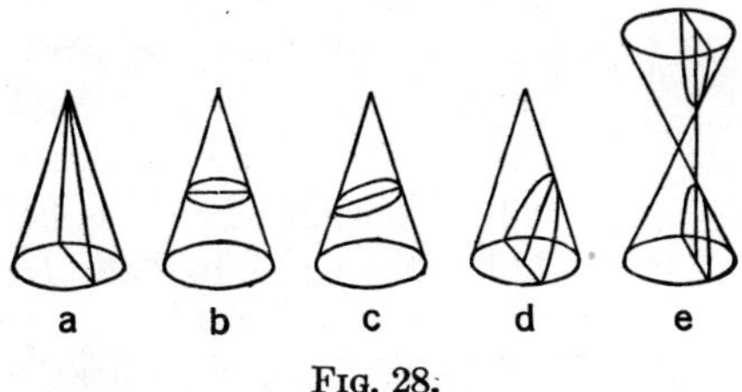

FIG. 28.

an ellipse, an hyperbola, and a parabola, only the last three of which are usually called *conic sections*.

If the plane passes through the vertex and any part of the base, the section will be a *triangle* as in Fig. 28*a*. When the plane cuts the cone parallel to the base, the section will be a *circle* as in Fig. 28*b*. When the cutting plane makes an angle with the base of less inclination than the side of the cone, as in Fig. 28*c*, the section will be an *ellipse*. When the cutting plane and the side of the cone make equal angles with the base, the section will be a *parabola* as in Fig. 28*d*. The section is a *hyperbola* when the cutting plane makes a greater angle with the base than the side of the cone makes, Fig. 28*e*. If the sides of the cone be continued through the vertex, forming an opposite equal cone, and the plane also continued to cut the opposite cone, this latter section will be the opposite hyperbola to the former.

Conic sections have considerable practical usefulness. Reinforced concrete arch bridges are often elliptical, parabolic, or even hyperbolic in section. Where curves with large diameters are needed such as for the cross-section of a pavement, the camber of a bridge, or the upper chord of a truss bridge, a parabolic curve is usually used instead of a circular curve because it is more readily computed and laid out. If a source of rays is placed at a certain point called a focus within a parabolic surface, these rays will be reflected in parallel lines. This principle is made use of in heat and light reflectors.

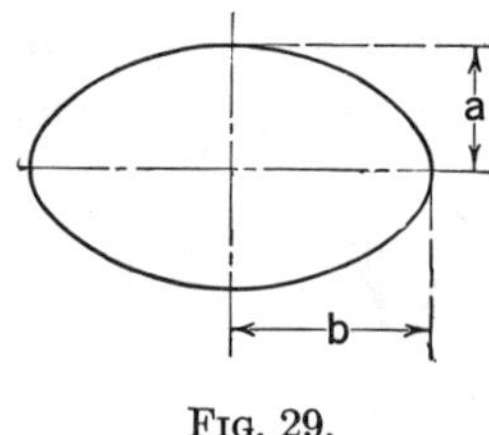

Fig. 29.

The subject of conic sections belongs to the study of analytical geometry which cannot be covered in this book.

Circumference and Area of an Ellipse.—The approximate circumference of an ellipse may be found by the following equation when a is half the smallest diameter and b half the largest diameter:

$$\text{Circumference} = \pi\sqrt{2(a^2 + b^2)}$$

The area of an ellipse is given by

$$\text{Area} = \pi \times a \times b$$

Geometrical Drawing.—Euclidean geometry is based on constructions using as the only tools a pencil, a pair of compasses, and a straight-edge or ruler. These constructions are simple and very useful. For instance, a building foreman may be confronted with the problem of laying out a line perpendicular to another line and of lengths too great for the effective use of the carpenter's square. Then, knowing the principles of geometrical construction and using a string for compasses, a sight-line between two nails, or a board, for a straight-edge, and a pencil, he can erect the perpendicular just as readily as it can be drawn on paper.

The following are the more important constructions:

To divide a straight line into a given number of equal parts. (See Fig. 30.)

Given line *a b*, which is to be divided into a given number of equal parts. Draw the line *b c*, of indefinite length, and point off from *b* the required number of equal parts, as *h*, *g*, *f*, *e*, *d*, *c′*; join *c′* and *a*, and draw the other lines parallel to *c′ a*.

To erect a perpendicular at a given point on a straight line. (See Fig. 31.)

Given line *a b* and the point *x*. The required perpendicular is *x y*.

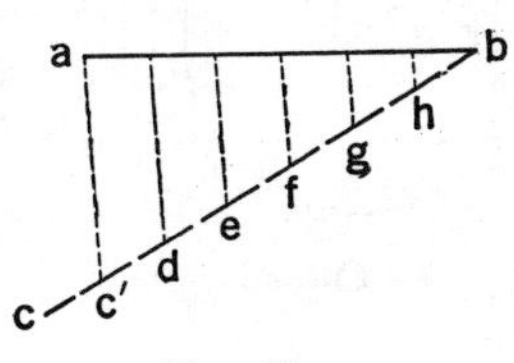

FIG. 30.

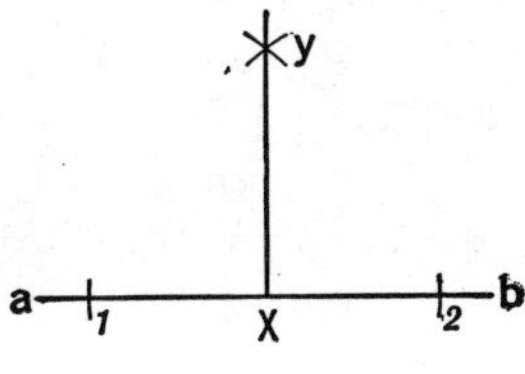

FIG. 31.

SOLUTION:

With *x* as center and any radius, as *x* 1, cut the line *a b* at 1 and 2. With 1 and 2 as centers and with a radius somewhat greater than 1 to *x*, describe arcs intersecting each other at *y*. Draw *x y*. This will be the required perpendicular.

From a given point without a straight line to draw a perpendicular to the line. (See Fig. 32.)

Given line *a b* and the point *c*. The required perpendicular is *x*.

SOLUTION:

With the point *c* as center and any radius as *c* 1, strike the arc 1 to 2. With 1 and 2 as centers and any suitable radius, describe arcs intersecting each other at *n*, lay the straight-edge through points *n* and *c* and draw the perpendicular *x*.

To erect a perpendicular at the extremity of a straight line. (See Fig. 33.)

Given line $a\ b$. The required perpendicular is x.

SOLUTION:

From any point, as c, with radius as $a\ c$, draw the circle. From point of intersection, n, through center, c, draw the diameter $n\ p$. From the point a, through the point of intersection at p, draw the perpendicular x.

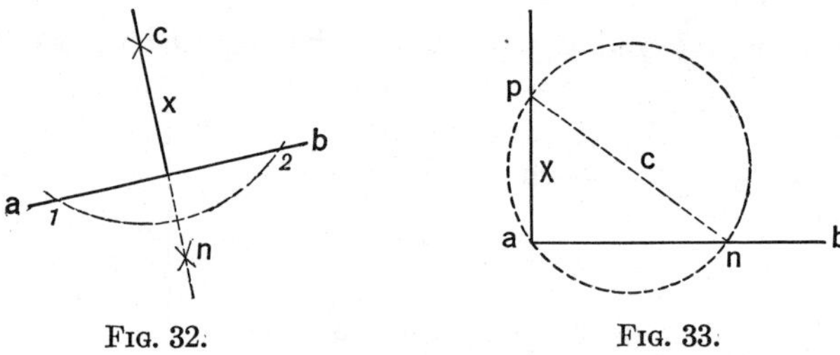

FIG. 32. FIG. 33.

The correctness of this construction is founded on the principle that inside a half circle no other angle but an angle of 90° can simultaneously touch three points in the circumference when two of these points are in the point of intersection with the diameter and the circumference and the third one anywhere on the circumference of the half circle. The pattern maker is making

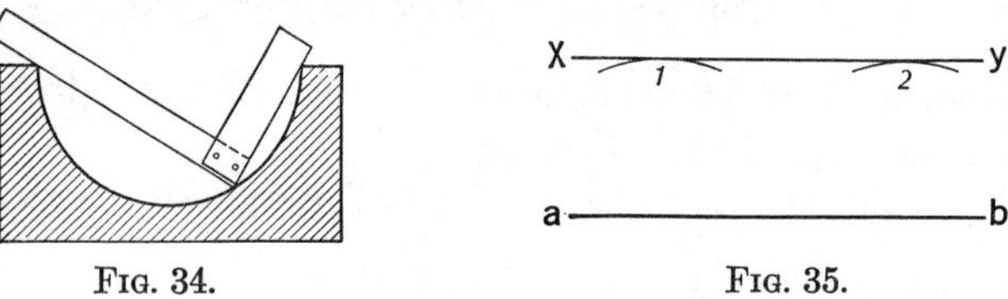

FIG. 34. FIG. 35.

practical use of this geometrical principle, when he by a common carpenter's square is trying the correctness of a semi-circular core box, as shown in Fig. 34.

Draw a line parallel to a given line. (See Fig. 35.)

Given line $a\ b$. The required line $x\ y$.

SOLUTION:

Describe with the compass from the line $a\ b$, the arcs 1 and 2; draw line $x\ y$, touching these arcs.

To divide a given angle into two equal angles. (Fig. 36).

The given angle, $a\ b\ c$, is divided by the line $b\ d$.

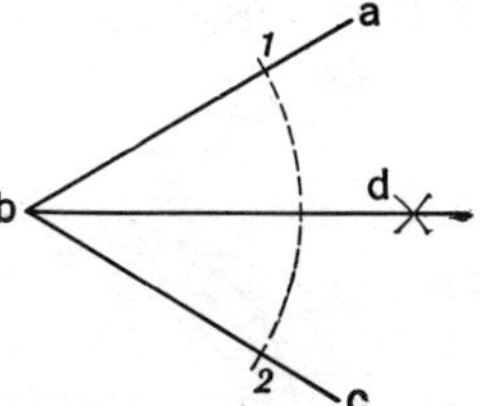

FIG 36.

SOLUTION:

With b as center and any radius, as b 1, describe the arc 1 to 2. With 1 and 2 as centers and any suitable radius, describe arcs cutting each other at d. Draw line $b\ d$, which will divide the angle into two equal parts.

To draw an angle equal to a given angle. (Fig. 37).

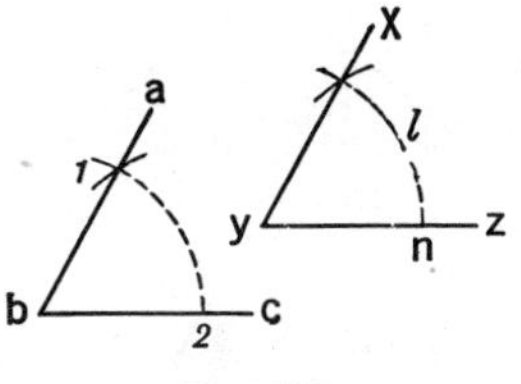

Given angle $a\ b\ c$. Construct angle $x\ y\ z$.

FIG. 37.

With b as center and any radius, as b 1, describe the arc 1 to 2, using y as center and without altering the compass describe the arc 1, intersecting $y\ z$. Measuring the distance from 2 to 1 on the given angle, transfer this measure to the arc 1, through the point of intersection. Draw the line $y\ x$, and this angle will be equal to the first angle.

NOTE.—Angles are usually measured by a tool called a protractor, looking somewhat like Fig. 38 or 39, usually made from metal, and supplied by dealers in drafting instruments. A protractor may also be constructed on paper and used for measuring angles, but it should then always be made on as large a scale as convenient.

To draw a protractor with a division of 5°. (See Fig. 39.)

Construct an angle of exactly 90°, divide the arc into nine

equal parts, then each part is 10°; divide each part into two equal parts and each is 5°.

Prove that the sum of the three angles in a triangle consists of 180°. (See Fig. 40.)

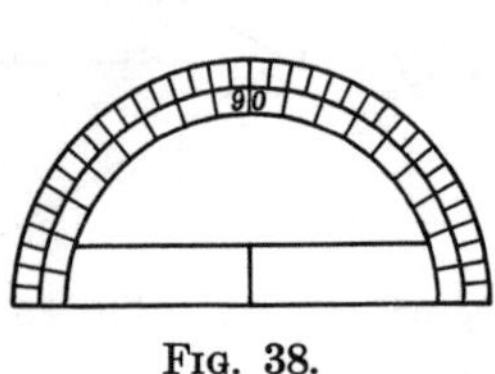

FIG. 38.

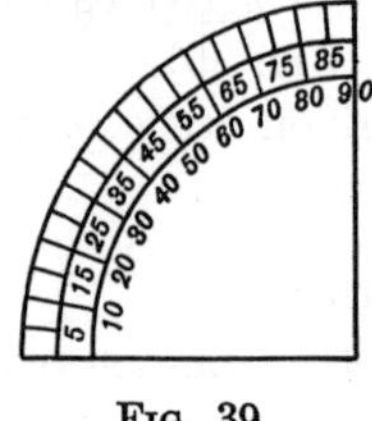

FIG. 39.

SOLUTION:

In the triangle $a\ b\ c$, extend the base line to i. Draw the line $o\ p$, parallel to the side $a\ b$, thereby the angle g will be equal to the angle d, and the angle h must be equal to the angle c. The angle f is one angle in the triangle and $f + g + h = 180°$.

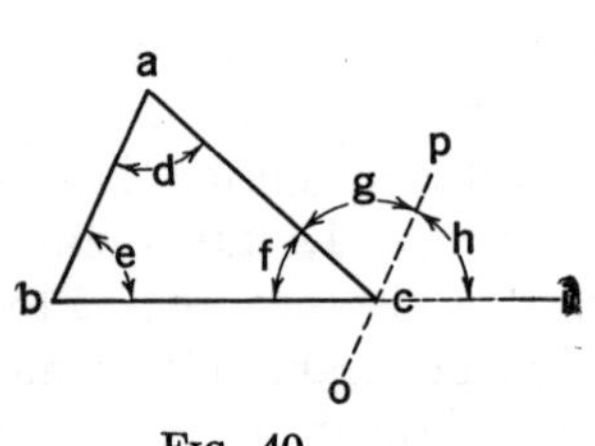

FIG. 40.

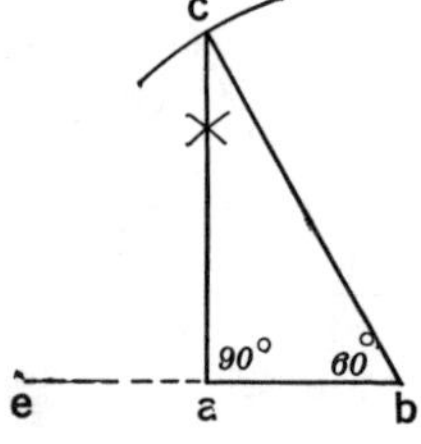

FIG. 41.

To draw on a given base line a triangle having angles 90°, 30°, and 60°. (See Fig. 41.)

Given line $a\ b$, required triangle is a, c, b.

SOLUTION:

Extend the line $a\ b$ to twice its length, to the point e. With e and b as centers strike arcs intersecting each other and erect the perpendicular $a\ c$. With b as center and a radius be draw an arc intersecting ac at c. Connect b and c. This will complete the triangle.

To draw a square inside a given circle. (See Fig. 42.)

Solution:

Draw the line *a b* through the center of the circle. From points of intersection at *a* and *b*, describe with any suitable radius arcs intersecting at *n* and *m*. Draw through the points the line *c d*. Connect the points of intersection on the circle, and the required square is constructed.

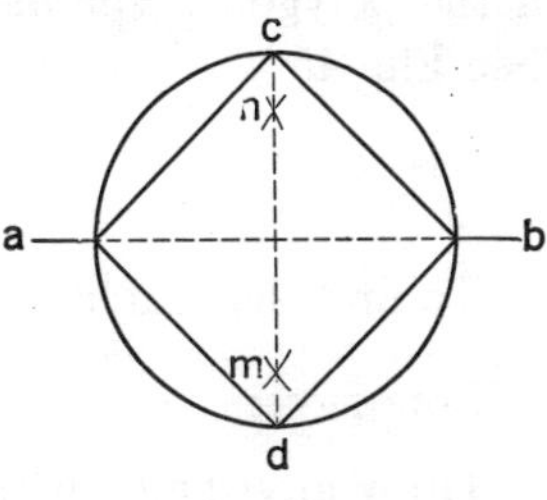

Fig. 42.

To draw a square outside a given circle. (See Fig. 43.)

Solution:

Draw lines *a b* and *c d*, and from points of intersection at *b* and *c*, describe half circles; their points of intersection determine the sides of the square.

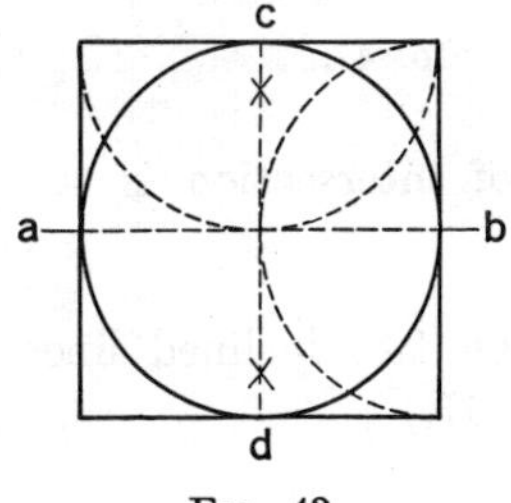

Fig. 43.

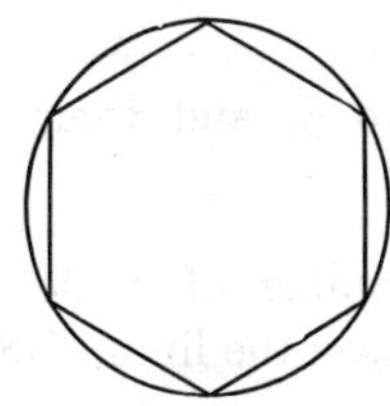

Fig. 44.

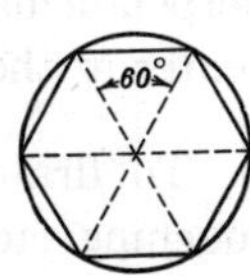

Fig. 45.

To draw a hexagon within a given circle. (See Fig. 44.)

Apply the radius as a chord successively about the circle; the resulting figure will be a hexagon.

To inscribe in a circle a regular polygon of any given number of sides.

Solution:

Divide 360 by the number of sides, and the quotient is the number of degrees, minutes, and seconds contained in the center angle of a triangle, of which one side will make one of the sides in the polygon. For instance, draw a hexagon by this method. (See Fig. 45.)

$$\frac{360}{6} = 60°$$

To find the center in a given circle. (See Fig. 46.)

SOLUTION:

Draw anywhere on the circumference of the circle two chords at approximately right angles to each other; bisect these by the

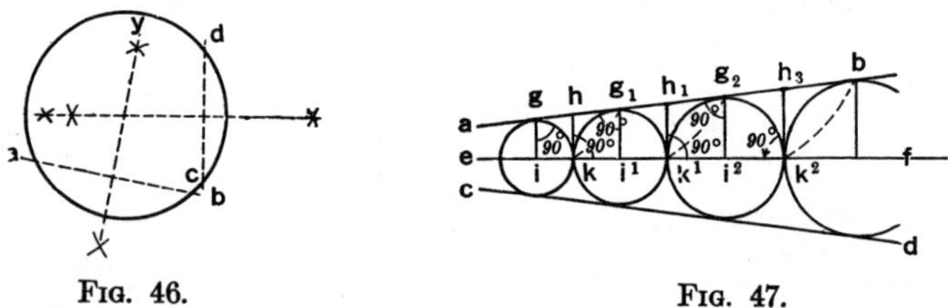

FIG. 46. FIG. 47.

perpendiculars x and y, and their point of intersection is the center of the circle.

To draw any number of circles between two inclined lines touching each other and the lines. (See Fig. 47.)

SOLUTION:

Bisect the inclination of the given lines $a\,b$, $c\,d$ by the line $e\,j$ From a point i in this line draw the perpendicular $i\,g$ to the line $a\,b$ and at i describe the circle $g\,e$ touching the lines and cutting the center line at k. From k draw $k\,h$ perpendicular to the center line and cutting $a\,b$ at h and from h describe an arc $k\,g'$ cutting $a\,b$ at $g'\,l''$ parallel to $g\,i$ the center of the next circle to be described with radius $k\,i'$ and so on for the next.

To draw a circle through three given points. (See Fig. 48.) The given points are a, b, and c.

SOLUTION:

From *a* and *b* as centers with suitable radius, describe arcs intersecting at *e e*. Draw a line through these points. From *b* and *c* as centers, describe arcs intersecting at *d d*; draw a line through these points. The point where these two lines intersect is the center of the circle.

To draw two tangents to a circle from a given point without same circle. (See Fig. 49.)

Given point *a*, and the circle with the center *n*. The required tangents are *a d* and *a b*.

SOLUTION:

Bisect line *n a*. With *c* as center and radius *a c*, describe the arc *b d* through the center of the circle. The points of intersection

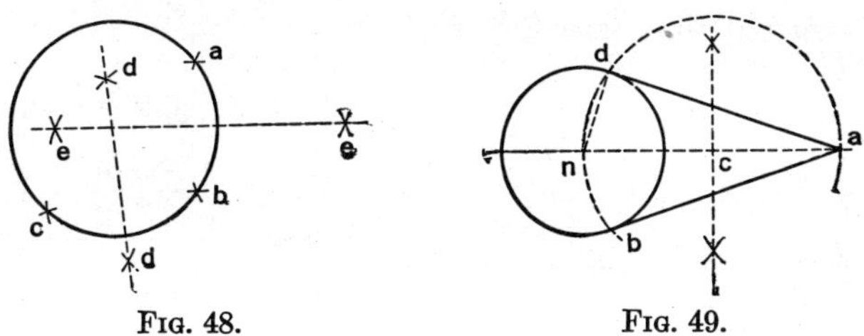

FIG. 48. FIG. 49.

at *b* and *d* are the points where the required tangents *a b* and *a d* will touch the circle.

To draw a tangent to a given point in a given circle. (See Fig. 50.)

Given circle and the point *h*, *x y* is required.

SOLUTION:

The radius is drawn to the point *h* and a line constructed perpendicular to it at the point *h*. This perpendicular, touching the circle at *h*, is called a *tangent*.

To draw a circle of a certain size that will touch the periphery of two given circles. (See Fig. 51.)

Given the diameter of circles *a*, *b*, and *c*. Locate the center for circle *c*, when centers for *a* and *b* are given.

SOLUTION:

From center of *a*, describe an arc with a radius equal to the sum of radii of *a* and *c*. From *b* as center, describe another arc using a radius equal to the sum of the radii of *b* and *c*. The point of intersection of those two arcs is the center of the circle *c*.

NOTE.—This construction is useful when locating the center for an intermediate gear. For instance, if *a* and *b* are the pitch circles of two gears, *c* would be the pitch circle located in correct position to connect *a* and *b*.

To draw an ellipse, the longest and shortest diameter being given. The diameters *a b* and *c d* are given. The required ellipse is constructed thus (see Fig. 52):

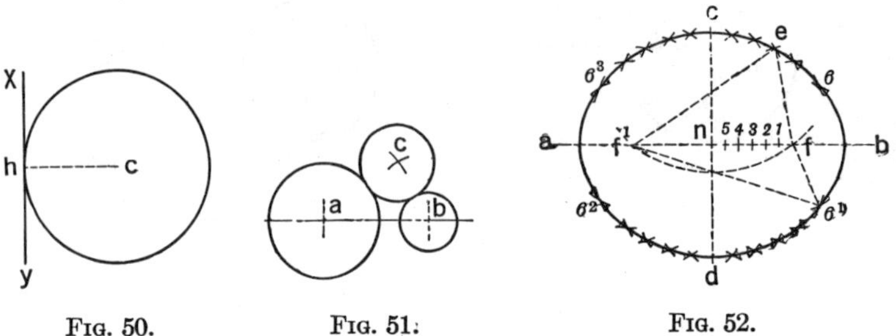

FIG. 50. FIG. 51. FIG. 52.

From *c* as center with a radius *a n*, describe an arc $f^1 f$. The points where this arc intersect *a b* are foci. The distance *f n* is divided into any number of parts, as 1, 2, 3, 4, 5. With radius 1 to *b*, and the focus *f* as center, describe arcs 6 and 6^1 with the same radius and with f^1 as center describe arcs 6^2 and 6^3. With radius 1 to *a* and f^1 as center, describe arcs intersecting at 6 and 6^1; with the same radius and with *f* as center, describe arcs intersecting at 6^2 and 6^3. Continue this operation for points 2, 3, etc., and when all the points for the circumference are in this way marked out, draw the ellipse by using a scroll. It is a property with ellipses that the sum of any two lines drawn from the foci to any point in the circumference is equal to the largest diameter. For instance:

$$f^1 e + f e, \quad = a\,b, \quad \text{or} \quad f\,6^1 + f^1\,6^1, \quad = a\,b.$$

Cycloids.—Suppose that a round disc, *c*, rolls on a straight line, *a*, *b*, and that a lead pencil is fastened at the point *r*; it will then describe a curved line, *a*, *l*, *r*, *n*, *b*. This line is called a *cycloid*. (See Fig. 53.)

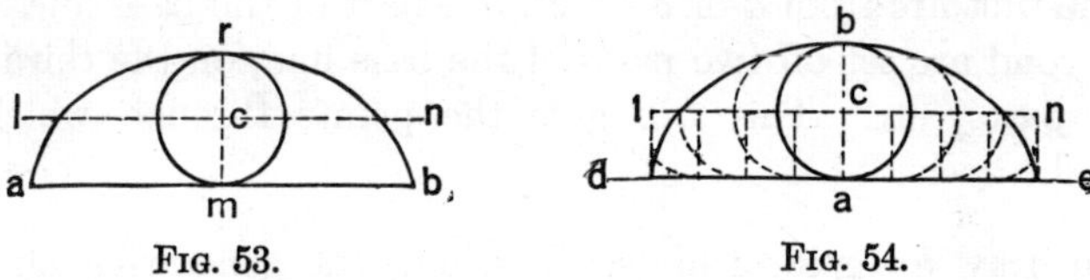

FIG. 53. FIG. 54.

This supposed disc is usually called the *generating circle.* The line *a b* is the base line of the cycloid and is equal in length to π times *m r*, or practically 3.1416 times the diameter of the generating circle. The length of the curved line *a*, *l*, *r*, *n*, *b* is four times *r m* (four times as long as the diameter of the generating circle).

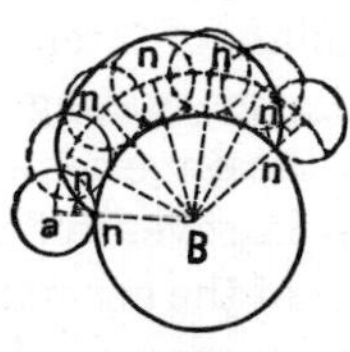

FIG. 55.

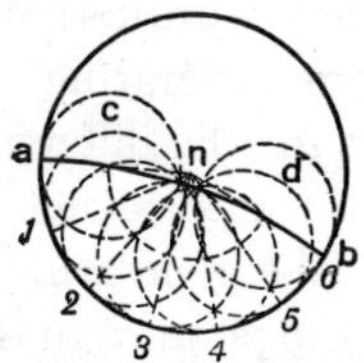

FIG. 56.

A circle rolling on a straight line generates a cycloid. (See Figs. 53 and 54.)

A circle rolling upon another circle is generating an *epicycloid.* (See Fig. 55.)

A circle rolling within another circle generates a *hypocycloid.* (See Fig. 56.)

To draw a cycloid, the generating circle being given.

SOLUTION:

Divide the diameter of the rolling circle in 7 equal parts. Set off 11 of these parts on each side of *a* on the line *d e*. This will give a base line practically equal to the circumference. Divide the base line from the point *a* into any number of equal parts; erect the perpendiculars; with center-line as centers and a radius equal to the radius of the generating circle describe the arcs. On the first arc from *d* or *e* set off one part of the base line. On the second arc set off two parts of the base line; on the third arc, three parts, etc. This will give the points through which to draw the cycloid.

To draw an epicycloid (see Fig. 55), the generating circle *a* and the fundamental circle *B* being given.

SOLUTION:

Concentric with the circle *B*, describe an arc through the center of the generating circle. Divide the circumference of the generating circle into any number of equal parts and set this off on the circumference of the circle *B*. Through those points draw radial lines extending until they intersect the arc passing through the center of the generating circle. These points of intersection give the centers for the different positions of the generating circle, and for the rest, the construction is essentially the same as the cycloids. In Fig. 55 the generating circle is shown in seven different positions, and the point *n*, in the circumference of the generating circle, may be followed from the position at the extreme left for one full rotation to the position where it again touches the circle *B*.

To draw a hypocycloid. (See Fig. 56.)

The hypocycloid is the line generated by a point in a circle rolling within another larger circle, and is constructed thus (see Fig. 56):

Divide the circumference of the generating circle into any number of equal parts. Set off these on the circumference of the fundamental circle. From each point of division draw radial lines, 1, 2, 3, 4, 5, 6. From *n* as center describe an arc through the center of the generating circle, as the arc *c d*. The point of intersection between this arc and the radial lines are centers for the different positions of the generating circle. The distance from 1 to *a* on the fundamental circle is set off from 1 on the generating circle in its first new position; the distance 2 to *a* on the fundamental circle is set off from 2 on the generating circle in its second position, etc. For the rest, the construction is substantially the same as Figs. 54 and 55.

NOTE.—If the diameter of the generating circle is equal to the radius of the fundamental circle, the hypocycloid will be a straight line, which is the diameter of the fundamental circle.

Involute.—An involute is a curved line which may be assumed to be generated in the following manner: Suppose a string be placed

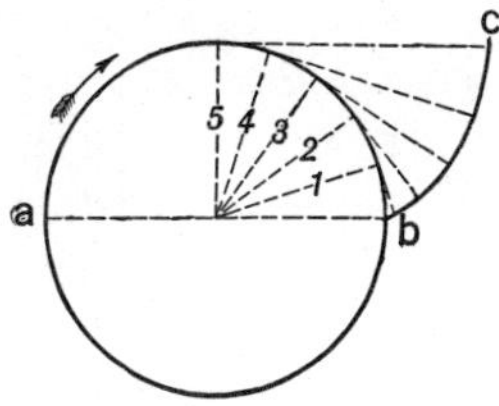

FIG. 57.

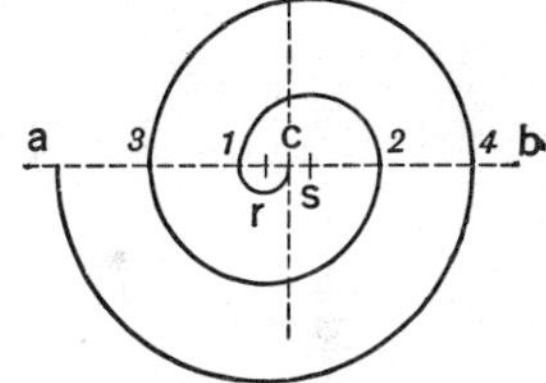

FIG. 58.

around a cylinder from *a* to *b*, in the direction of the arrow (see Fig. 57), and having a pencil attached at *b*; keep the string tight and move the pencil toward *c*, and the involute, *b c*, is generated.

To draw an involute.

SOLUTION:

From the point *b* (see Fig. 57) set off any number of radial lines at equal distances, as 1, 2, 3, 4, 5. From points of intersection draw the tangents (perpendicular to the radial lines). Set off on the first tangent the length of the arc 1 to *b*; on the second tangent the arc 2 to *b*, etc. This will give the points through which to draw the involute.

To draw a spiral from a given point, *c*.

SOLUTION:

Draw the line *a b* through the point *c*. Set off the centers *r* and *S*, one-fourth as far from *c* as the distance is to be between two lines in the spiral. Using *r* as center, describe the arc from *c* to 1; and using *S* as center, describe the arc from 1 to 2; using *r* as center, describe the arc from 2 to 3, etc.

V

Trigonometry

Trigonometry is that branch of geometry which deals with angles and with the solution of triangles by means of trigonometric functions.

Angles.—The opening between two straight intersecting lines is an *angle*. An angle may be designated in any one of several ways. Thus, in Fig. 1 we may speak of the angle B, the angle ABC, or the angle a, and refer in each instance to the same angle.

Angles are measured in *degrees*. One degree is $\frac{1}{360}$ of a whole angle, or angle describing a full circle. Then a 90-degree angle is one-quarter of a whole angle. It is called a right angle and the legs are perpendicular to each other. An angle of 180 degrees is equal to the sum of two right angles and is therefore a straight line. It is sometimes called a straight angle.

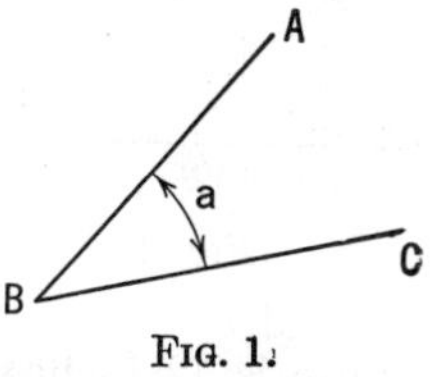

Fig. 1.

Trigonometric Functions.—If we have a right triangle whose acute angles are each 45 degrees and whose legs are each 1 unit long we know from geometry that the length of the hypotenuse is equal to the square root of the sum of the squares of the two sides. Then, in this case, the hypotenuse is equal to $\sqrt{2}$ units. Then, if we have *any* equilateral right triangle, the ratio of the length of legs to the length of the hypotenuse is $1 : \sqrt{2}$. This ratio may then be used to find the hypotenuse if the leg is given, and vice versa. Thus, if the hypotenuse of a 45-degree-angled

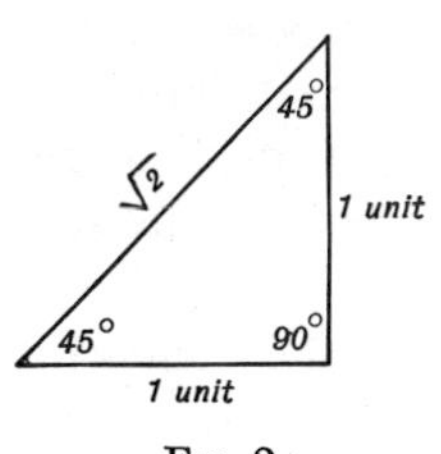

Fig. 2.

right triangle is 9 inches, the leg is $9 \times \frac{1}{\sqrt{2}}$ or 6.4 inches. Similarly, if the leg is given as 8 inches, the hypotenuse is $8 \times \frac{\sqrt{2}}{1}$ or 11.3 inches.

For a 45-degree-angled right triangle, the ratio of a side to the hypotenuse is *always* $\frac{1}{\sqrt{2}} = \frac{1}{1.414} = 0.707$, and the ratio of the hypotenuse to a side is *always* $\frac{\sqrt{2}}{1} = \frac{1.414}{1} = 1.414$.

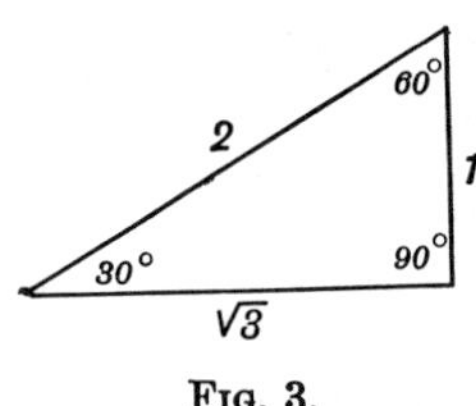

Fig. 3.

Let us now consider a right triangle whose angles are 30, 60, and 90 degrees. If the short side is 1 unit long, the hypotenuse is 2 units and the long leg $\sqrt{3}$ or 1.732 units long. Then, if we are given any 30-60-90 degree triangle and the length of one side, we can readily solve for the other sides. For example, if the hypotenuse is 12 inches, the short side is $12 \times \frac{1}{2}$ or 6 inches, and the long leg is $12 \times \frac{1.732}{2} = 10.4$ inches.

We have shown how the ratios of one side of a right triangle to another may be used in solving triangles. These ratios are called *trigonometric functions.* Not only are there definite ratios between the sides of right triangles with angles of 30 degrees, 45 degrees, and 60 degrees, as we have shown, but definite ratios exist for right triangles of *any* angle.

There are six fundamental trigometric functions known as (with abbreviations) *sine* (sin), *cosine* (cos), *tangent* (tan), *cotangent* (cot), *secant* (sec), and *cosecant* (csc).

The sine of an acute angle of a right triangle is the opposite side divided by the hypotenuse, or, in fractional form, opposite side over hypotenuse.

The cosine is the adjacent side over the hypotenuse.

The tangent is the opposite side over the adjacent side.

The cotangent is the adjacent side over the opposite side, or one over the tangent.

The secant is the hypotenuse over the adjacent side, or one over the cosine.

The cosecant is the hypotenuse over the opposite side, or one over the sine.

In Fig. 4 let a, b, and c represent the lengths of the sides of any right triangle, ABC. Then,

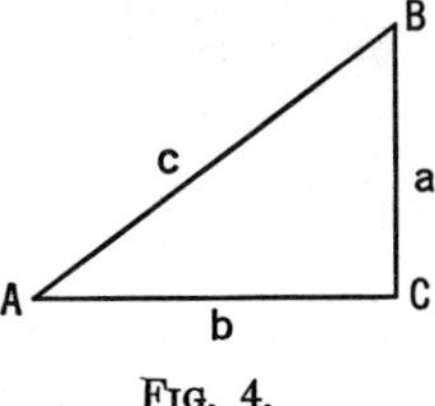

FIG. 4.

$$\sin A = \frac{a}{c} \qquad \cot A = \frac{b}{a}$$

$$\cos A = \frac{b}{c} \qquad \sec A = \frac{c}{b}$$

$$\tan A = \frac{a}{b} \qquad \csc A = \frac{c}{a}$$

Relations of Functions.—We notice that the cotangent, secant, and cosecant are reciprocals respectively of the tangent, cosine, and sine. Other relations between functions of one angle or of several angles, such as the functions of the sum of two angles, half an angle, twice an angle, etc., are very important and we give a few of them here:

Functions of one angle (A)

$$\sin^2 A + \cos^2 A = 1$$

$$\sec^2 A - \tan^2 A = 1$$

$$\csc^2 A - \cot^2 A = 1$$

Functions of the sum of two angles ($A + B$)

$$\sin (A + B) = \sin A \cos B + \cos A \sin B$$

$$\cos (A + B) = \cos A \cos B - \sin A \sin B$$

$$\tan (A + B) = \frac{\tan A + \tan B}{1 - \tan A \tan B}$$

$$\cot (A + B) = \frac{\cot A \cot B - 1}{\cot B + \cot A}$$

Functions of the difference of two angles $(A - B)$

$$\sin (A - B) = \sin A \cos B - \cos A \sin B$$

$$\cos (A - B) = \cos A \cos B + \sin A \sin B$$

$$\tan (A - B) = \frac{\tan A - \tan B}{1 + \tan A \tan B}$$

$$\cot (A - B) = \frac{\cot A \cot B + 1}{\cot B - \cot A}$$

Functions of one-half an angle $(\frac{1}{2}A)$

$$\sin \tfrac{1}{2}A = \frac{\sin A}{2 \cos \frac{1}{2}A} = \pm \sqrt{\frac{1 - \cos A}{2}}$$

$$\cos \tfrac{1}{2}A = \frac{\sin A}{2 \sin \frac{1}{2}A} = \pm \sqrt{\frac{1 + \cos A}{2}}$$

$$\tan \tfrac{1}{2}A = \frac{1 - \cos A}{\sin A} = \pm \sqrt{\frac{1 - \cos A}{1 + \cos A}}$$

$$\cot \tfrac{1}{2}A = \pm \sqrt{\frac{1 + \cos A}{1 - \cos A}}$$

Functions of twice an angle $(2A)$

$$\sin 2A = 2 \sin A \cos A = \frac{2 \tan A}{1 + \tan^2 A}$$

$$\cos 2A = \cos^2 A - \sin^2 A = 1 - 2 \sin^2 A$$

$$= 2 \cos^2 A - 1 = \frac{1 - \tan^2 A}{1 + \tan^2 A}$$

$$\tan 2A = \frac{2 \tan A}{1 - \tan^2 A} = \frac{\sin 3A - \sin A}{\cos 3A + \cos A}$$

$$\cot 2A = \frac{\cot^2 A - 1}{2 \cot A}$$

Functions of three times an angle $(3A)$

$$\sin 3A = 3 \sin A - 4 \sin^3 A$$

$$\cos 3A = 4 \cos^3 A - 3 \cos A$$

$$\tan 3A = \frac{3 \tan A - \tan^3 A}{1 - 3 \tan^2 A}$$

$$\cot 3A = \frac{\cot^3 A - 3 \cot A}{3 \cot^2 - 1}$$

Tables of Natural and Logarithmic Trigonometric Functions.— Tables for practical use need consist only of the values for sines, cosines, and tangents since the other functions can readily be obtained from these.

The *natural* functions are the actual values of the trigonometric functions themselves. The logarithms of these values are called the *logarithmetic* functions.

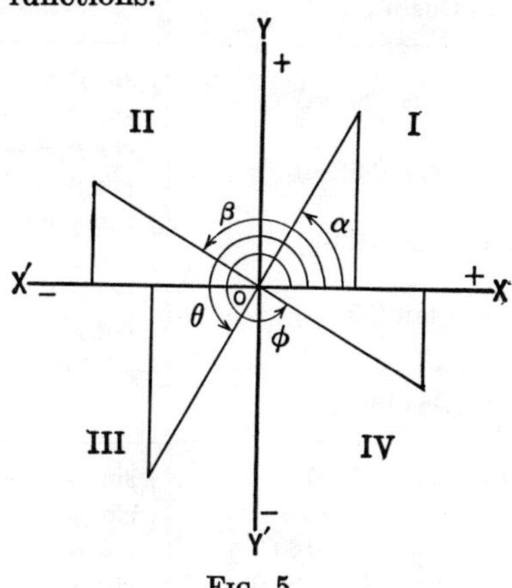

Fig. 5.

Table 2 contains the natural sines, tangents, cotangents and cosines. The functions from 0 degrees to 45 degrees are read *down* the page and the functions from 45 degrees to 90 degrees are read *up* the page.

The solution of problems with trigonometric functions often involves logarithmetic computations. A table giving directly the *logarithms* of the sines, cosines and tangents is a great convenience in such cases; these logarithmic functions are given in Table 3. The use of these tables will be illustrated later in the solution of triangles.

If a circle be imagined as divided into four quadrants and these numbered I, II, III, and IV as shown in Fig. 5, then an angle, such as α, which is less than 90 degrees is said to lie in the first quadrant. An angle between 90 degrees and 180 degrees, such as β, is said to lie in the second quadrant; an angle between 180 degrees and 270 degrees, in the third quadrant; and an angle between 270 degrees and 360 degrees in the fourth quadrant. The function of any angle may be reduced to the function of an angle not greater than 90 degrees by the use of Table 1 paying careful attention to signs.

TABLE 1

	1st Quadrant	2nd Quadrant
sin	$\sin\alpha = \cos(90°-\alpha)$	$\sin\beta = \sin(180°-\beta)$ $\sin\beta = \cos(\beta-90°)$
cos	$\cos\alpha = \sin(90°-\alpha)$	$\cos\beta = -\cos(180°-\beta)$ $\cos\beta = -\sin(\beta-90°)$
tan	$\tan\alpha = \cot(90°-\alpha)$	$\tan\beta = -\tan(180°-\beta)$ $\tan\beta = -\cot(\beta-90°)$
cot	$\cot\alpha = \tan(90°-\alpha)$	$\cot\beta = -\cot(180°-\beta)$ $\cot\beta = -\tan(\beta-90°)$
	3rd Quadrant	**4th Quadrant**
sin	$\sin\theta = -\sin(\theta-180°)$ $\sin\theta = -\cos(270°-\theta)$	$\sin\phi = -\sin(360°-\phi)$ $\sin\phi = -\cos(\phi-270°)$
cos	$\cos\theta = -\cos(\theta-180°)$ $\cos\theta = -\sin(270°-\theta)$	$\cos\phi = \cos(360°-\phi)$ $\cos\phi = \sin(\phi-270°)$
tan	$\tan\theta = \tan(\theta-180°)$ $\tan\theta = \cot(270°-\theta)$	$\tan\phi = -\tan(360°-\phi)$ $\tan\phi = -\cot(\phi-270°)$
cot	$\cot\theta = \cot(\theta-180°)$ $\cot\theta = \tan(270°-\theta)$	$\cot\phi = -\cot(360°-\phi)$ $\cot\phi = -\tan(\phi-270°)$

TABLE 2

Table of Natural Trigonometric Functions

2.—Natural Sines, Tangents, Cotangents, Cosines.

(Versed sine = 1 − cosine; coversed sine = 1 − sine.)

0° | 1°

′	Sine.	Tang.	Cotang.	Cosine.		′	Sine.	Tang.	Cotang.	Cosine.	
0	.0000000	.000000	Infinite	1.000000	60	0	.0174524	.017455	57.28996	.9998477	60
1	.0002909	.000291	3437.746	1.000000	59	1	.0177432	.017746	56.35059	.9998426	59
2	.0005818	.000582	1718.873	.9999998	58	2	.0180341	.018037	55.44151	.9998374	58
3	.0008727	.000872	1145.915	.9999996	57	3	.0183249	.018328	54.56130	.9998321	57
4	.0011636	.001163	859.4363	.9999993	56	4	.0186158	.018619	53.70858	.9998267	56
5	.0014544	.001454	687.5488	.9999989	55	5	.0189066	.018910	52.88211	.9998213	55
6	.0017453	.001745	572.9572	.9999985	54	6	.0191974	.019201	52.08067	.9998157	54
7	.0020362	.002036	491.1060	.9999979	53	7	.0194883	.019492	51.30315	.9998101	53
8	.0023271	.002327	429.7175	.9999973	52	8	.0197791	.019783	50.54850	.9998044	52
9	.0026180	.002618	381.9709	.9999966	51	9	.0200699	.020074	49.81572	.9997986	51
10	.0029089	.002908	343.7737	.9999958	50	10	.0203608	.020365	49.10388	.9997927	50
11	.0031998	.003199	312.5213	.9999949	49	11	.0206516	.020656	48.41208	.9997867	49
12	.0034907	.003490	286.4777	.9999939	48	12	.0209424	.020947	47.73950	.9997807	48
13	.0037815	.003781	264.4408	.9999928	47	13	.0212332	.021238	47.08534	.9997745	47
14	.0040724	.004072	245.5519	.9999917	46	14	.0215241	.021529	46.44886	.9997683	46
15	.0043633	.004363	229.1816	.9999905	45	15	.0218149	.021820	45.82935	.9997620	45
16	.0046542	.004654	214.8576	.9999892	44	16	.0221057	.022111	45.22614	.9997556	44
17	.0049451	.004945	202.2187	.9999878	43	17	.0223965	.022402	44.63859	.9997492	43
18	.0052360	.005236	190.9841	.9999863	42	18	.0226873	.022693	44.06611	.9997426	42
19	.0055268	.005526	180.9322	.9999847	41	19	.0229781	.022984	43.50812	.9997360	41
20	.0058177	.005817	171.8854	.9999831	40	20	.0232690	.023275	42.96407	.9997292	40
21	.0061086	.006108	163.7001	.9999813	39	21	.0235598	.023566	42.43346	.9997224	39
22	.0063995	.006399	156.2590	.9999795	38	22	.0238506	.023857	41.91579	.9997156	38
23	.0066904	.006690	149.4650	.9999776	37	23	.0241414	.024148	41.41058	.9997086	37
24	.0069813	.006981	143.2371	.9999756	36	24	.0244322	.024439	40.91741	.9997015	36
25	.0072721	.007272	137.5075	.9999736	35	25	.0247230	.024730	40.43583	.9996943	35
26	.0075630	.007563	132.2185	.9999714	34	26	.0250138	.025021	39.96546	.9996871	34
27	.0078539	.007854	127.3213	.9999692	33	27	.0253046	.025312	39.50589	.9996798	33
28	.0081448	.008145	122.7739	.9999668	32	28	.0255954	.025603	39.05677	.9996724	32
29	.0084357	.008436	118.5401	.9999644	31	29	.0258862	.025894	38.61773	.9996649	31
30	.0087265	.008726	114.5886	.9999619	30	30	.0261769	.026185	38.18845	.9996573	30
31	.0090174	.009017	110.8920	.9999593	29	31	.0264677	.026477	37.76861	.9996497	29
32	.0093083	.009308	107.4264	.9999567	28	32	.0267585	.026768	37.35789	.9996419	28
33	.0095992	.009599	104.1709	.9999539	27	33	.0270493	.027059	36.95600	.9996341	27
34	.0098900	.009890	101.1069	.9999511	26	34	.0273401	.027350	36.56265	.9996262	26
35	.0101809	.010181	98.21794	.9999482	25	35	.0276309	.027641	36.17759	.9996182	25
36	.0104718	.010472	95.48947	.9999452	24	36	.0279216	.027932	35.80055	.9996101	24
37	.0107627	.010763	92.90848	.9999421	23	37	.0282124	.028223	35.43128	.9996020	23
38	.0110535	.011054	90.46333	.9999389	22	38	.0285032	.028514	35.06954	.9995937	22
39	.0113444	.011345	88.14357	.9999357	21	39	.0287940	.028805	34.71511	.9995854	21
40	.0116353	.011636	85.93979	.9999323	20	40	.0290847	.029097	34.36777	.9995770	20
41	.0119261	.011927	83.84350	.9999289	19	41	.0293755	.029388	34.02730	.9995684	19
42	.0122170	.012217	81.84704	.9999254	18	42	.0296662	.029679	33.69350	.9995599	18
43	.0125079	.012508	79.94343	.9999218	17	43	.0299570	.029970	33.36619	.9995512	17
44	.0127987	.012799	78.12634	.9999181	16	44	.0302478	.030261	33.04517	.9995424	16
45	.0130896	.013090	76.39000	.9999143	15	45	.0305385	.030552	32.73026	.9995336	15
46	.0133805	.013381	74.72916	.9999105	14	46	.0308293	.030843	32.42129	.9995247	14
47	.0136713	.013672	73.13899	.9999065	13	47	.0311200	.031135	32.11809	.9995157	13
48	.0139622	.013963	71.61507	.9999025	12	48	.0314108	.031426	31.82051	.9995066	12
49	.0142530	.014254	70.15334	.9998984	11	49	.0317015	.031717	31.52839	.9994974	11
50	.0145439	.014545	68.75008	.9998942	10	50	.0319922	.032008	31.24157	.9994881	10
51	.0148348	.014836	67.40185	.9998900	9	51	.0322830	.032299	30.95992	.9994788	9
52	.0151256	.015127	66.10547	.9998856	8	52	.0325737	.032591	30.68330	.9994693	8
53	.0154165	.015418	64.85800	.9998812	7	53	.0328644	.032882	30.41158	.9994598	7
54	.0157073	.015709	63.65674	.9998766	6	54	.0331552	.033173	30.14461	.9994502	6
55	.0159982	.016000	62.49915	.9998720	5	55	.0334459	.033464	29.88229	.9994405	5
56	.0162890	.016291	61.38290	.9998673	4	56	.0337366	.033755	29.62449	.9994308	4
57	.0165799	.016582	60.30582	.9998625	3	57	.0340274	.034047	29.37110	.9994209	3
58	.0168707	.016873	59.26587	.9998577	2	58	.0343181	.034338	29.12200	.9994110	2
59	.0171616	.017164	58.26117	.9998527	1	59	.0346088	.034629	28.87708	.9994009	1
60	.0174524	.017455	57.28996	.9998477	0	60	.0348995	.034920	28.63625	.9993908	0
	Cosine.	Cotang	Tang.	Sine.	′		Cosine.	Cotang	Tang.	Sine.	′

89° | 88°

Note.—Secant = 1 ÷ cosine; cosecant = 1 ÷ sine.

2.—Natural Sines, Tangents, Cotangents, Cosines.—(Continued).

(Versed sine = 1 − cosine; coversed sine = 1 − sine.)

2°

′	Sine.	Tang.	Cotang.	Cosine.	
0	.0348995	.034920	28.63625	.9993908	60
1	.0351902	.035212	28.39939	.9993806	59
2	.0354809	.035503	28.16642	.9993704	58
3	.0357716	.035794	27.93723	.9993600	57
4	.0360623	.036085	27.71174	.9993495	56
5	.0363530	.036377	27.48985	.9993390	55
6	.0366437	.036668	27.27148	.9993284	54
7	.0369344	.036959	27.05655	.9993177	53
8	.0372251	.037250	26.84498	.9993069	52
9	.0375158	.037542	26.63669	.9992960	51
10	.0378065	.037833	26.43160	9992851	50
11	.0380971	.038124	26.22963	.9992740	49
12	.0383878	.038416	26 03073	.9992629	48
13	.0386785	.038707	25 83482	.9992517	47
14	.0389692	.038998	25.64183	.9992404	46
15	.0392598	.039290	25.45170	.9992290	45
16	.0395505	.039581	25.26436	.9992176	44
17	.0398411	.039872	25.07975	9992060	43
18	.0401318	.040164	24.89782	.9991944	42
19	.0404224	.040455	24.71851	.9991827	41
20	.0407131	.040746	24.54175	9991709	40
21	.0410037	.041038	24.36750	.9991590	39
22	.0412944	.041329	24 19571	.9991470	38
23	.0415850	.041621	24.02632	.9991350	37
24	.0418757	.041912	23.85927	.9991228	36
25	.0421663	.042203	23.69453	9991106	35
26	.0424569	.042495	23.53205	.9990983	34
27	.0427475	.042786	23.37177	9990859	33
28	.0430382	043078	23.21366	.9990734	32
29	.0433288	.043369	23 05767	.9990609	31
30	.0436194	.043660	22.90376	.9990482	30
31	.0439100	.043952	22.75189	.9990355	29
32	.0442006	.044243	22.60201	.9990227	28
33	.0444912	.044535	22.45409	.9990098	27
34	.0447818	.044826	22.30809	.9989968	26
35	.0450724	.045118	22.16398	.9989837	25
36	.0453630	.045409	22 02171	.9989706	24
37	.0456536	.045701	21.88125	.9989573	23
38	.0459442	.045992	21.74256	.9989440	22
39	.0462347	.046284	21.60563	.9989306	21
40	.0465253	046575	21.47040	9989171	20
41	.0468159	.046867	21.33685	.9989035	19
42	.0471065	.047158	21.20494	.9988899	18
43	.0473970	.047450	21.07466	.9988761	17
44	.0476876	.047741	20.94596	.9988623	16
45	.0479781	048033	20.81882	.9988484	15
46	.0482687	.048325	20.69322	.9988344	14
47	.0485592	.048616	20.56911	.9988203	13
48	.0488498	.048908	20.44648	.9988061	12
49	.0491403	.049199	20.32530	.9987919	11
50	.0494308	.049491	20.20555	.9987775	10
51	.0497214	.049782	20.08719	.9987631	9
52	.0500119	.050074	19.97021	.9987486	8
53	.0503024	.050366	19.85459	.9987340	7
54	.0505929	.050657	19 74029	.9987194	6
55	.0508835	.050949	19.62729	.9987046	5
56	.0511740	.051241	19.51558	.9986898	4
57	.0514645	.051532	19.40513	.9986748	3
58	.0517550	.051824	19.29592	.9986598	2
59	.0520455	.052116	19.18793	9986447	1
60	.0523360	052407	19 08113	9986295	0
	Cosine.	Cotang	Tang.	Sine.	′

87°

3°

′	Sine.	Tang.	Cotang.	Cosine.	
0	.0523360	.052407	19 08113	.9986295	60
1	.0526264	.052699	18.97552	.9986143	59
2	.0529169	.052991	18.87106	.9985989	58
3	.0532074	.053282	18.76775	.9985835	57
4	.0534979	.053574	18.66556	.9985680	56
5	.0537883	.053866	18.56447	.9985524	55
6	.0540788	.054158	18.46447	.9985367	54
7	.0543693	.054449	18.36553	.9985209	53
8	.0546597	054741	18.26765	.9985050	52
9	.0549502	.055033	18.17080	.9984891	51
10	.0552406	.055325	18.07497	.9984731	50
11	0555311	.055616	17.98015	.9984570	49
12	0558215	.055908	17.88631	.9984408	48
13	.0561119	.056200	17.79344	.9984245	47
14	.0564024	.056492	17.70152	.9984081	46
15	.0566928	.056784	17.61055	.9983917	45
16	.0569832	.057075	17.52051	.9983751	44
17	.0572736	.057367	17.43138	.9983585	43
18	.0575640	.057659	17.34315	.9983418	42
19	.0578544	.057951	17.25580	.9983250	41
20	.0581448	.058243	17.16933	.9983082	40
21	.0584352	.058535	17.08372	.9982912	39
22	.0587256	.058827	16.99895	.9982742	38
23	.0590160	.059119	16.91502	.9982570	37
24	.0593064	.059410	16.83191	.9982398	36
25	.0595967	059702	16.74961	.9982225	35
26	.0598871	.059994	16.66811	.9982052	34
27	.0601775	.060286	16.58739	.9981877	33
28	.0604678	.060578	16.50745	.9981701	32
29	.0607582	.060870	16.42827	.9981525	31
30	.0610485	.061162	16.34985	.9981348	30
31	.0613389	061454	16.27217	.9981170	29
32	.0616292	.061746	16.19522	.9980991	28
33	.0619196	.062038	16.11899	.9980811	27
34	.0622099	.062330	16.04348	9980631	26
35	.0625002	.062622	15.96866	.9980450	25
36	.0627905	.062914	15.89454	9980267	24
37	.0630808	.063206	15.82110	.9980084	23
38	.0633711	.063498	15.74833	.9979900	22
39	.0636614	.063790	15.67623	.9979716	21
40	.0639517	.064082	15.60478	9979530	20
41	.0642420	.064375	15.53398	.9979343	19
42	.0645323	.064667	15.46381	.9979156	18
43	.0648226	.064959	15.39427	.9978968	17
44	.0651129	.065251	15.32535	.9978779	16
45	.0654031	.065543	15.25705	.9978589	15
46	.0656934	.065835	15.18934	.9978399	14
47	.0659836	.066127	15.12224	.9978207	13
48	.0662739	.066419	15.05572	.9978015	12
49	.0665641	.066712	14.98978	.9977821	11
50	.0668544	.067004	14.92441	.9977627	10
51	.0671446	.067296	14.85961	.9977433	9
52	.0674349	.067588	14.79537	.9977237	8
53	.0677251	.067880	14.73167	.9977040	7
54	.0680153	.068173	14.66852	.9976843	6
55	.0683055	.068465	14.60591	.9976645	5
56	.0685957	.068757	14.54383	.9976445	4
57	.0688859	.069049	14.48227	.9976245	3
58	.0691761	.069342	14.42123	.9976045	2
59	.0694663	.069634	14 36069	.9975843	1
60	.0697565	069926	14.30066	.9975641	0
	Cosine.	Cotang	Tang.	Sine.	′

86°

Note.—Secant = 1 + cosine. Cosecant = 1 + sine.

2. —**Natural Sines,** TANGENTS, COTANGENTS, COSINES.—(Continued)

(Versed sine = 1 − cosine; coversed sine = 1 − sine.)

4°

′	Sine.	Tang.	Cotang.	Cosine.	
0	.0697565	.069926	14.30066	.9975641	**60**
1	.0700467	.070219	14.24113	.9975437	59
2	.0703368	.070511	14.18209	.9975233	58
3	.0706270	.070803	14.12353	.9975028	57
4	.0709171	.071096	14.06545	.9974822	56
5	.0712073	.071388	14.00785	.9974615	**55**
6	.0714974	.071680	13.95071	.9974408	54
7	.0717876	.071973	13.89404	.9974199	53
8	.0720777	.072265	13.83782	.9973990	52
9	.0723678	.072558	13.78206	.9973780	51
10	.0726580	.072850	13.72673	.9973569	**50**
11	.0729481	.073143	13.67185	.9973357	49
12	.0732382	.073435	13.61740	.9973145	48
13	.0735283	.073727	13.56339	.9972931	47
14	.0738184	.074020	13.50979	.9972717	46
15	.0741085	.074312	13.45662	.9972502	**45**
16	.0743986	.074605	13.40386	.9972286	44
17	.0746887	.074897	13.35151	.9972069	43
18	.0749787	.075190	13.29957	.9971851	44
19	.0752688	.075482	13.24803	.9971633	41
20	.0755589	.075775	13.19688	.9971413	**40**
21	.0758489	.076068	13.14612	.9971193	39
22	.0761390	.076360	13.09575	.9970972	38
23	.0764290	.076653	13.04576	.9970750	37
24	.0767190	.076945	12.99616	.9970528	36
25	0770091	.077238	12.94692	.9970304	**35**
26	.0772991	.077531	12.89805	.9970080	34
27	.0775891	.077823	12.84955	.9969854	33
28	.0778791	.078116	12.80141	.9969628	32
29	.0781691	.078409	12.75363	.9969401	31
30	.0784591	.078701	12.70620	.9969173	**30**
31	.0787491	.078994	12.65912	.9968945	29
32	.0790391	.079287	12.61239	.9968715	28
33	.0793290	.079579	12.56599	.9968485	27
34	.0796190	.079872	12.51994	.9968254	26
35	.0799090	.080165	12.47422	.9968022	**25**
36	.0801989	.080458	12.42883	.9967789	24
37	.0804889	.080750	12.38376	.9967555	23
38	.0807788	.081043	12.33902	.9967321	22
39	.0810687	.081336	12.29460	.9967085	21
40	.0813587	.081629	12.25050	.9966849	**20**
41	.0816486	.081922	12.20671	.9966612	19
42	.0819385	.082215	12.16323	.9966374	18
43	.0822284	.082507	12.12006	.9966135	17
44	.0825183	.082800	12.07719	.9965895	16
45	.0828082	.083093	12.03462	.9965655	**15**
46	.0830981	.083386	11.99234	.9965414	14
47	.0833880	.083679	11.95037	.9965172	13
48	.0836778	.083972	11.90868	.9964929	12
49	.0839677	.084265	11.86728	.9964685	11
50	.0842576	.084558	11.82616	.9964440	**10**
51	.0845474	.084851	11.78533	.9964195	9
52	.0848373	.085144	11.74477	.9963948	8
53	.0851271	.085437	11.70450	.9963701	7
54	.0854169	.085730	11.66449	.9963453	6
55	.0857067	.086023	11.62476	.9963204	**5**
56	.0859966	.086316	11.58529	.9962954	4
57	.0862864	.086609	11.54609	.9962704	3
58	.0865762	.086902	11.50715	.9962452	2
59	.0868660	.087195	11.46847	.9962200	1
60	.0871557	.087488	11.43005	.9961947	**0**
	Cosine.	Cotang	Tang.	Sine.	′

85°

5°

′	Sine.	Tang.	Cotang.	Cosine.	
0	.0871557	.087488	11.43005	.9961947	**60**
1	.0874455	.087781	11.39188	.9961693	59
2	.0877353	.088074	11.35397	.9961438	58
3	.0880251	.088368	11.31630	.9961183	57
4	.0883148	.088661	11.27888	.9960926	56
5	.0886046	.088954	11.24171	.9960669	**55**
6	.0888943	.089247	11.20478	.9960411	54
7	.0891840	.089540	11.16808	.9960152	53
8	.0894738	.089834	11.13163	.9959892	52
9	.0897635	.090127	11.09541	.9959631	51
10	.0900532	.090420	11.05943	.9959370	**50**
11	.0903429	.090713	11.02367	.9959107	49
12	.0906326	.091007	10.98815	.9958844	48
13	.0909223	.091300	10.95285	.9958580	47
14	.0912119	.091593	10.91777	.9958315	46
15	.0915016	.091887	10.88292	.9958049	**45**
16	.0917913	.092180	10.84828	.9957783	44
17	.0920809	.092473	10.81387	.9957515	43
18	.0923706	.092767	10.77967	.9957247	42
19	.0926602	.093060	10.74568	.9956978	41
20	.0929499	.093354	10.71191	.9956708	**40**
21	.0932395	.093647	10.67834	.9956437	39
22	.0935291	.093940	10.64499	.9956165	38
23	.0938187	.094234	10.61184	.9955892	37
24	.0941083	.094527	10.57889	.9955620	36
25	.0943979	.094821	10.54615	.9955345	**35**
26	.0946875	.095114	10.51360	.9955070	34
27	.0949771	.095408	10.48126	.9954794	33
28	.0952666	.095701	10.44911	.9954517	32
29	.0955562	.095995	10.41715	.9954240	31
30	.0958458	.096289	10.38539	.9953962	**30**
31	.0961353	.096582	10.35382	.9953683	29
32	.0964248	.096876	10.32244	.9953403	28
33	.0967144	.097169	10.29125	.9953122	27
34	.0970039	.097463	10.26024	.9952840	26
35	.0972934	.097757	10.22942	.9952557	**25**
36	.0975829	.098050	10.19878	.9952274	24
37	.0978724	.098344	10.16833	.9951990	23
38	.0981619	.098638	10.13805	.9951705	22
39	.0984514	.098932	10.10795	.9951419	21
40	.0987408	.099225	10.07803	.9951132	**20**
41	.0990303	.099519	10.04828	.9950844	19
42	.0993197	.099813	10.01871	.9950556	18
43	.0996092	.100107	9.989305	.9950266	17
44	.0998986	.100400	9.960072	.9949976	16
45	.1001881	.100694	9.931008	.9949685	**15**
46	.1004775	.100988	9.902112	.9949393	14
47	.1007669	.101282	9.873382	.9949101	13
48	.1010563	.101576	9.844816	.9948807	12
49	.1013457	.101870	9.816414	.9948513	11
50	.1016351	.102164	9.788173	.9948217	**10**
51	.1019245	.102458	9.760092	.9947921	9
52	.1022138	.102752	9.732171	.9947625	8
53	.1025032	.103046	9.704407	.9947327	7
54	.1027925	.103339	9.676800	.9947028	6
55	.1030819	.103634	9.649347	.9946729	**5**
56	.1033712	.103928	9.622048	.9946428	4
57	.1036605	.104222	9.594902	.9946127	3
58	.1039499	.104516	9.567906	.9945825	2
59	.1042392	.104810	9.541061	.9945523	1
60	.1045285	.105104	9.514364	.9945219	**0**
	Cosine.	Cotang	Tang.	Sine.	′

84°

Note.—Secant = 1 ÷ cosine. Cosecant = 1 ÷ sine.

2. —Natural Sines, TANGENTS, COTANGENTS, COSINES.—(Continued).

(Versed sine = 1 − cosine; coversed sine = 1 − sine.)

6° ′	Sine.	Tang.	Cotang.	Cosine.		7° ′	Sine.	Tang.	Cotang.	Cosine.	
0	.1045285	.105104	9.514364	.9945219	60	0	.1218693	.122784	8.144346	.9925462	60
1	.1048178	.105398	9.487814	.9944914	59	1	.1221581	.123079	8.124807	.9925107	59
2	.1051070	.105692	9.461411	.9944609	58	2	.1224468	.123375	8.105359	.9924751	58
3	.1053963	.105986	9.435153	.9944303	57	3	.1227355	.123670	8.086004	.9924394	57
4	.1056856	.106280	9.409038	.9943996	56	4	.1230241	.123965	8.066739	.9924037	56
5	.1059748	.106575	9.383066	.9943688	55	5	.1233128	.124261	8.047564	.9923679	55
6	.1062641	.106869	9.357235	.9943379	54	6	.1236015	.124556	8.028479	.9923319	54
7	.1065533	.107163	9.331545	.9943070	53	7	.1238901	.124852	8.009483	.9922959	53
8	.1068425	.107457	9.305993	.9942760	52	8	.1241788	.125147	7.990575	.9922599	52
9	.1071318	.107751	9.280580	.9942448	51	9	.1244674	.125442	7.971755	.9922237	51
10	.1074210	.108046	9.255303	.9942136	50	10	.1247560	.125738	7.953022	.9921874	50
11	.1077102	.108340	9.230162	.9941823	49	11	.1250446	.126033	7.934375	.9921511	49
12	.1079994	.108634	9.205156	.9941510	48	12	.1253332	.126329	7.915815	.9921147	48
13	.1082885	.108929	9.180283	.9941195	47	13	.1256218	.126624	7.897339	.9920782	47
14	.1085777	.109223	9.155543	.9940880	46	14	.1259104	.126920	7.878948	.9920416	46
15	.1088669	.109517	9.130934	.9940563	45	15	.1261990	.127216	7.860642	.9920049	45
16	.1091560	.109812	9.106456	.9940246	44	16	.1264875	.127511	7.842419	.9919682	44
17	.1094452	.110106	9.082107	.9939928	43	17	.1267761	.127807	7.824279	.9919314	43
18	.1097343	.110401	9.057886	.9939610	42	18	.1270646	.128103	7.806221	.9918944	42
19	.1100234	.110695	9.033793	.9939290	41	19	.1273531	.128398	7.788245	.9918574	41
20	.1103126	.110989	9.009826	.9938969	40	20	.1276416	.128694	7.770350	.9918204	40
21	.1106017	.111284	8.985984	.9938648	39	21	.1279302	.128990	7.752536	.9917832	39
22	.1108908	.111578	8.962266	.9938326	38	22	.1282186	.129285	7.734802	.9917459	38
23	.1111799	.111873	8.938672	.9938003	37	23	.1285071	.129581	7.717148	.9917086	37
24	.1114689	.112168	8.915200	.9937679	36	24	.1287956	.129877	7.699573	.9916712	36
25	.1117580	.112462	8.891850	.9937355	35	25	.1290841	.130173	7.682076	.9916337	35
26	.1120471	.112757	8.868620	.9937029	34	26	.1293725	.130469	7.664658	.9915961	34
27	.1123361	.113051	8.845510	.9936703	33	27	.1296609	.130764	7.647317	.9915584	33
28	.1126252	.113346	8.822518	.9936375	32	28	.1299494	.131060	7.630053	.9915206	52
29	.1129142	.113641	8.799644	.9936047	31	29	.1302378	.131356	7.612865	.9914828	31
30	.1132032	.113935	8.776887	.9935719	30	30	.1305262	.131652	7.595754	.9914449	30
31	.1134922	.114230	8.754246	.9935389	29	31	.1308146	.131948	7.578717	.9914069	29
32	.1137812	.114525	8.731719	.9935058	28	32	.1311030	.132244	7.561756	.9913688	28
33	.1140702	.114819	8.709307	.9934727	27	33	.1313913	.132540	7.544869	.9913306	27
34	.1143592	.115114	8.687008	.9934395	26	34	.1316797	.132836	7.528057	.9912923	26
35	.1146482	.115409	8.664822	.9934062	25	35	.1319681	.133132	7.511317	.9912540	25
36	.1149372	.115703	8.642747	.9933728	24	36	.1322564	.133428	7.494651	.9912155	24
37	.1152261	.115998	8.620783	.9933393	23	37	.1325447	.133724	7.478057	.9911770	23
38	.1155151	.116293	8.598929	.9933057	22	38	.1328330	.134020	7.461535	.9911384	22
39	.1158040	.116588	8.577183	.9932721	21	39	.1331213	.134316	7.445085	.9910997	21
40	.1160929	.116883	8.555546	.9932384	20	40	.1334096	.134612	7.428706	.9910610	20
41	.1163818	.117178	8.534017	.9932045	19	41	.1336979	.134909	7.412397	.9910221	19
42	.1166707	.117473	8.512594	.9931706	18	42	.1339862	.135205	7.396159	.9909832	18
43	.1169596	.117767	8.491277	.9931367	17	43	.1342744	.135501	7.379990	.9909442	17
44	.1172485	.118062	8.470065	.9931026	16	44	.1345627	.135797	7.363891	.9909051	16
45	.1175374	.118357	8.448957	.9930685	15	45	.1348509	.136094	7.347861	.9908659	15
46	.1178263	.118652	8.427953	.9930342	14	46	.1351392	.136390	7.331898	.9908266	14
47	.1181151	.118947	8.407051	.9929999	13	47	.1354274	.136686	7.316004	.9907873	13
48	.1184040	.119242	8.386251	.9929655	12	48	.1357156	.136983	7.300178	.9907478	12
49	.1186928	.119537	8.365553	.9929310	11	49	.1360038	.137279	7.284418	.9907083	11
50	.1189816	.119832	8.344955	.9928965	10	50	.1362919	.137575	7.268725	.9906687	10
51	.1192704	.120127	8.324457	.9928618	9	51	.1365801	.137872	7.253098	.9906290	9
52	.1195593	.120423	8.304058	.9928271	8	52	.1368683	.138168	7.237537	.9905893	8
53	.1198481	.120718	8.283757	.9927922	7	53	.1371564	.138465	7.222042	.9905494	7
54	.1201368	.121013	8.263554	.9927573	6	54	.1374445	.138761	7.206611	.9905095	6
55	.1204256	.121308	8.243448	.9927224	5	55	.1377327	.139058	7.191245	.9904694	5
56	.1207144	.121603	8.223438	.9926873	4	56	.1380208	.139354	7.175943	.9904293	4
57	.1210031	.121898	8.203523	.9926521	3	57	.1383089	.139651	7.160705	.9903891	3
58	.1212919	.122194	8.183704	.9926169	2	58	.1385970	.139947	7.145530	.9903489	2
59	.1215806	.122489	8.163978	.9925816	1	59	.1388850	.140244	7.130419	.9903085	1
60	.1218693	.122784	8.144346	.9925462	0	60	.1391731	.140540	7.115369	.9902681	0
	Cosine.	Cotang	Tang.	Sine.	′ 83°		Cosine.	Cotang	Tang.	Sine.	′ 82°

Note.—Secant = 1 ÷ cosine. Cosecant = 1 ÷ sine.

2.—Natural Sines, Tangents, Cotangents, Cosines.—(Continued.)

(Versed sine = 1 − cosine; coversed sine = 1 − sine.)

8°

′	Sine.	Tang.	Cotang.	Cosine.	
0	.1391731	.140540	7.115369	.9902681	60
1	.1394612	.140837	7.100382	.9902275	59
2	.1397492	.141134	7.085457	.9901869	58
3	.1400372	.141430	7.070593	.9901462	57
4	.1403252	.141727	7.055790	.9901055	56
5	.1406132	.142024	7.041048	.9900646	55
6	.1409012	.142321	7.026366	.9900237	54
7	.1411892	.142617	7.011744	.9899826	53
8	.1414772	.142914	6.997180	.9899415	52
9	.1417651	.143211	6.982678	.9899003	51
10	.1420531	.143508	6.968233	.9898590	50
11	.1423410	.143805	6.953847	.9898177	49
12	.1426289	.144102	6.939519	.9897762	48
13	.1429168	.144399	6.925248	.9897347	47
14	.1432047	.144696	6.911035	.9896931	46
15	.1434926	.144993	6.896879	.9896514	45
16	.1437805	.145290	6.882780	.9896096	44
17	.1440684	.145587	6.868737	.9895677	43
18	.1443562	.145884	6.854750	.9895258	42
19	.1446440	.146181	6.840819	.9894838	41
20	.1449319	.146478	6.826943	.9894416	40
21	.1452197	.146775	6.813122	.9893994	39
22	.1455075	.147072	6.799356	.9893572	38
23	.1457953	.147369	6.785644	.9893148	37
24	.1460830	.147667	6.771986	.9892723	36
25	.1463708	.147964	6.758382	.9892298	35
26	.1466585	.148261	6.744831	.9891872	34
27	.1469463	.149559	6.731334	.9891445	33
28	.1472340	.148856	6.717889	.9891017	32
29	.1475217	.149153	6.704496	.9890588	31
30	.1478094	.149451	6.691156	.9890159	30
31	.1480971	.149748	6.677867	.9889728	29
32	.1483848	.150045	6.664630	.9889297	28
33	.1486724	.150343	6.651444	.9888865	27
34	.1489601	.150640	6.638310	.9888432	26
35	.1492477	.150938	6.625225	.9887998	25
36	.1495353	.151235	6.612191	.9887564	24
37	.1498230	.151533	6.599208	.9887128	23
38	.1501106	.151830	6.586273	.9886692	22
39	.1503981	.152128	6.573389	.9886255	21
40	.1506857	.152426	6.560553	.9885817	20
41	.1509733	.152723	6.547767	.9885378	19
42	.1512608	.153021	6.535029	.9884939	18
43	.1515484	.153319	6.522339	.9884498	17
44	.1518359	.153617	6.509698	.9884057	16
45	.1521234	.153914	6.497104	.9883615	15
46	.1524109	.154212	6.484558	.9883172	14
47	.1526984	.154510	6.472059	.9882728	13
48	.1529858	.154808	6.459607	.9882284	12
49	.1532733	.155106	6.447201	.9881838	11
50	.1535607	.155404	6.434842	.9881392	10
51	.1538482	.155701	6.422530	.9880945	9
52	.1541356	.155999	6.410263	.9880497	8
53	.1544230	.156297	6.398042	.9880048	7
54	.1547104	.156595	6.385866	.9879599	6
55	.1549978	.156893	6.373735	.9879148	5
56	.1552851	.157191	6.361650	.9878697	4
57	.1555725	.157490	6.349609	.9878245	3
58	.1558598	.157788	6.337612	.9877792	2
59	.1561472	.158086	6.325660	.9877338	1
60	.1564345	.158384	6.313751	.9876883	0
	Cosine.	Cotang	Tang.	Sine.	′

81°

9°

′	Sine.	Tang.	Cotang.	Cosine.	
0	.1564345	.158384	6.313751	.9876883	60
1	.1567218	.158682	6.301886	.9876428	59
2	.1570091	.158980	6.290065	.9875972	58
3	.1572963	.159279	6.278286	.9875514	57
4	.1575836	.159577	6.266551	.9875057	56
5	.1578708	.159875	6.254858	.9874598	55
6	.1581581	.160174	6.243208	.9874138	54
7	.1584453	.160472	6.231600	.9873678	53
8	.1587325	.160770	6.220034	.9873216	52
9	.1590197	.161069	6.208510	.9872754	51
10	.1593069	.161367	6.197027	.9872291	50
11	.1595940	.161666	6.185586	.9871827	49
12	.1598812	.161964	6.174186	.9871363	48
13	.1601683	.162263	6.162827	.9870897	47
14	.1604555	.162561	6.151508	.9870431	46
15	.1607426	.162860	6.140230	.9869964	45
16	.1610297	.163159	6.128992	.9869496	44
17	.1613167	.163457	6.117794	.9869027	43
18	.1616038	.163756	6.106636	.9868557	42
19	.1618909	.164055	6.095517	.9868087	41
20	.1621779	.164353	6.084438	.9867615	40
21	.1624650	.164652	6.073397	.9867143	39
22	.1627520	.164951	6.062396	.9866670	38
23	.1630390	.165250	6.051434	.9866196	37
24	.1633260	.165548	6.040510	.9865722	36
25	.1636129	.165847	6.029624	.9865246	35
26	.1638999	.166146	6.018777	.9864770	34
27	.1641868	.166445	6.007967	.9864293	33
28	.1644738	.166744	5.997195	.9863815	32
29	.1647607	.167043	5.986461	.9863336	31
30	.1650476	.167342	5.975764	.9862856	30
31	.1653345	.167641	5.965104	.9862375	29
32	.1656214	.167940	5.954481	.9861894	28
33	.1659082	.168239	5.943895	.9861412	27
34	.1661951	.168539	5.933345	.9860929	26
35	.1664819	.168838	5.922832	.9860445	25
36	.1667687	.169137	5.912355	.9859960	24
37	.1670556	.169436	5.901913	.9859475	23
38	.1673423	.169735	5.891508	.9858988	22
39	.1676291	.170035	5.881138	.9858501	21
40	.1679159	.170334	5.870804	.9858013	20
41	.1682026	.170633	5.860505	.9857524	19
42	.1684894	.170933	5.850241	.9857035	18
43	.1687761	.171232	5.840011	.9856544	17
44	.0690628	.171532	5.829817	.9856053	16
45	.1693495	.171831	5.819657	.9855561	15
46	.1696362	.172130	5.809531	.9855068	14
47	.1699228	.172430	5.799440	.9854574	13
48	.1702095	.172730	5.789382	.9854079	12
49	.1704961	.173029	5.779358	.9853583	11
50	.1707828	.173329	5.769368	.9853087	10
51	.1710694	.173628	5.759412	.9852590	9
52	.1713560	.173928	5.749488	.9852092	8
53	.1716425	.174228	5.739598	.9851593	7
54	.1719291	.174527	5.729741	.9851093	6
55	.1722156	.174827	5.719917	.9850593	5
56	.1725022	.175127	5.710125	.9850091	4
57	.1727887	.175427	5.700366	.9849589	3
58	.1730752	.175727	5.690639	.9849086	2
59	.1733617	.176027	5.680944	.9848582	1
60	.1736482	.176327	5.671281	.9848078	0
	Cosine.	Cotang	Tang.	Sine.	′

80°

Note.—Secant = 1 ÷ cosine. Cosecant = 1 ÷ sine.

2.—Natural Sines, Tangents, Cotangents, Cosines.—(Continued.)

(Versed sine = 1 − cosine; coversed sine = 1 − sine).

10° | 11°

′	Sine.	Tang.	Cotang.	Cosine.			Sine.	Tang.	Cotang.	Cosine.	
0	.1736482	.176327	5.671281	.9848078	60	0	.1908090	.194380	5.144554	.9816272	60
1	.1739346	.176626	5.661650	.9847572	59	1	.1910945	.194682	5.136576	.9815716	59
2	.1742211	.176926	5.652051	.9847066	58	2	.1913801	.194984	5.128622	.9815160	58
3	1745075	.177226	5.642483	.9846558	57	3	.1916656	.195286	5.120692	.9814603	57
4	.1747939	.177527	5.632947	.9846050	56	4	.1919510	.195588	5.112785	.9814045	56
5	.1750803	.177827	5.623442	.9845542	55	5	.1922365	.195890	5.104902	.9813486	55
6	.1753667	.178127	5.613968	.9845032	54	6	.1925220	.196192	5.097042	.9812927	54
7	.1756531	.178427	5.604524	.9844521	53	7	.1928074	.196494	5.089206	.9812366	53
8	.1759395	.178727	5.595112	.9844010	52	8	.1930928	.196796	5.081392	.9811805	52
9	.1762258	.179027	5.585730	.9843498	51	9	.1933782	.197098	5.073602	.9811243	51
10	.1765121	.179327	5.576378	.9842985	50	10	.1936636	.197400	5.065835	.9810680	50
11	.1767984	.179628	5.567057	.9842471	49	11	.1939490	.197703	5.058090	.9810116	49
12	.1770847	.179928	5.557766	.9841956	48	12	.1942344	.198005	5.050369	.9809552	48
13	.1773710	.180228	5.548505	.9841441	47	13	.1945197	.198307	5.042670	.9008986	47
14	.1776573	.180529	5.539274	.9840924	46	14	.1948050	.198610	5.034993	.9808420	46
15	.1779435	.180829	5.530072	.9840407	45	15	.1950903	.198912	5.027339	.9807853	45
16	.1782298	.181129	5.520900	.9839889	44	16	.1953756	.199214	5.019707	.9807285	44
17	.1785160	.181430	5.511757	.9839370	43	17	.1956609	.199517	5.012098	.9806716	43
18	.1788022	.181730	5.502644	.9838850	42	18	.1959461	.199819	5.004511	.9806147	42
19	.1790884	.182031	5.493560	.9838330	41	19	.1962314	.200122	4.996945	.9805576	41
20	.1793746	.182331	5.484505	.9837808	40	20	.1965166	.200424	4.989402	.9805005	40
21	.1796607	.182632	5.475478	.9837286	39	21	.1968018	.200727	4.981881	.9804433	39
22	.1799469	.182933	5.466481	.9836763	38	22	.1970870	.201030	4.974381	.9803860	38
23	.1802330	.183233	5.457512	.9836239	37	23	.1973722	.201332	4.966903	.9803286	37
24	.1805191	.183534	5.448571	.9835715	36	24	.1976573	.201635	4.959447	.9802712	36
25	.1808052	.183835	5.439659	.9835189	35	25	.1979425	.201938	4.952012	.9802136	35
26	.1810913	.184135	5.430775	.9834663	34	26	.1982276	.202240	4.944599	.9801560	34
27	.1813774	.184436	5.421918	.9834136	33	27	.1985127	.202543	4.937206	.9800983	33
28	.1816635	.184737	5.413090	.9833608	32	28	.1987978	.202846	4.929835	.9800405	32
29	.1819495	.185038	5.404290	.9833079	31	29	.1990829	.203149	4.922485	.9799827	31
30	.1822355	.185339	5.395517	.9832549	30	30	.1993679	.203452	4.915157	.9799247	30
31	.1825215	.185639	5.386771	.9832019	29	31	.1996530	.203755	4.907849	.9798667	29
32	.1828075	.185940	5.378053	.9831487	28	32	.1999380	.204058	4.900562	.9798086	28
33	.1830935	.186241	5.369363	.9830955	27	33	.2002230	.204361	4.893295	.9797504	27
34	.1833795	.186542	5.360699	.9830422	26	34	.2005080	.204664	4.886049	.9796921	26
35	.1836654	.186843	5.352062	.9829888	25	35	.2007930	.204967	4.878824	.9796337	25
36	.1839514	.187144	5.343452	.9829353	24	36	.2010779	.205270	4.871620	.9795752	24
37	.1842373	.187446	5.334869	.9828818	23	37	.2013629	.205573	4.864435	.9795167	23
38	.1845232	.187747	5.326313	.9828282	22	38	.2016478	.205876	4.857271	.9794581	22
39	.1848091	.188048	5.317783	.9827744	21	39	.2019327	.206180	4.850128	.9793994	21
40	.1850949	.188349	5.309279	.9827206	20	40	.2022176	.206483	4.843004	.9793406	20
41	.1853808	.188650	5.300801	.9826668	19	41	.2025024	.206786	4.835901	.9792818	19
42	.1856666	.188952	5.292350	.9826128	18	42	.2027873	.207090	4.828817	.9792228	18
43	.1859524	.189253	5.283925	.9825587	17	43	.2030721	.207393	4.821753	.9791638	17
44	.1862382	.189554	5.275525	.9825046	16	44	.2033569	.207696	4.814709	.9791047	16
45	.1865240	.189855	5.267151	.9824504	15	45	.2036418	.208000	4.807685	.9790455	15
46	.1868098	.190157	5.258803	.9823961	14	46	.2039265	.208303	4.800680	.9789862	14
47	.1870956	.190458	5.250480	.9823417	13	47	.2042113	.208607	4.793695	.9789268	13
48	.1873813	.190760	5.242183	.9822873	12	48	.2044961	.208910	4.736730	.9788674	12
49	.1876670	.191061	5.233911	.9822327	11	49	.2047808	.209214	4.779783	.9788079	11
50	.1879528	.191363	5.225664	.9821781	10	50	.2050655	.209518	4.772856	.9787483	10
51	.1882385	.191664	5.217442	.9821234	9	51	.2053502	.209821	4.765949	.9786886	9
52	.1885241	.191966	5.209245	.9820686	8	52	.2056349	.210125	4.759060	.9786288	8
53	.1880898	.192268	5.201073	.9820137	7	53	.2059195	.210429	4.752190	.9785689	7
54	.1890954	.192569	5.192926	9819587	6	54	.2062042	.210733	4.745340	.9785090	6
55	.1893811	.192871	5.184803	.9819037	5	55	.2064888	.211036	4.738508	.9784490	5
56	.1896667	.193173	5.176705	.9818485	4	56	.2067734	.211340	4.731695	.9783889	4
57	.1899523	.193474	5.168631	.9817933	3	57	.2070580	.211644	4.724901	.9783287	3
58	.1902379	.193776	5.160581	.9817380	2	58	.2073426	.211948	4.718125	.9782684	2
59	.1905234	.194078	5.152555	.9816826	1	59	.2076272	.212252	4.711368	.9782080	1
60	.1908090	.194380	5.144554	.9816272	0	60	.2079117	.212556	4.704630	.9781476	0
	Cosine.	Cotang	Tang.	Sine.	′		Cosine.	Cotang	Tang.	Sine.	

79° | 78°

Note.—Secant = 1 ÷ cosine. Cosecant = 1 ÷ sine.

2.—Natural Sines, Tangents, Cotangents, Cosines.—(Continued.)

(Versed sine = 1 − cosine; coversed sine = 1 − sine.)

12°							13°					
′	Sine.	Tang.	Cotang.	Cosine.		′	Sine.	Tang.	Cotang.	Cosine.	
0	.2079117	.212556	4.704630	.9781476	60	0	.2249511	.230868	4.331475	.9743701	60
1	.2081962	.212860	4.697910	.9780871	59	1	.2252345	.231174	4.325734	.9743046	59
2	.2084807	.213164	4.691208	.9780265	58	2	.2255179	.231481	4.320007	.9742390	58
3	.2087652	.213468	4.684524	.9779658	57	3	.2258013	.231787	4.314295	.9741734	57
4	.2090497	.213773	4.677859	.9779050	56	4	.2260846	.232094	4,308597	.9741077	56
5	.2093341	.214077	4.671212	.9778441	55	5	.2263680	.232400	4.302913	.9740419	55
6	.2096186	.214381	4.664583	.9777832	54	6	.2266513	.232707	4.297244	.9739760	54
7	.2099030	.214685	4.657972	.9777222	53	7	.2269346	.233014	4.291588	.9739100	53
8	.2101874	.214990	4.651378	.9776611	52	8	.2272179	.233320	4.285947	.9738439	52
9	.2104718	.215294	4.644803	.9775999	51	9	.2275012	.233627	4.280319	.9737778	51
10	.2107561	.215598	4.638245	.9775386	50	10	.2277844	.233934	4.274706	.9737116	50
11	.2110405	.215903	4.631705	.9774773	49	11	.2280677	.234241	4.269107	.9736453	49
12	.2113248	.216207	4.625183	.9774159	48	12	.2283509	.234547	4.263521	.9735789	48
13	.2116091	.216512	4.618678	.9773544	47	13	.2286341	.234854	4.257950	.9735124	47
14	.2118934	.216816	4.612190	.9772928	46	14	.2289172	.235161	4.252392	.9734458	46
15	.2121777	.217121	4.605720	.9772311	45	15	.2292004	.235468	4.246848	.9733792	45
16	.2124619	.217425	4.599268	.9771693	44	16	.2294835	.235775	4.241317	.9733125	44
17	.2127462	.217730	4.592832	.9771075	43	17	.2297666	.236082	4.235800	.9732457	43
18	.2130304	.218035	4.586414	.9770456	42	18	.2300497	.236390	4.230297	.9731789	42
19	.2133146	.218340	4.580012	.9769836	41	19	.2303328	.236697	4.224808	.9731119	41
20	.2135988	.218644	4.573628	.9769215	40	20	.2306159	.237004	4.219331	.9730449	40
21	.2138829	.218949	4.567261	.9768593	39	21	.2308989	.237311	4.213869	.9729777	39
22	.2141671	.219254	4.560911	.9767970	38	22	.2311819	.237618	4.208419	.9729105	38
23	.2144512	.219559	4.554577	.9767347	37	23	.2314649	.237926	4.202983	.9728432	37
24	.2147353	.219864	4.548260	.9766723	36	24	.2317479	.238233	4.197560	.9727759	36
25	.2150194	.220169	4.541960	.9766098	35	25	.2320309	.238541	4.192151	.9727084	35
26	.2153035	.220474	4.535677	.9765472	34	26	.2323138	.238848	4.186754	.9726409	34
27	.2155876	.220779	4.529410	.9764845	33	27	.2325967	.239156	4.181371	.9725733	33
28	.2158716	.221084	4.523160	.9764217	32	28	.2328796	.239463	4.176001	.9725056	32
29	.2161556	.221389	4.516926	.9763589	31	29	.2331625	.239771	4.170644	.9724378	31
30	.2164396	.221694	4.510708	.9762960	30	30	.2334454	.240078	4.165299	.9723699	30
31	.2167236	.221999	4.504507	.9762330	29	31	.2337282	.240386	4.159968	.9723020	29
32	.2170076	.222305	4.498322	.9761699	28	32	.2340110	.240694	4.154650	.9722339	28
33	.2172915	.222610	4.492153	.9761067	27	33	.2342938	.241001	4.149344	.9721658	27
34	.2175754	.222915	4.486000	.9760435	26	34	.2345766	.241309	4.144051	.9720976	26
35	.2178593	.223221	4.479863	.9759802	25	35	.2348594	.241617	4.138771	.9720294	25
36	.2181432	.223526	4.473742	.9759168	24	36	.3351421	.241925	4.133504	.9719610	24
37	.2184271	.223831	4.467637	.9758533	23	37	.2354248	.242233	4.128249	.9718926	23
38	.2187110	.224137	4.461548	.9757897	22	38	.2357075	.242541	4.123007	.9718240	22
39	.2189948	.224442	4.455475	.9757260	21	38	.2359902	.242849	4.117778	.9717554	21
40	.2192786	.224748	4.449418	.9756623	20	40	.2362729	.243157	4.112561	.9716867	20
41	.2195624	.225054	4.443376	.9755985	19	41	.2365555	.243465	4.107356	.9716180	19
42	.2198462	.225359	4.437350	.9755345	18	42	.2368381	.243773	4.102164	.9715491	18
43	.2201300	.225665	4.431339	.9754706	17	43	.2371207	.244081	4.096985	.9714802	17
44	.2204137	.225971	4.425343	.9754065	16	44	.2374033	.244390	4.091817	.9714112	16
45	.2206974	.226276	4.419364	.9753423	15	45	.2376859	.244698	4.086662	.9713421	15
46	.2209811	.226582	4.413399	.9752781	14	46	.2379684	.245006	4.081519	.9712729	14
47	.2212648	.226888	4.407450	.9752138	13	47	.2382510	.245315	4.076389	.9712036	13
48	.2215485	.227194	4.401516	.9751494	12	48	.2385335	.245623	4.071270	.9711343	12
49	.2218321	.227500	4.395597	.9750849	11	49	.2388159	.245932	4.066164	.9710649	11
50	.2221158	.227806	4.389694	.9750203	10	50	.2390984	.246240	4.061070	.9709953	10
51	.2223994	.228112	4.383805	.9749556	9	51	.2393808	.246549	4.055987	.9709258	9
52	.2226830	.228418	4.377931	.9748909	8	52	.2396633	.246857	4.050917	.9708561	8
53	.2229666	.228724	4.372073	.9748261	7	53	.2399457	.247166	4.045859	.9707863	7
54	.2232501	.229030	4.366229	.9747612	6	54	.2402280	.247475	4.040812	.9707165	6
55	.2235337	.229336	4.360400	.9746962	5	55	.2405104	.247783	4.035777	.9706466	5
56	.2238172	.229642	4.354586	.9746311	4	56	.2407927	.248092	4.030755	.9705766	4
57	.2241007	.229949	4.348786	.9745660	3	57	.2410751	.248401	4.025744	.9705065	3
58	.2243842	.230255	4.343001	.9745008	2	58	.2413574	.248710	4.020744	.9704363	2
59	.2246676	.230561	4.337231	.9744355	1	59	.2416396	.249019	4.015757	.9703660	1
60	.2249511	.230868	4.331475	.9743701	0	60	.2419219	.249328	4.010780	.9702957	0
	Cosine.	Cotang	Tang.	Sine.			Cosine.	Cotang	Tang.	Sine.	′
					77°						76°

Note.—Secant = 1 ÷ cosine. Cosecant = 1 ÷ sine.

2.—Natural Sines, Tangents, Cotangents, Cosines.—(Continued.)

(Versed sine = 1 − cosine; coversed sine = 1 − sine.)

14° 15°

′	Sine.	Tang.	Cotang.	Cosine.		′	Sine.	Tang.	Cotang.	Cosine.	
0	.2419219	.249328	4.010780	.9702957	60	0	.2588190	.267949	3.732050	.9659258	60
1	.2422041	.249637	4.005816	.9702253	59	1	.2591000	.268261	3.727713	.9658505	59
2	.2424863	.249946	4.000863	.9701548	58	2	.2593810	.268572	3.723384	.9657751	58
3	.2427685	.250255	3.995922	.9700842	57	3	.2596619	.268884	3.719065	.9656996	57
4	.2430507	.250564	3.990992	.9700135	56	4	.2599428	.269196	3.714756	.9656240	56
5	.2433329	.250873	3.986073	.9699428	55	5	.2602237	.269508	3.710455	.9655484	55
6	.2436150	.251182	3.981166	.9698720	54	6	.2605045	.269820	3.706164	.9654726	54
7	.2438971	.251491	3.976271	.9698011	53	7	.2607853	.270132	3.701883	.9653968	53
8	.2441792	.251801	3.971386	.9697301	52	8	.2610662	.270444	3.697610	.9653209	52
9	.2444613	.252110	3.966513	.9696591	51	9	.2613469	.270757	3.693346	.9652449	51
10	.2447433	.252420	3.961651	.9695879	50	10	.2616277	.271069	3.689092	.9651689	50
11	.2450254	.252729	3.956801	.9695167	49	11	.2619085	.271381	3.684847	.9650927	49
12	.2453074	.253038	3.951961	.9694453	48	12	.2621892	.271694	3.680611	.9650165	48
13	.2455894	.253348	3.947133	.9693740	47	13	.2624699	.272006	3.676384	.9649402	47
14	.2458713	.253658	3.942315	.9693025	46	14	.2627506	.272318	3.672166	.9648638	46
15	.2461533	.253967	3.937509	.9692309	45	15	.2630312	.272631	3.667957	.9647873	45
16	.2464352	.254277	3.932714	.9691593	44	16	.2633118	.272943	3.663757	.9647108	44
17	.2467171	.254587	3.927929	.9690875	43	17	.2635925	.273256	3.659566	.9646341	43
18	.2469990	.254896	3.923156	.9690157	42	18	.2638730	.273569	3.655384	.9645574	42
19	.2472809	.255206	3.918393	.9689438	41	19	.2641536	.273881	3.651211	.9644806	41
20	.2475627	.255516	3.913642	.9688719	40	20	.2644342	.274194	3.647046	.9644037	40
21	.2478445	.255826	3.908901	.9687998	39	21	.2647147	.274507	3.642891	.9643268	39
22	.2481263	.256136	3.904171	.9687277	38	22	.2649952	.274820	3.638744	.9642497	38
23	.2484081	.256446	3.899451	.9686555	37	23	.2652757	.275133	3.634606	.9641726	37
24	.2486899	.256756	3.894742	.9685832	36	24	.2655561	.275445	3.630477	.9640954	36
25	.2489716	.257066	3.890044	.9685108	35	25	.2658366	.275758	3.626356	.9640181	35
26	.2492533	.257376	3.885357	.9684383	34	26	.2661170	.276071	3.622244	.9639407	34
27	.2495350	.257686	3.880680	.9683658	33	27	.2663973	.276385	3.618141	.9638633	33
28	.2498167	.257997	3.876014	.9682931	32	28	.2666777	.276698	3.614046	.9637858	32
29	.2500984	.258307	3.871358	.9682204	31	29	.2669581	.277011	3.609960	.9637081	31
30	.2503800	.258617	3.866713	.9681476	30	30	.2672384	.277324	3.605883	.9636305	30
31	.2506616	.258928	3.862078	.9680748	29	31	.2677187	.277637	3.601814	.9635527	29
32	.2509432	.259238	3.857453	.9680018	28	32	.2677989	.277951	3.597754	.9634748	28
33	.2512248	.259548	3.852839	.9679288	27	33	.2680792	.278264	3.593702	.9633969	27
34	.2515063	.259859	3.848235	.9678557	26	34	.2683594	.278578	3.589659	.9633189	26
35	.2517879	.260169	3.843642	.9677825	25	35	.2686396	.278891	3.585624	.9632408	25
36	.2520694	.260480	3.839059	.9677092	24	36	.2689198	.279205	3.581597	.9631626	24
37	.2523508	.260791	3.834486	.9676358	23	37	.2692000	.279518	3.577579	.9630843	23
38	.2526323	.261101	3.829923	.9675624	22	38	.2694801	.279832	3.573569	.9630060	22
39	.2529137	.261412	3.825370	.9674888	21	39	.2697602	.280145	3.569568	.9629275	21
40	.2531952	.261723	3.820828	.9674152	20	40	.2700403	.280459	3.565574	.9628490	20
41	.2534766	.262034	3.816295	.9673415	19	41	.2703204	.280773	3.561590	.9627704	19
42	.2537579	.262345	3.811773	.9672678	18	42	.2706004	.281087	3.557613	.9626917	18
43	.2540393	.262656	3.807260	.9671939	17	43	.2708805	.281401	3.553644	.9626130	17
44	.2543206	.262967	3.802758	.9671200	16	44	.2711605	.281715	3.549684	.9625342	16
45	.2546019	.263278	3.798266	.9670459	15	45	.2714404	.282029	3.545732	.9624552	15
46	.2548832	.263589	3.793783	.9669718	14	46	.2717204	.282343	3.541788	.9623762	14
47	.2551645	.263900	3.789310	.9668977	13	47	.2720003	.282657	3.537852	.9622972	13
48	.2554458	.264211	3.784848	.9668234	12	48	.2722802	.282971	3.533925	.9622180	12
49	.2557270	.264522	3.780395	.9667490	11	49	.2725601	.283285	3.530005	.9621387	11
50	.2560082	.264833	3.775951	.9666746	10	50	.2728400	.283599	3.526093	.9620594	10
51	.2562894	.265145	3.771518	.9666001	9	51	.2731198	.283914	3.522190	.9619800	9
52	.2565705	.265456	3.767094	.9665255	8	52	.2733997	.284228	3.518294	.9619005	8
53	.2568517	.265768	3.762680	.9664508	7	53	.2736794	.284543	3.514407	.9618210	7
54	.2571328	.266079	3.758276	.9663761	6	54	.2739592	.284857	3.510527	.9617413	6
55	.2574139	.266390	3.753881	.9663012	5	55	.2742390	.285172	3.506655	.9616616	5
56	.2576950	.266702	3.749496	.9662263	4	56	.2745187	.285486	3.502791	.9615818	4
57	.2579760	.267014	3.745120	.9661513	3	57	.2747984	.285801	3.498935	.9615019	3
58	.2582570	.267325	3.740754	.9660762	2	58	.2750781	.286115	3.495087	.9614219	2
59	.2585381	.267637	3.736398	.9660011	1	59	.2753577	.286430	3.491247	.9613418	1
60	.2588190	.267949	3.732050	.9659258	0	60	.2756374	.286745	3.487414	.9612617	0
	Cosine.	Cotang	Tang.	Sine.			Cosine.	Cotang	Tang.	Sine.	′

75° 74°

Note.—Secant = 1 ÷ cosine. Cosecant = 1 ÷ sine.

2.–Natural Sines. TANGENTS, COTANGENTS, COSINES.—**(Continued.)**

(Versed sine = 1 – cosine; coversed sine = 1 – sine.)

16° **17°**

	Sine.	Tang.	Cotang.	Cosine.			Sine.	Tang.	Cotang.	Cosine.	
0	.2756374	.286745	3.487414	.9612617	**60**	**0**	.2923717	.305730	3.270852	.9563048	**60**
1	.2759170	.287060	3.483589	.9611815	59	1	.2926499	.306048	3.267452	.9562197	59
2	.2761965	.287375	3.479772	.9611012	58	2	.2929280	.306367	3.264059	.9561345	58
3	.2764761	.287690	3.475963	.9610208	57	3	.2932061	.306685	3.260672	.9560492	57
4	.2767556	.288005	3.472161	.9609403	56	4	.2934842	.307003	3.257292	.9559639	56
5	.2770352	.288320	3.468367	.9608598	**55**	**5**	.2937623	.307321	3.253918	.9558785	**55**
6	.2773147	.288635	3.464581	.9607792	54	6	.2940403	.307640	3.250550	.9557930	54
7	.2775941	.288950	3.460802	.9606984	53	7	.2943183	.307958	3.247189	.9557074	53
8	.2778736	.289265	3.457031	.9606177	52	8	.2945963	.308277	3.243834	.9556218	52
9	.2781530	.289580	3.453267	.9605368	51	9	.2948743	.308595	3.240486	.9555361	51
10	.2784324	.289896	3.449512	.9604558	**50**	**10**	.2951522	.308914	3.237143	.9554502	**50**
11	.2787118	.290211	3.445763	.9603748	49	11	.2954302	.309233	3.233807	.9553643	49
12	.2789911	.290526	3.442022	.9602937	48	12	.2957081	.309551	3.230478	.9552784	48
13	.2792704	.290842	3.438289	.9602125	47	13	.2959859	.309870	3.227154	.9551923	47
14	.2795497	.291157	3.434563	.9601312	46	14	.2962638	.310189	3.223837	.9551062	46
15	.2798290	.291473	3.430844	.9600499	**45**	**15**	.2965416	.310508	3.220526	.9550199	**45**
16	.2801083	.291789	3.427133	.9599684	44	16	.2968194	.310827	3.217221	.9549336	44
17	.2803875	.292104	3.423429	.9598869	43	17	.2970971	.311146	3.213922	.9548473	43
18	.2806667	.292420	3.419733	.9598053	42	18	.2973749	.311465	3.210630	.9547608	42
19	.2809459	.292736	3.416044	.9597236	41	19	.2976526	.311784	3.207344	.9546743	41
20	.2812251	.293052	3.412362	.9596418	**40**	**20**	.2979303	.312103	3.204063	.9545876	**40**
21	.2815042	.293368	3.408688	.9595600	39	21	.2982079	.312422	3.200789	.9545009	39
22	.2817833	.293683	3.405021	.9594781	38	22	.2984856	.312742	3.197521	.9544141	38
23	.2820624	.293999	3.401361	.9593961	37	23	.2987632	.313061	3.194259	.9543273	37
24	.2823415	.294316	3.397708	.9593140	36	24	.2990408	.313381	3.191003	.9542403	36
25	.2826205	.294632	3.394063	.9592318	**35**	**25**	.2993184	.313700	3.187754	.9541533	**35**
26	.2828995	.294948	3.390424	.9591496	34	26	.2995959	.314020	3.184510	.9540662	34
27	.2831785	.295264	3.386793	.9590672	33	27	.2998734	.314339	3.181272	.9539790	33
28	.2834575	.295580	3.383169	.9589848	32	28	.3001509	.314659	3.178040	.9538917	32
29	.2837364	.295897	3.379553	.9589023	31	29	.3004284	.314979	3.174814	.9538044	31
30	.2840153	.296213	3.375943	.9588197	**30**	**30**	.3007058	.315298	3.171594	.9537170	**30**
31	.2842942	.296529	3.372340	.9587371	29	31	.3009832	.315618	3.168380	.9536294	29
32	.2845731	.296846	3.368745	.9586543	28	32	.3012606	.315938	3.165172	.9535418	28
33	.2848520	.297163	3.365156	.9585715	27	33	.3015380	.316258	3.161970	.9534542	27
34	.2851308	.297479	3.361575	.9584886	26	34	.3018153	.316578	3.158774	.9533664	26
35	.2854096	.297796	3.358000	.9584056	**25**	**35**	.3020926	.316898	3.155584	.9532786	**25**
36	.2856884	.298112	3.354433	.9583226	24	36	.3023699	.317218	3.152399	.9531907	24
37	.2859671	.298429	3.350872	.9582394	23	37	.3026471	.317538	3.149220	.9531027	23
38	.2862458	.298746	3.347319	.9581562	22	38	.3029244	.317859	3.146047	.9530146	22
39	.2865246	.299063	3.343772	.9580729	21	39	.3032016	.318179	3.142880	.9529264	21
40	.2868032	.299380	3.340232	.9579895	**20**	**40**	.3034788	.318499	3.139719	.9528382	**20**
41	.2870819	.299697	3.336699	.9579060	19	41	.3037559	.318820	3.136563	.9527499	19
42	.2873605	.300014	3.333173	.9578225	18	42	.3040331	.319140	3.133414	.9526615	18
43	.2876391	.300331	3.329654	.9577389	17	43	.3043102	.319461	3.130270	.9525730	17
44	.2879177	.300648	3.326141	.9576552	16	44	.3045872	.319781	3.127131	.9524844	16
45	.2881963	.300965	3.322636	.9575714	**15**	**45**	.3048643	.320102	3.123999	.9523958	**15**
46	.2884748	.301283	3.319137	.9574875	14	46	.3051413	.320423	3.120872	.9523071	14
47	.2887533	.301600	3.315645	.9574035	13	47	.3054183	.320744	3.117750	.9522183	13
48	.2890318	.301917	3.312159	.9573195	12	48	.3056953	.321064	3.114635	.9521294	12
49	.2893103	.302235	3.308681	.9572354	11	49	.3059723	.321385	3.111525	.9520404	11
50	.2895887	.302552	3.305209	.9571512	**10**	**50**	.3062492	.321706	3.108421	.9519514	**10**
51	.2898671	.302870	3.310743	.9570669	9	51	.3065261	.322027	3.105322	.9518623	9
52	.2901455	.303187	3.298285	.9569825	8	52	.3068030	.322348	3.102229	.9517731	8
53	.2904239	.303505	3.294833	.9568981	7	53	.3070798	.322670	3.099141	.9516838	7
54	.2907022	.303823	3.291387	.9568136	6	54	.3073566	.322991	3.096059	.9515944	6
55	.2909805	.304141	3.287948	.9567290	**5**	**55**	.3076334	.323312	3.092983	.9515050	**5**
56	.2912588	.304458	3.284516	.9566443	4	56	.3079102	.323633	3.089912	.9514154	4
57	.2915371	.304776	3.281090	.9565595	3	57	.3081869	.323955	3.086846	.9513258	3
58	.2918153	.305094	3.277671	.9564747	2	58	.3084636	.324276	3.083786	.9512361	2
59	.2920935	.305412	3.274258	.9563898	1	59	.3087403	.324598	3.080732	.9511464	1
60	.2923717	305730	3.270852	9563048	**0**	**60**	3090170	324919	3 077683	.9510565	**0**
	Cosine.	Cotang	Tang.	Sine.			Cosine.	Cotang	Tang.	Sine.	

73° **72°**

Note.—Secant = 1 ÷ cosine. Cosecant = 1 ÷ sine.

2.—**Natural Sines,** TANGENTS, COTANGENTS, COSINES.—(Continued.)

(Versed sine = 1 − cosine; coversed sine = 1 − sine.)

18°

'	Sine.	Tang.	Cotang.	Cosine.	
0	.3090170	.324919	3.077683	.9510565	**60**
1	.3092936	.325241	3.074640	.9509666	59
2	.3095702	.325563	3.071602	.9508766	58
3	.3098468	.325884	3.068569	.9507865	57
4	.3101234	.326206	3.065542	.9506963	56
5	.3103999	.326528	3.062520	.9506061	**55**
6	.3106764	.326850	3.059503	.9505157	54
7	.3109529	.327172	3.056492	.9504253	53
8	.3112294	.327494	3.053487	.9503348	52
9	.3115058	.327816	3.050486	.9502443	51
10	.3117822	.328138	3.047491	.9501536	**50**
11	.3120586	.328461	3.044501	.9500629	49
12	.3123349	.328783	3.041517	.9499721	48
13	.3126112	.329105	3.038538	.9498812	47
14	.3128875	.329428	3.035564	.9497902	46
15	.3131638	.329750	3.032595	.9496991	**45**
16	.3134400	.330073	3.029632	.9496080	44
17	.3137163	.330395	3.026673	.9495168	43
18	.3139925	.330718	3.023720	.9494255	42
19	.3142686	.331041	3.020772	.9493341	41
20	.3145448	.331363	3.017830	.9492426	**40**
21	.3148209	.331686	3.014892	.9491511	39
22	.3150969	.332009	3.011960	.9490595	38
23	.3153730	.332332	3.009033	.9489678	37
24	.3156490	.332655	3.006110	.9488760	36
25	.3159250	.332978	3.003193	.9487842	**35**
26	.3162010	.333302	3.000282	.9486922	34
27	.3164770	.333625	2.997375	.9486002	33
28	.3167529	.333948	2.994473	.9485081	32
29	.3170288	.334271	2.991576	.9484159	31
30	.3173047	.334595	2.988685	.9483237	**30**
31	.3175805	.334918	2.985798	.9482313	29
32	.3178563	.335242	2.982916	.9481389	28
33	.3181321	.335566	2.980040	.9480464	27
34	.3184079	.335889	2.977168	.9479538	26
35	.3126836	.336213	2.974301	.9478612	**25**
36	.3189593	.336537	2.971439	.9477684	24
37	.3192350	.336861	2.968583	.9476756	23
38	.3195106	.337185	2.965731	.9475827	22
39	.3197863	.337509	2.962884	.9474897	21
40	.3200619	.337833	2.960042	.9473966	**20**
41	.3203374	.338157	2.957205	.9473035	19
42	.3206130	.338481	2.954372	.9472103	18
43	.3208885	.338805	2.951545	.9471170	17
44	.3211640	.339129	2.948722	.9470236	16
45	.3214395	.339454	2.945905	.9469301	**15**
46	.3217149	.339778	2.943092	.9468366	14
47	.3219903	.340103	2.940284	.9467430	13
48	.3222657	.340427	2.937480	.9466493	12
49	.3225411	.340752	2.934682	.9465555	11
50	.3228164	.341077	2.931888	.9464616	**10**
51	.3230917	.341401	2.929099	.9463677	9
52	.3233670	.341726	2.926315	.9462736	8
53	.3236422	.342051	2.923535	.9461795	7
54	.3239174	.342376	2.920761	.9460854	6
55	.3241926	.342701	2.917990	.9459911	**5**
56	.3244678	.343026	2.915225	.9458968	4
57	.3247429	.343351	2.912464	.9458023	3
58	.3250180	.343677	2.909708	.9457078	2
59	.3252931	.344002	2.906957	.9456132	1
60	.3255682	.344327	2.904210	.9455186	**0**
	Cosine.	Cotang	Tang.	Sine.	'

71°

19°

'	Sine.	Tang.	Cotang.	Cosine.	
0	.3255682	.344327	2.904210	.9455186	**60**
1	.3258432	.344653	2.901468	.9454238	59
2	.3261182	.344978	2.898731	.9453290	58
3	.3263932	.345304	2.895998	.9452341	57
4	.3266681	.345629	2.893270	.9451391	56
5	.3269430	.345955	2.890546	.9450441	**55**
6	.3272179	.346281	2.887827	.9449489	54
7	.3274928	.346606	2.885113	.9448537	53
8	.3277676	.346932	2.882403	.9447584	52
9	.3280424	.347258	2.879697	.9446630	51
10	.3283172	.347584	2.876997	.9445675	**50**
11	.3285919	.347910	2.874300	.9444720	49
12	.3288666	.348236	2.871608	.9443764	48
13	.3291413	.348563	2.868921	.9442807	47
14	.3294160	.348889	2.866238	.9441849	46
15	.3296906	.349215	2.863560	.9440890	**45**
16	.3299653	.349542	2.860886	.9439931	44
17	.3302398	.349868	2.858216	.9438971	43
18	.3305144	.350195	2.855551	.9438010	42
19	.3307889	.350521	2.852891	.9437048	41
20	.3310634	.350848	2.850234	.9436085	**40**
21	.3313379	.351175	2.847583	.9435122	39
22	.3316123	.351501	2.844935	.9434157	38
23	.3318867	.351828	2.842292	.9433192	37
24	.3321611	.352155	2.839653	.9432227	36
25	.3324355	.352482	2.837019	.9431260	**35**
26	.3327098	.352809	2.834389	.9430293	34
27	.3329841	.353136	2.831763	.9429324	33
28	.3332584	.353464	2.829142	.9428355	32
29	.3335326	.353791	2.826525	.9427386	31
30	.3338069	.354118	2.823912	.9426415	**30**
31	.3340810	.354446	2.821304	.9425444	29
32	.3343552	.354773	2.818700	.9424471	28
33	.3346293	.355101	2.816100	.9423498	27
34	.3349034	.355428	2.813504	.9422525	26
35	.3351775	.355756	2.810913	.9421550	**25**
36	.3354516	.356084	2.808326	.9420575	24
37	.3357256	.356411	2.805743	.9419598	23
38	.3359996	.356739	2.803164	.9418621	22
39	.3362735	.357067	2.800590	.9417644	21
40	.3365475	.357395	2.798019	.9416665	**20**
41	.3368214	.357723	2.795453	.9415686	19
42	.3370953	.358051	2.792891	.9414705	18
43	.3373691	.358380	2.790333	.9413724	17
44	.3376429	.358708	2.787780	.9412743	16
45	.3379167	.359036	2.785230	.9411760	**15**
46	.3381905	.359365	2.782685	.9410777	14
47	.3384642	.359693	2.780144	.9409793	13
48	.3387379	.360022	2.777606	.9408808	12
49	.3390116	.360350	2.775073	.9407822	11
50	.3392852	.360679	2.772544	.9406835	**10**
51	.3395589	.361008	2.770019	.9405848	9
52	.3398325	.361337	2.767499	.9404860	8
53	.3401060	.361666	2.764982	.9403871	7
54	.3403796	.361994	2.762469	.9402881	6
55	.3406531	.362324	2.759960	.9401891	**5**
56	.3409265	.362653	2.757456	.9400899	4
57	.3412000	.362982	2.754955	.9399907	3
58	.3414734	.363311	2.752458	.9398914	2
59	.3417468	.363640	2.749966	.9397921	1
60	.3420201	.363970	2.747477	.9396926	**0**
	Cosine.	Cotang	Tang.	Sine.	'

70°

Note.—Secant = 1 ÷ cosine. Cosecant = 1 ÷ sine.

2.—Natural Sines, Tangents, Cotangents, Cosines.—(Continued.)

(Versed sine = 1 − cosine; coversed sine = 1 − sine.)

20°

	Sine.	Tang.	Cotang.	Cosine.	
0	.3420201	.363970	2.747477	.9396926	60
1	.3422935	.364299	2.744992	.9395931	59
2	.3425668	.364629	2.742512	.9394935	58
3	.3428400	.364958	2.740035	.9393938	57
4	.3431133	.365288	2.737562	.9392940	56
5	.3433865	.365618	2.735093	.9391942	55
6	.3436597	.365948	2.732628	.9390943	54
7	.3439329	.366277	2.730167	.9389943	53
8	.3442060	.366607	2.727710	.9388942	52
9	.3444791	.366937	2.725256	.9387940	51
10	.3447521	.367268	2.722807	.9386938	50
11	.3450252	.367598	2.720362	.9385934	49
12	.3452982	.367928	2.717920	.9384930	48
13	.3455712	.368258	2.715482	.9383925	47
14	.3458441	.368589	2.713048	.9382920	46
15	.3461171	.368919	2.710618	.9381913	45
16	.3463900	.369250	2.708192	.9380906	44
17	.3466628	.369580	2.705769	.9379898	43
18	.3469357	.369911	2.703351	.9378889	42
19	.3472085	.370242	2.700936	.9377880	41
20	.3474812	.370572	2.698525	.9376869	40
21	.3477540	.370903	2.696118	.9375858	39
22	.3480267	.371234	2.693714	.9374846	38
23	.3482994	.371565	2.691314	.9373833	37
24	.3485720	.371896	2.688919	.9372820	36
25	.3488447	.372227	2.686526	.9371806	35
26	.3491173	.372559	2.684138	.9370790	34
27	.3493898	.372890	2.681753	.9369774	33
28	.3496624	.373221	2.679372	.9368758	32
29	.3499349	.373553	2.676995	.9367740	31
30	.3502074	.373884	2.674621	.9366722	30
31	.3504798	.374216	2.672251	.9365703	29
32	.3507523	.374547	2.669885	.9364683	28
33	.3510246	.374879	2.667522	.9363662	27
34	.3512970	.375211	2.665163	.9362641	26
35	.3515693	.375543	2.662808	.9361618	25
36	.3518416	.375875	2.660456	.9360595	24
37	.3521139	.376207	2.658108	.9359571	23
38	.3523862	.376539	2.655764	.9358547	22
39	.3526584	.376871	2.653423	.9357521	21
40	.3529306	.377203	2.651086	.9356495	20
41	.3532027	.377536	2.648753	.9355468	19
42	.3534748	.377868	2.646423	.9354440	18
43	.3537469	.378201	2.644096	.9353412	17
44	.3540190	.378533	2.641774	.9352382	16
45	.3542910	.378866	2.639454	.9351352	15
46	.3545630	.379198	2.637139	.9350321	14
47	.3548350	.379531	2.634827	.9349289	13
48	.3551070	.379864	2.632518	.9348257	12
49	.3553789	.380197	2.630213	.9347223	11
50	.3556508	.380530	2.627912	.9346189	10
51	.3559226	.380863	2.625614	.9345154	9
52	.3561944	.381196	2.623319	.9344119	8
53	.3564662	.381529	2.621028	.9343082	7
54	.3567380	.381862	2.618741	.9342045	6
55	.3570097	.382196	2.616457	.9341007	5
56	.3572814	.382529	2.614176	.9339968	4
57	.3575531	.382863	2.611899	.9338928	3
58	.3578248	.383196	2.609625	.9337888	2
59	.3580964	.383530	2.607355	.9336846	1
60	.3583679	.383864	2.605089	.9335804	0
	Cosine.	Cotang	Tang.	Sine.	

69°

21°

	Sine.	Tang.	Cotang.	Cosine.	
0	.3583679	.383864	2.605089	.9335804	60
1	.3586395	.384197	2.602825	.9334761	59
2	.3589110	.384531	2.600565	.9333718	58
3	.3591825	.384865	2.598309	.9332673	57
4	.3594540	.385199	2.596056	.9331628	56
5	.3597254	.385533	2.593806	.9330582	55
6	.3599968	.385867	2.591560	.9329535	54
7	.3602682	.386202	2.589317	.9328488	53
8	.3605395	.386536	2.587078	.9327439	52
9	.3608108	.386870	2.584842	.9326390	51
10	.3610821	.387205	2.582609	.9325340	50
11	.3613534	.387539	2.580380	.9324290	49
12	.3616246	.387874	2.578153	.9323238	48
13	.3618958	.388209	2.575931	.9322186	47
14	.3621669	.388543	2.573711	.9321133	46
15	.3624380	.388878	2.571495	.9320079	45
16	.3627091	.389213	2.569283	.9319024	44
17	.3629802	.389548	2.567073	.9317969	43
18	.3632512	.389883	2.564867	.9316912	42
19	.3635222	.390218	2.562664	.9315855	41
20	.3637932	.390554	2.560464	.9314797	40
21	.3640641	.390889	2.558268	.9313739	39
22	.3643351	.391224	2.556075	.9312679	38
23	.3646059	.391560	2.553885	.9311619	37
24	.3648768	.391895	2.551699	.9310558	36
25	.3651476	.392231	2.549516	.9309496	35
26	.3654184	.392567	2.547335	.9308434	34
27	.3656891	.392902	2.545159	.9307370	33
28	.3659599	.393238	2.542985	.9306306	32
29	.3662306	.393574	2.540815	.9305241	31
30	.3665012	.393910	2.538647	.9304176	30
31	.3667719	.394246	2.536483	.9303109	29
32	.3670425	.394582	2.534323	.9302042	28
33	.3673130	.394918	2.532165	.9300974	27
34	.3675836	.395255	2.530011	.9299905	26
35	.3678541	.395591	2.527859	.9298835	25
36	.3681246	.395928	2.525711	.9297765	24
37	.3683950	.396264	2.523566	.9296694	23
38	.3686654	.396601	2.521424	.9295622	22
39	.3689358	.396937	2.519286	.9294549	21
40	.3692061	.397274	2.517150	.9293475	20
41	.3694765	.397611	2.515018	.9292401	19
42	.3697468	.397948	2.512889	.9291326	18
43	.3700170	.398285	2.510762	.9290250	17
44	.3702872	.398622	2.508639	.9289173	16
45	.3705574	.398959	2.506519	.9288096	15
46	.3708276	.399296	2.504403	.9287017	14
47	.3710977	.399634	2.502289	.9285938	13
48	.3713678	.399971	2.500178	.9284858	12
49	.3716379	.400308	2.498070	.9283778	11
50	.3719079	.400646	2.495966	.9282696	10
51	.3721780	.400984	2.493864	.9281614	9
52	.3724479	.401321	2.491766	.9280531	8
53	.3727179	.401659	2.489670	.9279447	7
54	.3729878	.401997	2.487578	.9278363	6
55	.3732577	.402335	2.485488	.9277277	5
56	.3735275	.402673	2.483402	.9276191	4
57	.3737973	.403011	2.481319	.9275104	3
58	.3740671	.403349	2.479238	.9274016	2
59	.3743369	.403687	2.477161	.9272928	1
60	.3746066	.404026	2.475086	.9271839	0
	Cosine.	Cotang	Tang.	Sine.	

68°

Note.—Secant = 1 ÷ cosine. **Cosecart = 1 ÷ sine.**

2.—Natural Sines, Tangents, Cotangents, Cosines.—(Continued.)

(Versed sine = 1 − cosine; coversed sine = 1 − sine.)

22°

′	Sine.	Tang.	Cotang.	Cosine.	
0	.3746066	.404026	2.475086	.9271839	60
1	.3748763	.404364	2.473015	.9270748	59
2	.3751459	.404703	2.470947	.9269658	58
3	.3754156	.405041	2.468881	.9268566	57
4	.3756852	.405380	2.466819	.9267474	56
5	.3759547	.405719	2.464759	.9266380	55
6	.3762243	.406057	2.462703	.9265286	54
7	.3764938	.406396	2.460649	.9264192	53
8	.3767632	.406735	2.458598	.9263096	52
9	.3770327	.407074	2.456551	.9262000	51
10	.3773021	.407413	2.454506	.9260902	50
11	.3775714	.407753	2.452464	.9259805	49
12	.3778408	.408092	2.450425	.9258706	48
13	.3781101	.408431	2.448389	.9257606	47
14	.3783794	.408771	2.446355	.9256506	46
15	.3786486	.409110	2.444325	.9255405	45
16	.3789178	.409450	2.442298	.9254303	44
17	.3791870	.409790	2.440273	.9253201	43
18	.3794562	.410129	2.438251	.9252097	42
19	.3797253	.410469	2.436233	.9250993	41
20	.3799944	.410809	2.434217	.9249888	40
21	.3802634	.411149	2.432204	.9248782	39
22	.3805324	.411489	2.430193	.9247676	38
23	.3808014	.411830	2.428186	.9246568	37
24	.3810704	.412170	2.426181	.9245460	36
25	.3813393	.412510	2.424180	.9244351	35
26	.3816082	.412851	2.422181	.9243242	34
27	.3818770	.413191	2.420185	.9242131	33
28	.3821459	.413532	2.418191	.9241020	32
29	.3824147	.413872	2.416201	.9239908	31
30	.3826834	.414213	2.414213	.9238795	30
31	.3829522	.414554	2.412228	.9237682	29
32	.3832209	.414895	2.410246	.9236567	28
33	.3834895	.415236	2.408267	.9235452	27
34	.3837582	.415577	2.406290	.9234336	26
35	.3840268	.415918	2.404316	.9233220	25
36	.3842953	.416259	2.402345	.9232102	24
37	.3845639	.416601	2.400377	.9230984	23
38	.3848324	.416942	2.398411	.9229865	22
39	.3851008	.417284	2.396449	.9228745	21
40	.3853693	.417625	2.394488	.9227624	20
41	.3856377	.417967	2.392531	.9226503	19
42	.3859060	.418309	2.390576	.9225381	18
43	.3861744	.418650	2.388625	.9224258	17
44	.3864427	.418992	2.386675	.9223134	16
45	.3867110	.419334	2.384729	.9222010	15
46	.3869792	.419676	2.382785	.9220884	14
47	.3872474	.420019	2.380844	.9219758	13
48	.3875156	.420361	2.378906	.9218632	12
49	.3877837	.420703	2.376970	.9217504	11
50	.3880518	.421046	2.375037	.9216375	10
51	.3883199	.421388	2.373106	.9215246	9
52	.3885880	.421731	2.371179	.9214116	8
53	.3888560	.422073	2.369254	.9212986	7
54	.3891240	.422416	2.367331	.9211854	6
55	.3893919	.422759	2.365411	.9210722	5
56	.3896598	.423102	2.363494	.9209589	4
57	.3899277	.423445	2.361580	.9208455	3
58	.3901955	.423788	2.359668	.9207320	2
59	.3904633	.424131	2.357759	.9206185	1
60	.3907311	.424474	2.355852	.9205049	0
	Cosine.	Cotang.	Tang.	Sine.	′

67°

23°

′	Sine.	Tang.	Cotang.	Cosine.	
0	.3907311	.424474	2.355852	.9205049	60
1	.3909989	.424818	2.353948	.9203912	59
2	.3912666	.425161	2.352046	.9202774	58
3	.3915343	.425505	2.350148	.9201635	57
4	.3918019	.425848	2.348251	.9200496	56
5	.3920695	.426192	2.346358	.9199356	55
6	.3923371	.426536	2.344467	.9198215	54
7	.3926047	.426880	2.342578	.9197073	53
8	.3928722	.427223	2.340692	.9195931	52
9	.3931397	.427568	2.338809	.9194788	51
10	.3934071	.427912	2.336928	.9193644	50
11	.3936745	.428256	2.335050	.9192499	49
12	.3939419	.428600	2.333174	.9191353	48
13	.3942093	.428944	2.331301	.9190207	47
14	.3944766	.429289	2.329431	.9189060	46
15	.3947439	.429633	2.327563	.9187912	45
16	.3950111	.429978	2.325697	.9186763	44
17	.3952783	.430323	2.323834	.9185614	43
18	.3955455	.430668	2.321974	.9184464	42
19	.3958127	.431012	2.320116	.9183313	41
20	.3960798	.431357	2.318260	.9182161	40
21	.3963468	.431703	2.316407	.9181009	39
22	.3966139	.432048	2.314557	.9179855	38
23	.3968809	.432393	2.312709	.9178701	37
24	.3971479	.432738	2.310863	.9177546	36
25	.3974148	.433084	2.309020	.9176391	35
26	.3976818	.433429	2.307180	.9175234	34
27	.3979486	.433775	2.305342	.9174077	33
28	.3982155	.434120	2.303506	.9172919	32
29	.3984823	.434466	2.301673	.9171760	31
30	.3987491	.434812	2.299842	.9170601	30
31	.3990158	.435158	2.298014	.9169440	29
32	.3992825	.435504	2.296188	.9168279	28
33	.3995492	.435850	2.294365	.9167118	27
34	.3998158	.436196	2.292544	.9165955	26
35	.4000825	.436542	2.290725	.9164791	25
36	.4003490	.436889	2.288909	.9163627	24
37	.4006156	.437235	2.287095	.9162462	23
38	.4008821	.437582	2.285284	.9161297	22
39	.4011486	.437928	2.283475	.9160130	21
40	.4014150	.438275	2.281669	.9158963	20
41	.4016814	.438622	2.279865	.9157795	19
42	.4019478	.438969	2.278063	.9156626	18
43	.4022141	.439316	2.276264	.9155456	17
44	.4024804	.439663	2.274467	.9154286	16
45	.4027467	.440010	2.272672	.9153115	15
46	.4030129	.440357	2.270880	.9151943	14
47	.4032791	.440705	2.269090	.9150770	13
48	.4035453	.441052	2.267303	.9149597	12
49	.4038114	.441400	2.265518	.9148422	11
50	.4040775	.441747	2.263735	.9147247	10
51	.4043436	.442095	2.261955	.9146072	9
52	.4046096	.442443	2.260177	.9144895	8
53	.4048756	.442791	2.258401	.9143718	7
54	.4051416	.443139	2.256628	.9142540	6
55	.4054075	.443487	2.254857	.9141361	5
56	.4056734	.443835	2.253088	.9140181	4
57	.4059393	.444183	2.251322	.9139001	3
58	.4062051	.444531	2.249558	.9137819	2
59	.4064709	.444880	2.247796	.9136637	1
60	.4067366	.445228	2.246036	.9135455	0
	Cosine.	Cotang.	Tang.	Sine.	′

66°

Note.—Secant = 1 ÷ cosine. Cosecant = 1 ÷ sine.

2.—**Natural Sines,** Tangents, Cotangents, Cosines.—(Continued.)

(Versed sine = 1 − cosine; coversed sine = 1 − sine.)

24°

′	Sine.	Tang.	Cotang.	Cosine.	
0	.4067366	.445228	2.246036	.9135455	**60**
1	.4070024	.445577	2.244279	.9134271	59
2	.4072681	.445926	2.242524	.9133087	58
3	.4075337	.446274	2.240772	.9131902	57
4	.4077993	.446623	2.239021	.9130716	56
5	.4080649	.446972	2.237273	.9129529	**55**
6	.4083305	.447321	2.235528	.9128342	54
7	.4085960	.447670	2.233784	.9127154	53
8	.4088615	.448020	2.232043	.9125965	52
9	.4091269	.448369	2.230304	.9124775	51
10	.4093923	.448718	2.228567	.9123584	**50**
11	.4096577	.449068	2.226833	.9122393	49
12	.4099230	.449417	2.225100	.9121201	48
13	.4101883	.449767	2.223370	.9120008	47
14	.4104536	.450117	2.221643	.9118815	46
15	.4107189	.450467	2.219917	.9117620	**45**
16	.4109841	.450817	2.218194	.9116425	44
17	.4112492	.451167	2.216473	.9115229	43
18	.4115144	.451517	2.214754	.9114033	42
19	.4117795	.451867	2.213037	.9112835	41
20	.4120445	.452217	2.211323	.9111637	**40**
21	.4123096	.452568	2.209611	.9110438	39
22	.4125745	.452918	2.207901	.9109238	38
23	.4128395	.453269	2.206193	.9108038	37
24	.4131044	.453620	2.204487	.9106837	36
25	.4133693	.453970	2.202784	.9105635	**35**
26	.4136342	.454321	2.201083	.9104432	34
27	.4138990	.454672	2.199384	.9103228	33
28	.4141638	.455023	2.197687	.9102024	32
29	.4144285	.455375	2.195992	.9100819	31
30	.4146932	.455726	2.194299	.9099613	**30**
31	.4149579	.456077	2.192609	.9098406	29
32	.4152226	.456429	2.190921	.9097199	28
33	.4154872	.456780	2.189234	.9095990	27
34	.4157517	.457132	2.187551	.9094781	26
35	.4160163	.457483	2.185869	.9093572	**25**
36	.4162808	.457835	2.184189	.9092361	24
37	.4165453	.458187	2.182511	.9091150	23
38	.4168097	.458539	2.180836	.9089938	22
39	.4170741	.458891	2.179163	.9088725	21
40	.4173385	.459243	2.177492	.9087511	**20**
41	.4176028	.459596	2.175822	.9086297	19
42	.4178671	.459948	2.174155	.9085082	18
43	.4181313	.460301	2.172491	.9083866	17
44	.4183956	.460653	2.170828	.9082649	16
45	.4186597	.461006	2.169167	.9081432	**15**
46	.4189239	.461359	2.167509	.9080214	14
47	.4191880	.461711	2.165852	.9078995	13
48	.4194521	.462064	2.164198	.9077775	12
49	.4197161	.462417	2.162546	.9076554	11
50	.4199801	.462771	2.160895	.9075333	**10**
51	.4202441	.463124	2.159247	.9074111	9
52	.4205080	.463477	2.157601	.9072888	8
53	.4207719	.463831	2.155957	.9071665	7
54	.4210358	.464184	2.154315	.9070440	6
55	.4212996	.464538	2.152675	.9069215	**5**
56	.4215634	.464891	2.151037	.9067989	4
57	.4218272	.465245	2.149402	.9066762	3
58	.4220909	.465599	2.147768	.9065535	2
59	.4223546	.465953	2.146136	.9064307	1
60	.4226183	.466307	2.144506	.9063078	**0**
	Cosine.	Cotang.	Tang.	Sine.	′

65°

25°

′	Sine.	Tang.	Cotang.	Cosine.	
0	.4226183	.466307	2.144506	.9063078	**60**
1	.4228819	.466661	2.142879	.9061848	59
2	.4231455	.467016	2.141253	.9060618	58
3	.4234090	.467370	2.139630	.9059386	57
4	.4236725	.467725	2.138008	.9058154	56
5	.4239360	.468079	2.136389	.9056922	**55**
6	.4241994	.468434	2.134771	.9055688	54
7	.4244628	.468789	2.133155	.9054454	53
8	.4247262	.469143	2.131542	.9053219	52
9	.4249895	.469498	2.129930	.9051983	51
10	.4252528	.469853	2.128321	.9050746	**50**
11	.4255161	.470209	2.126713	.9049509	49
12	.4257793	.470564	2.125108	.9048271	48
13	.4260425	.470919	2.123504	.9047032	47
14	.4263056	.471275	2.121903	.9045792	46
15	.4265687	.471630	2.120303	.9044551	**45**
16	.4268318	.471986	2.118705	.9043310	44
17	.4270949	.472342	2.117110	.9042068	43
18	.4273579	.472697	2.115516	.9040825	42
19	.4276208	.473053	2.113924	.9039582	41
20	.4278838	.473409	2.112334	.9038338	**40**
21	.4281467	.473765	2.110747	.9037093	39
22	.4284095	.474122	2.109161	.9035847	38
23	.4286723	.474478	2.107577	.9034600	37
24	.4289351	.474834	2.105995	.9033353	36
25	.4291979	.475191	2.104415	.9032105	**35**
26	.4294606	.475548	2.102836	.9030856	34
27	.4297233	.475904	2.101260	.9029606	33
28	.4299859	.476261	2.099686	.9028356	32
29	.4302485	.476618	2.098114	.9027105	31
30	.4305111	.476975	2.096543	.9025853	**30**
31	.4307736	.477332	2.094975	.9024600	29
32	.4310361	.477689	2.093408	.9023347	28
33	.4312986	.478047	2.091843	.9022092	27
34	.4315610	.478404	2.090280	.9020838	26
35	.4318234	.478762	2.088720	.9019582	**25**
36	.4320857	.479119	2.087161	.9018325	24
37	.4323481	.479477	2.085603	.9017068	23
38	.4326103	.479835	2.084048	.9015810	22
39	.4328726	.480193	2.082495	.9014551	21
40	.4331348	.480551	2.080943	.9013292	**20**
41	.4333970	.480909	2.079394	.9012031	19
42	.4336591	.481267	2.077846	.9010770	18
43	.4339212	.481625	2.076300	.9009508	17
44	.4341832	.481984	2.074756	.9008246	16
45	.4344453	.482342	2.073214	.9006982	**15**
46	.4347072	.482701	2.071674	.9005718	14
47	.4349692	.483060	2.070135	.9004453	13
48	.4352311	.483418	2.068599	.9003188	12
49	.4354930	.483777	2.067064	.9001921	11
50	.4357548	.484136	2.065531	.9000654	**10**
51	.4360166	.484495	2.064000	.8999386	9
52	.4362784	.484855	2.062471	.8998117	8
53	.4365401	.485214	2.060944	.8996848	7
54	.4368018	.485573	2.059418	.8995578	6
55	.4370634	.485933	2.057895	.8994307	**5**
56	.4373251	.486293	2.056373	.8993035	4
57	.4375866	.486652	2.054853	.8991763	3
58	.4378482	.487012	2.053334	.8990489	2
59	.4381097	.487372	2.051818	.8989215	1
60	.4383711	.487732	2.050303	.8987940	**0**
	Cosine.	Cotang.	Tang.	Sine.	′

64°

Note.—Secant = 1 ÷ cosine. Cosecant = 1 ÷ sine.

2.—Natural Sines, Tangents, Cotangents, Cosines.—(Continued.)

(Versed sine = 1 − cosine; coversed sine = 1 − sine.)

26°

′	Sine.	Tang.	Cotang.	Cosine.	
0	.4383711	.487732	2.050303	.8987940	**60**
1	.4386326	.488092	2.048791	.8986665	59
2	.4388940	.488453	2.047280	.8985389	58
3	.4391553	.488813	2.045770	.8984112	57
4	.4394166	.489173	2.044263	.8982834	56
5	.4396779	.489534	2.042757	.8981555	**55**
6	.4399392	.489894	2.041254	.8980276	54
7	.4402004	.490255	2.039751	.8978996	53
8	.4404615	.490616	2.038251	.8977715	52
9	.4407227	.490977	2.036753	.8976433	51
10	.4409838	.491338	2.035256	.8975151	**50**
11	.4412448	.491699	2.033761	.8973868	49
12	.4415059	.492061	2.032268	.8972584	48
13	.4417668	.492422	2.030776	.8971299	47
14	.4420278	.492783	2.029287	.8970014	46
15	.4422887	.493145	2.027799	.8968727	**45**
16	.4425496	.493507	2.026313	.8967440	44
17	.4428104	.493868	2.024828	.8966153	43
18	.4430712	.494230	2.023346	.8964864	42
19	.4433319	.494592	2.021865	.8963575	41
20	.4435927	.494954	2.020386	.8962285	**40**
21	.4438534	.495317	2.018908	.8960994	39
22	.4441140	.495679	2.017433	.8959703	38
23	.4443746	.496041	2.015959	.8958411	37
24	.4446352	.496404	2.014486	.8957118	36
25	.4448957	.496766	2.013016	.8955824	**35**
26	.4451562	.497129	2.011547	.8954529	34
27	.4454167	.497492	2.010080	.8953234	33
28	.4456771	.497855	2.008615	.8951938	32
29	.4459375	.498218	2.007151	.8950641	31
30	.4461978	.498581	2.005689	.8949344	**30**
31	.4464581	.498944	2.004229	.8948045	29
32	.4467184	.499308	2.002771	.8946746	28
33	.4469786	.499671	2.001314	.8945446	27
34	.4472388	.500035	1.999859	.8944146	26
35	.4474990	.500398	1.998405	.8942844	**25**
36	.4477591	.500762	1.996953	.8941542	24
37	.4480192	.501126	1.995503	.8940240	23
38	.4482792	.501490	1.994055	.8938936	22
39	.4485392	.501854	1.992608	.8937632	21
40	.4487992	.502218	1.991163	.8936326	**20**
41	.4490591	.502583	1.989720	.8935021	19
42	.4493190	.502947	1.988278	.8933714	18
43	.4495789	.503312	1.986838	.8932406	17
44	.4498387	.503676	1.985400	.8931098	16
45	.4500984	.504041	1.983963	.8929789	**15**
46	.4503582	.504406	1.982528	.8928480	14
47	.4506179	.504771	1.981095	.8927169	13
48	.4508775	.505136	1.979663	.8925858	12
49	.4511372	.505501	1.978233	.8924546	11
50	.4513967	.505866	1.976805	.8923234	**10**
51	.4516563	.506232	1.975378	.8921920	9
52	.4519158	.506597	1.973953	.8920606	8
53	.4521753	.506963	1.972529	.8919291	7
54	.4524347	.507329	1.971107	.8917975	6
55	.4526941	.507694	1.969687	.8916659	**5**
56	.4529535	.508060	1.968268	.8915342	4
57	.4532128	.508426	1.966851	.8914024	3
58	.4534721	.508792	1.965436	.8912705	2
59	.4537313	.509159	1.964022	.8911385	1
60	.4539905	.509525	1.962610	.8910065	**0**
	Cosine.	Cotang	Tang.	Sine.	′

63°

27°

′	Sine.	Tang.	Cotang.	Cosine.	
0	.4539905	.509525	1.962610	.8910065	**60**
1	.4542497	.509891	1.961200	.8908744	59
2	.4545088	.510258	1.959791	.8907423	58
3	.4547679	.510625	1.958383	.8906100	57
4	.4550269	.510991	1.956978	.8904777	56
5	.4552859	.511358	1.955573	.8903453	**55**
6	.4555449	.511725	1.954171	.8902128	54
7	.4558038	.512093	1.952770	.8900803	53
8	.4560627	.512460	1.951371	.8899476	52
9	.4563216	.512827	1.949973	.8898149	51
10	.4565804	.513195	1.948577	.8896822	**50**
11	.4568392	.513562	1.947112	.8895493	49
12	.4570979	.513930	1.945789	.8894164	48
13	.4573566	.514298	1.944398	.8892834	47
14	.4576153	.514665	1.943008	.8891503	46
15	.4578739	.515033	1.941620	.8890171	**45**
16	.4581325	.515401	1.940233	.8888839	44
17	.4583910	.515770	1.938848	.8887506	43
18	.4586496	.516138	1.937464	.8886172	42
19	.4589080	.516506	1.936082	.8884838	41
20	.4591665	.516875	1.934702	.8883503	**40**
21	.4594248	.517244	1.933323	.8882166	39
22	.4596832	.517612	1.931945	.8880830	38
23	.4599415	.517981	1.930569	.8879492	37
24	.4601998	.518350	1.929195	.8878154	36
25	.4604580	.518719	1.927822	.8876815	**35**
26	.4607162	.519089	1.926451	.8875475	34
27	.4609744	.519458	1.925081	.8874134	33
28	.4612325	.519827	1.923713	.8872793	32
29	.4614906	.520197	1.922347	.8871451	31
30	.4617486	.520567	1.920982	.8870108	**30**
31	.4620066	.520936	1.919618	.8868765	29
32	.4622646	.521306	1.918256	.8867420	28
33	.4625225	.521676	1.916896	.8866075	27
34	.4627804	.522046	1.915537	.8864730	26
35	.4630382	.522417	1.914179	.8863383	**25**
36	.4632960	.522787	1.912823	.8862036	24
37	.4635538	.523157	1.911469	.8860688	23
38	.4638115	.523528	1.910116	.8859339	22
39	.4640692	.523899	1.908764	.8857989	21
40	.4643269	.524269	1.907414	.8856639	**20**
41	.4645845	.524640	1.906066	.8855288	19
42	.4648420	.525011	1.904719	.8853936	18
43	.4650996	.525382	1.903373	.8852584	17
44	.4653571	.525754	1.902029	.8851230	16
45	.4656145	.526125	1.900687	.8849876	**15**
46	.4658719	.526496	1.899346	.8848522	14
47	.4661293	.526868	1.898006	.8847166	13
48	.4663866	.527240	1.896668	.8845810	12
49	.4666439	.527612	1.895332	.8844453	11
50	.4669012	.527983	1.893997	.8843095	**10**
51	.4671584	.528356	1.892663	.8841736	9
52	.4674156	.528728	1.891331	.8840377	8
53	.4676727	.529100	1.890000	.8839017	7
54	.4679298	.529472	1.888671	.8837656	6
55	.4681869	.529845	1.887343	.8836295	**5**
56	.4684839	.530217	1.886017	.8834933	4
57	.4687009	.530590	1.884692	.8833569	3
58	.4689578	.530963	1.883369	.8832206	2
59	.4692147	.531336	1.882047	.8830841	1
60	.4694716	.531709	1.880726	.8829476	**0**
	Cosine.	Cotang	Tang.	Sine.	′

62°

Note.—Secant = 1 ÷ cosine. Cosecant = 1 ÷ sine.

2.—**Natural Sines,** TANGENTS, COTANGENTS, COSINES.—**(Continued.)**

(Versed sine = 1 − cosine; coversed sine = 1 − sine.)

28°

′	Sine.	Tang.	Cotang.	Cosine.	
0	.4694716	.531709	1.880726	.8829476	**60**
1	.4697284	.532082	1.879407	.8828110	59
2	.4699852	.532455	1.878089	.8826743	58
3	.4702419	.532829	1.876773	.8825376	57
4	.4704986	.533202	1.875458	.8824007	56
5	.4707553	.533576	1.874145	.8822638	**55**
6	.4710119	.533950	1.872833	.8821269	54
7	.4712685	.534324	1.871523	.8819898	53
8	.4715250	.534698	1.870214	.8818527	52
9	.4717815	.535072	1.868906	.8817155	51
10	.4720380	.535446	1.867600	.8815782	**50**
11	.4722944	.535820	1.866295	.8814409	49
12	.4725508	.536195	1.864992	.8813035	48
13	.4728071	.536569	1.863690	.8811660	47
14	.4730634	.536944	1.862389	.8810284	46
15	.4733197	.537319	1.861090	.8808907	**45**
16	.4735759	.537694	1.859792	.8807530	44
17	.4738321	.538069	1.858496	.8806152	43
18	.4740882	.538444	1.857201	.8804774	42
19	.4743443	.538819	1.855908	.8803394	41
20	.4746004	.539195	1.854615	.8802014	**40**
21	.4748564	.539570	1.853325	.8800633	39
22	.4751124	.539946	1.852035	.8799251	38
23	.4753683	.540322	1.850747	.8797869	37
24	.4756242	.540698	1.849461	.8796486	36
25	.4758801	.541074	1.848176	.8795102	**35**
26	.4761359	.541450	1.846892	.8793717	34
27	.4763917	.541826	1.845609	.8792332	33
28	.4766474	.542202	1.844328	.8790946	32
29	.4769031	.542579	1.843049	.8789559	31
30	.4771588	.542955	1.841770	.8788171	**30**
31	.4774144	.543332	1.840494	.8786783	29
32	.4776700	.543709	1.839218	.8785394	28
33	.4779255	.544086	1.837944	.8784004	27
34	.4781810	.544463	1.836671	.8782613	26
35	.4784364	.544840	1.835399	.8781222	**25**
36	.4786919	.545217	1.834129	.8779830	24
37	.4789472	.545595	1.832861	.8778437	23
38	.4792026	.545972	1.831593	.8777043	22
39	.4794579	.546350	1.830327	.8775649	21
40	.4797131	.546728	1.829062	.8774254	**20**
41	.4799683	.547106	1.827799	.8772858	19
42	.4802235	.547484	1.826537	.8771462	18
43	.4804786	.547862	1.825276	.8770064	17
44	.4807337	.548240	1.824017	.8768666	16
45	.4809888	.548618	1.822759	.8767268	**15**
46	.4812438	.548997	1.821502	.8765868	14
47	.4814987	.549375	1.820247	.8764468	13
48	.4817537	.549754	1.818993	.8763067	12
49	.4820086	.550133	1.817740	.8761665	11
50	.4822634	.550512	1.816489	.8760263	**10**
51	.4825182	.550891	1.815239	.8758859	9
52	.4827730	.551270	1.813990	.8757455	8
53	.4830277	.551650	1.812743	.8756051	7
54	.4832824	.552029	1.811496	.8754645	6
55	.4835370	.552409	1.810252	.8753239	**5**
56	.4837916	.552789	1.809008	.8751832	4
57	.4840462	.553168	1.807766	.8750425	3
58	.4843007	.553548	1.806525	.8749016	2
59	.4845552	.553928	1.805286	.8747607	1
60	.4848096	.554309	1.804047	.8746197	**0**
	Cosine.	Cotang	Tang.	Sine.	

61°

29°

′	Sine.	Tang.	Cotang.	Cosine.	
0	.4848096	.554309	1.804047	.8746197	**60**
1	.4850640	.554689	1.802810	.8744786	59
2	.4853184	.555069	1.801575	.8743375	58
3	.4855727	.555450	1.800340	.8741963	57
4	.4858270	.555831	1.799107	.8740550	56
5	.4860812	.556211	1.797875	.8739137	**55**
6	.4863354	.556592	1.796645	.8737722	54
7	.4865895	.556973	1.795416	.8736307	53
8	.4868436	.557355	1.794188	.8734891	52
9	.4870977	.557736	1.792961	.8733475	51
10	.4873517	.558117	1.791736	.8732058	**50**
11	.4876057	.558499	1.790512	.8730640	49
12	.4878597	.558881	1.789289	.8729221	48
13	.4881136	.559262	1.788067	.8727801	47
14	.4883674	.559644	1.786847	.8726381	46
15	.4886212	.560026	1.785628	.8724960	**45**
16	.4888750	.560409	1.784410	.8723538	44
17	.4891288	.560791	1.783194	.8722116	43
18	.4893825	.561173	1.781979	.8720693	42
19	.4896361	.561556	1.780765	.8719269	41
20	.4898897	.561939	1.779552	.8717844	**40**
21	.4901433	.562321	1.778340	.8716419	39
22	.4903968	.562704	1.777130	.8714993	38
23	.4906503	.563087	1.775921	.8713566	37
24	.4909038	.563471	1.774714	.8712138	36
25	.4911572	.563854	1.773507	.8710710	**35**
26	.4914105	.564237	1.772302	.8709281	34
27	.4916638	.564621	1.771098	.8707851	33
28	.4919171	.565005	1.769895	.8706420	32
29	.4921704	.565388	1.768694	.8704989	31
30	.4924236	.565772	1.767494	.8703557	**30**
31	.4926767	.566156	1.766295	.8702124	29
32	.4929298	.566541	1.765097	.8700691	28
33	.4931829	.566925	1.763900	.8699256	27
34	.4934359	.567309	1.762705	.8697821	26
35	.4936889	.567694	1.761511	.8696386	**25**
36	.4939419	.568079	1.760318	.8694949	24
37	.4941948	.568463	1.759126	.8693512	23
38	.4944476	.568848	1.757936	.8692074	22
39	.4947005	.569233	1.756747	.8690636	21
40	.4949532	.569619	1.755559	.8689196	**20**
41	.4952060	.570004	1.754372	.8687756	19
42	.4954587	.570389	1.753186	.8686315	18
43	.4957113	.570775	1.752002	.8684874	17
44	.4959639	.571161	1.750819	.8683431	16
45	.4962165	.571547	1.749637	.8681988	**15**
46	.4964690	.571933	1.748456	.8680544	14
47	.4967215	.572319	1.747276	.8679100	13
48	.4969740	.572705	1.746098	.8677655	12
49	.4972264	.573091	1.744921	.8676209	11
50	.4974787	.573478	1.743745	.8674762	**10**
51	.4977310	.573864	1.742570	.8673314	9
52	.4979833	.574251	1.741396	.8671866	8
53	.4982355	.574638	1.740224	.8670417	7
54	.4984877	.575025	1.739053	.8668967	6
55	.4987399	.575412	1.737883	.8667517	**5**
56	.4989920	.575799	1.736714	.8666066	4
57	.4992441	.576187	1.735546	.8664614	3
58	.4994961	.576574	1.734380	.8663161	2
59	.4997481	.576962	1.733214	.8661708	1
60	.5000000	.577350	1.732050	.8660254	**0**
	Cosine.	Cotang	Tang.	Sine.	′

60°

Note.—Secant = 1 ÷ cosine. Cosecant = 1 ÷ sine.

2.—Natural Sines, Tangents, Cotangents, Cosines.—(Continued.)

(Versed sine = 1 − cosine; coversed sine = 1 − sine.)

30°

′	Sine.	Tang.	Cotang.	Cosine.	
0	.5000000	.577350	1.732050	.8660254	60
1	.5002519	.577738	1.730887	.8658799	59
2	.5005037	.578126	1.729726	.8657344	58
3	.5007556	.578514	1.728565	.8655887	57
4	.5010073	.578902	1.727406	.8654430	56
5	.5012591	.579291	1.726247	.8652973	55
6	.5015107	.579679	1.725090	.8651514	54
7	.5017624	.580068	1.723934	.8650055	53
8	.5020140	.580457	1.722779	.8648595	52
9	.5022655	.580846	1.721626	.8647134	51
10	.5025170	.581235	1.720473	.8645673	50
11	.5027685	.581624	1.719322	.8644211	49
12	.5030199	.582013	1.718172	.8642748	48
13	.5032713	.582403	1.717023	.8641284	47
14	.5035227	.582793	1.715875	.8639820	46
15	.5037740	.583182	1.714728	.8638355	45
16	.5040252	.583572	1.713582	.8636889	44
17	.5042765	.583962	1.712438	.8635423	43
18	.5045276	.584352	1.711294	.8633956	42
19	.5047788	.584743	1.710152	.8632488	41
20	.5050298	.585133	1.709011	.8631019	40
21	.5052809	.585524	1.707871	.8629549	39
22	.5055319	.585914	1.706732	.8628079	38
23	.5057828	.586305	1.705595	.8626608	37
24	.5060338	.586696	1.704458	.8625137	36
25	.5062846	.587087	1.703323	.8623664	35
26	.5065355	.587478	1.702189	.8622191	34
27	.5067863	.587870	1.701055	.8620717	33
28	.5070370	.588261	1.699923	.8619243	32
29	.5072877	.588653	1.698792	.8617768	31
30	.5075384	.589045	1.697663	.8616292	30
31	.5077890	.589436	1.696534	.8614815	29
32	.5080396	.589828	1.695406	.8613337	28
33	.5082901	.590221	1.694280	.8611859	27
34	.5085406	.590613	1.693155	.8610380	26
35	.5087910	.591005	1.692030	.8608901	25
36	.5090414	.591398	1.690907	.8607420	24
37	.5092918	.591791	1.689785	.8605939	23
38	.5095421	.592183	1.688664	.8604457	22
39	.5097924	.592576	1.687544	.8602975	21
40	.5100426	.592969	1.686426	.8601491	20
41	.5102928	.593363	1.685308	.8600007	19
42	.5105429	.593756	1.684191	.8593523	18
43	.5107930	.594150	1.683076	.8597037	17
44	.5110431	.594543	1.681962	.8595551	16
45	.5112931	.594937	1.680848	.8594064	15
46	.5115431	.595331	1.679736	.8592576	14
47	.5117930	.595725	1.678625	.8591088	13
48	.5120429	.596119	1.677515	.8589599	12
49	.5122927	.596514	1.676406	.8588109	11
50	.5125425	.596908	1.675298	.8586619	10
51	.5127923	.597303	1.674192	.8585127	9
52	.5130420	.597697	1.673086	.8583635	8
53	.5132916	.598092	1.671981	.8582143	7
54	.5135413	.598487	1.670878	.8580649	6
55	.5137908	.598882	1.669775	.8579155	5
56	.5140404	.599278	1.668674	.8577660	4
57	.5142899	.599673	1.667574	.8576164	3
58	.5145393	.600069	1.666474	.8574668	2
59	.5147887	.600464	1.665376	.8573171	1
60	.5150381	.600860	1.664279	.8571673	0
	Cosine.	Cotang	Tang.	Sine.	′

59°

31°

′	Sine.	Tang.	Cotang.	Cosine.	
0	.5150381	.600860	1.664279	.8571673	60
1	.5152874	.601256	1.663183	.8570174	59
2	.5155367	.601652	1.662088	.8568675	58
3	.5157859	.602049	1.660994	.8567175	57
4	.5160351	.602445	1.659901	.8565674	56
5	.5162842	.602841	1.658809	.8564173	55
6	.5165333	.603238	1.657718	.8562671	54
7	.5167824	.603635	1.656629	.8561168	53
8	.5170314	.604032	1.655540	.8559664	52
9	.5172804	.604429	1.654452	.8558160	51
10	.5175293	.604826	1.653366	.8556655	50
11	.5177782	.605224	1.652280	.8555149	49
12	.5180270	.605621	1.651196	.8553643	48
13	.5182758	.606019	1.650112	.8552135	47
14	.5185246	.606417	1.649030	.8550627	46
15	.5187733	.606814	1.647949	.8549119	45
16	.5190219	.607213	1.646868	.8547609	44
17	.5192705	.607611	1.645789	.8546099	43
18	.5195191	.608009	1.644711	.8544588	42
19	.5197676	.608408	1.643633	.8543077	41
20	.5200161	.608806	1.642557	.8541564	40
21	.5202646	.609205	1.641482	.8540051	39
22	.5205130	.609604	1.640408	.8538538	38
23	.5207613	.610003	1.639335	.8537023	37
24	.5210096	.610402	1.638263	.8535508	36
25	.5212579	.610801	1.637191	.8533992	35
26	.5215061	.611201	1.636121	.8532475	34
27	.5217543	.611601	1.635052	.8530958	33
28	.5220024	.612000	1.633984	.8529440	32
29	.5222505	.612400	1.632917	.8527921	31
30	.5224986	.612800	1.631851	.8526402	30
31	.5227466	.613201	1.630786	.8524881	29
32	.5229945	.613601	1.629722	.8523360	28
33	.5232424	.614001	1.628659	.8521839	27
34	.5234903	.614402	1.627597	.8520316	26
35	.5237381	.614803	1.626536	.8518793	25
36	.5239859	.615204	1.625476	.8517269	24
37	.5242336	.615605	1.624417	.8515745	23
38	.5244813	.616006	1.623359	.8514219	22
39	.5247290	.616407	1.622302	.8512693	21
40	.5249766	.616809	1.621246	.8511167	20
41	.5252241	.617210	1.620192	.8509639	19
42	.5254717	.617612	1.619138	.8508111	18
43	.5257191	.618014	1.618085	.8506582	17
44	.5259665	.618416	1.617033	.8505053	16
45	.5262139	.618818	1.615982	.8503522	15
46	.5264613	.619221	1.614932	.8501991	14
47	.5267085	.619623	1.613882	.8500459	13
48	.5269558	.620026	1.612834	.8498927	12
49	.5272030	.620429	1.611787	.8497394	11
50	.5274502	.620832	1.610741	.8495860	10
51	.5276973	.621235	1.609696	.8494325	9
52	.5279443	.621638	1.608652	.8492790	8
53	.5281914	.622041	1.607609	.8491254	7
54	.5284383	.622445	1.606567	.8489717	6
55	.5286853	.622848	1.605526	.8488179	5
56	.5289322	.623252	1.604485	.8486641	4
57	.5291790	.623656	1.603446	.8485102	3
58	.5294258	.624060	1.602408	.8483562	2
59	.5296726	.624465	1.601370	.8482022	1
60	.5299193	.624869	1.600334	.8480481	0
	Cosine.	Cotang	Tang.	Sine.	′

58°

Note.—Secant = 1 ÷ cosine. Cosecant = 1 ÷ sine.

2.—Natural Sines, Tangents, Cotangents, Cosines.—(Continued.)

(Versed sine = 1 − cosine; coversed sine = 1 − sine.)

32°

′	Sine.	Tang.	Cotang.	Cosine.	
0	.5299193	.624869	1.600334	.8480481	**60**
1	.5301659	.625273	1.599299	.8478939	59
2	.5304125	.625678	1.598264	.8477397	58
3	.5306591	.626083	1.597231	.8475853	57
4	.5309057	.626488	1.596198	.8474309	56
5	.5311521	.626893	1.595167	.8472765	**55**
6	.5313986	.627298	1.594136	.8471219	54
7	.5316450	.627704	1.593107	.8469673	53
8	.5318913	.628109	1.592078	.8468126	52
9	.5321376	.628515	1.591050	.8466579	51
10	.5323839	.628921	1.590023	.8465030	**50**
11	.5326301	.629327	1.588997	.8463481	49
12	5328763	.629733	1.587973	.8461932	48
13	.5331224	.630139	1.586949	.8460381	47
14	.5333685	.630546	1.585926	.8458830	46
15	.5336145	.630953	1.584904	.8457278	**45**
16	.5338605	.631359	1.583883	.8455726	44
17	.5341065	.631766	1.582862	.8454172	43
18	.5343523	.632173	1.581843	.8452618	42
19	.5345982	.632581	1.580825	.8451064	41
20	.5348440	.632988	1.579807	.8449508	**40**
21	.5350898	.633395	1.578791	.8447952	39
22	.5353355	.633803	1.577776	.8446395	38
23	.5355812	.634211	1.576761	.8444838	37
24	.5358268	.634619	1.575747	.8443279	36
25	.5360724	.635027	1.574735	.8441720	**35**
26	.5363179	.635435	1.573723	.8440161	34
27	.5365634	.635844	1.572712	.8438600	33
28	.5368089	.636252	1.571702	.8437039	32
29	.5370543	.636661	1.570693	.8435477	31
30	.5372996	.637070	1.569685	.8433914	**30**
31	.5375449	.637479	1.568678	.8432351	29
32	.5377902	.637888	1.567672	.8430787	28
33	.5380354	.638297	1.566666	.8429222	27
34	.5382806	.638707	1.565662	.8427657	26
35	.5385257	.639116	1.564659	.8426091	**25**
36	.5387708	.639526	1.563656	.8424524	24
37	.5390158	.639936	1.562654	.8422956	23
38	.5392608	.640346	1.561654	.8421388	22
39	.5395058	.640756	1.560654	.8419819	21
40	.5397507	.641167	1.559655	.8418249	**20**
41	.5399955	.641577	1.558657	.8416679	19
42	.5402403	.641988	1.557660	.8415108	18
43	.5404851	.642399	1.556663	.8413536	17
44	.5407298	.642810	1.555668	.8411963	16
45	.5409745	.643221	1.554674	.8410390	**15**
46	.5412191	.643632	1.553680	.8408816	14
47	.5414637	.644044	1.552688	.8407241	13
48	.5417082	.644456	1.551696	.8405666	12
49	.5419527	.644867	1.550705	.8404090	11
50	.5421971	.645279	1.549715	.8402513	**10**
51	.5424415	.645691	1.548726	.8400936	9
52	.5426859	.646104	1.547738	.8399357	8
53	.5429302	.646516	1.546751	.8397778	7
54	.5431744	.646929	1.545764	.8396199	6
55	.5434187	.647341	1.544779	.8394618	**5**
56	.5436628	.647754	1.543794	.8393037	4
57	.5439069	.648167	1.542810	.8391455	3
58	.5441510	.648580	1.541828	.8389873	2
59	.5443951	.648994	1.540846	.8388290	1
60	.5446390	.649407	1.539865	.8386706	**0**
	Cosine.	Cotang	Tang.	Sine.	′

57°

33°

′	Sine.	Tang.	Cotang.	Cosine.	
0	.5446390	.649407	1.539865	.8386706	**60**
1	.5448830	.649821	1.538884	.8385121	59
2	.5451269	.650235	1.537905	.8383536	58
3	.5453707	.650649	1.536927	.8381950	57
4	.5456145	.651063	1.535949	.8380363	56
5	.5458583	.651477	1.534972	.8378775	**55**
6	.5461020	.651891	1.533996	.8377187	54
7	.5463456	.652306	1.533021	.8375598	53
8	.5465892	.652721	1.532047	.8374009	52
9	.5468328	.653136	1.531074	.8372418	51
10	.5470763	.653551	1.530102	.8370827	**50**
11	.5473198	.653996	1.529130	.8369236	49
12	.5475632	.654381	1.528160	.8367643	48
13	.5478066	.654797	1.527190	.8366050	47
14	5480499	.655212	1.526221	.8364456	46
15	.5482932	.655628	1.525253	.8362862	**45**
16	.5485365	.656044	1.524286	.8361266	44
17	.5487797	.656460	1.523320	.8359670	43
18	.5490228	.656877	1.522354	.8358074	42
19	.5492659	.657293	1 521389	.8356476	41
20	.5495090	.657710	1.520426	.8354878	**40**
21	.5497520	.658127	1.519463	.8353279	39
22	.5499950	.658544	1 518501	.8351680	38
23	.5502379	.658961	1.517540	.8350080	37
24	.5504807	.659378	1 516579	.8348479	36
25	.5507236	.659796	1.515620	.8346877	**35**
26	.5509663	.660213	1.514661	.8345275	34
27	.5512091	.660631	1.513703	.8343672	33
28	.5514518	.661049	1 512746	.8342068	32
29	.5516944	.661467	1.511790	.8340463	31
30	.5519370	.661885	1.510835	.8338858	**30**
31	.5521795	.662304	1.509880	.8337252	29
32	.5524220	.662722	1.508927	.8335646	28
33	.5526645	.663141	1.507974	.8334038	27
34	.5529069	.663560	1.507022	.8332430	26
35	.5531492	.663979	1.506071	.8330822	**25**
36	.5533915	.664398	1.505121	.8329212	24
37	.5536338	.664817	1.504171	.8327602	23
38	.5538760	.665237	1 503222	.8325991	22
39	.5541182	.665657	1.502275	.8324380	21
40	.5543603	.666076	1.501328	.8322768	**20**
41	.5546024	.666496	1.500382	.8321155	19
42	.5548444	.666917	1.499436	.8319541	18
43	.5550864	.667337	1.498492	.8317927	17
44	.5553283	.667758	1.497548	.8316312	16
45	.5555702	.668178	1.496605	.8314696	**15**
46	.5558121	.668599	1.495663	.8313080	14
47	.5560539	.669020	1.494722	.8311463	13
48	.5562956	.669441	1.493782	.8309845	12
49	.5565373	.669863	1.492842	.8308226	11
50	.5567790	.670284	1.491903	.8306607	**10**
51	.5570206	.670706	1.490965	.8304987	9
52	.5572621	.671128	1.490028	.8303366	8
53	.5575036	.671550	1.489092	.8301745	7
54	.5577451	.671972	1.488157	.8300123	6
55	.5579865	.672394	1.487222	.8298500	**5**
56	.5582279	.672816	1.486288	.8296877	4
57	.5584692	.673239	1.485355	.8295252	3
58	.5587105	.673662	1.484423	.8293628	2
59	.5589517	.674085	1.483491	.8292002	1
60	.5591929	.674508	1.482561	.8290376	**0**
	Cosine.	Cotang	Tang.	Sine.	′

56°

Note.—Secant = 1 ÷ cosine. Cosecant = 1 ÷ sine.

2.—Natural Sines, Tangents, Cotangents, Cosines.—(Continued.)

(Versed sine = 1 − cosine; coversed sine = 1 − sine.)

34° 35°

′	Sine.	Tang.	Cotang.	Cosine.		′	Sine.	Tang.	Cotang.	Cosine.	
0	.5591929	.674508	1.482561	.8290376	**60**	**0**	.5735764	.700207	1.428148	.8191520	**60**
1	.5594340	.674931	1.481631	.8288749	59	1	.5738147	.700641	1.427264	.8189852	59
2	.5596751	.675355	1.480702	.8287121	58	2	.5740529	.701074	1.426381	.8188182	58
3	.5599162	.675779	1.479773	.8285493	57	3	.5742911	.701508	1.425498	.8186512	57
4	.5601572	.676202	1.478846	.8283864	56	4	.5745292	.701943	1.424617	.8184841	56
5	.5603981	.676626	1.477919	.8282234	**55**	**5**	.5747672	.702377	1.423736	.8183169	**55**
6	.5606390	.677050	1.476993	.8280603	54	6	.5750053	.702811	1.422856	.8181497	54
7	.5608798	.677475	1.476068	.8278972	53	7	.5752432	.703246	1.421976	.8179824	53
8	.5611206	.677899	1.475144	.8277340	52	8	.5754811	.703681	1.421097	.8178151	52
9	.5613614	.678324	1.474221	.8275708	51	9	.5757190	.704116	1.420220	.8176476	51
10	.5616021	.678749	1.473298	.8274074	**50**	**10**	.5759568	.704551	1.419342	.8174801	**50**
11	.5618428	.679174	1.472376	.8272440	49	11	.5761946	.704986	1.418466	.8173125	49
12	.5620834	.679599	1.471455	.8270806	48	12	.5764323	.705422	1.417590	.8171449	48
13	.5623239	.680024	1.470535	.8269170	47	13	.5766700	.705858	1.416715	.8169772	47
14	.5625645	.680450	1.469615	.8267534	46	14	.5769076	.706294	1.415840	.8168094	46
15	.5628049	.680875	1.468696	.8265897	**45**	**15**	.5771452	.706730	1.414967	.8166416	**45**
16	.5630453	.681301	1.467778	.8264260	44	16	.7773827	.707166	1 414094	.8164736	44
17	.5632857	.681727	1.466861	.8262622	43	17	.5776202	.707602	1.413222	8163056	43
18	.5635260	.682153	1.465945	.8260983	42	18	.5778576	.708039	1 412350	8161376	42
19	.5637663	.682580	1.465029	.8259343	41	19	.5780950	.708476	1.411479	.8159695	41
20	.5640066	.683006	1.464114	.8257703	**40**	**20**	.5783323	.708913	1.410609	.8158013	**40**
21	.5642467	.683433	1.463200	.8256062	39	21	.5785696	.709350	1.409740	.8156330	39
22	.5644869	.683860	1.462287	.8254420	38	22	.5788069	.709787	1.408871	.8154647	38
23	.5647270	.684287	1.461374	.8252778	37	23	.5790440	.710225	1.408003	.8152963	37
24	.5649670	.684714	1.460463	.8251135	36	24	.5792812	.710663	1.407136	.8151278	36
25	.5652070	.685141	1.459552	.8249491	**35**	**25**	.5795183	.711100	1.406270	.8149593	**35**
26	.5654469	.685569	1.458642	.8247847	34	26	.5797553	.711539	1 405404	8147906	34
27	.5656868	.685996	1.457732	.8246202	33	27	.5799923	.711977	1.404539	.8146220	33
28	.5659267	.686424	1.456824	.8244556	32	28	.5802292	.712415	1.403674	.8144532	32
29	.5661665	.686852	1.455916	.8242909	31	29	.5804661	.712854	1.402811	.8142844	31
30	.5664062	.687281	1.455009	.8241262	**30**	**30**	.5807030	.713293	1.401948	.8141155	**30**
31	.5666459	.687709	1.454102	.8239614	29	31	.5809397	.713732	1.401086	.8139466	29
32	.5668856	.688137	1.453197	.8237965	28	32	.5811765	.714171	1.400224	.8137775	28
33	.5671252	.688566	1.452292	.8236316	27	33	.5814132	.714610	1.399363	.8136084	27
34	.5673648	.688995	1.451388	.8234666	26	34	.5816498	.715050	1.398503	.8134393	26
35	.5676043	.689424	1.450485	.8233015	**25**	**35**	.5818864	.715490	1.397644	.8132701	**25**
36	.5678437	.689853	1.449582	.8231364	24	36	.5821230	.715929	1.396785	.8131008	24
37	.5680832	.690283	1.448680	.8229712	23	37	5823595	.716369	1.395927	.8129314	23
38	.5683225	.690712	1.447779	.8228059	22	38	.5825959	.716810	1.395069	.8127620	22
39	.5685619	.691142	1.446879	.8226405	21	39	.5828323	717250	1.394213	.8125925	21
40	.5688011	.691572	1.445980	.8224751	**20**	**40**	.5830687	.717691	1.393357	8124229	**20**
41	.5690403	.692002	1.445081	.8223096	19	41	.5833050	718131	1.392501	8122532	19
42	.5692795	.692432	1.444183	.8221440	18	42	.5835412	.718572	1.391647	.8120835	18
43	5695187	.692863	1.443286	.8219784	17	43	.5837774	.719014	1.390793	8119137	17
44	.5697577	.693293	1.442389	.8218127	16	44	.5840136	719455	1.389940	8117439	16
45	.5699968	.693724	1.441494	.8216469	**15**	**45**	.5842497	719897	1.389087	.8115740	**15**
46	.5702357	.694155	1.440599	.8214811	14	46	.5844857	.720338	1 388235	.8114040	14
47	.5704747	.694586	1.439704	.8213152	13	47	.5847217	.720780	1.387384	8112339	13
48	.5707136	.695018	1.438811	.8211492	12	48	.5849577	.721222	1.386534	8110638	12
49	.5709524	.695449	1.437918	.8209832	11	49	.5851936	721665	1.385684	.8108936	11
50	.5711912	.695881	1.437026	.8208170	**10**	**50**	.5854294	.722107	1 384835	8107234	**10**
51	.5714299	.696313	1.436135	.8206509	9	51	.5856652	.722550	1.383986	.8105530	9
52	.5716686	.696745	1.435245	.8204846	8	52	.5859010	.722993	1.383139	.8103826	8
53	.5719073	.697177	1.434355	.8203183	7	53	.5861367	.723436	1.382292	.8102122	7
54	.5721459	.697609	1.433466	.8201519	6	54	.5863724	723879	1.381445	8100416	6
55	.5723844	.698042	1.432578	.8199854	**5**	**55**	.5866080	.724322	1.380600	8098710	**5**
56	.5726229	.698474	1.431690	.8198189	4	56	.5868435	.724766	1.379755	.8097004	4
57	.5728614	.698907	1.430803	.8196523	3	57	.5870790	725210	1.378910	.8095296	3
58	.5730998	.699340	1.429917	8194856	2	58	.5873145	.725654	1 378067	.8093588	2
59	.5733381	.699774	1.429032	.8193189	1	59	.5875499	726098	1.377224	.8091879	1
60	.5735764	.700207	1.428148	.8191520	**0**	**60**	.5877853	726542	1.376381	.8090170	**0**
	Cosine.	Cotang	Tang.	Sine.	′		Cosine.	Cotang	Tang.	Sine.	′

55° 54°

Note.—Secant = 1 ÷ cosine. Cosecant = 1 ÷ sine.

2.–Natural Sines, TANGENTS, COTANGENTS, COSINES.—(Continued.)

(Versed sine = 1 − cosine; coversed sine = 1 − sine.)

36° | 37°

′	Sine.	Tang.	Cotang.	Cosine.		′	Sine.	Tang.	Cotang.	Cosine.	
0	.5877853	.726542	1.376381	.8090170	**60**	**0**	.6018150	.753554	1.327044	.7986355	**60**
1	.5880206	.726987	1.375540	.8088460	59	1	.6020473	.754010	1.326242	.7984604	59
2	.5882558	.727431	1.374699	.8086749	58	2	.6022795	.754466	1.325439	.7982853	58
3	.5884910	.727876	1.373859	.8085037	57	3	.6025117	.754923	1.324638	.7981100	57
4	.5887262	.728321	1.373019	.8083325	56	4	.6027439	.755379	1.323837	.7979347	56
5	.5889613	.728767	1.372180	.8081612	**55**	**5**	.6029760	.755836	1.323036	.7977594	**55**
6	.5891964	.729212	1.371342	.8079899	54	6	.6032080	.756294	1.322237	.7975839	54
7	.5894314	.729658	1.370504	.8078185	53	7	.6034400	.756751	1.321437	.7974084	53
8	.5896663	.730104	1.369667	.8076470	52	8	.6036719	.757209	1.320639	.7972329	52
9	.5899012	.730550	1.368831	.8074754	51	9	.6039038	.757666	1.319841	.7970572	51
10	.5901361	.730996	1.367995	.8073038	**50**	**10**	.6041356	.758124	1.319044	.7968815	**50**
11	.5903709	.731442	1.367161	.8071321	49	11	.6043674	.758582	1.318247	.7967058	49
12	.5906057	.731889	1.366326	.8069603	48	12	.6045991	.759041	1.317451	.7965299	48
13	.5908404	.732336	1.365493	.8067885	47	13	.6048308	.759499	1.316655	.7963540	47
14	.5910750	.732783	1.364660	.8066166	46	14	.6050624	.759958	1.315861	.7961780	46
15	.5913096	.733230	1.363827	.8064446	**45**	**15**	.6052940	.760417	1.315066	.7960020	**45**
16	.5915442	.733677	1.362996	.8062726	44	16	.6055255	.760876	1.314273	.7958259	44
17	.5917787	.734125	1.362165	.8061005	43	17	.6057570	.761336	1.313480	.7956497	43
18	.5920132	.734573	1.361335	.8059283	42	18	.6059884	.761795	1.312687	.7954735	42
19	.5922476	.735021	1.360505	.8057560	41	19	.6062198	.762255	1.311895	.7952972	41
20	.5924819	.735469	1.359676	.8055837	**40**	**20**	.6064511	.762715	1.311104	.7951208	**40**
21	.5927163	.735917	1.358848	.8054113	39	21	.6066824	.763175	1.310314	.7949444	39
22	.5929505	.736366	1.358020	.8052389	38	22	.6069136	.763636	1.309523	.7947678	38
23	.5931847	.736814	1.357193	.8050664	37	23	.6071447	.764096	1.308734	.7945913	37
24	.5934189	.737263	1.356367	.8048938	36	24	.6073758	.764557	1.307945	.7944146	36
25	.5936530	.737712	1.355541	.8047211	**35**	**25**	.6076069	.765018	1.307157	.7942379	**35**
26	.5938871	.738162	1.354716	.8045484	34	26	.6078379	.765480	1.306369	.7940611	34
27	.5941211	.738611	1.353891	.8043756	33	27	.6080689	.765941	1.305582	.7938843	33
28	.5943550	.739061	1.353068	.8042028	32	28	.6082998	.766403	1.304796	.7937074	32
29	.5945889	.739511	1.352244	.8040299	31	29	.6085306	.766864	1.304010	.7935304	31
30	.5948228	.739961	1.351422	.8038569	**30**	**30**	.6087614	.767327	1.303225	.7933533	**30**
31	.5950566	.740411	1.350600	.8036838	29	31	.6089922	.767789	1.302440	.7931762	29
32	.5952904	.740861	1.349779	.8035107	28	32	.6092229	.768251	1.301656	.7929990	28
33	.5955241	.741312	1.348958	.8033375	27	33	.6094535	.768714	1.300873	.7928218	27
34	.5957577	.741763	1.348139	.8031642	26	34	.6096841	.769177	1.300090	.7926445	26
35	.5959913	.742214	1.347319	.8029909	**25**	**35**	.6099147	.769640	1.299308	.7924671	**25**
36	.5962249	.742665	1.346501	.8028175	24	36	.6101452	.770103	1.298526	.7922896	24
37	.5964584	.743117	1.345683	.8026440	23	37	.6103756	.770567	1.297745	.7921121	23
38	.5966918	.743568	1.344865	.8024705	22	38	.6106060	.771030	1.296964	.7919345	22
39	.5969252	.744020	1.344049	.8022969	21	39	.6108363	.771494	1.296185	.7917569	21
40	.5971586	.744472	1.343233	.8021232	**20**	**40**	.6110666	.771958	1.295405	.7915792	**20**
41	.5973919	.744924	1.342417	.8019495	19	41	.6112969	.772423	1.294627	.7914014	19
42	.5976251	.745377	1.341602	.8017756	18	42	.6115270	.772887	1.293848	.7912235	18
43	.5978583	.745829	1.340788	.8016018	17	43	.6117572	.773352	1.293071	.7910456	17
44	.5980915	.746282	1.339975	.8014278	16	44	.6119873	.773817	1.292294	.7908676	16
45	.5983246	.746735	1.339162	.8012538	**15**	**45**	.6122173	.774282	1.291517	.7906896	**15**
46	.5985577	.747188	1.338350	.8010797	14	46	.6124473	.774748	1.290742	.7905115	14
47	.5987906	.747642	1.337538	.8009056	13	47	.6126772	.775213	1.289966	.7903333	13
48	.5990236	.748095	1.336727	.8007314	12	48	.6129071	.775679	1.289192	.7901550	12
49	.5992565	.748549	1.335917	.8005571	11	49	.6131369	.776145	1.288418	.7899767	11
50	.5994893	.749003	1.335107	.8003827	**10**	**50**	.6133666	.776611	1.287644	.7897983	**10**
51	.5997221	.749457	1.334298	.8002083	9	51	.6135964	.777078	1.286871	.7896198	9
52	.5999549	.749911	1.333490	.8000338	8	52	.6138260	.777544	1.286099	.7894413	8
53	.6001876	.750366	1.332682	.7998593	7	53	.6140556	.778011	1.285327	.7892627	7
54	.6004202	.750821	1.331875	.7996847	6	54	.6142852	.778478	1.284556	.7890841	6
55	.6006528	.751276	1.331068	.7995100	**5**	**55**	.6145147	.778946	1.283786	.7889054	**5**
56	.6008854	.751731	1.330262	.7993352	4	56	.6147442	.779413	1.283016	.7887266	4
57	.6011179	.752186	1.329457	.7991604	3	57	.6149736	.779881	1.282246	.7885477	3
58	.6013503	.752642	1.328652	.7989855	2	58	.6152029	.780349	1.281477	.7883688	2
59	.6015827	.753098	1.327848	.7988105	1	59	.6154322	.780817	1.280709	.7881898	1
60	.6018150	.753554	1.327044	.7986355	**0**	**60**	.6156615	.781285	1.279941	.7880108	**0**
	Cosine.	Cotang	Tang.	Sine.	′		Cosine.	Cotang	Tang.	Sine.	′

53° | 52°

Note.—Secant = 1 ÷ cosine. Cosecant = 1 ÷ sine.

2.—**Natural Sines,** TANGENTS, COTANGENTS, COSINES.—(Continued.)

(Versed sine = 1 − cosine; coversed sine = 1 − sine.)

38°

'	Sine.	Tang.	Cotang.	Cosine.	
0	.6156615	.781285	1.279941	.7880108	**60**
1	.6158907	.781754	1.279174	.7878316	59
2	.6161198	.782222	1.278407	.7876524	58
3	.6163489	.782691	1.277641	.7874732	57
4	.6165780	.783161	1.276876	.7872939	56
5	.6168069	.783630	1.276111	.7871145	**55**
6	.6170359	.784100	1.275347	7869350	54
7	.6172648	.784570	1 274583	7867555	53
8	.6174936	.785040	1 273820	.7865759	52
9	.6177224	.785510	1.273057	.7863963	51
10	6179511	.785980	1.272295	.7862165	**50**
11	.6181798	.786451	1.271534	.7860367	49
12	.6184084	786922	1.270773	.7858569	48
13	.6186370	.787393	1 270013	.7856770	47
14	.6188655	.787864	1.269253	.7854970	46
15	.6190939	.788336	1.268494	.7853169	**45**
16	.6193224	.788808	1 267735	.7851368	44
17	.6195507	.789280	1.266977	.7849566	43
18	.6197790	.789752	1.266219	.7847764	42
19	.6200073	.790224	1.265462	.7845961	41
20	.6202355	.790697	1.264706	.7844157	**40**
21	.6204636	.791170	1.263950	.7842352	39
22	.6206917	.791643	1.263195	.7840547	38
23	.6209198	.792116	1.262440	.7838741	37
24	.6211478	.792590	1.261686	.7836935	36
25	.6213757	.793064	1.260932	7835127	**35**
26	.6216036	.793537	1.260179	.7833320	34
27	.6218314	.794012	1.259426	.7831511	33
28	.6220592	.794486	1.258674	.7829702	32
29	.6222870	.794961	1.257923	.7827892	31
30	.6225146	.795435	1.257172	.7826082	**30**
31	.6227423	.795911	1.256421	.7824270	29
32	.6229698	.796386	1.255672	.7822459	28
33	.6231974	.796861	1.254922	.7820646	27
34	.6234248	.797337	1.254174	.7818833	26
35	.6236522	.797813	1.253426	.7817019	**25**
36	.6238796	.798289	1.252678	.7815205	24
37	.6241069	.798765	1.251931	.7813390	23
38	.6243342	.799242	1.251184	.7811574	22
39	.6245614	.799719	1.250438	.7809757	21
40	.6247885	.800196	1.249693	.7807940	**20**
41	.6250156	.800673	1.248948	.7806123	19
42	.6252427	.801151	1.248204	.7804304	18
43	6254696	.801628	1.247460	.7802485	17
44	.6256966	.802106	1.246716	.7800665	16
45	.6259235	.802584	1.245974	.7798845	**15**
46	.6261503	.803063	1.245232	.7797024	14
47	.6263771	.803541	1.244490	.7795202	13
48	.6266038	.804020	1 243749	.7793380	12
49	.6268305	.804499	1.243008	.7791557	11
50	.6270571	804979	1 242268	.7789733	**10**
51	.6272837	.805458	1 241529	.7787909	9
52	6275102	.805938	1 240790	.7786084	8
53	.6277366	.806418	1.240051	7784258	7
54	.6279631	.806898	1 239313	.7782431	6
55	.6281894	.807378	1.238576	7780604	**5**
56	.6284157	.807859	1 237839	.7778777	4
57	.6286420	.808340	1.237103	.7776949	3
58	.6288682	.808821	1.236367	.7775120	2
59	.6290943	.809302	1.235631	.7773290	1
60	.6293204	809784	1.234897	.7771460	**0**
	Cosine.	Cotang	Tang.	Sine.	'

51°

39°

'	Sine.	Tang.	Cotang.	Cosine.	
0	.6293204	.809784	1.234897	.7771460	**60**
1	.6295464	.810265	1.234162	.7769629	59
2	.6297724	.810747	1.233429	.7767797	58
3	.6299983	.811230	1.232696	.7765965	57
4	.6302242	.811712	1.231963	.7764132	56
5	.6304500	.812195	1.231231	.7762298	**55**
6	6306758	.812678	1.230499	.7760464	54
7	.6309015	.813161	1.229768	.7758629	53
8	.6311272	.813644	1.229038	.7756794	52
9	.6313528	.814128	1.228308	.7754957	51
10	.6315784	.814611	1.227578	.7753121	**50**
11	.6318039	.815095	1.226849	.7751283	49
12	.6320293	.815580	1.226121	.7749445	48
13	.6322547	.816064	1.225393	.7747606	47
14	.6324800	.816549	1.224665	.7745767	46
15	.6327053	.817034	1.223938	.7743926	**45**
16	.6329306	.817519	1.223212	.7742086	44
17	.6331557	.818004	1.222486	.7740244	43
18	.6333809	.818490	1.221761	.7738402	42
19	.6336059	.818976	1.221036	.7736559	41
20	.6338310	.819462	1.220312	.7734716	**40**
21	.6340559	.819948	1.219588	.7732872	39
22	.6342808	.820435	1.218865	.7731027	38
23	.6345057	.820922	1.218142	.7729182	37
24	.6347305	.821409	1.217419	.7727336	36
25	.6349553	.821896	1.216698	.7725489	**35**
26	.6351800	.822384	1.215976	.7723642	34
27	.6354046	.822871	1.215256	.7721794	33
28	.6356292	.823359	1.214535	.7719945	32
29	.6358537	.823847	1.213816	.7718096	31
30	.6360782	.824336	1.213097	.7716246	**30**
31	.6363026	824825	1.212378	.7714395	29
32	.6365270	.825314	1.211660	.7712544	28
33	.6367513	.825803	1.210942	.7710692	27
34	.6369756	.826292	1.210225	.7708840	26
35	.6371998	.826782	1.209508	.7706986	**25**
36	.6374240	.827271	1.208792	.7705132	24
37	.6376481	.827762	1.208076	.7703278	23
38	.6378721	.828252	1.207361	.7701423	22
39	.6380961	.828742	1 206646	.7699567	21
40	.6383201	.829233	1.205932	.7697710	**20**
41	.6385440	.829724	1.205219	7695853	19
42	.6387678	.830216	1.204505	.7693996	18
43	.6389916	.830707	1.203793	.7692137	17
44	.6392153	.831199	1.203081	.7690278	16
45	.6394390	.831691	1.202369	.7688418	**15**
46	.6396626	.832183	1.201658	.7686558	14
47	.6398862	.832675	1.200947	.7684697	13
48	.6401097	.833168	1.200237	.7682835	12
49	.6403332	.833661	1 199527	.7680973	11
50	.6405566	834154	1 198818	.7679110	**10**
51	.6407799	.834648	1 198109	.7677246	9
52	.6410032	835141	1 197401	7675382	8
53	.6412264	835635	1.196693	.7673517	7
54	6414496	.836129	1 195986	.7671652	6
55	.6416728	.836624	1.195279	7669785	**5**
56	.6418958	.837118	1.194573	.7667918	4
57	.6421189	837613	1 193867	7666051	3
58	.6423418	838108	1.193162	7664183	2
59	.6425647	.838604	1 192457	7662314	1
60	.6427876	839099	1.191753	.7660444	**0**
	Cosine.	Cotang	Tang.	Sine.	'

50°

Note.—Secant = 1 ÷ cosine. Cosecant = 1 ÷ sine

2.—**Natural Sines,** Tangents, Cotangents, Cosines.—**(Continued.)**

(Versed sine = 1 − cosine; coversed sine = 1 − sine.)

40°						41°					
′	Sine.	Tang.	Cotang.	Cosine.		′	Sine.	Tang.	Cotang.	Cosine.	
0	.6427876	.839099	1.191753	.7660444	**60**	**0**	.6560590	.869286	1.150368	.7547096	**60**
1	.6430104	.839595	1.191049	.7658574	59	1	.6562785	.869797	1.149692	.7545187	59
2	.6432332	.840091	1.190346	.7656704	58	2	.6564980	.870308	1.149017	.7543278	58
3	.6434559	.840587	1.189643	.7654832	57	3	.6567174	.870820	1.148342	.7541368	57
4	.6436785	.841084	1.188941	.7652960	56	4	.6569367	.871331	1.147668	.7539457	56
5	.6439011	.841581	1.188239	.7651087	**55**	**5**	.6571560	.871843	1.146994	.7537546	**55**
6	.6441236	.842078	1.187538	.7649214	54	6	.6573752	.872355	1.146321	.7535634	54
7	.6443461	.842575	1.186837	.7647340	53	7	.6575944	.872868	1.145648	.7533721	53
8	.6445685	.843073	1.186136	.7645465	52	8	.6578135	.873380	1.144976	.7531808	52
9	.6447909	.843570	1.185437	.7643590	51	9	.6580326	.873893	1.144304	.7529894	51
10	.6450132	.844068	1.184737	.7641714	**50**	**10**	.6582516	.874406	1.143632	.7527980	**50**
11	.6452355	.844567	1.184038	.7639838	49	11	.6584706	.874920	1.142961	.7526065	49
12	.6454577	.845065	1.183340	.7637960	48	12	.6586895	.875433	1.142290	.7524149	48
13	.6456798	.845564	1.182642	.7636082	47	13	.6589083	.875947	1.141620	.7522233	47
14	.6459019	.846063	1.181944	.7634204	46	14	.6591271	.876462	1.140950	.7520316	46
15	.6461240	.846562	1.181247	.7632325	**45**	**15**	.6593458	.876976	1.140281	.7518398	**45**
16	.6463460	.847062	1.180551	.7630445	44	16	.6595645	.877491	1.139612	.7516480	44
17	.6465679	.847561	1.179855	.7628561	43	17	.6597831	.878006	1.138944	.7514561	43
18	.6467898	.848061	1.179159	.7626683	42	18	.6600017	.878521	1.138276	.7512641	42
19	.6470116	.848561	1.178464	.7624802	41	19	.6602202	.879037	1.137608	.7510721	41
20	.6472334	.849062	1.177769	.7622919	**40**	**20**	.6604386	.879552	1.136941	.7508800	**40**
21	.6474551	.849563	1.177075	.7621036	39	21	.6606570	.880068	1.136274	.7506879	39
22	.6476767	.850064	1.176382	.7619152	38	22	.6608754	.880585	1.135608	.7504957	38
23	.6478984	.850565	1.175688	.7617268	37	23	.6610936	.881101	1.134942	.7503034	37
24	.6481199	.851066	1.174996	.7615383	36	24	.6613:19	.881618	1.134277	.7501111	36
25	.6483414	.851568	1.174303	.7613497	**35**	**25**	.6615300	.882135	1.133612	.7499187	**35**
26	.6485628	.852070	1.173612	.7611611	34	26	.6617482	.882653	1.132947	.7497262	34
27	.6487842	.852572	1.172920	.7609724	33	27	.6619662	.883170	1.132283	.7495337	33
28	.6490056	.853075	1.172229	.7607837	32	28	.6621842	.883688	1.131620	.7493411	32
29	.6492268	.853577	1.171539	.7605949	31	29	.6624022	.884206	1.130957	.7491484	31
30	.6494480	.854080	1.170849	.7604060	**30**	**30**	.6626200	.884725	1.130294	.7489557	**30**
31	.6496692	.854583	1.170160	.7602170	29	31	.6628379	.885244	1.129632	.7487629	29
32	.6498903	.855087	1.169471	.7600280	28	32	.6630557	.885763	1.128970	.7485701	28
33	.6501114	.855591	1.168782	.7598389	27	33	.6632734	.886282	1.128308	.7483772	27
34	.6503324	.856095	1.168094	.7596498	26	34	.6634910	.886801	1.127647	.7481842	26
35	.6505533	.856599	1.167407	.7594606	**25**	**35**	.6637087	.887321	1.126987	.7479912	**25**
36	.6507742	.857103	1.166720	.7592713	24	36	.6639262	.887841	1.126327	.7477981	24
37	.6509951	.857608	1.166033	.7590820	23	37	.6641437	.888361	1.125667	.7476049	23
38	.6512158	.858113	1.165347	.7588926	22	38	.6643612	.888882	1.125008	.7474117	22
39	.6514366	.858618	1.164661	.7587031	21	39	.8645785	.889403	1.124349	.7472184	21
40	.6516572	.859124	1.163976	.7585136	**20**	**40**	.6647959	.889924	1.123690	.7470251	**20**
41	.6518778	.859629	1.163291	.7583240	19	41	.6650131	.890445	1.123032	.7468317	19
42	.6520984	.860135	1.162607	.7581343	18	42	.6652304	.890967	1.122375	.7466382	18
43	.6523189	.860641	1.161923	.7579446	17	43	.6654475	.891489	1.121718	.7464446	17
44	.6525394	.861148	1.161240	.7577548	16	44	.6656646	.892011	1.121061	.7462510	16
45	.6527598	.861655	1.160557	.7575650	**15**	**45**	.6658817	.892534	1.120405	.7460574	**15**
46	.6529801	.862162	1.159874	.7573751	14	46	.6660987	.893056	1.119749	.7458636	14
47	.6532004	.862669	1.159192	.7571851	13	47	.6663156	.893579	1.119094	.7456699	13
48	.6534206	.863176	1.158511	.7569951	12	48	.6665325	.894103	1.118439	.7454760	12
49	.6536408	.863684	1.157830	.7568050	11	49	.6667493	.894626	1.117784	.7452821	11
50	.6538609	.864192	1.157149	.7566148	**10**	**50**	.6669661	.895150	1.117130	.7450881	**10**
51	.6540810	.864700	1.156469	.7564246	9	51	.6671828	.895674	1.116476	.7448941	9
52	.6543010	.865209	1.155789	.7562343	8	52	.6673994	.896199	1.115823	.7446999	8
53	.6545209	.865718	1.155110	.7560439	7	53	.6676160	.896723	1.115170	.7445058	7
54	.6547408	.866227	1.154431	.7558535	6	54	.6678326	.897248	1.114518	.7443115	6
55	.6549607	.866736	1.153753	.7556630	**5**	**55**	.6680490	.897773	1.113866	.7441173	**5**
56	.6551804	.867246	1.153075	.7554724	4	56	.6682655	.898299	1.113214	.7439229	4
57	.6554002	.867755	1.152397	.7552818	3	57	.6684818	.898825	1.112563	.7437285	3
58	.6556198	.868265	1.151721	.7550911	2	58	.6686981	.899351	1.111912	.7435340	2
59	.6558395	.868776	1.151044	.7549004	1	59	.6889144	.899877	1.111262	.7433394	1
60	.6560590	.869286	1.150368	.7547096	**0**	**60**	.6691306	.900404	1.110612	.7431448	**0**
	Cosine.	Cotang	Tang.	Sine.	′		Cosine.	Cotang	Tang.	Sine.	′
					49°						48°

Note.—Secant = 1 ÷ cosine. Cosecant = 1 ÷ sine.

2.—Natural Sines, Tangents, Cotangents, Cosines.—(Continued.)

(Versed sine = 1 − cosine; coversed sine = 1 − sine.)

42°

′	Sine.	Tang.	Cotang.	Cosine.	
0	.6691306	.900404	1.110612	.7431448	**60**
1	.6693468	.900930	1.109963	.7429502	59
2	.6695628	.901458	1.109314	.7427554	58
3	.6697789	.901985	1.108665	.7425606	57
4	.6699948	.902513	1.108017	.7423658	56
5	.6702108	.903041	1.107369	.7421708	**55**
6	.6704266	.903569	1.106721	.7419758	54
7	.6706424	.904097	1.106075	.7417808	53
8	.6708582	.904626	1.105428	.7415857	52
9	.6710739	.905155	1.104782	.7413905	51
10	.6712895	.905685	1.104136	.7411953	**50**
11	.6715051	.906214	1.103491	.7410000	49
12	.6717206	.906744	1.102846	.7408046	48
13	.6719361	.907274	1.102201	.7406092	47
14	.6721515	.907805	1.101557	.7404137	46
15	.6723668	.908336	1.100914	.7402181	**45**
16	.6725821	.908867	1.100270	.7400225	44
17	.6727973	.909398	1.099628	.7398268	43
18	.6730125	.909930	1.098985	.7396311	42
19	.6732276	.910461	1.098343	.7394353	41
20	.6734427	.910994	1.097702	.7392394	**40**
21	.6736577	.911526	1.097060	.7390435	39
22	.6738727	.912059	1.096420	.7388475	38
23	.6740876	.912592	1.095779	.7386515	37
24	.6743024	.913125	1.095139	.7384553	36
25	.6745172	.913659	1.094500	.7382592	**35**
26	.6747319	.914192	1.093861	.7380629	34
27	.6749466	.914727	1.093222	.7378666	33
28	.6751612	.915261	1.092584	.7376703	32
29	.6753757	.915796	1.091946	.7374738	31
30	.6755902	.916331	1.091308	.7372773	**30**
31	.6758046	.916866	1.090671	.7370808	29
32	.6760190	.917402	1.090034	.7368842	28
33	.6762333	.917937	1.089398	.7366875	27
34	.6764476	.918474	1.088762	.7364908	26
35	.6766618	.919010	1.088126	.7362940	**25**
36	.6768760	.919547	1.087491	.7360971	24
37	.6770901	.920084	1.086857	.7359002	23
38	.6773041	.920621	1.086222	.7357032	22
39	.6775181	.921159	1.085588	.7355061	21
40	.6777320	.921696	1.084955	.7353090	**20**
41	.6779459	.922235	1.084322	.7351118	19
42	.6781597	.922773	1.083689	.7349146	18
43	.6783734	.923312	1.083057	.7347173	17
44	.6785871	.923851	1.082425	.7345199	16
45	.6788007	.924391	1.081793	.7343225	**15**
46	.6790143	.924930	1.081162	.7341250	14
47	.6792278	.925470	1.080532	.7339275	13
48	.6794413	.926010	1.079901	.7337299	12
49	.6796547	.926550	1.079271	.7335322	11
50	.6798681	.927091	1.078642	.7333345	**10**
51	.6800813	.927632	1.078013	.7331367	9
52	.6802946	.928173	1.077384	.7329388	8
53	.6805078	.928715	1.076756	.7327409	7
54	.6807209	.929257	1.076128	.7325429	6
55	.6809339	.929799	1.075500	.7323449	**5**
56	.6811469	.930342	1.074873	.7321467	4
57	.6813599	.930884	1.074246	.7319486	3
58	.6815728	.931428	1.073620	.7317503	2
59	.6817856	.931971	1.072994	.7315521	1
60	.6819984	.932515	1.072368	.7313537	**0**
	Cosine.	Cotang	Tang.	Sine.	′

47°

43°

′	Sine.	Tang.	Cotang.	Cosine.	
0	.6819984	.932515	1.072368	.7313537	**60**
1	.6822111	.933059	1.071743	.7311553	59
2	.6824237	.933603	1.071118	.7309568	58
3	.6826363	.934147	1.070494	.7307583	57
4	.6828489	.934692	1.069870	.7305597	56
5	.6830613	.935238	1.069246	.7303610	**55**
6	.6832738	.935783	1.068623	.7301623	54
7	.6834861	.936329	1.068000	.7299635	53
8	.6836984	.936875	1.067377	.7297646	52
9	.6839107	.937421	1.066755	.7295657	51
10	.6841229	.937968	1.066134	.7293668	**50**
11	.6843350	.938515	1.065512	.7291677	49
12	.6845471	.939062	1.064891	.7289686	48
13	.6847591	.939610	1.064271	.7287695	47
14	.6849711	.940157	1.063651	.7285703	46
15	.6851830	.940706	1.063031	.7283710	**45**
16	.6853948	.941254	1.062411	.7281716	44
17	.6856066	.941803	1.061792	.7279722	43
18	.6858184	.942352	1.061174	.7277728	42
19	.6860300	.942901	1.060556	.7275732	41
20	.6862416	.943451	1.059938	.7273736	**40**
21	.6864532	.944001	1.059320	.7271740	39
22	.6866647	.944551	1.058703	.7269743	38
23	.6868761	.945102	1.058086	.7267745	37
24	.6870875	.945653	1.057470	.7265747	36
25	.6872988	.946204	1.056854	.7263748	**35**
26	.6875101	.946755	1.056238	.7261748	34
27	.6877213	.947307	1.055623	.7259748	33
28	.6879325	.947859	1.055008	.7257747	32
29	.6881435	.948411	1.054394	.7255746	31
30	.6883546	.948964	1.053780	.7253744	**30**
31	.6885655	.949517	1.053166	.7251741	29
32	.6887765	.950070	1.052553	.7249738	28
33	.6889873	.950624	1.051940	.7247734	27
34	.6891981	.951178	1.051327	.7245729	26
35	.6894089	.951732	1.050715	.7243724	**25**
36	.6896195	.952287	1.050103	.7241719	24
37	.6898302	.952842	1.049492	.7239712	23
38	.6900407	.953397	1.048880	.7237705	22
39	.6902512	.953952	1.048270	.7235698	21
40	.6904617	.954508	1.047659	.7233690	**20**
41	.6906721	.955064	1.047049	.7231681	19
42	.6908824	.955620	1.046440	.7229671	18
43	.6910927	.956177	1.045831	.7227661	17
44	.6913029	.956734	1.045222	.7225651	16
45	.6915131	.957291	1.044613	.7223640	**15**
46	.6917232	.957849	1.044005	.7221628	14
47	.6919332	.958407	1.043397	.7219615	13
48	.6921432	.958965	1.042790	.7217602	12
49	.6923531	.959524	1.042183	.7215589	11
50	.6925630	.960082	1.041576	.7213574	**10**
51	.6927728	.960642	1.040970	.7211559	9
52	.6929825	.961201	1.040364	.7209544	8
53	.6931922	.961761	1.039758	.7207528	7
54	.6934018	.962321	1.039153	.7205511	6
55	.6936114	.962881	1.038548	.7203494	**5**
56	.6938209	.963442	1.037944	.7201476	4
57	.6940304	.964003	1.037340	.7199457	3
58	.6942398	.964565	1.036736	.7197438	2
59	.6944491	.965126	1.036133	.7195418	1
60	.6946584	.965688	1.035530	.7193398	**0**
	Cosine.	Cotang	Tang.	Sine.	′

46°

Note.—Secant = 1 ÷ cosine. Cosecant = 1 ÷ sine.

2.—**Natural Sines,** TANGENTS, COTANGENTS, COSINES.—(Concluded.)

(Versed sine = 1 − cosine; coversed sine = 1 − sine.)

44° 44°

′	Sine.	Tang.	Cotang.	Cosine.		′	Sine.	Tang.	Cotang.	Cosine.	
0	.6946584	.965688	1.035530	.7193398	**60**	**30**	.7009093	.982697	1.017607	.7132504	**30**
1	.6948676	.966251	1.034927	.7191377	59	31	.7011167	.983269	1.017015	.7130465	29
2	.6950767	.966813	1.034325	.7189355	58	32	.7013241	.983841	1.016423	.7128426	28
3	.6952858	.967376	1.033723	.7187333	57	33	.7015314	.984414	1.015832	7126385	27
4	.6954949	.967939	1.033122	.7185310	56	34	.7017387	.984987	1.015241	.7124344	26
5	.6957039	.968503	1.032520	.7183287	**55**	**35**	.7019459	.985560	1 014651	7122303	**25**
6	.6959128	.969067	1.031919	.7181263	54	36	.7021531	.986133	1 014061	7120260	24
7	.6961217	.969631	1.031319	.7179238	53	37	.7023601	.986707	1 013471	.7118218	23
8	.6963305	.970196	1 030719	.7177213	52	38	.7025672	.987282	1 012881	.7116174	22
9	.6965392	.970761	1.030119	.7175187	51	39	.7027741	.987856	1 012292	7114130	21
10	.6967479	.971326	1.029520	.7173161	**50**	**40**	7029811	.988431	1 011703	.7112086	**20**
11	.6969565	.971891	1.028921	.7171134	49	41	.7031879	.989006	1 011115	7110041	19
12	6971651	.972457	1.028322	.7169106	48	42	7033947	.989582	1 010527	7107995	18
13	.6973736	.973023	1.027724	.7167078	47	43	.7036014	.990158	1 009939	7105948	17
14	.6975821	.973590	1.027126	.7165049	46	44	.7038081	.990734	1 009352	.7103901	16
15	.6977905	.974156	1.026528	.7163019	**45**	**45**	7040147	.991311	1 008764	.7101854	**15**
16	.6979988	.974724	1.025931	.7160989	44	46	.7042213	.991888	1 008178	.7099806	14
17	.6982071	.975291	1.025334	.7158959	43	47	.7044278	.992465	1 007591	.7097757	13
18	.6984153	.975859	1.024738	.7156927	42	48	.7046342	.993042	1 007005	7095707	12
19	.6986234	.976427	1.024141	.7154895	41	49	.7048406	.993620	1 006420	.7093657	11
20	.6988315	.976995	1.023546	.7152863	**40**	**50**	.7050469	.994199	1 005834	.7091607	**10**
21	.6990396	.977564	1.022950	.7150830	39	51	.7052532	.994777	1 005249	.7089556	9
22	.6992476	.978133	1.022355	.7148796	38	52	.7054594	.995356	1 004665	7087504	8
23	.6994555	.978702	1.021760	.7146762	37	53	.7056655	.995935	1 004080	.7085451	7
24	.6996633	.979272	1.021166	.7144727	36	54	.7058716	.996515	1.003496	.7083398	6
25	.6998711	.979842	1.020572	.7142691	**35**	**55**	.7060776	.997095	1.002913	.7081345	**5**
26	.7000789	.980412	1.019978	.7140655	34	56	.7062835	.997675	1.002329	.7079291	4
27	.7002866	.980983	1.019385	.7138618	33	57	.7064894	.998256	1.001746	.7077236	3
28	.7004942	.981554	1.018792	.7136581	32	58	.7066953	.998837	1.001164	.7075180	2
29	.7007018	.982125	1.018199	.7134543	31	59	.7069011	.999418	1 000581	.7073124	1
30	.7009093	.982697	1.017607	.7132504	**30**	**60**	.7071068	1.00000	1.000000	.7071068	**0**
	Cosine.	Cotang	Tang.	Sine.	′		Cosine.	Cotang	Tang.	Sine.	′

45° 45°

Note.—Secant = 1 ÷ cosine. Cosecant = 1 ÷ sine.

TABLE 3

Table of Logarithmic Sines

3.—Logarithmic Sines, Tangents, Cotangents, Cosines. (Secants, Cosecants.)*

0°						1°					
'	Sine.	Tang.	Cotang.	Cosine.		'	Sine.	Tang.	Cotang.	Cosine.	
0	Inf. Neg.	Inf. Neg.	Infinite.	10.00000	**60**	**0**	8.24186	8.24192	11.75808	9.99993	**60**
1	6.46373	6.46373	13.53627	.00000	59	1	.24903	.24910	.75090	.99993	59
2	.76476	.76476	.23524	.00000	59	2	.25609	.25616	.74384	.99993	58
3	6.94085	6.94085	13.05915	.00000	57	3	.26304	.26312	73688	.99993	57
4	7.06579	7.06579	12.93421	.00000	56	4	.26988	.26996	.73004	.99992	56
5	7.16270	7.16270	12.83730	10.00000	**55**	**5**	8.27661	8.27669	11.72331	9.99992	**55**
6	.24188	24188	.75812	.00000	54	6	.28324	.28332	.71668	.99992	54
7	.30882	30882	69118	.00000	53	7	.28977	.28986	.71014	.99992	53
8	.36682	.36682	.63318	.00000	52	8	.29621	.29629	.70371	.99992	52
9	.41797	.41797	.58203	.00000	51	9	.30255	.30263	.69737	.99991	51
10	7.46373	7.46373	12.53627	10.00000	**50**	**10**	8.30879	8.30888	11.69112	9.99991	**50**
11	.50512	.50512	.49488	.00000	49	11	.31495	.31505	.68495	.99991	49
12	54291	54291	.45709	00000	48	12	.32103	.32112	.67888	.99990	48
13	.57767	.57767	.42233	.00000	47	13	.32702	.32711	.67289	.99990	47
14	.60985	.60986	.39014	.00000	46	14	.33292	.33302	.66698	.99990	46
15	7.63982	7.63982	12.36018	10.00000	**45**	**15**	8.33875	8.33886	11.66114	9.99990	**45**
16	.66784	.66785	.33215	10.00000	44	16	.34450	.34461	.65539	.99989	44
17	.69417	.69418	.30582	9.99999	43	17	.35018	.35029	.64971	.99989	43
18	.71900	.71900	.28100	.99999	42	18	.35578	.35590	.64410	.99989	42
19	.74248	.74248	.25752	.99999	41	19	.36131	.36143	.63857	.99989	41
20	7.76475	7.76476	12.23524	9.99999	**40**	**20**	8.36678	8.36689	11.63311	9.99988	**40**
21	.78594	.78595	.21405	.99999	39	21	.37217	.37229	.62771	.99988	39
22	.80615	.80615	.19385	.99999	38	22	.37750	.37762	.62238	.99988	38
23	.82545	.82546	.17454	.99999	37	23	.38276	.38289	.61711	.99987	37
24	.84393	.84394	.15606	.99999	36	24	.38796	.38809	.61191	.99987	36
25	7.86166	7.86167	12.13833	9.99999	**35**	**25**	8.39310	8.39323	11.60677	9.99987	**35**
26	.87870	.87871	.12129	.99999	34	26	.39818	.39832	.60168	.99986	34
27	.89509	.89510	.10490	.99999	33	27	.40320	.40334	.59666	.99986	33
28	.91088	.91089	.08911	.99999	32	28	.40816	.40830	.59170	.99986	32
29	.92612	.92613	.07387	.99998	31	29	.41307	.41321	.58679	.99985	31
30	7.94084	7.94086	12.05914	9.99998	**30**	**30**	8.41792	8.41807	11.58193	9.99985	**30**
31	.95508	.95510	.04490	.99998	29	31	.42272	.42287	.57713	.99985	29
32	.96887	.96889	.03111	.99998	28	32	.42746	.42762	.57238	.99984	28
33	.98223	.98225	.01775	.99998	27	33	.43216	.43232	.56768	.99984	27
34	7.99520	7.99522	12.00478	.99998	26	34	.43680	.43696	.56304	.99984	26
35	8.00779	8.00781	11.99219	9.99998	**25**	**35**	8.44139	8.44156	11.55844	9.99983	**25**
36	.02002	.02004	.97996	.99998	24	36	.44594	.44611	.55389	.99983	24
37	.03192	.03194	.96806	.99997	23	37	.45044	.45061	.54939	.99983	23
38	.04350	.04353	95647	.99997	22	38	.45489	.45507	.54493	.99982	22
39	.05478	.05481	.94519	.99997	21	39	.45930	.45948	.54052	.99982	21
40	8.06578	8.06581	11.93419	9.99997	**20**	**40**	8.46366	8.46385	11.53615	9.99982	**20**
41	.07650	.07653	.92347	9°997	19	41	.46799	.46817	.53183	.99981	19
42	.08696	.08700	.91300	.99997	18	42	.47226	.47245	.52755	.99981	18
43	.09718	.09722	.90278	.99997	17	43	.47650	.47669	.52331	.99981	17
44	.10717	.10720	.89280	.99996	16	44	.48069	.48089	.51911	.99980	16
45	8.11693	8.11696	11.88304	9.99996	**15**	**45**	8.48485	8.48505	11.51495	9.99980	**15**
46	.12647	.12651	.87349	.99996	14	46	.48896	.48917	.51083	.99979	14
47	.13581	.13585	.86415	.99996	13	46	.49304	.49325	.50675	.99979	13
48	.14495	.14500	.85500	.99996	12	48	.49708	.49729	.50271	.99979	12
49	.15391	.15395	.84605	.99996	11	49	.50108	.50130	.49870	.99978	11
50	8.16268	8.16273	11.83727	9.99995	**10**	**50**	8.50504	8.50527	11.49473	9.99978	**10**
51	.17128	.17133	.82867	.99995	9	51	.50897	.50920	.49080	.99977	9
52	.17971	.17976	82024	.99995	8	52	.51287	.51310	.48690	.99977	8
53	.18798	.18804	.81196	.99995	7	53	.51673	.51696	.48304	.99977	7
54	.19610	.19616	80384	.99995	6	54	.52055	.52079	.47921	.99976	6
55	8.20407	8.20413	11.79587	9.99994	**5**	**55**	8.52434	8.52459	11.47541	9.99976	**5**
56	.21189	.21195	.78805	.99994	4	56	.52810	.52835	.47165	.99975	4
57	.21958	21964	.78036	.99994	3	57	.53183	.53208	.46792	.99975	3
58	.22713	.22720	.77280	.99994	2	58	.53552	.53578	.46422	.99974	2
59	.23456	.23462	76538	.99994	1	59	.53919	.53945	.46055	.99974	1
60	8.24186	8.24192	11.75808	9.99993	**0**	**60**	8.54282	8.54308	11.45692	9.99974	**0**
	Cosine.	Cotang	Tang.	Sine.	'		Cosine.	Cotang.	Tang.	Sine.	'
					89°						88°

*Log secant = colog cosine = 1 − log cosine; log cosecant = colog sine = 1 − log sine.

Ex.—Log sec 0°− 30′ = 10.00002. *Ex.*—Log cosec 0°− 30′ = 12.05916.

3.—Logarithmic Sines, Tangents, Cotangents, Cosines.
(Secants, Cosecants.) *—(Cont'd.)

2°

′	Sine.	Tang.	Cotang.	Cosine.	
0	8.54282	8.54308	11.45692	9.99974	**60**
1	.54642	.54669	.45331	.99973	59
2	.54999	.55027	.44973	.99973	58
3	.55354	.55382	.44618	.99972	57
4	.55705	.55734	.44266	.99972	56
5	8.56054	8.56083	11.43917	9.99971	**55**
6	.56400	.56429	.43571	.99971	54
7	.56743	.56773	.43227	.99970	53
8	.57084	.57114	.42886	.99970	52
9	.57421	.57452	.42548	.99969	51
10	8.57757	8.57788	11.42212	9.99969	**50**
11	.58089	.58121	.41879	.99968	49
12	.58419	.58451	.41549	.99968	48
13	.58747	.58779	.41221	.99967	47
14	.59072	.59105	.40895	.99967	46
15	8.59395	8.59428	11.40572	9.99967	**45**
16	.59715	.59749	.40251	.99966	44
17	.60033	.60068	.39932	.99966	43
18	.60349	.60384	.39616	.99965	42
19	.60662	.60698	.39302	.99964	41
20	8.60973	8.61009	11.38991	9.99964	**40**
21	.61282	.61319	.38681	.99963	39
22	.61589	.61626	.38374	.99963	38
23	.61894	.61931	.38069	.99962	37
24	.62196	.62234	.37766	.99962	36
25	8.62497	8.62535	11.37465	9.99961	**35**
26	.62795	.62834	.37166	.99961	34
27	.63091	.63131	.36869	.99960	33
28	.63385	.63426	.36574	.99960	32
29	.63678	.63718	.36282	.99959	31
30	8.63968	8.64009	11.35991	9.99959	**30**
31	.64256	.64298	.35702	.99958	29
32	.64543	.64585	.35415	.99958	28
33	.64827	.64870	.35130	.99957	27
34	.65110	.65154	.34846	.99956	26
35	8.65391	8.65435	11.34565	9.99956	**25**
36	.65670	.65715	.34285	.99955	24
37	.65947	.65993	.34007	.99955	23
38	.66223	.66269	.33731	.99954	22
39	.66497	.66543	.33457	.99954	21
40	8.66769	8.66816	11.33184	9.99953	**20**
41	.67039	.67087	.32913	.99952	19
42	.67308	.67356	.32644	.99952	18
43	.67575	.67624	.32376	.99951	17
44	.67841	.67890	.32110	.99951	16
45	8.68104	8.68154	11.31846	9.99950	**15**
46	.68367	.68417	.31583	.99949	14
47	.68627	.68678	.31322	.99949	13
48	.68886	.68938	.31062	.99948	12
49	.69144	.69196	.30804	.99948	11
50	8.69400	8.69453	11.30547	9.99947	**10**
51	.69654	.69708	.30292	.99946	9
52	.69907	.69962	.30038	.99946	8
53	70159	.70214	.29786	.99945	7
54	.70409	.70465	.29535	.99944	6
55	8.70658	8.70714	11.29286	9.99944	**5**
56	.70905	.70962	.29038	.99943	4
57	.71151	.71208	.28792	.99942	3
58	.71395	.71453	.28547	.99942	2
59	.71638	.71697	.28303	.99941	1
60	8.71880	8.71940	11.28060	9.99940	**0**
	Cosine.	Cotang.	Tang.	Sine.	′

87°

3°

′	Sine.	Tang.	Cotang.	Cosine.	
0	8.71880	8.71940	11.28060	9.99940	**60**
1	.72120	.72181	.27819	.99940	59
2	.72359	.72420	.27580	.99939	58
3	.72597	.72659	.27341	.99938	57
4	.72834	.72896	.27104	.99938	56
5	8.73069	8.73132	11.26868	9.99937	**55**
6	.73303	.73366	.26634	.99936	54
7	.73535	.73600	.26400	.99936	53
8	.73767	.73832	.26168	.99935	52
9	.73997	.74063	.25937	.99934	51
10	8.74226	8.74292	11.25708	9.99934	**50**
11	.74454	.74521	.25479	.99933	49
12	.74680	.74748	.25252	.99932	48
13	.74906	.74974	.25026	.99932	47
14	.75130	.75199	.24801	.99931	46
15	8.75353	8.75423	11.24577	9.99930	**45**
16	.75575	.75645	.24355	.99929	44
17	.75795	.75867	.24133	.99929	43
18	.76015	.76087	.23913	.99928	42
19	.76234	.76306	.23694	.99927	41
20	8.76451	8.76525	11.23475	9.99926	**40**
21	.76667	.76742	.23258	.99926	39
22	.76883	.76958	.23042	.99925	38
23	.77097	.77173	.22827	.99924	37
24	.77310	.77387	.22613	.99923	36
25	8.77522	8.77600	11.22400	9.99923	**35**
26	.77733	.77811	.22189	.99922	34
27	.77943	.78022	.21978	.99921	33
28	.78152	.78232	.21768	.99920	32
29	.78360	.78441	.21559	.99920	31
30	8.78568	8.78649	11.21351	9.99919	**30**
31	.78774	.78855	.21145	.99918	29
32	.78979	.79061	.20939	.99917	28
33	.79183	.79266	.20734	.99917	27
34	.79386	.79470	.20530	.99916	26
35	8.79588	8.79673	11.20327	9.99915	**25**
36	.79789	.79875	.20125	.99914	24
37	.79990	.80076	.19924	.99913	23
38	.80189	.80277	.19723	.99913	22
39	.80388	.80476	.19524	.99912	21
40	8.80585	8.80674	11.19326	9.99911	**20**
41	.80782	.80872	.19128	.99910	19
42	.80978	.81068	.18932	.99909	18
43	.81173	.81264	.18736	.99909	17
44	.81367	.81459	.18541	.99908	16
45	8.81560	8.81653	11.18347	9.99907	**15**
46	.81752	.81846	.18154	.99906	14
47	.81944	.82038	.17962	99905	13
48	.82134	.82230	.17770	.99904	12
49	.82324	.82420	.17580	.99904	11
50	8.82513	8.82610	11.17390	9.99903	**10**
51	.82701	.82799	17201	.99902	9
52	.82888	.82987	.17013	.99901	8
53	.83075	.83175	.16825	.99900	7
54	.83261	.83361	.16639	99899	6
55	8.83446	8.83547	11.16453	9.99898	**5**
56	.83630	.83732	.16268	.99898	4
57	.83813	.83916	.16084	.99897	3
58	.83996	.84100	15900	.99896	2
59	.84177	.84282	.15718	.99895	1
60	8.84358	8.84464	11.15536	9.99894	**0**
	Cosine.	Cotang.	Tang.	Sine.	

86°

*Log secant = colog cosine = 1 − log cosine; log cosecant = colog sine = 1 − log sine.

Ex.—Log sec 2°− 30′ = **10.00041** *Ex.*—Log cosec 2°− 30′ = **11.36032.**

3. —Logarithmic Sines, Tangents, Cotangents, Cosines.—(Cont'd.) (Secants, Cosecants.)*

4°

′	Sine.	Tang.	Cotang.	Cosine.	
0	8.84358	8.84464	11.15536	9.99894	**60**
1	.84539	.84646	.15354	.99893	59
2	.84718	.84826	.15174	.99892	58
3	.84897	.85006	.14994	.99891	57
4	.85075	.85185	.14815	.99891	56
5	8.85252	8.85363	11.14637	9.99890	**55**
6	.85429	.85540	.14460	.99889	54
7	.85605	.85717	.14283	.99888	53
8	.85780	.85893	.14107	.99887	52
9	.85955	.86069	.13931	.99886	51
10	8.86128	8.86243	11.13757	9.99885	**50**
11	.86301	.86417	.13583	.99884	49
12	.86474	.86591	.13409	.99883	48
13	.86645	.86763	.13237	.99882	47
14	.86816	.86935	.13065	.99881	46
15	8.86987	8.87106	11.12894	9.99880	**45**
16	.87156	.87277	.12723	.99879	44
17	.87325	.87447	.12553	.99879	43
18	.87494	.87616	.12384	.99878	42
19	.87661	.87785	.12215	.99877	41
20	8.87829	8.87953	11.12047	9.99876	**40**
21	.87995	.88120	.11880	.99875	39
22	.88161	.88287	.11713	.99874	38
23	.88326	.88453	.11547	.99873	37
24	.88490	.88618	.11382	.99872	36
25	8.88654	8.88783	11.11217	9.99871	**35**
26	.88817	.88948	.11052	.99870	34
27	.88980	.89111	.10889	.99869	33
28	.89142	.89274	.10726	.99868	32
29	.89304	.89437	.10563	.99867	31
30	8.89464	8.89598	11.10402	9.99866	**30**
31	.89625	.89760	.10240	.99865	29
32	.89784	.89920	.10080	.99864	28
33	.89943	.90080	.09920	.99863	27
34	.90102	.90240	.09760	.99862	26
35	8.90260	8.90399	11.09601	9.99861	**25**
36	.90417	.90557	.09443	.99860	24
37	.90574	.90715	.09285	.99859	23
38	.90730	.90872	.09128	.99858	22
39	.90885	.91029	.08971	.99857	21
40	8.91040	8.91185	11.08815	9.99856	**20**
41	.91195	.91340	.08660	.99855	19
42	.91349	.91495	.08505	.99854	18
43	.91502	.91650	.08350	.99853	17
44	.91655	.91803	.08197	.99852	16
45	8.91807	8.91957	11.08043	9.99851	**15**
46	.91959	.92110	.07890	.99850	14
47	.92110	.92262	.07738	.99848	13
48	.92261	.92414	.07586	.99847	12
49	.92411	.92565	.07435	.99846	11
50	8.92561	8.92716	11.07284	9.99845	**10**
51	.92710	.92866	.07134	.99844	9
52	.92859	.93016	.06984	.99843	8
53	.93007	.93165	.06835	.99842	7
54	.93154	.93313	.06687	.99841	6
55	8.93301	8.93462	11.06538	9.99840	**5**
56	.93448	.93609	.06391	.99839	4
57	.93594	.93756	.06244	.99838	3
58	.93740	.93903	.06097	.99837	2
59	.93885	.94049	.05951	.99836	1
60	8.94030	8.94195	11.05805	9.99834	**0**
	Cosine.	Cotang.	Tang.	Sine.	′

85°

5°

′	Sine.	Tang.	Cotang.	Cosine.	
0	8.94030	8.94195	11.05805	9.99834	**60**
1	.94174	.94340	.05660	.99833	59
2	.94317	.94485	.05515	.99832	58
3	.94461	.94630	.05370	.99831	57
4	.94603	.94773	.05227	.99830	56
5	8.94746	8.94917	11.05083	9.99829	**55**
6	.94887	.95060	.04940	.99828	54
7	.95029	.95202	.04798	.99827	53
8	.95170	.95344	.04656	.99825	52
9	.95310	.95486	.04514	.99824	51
10	8.95450	8.95627	11.04373	9.99823	**50**
11	.95589	.95767	.04233	.99822	49
12	.95728	.95908	.04092	.99821	48
13	.95867	.96047	.03953	.99820	47
14	.96005	.96187	.03813	.99819	46
15	8.96143	8.96325	11.03675	9.99817	**45**
16	.96280	.96464	.03536	.99816	44
17	.96417	.96602	.03398	.99815	43
18	.96553	.96739	.03261	.99814	42
19	.96689	.96877	.03123	.99813	41
20	8.96825	8.97013	11.02987	9.99812	**40**
21	.96960	.97150	.02850	.99810	39
22	.97095	.97285	.02715	.99809	38
23	.97229	.97421	.02579	.99808	37
24	.97363	.97556	.02444	.99807	36
25	8.97496	8.97691	11.02309	9.99806	**35**
26	.97629	.97825	.02175	.99804	34
27	.97762	.97959	.02041	.99803	33
28	.97894	.98092	.01908	.99802	32
29	.98026	.98225	.01775	.99801	31
30	8.98157	8.98358	11.01642	9.99800	**30**
31	.98288	.98490	.01510	.99798	29
32	.98419	.98622	.01378	.99797	28
33	.98549	.98753	.01247	.99796	27
34	.98679	.98884	.01116	.99795	26
35	8.98808	8.99015	11.00985	9.99793	**25**
36	.98937	.99145	.00855	.99792	24
37	.99066	.99275	.00725	.99791	23
38	.99194	.99405	.00595	.99790	22
39	.99322	.99534	.00466	.99788	21
40	8.99450	8.99662	11.00338	9.99787	**20**
41	.99577	.99791	.00209	.99786	19
42	.99704	8.99919	11.00081	.99785	18
43	.99830	9.00046	10.99954	.99783	17
44	8.99956	.00174	.99826	.99782	16
45	9.00082	9.00301	10.99699	9.99781	**15**
46	.00207	.00427	.99573	.99780	14
47	.00332	.00553	.99447	.99778	13
48	.00456	.00679	.99321	.99777	12
49	.00581	.00805	.99195	.99776	11
50	9.00704	9.00930	10.99070	9.99775	**10**
51	.00828	.01055	.98945	.99773	9
52	.00951	.01179	.98821	.99772	8
53	.01074	.01303	.98697	.99771	7
54	.01196	.01427	.98573	.99769	6
55	9.01318	9.01550	10.98450	9.99768	**5**
56	.01440	.01673	.98327	.99767	4
57	.01561	.01796	.98204	.99765	3
58	.01682	.01918	.98082	.99764	2
59	.01803	.02040	.97960	.99763	1
60	9.01923	9.02162	10.97838	9.99761	**0**
	Cosine.	Cotang.	Tang.	Sine.	′

84°

*Log secant = colog cosine = 1 − log cosine; log cosecant = colog sine = 1 − log sine.

Ex.—Log sec 4°- 30′ = 10.00134. *Ex.*—Log cosec 4°- 30′ = 11.10536.

3.—Logarithmic Sines, Tangents, Cotangents, Cosines.—(Cont'd.)
(Secants, Cosecants.)*

6°

′	Sine.	Tang.	Cotang.	Cosine.	
0	9.01923	9.02162	10.97838	9.99761	**60**
1	.02043	.02283	.97717	.99760	59
2	.02163	.02404	.97596	.99759	58
3	.02283	.02525	.97475	.99757	57
4	.02402	.02645	.97355	.99756	56
5	9.02520	9.02766	10.97234	9.99755	**55**
6	.02639	.02885	.97115	.99753	54
7	.02757	.03005	.96995	.99752	53
8	.02874	.03124	.96876	.99751	52
9	.02992	.03242	.96758	.99749	51
10	9.03109	9.03361	10.96639	9.99748	**50**
11	.03226	.03479	.96521	.99747	49
12	.03342	.03597	.96403	.99745	48
13	.03458	.03714	.96286	.99744	47
14	.03574	.03832	.96168	.99742	46
15	9.03690	9.03948	10.96052	9.99741	**45**
16	.03805	.04065	95935	.99740	44
17	.03920	.04181	.95819	.99738	43
18	.04034	.04297	.95703	.99737	42
19	.04149	.04413	.95587	.99736	41
20	9.04262	9.04528	10.95472	9.99734	**40**
21	.04376	.04643	.95357	.99733	39
22	.04490	.04758	.95242	.99731	38
23	.04603	.04873	.95127	.99730	37
24	.04715	.04987	.95013	.99728	36
25	9.04828	9.05101	10.94899	9.99727	**35**
26	.04940	.05214	.94786	.99726	34
27	.05052	.05328	.94672	.99724	33
28	.05164	.05441	.94559	.99723	32
29	.05275	.05553	.94447	.99721	31
30	9.05386	9.05666	10.94334	9.99720	**30**
31	.05497	.05778	.94222	.99718	29
32	.05607	.05890	.94110	.99717	28
33	.05717	.06002	.93998	.99716	27
34	.05827	.06113	.93887	.99714	26
35	9.05937	9.06224	10.93776	9.99713	**25**
36	.06046	.06335	.93665	.99711	24
37	.06155	.06445	.93555	.99710	23
38	.06264	.06556	.93444	.99708	22
39	.06372	.06666	.93334	.99707	21
40	9.06481	9.06775	10.93225	9.99705	**20**
41	.06589	.06885	.93115	.99704	19
42	.06696	.06994	.93006	.99702	18
43	.06804	.07103	.92897	.99701	17
44	.06911	.07211	.92789	.99699	16
45	9.07018	9.07320	10.92680	9.99698	**15**
46	.07124	.07428	.92572	.99696	14
47	.07231	.07536	.92464	.99695	13
48	.07337	.07643	.92357	.99693	12
49	.07442	.07751	.92249	.99692	11
50	9.07548	9.07858	10.92142	9.99690	**10**
51	.07653	.07964	.92036	.99689	9
52	.07758	.08071	.91929	.99687	8
53	.07863	.08177	.91823	.99686	7
54	.07968	.08283	.91717	.99684	6
55	9.08072	9.08389	10.91611	9.99683	**5**
56	.08176	.08495	.91505	.99681	4
57	.08280	.08600	.91400	.99680	3
58	.08383	.08705	.91295	.99678	2
59	.08486	.08810	.91190	.99677	1
60	9.08589	9.08914	10.91086	9.99675	**0**
	Cosine.	Cotang.	Tang.	Sine.	′
					83°

7°

′	Sine.	Tang.	Cotang.	Cosine.	
0	9.08589	9.08914	10.91086	9.99675	**60**
1	.08692	.09019	.90981	.99674	59
2	.08795	.09123	.90877	.99672	58
3	.08897	.09227	.90773	.99670	57
4	.08999	.09330	.90670	.99669	56
5	9.09101	9.09434	10.90566	9.99667	**55**
6	.09202	.09537	.90463	99666	54
7	.09304	.09640	.90360	.99664	53
8	.09405	.09742	.90258	.99663	52
9	.09506	.09845	.90155	.99661	51
10	9.09606	9.09947	10.90053	9.99659	**50**
11	.09707	.10049	.89951	.99658	49
12	.09807	.10150	.89850	.99656	48
13	.09907	.10252	.89748	.99655	47
14	.10006	.10353	.89647	.99653	46
15	9.10106	9.10454	10.89546	9.99651	**45**
16	.10205	.10555	.89445	.99650	44
17	.10304	.10656	.89344	.99648	43
18	.10402	.10756	.89244	.99647	42
19	.10501	.10856	.89144	.99845	41
20	9.10599	9.10956	10.89044	9.99643	**40**
21	.10697	.11056	.88944	.99642	39
22	.10795	.11155	.88845	.99640	38
23	.10893	.11254	.88746	.99638	37
24	.10990	.11353	.88647	.99637	36
25	9.11087	9.11452	10.88548	9.99635	**35**
26	.11184	.11551	.88449	.99633	34
27	.11281	.11649	.88351	.99632	33
28	.11377	.11747	.88253	.99630	32
29	.11474	.11845	.88155	.99629	31
30	9.11570	9.11943	10.88057	9.99627	**30**
31	.11666	.12040	.87960	.99625	29
32	.11761	.12138	.87862	.99624	28
33	.11857	.12235	.87765	.99622	27
34	.11952	.12332	.87668	.99620	26
35	9.12047	9.12428	10.87572	9.99618	**25**
36	.12142	.12525	.87475	.99617	24
37	.12236	.12621	.87379	.99615	23
38	.12331	.12717	.87283	.99613	22
39	.12425	.12813	.87187	.99612	21
40	9.12519	9.12909	10.87091	9.99610	**20**
41	.12612	.13004	.86996	.99608	19
42	.12706	.13099	.86901	.99607	18
43	.12799	.13194	.86806	.99605	17
44	.12892	.13289	.86711	.99603	16
45	9.12985	9.13384	10.86616	9.99601	**15**
46	.13078	.13478	.86522	.99600	14
47	.13171	.13573	.86427	.99598	13
48	.13263	.13667	.86333	.99596	12
49	.13355	.13761	.86239	.99595	11
50	9.13447	9.13854	10.86146	9.99593	**10**
51	.13539	.13948	.86052	.99591	9
52	.13630	.14041	.85959	.99589	8
53	.13722	.14134	.85866	.99588	7
54	.13813	.14227	.85773	.99586	6
55	9.13904	9.14320	10.85680	9.99584	**5**
56	13994	.14412	.85588	.99582	4
57	.14085	.14504	.85496	.99581	3
58	.14175	.14597	.85403	.99579	2
59	.14266	.14688	.85312	.99577	1
60	9.14356	9.14780	10.85220	9.99575	**0**
	Cosine.	Cotang.	Tang.	Sine.	′
					82°

*Log secant = colog cosine = 1 − log cosine; log cosecant = colog sine = 1 − log sine.

Ex.—Log sec 6° − 30′ = 10.00280. *Ex.*—Log cosec 6° − 30′ = 10.94614.

3. —Logarithmic Sines, Tangents, Cotangents, Cosines.—(Cont'd.) (Secants, Cosecants.)*

8°

'	Sine.	Tang.	Cotang.	Cosine.	
0	9.14356	9.14780	10.85220	9.99575	60
1	.14445	.14872	.85128	.99574	59
2	.14535	.14963	.85037	.99572	58
3	.14624	.15054	.84946	.99570	57
4	.14714	.15145	.84855	.99568	56
5	9.14803	9.15236	10.84764	9.99566	55
6	.14891	.15327	.84673	.99565	54
7	.14980	.15417	.84583	.99563	53
8	.15069	.15508	.84492	.99561	52
9	.15157	.15598	.84402	.99559	51
10	9.15245	9.15688	10.84312	9.99557	50
11	.15333	.15777	.84223	.99556	49
12	.15421	.15867	.84133	.99554	48
13	.15508	.15956	.84044	.99552	47
14	.15596	.16046	.83954	.99550	46
15	9.15683	9.16135	10.83865	9.99548	45
16	.15770	.16224	.83776	.99546	44
17	.15857	.16312	.83688	.99545	43
18	.15944	.16401	.83599	.99543	42
19	.16030	.16489	.83511	.99541	41
20	9.16116	9.16577	10.83423	9.99539	40
21	.16203	.16665	.83335	.99537	39
22	.16289	.16753	.83247	.99535	38
23	.16374	.16841	.83159	.99533	37
24	.16460	.16928	.83072	.99532	36
25	9.16545	9.17016	10.82984	9.99530	35
26	.16631	.17103	.82897	.99528	34
27	.16716	.17190	.82810	.99526	33
28	.16801	.17277	.82723	.99524	32
29	.16886	.17363	.82637	.99522	31
30	9.16970	9.17450	10.82550	9.99520	30
31	.17055	.17536	.82464	.99518	29
32	.17139	.17622	.82378	.99517	28
33	.17223	.17708	.82292	.99515	27
34	.17307	.17794	.82206	.99513	26
35	9.17391	9.17880	10.82120	9.99511	25
36	.17474	.17965	.82035	.99509	24
37	.17558	.18051	.81949	.99507	23
38	.17641	.18136	.81864	.99505	22
39	.17724	.18221	.81779	.99503	21
40	9.17807	9.18306	10.81694	9.99501	20
41	.17890	.18391	.81609	.99499	19
42	.17973	.18475	.81525	.99497	18
43	.18055	.18560	.81440	.99495	17
44	.18137	.18644	.81356	.99494	16
45	9.18220	9.18728	10.81272	9.99492	15
46	.18302	.18812	.81188	.99490	14
47	.18383	.18896	.81104	.99488	13
48	.18465	.18979	.81021	.99486	12
49	.18547	.19063	.80937	.99484	11
50	9.18628	9.19146	10.80854	9.99482	10
51	.18709	.19229	.80771	.99480	9
52	.18790	.19312	.80688	.99478	8
53	.18871	.19395	.80605	.99476	7
54	.18952	.19478	.80522	.99474	6
55	9.19033	9.19561	10.80439	9.99472	5
56	.19113	.19643	.80357	.99470	4
57	.19193	.19725	.80275	.99468	3
58	.19273	.19807	.80193	.99466	2
59	.19353	.19889	.80111	.99464	1
60	9.19433	9.19971	10.80029	9.99462	0
	Cosine.	Cotang.	Tang.	Sine.	'

81°

9°

'	Sine.	Tang.	Cotang.	Cosine.	
0	9.19433	9.19971	10.80029	9.99462	60
1	.19513	.20053	.79947	.99460	59
2	.19592	.20134	.79866	.99458	58
3	.19672	.20216	.79784	.99456	57
4	.19751	.20297	.79703	.99454	56
5	9.19830	9.20378	10.79622	9.99452	55
6	.19909	.20459	.79541	.99450	54
7	.19988	.20540	.79460	.99448	53
8	.20067	.20621	.79379	.99446	52
9	.20145	.20701	.79299	.99444	51
10	9.20223	9.20782	10.79218	9.99442	50
11	.20302	.20862	.79138	.99440	49
12	.20380	.20942	.79058	.99438	48
13	.20458	.21022	.78978	.99436	47
14	.20535	.21102	.78898	.99434	46
15	9.20613	9.21182	10.78818	9.99432	45
16	.20691	.21261	.78739	.99429	44
17	.20768	.21341	.78659	.99427	43
18	.20845	.21420	.78580	.99425	42
19	.20922	.21499	.78501	.99423	41
20	9.20999	9.21578	10.78422	9.99421	40
21	.21076	.21657	.78343	.99419	39
22	.21153	.21736	.78264	.99417	38
23	.21229	.21814	.78186	.99415	37
24	.21306	.21893	.78107	.99413	36
25	9.21382	9.21971	10.78029	9.99411	35
26	.21458	.22049	.77951	.99409	34
27	.21534	.22127	.77873	.99407	33
28	.21610	.22205	.77795	.99404	32
29	.21685	.22283	.77717	.99402	31
30	9.21761	9.22361	10.77639	9.99400	30
31	.21836	.22438	.77562	.99398	29
32	.21912	.22516	.77484	.99396	28
33	.21987	.22593	.77407	.99394	27
34	.22062	.22670	.77330	.99392	26
35	9.22137	9.22747	10.77253	9.99390	25
36	.22211	.22824	.77176	.99388	24
37	.22286	.22901	.77099	.99385	23
38	.22361	.22977	.77023	.99383	22
39	.22435	.23054	.76946	.99381	21
40	9.22509	9.23130	10.76870	9.99379	20
41	.22583	.23206	.76794	.99377	19
42	.22657	.23283	.76717	.99375	18
43	.22731	.23359	.76641	.99372	17
44	.22805	.23435	.76565	.99370	16
45	9.22878	9.23510	10.76490	9.99368	15
46	.22952	.23586	.76414	.99366	14
47	.23025	.23661	.76339	.99364	13
48	.23098	.23737	.76263	.99362	12
49	.23171	.23812	.76188	.99359	11
50	9.23244	9.23887	10.76113	9.99357	10
51	.23317	.23962	.76038	.99355	9
52	.23390	.24037	.75963	.99353	8
53	.23462	.24112	.75888	.99351	7
54	.23535	.24186	.75814	.99348	6
55	9.23607	9.24261	10.75739	9.99346	5
56	.23679	.24335	.75665	.99344	4
57	.23752	.24410	.75590	.99342	3
58	.23823	.24484	.75516	.99340	2
59	.23895	.24558	.75442	.99337	1
60	9.23967	9.24632	10.75368	9.99335	0
	Cosine.	Cotang.	Tang.	Sine.	'

80°

*Log secant = colog cosine = 1 − log cosine; log cosecant = colog sine = 1 − log sine.

Ex.—Log sec 8°− 30′ = 10.00480. *Ex.*—Log cosec 8°− 30′ = 10.83030.

3. —Logarithmic Sines, TANGENTS, COTANGENTS, COSINES.—(Cont'd.)
(SECANTS, COSECANTS.)*

10°						11°					
'	Sine.	Tang.	Cotang.	Cosine.		'	Sine.	Tang.	Cotang.	Cosine.	
0	9.23967	9.24632	10.75368	9.99335	**60**	**0**	9.28060	9.28865	10.71135	9.99195	**60**
1	.24039	.24706	.75294	.99333	59	1	.28125	.28933	.71067	.99192	59
2	.24110	.24779	.75221	.99331	58	2	.28190	.29000	.71000	.99190	58
3	.24181	.24853	.75147	.99328	57	3	.28254	.29067	.70933	.99187	57
4	.24253	.24926	.75074	.99326	56	4	.28319	.29134	.70866	.99185	56
5	9.24324	9.25000	10.75000	9.99324	**55**	**5**	9.28384	9.29201	10.70799	9.99182	**55**
6	.24395	.25073	.74927	.99322	54	6	.28448	.29268	.70732	.99180	54
7	.24466	.25146	.74854	.99319	53	7	.28512	.29335	.70665	.99177	53
8	.24536	.25219	.74781	.99317	52	8	.28577	.29402	.70598	.99175	52
9	.24607	.25292	.74708	.99315	51	9	.28641	.29468	.70532	.99172	51
10	9.24677	9.25365	10.74635	9.99313	**50**	**10**	9.28705	9.29535	10.70465	9.99170	**50**
11	.24748	.25437	.74563	.99310	49	11	.28769	.29601	.70399	.99167	49
12	.24818	.25510	.74490	.99308	48	12	.28833	.29668	.70332	.99165	48
13	.24888	.25582	.74418	.99306	47	13	.28896	.29734	.70266	.99162	47
14	.24958	.25655	.74345	.99304	46	14	.28960	.29800	.70200	.99160	46
15	9.25028	9.25727	10.74273	9.99301	**45**	**15**	9.29024	9.29866	10.70134	9.99157	**45**
16	.25098	.25799	.74201	.99299	44	16	.29087	.29932	.70068	.99155	44
17	.25168	.25871	.74129	.99297	43	17	.29150	.29998	.70002	.99152	43
18	.25237	.25943	.74057	.99294	42	18	.29214	30064	.69936	.99150	42
19	.25307	.26015	.73985	.99292	41	19	.29277	.30130	.69870	.99147	41
20	9.25376	9.26086	10.73914	9.99290	**40**	**20**	9.29340	9.30195	10.69805	9.99145	**40**
21	.25445	.26158	.73842	.99288	39	21	.29403	.30261	.69739	.99142	39
22	.25514	.26229	.73771	.99285	38	22	.29466	.30326	.69674	.99140	38
23	.25583	.26301	.73699	.99283	37	23	.29529	.30391	.69609	.99137	37
24	.25652	.26372	.73628	.99281	36	24	.29591	.30457	.69543	.99135	36
25	9.25721	9.26443	10.73557	9.99278	**35**	**25**	9.29654	9.30522	10.69478	9.99132	**35**
26	.25790	.26514	.73486	.99276	34	26	.29716	.30587	.69413	.99130	34
27	.25858	.26585	.73415	.99274	33	27	.29779	.30652	.69348	.99127	33
28	.25927	.26655	.73345	.99271	32	28	.29841	.30717	.69283	.99124	32
29	.25995	.26726	.73274	.99269	31	29	.29903	.30782	.69218	.99122	31
30	9.26063	9.26797	10.73203	9.99267	**30**	**30**	9.29966	9.30846	10.69154	9.99119	**30**
31	.26131	.26867	.73133	.99264	29	31	.30028	.30911	.69089	.99117	29
32	.26199	.26937	.73063	.99262	28	32	.30090	.30975	.69025	.99114	28
33	.26267	.27008	.72992	.99260	27	33	.30151	.31040	.68960	.99112	27
34	.26335	.27078	.72922	.99257	26	34	.30213	.31104	.68896	.99109	26
35	9.26403	9.27148	10.72852	9.99255	**25**	**35**	9.30275	9.31168	10.68832	9.99106	**25**
36	.26470	.27218	.72782	.99252	24	36	.30336	.31233	.68767	.99104	24
37	.26538	.27288	.72712	.99250	23	37	.30398	.31297	.68703	.99101	23
38	.26605	.27357	.72643	.99248	22	38	.30459	.31361	.68639	.99099	22
39	.26672	.27427	.72573	.99245	21	39	.30521	.31425	.68575	.99096	21
40	9.26739	9.27496	10.72504	9.99243	**20**	**40**	9.30582	9.31489	10.68511	9.99093	**20**
41	.26806	.27566	.72434	.99241	19	41	.30643	.31552	.68448	.99091	19
42	.26873	.27635	.72365	.99238	18	42	.30704	.31616	.68384	.99088	18
43	.26940	.27704	.72296	.99236	17	43	.30765	.31679	.68321	.99086	17
44	.27007	.27773	.72227	.99233	16	44	.30826	.31743	.68257	.99083	16
45	9.27073	9.27842	10.72158	9.99231	**15**	**45**	9.30887	9.31806	10.68194	9.99080	**15**
46	.27140	.27911	.72089	.99229	14	46	.30947	.31870	.68130	99078	14
47	.27206	.27980	.72020	.99226	13	47	.31008	.31933	.68067	.99075	13
48	.27273	.28049	.71951	.99224	12	48	.31068	.31996	.68004	.99072	12
49	.27339	.28117	.71883	.99221	11	49	.31129	.32059	.67941	.99070	11
50	9.27405	9.28186	10.71814	9.99219	**10**	**50**	9.31189	9.32122	10.67878	9.99067	**10**
51	.27471	.28254	.71746	.99217	9	51	.31250	.32185	.67815	.99064	9
52	.27537	.28323	.71677	.99214	8	52	.31310	.32248	.67752	.99062	8
53	.27602	.28391	.71609	.99212	7	53	.31370	.32311	.67689	.99059	7
54	.27668	.28459	.71541	.99209	6	54	.31430	.32373	.67627	.99056	6
55	9.27734	9.28527	10.71473	9.99207	**5**	**55**	9.31490	9.32436	10.67564	9.99054	**5**
56	.27799	.28595	.71405	.99204	4	56	.31549	.32498	.67502	.99051	4
57	.27864	.28662	.71338	.99202	3	57	.31609	.32561	.67439	.99048	3
58	.27930	.28730	.71270	.99200	2	58	.31669	.32623	.67377	.99046	2
59	.27995	.28798	.71202	.99197	1	59	.31728	.32685	.67315	.99043	1
60	9.28060	9.28865	10.71135	9.99195	**0**	**60**	9.31788	9.32747	10.67253	9.99040	**0**
	Cosine.	Cotang.	Tang.	Sine.	'		Cosine.	Cotang.	Tang.	Sine.	'
					79°						78°

*Log secant = colog cosine = 1 − log cosine; log cosecant = colog sine = 1 − log sine.

Ex.—Log sec 10°- 30′ = 10.00733. *Ex.*—Log cosec 10°- 30′ = 10.73937.

3.—**Logarithmic Sines,** TANGENTS, COTANGENTS, COSINES.—(Cont'd.)
(SECANTS, COSECANTS.)*

12°						13°					
′	Sine.	Tang.	Cotang.	Cosine.		′	Sine.	Tang.	Cotang.	Cosine.	
0	9.31788	9.32747	10.67253	9.99040	**60**	**0**	9.35209	9.36336	10.63664	9.98872	**60**
1	.31847	.32810	.67190	.99038	59	1	.35263	.36394	.63606	.98869	59
2	.31907	.32872	.67128	.99035	58	2	.35318	.36452	.63548	.98867	58
3	.31966	.32933	.67067	.99032	57	3	.35373	.36509	.63491	.98864	57
4	.32025	.32995	.67005	.99030	56	4	.35427	.36566	.63434	.98861	56
5	9.32084	9.33057	10.66943	9.99027	**55**	**5**	9.35481	9.36624	10.63376	9.98858	**55**
6	.32143	.33119	.66881	.99024	54	6	.35536	.36681	.63319	.98855	54
7	.32202	.33180	.66820	.99022	53	7	.35590	.36738	.63262	.98852	53
8	.32261	.33242	.66758	.99019	52	8	.35644	.36759	.63205	.98849	52
9	.32319	.33303	.66697	.99016	51	9	.35698	.36852	.63148	.98846	51
10	9.32378	9.33365	10.66635	9.99013	**50**	**10**	9.35752	9.36909	10.63091	9.98843	**50**
11	.32437	.33426	.66574	.99011	49	11	.35806	.36966	.63034	.98840	49
12	.32495	.33487	.66513	.99008	48	12	.35860	.37023	.62977	.98837	48
13	.32553	.33548	.66452	.99005	47	13	.35914	.37080	.62920	.98834	47
14	.32612	.33609	.66391	.99002	46	14	.35968	.37137	.62863	.98831	46
15	9.32670	9.33670	10.66330	9.99000	**45**	**15**	9.36022	9.37193	10.62807	9.98828	**45**
16	.32728	.33731	.66269	.98997	44	16	.36075	.37250	.62750	.98825	44
17	.32786	.33792	.66208	.98994	43	17	.36129	.37306	.62694	.98822	43
18	.32844	.33853	.66147	.98991	42	18	.36182	.37363	.62637	.98819	42
19	.32902	.33913	.66087	.98989	41	19	.36236	.37419	.62581	.98816	41
20	9.32960	9.33974	10.66026	9.98986	**40**	**20**	9.36289	9.37476	10.62524	9.98813	**40**
21	.33018	.34034	.65966	.98983	39	21	.36342	.37532	.62468	.98810	39
22	.33075	.34095	.65905	.98980	38	22	.36395	.37588	.62412	.98807	38
23	.33133	.34155	.65845	.98978	37	23	.36449	.37644	.62356	.98804	37
24	.33190	.34215	.65785	.98975	36	24	.36502	.37700	.62300	.98801	36
25	9.33248	9.34276	10.65724	9.98972	**35**	**25**	9.36555	9.37756	10.62244	9.98798	**35**
26	.33305	.34336	.65664	.98969	34	26	.36608	.37812	.62188	.98795	34
27	.33362	.34396	.65604	.98967	33	27	.36660	.37868	.62132	.98792	33
28	.33420	.34456	.65544	.98964	32	28	.36713	.37924	.62076	.98789	32
29	.33477	.34516	.65484	.98961	31	29	.36766	.37980	.62020	98786	31
30	9.33534	9.34576	10.65424	9.98958	**30**	**30**	9.36819	9.38035	10.61965	9.98783	**30**
31	.33591	.34635	.65365	.98955	29	31	.36871	.38091	.61909	.98780	29
32	.33647	.34695	.65305	.98953	28	32	.36924	.38147	.61853	.98777	28
33	.33704	.34755	.65245	.98950	27	33	.36976	.38202	.61798	.98774	27
34	.33761	.34814	.65186	.98947	26	34	.37028	.38257	.61743	.98771	26
35	9.33818	9.34874	10.65126	9.98944	**25**	**35**	9.37081	9.38313	10.61687	9.98768	**25**
36	.33874	.34933	.65067	.98941	24	36	.37133	.38368	.61632	.98765	24
37	.33931	.34992	.65008	.98938	23	37	.37185	.38423	.61577	.98762	23
38	.33987	.35051	.64949	.98936	22	38	.37237	.38479	.61521	.98759	22
39	.34043	.35111	.64889	.98933	21	39	.37289	.38534	.61466	.98756	21
40	9.34100	9.35170	10.64830	9.98930	**20**	**40**	9.37341	9.38589	10.61411	9.98753	**20**
41	.34156	.35229	.64771	.98927	19	41	.37393	.38644	.61356	.98750	19
42	.34212	.35288	.64712	.98924	18	42	.37445	.38699	.61301	.98746	18
43	.34268	.35347	.64653	.98921	17	43	.37497	.38754	.61246	.98743	17
44	.34324	.35405	.64595	.98919	16	44	.37549	.38808	.61192	.98740	16
45	9.34380	9.35464	10.64536	9.98916	**15**	**45**	9.37600	9.38863	10.61137	9.98737	**15**
46	.34436	.35523	.64477	.98913	14	46	.37652	.38918	.61082	.98734	14
47	.34491	.35581	.64419	.98910	13	47	.37703	.38972	.61028	.98731	13
48	.34547	.35640	.64360	.98907	12	48	.37755	.39027	.60973	.98728	12
49	.34602	.35698	.64302	.98904	11	49	.37806	.39082	.60918	.98725	11
50	9.34658	9.35757	10.64243	9.98901	**10**	**50**	9.37858	9.39136	10.60864	9.98722	**10**
51	.34713	.35815	.64185	.98898	9	51	.37909	.39190	.60810	.98719	9
52	.34769	.35873	.64127	.98896	8	52	.37960	.39245	.60755	.98517	8
53	.34824	.35931	.64069	.98893	7	53	.38011	.39299	.60701	.98712	7
54	.34879	.35989	.64011	.98890	6	54	.38062	.39353	.60647	.98709	6
55	9.34934	9.36047	10.63953	9.98887	**5**	**55**	9.38113	9.39407	10.60593	9.98706	**5**
56	.34989	.36105	.63895	.98884	4	56	.38164	.39461	.60539	.98703	4
57	.35044	.36163	.63837	.98881	3	57	.38215	.39515	.60485	.98700	3
58	.35099	.36221	.63779	.98878	2	58	.38266	.39569	.60431	.98697	2
59	.35154	.36279	.63721	.98875	1	59	.38317	.39623	.60377	.98694	1
60	9.35209	9.36336	10.63664	9.98872	**0**	**60**	9.38368	9.39677	10.60323	9.98690	**0**
	Cosine.	Cotang.	Tang.	Sine.	′		Cosine.	Cotang.	Tang.	Sine.	′
					77°						76°

*Log secant = colog cosine = 1 − log cosine; log cosecant = colog sine = 1 − log sine.

Ex.—Log sec 12°- 30′ = 10.01042. *Ex.*—Log cosec 12°- 30′ = 10.66466.

3. –Logarithmic Sines, Tangents, Cotangents, Cosines.—(Cont'd.)
(Secants, Cosecants.)*

14°						15°					
′	Sine.	Tang.	Cotang.	Cosine.	′	′	Sine.	Tang.	Cotang.	Cosine.	′
0	9.38368	9.39677	10.60323	9.98690	**60**	**0**	9.41300	9.42805	10.57195	9.98494	**60**
1	.38418	.39731	.60269	.98687	59	1	.41347	.42856	.57144	.98491	59
2	.38469	.39785	.60215	.98684	58	2	.41394	.42906	.57094	.98488	58
3	.38519	.39838	.60162	.98681	57	3	.41441	.42957	.57043	.98484	57
4	.38570	.39892	.60108	.98678	56	4	.41488	.43007	.56993	.98481	56
5	9.38620	9.39945	10.60055	9.98675	**55**	**5**	9.41535	9.43057	10.56943	9.98477	**55**
6	.38670	.39999	.60001	.98671	54	6	.41582	.43108	.56892	.98474	54
7	.38721	.40052	.59948	.98668	53	7	.41628	.43158	.56842	.98471	53
8	.38771	.40106	.59894	.98665	52	8	.41675	.43208	.56792	.98467	52
9	.38821	.40159	.59841	.98662	51	9	.41722	.43258	.56742	.98464	51
10	9.38871	9.40212	10.59788	9.98659	**50**	**10**	9.41768	9.43308	10.56692	9.98460	**50**
11	.38921	.40266	.59734	.98656	49	11	.41815	.43358	.56642	.98457	49
12	.38971	.40319	.59681	.98652	48	12	.41861	.43408	.56592	.98453	48
13	.39021	.40372	.59628	.98649	47	13	.41908	.43458	.56542	.98450	47
14	.39071	.40425	.59575	.98646	46	14	.41954	.43508	.56492	.98447	46
15	9.39121	9.40478	10.59522	9.98643	**45**	**15**	9.42001	9.43558	10.56442	9.98443	**45**
16	.39170	.40531	.59469	.98640	44	16	.42047	.43607	.56393	.98440	44
17	.39220	.40584	.59416	.98636	43	17	.42093	.43657	.56343	.98436	43
18	.39270	.40636	.59364	.98633	42	18	.42140	.43707	.56293	.98433	42
19	.39319	.40689	.59311	.98630	41	19	.42186	.43756	.56244	.98429	41
20	9.39369	9.40742	10.59258	9.98627	**40**	**20**	9.42232	9.43806	10.56194	9.98426	**40**
21	.39418	.40795	.59205	.98623	39	21	.42278	.43855	.56145	.98422	39
22	.39467	.40847	.59153	.98620	38	22	.42324	.43905	.56095	.98419	38
23	.39517	.40900	.59100	.98617	37	23	.42370	.43954	.56046	.98415	37
24	.39566	.40952	.59048	.98614	36	24	.42416	.44004	.55996	.98412	36
25	9.39615	9.41005	10.58995	9.98610	**35**	**25**	9.42461	9.44053	10.55947	9.98409	**35**
26	.39664	.41057	.58943	.98607	34	26	.42507	.44102	.55898	.98405	34
27	.39713	.41109	.58891	.98604	33	27	.42553	.44151	.55849	.98402	33
28	.39762	.41161	.58839	.98601	32	28	.42599	.44201	.55799	.98398	32
29	.39811	.41214	.58786	.98597	31	29	.42644	.44250	.55750	.98395	31
30	9.39860	9.41266	10.58734	9.98594	**30**	**30**	9.42690	9.44299	10.55701	9.98391	**30**
31	.39909	.41318	.58682	.98591	29	31	.42735	.44348	.55652	.98388	29
32	.39958	.41370	.58630	.98588	28	32	.42781	.44397	.55603	.98384	28
33	.40006	.41422	.58578	.98584	27	33	.42826	.44446	.55554	.98381	27
34	.40055	.41474	.58526	.98581	26	34	.42872	.44495	.55505	.98377	26
35	9.40103	9.41526	10.58474	9.98578	**25**	**35**	9.42917	9.44544	10.55456	9.98373	**25**
36	.40152	.41578	.58422	.98574	24	36	.42962	.44592	.55408	.98370	24
37	.40200	.41629	.58371	.98571	23	37	.43008	.44641	.55359	.98366	23
38	.40249	.41681	.58319	.98568	22	38	.43053	.44690	.55310	.98363	22
39	.40297	.41733	.58267	.98565	21	39	.43098	.44738	.55262	.98359	21
40	9.40346	9.41784	10.58216	9.98561	**20**	**40**	9.43143	9.44787	10.55213	9.98356	**20**
41	.40394	.41836	.58164	.98558	19	41	.43188	.44836	.55164	.98352	19
42	.40442	.41887	.58113	.98555	18	42	.43233	.44884	.55116	.98349	18
43	.40490	.41939	.58061	.98551	17	43	.43278	.44933	.55067	.98345	17
44	.40538	.41990	.58010	.98548	16	44	.43323	.44981	.55019	.98342	16
45	9.40586	9.42041	10.57959	9.98545	**15**	**45**	9.43367	9.45029	10.54971	9.98338	**15**
46	.40634	.42093	.57907	.98541	14	46	.43412	.45078	.54922	.98334	14
47	.40682	.42144	.57856	.98538	13	47	.43457	.45126	.54874	.98331	13
48	.40730	.42195	.57805	.98535	12	48	.43502	.45174	.54826	.98327	12
49	.40778	.42246	.57754	.98531	11	49	.43546	.45222	.54778	.98324	11
50	9.40825	9.42297	10.57703	9.98528	**10**	**50**	9.43591	9.45271	10.54729	9.98320	**10**
51	.40873	.42348	.57652	.98525	9	51	.43635	.45319	.54681	.98317	9
52	.40921	.42399	.57601	.98521	8	52	.43680	.45367	.54633	.98313	8
53	.40968	.42450	.57550	.98518	7	53	.43724	.45415	.54585	.98309	7
54	.41016	.42501	.57499	.98515	6	54	.43769	.45463	.54537	.98306	6
55	9.41063	9.42552	10.57448	9.98511	**5**	**55**	9.43813	9.45511	10.54489	9.98302	**5**
56	.41111	.42603	.57397	.98508	4	56	.43857	.45559	.54441	.98299	4
57	.41158	.42653	.57347	.98505	3	57	.43901	.45606	.54394	.98295	3
58	.41205	.42704	.57296	.98501	2	58	.43946	.45654	.54346	.98291	2
59	.41252	.42755	.57245	.98498	1	59	.43990	.45702	.54298	.98288	1
60	9.41300	9.42805	10.57195	9.98494	**0**	**60**	9.44034	9.45750	10.54250	9.98284	**0**
	Cosine.	Cotang.	Tang.	Sine.	′		Cosine.	Cotang.	Tang.	Sine.	′
					75°						74°

*Log secant = colog cosine = 1 − log cosine. log cosecant = colog sine = 1 − log sine.
Ex.—Log sec 14°- 30′ = 10.01406. *Ex.*—Log cosec 14°- 30′ = 10.60140.

3.—Logarithmic Sines, Tangents, Cotangents, Cosines.—(Cont'd.)
(Secants, Cosecants.)*

16°

′	Sine.	Tang.	Cotang.	Cosine.	
0	9.44034	9.45750	10.54250	9.98284	60
1	.44078	.45797	.54203	.98281	59
2	.44122	.45845	.54155	.98277	58
3	.44166	.45892	.54108	.98273	57
4	.44210	.45940	54060	.98270	56
5	9.44253	9.45987	10.54013	9.98266	55
6	.44297	.46035	.53965	.98262	54
7	.44341	.46082	.53918	.98259	53
8	.44385	.46130	.53870	.98255	52
9	.44428	.46177	.53823	.98251	51
10	9.44472	9.46224	10.53776	9.98248	50
11	.44516	.46271	.53729	.98244	49
12	.44559	.46319	.53681	.98240	48
13	.44602	.46366	.53634	.98237	47
14	.44646	.46413	.53587	.98233	46
15	9.44689	9.46460	10.53540	9.98229	45
16	.44733	.46507	.53493	.98226	44
17	.44776	.46554	.53446	.98222	43
18	.44819	.46601	.53399	.98218	42
19	.44862	.46648	.53352	.98215	41
20	9.44905	9.46694	10.53306	9.98211	40
21	.44948	.46741	.53259	.98207	39
22	.44992	.46788	.53212	.98204	38
23	.45035	.46835	.53165	.98200	37
24	.45077	.46881	.53119	.98196	36
25	9.45120	9.46928	10.53072	9.98192	35
26	.45163	.46975	.53025	.98189	34
27	.45206	.47021	.52979	.98185	33
28	.45249	.47068	.52932	.98181	32
29	.45292	.47114	.52886	.98177	31
30	9.45334	9.47160	10.52840	9.98174	30
31	.45377	.47207	.52793	.98170	29
32	.45419	.47253	.52747	.98166	28
33	.45462	.47299	.52701	.98162	27
34	.45504	.47346	.52654	.98159	26
35	9.45547	9.47392	10.52608	9.98155	25
36	.45589	.47438	.52562	.98151	24
37	.45632	.47484	.52516	.98147	23
38	.45674	.47530	.52470	.98144	22
39	.45716	.47576	.52424	.98140	21
40	9.45758	9.47622	10.52378	9.98136	20
41	.45801	.47668	.52332	.98132	19
42	.45843	.47714	.52286	.98129	18
43	.45885	.47760	.52240	.98125	17
44	.45927	.47806	.52194	.98121	16
45	9.45969	9.47852	10.52148	9.98117	15
46	.46011	.47897	.52103	.98113	14
47	.46053	.47943	.52057	.98110	13
48	.46095	.47989	.52011	.98106	12
49	.46136	.48035	.51965	.98102	11
50	9.46178	9.48080	10.51920	9.98098	10
51	.46220	.48126	.51874	.98094	9
52	.46262	.48171	.51829	.98090	8
53	.46303	.48217	.51783	.98087	7
54	.46345	.48262	.51738	.98083	6
55	9.46386	9.48307	10.51693	9.98079	5
56	.46428	.48353	.51647	.98075	4
57	.46469	.48398	.51602	.98071	3
58	.46511	.48443	.51557	.98067	2
59	.46552	.48489	.51511	.98063	1
60	9.46594	9.48534	10.51466	9.98060	0
	Cosine.	Cotang.	Tang.	Sine.	′

73°

17°

′	Sine.	Tang.	Cotang.	Cosine.	
0	9.46594	9.48534	10.51466	9.98060	60
1	.46635	.48579	.51421	.98056	59
2	.46676	.48624	.51376	.98052	58
3	.46717	.48669	.51331	.98048	57
4	.46758	.48714	.51286	.98044	56
5	9.46800	9.48759	10.51241	9.98040	55
6	.46841	.48804	.51196	.98036	54
7	.46882	.48849	.51151	.98032	53
8	.46923	.48894	.51106	.98029	52
9	.46964	.48939	.51061	.98025	51
10	9.47005	9.48984	10.51016	9.98021	50
11	.47045	.49029	.50971	.98017	49
12	.47086	.49073	.50927	.98013	48
13	.47127	.49118	.50882	.98009	47
14	.47168	.49163	.50837	.98005	46
15	9.47209	9.49207	10.50793	9.98001	45
16	.47249	.49252	.50748	.97997	44
17	.47290	.49296	.50704	.97993	43
18	.47330	.49341	.50659	.97989	42
19	.47371	.49385	.50615	.97986	41
20	9.47411	9.49430	10.50570	9.97982	40
21	.47452	.49474	.50526	.97978	39
22	.47492	.49519	.50481	.97974	38
23	.47533	.49563	.50437	.97970	37
24	.47573	.49607	.50393	.97966	36
25	9.47613	9.49652	10.50348	9.97962	35
26	.47654	.49696	.50304	.97958	34
27	.47694	.49740	.50260	.97954	33
28	.47734	.49784	.50216	.97950	32
29	.47774	.49828	.50172	.97946	31
30	9.47814	9.49872	10.50128	9.97942	30
31	.47854	.49916	.50084	.97938	29
32	.47894	.49960	.50040	.97934	28
33	.47934	.50004	.49996	.97930	27
34	.47974	.50048	.49952	.97926	26
35	9.48014	9.50092	10.49908	9.97922	25
36	.48054	.50136	.49864	.97918	24
37	.48094	.50180	.49820	.97914	23
38	.48133	.50223	.49777	.97910	22
39	.48173	.50267	.49733	.97906	21
40	9.48213	9.50311	10.49689	9.97902	20
41	.48252	.50355	.49645	.97898	19
42	.48292	.50398	.49602	.97894	18
43	.48332	.50442	.49558	.97890	17
44	.48371	.50485	.49515	.97886	16
45	9.48411	9.50529	10.49471	9.97882	15
46	.48450	.50572	.49428	.97878	14
47	.48490	.50616	.49384	.97874	13
48	.48529	.50659	.49341	.97870	12
49	.48568	.50703	.49297	.97866	11
50	9.48607	9.50746	10.49254	9.97861	10
51	.48647	.50789	.49211	.97857	9
52	.48686	.50833	.49167	.97853	8
53	.48725	.50876	.49124	.97849	7
54	.48764	.50919	.49081	.97845	6
55	9.48803	9.50962	10.49038	9.97841	5
56	.48842	.51005	.48995	.97837	4
57	.48881	.51048	.48952	.97833	3
58	.48920	.51092	.48908	.97829	2
59	.48959	.51135	.48865	.97825	1
60	9.48998	9.51178	10.48822	9.97821	0
	Cosine.	Cotang.	Tang.	Sine.	′

72°

*Log secant = colog cosine = 1 − log cosine; log cosecant = colog sine = 1 − log sine.

Ex.—Log sec 16°- 30′ = 10.01826. *Ex.*—Log cosec 16°- 30′ = 10.54666.

3.—Logarithmic Sines, Tangents, Cotangents, Cosines.—(Cont'd.) (Secants, Cosecants.)*

18°						19°					
'	Sine.	Tang.	Cotang.	Cosine.		'	Sine.	Tang.	Cotang.	Cosine.	
0	9.48998	9.51178	10.48822	9.97821	**60**	**0**	9.51264	9.53697	10.46303	9.97567	**60**
1	.49037	.51221	.48779	.97817	59	1	.51301	.53738	.46262	.97563	59
2	.49076	.51264	.48736	.97812	58	2	.51338	.53779	.46221	.97558	58
3	.49115	.51306	.48694	.97808	57	3	.51374	.53820	.46180	.97554	57
4	.49153	.51349	.48651	.97804	56	4	.51411	.53861	.46139	.97550	56
5	9.49192	9.51392	10.48608	9.97800	**55**	**5**	9.51447	9.53902	10.46098	9.97545	**55**
6	.49231	.51435	.48565	.97796	54	6	.51484	.53943	.46057	.97541	54
7	.49269	.51478	.48522	.97792	53	7	.51520	.53984	.46016	.97536	53
8	.49308	.51520	.48480	.97788	52	8	.51557	.54025	.45975	.97532	52
9	.49347	.51563	.48437	.97784	51	9	.51593	.54065	.45935	.97528	51
10	9.49385	9.51606	10.48394	9.97779	**50**	**10**	9.51629	9.54106	10.45894	9.97523	**50**
11	.49424	.51648	.48352	.97775	49	11	.51666	.54147	.45853	.97519	49
12	.49462	.51691	.48309	.97771	48	12	.51702	.54187	.45813	.97515	48
13	.49500	.51734	.48266	.97767	47	13	.51738	.54228	.45772	.97510	47
14	.49539	.51776	.48224	.97763	46	14	.51774	.54269	.45731	.97506	46
15	9.49577	9.51819	10.48181	9.97759	**45**	**15**	9.51811	9.54309	10.45691	9.97501	**45**
16	.49615	.51861	.48139	.97754	44	16	.51847	.54350	.45650	.97497	44
17	.49654	.51903	.48097	.97750	43	17	.51883	.54390	.45610	.97492	43
18	.49692	.51946	.48054	.97746	42	18	.51919	.54431	.45569	.97488	42
19	.49730	.51988	.48012	.97742	41	19	.51955	.54471	.45529	.97484	41
20	9.49768	9.52031	10.47969	9.97738	**40**	**20**	9.51991	9.54512	10.45488	9.97479	**40**
21	.49806	.52073	.47927	.97734	39	21	.52027	.54552	.45448	.97475	39
22	.49844	.52115	.47885	.97729	38	22	.52063	.54593	.45407	.97470	38
23	.49882	.52157	.47843	.97725	37	23	.52099	.54633	.45367	.97466	37
24	.49920	.52200	.47800	.97721	36	24	.52135	.54673	.45327	.97461	36
25	9.49958	9.52242	10.47758	9.97717	**35**	**25**	9.52171	9.54714	10.45286	9.97457	**35**
26	.49996	.52284	.47716	.97713	34	26	.52207	.54754	.45246	.97453	34
27	.50034	.52326	.47674	.97708	33	27	.52242	.54794	.45206	.97448	33
28	.50072	.52368	.47632	.97704	32	28	.52278	.54835	.45165	.97444	32
29	.50110	.52410	.47590	.97700	31	29	.52314	.54875	.45125	.97439	31
30	9.50148	9.52452	10.47548	9.97696	**30**	**30**	9.52350	9.54915	10.45085	9.97435	**30**
31	.50185	.52494	.47506	.97691	29	31	.52385	.54955	.45045	.97430	29
32	.50223	.52536	.47464	.97687	28	32	.52421	.54995	.45005	.97426	28
33	.50261	.52578	.47422	.97683	27	33	.52456	.55035	.44965	.97421	27
34	.50298	.52620	.47380	.97679	26	34	.52492	.55075	.44925	.97417	26
35	9.50336	9.52661	10.47339	9.97674	**25**	**35**	9.52527	9.55115	10.44885	9.97412	**25**
36	.50374	.52703	.47297	.97670	24	36	.52563	.55155	.44845	.97408	24
37	.50411	.52745	.47255	.97666	23	37	.52598	.55195	.44805	.97403	23
38	.50449	.52787	.47213	.97662	22	38	.52634	.55235	.44765	.97399	22
39	.50486	.52829	.47171	.97657	21	39	.52669	.55275	.44725	.97394	21
40	9.50523	9.52870	10.47130	9.97653	**20**	**40**	9.52705	9.55315	10.44685	9.97390	**20**
41	.50561	.52912	.47088	.97649	19	41	.52740	.55355	.44645	.97385	19
42	.50598	.52953	.47047	.97645	18	42	.52775	.55395	.44605	.97381	18
43	.50635	.52995	.47005	.97640	17	43	.52811	.55434	.44566	.97376	17
44	.50673	.53037	.46963	.97636	16	44	.52846	.55474	.44526	.97372	16
45	9.50710	9.53078	10.46922	9.97632	**15**	**45**	9.52881	9.55514	10.44486	9.97367	**15**
46	.50747	.53120	.46880	.97628	14	46	.52916	.55554	.44446	.97363	14
47	.50784	.53161	.46839	.97623	13	47	.52951	.55593	.44407	.97358	13
48	.50821	.53202	.46798	.97619	12	48	.52986	.55633	.44367	.97353	12
49	.50858	.53244	.46756	.97615	11	49	.53021	.55673	.44327	.97349	11
50	9.50896	9.53285	10.46715	9.97610	**10**	**50**	9.53056	9.55712	10.44288	9.97344	**10**
51	.50933	.53327	.46673	.97606	9	51	.53092	.55752	.44248	.97340	9
52	.50970	.53368	.46632	.97602	8	52	.53126	.55791	.44209	.97335	8
53	.51007	.53409	.46591	.97597	7	53	.53161	.55831	.44169	.97331	7
54	.51043	.53450	.46550	.97593	6	54	.53196	.55870	.44130	.97326	6
55	9.51080	9.53492	10.46508	9.97589	**5**	**55**	9.53231	9.55910	10.44090	9.97322	**5**
56	.51117	.53533	.46467	.97584	4	56	.53266	.55949	.44051	.97317	4
57	.51154	.53574	.46426	.97580	3	57	.53301	.55989	.44011	.97312	3
58	.51191	.53615	.46385	.97576	2	58	.53336	.56028	.43972	.97308	2
59	.51227	.53656	.46344	.97571	1	59	.53370	.56067	.43933	.97303	1
60	9.51264	9.53697	10.46303	9.97567	**0**	**60**	9.53405	9.56107	10.43893	9.97299	**0**
	Cosine.	Cotang.	Tang.	Sine.	'		Cosine.	Cotang.	Tang.	Sine.	'
					71°						70°

*Log secant = colog cosine = 1 − log cosine; log cosecant = colog sine = 1 − log sine.

Ex.—Log sec 18°- 30′ = **10.02304**. *Ex.*—Log cosec 18°- 30′ = **10.49852**.

3.—Logarithmic Sines, Tangents, Cotangents, Cosines.—(Cont'd.) (Secants, Cosecants.)*

20°

'	Sine.	Tang.	Cotang.	Cosine.	
0	9.53405	9.56107	10.43893	9.97299	60
1	.53440	.56146	.43854	.97294	59
2	.53475	.56185	.43815	.97289	58
3	.53509	.56224	.43776	.97285	57
4	.53544	.56264	43736	.97280	56
5	9.53578	9.56303	10.43697	9.97276	55
6	.53613	.56342	.43658	.97271	54
7	.53647	.56381	.43619	.97266	53
8	.53682	.56420	.43580	.97262	52
9	.53716	.56459	.43541	.97257	51
10	9.53751	9.56498	10.43502	9.97252	50
11	.53785	.56537	.43463	.97248	49
12	.53819	.56576	.43424	.97243	48
13	.53854	.56615	.43385	.97238	47
14	.53888	.56654	.43346	.97234	46
15	9.53922	9.56693	10.43307	9.97229	45
16	.53957	.56732	.43268	.97224	44
17	.53991	.56771	.43229	.97220	43
18	.54025	.56810	.43190	.97215	42
19	.54059	.56849	.43151	.97210	41
20	9.54093	9.56887	10.43113	9.97206	40
21	.54127	.56926	.43074	.97201	39
22	.54161	.56965	.43035	.97196	38
23	.54195	.57004	.42996	.97192	37
24	.54229	.57042	.42958	.97187	36
25	9.54263	9.57081	10.42919	9.97182	35
26	.54297	.57120	.42880	.97178	34
27	.54331	.57158	.42842	.97173	33
28	.54365	.57197	.42803	.97168	32
29	.54399	.57235	.42765	.97163	31
30	9.54433	9.57274	10.42726	9.97159	30
31	.54466	.57312	.42688	.97154	29
32	.54500	.57351	.42649	.97149	28
33	.54534	.57389	.42611	.97145	27
34	.54567	.57428	.42572	.97140	26
35	9.54601	9.57466	10.42534	9.97135	25
36	.54635	.57504	.42496	.97130	24
37	.54668	.57543	.42457	.97126	23
38	.54702	.57581	.42419	.97121	22
39	.54735	.57619	.42381	.97116	21
40	9.54769	9.57658	10.42342	9.97111	20
41	.54802	.57696	.42304	.97107	19
42	.54836	.57734	.42266	.97102	18
43	.54869	.57772	.42228	.97097	17
44	.54903	.57810	.42190	.97092	16
45	9.54936	9.57849	10.42151	9.97087	15
46	.54969	.57887	.42113	.97083	14
47	.55003	.57925	.42075	.97078	13
48	.55036	.57963	.42037	.97073	12
49	.55069	.58001	.41999	.97068	11
50	9.55102	9.58039	10.41961	9.97063	10
51	.55136	.58077	.41923	.97059	9
52	.55169	.58115	.41885	.97054	8
53	.55202	.58153	.41847	.97049	7
54	.55235	.58191	.41809	.97044	6
55	9.55268	9.58229	10.41771	9.97039	5
56	.55301	.58267	.41733	.97035	4
57	.55334	.58304	.41696	.97030	3
58	.55367	.58342	.41658	.97025	2
59	.55400	.58380	.41620	.97020	1
60	9.55433	9.58418	10.41582	9.97015	0
	Cosine.	Cotang.	Tang.	Sine.	'

69°

21°

'	Sine.	Tang.	Cotang.	Cosine.	
0	9.55433	9.58418	10.41582	9.97015	60
1	.55466	.58455	.41545	.97010	59
2	.55499	.58493	.41507	.97005	58
3	.55532	.58531	.41469	.97001	57
4	.55564	.58569	.41431	.96996	56
5	9.55597	9.58606	10.41394	9.96991	55
6	.55630	.58644	.41356	.96986	54
7	.55663	.58681	.41319	.96981	53
8	.55695	.58719	.41281	.96976	52
9	.55728	.58757	.41243	.96971	51
10	9.55761	9.58794	10.41206	9.96966	50
11	.55793	.58832	.41168	.96962	49
12	.55826	.58869	.41131	.96957	48
13	.55858	.58907	.41093	.96952	47
14	.55891	.58944	.41056	.96947	46
15	9.55923	9.58981	10.41019	9.96942	45
16	.55956	.59019	.40981	.96937	44
17	.55988	.59056	.40944	.96932	43
18	.56021	.59094	.40906	.96927	42
19	.56053	.59131	.40869	.96922	41
20	9.56085	9.59168	10.40832	9.96917	40
21	.56118	.59205	.40795	.96912	39
22	.56150	.59243	.40757	.96907	38
23	.56182	.59280	.40720	.96903	37
24	.56215	.59317	.40683	.96898	36
25	9.56247	9.59354	10.40646	9.96893	35
26	.56279	.59391	.40609	.96888	34
27	.56311	.59429	40571	.96883	33
28	.56343	.59466	.40534	.96878	32
29	.56375	.59503	.40497	.96873	31
30	9.56408	9.59540	10.40460	9.96868	30
31	.56440	.59577	.40423	.96863	29
32	.56472	.59614	.40386	.96858	28
33	.56504	.59651	.40349	.96853	27
34	.56536	.59688	.40312	.96848	26
35	9.56568	9.59725	10.40275	9.96843	25
36	.56599	.59762	.40238	.96838	24
37	.56631	.59799	.40201	.96833	23
38	.56663	.59835	.40165	.96828	22
39	.56695	.59872	.40128	.96823	21
40	9.56727	9.59909	10.40091	9.96818	20
41	.56759	.59946	.40054	.96813	19
42	.56790	.59983	.40017	.96808	18
43	.56822	.60019	.39981	.96803	17
44	.56854	.60056	.39944	.96798	16
45	9.56886	9.60093	10.39907	9.96793	15
46	.56917	.60130	.39870	.96788	14
47	.56949	.60166	.39834	.96783	13
48	.56980	.60203	.39797	.96778	12
49	.57012	.60240	.39760	.96772	11
50	9.57044	9.60276	10.39724	9.96767	10
51	.57075	.60313	.39687	.96762	9
52	.57107	.60349	.39651	.96757	8
53	.57138	.60386	.39614	.96752	7
54	.57169	.60422	.39578	.96747	6
55	9.57201	9.60459	10.39541	9.96742	5
56	.57232	.60495	.39505	.96737	4
57	.57264	.60532	.39468	.96732	3
58	.57295	.60568	.39432	.96727	2
59	.57326	.60605	.39395	.96722	1
60	9.57358	9.60641	10.39359	9.96717	0
	Cosine.	Cotang.	Tang.	Sine.	'

68°

*Log secant = colog cosine = 1 − log cosine; log cosecant = colog sine = 1 − log sine.

Ex.—Log sec 20°- 30′ = 10.02841. *Ex.*—Log cosec 20°- 30′ = 10.45567.

3.—**Logarithmic Sines,** TANGENTS, COTANGENTS, COSINES.—(Cont'd.)
(SECANTS, COSECANTS.)*

22°						23°					
'	Sine.	Tang.	Cotang.	Cosine.		'	Sine.	Tang.	Cotang.	Cosine.	
0	9.57358	9.60641	10.39359	9.96717	**60**	**0**	9.59188	9.62785	10.37215	9.96403	**60**
1	.57389	.60677	.39323	.96711	59	1	.59218	.62820	.37180	.96397	59
2	.57420	.60714	.39286	.96706	58	2	.59247	.62855	.37145	.96392	58
3	.57451	.60750	.39250	.96701	57	3	.59277	.62890	.37110	.96387	57
4	.57482	.60786	.39214	.96696	56	4	.59307	.62926	.37074	.96381	56
5	9.57514	9.60823	10.39177	9.96691	**55**	**5**	9.59336	9.62961	10.37039	9.96376	**55**
6	.57545	.60859	.39141	.96686	54	6	.59366	.62996	.37004	.96370	54
7	.57576	.60895	.39105	.96681	53	7	.59396	.63031	.36969	.96365	53
8	.57607	.60931	.39069	.96676	52	8	.59425	.63066	.36934	.96360	52
9	.57638	.60967	.39033	.96670	51	9	.59455	.63101	.36899	.96354	51
10	9.57669	9.61004	10.38996	9.96665	**50**	**10**	9.59484	9.63135	10.36865	9.96349	**50**
11	.57700	.61040	.38960	.96660	49	11	.59514	.63170	.36830	.96343	49
12	.57731	.61076	.38924	.96655	48	12	.59543	.63205	.36795	.96338	48
13	.57762	.61112	.38888	.96650	47	13	.59573	.63240	.36760	.96333	47
14	.57793	.61148	.38852	.96645	46	14	.59602	.63275	.36725	.96327	46
15	9.57824	9.61184	10.38816	9.96640	**45**	**15**	9.59632	9.63310	10.36690	9.96322	**45**
16	.57855	.61220	.38780	.96634	44	16	.59661	.63345	.36655	.96316	44
17	.57885	.61256	.38744	.96629	43	17	.59690	.63379	.36621	.96311	43
18	.57916	.61292	.38708	96624	42	18	.59720	.63414	.36586	.96305	42
19	.57947	.61328	.38672	.96619	41	19	.59749	.63449	.36551	.96300	41
20	9.57978	9.61364	10.38636	9.96614	**40**	**20**	9.59778	9.63484	10.36516	9.96294	**40**
21	.58008	.61400	.38600	.96608	39	21	.59808	.63519	.36481	.96289	39
22	.58039	.61436	.38564	.96603	38	22	.59837	.63553	.36447	.96284	38
23	.58070	.61472	.38528	.96598	37	23	.59866	.63588	.36412	.96278	37
24	.58101	.61508	.38492	.96593	36	24	.59895	.63623	.36377	.96273	36
25	9.58131	9.61544	10.38456	9.96588	**35**	**25**	9.59924	9.63657	10.36343	9.96267	**35**
26	.58162	.61579	.38421	.96582	34	26	.59954	63692	.36308	.96262	34
27	.58192	.61615	.38385	.96577	33	27	.59983	.63726	.36274	.96256	33
28	.58223	.61651	.38349	.96572	32	28	.60012	.63761	.36239	.96251	32
29	.58253	.61687	.38313	.96567	31	29	.60041	.63796	.36204	.96245	31
30	9.58284	9.61722	10.38278	9.96562	**30**	**30**	9.60070	9.63830	10.36170	9.96240	**30**
31	.58314	.61758	.38242	.96556	29	31	.60099	.63865	.36135	.96234	29
32	.58345	.61794	.38206	.96551	28	32	.60128	.63899	.36101	.96229	28
33	.58375	.61830	.38170	.96546	27	33	.60157	.63934	.36066	.96223	27
34	.58406	.61865	.38135	.96541	26	34	.60186	.63968	.36032	.96218	26
35	9.58436	9.61901	10.38099	9.96535	**25**	**35**	9.60215	9.64003	10.35997	9.96212	**25**
36	.58467	.61936	.38064	.96530	24	36	.60244	.64037	.35963	.96207	24
37	.58497	.61972	.38028	.96525	23	37	.60273	.64072	.35928	.96201	23
38	.58527	.62008	.37992	.96520	22	38	.60302	.64106	.35894	.96196	22
39	.58557	.62043	.37957	.96514	21	39	.60331	.64140	.35860	.96190	21
40	9.58588	9.62079	10.37921	9.96509	**20**	**40**	9.60359	9.64175	10.35825	9.96185	**20**
41	.58618	.62114	.37886	.96504	19	41	.60388	.64209	.35791	.96179	19
42	.58648	.62150	.37850	.96498	18	42	.60417	.64243	.35757	.96174	18
43	.58678	.62185	.37815	.96493	17	43	.60446	.64278	.35722	.96168	17
44	.58709	.62221	.37779	.96488	16	44	.60474	.64312	.35688	.96162	16
45	9.58739	9.62256	10.37744	9.96483	**15**	**45**	9.60503	9.64346	10.35654	9.96157	**15**
46	.58769	.62292	.37708	.96477	14	46	.60532	.64381	.35619	.96151	14
47	.58799	.62327	.37673	.96472	13	47	.60561	.64415	.35585	.96146	13
48	.58829	.62362	.37638	.96467	12	48	.60589	.64449	.35551	.96140	12
49	.58859	.62398	.37602	.96461	11	49	.60618	.64483	.35517	.96135	11
50	9.58889	9.62433	10.37567	9.96456	**10**	**50**	9.60646	9.64517	10.35483	9.96129	**10**
51	.58919	.62468	.37532	.96451	9	51	.60675	.64552	.35448	.96123	9
52	.58949	.62504	.37496	.96445	8	52	.60704	.64586	.35414	.96118	8
53	.58979	.62539	.37461	.96440	7	53	.60732	.64620	.35380	.96112	7
54	.59009	.62574	.37426	.96435	6	54	.60761	.64654	.35346	.96107	6
55	9.59039	9.62609	10.37391	9.96429	**5**	**55**	9.60789	9.64688	10.35312	9.96101	**5**
56	.59069	.62645	.37355	.96424	4	56	.60818	.64722	.35278	.96095	4
57	.59098	.62680	.37320	.96419	3	57	.60846	.64756	.35244	.96090	3
58	.59128	.62715	.37285	.96413	2	58	.60875	.64790	.35210	.96084	2
59	.59158	.62750	.37250	.96408	1	59	.60903	.64824	.35176	.96079	1
60	9.59188	9.62785	10.37215	9.96403	**0**	**60**	9.60931	9.64858	10.35142	9.96073	**0**
	Cosine.	Cotang.	Tang.	Sine.	'		Cosine.	Cotang.	Tang.	Sine.	'
					67°						66°

*Log secant = colog cosine = 1 − log cosine; log cosecant = colog sine = 1 − log sine.

Ex.—Log sec 22°- 30′ = 10.03438. *Ex.*—Log cosec 22°- 30′ = **10.41716.**

3.—Logarithmic Sines, TANGENTS, COTANGENTS, COSINES.—(Cont'd.)
(SECANTS, COSECANTS.)*

24°

'	Sine.	Tang.	Cotang.	Cosine.	
0	9.60931	9.64858	10.35142	9.96073	**60**
1	.60960	.64892	.35108	.96067	59
2	.60988	.64926	.35074	.96062	58
3	.61016	.64960	.35040	.96056	57
4	.61045	.64994	.35006	.96050	56
5	9.61073	9.65028	10.34972	9.96045	**55**
6	.61101	.65062	.34938	.96039	54
7	.61129	.65096	.34904	.96034	53
8	.61158	.65130	.34870	.96028	52
9	.61186	.65164	.34836	.96022	51
10	9.61214	9.65197	10.34803	9.96017	**50**
11	.61242	.65231	.34769	.96011	49
12	.61270	.65265	.34735	.96005	48
13	.61298	.65299	.34701	.96000	47
14	.61326	.65333	.34667	.95994	46
15	9.61354	9.65366	10.34634	9.95988	**45**
16	.61382	.65400	.34600	.95982	44
17	.61411	.65434	.34566	.95977	43
18	.61438	.65467	.34533	.95971	42
19	.61466	.65501	.34499	.95965	41
20	9.61494	9.65535	10.34465	9.95960	**40**
21	.61522	.65568	.34432	.95954	39
22	.61550	.65602	.34398	.95948	38
23	.61578	.65636	.34634	.95942	37
24	.61606	.65669	.34331	.95937	36
25	9.61634	9.65703	10.34297	9.95931	**35**
26	.61662	.65736	.34264	.95925	34
27	.61689	.65770	.34230	.95920	33
28	.61717	.65803	.34197	.95914	32
29	.61745	.65837	.34163	.95908	31
30	9.61773	9.65870	10.34130	9.95902	**30**
31	.61800	.65904	.34096	.95897	29
32	.61828	.65937	.34063	.95891	28
33	.61856	.65971	.34029	.93885	27
34	.61883	.66004	.33996	.95879	26
35	9.61911	9.66038	10.33962	9.95873	**25**
36	.61939	.66071	.33929	.95868	24
37	.61966	.66104	.33896	.95862	23
38	.61994	.66138	.33862	.95856	22
39	.62021	.66171	.33829	.95850	21
40	9.62049	9.66204	10.33796	9.95844	**20**
41	.62076	.66238	.33762	.95839	19
42	.62104	.66271	.33729	.95833	18
43	.62131	.66304	.33696	.95827	17
44	.62159	.66337	.33663	.95821	16
45	9.62186	9.66371	10.33629	9.95815	**15**
46	.62214	.66404	.33596	.95810	14
47	.62241	.66437	.33563	.95804	13
48	.62268	.66470	.33530	.95798	12
49	.62296	.66503	.33497	.95792	11
50	9.62323	9.66537	10.33463	9.95786	**10**
51	.62350	.66570	.33430	.95780	9
52	.62377	.66603	.33397	.95775	8
53	.62405	.66636	.33364	.95769	7
54	.62432	.66669	.33331	.95763	6
55	9.62459	9.66702	10.33298	9.95757	**5**
56	.62486	.66735	.33265	.95751	4
57	.62513	.66768	.33232	.95745	3
58	.62541	.66801	.33199	.95739	2
59	.62568	.66834	.33166	.95733	1
60	9.62595	9.66867	10.33133	9.95728	**0**
	Cosine.	Cotang.	Tang.	Sine.	'

65°

25°

'	Sine.	Tang.	Cotang.	Cosine.	
0	9.62595	9.66867	10.33133	9.95728	**60**
1	.62622	.66900	.33100	.95722	59
2	.62649	.66933	.33067	.95716	58
3	.62676	.66966	.33034	.95710	57
4	.62703	.66999	.33001	.95704	56
5	9.62730	9.67032	10.32968	9.95698	**55**
6	.62757	.67065	.32935	.95692	54
7	.62784	.67098	.32902	.95686	53
8	.62811	.67131	.32869	.95680	52
9	.62838	.67163	.32837	.95674	51
10	9.62865	9.67196	10.32804	9.95668	**50**
11	.62892	.67229	.32771	.95663	49
12	.62918	.67262	.32738	.95657	48
13	.62945	.67295	.32705	.95651	47
14	.62972	.67327	.32673	.95645	46
15	9.62999	9.67360	10.32640	9.95639	**45**
16	.63026	.67393	.32607	.95633	44
17	.63052	.67426	.32574	.95627	43
18	.63079	.67458	.32542	.95621	42
19	.63106	.67491	.32509	.95615	41
20	9.63133	9.67524	10.32476	9.95609	**40**
21	.63159	.67556	.32444	.95603	39
22	.63186	.67589	.32411	.95597	38
23	.63213	.67622	.32378	.95591	37
24	.63239	.67654	.32346	.95585	36
25	9.63266	9.67687	10.32313	9.95579	**35**
26	.63292	.67719	.32281	.95573	34
27	63319	.67752	.32248	.95567	33
28	.63345	.67785	.32215	.95561	32
29	.63372	.67817	.32183	.95555	31
30	9.63398	9.67850	10.32150	9.95549	**30**
31	.63425	.67882	.32118	.95543	29
32	.63451	.67915	.32085	.95537	28
33	.63478	.67947	.32053	.95531	27
34	.63504	.67980	.32020	.95525	26
35	9.63531	9.68012	10.31988	9.95519	**25**
36	.63557	.68044	.31956	.95513	24
37	.63583	.68077	.31923	.95507	23
38	.63610	.68109	.31891	.95500	22
39	.63636	.68142	.31858	.95494	21
40	9.63662	9.68174	10.31826	9.95488	**20**
41	.63689	.68206	.31794	.95482	19
42	.63715	.68239	.31761	.95476	18
43	.63741	.68271	.31729	.95470	17
44	.63767	.68303	.31697	.95464	16
45	9.63794	9.68336	10.31664	9.95458	**15**
46	.63820	.68368	.31632	.95452	14
47	.63846	.68400	.31600	.95446	13
48	.63872	.68432	.31568	.95440	12
49	.63898	.68465	.31535	.95434	11
50	9.63924	9.68497	10.31503	9.95427	**10**
51	.63950	.68529	.31471	.95421	9
52	.63976	.68561	.31439	.95415	8
53	.64002	.68593	.31407	.95409	7
54	.64028	.68626	.31374	.95403	6
55	9.64054	9.68658	10.31342	9.95397	**5**
56	.64080	.68690	.31310	.95391	4
57	.64106	.68722	.31278	.95384	3
58	.64132	.68754	.31246	.95378	2
59	.64158	.68786	.31214	.95372	1
60	9.64184	9.68818	10.31182	9.95366	**0**
	Cosine.	Cotang.	Tang.	Sine.	'

64°

*Log secant = colog cosine = 1 − log cosine; log cosecant = colog sine = 1 − log sine.

Ex.—Log sec 24°- 30′ = 10.04098. *Ex.*—Log cosec 24°- 30′ = 10.38227.

3.—Logarithmic Sines, Tangents, Cotangents, Cosines.—(Cont'd.) (Secants, Cosecants.)*

26°						27°					
′	Sine.	Tang.	Cotang.	Cosine.		′	Sine.	Tang.	Cotang.	Cosine.	
0	9.64184	9.68818	10.31182	9.95366	**60**	**0**	9.65705	9.70717	10.29283	9.94988	**60**
1	.64210	.68850	.31150	.95360	59	1	.65729	.70748	.29252	.94982	59
2	.64236	.68882	.31118	.95354	58	2	.65754	.70779	.29221	.94975	58
3	.64262	.68914	.31086	.95348	57	3	.65779	.70810	.29190	.94969	57
4	.64288	.68946	.31054	.95341	56	4	.65804	.70841	.29159	.94962	56
5	9.64313	9.68978	10.31022	9.95335	**55**	**5**	9.65828	9.70873	10.29127	9.94956	**55**
6	.64339	.69010	.30990	.95329	54	6	.65853	.70904	.29096	.94949	54
7	.64365	.69042	.30958	.95323	53	7	.65878	.70935	.29065	.94943	53
8	.64391	.69074	.30926	.95317	52	8	.65902	.70966	.29034	.94936	52
9	.64417	.69106	.30894	.95310	51	9	.65927	.70997	.29003	.94930	51
10	9.64442	9.69138	10.30862	9.95304	**50**	**10**	9.65952	9.71028	10.28972	9.94923	**50**
11	.64468	.69170	.30830	.95298	49	11	.65976	.71059	.28941	.94917	49
12	.64494	.69202	.30798	.95292	48	12	.66001	.71090	.28910	.94911	48
13	.64519	.69234	.30766	.95286	47	13	.66025	.71121	.28879	.94904	47
14	.64545	.69266	.30734	.95279	46	14	.66050	.71153	.28847	.94898	46
15	9.64571	9.69298	10.30702	9.95273	**45**	**15**	9.66075	9.71184	10.28816	9.94891	**45**
16	.64596	.69329	.30671	.95267	44	16	.66099	.71215	.28785	.94885	44
17	.64622	.69361	.30639	.95261	43	17	.66124	.71246	.28754	.94878	43
18	.64647	.69393	.30607	.95254	42	18	.66148	.71277	.28723	.94871	42
19	.64673	.69425	.30575	.95248	41	19	.66173	.71308	.28692	.94865	41
20	9.64698	9.69457	10.30543	9.95242	**40**	**20**	9.66197	9.71339	10.28661	9.94858	**40**
21	.64724	.69488	.30512	.95236	39	21	.66221	.71370	.28630	.94852	39
22	.62749	.69520	.30480	.95229	38	22	.66246	.71401	.28599	.94845	38
23	.64775	.69552	.30448	.95223	37	23	.66270	.71431	.28569	.94839	37
24	.64800	.69584	.30416	.95217	36	24	.66295	.71462	.28538	.94832	36
25	9.64826	9.69615	10.30385	9.95211	**35**	**25**	9.66319	9.71493	10.28507	9.94826	**35**
26	.64851	.69647	.30353	.95204	34	26	.66343	.71524	.28476	.94819	34
27	.64877	.69679	.30321	.95198	33	27	.66368	.71555	.28445	.94813	33
28	.64902	.69710	.30290	.95192	32	28	.66392	.71586	.28414	.94806	32
29	.64927	.69742	.30258	.95185	31	29	.66416	.71617	.28383	.94799	31
30	9.64953	9.69774	10.30226	9.95179	**30**	**30**	9.66441	9.71648	10.28352	9.94793	**30**
31	.64978	.69805	.30195	.95173	29	31	.66465	.71679	.28321	.94786	29
32	.65003	.69837	.30163	.95167	28	32	.66489	.71709	.28291	.94780	28
33	.65029	.69868	.30132	.95160	27	33	.66513	.71740	.28260	.94773	27
34	.65054	.69900	.30100	.95154	26	34	.66537	.71771	.28229	.94767	26
35	9.65079	9.69932	10.30068	9.95148	**25**	**35**	9.66562	9.71802	10.28198	9.94760	**25**
36	.65104	.69963	.30037	.95141	24	36	.66586	.71833	.28167	.94753	24
37	.65130	.69995	.30005	.95135	23	37	.66610	.71863	.28137	.94747	23
38	.65155	.70026	.29974	.95129	22	38	.66634	.71894	.28106	.94740	22
39	.65180	.70058	.29942	.95122	21	39	.66658	.71925	.28075	.94734	21
40	9.65205	9.70089	10.29911	9.95116	**20**	**40**	9.66682	9.71955	10.28045	9.94727	**20**
41	.65230	.70121	.29879	.95110	19	41	.66706	.71986	.28014	.94720	19
42	.65255	.70152	.29848	.95103	18	42	.66731	.72017	.27983	.94714	18
43	.65281	.70184	.29816	.95097	17	43	.66755	.72048	.27952	.94707	17
44	.65306	.70215	.29785	.95090	16	44	.66779	.72078	.27922	.94700	16
45	9.65331	9.70247	10.29753	9.95084	**15**	**45**	9.66803	9.72109	10.27891	9.94694	**15**
46	.65356	.70278	.29722	.95078	14	46	.66827	.72140	.27860	.94687	14
47	.65381	.70309	.29691	.95071	13	47	.66851	.72170	.27830	.94680	13
48	.65406	.70341	.29659	.95065	12	48	.66875	.72201	.27799	.94674	12
49	.65431	.70372	.29628	.95059	11	49	.66899	.72231	.27769	.94667	11
50	9.65456	9.70404	10.29596	9.95052	**10**	**50**	9.66922	9.72262	10.27738	9.94660	**10**
51	.65481	.70435	.29565	.95046	9	51	.66946	.72293	.27707	.94654	9
52	.65506	.70466	.29534	.95039	8	52	.66970	.72323	.27677	.94647	8
53	.65531	.70498	.29502	.95033	7	53	.66994	.72354	.27646	.94640	7
54	.65556	.70529	.29471	.95027	6	54	.67018	.72384	.27616	.94634	6
55	9.65580	9.70560	10.29440	9.95020	**5**	**55**	9.67042	9.72415	10.27585	9.94627	**5**
56	.65605	.70592	.29408	.95014	4	56	.67066	.72445	.27555	.94620	4
57	.65630	.70623	.29377	.95007	3	57	.67090	.72476	.27524	.94614	3
58	.65655	.70654	.29346	.95001	2	58	.67113	.72506	.27494	.94607	2
59	.65680	.70685	.29315	.94995	1	59	.67137	.72537	.27463	.94600	1
60	9.65705	9.70717	10.29283	9.94988	**0**	**60**	9.67161	9.72567	10.27433	9.94593	**0**
	Cosine.	Cotang.	Tang.	Sine.	′		Cosine.	Cotang.	Tang.	Sine.	′
					63°						62°

*Log secant = colog cosine = 1 − log cosine; log cosecant = colog sine = 1 − log sine.

Ex.—Log sec 26°- 30′ = 10.04821. *Ex.*—Log cosec 26°- 30′ = 10.35047.

3.—**Logarithmic Sines,** TANGENTS, COTANGENTS, COSINES.—(Cont'd.)
(SECANTS, COSECANTS.)*

28°						29°					
′	Sine.	Tang.	Cotang.	Cosine.		′	Sine.	Tang.	Cotang.	Cosine.	
0	9.67161	9.72567	10.27433	9.94593	**60**	**0**	9.68557	9.74375	10.25625	9.94182	**60**
1	.67185	.72598	.27402	.94587	59	1	.68580	.74405	.25595	.94175	59
2	.67208	.72628	.27372	.94580	58	2	.68603	.74435	.25565	.94168	58
3	.67232	.72659	.27341	.94573	57	3	.68625	.74465	.25535	.94161	57
4	.67256	.72689	.27311	.94567	56	4	.68648	.74494	.25506	.94154	56
5	9.67280	9.72720	10.27280	9.94560	**55**	**5**	9.68671	9.74524	10.25476	9.94147	**55**
6	.67303	.72750	.27250	.94553	54	6	.68694	74554	.25446	.94140	54
7	.67327	.72780	.27220	.94546	53	7	.68716	.74583	.25417	.94133	53
8	.67350	.72811	.27189	.94540	52	8	.68739	.74613	.25387	94126	52
9	.67374	.72841	.27159	.94533	51	9	.68762	.74643	.25357	.94119	51
10	9.67398	9.72872	10.27128	9.94526	**50**	**10**	9.68784	9.74673	10.25327	9.94112	**50**
11	.67421	.72902	.27098	.94519	49	11	.68807	.74702	.25298	.94105	49
12	.67445	.72932	.27068	.94513	48	12	.68829	.74732	.25268	.94098	48
13	.67468	.72963	.27037	.94506	47	13	.68852	.74762	.25238	94090	47
14	.67492	.72993	.27007	.94499	46	14	.68875	74791	.25209	.94083	46
15	9.67515	9.73023	10.26977	9.94492	**45**	**15**	9.68897	9.74821	10.25179	9.94076	**45**
16	.67539	.73054	.26946	.94485	44	16	.68920	.74851	.25149	.94069	44
17	.67562	.73084	.26916	.94479	43	17	.68942	.74880	.25120	.94062	43
18	.67586	.73114	.26886	.94472	42	18	.68965	.74910	.25090	.94055	42
19	.67609	.73144	.26856	.94465	41	19	.68987	.74939	.25061	.94048	41
20	9.67633	9.73175	10.26825	9.94458	**40**	**20**	9.69010	9.74969	10.25031	9.94041	**40**
21	.67656	.73205	.26795	.94451	39	21	.69032	.74998	.25002	.94034	39
22	.67680	.73235	.26765	.94445	38	22	.69055	.75028	.24972	.94027	38
23	.67703	.73265	.26735	.94438	37	23	.69077	.75058	.24942	.94020	37
24	.67726	.73295	.26705	.94431	36	24	.69100	.75087	.24913	.94012	36
25	9.67750	9.73326	10.26674	9.94424	**35**	**25**	9.69122	9.75117	10.24883	9.94005	**35**
26	.67773	.73356	.26644	.94417	34	26	.69144	.75146	.24854	.93998	34
27	.67796	.73386	.26614	.94410	33	27	.69167	.75176	.24824	.93991	33
28	.67820	.73416	.26584	.94404	32	28	.69189	.75205	.24795	.93984	32
29	.67843	.73446	.26554	.94397	31	29	.69212	.75235	.24765	.93977	31
30	9.67866	9.73476	10.26524	9.94390	**30**	**30**	9.69234	9.75264	10.24736	9.93970	**30**
31	.67890	.73507	.26493	.94383	29	31	.69256	.75294	.24706	.93963	29
32	.67913	.73537	.26463	.94376	28	32	.69279	75323	.24677	.93955	28
33	.67936	.73567	.26433	.94369	27	33	.69301	.75353	.24647	.93948	27
34	.67959	.73597	.26403	.94362	26	34	.69323	.75382	.24618	.93941	26
35	9.67982	9.73627	10.26373	9.94355	**25**	**35**	9.69345	9.75411	10.24589	9.93934	**25**
36	.68006	.73657	.26343	.94349	24	36	.69368	.75441	.24559	.93927	24
37	.68029	.73687	.26313	.94342	23	37	.69390	75470	.24530	93920	23
38	.68052	.73717	.26283	.94335	22	38	.69412	75500	.24500	.93912	22
39	.68075	.73747	.26253	.94328	21	39	.69434	.75529	.24471	.93905	21
40	9.68098	9.73777	10.26223	9.94321	**20**	**40**	9.69456	9.75558	10.24442	9.93898	**20**
41	.68121	.73807	.26193	.94314	19	41	.69479	.75588	.24412	.93891	19
42	.68144	.73837	.26163	.94307	18	42	.69501	.75617	.24383	.93884	18
43	.68167	.73867	.26133	.94300	17	43	.69523	.75647	.24353	.93876	17
44	.68190	.73897	.26103	.94293	16	44	.69545	.75676	.24324	.93869	16
45	9.68213	9.73927	10.26073	9.94286	**15**	**45**	9.69567	9.75705	10.24295	9.93862	**15**
46	.68237	.73957	.26043	.94279	14	46	.69589	.75735	.24265	.93855	14
47	.68260	.73987	.26013	.94273	13	47	.69611	.75764	.24236	.93847	13
48	.68283	.74017	.25983	.94266	12	48	.69633	.75793	.24207	.93840	12
49	.68305	.74047	.25953	.94259	11	49	.69655	.75822	.24178	.93833	11
50	9.68328	9.74077	10.25923	9.94252	**10**	**50**	9.69677	9.75852	10.24148	9.93826	**10**
51	.68351	.74107	.25893	.94245	9	51	.69699	.75881	.24119	.93819	9
52	.68374	.74137	.25863	.94238	8	52	.69721	.75910	.24090	.93811	8
53	.68397	.74166	.25834	.94231	7	53	.69743	.75939	.24061	.93804	7
54	.68420	.74196	.25804	.94224	6	54	.69765	.75969	.24031	.93797	6
55	9.68443	9.74226	10.25774	9.94217	**5**	**55**	9.69787	9.75998	10.24002	9.93789	**5**
56	.68466	.74256	.25744	.94210	4	56	.69809	.76027	.23973	.93782	4
57	.68489	.74286	.25714	.94203	3	57	.69831	.76056	.23944	.93775	3
58	.68512	.74316	.25684	.94196	2	58	.69853	.76086	.23914	.93768	2
59	.68534	.74345	.25655	.94189	1	59	.69875	.76115	.23885	.93760	1
60	9.68557	9.74375	10.25625	9.94182	**0**	**60**	9.69897	9.76144	10.23856	9.93753	**0**
	Cosine.	Cotang.	Tang.	Sine.	′		Cosine.	Cotang.	Tang.	Sine.	′
					61°						60°

*Log secant = colog cosine = 1 − log cosine; log cosecant = colog sine = 1 − log sine.

Ex.—Log sec 28°- 30′ = 10.05610. *Ex.*—Log cosec 28°- 30′ = 10.32134

3.—Logarithmic Sines, TANGENTS, COTANGENTS, COSINES.—(Cont'd.)
(SECANTS, COSECANTS.)*

30°							31°				
′	Sine.	Tang.	Cotang.	Cosine.		′	Sine.	Tang.	Cotang.	Cosine.	
0	9.69897	9.76144	10.23856	9.93753	**60**	**0**	9.71184	9.77877	10.22123	9.93307	**60**
1	.69919	.76173	.23827	.93746	59	1	.71205	.77906	.22094	.93299	59
2	.69941	.76202	.23798	.93738	58	2	.71226	.77935	.22065	.93291	58
3	.69963	.76231	.23769	.93731	57	3	.71247	.77963	.22037	.93284	57
4	.69984	.76261	.23739	.93724	56	4	.71268	.77992	.22008	.93276	56
5	9.70006	9.76290	10.23710	9.93717	**55**	**5**	9.71289	9.78020	10.21980	9.93269	**55**
6	.70028	.76319	.23681	.93709	54	6	.71310	.78049	.21951	.93261	54
7	.70050	.76348	.23652	.93702	53	7	.71331	.78077	.21923	.93253	53
8	.70072	.76377	.23623	.93695	52	8	.71352	.78106	.21894	.93246	52
9	.70093	.76406	.23594	.93687	51	9	.71373	.78135	.21865	.93238	51
10	9.70115	9.76435	10.23565	9.93680	**50**	**10**	9.71393	9.78163	10.21837	9.93230	**50**
11	.70137	.76464	.23536	.93673	49	11	.71414	.78192	.21808	.93223	49
12	.70159	.76493	.23507	.93665	48	12	.71435	.78220	.21780	.93215	48
13	.70180	.76522	.23478	.93658	47	13	.71456	.78249	.21751	.93207	47
14	.70202	.76551	.23449	.93650	46	14	.71477	.78277	.21723	.93200	46
15	9.70224	9.76580	10.23420	9.93643	**45**	**15**	9.71498	9.78306	10.21694	9.93192	**45**
16	.70245	.76609	.23391	.93636	44	16	.71519	.78334	.21666	.93184	44
17	.70267	.76639	.23361	.93628	43	17	.71539	.78363	.21637	.93177	43
18	.70288	.76668	.23332	.93621	42	18	.71560	.78391	.21609	.93169	42
19	.70310	.76697	.23303	.93614	41	19	.71581	.78419	.21581	.93161	41
20	9.70332	9.76725	10.23275	9.93606	**40**	**20**	9.71602	9.78448	10.21552	9.93154	**40**
21	.70353	.76754	.23246	.93599	39	21	.71622	.78476	.21524	.93146	39
22	.70375	.76783	.23217	.93591	38	22	.71643	.78505	.21495	.93138	38
23	.70396	.76812	.23188	.93584	37	23	.71664	.78533	.21467	.93131	37
24	.70418	.76841	.23159	.93577	36	24	.71685	.78562	.21438	.93123	36
25	9.70439	9.76870	10.23130	9.93569	**35**	**25**	9.71705	9.78590	10.21410	9.93115	**35**
26	.70461	.76899	.23101	.93562	34	26	.71726	.78618	.21382	.93107	34
27	.70482	.76928	.23072	.93554	33	27	.71747	.78647	.21353	.93100	33
28	.70504	.76957	.23043	.93547	32	28	.71767	.78675	.21325	.93092	32
29	.70525	.76986	.23014	.93539	31	29	.71788	.78704	.21296	.93084	31
30	9.70547	9.77015	10.22985	9.93532	**30**	**30**	9.71809	9.78732	10.21268	9.93077	**30**
31	.70568	.77044	.22956	.93525	29	31	.71829	.78760	.21240	.93069	29
32	.70590	.77073	.22927	.93517	28	32	.71850	.78789	.21211	.93061	28
33	.70611	.77101	.22899	.93510	27	33	.71870	.78817	.21183	.93053	27
34	.70633	.77130	.22870	.93502	26	34	.71891	.78845	.21155	.93046	26
35	9.70654	9.77159	10.22841	9.93495	**25**	**35**	9.71911	9.78874	10.21126	9.93038	**25**
36	.70675	.77188	.22812	.93487	24	36	.71932	.78902	.21098	.93030	24
37	.70697	.77217	.22783	.93480	23	37	.71952	.78930	.21070	.93022	23
38	.70718	.77246	.22754	.93472	22	38	.71973	.78959	.21041	.93014	22
39	.70739	.77274	.22726	.93465	21	39	.71994	.78987	.21013	.93007	21
40	9.70761	9.77303	10.22697	9.93457	**20**	**40**	9.72014	9.79015	10.20985	9.92999	**20**
41	.70782	.77332	.22668	.93450	19	41	.72034	.79043	.20957	.92991	19
42	.70803	.77361	.22639	.93442	18	42	.72055	.79072	.20928	.92983	18
43	.70824	.77390	.22610	.93435	17	43	.72075	.79100	.20900	.92976	17
44	.70846	.77418	.22582	.93427	16	44	.72096	.79128	.20872	.92968	16
45	9.70867	9.77447	10.22553	9.93420	**15**	**45**	9.72116	9.79156	10.20844	9.92960	**15**
46	.70888	.77476	.22524	.93412	14	46	.72137	.79185	.20815	.92952	14
47	.70909	.77505	.22495	.93405	13	47	.72157	.79213	.20787	.92944	13
48	.70931	.77533	.22467	.93397	12	48	.72177	.79241	.20759	.92936	12
49	.70952	.77562	.22438	.93390	11	49	.72198	.79269	.20731	.92929	11
50	9.70973	9.77591	10.22409	9.93382	**10**	**50**	9.72218	9.79297	10.20703	9.92921	**10**
51	.70994	.77619	.22381	.93375	9	51	.72238	.79326	.20674	.92913	9
52	.71015	.77648	.22352	.93367	8	52	.72259	.79354	.20646	.92905	8
53	.71036	.77677	.22323	.93360	7	53	.72279	.79382	.20618	.92897	7
54	.71058	.77706	.22294	.93352	6	54	.72299	.79410	.20590	.92889	6
55	9.71079	9.77734	10.22266	9.93344	**5**	**55**	9.72320	9.79438	10.20562	9.92881	**5**
56	.71100	.77763	.22237	.93337	4	56	.72340	.79466	.20534	.92874	4
57	.71121	.77791	.22209	.93329	3	57	.72360	.79495	.20505	.92866	3
58	.71142	.77820	.22180	.93322	2	58	.72381	.79523	.20477	.92858	2
59	.71163	.77849	.22151	.93314	1	59	.72401	.79551	.20449	.92850	1
60	9.71184	9.77877	10.22123	9.93307	**0**	**60**	9.72421	9.79579	10.20421	9.92842	**0**
	Cosine.	Cotang.	Tang.	Sine.	′		Cosine.	Cotang.	Tang.	Sine.	′
					59°						58°

*Log secant = colog cosine = 1 − log cosine; log cosecant = colog sine = 1 − log sine.

Ex.—Log sec 30°− 30′ = 10.06468. *Ex.*—Log cosec 30°− 30′ = 10.29453.

3. —**Logarithmic Sines,** TANGENTS, COTANGENTS, COSINES.—(Cont'd.)
(SECANTS, COSECANTS.)*

32°						33°					
′	Sine.	Tang.	Cotang.	Cosine.		′	Sine.	Tang.	Cotang.	Cosine.	
0	9.72421	9.79579	10.20421	9.92842	**60**	**0**	9.73611	9.81252	10.18748	9.92359	**60**
1	.72441	.79607	.20393	.92834	59	1	.73630	.81279	.18721	.92351	59
2	.72461	.79635	.20365	.92826	58	2	.73650	.81307	.18693	.92343	58
3	.72482	.79663	.20337	.92818	57	3	.73669	.81335	.18665	.92335	57
4	.72502	.79691	.20309	.92810	56	4	.73689	.81362	.18638	.92326	56
5	9.72522	9.79719	10.20281	9.92803	**55**	**5**	9.73708	9.81390	10.18610	9.92318	**55**
6	.72542	.79747	.20253	.92795	54	6	.73727	.81418	.18582	.92310	54
7	.72562	.79776	.20224	.92787	53	7	.73747	.81445	.18555	.92302	53
8	.72582	.79804	.20196	.92779	52	8	.73766	.81473	.18527	.92293	52
9	.72602	.79832	.20168	.92771	51	9	.73785	.81500	.18500	.92285	51
10	9.72622	9.79860	10.20140	9.92763	**50**	**10**	9.73805	9.81528	10.18472	9.92277	**50**
11	.72643	.79888	.20112	.92755	49	11	.73824	.81556	.18444	.92269	49
12	.72663	.79916	.20084	.92747	48	12	.73843	.81583	.18417	.92260	48
13	.72683	.79944	.20056	.92739	47	13	.73863	.81611	.18389	.92252	47
14	.72703	.79972	.20028	.92731	46	14	.73882	.81638	.18362	.92244	46
15	9.72723	9.80000	10.20000	9.92723	**45**	**15**	9.73901	9.81666	10.18334	9.92235	**45**
16	.72743	.80028	.19972	.92715	44	16	.73921	.81693	.18307	.92227	44
17	.72763	.80056	.19944	.92707	43	17	.73940	.81721	.18279	.92219	43
18	.72783	.80084	.19916	.92699	42	18	.73959	.81748	.18252	.92211	42
19	.72803	.80112	.19888	.92691	41	19	.73978	.81776	.18224	.92202	41
20	9.72823	9.80140	10.19860	9.92683	**40**	**20**	9.73997	9.81803	10.18197	9.92194	**40**
21	.72843	.80168	.19832	.92675	39	21	.74017	.81831	.18169	.92186	39
22	.72863	.80195	.19805	.92667	38	22	.74036	.81858	.18142	.92177	38
23	.72883	.80223	.19777	.92659	37	23	.74055	.81886	.18114	.92169	37
24	.72902	.80251	.19749	.92651	36	24	.74074	.81913	.18087	.92161	36
25	9.72922	9.80279	10.19721	9.92643	**35**	**25**	9.74093	9.81941	10.18059	9.92152	**35**
26	.72942	.80307	.19693	.92635	34	26	.74113	.81968	.18032	.92144	34
27	.72962	.80335	.19665	.92627	33	27	.74132	.81996	.18004	.92136	33
28	.72982	.80363	.19637	.92619	32	28	.74151	.82023	.17977	.92127	32
29	.73002	.80391	.19609	.92611	31	29	.74170	.82051	.17949	.92119	31
30	9.73022	9.80419	10.19581	9.92603	**30**	**30**	9.74189	9.82078	10.17922	9.92111	**30**
31	.73041	.80447	.19553	.92595	29	31	.74208	.82106	.17894	.92102	29
32	.73061	.80474	.19526	.92587	28	32	.74227	.82133	.17867	.92094	28
33	.73081	.80502	.19498	.92579	27	33	.74246	.82161	.17839	.92086	27
34	.73101	.80530	.19470	.92571	26	34	.74265	.82188	.17812	.92077	26
35	9.73121	9.80558	10.19442	9.92563	**25**	**35**	9.74284	9.82215	10.17785	9.92069	**25**
36	.73140	.80586	.19414	.92555	24	36	.74303	.82243	.17757	.92060	24
37	.73160	.80614	.19386	.92546	23	37	.74322	.82270	.17730	.92052	23
38	.73180	.80642	.19358	.92538	22	38	.74341	.82298	.17702	.92044	22
39	.73200	.80669	.19331	.92530	21	39	.74360	.92325	.17675	.92035	21
40	9.73219	9.80697	10.19303	9.92522	**20**	**40**	9.74379	9.82352	10.17648	9.92027	**20**
41	.73239	.80725	.19275	.92514	19	41	.74398	.82380	.17620	.92018	19
42	.73259	.80753	.19247	.92506	18	42	.74417	.82407	.17593	.92010	18
43	.73278	.80781	.19219	.92498	17	43	.74436	.82435	.17565	.92002	17
44	.73298	.80808	.19192	.92490	16	44	.74455	.82462	.17538	.91993	16
45	9.73318	9.80836	10.19164	9.92482	**15**	**45**	9.74474	9.82489	10.17511	9.91985	**15**
46	.73337	.80864	.19136	.92473	14	46	.74493	.82517	.17483	.91976	14
47	.73357	.80892	.19108	.92465	13	47	.74512	.82544	.17456	.91968	13
48	.73377	.80919	.19081	.92457	12	48	.74531	.82571	.17429	.91959	12
49	.73396	.80947	.19053	.92449	11	49	.74549	.82599	.17401	.91951	11
50	9.73416	9.80975	10.19025	9.92441	**10**	**50**	9.74568	9.82626	10.17374	9.91942	**10**
51	.73435	.81003	.18997	.92433	9	51	.74587	.82653	.17347	.91934	9
52	.73455	.81030	.18970	.92425	8	52	.74606	.82681	.17319	.91925	8
53	.73474	.81058	.18942	.92416	7	53	.74625	.82708	.17292	.91917	7
54	.73494	.81086	.18914	.92408	6	54	.74644	.82735	.17265	.91908	6
55	9.73513	9.81113	10.18887	9.92400	**5**	**55**	9.74662	9.82762	10.17238	9.91900	**5**
56	.73533	.81141	.18859	.92392	4	56	.74681	.82790	.17210	.91891	4
57	.73552	.81169	.18831	.92384	3	57	.74700	.82817	.17183	.91883	3
58	.73572	.81196	.18804	.92376	2	58	.74719	.82844	.17156	.91874	2
59	.73591	.81224	.18776	.92367	1	59	.74737	.82871	.17129	.91866	1
60	9.73611	9.81252	10.18748	9.92359	**0**	**60**	9.74756	9.82899	10.17101	9.91857	**0**
	Cosine.	Cotang.	Tang.	Sine.	′		Cosine.	Cotang.	Tang.	Sine.	
					57°						56°

*Log secant = colog cosine = 1 − log cosine; log cosecant = colog sine = 1 − log sine.
Ex.—Log sec 32°- 30′ = 10.07397. *Ex.*—Log cosec 32°- 30′ = 10.26978.

3.—Logarithmic Sines, Tangents, Cotangents, Cosines.—(Cont'd.)
(Secants, Cosecants.)*

34°

'	Sine.	Tang.	Cotang.	Cosine.	
0	9.74756	9.82899	10.17101	9.91857	60
1	.74775	.82926	.17074	.91849	59
2	.74794	.82953	.17047	.91840	58
3	.74812	.82980	.17020	.91832	57
4	.74831	.83008	.16992	.91823	56
5	9.74850	9.83035	10.16965	9.91815	55
6	.74868	.83062	.16938	.91806	54
7	.74887	.83089	.16911	.91798	53
8	.74906	.83117	.16883	.91789	52
9	.74924	.83144	.16856	.91781	51
10	9.74943	9.83171	10.16829	9.91772	50
11	.74961	.83198	.16802	.91763	49
12	.74980	.83225	.16775	.91755	48
13	.74999	.83252	.16748	.91746	47
14	.75017	.83280	.16720	.91738	46
15	9.75036	9.83307	10.16693	9.91729	45
16	.75054	.83334	.16666	.91720	44
17	.75073	.83361	.16639	.91712	43
18	.75091	.83388	.16612	.91703	42
19	.75110	.83415	.16585	.91695	41
20	9.75128	9.83442	10.16558	9.91686	40
21	.75147	.83470	.16530	.91677	39
22	.75165	.83497	.16503	.91669	38
23	.75184	.83524	.16476	.91660	37
24	.75202	.83551	.16449	.91651	36
25	9.75221	9.83578	10.16422	9.91643	35
26	.75239	.83605	.16395	.91634	34
27	.75258	.83632	.16368	.91625	33
28	.75276	.83659	.16341	.91617	32
29	.75294	.83686	.16314	.91608	31
30	9.75313	9.83713	10.16287	9.91599	30
31	.75331	.83740	.16260	.91591	29
32	.75350	.83768	.16232	.91582	28
33	.75368	.83795	.16205	.91573	27
34	.75386	.83822	.16178	.91565	26
35	9.75405	9.83849	10.16151	9.91556	25
36	.75423	.83876	.16124	.91547	24
37	.75441	.83903	.16097	.91538	23
38	.75459	.83930	.16070	.91530	22
39	.75478	.83957	.16043	.91521	21
40	9.75496	9.83984	10.16016	9.91512	20
41	.75514	.84011	.15989	.91504	19
42	.75533	.84038	.15962	.91495	18
43	.75551	.84065	.15935	.91486	17
44	.75569	.84092	.15908	.91477	16
45	9.75587	9.84119	10.15881	9.91469	15
46	.75605	.84146	.15854	.91460	14
47	.75624	.84173	.15827	.91451	13
48	.75642	.84200	.15800	.91442	12
49	.75660	.84227	.15773	.91433	11
50	9.75678	9.84254	10.15746	9.91425	10
51	.75696	.84280	.15720	.91416	9
52	.75714	.84307	.15693	.91407	8
53	.75733	.84334	.15666	.91398	7
54	.75751	.84361	.15639	.91389	6
55	9.75769	9.84388	10.15612	9.91381	5
56	.75787	.84415	.15585	.91372	4
57	.75805	.84442	.15558	.91363	3
58	.75823	.84469	.15531	.91354	2
59	.75841	.84496	.15504	.91345	1
60	9.75859	9.84523	10.15477	9.91336	0
	Cosine.	Cotang.	Tang.	Sine.	'

55°

35°

'	Sine.	Tang.	Cotang.	Cosine.	
0	9.75859	9.84523	10.15477	9.91336	60
1	.75877	.84550	.15450	.91328	59
2	.75895	.84576	.15424	.91319	58
3	.75913	.84603	.15397	.91310	57
4	.75931	.84630	.15370	.91301	56
5	9.75949	9.84657	10.15343	9.91292	55
6	.75967	.84684	.15316	.91283	54
7	.75985	.84711	.15289	.91274	53
8	.76003	.84738	.15262	.91266	52
9	.76021	.84764	.15236	.91257	51
10	9.76039	9.84791	10.15209	9.91248	50
11	.76057	.84818	.15182	.91239	49
12	.76075	.84845	.15155	.91230	48
13	.76093	.84872	.15128	.91221	47
14	.76111	.84899	.15101	.91212	46
15	9.76129	9.84925	10.15075	9.91203	45
16	.76146	.84952	.15048	.91194	44
17	.76164	.84979	.15021	.91185	43
18	.76182	.85006	.14994	.91176	42
19	.76200	.85033	.14967	.91167	41
20	9.76218	9.85059	10.14941	9.91158	40
21	.76236	.85086	.14914	.91149	39
22	.76253	.85113	.14887	.91141	38
23	.76271	.85140	.14860	.91132	37
24	.76289	.85166	.14834	.91123	36
25	9.76307	9.85193	10.14807	9.91114	35
26	.76324	.85220	.14780	.91105	34
27	.76342	.85247	.14753	.91096	33
28	.76360	.85273	.14727	.91087	32
29	.76378	.85300	.14700	.91078	31
30	9.76395	9.85327	10.14673	9.91069	30
31	.76413	.85354	.14646	.91060	29
32	.76431	.85380	.14620	.91051	28
33	.76448	.85407	.14593	.91042	27
34	.76466	.85434	.14566	.91033	26
35	9.76484	9.85460	10.14540	9.91023	25
36	.76501	.85487	.14513	.91014	24
37	.76519	.85514	.14486	.91005	23
38	.76537	.85540	.14460	.90996	22
39	.76554	.85567	.14433	.90987	21
40	9.76572	9.85594	10.14406	9.90978	20
41	.76590	.85620	.14380	.90969	19
42	.76607	.85647	.14353	.90960	18
43	.76625	.85674	.14326	.90951	17
44	.76642	.85700	.14300	.90942	16
45	9.76660	9.85727	10.14273	9.90933	15
46	.76677	.85754	.14246	.90924	14
47	.76695	.85780	.14220	.90915	13
48	.76712	.85807	.14193	.90906	12
49	.76730	.85834	.14166	.90896	11
50	9.76747	9.85860	10.14140	9.90887	10
51	.76765	.85887	.14113	.90878	9
52	.76782	.85913	.14087	.90869	8
53	.76800	.85940	.14060	.90860	7
54	.76817	.85967	.14033	.90851	6
55	9.76835	9.85993	10.14007	9.90842	5
56	.76852	.86020	.13980	.90832	4
57	.76870	.86046	.13954	.90823	3
58	.76887	.86073	.13927	.90814	2
59	.76904	.86100	.13900	.90805	1
60	9.76922	9.86126	10.13874	9.90796	0
	Cosine.	Cotang.	Tang.	Sine.	'

54°

*Log secant = colog cosine = 1 − log cosine; log cosecant = colog sine = 1 − log sine.

Ex.—Log sec 34°- 30′ = 10.08401. *Ex.*—Log cosec 34°- 30′ = 10.24687.

3.—Logarithmic Sines, Tangents, Cotangents, Cosines—(Cont'd.) (Secants, Cosecants.)*

36°

′	Sine.	Tang.	Cotang.	Cosine.	
0	9.76922	9.86126	10.13874	9.90796	60
1	.76939	.86153	.13847	.90787	59
2	.76957	.86179	.13821	.90777	58
3	.76974	.86206	.13794	.90768	57
4	.76991	.86232	.13768	.90759	56
5	9.77009	9.86259	10.13741	9.90750	55
6	.77026	.86285	.13715	.90741	54
7	.77043	.86312	.13688	.90731	53
8	.77061	.86338	.13662	.90722	52
9	.77078	.86365	.13635	.90713	51
10	9.77095	9.86392	10.13608	9.90704	50
11	.77112	.86418	.13582	.90694	49
12	.77130	.86445	.13555	.90685	48
13	.77147	.86471	.13529	.90676	47
14	.77164	.86498	.13502	.90667	46
15	9.77181	9.86524	10.13476	9.90657	45
16	.77199	.86551	.13449	.90648	44
17	.77216	.86577	.13423	.90639	43
18	.77233	.86603	.13397	.90630	42
19	.77250	.86630	.13370	.90620	41
20	9.77268	9.86656	10.13344	9.90611	40
21	.77285	.86683	.13317	.90602	39
22	.77302	.86709	.13291	.90592	38
23	.77319	.86736	.13264	.90583	37
24	.77336	.86762	.13238	.90574	36
25	9.77353	9.86789	10.13211	9.90565	35
26	.77370	.86815	.13185	.90555	34
27	.77387	.86842	.13158	.90546	33
28	.77405	.86868	.13132	.90537	32
29	.77422	.86894	.13106	.90527	31
30	9.77439	9.86921	10.13079	9.90518	30
31	.77456	.86947	.13053	.90509	29
32	.77473	.86974	.13026	.90499	28
33	.77490	.87000	.13000	.90490	27
34	.77507	.87027	.12973	.90480	26
35	9.77524	9.87053	10.12947	9.90471	25
36	.77541	.87079	.12921	.90462	24
37	.77558	.87106	.12894	.90452	23
38	.77575	.87132	.12868	.90443	22
39	.77592	.87158	.12842	.90434	21
40	9.77609	9.87185	10.12815	9.90424	20
41	.77626	.87211	.12789	.90415	19
42	.77643	.87238	.12762	.90405	18
43	.77660	.87264	.12736	.90396	17
44	.77677	.87290	.12710	.90386	16
45	9.77694	9.87317	10.12683	9.90377	15
46	.77711	.87343	.12657	.90368	14
47	.77728	.87369	.12631	.90358	13
48	.77744	.87396	.12604	.90349	12
49	.77761	.87422	.12578	.90339	11
50	9.77778	9.87448	10.12552	9.90330	10
51	.77795	.87475	.12525	.90320	9
52	.77812	.87501	.12499	.90311	8
53	.77829	.87527	.12473	.90301	7
54	.77846	.87554	.12446	.90292	6
55	9.77862	9.87580	10.12420	9.90282	5
56	.77879	.87606	.12394	.90273	4
57	.77896	.87633	.12367	.90263	3
58	.77913	.87659	.12341	.90254	2
59	.77930	.87685	.12315	.90244	1
60	9.77946	9.87711	10.12289	9.90235	0
	Cosine.	Cotang.	Tang.	Sine.	′

53°

37°

′	Sine.	Tang.	Cotang.	Cosine.	
0	9.77946	9.87711	10.12289	9.90235	60
1	.77963	.87738	12262	.90225	59
2	.77980	87764	.12236	.90216	58
3	.77997	.87790	.12210	.90206	57
4	.78013	.87817	.12183	.90197	56
5	9.78030	9.87843	10.12157	9.90187	55
6	.78047	.87869	.12131	.90178	54
7	.78063	.87895	.12105	.90168	53
8	.78080	.87922	.12078	.90159	52
9	.78097	.87948	.12052	.90149	51
10	9.78113	9.87974	10.12026	9.90139	50
11	.78130	.88000	.12000	.90130	49
12	.78147	.88027	.11973	.90120	48
13	.78163	.88053	.11947	.90111	47
14	.78180	.88079	.11921	.90101	46
15	9.78197	9.88105	10.11895	9.90091	45
16	.78213	.88131	.11869	.90082	44
17	.78230	.88158	.11842	.90072	43
18	.78246	.88184	.11816	.90063	42
19	.78263	.88210	.11790	.90053	41
20	9.78280	9.88236	10.11764	9.90043	40
21	.78296	.88262	.11738	.90034	39
22	.78313	.88289	.11711	.90024	38
23	.78329	.88315	.11685	.90014	37
24	.78346	.88341	.11659	.90005	36
25	9.78362	9.88367	10.11633	9.89995	35
26	.78379	.88393	.11607	.89985	34
27	.78395	.88420	.11580	.89976	33
28	.78412	.88446	.11554	.89966	32
29	.78428	.88472	.11528	.89956	31
30	9.78445	9.88498	10.11502	9.89947	30
31	.78461	.88524	.11476	.89937	29
32	.78478	.88550	.11450	.89927	28
33	.78494	.88577	.11423	.89918	27
34	.78510	.88603	.11397	.89908	26
35	9.78527	9.88629	10.11371	9.89898	25
36	.78543	.88655	.11345	.89888	24
37	.78560	.88681	.11319	.89879	23
38	.78576	.88707	.11293	.89869	22
39	.78592	.88733	.11267	.89859	21
40	9.78609	9.88759	10.11241	9.89849	20
41	.78625	.88786	.11214	.89840	19
42	.78642	.88812	.11188	.89830	18
43	.78658	.88838	.11162	.89820	17
44	.78674	.88864	.11136	.89810	16
45	9.78691	9.88890	10.11110	9.89801	15
46	.78707	.88916	.11084	.89791	14
47	.78723	.88942	.11058	.89781	13
48	.78739	.88968	.11032	.89771	12
49	.78756	.88994	.11006	.89761	11
50	9.78772	9.89020	10.10980	9.89752	10
51	.78788	.89046	.10954	.89742	9
52	.78805	.89073	.10927	.89732	8
53	.78821	.89099	.10901	.89722	7
54	.78837	.89125	.10875	.89712	6
55	9.78853	9.89151	10.10849	9.89702	5
56	.78869	.89177	.10823	.89693	4
57	.78886	.89203	.10797	.89683	3
58	.78902	.89229	.10771	.89673	2
59	.78918	.89255	.10745	.89663	1
60	9.78934	9.89281	10.10719	9.89653	0
	Cosine.	Cotang.	Tang.	Sine.	′

52°

*Log secant = colog cosine = 1 − log cosine; log cosecant = colog sine = 1 − log sine.

Ex.—Log sec 36°− 30′ = 10.09482. *Ex.*—Log cosec 36°− 30′ = 10.22561

3.—Logarithmic Sines, Tangents, Cotangents, Cosines.—(Cont'd.) (Secants, Cosecants.)*

38°						39°					
′	Sine.	Tang.	Cotang.	Cosine.		′	Sine.	Tang.	Cotang.	Cosine.	
0	9.78934	9.89281	10.10719	9.89653	**60**	**0**	9.79887	9.90837	10.09163	9.89050	**60**
1	.78950	.89307	.10693	.89643	59	1	.79903	.90863	.09137	.89040	59
2	.78967	.89333	.10667	.89633	58	2	.79918	.90889	.09111	.89030	58
3	.78983	.89359	.10641	.89624	57	3	.79934	.90914	.09086	.89020	57
4	.78999	.89385	.10615	.89614	56	4	.79950	.90940	.09060	.89009	56
5	9.79015	9.89411	10.10589	9.89604	**55**	**5**	9.79965	9.90966	10.09034	9.88999	**55**
6	.79031	.89437	.10563	.89594	54	6	.79981	.90992	.09008	.88989	54
7	.79047	.89463	.10537	.89584	53	7	.79996	.91018	.08982	.88978	53
8	.79063	.89489	.10511	.89574	52	8	.80012	.91043	.08957	.88968	52
9	.79079	.89515	.10485	.89564	51	9	.80027	.91069	.08931	.88958	51
10	9.79095	9.89541	10.10459	9.89554	**50**	**10**	9.80043	9.91095	10.08905	9.88948	**50**
11	.79111	.89567	.10433	.89544	49	11	.80058	.91121	.08879	.88937	49
12	.79128	.89593	.10407	.89534	48	12	.80074	.91147	.08853	.88927	48
13	.79144	.89619	.10381	.89524	47	13	.80089	.91172	.08828	.88917	47
14	.79160	.89645	.10355	.89514	46	14	.80105	.91198	.08802	.88906	46
15	9.79176	9.89671	10.10329	9.89504	**45**	**15**	9.80120	9.91224	10.08776	9.88896	**45**
16	.79192	.89697	.10303	.89495	44	16	.80136	.91250	.08750	.88886	44
17	.79208	.89723	.10277	.89485	43	17	.80151	.91276	.08724	.88875	43
18	.79224	.89749	.10251	.89475	42	18	.80166	.91301	.08699	.88865	42
19	.79240	.89775	.10225	.89465	41	19	.80182	.91327	.08673	.88855	41
20	9.79256	9.89801	10.10199	9.89455	**40**	**20**	9.80197	9.91353	10.08647	9.88844	**40**
21	.79272	.89827	.10173	.89445	39	21	.80213	.91379	.08621	.88834	39
22	.79288	.89853	.10147	.89435	38	22	.80228	.91404	.08596	.88824	38
23	.79304	.89879	.10121	.89425	37	23	.80244	.91430	.08570	.88813	37
24	.79319	.89905	.10095	.89415	36	24	.80259	.91456	.08544	.88803	36
25	9.79335	9.89931	10.10069	9.89405	**35**	**25**	9.80274	9.91482	10.08518	9.88793	**35**
26	.79351	.89957	.10043	.89395	34	26	.80290	.91507	.08493	.88782	34
27	.79367	.89983	.10017	.89385	33	27	.80305	.91533	.08467	.88772	33
28	.79383	.90009	.09991	.89375	32	28	.80320	.91559	.08441	.88761	32
29	.79399	.90035	.09965	.89364	31	29	.80336	.91585	.08415	.88751	31
30	9.79415	9.90061	10.09939	9.89354	**30**	**30**	9.80351	9.91610	10.08390	9.88741	**30**
31	.79431	.90086	.09914	.89344	29	31	.80366	.91636	.08364	.88730	29
32	.79447	.90112	.09888	.89334	28	32	.80382	.91662	.08338	.88720	28
33	.79463	.90138	.09862	.89324	27	33	.80397	.91688	.08312	.88709	27
34	.79478	.90164	.09836	.89314	26	34	.80412	.91713	.08287	.88699	26
35	9.79494	9.90190	10.09810	9.89304	**25**	**35**	9.80428	9.91739	10.08261	9.88688	**25**
36	.79510	.90216	.09784	.89294	24	36	.80443	.91765	.08235	.88678	24
37	.79526	.90242	.09758	.89284	23	37	.80458	.91791	.08209	.88668	23
38	.79542	.90268	.09732	.89274	22	38	.80473	.91816	.08184	.88657	22
39	.79558	.90294	.09706	.89264	21	39	.80489	.91842	.08158	.88647	21
40	9.79573	9.90320	10.09680	9.89254	**20**	**40**	9.80504	9.91868	10.08132	9.88636	**20**
41	.79589	.90346	.09654	.89244	19	41	.80519	.91893	.08107	.88626	19
42	.79605	.90371	.09629	.89233	18	42	.80534	.91919	.08081	.88615	18
43	.79621	.90397	.09603	.89223	17	43	.80550	.91945	.08055	.88605	17
44	.79636	.90423	.09577	.89213	16	44	.80565	.91971	.08029	.88594	16
45	9.79652	9.90449	10.09551	9.89203	**15**	**45**	9.80580	9.91996	10.08004	9.88584	**15**
46	.79668	.90475	.09525	.89193	14	46	.80595	.92022	.07978	.88573	14
47	.79684	.90501	.09499	.89183	13	47	.80610	.92048	.07952	.88563	13
48	.79699	.90527	.09473	.89173	12	48	.80625	.92073	.07927	.88552	12
49	.79715	.90553	.09447	.89162	11	49	.80641	.92099	.07901	.88542	11
50	9.79731	9.90578	10.09422	9.89152	**10**	**50**	9.80656	9.92125	10.07875	9.88531	**10**
51	.79746	.90604	.09396	.89142	9	51	.80671	.92150	.07850	.88521	9
52	.79762	.90630	.09370	.89132	8	52	.80686	.92176	.07824	.88510	8
53	.79778	.90656	.09344	.89122	7	53	.80701	.92202	.07798	.88499	7
54	.79793	.90682	.09318	.89112	6	54	.80716	.92227	.07773	.88489	6
55	9.79809	9.90708	10.09292	9.89101	**5**	**55**	9.80731	9.92253	10.07747	9.88478	**5**
56	.79825	.90734	.09266	.89091	4	56	.80746	.92279	.07721	.88468	4
57	.79840	.90759	.09241	.89081	3	57	.80762	.92304	.07696	.88457	3
58	.79856	.90785	.09215	.89071	2	58	.80777	.92330	.07670	.88447	2
59	.79872	.90811	.09189	.89060	1	59	.80792	.92356	.07644	.88436	1
60	9.79887	9.90837	10.09163	9.89050	**0**	**60**	9.80807	9.92381	10.07619	9.88425	**0**
	Cosine.	Cotang.	Tang.	Sine.	′		Cosine.	Cotang.	Tang.	Sine.	′
					51°						50°

*Log secant = colog cosine = 1 − log cosine; log cosecant = colog sine = 1 − log sine.

Ex.—Log sec 38°− 30′ = 10 10646. *Ex.*—Log cosec 38°− 30′ = 10.20585.

3. —Logarithmic Sines, TANGENTS, COTANGENTS, COSINES.—(Cont'd.)
(SECANTS, COSECANTS.)*

40°						41°					
′	Sine.	Tang.	Cotang.	Cosine.		′	Sine.	Tang.	Cotang.	Cosine.	
0	9.80807	9.92381	10.07619	9.88425	**60**	**0**	9.81694	9.93916	10.06084	9.87778	**60**
1	.80822	.92407	.07593	.88415	59	1	.81709	.93942	.06058	.87767	59
2	.80837	.92433	.07567	.88404	58	2	.81723	.93967	.06033	.87756	58
3	.80852	.92458	.07542	.88394	57	3	.81738	.93993	.06007	.87745	57
4	.80867	.92484	.07516	.88383	56	4	.81752	.94018	.05982	.87734	56
5	9.80882	9.92510	10.07490	9.88372	**55**	**5**	9.81767	9.94044	10.05956	9.87723	**55**
6	.80897	.92535	.07465	.88362	54	6	.81781	.94069	.05931	.87712	54
7	.80912	.92561	.07439	.88351	53	7	.81796	.94095	.05905	.87701	53
8	.80927	.92587	.07413	.88340	52	8	.81810	.94120	.05880	.87690	52
9	.80942	.92612	.07388	.88330	51	9	.81825	.94146	.05854	.87679	51
10	9.80957	9.92638	10.07362	9.88319	**50**	**10**	9.81839	9.94171	10.05829	9.87668	**50**
11	.80972	.92663	.07337	.88308	49	11	.81854	.94197	.05803	.87657	49
12	.80987	.92689	.07311	.88298	48	12	.81868	.94222	.05778	.87646	48
13	.81002	.92715	.07285	.88287	47	13	.81882	.94248	.05752	.87635	47
14	.81017	.92740	.07260	.88276	46	14	.81897	.94273	.05727	.87624	46
15	9.81032	9.92766	10.07234	9.88266	**45**	**15**	9.81911	9.94299	10.05701	9.87613	**45**
16	.81047	.92792	.07208	.88255	44	16	.81926	.94324	.05676	.87601	44
17	.81061	.92817	.07183	.88244	43	17	.81940	.94350	.05650	.87590	43
18	.81076	.92843	.07157	.88234	42	18	.81955	.94375	.05625	.87579	42
19	.81091	.92868	.07132	.88223	41	19	.81969	.94401	.05599	.87568	41
20	9.81106	9.92894	10.07106	9.88212	**40**	**20**	9.81983	9.94426	10.05574	9.87557	**40**
21	.81121	.92920	.07080	.88201	39	21	.81998	.94452	.05548	.87546	39
22	.81136	.92945	.07055	.88191	38	22	.82012	.94477	.05523	.87535	38
23	.81151	.92971	.07029	.88180	37	23	.82026	.94503	.05497	.87524	37
24	.81166	.92996	.07004	.88169	36	24	.82041	.94528	.05472	.87513	36
25	9.81180	9.93022	10.06978	9.88158	**35**	**25**	9.82055	9.94554	10.05446	9.87501	**35**
26	.81195	.93048	.06952	.88148	34	26	.82069	.94579	.05421	.87490	34
27	.81210	.93073	.06927	.88137	33	27	.82084	.94604	.05396	.87479	33
28	.81225	.93099	.06901	.88126	32	28	.82098	.94630	.05370	.87468	32
29	.81240	.93124	.06876	.88115	31	29	.82112	.94655	.05345	.87457	31
30	9.81254	9.93150	10.06850	9.88105	**30**	**30**	9.82126	9.94681	10.05319	9.87446	**30**
31	.81269	.93175	.06825	.88094	29	31	.82141	.94706	.05294	.87434	29
32	.81284	.93201	.06799	.88083	28	32	.82155	.94732	.05268	.87423	28
33	.81299	.93227	.06773	.88072	27	33	.82169	.94757	.05243	.87412	27
34	.81314	.93252	.06748	.88061	26	34	.82184	.94783	.05217	.87401	26
35	9.81328	9.93278	10.06722	9.88051	**25**	**35**	9.82198	9.94808	10.05192	9.87390	**25**
36	.81343	.93303	.06697	.88040	24	36	.82212	.94834	.05166	.87378	24
37	.81358	.93329	.06671	.88029	23	37	.82226	.94859	.05141	.87367	23
38	.81372	.93354	.06646	.88018	22	38	.82240	.94884	.05116	.87356	22
39	.81387	.93380	.06620	.88007	21	39	.82255	.94910	.05090	.87345	21
40	9.81402	9.93406	10.06594	9.87996	**20**	**40**	9.82269	9.94935	10.05065	9.87334	**20**
41	.81417	.93431	.06569	.87985	19	41	.82283	.94961	.05039	.87322	19
42	.81431	.93457	.06543	.87975	18	42	.82297	.94986	.05014	.87311	18
43	.81446	.93482	.06518	.87964	17	43	.82311	.95012	.04988	.87300	17
44	.81461	.93508	.06492	.87953	16	44	.82326	.95037	.04963	.87288	16
45	9.81475	9.93533	10.06467	9.87942	**15**	**45**	9.82340	9.95062	10.04938	9.87277	**15**
46	.81490	.93559	.06441	.87931	14	46	.82354	.95088	.04912	.87266	14
47	.81505	.93584	.06416	.87920	13	47	.82368	.95113	.04887	.87255	13
48	.81519	.93610	.06390	.87909	12	48	.82382	.95139	.04861	.87243	12
49	.81534	.93636	.06364	.87898	11	49	.82396	.95164	.04836	.87232	11
50	9.81549	9.93661	10.06339	9.87887	**10**	**50**	9.82410	9.95190	10.04810	9.87221	**10**
51	.81563	.93687	.06313	.87877	9	51	.82424	.95215	.04785	.87209	9
52	.81578	.93712	.06288	.87866	8	52	.82439	.95240	.04760	.87198	8
53	.81592	.93738	.06262	.87855	7	53	.82453	.95266	.04734	.87187	7
54	.81607	.93763	.06237	.87844	6	54	.82467	.95291	.04709	.87175	6
55	9.81622	9.93789	10.06211	9.87833	**5**	**55**	9.82481	9.95317	10.04683	9.87164	**5**
56	.81636	.93814	.06186	.87822	4	56	.82495	.95342	.04658	.87153	4
57	.81651	.93840	.06160	.87811	3	57	.82509	.95368	.04632	.87141	3
58	.81665	.93865	.06135	.87800	2	58	.82523	.95393	.04607	.87130	2
59	.81680	.93891	.06109	.87789	1	59	.82537	.95418	.04582	.87119	1
60	9.81694	9.93916	10.06084	9.87778	**0**	**60**	9.82551	9.95444	10.04556	9.87107	**0**
	Cosine.	Cotang.	Tang.	Sine.	′		Cosine.	Cotang.	Tang.	Sine.	′
					49°						48°

*Log secant = colog cosine = 1 − log cosine; log cosecant = colog sine = 1 − log sine.

Ex.—Log sec 40°− 30′ = 10.11895. *Ex.*—Log cosec 40°− 30′ = 10.18746.

3. –Logarithmic Sines, TANGENTS, COTANGENTS, COSINES.—(Cont'd.)
(SECANTS, COSECANTS.)*

42°							43°				
′	Sine.	Tang.	Cotang.	Cosine.		′	Sine.	Tang.	Cotang.	Cosine.	
0	9.82551	9.95444	10.04556	9.87107	**60**	**0**	9.83378	9.96966	10.03034	9.86413	**60**
1	.82565	.95469	.04531	.87096	59	1	.83392	.96991	.03009	.86401	59
2	.82579	.95495	.04505	.87085	58	2	.83405	.97016	.02984	.86389	58
3	.82593	.95520	.04480	.87073	57	3	.83419	.97042	.02958	.86377	57
4	.82607	.95545	.04455	.87062	56	4	.83432	.97067	.02933	.86366	56
5	9.82621	9.95571	10.04429	9.87050	**55**	**5**	9.83446	9.97092	10.02908	9.86354	**55**
6	.82635	.95596	.04404	.87039	54	6	.83459	.97118	.02882	.86342	54
7	.82649	.95622	.04378	.87028	53	7	.83473	.97143	.02857	.86330	53
8	.82663	.95647	.04353	.87016	52	8	.83486	.97168	.02832	.86318	52
9	.82677	.95672	.04328	.87005	51	9	.83500	.97193	.02807	.86306	51
10	9.82691	9.95698	10.04302	9.86993	**50**	**10**	9.83513	9.97219	10.02781	9.86295	**50**
11	.82705	.95723	.04277	.86982	49	11	.83527	.97244	.02756	.86283	49
12	.82719	.95748	.04252	.86970	48	12	.83540	.97269	.02731	.86271	48
13	.82733	.95774	.04226	.86959	47	13	.83554	.97295	.02705	.86259	47
14	.82747	.95799	.04201	.86947	46	14	.83567	.97320	.02680	.86247	46
15	9.82761	9.95825	10.04175	9.86936	**45**	**15**	9.83581	9.97345	10.02655	9.86235	**45**
16	.82775	.95850	.04150	.86924	44	16	.83594	.97371	.02629	.86223	44
17	.82788	.95875	.04125	.86913	43	17	.83608	.97396	.02604	.86211	43
18	.82802	.95901	.04099	.86902	42	18	.83621	.97421	.02579	.86200	42
19	.82816	.95926	.04074	.86890	41	19	.83634	.97447	.02553	.86188	41
20	9.82830	9.95952	10.04048	9.86879	**40**	**20**	9.83648	9.97472	10.02528	9.86176	**40**
21	.82844	.95977	.04023	.86867	39	21	.83661	.97497	.02503	.86164	39
22	.82858	.96002	.03998	.86855	38	22	.83674	.97523	.02477	.86152	38
23	.82872	.96028	.03972	.86844	37	23	.83688	.97548	.02452	.86140	37
24	.82885	.96053	.03947	.86832	36	24	.83701	.97573	.02427	.86128	36
25	9.82899	9.96078	10.03922	9.86821	**35**	**25**	9.83715	9.97598	10.02402	9.86116	**35**
26	.82913	.96104	.03896	.86809	34	26	.83728	.97624	.02376	.86104	34
27	.82927	.96129	.03871	.86798	33	27	.83741	.97649	.02351	.86092	33
28	.82941	.96155	.03845	.86786	32	28	.83755	.97674	.02326	.86080	32
29	.82955	.96180	.03820	.86775	31	29	.83768	.97700	.02300	.86068	31
30	9.82968	9.96205	10.03795	9.86763	**30**	**30**	9.83781	9.97725	10.02275	9.86056	**30**
31	.82982	.96231	.03769	.86752	29	31	.83795	.97750	.02250	.86044	29
32	.82996	.96256	.03744	.86740	28	32	.83808	.97776	.02224	.86032	28
33	.83010	.96281	.03719	.86728	27	33	.83821	.97801	.02199	.86020	27
34	.83023	.96307	.03693	.86717	26	34	.83834	.97826	.02174	.86008	26
35	9.83037	9.96332	10.03668	9.86705	**25**	**35**	9.83848	9.97851	10.02149	9.85996	**25**
36	.83051	.96357	.03643	.86694	24	36	.83861	.97877	.02123	.85984	24
37	.83065	.96383	.03617	.86682	23	37	.83874	.97902	.02098	.85972	23
38	.83078	.96408	.03592	.86670	22	38	.83887	.97927	.02073	.85960	22
39	.83092	.96433	.03567	.86659	21	39	.83901	.97953	.02047	.85948	21
40	9.83106	9.96459	10.03541	9.86647	**20**	**40**	9.83914	9.97978	10.02022	9.85936	**20**
41	.83120	.96484	.03516	.86635	19	41	.83927	.98003	.01997	.85924	19
42	.83133	.96510	.03490	.86624	18	42	.83940	.98029	.01971	.85912	18
43	.83147	.96535	.03465	.86612	17	43	.83954	.98054	.01946	.85900	17
44	.83161	.96560	.03440	.86600	16	44	.83967	.98079	.01921	.85888	16
45	9.83174	9.96586	10.03414	9.86589	**15**	**45**	9.83980	9.98104	10.01896	9.85876	**15**
46	.83188	.96611	.03389	.86577	14	46	.83993	.98130	.01870	.85864	14
47	.83202	.96636	.03364	.86565	13	47	.84006	.98155	.01845	.85851	13
48	.83215	.96662	.03338	.86554	12	48	.84020	.98180	.01820	.85839	12
49	.83229	.96687	.03313	.86542	11	49	.84033	.98206	.01794	.85827	11
50	9.83242	9.96712	10.03288	9.86530	**10**	**50**	9.84046	9.98231	10.01769	9.85815	**10**
51	.83256	.96738	.03262	.86518	9	51	.84059	.98256	.01744	.85803	9
52	.83270	.96763	.03237	.86507	8	52	.84072	.98281	.01719	.85791	8
53	.83283	.96788	.03212	.86495	7	53	.84085	.98307	.01693	.85779	7
54	.83297	.96814	.03186	.86483	6	54	.84098	.98332	.01668	.85766	6
55	9.83310	9.96839	10.03161	9.86472	**5**	**55**	9.84112	9.98357	10.01643	9.85754	**5**
56	.83324	.96864	.03136	.86460	4	56	.84125	.98383	.01617	.85742	4
57	.83338	.96890	.03110	.86448	3	57	.84138	.98408	.01592	.85730	3
58	.83351	.96915	.03085	.86436	2	58	.84151	.98433	.01567	.85718	2
59	.83365	.96940	.03060	.86425	1	59	.84164	.98458	.01542	.85706	1
60	9.83378	9.96966	10.03034	9.86413	**0**	**60**	9.84177	9.98484	10.01516	9.85693	**0**
	Cosine.	Cotang.	Tang.	Sine.	′		Cosine.	Cotang.	Tang.	Sine.	
					47°						46°

*Log secant = colog cosine = 1 − log cosine; log cosecant = colog sine = 1 − log sine.
Ex.—Log sec 42°– 30′ = 10.13237. *Ex.*—Log cosec 42°– 30′ = 10.17032.

3.—Logarithmic Sines, Tangents, Cotangents, Cosines.—(Concl'd.) (Secants, Cosecants.)

44° | 44°

'	Sine.	Tang.	Cotang.	Cosine.		'	Sine.	Tang.	Cotang.	Cosine.	
0	9.84177	9.98484	10.01516	9.85693	**60**	**30**	9.84566	9.99242	10.00758	9.85324	**30**
1	.84190	.98509	.01491	.85681	59	31	.84579	.99267	.00733	.85312	29
2	.84203	.98534	.01466	.85669	58	32	.84592	.99293	.00707	.85299	28
3	.84216	.98560	.01440	.85657	57	33	.84605	.99318	.00682	.85287	27
4	.84229	.98585	.01415	.85645	56	34	84618	.99343	.00657	.85274	26
5	9.84242	9.98610	10.01390	9.85632	**55**	**35**	9.84630	9.99368	10.00632	9.85262	**25**
6	.84255	.98635	.01365	.85620	54	36	.84643	.99394	.00606	.85250	24
7	.84269	.98661	.01339	.85608	53	37	.84656	.99419	.00581	.85237	23
8	.84282	.98686	.01314	.85596	52	38	.84669	.99444	.00556	.85225	22
9	.84295	.98711	.01289	.85583	51	39	.84682	.99469	.00531	.85212	21
10	9.84308	9.98737	10.01263	9.85571	**50**	**40**	9.84694	9.99495	10.00505	9.85200	**20**
11	.84321	.98762	.01238	.85559	49	41	.84707	.99520	.00480	.85187	19
12	.84334	.98787	.01213	.85547	48	42	.84720	.99545	.00455	.85175	18
13	.84347	.98812	.01188	.85534	47	43	.84733	.99570	.00430	.85162	17
14	.84360	.98838	.01162	.85522	46	44	.84745	.99596	.00404	.85150	16
15	9.84373	9.98863	10.01137	9.85510	**45**	**45**	9.84758	9.99621	10.00379	9.85137	**15**
16	.84385	.98888	.01112	.85497	44	46	.84771	.99646	.00354	.85125	14
17	.84398	.98913	.01087	.85485	43	47	.84784	.99672	.00328	.85112	13
18	.84411	.98939	.01061	.85473	42	48	.84796	.99697	.00303	.85100	12
19	.84424	.98964	.01036	.85460	41	49	.84809	.99722	.00278	.85087	11
20	9.84437	9.98989	10.01011	9.85448	**40**	**50**	9.84822	9.99747	10.00253	9.85074	**10**
21	.84450	.99015	.00985	.85436	39	51	.84835	.99773	.00227	.85062	9
22	.84463	.99040	.00960	.85423	38	52	.84847	.99798	.00202	.85049	8
23	.84476	.99065	.00935	.85411	37	53	.84860	.99823	.00177	.85037	7
24	.84489	.99090	.00910	.85399	36	54	.84873	.99848	.00152	.85024	6
25	9.84502	9.99116	10.00884	9.85386	**35**	**55**	9.84885	9.99874	10.00126	9.85012	**5**
26	.84515	.99141	.00859	.85374	34	56	.84898	.99899	.00101	.84999	4
27	.84528	.99166	.00834	.85361	33	57	.84911	.99924	.00076	.84986	3
28	.84540	.99191	.00809	.85349	32	58	.84923	.99949	.00051	.84974	2
29	.84553	.99217	.00783	.85337	31	59	.84936	.99975	.00025	.84961	1
30	9.84566	9.99242	10.00758	9.85324	**30**	**60**	9.84949	10.00000	10.00000	9.84949	**0**
	Cosine.	Cotang.	Tang.	Sine.	'		Cosine.	Cotang.	Tang.	Sine.	'

45° | 45°

3a.—Table for Finding the Logarithmic Sines and Tangents of Small Angles.

[Values of S and T in Formulas Below.*]

A.	A(sec.).	S.	A.	A(sec.).	T.	A.	A(sec.).	T.
0°00′00″	0000″	4.68557	0°00′00″	000″	4.68557	1°25′40″	5140″	4.68566
0°40′10″	2410″	57	0°03′00″	180″	57	1°25′50″	5150″	67
0°40′20″	2420″	56	0°03′20″	200″	58	1°30′20″	5420″	67
0°57′00″	3420″	56	0°28′40″	1720″	58	1°30′30″	5430″	68
0°57′10″	3430″	55	0°28′50″	1730″	59	1°34′40″	5680″	68
1°09′50″	4190″	55	0°40′20″	2420″	59	1°34′50″	5690″	69
1°10′00″	4200″	54	0°40′30″	2430″	60	1°39′00″	5940″	69
1°20′40″	4840″	54	0°49′30″	2970″	60	1°39′10″	5950″	70
1°20′50″	4850″	53	0°49′40″	2980″	61	1°43′00″	6180″	70
1°30′10″	5410″	53	0°57′10″	3430″	61	1°43′10″	6190″	71
1°30′20″	5420″	52	0°57′20″	3440″	62	1°46′50″	6410″	71
1°38′50″	5930″	52	1°03′50″	3830″	62	1°47′00″	6420″	72
1°39′00″	5940″	51	1°04′00″	3840″	63	1°50′40″	6640″	72
1°46′50″	6410″	51	1°10′00″	4200″	63	1°50′50″	6650″	73
1°47′00″	6420″	50	1°10′10″	4210″	64	1°54′10″	6850″	73
1°54′10″	6850″	50	1°15′30″	4530″	64	1°54′20″	6860″	74
1°54′20″	6860″	49	1°15′40″	4540″	65	1°57′40″	7060″	74
2°01′00″	7260″	49	1°20′50″	4850″	65	1°57′50″	7070″	75
2°01′10″	7270″	48	1°21′00″	4860″	66	2°01′10″	7270″	75

* Log sin $A = \log A$ (seconds)$+S$. Log tan $A = \log A$ (seconds)$+T$.

The Solution of Right-Angled Triangles.—Let triangle $A\ B\ C$ of Fig. 4 (p. 141) represent any right-angled triangle and a, b, and c, the lengths of its sides. Then, with any two sides, or any one side and one acute angle known, the missing information can be obtained by the following formulas:

TABLE 4

SOLUTION OF RIGHT-ANGLED TRIANGLES

Sides and Angles Known	Formulas for Sides and Angles to be Found		
Sides c and a	$b = \sqrt{c^2 - a^2}$	$\sin A = \frac{a}{c}$	$B = 90° - A$
Sides c and b	$a = \sqrt{c^2 - b^2}$	$\sin B = \frac{b}{c}$	$A = 90° - B$
Sides a and b	$c = \sqrt{a^2 + b^2}$	$\tan A = \frac{a}{b}$	$B = 90° - A$
Side c Ang. A	$a = c \times \sin A$	$b = c \times \cos A$	$B = 90° - A$
Side c Ang. B	$a = c \times \cos B$	$b = c \times \sin B$	$A = 90° - B$
Side a Ang. A	$c = \frac{a}{\sin A}$	$b = a \times \cot A$	$B = 90° - A$
Side a Ang. B	$c = \frac{a}{\cos B}$	$b = a \times \tan B$	$A = 90° - B$
Side b Ang. A	$c = \frac{b}{\cos A}$	$a = b \times \tan A$	$B = 90° - A$
Side b Ang. B	$c = \frac{b}{\sin B}$	$a = b \times \cot B$	$A = 90° - B$

ILLUSTRATION: A gabled roof has a pitch of 45 degrees. What is the length of the rafters if the span is 20 feet?

In this case A and B $= 45$ degrees, $C = 20$ feet.

Length of rafter $= a = c \times \sin A$

$a = 20 \times .707 = 14.14$ feet (Ans.)

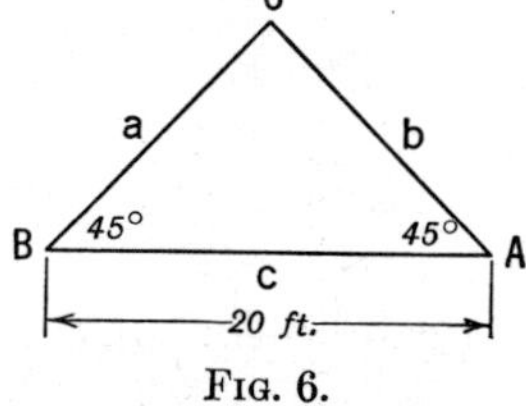

FIG. 6.

ILLUSTRATION: Figure 7 shows a method used to measure the distance CB across a river. A surveying party sets points A and C, then with a transit at C, a right angle is turned and point B set. The distance b is measured and also the angle A. What is the distance across the river (a) if b is 487.32 feet and $\angle A$ is 35°17′?

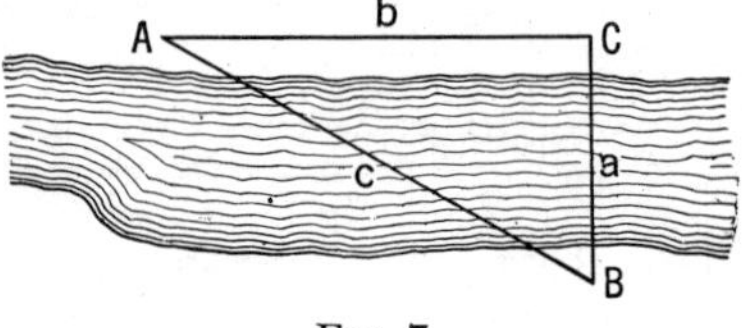

FIG. 7.

$$\begin{aligned} a &= b \times \tan A \\ \log 487.32 &= 2.68782 \\ \log \tan 35°17' &= 9.84979 \\ \hline \log a &= 2.53761 \\ a &= 344.85 \text{ feet (Ans.)} \end{aligned}$$

Solution of Any Plane Triangle.—Not only right triangles but any plane triangle may be solved by trigonometric formulas if two sides and an angle, or two angles and a side, or three sides are given. Four cases will be considered.

CASE I. Given any two sides b and c and their included angle A. Use any one of the following sets of formulas:

(1) $$\tfrac{1}{2}(B + C) = 90° - \tfrac{1}{2}A$$

$$\tan \tfrac{1}{2}(B - C) = \frac{b - c}{b + c} \tan \tfrac{1}{2}(B + C)$$

$$B = \tfrac{1}{2}(B + C) + \tfrac{1}{2}(B - C)$$

$$C = \tfrac{1}{2}(B + C) - \tfrac{1}{2}(B - C)$$

$$a = \frac{b \sin A}{\sin B}$$

(2) $$\tan C = \frac{c \sin A}{b - c \cos A}$$

$$B = 180° - (A + C)$$

$$a = \frac{c \sin A}{\sin C}$$

(3) $$a = \sqrt{b^2 + c^2 - 2bc \cos A}$$

$$\sin B = \frac{b \sin A}{a}$$

$$C = 180° - (A + B)$$

Case II. Given any two angles A and B and any side c.

$$C = 180° - (A + B)$$

$$a = \frac{c \sin A}{\sin C}$$

$$b = \frac{c \sin B}{\sin C}$$

Case III. Given the three sides a, b, and c. Use either of the following sets of formulas:

(1) $$\cos A = \frac{b^2 + c^2 - a^2}{2bc}$$

$$\cos B = \frac{a^2 + c^2 - b^2}{2ac}$$

$$C = 180° - (A + B)$$

(2) Let $$s = \tfrac{1}{2}(a + b + c)$$

$$r = \sqrt{\frac{(s - a)(s - b)(s - c)}{s}}$$

$$\tan \tfrac{1}{2}A = \frac{r}{s - a}$$

$$\tan \tfrac{1}{2}B = \frac{r}{s - b}$$

$$\tan \tfrac{1}{2}C = \frac{r}{s - c}$$

(3) Following also comment for case III, let c be longest side and $a > b$. Then (see Fig. 8) or similarly for any triangle:

$$g = \tfrac{1}{2}\left[\frac{(a+b)(a-b)}{c} + c\right]$$

$$s = c - g$$

$$\cos A = \frac{s}{b}$$

$$\cos B = \frac{g}{a}$$

$$C = 180° - (A + B)$$

CASE IV. Given any two sides a and b and an angle A opposite either one of these

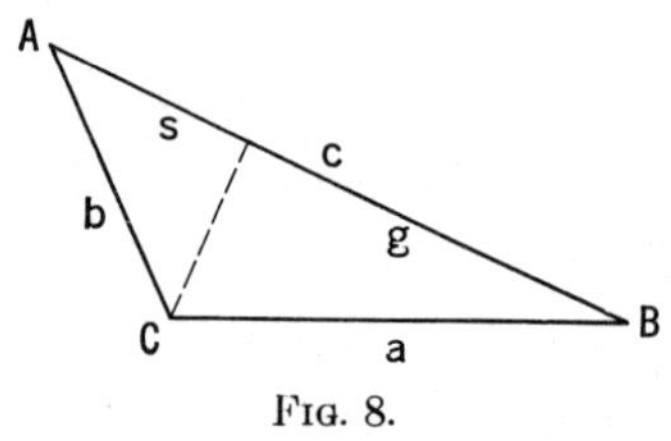

FIG. 8.

$$\sin C = \frac{c \sin A}{a}$$

$$B = 180° - (A + C)$$

$$b = \frac{a \sin B}{\sin A}$$

NOTE. There may be two values for the angle C. If, however, one solution is such that $A + C > 180°$, use only the other value.

Reference.—An excellent treatise on trigonometry is also contained in the set of mathematics books by J. E. Thompson. This is entitled TRIGONOMETRY FOR THE PRACTICAL MAN and is published by the D. Van Nostrand Company.

VI

Differential Calculus

In arithmetic we deal with numbers that represent fixed quantities or *constants*. In algebra, the numbers may be constants or they may vary (variables), but in any given problem, the numbers remain constant throughout the consideration of that particular problem.

Other problems often arise in which the quantities involved, or the symbol expressing these quantities, is continually changing and therefore cannot be solved by arithmetic or algebra. For example, if a weight is dropped and allowed to fall freely, its speed steadily increases. If thrown directly upward, it moves slower and slower as it rises until it stops. Then it begins to fall and moves more and more rapidly until it hits the ground. Many other such examples could be cited.

The branch of mathematics which is very helpful in understanding these phenomena is known as the calculus. Calculus is concerned with the study of rates of change of quantities. In this chapter, some of the basic ideas underlying calculus are explained and a few practical applications are cited.

Differentiation.—If a straight line of length x is divided into an enormous number of parts, then the length of each of these infinitesimal parts is represented by the symbol dx. This expression is known as the *differential* of x. Thus the differential of a line x is dx. (Note that dx does not signify d times x, but is a single symbol in itself.)

Now suppose a square is formed whose sides are x units long. The area (x times x) will be x^2. If the length of each side is increased by an infinitesimally small amount dx, then the length of each side becomes $x + dx$, and the new area is $(x + dx)(x + dx)$ or $x^2 + 2xdx + (dx)^2$. The first term is the original

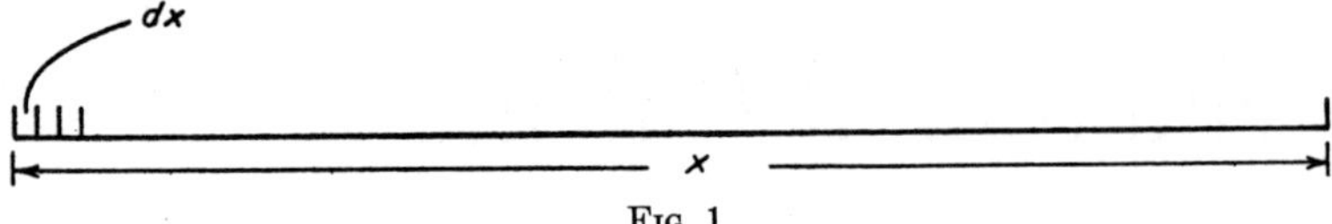

FIG. 1.

area, and the second and third terms represent the additional area formed by increasing the two sides of the square by the amount dx.

For practical purposes we can neglect $(dx)^2$ because its magnitude is negligible (for example, if dx equals 0.001 inch, then $(dx)^2$ equals 0.000001 inch and since dx actually is very much smaller than 0.001, when squared, it becomes negligible). The remaining terms represent the original area (x^2) plus the increase, or change, in the area $(2xdx)$.

The expression $2xdx$ is known as the *differential* of x^2. Visually (see Fig. 2), $2xdx$ is the sum of each area strip, neglecting

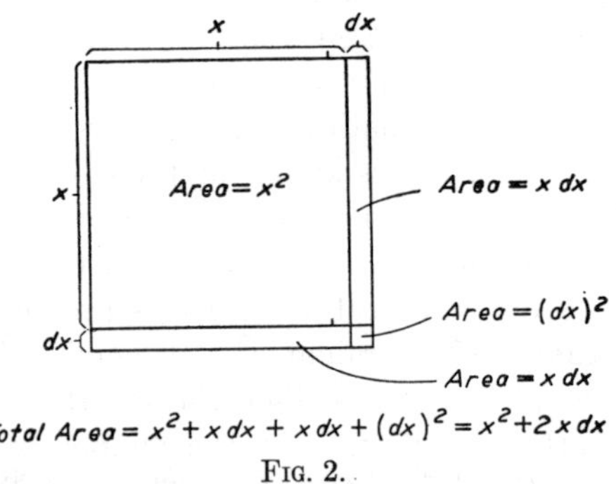

FIG. 2.

the corner $(dx)^2$, formed by changing each side of the square by an amount dx. Similarly the differential of x^3 (representing, for example, the volume of a cube whose sides are each x units long) can be shown to be $3x^2dx$. Thus, we find a pattern forming:

the differential of x is dx
the differential of x^2 is $2\ xdx$
the differential of x^3 is $3x^2dx$
the differential of x^4 is $4x^3dx$
etc.

We now can devise a formula which will tell us the differential of x raised to any power n. It can be written:

the differential of x^n equals $nx^{n-1}dx$

This formula can be applied to the area problem shown in Fig. 2. Given a square with sides x and area A, what happens to the area if each side is increased by an amount dx?

$$A = x^2$$

Taking the differential of both sides of the above equation we get

$$\frac{d(A)}{dx} = \frac{d(x^2)}{dx} = 2xdx$$

The expression d/dx means the differential of a variable with respect to x. For example, dy/dx would be the differential of y with respect to x, or the change in y caused by a change in x. This is referred to as the derivative of y with respect to x.

Differentiation of Constants.—A constant quantity, having no rate of change, cannot be differentiated; therefore its differential is zero. If the constant is a coefficient of a variable, as 6 in the expression $6x$, then the constant which multiplies the variable remains unchanged in the differentiating process. Thus:

the differential of $6 =$ zero
the differential of $6x = 6dx$
the differential of $6x^2 = 6(2xdx) = 12xdx$

Differential of Sum or Difference.—To differentiate an algebraic expression consisting of several terms with positive or negative signs before them such as:

$$x^2 - 2x + 6 + 3x^4$$

each term must be differentiated separately. This is so because each term is separate and distinct from the other terms and therefore its differential or rate of growth will be distinct or separate from the other terms. Thus·

the differential of $x^2 - 2x + 6 + 3x^4 = 2xdx - 2dx + 0 + 12x^3dx$

Differentiation of a Product.—To illustrate how the product of two variables is differentiated, consider a rectangle whose sides are x and y, such as the one shown in Fig. 3.

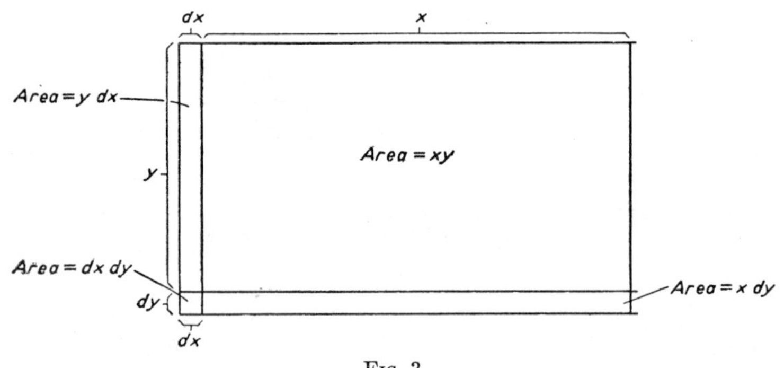

FIG. 3.

Its area is equal to the product xy. If the sides are increased by the amounts dx and dy respectively, then:

$$\text{New area} = xy \text{ (old area)} + ydx + xdy + dydx$$

The last term $dydx$ is a very small quantity and therefore can be safely neglected. Therefore, the differential of the original area xy is $xdy + ydx$. This can be generalized for every case:

> The differential of the product of two variables is equal to the first multiplied by the differential of the second, plus the second multiplied by the differential of the first. Thus, the differential of $x^2y = x^2dy + y(2xdx) = x^2dy + 2yxdx$

This law also holds for any number of variables. For example, the differential of $xyz = xydz + xzdy + yzdx$.

Differential of a Fraction.—To differentiate a fraction, say x/y, the expression is changed to a product by writing it in the form xy^{-1}, using the negative exponent to eliminate the fraction. Then the product is differentiated in the manner just described. Thus:

$$\text{differential } xy^{-1} = -xy^{-2}dy + y^{-1}dx$$
$$= -xdy/y^2 + dx/y$$

reducing to a common denominator:

$$= \frac{(ydx - xdy)}{y^2}$$

This can be expressed in the following way:

The differential of a fraction is equal to the denominator times the differential of the numerator, minus the numerator times the differential of the denominator, all divided by the square of the denominator.

By further mathematical analysis the differentials for other expressions can be found. The following table summarizes those we have considered and also includes several others:

Expression	*Differential*	*Derivative*
$x + y + z + \cdots$	$dx + dy + dz + \cdots$	
$y = C$	$dy = 0$	$dy/dx = 0$
$y = x + C$	$dy = dx + 0$	$dy/dx = 1$
$y = \pm x$	$dy = \pm dx$	$dy/dx = \pm 1$
$y = Cx$	$dy = Cdx$	$dy/dx = C$
$y = x^n$	$dy = nx^{n-1}dx$	$dy/dx = nx^{n-1}$
$y = x^{1/n}$	$dy = (1/n)x^{(1/n)-1}dx$	$dy/dx = (1/n)x^{(1/n)-1}$
$z = xy$	$dz = xdy + ydx$	
$z = x/y$	$dz = \dfrac{ydx - xdy}{y^2}$	

It should be noted that differentials can also be found for trigonometric functions and logarithmic functions.

Maximum and Minimum Problems.—One of the most important and practical applications of differential calculus is the solution of problems involving maximum and minimum values. Suppose we had a hollow box into which we placed an uninflated balloon. If we inflate the balloon, its volume V changes at every instant with respect to the box volume B. The rate of change of the volume occupied by the balloon with respect to the volume of the box is dV/dB and it has some significant value so long as the balloon continues to be inflated. When the balloon completely fills the box, reaching its maximum possible size, the volume V no longer changes and at this point dV/dB is zero, and V is now at its maximum value.

Then, if we let the air out of the balloon, V again changes (now decreasing with respect to B) and dV/dB again has a significant value. When all the air has escaped from the balloon, dV/dB again becomes zero, for once again V is no longer changing with respect to B, now having reached its minimum value.

This example illustrates one of the most important rules in calculus: When the derivative of an expression is zero, the value of the expression is at a maximum or a minimum. In the example just cited, when the balloon filled the box, dV/dB was zero and the volume of the balloon was at its maximum. Similarly, when the balloon was deflated, dV/dB again became zero and its volume was now at its minimum value.

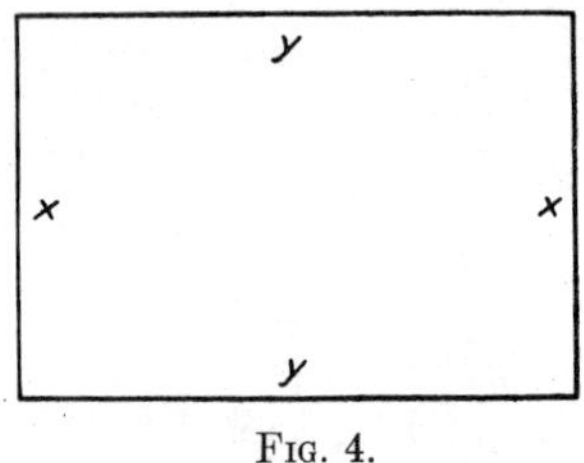

FIG. 4.

Maximum and Minimum Examples.—The following two practical problems further illustrate the application of maximum and minimum rules.

EXAMPLE 1. What are the dimensions of the largest possible rectangle which can be enclosed by a fence 600 ft long (Fig. 4)?

Perimeter of rectangle $= 2x + 2y = 600$ ft; area of rectangle $= xy = A$.

$$2x = 600 - 2y \qquad A = xy = (300 - y)y$$
$$x = 300 - y \qquad A = -y^2 + 300y$$

To find maximum area, differentiate the equation:

$$A = -y^2 + 300y$$

$$\frac{dA}{dy} = -2y + 300$$

Now set the derivative equal to zero to find the value of y which gives a maximum A (the minimum value in this case is zero).

$$-2y + 300 = 0$$

$$2y = 300$$

$$y = 150$$

Substituting in the perimeter equation,

$$x = 300 - y$$

we obtain

$$x = 300 - 150 = 150$$

Therefore, to enclose the maximum area, the rectangle measures 150 ft $\times$ 150 ft, or 22,500 sq ft. Note that, for a given perimeter, the rectangle which will enclose the maximum area turns out to be a square.

EXAMPLE 2. Transmitting electricity from the powerhouse to the consumer involves the cost of the heat loss and the cost of the copper wire used to carry the electrical energy. The heat loss is inversely proportional to the cross-section of the wire. The cost of the copper wire is directly proportional to the cross-section of the wire. In a certain electrical system, the cost C turns out to be

$$C = \frac{2}{A} + \frac{2A}{3}$$

where C is the cost and A is the cross-section of the wire. Find the area A which makes the cost C a minimum.

To find minimum C differentiate the expression:

$$C = \frac{2}{A} + \frac{2A}{3}.$$

$$\frac{dC}{dA} = \frac{-2}{A^2} + \frac{2}{3}$$

Now set the derivative equal to zero and solve for A.

$$\frac{-2}{A^2} + \frac{2}{3} = 0$$

$$\frac{A^2}{2} = \frac{3}{2}$$

$$A^2 = 3$$

$$A = \sqrt{3} = 1.732 \text{ sq. inches}$$

In this chapter we have presented some of the basic ideas underlying the calculus. Understandably we have only scratched the surface of this fascinating and useful branch of mathematics. For a further treatment of the subject (including *integration* which is the exact opposite of differentiation) the reader is referred to *The Calculus for the Practical Man* by J. E. Thompson, D. Van Nostrand and Co., New York, N. Y.

VII

Mechanics

Mechanics.—Mechanics is a science which treats of the action of forces and their effect upon bodies. A force is defined as a push or pull which tends to change the velocity or direction of a body's motion. The units by which a force is measured are pounds or tons. Distance measured in linear units and time expressed in seconds, minutes, etc., are two other elementary quantities in mechanics from which numerous compound quantities are derived.

Work is the product of force by distance. The units for measuring work are derived from the units of force and distance. In the British system, the unit of work is the foot-pound.

Power is the time rate of doing work. In mechanics it is the product of force by distance divided by time. Power is commonly expressed as inch-pounds per minute, foot-pounds per minute or second, etc. Horsepower, H.P., is the unit of power adopted for engineering work. One horsepower = 33,000 foot-pounds per minute = 550 foot-pounds per second.

Velocity is the time rate of motion. It is distance divided by time, and is expressed in feet per minute, miles per hour, etc.

Stress and Strain.—An external force applied to a body, so as to pull it apart, is resisted by an internal force, or resistance, and the action of these forces cause a displacement of the molecules, or deformation. The external forces are called stresses while the alteration produced by the stresses is called by the term strain. For example, a load on a steel column tends to compress or crush the column. At the same time, the column reacts against the tendency of the load to crush it and exerts a force opposite to the load. The external force or the tendency of the outside load to change the shape of the column is called stress. The internal force or the resistance of the column to the tendency of the outside load to change its shape is called strain.

There are five kinds of stresses:

1. Tensile stress, or pull, is a force which tends to elongate a piece of material.

2. Compressive stress, or push, is a force which tends to shorten a piece of material.

3. Shearing stress is a force which tends to force one part of a piece of material to slide over an adjacent part.

4. Torsional stress, a form of shearing stress, is a force which tends to twist a piece of material.

5. Transverse stress, a combination of tension and compression, is a force which tends to bend a piece of material.

All stresses to which a material is subjected cause a deformation in it. If the stress is not too great, however, the material will return to its original shape and dimensions when the external stress is removed. The property which enables a material to return to its original shape and dimensions is called its elasticity.

The elastic limit is the unit stress beyond which the material will not return to its original shape when the load is removed.

There is a law, called Hooke's Law, which expresses the relation between the amount of stress applied to a body and the amount of strain it produces.

Hooke's Law. The amount of change in the shape of an elastic body is proportional to the force applied, provided that the elastic limit is not exceeded. In other words the strain is directly proportional to the stress.

For different stresses the rule becomes:

Tensile stress, the stretch is proportional to the force applied

Torsional stress, the twist is proportional to the stress causing it.

Transverse stress, the deflections are proportioned to the loads causing them.

ILLUSTRATION. If a weight of one pound is hung on a spring it lengthens the spring 1.5 inch; what weight would lengthen it 0.75 inch?

$$x : 1 = 0.75 : 1.5$$

$$x = \frac{1 \times 0.75}{1.5} = 0.5$$

Therefore, $\frac{1}{2}$ pound weight would lengthen the spring $\frac{3}{4}$ inch.

Modulus of Elasticity.—The modulus of elasticity is a term expressing the relation between the amount of extension or compression of a material and the load producing that extension or compression. It is defined as the load per unit of section divided by the extension per unit of length.

The following table gives the moduli of elasticity for various materials.

Material	Modulus	Material	Modulus
Brass, cast......	9,170,000	Tin, cast........	4,600,000
Copper..........	15,000,000	Iron, cast........	12,000,000
Lead............	1,000,000	Steel............	28,000,000

The following rule may be used to find the modulus of elasticity, commonly designated by E.

Divide the stress per square inch by the elongation in one inch caused by this stress. Expressed as a formula:

$$E = \frac{P}{e}$$

where E = modulus of elasticity in pounds
P = stress
e = elongation in inches

Illustration: If the elongation of 0.02 inch is produced in a bar 10 inches long by a load of 48,000 pounds per square inch of cross section of the bar, find the modulus of elasticity.

$$E = \frac{P}{e}$$

$$= \frac{48{,}000 \times 10}{0.02}$$

$$= 24{,}000{,}000$$

Therefore, the modulus of elasticity is 24,000,000 pounds.

Graphical Representation of Forces.—Forces may be represented geometrically by straight lines, proportional to the forces. The three characteristics which, when known, determine a force are (1) direction, (2) place of application, and (3) magnitude. These three are defined as follows:

1. The direction of a force is the direction in which it tends to move the body upon which it acts.

2. The place of application is usually assumed to be a point such as the center of gravity.

3. The magnitude is measured in pounds.

Composition of Forces.—The operation of finding a single force whose effect is the same as that of two or more given forces is called the composition of forces. This single force is called the resultant of the given forces. The separate forces which can be so combined are called the components.

Resolution of Forces.—The operation of finding two or more components of a given force is called the resolution of forces.

Straight lines, drawn to a convenient scale, may be used to represent the forces and arrowheads the direction of the force, the length of the line being its magnitude. The point of application may be any point on the line, although usually it is more convenient to assume the point to be at one end.

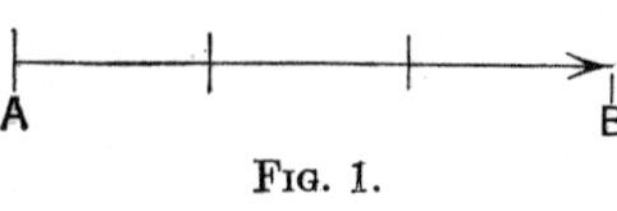

Fig. 1.

In the sketch at the left a force is supposed to act along $A\ B$ in a direction from left to right.

ILLUSTRATION: In the above sketch if A is assumed to be the point of application, the force is exerted as a pull; but if point B is assumed to be the point of application, it would indicate that the force is exerted as a push. If the line is 3 units long and if each unit represents 5 pounds, the line $A\ B$ represents a force of fifteen pounds applied at A.

Composition and Resolution of Forces.—The following rules may be used in the composition and resolution of forces:

1. The resultant of two forces acting in the same direction, is equal to the sum of the forces.

ILLUSTRATION: Two forces $A\ B$ equal to two pounds and $A\ C$ equal to four pounds are both applied at point A. Find the resultant $A\ D$.

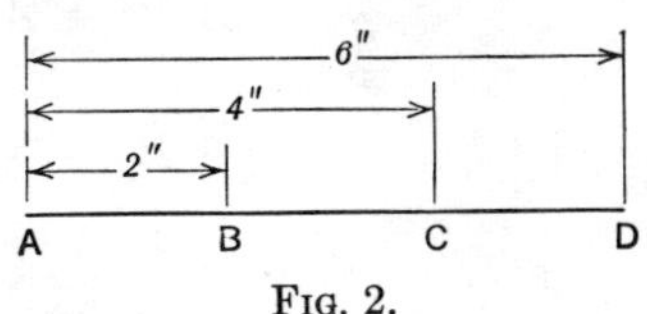

FIG. 2.

$$A\ D = \text{Sum of the forces}$$
$$= 2 + 4 = 6$$

Therefore, the resultant equals 6 pounds.

2. If two forces act in opposite directions, then their resultant is equal to their difference, and the direction of the resultant is the same as the direction of the greater of the two forces.

ILLUSTRATION: Two forces one $A\ B$ equal to 3 pounds and one $A\ C$ equal to 5 pounds are both applied at A. Find the resultant.

$$A\ D = \text{Difference of two forces}$$
$$= 5 - 3 = 2$$

Therefore, the resultant is 2 pounds and acts in the direction of $A\ C$.

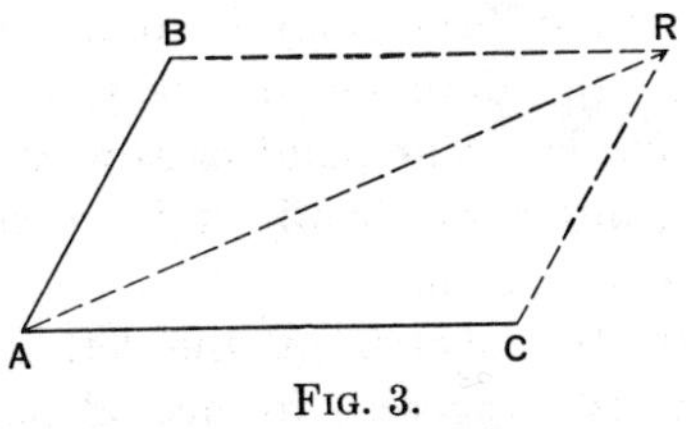

FIG. 3.

Parallelogram of Forces.— If two forces acting on a point are represented in magnitude and direction by the adjacent sides of a parallelogram $A\ B$ and $A\ C$ in the sketch on the left, the resultant will be represented in magnitude and direction by the diagonal $A\ R$ drawn from the intersection of the two component forces.

ILLUSTRATION: If in the figure at the left two forces, one $A\ C$ of 4 pounds acting in the direction of the arrow, and, one $A\ B$ of 3 pounds acting in the direction of the arrow, are both applied at A, find the resultant $A\ R$.

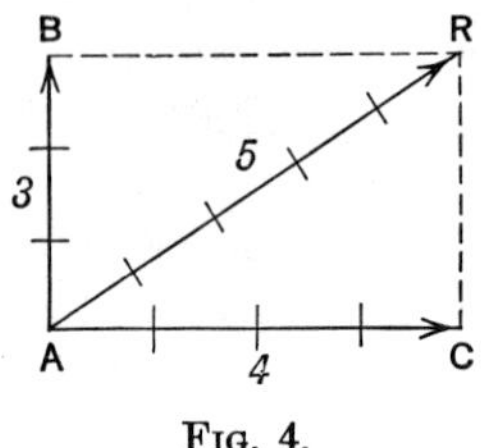

FIG. 4.

Use the geometrical proposition relative to the right triangle i.e., the square on the hypotenuse is equal to the sum of the squares on the other two sides. Expressed as a formula:

$$\overline{A\ R}^2 = \overline{A\ C}^2 + \overline{A\ B}^2$$

$$\overline{A\ R} = \sqrt{\overline{A\ C}^2 + \overline{A\ B}^2}$$

$$= \sqrt{(4 \times 4) + (3 \times 3)}$$

$$= \sqrt{25} = 5$$

Therefore, the resultant is equal to 5 pounds.

Factor of Safety.—A factor of safety is defined as the ratio in which the load that is just sufficient to overcome instantly the strength of a piece of material is greater than the greatest safe ordinary working load. The character of the loading determines in a large degree the margin that should be left for safety. The following table gives the factor of safety for some metals which have been determined by an analytical method:

Cast-iron and other castings..........	4
Wrought iron or mild steel...........	3
Oil-tempered or nickel steel	$2\frac{1}{4}$
Hardened steel......................	3
Bronze or brass, rolled or forged......	3

TABLE 1

AVERAGE ULTIMATE STRENGTH OF COMMON METALS; POUNDS PER SQUARE INCH

Material	Tension	Compression	Shear	Modulus of Elasticity
Cast iron	15,000	80,000	18,000	12,000,000
Wrought iron	48,000	46,000	40,000	27,000,000
Steel castings	70,000	70,000	60,000	30,000,000
Steel structural	60,000	60,000	50,000	29,000,000
Cast brass	24,000	30,000	36,000	9,000,000

TABLE 2

GENERAL FACTORS OF SAFETY

Material	Steady load	Load varying from zero to maximum in one direction	Load varying from zero to maximum in both directions	Suddenly varying loads
Cast iron	6	10	15	20
Wrought iron	4	6	8	12
Steel	5	6	8	12

Symbols and Formulas for the Strength of Materials.—The following symbols are commonly used in the formulas:

A = area of cross section of material in square inches;
E = modulus of elasticity;
I = moment of inertia of section about an axis passing through the center of gravity;
I_p = polar moment of inertia of section;
M_b = maximum bending moment in inch-pounds;
M_t = moment of force tending to twist (torsional moment) in inch-pounds;
P = total stress in pounds;

Y = distance from center of gravity to most remote fiber;
S = permissible working stress in pounds per square inch;
Z = section modulus for bending (moment of resistance);
Z_p = section modulus for torsion;
e = elongation or shortening in inches;
l = length in inches.

These formulas may be used to calculate strength of materials:

For tension and compression: $P = A \times S; \quad e = \frac{Pl}{AE}$

For shear: $P = A \times S$

For torsion: $M_t = \frac{SI_p}{Y} = SZ_p$

For bending: $M_b = \frac{SI}{Y} = SZ$

Combined bending and torsion:

$$\text{Combined moment} = \sqrt{M_b^2 + M_t^2} = SZ$$

The following group of illustrative problems shows how these formulas are employed in practice. The calculations are simple and straightforward, and the reader should have no difficulty in following the steps involved.

ILLUSTRATIVE PROBLEMS

(1) Find the diameter of a wrought iron bar which is to support (in tension) a load of 32,000 pounds if the load is gradually applied and after reaching its maximum value gradually removed.

$$48{,}000 \div 6 = 8{,}000$$

$$P = A \times S$$

$$A = \frac{P}{S} = \frac{32{,}000}{8{,}000} = 4 \text{ sq. in.}$$

$$\text{Diameter} = 1.128\sqrt{4} = 2.256 \text{ inches}$$

Divide 48,000 obtained from the ultimate strength table on page 209 by 6, the factor of safety obtained from table on same page. 48,000 ÷ 6 = 8,000 pounds per square inch. Then dividing 32,000 by 8,000 obtain the answer of 4 square inches. The diameter of a circle of this area is $2\frac{1}{4}$ inches approx.

(2) In the above problem what would be the total elongation of the bar under full load if the bar were 6 feet long?

Multiply 32,000, the load in the above problem, by 6, the length of the bar, and then by 12, the number of inches in one foot. Divide this product by the product of the area 4 and the modulus of elasticity, 27,000,000, obtained from the table on page 209. The quotient, 0.021 inch, is the total elongation.

$$e = \frac{P}{AE}$$

$$= \frac{32{,}000 \times 6 \times 12}{4 \times 27{,}000{,}000}$$

$$= 0.021 \text{ in.}$$

(3) A square bar 3 feet long firmly fixed at one end, is supporting a load of 4,000 pounds at the outer free end. If the bar is to be made of structural steel and the load is steady, find the size of bar required for safe loading.

$$4{,}000 \times 36 = 144.000$$

$$60.000 \div 5 = 12{,}000$$

$$M_b = \frac{SI}{Y}$$

$$144{,}000 = \frac{12.000 \times s^4}{12 \times \frac{1}{2}s} = \frac{12{,}000s^3}{6}$$

$$s^3 = 72, \qquad s = 4.16$$

M_b = load × lever arm in inches, in this case 4,000 × 36 or 144,000 inch-pounds.

S = safe stress = 60,000 obtained from the table divided by 5, the factor of safety, for a safe load for steel, or 12,000;
$I = s^4 \div 12$ for a square, if s = side of square;
$Y = s \div 2$ in this case.

Substituting these values in the equation find size of bar to be 4.16 inches.

(4) A square bar made of structural steel is subjected to a steady torsional moment of 80,000 inch-pounds. Find the size of bar required for safe loading.

$$M_t = 80{,}000$$

Ultimate strength in shear for torsion = $60{,}000 \times \frac{4}{5} = 48{,}000$.

$$S = 48{,}000 \div 5 \text{ (factor of safety for steel)} = 9{,}600$$

$$Z_p = \tfrac{2}{9}s^3 \text{ for a square, if } s = \text{side of square}$$

Substituting the above values in the equation for M and evaluating find $s^3 = 32.5$ and $s = 3.17$ in.

$$M_t = SZ_p$$

$$80{,}000 = 9{,}600 \times \tfrac{2}{9}s^3$$

$$s^3 = 32.5, \quad s = 3.17$$

(5) If the bar in the two previous problems is subjected to combined bending and torsion find the size of square bar required to withstand the combined moment safely.

$$M_b = 144{,}000 \text{ inch-pounds in (3)}$$

$$M_t = 80{,}000 \text{ inch-pounds in (4)}$$

$$\text{Combined Moment} = \sqrt{144{,}000^2 + 80{,}000^2}$$

$$= 165{,}000 \text{ approx.}$$

$$165{,}000 = SZ$$

thus

$$165{,}000 = 12{,}000 \times \frac{s^3}{6}$$

and

$$s^3 = 82.5, \quad s = 4.35 \text{ in.}$$

Thus 12,000 obtained from (3) is the safe load for steel. $\frac{s^3}{6}$ is the formula for section modulus. Substituting these values in the equation obtain 4.35 in., the size of the required bar.

Simple Machines.—A machine is a device by which useful work is done in such a way that the operator gains in effort, speed or convenience. A machine is a simple one when it contains but one moving part. The six fundamental *simple machines* are the *lever*, the *pulley*, the *screw*, the *inclined plane*, the *wedge*, and the *wheel and axle*. Practically all of these machines are used in the machine shop in some form or other. It is, therefore, desirable to know their properties.

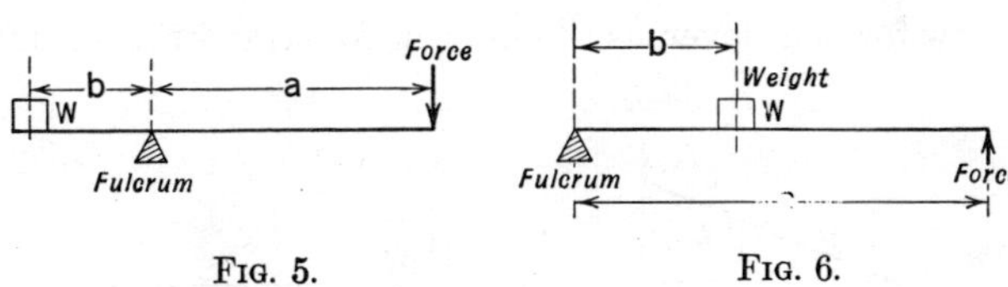

Fig. 5. Fig. 6.

Levers.—A lever is an inflexible rod capable of motion about a fixed point, called a *fulcrum*. The rod may be straight, curved or bent at any angle.

There are three kinds or classes of levers which differ in the respective locations of the applied force, the moved weight and the fulcrum.

In the *lever of the first class*, the fulcrum lies between the points at which the force and the load act (Fig. 5). An example of this type of lever is a claw hammer pulling a nail.

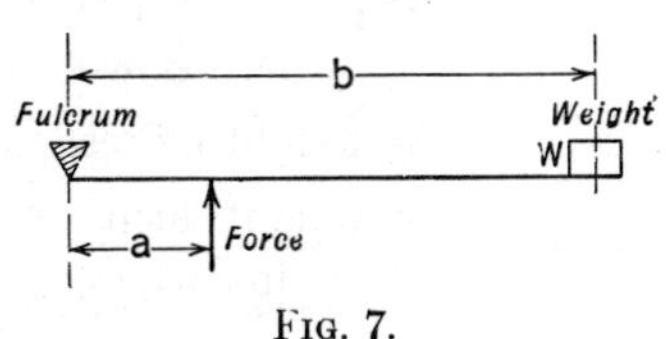

Fig. 7.

In the *lever of the second class*, the load acts at a point between the fulcrum and the force (Fig. 6). An example of this type of lever is a wheelbarrow in which the wheel axle is the fulcrum.

In the *lever of the third class* the force acts between the fulcrum and the load.

Levers are usually used to gain power at the expense of time or motion. Thus, in a first class lever, if the distance from the fulcrum to the force is five times the distance from the fulcrum to the weight, it will give five times the power, but the force will have to move a distance five times greater than the weight.

Levers of the third class involve a mechanical disadvantage as the power must always be greater than the weight. However, there is a gain in motion.

Law of the Lever.—The force multiplied by its distance to the fulcrum is equal to the weight multiplied by its distance to the fulcrum.

The law for bent levers is the same as for straight levers but the

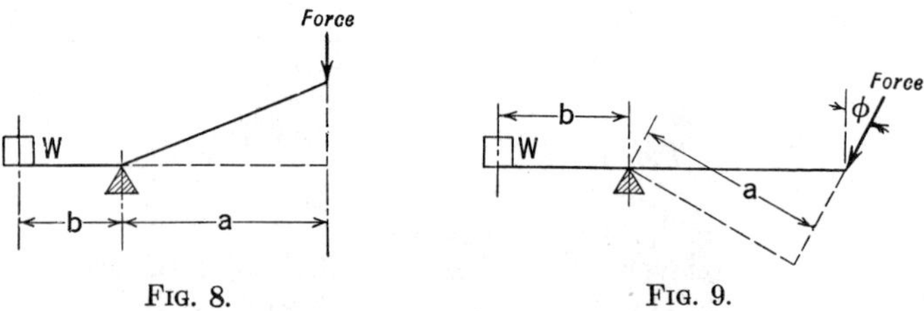

FIG. 8. FIG. 9.

length of arms is computed on lines from the fulcrum at right angles to the direction in which the power and weight act. (See Figs. 8 and 9).

Letting P = power or force;

a = power arm or distance from the fulcrum to the point where power is applied;

W = weight or resistance;

b = weight arm or distance from the fulcrum to the point where the weight or resistance is applied;

the law may then be stated as follows:

$$P \times a = W \times b$$

From this the following relations may be obtained by transposition.

$$P = \frac{W \times b}{a}, \qquad W = \frac{P \times a}{b},$$

$$a = \frac{W \times b}{P}, \qquad b = \frac{P \times a}{W}.$$

ILLUSTRATION: What force in pounds is applied at the brake shoe shown in Fig. 10 if a force of 50 pounds is exerted on the pedal?

In this case, $P = 50$ lb, $a = 14$ in., and $b = 5$ in.

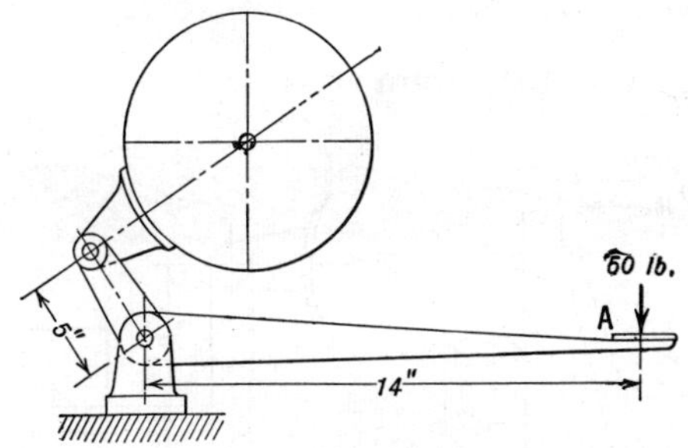

FIG. 10.

$$\text{Then } W = \frac{P \times a}{b} = \frac{50 \times 14}{5} = 140 \text{ lb} \quad \text{(Ans.)}$$

This is an example of a lever of the first class.

ILLUSTRATION: What force will be exerted by the rod "A" in Fig. 11 if a force of 40 pounds is exerted at the handle of the lever?

Here, $P = 40$ lb, $a = 6 + 20 = 26$ in., $b = 6$ in.

$$\text{Then, } W = \frac{P \times a}{b} = \frac{40 \times 26}{6} = 173 \text{ lb} \quad \text{(Ans.)}$$

This is a lever of the second class.

ILLUSTRATION: Figure 12 shows an air brake layout. If the piston in the air cylinder is 10 inches in diameter and the air pressure is 100 pounds per square inch, what is the force on the brake shoe?

$$\text{Area of piston} = \pi r^2 = \pi \times 5 \times 5 = 25\pi \text{ sq. in.}$$

$$\text{Force on brake rod} = P = 100 \times 25\pi = 2500\pi \text{ lb.}$$

$$a = 12 \text{ in.} \quad b = 12 + 8 = 20 \text{ in.}$$

$$\text{Then } W = \frac{P \times a}{b} = \frac{2500\pi \times 12}{20} = 1500\pi = 4712 \text{ lb.} \quad \text{(Ans.)}$$

This is the lever of the third class.

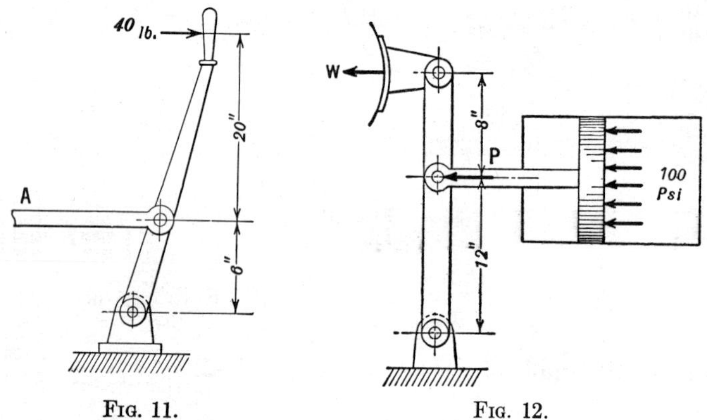

FIG. 11.

FIG. 12.

Wheel and Axle.—This is simply an application of the lever of the first order so that the power and resistance may act through greater distances; the radius of the wheel is the lever arm of the power and that of the axle at the bearing, the lever arm of the resistance. The hoist on a derrick, the capstan on a ship, and the dumbwaiter hoist are common examples of this type of machine.

In considering the wheel and axle, the same formulas are used, the radius of the wheel, R, and the radius of the axle, r, being used for power arm and weight arm. Then

$$P : W = r : R$$

and $$P = \frac{W \times r}{R}$$

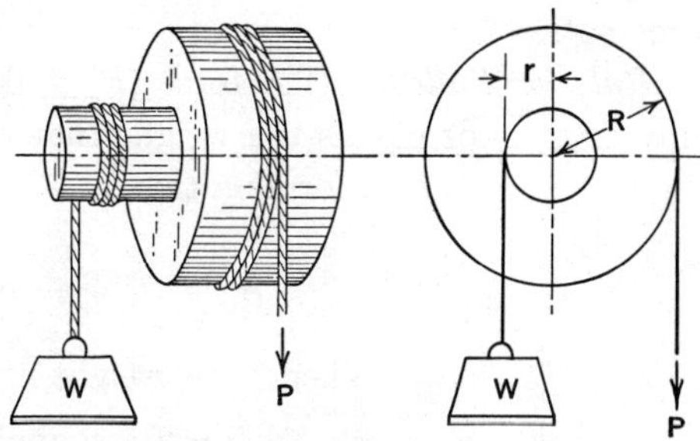

FIG. 13.

ILLUSTRATION: If the radius of a drum on which is wound the lifting rope of a windlass is 2 inches, find the power that must be exerted at the periphery of a wheel 20 inches in diameter when mounted on the same shaft as the drum and transmitting power to it if 1800 pounds is to be lifted.

$$P = \frac{1800 \times 2}{10} = 360 \text{ lb. (Ans.)}$$

Pulleys.—A pulley is a wheel mounted to revolve on an axle and has a grooved rim in which a cord, band or chain is passed to transmit the force applied in another direction. A pulley block is a device for holding one or more pulleys as a unit.

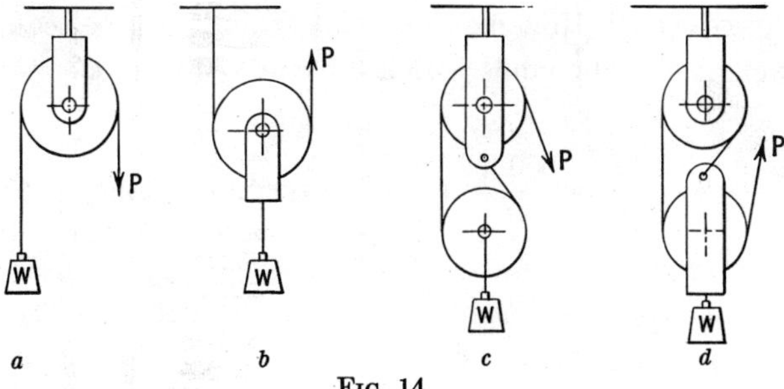

FIG. 14.

Pulleys are either fixed or movable, depending on whether they are held in a fixed position or move with the load. Fig. 14*a* shows a fixed pulley and Fig. 14*b* a movable pulley. In the case of the former the only mechanical advantage is the change in the direction of the applied force.

Figure 14*c* and *d* shows two combinations of fixed and movable pulleys. In each of these arrangements, the weight will move through half the distance through which the pulling force acts.

Rule for Pulleys.—The force (P) multiplied by the number of moving strands equals the weight that can be raised. Stated as a formula this is,

$$W = P \times n \quad \text{or} \quad P = \frac{W}{n} \quad \text{or} \quad n = \frac{W}{P}$$

When W = weight lifted;

P = force applied on free strand;

n = number of moving strands.

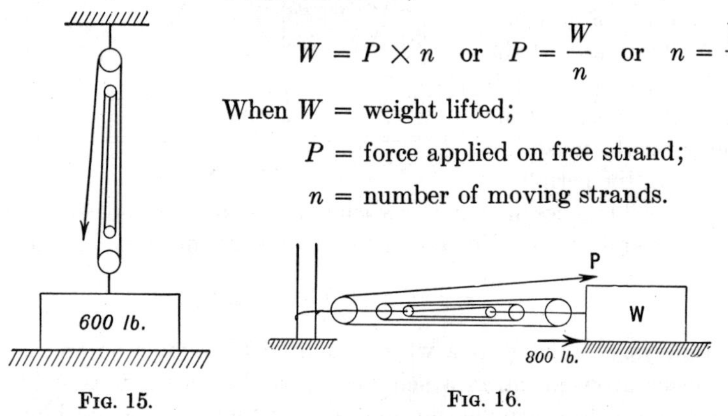

FIG. 15. FIG. 16.

ILLUSTRATION: How many moving strands will be required to lift a weight of 600 pounds with a force of 150 pounds?

$$n = \frac{W}{P} = \frac{600}{150} = 4 \text{ moving strands.} \quad \text{(Ans.)}$$

ILLUSTRATION: A weight offers a resistance of 800 pounds to being pulled along a floor. What force will be required to pull it if a block and tackle with six moving strands is attached? (Fig. 16).

$$P = \frac{W}{n} = \frac{800}{6} = 133 \text{ lb.} \quad \text{(Ans.)}$$

Differential Pulley.—Figure 17 shows a differential pulley which has great general usefulness. In this device an endless chain sprocketed to the pulley wheels replaces the rope. The two pulleys at the top are slightly different diameters and are attached so that they rotate as a unit.

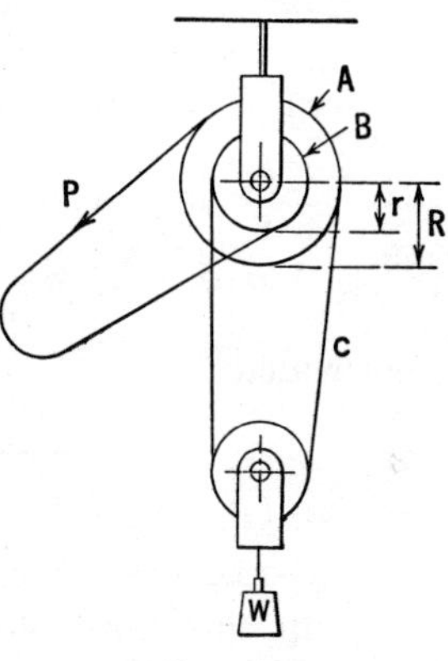

FIG. 17.

When the chain is drawn over the larger pulley it passes around the lower pulley and up over the small wheel from which it is unwound, causing the loop in which the movable pulley rests to shorten by an amount equal to the difference in circumference of the two upper wheels, when they have made one revolution. The weight is moved by an amount equal to one-half this difference.

This may be condensed into a formula as follows:

$$P = \frac{W(R - r)}{2R} \quad \text{or} \quad W = \frac{2PR}{R - r}$$

ILLUSTRATION: A weight of 800 pounds is to be lifted by a differential pulley whose upper wheels are 16 inches and 15 inches in diameter, respectively. What pull or force will be required?

$$P = \frac{W(R - r)}{2R} = \frac{800(\frac{16}{2} - \frac{15}{2})}{2 \times \frac{16}{2}} = \frac{800}{32} = 25 \text{ lb.} \quad \text{(Ans.)}$$

What is the ratio of load to power in this illustration?

$$\text{Ratio} = \frac{\text{Load}}{\text{Force}} = \frac{800}{25} = \frac{32}{1} \quad \text{(Ans.)}$$

Inclined Planes.—An inclined plane is a flat surface sloping or inclined from the horizontal. A body moving up an inclined plane is opposed both by gravity and friction, while one moving down an inclined plane is assisted by gravity and opposed by only friction.

When the force which is being applied is exerted in a direction parallel to the inclined surface as in Fig. 18, it is evident that the power must move through the distance equal to the length of the incline in order to raise the weight through the distance H. The gain in power will then be equal to the length of the incline divided by the height, or

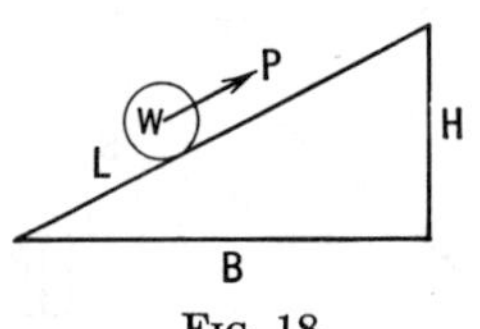

FIG. 18.

$$\frac{P}{W} = \frac{H}{L}$$

from which

$$P = \frac{W \times H}{L}, \quad \text{and} \quad W = \frac{P \times L}{H}.$$

ILLUSTRATION: A roll of paper weighing 500 pounds is to be rolled up onto a 3-foot loading platform by the use of an incline 12 feet long. What force will be required if it acts parallel to the incline? (Fig. 18.)

$$P = \frac{W \times H}{L} = \frac{500 \times 3}{12} = 125 \text{ lb.} \quad \text{(Ans.)}$$

If a force acts along a line parallel to the base as in Fig. 19 then

$$\frac{P}{W} = \frac{H}{B} \quad \text{and} \quad P = \frac{W \times H}{B}, \quad \text{and} \quad W = \frac{P \times B}{H}$$

ILLUSTRATION: What force will be required in the above problem if the force moving the roll of paper acts horizontally? (Fig. 19).

If $L = 12$ and $H = 3$, then by the law of right triangles, B is the square root of the differences of the squares of the hypotenuse and the opposite side, or

$$B = \sqrt{12^2 - 3^2} = \sqrt{144 - 9} = \sqrt{135} = 11.62 \text{ ft.}$$

Then, $$P = \frac{W \times H}{B} = \frac{500 \times 3}{11.62} = 129.1 \text{ lb.} \quad \text{(Ans.)}$$

If a force acts at any angle to the plane as X in Fig. 20 and the angle of the incline makes Y degrees with the horizontal, then

$$\frac{P}{W} = \frac{\sin Y}{\cos X}$$

From which

$$P = \frac{W \times \sin Y}{\cos X}, \quad W = \frac{P \times \cos X}{\sin Y} \quad \text{and} \quad \cos X = \frac{W \times \sin Y}{P}$$

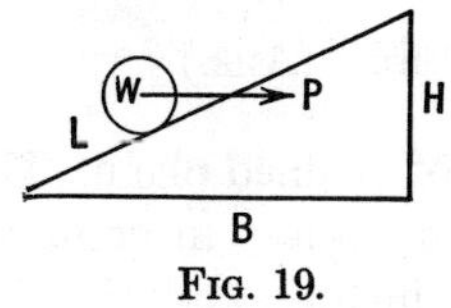

Fig. 19.

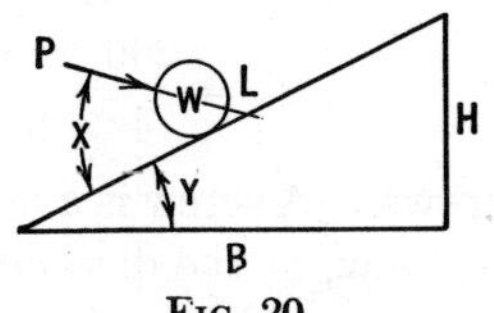

Fig. 20.

Illustration: A boiler drum weighing one ton is to be rolled up a 10-degree incline. What force will be required (ignoring friction) if X is 20 degrees?

$$\sin Y = \sin 10° = 0.1736$$
$$\cos X = \cos 20° = 0.9397$$

Then,

$$P = \frac{W \times \sin Y}{\cos X} = \frac{2000 \times 0.1736}{0.9397} = \frac{347.2}{0.9397} = 369.5 \text{ lb.} \quad \text{(Ans.)}$$

Wedges.—A wedge is a pair of inclined planes united at their bases. The power is usually applied by a blow of a heavy body or by pressure. Wedges are used for splitting logs and stones and raising heavy weights short distances. Due to excessive friction, they are not very efficient.

Ignoring friction, the relations of weight and force may be expressed,

$$\frac{P}{W} = \frac{T}{L}, \quad \text{or} \quad P = \frac{W \times T}{L}$$

when, P = power applied;
W = weight or resistance;
T = thickness of wedge at base;
L = length of wedge.

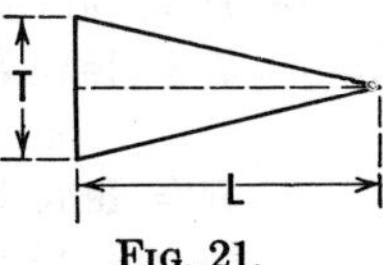

Fig. 21.

This may be expressed in the following forms.

$$W = \frac{P \times L}{T}, \quad L = \frac{W \times T}{P}, \quad \text{and} \quad T = \frac{L \times P}{W}$$

Illustration: What force will be required to drive a wedge 4 inches long and $\frac{3}{8}$ inch thick to raise a 200-pound casting?

$$P = \frac{W \times T}{L} = \frac{200 \times \frac{3}{8}}{4}$$

$$P = \frac{200 \times 3}{4 \times 8} = \frac{600}{32} = 19 \text{ lb.} \quad \text{(Ans.)}$$

Screws.—A screw is a modified form of inclined plane. The lead of the screw, or the distance the thread advances in going around once, corresponds to the height of the incline, and the distance around the screw measured on the thread is the length of the incline.

When a force is applied to raise a weight or overcome resistance by means of a screw or nut, either the screw or nut may be fixed, the other being movable. The force is generally applied at the end of a wrench or lever arm or at the circumference of a wheel. The ratio of the power to weight is independent of the diameter of the screw. In actual work, a considerable proportion of the power transmitted is lost through friction.

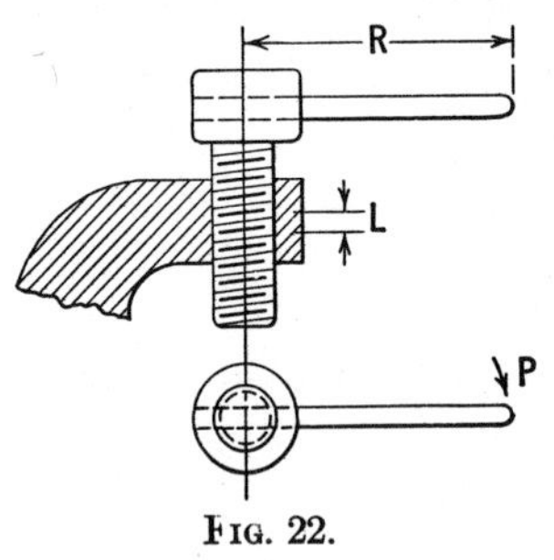

Fig. 22.

Ignoring friction, the force multiplied by the circumference of the circle through which the force arm moves, equals the weight or resulting force multiplied by the lead of the screw. This may be expressed as an equation:

$$\frac{P}{W} = \frac{L}{2\pi R}$$

When P = power applied;

L = lead of screw; in single threads the lead is equal to the pitch, in double threads the lead is twice the pitch, etc.

R = length of bar, wrench, or radius of hand wheel used to operate screw;

W = resulting force or weight moved.

NOTE. All lengths must be expressed in the same unit and all forces in one unit.

The equation may also be expressed in the following forms:

$$P = \frac{W \times L}{2\pi R}, \quad \text{and} \quad W = \frac{P \times 2\pi R}{L}$$

ILLUSTRATION: What is the pressure produced in a milling machine vise if the screw has six single threads per inch, the handle a length of 10 inches, if a pressure of 50 pounds is applied and the loss through friction is 40 per cent?

If 40 per cent of the power is lost in friction only $50 - (50 \times 0.40) = 30$ pounds of pressure remains for useful work.

Since there are six single threads per inch, the lead (L) is $\frac{1}{6}$ inch.

Then, $$W = \frac{P \times 2\pi R}{L} = \frac{30 \times 2 \times 10}{\frac{1}{6}} \times 3.14$$

$$W = 300 \times 6 \times 2 \times 3.14 = 3{,}600 \times 3.14 = 11{,}310 \text{ lb.} \quad \text{(Ans.)}$$

Mechanical Advantage.—The mechanical advantage of a perfect machine is the number obtained by dividing the resistance by the effort. Expressed as a formula:

$$\text{Mechanical advantage} = \frac{\text{resistance}}{\text{effort}}$$

Many machine problems may be solved by using the principle of mechanical advantage. If a machine has a mechanical advantage of 5, an effort of 20 pounds will lift 5 times as much weight, or 100 pounds.

In the lever the mechanical advantage is found by dividing the length of effort arm by the length of resistance arm; in the wheel and axle by dividing the radius of the wheel by the radius of the axle. The mechanical advantage of a fixed pulley is 1, of a single movable pulley is 2.

Mechanical Efficiency.—The efficiency of a machine is a fraction expressing the ratio of the useful work to the whole work performed which is equal to the energy expended.

$$\text{Efficiency of a machine} = \frac{\text{useful output}}{\text{input}}$$

Efficiency in machines is always expressed by a percent. Thus, if 100 units of work are put into a machine and only 95 units are gotten out, the efficiency of the machine is $\frac{95}{100}$ or 95%.

Friction is the chief cause of the loss of efficiency in most machines.

Pages 228–235 contain tables which give moments of inertia and other properties of different cross sections (of such outlines) frequently met with in structural steel shapes and in cast iron designs.

The Moment of Inertia.—The moment of inertia of any cross-section may be defined as the sum of the products obtained by multiplying each of the elementary areas of which the section is composed by the square of the distance of the center of gravity of the elementary area to the neutral axis of the section. The moment of inertia varies, in the same body, according to the position of the axis. It is the least possible when the axis passes through the center of gravity.

The Section-Modulus or Section-Factor.—The strength of sections to resist strains either as girders or as columns, depends not only on the area but also on the form of the section, and the property which forms the basis of the constants used in the formulas for the strength of girders and columns to express the effect of form, is the moment of inertia about its neutral axis. The modulus of resistance of any section to transverse bending is its moment of inertia divided by the normal distance of the extreme fiber from the neutral axis.

Radius of Gyration.—The effect of the form of the cross-

TABLE 3

Coefficients of Deflection of Steel Beams for Uniformly Distributed Loads

Span in feet	Fiber Stress, pounds per square inch		Span in feet	Fiber Stress, pounds per square inch	
	16,000	12,500		16,000	12,500
1	0.017	0.013	21	7.299	5.703
2	0.066	0.052	22	8.011	6.259
3	0.149	0.166	23	8.756	6.841
4	0.265	0.207	24	9.534	7.448
5	0.414	0.323	25	10.345	8.082
6	0.596	0.466	26	11.189	8.741
7	0.811	0.634	27	12.066	9.427
8	1.059	0.828	28	12.977	10.138
9	1.341	1.047	29	13.920	10.875
10	1.655	1.293	30	14.897	11.638
11	2.003	1.565	31	15.906	12.427
12	2.383	1.862	32	16.949	13.241
13	2.797	2.185	33	18.025	14.082
14	3.244	2.534	34	19.134	14.948
15	3.724	2.909	35	20.276	15.841
16	4.237	3.310	36	21.451	16.759
17	4.783	3.737	37	22.659	17.703
18	5.363	4.190	38	23.901	18.672
19	5.975	4.668	39	25.175	19.668
20	6.621	5.172	40	26.483	20.690

section of a column on its strength is determined by a quantity called the radius of gyration, which is the normal distance from the neutral axis to the center of gyration. The center of gyration is defined as the point where the entire area might be concentrated and have the same moment of inertia as the actual distributed area.

The following notation is used:

A = the area of section in square inches;
d = the depth of cross-section in inches;
I = the moment of inertia in inches;
r = the radius of gyration in inches;
S = the section modulus in inches;
X_1z = the distance of the center of gravity of section from extreme fiber in inches.

Deflection of Steel Beams.—To find the deflection in inches of a section symmetrical about the neutral axis, such as the section of an I beam, channel, zee, etc., divide the coefficient in the table corresponding to the given span and fiber-stress by the depth of the section in inches.

ILLUSTRATION: Find the deflection in a 10-inch 25-pound beam of a 10-foot span, under its maximum distributed load of 13 tons, the fiber-stress being taken at 12,500 pounds per square inch.

The table of coefficients, page 225, gives the deflection of a 10-foot span as 1.293 for a fiber stress of 12,500. Therefore, $1.293 \div 10 = 0.1293$ the deflection at the middle.

Another method of calculating the deflection of steel beams is to use Table 3A on page 227. When given the length and depth of a beam, girder, or truss, and the design unit stress; then the corresponding factor from Table 3A, multiplied by the length in feet, will give the center deflection in inches.

For unit stress values not shown in Table 3A, multiply the factor for 10,000 psi by the ratio of the design unit stress to 10,000. For example, if the design unit stress is 13,000 psi, mul-

TABLE 3A*

COEFFICIENTS OF DEFLECTION OF STEEL BEAMS

Ratio of $\frac{\text{Depth}}{\text{Span}}$	Maximum Fibre Stress in Lbs. per Sq. In.					
	10,000	12,000	14,000	16,000	18,000	20,000
1/4	.0034	.0041	.0048	.0054	.0061	.0068
1/5	.0043	.0051	.0060	.0068	.0077	.0085
1/6	.0051	.0061	.0072	.0082	.0092	.0102
1/7	.0060	.0072	.0084	.0096	.0107	.0119
1/8	.0068	.0082	.0095	.0109	.0123	.0136
1/9	.0077	.0092	.0107	.0123	.0138	.0153
1/10	.0085	.0102	.0119	.0136	.0153	.0170
1/11	.0094	.0112	.0131	.0150	.0169	.0187
1/12	.0102	.0122	.0143	.0164	.0184	.0204

* Courtesy American Institute of Steel Construction.

tiply the factors listed in the column headed "10,000" by 13,000/10,000 or 1.3.

Table 3A assumes uniformly distributed loading. For a single load at center, multiply these factors by 0.8; for two equal loads at the third points, by 1.02. The factors shown are strictly correct for beams of constant section; close for cover-plated beams and girders; and reasonably approximate for trusses.

The following pages give tables of the moments of inertia and other properties of different cross-section of such outlines as are most frequently met with in structural steel shapes, together with the formulas used to determine bending moments and deflections for steel beams. From these the total safe load may be determined and the proper size beam may be selected from tables in the *Steel Construction Manual* of the American Institute of Steel Construction, New York, N. Y.

TABLE 4

PROPERTIES OF VARIOUS SECTIONS

Sections.	Area of Section. A	Distance from Neutral Axis to Extremities of Section. x and x_1
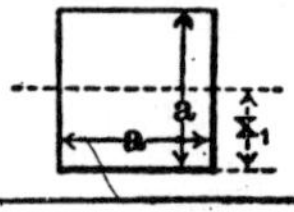	a^2	$x_1 = \frac{a}{2}$
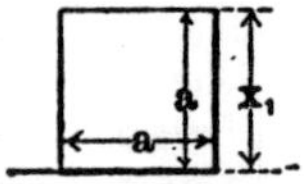	a^2	$x_1 = a$
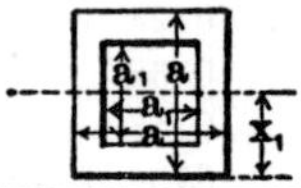	$a^2 - a_1^2$	$x_1 = \frac{a}{2}$
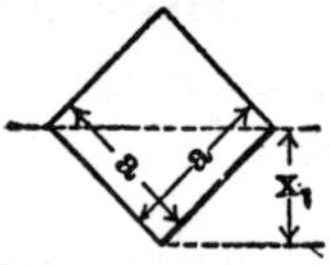	a^2	$x_1 = \frac{a}{\sqrt{2}} = .707a$
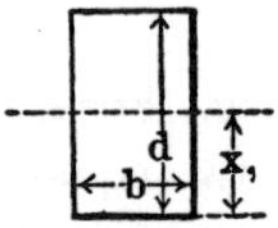	bd	$x_1 = \frac{d}{2}$
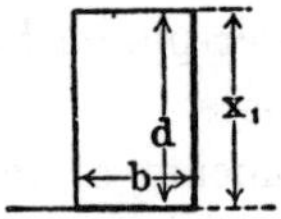	bd	$x_1 = d$
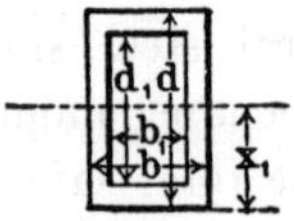	$bd - b_1d_1$	$x_1 = \frac{d}{2}$
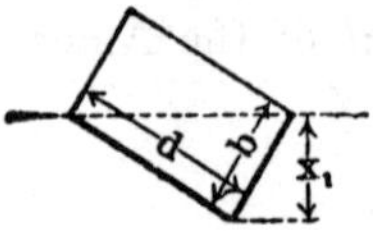	bd	$x_1 = \frac{b\,d}{\sqrt{b^2 + d^2}}$

TABLE 4.—PROPERTIES OF VARIOUS SECTIONS—*Continued*

Moment of Inertia. I	**Section Modulus.** $S = \frac{I}{x_1}$.	**Radius of Gyration.** $r = \sqrt{\frac{I}{A}}$.
$\frac{a^4}{12}$	$\frac{a^3}{6}$	$\frac{a}{\sqrt{12}} = .289a$
$\frac{a^4}{3}$	$\frac{a^3}{3}$	$\frac{a}{\sqrt{3}} = .577a$
$\frac{a^4 - a_1^4}{12}$	$\frac{a^4 - a_1^4}{6a}$	$\sqrt{\frac{a^2 + a_1^2}{12}}$
$\frac{a^4}{12}$	$\frac{a^3}{6\sqrt{2}} = .118a^3$	$\frac{a}{\sqrt{12}} = .289a$
$\frac{bd^3}{12}$	$\frac{bd^2}{6}$	$\frac{d}{\sqrt{12}} = .289d$
$\frac{bd^3}{3}$	$\frac{bd^2}{3}$	$\frac{d}{\sqrt{3}} = .577d$
$\frac{bd^3 - b_1d_1^3}{12}$	$\frac{bd^3 - b_1d_1^3}{6d}$	$\sqrt{\frac{bd^3 - b_1d_1^3}{12\,(bd - b_1d_1)}}$
$\frac{b^3d^3}{6\,(b^2 + d^2)}$	$\frac{b^2d^2}{6\sqrt{b^2 + d^2}}$	$\frac{bd}{\sqrt{6\,(b^2 + d^2)}}$

TABLE 4.—Properties of Various Sections—*Continued*

Sections.	Area of Section. A	Distance from Neutral Axis to Extremities of Section. x and x_1
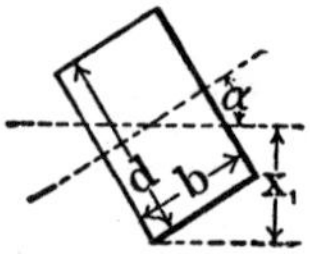	bd	$x_1 = \frac{d \cos \alpha + b \sin \alpha}{2}$
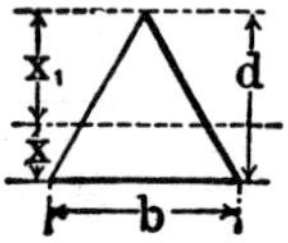	$\frac{bd}{2}$	$x = \frac{d}{3}$ $x_1 = \frac{2d}{3}$
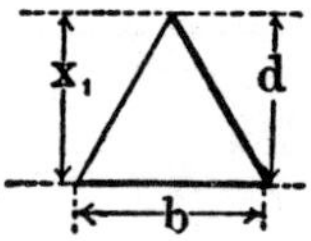	$\frac{bd}{2}$	$x_1 = d$
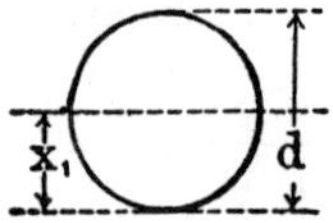	$\frac{\pi d^2}{4} = .785d^2$	$x_1 = \frac{d}{2}$
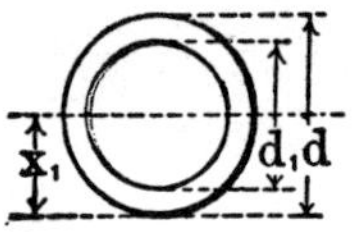	$\frac{\pi (d^2 - d_1^2)}{4} = .785 (d^2 - d_1^2)$	$x_1 = \frac{d}{2}$
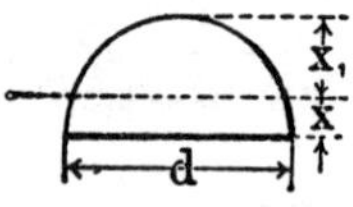	$\frac{\pi d^2}{8} = .393d^2$	$x = \frac{2d}{3\pi} = .212d$ $x_1 = \frac{(3\pi - 4) d}{6\pi} = .288d$
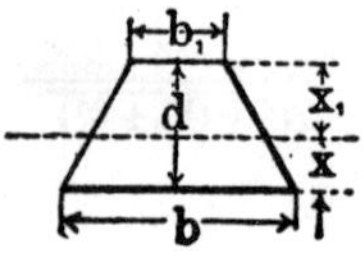	$\frac{b + b_1}{2} \cdot d$	$x = \frac{b + 2b_1}{b + b_1} \cdot \frac{d}{3}$ $x_1 = \frac{b_1 + 2b}{b + b_1} \cdot \frac{d}{3}$

TABLE 4.—Properties of Various Sections—*Continued*

Moment of Inertia. I	**Section Modulus.** $S = \frac{I}{x_1}$.	**Radius of Gyration.** $r = \sqrt{\frac{I}{A}}$
$\frac{bd}{12}(d^2\cos^2 a + b^2\sin^2 a)$	$\frac{db}{6}\left(\frac{d^2\cos^2 a + b^2\sin^2 a}{d\cos a + b\sin a}\right)$	$\sqrt{\frac{d^2\cos^2 a + b^2\sin^2 a}{12}}$
$\frac{bd^3}{36}$	$\frac{bd^2}{24}$	$\frac{d}{\sqrt{18}} = .236d$
$\frac{bd^3}{12}$	$\frac{bd^2}{12}$	$\frac{d}{\sqrt{6}} = .408d$
$\frac{\pi d^4}{64} = .049d^4$	$\frac{\pi d^3}{32} = .098d^3$	$\frac{d}{4}$
$\frac{\pi(d^4 - d_1^4)}{64} = .049(d^4 - d_1^4)$	$\frac{\pi}{32}\frac{(d^4 - d_1^4)}{d} = .098\frac{(d^4 - d_1^4)}{d}$	$\frac{\sqrt{d^2 + d_1^2}}{4}$
$\frac{9\pi^2 - 64}{1152\pi}\cdot d^4 = .007d^4$	$\frac{9\pi^2 - 64}{192(3\pi - 4)}\cdot d^3 = .024d^3$	$\frac{\sqrt{9\pi^2 - 64}}{12\pi}\cdot d = .132d$
$\frac{b^2 + 4bb_1 + b_1^2}{36(b + b_1)}\cdot d^3$	$\frac{b^2 + 4bb_1 + b_1^2}{12(b_1 + 2b)}\cdot d^2$	$\frac{d}{6(b+b_1)}\sqrt{2(b^2+4bb_1+b_1^2)}$

TABLE 4.—PROPERTIES OF VARIOUS SECTIONS—*Continued*

Sections.	Area of Section. A	Distance from Neutral Axis to Extremities of Section. x and x_1
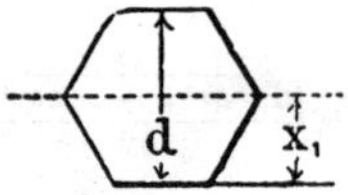 d, x_1	$\frac{3}{2}d^2 \tan. 30° = .866d^2$	$x_1 = \frac{d}{2}$
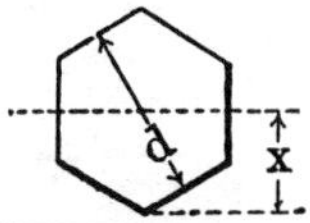 d, x	$\frac{3}{2}d^2 \tan. 30° = .866d^2$	$x_1 = \frac{d}{2 \cos 30°} = .577d$
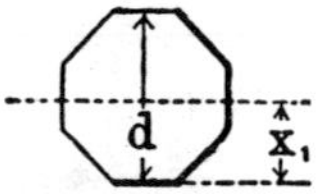 d, x_1	$2d^2 \tan. 22\frac{1}{2}° = .828d^2$	$x_1 = \frac{d}{2}$
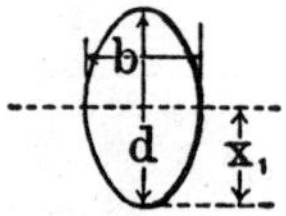 b, d, x_1	$\frac{\pi bd}{4} = .785\ bd$	$x_1 = \frac{d}{2}$
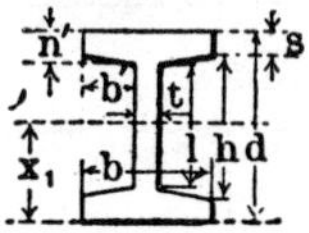 n′, s, b′, t, b, l, h, d, x_1	$td + 2b'(s + n')$	$x_1 = \frac{d}{2}$
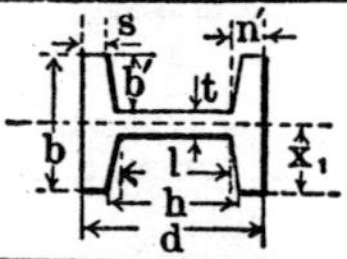 s, n′, b′, t, b, l, h, d, x_1	$td + 2b'(s + n')$	$x_1 = \frac{b}{2}$
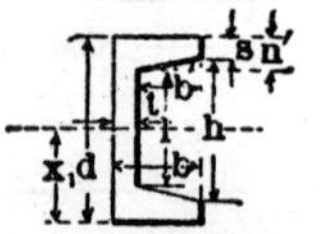 s, n′, t, b′, l, h, b, d, x_1	$td + b'(s + n')$	$x_1 = \frac{d}{2}$
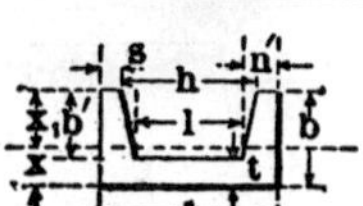 s, h, n′, x_1, b′, l, b, x, t, d	$td + b'(s + n')$	$x = \left[b^2s + \frac{ht^2}{2} + \frac{g}{3}(b-t)^2 (b + 2t)\right] + A$ $x_1 = b - x$

TABLE 4.—PROPERTIES OF VARIOUS SECTIONS—*Continued*

Moment of Inertia. I	Section Modulus. $S = \frac{I}{x_1}$	Radius of Gyration. $r = \sqrt{\frac{I}{A}}$
$\frac{A}{12}\left[\frac{d^2(1+2\cos^2 30^\circ)}{4\cos^2 30^\circ}\right]$ $= .06d^4$	$\frac{A}{6}\left[\frac{d(1+2\cos^2 30^\circ)}{4\cos^2 30^\circ}\right] = .12d^3$	$\frac{d}{4\cos 30^\circ}\sqrt{\frac{1+2\cos^2 30^\circ}{3}}$ $= .261d$
$\frac{A}{12}\left[\frac{d^2(1+2\cos^2 30^\circ)}{4\cos^2 30^\circ}\right]$ $= .06d^4$	$\frac{A}{6}\left[\frac{d(1+2\cos^2 30^\circ)}{4\cos 30^\circ}\right]$ $= .104d^3$	$\frac{d}{4\cos 30^\circ}\sqrt{\frac{1+2\cos^2 30^\circ}{3}}$ $= .261d$
$\frac{A}{12}\left[\frac{d^2(1+2\cos^2 22\frac{1}{2}^\circ)}{4\cos^2 22\frac{1}{2}^\circ}\right]$ $= .055d^4$	$\frac{A}{6}\left[\frac{d(1+2\cos^2 22\frac{1}{2}^\circ)}{4\cos 22\frac{1}{2}^\circ}\right]$ $= .109d^3$	$\frac{d}{4\cos 22\frac{1}{2}^\circ}\sqrt{\frac{1+2\cos^2 22\frac{1}{2}^\circ}{3}}$ $= .257d$
$\frac{\pi bd^3}{64} = .049bd^3$	$\frac{\pi bd^2}{32} = .098bd^2$	$\frac{d}{4}$
$\frac{1}{12}\left[bd^3 - \frac{1}{4g}(h^4 - l^4)\right]$	$\frac{2I}{d}$	$r = \sqrt{\frac{I}{A}}$
$\frac{1}{12}\left[b^3(d-h) + lt^3 + \frac{g}{4}(b^4 - t^4)\right]$	$\frac{2I}{b}$	$r = \sqrt{\frac{I}{A}}$
$\frac{1}{12}\left[bd^3 - \frac{1}{8g}(h^4 - l^4)\right]$	$\frac{2I}{d}$	$r = \sqrt{\frac{I}{A}}$
$\frac{1}{3}\left[2sb^3 + lt^3 + \frac{g}{2}(b^4 - t^4)\right]$ $- Ax^2$	$\frac{I}{b - x}$	$r = \sqrt{\frac{I}{A}}$

TABLE 4.—PROPERTIES OF VARIOUS SECTIONS—*Continued*

Sections.	Area of Section. A	Distance from Neutral Axis to Extremities of Section. x and x_1
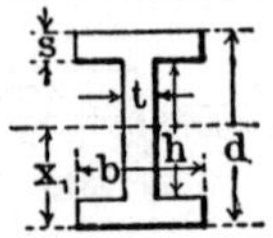	$bd - h(b - t)$	$x_1 = \frac{d}{2}$
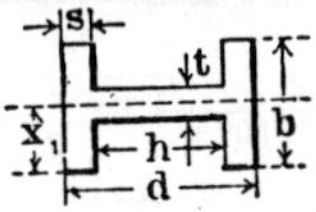	$bd - h(b - t)$	$x_1 = \frac{b}{2}$
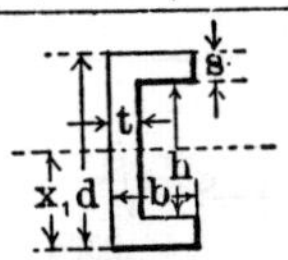	$bd - h(b - t)$	$x_1 = \frac{d}{2}$
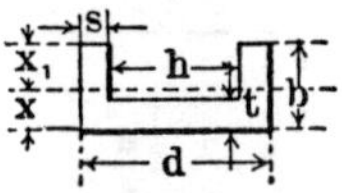	$bd - h(b - t)$	$x = \frac{2b^2s + ht^2}{2A}$ $x_1 = b - x$
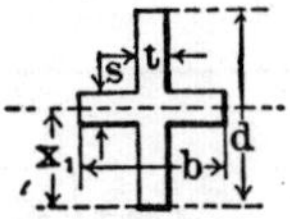	$td + s(b - t)$	$x_1 = \frac{d}{2}$
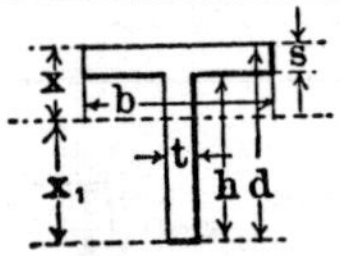	$bs + ht$	$x = \frac{d^2t + s^2(b - t)}{2A}$ $x_1 = d - x$
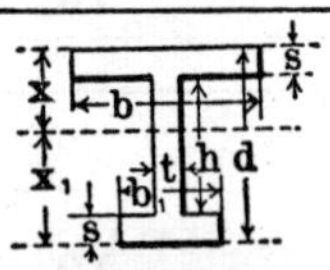	$bs + ht + b_1s$	$x = \frac{td^2 + s^2(b - t) + s(b_1 - t)(2d - s)}{2A}$ $x_1 = d - x$
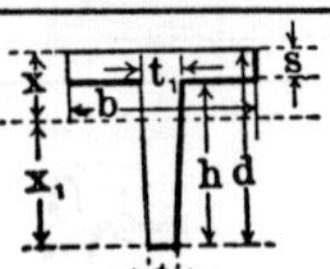	$bs + \frac{h(t + t_1)}{2}$	$x = \frac{3bs^2 + 3th(d + s) + h(t_1 - t)(h + 3s)}{6A}$ $x_1 = d - x$

TABLE 4.—Properties of Various Sections—*Concluded*

Moment of Inertia. I	Section Modulus. $S = \frac{I}{x_1}$	Radius of Gyration. $r = \sqrt{\frac{I}{A}}$
$\frac{bd^3 - h^3(b-t)}{12}$	$\frac{bd^3 - h^3(b-t)}{6d}$	$\sqrt{\frac{bd^3 - h^3(b-t)}{12\,[bd - h(b-t)]}}$
$\frac{2sb^3 + ht^3}{12}$	$\frac{2sb^3 + ht^3}{6b}$	$\sqrt{\frac{2sb^3 + ht^3}{12\,[bd - h(b-t)]}}$
$\frac{bd^3 - h^3(b-t)}{12}$	$\frac{bd^3 - h^3(b-t)}{6d}$	$\sqrt{\frac{bd^3 - h^3(b-t)}{12\,[bd - h(b-t)]}}$
$\frac{2sb^3 + ht^3}{3} - Ax^2$	$\frac{I}{b-x}$	$\sqrt{\frac{I}{A}}$
$\frac{td^3 + s^3(b-t)}{12}$	$\frac{td^3 + s^3(b-t)}{6d}$	$\sqrt{\frac{td^3 + s^3(b-t)}{12\,[td + s(b-t)]}}$
$\frac{tx_1^3 + bx^3 - (b-t)(x-s)^3}{3}$	$\frac{I}{d-x}$	$\sqrt{\frac{tx_1^3 + bx^3 - (b-t)(x-s)^3}{3(bs+ht)}}$
$\frac{bx^3 + b_1x_1^3 - (b-t)(x-s)^3}{3} - \frac{(b_1-t)(x_1-s)^3}{3}$	$\frac{I}{d-x}$	$\left[\frac{bx^3 + b_1x_1^3 - (b-t)(x-s)^3}{3(bs+ht+b_1s)} - \frac{(b_1-t)(x_1-s)^3}{3(bs+ht+b_1s)}\right]^{\frac{1}{2}}$
$\frac{4bs^3 + h^3(3t+t_1)}{12} - A(x-s)^2$	$\frac{I}{d-x}$	$\sqrt{\frac{I}{A}}$

TABLE 5

Bending Moments and Deflections for Beams of Uniform Section

W = Total Load, in lbs., uniformly distributed, including the weight of beam.

W_1 = Total Superimposed or Live Load, in lbs., uniformly distributed.

W_2 = Total Weight of Beam or Dead Load, in lbs., uniformly distributed.

P, P_1, P_2, P_3 = Loads, in lbs., concentrated at any points.

M = Total Bending Moment, in inch-lbs.

M_{w1}, M_p = Bending Moments, in inch-lbs., due to Weights W_1 and P respectively.

I = Moment of Intertia, in inches^4.

l = Length of Span, in inches.

E = Modulus of Elasticity, in lbs. per square inch = 29 000 000 for steel.

W_s = Total Safe Load, in lbs., uniformly distributed, including weight of beam = Total Safe Load of Tables.

The ordinates in diagrams give the bending moments for corresponding points on beam. For superimposed load only, make W_2 in formulæ equal to zero.

---

(1) Beam Supported at both ends and Uniformly Loaded.

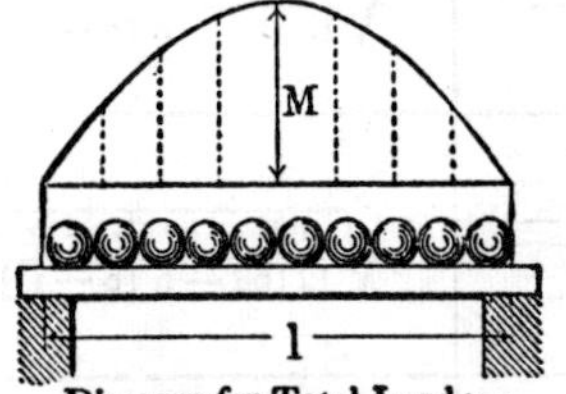

Diagram for Total Load:—

Draw parabola having $M = \frac{Wl}{8}$

Safe Superimposed Load, in lbs., uniformly distributed, $W'_s = W_s - W_2$.

Maximum Bending Moment at middle of beam $= M = \frac{Wl}{8} = \frac{(W_1 + W_2)\,l}{8}$.

Maximum Shear at points of support $= \frac{W}{2} = \frac{W_1 + W_2}{2}$.

Maximum Deflection $= \frac{5}{384}\,\frac{Wl^3}{EI} = \frac{5}{384}\,\frac{(W_1 + W_2)\,l^3}{EI}$.

---

(2) Beam Supported at both ends with Load Concentrated at the Middle.

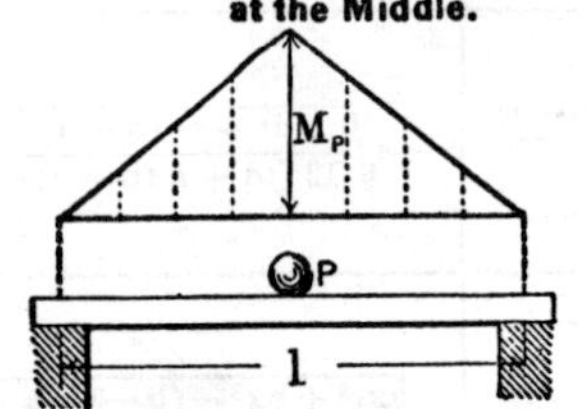

Diagram for Superimposed Load:—

Draw triangle having $M_p = \frac{Pl}{4}$

Diagram for Dead Load similar to Case(1)

Safe Superimposed Load, in lbs., concentrated, $P_s = \frac{W_s - W_2}{2}$.

Maximum Bending Moment at middle of beam $= M = \frac{Pl}{4} + \frac{W_2 l}{8}$.

Maximum Shear at points of support $= \frac{P + W_2}{2}$.

Max. Deflection $= \frac{Pl^3}{48EI} + \frac{5}{384}\,\frac{W_2\,l^3}{EI}$.

---

(3) Beam fixed at one end, Unsupported at the other and Uniformly Loaded.

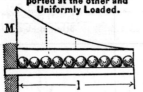

M

l

Diagram for Total Load:—

Draw Parabola having $M = \frac{Wl}{2}$

Safe Superimposed Load, in lbs., uniformly distributed, $W'_s = \frac{W_s}{4} - W_2$.

Maximum Bending Moment at point of support $= \frac{Wl}{2} = \frac{(W_1 + W_2)\,l}{2}$.

Maximum Shear at point of support $= W = W_1 + W_2$.

Max. Deflection $= \frac{Wl^3}{8EI} = \frac{(W_1 + W_2)l^3}{8EI}$.

TABLE 5.—BENDING MOMENTS AND DEFLECTIONS FOR BEAMS OF UNIFORM SECTION.—*Continued*

W = Total Load, in lbs., uniformly distributed, including the weight of beam.

W_1 = Total Superimposed or Live Load, in lbs., uniformly distributed.

W_2 = Total Weight of Beam or Dead Load, in lbs., uniformly distributed.

P, P_1, P_2, P_3 = Loads, in lbs., concentrated at any points.

M = Total Bending Moment, in inch-lbs.

M_{w1}, M_p = Bending Moments, in inch-lbs., due to Weights W_1 and P respectively.

I = Moment of Inertia, in inches[4].

l = Length of Span, in inches.

E = Modulus of Elasticity, in lbs. per square inch = 29 000 000 for steel.

W_s = Total Safe Load, in lbs., uniformly distributed, including weight of beam = Total Safe Load of Tables.

The ordinates in diagrams give the bending moments for corresponding points on beam. For superimposed load only, make W_2 in formulæ equal to zero.

(4) Beam fixed at one end, and Unsupported at the other, with Load Concentrated at the free end.

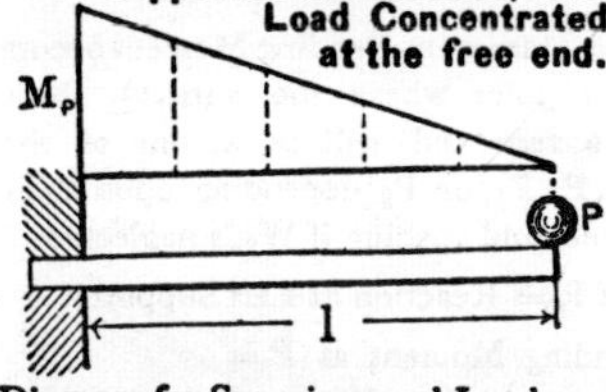

Diagram for Superimposed Load:—
Draw triangle having $M_p = Pl$.
Diagram for Dead Load similar to Case(3)

Safe Superimposed Load, in lbs., concentrated, $P_s = \frac{W_s - 4W_2}{8}$.

Maximum Bending Moment at point of support $= Pl + \frac{W_2 l}{2}$.

Maximum Shear at point of support $= P + W_2$.

Maximum Deflection $= \frac{Pl^3}{3EI} + \frac{W_2 l^3}{8EI}$.

(5) Beam Supported at both ends with Load Concentrated at any point.

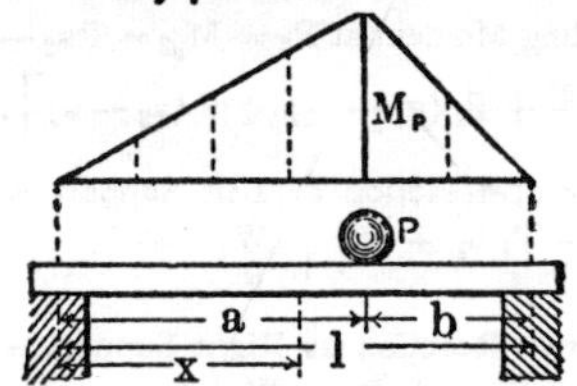

Diagram for Superimposed Load:—

Draw triangle having $M_p = \frac{Pab}{l}$.

Diagram for Dead Load similar to Case (1)

Safe Superimposed Load, in lbs., concentrated, $P_s = \frac{W_s l^2 - 4a\,W_2\,(l - a)}{8ab}$.

Maximum Bending Moment under load $= \frac{a\,(2\,Pb + W_2 l - W_2 a)}{2l}$.

Max. Shear at Sup. near a $= \frac{Pb}{l} + \frac{W_2}{2}$.

Max. Shear at Sup. near b $= \frac{Pa}{l} + \frac{W_2}{2}$.

Deflection at distance x from left support $= \frac{1}{3lEI}\left[\frac{2al - a^2}{3}\right]^{\frac{3}{2}}$
$\left[Pb + \frac{W_2}{8}\left(2l - \sqrt{\frac{2al - a^2}{3}} - \frac{3l^3}{2al - a^2}\right)\right]$

$x = \sqrt{\frac{2al - a^2}{3}}$ = Distance, from left support, of point of maximum deflection for superimposed load.

(6) Beam Supported at both ends with two Symmetrical Loads.

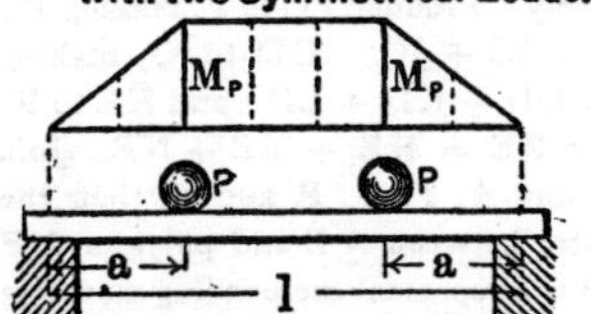

Diagram for Superimposed Load:—
Draw trapezoid having $M_p = Pa$.
Diagram for Dead Load similar to Case(1)

Safe Superimposed Load, in lbs., concentrated, each, $P_s = \frac{W_s l - W_2 l}{8a}$.

Maximum Bending Moment at center of beam $= Pa + \frac{W_2 l}{8}$.

Maximum Shear at points of support $= \frac{2P + W_2}{2}$.

Maximum Deflection $= \frac{Pa}{24EI}\left(3l^2 - 4a^2\right) + \frac{5}{384}\frac{W_2 l^3}{EI}$.

TABLE 5.—BENDING MOMENTS AND DEFLECTIONS FOR BEAMS OF UNIFORM SECTION.—*Continued*

W = Total Load, in lbs., uniformly distributed, including the weight of beam.

W_1 = Total Superimposed or Live Load, in lbs., uniformly distributed.

W_2 = Total Weight of Beam or Dead Load, in lbs., uniformly distributed.

P, P_1, P_2, P_3 = Loads, in lbs., concentrated at any points.

M = Total Bending Moment, in inch-lbs.

M_{w1}, M_p = Bending Moments, in inch-lbs., due to Weights W_1 and P respectively.

I = Moment of Inertia, in inches⁴.

l = Length of Span, in inches.

E = Modulus of Elasticity, in lbs., per square inch = 29 000 000 for steel.

W_s = Total Safe Load, in lbs., uniformly distributed, including the weight of beam = Total Safe Load of Tables.

The ordinates in diagrams give the bending moments for corresponding points on beam. For superimposed load only, make W_2 in formulæ equal to zero.

(7) Beam Supported at both ends with Loads Concentrated at various Points.

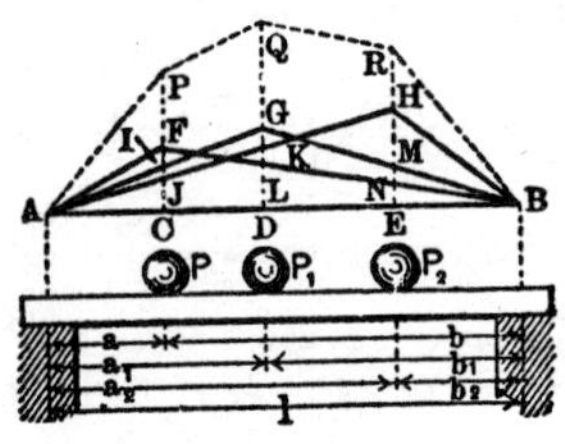

The total bending moment at any point produced by all the weights is equal to the sum of the moments at that point produced by each of the weights separately.

Diagram for Dead Load similar to Case (1)

The Maximum Bending Moment occurs at the point where the vertical shear equals zero and will be at one of the loads P, P_1, or P_2 depending upon their amounts and spacing if W_2 is neglected.

Let R = Reaction at Left Support.

Bending Moment at P =

$$M_p = Ra - \frac{W_2\, a^2}{2l}.$$

Bending Moment at P_1 =

$$M_{p1} = Ra_1 - \left[\frac{W_2\, a_1^2}{2l} + P(a_1 - a)\right].$$

Bending Moment at $P_2 = M_{p2} = Ra_2 -$

$$\left[\frac{W_2\, a_2^2}{2l} + P_1(a_2 - a_1) + P(a_2 - a)\right].$$

Shear or Reaction at Left Support =

$$\frac{P_2 b_2 + P_1 b_1 + Pb}{l} + \frac{W_2}{2}.$$

Shear or Reaction at Right Support =

$$\frac{P_2 a_2 + P_1 a_1 + Pa}{l} + \frac{W_2}{2}.$$

Diagram for Superimposed Load:—Draw as in Case (5) the Ordinates FC, GD and HE representing the bending moments due to loads P, P_1 and P_2 respectively. Produce FC to P, making PC = FC + IC + JC; GD to Q, making QD = GD + KD + LD; and HE to R, making RE = HE + ME + NE. Join the points A, P, Q, R and B, then the ordinates between A B and polygon A P Q R B will represent the bending moments for corresponding points on beam.

TABLE 5.—BENDING MOMENTS AND DEFLECTIONS FOR BEAMS OF UNIFORM SECTION.—*Concluded*

W = Total Load, in lbs., uniformly distributed, including the weight of beam. W_1 = Total Superimposed or Live Load, in lbs., uniformly distributed. W_2 = Total Weight of Beam or Dead Load, in lbs., uniformly distributed. P, P_1, P_2, P_3 = Loads, in lbs., concentrated at any points.	M = Total Bending Moment, in inch-lbs. M_{w1}, M_p = Bending Moments, in inch-lbs., due to Weights W_1 and P respectively. I = Moment of Inertia, in inches⁴. l = Length of Span, in inches. E = Modulus of Elasticity, in lbs., per square inch = 29 000 000 for steel. W_s = Total Safe Load, in lbs., uniformly distributed, including the weight of beam = Total Safe Load of Tables.

The ordinates in diagrams give the bending moments for corresponding points on beam. For superimposed load only, make W_2 in formulæ equal to zero.

(8) Beam Fixed at both ends and Uniformly Loaded.

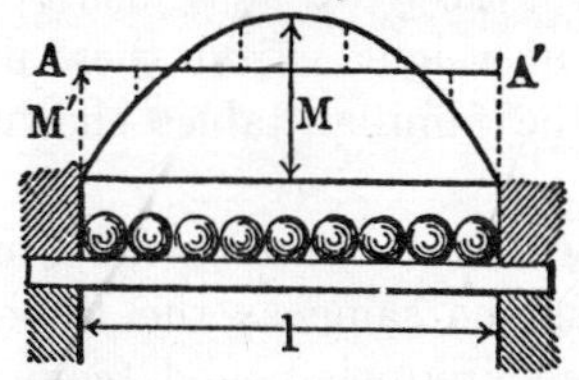

Diagram for Total Load:—Draw parabola having $M = \frac{Wl}{8}$. Also A A' parallel to base and at a distance $M' = \frac{Wl}{12}$. The Vertical distances between the parabola and line A A' are the moments for corresponding points on beam.

Safe Superimposed Load, in lbs., uniformly distributed, $W'_s = \frac{2}{3} W_s - W_2$.

Distance of points of contra-flexure from supports = .2113l.

Maximum Bending Moment at points of support $= \frac{Wl}{12} = \frac{(W_1 + W_2)\,l}{12}$.

Bending Moment at middle of beam $= \frac{Wl}{24} = \frac{(W_1 + W_2)\,l}{24}$.

Maximum Shear at points of support $= \frac{W_1 + W_2}{2}$.

Maximum Deflection $= \frac{Wl^3}{384EI} = \frac{(W_1 + W_2)\,l^3}{384EI}$.

(9) Beam Fixed at both ends with Load Concentrated at the Middle.

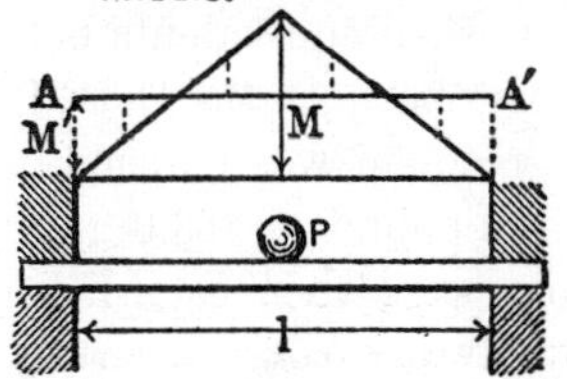

Diagram for Superimposed Load:—Draw triangle having $M = \frac{Pl}{4}$. Also A A' parallel to base and at a distance $M' = \frac{Pl}{8}$. The Vertical distances between the triangle and line A A' are the moments for corresponding points on beam.

Diagram for Dead Load similar to Case (8)

Safe Superimposed Load, in lbs., concentrated, $P_s = W_s - \frac{2}{3} W_2$.

Distance of points of contra-flexure from supports = ¼l.

Maximum Bending Moment at points of support $= \frac{Pl}{8} + \frac{W_2 l}{12}$.

Bending Moment at middle of beam $= \frac{Pl}{8} + \frac{W_2 l}{24}$.

Maximum Shear at points of support $= \frac{P + W_2}{2}$.

Maximum Deflection $= \frac{Pl^3}{192EI} + \frac{W_2 l^3}{384EI}$.

VIII

Weights and Measures

In the United States and other English-speaking countries a customary system of weights and measures has been in use for many hundreds of years. Its basis is the yard as a measure of length and the pound as a measure of weight. From these units many other units are derived by the familiar tables shown in the following pages.

The troy pound, which is accepted as standard in the United States, contains 5760 grains, and is the same as the Imperial troy pound of Great Britain. The avoirdupois pound (commercial) of the United States contains 7000 grains, and agrees with the British avoirdupois pound within 0.001 grain.

The U. S. yard differs from the British yard by less than 2 parts per million. However, certain derived measures, such as the liquid and dry measures differ widely. Therefore there is widespread use in both countries of the metric system, which was made a permissive system for use in the U. S. by an Act of Congress of 1866. That Act also established equivalents between the two systems. After some conferences between the U. S., Great Britain, Canada, and Australia, an agreement was reached defining the inch as 25.4 millimeters and the pound as 453.6 grams in all these countries. These equivalents were then established in the U. S. by proclamation, but not by law.

The name units of length and mass in the metric system are the meter and the gram; all larger and smaller units are multiples of ten, their names being formed by using prefixes. Thus deka- is 10 times, hecto- is 100 times, kilo- is 1000 times, deci- is $1/10$, centi- is $1/100$, milli- is $1/1000$, etc. Examples are kilogram and millimeter.

The legal equivalent of the meter as established by Act of Congress is 39.37 inches = 3.28083 feet = 1.093611 yards.

Long Measure—Measures of Length

12 inches (in.)	= 1 foot (ft.)
3 feet	= 1 yard (yd.)
1760 yards, or 5280 feet	= 1 mile (mi.)

Additional measures of length occasionally used are:

1000 mils = 1 inch; 3 inches = 1 palm; 4 inches = 1 hand
9 inches = 1 span; $2\frac{1}{2}$ feet = 1 military space
$5\frac{1}{2}$ yards or $16\frac{1}{2}$ feet = 1 rod; 2 yards = 1 fathom;
a cable length = 120 fathoms = 720 feet;
1 inch = 0.0001157 cable length = 0.013889 fathom = 0.111111 span.

Old Land or Surveyors' Measure*

7.92 inches = 1 link (l.)
100 links, or 66 feet, or 4 rods = 1 chain (ch.)
10 chains or 220 yards = 1 furlong
8 furlongs or 80 chains = 1 mile (mi.)

Nautical Measure

6080.26 feet or 1.15156 statute miles = 1 nautical mile or knot †
3 nautical miles = 1 league
60 nautical miles, or 69.169 statute miles = 1 degree at the equator
360 degrees = circumference of the earth at the equator

* Sometimes called Gunter's Chain.
† The value varies according to different measures of the earth's diameter.

Square Measure—Measures of Surface*

144 square inches (sq. in.)	= 1 square foot (sq. ft.)
9 square feet	= 1 square yard (sq. yd.)
$30\frac{1}{4}$ square yards or $272\frac{1}{4}$ square feet	= 1 square rod (sq. rd.)
160 square rods or 43,560 square feet	= 1 acre (A.)
640 acres	= 1 square mile (sq. mi.)

Surveyors' Measure

16 square rods	= 1 square chain (sq. ch.)
10 square chains	= 1 acre (A.)
640 acres	= 1 square mile (sq. mi.)
1 square mile	= 1 section (sec.)
36 sections	= 1 township (tp.)

Measures used for Diameters and Areas of Electric Wires

Circular inch: a circular inch is the area of a circle 1 inch in diameter.

1 circular inch = 0.7854 square inch

1 square inch = 1.2732 circular inches

1 circular inch = 1,000,000 circular mils = 1000 MCM

Circular mil: a circular mil is the area of a circle one mil, or 0.001 inch in diameter. In larger cable sizes, 1000 circular mils usually is written 1 MCM.

* Square measures are used in computing area or surfaces, as land, lumber painting, etc.

Solid or Cubic Measure—Measures of Volume*

1728 cubic inches (cu. in.) = 1 cubic foot (cu. ft.)
27 cubic feet = 1 cubic yard (cu. yd.)

The following measures are also used for wood and masonry.

1 cord of wood = a pile, 4 × 4 × 8 feet = 128 cubic feet
1 perch of masonry = $16\frac{1}{2} \times 1\frac{1}{2} \times 1$ foot = $24\frac{3}{4}$ cubic feet

Shipping Measure

Register Ton—For register tonnage or for measuring entire internal capacity of a ship or vessel:

100 cubic feet = 1 register ton

Shipping Ton—For the measurement of cargo.

40 cubic feet = 1 United States shipping ton = 32.143 U. S. bushels

42 cubic feet = 1 British shipping ton = 32.719 imperial bushels.

Carpenter's Rule—To find the weight a vessel will carry multiply the length of keel by the breadth at main beam by the depth of the hold in feet and divide by 95 (the cubic feet allowed for a ton). The result will be the tonnage.

Dry Measure—United States†

2 pints (pt.) = 1 quart (qt.)
8 quarts = 1 peck (pk.)
4 pecks = 1 bushel (bu.)

* This table is used in measuring bodies having three dimensions; length, breadth, and height or depth.

† This measure is used in measuring grain, fruit and other articles not liquid. The standard U. S. bushel is the Winchester bushel, which is, in cylinder form $18\frac{1}{2}$ inches in diameter and 8 inches deep and contains 2150 42 cubic inches. A struck bushel contains 2150.42 cubic inches = 1.2445 cubic feet; 1 cubic foot = 0.80356 struck bushel.

The British Imperial bushel = 8 imperial gallons or 2,219.360 cubic inches = 1.2837 cubic feet. The British quarter = 8 imperial bushels.

Liquid Measure

4 gills (gi.)	= 1 pint (pt.)
2 pints	= 1 quart (qt.)
4 quarts	= 1 gallon (gal.) {U. S. 231 cubic inches; British 277.274 cubic inches}
1 cubic foot	= 7.48 U. S. gallons

Old Liquid Measure

31½ gallons	= 1 barrel (bbl.)
42 gallons	= 1 tierce
2 barrels or 63 gallons	= 1 hogshead (hhd.)
84 gallons or 2 tierces	= 1 puncheon
2 hogsheads or 4 barrels or 126 gallons	= 1 pipe or butt
2 pipes or 3 puncheons	= 1 tun

Apothecaries' Fluid Measure

60 minims = 1 fluid drachm; 8 drachms = 1 fluid ounce

1 U. S. fluid ounce = 8 drachms = 1.805 cubic inch = $\frac{1}{128}$ U. S. gallon. The fluid ounce in Great Britian is 1.732 cubic inches.

MEASURES OF WEIGHT

Avoirdupois or Commercial Weight

16 drachms or 437.5 grains	= 1 ounce (oz.)
16 ounces or 7000 grains	= 1 pound (lb.)
2000 pounds	= 1 net or short ton
2240 pounds	= 1 gross or long ton
2204.6 pounds	= 1 metric ton

Measures of weight occasionally used in collecting duties on foreign goods at U. S. custom houses and in freighting coal are:

1 hundredweight = 4 quarters = 112 pounds (1 gross or long ton = 20 hundredweight); 1 quarter = 28 pounds; 1 stone = 14 pounds; 1 quintal = 100 pounds.

Troy Weight *

24 grains	= 1 pennyweight (pwt.)
20 pennyweights	= 1 ounce (oz.)
12 ounces or 5760 grains	= 1 pound (lb.)

A carat of the jewelers, for precious stones = 3.2 grains in the United States. The International carat = 3.168 grains or 200 milligrams. In avoirdupois, apothecaries' and troy weights, the grain is the same, 1 pound troy = 0.82286 pound avoirdupois.

Apothecaries' Weight †

20 grains (gr.)	= 1 scruple (℈)
3 scruples	= 1 drachm (ʒ)
8 drachms	= 1 ounce (℥)
12 ounces	= 1 pound troy (lb.)

Measures of Time

one millionth of a second	= 1 microsecond (μsec.)
one thousandth of a second	= 1 millisecond (msec.)
1/3600 hour	= 1 second (sec.)
60 seconds (sec.)	= 1 minute (min.)
60 minutes	= 1 hour (hr.)
24 hours	= 1 day (da.)
7 days	= 1 week (wk.)
365 days	= 1 solar year (yr.)
366 days	= 1 leap-year (every four years)
100 years	= 1 century

By the Gregorian calender every year whose number is divisible by 4 is a leap year except that the centesimal years (each 100 years: 1800, 1900, 2000, etc.) are leap-years only when the number of the year is divisible by 400.

* Used for weighing gold, silver, jewels, etc.

† This table is used in compounding medicines and prescriptions.

A solar day is measured by the rotation of the earth upon its axis, with respect to the sun. In astronomical calculations and in nautical time the day begins at noon.

In civil calculations the day commences at midnight, and is divided into two parts of 12 hours each. A mean lunar month, or lunation of the moon, is 29 days, 12 hours, 44 minutes, 2 seconds, and 5.24 thirds (29.53 days).

In one hour a point on the earth's surface describes $\frac{1}{24}$ of $360° = 15°$, in one minute $\frac{1}{60}$ of $15° = 15'$, and in one second $\frac{1}{60}$ of $15' = 15''$.

In military calculations, the day begins at midnight and the hours are numbered around the clock from 0000 to 2400. Thus 8:46 A.M. is written as 0846 hours, 4:36 P.M. is written as 1636 hours and 10:32 P.M. is written as 2232 hours. The United States Army adopted this 24-hour clock system July 1, 1942. This method of time notation is easily used by noting that all times prior to noon are written directly (8:46 A.M. = 0846 hours), and all times past noon are written by adding 12 to the hour involved (4:36 P.M. = 12 + 436 = 1636 hours). No colon is used in this notation.

Circular and Angular Measures *

60 seconds (″)	=	1 minute (′)
60 minutes	=	1 degree (°)
90 degrees	=	1 quadrant
360 degrees	=	1 circumference

A second is usually sub-divided into tenths and hundreths. A minute of the earth's circumference is a geographical mile.

* This table is used for measuring angles and arcs, and for determining latitude and longitude.

TABLE 1

Rectangular Tanks

Capacity in U. S. Gallons Per Foot of Depth

Widths, Feet	Length of Tank—in Feet						
	2	2½	3	3½	4	4½	5
2	29.92	37.40	44.88	52.36	59.84	67.32	74.81
2½	—	46.75	56.10	65.45	74.81	84.16	93.51
3	—	—	67.32	78.55	89.77	101.0	112.2
3½	—	—	—	91.64	104.7	117.8	130.9
4	—	—	—	—	119.7	134.6	149.6
4½	—	—	—	—	—	151.5	168.3
5	—	—	—	—	—	—	187.0

Widths, Feet	5½	6	6½	7	7½	8	8½
2	82.29	89.77	97.25	104.7	112.2	119.7	127.2
2½	102.9	112.2	121.6	130.9	140.3	149.6	159.0
3	123.4	134.6	145.9	157.1	168.3	179.5	190.8
3½	144.0	157.1	170.2	183.3	196.4	209.5	222.5
4	164.6	179.5	194.5	209.5	224.4	239.4	254.3
4½	185.1	202.0	218.8	235.6	252.5	269.3	286.1
5	205.7	224.4	243.1	261.8	280.5	299.2	317.9
5½	226.3	246.9	267.4	288.0	308.6	329.1	349.7
6	—	269.3	291.7	314.2	336.6	359.1	381.5
6½	—	—	316.1	340.4	364.7	389.0	413.3
7	—	—	—	366.5	392.7	418.9	445.1
7½	—	—	—	—	420.8	448.8	476.9
8	—	—	—	—	—	478.8	508.7
8½	—	—	—	—	—	—	540.5

Widths, Feet	9	9½	10	10½	11	11½	12
2	134.6	142.1	149.6	157.1	164.6	172.1	179.5
2½	168.3	177.7	187.0	196.4	205.7	215.1	224.4
3	202.0	213.2	224.4	235.6	246.9	258.1	269.3
3½	235.6	248.7	261.8	274.9	288.0	301.1	314.2
4	269.3	284.3	299.2	314.2	329.1	344.1	359.1
4½	303.0	319.3	336.6	353.5	370.3	387.1	403.9
5	336.6	355.3	374.0	392.7	411.4	430.1	448.8
5½	370.3	390.9	411.4	432.0	452.6	473.1	493.7
6	403.9	426.4	448.8	471.3	493.7	516.2	538.6
6½	437.6	461.9	486.2	510.5	534.9	559.2	583.5
7	471.3	497.5	523.6	549.8	576.0	602.2	628.4
7½	504.9	533.0	561.0	589.1	617.1	645.2	673.2
8	538.6	568.5	598.4	628.4	658.3	688.2	718.1
8½	572.3	604.1	635.8	667.6	699.4	731.2	763.0
9	605.9	639.6	673.2	706.9	740.6	774.2	807.9
9½	—	675.1	710.6	746.2	781.7	817.2	852.3
10	—	—	748.1	785.5	822.9	860.3	897.7
10½	—	—	—	824.7	864.0	903.3	942.5
11	—	—	—	—	905.1	946.3	987.4
11½	—	—	—	—	—	989.3	1032
12	—	—	—	—	—	—	1077

U. S. Gallon of water weighs 8.34523 Pounds Avoirdupois at 4° C.

TABLE 2

CIRCULAR TANKS

Capacity in U. S. Gallons Per Foot of Depth

Diam., Ft. In.	Gallons	Diam., Ft. In.	Gallons	Diam., Ft. In.	Gallons
1	5.875	3 6	71.97	5 11	205.7
1 1	6.895	3 7	75.44	6	211.5
1 2	7.997	3 8	78.99	6 3	229.5
1 3	9.180	3 9	82.62	6 6	248.2
1 4	10.44	3 10	86.33	6 9	267.7
1 5	11.79	3 11	90.13	7	287.9
1 6	13.22	4	94.00	7 3	308.8
1 7	14.73	4 1	97.96	7 6	330.5
1 8	16.32	4 2	102.0	7 9	352.9
1 9	17.99	4 3	106.1	8	376.0
1 10	19.75	4 4	110.3	8 3	399.9
1 11	21.58	4 5	114.6	8 6	424.5
2	23.50	4 6	119.0	8 9	449.8
2 1	25.50	4 7	123.4	9	475.9
2 2	27.58	4 8	127.9	9 3	502.7
2 3	29.74	4 9	132.6	9 6	530.2
2 4	31.99	4 10	137.3	9 9	558.5
2 5	34.31	4 11	142.0	10	587.5
2 6	36.72	5	146.9	10 3	617.3
2 7	39.21	5 1	151.8	10 6	647.7
2 8	41.78	5 2	156.8	10 9	679.0
2 9	44.43	5 3	161.9	11	710.9
2 10	47.16	5 4	167.1	11 3	743.6
2 11	49.98	5 5	172.4	11 6	777.0
3	52.88	5 6	177.7	11 9	811.1
3 1	55.86	5 7	183.2	12	846.0
3 2	58.92	5 8	188.7	12 3	881.6
3 3	62.06	5 9	194.2	12 6	918.0
3 4	65.28	5 10	199.9	12 9	955.1
3 5	68.58				

U. S. Gallon of water weighs 8.34523 Pounds Avoirdupois at 4° C.

Water Conversion Factors

U. S. gallons	×	8.33	= pounds
U. S. gallons	×	0.13368	= cubic feet
U. S. gallons	×	231	= cubic inches
U. S. gallons	×	0.83	= English gallons
U. S. gallons	×	3.78	= liters
English gallons (Imperial)	×	10	= pounds
English gallons (Imperial)	×	0.16	= cubic feet
English gallons (Imperial)	×	277.274	= cubic inches
English gallons (Imperial)	×	1.2	= U. S. gallons
English gallons (Imperial)	×	4.537	= liters
Cubic inches of water (39.1°)	×	0.036024	= pounds
Cubic inches of water (39.1°)	×	0.004329	= U. S. gallons
Cubic inches of water (39.1°)	×	0.003607	= English gallons
Cubic inches of water (39.1°)	×	0.576384	= ounces
Cubic feet (of water) (39.1°)	×	62.425	= pounds
Cubic feet (of water) (39.1°)	×	7.48	= U. S. gallons
Cubic feet (of water) (39.1°)	×	6.232	= English gallons
Cubic feet (of water) (39.1°)	×	0.028	= tons
Pounds of water	×	27.72	= cubic inches
Pounds of water	×	0.01602	= cubic feet
Pounds of water	×	0.12	= U. S. gallons
Pounds of water	×	0.10	= English gallons

Miscellaneous Tables

Numbers

12 units = 1 dozen
12 dozen = 1 gross
12 gross = 1 great gross
20 units = 1 score

Paper

24 sheets = 1 quire
20 quires = 1 ream
2 reams = 1 bundle
5 bundles = 1 bale

Books

A book of sheets folded in:
2 leaves is a folio
4 leaves is a quarto
8 leaves is an octavo
12 leaves is a duodecimo
16 leaves is a 16mo.

The Metric System

The metric system is a system of weights and measures based upon a unit called a meter and expressed in the decimal scale. The meter was intended to be one ten millionth of the distance from the equator to either pole, but more careful measurements show that this distance is 10,001,887 meters. The value of the meter, as authorized by the United States Government, is 39.37 inches.

The names of derived metric denominations are formed by prefixing to the name of the primary unit of measure:

Micro, a millionth $= \frac{1}{1,000,000}$
Milli, a thousandth $= \frac{1}{1000}$
Centi, a hundredth $= \frac{1}{100}$
Deci, a tenth $= \frac{1}{10}$
Deca, ten = 10
Hecto, one hundred = 100
Kilo, one thousand = 1000
Myria, ten thousand = 10,000
Mega, one million = 1,000,000

The principal units of the metric system are:

The meter for lengths
The square meter for surfaces
The cubic meter for large volumes
The liter for small volumes
The gram for weights

Measures of Length

10 millimeters (mm.)	= 1 centimeter (cm.)
10 centimeters	= 1 decimeter (dm.)
10 decimeters	= 1 meter (m.)
10 meters	= 1 decameter (Dm.)
10 decameters	= 1 hectometer (Hm.)
10 hectometers	= 1 kilometer (Km.)
10 kilometers	= 1 myriameter

A meter is used in ordinary measurements; the centimeter or millimeter in calculating very small distances; and the kilometer for long distances.

Square Measures—Measures of Surface

100 square millimeters ($mm.^2$)	= 1 square centimeter ($cm.^2$)
100 square centimeters	= 1 square decimeter ($dm.^2$)
100 square decimeters	= 1 square meter ($m.^2$)
100 centiares, or square meters	= 1 are (a.)
100 ares	= 1 hectare (ha.)

The square meter is used for ordinary surfaces; the are, a square, each of whose sides is 10 meters, is the unit of land measure.

Cubic Measure—Measures of Volume

1000 cubic millimeters ($mm.^3$)	= 1 cubic centimeter ($cm.^3$ or cc.)
1000 cubic centimeters	= 1 cubic decimeter ($dm.^3$)
1000 cubic decimeters	= 1 cubic meter ($m.^3$)

The term stere is used to designate the cubic meter in measuring wood and timber. A tenth of a stere is a decistere, and ten steres are a decastere.

Liquid and Dry Measures—Measures of Capacity

10 milliliters (ml.)	= 1 centiliter (cl.)
10 centiliters	= 1 deciliter (dl.)
10 deciliters	= 1 liter (l.)
10 liters	= 1 decaliter (Dl.)
10 decaliters	= 1 hectoliter (Hl.)
10 hectoliters	= 1 kiloliter (Kl.)

The liter, which is a cube each of whose edges is $\frac{1}{10}$ of a meter in length, is the principal unit of measures of capacity. The hectoliter is the unit that is used in measuring large quantities of grain, fruits, roots, and liquids.

Measures of Weight

10 milligrams (mg.)	= 1 centigram (cg.)
10 centigrams	= 1 decigram (dg.)
10 decigrams	= 1 gram (g.)
10 grams	= 1 decagram (Dg.)
10 decagrams	= 1 hectogram (Hg.)
10 hectograms	= 1 kilogram (Kg.)
1000 kilograms	= 1 (metric) ton (T.)

The gram, which is the primary unit of weights, is the weight of one cubic centimeter of pure distilled water at a temperature of 39.2° F., the kilogram is the weight of 1 liter of water; the ton is the weight of 1 cubic meter of water. The gram is used in weighing gold, jewels, and small quantities of things. The kilogram, commonly called kilo for brevity, is used by grocers; the ton is used for weighing heavy articles.

Heat and Power Equivalents

1 Horsepower =	746 watts
	0.746 kilowatt
	33,000 foot pounds per minute
	550 foot pounds per second
	2546.5 heat units per hour
	42.4 heat units per minute
	0.707 heat unit per second
	0.175 pound carbon oxidized per hour
	2.64 pounds of water evaporated per hour from and at 212° F.
1 Heat unit (British thermal unit) =	778 foot pounds
	1,055 watt second
	0.000293 kilowatt hour
	0.000393 horsepower hour
	0.001036 pound water evaporated from or at 212° F.
	107.6 kilogram meters
Heat unit per square foot per minute =	0.122 watt per square inch
	0.0176 kilowatt per square foot
	0.0236 horsepower per square foot
1 Horsepower-hour =	0.746 kilowatt hour
	1,980,000 foot pounds
	2546.5 heat units
	2.64 pounds water evaporated from and at 212° F.
	17.0 pounds water raised from 62° F. to 212° F.
1 Pound of water evaporated from and at 212° F =	0.283 kilowatt hour
	0.379 horsepower hour
	965.2 heat units
	1,019,000 joules
	751,300 foot pounds

Measures of Pressure

1 Pound per square inch =	144 pounds per square foot 0.068 atmosphere 2.042 inches of mercury at 62° F. 27.7 inches of water at 62° F. 2.31 feet of water at 62° F.
1 Atmosphere =	30 inches of mercury at 62° F. 14.7 pounds per square inch 2116.3 pounds per square foot 33.95 feet of water at 62° F.
1 Foot of water at 62° F. =	62.355 pounds per square foot 0.433 pound per square inch
1 Inch of mercury at 62° F.	1.132 foot of water 13.58 inches of water 0.491 pound per square inch

Metric and English Conversion Table

Measures of Length

1 millimeter	=	0.03937 inch
1 centimeter	=	0.3937 inch
1 meter	=	39.37 inches 3.2808 feet 1.0936 yards
1 kilometer	=	0.6214 mile
1 inch	=	25.4 millimeters 2.54 centimeters
1 foot	=	304.8 millimeters 0.3048 meter
1 yard	=	0.9144 meter
1 mile	=	1.609 kilometer

Square Measure—Measures of Surface

1 square millimeter	=	0.00155 square inch
1 square centimeter	=	0.155 square inch
1 square meter	=	10.764 square feet 1.196 square yard
1 are	=	0.0247 acre 1076.4 square feet
1 hectare	=	2.471 acres 107,640 square feet
1 square kilometer	=	0.3861 square mile 247.1 acres
1 square inch	=	6.452 square centimeters 645.2 square millimeters
1 square foot	=	0.0929 square meter 9.290 square centimeters
1 square yard	=	0.836 square meter
1 acre	=	0.4047 hectare 40.47 ares
1 square mile	=	2.5899 square kilometers

Cubic Measure—Measures of Volume and Capacity

1 cubic centimeter	=	0.061 cubic inch
1 cubic decimeter	=	61.023 cubic inches 0.0353 cubic foot
1 cubic meter	=	35.314 cubic feet 1.308 cubic yards 264.2 U. S. gallons

1 liter	=	1 cubic decimeter 61.023 cubic inches 0.0353 cubic foot 1.0567 U. S. quarts 0.2642 U. S. gallons 2.202 lbs. of water at 62° F.
1 cubic inch	=	16.383 cubic centimeters
1 cubic foot	=	0.02832 cubic meter 28.317 cubic decimeters 28.317 liters
1 cubic yard	=	0.7645 cubic meter
1 gallon U. S.	=	3.785 liters
1 gallon British	=	4.543 liters

Measures of Weight

1 gram	=	0.03216 ounce troy 0.03527 ounce avoirdupois 15.432 grains
1 kilogram	=	2.2046 pounds avoirdupois 35.274 ounces avoirdupois
1 metric ton	=	0.9842 ton of 2,240 pounds 19.68 hundredweight 2204.6 pounds 1.1023 tons of 2,000 pounds

---

1 grain	=	0.0648 gram
1 ounce troy	=	31.103 grams
1 ounce avoirdupois	=	28.35 grams
1 pound	=	0.4536 kilogram 453.6 grams
1 ton of 2240 pounds	=	1.016 metric tons 1016 kilograms

TABLE 3

Inches and Equivalents in Millimeters

Inches	MM	Inches	MM	Inches	MM
1/64	.397	45/64	17.859	26	660.4
1/32	.794	23/32	18.256	27	685.8
3/64	1.191	47/64	18.653	28	711.2
1/16	1.588	3/4	19.050	29	637.6
5/64	1.984	49/64	19.447	30	762.0
3/32	2.381	25/32	19.844	31	787.4
7/64	2.778	51/64	20.241	32	812.8
1/8	3.175	13/16	20.638	33	838.2
9/64	3.572	53/64	21.034	34	863.6
5/32	3.969	27/32	21.431	35	889.0
11/64	4.366	55/64	21.828	36	914.4
3/16	4.763	7/8	22.225	37	939.8
13/64	5.159	57/64	22.622	38	965.2
7/32	5.556	29/32	23.019	39	990.6
15/64	5.953	59/64	23.416	40	1016.0
1/4	6.350	15/16	23.813	41	1041.4
17/64	6.747	61/64	24.209	42	1066.8
9/32	7.144	31/32	24.606	43	1092.2
19/64	7.540	63/64	25.003	44	1117.6
5/16	7.938	1	25.400	45	1143.0
21/64	8.334	2	50.8	46	1168.4
11/32	8.731	3	76.2	47	1193.8
23/64	9.128	4	101.6	48	1219.2
3/8	9.525	5	127.0	49	1244.6
25/64	9.922	6	152.4	50	1270.0
13/32	10.319	7	177.8	51	1295.4
27/64	10.716	8	203.2	52	1320.8
7/16	11.113	9	228.6	53	1346.2
29/64	11.509	10	254.0	54	1371.6
15/32	11.906	11	279.4	55	1397.0
31/64	12.303	12	304.8	56	1422.4
1/2	12.700	13	330.2	57	1447.8
33/64	13.097	14	355.6	58	1473.2
17/32	13.494	15	381.0	59	1498.6
35/64	13.891	16	406.4	60	1524.0
9/16	14.288	17	431.8	61	1549.4
37/64	14.684	18	457.2	62	1574.8
19/32	15.081	19	482.6	63	1600.2
39/64	15.478	20	508.0	64	1625.6
5/8	15.875	21	533.4	65	1651.0
41/64	16.272	22	558.8	66	1676.4
21/32	16.669	23	584.2	67	1701.8
43/64	17.066	24	609.6	68	1727.2
11/16	17.463	25	635.0	69	1752.6

3.—INCHES AND EQUIVALENTS IN MILLIMETERS—*Continued*

Inches	MM	Inches	MM	Inches	MM
70	1778.0	114	2895.6	158	4013.2
71	1803.4	115	2921.0	159	4038.6
72	1828.8	116	2946.4	160	4064.0
73	1854.2	117	2971.8	161	4089.4
74	1879.6	118	2997.2	162	4114.8
75	1905.0	119	3022.6	163	4140.2
76	1930.4	120	3048.0	164	4165.6
77	1955.8	121	3073.4	165	4191.0
78	1981.2	122	3098.8	166	4216.4
79	2006.6	123	3124.2	167	4241.8
80	2032.0	124	3149.6	168	4267.2
81	2057.4	125	3175.0	169	4292.6
82	2082.8	126	3200.4	170	4318.0
83	2108.2	127	3225.8	171	4343.4
84	2133.6	128	3251.2	172	4368.8
85	2159.0	129	3276.6	173	4394.2
86	2184.4	130	3302.0	174	4419.6
87	2209.8	131	3327.4	175	4445.0
88	2235.2	132	3352.8	176	4470.4
89	2260.6	133	3378.2	177	4495.8
90	2286.0	134	3403.6	178	4521.2
91	2311.4	135	3429.0	179	4546.6
92	2336.8	136	3454.4	180	4572.0
93	2362.2	137	3479.8	181	4597.4
94	2387.6	138	3505.2	182	4622.8
95	2413.0	139	3530.6	183	4648.2
96	2438.4	140	3556.0	184	4673.6
97	2463.8	141	3581.4	185	4699.0
98	2489.2	142	3606.8	186	4724.4
99	2514.6	143	3632.2	187	4749.8
100	2540.0	144	3657.6	188	4775.2
101	2565.4	145	3683.0	189	4800.6
102	2590.8	146	3708.4	190	4826.0
103	2616.2	147	3733.8	191	4851.4
104	2641.6	148	3759.2	192	4876.8
105	2667.0	149	3784.6	193	4902.2
106	2692.4	150	3810.0	194	4927.6
107	2717.8	151	3835.4	195	4953.0
108	2743.2	152	3860.8	196	4978.4
109	2768.6	153	3886.2	197	5003.8
110	2794.0	154	3911.6	198	5029.2
111	2819.4	155	3937.0	199	5054.6
112	2844.8	156	3962.4	200	5080.0
113	2870.2	157	3987.8		

TABLE 4

MILLIMETERS AND EQUIVALENTS IN INCHES

MM	Inches	MM	Inches	MM	Inches
1/100	.0004	45/100	.0177	89/100	.0350
2/100	.0008	46/100	.0181	90/100	.0354
3/100	.0012	47/100	.0185	91/100	.0358
4/100	.0016	48/100	.0189	92/100	.0362
5/100	.0020	49/100	.0193	93/100	.0366
6/100	.0024	50/100	.0197	94/100	.0370
7/100	.0028	51/100	.0201	95/100	.0374
8/100	.0031	52/100	.0205	96/100	.0378
9/100	.0035	53/100	.0209	97/100	.0382
10/100	.0039	54/100	.0213	98/100	.0386
11/100	.0043	55/100	.0217	99/100	.0390
12/100	.0047	56/100	.0221	1	.0394
13/100	.0051	57/100	.0224	2	.0787
14/100	.0055	58/100	.0228	3	.1181
15/100	.0059	59/100	.0232	4	.1575
16/100	.0063	60/100	.0236	5	.1969
17/100	.0067	61/100	.0240	6	.2362
18/100	.0071	62/100	.0244	7	.2756
19/100	.0075	63/100	.0248	8	.3150
20/100	.0079	64/100	.0252	9	.3543
21/100	.0083	65/100	.0256	10	.3937
22/100	.0087	66/100	.0260	11	.4331
23/100	.0091	67/100	.0264	12	.4724
24/100	.0094	68/100	.0268	13	.5118
25/100	.0098	69/100	.0272	14	.5512
26/100	.0102	70/100	.0276	15	.5906
27/100	.0106	71/100	.0280	16	.6299
28/100	.0110	72/100	.0284	17	.6693
29/100	.0114	73/100	.0287	18	.7087
30/100	.0118	74/100	.0291	19	.7480
31/100	.0122	75/100	.0295	20	.7874
32/100	.0126	76/100	.0299	21	.8268
33/100	.0130	77/100	.0303	22	.8661
34/100	.0134	78/100	.0307	23	.9055
35/100	.0138	79/100	.0311	24	.9449
36/100	.0142	80/100	.0315	25	.9843
37/100	.0146	81/100	.0319	26	1.0236
38/100	.0150	82/100	.0323	27	1.0630
39/100	.0154	83/100	.0327	28	1.1024
40/100	.0158	84/100	.0331	29	1.1417
41/100	.0161	85/100	.0335	30	1.1811
42/100	.0165	86/100	.0339	31	1.2205
43/100	.0169	87/100	.0343	32	1.2598
44/100	.0173	88/100	.0347	33	1.2992

4.—MILLIMETERS AND EQUIVALENTS IN INCHES—*Continued*

MM	Inches	MM	Inches	MM	Inches
34	1.3386	78	3.0709	122	4.8031
35	1.3780	79	3.1102	123	4.8425
36	1.4173	80	3.1496	124	4.8819
37	1.4567	81	3.1890	125	4.9213
38	1.4961	82	3.2283	126	4.9606
39	1.5354	83	3.2677	127	5.0000
40	1.5748	84	3.3071	128	5.0394
41	1.6142	85	3.3465	129	5.0787
42	1.6535	86	3.3858	130	5.1181
43	1.6929	87	3.4252	131	5.1575
44	1.7323	88	3.4646	132	5.1968
45	1.7717	89	3.5039	133	5.2362
46	1.8110	90	3.5433	134	5.2756
47	1.8504	91	3.5827	135	5.3150
48	1.8898	92	3.6220	136	5.3543
49	1.9291	93	3.6614	137	5.3937
50	1.9685	94	3.7008	138	5.4331
51	2.0079	95	3.7402	139	5.4724
52	2.0472	96	3.7795	140	5.5118
53	2.0866	97	3.8189	141	5.5512
54	2.1260	98	3.8583	142	5.5905
55	2.1654	99	3.8976	143	5.6299
56	2.2047	100	3.9370	144	5.6693
57	2.2441	101	3.9764	145	5.7087
58	2.2835	102	4.0157	146	5.7480
59	2.3228	103	4.0551	147	5.7874
60	2.3622	104	4.0945	148	5.8268
61	2.4016	105	4.1339	149	5.8661
62	2.4409	106	4.1732	150	5.9055
63	2.4803	107	4.2126	151	5.9449
64	2.5197	108	4.2520	152	5.9842
65	2.5591	109	4.2913	153	6.0236
66	2.5984	110	4.3307	154	6.0630
67	2.6378	111	4.3701	155	6.1024
68	2.6772	112	4.4094	156	6.1417
69	2.7165	113	4.4488	157	6.1811
70	2.7559	114	4.4882	158	6.2205
71	2.7953	115	4.5276	159	6.2598
72	2.8346	116	4.5669	160	6.2992
73	2.8740	117	4.6063	161	6.3386
74	2.9134	118	4.6457	162	6.3779
75	2.9528	119	4.6850	163	6.4173
76	2.9921	120	4.7244	164	6.4567
77	3.0315	121	4.7638	165	6.4961

4.—Millimeters and Equivalents in Inches—*Concluded*

MM	Inches	MM	Inches	MM	Inches
166	6.5354	211	8.3071	256	10.079
167	6.5748	212	8.3464	257	10.118
168	6.6142	213	8.3858	258	10.157
169	6.6535	214	8.4252	259	10.197
170	6.6929	215	8.4646	260	10.236
171	6.7323	216	8.5039	261	10.276
172	6.7716	217	8.5433	262	10.315
173	6.8110	218	8.5827	263	10.354
174	6.8504	219	8.6220	264	10.394
175	6.8898	220	8.6614	265	10.433
176	6.9291	221	8.7008	266	10.472
177	6.9685	222	8.7401	267	10.512
178	7.0079	223	8.7795	268	10.551
179	7.0472	224	8.8189	269	10.591
180	7.0866	225	8.8583	270	10.630
181	7.1260	226	8.8976	271	10.669
182	7.1653	227	8.9370	272	10.709
183	7.2047	228	8.9764	273	10.748
184	7.2441	229	9.0157	274	10.787
185	7.2835	230	9.0551	275	10.827
186	7.3228	231	9.0945	276	10.866
187	7.3622	232	9.1338	277	10.905
188	7.4016	233	9.1732	278	10.945
189	7.4409	234	9.2126	279	10.984
190	7.4803	235	9.2520	280	11.024
191	7.5197	236	9.2913	281	11.063
192	7.5590	237	9.3307	282	11.102
193	7.5984	238	9.3701	283	11.142
194	7.6378	239	9.4094	284	11.181
195	7.6772	240	9.4488	285	11.220
196	7.7165	241	9.4882	286	11.260
197	7.7559	242	9.5275	287	11.299
198	7.7953	243	9.5669	288	11.339
199	7.8346	244	9.6063	289	11.378
200	7.8740	245	9.6457	290	11.417
201	7.9134	246	9.6850	291	11.457
202	7.9527	247	9.7244	292	11.496
203	7.9921	248	9.7638	293	11.535
204	8.0315	249	9.8031	294	11.575
205	8.0709	250	9.8425	295	11.614
206	8.1102	251	9.8819	296	11.654
207	8.1496	252	9.9212	297	11.693
208	8.1890	253	9.9606	298	11.732
209	8.2283	254	10.000	299	11.772
210	8.2677	255	10.039		

USEFUL FACTORS, ENGLISH MEASURES

Inches............	×	0.08333	= feet
"	×	0.02778	= yards
"	×	0.00001578	= miles
Square inches......	×	0.00695	= square feet
" "	×	0.0007716	= square yards
Cubic inches.......	×	0.00058	= cubic feet
" "	×	0.0000214	= cubic yards
" "	×	0.004329	= U. S. gallons
Feet..............	×	0.3334	= yards
"	×	0.00019	= miles
Square feet........	×	144.0	= square inches
" "	×	0.1112	= square yards
Cubic feet.........	×	1,728	= cubic inches
" "	×	0.03704	= cubic yards
" "	×	7.48	= U. S. gallons
Yards.............	×	36	= inches
"	×	3	= feet
"	×	0.0005681	= miles
Square yards.......	×	1,296	= square inches
" "	×	9	= square feet
Cubic yards........	×	46,656	= cubic inches
" "	×	27	= cubic feet
Miles.............	×	63,360	= inches
"	×	5,280	= feet
"	×	1,760	= yards
Avoirdupois ounces.	×	0.0625	= pounds
" " .	×	0.00003125	= tons
" pounds.	×	16	= ounces
" " .	×	.001	= hundredweight
" " .	×	.0005	= tons
" " .	×	27.681	= cubic inches of water at 39.2° F
" tons...	×	32,000	= ounces
" " ...	×	2,000	= pounds
Watts.............	×	0.00134	= horse power
Horse power.......	×	746	= watts

Weight of round iron per foot = square of diameter in quarter inches ÷ 6.
Weight of flat iron per foot = width × thickness× 10/3.
Weight of flat plates per square foot = 5 pounds for each 1/8 inch thickness.

Useful Factors, Metric Measures

Millimeters × 0.03937	= inches
Millimeters ÷ 25.4	= inches
Centimeters × 0.3937	= inches
Centimeters ÷ 2.54	= inches
Meters × 39.37	= inches
Meters × 3.281	= feet
Meters × 1.094	= yards
Kilometers × 0.621	= miles
Kilometers ÷ 1.6093	= miles
Kilometers × 3280.7	= feet
Square millimeters × 0.0155	= square inches
Square millimeters ÷ 645.1	= square inches
Square centimeters × 0.155	= square inches
Square centimeters ÷ 6.451	= square inches
Square meters × 10.764	= square feet
Square kilometers × 247.1	= acres
Hectares × 2.471	= acres
Cubic centimeters ÷ 16.385	= cubic inches
Cubic centimeters ÷ 3.69	= fluid drachms, U. S. Pharmacopœia
Cubic centimeters ÷ 29.57	= fluid ounce U. S. Pharmacopœia
Cubic meters × 35.315	= cubic feet
Cubic meters × 1.038	= cubic yards
Cubic meters × 264.2	= gallons, United States
Liters × 61.022	= cubic inches
Liters × 33.84	= fluid ounces
Liters × 0.2642	= gallons, United States

Liters ÷ 3.78	= gallons, United States
Liters ÷ 28.316	= cubic feet
Hectoliters × 3.531	= cubic feet
Hectoliters × 2.84	= bushels, United States
Hectoliters × 0.131	= cubic yards
Hectoliters × 26.42	= gallons, United States
Grams × 15.432	= grains
Grams (water) ÷ 29.57	= fluid ounces
Grams ÷ 28.35	= ounces, avoirdupois
Kilograms × 2.2046	= pounds
Kilograms × 35.3	= ounces, avoirdupois
Kilograms ÷ 1102.3	= tons, 2000 pounds

Specific Gravity

The relative heaviness of substances is of much practical importance to the industrial world. In the metal industry research workers are constantly seeking for relatively light materials that possess great strength.

Weight measures the earth's pull upon body, and depends upon the body's mass. But substances which are equal in volume vary in heaviness. Thus, it is evident that the pull of gravity is stronger on some substances than on others. As the weight of a body is the measure of the pull between all bodies and the earth, or gravity, the specific gravity of a substance is found by comparing the weight of a certain volume of that substance with the weight of an equal volume of another substance taken as a standard.

The specific gravity of a substance is its weight as compared with the weight of an equal bulk of pure water.

Rule.—To calculate the specific gravity of a substance, find the weight of the body in air and divide by the difference of the weight of the body in air and the weight of the body submerged in water.

Expressed as a formula:

$$\text{Specific gravity} = \frac{W}{W - w}$$

where W = weight of body in air
w = weight of body submerged in water

ILLUSTRATION: Find the specific gravity of a lump of coal that weighs 150 grams in air and 60 grams immersed in water.

$$\text{Specific gravity} = \frac{W}{W - w}$$

$$= \frac{150}{150 - 60}$$

$$= \frac{150}{90} = 1.66$$

Specific gravity determinations are usually referred to the standard of the weight of water at 62° F., 62.355 pounds per cubic feet. The formula becomes:

$$\text{Specific gravity} = \frac{\text{weight of solid}}{\text{weight of equal volume of water}}$$

ILLUSTRATION: Find the specific gravity of a cube of steel 1 foot on a side and weighing 489.6 pounds per cubic foot.

$$\text{Specific gravity} = \frac{\text{weight of solid}}{\text{weight of equal volume of water}}$$

$$= \frac{489.6}{62.355} = 7.85$$

The following tables give the specific gravities and weights of various substances.

TABLE 5

SPECIFIC GRAVITIES AND WEIGHTS OF VARIOUS SUBSTANCES

The Basis for Specific Gravities is Pure Water at 62 Degrees Fah., Barometer 30 Inches. Weight of One Cubic Foot, 62.355 Pounds.	Average Specific Gravity. Water = 1.	Average Weight of One Cubic Foot. Pounds.
Air, atmospheric at 60 degrees F., under pressure of one atmosphere, or 14.7 pounds per square inch, weighs $\frac{1}{815}$th as much as water	.00123	.0765
Aluminum	2.6	162
Anthracite, 1.3 to 1.84; of Penna., 1.3 to 1.7.	1.5	93.5
" broken, of any size, loose	....	52 to 56
" " moderately shaken	....	56 to 60
" " heaped bushel, loose, 77 to 83 pounds	....	....
" " a ton loose occupies 40 to 43 cubic feet	....	....
Antimony, cast	6.70	418
" native	6.67	416
Ash, perfectly dry	.752	47
" American White, dry	.61	38
Ashes of soft coal, solidly packed		40 to 45
Asphaltum, 1 to 1.8	1.4	87.3
Brass (copper and zinc), cast, 7.8 to 8.4	8.1	504
" rolled	8.4	524
Brick, best pressed	....	150
" common and hard	....	125
" soft inferior	....	100
Brickwork, pressed brick, fine joints	....	140
" medium quality	....	125
" coarse, inferior, soft	....	100
" at 125 pounds per cubic foot, 1 cubic yard equals 1.507 tons, and 17.92 cubic feet equal 1 ton	....	....
Bronze, copper 8, tin 1 (gun metal)	8.5	529
Cement, hydraulic. American, Rosendale, ground and loose	....	56
" hydraulic. American, Rosendale, U. S. struck bush., 70 pounds	....	....
" hydraulic. American, Rosendale, Louisville bushel, 62 pounds	....	....
" hydraulic. American, Cumberland, ground, loose	....	65
" hydraulic. American, Cumberland, ground, thoroughly shaken	....	85
" hydraulic. English Portland (U.S. struck bushel, 100 to 128)	....	81 to 102

TABLE 5.—Specific Gravities and Weights of Various Substances—*Continued*

The Basis for Specific Gravities is Pure Water at 62 Degrees Fah., Barometer 30 Inches. Weight of One Cubic Foot, 62.355 Pounds.	Average Specific Gravity. Water = 1.	Average Weight of One Cubic Foot. Pounds.
Cement, hydraulic. English Portland, a barrel, 400 to 430 pounds	....	....
" hydraulic. American Portland, loose	....	88
" hydraulic. American Portland, thoroughly shaken	....	110
Charcoal of pines and oaks	....	15 to 30
Chalk	2.5	156
Cherry, perfectly dry	.672	42
Clay, potters', dry, 1.8 to 2.1	1.9	119
" dry in lump, loose	....	63
Coal, bituminous, solid, 1.2 to 1.5	1.35	84
" bituminous, solid, Cambria Co., Pa., 1.27–1.34	....	79 to 84
" bituminous, broken, of any size, loose ..	....	47 to 52
" bituminous, moderately shaken	....	51 to 56
" bituminous, a heaped bushel, loose, 70 to 78	....	....
" bituminous, 1 ton occupies 43 to 48 cubic feet	....	....
Coke, loose, good quality	....	23 to 32
" loose, a heaped bushel, 35 to 42	....	....
" 1 ton occupies 80 to 97 cubic feet	....	....
Corundum, pure, 3.8 to 4	3.9	....
Copper, cast, 8.6 to 8.8	8.7	542
" rolled, 8.8 to 9	8.9	555
Cork, dry	.24	15
Earth, common loam, perfectly dry, loose ...	....	72 to 80
" " " perfectly dry, shaken ..	....	82 to 92
" " " perfectly dry, rammed.	....	90 to 100
" " " slightly moist, loose ..	....	70 to 76
" " " more moist, loose	....	66 to 68
" " " more moist, shaken ...	....	75 to 90
" " " more moist, packed ...	....	90 to 100
" " " as soft flowing mud ...	....	104 to 112
" " " as soft flowing mud well pressed	....	110 to 120
Elm, perfectly dry	.56	35
Flint	2.6	162
Glass, 2.5 to 3.45	2.98	186
" common window	2.52	157
Gneiss, common, 2.62 to 2.76	2.69	168

TABLE 5.—Specific Gravities and Weights of Various Substances—*Continued*

The Basis for Specific Gravities is Pure Water at 62 Degrees Fah., Barometer 30 Inches. Weight of One Cubic Foot, 62.355 Pounds.	Average Specific Gravity. Water = 1.	Average Weight of One Cubic Foot. Pounds.
Gneiss, in loose piles	....	96
Gold, cast, pure or 24 karat	19.258	1204
" pure, hammered	19.5	1217
Granite, 2.56 to 2.88	2.72	170
Greenstone, trap, 2.8 to 3.2	3.00	187
Gypsum, plaster of Paris, 2.24 to 2.30	2.27	141.6
Hickory, perfectly dry	.85	53
Ice, .917 to .922	.92	57.4
Iron, cast, 6.9 to 7.4	7.15	446
" grey foundry, cold	7.21	450
" " molten	6.94	433
" wrought	7.69	480
Lead, commercial	11.38	709.6
Lignumvitæ (dry)	.65–1.33	41 to 83
Limestone and marble	2.6	164.4
Lime, quick	1.5	95
" quick, ground, well shaken, per struck bushel 80 pounds	....	64
" quick, ground, thoroughly shaken, per struck bushel 93¾ pounds	....	75
Locust, dry	.71	44
Mahogany, Spanish, dry	.85	53
" Honduras, dry	.56	35
Maple, dry	.79	49
Marble (see Limestone).		
Masonry of granite or limestone, well-dressed	....	165
" of granite, well-scabbled mortar rubble, about ⅕ of mass will be mortar	....	154
" of granite, well-scabbled dry rubble	....	138
" of granite, roughly scabbled mortar rubble, about ¼ to ⅓ of mass will be mortar	....	150
" of granite, scabbled dry rubble	....	125
" of sandstone, ⅛ less than granite	....	....
Masonry of brickwork		
Mercury, at 32 degrees Fah	13.62	849
Mica, 2.75 to 3.1	2.93	183
Mortar, hardened, 1.4 to 1.9	1.65	103
Mud, dry, close	....	80 to 110
" wet, moderately pressed	....	110 to 130
" " fluid	....	104 to 120

TABLE 5.—Specific Gravities and Weights of Various Substances—*Concluded*

The Basis for Specific Gravities is Pure Water at 62 Degrees Fah., Barometer 30 Inches. Weight of One Cubic Foot, 62.355 Pounds.	Average Specific Gravity. Water = 1.	Average Weight of One Cubic Foot. Pounds.
Oak, live, perfectly dry, .88–1.02 (see note below)	.95	59.3
" Red, Black, perfectly dry	....	32 to 45
Petroleum	.878	54.8
Pitch	1.15	71.7
Poplar, dry (see note below)	.47	29
Platinum	21.5	1342
Quartz	2.65	165
Rosin	1.10	68.6
Salt, coarse, (per struck bushel, Syracuse, N. Y., 56 pounds)	....	45
Sand, of pure quartz, perfectly dry and loose	....	90 to 106
" " " voids full of water	....	118 to 129
" " " very large and small grains, dry	....	117
Sandstone, 2.1 to 2.73, 131 to 171	2.41	151
" quarried and piled, 1 measure solid makes 1¾ (about) piled	....	86
Snow, fresh fallen	....	5 to 12
" moistened, compacted by rain	....	15 to 50
Sycamore, perfectly dry (see note below)	.59	37
Shales, red or black, 2.4 to 2.8	2.6	162
Silver	10.5	655
Slate, 2.7 to 2.9	2.8	175
Soapstone, 2.65 to 2.8	2.73	170
Steel	7.85	490
Sulphur	2.00	125
Tallow	.94	58.6
Tar	1	62.355
Tin, cast, 7.2 to 7.5	7.35	459
Walnut, Black, perfectly dry (see note below)	.61	38
Water, pure rain, distilled, at 32 degrees F., Bar. 30 inches.	....	62.417
" " " at 62 degrees F., Bar. 30 inches.	1	62.355
" " " at 212 degrees F., Bar. 30 inches	....	59.7
" sea, 1.026 to 1.030	1.028	64.08
Zinc or spelter, 6.8 to 7.2	7.00	437.5

Note.—Green timbers usually weigh from one-fifth to nearly one-half more than dry; ordinary building timbers, tolerably seasoned, one-sixth more.

When the specific gravity of a substance is known the weight per cubic foot of the substance can be found by multiplying the specific gravity by 62.355; the weight of one cubic inch by multiplying the specific gravity by 0.0361 the weight of one cubic inch of pure water at 62° F.

ILLUSTRATION: From the table, page 268, the specific gravity of cast iron is given as 7.2. Find the weight of 6 cubic inches of cast iron.

$$7.2 \times 0.0361 \times 6 = 1.5586 \text{ pounds}$$

If the weight per cubic foot of a substance is known, the specific gravity can be calculated by multiplying this weight by 0.01604.

ILLUSTRATION: Find the specific gravity of a cubic foot of cast tin that weighs 455 pounds.

$$455 \times 0.01604 = 7.29$$

Specific Gravity of Liquids. The specific gravity of liquids is the number which indicates how much a certain volume of the liquid weighs compared with an equal volume of water.

TABLE 6

Specific Gravity of Liquids

Liquid	Sp. Gr.	Liquid	Sp. Gr.	Liquid	Sp. Gr.
Acetic acid	1.06	Fluoric acid	1.50	Petroleum oil	0.82
Alcohol, commerical	0.83	Gasoline	0.70	Phosphoric acid	1.78
Alcohol, pure	0.79	Kerosene	0.80	Rape oil	0.92
Ammonia	0.89	Linseed oil	0.94	Sulphuric acid	1.84
Benzine	0.69	Mineral oil	0.92	Tar	1.00
Bromine	2.97	Muriatic acid	1.20	Turpentine oil	0.87
Carbolic acid	0.96	Naphtha	0.76	Vinegar	1.08
Carbon disulphide	1.26	Nitric acid	1.22	Water	1.00
Cotton-seed oil	0.93	Olive oil	0.92	Water, sea	1.03
Ether, sulphuric	0.72	Palm oil	0.97	Whale oil	0.92

There are three methods of determining the specific gravity of liquids:

(1) Hydrometer method, in which the specific gravity of the liquid tested is read as the scale division marking the liquid level on the stem of the hydrometer.

(2) Bottle method, in which the specific gravity

$$= \frac{\text{weight of liquid in a full bottle}}{\text{weight of water in a full bottle}}$$

(3) Displacement method in which specific gravity

$$= \frac{\text{weight of liquid displaced by a body}}{\text{weight of equal volume of water displaced by the body}}$$

Specific Gravity of Gases.—The specific gravity of gases is the number which indicates their weight in comparison with that of an equal volume of air. The specific gravity of air is 1, and the comparison is made at 32° F.

TABLE 7

Specific Gravity of Gases at 32 degrees F.

Gas	Sp. Gr.	Gas	Sp. Gr.	Gas	Sp. Gr.
Air	1.000	Ether vapor	2.586	Marsh gas	0.555
Acetylene	0.920	Ethylene	0.967	Nitrogen	0.971
Alcohol vapor	1.601	Hydrofluoric acid	2.370	Nitric oxide	1.039
Ammonia	0.592	Hydrochloric acid	1.261	Nitrous oxide	1.527
Carbon dioxide	1.520	Hydrogen	0.069	Oxygen	1.106
Carbon monoxide	0.967	Illuminating gas	0.400	Sulphur dioxide	2.250
Chlorine	2.423	Mercury vapor	6.940	Water vapor	0.623

1 cubic foot of air at 32 degrees F. and atmospheric pressure weighs 0.0807 pound.

Weights of Materials

The weight of any object may be found by calculating its volume in cubic inches or cubic feet and multiplying this volume by the unit of weight, that is, the weight per cubic foot or cubic inch of the material of which the object is made.

Weight of Square Bars:

ILLUSTRATION: (1) Find the weight of a wrought iron bar 1 foot long and 1 inch square if one cubic inch weighs 0.2778 pound.

$$0.2778 \times 12 = 3.33$$

Therefore, a wrought iron bar 1 inch square and one foot long weighs 3.33 pounds.

(2) Find the weight of a steel bar 1 foot long and 2 inches square if one cubic inch weighs 0.2835 pound.

$$0.2835 \times (2 \times 2) \times 12 = 13.63$$

Therefore, a steel bar 2 inches square, the cross section area 4 sq in., and 1 foot long weighs 13.63 pounds.

Weight of Sheet Metal.—The weight of one square foot of sheet iron equals 40 × thickness in thousandths of an inch. A square sheet of iron plate 1 inch thick and measuring 1 foot on each side contains:

$$12 \times 12 \times 1 = 144 \text{ cubic inches}$$

$$144 \times 0.2778 \text{ (the weight of 1 cubic inch of iron)} = 40$$

Therefore, the weight of a sheet of iron plate 1 inch thick and 1 foot on each side weighs 40 pounds.

Illustration: What is the weight of 1 sq. ft. of sheet iron, No. 20 gage, i.e., 0.032 inch thick.

$$40 \times 0.032 = 1.28$$

Therefore the weight is 1.28 pounds.

Illustration: Find the weight of a sheet of steel 6 feet 8 inches long, 2 feet 6 inches wide and No. 2 gage, i.e., 0.2576 inch thick.

$$6 \text{ feet } 8 \text{ inches} = 80 \text{ inches}, \quad 2 \text{ feet } 6 \text{ inches} = 30 \text{ inches}$$

$$80 \times 30 \times 0.2576 \times 0.2835 = 165.26$$

Therefore, the weight of the bar is 165.26 pounds.

Weight of Round Bars.—The weight of round bars are found by a similar method used in square bars, the only difference being that the area of the end of the bar is the area of a circle whose diameter is given.

Illustration: Find the weight of a steel bar 1 inch in diameter and 1 foot long.

$$0.2835 \times (1^2 \times 0.7854) \times 12 = 2.67$$

Therefore a round steel bar 1 inch in diameter and 1 foot long weighs 2.67 pounds.

Table 6 may also be used to calculate the weights of round, square and hexagon steel bars.

Illustration: Find the weight of a steel bar 1 inch in diameter and 1 foot long.

From the table, weight per inch of a 1 inch round bar is 0.2227 lb. Therefore, $12 \times 0.2227 = 2.67$ pounds.

TABLE 8

Weights and Areas of Round, Square and Hexagon Steel

Weight of one cubic inch = 0.2836 lb
Weight of one cubic foot = 490 lb

Thickness or Diameter	Area = Diam.² × 0.7854			Area = Side² × 1		Area = Diam.² × 0.866	
	Round			Square		Hexagon	
	Weight Per Inch	Area Square Inches	Circumference Inches	Weight Per Inch	Area Square Inches	Weight Per Inch	Area Square Inches
1/32	0.0002	0.0008	0.0981	0.0003	0.0010	0.0002	0.0008
1/16	.0009	.0031	.1963	.0011	.0039	.0010	.0034
3/32	.0020	.0069	.2995	.0025	.0088	.0022	.0076
1/8	.0035	.0123	.3927	.0044	.0156	.0038	.0135
5/32	.0054	.0192	.4908	.0069	.0244	.0060	.0211
3/16	.0078	.0276	.5890	.0101	.0352	.0086	.0304
7/32	.0107	.0376	.6872	.0136	.0479	.0118	.0414
1/4	.0139	.0491	.7854	.0177	.0625	.0154	.0540
9/32	.0176	.0621	.8835	.0224	.0791	.0194	.0686
5/16	.0218	.0767	.9817	.0277	.0977	.0240	.0846
11/32	.0263	.0928	1.0799	.0335	.1182	.0290	.1023
3/8	.0313	.1104	1.1781	.0405	.1406	.0345	.1218
13/32	.0368	.1296	1.2762	.0466	.1651	.0405	.1428
7/16	.0426	.1503	1.3744	.0543	.1914	.0470	.1658
15/32	.0489	.1726	1.4726	.0623	.2197	.0540	.1903
1/2	.0557	.1963	1.5708	.0709	.2500	.0614	.2161
17/32	.0629	.2217	1.6689	.0800	.2822	.0693	.2444
9/16	.0705	.2485	1.7671	.0897	.3164	.0777	.2743
19/32	.0785	.2769	1.8653	.1036	.3526	.0866	.3053
5/8	.0870	.3068	1.9635	.1108	.3906	.0959	.3383
21/32	.0959	.3382	2.0616	.1221	.4307	.1058	.3730
11/16	.1053	.3712	2.1598	.1340	.4727	.1161	.4093
23/32	.1151	.4057	2.2580	.1465	.5166	.1270	.4474
3/4	.1253	.4418	2.3562	.1622	.5625	.1382	.4871
25/32	.1359	.4794	2.4543	.1732	.6103	.1499	.5286
13/16	.1470	.5185	2.5525	.1872	.6602	.1620	.5712
27/32	.1586	.5591	2.6507	.2019	.7119	.1749	.6165
7/8	.1705	.6013	2.7489	.2171	.7656	.1880	.6631

TABLE 8—(*Continued*)

Thickness or Diameter	Area = Diam.² × 0.7854			Area = Side² × 1		Area = Diam.² × 0.866	
	Round			Square		Hexagon	
	Weight Per Inch	Area Square Inches	Circumference Inches	Weight Per Inch	Area Square Inches	Weight Per Inch	Area Square Inches
29/32	0.1829	0.6450	2.8470	0.2329	0.8213	0.2015	0.7112
15/16	.1958	.6903	2.9452	.2492	.8789	.2159	.7612
31/32	.2090	.7371	3.0434	.2661	.9384	.2305	.8127
1	.2227	.7854	3.1416	.2836	1.0000	.2456	.8643
1 1/16	.2515	.8866	3.3379	.3201	1.1289	.2773	.9776
1 1/8	.2819	.9940	3.5343	.3589	1.2656	.3109	1.0973
1 3/16	.3141	1.1075	3.7306	.4142	1.4102	.3464	1.2212
1 1/4	.3480	1.2272	3.9270	.4431	1.5625	.3838	1.3531
1 5/16	.3837	1.3530	4.1233	.4885	1.7227	.4231	1.4919
1 3/8	.4211	1.4849	4.3197	.5362	1.8906	.4643	1.6373
1 7/16	.4603	1.6230	4.5160	.5860	2.0664	.5076	1.7898
1 1/2	.5012	1.7671	4.7124	.6487	2.2500	.5526	1.9485
1 9/16	.5438	1.9175	4.9087	.6930	2.4414	.5996	2.1143
1 5/8	.5882	2.0739	5.1051	.7489	2.6406	.6480	2.2847
1 11/16	.6343	2.2365	5.3014	.8076	2.8477	.6994	2.4662
1 3/4	.6821	2.4053	5.4978	.8685	3.0625	.7521	2.6522
1 13/16	.7317	2.5802	5.6941	.9316	3.2852	.8069	2.8450
1 7/8	.7831	2.7612	5.8905	.9970	3.5156	.8635	3.0446
1 15/16	.8361	2.9483	6.0868	1.0646	3.7539	.9220	3.2509
2	.8910	3.1416	6.2832	1.1342	4.0000	.9825	3.4573
2 1/16	.9475	3.3410	6.4795	1.2064	4.2539	1.0448	3.6840
2 1/8	1.0058	3.5466	6.6759	1.2806	4.5156	1.1091	3.9106
2 3/16	1.0658	3.7583	6.8722	1.3570	4.7852	1.1753	4.1440
2 1/4	1.1276	3.9761	7.0686	1.4357	5.0625	1.2434	4.3892
2 5/16	1.1911	4.2000	7.2649	1.5165	5.3477	1.3135	4.6312
2 3/8	1.2564	4.4301	7.4613	1.6569	5.6406	1.3854	4.8849
2 7/16	1.3234	4.6664	7.6575	1.6849	5.9414	1.4593	5.1454
2 1/2	1.3921	4.9087	7.8540	1.7724	6.2500	1.5351	5.4126
2 5/8	1.5348	5.4119	8.2467	1.9541	6.8906	1.6924	5.9674
2 3/4	1.6845	5.9396	8.6394	2.1446	7.5625	1.8574	6.5493
2 7/8	1.8411	6.4918	9.0321	2.3441	8.2656	2.0304	7.1590
3	2.0046	7.0686	9.4248	2.5548	9.0000	2.2105	7.7941

TABLE 8—(*Concluded*)

Thickness or Diameter	Area = Diam.² × 0.7854			Area = Side² × 1		Area = Diam.² × 0.866	
	Round			Square		Hexagon	
	Weight Per Inch	Area Square Inches	Circumference Inches	Weight Per Inch	Area Square Inches	Weight Per Inch	Area Square Inches
3⅛	2.1752	7.6699	9.8175	2.7719	9.7656	2.3986	8.4573
3¼	2.3527	8.2958	10.2102	2.9954	10.5625	2.5918	9.1387
3⅜	2.5371	8.9462	10.6029	3.2303	11.3906	2.7977	9.8646
3½	2.7286	9.6211	10.9956	3.4740	12.2500	3.0083	10.6089
3⅝	2.9269	10.3206	11.3883	3.7265	13.1407	3.2275	11.3798
3¾	3.1323	11.0447	11.7810	3.9880	14.0625	3.4539	12.1785
3⅞	3.3446	11.7932	12.1737	4.2582	15.0156	3.6880	13.0035
4	3.5638	12.5664	12.5664	4.5374	16.0000	3.9298	13.8292
4⅛	3.7900	13.3640	12.9591	4.8254	17.0156	4.1792	14.7359
4¼	4.0232	14.1863	13.3518	5.1223	18.0625	4.4364	15.6424
4⅜	4.2634	15.0332	13.7445	5.4280	19.1406	4.7011	16.5761
4½	4.5105	15.9043	14.1372	5.7426	20.2500	4.9736	17.5569
4⅝	4.7645	16.8002	14.5299	6.0662	21.3906	5.2538	18.5249
4¾	5.0255	17.7205	14.9226	6.6276	22.5625	5.5416	19.5397
4⅞	5.2935	18.6655	15.3153	6.7397	23.7656	5.8371	20.5816
5	5.5685	19.6350	15.7080	7.0897	25.0000	6.1403	21.6503
5⅛	5.8504	20.6290	16.1007	7.4496	26.2656	6.4511	22.7456
5¼	6.1392	21.6475	16.4934	7.8164	27.5624	6.7697	23.8696
5⅜	6.4351	22.6905	16.8861	8.1930	28.8906	7.0959	25.0198
5½	6.7379	23.7583	17.2788	8.5786	30.2500	7.4298	26.1971
5⅝	7.0476	24.8505	17.6715	8.9729	31.6406	7.7713	27.4013
5¾	7.3643	25.9672	18.0642	9.3762	33.0625	8.1214	28.6361
5⅞	7.6880	27.1085	18.4569	9.7883	34.5156	8.4774	29.8913
6	8.0186	28.2743	18.8496	10.2192	36.0000	8.8420	31.1765
6¼	8.7007	30.6796	19.6350	11.0877	39.0625	9.5943	33.8291
6½	9.4107	33.1831	20.4204	11.9817	42.2500	10.3673	36.5547
6¾	10.1485	35.7847	21.2058	12.9211	45.5625	11.1908	39.4584
7	10.9142	38.4845	21.9912	13.8960	49.0000	12.0351	42.4354
7½	12.5291	44.1786	23.5620	15.9520	56.2500	13.8158	48.7142
8	14.2553	50.2655	25.1328	18.1497	64.0000	15.7192	55.3169

Multiply above weights by 0.993 for wrought iron, 0.918 for cast iron 1.0331 for cast brass, 1.1209 for copper, 1.1748 for phos. bronze, and 0.3265 for aluminum.

IX

Excavation and Foundations

Excavation.—Excavation of earth and rock involves three or four general operations on the excavated material; viz., (*a*) loosening, (*b*) loading, (*c*) hauling, and (*d*) dumping. Rock, hardpan, and frozen ground may be loosened most economically with explosives, although pneumatic spades may be used on the latter two where explosives are not permitted.

In soft ground, loosening and loading become one operation. On small work, picks and shovels are used to break up the ground

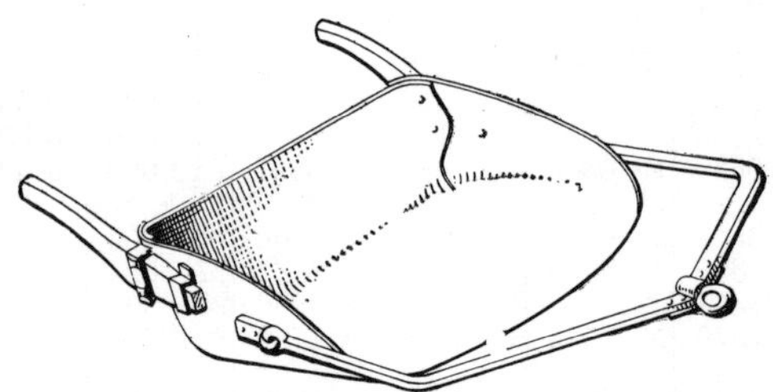

Fig. 1.—Western Slip or Drag Scraper.

and load it into dump wagons. Drag scrapers such as shown in Fig. 1 are also widely used on small building excavation, particularly when the dirt may be disposed of close at hand. In the case of excavations for larger buildings, diesel or gasoline shovels are generally used. These dump into trucks which have access to the hole by ramps or elevators.

Three special types of excavations will be considered in the following paragraphs; namely, foundation, right-of-way cut, and borrow pit excavations.

Laying Out a Foundation.—The first step preparatory to excavating for the foundation of a small building is to set stakes into the ground on the lines of the excavation and some distance back from the corners as shown in Fig. 2. These stakes should be set by an engineer or surveyor. When lines are stretched between the stakes, the diagonals between the corners are equal when the excavation is rectangular. When the corners are supposed to be square, the angle may be checked by laying off a distance of 6 feet from the corner along one line and 8 feet from the corner on the other. The distance between these two points should measure 10 feet.

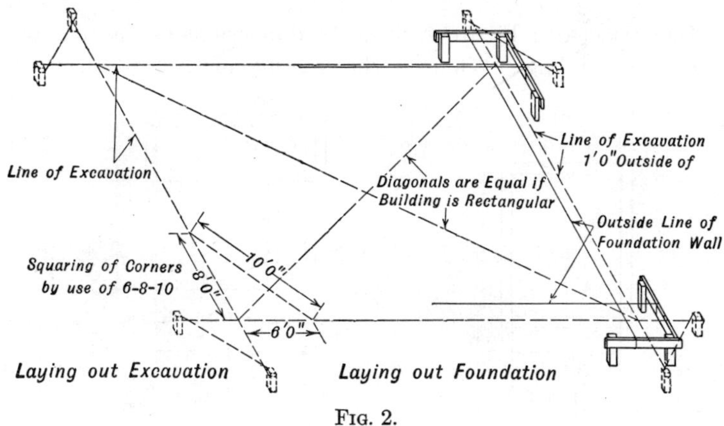

FIG. 2.

Excavation lines should be set 1 foot outside of the foundation lines to allow sufficient working space.

Estimating Quantity of Excavated Material—Material removed from an excavation is measured by cubic yards " in place." That is, it is measured as solid ground and not as the loose material which is hauled away and dumped. The reason for this is that the latter occupies a volume about 25 percent greater than its original volume. The problem of measuring the amount of material excavated becomes then a case of determining the volume of the resulting hole. When the ground is level and the figure regular, the computation is quite simple. It can best be illustrated by a few examples.

ILLUSTRATION: Figure 3 shows the plan of a building whose outside dimensions are 20 feet and 32 feet. What is the volume of excavation if the depth is uniformly eight feet and the lines of excavation are one foot outside the building lines?

The dimensions of the hole are 8 ft. $\times$ 22 ft. $\times$ 34 ft.

Volume $= 8 \times 22 \times 34 = 5984$ cu. ft.

Changing to cu. yd., volume $= \dfrac{5984}{27} = 222$ cu. yd. (Ans.)

ILLUSTRATION: Figure 4 gives the dimensions of the plan of a T-shaped building. What is the volume of excavation if the

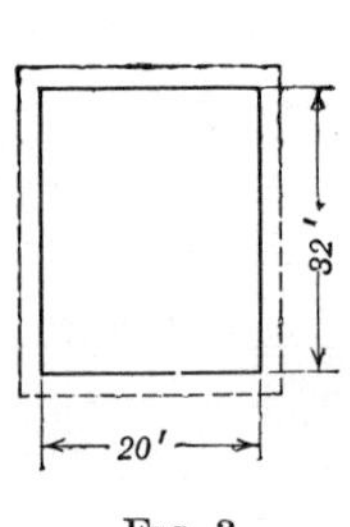

FIG. 3.

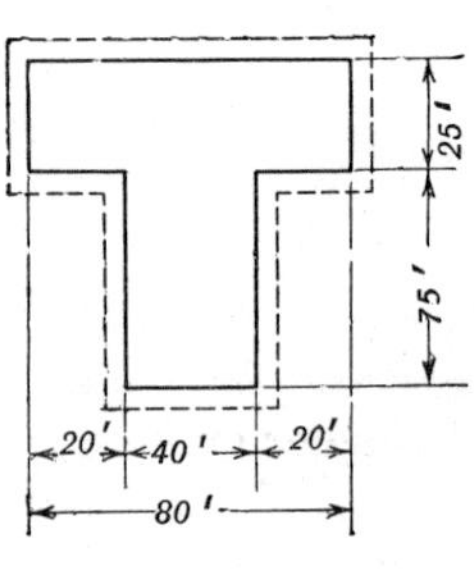

FIG. 4.

excavating line is one foot outside the building line and the depth is nine feet?

The area of the excavation can be computed most readily by mentally dividing its plan into two rectangles, one 82 feet by 27 feet and the other 75 feet by 42 feet. The areas of these are

$$82 \times 27 = 2214 \text{ sq. ft.}$$
$$75 \times 42 = 3150 \text{ sq. ft.}$$
$$\text{Total } 5364 \text{ sq. ft.}$$

The volume in cubic feet is then the total area times the depth of 9 feet. This is changed to cubic yards by dividing by 27. If these operations are set up together, the computation may be completed mentally:

$$\frac{5364 \times \cancel{9}}{\underset{3}{\cancel{27}}} = 1788 \text{ cu. yd. (Ans.)}$$

ILLUSTRATION: A building of the dimensions shown in Fig. 5 is to be built on a triangular lot. If the excavation is eight feet deep and one foot outside the building line, what volume of earth will have to be removed?

The problem gives us the three sides of an oblique-angled triangle, but we do not know any of the dimensions of the larger triangle represented by the excavation line. Determining these dimensions would be a tedious operation not warranted by this problem. A practical solution is to solve for the area of the triangle represented by the building line and add to this the area of a strip one foot wide and slightly longer than the perimeter of the triangle.

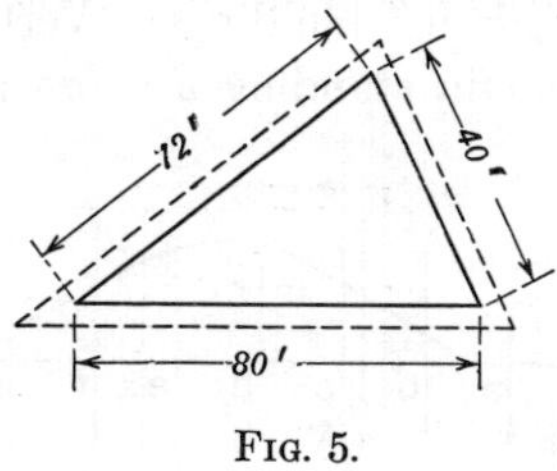

FIG. 5.

From geometry we know that the area of any triangle, whose three sides are represented by a, b, and c, is

$$\sqrt{S(S-a)(S-b)(S-c)}$$

when $S = \frac{1}{2}(a + b + c)$. Using this, we proceed to find the area of the inner triangle.

$$S = \tfrac{1}{2}(a + b + c) = \tfrac{1}{2}(72 + 40 + 80) = 96$$

$$\text{Area} = \sqrt{96(96 - 72)(96 - 40)(96 - 80)}$$

$$= \sqrt{96 \times 24 \times 56 \times 16} = 1437 \text{ sq. ft.}$$

In computing the area of the one-foot strip around this triangle, let us arbitrarily add 3 feet to the sum of the lengths of the sides of the foundation wall. Then the area is

$$\text{area of strip} = 1 \times (72 + 40 + 80 + 3) = 195 \text{ sq. ft.}$$

Adding this to the area of the triangle we obtain, 1437 + 195 = 1632 sq ft. The volume of the excavation is then this area times the depth, 8 feet, and divided by 27 to change to cubic yards, or

$$\text{Volume} = \frac{1632 \times 8}{27} = 483 \text{ cu. yd. (Ans.)}$$

Average End Area Method of Estimating Earthwork.—The preceding paragraphs have considered only excavations regular in shape and with vertical sides such as are common in foundation work for buildings. Vertical faces of earth will, however, only remain standing a short time and when a permanent depression

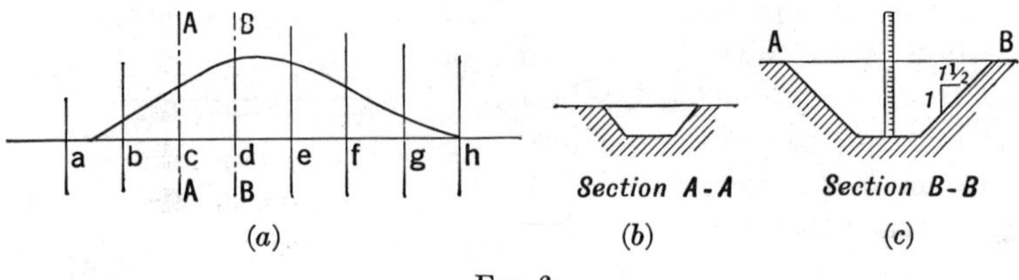

Fig. 6.

in earth is desired without retaining walls, the sides of the excavation must be sloped.

The slope which a loose material will naturally assume and at which it will remain stable, is called the *angle of repose*, referred to the horizontal. Sand has an angle of repose of about 34 degrees, a mixture of sand, gravel, and clay, an angle of about 45 degrees, while sound rock will stand vertical or at an angle of 90 degrees.

An irregular excavation or a uniform excavation through irregular ground is usually measured by dividing the total volume into small prisms and arriving at the sum of the volumes of these prisms. For example, let Fig. 6 (*a*) represent the profile of a hill through which a driveway is to be cut, and Fig. 6 (*b*) a cross-section at *A–A* while (*c*) is a cross-section at *B–B*. The volume to be excavated between sections *A–A* and *B–B* is a six-sided prism whose shape is approximately as shown in Fig. 7. In the average-end-area method of computing this volume, the area of *ABCD* is averaged with the area of *EFGH*, resulting in the area of the mid-section *IJKL*. This is multiplied by the distance between the cross-sections (*CG* or *DE*) to obtain the volume. If then, in Fig. 6 (*a*) we average the areas of the sections, *a* and *b*, *b* and *c*, *c* and *d*, *d* and *e*, *e* and *f*, *f* and *g*, *g* and *h*, and multiply each average by the distance between its respective end areas, we will obtain the volume of the entire excavation.

It is to be noted that the result is only approximately correct and that the error increases as the difference in areas of the end sections increases. However, the method represents accepted practice in engineering work.

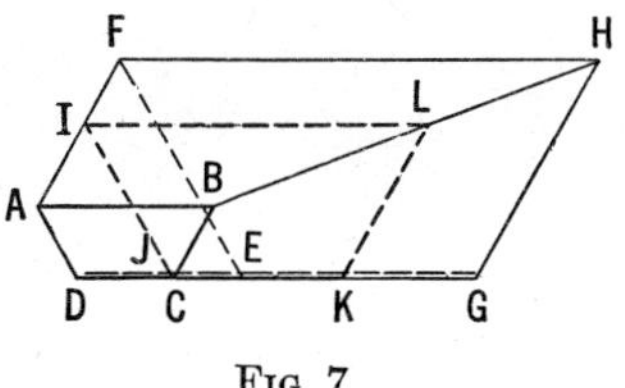

FIG. 7.

Right-of-Way Excavations.—The method outlined in the preceding paragraphs is equally applicable to cuts for driveways, highways, railways, canals, etc., which we shall call right-of-way excavations for want of a more descriptive term. These excavations, or "cuts," as they are called, have the common property of being generally uniform in shape of cross-section, the only major variation being the depth of the cut. This being true, it has been possible to develop tables so that the volumes can be estimated with a minimum of computation and without the use of surveying instruments.

Whether the tables or direct computation are used, a longitudinal line is first laid out along the centerline of the work and stakes or markers are set at horizontal intervals of 100 feet along this line. These points are called *stations*. If the ground is very irregular, the intervals may be only 50 feet, and for rock excavation the interval is often only 25 feet.

The use of the tables requires a knowledge of the width of the base of a roadway, the slope of the sides, and the depth of the cut

TABLE 1

Level Sections (Earthwork); Height, 0–60 Ft.
Base of Roadway, 16 Ft., Side Slope, 1 to 1

Note.—The last two columns enable us to use any other base than 14 ft.: Ex.—Given height, 20.3 ft.; roadway 14 ft. Then we have, 2729.2−(148.15+2.22)=2578.8 cu. yds.

[Cu. Yds. per 100-Ft. Station.]

Ht. Ft.	.0	.1	.2	.3	.4	.5	.6	.7	.8	.9	Width of 2 Ft. Cu.Yds
0	......	6.0	12.0	18.1	24.3	30.6	36.9	43.3	49 8	56 3	....
1	63.0	69.7	76.4	83.3	90.2	97.2	104.3	111.4	118.7	126 0	7.41
2	133.3	140.8	148.3	155.9	163.6	171.3	179.1	187.0	195.0	203.0	14.81
3	211.1	219.3	227.6	235.9	244.3	252.8	261.3	270.0	278.7	287.4	22.22
4	296.3	305.2	314.2	323.3	332.4	341.7	351.0	360.3	369.8	379.3	29.63
5	388.9	398.6	408.3	418.1	428.0	438.0	448.0	458.1	468.3	478.6	37.04
6	488.9	499.3	509.8	520.3	531.0	541.7	552.4	563.3	574.2	585.2	44.44
7	596.3	607.4	618.7	630.0	641.3	652.8	664.3	675.9	687.6	699.3	51.85
8	711.1	723.0	735.0	747.0	759.1	771.3	783.6	795.9	808.3	820.8	59.26
9	833.3	846.0	858.7	871.4	884.3	897.2	910.2	923.3	936.4	949.7	66.67
10	963.0	976.3	898.8	1003.3	1016.9	1030.6	1044.3	1058.1	1072.0	1086.0	74.07
11	1100.0	1114.1	1128.3	1142.6	1156.9	1171.3	1185.8	1200.3	1215.0	1229.7	81.48
12	1244.4	1259.3	1274.2	1289.2	1304.3	1319.4	1334.7	1350.0	1365.3	1380.8	88.89
13	1396.3	1411.9	1427.6	1443.3	1459.1	1475.0	1491.0	1507.0	1523.1	1539.3	96.30
14	1555.6	1571.9	1588.3	1604.8	1621.3	1638.0	1654.7	1671.4	1688.3	1705.2	103.70
15	1722.2	1739.3	1756.4	1773.7	1791.0	1808.3	1825.8	1843.3	1860.9	1878.6	111.11
16	1896.3	1914.1	1932.0	1950.0	1968.0	1986.1	2004.3	2022.6	2040.9	2059.3	118.52
17	2077.8	2096.3	2115.0	2133.7	2152.4	2171.3	2190.2	2209.2	2228.3	2247.4	125.93
18	2266.7	2286.0	2305.3	2324.8	2344.3	2363.9	2383.6	2403.3	2423.1	2443.0	133.33
19	2463.0	2483.0	2503.1	2523.3	2543.6	2563.9	2584.3	2604.8	2625.3	2646.0	140.74
20	2666.7	2687.4	2708.3	2729.2	2750.2	2771.3	2792.4	2813.7	2835.0	2856.3	148.15
21	2877.8	2899.3	2920.9	2942.6	2964.3	2986.1	3008.0	3030.0	3052.0	3074.1	156.56
22	3096.3	3118.6	3140.9	3163.3	3185.8	3208.3	3231.0	3253.7	3276.4	3299.3	162.96
23	3322.2	3345.2	3368.3	3391.4	3414.7	3438.0	3461.3	3484.8	3508.3	3531.9	170.37
24	3555.6	3579.3	3603.1	3627.0	3651.0	3675.0	3699.1	3723.3	3747.6	3771.9	177.78
25	3796.3	3820.8	3845.3	3870.0	3894.7	3919.4	3944.3	3969.2	3994.2	4019.3	185.19
26	4044.4	4069.7	4095.0	4120.3	4145.8	4171.3	4196.9	4222.6	4248.3	4274.1	192.59
27	4300.0	4326.0	4352.0	4378.1	4404.3	4430.6	4456.9	4483.3	4509.8	4536.3	200.00
28	4563.0	4589.7	4616.4	4643.3	4670.2	4697.2	4724.3	4751.4	4778.7	4806.0	207.41
29	4833.3	4860.8	4888.3	4915.9	4943.6	4971.3	4999.1	5027.0	5055.0	5083.0	214.81
30	5111.1	5139.3	5167.6	5195.9	5224.3	5252.8	5281.3	5310.0	5338.7	5367.4	222.22
31	5396.3	5425.2	5454.2	5483.3	5512.4	5541.7	5571.0	5600.3	5629.8	5659.3	229.63
32	5688.9	5718.6	5748.3	5778.1	5808.0	5838.0	5868.0	5898.1	5928.3	5958.6	237.04
33	5988.9	6019.3	6049.8	6080.3	6111.0	6141.7	6172.4	6203.3	6234.2	6265.2	244.44
34	6296.3	6327.4	6358.7	6390.0	6421.3	6452.8	6484.3	6515.9	6547.6	6579.3	251.85
35	6611.1	6643.0	6675.0	6707.0	6738.1	6771.3	6803.6	6835.9	6868.3	6900.8	259.26
36	6933.3	6966.0	6998.7	7031.4	7064.3	7097.2	7130.2	7163.3	7196.4	7229.7	266.67
37	7263.0	7296.3	7329.8	7363.3	7396.9	7430.6	7464.3	7498.1	7532.0	7566.0	274.07
38	7600.0	7634.1	7668.3	7702.6	7736.9	7771.3	7805.8	7840.3	7875.0	7909.7	281.48
39	7944.4	7979.3	8014.2	8049.2	8084.3	8119.4	8154.7	8190.0	8225.3	8260 8	288.89
40	8296.3	8331.9	8367.6	8403.3	8439.1	8475.0	8511.0	8547.0	8583.1	8619.3	296.30
41	8655.6	8691.9	8728.3	8764.8	8801.3	8838.0	8874.7	8911.4	8948.3	8985.2	303.70
42	9022.2	9059.3	9096.4	9133.7	9171.0	9208.3	9245.8	9283.3	9320.9	9358.6	311.11
43	9396.3	9434.1	9472.0	9510.0	9548.0	9586.1	9624.3	9662.6	9700.9	9739.3	318.52
44	9777.8	9816.3	9855.0	9893.7	9932.4	9971.3	10010	10049	10088	10127	325.93
45	10167	10206	10245	10285	10324	10364	10404	10443	10483	10523	333.33
46	10563	10603	10643	10683	10724	10764	10804	10845	10885	10926	340.74
47	10967	11007	11048	11089	11130	11171	11212	11254	11295	11336	348.15
48	11378	11419	11461	11503	11544	11586	11628	11670	11712	11754	355.56
49	11796	11839	11881	11923	11966	12008	12051	12094	12136	12179	362.96
50	12222	12265	12308	12351	12395	12438	12481	12525	12568	12612	370.37
51	12656	12699	12743	12787	12831	12875	12919	12963	13008	13052	377.78
52	13096	13141	13185	13230	13275	13319	13364	13409	13454	13499	385.19
53	13544	13590	13635	13680	13726	13771	13817	13863	13908	13954	392.59
54	14000	14046	14092	14138	14184	14231	14277	14323	14370	14416	400.00
55	14463	14510	14556	14603	14650	14697	14744	14791	14839	14886	407.41
56	14933	14981	15028	15076	15124	15171	15219	15267	15315	15363	414.81
57	15411	15459	15508	15556	15604	15653	15701	15750	15799	15847	422.22
58	15896	15945	15994	16043	16092	16142	16191	16240	16290	16339	429.63
59	16389	16439	16488	16538	16588	16638	16688	16738	16788	16839	437.04
60	16889	16939	16990	17040	17091	17142	17192	17243	17294	17345	444.44

Note that Base, Slope, and Cu. Yds. in this table may all be multiplied by the same factor; thus, using factor of ¾ for height of 27.2 ft., we have, 3264 cu. yds. for base of 12 ft. and slopes ¾ to 1.

Add for Tenths of Feet in Height.

P. P.	7.41
1	.74
2	1.48
3	2.22
4	2.96
5	3.70
6	4.44
7	5.19
8	5.93
9	6.67

Use with preceding column only.

TABLE 2

Level Sections (Earthwork); Height, 0–60 Ft. Base of Roadway, 16 Ft., Side Slopes, 1½ to 1

Note.—The last two columns enable us to use any other base than 16 ft.: Ex.—Given height, 39.7 ft.; roadway 14 ft. Then we have, **11109−(288.89+5.19)=10815** cu. yds.

[Cu. Yds. per 100-Ft. Station.]

Note that Base, Slope, and Cu. Yds. in this table may all be multiplied by the same factor; thus, using factor of 1½ for height of 39.1 ft., we have, 16215 cu. yds. for base of 24 ft. and slopes 2¼ to 1.

Ht. Ft.	.0	.1	.2	.3	.4	.5	.6	.7	.8	.9	Width of 2 Ft. Cu. Yds
0	......	6.0	12.1	18.3	24.6	31.0	37.6	44.2	51.0	57.8	....
1	64.8	71.9	79.1	86.4	93.9	101.4	109.0	116.8	124.7	132.6	7.41
2	140.7	148.9	157.3	165.7	174.2	182.9	191.6	200.5	209.5	218.6	14.81
3	227.8	237.1	246.5	256.1	265.7	275.5	285.3	295.3	305.4	315.6	22.22
4	325.9	336.4	346.9	357.5	368.3	379.2	390.1	401.2	412.4	423.8	29.63
5	435.2	446.7	458.4	470.1	482.0	494.0	506.1	518.3	530.6	543.0	37.04
6	555.6	568.2	581.0	593.8	606.8	619.9	633.1	646.4	659.9	673.4	44.44
7	687.0	700.8	714.7	728.6	742.7	756.9	771.3	785.7	800.2	814.9	51.85
8	829.6	844.5	859.5	874.6	889.8	905.1	920.5	936.1	951.7	967.5	59.26
9	983.3	999.3	1015.4	1031.6	1047.9	1064.4	1080.9	1097.5	1114.3	1131.2	66.67
10	1148.1	1165.2	1182.4	1199.8	1217.2	1234.7	1252.4	1270.1	1288.0	1306.0	74.07
11	1324.1	1342.3	1360.6	1379.0	1397.6	1416.2	1435.0	1453.8	1472.8	1491.9	81.48
12	1511.1	1530.4	1549.9	1569.4	1589.0	1608.8	1628.7	1648.6	1668.7	1688.9	88.89
13	1709.3	1729.7	1750.2	1770.9	1791.6	1812.5	1833.5	1854.6	1875.8	1897.1	96.30
14	1918.5	1940.1	1961.7	1983.5	2005.3	2027.3	2049.4	2071.6	2093.9	2116.4	103.70
15	2138.9	2161.5	2184.3	2207.2	2230.1	2253.2	2276.4	2299.8	2323.2	2346.7	111.11
16	2370.4	2394.1	2418.0	2442.0	2466.1	2490.3	2514.6	2539.0	2563.6	2588.2	118.52
17	2613.0	2637.8	2662.8	2687.9	2713.1	2738.4	2763.9	2789.4	2815.0	2840.8	125.93
18	2866.7	2892.6	2918.7	2944.9	2971.3	2997.7	3024.2	3050.9	3077.6	3104.5	133.33
19	3131.5	3158.6	3185.8	3213.1	3240.5	3268.1	3295.7	3323.5	3351.3	3379.3	140.74
20	3407.4	3435.6	3463.9	3492.4	3520.9	3549.5	3578.3	3607.2	3636.1	3665.2	148.15
21	3694.4	3723.8	3753.2	3782.7	3812.4	3842.1	3872.0	3902.0	3932.1	3962.3	156.56
22	3992.6	4023.0	4053.6	4084.2	4115.0	4145.8	4176.8	4207.9	4239.1	4270.4	162.96
23	4301.9	4333.4	4365.0	4396.8	4428.7	4460.6	4492.7	4524.9	4557.3	4589.7	170.37
24	4622.2	4654.9	4687.6	4720.5	4753.5	4786.6	4819.8	4853.1	4886.5	4920.1	177.78
25	4953.7	4987.5	5021.3	5055.3	5089.4	5123.6	5157.9	5192.4	5226.9	5261.5	185.19
26	5296.3	5331.2	5366.1	5401.2	5436.4	5471.8	5507.2	5542.7	5578.4	5614.1	192.59
27	5650.0	5686.0	5722.1	5758.3	5794.6	5831.0	5867.6	5904.2	5941.0	5977.8	200.00
28	6014.8	6051.9	6089.1	6126.4	6163.9	6201.4	6239.0	6276.8	6314.7	6352.6	207.41
29	6390.7	6428.9	6467.3	6505.7	6544.2	6582.9	6621.6	6660.5	6699.5	6738.6	214.81
30	6777.8	6817.1	6856.5	6896.1	6935.7	6975.5	7015.3	7055.3	7095.4	7135.6	222.22
31	7175.9	7216.4	7256.9	7297.5	7338.3	7379.2	7420.1	7461.2	7502.4	7543.8	229.63
32	7585.2	7626.7	7668.4	7710.1	7752.0	7794.0	7836.1	7878.3	7920.6	7963.0	237.04
33	8005.6	8048.2	8091.0	8133.8	8176.8	8219.9	8263.1	8306.4	8349.9	8393.4	244.44
34	8437.0	8480.8	8524.7	8568.6	8612.7	8656.9	8701.3	8745.7	8790.2	8834.9	251.85
35	8879.6	8924.5	8969.5	9014.6	9059.8	9105.1	9150.5	9196.1	9241.7	9287.5	259.26
36	9333.3	9379.3	9425.4	9471.6	9517.9	9564.4	9610.9	9657.5	9704.3	9751.2	266.67
37	9798.1	9845.2	9892.4	9939.8	9987.2	10035	10082	10130	10178	10226	274.07
38	10274	10322	10371	10419	10468	10516	10565	10614	10663	10712	281.48
39	10761	10810	10860	10909	10959	11009	11059	11109	11159	11209	288.89
40	11259	11310	11360	11411	11462	11513	11563	11615	11666	11717	296.30
41	11769	11820	11872	11923	11975	12027	12079	12132	12184	12236	303.70
42	12289	12342	12394	12447	12500	12553	12606	12660	12713	12767	311.11
43	12820	12874	12928	12982	13036	13090	13145	13199	13254	13308	318.52
44	13363	13418	13473	13528	13583	13638	13694	13749	13805	13861	325.93
45	13917	13973	14029	14085	14141	14198	14254	14311	14368	14425	333.33
46	14481	14539	14596	14653	14711	14768	14826	14883	14941	14999	340.74
47	15057	15116	15174	15232	15291	15350	15408	15467	15526	15585	348.15
48	15644	15704	15763	15823	15882	15942	16002	16062	16122	16182	355.56
49	16243	16303	16364	16424	16485	16546	16607	16668	16729	16790	362.96
50	16852	16913	16975	17037	17099	17161	17223	17285	17347	17410	370.37
51	17472	17535	17598	17661	17723	17787	17850	17913	17977	18040	377.78
52	18104	18167	18231	18295	18359	18424	18488	18552	18617	18682	385.19
53	18746	18811	18876	18941	19006	19072	19137	19203	19268	19334	392.59
54	19400	19466	19532	19598	19665	19731	19798	19864	19931	19998	400.00
55	20065	20132	20199	20266	20334	20401	20469	20537	20605	20673	407.41
56	20741	20809	20877	20946	21014	21083	21152	22221	21289	21359	414.81
57	21428	21497	21567	21636	21706	21775	21845	21915	21985	22056	422.22
58	22126	22196	22267	22338	22408	22479	22550	22621	22692	22764	429.63
59	22835	22907	22978	23050	23122	23194	23266	23338	23411	23483	437.04
60	23556	23628	23701	23774	23847	23920	23993	24066	24140	24213	444.44

Add for Tenths of Feet in Height.

P. P.	7.41
1	.74
2	1.48
3	2.22
4	2.96
5	3.70
6	4.44
7	5.19
8	5.93
9	6.67

Use with preceding column only.

TABLE 3

Level Sections (Earthwork); Height, 0–60 Ft.

Base of Roadway, 28 Ft., Side Slopes, 1 to 1

Note.—The last two columns enable us to use any other base than 28 ft.: Ex.—Given height, 57.5 ft.; roadway 26 ft. Then we have, 18208−(422.22+3.70)=17782 cu. yds.

[Cu. Yds. per 100-Ft. Station.]

Ht. Ft.	.0	.1	.2	.3	.4	.5	.6	7	.8	.9	Width of 2 Ft. Cu. Yds
0	......	10.4	20.9	31.4	42.1	52.8	63.6	74.4	85.3	96.3	....
1	107.4	118.6	129.8	141.1	152.4	163.9	175.4	187.0	198.7	210.4	7.41
2	222.2	234.1	246.1	258.1	270.2	282.4	294.7	307.0	319.4	331.9	14.81
3	344.4	357.1	369.8	382.6	395.4	408.3	421.3	434.4	447.6	460.8	22.22
4	474.1	487.4	500.9	514.4	528.0	541.7	555.4	569.2	583.1	597.1	29.63
5	611.1	625.2	639.4	653.7	668.0	682.4	696.9	711.4	726.1	740.8	37.04
6	755.6	770.4	785.4	800.4	815.5	830.6	845.8	861.1	876.5	891.9	44.44
7	907.5	923.0	938.7	954.5	970.3	986.1	1002.1	1018.1	1034.2	1050.4	51.85
8	1066.7	1083.0	1099.4	1115.9	1132.4	1149.1	1165.8	1182.6	1199.4	1216.3	59.26
9	1233.3	1250.4	1267.6	1284.8	1302.1	1319.4	1336.9	1354.4	1372.0	1389.7	66.67
10	1407.4	1425.2	1443.1	1461.1	1479.1	1497.2	1515.4	1533.7	1552.0	1570.4	74.07
11	1588.9	1607.4	1626.1	1644.8	1663.6	1682.4	1701.3	1720.3	1739.4	1758.6	81.48
12	1777.8	1797.1	1816.4	1835.9	1855.4	1875.0	1894.7	1914.4	1934.2	1954.1	88.89
13	1974.1	1994.1	2014.2	2034.4	2054.7	2075.0	2095.4	2115.9	2136.4	2157.1	96.30
14	2177.8	2198.6	2219.4	2240.3	2261.3	2282.4	2303.6	2324.8	2346.1	2367.4	103.70
15	2388.9	2410.4	2432.0	2453.7	2475.4	2497.2	2519.1	2541.1	2563.1	2585.2	111.11
16	2607.4	2629.7	2652.0	2674.4	2696.9	2719.4	2742.1	2764.8	2787.6	2810.4	118.52
17	2833.3	2856.3	2879.4	2902.6	2925.8	2949.1	2972.4	2995.9	3019.4	3043.0	125.93
18	3066.7	3090.4	3114.2	3138.1	3162.1	3186.1	3210.2	3234.4	3258.7	3283.0	133.33
19	3307.4	3331.9	3356.4	3381.1	3405.8	3430.6	3455.4	3480.3	3505.3	3530.4	140.74
20	3555.6	3580.8	3606.1	3631.4	3656.9	3682.4	3708.0	3733.7	3759.4	3785.2	148.15
21	3811.1	3837.1	3863.1	3889.2	3915.4	3941.7	3968.0	3994.4	4020.9	4047.4	155.56
22	4074.1	4100.8	4127.6	4154.4	4181.3	4208.3	4235.4	4262.6	4289.8	4317.1	162.96
23	4344.4	4371.9	4399.4	4427.0	4454.7	4482.4	4510.2	4538.1	4566.1	4594.1	170.37
24	4622.2	4650.4	4678.7	4707.0	4735.4	4763.9	4792.4	4821.1	4849.8	4878.6	177.78
25	4907.4	4936.3	4965.3	4994.4	5023.6	5052.8	5082.1	5111.4	5140.9	5170.4	185.19
26	5200.0	5229.7	5259.4	5289.2	5319.1	5349.1	5379.1	5409.2	5439.4	5469.7	192.59
27	5500.0	5530.4	5560.9	5591.4	5622.1	5652.8	5683.6	5714.4	5745.3	5776.3	200.00
28	5807.4	5838.6	5869.8	5901.1	5932.4	5963.9	5995.4	6127.0	6158.7	6190.4	207.41
29	6122.2	6154.1	6186.1	6218.1	6250.2	6282.4	6314.7	6347.0	6379.4	6411.9	214.81
30	6444.4	6477.1	6509.8	6542.6	6575.4	6608.3	6641.3	6674.4	6707.6	6740.8	222.22
31	6774.1	6807.4	6840.9	6874.4	6908.0	6941.7	6975.4	7009.2	7043.1	7077.1	229.63
32	7111.1	7145.2	7179.4	7213.7	7248.0	7282.4	7316.9	7351.4	7386.1	7420.8	237.04
33	7455.6	7490.4	7525.3	7560.3	7595.4	7630.6	7665.8	7701.1	7736.4	7771.9	244.44
34	7807.4	7843.0	7878.7	7914.4	7950.2	7986.1	8022.1	8058.1	8094.2	8130.4	251.85
35	8166.7	8203.0	8239.4	8275.9	8212.4	8349.1	8385.8	8422.6	8459.4	8496.3	259.26
36	8533.3	8570.4	8607.6	8644.8	8682.1	8719.4	8756.9	8794.4	8832.0	8869.7	266.67
37	8907.4	8945.2	8983.1	9021.1	9059.1	9097.2	9135.4	9173.7	9212.0	9250.4	274.07
38	9288.9	9327.4	9366.1	9404.8	9443.6	9482.4	9521.3	9560.3	9599.4	9638.6	281.48
39	9677.8	9717.1	9756.4	9795.9	9835.4	9875.0	9914.7	9954.4	9994.2	10034	288.89
40	10074	10114	10154	10194	10235	10275	10315	10356	10396	10437	296.30
41	10478	10519	10559	10600	10641	10682	10724	10765	10806	10847	303.70
42	10889	10930	10972	11014	11055	11097	11139	11181	11223	11265	311.11
43	11307	11350	11392	11434	11477	11519	11562	11605	11648	11690	318.52
44	11733	11776	11819	11863	11906	11949	11992	12036	12079	12123	325.93
45	12167	12210	12254	12298	12342	12386	12430	12474	12519	12563	333.33
46	12607	12652	12696	12741	12786	12831	12875	12920	12965	13010	340.74
47	13056	13101	13146	13191	13237	13282	13328	13374	13419	13465	348.15
48	13511	13587	13603	13649	13695	13742	13788	13834	13881	13927	355.56
49	13974	14021	14068	14114	14161	14208	14255	14303	14350	14397	362.96
50	14444	14492	14539	14587	14635	14682	14730	14778	14826	14874	370.37
51	14922	14970	15019	15067	15115	15164	15212	15261	15310	15359	377.78
52	15407	15456	15505	15554	15604	15653	15702	15751	15801	15850	385.19
53	15900	15950	15999	16049	16099	16149	16199	16249	16299	16350	392.59
54	16400	16450	16501	16551	16602	16653	16704	16754	16805	16856	400.00
55	16907	16959	17010	17061	17112	17164	17215	17267	17319	17370	407.41
56	17422	17474	17526	17578	17630	17682	17735	17787	17839	17892	414.81
57	17944	17997	18050	18103	18155	18208	18261	18314	18368	18421	422.22
58	18474	18527	18581	18634	18688	18742	18795	18849	18903	18957	429.63
59	19011	19065	19119	19174	19228	19282	19337	19391	19446	19501	437.04
60	19556	19610	19665	19720	19775	19831	19886	19941	19996	20052	444.44

Note that Base, Slope, and Cu. Yds. in this table may all be multiplied by the same factor; thus, using factor of ½ for height of 45.1 ft., we have, 6105 cu. yds. for base of 14 ft. and slopes ½ to 1.

Add for Tenths of Feet in Height.

P. P.	7.41
1	.74
2	1.48
3	2.22
4	2.96
5	3.70
6	4.44
7	5.19
8	5.93
9	6.67

Use with preceding column only.

TABLE 4

Level Sections (Earthwork); Height, 0–60 Ft.
Base of Roadway, 28 Ft., Side Slopes, 1½ to 1

Note.—The last two columns enable us to use any other base than 28 ft.: Ex.—Given height, 33.6 ft.; roadway 30 ft. Then we have, **9756.4+ (244.44+4.44)=10005.3** cu. yds.

[Cu. Yds. per **100**-Ft. Station.]

Ht. Ft.	.0	.1	.2	.3	.4	.5	.6	.7	.8	.9	Width of 2 Ft. Cu. Yds	P. P. 7.41
0	......	10.4	21.0	31.6	42.4	53.2	64.2	75.3	86.5	97.9	....	
1	109.3	120 8	132.5	144.3	156.1	168.1	180.2	192.4	204.8	217.2	7.41	
2	229.6	242.3	255.0	267.9	280.9	294.0	307.2	320.5	334.0	347.5	14.81	
3	361.2	374.9	388.8	402.8	416.9	431.1	445.4	459.9	474.4	489.1	22.22	
4	503.7	518.6	533.6	548.6	563.9	579.3	594.7	610.2	625.8	641.6	29.63	
5	657.5	673.4	689.5	705.7	722.1	738.5	755.0	771.7	788.4	805.3	37.04	
6	822.2	839.3	856.5	873.8	891.2	908.8	926.4	944.2	962.0	980.0	44.44	
7	998.1	1016.4	1034.7	1053.1	1071.6	1090.3	1109.0	1127.9	1146.9	1166.0	51.85	
8	1185.2	1204.5	1223.9	1243.5	1263.1	1282.9	1302.7	1322.7	1342.8	1363.0	59.26	
9	1383.3	1403.8	1424.3	1444.9	1465.7	1486.6	1507.6	1528.6	1549.8	1571.2	66.67	
10	1592.6	1614.1	1635.8	1657.5	1679.4	1701.4	1723.5	1745.7	1768.0	1790.4	74.07	
11	1813.0	1835.6	1858.4	1881.2	1904.2	1927.3	1950.5	1973.8	1997.3	2020.8	81.48	
12	2044.4	2068.2	2092.1	2116.1	2040 1	2164.3	2188.7	2213.1	2237.6	2262.3	88.89	
13	2287.0	2311.9	2336.9	2362.0	2387.2	2412.5	2437.9	2463.5	2489.1	2514.9	96.30	
14	2540.7	2566.7	2592.8	2619.0	2645.3	2671.8	2698.3	2724.9	2751.7	2778.6	103.70	
15	2805.6	2832.6	2859.9	2887.2	2914.6	2942.1	2969.8	2997.5	3025.4	3053.4	111.11	
16	3081.5	3109.7	3138.0	3166.4	3195.0	3223.6	3252.4	3281.2	3310.2	3339.3	118.52	
17	3368.5	3397.8	3427.3	3456.8	3486.4	3516.2	3546.1	3576.1	3606.1	3636.4	125.93	
18	3666.7	3697.1	3727.6	3758.3	3789.0	3819.9	3850.9	3882.0	3913.2	3944.5	133.33	
19	3975.9	4007.5	4039.1	4070.9	4102.7	4134.7	4166.8	4199.0	4231.3	4263.8	140.74	
20	4296.3	4328.9	4361.7	4394.6	4427.6	4460.6	4493.9	4527.2	4560.6	4594.1	148.15	
21	4627.8	4661.5	4695.4	4729.4	4763.5	4797.7	4832.0	4866.4	4901.0	4935.6	155.56	
22	4970.4	5005.2	5040.2	5075.3	5110.5	5145.8	5181.3	5216.8	5252.4	5288.2	162.96	
23	5324.1	5360.1	5396.1	5432.4	5468.7	5505.1	5541.6	5578.3	5615.0	5651.9	170.37	
24	5688.9	5726.0	5763.2	5800.5	5837.9	5875.5	5913.1	5950.9	5988.7	6026.7	177.78	
25	6064.8	6103.0	6141.3	6179.8	6218.3	6256.9	6295.7	6334.6	6373.6	6412.6	185.19	
26	6451.9	6491.2	6530.0	6570.1	6609.8	6649.5	6689.4	6729.4	6769.5	6809.7	192.59	
27	6850.0	6890.4	6931.0	6971.6	7012.4	7053.2	7094.2	7135.3	7176.5	7217.8	200.00	1 .74
28	7259.3	7300.8	7342.4	7384.2	7426.1	7463.1	7510.1	7552.4	7594.7	7637.1	207.41	2 1.48
29	7679.6	7722.3	7765.0	7807.9	7850.9	7894.0	7937.2	7980.5	8023.9	8067.5	214.81	3 2.22
30	8111.1	8154.9	8198.7	8242.7	8236.8	8331.0	8375.3	8419.8	8464.3	8508.9	222.22	4 2.96
31	8553.7	8598.6	8643.6	8688.6	8733.9	8779.2	8824.6	8870.1	8915.8	8961.5	229.63	5 3.70
32	9007.4	9053.4	9099.5	9145.7	9192.0	9238.4	9285.0	9331.6	9378.4	9425.2	237.04	6 4.44
33	9472.2	9519.3	9566.5	9613.8	9661.3	9708.8	9756.4	9804.2	9852.1	9900.1	244.44	7 5.19
34	9948.1	9996.4	10045	10093	10142	10190	10239	10288	10337	10386	251.85	8 5.93
35	10435	10484	10534	10583	10633	10683	10732	10782	10832	10882	259.26	9 6.67
36	10933	10983	11034	11084	11135	11186	11237	11288	11339	11391	266.67	
37	11443	11494	11546	11598	11649	11701	11753	11806	11858	11910	274.07	
38	11963	12016	12068	12121	12174	12227	12281	12334	12387	12441	281.48	
39	12494	12548	12602	12656	12710	12764	12819	12873	12928	12982	288.89	
40	13037	13092	13147	13202	13257	13312	13368	13423	13479	13535	296.30	
41	13591	13647	13703	13759	13815	13872	13928	13985	14042	14099	303.70	
42	14156	14213	14270	14327	14385	14442	14500	14558	14615	14673	311.11	
43	14731	14790	14848	14906	14965	15024	15082	15141	15200	15259	318.52	
44	15318	15378	15437	15497	15556	15616	15676	15736	15796	15856	325.93	
45	15917	15977	16038	16098	16159	16220	16281	16342	16403	16465	333.33	
46	16526	16587	16649	16711	16773	16835	16897	16959	17021	17084	340.74	
47	17146	17209	17272	17335	17398	17461	17524	17587	17651	17714	348.15	
48	17778	17842	17905	17969	18033	18098	18162	18226	18291	18356	355.56	
49	18420	18485	18550	18615	18680	18746	18811	18877	18942	19008	362.96	
50	19074	19140	19206	19272	19339	19405	19472	19538	19605	19672	370.37	
51	19739	19806	19873	19940	20008	20075	20143	20211	20279	20347	377.78	
52	20415	20483	20551	20620	20688	20757	20826	20894	20963	21032	385.19	
53	21102	21171	21241	21310	21380	21450	21519	21589	21659	21730	392.59	
54	21800	21870	21941	22012	22082	22153	22224	22295	22366	22438	400.00	
55	22509	22581	22652	22724	22796	22868	22940	23012	23085	23157	407.41	
56	23230	23302	23375	23448	23521	23594	23667	23741	23814	23888	414.81	
57	23961	24035	24109	24183	24257	24331	24405	24480	24554	24629	422.22	
58	24704	24779	24854	24929	25004	25079	25155	25230	25306	25381	429.63	
59	25457	25533	25609	25686	25762	25838	25915	25992	26068	26145	437.04	
60	26222	26299	26376	26454	26531	26609	26686	26764	26842	26920	444.44	

Note that Base, Slope, and Cu. Yds. in this table may all be multiplied by the same factor; thus, using factor of ½ for height of 51.4 ft., we have, 10004 cu. yds. for base of 14 ft. and slopes ¾ to 1.

Add for Tenths of Feet in Height.

Use with preceding column only.

at the centerline of each station. The latter may be obtained by scaling the depth on a profile drawing or sighting on a graduated rod as from A to B in Fig. 6 (c). Side slopes of an earth excavation are usually about 45 degrees and the slope is given on the drawings as $1\frac{1}{2}$ to 1, 1 to 1, etc., which means " $1\frac{1}{2}$ foot horizontal to 1 foot vertical," "1 foot horizontal to 1 foot vertical," etc. See Fig. 6 (c).

Tables 1 and 2 give the cubic yards of excavation per 100-foot stations for a 16-foot roadway and side slopes of 1 to 1 and $1\frac{1}{2}$ to 1, respectively. Tables 3 and 4 give the corresponding data on roadways 28 feet wide. These tables are also applicable to the determination of the volumes of fills, since the inverted cross-section of a typical cut is the cross-section of a typical fill.

Illustration: How many cubic yards of excavation are involved in the cut shown in Fig. 6 (a) if the roadway is 16 feet wide, the side slopes $1\frac{1}{2}$ to 1 and the centerline depth in feet at the various stations 100 feet apart as follows: a, 0.0; b, 4.7; c, 10.4; d, 15.3; e, 14.7; f, 12.1; g, 6.2; h, 1.2?

Table 2 applies to the conditions of this problem. Taking the values from this table for the depths (or heights) corresponding to each station, we obtain the following total:

Station	Height Feet	Cubic Yards per 100-Ft Station
a	0.0	0.0
b	4.7	401.2
c	10.4	1217.2
d	15.3	2207.2
e	14.7	2071.6
f	12.1	1530.4
g	6.2	581.0
h	1.2	79.1
		8087.7 cu. yd. (Ans.)

This would be given in an estimate as 8100 cu. yd.

Borrow Pit Excavation.—When a fill of earth is to be made, the material is taken from what is called a "borrow pit." It is often necessary to measure the amount of material which has been removed from such a pit, and since its shape is generally irregular, tables cannot be used and the average-end-area method is often applied.

As in the case of the measurement of right-of-way excavations, the determination of the volume of a borrow pit requires the use of a base line and a determination of the profiles of the ground before and after excavation at right angles to and at regular intervals along the base line. Figure 8 shows the plan of a borrow pit with the base line and stations. In practice, the setting of the base lines and the measurement of the profiles is the work of a surveying party and this phase is beyond the scope of this book. We only propose to show how the volume of excavation is computed after the surveying notes have been made and plotted. For the sake of simplicity only a minimum number of cross-sections will be used and each of these as elementary as possible.

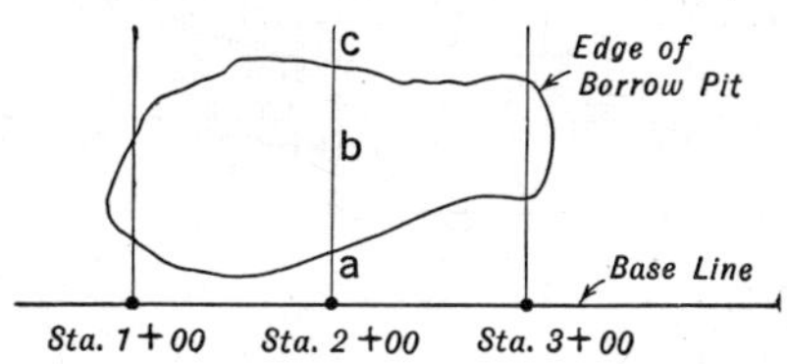

FIG. 8.

In Fig. 9 let *abc* represent a profile of the original ground surface of a borrow pit at a point such as at Sta. 2 + 00 in Fig. 8, and let *adec* represent the final ground surface. The two form a cross-section. It will be noted that by referring the points in this figure to a reference line such as *fh* and dropping perpendiculars, a number of trapezoids are formed. From geometry we know that the area of trapezoid *abgf* is

$$\frac{(af + \overline{bg})}{2} \times \overline{fg} \quad \text{or} \quad \tfrac{1}{2}(\overline{af} \times \overline{fg} + \overline{bg} \times \overline{fg})$$

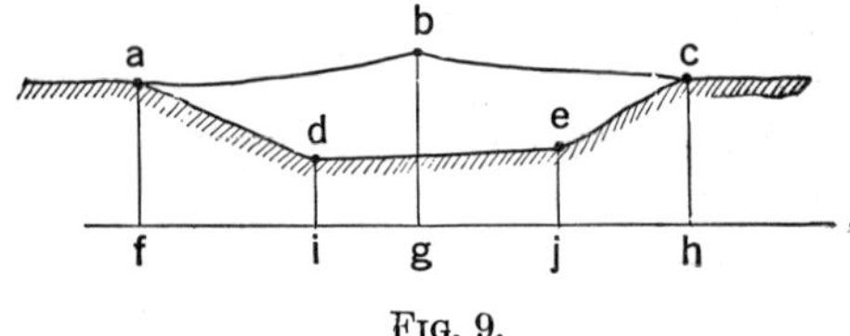

b
a
c
d
e
f
i
g
j
h

Fig. 9.

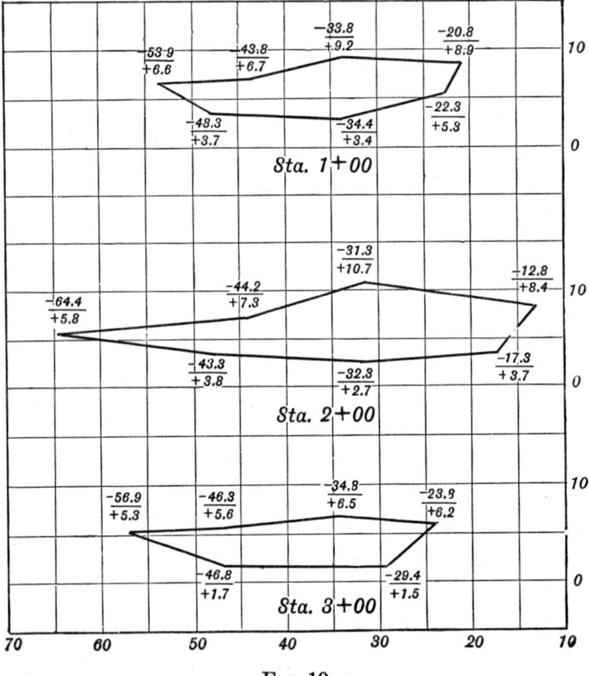

−33.8
+9.2
−20.8
+8.9
−53.9
+6.6
−43.8
+6.7
10
−22.3
+5.3
−48.3
+3.7
−34.4
+3.4
0
Sta. 1+00
−31.3
+10.7
−12.8
+8.4
−44.2
+7.3
10
−64.4
+5.8
−43.3
+3.8
−32.3
+2.7
−17.3
+3.7
0
Sta. 2+00
−34.8
+6.5
10
−56.9
+5.3
−46.3
+5.6
−23.3
+6.2
−46.8
+1.7
−29.4
+1.5
0
Sta. 3+00
70
60
50
40
30
20
10

Fig. 10.

We can similarly find the area of each of the other trapezoids in the figure by multiplying half the sum of the two sides by the base. Then, if we subtract the sum of the areas *adif*, *deji*, and *echj*, from the sum of the areas of *abgf* and *bchg*, it is obvious that the remainder will be the area sought, *abced*. This is the principle on which is based the method of computation described in the next paragraph.

Figure 10 represents three cross-sections such as might be obtained from an excavation such as is shown in plan in Fig. 8. Each " break " in the ground level is represented by a point on the plotted cross-section. At each point the figure above the line is the distance from the base line (distances to the left are marked as − and those to the right as +) while the figure below the line is the elevation above or below an arbitrarily selected grade (marked + if above and − if below). In this case the base line is represented by the right-hand margin. Beginning at any point, proceed clockwise around the figure multiplying each elevation by the distance for the point next in advance minus the distance for the preceding point, with due observance of algebraic signs. The algebraic sum of these products divided by 2 is the area.

ILLUSTRATION (Station 1 + 00, Fig. 10)

	+	−
$+6.6[-43.8-(-48.3)] =$	29.7	
$+6.7[-33.8-(-53.9)] =$	134.7	
$+9.2[-20.8-(-43.8)] =$	211.6	
$+8.9[-22.3-(-33.8)] =$	102.4	
$+5.3[-34.4-(-20.8)] =$		72.1
$+3.4[-48.3-(-22.3)] =$		88.4
$+3.7[-53.9-(-34.4)] =$		72.2
	+ 478.4	− 232.7
	− 232.7	
	2)245.7	
	122.8 sq. ft.	

By carrying out a similar computation for the areas of the cross-sections at stations 2 + 00 and 3 + 00 we find that these are 220.6 sq ft and 112.4 sq. ft. respectively. Then, by the average-end-area method, the volume is the product of the average of the areas of two adjacent cross-sections and the distance between them. In this case the distance between cross-sections is 100 feet. In actual practice these computations involve many cross-sections and they can be handled most conveniently in tabular form as follows:

Station	Area, Square Feet	Average Area, Square Feet	Distance, Feet	Volume, Cubic Feet
1+00	122.8			
		171.7	100	17,170
2+00	220.6			
		166.5	100	16,650
3+00	112.4			
				27)33,820
				1,253 cu. yd. (Ans.)

Planimeter Measurements.—If cross-sections are plotted on coordinate paper to a scale of 1 in. = 10 ft, then 1 sq in. on the paper represents 100 sq ft. An instrument known as a planimeter,

Fig. 11.

of which one form is shown in Fig. 11, is a convenient and fairly accurate device which may be used for measuring directly the areas plotted on paper. It consists of a point P which is held

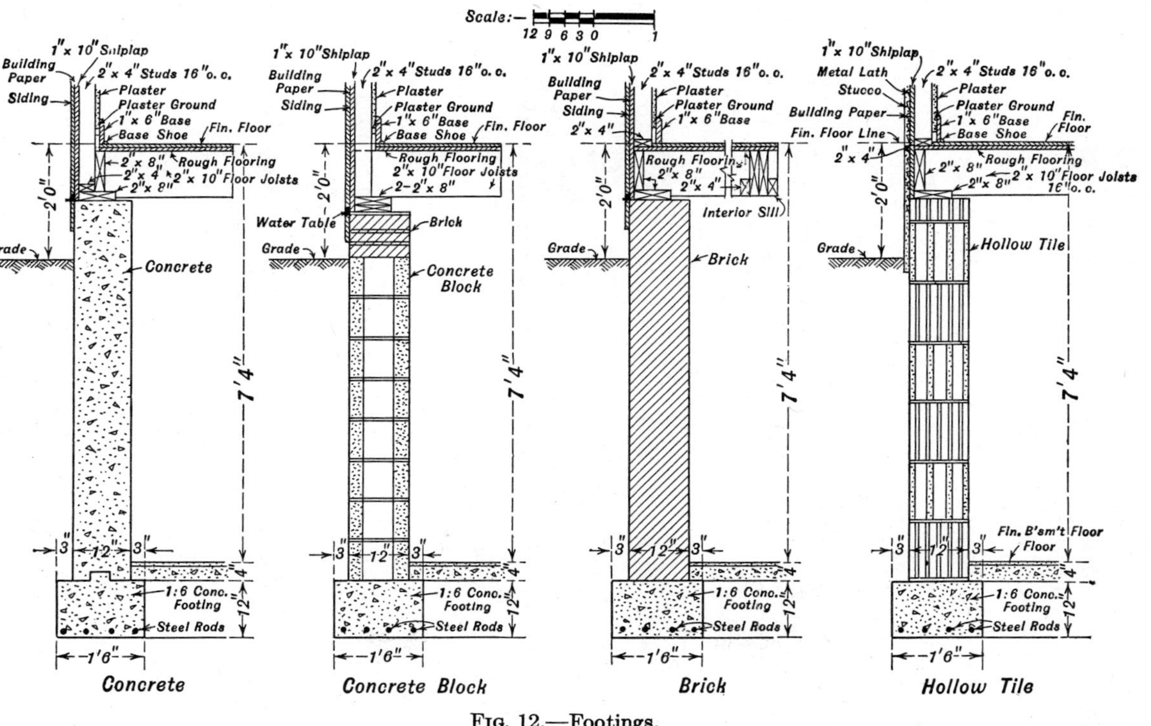

FIG. 12.—Footings.

stationary on the paper by a weight, a point T with which the outline of an area to be measured is traced in a clockwise direction, and the roller R which slides over the paper and records the area. With the vernier, the area may be read to hundredths of a square inch.

In operating a planimeter, the instrument is set down on the paper with the point P well outside the figure to be measured.

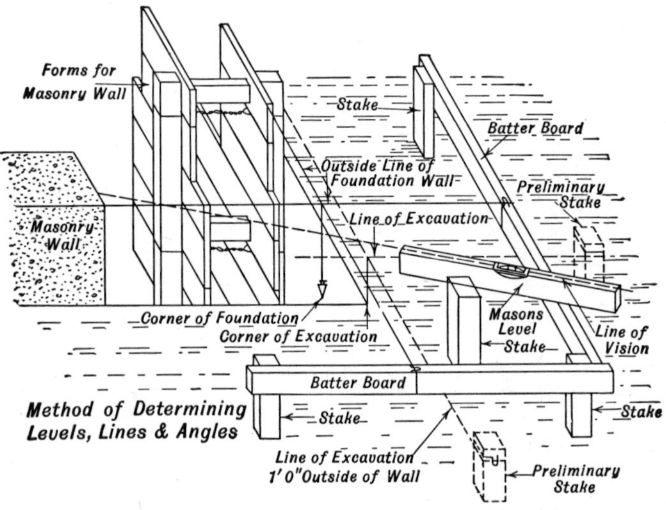

Method of Staking and Laying out the Foundation Walls

Fig. 13.

Then point T is moved to a starting point on the outline of the figure. Next, the reading on the roller is taken and recorded. Then the handle above T is gripped lightly and the point moved slowly and uniformly along the outline in a clockwise direction until the original starting point is reached. With T remaining on the starting point, the reading on the wheel is again taken and recorded. The difference between the first and second readings is the area outlined in square inches. The area of the section is then converted to square feet by multiplying by the scale factor.

Dwelling Foundations.—The type of foundation selected for a structure depends on the weight of the structure and the allowable bearing capacity of the soil. In the case of ordinary dwellings, however, the weight of the building distributed over the foundation wall results in a unit pressure on the soil so low that no special treatment is necessary for adequate support. Of more concern in this case is the matter of even settlement to prevent cracks in walls and plaster. It is therefore a rather general practice to build a footing somewhat wider than the foundation wall as shown in Fig. 12. The use of steel rods at the bottom makes the footing act as a beam and results in better distribution of pressures.

The foundation is staked out on the ground after the excavation has been completed. This consists of setting batter boards as shown in Fig. 13 and stretching lines between them when needed to align the walls. Here again a right angle formed by the lines can be checked by the method discussed on page 280. If a grade stake has been set by an instrument man and properly preserved, elevations may be checked by the use of a mason's level as shown in Fig. 13.

The footing, if concrete, may be poured directly into the excavation made for it without form work. It then serves the useful function of being a solid and level support for concrete forms which may be erected on it as shown in Fig. 14.

Heavy Foundations.—Foundations which must carry heavy loads require careful attention to the bearing capacity of the soil. Safe bearing capacities in tons per square foot are:

Material	Allowable Bearing Tons per Sq. Ft.
Quicksand and alluvial soil	½
Soft clay	1
Moderately dry clay, fine sand	2
Firm and dry loam or clay	3
Compact coarse sand or stiff gravel	4
Coarse gravel	6
Gravel and sand, well-cemented	8
Good hardpan or hard shale	10
Very hard native bedrock	20
Rock under caissons	25

The unit soil pressure exerted by a foundation is computed by dividing the total weight by the bearing area.

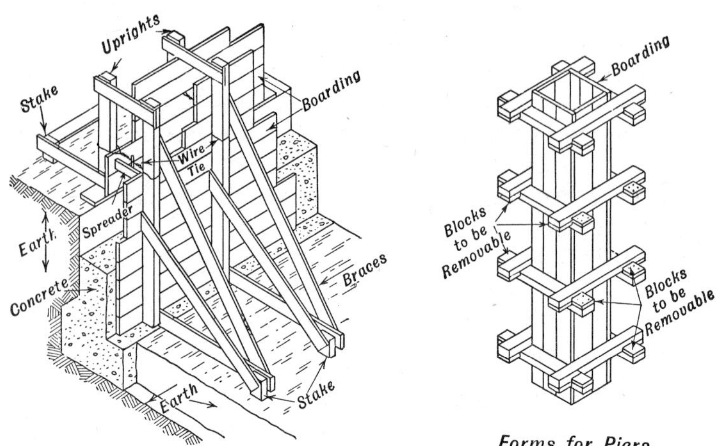

Forms for Walls in Solid Earth

Forms for Piers

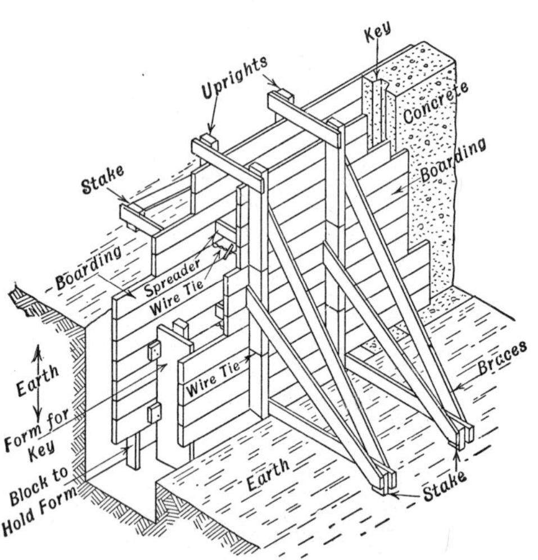

Forms for Wall in Soft Earth and Method of Keying Wall for a Halt in Concrete

FIG. 14.

ILLUSTRATION: A building weighing 100 tons rests on a footing 18 inches wide and of a total length of 84 feet. What is the unit soil pressure?

Bearing area $= 1.5 \times 84 = 126$ sq. ft.

Unit soil pressure $= \dfrac{100}{126} = 0.8$ ton per sq. ft. (Ans.)

ILLUSTRATION: A spread footing 8 feet square carries a column load of 55 tons. What is the unit soil pressure?

Bearing area $= 8 \times 8 = 64$ sq. ft.

Unit soil pressure $= \dfrac{55}{64} = 0.86$ ton per sq. ft. (Ans.)

ILLUSTRATION: What area of bearing is required to support a load of 72 tons on coarse gravel?

The safe bearing on coarse gravel is 6 tons per square foot. Then, required bearing area is

$$\frac{72}{6} = 12 \text{ sq. ft. (Ans.)}$$

ILLUSTRATION: What area of bearing is required for a load of 536,000 pounds on firm clay?

$$\text{Changing to tons, } \frac{536{,}000}{2000} = 268 \text{ tons}$$

Safe bearing on firm clay is 3 tons per square foot. Then, bearing area required is,

$$\frac{268}{3} = 89\tfrac{1}{3} \text{ sq. ft. (Ans.)}$$

Weights of Structures.—It is obvious from these illustrations that the determination of the weights of structures is a necessary preliminary to the determination of bearing pressures. Average unit weights of building materials in pounds per cubic foot which may be used for this purpose, are shown in Table 5:

TABLE 5

AVERAGE WEIGHTS OF VARIOUS MATERIALS

Material	Weight, lb. per cu. ft.	Material	Weight, lb. per cu. ft.
MASONRY CONSTRUCTION		TIMBER, U. S. SEASONED (Moisture, 15–20 per cent)	
Ashlar, granite	165	Ash	45
Ashlar, limestone, marble	160	Beech	46
Ashlar, sandstone, bluestone	140	Birch	32
Dry rubble, granite	130	Butternut	26
Dry rubble, limestone, marble	125	Cedar	39
Dry rubble, sandstone, bluestone	110	Cherry	44
Mortar rubble, granite	155	Chestnut	41
Mortar rubble, limestone, marble	150	Cypress	30
Mortar rubble, sandstone, bluestone	130	Elm	40
Brick masonry, common brick	120	Fir, Douglas Spruce	34
Brick masonry, pressed brick	140	Hemlock	29
Concrete masonry:		Hickory	48
cinder concrete	115	Mahogany	44
slag concrete	130	Maple, hard	42
stone or gravel concrete, plain	145	Maple, white	33
stone or gravel concrete, reinforced	150	Oak, red, black	42
		Oak, white	46
STONE, QUARRIED, PILED		Pine, Oregon	32
Basalt, granite, gneiss	96	Pine, red	30
Limestone, marble, quartz	95	Pine, white	28
Sandstone	82	Pine, yellow, long leaf	44
		Pine, yellow, short leaf	38
		Poplar	27
		Redwood	28
		Spruce	27

A common building material omitted from this list is steel, which has a unit weight of 490 lb per cu ft. The reason for the omission is that its weight in structures is not estimated on a volume basis. The weight of each member is determined separately from the tables of weights per linear foot of structural steel which are found in such handbooks as the AISC Steel Construction Manual.

In estimating the weights of buildings and other structures for the purpose of determining bearing pressures, the quantity of each material must often be computed separately. This divided by the unit weight in the above table gives the total weight of that material. Particular attention must be paid to the *distribution* of weights so that the computed bearing on a pedestal footing, for example, will represent only the load transmitted to that particular footing.

The weight of walls, floors, roofs, partitions, and all permanent construction is called *dead load.* All other loads are variable loads or *live loads.* Live loads per square foot of floor should be figured with the following *minimum* values:

Type of Structure	Live Load, Pounds per Square Foot
Dwelling, apartment, hotel	40
Hospitals, operating rooms	60
Office building, first floor	100
" " , all floors above the first	80
School, classrooms	40
Assembly halls, fixed seats	60
Ordinary stores, light manufacturing, light storage	125
Dining rooms, public	100
Theaters, orchestra and balconies	60

When computing bearing pressures, both the dead loads and the live loads must be taken into consideration in the following relation: For warehouses and factories, full dead and full live loads; for stores, light factories, churches, school houses, and places of public amusement or assembly, full dead and 75 percent of live loads; for office buildings, hotels, dwellings, apartment houses, full dead load and 60 percent of live load.

Pile Foundations.—When a ground surface does not have sufficient bearing power to support a structure but is underlaid by a stratum of satisfactory material, it is common practice to support the structure on piles. Wooden piles should be sound straight timbers, not less than six inches in diameter at the point or less than twelve inches at the butt. Piles are driven point downward by the successive blows of a hammer which is either permitted to fall a considerable distance or is actuated by steam pressure through short strokes.

When piles are driven to rock or hardpan, that is, driven to "refusal," the safe sustaining power is that of the pile as a column, provided that the maximum load on any pile should not exceed 20 tons. When a pile is not driven to refusal, its bearing capacity may be determined by the following formulas, known as the ENGINEERING NEWS formulas:

The safe load in pounds when a drop hammer is used is,

$$p = \frac{2wh}{s+1}$$

when w = weight of hammer in pounds
h = fall of the hammer in feet
s = penetration in inches under the last blow
p = safe load in pounds

ILLUSTRATION: What is the safe bearing power of a pile which penetrates $\frac{3}{4}$ inch under the last blow of a 3000-lb hammer dropping 20 feet?

$$p = \frac{2wh}{s+1} = \frac{2 \times 3000 \times 20}{\frac{3}{4}+1}$$

$$p = \frac{120\,000}{1.75} = 68{,}000 \text{ lb} = 34.3 \text{ tons}$$

This exceeds the maximum allowable of 20 tons, so the pile cannot be counted on to carry more than 20 tons. (Ans.)

When a steam hammer is used, the safe load P per pile may be computed from the following formula:

$$P = \frac{2h(w + am - b)}{s + k}$$

in which w = weight of striking part of hammer, in lb.
h = stroke, or height of fall, in ft.
a = effective area of piston, in sq. in.
m = mean effective pressure of the steam on the downward stroke, in lb. per sq. in.
b = total back pressure, in lb.
k = a constant, sometimes taken at from 0.1 to 0.3. The Navy Department uses $k = 0.3W/w$, in which W is the weight of the pile, in lb., and w is the same as above.

X

Concrete

Definitions.—The word "concrete" now generally refers to masonry material which is made from Portland cement, water, sand, and stone or gravel. *Portland Cement* is the product formed by pulverizing the clinker produced by heating to incipient fusion a properly proportioned mixture of siliceous, argillaceous, and calcareous material. *Water* for concrete must be fresh water and free from injurious salts or organic material. *Sand,* for concrete, also referred to as *fine aggregate,* must be graded, clean, and free from clay, loam, or organic impurities. The *coarse aggregate* is gravel or crushed rock. This should be made up of strong and unlaminated particles no larger in size than half the thickness of the thinnest section of concrete to be poured.

The *proportion* of other material to concrete is usually given by volume. Cement is measured by the bag, which contains 1 cu. ft., and the aggregates in cubic feet. It is standard practice to designate a concrete mix by three numbers, such as 1 : 3.75 : 5. The first number always refers to the number of bags of cement; the second number is the number of cubic feet of fine aggregate, usually sand, and the third number is the number of cubic feet of coarse aggregate, such as pebbles, crushed stone, steam cinders, etc.

Strength of Concrete.—In many concrete structures, the concrete is called upon to carry a certain amount of load. In other structures such as dwelling foundations, strength is a secondary factor, but perviousness or water-tightness is an important consideration. Whether density or strength is desired, both are arrived at simultaneously by proper proportioning of the ingredients of the mix.

Field and laboratory experiments have determined that with concrete of a given plastic consistency (not so stiff as to be harsh and not soft enough to permit the aggregates to separate) the strength will depend upon the ratio of water to cement in the mixture; the smaller this ratio, the stronger the resulting concrete.

In heavy construction work such as hydroelectric dams, bridge foundations, and large buildings, the water-cement ratio is calculated from engineering data and carefully maintained throughout the job. Climatic conditions, exposure to sea water, and other environmental factors are all taken into account in determining the correct water-cement ratio.

Research and practical experience under actual job conditions have produced concrete mixes suitable for a number of ordinary types of jobs. Several of these mixes are shown in Table 1 which includes the recommended quantities of water for different classes of work as well as the proportions of cement to sand and coarse aggregate that have proved successful for each type of construction.

A trial batch should be mixed and tested for workability before the final proportions of a particular mix are fixed. It may be found that the exact proportion of water given in Table 1 results in a mixture that is too stiff, too wet, or lacking in smoothness and workability. This situation may be remedied easily by changing the amount of the aggregates slightly, but *not the water*. If the mix is too wet, add sand and pebbles slowly until the desired stiffness is obtained. If the mix is too stiff, cut down the proportion of the sand and coarse aggregate *slightly* in another trial batch.

In the selection of trial proportions of the aggregates, it should be remembered that increasing the proportion of coarse aggregate up to a certain point reduces the cement factor. Beyond this point the saving in cement is very slight, while the deficiency in mortar increases the labor cost of placing and finishing. An excess

TABLE 1

QUANTITIES OF CEMENT AND AGGREGATES FOR VARIOUS JOBS

Kind of Concrete Work	Mix by Vol. Materials Cubic Feet			Workability of the Mix	Gals. Water per Bag When Mixing	One Bag Batch Makes This Much Concrete — Cu. Ft.	Materials for One Cubic Yard of Concrete			
	Cement Bags	Sand Cu. Ft.	Stone Gravel Cu. Ft.				Cement Bags	Sand Cu. Ft.	Stone Gravel Cu. Ft.	Gals. Water When Mixing
Footings, Heavy Foundations	1	3.75	5	Stiff	6.4	6.2	4.3	16.3	21.7	27.6
Watertight Construction, Walls	1	2.5	3.5	Med.	5	4.5	6.0	15.0	21.0	30
Driveways, Floors, Walks } One Course	1	2.5	3	Stiff	4.4	4.1	6.5	16.3	19.5	28.7
Driveways, Floors, Walks } Two Courses	1	Top 2	0	Stiff	3.6	2.14	12.6	25.2	0	45.3
	1	Bottom 2.5	4	Stiff	4.9	4.8	5.7	14.2	22.8	27.8
Pavements	1	2.2	3.5	Stiff	4.3	4.2	6.4	14.1	22.4	27.5
Watertight Construction for Tanks, Wells, Cast units like posts, slabs	1	2	3	Med.	4.1	3.8	7.1	14.2	21.3	29.3
				Wet	4.9	3.9	6.9	13.8	20.7	33.7
Heavy Duty Floor, Barns, Shops, etc.	1	1.25	2	Stiff	3.4	2.8	9.8	12.3	19.6	33.9
Mortar for Brick, Concrete Blocks	1	6	1 Sack 50 lb. Hydrated Lime	Med.	12.5	5.5	4.9	29.4	5	61.2

of coarse material will produce mixtures that are undersanded and harsh. (See Fig. 1.)

Estimating Quantities. Table 1 may be used for estimating the quantities of cement and aggregates needed for a job, and hence the cost of materials, based on local prices. For example,

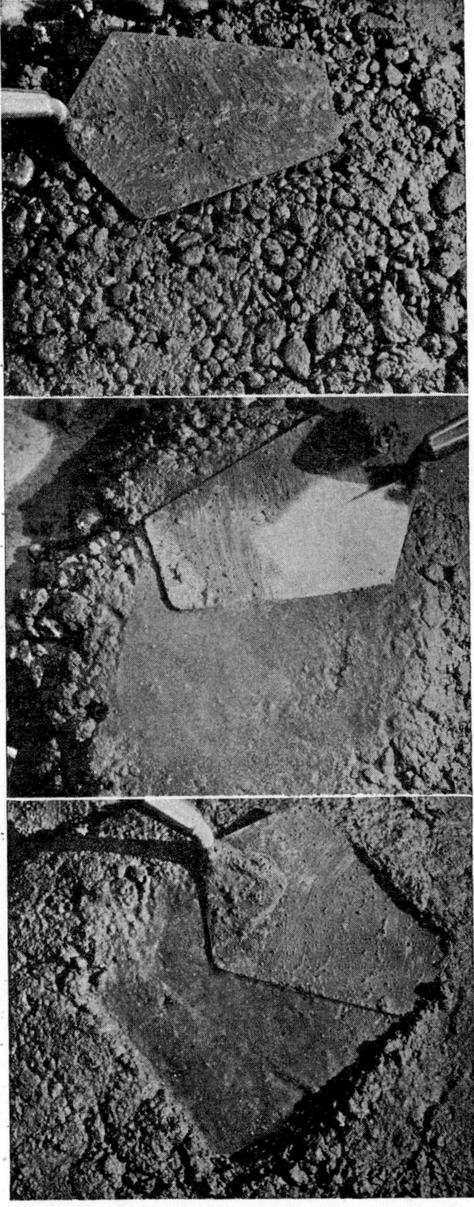

A concrete mixture in which there is not sufficient cement-sand mortar to fill spaces between pebbles. Such a mixture will be hard to work and will result in rough, honeycombed surfaces.

A concrete mixture which contains correct amount of cement-sand mortar. With light troweling all spaces between pebbles are filled with mortar. Note appearance on edges of pile. This is a good workable mixture and will give maximum yield of concrete with a given amount of cement.

A concrete mixture in which there is an excess of cement-sand mortar. While such a mixture is plastic and workable and will produce smooth surfaces, the yield of concrete will be low. Such concrete is also likely to be porous.

FIG. 1

a footing 27′ long, 2′ wide, and 1′ thick requires 54 cu. ft. or 2 cu. yds. of concrete. A 1 : 3.75 : 5 mix is recommended for this type of work by Table 1. To make 1 cu. yd. of this mix it is necessary to use 4.3 bags of cement, 16.3 cu. ft. of sharp sand, and 21.7 cu. ft. of gravel or crushed stone, and 27.6 gal. of clean water. Although the total volume of the materials when dry is much greater than 1 cu. yd., the addition of water reduces the net volume by causing the sand and cement to fill in the voids in the coarse aggregate. To obtain the 2 cu. yds. of concrete required for the footing it is necessary to double the quantities just given for one cubic yard. No allowance for waste is included in Table 1 and this factor cannot be ignored. It is customary to allow about 5% to 10% for material waste, on the job, so the following quantities should be purchased for the footing: cement, 9 bags; sand, 1⅓ cu. yds.; gravel or crushed stone, 1¾ cu. yds.

Mixing concrete or mortar involves several simple operations and requires some forethought. Excellent concrete can be mixed by hand, although machine mixing is preferred because better quality concrete results.

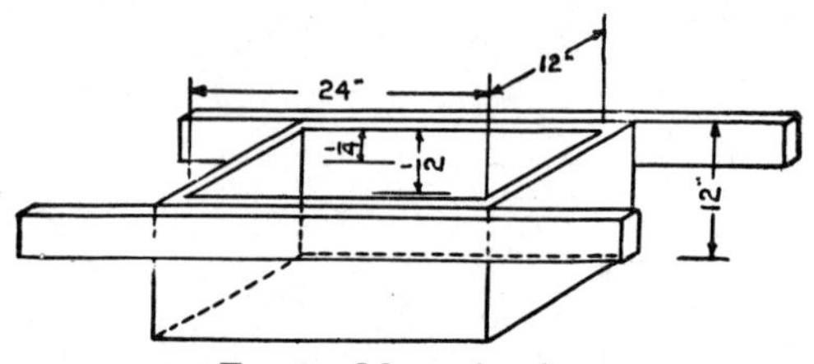

FIG. 2.—Measuring box.

Measuring Box. A simple and convenient measure for proportioning aggregates is a bottomless measuring box, shown in Fig. 2. This measure is simply a strong box with no bottom, made to exact *inside* measurements. The handles help in lifting or moving the box. As shown it has a capacity of 2 cu. ft., with marks on the inside of the box for 1 cu. ft. and for ½ cu. ft.

Selection of Consistency.—Consistency relates to the fluidity of the concrete. Figure 3 illustrates three samples of concrete with consistencies varying from "stiff" to "wet." It is important that the consistency be suited to the work to be done. Stiff

FIG. 3.—Showing Stiff, Medium, and Wet Mixtures of Concrete.

The stiff consistency is recommended for footings, walls, and pavements. The medium mix is suitable for tank walls, floors, slabs, beams, etc. The wet mix is suitable for thin walls and columns.

concrete is best suited for foundations, pavements, and massive walls. When poured into forms, it may require considerable spading to obtain smooth faces. Concrete of medium consistency is recommended for beams, slabs, and walls, where a smooth face and good bond with reinforcing steel is essential. A wet consistency should only be used in thin walls or columns or where the reinforcing is so heavy that spading is difficult or impossible.

Mixing the Concrete.—Except for small quantities mixed by hand, practically all concrete is mixed in mixing drums of the batch type. Portable mixers are usually provided with skips for charging the drum and one of the problems is that of measuring the aggregates to be dumped into the skip. In highway work a batching plant is often used and the properly proportioned aggregates are dumped into the skip from batch boxes. In charging from open stock piles, the wheelbarrow serves the double purpose of conveying the material and measuring it at the same time. The ordinary wheelbarrow used in construction work holds two cubic feet when struck off and three cubic feet when heaped. If there is any doubt as to its capacity, a wheelbarrow can readily be calibrated by filling it with sand from a box with 12 inches for each internal dimension.

On construction projects with central mixing plants the aggregates are measured in hoppers which must be calibrated by computing the internal volumes if they are of the fixed type, or adjusted according to the manufacturer's rating if they are of the patented automatic type.

A mixer should not be charged with material in excess of its rated capacity. Not only does overloading prevent the aggregates from mixing properly, but rich mortar is apt to be lost by slopping out. There is the additional disadvantage that an overloaded mixer may stall its driving engine and the mixer can then usually be started again only after most of its load has been laboriously shoveled out.

Placing Concrete.—Concrete should be deposited into a form at several points so that no appreciable flow results within the form. Such flow causes segregation of the materials and results in porous concrete. The concrete should not be allowed to fall more than a few feet into a form. It should be spaded the minimum amount necessary to make it flow around reinforcing bars and into corners of the form. If water accumulates in a form it should be drained off carefully so that cement will not be lost. A small hole drilled into the form may amply serve this purpose.

Concreting operations should not be undertaken in freezing weather unless provision is made for heating the aggregates and for protecting the newly placed concrete from frost for seventy-two hours.

Concrete will develop greater strength and wearing qualities if it is " cured " by being kept moist for a week or ten days after pouring. This may effectively be accomplished by covering with burlap and sprinkling or, in the case of pavements, by building earth dikes and permitting pools of water to stand on the concrete.

Practical Project—Concrete Stoop.—In order to show how the previous information is used in practice, a practical project will be considered—that of constructing a concrete stoop and steps such as the one shown in Fig. 4. It is a stoop consisting of three steps and is intended for installation at the rear or side entrance of a small house.*

The platform, or upper step, of this stoop is 3′ square. Each step is 3′ wide, and has an 8″ riser, and a 12″ tread. All the dimensions of this project are shown in the accompanying illustra-

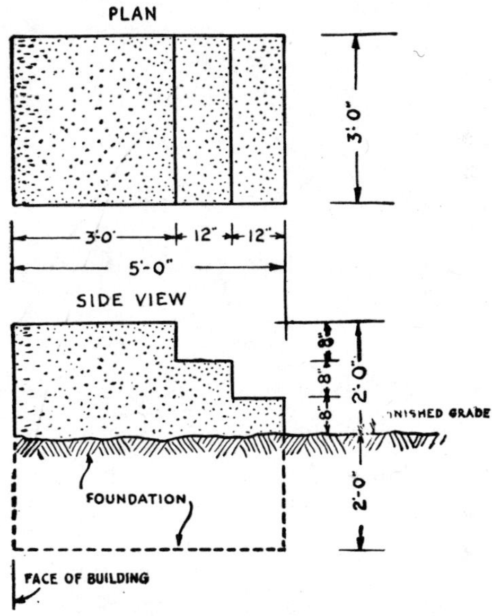

Fig. 4.—Dimensions of Stoop and Steps.

tion (Fig. 4). Note that the foundation extends 2′ below grade (ground level).

The materials needed to build this project consist of the lumber for the forms, shown in Fig. 5, and the materials for making the concrete. The quantities of these materials are listed in Table 2.

* This project is taken from *Masonry* by K. H. Bailey, D. Van Nostrand Co., Inc., New York.

TABLE 2

BILL OF MATERIAL

Number Required	Part	Size	Kind of Material
2	Sides of Top Step	3′ 1″ × 8″ × 1″	Wood
2	Sides of Middle Step	4′ 1″ × 8″ × 1″	"
2	Sides of Bottom Step	5′ 1″ × 8″ × 1″	"
3	Front of Steps (risers)	3′ × 8″ × 1″	"
1	Brace	3′ × 4″ × 2″	"
6	Stakes for Holding Form	1′ 6″ × 2″ × 4″	"
12	Cleats	Various	"
9		Bags	Cement
36		Cubic Feet	Sand
45		" "	Gravel & Crushed Stone
56		Gallons	Water

These quantities have been computed by the use of Table 1, page 292, which shows that a 1 : 3.75 : 5 mix is suitable for this kind of work. A batch amounting to 2 cu. yds. of mixed concrete will be obtained.

The Excavation. The first step in constructing this stoop is the excavation. It should be made at the point where the steps are to be installed. It should be 2′ deep and its dimensions are 5′ × 3′. When digging, take care to cut the sides of the hole so that no dirt or stones fall into the space, because these earth sides are used to hold the concrete, so that no form is necessary below grade. The foundation of the building forms one end of the hole, and must be well cleaned of all dirt so that the concrete can bond with it. Tamp the earth in the bottom of the hole, unless it is already firm, to produce a sound base for the concrete.

The Form. When the excavation is complete, work can begin on the form. Naturally, it must be made to the exact dimensions given, with all measurements made on the inside. It is usually convenient to build the form in a shop or workroom and place it on the foundation afterwards. This is easy to do, if it is nailed together securely, as shown in Fig. 5, before it is moved.

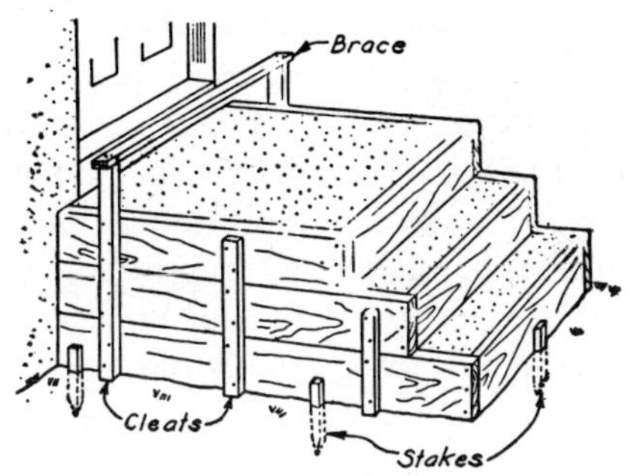

FIG. 5.—The Form.

In building the form, nail the three side pieces together first, using the cleats for that purpose, as shown in Fig. 5. Then make the other side in exactly the same way. Finally, the two sides are nailed to the front pieces, the risers, which must be carefully cut to size. The joints should be made as close as possible to prevent leaks. The last step in making the form is to nail on the brace that appears in Fig. 5. This is the piece of 2″ × 4″ exactly 3′ long, which acts as a spreader to keep the sides of the form at the proper distance apart until some concrete is poured. Be sure to remove it before the concrete is high enough to touch it.

Pouring the Foundation. The first step in concrete work is to pour the foundation, the part below ground level, and let it set for at least a day before doing the work above grade. Then the form is set in place and the job is finished. To mix the concrete for the foundation, take 60% of the total materials given, that is, 5½ bags of cement, 21½ cu. ft. of sand, 27 cu. ft. of gravel or crushed stone, and 33½ gals. of water. Mix these materials, and pour the mixture into the excavation, following the methods explained earlier in this section. Remember that thorough mixing is most important. More troubles result from too little mixing than from too much, especially with stiff concrete such as is used for this job.

After this foundation has set for at least one day, set up the form on it. The form must be placed so that it is level and square with the house before it is fastened and braced firmly into place. It is secured by means of the 2″ × 4″ stakes illustrated, and also by 3 or 4 diagonal braces (not illustrated) running from the top edge of the sides to stakes in the ground outside. Put these wherever the form needs support to prevent moving, or spreading under weight of the concrete. Chips of brick or slate should be used to wedge up the form by inserting them between the bottom edges of the side pieces and foundation. As a final check on the form after it has been placed in position, but before the stakes and braces are secured, measure the diagonals across the square that will form the stoop. If both are the same length the sides are square, parallel, and ready to be fastened into position as rigidly as possible.

Platform and Steps. When the form has been properly placed, the next step is to spread a thick cement paste over the concrete foundation to bond it with the new concrete for the steps. To make this, mix ½ bag of cement with 3 gals. of water. Then spread the paste evenly over the top of the foundation. This is done immediately before pouring the final batch of concrete, which is prepared by mixing the remainder of the materials, 3½ bags of cement, 14½ cu. ft. of sand, 18 cu. ft. of gravel or crushed stone, and 22½ gals. of water. This pouring must be done very carefully to avoid hitting or jarring the form out of position. After the operation is complete, the form should be filled to the top of the steps and platform. If water collects on these flat surfaces it should be removed with a sponge or brush, so that the concrete is left moist, but never puddled.

As soon as the concrete has set a little, that is, after a few hours, the platform and steps may be finished with a wood float.

This produces an even, yet gritty surface that is not slippery in wet weather. A small trowel or dull knife blade should be run along the inside edge of the form between the boards and concrete to make a smooth and slightly rounded edge on all exposed surfaces.

Leave the form in position for a week. Then remove it, and cure the concrete by covering it with straw, or burlap bags, and keep damp for a week or 10 days. After this curing, the stoop is ready for use. If desired, an additional finish may be applied by stripping the forms after 24 hours, and trowelling a thick cement paste onto the damp concrete with a wood float. After this finish has had time to harden, the work is cured in the usual manner.

Transit-Mixed Concrete.—The work of installing concrete can be materially reduced by ordering ready mixed concrete to be delivered by truck. This service is available even for relatively small jobs, since most of the suppliers deliver quantities as small as 1 cubic yard. However, the truck cannot "stand by," so that the job must be ready to receive the entire quantity ordered at one pouring. Therefore, if the site is not on a road or driveway, the best plan is to place planking (which can be rented in some localities) so that the truck can drive to the site. Of course, arrangements might be made to have enough men and wheelbarrows on hand to transfer the concrete from the truck to the site, but this plan is seldom desirable for smaller jobs.

In any case, the grade and quantity of the concrete must be carefully determined in advance. The suppliers of transit-mixed concrete do not usually classify their mixes according to the proportions of cement, fine aggregate and coarse aggregate, as explained earlier in this chapter, but according to bearing strength in pounds per square foot. Grades usually offered are 1000, 1500, 2000, 2500, 3000, 3500, 4000, 4500 and 5000 pound strengths. These figures are not, however, actual test results, but include safety factors. Therefore, the supplier should be consulted in choosing the grade for a particular job. One should note, however, that these grades increase in density as they increase in strength—for that reason the higher grades are more resistant to surface abrasion. Thus for footings for residential structures, i.e., homes, garages, etc., the 2000 pound grade is usually considered adequate, while for garage and cellar floors, walks, etc., the 3000 pound grade is usually the minimum specified, even though the recommended practice of using reinforcement to prevent temperature and tension cracking is followed. (Concrete is much weaker in tension than in compression.)

In ordering concrete for transit-delivery, the cubical contents of the volume to be filled are calculated, and an additional percentage added to compensate for any spillage that may occur. The percentage depends upon the method of placement and the bounding surfaces of the volume to be filled. If the surfaces bounding the volume consist entirely of wooden forms, and if the arrangements for pouring do not indicate that much loss is likely, the excess figure can be held to 10%. On the other hand, if one bounding surface is the coarse aggregate foundation for a cellar or garage floor, the increase should be 20 or 25%.

Reinforcement for concrete is of steel, which has the high tensile strength necessary to balance the weak tensile strength of concrete. The two important types of reinforcement are rods, used, for example, in beams and columns, and mesh, used in floors and walks; in roads rods are often used as well as mesh, to carry the tension between sections.

The correct placement of reinforcing is based upon the principle, established by stress analysis, that the greatest tensile stresses occur close to the outside surfaces. The application to the placement of reinforcing bars in columns and beams, and to reinforcing mesh is floors and roads, is shown in Fig. 6.

The use of additives in concrete is often found to improve special properties. The incorporation of asphalt and pitch emulsions, while producing a dark color, gives improved resistance to abrasion and thermal cracking. Waterproofing can also be effected by additives, in addition to the membrane method shown in the road sketched in Fig. 6. A representative formulation for a waterproofing additive is calcium chloride 25%, sodium silicate

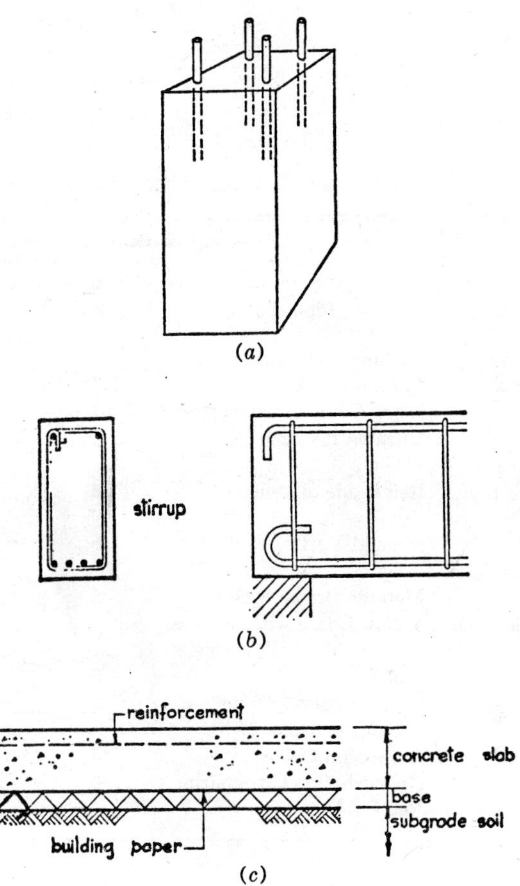

Fig. 6.—Steel Reinforcing for Concrete as Placed in (*a*) Columns; (*b*) Beams; and (*c*) Roads and Walks.

3%, ground silica 3%, and water 69%. One quart of this mixture is added to the concrete per bag of cement used. A 2% solution of ammonium stearate is also used for this purpose. Additives are also used for coloring concrete, according to the color desired as shown in Table 3.

TABLE 3

FORMULAS FOR COLORING CONCRETE

Desired Color	Pigment	Pounds per Bag of Cement: Light Shade	Pounds per Bag of Cement: Medium Shade
Black, Blue-black and Grays	Germantown Lampblack	½	1
	Carbon Black	½	1
	Black Oxide of Manganese	1	2
Blue	Ultramarine Blue	5	9
Brownish Red to Dull Brick Red	Red Oxide of Iron	5	9
Bright Red to Vermilion	Mineral Turkey Red	5	9
Brown to Reddish Brown	Metallic Brown (oxide)	5	9
Buff, Colonial Tint	Yellow Ochre with not less than 15% Yellow Oxide of Iron	5	9
Cream	Small Quantity of Yellow Oxide of Iron		
Green	Chromium Oxide	5	9
	Greenish-blue Ultramarine	6	
Pink	Red Oxide of Iron	4	8

XI

Brickwork

Uses of Brickwork.—Brickwork is well adapted to many kinds of masonry construction and is used extensively in building walls, tunnel linings, small arches, culverts, street paving, sewers, etc. The convenience in handling and laying brick, in forming arches and "rounding" corners makes it particularly useful in these classes of construction. In fire-resisting qualities it is superior to most natural building stone and in general durability it has a high rating.

Bond.—Bricks laid longitudinally in a wall are called *stretchers*. Bricks laid across the wall are called *headers*. (See Fig. 1.)

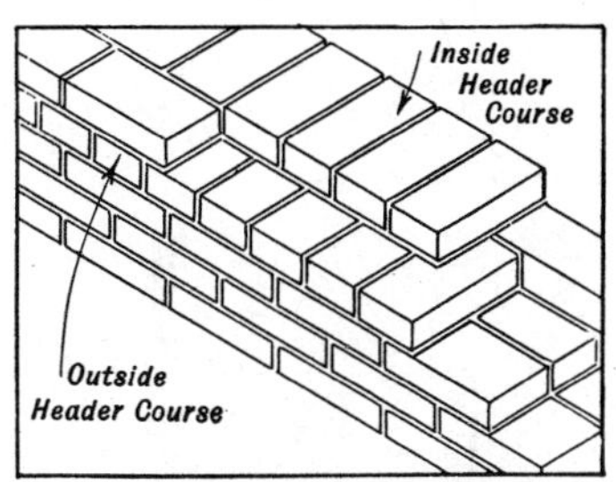

Fig. 1.—Method of lapping inside and outside header courses in 12″ solid brick walls, common bond.

The method of arrangement of the stretchers and headers in the same or adjacent courses of a wall is spoken of as the *bond.* Cost and appearance are the chief considerations in the selection of style of bond.

Common Bond.—This consists of one course of headers to every four to six courses of stretchers. Local building codes usually specify how many stretcher courses are permitted to each header course, but placing a header course at every sixth course is a safe rule. The header course may be either plain or "Flemish."

English Bond.—This bond is composed of alternate courses of headers and stretchers, the headers centering on the stretchers or the joints between them. This is considered the strongest bond.

Flemish Bond.—Each course is made up of alternate headers and stretchers. This bond gives a strong construction and a pleasing appearance.

Types of Joints.—There are four types of joints commonly used in brickwork: (1) shoved joints, (2) grouted joints, (3) open joints, and (4) dry joints.

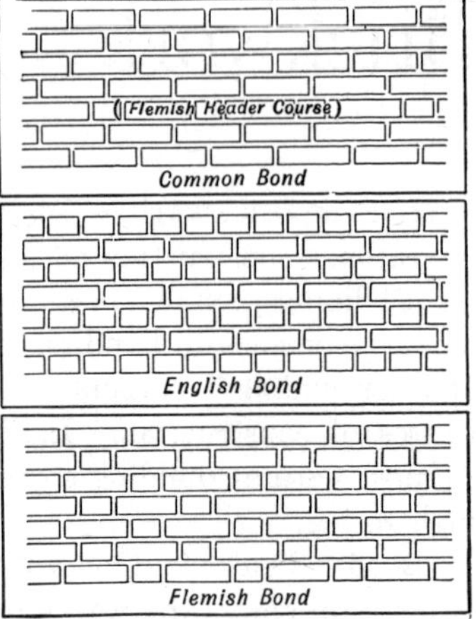

Fig. 2.

Shoved Joints.—The brick is laid on a bed of mortar a little thicker than the finished joint will be. It is then pressed downward and sideways, the soft mortar rising and filling the vertical joints. This joint produces strong and watertight masonry. (See Fig. 3.)

Grouted Joint.—The brick is laid on a level bed of mortar and the vertical joints are filled with mortar to which water has been added till it is of a " soupy " consistency.

Fig. 3.—Method of Forming Shoved Joint.

Shoved and grouted joints constitute the two types of *filled joints.* This type construction is used in fire, party and division walls, also in chimneys and piers or walls designed for heavy loads.

Open Joints.—This type of joint is often permitted above ground in dwelling construction. It is used principally with common bond and consists of laying the stretcher course on flat beds of mortar and leaving the middle vertical joint unfilled. Each header course has, however, filled joints.

Dry Joints.—This consists of laying a course of bricks directly on top of the lower course with no mortar in between. It is sometimes used on the interior face of every sixth course in cheap construction work, but its use is not recommended.

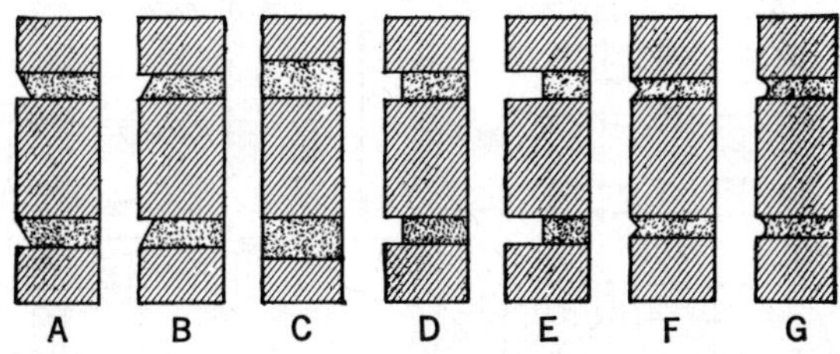

Fig. 4.—Common Types of Joints. (a) Struck joint, (b) weathered joint, (c) flush or plain cut joint, (d) raked joint, (e) stripped joint, (f) "V" joint, (g) concave joint.

Types of Exposed Joints.—Exposed joints are finished in a variety of ways for aesthetic effect or structural strength. The various types of exposed joints are illustrated in Fig. 4. Some can be made with the trowel and others require special implements.

Fireplaces.—A fireplace that will burn fuel properly and radiate maximum heat can be designed based on proportions gained from experience. The width of the opening should not be too great for moderate size rooms. For a living room with 300 sq. ft. floor space, the width should be 30 to 36 in. The height of the opening should be from 30 to 34 in., regardless of the width.

The combustion chamber must be properly proportioned for

Dimensions for various fireplace openings.

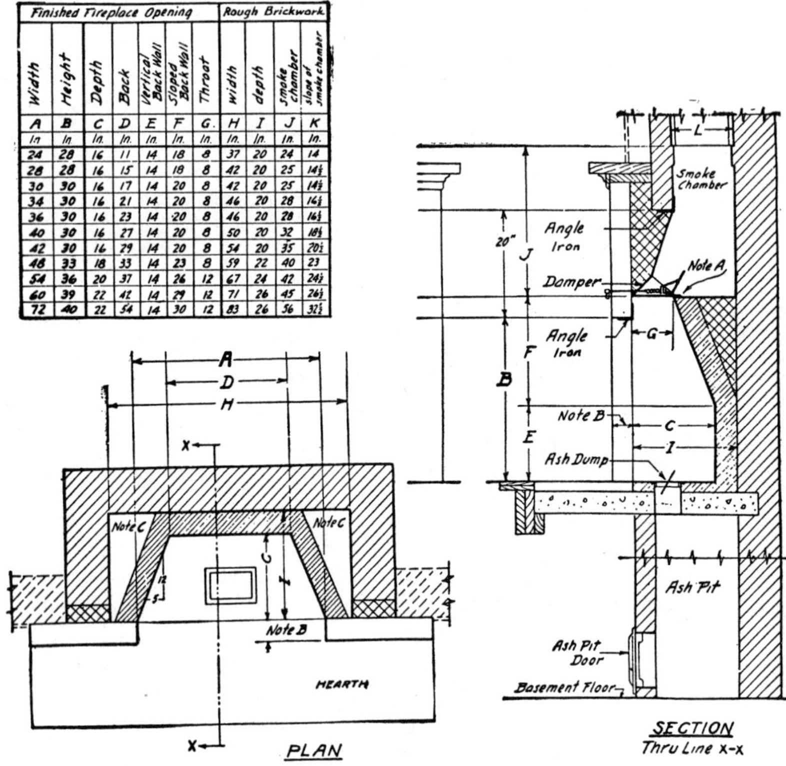

Finished Fireplace Opening							Rough Brickwork			
Width	Height	Depth	Back	Vertical Back Wall	Sloped Back Wall	Throat	width	depth	smoke chamber	slope of smoke chamber
A	B	C	D	E	F	G.	H	I	J	K
In.	In.	In.	In.	In.	In.	In.	In.	In.	In.	In.
24	28	16	11	14	18	8	37	20	24	14
28	28	16	15	14	18	8	42	20	25	14½
30	30	16	17	14	20	8	42	20	25	14½
34	30	16	21	14	20	8	46	20	28	16½
36	30	16	23	14	20	8	46	20	28	16½
40	30	16	27	14	20	8	50	20	32	18½
42	30	16	29	14	20	8	54	20	35	20½
48	33	18	33	14	23	8	59	22	40	23
54	36	20	37	14	26	12	67	24	42	24½
60	39	22	42	14	29	12	71	26	45	26½
72	40	22	54	14	30	12	83	26	56	32½

Fig. 5.—Construction Sketch of Fireplace.

Note A. The back flange of the damper must be protected from intense heat, by being fully supported and shielded by the back wall masonry.

Note B. The drawing shows the brick fireplace front as 4″ thick. No dimensions are given for this particular part of the fireplace because of the variety of materials that may be used to face the opening, such as marble, stone, tile, etc., all of which have different thicknesses.

Note C. These hollow spaces should be filled with masonry to form a solid backing. If it is necessary to include one or more furnace flues in either space beside the fireplace, the overall dimensions of the chimney must be increased to allow at least 8″ of solid brickwork on all sides of each flue.

the sake of both proper draft and heat radiation. The upper part on all sides should slope in gently to the size of the throat. This slope should preferably be no greater than about 30 degrees from the vertical or to a ratio of approximately 3 inches horizontal to 5 inches vertical. The slope should start from a point a little less than halfway up from the hearth to the throat. Not only should the sides slope toward the center, but they should be splayed toward the back. The amount of splay which gives the maximum radiation is about 5 inches for each 12 inches of depth.

A smoke shelf is an essential part of a fireplace. This performs the function of deflecting upward the downward air currents in the chimney. A damper placed at the throat is a further aid to keep smoke from being blown back into the room. A combination metal throat and damper is on the market and is a valuable aid in fireplace construction.

Flue Sizes for Fireplaces.—It is desirable to have a relatively high velocity of the gases through the throat of a fireplace. This

TABLE 1

Fireplace Dimensions			Flue Sizes			
			Rectangular		Circular	
Width of opening, inches	Approximate height, inches	Depth of opening, inches	Outside dimensions, inches	Effective area, square inches	Diameter, inches	Effective area, square inches
24	28	17—20	8½×8½	41	10	78
28	28	17—20	8½×13	70	10	78
30	30	17—21	8½×13	70	12	113
34	30	17—21	8½×13	70	12	113
36	30	21	8½×18	97	12	113
40	30	21—24	8½×18	97	15	177
42	30	21—25	8½×18	97	15	177
48	32	21—26	13 ×13	100	15	177

requires a flue of proper proportions. For a chimney 30 feet or more in height, the flue area should be about one-twelfth the area of the fireplace opening; where the chimney is 20 feet high or less, one-tenth the area is more desirable. In a flue of rectangular cross-section the gases travel up only in the center so that the *effective area* is considerably less than the cross-sectional area. Table 1 shows the sizes of flues required for various sizes of fireplaces with proper reductions made to obtain effective area.

Chimney Construction.—A chimney should be built up from a footing in the basement. Not more than two flues should occupy the same chimney space. Where three or more flues are necessary, a 4-inch partition called a *withe* should be incorporated. Every fireplace, stove, or furnace should have a separate flue.

A chimney should extend at least three feet above the highest point at which it comes into contact with the roof and at least two feet higher than any ridge within ten feet of the chimney.

The thickness of the chimney wall may be 4 inches when a flue lining of fire clay is used, but must be at least 8 inches thick with joints carefully pointed if no flue lining is used.

Mortar for Brickwork.—The mortar used in brickwork may vary considerably depending upon the class of work. For instance, the National Board of Fire Underwriters' Chimney Ordinance requires that chimney mortar be composed of two bags of Portland cement and one bag hydrated lime mixed dry and added to three times its volume of clean sharp sand, and mortar where a structure is exposed to a considerable amount of stress should be composed of one part Portland cement and two parts sand, while for most ordinary brickwork, one part fresh, well-slaked lime to $2\frac{1}{2}$ to 3 parts sand will answer. Between these limits there are various mixtures of Portland cement, natural cement, common lime, and sand. Table 2 gives a few of these mixtures.

Estimating Brickwork.—The standard size of a common brick is 8 in. $\times$ $2\frac{1}{4}$ in. $\times$ $3\frac{3}{4}$ in. and the most common thickness of joint is $\frac{1}{2}$ inch. It is readily seen that the number of bricks in a wall depends on the thickness of the wall, the thickness of the

TABLE 2

Class	Portland Cement	Natural Cement	Common Lime	Clean, Sharp Sand
A	1	....	....	2
A_1	1	....	....	2½
A_2	1	....	....	3
B	....	1	....	2
B_1	....	1	....	2½
B_2	....	1	....	3
C	....	....	1	2
C_1	....	....	1	2½
C_2	....	....	1	3
D	1	1	....	4
D_1	1	1	....	5
D_2	1	1	....	6
			Lime Paste	
E, etc.	1	....	1	2
F, etc.	....	1	1	2

Class A is used in superior building construction, for railroad masonry in general, tunnel lining and sewers; class E for building work of the highest class; class C for common brickwork as in buildings.

joints and the type of the bond. Estimating the quantities for brickwork is greatly facilitated by the use of tables which are here reproduced by courtesy of the Common Brick Manufacturers Association. Table 4 gives the number of courses in brick walls of various heights. Table 3 gives the quantities of brick and mortar in footings, piers, and chimneys of various proportions. Table 5 gives the weights of solid brick walls of unit areas. Table 6 gives the number of bricks in solid walls of various thicknesses. Table 7 gives the quantities of material needed for the mortar.

Use of the Tables—Estimating Problem.—Figure 6 shows a simple brick dwelling with 12-inch walls resting on footings and reaching to the first floor level. From the first floor to the roof the walls are 8 inches thick. The problem is to determine the approximate quantity of brick and mortar needed. Regard the remote sides of the house as having door and window openings equivalent to those shown.

Since the footing is symmetrical about the center line of each wall, the actual length of the footing will be the sum of the distances from center to center of each wall. Since the walls at the base are 12 inches thick, we can obtain the distances from center

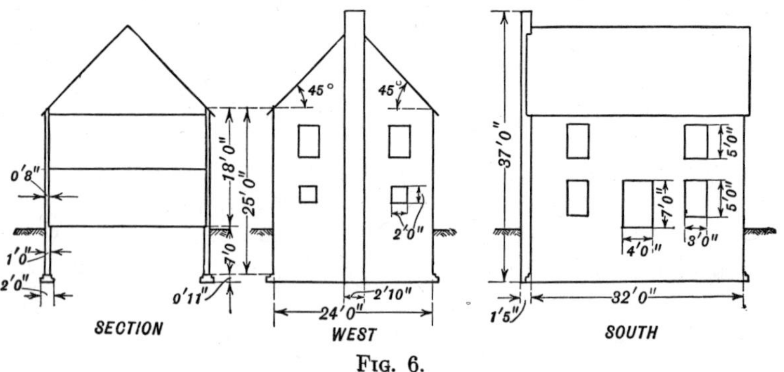

FIG. 6.

to center of each wall by subtracting 6 inches or $\frac{1}{2}$ foot from each end. Then the length of the footing for the south side of the house is $32 - \frac{1}{2} - \frac{1}{2} = 31$ feet. Similarly, the length of the west footing is 23 feet. The total length of footings all around the house is then $31 + 31 + 23 + 23 = 108$ feet. Referring to Table 3 we note that the quantities for footings are given in terms of 100 feet. In this example we have $\frac{108}{100} = 1.08$ hundreds of feet. The quantities given in the table for footings for a 12-inch wall are 2812 bricks and 48 cubic feet of mortar. We multiply these two figures by 1.08 and obtain

$2812 \times 1.08 = 3037$ bricks for the footings

$48 \times 1.08 = 51.8$ cubic feet of mortar for the footings.

TABLE 3

QUANTITIES OF BRICK AND MORTAR IN FOOTINGS, PIERS, AND CHIMNEYS*

Construction	Number of Brick	Mortar Cu. Ft.
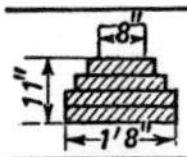 8" Wall	2272	39
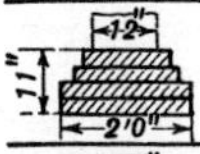 12" Wall	2812	48
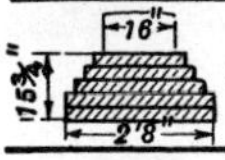 16" Wall	4592	78
PIERS—Quantities for 10 Ft. Height		
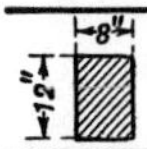 8" x 12" Solid	124	2¼
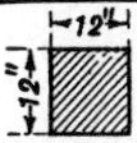 12" x 12" Solid	185	3¼
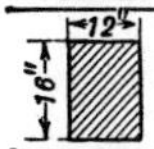 12" x 16" Solid	247	4½
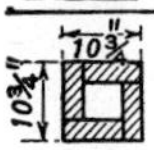 10¾" x 10¾" Hollow Brick Laid on Edge	113	1
CHIMNEYS—Quantities for 10 Ft. Height		
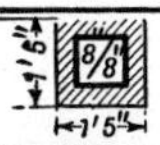 8" x 8" Flue	259	4½
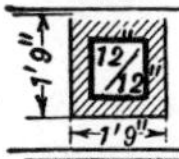 12" x 12" Flue	345	6
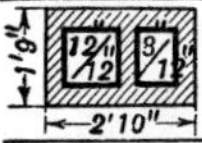 12" x 12" and 8" x 8" Flues	539	8½
 8" x 8" Flue	173	3
 12" x 12" Flue	238	4
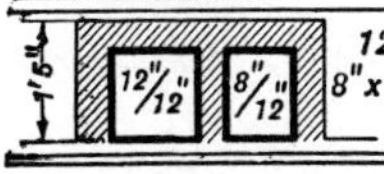 12" x 12" and 8" x 12" Flues	367	6½

*Quantities are for **100-foot** lengths of footing.

TABLE 4.

HEIGHT OF SOLID AND IDEAL BRICKWORK BY COURSES

Based on Standard Brick 2¼″×3¾″×8″

Height from Bottom of Mortar Joint to Bottom of Mortar Joint

Number of Courses	⅜″ Joints		½″ Joints		⅝″ Joints		¾″ Joints	Number of Courses
	Brick flat	Brick on edge	Brick flat	Brick on edge	Brick flat	Brick on edge	Brick flat	
1	2⅝″	4⅛″	2¾″	4¼″	2⅞″	4⅜″	3″	1
2	5¼″	8¼″	5½″	8½″	5¾″	8¾″	6″	2
3	7⅞″	1′— 0⅜″	8¼″	1′— 0¾″	8⅝″	1′— 1⅛″	9″	3
4	10½″	1′— 4½″	11″	1′— 5″	11½″	1′— 5½″	1′—0″	4
5	1′— 1⅛″	1′— 8⅝″	1′— 1¾″	1′— 9¼″	1′— 2⅜″	1′— 9⅞″	1′—3″	5
6	1′— 3¾″	2′— 0¾″	1′— 4½″	2′— 1½″	1′— 5¼″	2′— 2¼″	1′—6″	6
7	1′— 6⅜″	2′— 4⅞″	1′— 7¼″	2′— 5¾″	1′— 8⅛″	2′— 6⅝″	1′—9″	7
8	1′— 9″	2′— 9″	1′—10″	2′—10″	1′—11″	2′—11″	2′—0″	8
9	1′—11⅝″	3′— 1⅛″	2′— 0¾″	3′— 2¼″	2′— 1⅞″	3′— 3⅜″	2′—3″	9
10	2′— 2¼″	3′— 5¼″	2′— 3½″	3′— 6½″	2′— 4¾″	3′— 7¾″	2′—6″	10
11	2′— 4⅞″	3′— 9⅜″	2′— 6¼″	3′—10¾″	2′— 7⅝″	4′— 0⅛″	2′—9″	11
12	2′— 7½″	4′— 1½″	2′— 9″	4′— 3″	2′—10½″	4′— 4½″	3′—0″	12
13	2′—10⅛″	4′— 5⅝″	2′—11¾″	4′— 7¼″	3′— 1⅜″	4′— 8⅞″	3′—3″	13
14	3′— 0¾″	4′— 9¾″	3′— 2½″	4′—11½″	3′— 4¼″	5′— 1¼″	3′—6″	14
15	3′— 3⅜″	5′— 1⅞″	3′— 5¼″	5′— 3¾″	3′— 7⅛″	5′— 5⅝″	3′—9″	15
16	3′— 6″	5′— 6″	3′— 8″	5′— 8″	3′—10″	5′—10″	4′—0″	16
17	3′— 8⅝″	5′—10⅛″	3′—10¾″	6′— 0¼″	4′— 0⅞″	6′— 2⅜″	4′—3″	17
18	3′—11¼″	6′— 2¼″	4′— 1½″	6′— 4½″	4′— 3¾″	6′— 6¾″	4′—6″	18
19	4′— 1⅞″	6′— 6⅜″	4′— 4¼″	6′— 8¾″	4′— 6⅝″	6′—11⅛″	4′—9″	19
20	4′— 4½″	6′—10½″	4′— 7″	7′— 1″	4′— 9½″	7′— 3½″	5′—0″	20
21	4′— 7⅛″	7′— 2⅝″	4′— 9¾″	7′— 5¼″	5′— 0⅜″	7′— 7⅞″	5′—3″	21
22	4′— 9¾″	7′— 6¾″	5′— 0½″	7′— 9½″	5′— 3¼″	8′— 0¼″	5′—6″	22
23	5′— 0⅜″	7′—10⅞″	5′— 3¼″	8′— 1¾″	5′— 6⅛″	8′— 4⅝″	5′—9″	23

24	5′— 3″	8′— 3″	5′— 6″	8′— 6″	5′— 9″	8′— 9″	6′—0″	24
25	5′— 5⅝″	8′— 7⅛″	5′— 8¾″	8′—10¼″	5′—11⅞″	9′— 1⅜″	6′—3″	25
26	5′— 8¼″	8′—11¼″	5′—11½″	9′— 2½″	6′— 2¾″	9′— 5¾″	6′—6″	26
27	5′—10⅞″	9′— 3⅜″	6′— 2¼″	9′— 6¾″	6′— 5⅝″	9′—10⅛″	6′—9″	27
28	6′— 1½″	9′— 7½″	6′— 5″	9′—11″	6′— 8½″	10′— 2½″	7′—0″	28
29	6′— 4⅛″	9′—11⅝″	6′— 7¾″	10′— 3¼″	6′—11⅜″	10′— 6⅞″	7′—3″	29
30	6′— 6¾″	10′— 3¾″	6′—10½″	10′— 7½″	7′— 2¼″	10′—11¼″	7′—6″	30
31	6′— 9⅜″	10′— 7⅞″	7′— 1¼″	10′—11¾″	7′— 5⅛″	11′— 3⅝″	7′—9″	31
32	7′— 0″	11′— 0″	7′— 4″	11′— 4″	7′— 8″	11′— 8″	8′—0″	32
33	7′— 2⅝″	11′— 4⅛″	7′— 6¾″	11′— 8¼″	7′—10⅞″	12′— 0⅜″	8′—3″	33
34	7′— 5¼″	11′— 8¼″	7′— 9½″	12′— 0½″	8′— 1¾″	12′— 4¾″	8′—6″	34
35	7′— 7⅞″	12′— 0⅜″	8′— 0¼″	12′— 4¾″	8′— 4⅝″	12′— 9⅛″	8′—9″	35
36	7′—10½″	12′— 4½″	8′— 3″	12′— 9″	8′— 7½″	13′— 1½″	9′—0″	36
37	8′— 1⅛″	12′— 8⅝″	8′— 5¾″	13′— 1¼″	8′—10⅜″	13′— 5⅞″	9′—3″	37
38	8′— 3¾″	13′— 0¾″	8′— 8½″	13′— 5½″	9′— 1¼″	13′—10¼″	9′—6″	38
39	8′— 6⅜″	13′— 4⅞″	8′—11¼″	13′— 9¾″	9′— 4⅛″	14′— 2⅝″	9′—9″	39
40	8′— 9″	13′— 9″	9′— 2″	14′— 2″	9′— 7″	14′— 7″	10′—0″	40
41	8′—11⅝″	14′— 1⅛″	9′— 4¾″	14′— 6¼″	9′— 9⅞″	14′—11⅜″	10′—3″	41
42	9′— 2¼″	14′— 5¼″	9′— 7½″	14′—10½″	10′— 0¾″	15′— 3¾″	10′—6″	42
43	9′— 4⅞″	14′— 9⅜″	9′—10¼″	15′— 2¾″	10′— 3⅝″	15′— 8⅛″	10′—9″	43
44	9′— 7½″	15′— 1½″	10′— 1″	15′— 7″	10′— 6½″	16′— 0½″	11′—0″	44
45	9′—10⅛″	15′— 5⅝″	10′— 3¾″	15′—11¼″	10′— 9⅜″	16′— 4⅞″	11′—3″	45
46	10′— 0¾″	15′— 9¾″	10′— 6½″	16′— 3½″	11′— 0¼″	16′— 9¼″	11′—6″	46
47	10′— 3⅜″	16′— 1⅞″	10′— 9¼″	16′— 7¾″	11′— 3⅛″	17′— 1⅝″	11′—9″	47
48	10′— 6″	16′— 6″	11′— 0″	17′— 0″	11′— 6″	17′— 6″	12′—0″	48
49	10′— 8⅝″	16′—10⅛″	11′— 2¾″	17′— 4¼″	11′— 8⅞″	17′—10⅜″	12′—3″	49
50	10′—11¼″	17′— 2¼″	11′— 5½″	17′— 8½″	11′—11¾″	18′— 2¾″	12′—6″	50
60	13′— 1½″	20′— 7½″	13′—10″	21′— 3″	14′— 4½″	21′—10½″	15′—0″	60
70	15′— 3¾″	24′— 0¾″	16′— 0½″	24′— 9½″	16′— 9¼″	25′— 6¼″	17′—6″	70
80	17′— 6″	27′— 6″	18′— 4″	28′— 4″	19′— 2″	29′— 2″	20′—0″	80
90	19′— 8¼″	30′—11¼″	20′— 7½″	31′—10½″	21′— 6¾″	32′— 9¾″	22′—6″	90
100	21′—10½″	34′— 4½″	22′—11″	35′— 5″	23′—11½″	36′— 5½″	25′—0″	100

The lengths of the basement walls may also be regarded as running from center to center of each wall. The total length of the basement wall is, then, also 108 feet. The height of the basement wall is 7 feet. The area of this wall is then $108 \times 7 = 756$ square feet. Referring to Table 6 we find that for a 12-inch wall the quantities are,

for 700 sq. ft. = 12,937 bricks, 220 cu. ft. mortar
for 50 sq. ft. = 925 bricks, 16 cu. ft. mortar
for 6 sq. ft. = 6×1 = 111 bricks, 2 cu. ft. mortar

or a total for the basement wall of 13,973 bricks and 238 cubic feet mortar.

Turning our attention now to 8-inch walls, we subtract only 4 inches or one-third foot from each end to obtain the lengths of the sides from center to center of walls. The south wall is then $32 - \frac{1}{3} - \frac{1}{3} = 31\frac{1}{3}$ feet long. The height of this wall is 18 feet. The area is $31\frac{1}{3} \times 18 = 564$ square feet. From this figure must be subtracted the openings. There are four windows 3 feet by 5 feet and a door 4 feet by 7 feet. The total area of the window space is

$$3 \times 5 \times 4 = 60 \text{ sq. ft.}$$

$$\text{Area of door} = \quad 4 \times 7 = 28 \text{ sq. ft.}$$

$$\text{Total area } 88 \text{ sq. ft.}$$

The net area of the south wall is then $564 - 88 = 476$ square feet. The north wall has a like area.

The west and east 8-inch walls may be considered as consisting of a rectangle $23\frac{1}{3}$ feet by 18 feet in size and a 45-degree right triangle whose hypotenuse is $23\frac{1}{3}$ feet long. The area of the rectangle is $23\frac{1}{3} \times 18 = 420$ square feet. The areas of the window openings to be subtracted are

$$3 \times 5 \times 2 = 30 \text{ sq. ft.}$$

and

$$2 \times 2 \times 2 = 8 \text{ sq. ft.}$$

$$\text{Total } 38 \text{ sq. ft.}$$

TABLE 5

Average Weight of Solid Brick Walls

Brick Assumed to Weigh 4½ lb. each. ½″ Joints Filled with Mortar

Weight in Pounds per Square Foot of Wall Area		
4″ Wall	8″ Wall	12″ Wall
36.782	78.808	115.414

Thus from Table 5 a 125.5 square foot wall, 8 inches deep, weighs 125.5 × 78.808 or 9890.4 pounds.

TABLE 6

Bricks and Mortar for Solid Walls in All Bonds

Half-Inch Joints—All Joints Filled with Mortar

Square Feet Area of Wall	4-Inch Wall		8-Inch Wall		12-Inch Wall		16-Inch Wall		Square Feet Area of Wall
	Number of bricks	Cubic feet of mortar	Number of bricks	Cubic feet of mortar	Number of bricks	Cubic feet of mortar	Number of bricks	Cubic feet of mortar	
1	6.160	0.075	12.320	0.195	18.481	0.314	24.641	0.433	1
10	62	1	124	2	185	3½	247	4½	10
20	124	2	247	4	370	6½	493	9	20
30	185	2½	370	6	555	9½	740	13	30
40	247	3½	493	8	740	13	986	17½	40
50	309	4	617	10	925	16	1,233	22	50
60	370	5	740	12	1,109	19	1,479	26	60
70	432	5½	863	14	1,294	22	1,725	31	70
80	493	6½	986	16	1,479	25	1,972	35	80
90	555	7	1,109	18	1,664	28	2,218	39	90
100	617	8	1,233	20	1,849	32	2,465	44	100
200	1,233	15	2,465	39	3,697	63	4,929	87	200
300	1,849	23	3,697	59	5,545	94	7,393	130	300
400	2,465	30	4,929	78	7,393	126	9,857	173	400
500	3,081	38	6,161	98	9,241	157	12,321	217	500
600	3,697	46	7,393	117	11,089	189	14,786	260	600
700	4,313	53	8,625	137	12,937	220	17,250	303	700
800	4,929	61	9,857	156	14,786	251	19,714	347	800
900	5,545	68	11,089	175	16,634	283	22,178	390	900
1,000	6,161	76	12,321	195	18,482	314	24,642	433	1,000
2,000	12,321	151	24,642	390	36,963	628	49,284	866	2,000
3,000	18,482	227	36,963	584	55,444	942	73,926	1299	3,000
4,000	24,642	302	49,284	779	73,926	1255	98,567	1732	4,000
5,000	30,803	377	61,605	973	92,407	1569	123,209	2165	5,000
6,000	36,963	453	73,926	1168	110,888	1883	147,851	2599	6,000
7,000	43,124	528	86,247	1363	129,370	2197	172,493	3032	7,000
8,000	49,284	604	98,567	1557	147,851	2511	197,124	3465	8,000
9,000	55,444	679	110,888	1752	166,332	2825	221,776	3898	9,000
10,000	61,605	755	123,209	1947	184,813	3139	246,418	4331	10,000

TABLE 7

Quantities of Materials Required in Mortar

Cubic Feet of Mortar	Lime Mortar								Cubic Feet Mortar
	1 : 2½				1 : 3				
	180 lb barrels lump lime	or	50 lb sacks hydrated lime	Cubic yards sand	180 lb barrels lump lime	or	50 lb sacks hydrated lime	Cubic yard sand	
1	0.057	or	0.350	0.037	0.1	or	0.3	0.037	1
2	.1	or	.7	.1	.1	or	.6	.1	2
3	.2	or	1.1	.1	.1	or	.9	.1	3
4	.2	or	1.4	.1	.2	or	1.2	.1	4
5	.3	or	1.8	.2	.2	or	1.5	.2	5
6	.3	or	2.1	.2	.3	or	1.8	.2	6
7	.4	or	2.5	.3	.3	or	2.0	.3	7
8	.5	or	2.8	.3	.4	or	2.3	.3	8
9	.5	or	3.2	.3	.4	or	2.6	.3	9
10	.6	or	3.5	.4	.5	or	2.9	.4	10
11	.7	or	3.9	.4	.5	or	3.2	.4	11
12	.7	or	4.2	.4	.6	or	3.5	.4	12
13	.7	or	4.6	.5	.6	or	3.8	.5	13
14	.8	or	4.9	.5	.7	or	4.1	.5	14
15	.9	or	5.3	.6	.7	or	4.4	.6	15
16	.9	or	5.6	.6	.8	or	4.7	.6	16
17	1.0	or	6.0	.6	.8	or	5.0	.6	17
18	1.0	or	6.3	.7	.9	or	5.3	.7	18
19	1.1	or	6.7	.7	.9	or	5.5	.7	19
20	1.1	or	7.0	.7	.9	or	5.8	.7	20
27	1.5	or	9.5	1.0	1.3	or	7.8	1.0	27
30	1.7	or	10.5	1.1	1.4	or	8.7	1.1	30
40	2.3	or	14.0	1.5	1.9	or	11.7	1.5	40
50	2.8	or	17.5	1.9	2.4	or	14.6	1.9	50
60	3.4	or	21.0	2.2	2.8	or	17.5	2.2	60
70	3.9	or	24.5	2.6	3.3	or	20.4	2.6	70
80	4.6	or	28.0	3.0	3.8	or	23.3	3.0	80
90	5.1	or	31.5	3.3	4.3	or	26.3	3.3	90
100	6	or	35	4	5	or	29	4	100
200	11	or	70	7	9	or	58	7	200
300	17	or	105	11	14	or	88	11	300
400	23	or	140	15	19	or	117	15	400
500	29	or	175	19	24	or	146	19	500
600	34	or	210	22	28	or	175	22	600
700	40	or	245	26	33	or	204	26	700
800	46	or	280	30	38	or	233	30	800
900	51	or	315	33	43	or	263	33	900
1000	57	or	350	37	47	or	292	37	1000

Notes.—Quantities of lime are based on the use of good quality lime. Lime quantities are approximate and will vary with the grade of lime and the size of particles composing the sand. In the cement mortars, ¹⁄₁₀ of the cement by weight is replaced by dry hydrated lime or its equivalent in lump lime paste.

TABLE 7—*Continued*

Cubic Feet Mortar	Cement-Lime Mortar 1 : 1 : 6					Cement Mortar 1 : 2					Cubic Feet Mortar
	94 lbs net sacks cement	180 lb barrels lump lime	or	50 lb sacks hydrated lime	Cubic yards sand	94 lb net sacks cement	180 lb barrels lump lime	or	50 lb sacks hydrated lime	Cubic yards sand	
1	0.18	0.023	or	0.145	0.037	0.4	0.1	or	0.1	0.1	1
2	.4	.1	or	.3	.1	.9	.1	or	.2	.1	2
3	.5	.1	or	.4	.1	1.3	.1	or	.3	.1	3
4	.7	.1	or	.6	.1	1.8	.1	or	.4	.1	4
5	.9	.1	or	.7	.2	2.2	.1	or	.5	.2	5
6	1.1	.1	or	.9	.2	2.7	.1	or	.6	.2	6
7	1.3	.2	or	1.0	.3	3.1	.1	or	.6	.2	7
8	1.4	.2	or	1.2	.3	3.5	.1	or	.7	.3	8
9	1.6	.2	or	1.3	.3	3.9	.1	or	.8	.3	9
10	1.8	.2	or	1.5	.4	4.4	.2	or	.9	.3	10
11	2.0	.3	or	1.6	.4	4.9	.2	or	1.0	.4	11
12	2.2	.3	or	1.7	.4	5.3	.2	or	1.1	.4	12
13	2.3	.3	or	1.9	.5	5.7	.2	or	1.2	.4	13
14	2.5	.3	or	2.0	.5	6.2	.2	or	1.3	.5	14
15	2.7	.4	or	2.2	.6	6.6	.2	or	1.4	.5	15
16	2.9	.4	or	2.3	.6	7.1	.3	or	1.5	.5	16
17	3.1	.4	or	2.5	.6	7.5	.3	or	1.6	.6	17
18	3.2	.4	or	2.6	.7	8.0	.3	or	1.7	.6	18
19	3.4	.5	or	2.8	.7	8.4	.3	or	1.8	.6	19
20	3.6	.5	or	2.9	.7	8.9	.3	or	1.8	.7	20
27	4.1	.64	or	3.94	1.0	12.0	.4	or	2.5	.9	27
30	5.4	.7	or	4.4	1.1	13.3	.5	or	2.8	1.0	30
40	7.2	.9	or	5.8	1.5	17.7	.7	or	3.7	1.3	40
50	9.0	1.2	or	7.3	1.9	22.1	.8	or	4.6	1.7	50
60	10.8	1.4	or	8.8	2.2	26.6	1.0	or	5.5	2.1	60
70	12.6	1.7	or	10.2	2.6	31.0	1.1	or	6.5	2.4	70
80	14.4	1.9	or	11.7	3.0	35.4	1.3	or	7.4	2.8	80
90	16.2	2.1	or	13.1	3.3	39.8	1.5	or	8.3	3.1	90
100	18	2	or	15	4	44	2	or	9	3	100
200	36	5	or	29	7	88	3	or	18	7	200
300	54	7	or	44	11	132	5	or	28	10	300
400	72	10	or	58	15	177	7	or	37	14	400
500	90	12	or	73	19	221	8	or	46	17	500
600	108	14	or	88	22	265	10	or	55	21	600
700	126	17	or	102	26	310	11	or	65	24	700
800	144	19	or	117	30	354	13	or	74	28	800
900	162	21	or	131	33	398	15	or	83	31	900
1000	1 0	24	or	1[illegible]6	37	442	16	or	92	34	1000

NOTES.—Quantities of lime are based on the use of good quality lime. Lime quantities are approximate and will vary with the grade of lime and the size of particles composing the sand. In the cement mortars, 1/10 of the cement by weight is replaced by dry hydrated lime or its equivalent in lump lime paste.

TABLE 7—*Concluded*

Cubic Feet Mortar	Cement Mortar										Cubic Feet Mortar
	1 : 3					1 : 4					
	94 lb net sacks cement	180 lb barrels lump lime	or	50 lb sacks hydrated lime	Cubic yards sand	94 lb net sacks cement	180 lb barrels lump lime	or	50 lb sacks hydrated lime	Cubic yards sand	
1	0.3	0.1	or	0.1	0.1	0.3	0.1	or	0.1	0.1	1
2	.7	.1	or	.1	.1	.5	.1	or	.1	.1	2
3	1.0	.1	or	.2	.1	.8	.1	or	.2	.1	3
4	1.3	.1	or	.3	.2	1.1	.1	or	.2	.2	4
5	1.7	.1	or	.3	.2	1.3	.1	or	.3	.2	5
6	2.0	.1	or	.4	.2	1.6	.1	or	.3	.2	6
7	2.3	.1	or	.5	.3	1.8	.1	or	.4	.3	7
8	2.6	.1	or	.6	.3	2.1	.1	or	.4	.3	8
9	3.0	.1	or	.6	.3	2.4	.1	or	.5	.4	9
10	3.3	.1	or	.7	.4	2.6	.1	or	.6	.4	10
11	3.6	.1	or	.8	.4	2.9	.1	or	.6	.5	11
12	4.0	.1	or	.8	.5	3.2	.1	or	.7	.5	12
13	4.3	.2	or	.9	.5	3.4	.1	or	.7	.5	13
14	4.6	.2	or	.9	.5	3.7	.1	or	.8	.6	14
15	5.0	.2	or	1.0	.6	4.0	.2	or	.8	.6	15
16	5.3	.2	or	1.1	.6	4.2	.2	or	.9	.7	16
17	5.6	.2	or	1.2	.7	4.6	.2	or	.9	.7	17
18	6.0	.2	or	1.2	.7	4.8	.2	or	1.0	.7	18
19	6.3	.2	or	1.3	.7	5.0	.2	or	1.0	.8	19
20	6.6	.2	or	1.4	.8	5.3	.2	or	1.1	.8	20
27	8.9	.3	or	1.9	1.1	7.1	.3	or	1.5	1.1	27
30	9.9	.4	or	2.1	1.2	7.9	.3	or	1.7	1.2	30
40	13.2	.5	or	2.8	1.6	10.6	.4	or	2.2	1.6	40
50	16.5	.6	or	3.5	1.9	13.2	.5	or	2.8	2.1	50
60	19.8	.7	or	4.1	2.3	15.8	.6	or	3.3	2.5	60
70	23.1	.9	or	4.8	2.7	18.5	.7	or	3.9	2.9	70
80	26.4	1.0	or	5.5	3.1	21.1	.8	or	4.4	3.3	80
90	29.8	1.1	or	6.2	3.5	23.8	.9	or	5.0	3.7	90
100	33	1.2	or	7	4	26	1	or	6	4	100
200	66	2	or	14	8	53	2	or	11	8	200
300	99	4	or	21	12	79	3	or	17	12	300
400	132	5	or	28	16	106	4	or	22	16	400
500	165	6	or	35	19	132	5	or	28	21	500
600	198	7	or	41	23	158	6	or	33	25	600
700	231	9	or	48	27	184	7	or	39	29	700
800	265	10	or	55	31	211	8	or	44	33	800
900	298	11	or	62	35	238	9	or	50	37	900
1000	331	12	or	69	39	264	10	or	55	41	1000

NOTES.—Quantities of lime are based on the use of good quality lime. Lime quantities are approximate and will vary with the grade of lime and the size of particles composing the sand. In the cement mortars, 1/10 of the cement by weight is replaced by dry hydrated lime or its equivalent in lump lime paste.

This makes the net area of the rectangle 420 − 38 = 382 square feet.

In a 45-degree right triangle the sides bear a relation to the hypotenuse of $1 : \sqrt{2}$. In the case of the wall we know the hypotenuse to be $23\frac{1}{3}$ or 23.33 feet. Letting L represent the sloping side, we may set up the proportion,

$$\frac{L}{23.33} = \frac{1}{\sqrt{2}}$$

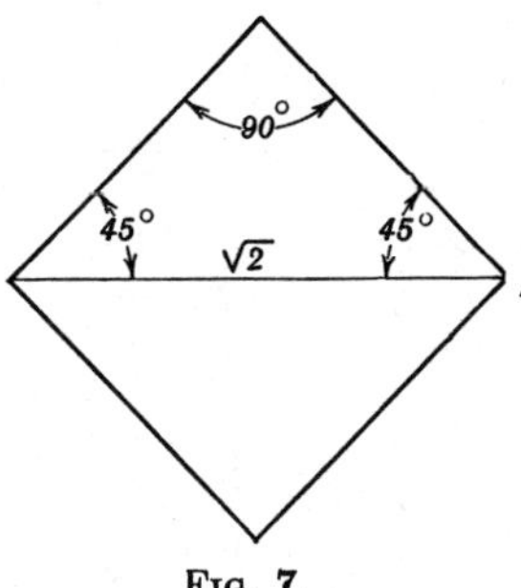

Fig. 7.

Transposing,

$$L = \frac{23.33}{\sqrt{2}} = \frac{23.33}{1.414} = 16.50 \text{ ft}$$

The area of the triangle is

$$\frac{16.5 \times 16.5}{2} = 136 \text{ sq ft}$$

This makes the area of each end wall 136 + 382 = 518 sq. ft.

We are now ready to add the areas of all the 8-inch walls:

South wall	=	476 sq. ft.
North wall	=	476 sq. ft.
East wall	=	518 sq. ft.
West wall	=	518 sq. ft.
Total area		1988 sq. ft.

Turning again to Table 6, but this time to the column for the 8-inch wall, we obtain the following quantities:

for 1000 sq. ft. =	12,321 bricks	195	cu. ft. mortar
900 sq. ft. =	11,089 bricks	175	cu. ft. mortar
80 sq. ft. =	986 bricks	16	cu. ft. mortar
8 sq. ft. =	[illegible] bricks	1.6	cu. ft. mortar

The totals for 8-inch wall are 24,495 bricks; 387.6 cu. ft. mortar.

This leaves only the chimney to be estimated. It will be noted that Table 4 gives the quantities for chimneys by 10-foot heights. Since the chimney in this problem is 37 feet high it is 3.7 10-foot lengths. The unit quantities given in the table for a chimney 1 foot 5 inches by 2 feet 10 inches in cross-section are 367 bricks and 6½ cubic feet mortar. Multiplying these quantities by 3.7 we obtain,

$$367 \times 3.7 = 1358 \text{ bricks for chimney}$$
$$6.5 \times 3.7 = 24.0 \text{ cu. ft. of mortar for chimney}$$

It remains only to make a recapitulation of all the quantities.

	Bricks	Mortar
Footings............	3,037	51.8
12-in walls..........	13,973	238.0
8-in walls...........	24,495	387.6
Chimney............	1,358	24.0
Totals...	42,863	701.4

With the quantity of mortar known, the amount of cement, lime, and sand required can readily be found from Table 7.

Concrete Blocks.—Portland cement concrete blocks are a comparatively cheap and easily handled building material. Because of the ease with which they may be manufactured, the blocks are available everywhere. Practically every building code in the country permits the use of Portland cement blocks, although some codes restrict their use to small buildings. A standard block is usually 8″ wide, 8″ high, and 16″ long. Half blocks, corners, and other shapes may be made to meet any need. The aggregate may be sand or steam cinders. The cinder block has several advantages such as lighter weight, porous surface for plaster, and excellent insulating qualities.

The standard concrete block produces a wall 8″ thick, in 8″ courses when laid in a single thickness. For 100 sq. ft. of wall area, one hundred and ten 8″ × 8″ × 16″ blocks are required. Concrete building tile are smaller and lighter than concrete blocks.

They are usually 12″ long and 8″ wide, but their height varies. The 3½″ and 5″ heights are the most common. Two hundred and twenty 5″, or three hundred, 3½″ tiles are required for 100 sq. ft. of wall area. Table 8 gives a complete tabulation of materials including mortar for joints, for estimating quantities of material per hundred square feet of wall area. It should be remembered that the table does not include wall opening such as doors and windows. These should be deducted before estimating the net wall area.

Reinforcement.—The increasing use of concrete block construction has extended not only to installations of wider and higher walls, but also to more severe external conditions (e.g., in hurricane and earthquake areas), and more troublesome structural conditions (e.g., settlement of footings, distortions due to expansion of floors and roofs). This situation has led to greater

TABLE 8

DATA ON VARIOUS CONCRETE MASONRY WALLS PER 100 SQ. FT. OF WALL AREA

Concrete Block

Description		Wall Thickness	Weight per Unit (lb.)	Number of Units (per 100 sq. ft. of wall area)	Mortar* (cu. ft.)	Weight, lb. (per 100 sq. ft. of wall area)
8″× 8″×16″	Hollow Partition Block	8″	50	110	3.25	5850
8″× 8″×12″		8″	38	146	3.50	6000
8″×12″×16″		12″	85	110	3.25	9700
8″× 3″×16″		3″	20	110	2.75	2600
9″× 3″×18″		3″	26	87	2.50	2500
12″× 3″×12″		3″	23	100	2.50	2550
8″× 3″×12″		3″	15	146	3.50	2550
8″× 4″×16″		4″	28	110	3.25	3450
9″× 4″×18″		4″	35	87	3.25	3350
12″× 4″×12″		4″	31	100	3.25	3450
8″× 4″×12″		4″	21	146	4.00	3500
8″× 6″×16″		6″	42	110	3.25	5000

* These figures are based on ⅜-inch mortar joints, 25% wastage included. Weight of mortar assumed at 103 pounds per cubic foot. For ½-inch mortar joints, use one-fourth more mortar than specified in the table.

TABLE 8—*Continued*

Concrete Tile

Description	Wall Thickness	Weight per Unit (lb.)	Number of Units (per 100 sq. ft. of wall area)	Mortar* (cu. ft.)	Weight, lb. (per 100 sq. ft. of wall area)
5″×8″×12″	8″	19.9	220	5.00	4900
5″×4″×12″	4″	9.9	220	5.00	2700
5″×6″×12″	6″	14.9	220	4.00	3800
3½″×8″×12″	8″	16.5	300	6.00	5550
3½″×4″×12″	4″	8.5	300	5.00	3050
3½″×6″×12″	6″	12.5	300	5.50	4300

Light Weight Concrete Block

Description	Wall Thickness	Weight per Unit (lb.)	Number of Units (per 100 sq. ft. of wall area)	Mortar* (cu. ft.)	Weight, lb. (per 100 sq. ft. of wall area)
8″× 8″×16″	8″	27–32	110	3.25	3300–3850
8″× 8″×12″	8″	21–24	146	3.50	3400–3850
8″×12″×16″	12″	46–54	110	3.25	5400–6300
8″× 3″×16″	3″	11–13	110	2.75	1500–1700
9″× 3″×18″	3″	14–17	87	2.50	1450–1700
12″× 3″×12″	3″	12–15	100	2.50	1450–1750
8″× 3″×12″	3″	8–10	146	3.50	1500–1800
8″× 4″×16″	4″	15–18	110	3.25	2000–2350
9″× 4″×18″	4″	19–22	87	3.25	2000–2250
12″× 4″×12″	4″	17–20	100	3.25	2050–2350
8″× 4″×12″	4″	11–13	146	4.00	2000–2300
8″× 6″×16″	6″	23–27	110	3.25	2900–3300

Light Weight Concrete Tile

Description	Wall Thickness	Weight per Unit (lb.)	Number of Units (per 100 sq. ft. of wall area)	Mortar* (cu. ft.)	Weight, lb. (per 100 sq. ft. of wall area)
5″×8″×12″	8″	10.8–12.7	220	5.00	2900–3300
5″×4″×12″	4″	5.4– 6.3	220	5.00	1700–1950
5″×6″×12″	6″	8.1– 9.5	220	4.00	2200–2500
3½″×8″×12″	8″	8.9–10.6	300	6.00	3300–3800
3½″×4″×12″	4″	4.6– 5.4	300	5.50	2000–2200
3½″×6″×12″	6″	6.8– 8.0	300	5.50	2600–3000

Concrete Brick

Description	Wall Thickness	Weight per Unit (lb.)	Number of Units (per 100 sq. ft. of wall area)	Mortar* (cu. ft.)	Weight, lb. (per 100 sq. ft. of wall area)
2¼″×3¾″×8″	8″	5	1300	18.10	8350

* These figures are based on ⅜-inch mortar joints, 25% wastage included. Weight of mortar assumed at 103 pounds per cubic foot. For ½-inch mortar joints, use one-fourth more mortar than specified in the table

use of reinforcing in concrete block construction. While the extent to which such reinforcing is necessary has no general answer, it is considered especially desirable in cases where plaster is to be applied directly to concrete block.

Therefore there is greater use of both rods and shaped sections for reinforcement. Fig. 8 shows an installation of vertical stabilization in a Vibrapac® pilaster block of the Besser Company, and Fig. 9 shows their unit designed to control the volume

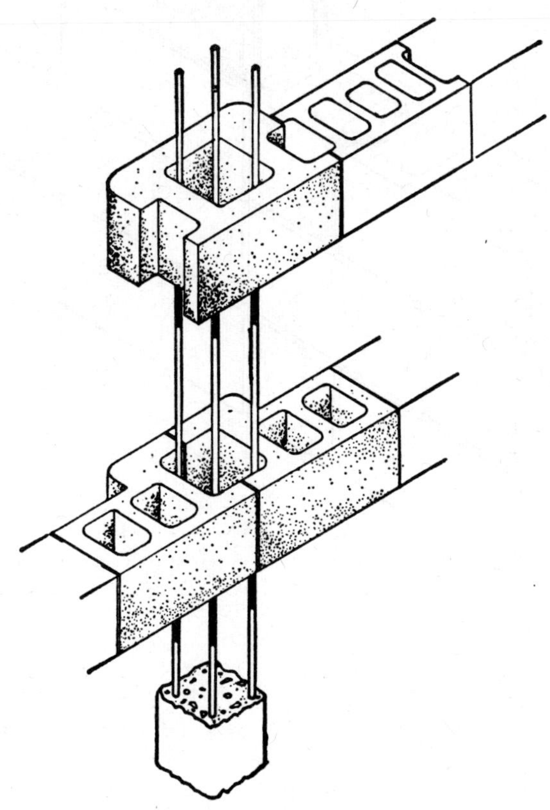

Fig. 8.

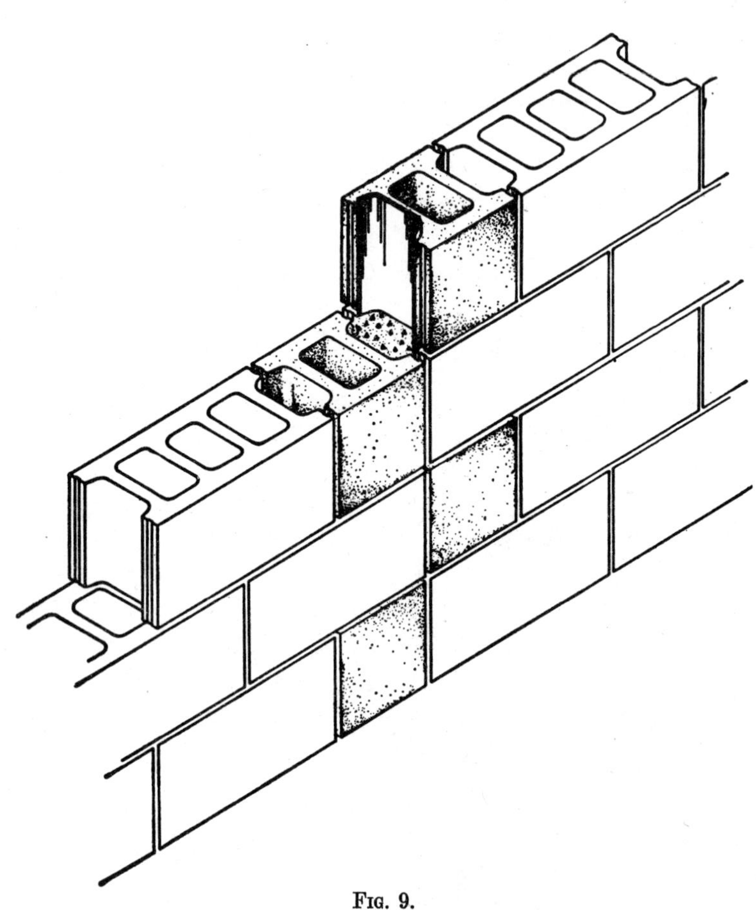

Fig. 9.

changes of a wall—this block being placed at specific intervals to relieve the volume changes in the wall. Figure 10 shows their block designed to permit construction of a 8″ to 16″ deep continuous reinforced concrete bond around entire building. This arrangement can be combined with vertical reinforced concrete column to attain masonry walls of great strength.

Figure 11 shows various installations of the Dur-O-Wal® reinforcement of the Dur-O-Wal Company. Fig. 11(*a*) shows single-side rods 16″ center to center with cross rods; Fig. 11(*b*) shows a similar installation with double side rods; while Fig. 11(*c*) shows an 8″ wall with a pilaster and control joint; and 11(*d*) shows a corner. Figure 12 shows the design of a unit of Dur-O-Wal reinforcing.

Painting Block Walls.—An attractive waterproof finish for exterior walls consists of two coats of Portland cement base paint, applied as follows:

Uniformly dampen wall and apply paint, mixed to creamy consistency. Scrub into surface with a fiber brush with 1½″ to 2″ bristles. Dampen painted surfaces the morning following application with a fog spray. Apply second coat as above to the dampened wall not less than two days after application of first coat. Dampen second coat with fog spray the morning after application.

Interior Walls.—While plaster is often applied directly to cement block walls, the opinion is quite general that, unless the wall is heavily reinforced, cracking is likely to occur. This situation is not nearly so likely if more porous blocks of cinders or slag are used, such as the Waylite® blocks, made by the Waylite Company. These blocks, having larger air spaces, are more effective insulators. Even for them, however, a metal lath installation is recommended if the wall is to be plastered. This is feasible for these blocks, because nails can be driven directly into them. Of course a wallboard interior surface provides the necessary insulating dead air space for all these types of building blocks.

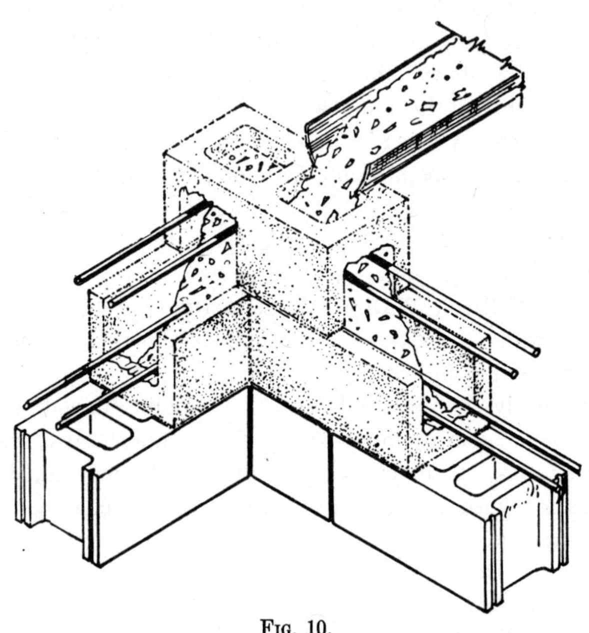

Fig. 10.

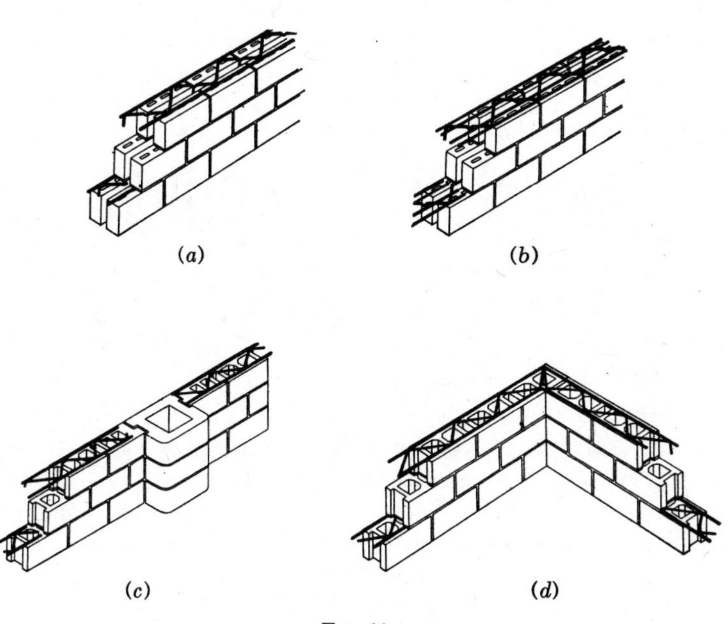

(a) (b)

(c) (d)

Fig. 11.

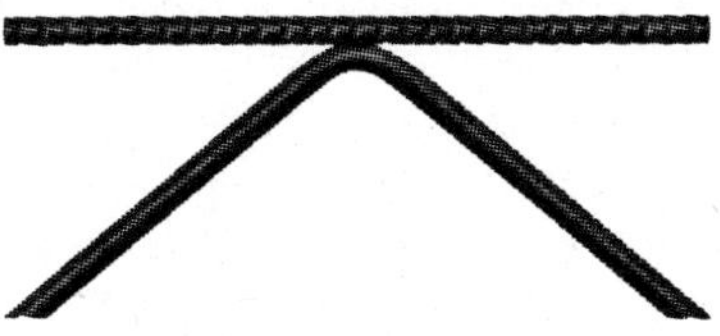

Fig. 12.

XII

Carpentry and Building

Carpentry finds its greatest expression in house building, and, although it is one of the oldest of the arts, its basic principles have changed but little with the passing of time. A man may build a house long or short, high or low, square or circular as his fancy dictates. For this reason practically every house built has its series of individual problems. Not all of these are fully solved when the building plans or working drawings reach the building foreman or the estimator. He must be able to interpret the plans in their proper light and independently find the missing information by computation or estimation.

The object of this section is to show how these computations are made and how estimates of material are arrived at. After a general discussion of board measure, it takes up house framing and surface covering, including walls, floors and roofs, all of which may be classified as *rough carpentry.* Interior trim and millwork, which may be called *finish carpentry,* does not present many problems in which mathematics is helpful.

Board Measure.—Lumber is measured in terms of *board feet,* abbreviated *fbm* (feet board measure). A board 12 inches wide, 12 inches long, and one inch thick contains one board foot of lumber. Similarly, a board 6 inches wide, 24 inches long, and 1 inch thick is also one board foot. The rule for determining the number of board feet in a piece of lumber may then be stated: *Multiply the length in feet by the width in inches and the thickness in inches* ($\frac{1}{2}$ *inch or over*) *and divide the product by twelve.* Stated as a formula, this is,

TABLE 1

Board Measure

Size	Length in feet							
	12	14	16	18	20	22	24	26
	Square feet							
1 × 8	8	9⅓	10⅔	12	13⅓	14⅔	16	17⅓
1 × 10	10	11⅔	13⅔	15	16⅔	18⅓	20	21⅔
1 × 12	12	14	16	18	20	22	24	26
1 × 14	14	16⅓	18⅔	21	23⅓	25⅔	28	30⅓
1 × 16	16	18⅔	21⅓	24	26⅔	29⅓	32	34⅔
2 × 3	6	7	8	9	10	11	12	13
2 × 4	8	9⅓	10⅔	12	13⅓	14⅔	16	17⅓
2 × 6	12	14	16	18	20	22	24	26
2 × 8	16	18⅔	21⅓	24	26⅔	29⅓	32	34⅔
2 × 10	20	23⅓	26⅔	30	33⅓	36⅔	40	43⅓
2 × 12	24	28	32	36	40	44	48	52
2 × 14	28	32⅔	37⅓	42	46⅔	51⅓	56	60⅔
2 × 16	32	37⅓	42⅔	48	53⅓	58⅔	64	69⅓
3 × 4	12	14	16	18	20	22	24	26
3 × 6	18	21	24	27	30	33	36	39
3 × 8	24	28	32	36	40	44	48	52
3 × 10	30	35	40	45	50	55	60	65
3 × 12	36	42	48	54	60	66	72	78
3 × 14	42	49	56	63	70	77	84	91
3 × 16	48	56	64	72	80	88	96	104
4 × 4	16	18⅔	21⅓	24	26⅔	29⅓	32	34⅔
4 × 6	24	28	32	36	40	44	48	52
4 × 8	32	37⅓	42⅔	48	53⅓	58⅔	64	69⅓
4 × 10	40	46⅔	53⅓	60	66⅔	73⅓	80	86⅔
4 × 12	48	56	64	72	80	88	96	104
4 × 14	56	65⅓	74⅔	84	93⅓	102⅔	112	121⅓
4 × 16	64	74⅔	85⅓	96	106⅔	117⅓	128	138⅔
6 × 6	36	42	48	54	60	66	72	78
6 × 8	48	56	64	72	80	88	96	104
6 × 10	60	70	80	90	100	110	120	130
6 × 12	72	84	96	108	120	132	144	156
6 × 14	84	98	112	126	140	154	168	182
6 × 16	96	112	128	144	160	176	192	208
8 × 8	64	74⅔	85⅓	96	106⅔	117⅓	128	138⅔
8 × 10	80	93⅓	106⅔	120	133⅓	146⅔	160	173⅓
8 × 12	96	112	128	144	160	176	192	208
8 × 14	112	130⅔	149⅓	168	186⅔	205⅓	224	242⅔
8 × 16	128	149⅓	170⅔	192	213⅓	234⅔	256	277⅓
10 × 10	100	116⅔	133⅓	150	166⅔	183⅓	200	216⅔
10 × 12	120	140	160	180	200	220	240	260
10 × 14	140	163⅓	186⅔	210	233⅓	256⅔	280	303⅓
10 × 16	160	186⅔	213⅓	240	266⅔	293⅓	320	346⅔
12 × 12	144	168	192	216	240	264	288	312
12 × 14	168	196	224	252	280	308	336	364
12 × 16	192	224	256	288	320	352	384	416
14 × 14	196	228⅔	261⅓	294	326⅔	359⅓	392	424⅔
14 × 16	224	261⅓	298⅔	336	373⅓	410⅔	448	485⅓
16 × 16	256	298⅔	341⅓	384	426⅔	469⅓	512	554⅔

TABLE 1

Board Measure — (*Continued*)

Size	Length in feet						
	28	30	32	34	36	38	40
	Square feet						
1×8	18⅔	20	21⅓	22⅔	24	25⅓	26⅔
1×10	23⅓	25	26⅔	28⅓	30	31⅔	33⅓
1×12	28	30	32	34	36	38	40
1×14	32⅔	35	37⅓	39⅔	42	44⅓	46⅔
1×16	37⅓	40	42⅔	45⅓	48	50⅔	53⅓
2×3	14	15	16	17	18	19	20
2×4	18⅔	20	21⅓	22⅔	24	25⅓	26⅔
2×6	28	30	32	34	36	38	40
2×8	37⅓	40	42⅔	45⅓	48	50⅔	53⅓
2×10	46⅔	50	53⅓	56⅔	60	63⅓	66⅔
2×12	56	60	64	68	72	76	80
2×14	65⅓	70	72⅔	79⅓	84	88⅔	93⅓
2×16	74⅔	80	85⅓	90⅔	96	101⅓	106⅔
3×4	28	30	32	34	36	38	40
3×6	42	45	48	51	54	57	60
3×8	56	60	64	68	72	76	80
3×10	70	75	80	85	90	95	100
3×12	84	90	96	102	108	114	120
3×14	98	105	112	119	126	133	140
3×16	112	120	128	136	144	152	160
4×4	37⅓	40	42⅔	45⅓	48	50⅔	53⅓
4×6	56	60	64	68	72	76	80
4×8	74⅔	80	85⅓	90⅔	96	101⅓	106⅔
4×10	93⅓	100	106⅔	113⅓	120	126⅔	133⅓
4×12	112	120	128	136	144	152	160
4×14	130⅔	140	149⅓	158⅔	168	177⅓	186⅔
4×16	149⅓	160	170⅔	181⅓	192	202⅔	213⅓
6×6	84	90	96	102	108	114	120
6×8	112	120	128	136	144	152	160
6×10	140	150	160	170	180	190	200
6×12	168	180	192	204	216	228	240
6×14	196	210	224	238	252	266	280
6×16	224	240	256	272	288	304	320
8×8	149⅓	160	170⅔	181⅓	192	202⅔	213⅓
8×10	186⅔	200	213⅓	226⅔	240	253⅓	266⅔
8×12	224	240	256	272	288	304	320
8×14	261⅓	280	298⅔	317⅓	336	354⅔	373⅓
8×16	298⅔	320	341⅓	362⅔	384	405⅓	426⅔
10×10	233⅓	250	266⅔	283⅓	300	316⅔	333⅓
10×12	280	300	320	340	360	380	400
10×14	326⅔	350	373⅓	396⅔	410	443⅓	466⅔
10×16	373⅓	400	426⅔	453⅓	480	506⅔	533⅓
12×12	336	360	384	408	432	456	480
12×14	392	420	448	476	504	532	560
12×16	448	480	512	544	576	608	640
14×14	457⅓	490	522⅔	555⅓	588	620⅔	653⅓
14×16	522⅓	560	597⅓	634⅔	672	709⅓	746⅔
16×16	597⅓	640	682⅔	725⅓	768	810⅔	853⅓

Note. — By simply multiplying or dividing the above amounts, the number of feet contained in other dimensions can be obtained.

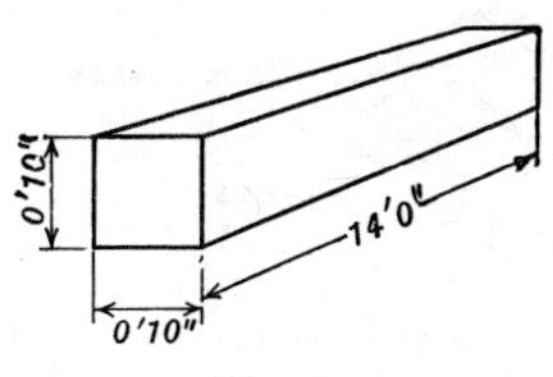

FIG. 1.

$$\text{fbm} = \frac{L \times w \times t}{12}$$

where L = length in feet;
w = width in inches;
t = thickness in inches.

ILLUSTRATION: What is the board measure of a timber 10 inches by 10 inches and 14 feet long?

$$\text{fbm} = \frac{L \times w \times t}{12} = \frac{14 \times 10 \times 10}{12} = 116.7 \text{ fbm} \quad \text{(Ans.)}$$

Lumber is measured on the basis of "rough stock." When lumber is "dressed" or planed, $\frac{1}{8}$ inch is taken off each side if the lumber is $1\frac{1}{2}$ inches or greater in thickness, and $\frac{1}{16}$ inch if the thickness is less than $1\frac{1}{2}$ inches. The purchaser pays, however, on the basis of its measurement before planing.

Thicknesses less than one inch are regarded as one inch in measuring lumber.

In measuring width of boards, fractions of an inch, one-half or greater are regarded as a whole inch, while fractions less than one-half inch are ignored. For example, a board $4\frac{1}{2}$ inches or $4\frac{3}{4}$ inches wide would be called 5 inches, while a board $4\frac{3}{8}$ inches wide would be measured as but 4 inches.

Building lumber is sold in standard lengths which are multiples of two feet from 10 to 24 feet, that is 10, 12, 14, etc. feet.

Lumber dealt with in large quantities is measured and sold by the thousand board feet (M fbm). Board feet are changed to thousand board feet by simply shifting the decimal point three places to the left. Thus, 28,500 fbm = 28.5 M fbm.

ILLUSTRATION: How many thousand board feet are there in 1200 pieces 2 inches × 4 inches and 18 feet long?

$$\text{fbm (one piece)} = \frac{L \times w \times t}{12} = \frac{18 \times 4 \times 2}{12} = 12 \text{ fbm}$$

$$1200 \times 12 = 14{,}400 \text{ fbm} = 14.4 \text{ M fbm} \quad \text{(Ans.)}$$

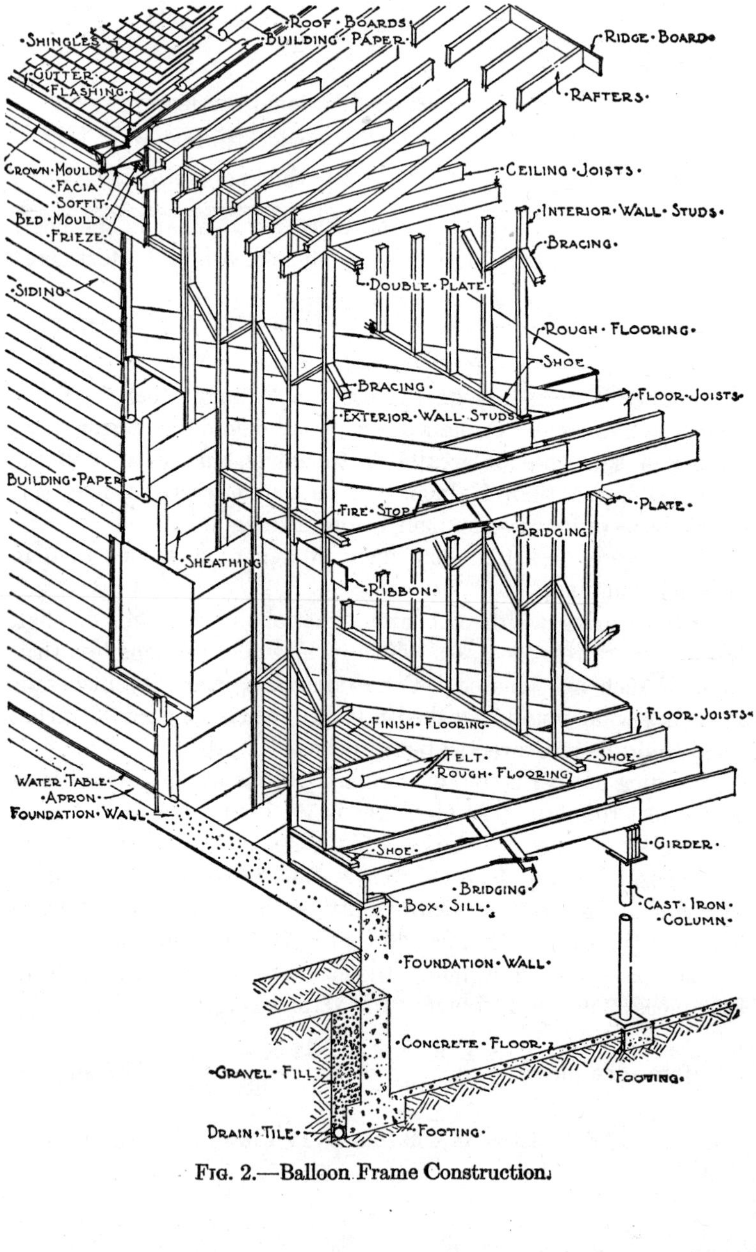

Fig. 2.—Balloon Frame Construction.

House Framing.—The details of frame dwelling construction have been so well standardized by building codes and convention that it is entirely feasible to make fairly accurate estimates of the quantities of material required from the general dimensions of the structure. In preparing orders for material for a building, it is well to bear in mind that the use of standard sizes is most economical and that a further saving is often effected by them in the elimination of unnecessary sawing and handling. When listing lumber, it is common practice to give the number of pieces first, then the width and thickness in inches and the length in feet or feet and inches. Thus, 24 pieces 2 × 4 in. by 16 ft 0 in.

This section will concern itself with a few typical details representing accepted standard practice. Figure 2 shows a corner of what is known as " balloon frame construction " and illustrates the terminology used in house framing and the general location of the various members. The following paragraphs will proceed to deal with the details separately.

Sills.—The first carpentry on a frame building usually begins after the completion of the foundation and consists of laying the sill. The sill may be either a solid timber, as a 4 in. by 6 in., or 4 in. by 8 in., or may be built up as from two 2 in. by 6 in., or 3 in. × 6 in. planks. The sill should be placed about an inch from the outer edge of the foundation, and should be bedded in mortar to secure even bearing and be securely anchored to the masonry. Joints at the corners are made by halving the sills as shown for both types in Fig. 3.

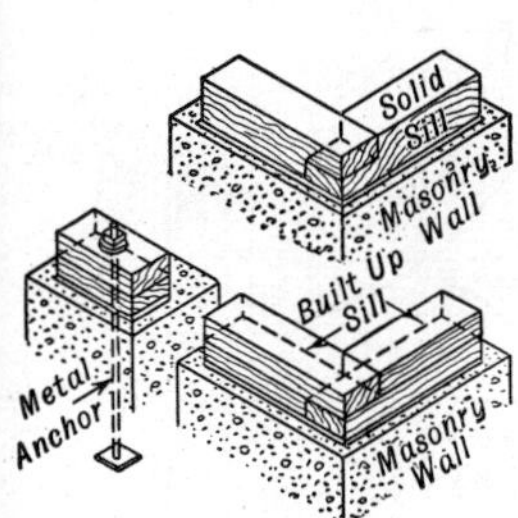

Fig. 3.

The length of sill required is, for practical purposes, the sum of the lengths of the outside walls, or the girth, plus an allowance of six inches in each length for splices. This will, of course, result in one-foot splices.

Illustration: If 4-inch by 6-inch timbers are to be used for the sills in the building shown in plan in Fig. 4, how many board

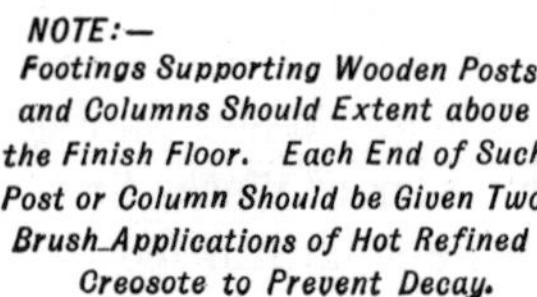

NOTE:— Footings Supporting Wooden Posts and Columns Should Extent above the Finish Floor. Each End of Such Post or Column Should be Given Two Brush Applications of Hot Refined Creosote to Prevent Decay.

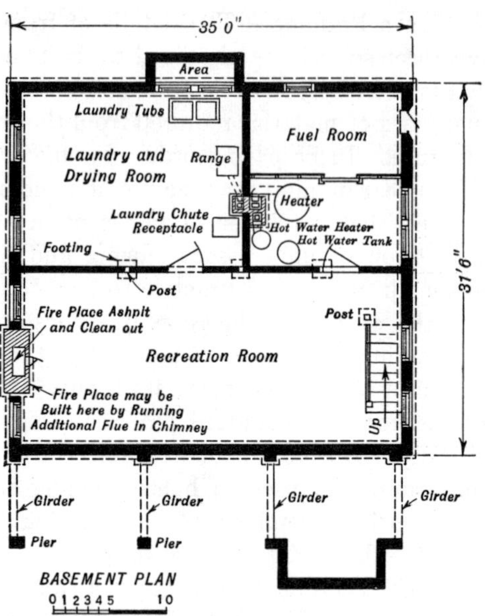

BASEMENT PLAN

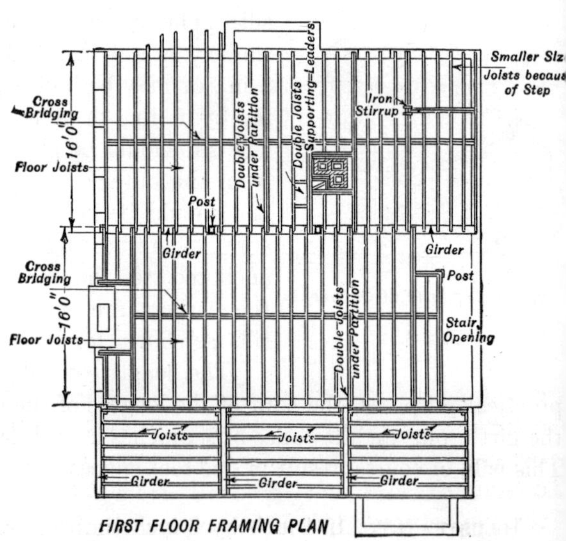

FIRST FLOOR FRAMING PLAN

Fig. 4.

feet will be required and what lengths of pieces can be used advantageously?

The girth of the building is, in round figures, 35 + 35 + 32 + 32 = 134 feet. One joint at about the middle of each wall will obviously be needed. This will add one-half foot to each of eight timbers, making the total length 134 + 4 = 138 feet. This allows nothing for waste and assumes that commercial lengths will fit.

Turning our attention to specific lengths of timbers needed for the house, we note that for the front and back, two timbers each $\frac{35}{2} + \frac{1}{2} = 18$ feet long will fit each of these walls without waste, or a total of 4 18-foot timbers. On the sides, $\frac{32}{2} + \frac{1}{2} = 16\frac{1}{2}$ feet, but the next larger commercial length is 18 feet. However, one 18-foot piece and one 16-foot piece will take care of each side nicely with a total waste of only about 2 feet. The bill of material for the sill would then read:

6 pieces 4 by 6 in. 18 ft long
2 pieces 4 by 6 in. 14 ft long } (Ans.)

The original estimate of 138 linear feet must now be revised by the addition of 2 feet to make a total of 140 feet. Converting this to board feet we obtain,

$$\text{fbm} = \frac{L \times w \times t}{12} = \frac{140 \times 6 \times 4}{12} = 280 \text{ fbm}$$

Floor Joists.—Floor joists form the support for the floor, as their name implies, and, in the case of those for the first floor, rest on edge on the sills. The joists may be anywhere from 2 in. by 6 in. to 3 in. by 14 in. in cross-section depending on the load, the span, and the extreme bending stress allowed by the building code for the kind and grade of lumber used. These factors also determine the spacing, which may be 12 in., 16 in., 20 in., or 24 in. center to center. Sixteen-inch spacing is the most common in dwelling construction because it conveniently connects up with the favored spacing of studding.

In the case of narrow buildings, joists span the entire width and rest on the side sills. In larger buildings where the span would be too great the joists have one end resting on wall sills and the other on girders supported by columns as shown in Fig. 4.

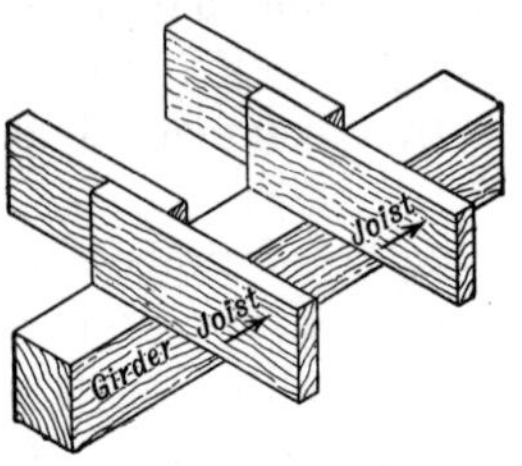

Joists Lapped on Top of Girder

Fig. 5.

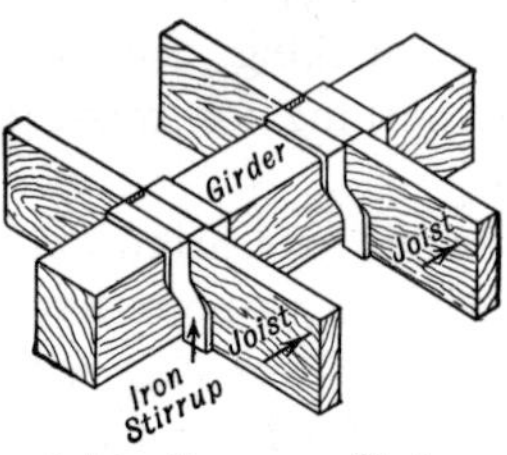

Joists Hung on Girder with Iron Stirrups

Fig. 6.

When sufficient basement headroom is available, they can be made to lap over the girder as shown in Fig. 5. This makes for a minimum amount of sawing since it is not material how far the end of the joist extends beyond the bearing on the girder.

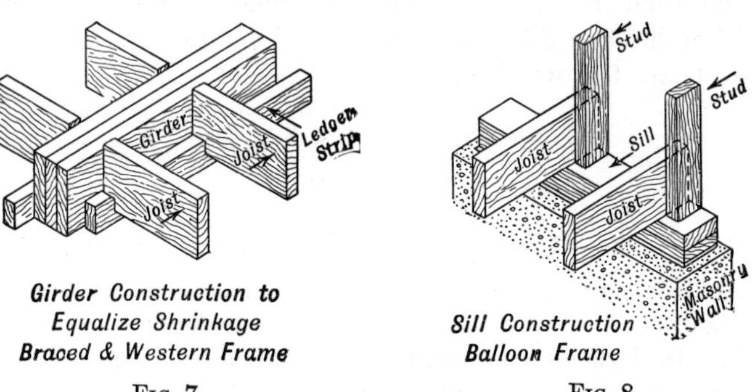

Girder Construction to Equalize Shrinkage Braced & Western Frame

Fig. 7.

Sill Construction Balloon Frame

Fig. 8.

Figures 6 and 7 show other girder connections which require less headroom.

At the wall bearing end, joists may either rest directly on the sills as shown in Fig. 8, or may be dapped a small amount as shown in Fig. 9 to bring their top surfaces to an absolutely level plane.

Under partitions and around floor openings, heavier members than the regular joists are required. These are called *trimmer beams*, but the required reinforcement is often accomplished by using double joists as shown in Fig. 4. The members around openings which are placed transverse to the direction of the joists are called *headers*, and these, too, are often made up of double joist timbers.

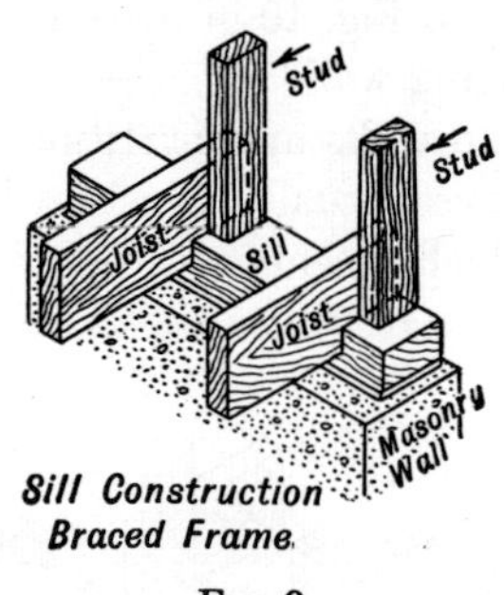

Sill Construction Braced Frame

Fig. 9.

When a joist spacing of twelve inches is used, the number of joists required will be equal to the length of the opening in feet plus one, plus one for each point at which the joists are doubled.

Illustration: A building with a floor space 17 feet wide and 60 feet long is to have joists spaced 12 inches center to center spanning the width. How many joists will be required for a floor if there are no floor openings but eight partitions to be supported?

Joists required = length in feet + 1 + number of partitions
Joists required = $60 + 1 + 8 = 69$ (Ans.)

The number of joists required when the spacing is 16 inches may be estimated by multiplying the distance of the opening across the joists in feet by $\frac{3}{4}$, adding 1 and adding further 1 for each doubling of joists.

Illustration: A floor 20 feet wide by 32 feet long is to have joists spaced 16 inches center to center transverse to the length of the house. How many joists will be required, if there are six points at which they must be doubled up?

Joists required = length in feet $\times \frac{3}{4} + 1 +$ no. of doublings
Joists required = $32 \times \frac{3}{4} + 1 + 6 = 24 + 7 = 31$ (Ans.)

Another method of estimating the number of floor joists is, of course, to count them from the plans. Thus in Fig. 4 it is an easy matter to determine that the equivalent of some 60 long joists will be required, with a slight addition for headers, and 24 short joists for the porch floor.

Floor joists are given lateral support by *cross bridging* consisting of $1\frac{1}{2}$ in. by 3 in. pieces nailed as shown in Fig. 10 in rows not more than 8 feet apart or from the supporting wall.

Studding.—The vertical members of the walls and partitions of a frame dwelling are called *studs*. These usually consist of 2 in. by 4 in. pieces of lumber spaced 16 inches center to center. In the outside walls they may be continuous from the sill to the

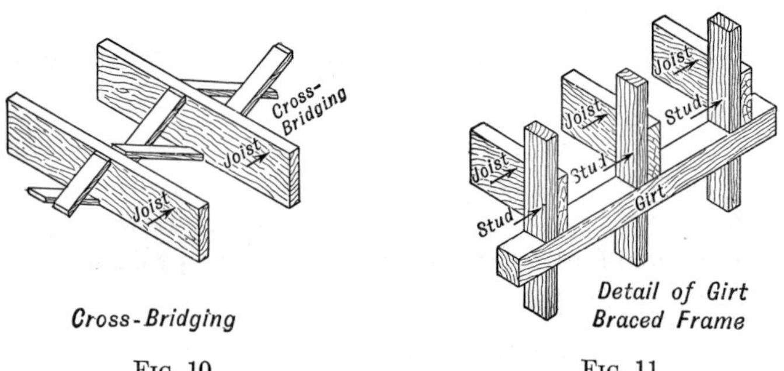

FIG. 10. FIG. 11.

roof plate as shown in Fig. 2, or they may terminate at the ceiling level and be capped by a plate or girt as shown in Fig. 11. Studding is doubled around openings and at corners although the construction at corners shown in Figs. 2 and 12 gives more convenient nailing surfaces for the lath. Studding is braced at the midpoint between floor and ceiling either by straight diagonal bridging or by herringbone bracing as shown in Fig. 2.

Studding spaced 16 inches center to center may be estimated by multiplying the lineal lengths of the walls and partitions by $\frac{3}{4}$ and adding one for each corner and opening. However, the more common and sufficiently accurate practice is to estimate one stud per lineal foot of walls and partitions, the surplus being sufficient for doubling at corners and openings.

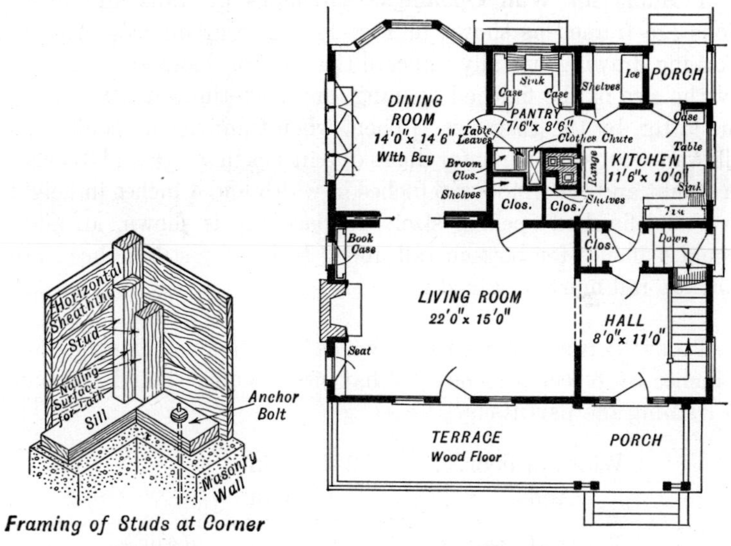

Framing of Studs at Corner

FIG. 12. FIG. 13.

ILLUSTRATION: The floor plan shown in Fig. 13 is the first floor plan of the same building as shown in Fig. 4. Estimate the approximate number of studs needed for the walls and partitions of this floor.

Length of outside walls = 35 + 35 + 32 + 32	=	134 feet
Center transverse partition..................	=	35 feet
Living room-hall partition..................	=	15 feet
Hall-stair partition = 8 + 6..............	=	14 feet
Dining room-pantry partition..............	=	14 feet
Pantry-kitchen partition..................	=	14 feet
Pantry-closet partition....................	=	8 feet
Total length of walls and partitions.......		234 feet

Then, if one stud is allowed for each lineal foot of wall and partition, the number required in this case will be 234. (Ans.)

Framing for Wall Openings.—Openings in walls and partitions are framed as shown in Fig. 14. The architect's plans or working drawings usually indicate the sizes of doors and windows by the size of the finished opening, and sometimes in the case of the latter, by the glass size. Then, when framing an opening, an allowance must be made for doors of 5 inches in width and 3 inches in height and for windows 6 inches in width and 4 inches in height over the finished opening size. If glass size is shown, an additional 4 inches for bottom rail and 2 inches for stiles, check rail and top rail must be added.

Illustration: Working drawings show door openings 2 feet 6 inches by 6 feet 6 inches. What size opening should be made in framing the partitions?

Width of door........	2 ft. 6 in.	
Add............	5 in.	
Width of opening.....	2 ft. 11 in.	(Ans.)
Height of door.......	6 ft. 6 in.	
Add............	3 in.	
Height of opening.....	6 ft. 9 in.	(Ans.)

Illustration: A working drawing shows a window opening 2 feet 4 inches by 4 feet 10 inches. What size opening should be provided in framing the wall?

Width of window....	2 ft. 4 in.	
Add............	6 in.	
Width of opening....	2 ft. 10 in.	
Height of window....	4 ft. 10 in.	
Add............	4 in.	
Height of opening....	5 ft. 2 in.	(Ans.)

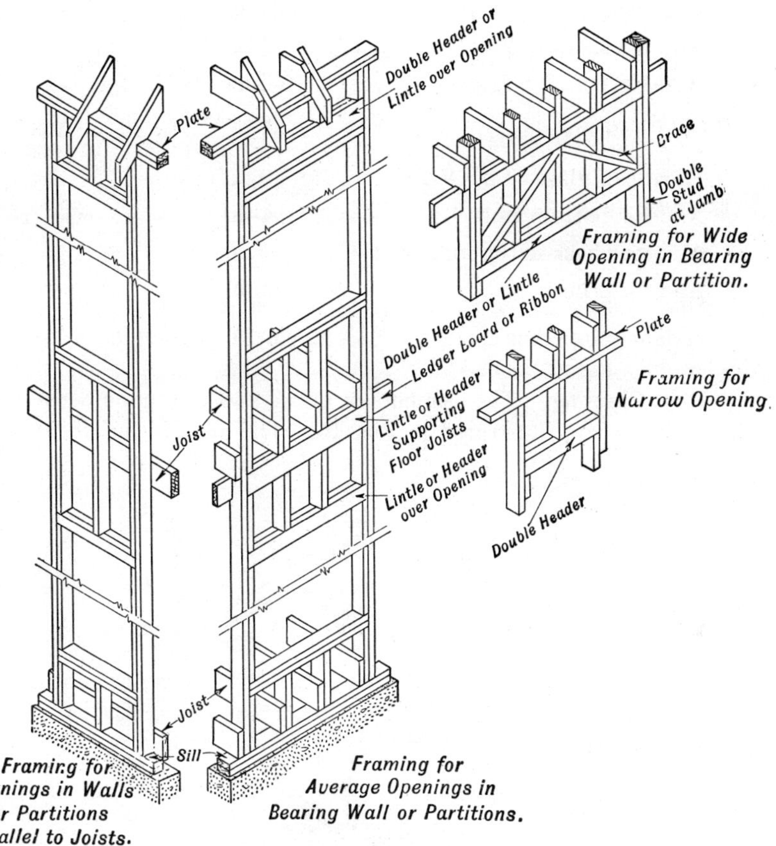

METHODS OF FRAMING AROUND OPENINGS IN WALLS AND PARTITIONS

FIG. 14.

ILLUSTRATION: Architect's drawings show a two-light window with glass sizes 24 inches by 20 inches. What size opening should be provided in framing the wall?

Glass width	24 in.	
Add for stiles = 2 + 2	4 in.	
Add for trim	6 in.	
Width of opening	34 in.	2 ft. 10 in. (Ans.)
Height of glass = 20 + 20	40 in.	
Add for bottom rail	4 in.	
Add for check & top rails = 2 + 2	4 in.	
Add for trim	4 in.	
Height of opening	52 in.	4 ft. 4 in. (Ans.)

Roof Framing.—The elements of a roof and the terms pertaining to them are illustrated in Fig. 15. The *span* is the distance

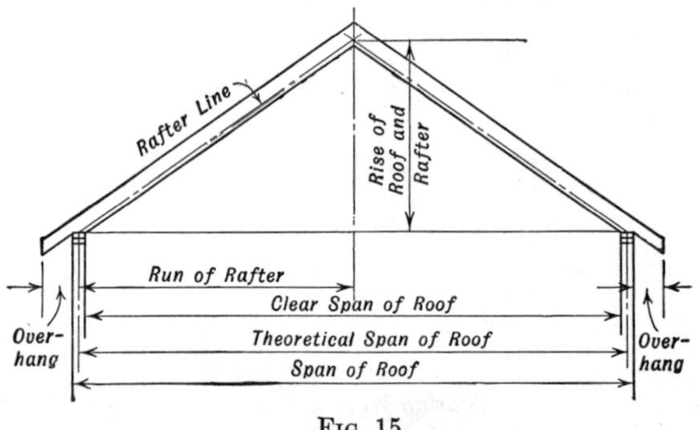

FIG. 15.

between the outer edges of the side walls supporting a roof. The *rise* is the vertical distance between the ridge and the plates supporting the roof. The *run* is the horizontal distance between the ridge and the outside edge of the plate supporting the roof.

The *pitch* of a roof is the slope of the rafters expressed as a ratio of the rise to the span. Thus, to find the pitch of a roof when the rise and span are given, merely substitute the known values in this equation,

$$\text{Pitch} = \frac{\text{rise}}{\text{span}}$$

ILLUSTRATION: What is the pitch of a roof whose rise is 6 feet and span 18 feet?

$$\text{Pitch} = \frac{\text{rise}}{\text{span}} = \frac{6}{18} = \frac{1}{3} \quad \text{(Ans.)}$$

To find the rise when the pitch and span are known, use the equation,

$$\text{Rise} = \text{pitch} \times \text{span}$$

ILLUSTRATION: What is the rise of a roof whose pitch is $\frac{2}{3}$ and span 24 feet?

$$\text{Rise} = \text{pitch} \times \text{span} = \tfrac{2}{3} \times 24 = 16$$

With these relationships in mind it is a simple matter to compute the length of the rafters by extracting the square root of the sum of the squares of the rise and the run, since these form a right triangle. Thus,

$$\text{Rafter length} = \sqrt{(\text{rise})^2 + (\text{run})^2}$$

The overhang for the eaves, if any, must then be added to this figure.

Another convenient method of determining the length of rafters is to let the inches on a steel square represent the rise and run in feet. Thus, in Fig. 16 the run of 20 feet is represented by 20 inches on the square and the rise of 10 feet is represented by 10 inches. Then the length of the diagonal in inches may be measured with a rule and this represents the length of the rafter in feet.

Flat Roof.—A flat roof or lean-to has but one pitch and is used widely on sheds, porches, dormers, etc. The slope is often just sufficient for drainage and the length of the rafters may be computed by either of the above methods.

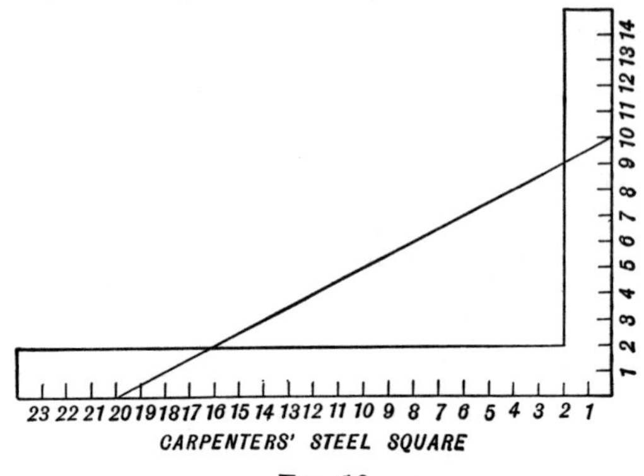

FIG. 16.

ILLUSTRATION: The roof shown in Fig. 17 has a rise of 18 inches and a run of 15 feet. How long must the rafters be if the overhang front and back is 8 inches?

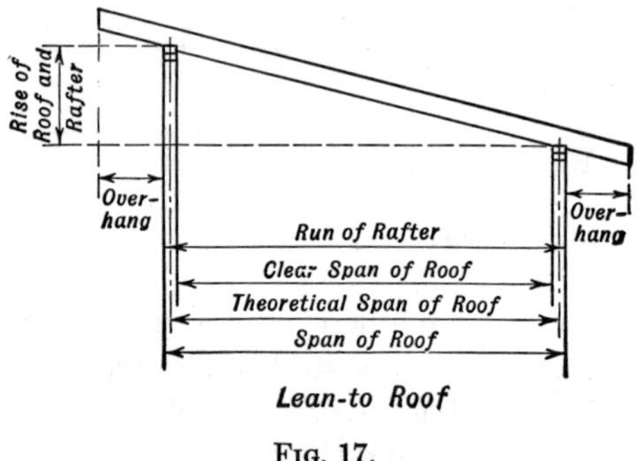

FIG. 17.

$$\text{Length of rafters} = \sqrt{(\text{rise})^2 + (\text{run})^2}$$

Length of rafters =

$$\sqrt{(1\tfrac{1}{2})^2 + (15)^2} = \sqrt{2.25 + 225} = \sqrt{227.25} = 15.075 \text{ ft}$$

Converting the decimal to inches and fractions of an inch by multiplying by 12 and referring to Table 1, page 19, we obtain a length of 15 ft. $0\frac{7}{8}$ in. To this must be added 16 inches for the overhangs.

	15 ft. $0\frac{7}{8}$ in.
	16 in.
Total length of rafters.......	16 ft. $4\frac{7}{8}$ in.

This problem illustrates that when the pitch is small, the length of the rafters will very nearly equal the run, so that in sheds and unimportant structures, where the exact amount of overhang is not of great concern, the overhang added to the run may be used for the length of the rafters. However, the calculation illustrated is important in the case of roofs of greater pitch and in dwelling construction.

Gable Roofs.—A gable roof has two sloping surfaces which meet at the ridge. Figure 18 shows an end view of such a roof.

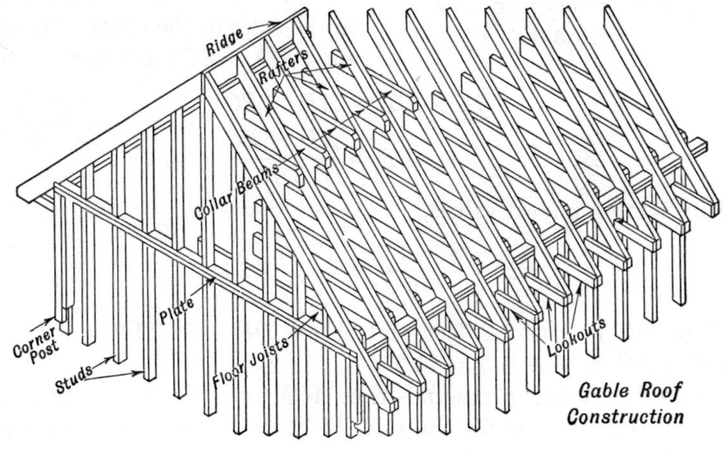

Fig. 18.

The length of the rafters is computed as for a flat roof except, of course, that an overhang occurs on only one end.

ILLUSTRATION: What is the length of the rafters of a roof which has a rise of 10 feet, a run of 12 feet and an overhang of 1 foot?

$$\text{Length of rafter} = \sqrt{(\text{rise})^2 + (\text{run})^2} = \sqrt{(10)^2 + (12)^2}$$
$$\sqrt{100 + 144} = \sqrt{244} = 15.62 \text{ ft.}$$

Changing the decimal 0.62 to inches by multiplying by 12, we obtain a length of 15 feet $7\frac{1}{2}$ inches to which must be added the overhang, making a total of 16 ft. $7\frac{1}{2}$ in. (Ans.)

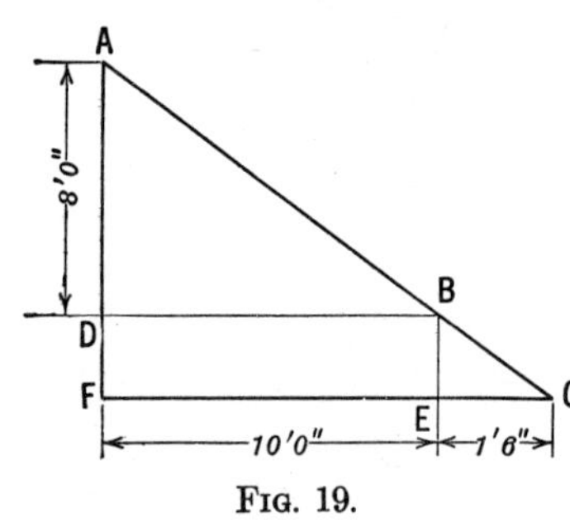

FIG. 19.

ILLUSTRATION: A roof has a rise of 8 feet and a run of 10 feet and eaves projecting a horizontal distance of 1 foot 6 inches. What is the length of the rafters?

From geometry we know that ABD and BCE (Fig. 19) are similar triangles and that therefore the sides of one are proportional to the sides of the other. Then, $BE : AD = EC : DB$ and

$$BE \times DB = AD \times EC$$
$$BE = \frac{AD \times EC}{DB}$$

Substituting known values,

$$BE = \frac{8 \times 1.5}{10} = \frac{12}{10} = 1.2 \text{ ft.}$$

Then DF is also 1.2 feet and $AD + DF = 8 + 1.2 = 9.2$ ft.; $FE + EC = 10 + 1.5 = 11.5$ ft.

We have then a new triangle, ACF, which can be solved in the regular manner for the side AC which represents the entire length of the rafter including overhang.

$$AC = \sqrt{(9.2)^2 + (11.5)^2} = \sqrt{84.64 + 132.25}$$
$$AC = \sqrt{216.89} = 14.727 = 14 \text{ ft. } 8\tfrac{3}{4} \text{ in.} \quad \text{(Ans.)}$$

Frequently two gable roofs will meet at right angles as shown in Fig. 20.

This construction calls for a *valley rafter* at the intersection of the roof surfaces. The valley rafter may be represented by the hypotenuse of a right triangle one of whose legs is the length of

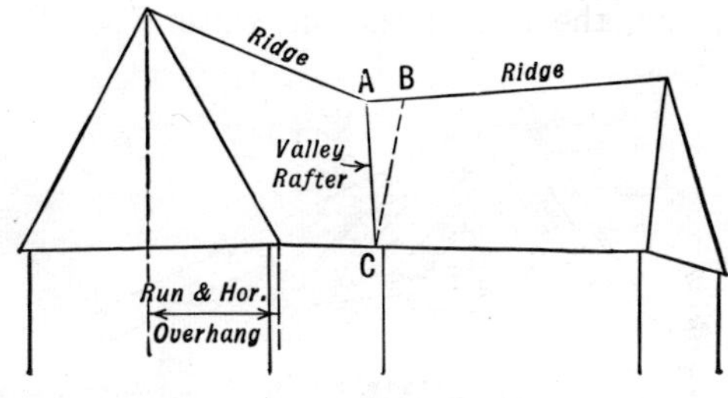

FIG. 20.

the common rafter BC, and the other leg the distance AB from the intersection of the ridges to a point in a plane with the extremeties of the rafters of the gable perpendicular to AB. The length of the valley rafter is then

$$AC = \sqrt{(AB)^2 + (BC)^2}$$

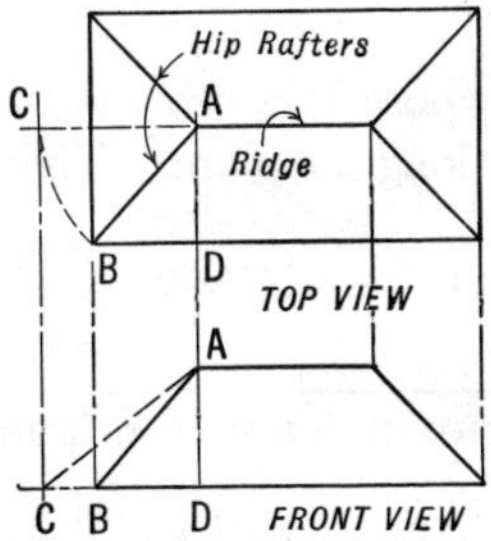

It will be noted that when two gables intersect at exactly right angles, the distance AB is equal to the run plus horizontal overhang of the intersecting gable.

Hip Roofs.—A hip roof has surfaces sloping toward all four walls as shown in Fig. 21. The only new problem which this involves is the caloulation of the length of the hip rafters.

FIG. 21.—Hipped Roof.

If a roof drawing is made to scale the length of the hip rafter can be found by scribing radius AB in Top View to point C on ridge center line. By dropping a vertical line to the line of plate in the Front View the actual length of hip rafter can be measured along AC.

The length of the hip rafter can be computed when the length of the common rafter AD and the distance BD are known. Then,

$$\text{Length of hip rafter} = \sqrt{(AD)^2 + (BD)^2}$$

When the pitches of the intersecting roof surfaces are equal, as they usually are, the run of the hip rafter (See Fig. 22) is the

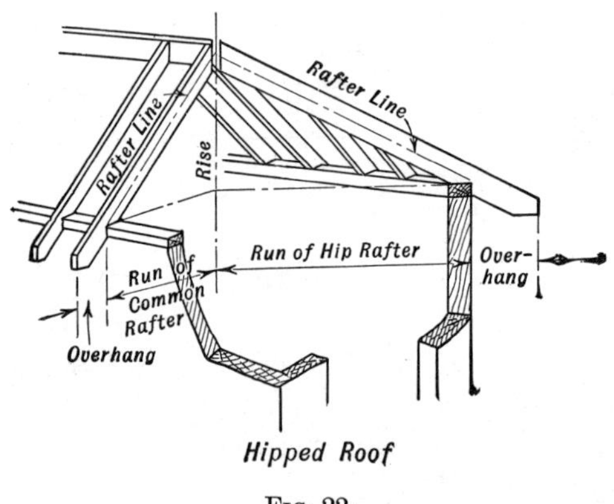

Fig. 22.

hypotenuse of the isosceles right triangle whose legs are the run of the common rafters and the distance BD along the plate. Then,

$$\text{Run of hip rafters} = \text{run of common rafters} \times \sqrt{2},$$

and,

$$\text{Length of hip rafters} = \sqrt{(\text{rise})^2 + 2(\text{run of common rafters})}$$

Illustration: A hip roof of equal pitch all around has a rise of 10 feet and a run of 14 feet. What is the length of the hip rafters?

$$\begin{aligned} \text{Length of hip rafters} &= \sqrt{(\text{rise})^2 + 2(\text{run})^2} \\ &= \sqrt{(10)^2 + 2(14)^2} = \sqrt{100 + 2 \times 196} \\ &= \sqrt{492} = 22.181 \text{ ft.} = 22 \text{ ft. } 2\tfrac{3}{16} \text{ in. (Ans.)} \end{aligned}$$

The length of the hip rafter can also be found without computation by scaling the distance AB on a plan or top view drawing, laying this distance off to a scale of 1 in. = 1 ft on one leg of a carpenters' square, as in Fig. 16, and laying the rise off on the other leg. Then the diagonal distance between these points is the scale length of the hip rafter.

Stair Construction.—The proportioning and construction of stairs present several nice problems of calculation. The elements of a stairway are shown in Fig. 23 and the details of framing in Fig. 24.

The ideal angle for a stairway is between 30 degrees and 35 degrees with the horizontal, although both steeper and flatter stairways are sometimes necessary. However, regardless of the angle of stair, a certain relationship between the rise and the run of each step must prevail. That is, the sum of the rise and the run shall not be less than 17 inches nor more than 18 inches. (It is to be noted that the run does not include the nosing.) Then, if a step has a rise of 7 inches, its run will be between 10 and 11 inches.

When the distance between two floors or the rise of the stair is known, and the approximate amount of the rise of each step has been determined, then the number of steps required may be found by dividing the rise of the stair by the rise of the step. If the quotient is not an even number, divide the rise of the stair by the nearest whole number of the quotient to obtain the exact rise of the step.

ILLUSTRATION: The distance between two floors is 12 feet 4 inches. How many steps will be required if the rise is to be about $7\frac{1}{4}$ inches?

$$12 \text{ ft. } 4 \text{ in.} = 148 \text{ in.} \qquad 148 \div 7.25 = 20.4$$

Then, since the quotient is not a whole number, divide the rise of the stair by 20. $148 \div 20 = 7.4$ or approximately $7\frac{13}{32}$.

The result shows that 20 steps each with a rise of $7\frac{13}{32}$ inches are required. (Ans.)

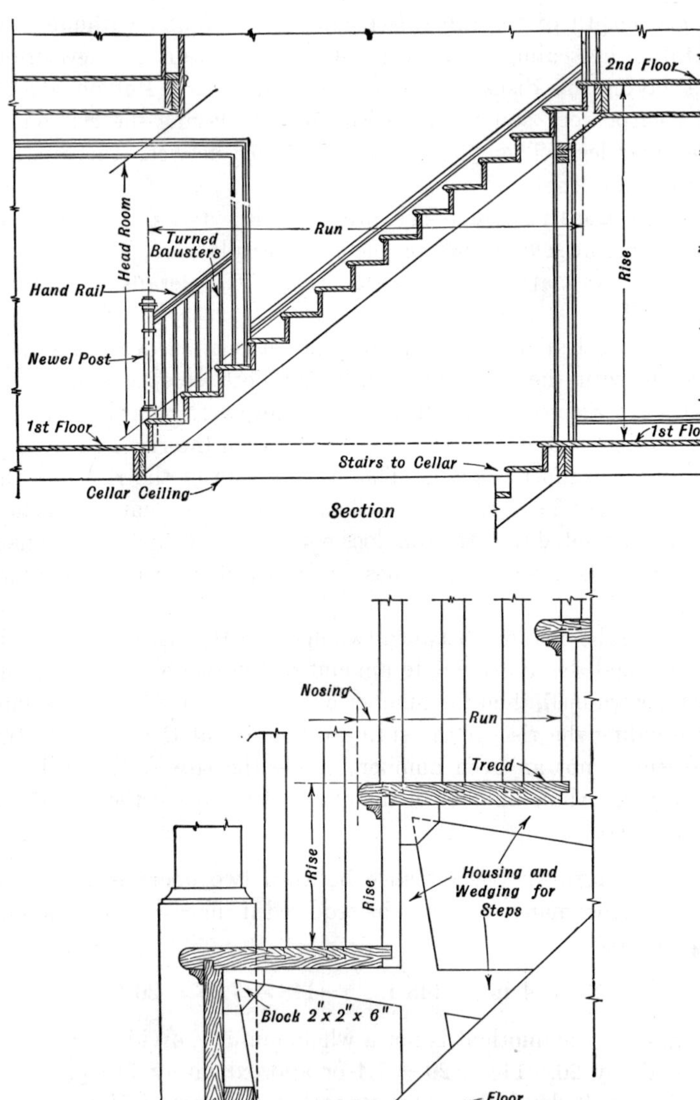

FIG. 23.—Stair Details.

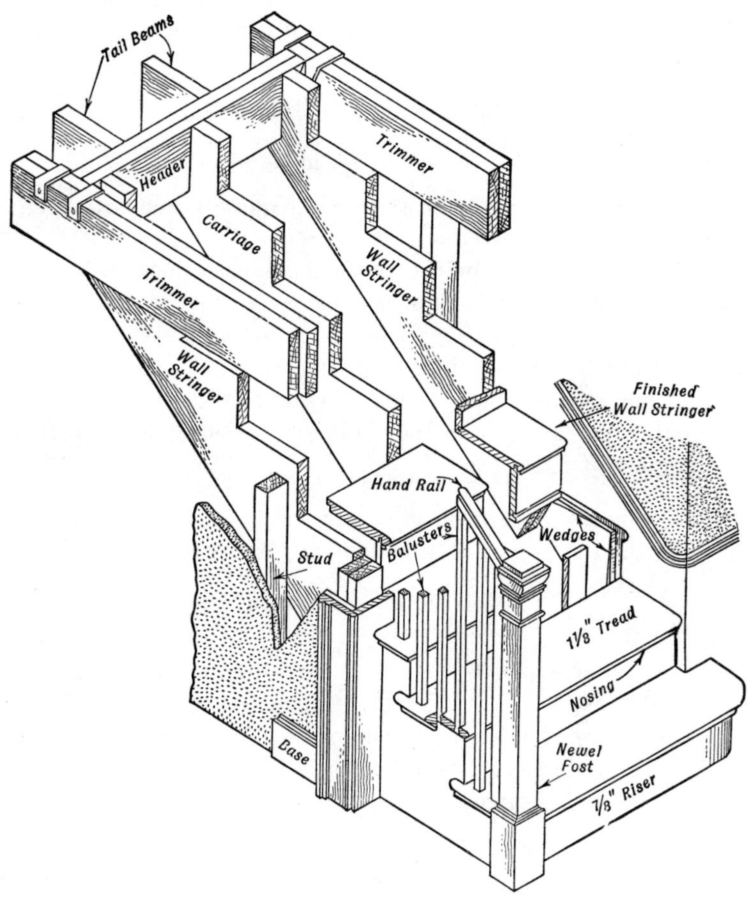

Front Elevation Frame of the Stairs

FIG. 24.

ILLUSTRATION: How many steps will be required between two floors with a difference in elevation of 9 feet 7 inches, if the rise is to be about 7 inches?

$$9 \text{ ft. } 7 \text{ in.} = 115 \text{ in.} \qquad \tfrac{115}{7} = 16.4$$

$$\tfrac{115}{16} = 7\tfrac{3}{16} \text{ in.}$$

The result shows that 16 steps are required, each with a rise of $7\frac{3}{16}$ inches. (Ans.)

The computations in the preceding illustrations instead of actually arriving at the number of steps, arrived at the number of risers. The top landing is not regarded as a step, and thus there is one less tread than riser in a stairway. Reference to Fig. 23 makes this clear. Then the width of the run of each step is equal to the total run of the stairway divided by one less than the number of risers.

ILLUSTRATION: The run of a stairway is 13 feet $1\frac{1}{2}$ inches. What is the run of each step if there are 16 risers?

$$13 \text{ ft. } 1\tfrac{1}{2} \text{ in.} = 157.5 \text{ in.}$$

$$\text{Run of step} = \frac{157.5}{16 - 1} = \frac{157.5}{15} = 10\tfrac{1}{2} \text{ in.} \quad \text{(Ans.)}$$

ILLUSTRATION: What is the run of each step if a stairway has 20 risers, a total rise of 12 feet 6 inches, and a slope of 35 degrees?

Since the length of the run of the stairway is lacking, it must be found by trigonometry. It is evident from the triangle in Fig. 25 that

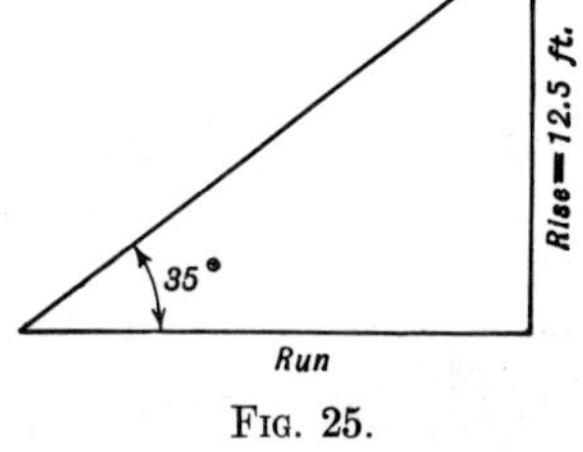

FIG. 25.

$$\frac{\text{run}}{\text{rise}} = \text{cotangent } 35°$$

Plywood, Wallboard, etc.—In recent years the use of laminated and specially treated large-surface covering materials, such as plywood and wallboard, has become increasingly popular. They are readily available and lend themselves to quick, easy installation on both interiors and exteriors.

Calculating the amount of material required is a relatively simple and straightforward procedure. Plywood can be furnished in panels $2\frac{1}{2}$ to 4 feet wide, by 5 to 12 feet long, in thicknesses from $\frac{3}{16}$ inch to $1\frac{1}{8}$ inches. To determine the number and size of panels required, calculate the square feet of area to be covered and divide it by the number of square feet in the panel to be used. The particular panel size selected is usually determined by the dimensions of the wall being sheathed.

ILLUSTRATION: A solid wall 8 feet high by 18 feet long with studs 16 inches on center to be sheathed with $\frac{5}{16}$-inch thick plywood. Find the number and size of panels required.

$$\text{Wall area} = 8 \times 18 = 144 \text{ sq ft}$$

Since the wall is 8 feet high, use panels 8 feet long. Since the studs are on 16-inch centers, use panels 4 feet wide.

Divide the wall area by the area of the panel selected (4×8, or 32 sq ft) to find the number of panels required.

$$\frac{144}{32} = 4\tfrac{1}{2} \text{ or 5 panels required}$$

ANSWER: 5 panels measuring 4×8 feet required.

Plywood is manufactured in two basic types (exterior and interior), and in several grades. The *type* refers to the type of bond between plies. Exterior-type panels are completely waterproof, and interior types are moisture resistant but not waterproof. The *grade* refers to the appearance of the wood veneer in the outer plies.

Standards for plywood have been set up by the American Plywood Association, a nonprofit organization of plywood manufacturers. Table 2 lists the different types, grades, and sizes of plywood commercially available. The trade names of the various grades are also defined below.

EXT-DFPA·A-A. Use where the appearance of both sides is important. Fences, built-ins, signs, boats, cabinets, commercial refrigerators, shipping containers, tote boxes and ducts.

EXT-DFPA·A-B. For uses similar to EXT·A-A panels, but where the appearance of one side is less important.

EXT-DFPA·A-C. Use where the appearance of only one side is important. Siding, soffits, fences, structural uses. Box car and truck lining and farm buildings.

EXT-DFPA·PLYFORM. Concrete form grade, with high re-use factor. Red painted edges, mill-oiled unless otherwise specified.

EXT-DFPA·B-C. An outdoor utility panel. For farm service and work buildings. Box car and truck linings, containers and agricultural equipment.

EXT-DFPA·C-C, PLUGGED, or EXT-DFPA·UNDERLAYMENT. Use as a base for resilient flooring, mosaic tile, carpeting, where unusual moisture conditions exist, such as bathrooms, utility rooms, kitchens. Use for conventional or single layer subfloor-underlayment. Other uses include pallets, pallet bins and farm service buildings.

EXT-DFPA·C-C. Unsanded grade with waterproof bond. Use for backing, rough construction, farm service buildings, crating, pallets and pallet bins.

INT-DFPA·A-A. For interior applications where both sides will be viewed. Built-ins, cabinets, furniture, and partitions.

INT-DFPA·A-B. For uses similar to INT·A-A panels, but where the appearance of one side is less important and two smooth surfaces are necessary.

INT-DFPA·A-D. Interior use where appearance of only one side is important. Paneling, built-ins, shelving, partitions and flow racks.

INT-DFPA·PLYFORM. Re-usable form plywood. Glue is moisture-resistant, not waterproof. Green painted edges, mill-oiled unless otherwise specified.

INT-DFPA·B-B. Interior utility panel used where two smooth sides are required.

INT-DFPA·B-D. Interior utility panel used where one smooth side is required. Good for backing, sides of built-ins. Industry: separator boards and bins.

INT-DFPA·UNDERLAYMENT. Base for resilient flooring and carpeting. A backing material where moisture is not a problem. Used for conventional or single layer subfloor-underlayment.

INT-DFPA·PLYSCORD. Unsanded structural grade panel for sheathing, subflooring, limited exposure crates, containers, pallets and dunnage.

INT-DFPA·PLYSCORD WITH EXTERIOR GLUE. Same as PlyScord, but with waterproof glue. Not a substitute for Exterior-type plywood.

TEXTURE 1-11, EXT-DFPA. Exterior type, sanded or unsanded, shiplapped edges with parallel grooves ¼″ deep, ⅜″ wide. Grooves 2″ or 4″ o.c. Available in 8′ and 10′ lengths and MD Overlay. Use for siding, gable ends, fences, interior paneling.

EXT-DFPA·303, SPECIALTY SIDING. Exterior siding or fencing panel with special surface treatment such as V-groove, channel groove, striated, brushed, Lauan One-Eleven, rough sawn and many other decorative surface patterns. Available in several wood face veneers including fir, red cedar, Philippine mahogany (lauan) and MD Overlay.

INT-DFPA·BRUSHED, INT-DFPA·EMBOSSED, INT-DFPA·STRIATED. Textured paneling, for accent walls, built-ins, counter facings, and displays. An effect of depth and dimension can be accomplished with the use of brushed, embossed or striated faced panels. For outdoor applications, use only Exterior-type.

EXT-DFPA·MD, OVERLAY. Exterior-type Medium Density Overlaid plywood with a smooth, fused, resin-fiber overlay. Especially good surface for painting. Exterior siding, soffits, kitchen cabinets, signs.

EXT-DFPA·HD, OVERLAY. Exterior-type High Density Overlaid plywood with hard, translucent, fused resin-fiber overlay. Slight grain pattern. Abrasion resistant. Painting not ordinarily required. Good for concrete forms, acid tanks, cabinets.

EXT-DFPA·PLYRON, INT-DFPA·PLYRON. Hardboard face on one or both sides. For concrete forms, counter tops, shelving, cabinet doors and built-ins.

INT-DFPA·N-N. Natural finish cabinet quality. Both sides select all heartwood veneer. Special jointed core construction. For furniture having a natural finish, cabinet doors, built-ins, etc. (5)

INT-DFPA·N-A. Same as N-N except one side is A faced for economy. Special jointed core construction. For furniture having a natural finish, cabinet doors, built-ins, etc. (5).

INT-DFPA·N-D. One side special select, all heartwood veneer. For wall paneling. (5)

MARINE·EXT-DFPA. Waterproof panel with limitations on core gaps and with special solid jointed core construction. Made especially for marine use. (2) Available also in overlay grades. Especially good for boat hulls.

DFPA·2·4·1, DFPA·2·4·1, WITH EXT. GLUE, DFPA·2·4·1 T & G, DFPA·2·4·1 T & G, WITH EXT. GLUE. Combination subfloor and underlayment. Base for resilient finish flooring, carpeting and wood strip flooring. Available in square edges; or with tongue and grooved sides or sides and ends. Use 2·4·1 with Exterior Glue in baths, kitchens or utility areas where moisture may be a problem (or when construction delays are anticipated). (1)

The thicknesses available of the first three types are ¼″, ⅜″, ½″, ⅝″, ¾″ and 1″. The next 11 types are available in all these thicknesses except, 1″. The other types are special veneers and other items whose available thicknesses must be checked with the supplier.

Standard size plywood panels are 4×8 feet. Other sizes are available upon order. King size panels 12′, 14′, 16′, 20′, and longer are also available. Panels ⅜″ and thinner have a minimum of 3 plies; ⅝″ to ¾″ have 5-ply minimum; thicker panels have 7-ply minimum.

Surface Covering

Up to this point, only structural members of buildings have been considered and the main concerns of these are strength and conformity with building regulations. The measurement of these members has generally been by the piece. Surface covering, on the other hand, while it is purchased by the board foot by nominal dimensions, covers areas only in proportion to its actual dimensions. Surface measure is made in square feet, or, for the sake of smaller figures, in *squares,* one square being a surface 10 feet by 10 feet or 100 square feet.

Certain factors pertaining to surface covering with common boards and strips are common to sheathing, rough flooring, and roof boarding. Thus, in any of these uses, a seven-inch board will cover a space less than seven inches wide and a ten-inch board will cover a space less than ten inches wide. When the area to be covered has been calculated, the following percentages must be added to make up for the scant widths:

Width of Board, Inches	Percentage to be Added	Width of Board, Inches	Percentage to be Added
3	14.39	8	6.66
4	10.34	9	5.88
5	8.11	10	5.26
6	6.66	11	4.76
7	5.66	12	4.35

This table does not provide for waste resulting from short ends. An additional 5 percent should be added for waste when sheathing is placed horizontally or when rough flooring is laid parallel to the walls. Ten percent should be added for waste when these coverings are laid diagonally.

Sheathing.—Sheathing may be nailed to the studding of a frame building either diagonally as shown in Fig. 2, or horizontally as shown in Fig. 12. It may be either matched or unmatched lumber $7/8$ inch thick and planed on at least one side.

In estimating the amount of lumber needed for sheathing the procedure is to calculate the net wall surfaces and add the proper percentage for waste and scant widths. The area of the triangular surface under the end of a gable roof is, by geometry, one-half the product of the rise and the span.

ILLUSTRATION: The bungalow shown in Fig. 26 is to be sheathed diagonally with 1-inch by 6-inch common boards. How many board feet of lumber will be required? (Assume door and window openings on far sides equal in area to those on the near sides.)

$$\begin{aligned}
\text{Area of side wall} &= 23 \text{ ft. } 10 \text{ in.} \times 10 \text{ ft. } 2 \text{ in.} - \text{openings} \\
&= 23.83 \times 10.17 - (3.17 \times 4.92 + 3.17 \times 5.42) \\
&= 242.35 - 32.78 = 180.58 \text{ sq. ft.}
\end{aligned}$$

$$\begin{aligned}
\text{Area of end wall} &= 18 \text{ ft. } 0 \text{ in.} \times 10 \text{ ft. } 2 \text{ in.} \\
&\quad + \tfrac{1}{2}(18 \text{ ft. } 0 \text{ in.} \times 6 \text{ ft. } 4 \text{ in.}) \\
&\quad - 2(5 \text{ ft. } 5 \text{ in.} \times 3 \text{ ft. } 2 \text{ in.}) \\
&\quad - 3 \text{ ft. } 4 \text{ in.} \times 8 \text{ ft. } 0 \text{ in.} \\
&\quad - 3 \text{ ft. } 4 \text{ in.} \times 1 \text{ ft. } 10 \text{ in.} \\
&= 18 \times 10.17 + \tfrac{1}{2}(18 \times 6.33) - 2(5.42 \times 3.17) \\
&\quad - 3.33 \times 8 - 3.33 \times 1.83 \\
&= 183.06 + 57 - 34.4 - 26.6 - 6.1 \\
&= 240.06 - 67.09 = 172.97 \text{ sq. ft.}
\end{aligned}$$

$$\begin{aligned}
\text{Total surface} &= 2 \text{ sides @ } 180.58 \text{ sq. ft.} = 361.16 \text{ sq. ft.} \\
&= 2 \text{ ends @ } 172.97 \text{ sq. ft.} = 345.94 \text{ sq. ft.} \\
&\qquad\qquad \text{Total} = 707.10 \text{ sq. ft.}
\end{aligned}$$

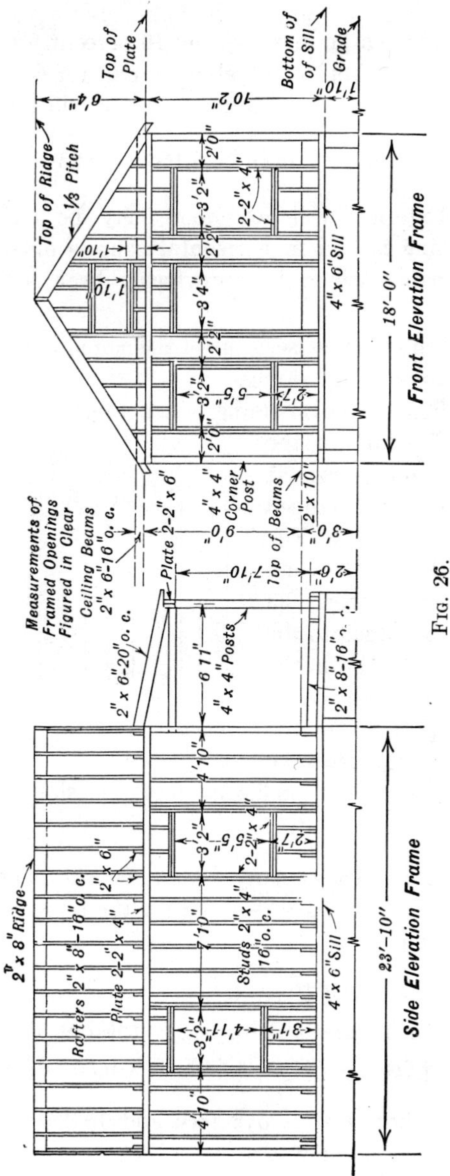
Top of Ridge
1/3 Pitch
6'4"
Top of Plate
10'2"
Bottom of Sill
1'10"
Grade
1'10"
1'10"
2'0"
3'2"
2-2" x 4"
2'2"
3'4"
2'2"
3'2"
5'5"
2'7"
2'0"
4" x 6" Sill
18'-0"
Front Elevation Frame
Measurements of Framed Openings Figured in Clear
Ceiling Beams 2" x 6"-16" o. c.
Plate 2-2" x 6"
4" x 4" Corner Post
9'0"
Top of Beams
2" x 10"
3'0"
Plate
7'10"
2'6"
2" x 6"-20" o. c.
6'11"
4" x 4" Posts
2" x 8"-16" o. c.
2" x 8" Ridge
Rafters 2" x 8"-16" o. c.
Plate 2-2" x 4"
2" x 6"
4'10"
3'2"
5'5"
2-2" x 4"
2'7"
7'10"
Studs 2" x 4" 16" o. c.
3'2"
4'11"
3'1"
4'10"
4" x 6" Sill
23'-10"
Side Elevation Frame

FIG. 26.

This area must then be increased by 10 percent for waste and by 6.66 percent (according to the above table) for scant widths or a total of 16.66 percent. The lumber needed will then be

$$707.10 + 707.10 \times 0.1666 = 825 \text{ fbm} \quad \text{(Ans.)}$$

Siding.—Exterior walls of wood may be either siding or shingles. Siding is laid in horizontal courses outside of a layer of building paper which has previously been attached to the sheathing.

Figure 27 shows cross-sections of the common bevel siding and several patterns of drop siding.

The usual size of bevel siding is a nominal width of 6 inches, a thickness of $\frac{1}{2}$ inch at the bottom edge and $\frac{1}{4}$ inch at the top edge. It is lapped on the wall as shown in Fig. 2. When laid with $4\frac{1}{2}$ inches exposed to the weather, 33 percent must be added to the area of the wall to obtain the area of siding required. With 4 inches exposed to the weather, 50 percent must be added. In both cases an additional 5 percent should be added for waste.

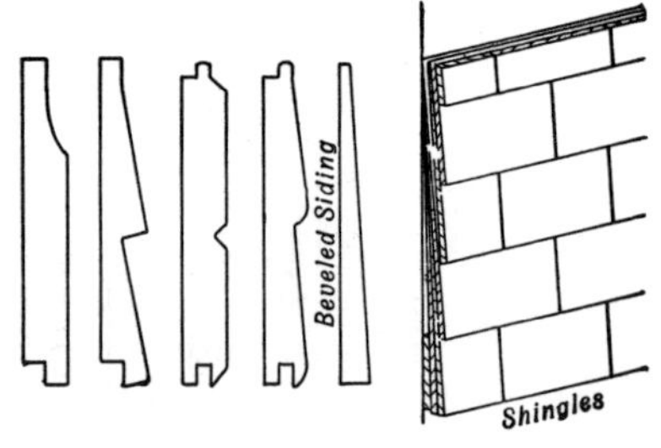

FIG. 27.—Siding.

ILLUSTRATION: How many board feet of bevel siding laid 4 inches to the weather are required for the bungalow in the previous illustration?

Net wall area = 707.10 sq. ft.

Add 50% for lap and 5% for waste; total of 55%

Lumber required = 707.10 + 707.10 × 0.55 = 1096 fbm (Ans.)

For drop siding with a $5\frac{3}{16}$ face add 16.3 percent for scant width and 5 percent for waste.

ILLUSTRATION: How many board feet of $5\frac{3}{16}$-inch drop siding would be required for the bungalow of the preceding exercises?

Net wall area = 707.10 square feet

Lumber required = 707.10 + 707.10 × 0.213 = 858 fbm. (Ans.)

Flooring.—Rough flooring should be laid diagonally on the floor joists. The lumber required is estimated in exactly the same manner as the sheathing.

ILLUSTRATION: How many board feet of lumber are required for a floor 26 feet by 28 feet if 7-inch common lumber is used and laid diagonally?

Area = 26 × 28 = 728 sq. ft.

Add for scant width	5.66%
Add for waste	10.00%
Total	15.66%

Lumber required = 728 + (728 × 0.1566) = 842 fbm (Ans.)

A finish flooring of hard maple, beech, birch or oak provides a substantial wearing surface. It is laid directly on top of the rough flooring at right angles to the direction of the floor joists, but never parallel to the rough flooring. It is nailed at intervals of 12 or 16 inches with 8-penny steel-cut flooring nails driven at an angle of 45 degrees and starting just above the tongue.

Hardwood flooring comes in thicknesses of $\frac{3}{8}$ in., $\frac{1}{2}$ in., $\frac{5}{8}$ in. and $\frac{25}{32}$ in. and in face widths of $1\frac{1}{2}$ in., 2 in., $2\frac{1}{4}$ in. and $3\frac{1}{4}$ in. The scant width loss due to the tongue and groove is considerable and the following percentages must be added when estimating the flooring required:

Face Width, Inches	Allowance, Percent
1½	50
2	37.5
2¼	33.3
3¼	24

An additional 3 to 5 percent must be added for waste in cutting and fitting.

ILLUSTRATION: How many board feet of flooring are required to lay 1252 square feet of $\frac{25}{32}$-in. by $2\frac{1}{4}$-in. flooring and allowing 5 percent for waste?

Scant width loss.........	= 33.3%	
Waste loss..............	= 5.0%	
Total loss...........	38.3%	= .383

Flooring required = 1252 + 1252 × 0.383 = 1732 fbm (Ans.)

Roofing.—The area of a gable roof is the sum of the two sloping surfaces. The area of one of these surfaces is equal to the product of the length of the roof and the slope length or the rafter length.

ILLUSTRATION: What is the area of a gable roof whose length is 35 feet and whose rafters are 18 feet long?

Area of $\frac{1}{2}$ of roof = 35 × 18 = 630 sq. ft.

Area of whole roof = 630 × 2 = 1260 sq. ft. = 12.6 squares (Ans.)

A hip roof has the same area as a gable roof of the same pitch, overhang and plate dimensions. Therefore, the area of a hip roof is equal to twice the product of the length of rafters on the long side and the length of the eaves on the long side.

A dormer having the same roof pitch as the main roof adds only the amount of the overhang to the area which would obtain if the dormer did not exist.

Roof rafters are covered with boarding as a support for the roof covering material. This boarding is usually tight sheathing as in Fig. 28 for slate or composition roofing.

Roof sheathing is estimated in the same manner as side-wall sheathing, the allowances for scant widths given at the head of this section being used, and 5 percent allowed for waste.

ILLUSTRATION: How many board feet of sheathing are required to cover a hip roof 35 feet long and a rafter length of 17 feet, if 1-inch by 6-inch boards are used?

Area of roof $= 2 \times 35 \times 17 = 1190$ sq. ft.

Add for scant widths.............	6.66%
Add for waste....................	5.00%
Total........................	11.66%

Lumber required for sheathing

$$= 1190 + 1190 \times 0.1166 = 1329 \text{ fbm} \quad \text{(Ans.)}$$

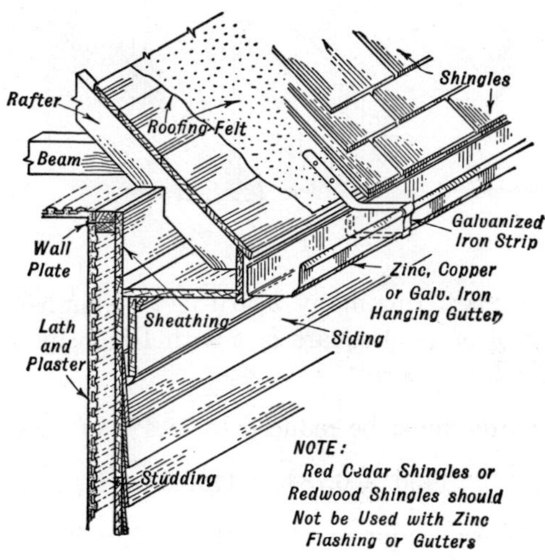

FIG. 28.

There is little unanimity on the question as to whether or not solid sheathing should be used under wood shingles. The alternative construction is the use of 1 in. bv 4 in. shingle lath spaced an inch apart, as shown in Fig. 29.

Since the actual width of a 4-inch board is $3\frac{5}{8}$ inches, if 1 inch is left open, only $\dfrac{3\frac{5}{8}}{1 + 3\frac{5}{8}} = \dfrac{3.625}{4.625} = 0.784 = 78.4$ percent of the roof area will be covered. When computing the lumber required for covering a roof with 1-in. by 4-in. shingle lath spaced 1 inch apart, only 78.4 percent of the actual area is considered. The usual factors for scant widths and waste still apply, however.

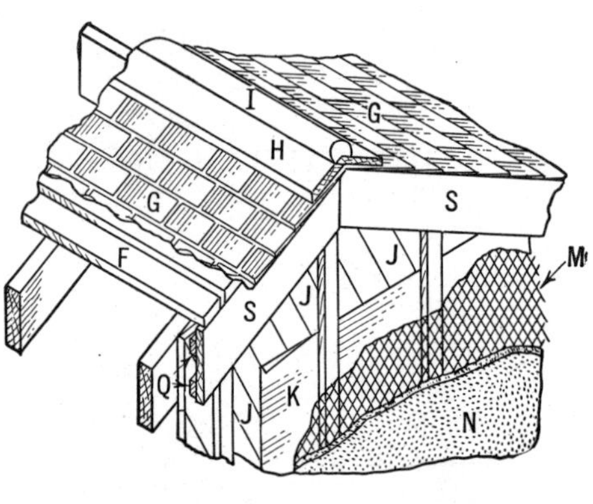

FIG. 29.

ILLUSTRATION: How many board feet of lumber are required to cover a roof of 1450 square feet with 1-inch by 4-inch shingle lath spaced 1 inch apart?

First the area must be reduced to 78.4% of its actual area.

$$1450 \times 0.784 = 1136.8 \text{ sq. ft.}$$

Allowance for scant widths.....	=	10.34%
Allowance for waste...........	=	5.00%
Total.....................		15.34%

Lumber required = 1136.8 + 1136.8 × 0.1534 = 1311 fbm (Ans.)

Shingles.—Cedar or cypress shingles form a roof covering of great durability. Shingles are sold in bundles which contain the equivalent of 250 shingles 4 inches wide. Actually they are of random widths. They come in lengths of 16, 18, and 24 inches and in butt thicknesses of from $\frac{5}{16}$ inch to $\frac{1}{2}$ inch. Shingles are listed in this fashion:

24-in. Royals, 4/2 in.

16-in. Perfects, 5/2 in.

The first figure gives the length of the shingle; (4/2 in.) means that 4 shingles measure 2 inches at the butts, and (5/2 in.) means that 5 shingles measure 2 inches at the butts.

The amount of roof surface which a bundle of shingles will cover depends on the amount exposed to the weather. Sixteen-inch roof shingles are laid 4 in., 4½ in., and 5 in. to the weather. Twenty-four-inch shingles are usually used for siding and laid 7½ in. or even 10 in. to the weather. The number of bundles of shingles required for each square of roof area including an allowance of 10 percent for waste is, for various exposures, as follows:

Exposure, Inches	Bundles per Square
4	4.0
4½	3.6
5	3.2
6	2.7
7½	2.1
10	1.6

ILLUSTRATION: How many bundles of shingles are required to cover a roof of 2240 square feet when 4½ inches are exposed to the weather?

2240 sq. ft. = 22.4 squares

$22.4 \times 3.6 = 81$ bundles (Ans.)

Nails Required.—The quantity of nails required for the various operations in the construction of a house may be obtained from Table 3.

ILLUSTRATION: What kind and how many pounds of nails are required for nailing 2400 fbm of 1-inch by 6-inches sheathing on 16 inches center to center studding?

The table shows 8d common to be the proper size and 32 pounds per 1000 fbm as the unit quantity. Then,

$$2.4 \times 32 = 77 \text{ lb for 2400 fbm} \quad \text{(Ans.)}$$

ILLUSTRATION: What kind and how many pounds of nails are required for nailing 1700 fbm of 1-inch by 8-inches drop siding nailed on 12-inch centers?

The table gives 8d casing as the proper size and 23 lb per 1000 fbm as the unit quantity. Then,

$$1.7 \times 23 = 39 \text{ lb for 1700 fbm} \quad \text{(Ans.)}$$

Interior Trim.—This work includes door jambs and trim, window frames, sash and trim, baseboards and mouldings. Frames and sash are seldom made up on the job these days, and dealers supply even door and window trim already cut and bundled. Baseboards, mouldings, etc., should be estimated to the nearest 100 feet in excess of the actual length wanted.

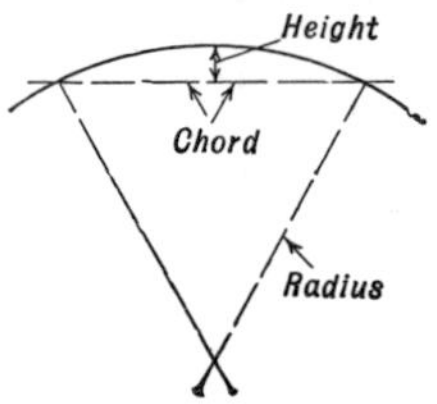

FIG. 30.

Determining Radius.—In making a bend as for a moulding or baseboard, of a known chord and height, the radius must be known so that a line can be struck to which to work.

To determine radius, add the square of half the chord to the square of the height and divide by twice the height. Thus, in Fig. 30, if the chord is 8 feet and the height 1 foot,

$$\text{Radius} = \frac{4^2 + 1^2}{2 \times 1} = 8.5 \text{ feet} \quad \text{(Ans.)}$$

A slight bend can be made in a board if soaked in hot water 30 minutes. Sharp bends can be made after wood has been cooked or steamed for at least 6 hours

TABLE 3

Wire Nails—Kinds and Quantities Required *

Length, in inches	Am. Steel & Wire Co.'s Steel Wire Gauge	Approx. No. to lbs.	Nailings	Sizes and Kinds of Material		Trade Names	Pounds per 1000 feet B. M. on center as follows: 12″	16″	20″	36″	48″
							Pounds				
2½	10¼	106	2	1 x 4	I. Used square edge, as platforms, floors, sheathing, or shiplap. II. When used D. & M., blind nailed, only ½ quantity named required.	8d common	60	48	37	23	20
2½	10¼	106	2	1 x 6		8d common	40	32	25	16	13
2½	10¼	106	2	1 x 8		8d common	31	27	20	12	10
2½	10¼	106	2	1 x 10		8d common	25	20	16	10	8
2½	10¼	106	3	1 x 12		8d common	31	24	20	12	10
4	6	31	2	2 x 4		20d common	105	80	65	60	33
4	6	31	2	2 x 6		20d common	70	54	43	27	22
4	6	31	2	2 x 8		20d common	53	40	53	21	17
4	6	31	3	2 x 10		20d common	60	50	40	25	20
4	6	31	3	2 x 12		20d common	52	41	33	21	17
6	2	11	2	3 x 4		60d common	197	150	122	76	61
6	2	11	2	3 x 6		60d common	131	97	82	52	42
6	2	11	2	3 x 8		60d common	100	76	61	38	34
6	2	11	3	3 x 10		60d common	178	137	110	70	55
6	2	11	3	3 x 12		60d common	145	115	92	58	46
2½	12½	189	2	Base, per 100 ft. lin.		8d finish	......	1	......	......	......
2½	10¼	106	2	Byrket lath		8d common	......	48	......	......	......

* Courtesy American Steel and Wire Company.

Wire Nails—Kind and Quantities Required—Cont.

Length, in inches	Am. Steel & Wire Co.'s Steel Wire Gauge	Approx. No. to lbs.	Nailings	Sizes and Kinds of Material	Trade Names	Pounds per 1000 feet B. M. on center as follows: 12″	16″	20″	36″	48″
						Pounds				
2½	12½	189	1	Ceiling, ¾ x 4	8d finish	18	14	......	......	......
2	13	309	1	Ceiling, ½ and ⅝	6d finish	11	8	......	......	......
2½	12½	189	2	Finish, ⅞	8d finish	25	12	......	......	......
3	11½	121	2	Finish, 1⅛	10d finish	12	10	......	......	......
2½	10	99	1	Flooring, 1 x 3	8d floor brads	42	32	......	......	......
2½	10	99	1	Flooring, 1 x 4	8d floor brads	32	26	......	......	......
2½	10	99	1	Flooring, 1 x 6	8d floor brads	22	18	......	......	......
4	6	31		Framing, 2x4 to 2x16	20d common	20	16	14	......	......
3½	8	49		requires 3 or more	16d common	10	10	8	......	......
3	9	69		sizes and vary greatly	10d common	8	6	5	......	......
6	2	11		Framing, 3x4 to 3x14	60d common	30	25	20	......	......
2½	11½	145	2	Siding, drop, 1 x 4	8d casing	45	35	......	......	......
2½	11½	145	2	Siding, drop, 1 x 6	8d casing	30	25	......	......	......
2½	11½	145	2	Siding, drop, 1 x 8	8d casing	23	18	......	......	......
2	13	309	1	Siding, bevel, ½ x 4	6d finish	23	18	......	......	......
2	13	309	1	Siding, bevel, ½ x 6	6d finish	15	13	......	......	......
2	13	309	1	Siding, bevel, ½ x 8	6d finish	12	10	......	......	......
				Casing, per opening	6d and 8d casing	About ½ pound per side.				

Wire Nails—Kinds and Quantities Required—Cont.

1¼	14	568	12″ o.c.	Flooring, ⅜ x 2....	3d brads.........	About 10 pounds per 1000 square feet.
1⅛	15	778	16″ o.c.	Lath, 48″..........	3d fine...........	6 pounds per 1000 pieces.
⅞	12	469	2″ o.c.	Ready roofing......	Barbed roofing.....	¾ of a pound to the square.
⅞	12	469	1″ o.c.	Ready roofing......	Barbed roofing.....	1½ pounds to the square.
⅞	12	180	2″ o.c.	Ready roofing...... (⅝ heads)	American felt roofing	1½ pounds to the square.
⅞	12	180	1″ o.c.	Ready roofing...... (⅝ heads)	American felt roofing	3 pounds to the square.
1¼	13	429	....	Shingles†.........	3d shingle........	4½ pounds; about 2 nails to each 4 inches.
1½	12	274	....	Shingles...........	4d shingle.........	7½ pounds; about 2 nails to each 4 inches.
⅞	12	180	4	Shingles...........	American felt roofing	12 lbs., 4 nails to shingle.
⅞	12	469	4	Shingles...........	Barbed roofing.....	4½ lbs., 4 nails to shingle.
1	16	1150	2″ o.c.	Wall board, around entire edge......	2d Barbed Berry, flat head	5 pounds, per 1,000 square feet.
1	15½	1010	3″ o.c.	Wall board, intermediate nailings..	2d casing or floor brad.....	2½ lbs., per 1.000 square feet.

†Wood shingles vary in width; asphalt are usually 8 inches wide. Regardless of width 1000 shingles are the equivalent of 1000 pieces 4 inches wide.

XIII

Lathing and Plastering

Laths form the supporting structure for plaster on walls and ceilings when the plaster cannot be applied directly to a firm base to which it will bind. Laths may be of either wood or metal and are nailed either to furring strips or to the studding of walls and partitions and to the under side of floor joists to form ceilings.

Wood Laths.—Wood laths are strips $1\frac{1}{2}$ in. wide, $\frac{1}{4}$ in. or $\frac{3}{8}$ in. thick, and 48 in. long sawed from pine, spruce, or hemlock. This length permits the lath to cover, without cutting, three spans between studs when these are placed on 16-inch centers. Laths for lime plaster are spaced $\frac{1}{4}$ in. or $\frac{3}{8}$ in. and closer for gypsum plaster. A bundle of 100 laths

spaced $\frac{1}{4}$ in., will cover 6.48 sq.yd.: equal to 1543 laths per 100 sq.yd.
spaced $\frac{3}{8}$ in., will cover 6.94 sq.yd.: equal to 1441 laths per 100 sq.yd.

About 10 pounds of fine lath nails are required per 100 square yards of lathing.

Illustration: How many bundles of laths will be required for lathing the walls and ceiling of a room 12 feet × 18 feet, ceiling 9 feet high, if the areas of the windows and doorways total 12 square yards and the spacing of the lath is $\frac{1}{4}$ inch? Allow 5% for waste.

Area of ceiling	$= 4 \times 6$	$=$	24 sq. yd.
Area of side walls	$= 3 \times 6 \times 2$	$=$	36 sq. yd.
Area of end walls	$= 3 \times 4 \times 2$	$=$	24 sq. yd.
		Total	84 sq. yd.

Total carried forward	84 sq. yd.
Area of openings	12
	72 sq. yd.
5% for waste	3.6
	75.6 sq. yd.

If one bundle covers 6.48 sq. yd., then the number of bundles required is $\frac{75.6}{6.48} = 11.7$ and the next larger whole number is, of course, 12 bundles. (Ans.)

Illustration: A room to be lathed has two window openings 2 ft. 10 in. by 5 ft. 2 in. and two door openings 3 ft. 0 in. by 7 ft. 0 in. What quantity of nails and how many bundles of lath will be required if the size of the room is 13 ft by 12 ft 6 in. and the height of the ceiling is 9 ft 6 in. and the spacing is $\frac{3}{8}$ inch? Allow 5% for waste.

In this problem it is more convenient to change the inches to tenths of a foot and compute the total area in square feet and reduce to square yards by dividing by 9.

Area of two windows	$= 2 \times 2.83 \times 5.17$	= 29.3 sq. ft.
Area of two doors	$= 2 \times 3 \times 7$	= 42.0 sq. ft.
	Total	71.3 sq. ft.
Area of ceiling	$= 13 \times 12.5$	= 162.5 sq. ft.
Area of end walls	$= 2 \times 9.5 \times 12.5$	= 237.5 sq. ft.
Area of side walls	$= 2 \times 9.5 \times 13.0$	= 247.0 sq. ft.
	Total	647.0 sq. ft.
	Area of openings	71.3 sq. ft.
		575.7 sq. ft.
	5% for waste	28.8 sq. ft.
		604.5 sq. ft.

Changing to square yards,

$$\text{area} = \frac{604.5}{9} = 67.17 \text{ sq. yd.}$$

If one bundle at $\frac{3}{8}$ in. spacing covers 6.94 sq. yd., then the number of bundles required will be

$$\frac{67.17}{6.94} = 9.7 \text{ or } 10 \text{ whole bundles} \quad \text{(Ans.)}$$

If 10 pounds of nails are required for 100 sq. yd., this room will require $10 \times \frac{67}{100} = 6.7$ pounds of nails. (Ans.)

Metal Lath.—Metal lath is manufactured in two general forms, as a wire mesh and as expanded metal (Figs. 1 and 2). Both forms are protected from corrosion by being painted, japanned or galvanized. Metal lath is not only a base for plaster but also serves as reinforcing. It is universally used in fireproof construction and is particularly adapted for thin partition walls and suspended ceilings.

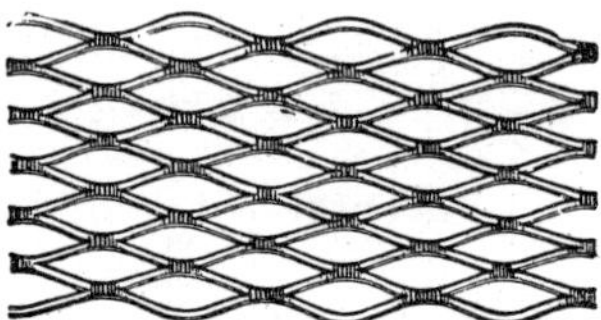

Fig. 1.—Expanded Metal Lath.

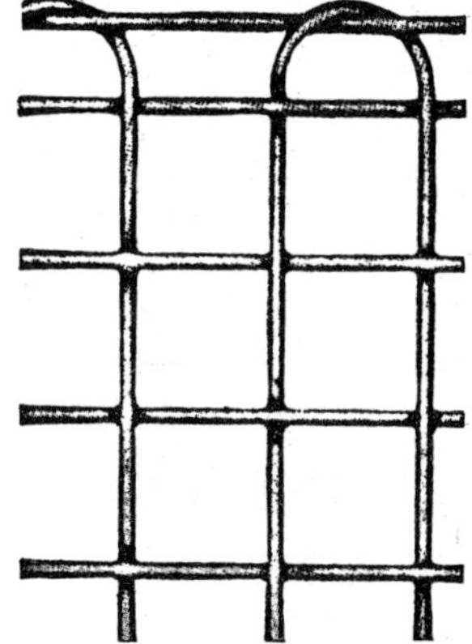

Fig. 2.—Wire Mesh Lath.*

Both wire lath and expanded metal lath are attached to steel furring with No. 18 gage annealed galvanized wire lacing and to wooden furring, studding, or floor joists with No. 13 gage galvanized wire staples spaced about six inches apart. The following are average quantities of lacing and staples required per 100 square yards of metal lath:*

* Courtesy Wickwire Spencer Steel Company.

Spacing of Furring, Inches, Center to Center	No. 18 Galvanized Wire Lacing, Pounds	1¼-In. No. 13 Galvanized Wire Staples, Pounds
12	6	9½
14	5	8
16	4½	7

Wire Lath.—Wire lath is woven from No. 18 to No. 21 Washburn & Moen gage wire with 2 and $2\frac{1}{2}$ meshes per lineal inch in each direction. Some forms have V-shaped metal stiffeners attached at intervals of 8 inches to provide the fabric with greater rigidity. The lath usually comes in rolls 150 feet long and 36 inches wide. Thus one roll will cover 50 square yards.

With 12-inch spacing of furring, a No. 19 gage plain wire lath is recommended, while the No. 18 gage is more suitable when the spacing of furring is 14 or 16 inches. If lath with V-stiffeners is used, a No. 20 gage wire is sufficient.

Illustration: An auditorium 50 feet by 100 feet with a 20-foot ceiling is to be lathed with wire lath on metal furring, 12 inches on centers. How many square yards of lath, how many rolls, and how many pounds of lacing will be required if the total area of doors and windows is 50 square yards?

Area of ceiling	= 50 × 100	=	5,000 sq. ft.
Area of end walls	= 2 × 20 × 50	=	2,000 sq. ft.
Area of side walls	= 2 × 20 × 100	=	4,000 sq. ft.
	Total area		11,000 sq. ft.

Reducing to square yards,

$$\frac{11{,}000}{9} = 1222 \text{ sq. yd.}$$

less openings 50 sq. yd.

Net area = 1172 sq. yd. of lath required (Ans.)

(*Courtesy Associated Metal Lath Manufacturers, Inc.*)

Fig. 3.—Most Advantageous Positions for Metal Lath for Fire Stops and Crack Prevention.

For Fire Stops—

(1) On all stud bearing partitions and walls and fire stops between studs. (Fire stops to be metal lath basket-shaped to fit between studs, coated with plaster or cement and filled with incombustible materials.)

(2) On ceilings under inhabited floors, especially over heating plants and coal bins.

(3) At chimney breasts, around flues and back of kitchen ranges.

(4) For stair-wells and under stairs.

(5) As a base and reinforcement for exterior stucco.

For Crack Prevention—

(a) On ceilings of prominent rooms.

(b) Lap 4 in. on either side of wall and partition angles, and around door bucks.

(c) Back of wainscots and tile mantels.

(d) Across plumbing pipes and heat ducts.

(e) Proper construction of exterior stud walls for successful stucco.

If each roll contains 50 square yards

$$\frac{1172}{50} = 23.4 \text{ or } 24 \text{ whole rolls required (Ans.)}$$

Wire lacing required at 6 pounds per 100 square yards is,

$$6 \times 11.72 = 70\tfrac{1}{4} \text{ pounds (Ans.)}$$

ILLUSTRATION: A ceiling is to be lathed on joists spaced 16 inches center to center. What size of plain or reinforced wire lath should be used?

No. 18 gage plain or No. 20 gage reinforced (Ans.)

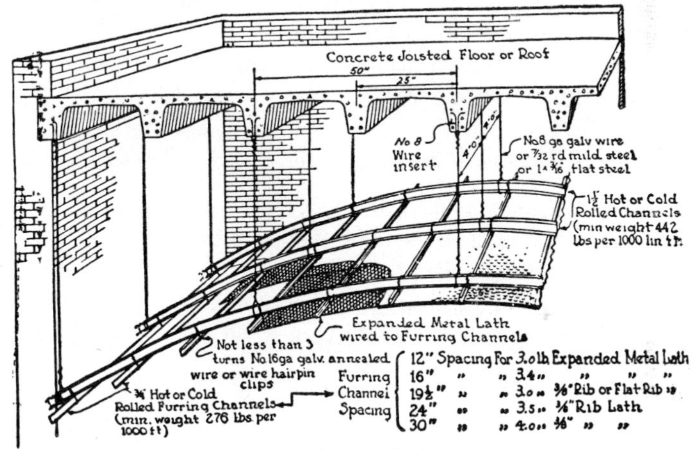

(*Courtesy Associated Metal Lath Manufacturers, Inc.*)

FIG. 4.—Metal Lath Used for Suspended Ceiling.

Expanded Metal Lath.—Expanded metal lath is made by punching and stamping sheet metal and then pulling it so that the punched slits open up as holes which hold the plaster. Ribs are quite frequently stamped into the metal to obtain greater rigidity.

The uses of expanded metal lath are illustrated in Fig. 3. It will be noted that not only is it used to support plaster by itself, but also in corners in combination with wood lath to prevent cracks. Fig. 4 shows the application to suspended ceiling.

Generally, the weight of the expanded metal per unit area is about one-half or less of the unit weight of the original sheet. The following are the minimum weights per square yard recommended for various uses:

Expanded Metal Lath for Interior Work

For vertical position attached to metal studs spaced not to exceed 12 in. on centers, 2.2 lb.

For vertical position attached to wood or metal studs not to exceed 16 in. on centers, 2.5 lb.

For horizontal position attached to metal supports spaced not to exceed 16 in. on centers, 3.4 lb.

For horizontal position attached to metal supports spaced not to exceed 12 in. on centers, 3.0 lb.

Expanded Metal Lath for Exterior Work

For any position attached to wood, metal, masonry, etc., 3.4 lb.

Expanded metal lath is manufactured in sheets of various dimensions, a common length being 8 feet, and widths ranging from 15 inches to 27 inches, with 24 inches as an average. It is sold in bundles of sheets which have a coverage of from 10 to 25 square yards per bundle.

ILLUSTRATION: A room 30 feet by 70 feet with a ceiling 18 feet high is to be lathed with expanded metal lath on metal furring on 12-inch centers. What total weight of lath will be required if the area of doors and windows is 34 square yards and a skylight 18 square yards?

$$\text{Area of ceiling} = 30 \times 70 = 2100 \text{ sq. ft.}$$

$$\frac{2100}{9} = 233 \text{ sq. yd.}$$

Subtracting skylight area,

$$233 - 18 = 215 \text{ sq. yd. (net area)}$$

The weight of lath required for a horizontal position on metal supports spaced 12 inches on centers is 3.0 pounds per square yard. Then the weight of lath required for ceiling is,

$$3.0 \times 215 = 645 \text{ lb.}$$

$$\text{Area of end walls} = 2 \times 18 \times 30 = 1080 \text{ sq. ft.}$$

$$\text{Area of side walls} = 2 \times 18 \times 70 = \underline{2520} \text{ sq. ft.}$$

$$\text{Total area } 3600 \text{ sq. ft.}$$

Reducing to square yards,

$$\text{Area} = \frac{3600}{9} = 400 \text{ sq. yd.}$$

$$\text{Net wall area} = 400 - 34 = 366 \text{ sq. yd.}$$

The weight of lath which may be used on this vertical surface is 2.2 pounds per square yard. Then the total weight required for the walls is,

$$2.2 \times 366 = 805 \text{ lb.}$$

The sum of the weights required for the ceiling and walls is

$$645 + 805 = 1450 \text{ lb. total weight of lath (Ans.)}$$

Plastering.—Plastering usually consists of three coats (Fig. 5), viz., (1) the rough or "scratch" coat which is applied directly to the wood or metal lath; (2) the "brown" coat which is floated onto the scratch coat, which has been scratched with a comb in order to roughen it so the brown coat will adhere better and (3) the finishing or "skim" coat which is applied to the brown coat after it has been finely scratched or roughened. When plaster is applied to a masonry wall, the scratch coat is often omitted, the brown coat being applied directly to the masonry. Plaster prepared in sheets and commonly known as plaster board or gypsum lath shipped ready for nailing is often substituted for the scratch coat and sometimes for both the scratch coat and the brown coat.

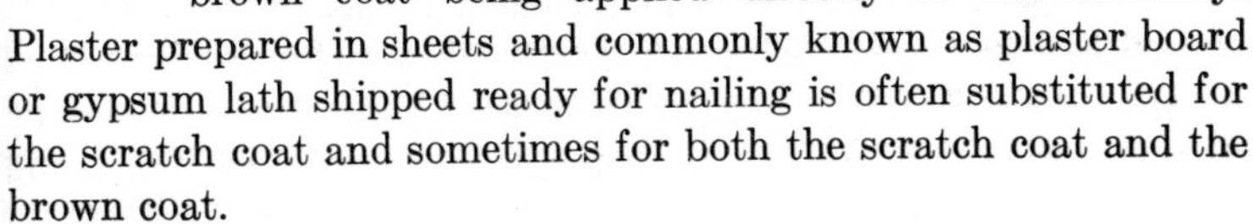

Fig. 5

Scratch Coat.—The scratch coat is applied with sufficient force to insure good key to the lath, and is composed of a mixture of slaked lime, clear river or pit sand free from salt and long cattle or goat hair (wood fiber, jute or asbestos is sometimes used instead of hair on cheap work). These are mixed in the proportions of one part lime paste to two parts sand, with $1\frac{1}{2}$ bushels of hair to each barrel of unslaked lime. Unslaked lime (quicklime) comes in lumps and is sold in barrels containing from 200 to 260 pounds. A barrel of Rockland, Me., lime weighs 220 pounds net, contains about $3\frac{1}{2}$ cubic feet and will make about 2.6 barrels or 9 cubic feet of paste. A barrel of 200 pounds will make about 8 cubic feet of paste. Approximately 9 cubic feet of lime paste, 18 cubic feet of sand, and 4 bushels of hair will cover about 40 square yards about $\frac{3}{8}$ inch thick on wooden laths and about 30 square yards on metal laths.

ILLUSTRATION: What quantities of materials will be required for the scratch coat in a building having 520 square yards of wood-lathed walls?

If one 220-pound barrel of lime, 18 cubic feet of sand, and 4 bushels of hair will cover 40 square yards, then $\frac{520}{40} = 13$ times these quantities will give the total amounts required.

$$13 \times 1 = 13 \quad \text{220-pound barrels of quicklime} \quad \text{(Ans.)}$$

$$\frac{13 \times 18}{27} = 8.7 \text{ cubic yards of sand} \quad \text{(Ans.)}$$

$$13 \times 4 = 52 \text{ bushels of hair} \quad \text{(Ans.)}$$

Quicklime must be slaked and aged before using. To obviate the delays incident to these operations, a *hydrated lime* may be used which has been slaked by the manufacturer and is marketed as a flocculent powder in 50-pound paper sacks. Hydrated lime is prepared for use by being sifted through a screen into an equal volume of water and permitted to soak undisturbed for 24 hours. This produces a putty or paste which is then mixed with the sand and hair.

The proportions of materials for the scratch coat using hydrated lime are: 1 sack (50 lb.) hydrated lime; 200 pounds of dry plastering sand; $\frac{1}{2}$ pound of hair or fiber. This will produce about 2.3 cu. ft. or 0.085 cu. yd. of plaster and will cover about $4\frac{1}{2}$ square yards on wood lath with a thickness of about $\frac{3}{8}$ inch, or $3\frac{1}{3}$ square yards on metal lath. The weight of a cubic foot of sand is about 100 pounds.

Illustration: What quantities of hydrated lime, sand and hair are required to apply a scratch coat on wood lath to 243 square yards of surface?

Since the quantities given in the statement of the proportions of materials produce a coverage of $4\frac{1}{2}$ square yards on wood lath, the factor obtained by dividing 243 by $4\frac{1}{2}$ when multiplied by these figures will give the total quantities required.

$$\frac{243}{4.5} = 54$$

Then,

$$54 \times 1 = 54 \text{ sacks of hydrated lime (Ans.)}$$

$$54 \times 200 = 10{,}800 \text{ lb. sand}$$

$$\frac{10{,}800}{100} = 108 \text{ cu. ft.} = \frac{108}{27} = 4 \text{ cu. yd. sand (Ans.)}$$

$$54 \times 0.5 = 27 \text{ lb. hair (Ans.)}$$

Brown Coat.—The brown coat is usually leaner in lime and has a smaller percentage of hair than the scratch coat. It is applied after the scratch coat has dried and is generally $\frac{1}{4}$ inch to $\frac{3}{8}$ inch thick. Considerable care is exercised in its application so that the surface produced will be straight and true and within about $\frac{1}{8}$ inch of the final finished surface or grounds.

When hydrated lime is used for the brown coat, the recommended proportions are: 1 sack (50 lb.) hydrated lime; 250 pounds of dry plastering sand and $\frac{1}{4}$ pound of hair. This will produce about 2.7 cubic feet or 0.1 cubic yard and will cover about 10 square yards to a thickness of $\frac{3}{8}$ inch.

Illustration: What quantities of material are required to cover 340 square yards of wall space with a brown coat of plaster $\frac{3}{8}$ inch thick?

$$\frac{340}{10} = 34$$

Then,

$$34 \times 1 = 34 \text{ sacks of hydrated lime (Ans.)}$$

$$34 \times 250 = 8500 \text{ lb. sand}$$

$$\frac{8500}{100 \times 27} = 3.15 \text{ cu. yd. sand (Ans.)}$$

$$34 \times \tfrac{1}{4} = 8\tfrac{1}{2} \text{ lb. hair (Ans.)}$$

Finish Coat.—The skim coat or finish coat is usually $\frac{1}{8}$ inch thick and contains no hair. It may be made with one part of slaked lime to two parts of clear white sand or marble dust. However, a harder finish may be obtained by using any of the patent plasters on the market. These are composed principally of plaster of Paris or gypsum. Hydrated lime is mixed with these to retard the time of set. The materials and proportions used depend on the type of finish desired.

White Smooth Finish.—This finish may be obtained by mixing 4 sacks (200 lb.) hydrated lime with 50 pounds of plaster of Paris. The resulting putty will cover about 45 square yards to a thickness of $\frac{1}{8}$ inch.

Sand Finish.—A mixture of $2\frac{1}{2}$ cubic feet each of lime, plaster of Paris, and white sand or marble dust will skim-coat about 100 square yards from $\frac{1}{16}$ in. to $\frac{1}{8}$ in. thick.

A coarser sand finish may be produced by mixing 2 sacks (100 lb.) of hydrated lime with 3 cubic feet (300 lb.) of plastering sand. This will cover about 65 square yards of surface.

Textured Finish.—A textured finish is made by first applying a sand finish coat and then a second heavier coat, and the texture desired worked in with tools or hands. This second or texture coat may be proportioned as follows: 3 sacks (150 lb.) of hydrated lime to 50 pounds of plaster of Paris.

ILLUSTRATION: What quantities of materials will be required for a white smooth finish coat of plaster on 355 square yards of surface?

Using the above proportions which yield a coverage of 45 square yards, we obtain,

$$\frac{355}{45} = 7.9 = \text{factor for multiplying ingredients in the mix.}$$

Then,

$$7.9 \times 4 = 31.6 = 32 \text{ whole bags of hydrated lime (Ans.)}$$

$$7.9 \times 50 = 395 \text{ lb. plaster of Paris (Ans.)}$$

Plaster of Paris is often sold in 100-pound bags. Four bags would be required in this case.

ILLUSTRATION: What quantity of materials would be required to make a finishing plaster composed of equal parts of lime, plaster of Paris, and sand to cover 1150 square yards of surface?

A mixture given above with ingredients in this proportion covers 100 square yards when $2\frac{1}{2}$ cu. ft. sand, $2\frac{1}{2}$ cu. ft. plaster of Paris, and $2\frac{1}{2}$ cu. ft. lime are mixed together.

Then,

$$\frac{1150}{100} = 11.5$$

and

$$11.5 \times 2.5 = 28.75 \text{ cu. ft. lime (Ans.)}$$

$$11.5 \times 2.5 = 28.75 \text{ cu. ft. plaster of Paris (Ans.)}$$

$$\frac{11.5 \times 2.5}{27} = 1.06 \text{ cu. yd. sand (Ans.)}$$

Thickness of Plaster.—The minimum total thickness of plaster on wood or metal lath should be $\frac{7}{8}$ inch from the face of the lath to the grounds divided as follows:

Scratch coat, average, $\frac{3}{8}$ inch

Brown coat, average, $\frac{3}{8}$ inch

Finish coat, average, $\frac{1}{8}$–$\frac{3}{8}$ inch according to finish

On brick, stone, hollow tile, concrete blocks or poured concrete, the minimum total thickness from the normal masonry line to the grounds should be $\frac{3}{4}$ inch for two-coat work divided as follows:

Brown coat, average, $\frac{3}{8}$ inch

Finish coat, average, $\frac{3}{8}$ inch

Stucco.—Plaster made with Portland cement is used in interior work only as a base coat to support bathroom, kitchen, or ornamental tile. In exterior work, however, such plaster, called *stucco*, is widely used in finishing buildings.

Stucco should always be supported on painted or galvanized metal lath on a wooden structure. It may be applied directly to masonry structures.

The first (scratch) and second (brown) coats each $\frac{3}{8}$ inch thick are usually composed of one part of Portland cement to three parts clean well-graded sand. Eight pounds of hydrated lime per sack of cement are often added to aid the plasticity of the mix. One sack of cement mixed with three cubic feet of sand and eight pounds of hydrated lime will cover about 11 square yards $\frac{3}{8}$ inch thick.

The same proportions or somewhat richer may be used for the finish coat, which may be from $\frac{1}{8}$ inch to $\frac{1}{4}$ inch thick depending on the finish. Smooth troweled, sand floated, rough trowel floated, rough cast, and pebble dash are some of the finishes effected.

Illustration: What quantities of materials are required for a three-coat stucco job, the finish coat being $\frac{1}{8}$ inch, smooth troweled and the total area of the houses to be stuccoed, 1400 square yards?

Since a scratch coat of one sack of cement, 3 cubic feet of sand and 8 pounds of hydrated lime will cover 11 square yards $\frac{3}{8}$ inch thick, then

$$\frac{1400}{11} = 127$$

and

$$127 \times 1 = 127 \text{ sacks of cement}$$

$$\frac{127 \times 3}{27} = 14.1 \text{ cu. yd. sand}$$

$$127 \times 8 = 1015 \text{ lb. hydrated lime}$$

The second coat will duplicate these quantities and the third coat will be one-third of these quantities. Then the total materials required are:

	Cement, Sacks	Sand, Cubic Yards	Hydrated Lime, Pounds
First coat	127	14.1	1015
Second coat	127	14.1	1015
Third coat	43	4.7	338
Totals	297	32.9	2368

Reducing these quantities to purchasable units, figuring 4 sacks of cement per barrel and 50 pounds of hydrated lime per bag, we have

$$\text{Cement, } \frac{297}{4} = 75 \text{ barrels (Ans.)}$$

Sand, 33 cubic yards (Ans.)

$$\text{Lime, } \frac{2368}{50} = 48 \text{ bags (Ans.)}$$

XIV

Painting, Paperhanging, Glazing

Paint and Its Uses.—Paint is a liquid medium having in suspension solid particles of various kinds known as pigments, and so formulated that it may be spread evenly and easily over a surface by brushes, rollers or sprays. After the liquid film on the surface dries, it should leave behind a coating that is uniform in appearance, and that will resist deterioration by its environment for a long period. A special requirement of exterior paints is that as they do undergo weathering over the years, they should do so by chalking (i.e., breaking down to form small particles) rather than by flaking or cracking. Paint is used to protect exterior wooden or metallic surfaces from decay or corrosion, and to enhance the appearance of all surfaces.

Composition of Paints.—The four major functional components of paints are the vehicle, the pigment or mixture of pigments, the drier and the thinner. The most general classification of paints is based upon differences in the vehicle. On this basis the major types are paints having oil-based vehicles, solvent-based vehicles and emulsion-based vehicles.

Oil Paints.—This class of paints includes the linseed oil paints which have been in use for hundreds of years, and owe their effectiveness to the ease with which linseed oil "dries," i.e., is oxidized by the air to form a hard adherent film. Other vegetable oils have been discovered which are equally or more effective as drying oils than linseed oil and many of them are used in present-day paints. They include soya, castor, safflower, coconut and tall oil, the latter being a product of the pine tree. The most important development of recent years, however, has not been the use of the new oils, but the production of oil vehicles containing in solution other substances to improve the durability and other properties of the paint film. Prominent in this group of additives are the synthetic resins, and paints based upon one of the most widely used types of resins, the alkyd resins, are found classified according to the percentage of resin used, as the long oil type, the medium oil type and the short oil type. These terms can be understood by considering that the more oil contained in the vehicle, the less its resin content, so that the long oil type is lowest in resin and highest in oil, while the short oil type being lowest in oil has the highest amount of resin. The long oil type is used where slower drying time and more even penetration is desired, while the short oil type is best for producing a hard finish on a metallic surface where rapid drying is not objectionable and may be an advantage, as for application by spraying.

Other resins besides the alkyd types are used in oil type paints, including especially the melamine and the urea resins. Moreover, these other resins are often used in combination with the alkyd resins to produce very hard finishes.

Solvent Based Paints.—In designating this group of paints by the word solvent, one must remember that the oils used in an oil paint are themselves solvents. The term solvent is used here to distinguish the other organic solvents from the oils. It should also be noted that it is a regular practice to add various organic solvents such as benzene, and toluene to oil based paints. What is meant here, however, are the products that contain solvents with very little oil. These are primarily the lacquer paints based on nitrocellulose. Such paints are used mostly for painting furniture and fixtures, and not for walls and similar surfaces, either interior or exterior. Nitrocellulose based paints vary widely in viscosity, to provide for differences in film thickness and methods of application. In addition to the various viscosity grades of nitrocellulose and differences in the solvents used, which are chiefly organic esters such as ethyl acetate, butyl acetate, amyl acetate and amyl butyrate, they include plasticizers as well. There are a great variety of these plasticizers and they range from simple vegetable oils such as castor oil, to high boiling solvents such as dibutyl or dioctyl phthalate and tricresyl phosphate. Many of them are themselves resins, such as the ester gums.

Emulsion Paints.—Emulsion paints are also called water based paints because they consist of a suspension in water of all the components of the paint. This suspension is so finely divided that it does not separate under any ordinary conditions. However, emulsion paints should never be exposed to freezing. The emulsion is often called by the word, "latex," though it is not usually a natural rubber but a synthetic resin with similar properties, such as the styrene-butadiene polymer. This resin is particularly useful where resistance to staining is important. However, it does tend to "yellow" and is therefore often replaced in formulas for emulsion paints by such other resins as polyvinyl acetate and acrylic resins. Rubber solutions and chlorinated rubber solutions are also used widely in emulsion paints.

Pigments.—With all of the types of vehicles discussed the types of pigments used are much the same. Their most important property is their covering power, and therefore the pigments of high covering power are called hiding pigments, while those of lower covering power are known as extender pigments. Pigments in the first class are the various types of titanium dioxide as well as zinc sulfide, zinc oxide and white lead, to mention only the white pigments. In the extender class would be such minerals as talc, clay, silica and gypsum. It should not be concluded that the extender pigments are merely used to cheapen the product; on the contrary, they may well make an important contribution to the hiding power, since this is not simply an additive property. They also may contribute greatly to other important properties of the finished coating, such as its mechanical strength and electrical resistance.

Driers.—Driers are substances used to speed up the reaction of a drying oil with the air, so that the film hardens more rapidly. It is important, however, to avoid adding too much drier, which would cause the paint film to crack or wrinkle, or to dry rapidly on the outside so that blisters would form from within.

Thinners.—The general purpose of thinners is to facilitate the application of the paint by reducing its viscosity. The aromatic hydrocarbons, such as benzene, toluene and xylene are excellent thinners for lacquer type paints For oil type paints turpentine is still the standard thinner. Emulsion type paints, especially those based on chlorinated rubber, tolerate very little addition of thinners and no such addition should be made other than in the manufacturing plant to this type of paint.

Varnishes.—In the past, varnishes were produced entirely by the high temperature processing of natural resins. However, at the present time the term is used for any finish which does not contain suspended pigments and which are based upon synthetic resins, oils and solvents. The phenolic resins have been used for years in high oil-content varnishes, because without high oil the phenolic resins give a finish that is too hard and brittle. They also tend to "yellow" and should only be used on very dark woods. More recently the melamine and epoxy resins have come into use in general purpose varnishes. Like oil paints varnishes are available in short, medium and long oil types which are described as "10 gallon varnish" "20 gallon varnish," etc., the numbers being the number of gallons of oil used per 100 pounds of synthetic resin. Of course, natural resins are still used in varnishes to some extent, especially to modify the properties of the synthetics.

Calcimine.—This material is a solution of chalk with glue as a binder and water as the thinner. It is an inexpensive material for interior walls and ceilings and cannot be cleaned by washing. It must be removed before redecorating. Formerly quite popular, it is being supplanted by more modern coatings which require no special surface preparation.

Shellac.—This versatile and widely used material consists of a natural resinous substance known as "lac" dissolved in alcohol which acts as the thinner. Sealing and finishing floors, sizing furniture or trim before painting or varnishing, and touching up "hot spots" on plaster are common applications of shellac.

Aluminum Paint.—This is a ready-mixed paint that has aluminum as its pigment. Three types of aluminum paint are commonly used: aluminum house paint for weather-exposed wood surfaces, metal and masonry aluminum paint for hard surfaces such as machinery, metal roofs, etc., and aluminum enamel for inside applications where high gloss or high heat-reflecting surfaces are desired. For increased roof surface protection and added heat reflection, special aluminum-asphalt paints are available.

Whitewash.—Common whitewash for sheds and barns can be made either by slaking one-half bushel (38 pounds) of common lime and straining, or by mixing one sack (50 pounds) of hydrated lime with water and adding a solution of 15 pounds of common salt in 7½ gallons of water and subsequently thinning with water as desired. If a disinfectant or insecticidal whitewash is desired, one or two quarts of crude carbolic acid should be added.

Spreading Rates.—The area over which a certain quantity of paint will spread depends on the nature and consistency of the paint and the porosity and roughness of the surface to which it is applied. Only approximate figures for average conditions can be given. Table 1, furnished through the courtesy of the National Paint, Varnish, and Lacquer Association, Inc., Washington, D. C., lists the average covering power of various types of paints.

Estimating Paint Requirements.—The quantity of paint required for a job may be estimated by dividing the area to be covered by the spreading rate of the particular paint for the kind of surface to be covered and the number of coats to be applied, as given in Table 1.

TABLE 1

SPREADING RATES OF VARIOUS PAINTS *

Surface and Product	Average Coverage in Square Feet per Gallon		
	1st (or Primer) Coat	2nd Coat	3rd Coat
Frame Siding			
Exterior House Paint	468	540	630
Aluminum Paint	550	600	
Trim (Exterior)			
Exterior Trim Paint	850	900	972
Porch Floors and Steps			
Porch and Deck Paint	378	540	576
Asbestos Wall Shingles			
Exterior House Paint	180	400	
Shingle Siding			
Exterior House Paint	342	423	
Shingle Stain	150	225	
Shingle Roofs			
Exterior Oil Paint	150	250	
Shingle Stain	120	200	
Brick (Exterior)			
Exterior Oil Paint	200	400	
Cement Water Paint	100	150	
Exterior Emulsion	215		
Cement and Cinder Block			
Cement Water Paint	100	140	
Exterior Oil Paint	180	240	
Medium Texture Stucco			
Exterior Oil Paint	153	360	360
Cement Water Paint	99	135	
Aluminum Paint	300	400	
Cement Floors and Steps (Exterior)			
Porch and Deck Paint	450	600	600
Color Stain and Finish	510	480	

TABLE 1 (*Continued*)

Spreading Rates of Various Paints *

Surface and Product	Average Coverage in Square Feet per Gallon		
	1st (or Primer) Coat	2nd Coat	3rd Coat
Doors and Windows (Interior)			
Enamel	603	405	504
Picture Molding, Chair Rails and Other Trim (Coverage per gal. in linear ft.)	1200	810	810
Floors, Hardwood (Interior)			
Oil Paint	540	450	
Shellac	540	675	765
Varnish	540	540	540
Linoleum			
Varnish	540	558	
Walls, Smooth-Finish Plaster			
Flat Oil Paint	630 Primer	540	630
Gloss or Semi-Gloss Oil Paint	630 Primer	540	540
Calcimine	720 Size	240	
Emulsion Paint (Latex Base Paints, Kemtone, etc.)	540	700	
Casein Water Paint	540	700	
Aluminum Paint	450	600	

* Courtesy National Paint, Varnish, and Lacquer Association, Inc., Washington, D. C.

Illustration: How many gallons of flat finish oil paint are required for two coats on the plaster walls of one room 14 feet by

22 feet and two rooms 13 feet by 15 feet if the ceilings are 9 feet high? Assume door and window openings to total 200 square feet.

Large room areas

$$\begin{aligned}
\text{End walls} &= 2 \times 14 \times 9 = 252 \text{ sq. ft.}\\
\text{Side walls} &= 2 \times 22 \times 9 = 396 \text{ sq. ft.}\\
\text{Ceiling} &= 14 \times 22 = 308 \text{ sq. ft.}
\end{aligned}$$

Two smaller room areas

$$\begin{aligned}
\text{End walls} &= 4 \times 13 \times 9 = 468 \text{ sq. ft.}\\
\text{Side walls} &= 4 \times 15 \times 9 = 540 \text{ sq. ft.}\\
\text{Ceiling} &= 2 \times 13 \times 15 = 390 \text{ sq. ft.}
\end{aligned}$$

Total area. . . . . 2354 sq. ft.
Area of openings. . . . . . 200 sq. ft.

2154 sq. ft.

Spreading rate per gallon for the first (or primer) coat is 630 sq. ft. from Table 1.

$$\text{Paint required for first coat} = \frac{2154}{630} = 3.4 \text{ gal. (approx.).}$$

Spreading rate per gallon for the second coat is 540 sq. ft. from Table 1.

$$\text{Paint required for second coat} = \frac{2154}{540} = 4.0 \text{ gal. (approx.).}$$

$$\begin{aligned}
\text{Total paint required for the two coats} &= 3.4 + 4.0\\
&= 7.4 = 8 \text{ gal.}
\end{aligned}$$

ILLUSTRATION: How much varnish is needed for two coats on a hardwood floor 60 feet by 40 feet?

$$\text{Area} = 60 \times 40 = 2400 \text{ sq. ft.}$$

Spreading rate per gallon for the first coat is 540 sq. ft. from Table 1.

$$\text{Varnish required for the first coat} = \tfrac{2400}{540} = 4.5 \text{ gal.}$$

Since the spreading rate is the same for the second coat, the same amount of varnish will be required for the second coat. Therefore the total amount of varnish required is $4.5 \times 2 = 9$ gal.

TABLE 2 *

SPREADING RATES OF ROOF PAINTS

Type of Roof	Square Feet Covered by One Gallon	
	Asbestos or Fibered Roof Coating	Roof or Metal Paint (Nonfibered)
Composition Roofing (Felt or Paper)	50–75	75–100
Concrete	50–75	50–75
Metal	75–100	100–250
Slag or Gravel	25–50	20–35

* Circular 736, National Paint, Varnish, and Lacquer Association, Inc. Washington, D. C.

ILLUSTRATION: How much aluminum paint is required for two coats on a wooden silo 12 feet in diameter and 30 feet high if it has a conical roof with a rise of 4 feet and an overhang of one foot?

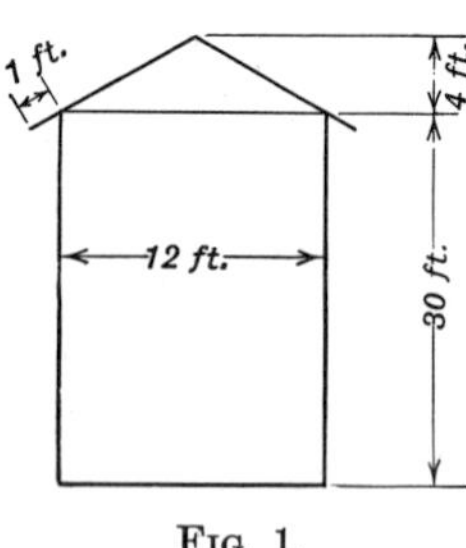

FIG. 1.

Computation of the roof area as a cone whose area is one-half of the product of the slant height and the circumference of the base, would be a refinement not warranted by the problem. It is sufficiently accurate to regard the roof as a disc 14 feet in diameter. Then,

$$\text{Area of roof} = \tfrac{1}{4}\pi D^2 = \frac{14 \times 14\pi}{4} = 49\pi = 154 \text{ sq. ft.}$$

$$\text{Area of cylinder} = \pi Dh = 12 \times 30 \times \pi = 1130 \text{ sq. ft.}$$

$$\text{Total} \quad 1284 \text{ sq. ft.}$$

Spreading rate first coat = 550 sq. ft. per gallon

$$\text{Paint required first coat} = \frac{1284}{550} = 2.3 \text{ gallons}$$

Spreading rate second coat = 600 sq. ft. per gallon

$$\text{Paint required second coat} = \frac{1284}{600} = 2.1 \text{ gallons}$$

Total pa'nt required = 2.3 + 2.1 = 4.4 = 5 gallons (Ans.)

ILLUSTRATION: A smooth hemispherical dome 32 feet in diameter is to be given one coat of asbestos roof paint. How many gallons of paint will be required?

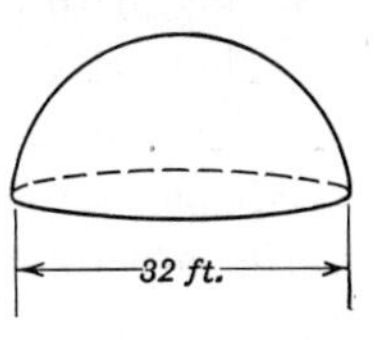

FIG. 2.

The area of a sphere is πD^2, then the area of a hemisphere is $\frac{\pi D^2}{2}$ and,

$$\text{Area of dome} = \frac{\pi D^2}{2} = \frac{\pi \times 32 \times 32}{2} = \pi \times 512 = 1610$$

Spreading rate = 100 sq. ft. per gallon (Table 2)

$$\text{Paint required} = \frac{1610}{100} = 16 \text{ gallons (Ans.)}$$

For a two-coat repainting job on the exterior of a house of moderate size and in good condition, it is fairly safe to estimate that as many gallons of paint as there are rooms in the house will be required. Half again as many gallons will be required for a three-coat job.

PAPER HANGING

Papers for the walls of rooms are printed with distemper color and with oil colors; the cheaper papers are made by machine, the more expensive are hand blocked.

Wall paper is usually made in rolls 18 inches wide and, single rolls, 8 yards long, double rolls, 16 yards long. A roll of border is the same length as a roll of wall paper.

Calculating the Number of Rolls.—There are several methods of figuring the numbers of rolls required to paper a room. Moreover, the methods of measurement vary in different localities. Some measure all the walls as solid, without deductions for the ordinary openings; others deduct one-half of a single roll for each ordinary door or window. Some do not deduct for openings less than 20 square feet in order to compensate for cutting and fitting; others add 15 percent to the area to allow for waste.

There is always waste in matching which must be allowed for; and the height of the room has a great deal to do with the number of strips that can be cut from a roll. Often a double roll cuts to better advantage than a single roll.

ILLUSTRATION: Find the number of rolls of paper for a room 9 feet in height, 15 feet long and 12 feet wide, if the room has one door and three windows each $3\frac{1}{2}$ feet wide.

First Method:

$$\begin{aligned} \text{Perimeter of room} &= 2 \times (12 \text{ ft.} + 15 \text{ ft.}) = 54 \text{ ft.} \\ \text{Width of door and windows} &= 4 \times 3\tfrac{1}{2} \text{ ft.} = 14 \text{ ft.} \\ \hline \text{Perimeter less door and windows} &= 40 \text{ ft.} \end{aligned}$$

Allowing one double roll or two single rolls for every seven feet,

$$40 \div 7 = 5\tfrac{5}{7}$$

Therefore, 6 double rolls will be required.

Second Method:

Perimeter of room = 54 ft.
Wall surface = 54 ft. × 9 ft. = 486 sq. ft.
Allowing 20 sq. ft. per opening = 4 × 20 = 80 sq. ft.
Area of single roll = 24 × $1\frac{1}{2}$ = 30 square feet = 406 sq. ft.

$$406 \div 30 = 13\frac{1}{2}$$

Therefore 14 single rolls will be required.

Third Method:

Perimeter of room in yards = 2 × (4 + 5) = 18 yards
Subtract width of doors and windows, approximately = $4\frac{1}{2}$ yards

$13\frac{1}{2}$ yards

Because a roll is $\frac{1}{2}$ yard wide, the number of strips = $13\frac{1}{2} \times 2 = 27$

Because the room is 9 feet high, each strip will be 9 feet or 3 yards long.

$27 \times 3 = 81$ yards required

$81 \div 16$ (the number of yards in a double roll) $= 5\frac{5}{16}$

Therefore 6 double rolls will be required.

Since the distance around the room is 54 feet or 18 yards, and a 2-strip roll of border contains 16 yards, $18 \div 16$ or $1\frac{1}{8}$ rolls of border are required.

The amount of wall paper needed for the ceiling is found by finding the area, 12 feet × 15 feet = 180 square feet.

Dividing the area of the ceiling in feet by the area of 1 roll in feet, $24 \times 1\frac{1}{2} = 36$. Then, $180 \div 36 = 5$ rolls.

Allowing 1 roll for trimming and matching, 6 single rolls would be required.

TABLE 3

ROLLS OF WALL PAPER AND BORDER REQUIRED FOR VARIOUS-SIZED ROOMS *

Dimensions of Room in Feet	Height of Ceiling in Feet	Number of Doors	Number of Windows	Rolls of Paper	Yards of Border
7×9	9	1	1	7	11
7×9	10	1	1	8	11
8×10	9	1	1	8	12
8×10	10	1	1	9	12
9×11	9	1	1	10	14
9×11	10	1	1	11	14
10×12	9	1	1	10	15
10×12	10	1	1	11	15
11×12	9	2	2	9	16
11×12	10	2	2	10	16
12×13	9	2	2	10	17
12×13	10	2	2	11	17
12×15 or 13×14	9	2	2	11	18
12×15 or 13×14	10	2	2	13	18
13×15	9	2	2	11	19
13×15	10	2	2	13	19
14×16	9	2	2	12	20
14×16	10	2	2	14	20
14×18	9	2	2	13	22
14×18	10	2	2	15	22
15×16	10	2	2	15	21
15×17	12	2	2	19	22

* 18″ rolls; papering of ceilings not included.

OTHER SURFACE COVERINGS

In addition to painting and paperhanging, walls, ceilings, and floors can be covered by a wide variety of other materials. These include wood, plastic, metal, and special compositions which come in many sizes. A number of the commercially available types are listed in Tables 4 and 5.

TABLE 4

FLOOR COVERINGS

Tiles	Tile Sizes in Inches								
	4 × 4	6 × 6	6 × 12	9 × 9	9 × 18	12 × 12	12 × 24	18 × 24	18 × 36
Asphalt			x	x		x	x	x	
Cork		x	x	x		x			
Linoleum				x					
Plastic		x	x	x		x	x	x	
Rubber	x	x	x	x	x	x			x

CONTINUOUS COVERINGS

Linoleum and Vinyl floor coverings are available in continuous rolls in several standard widths, some of which are:

Linoleum: 2′, 3′, 6′, 9′, and 12′
Vinyl: 6′ and 9′

TABLE 5

CEILING AND WALL COVERINGS

Acoustic Materials		Tile Sizes in Inches			
Type	Thickness	6 × 12	12 × 12	12 × 24	24 × 24
"Cushiontone" *	½″, ¾″, 1″	x	x	x	x
"Travertone" *	11/16″, 13/16″	x	x		
"Arrestone" *	2½″		x	x	
"Corkoustic" *	1¼″	x	x		
"Celotex" †					
Perf. Asbestos Bd.	3/16″, ⅛″		x	x	x
Perf. Acoustic Tile	⅝″		x		
Perf. Acoustic Tile	1″		x	x	

Paneling		Panel Length in Feet						
Type	Width	5	6	7	8	9	10	12
Wall Plank	8″, 10″, 12″, 16″				x		x	x
½″ Building Board	4′		x	x	x	x	x	x
1″ Building Board	4′				x		x	x
3/16″–¾″ Plywood	2½′, 3′, 3½′, 4′	x	x	x	x	x	x	x
1/85″ "Flexwood"	18″, 24″				x		x	x

WALL TILE

Aluminum—4¼″ × 4¼″, 5″ × 5″, 5″ × 10″, 10″ × 10″
Plastic—4¼″ × 4¼″, 4¼″ × 8½″, 8½″ × 8½″
Plastic Asbestos—4″ × 12″, 6″ × 6″, 6″ × 12″, 9″ × 18″, 12″ × 12″, 18″ × 24″
"Cork Wall"—6″ × 12″, 9″ × 9″, 12″ × 24″, 24″ × 48″
Steel, Aluminum, Stainless Steel—4¼″ × 4¼″, 6″ × 6″, 8½″ × 8½″

* Armstrong Cork Company.
† Johns Manville.

Generally speaking, the calculations involved in estimating the amount of material needed to cover a given surface are simple and straightforward. Areas of surfaces are calculated in a manner similar to that described for painting and paperhanging.

Allowances are made for openings, and the net area to be covered is divided by the area of the tile, panel, or strip being used.

For example, if it is desired to cover a 10′ × 15′ floor with 9″ × 9″ tiles the problem is solved as follows:

$$\frac{\text{Area of floor in square feet}}{\text{Area of tile in square feet}} = \frac{10 \times 15}{\frac{9}{12} \times \frac{9}{12}} = \frac{150}{0.75 \times 0.75} = \frac{150}{0.5625}$$

$$= 266.67 \quad = 267 \text{ tiles required.}$$

To allow for wastage and trim it is wise to add approximately 5% to the calculated figure, which would bring the total number of tiles required to about 280.

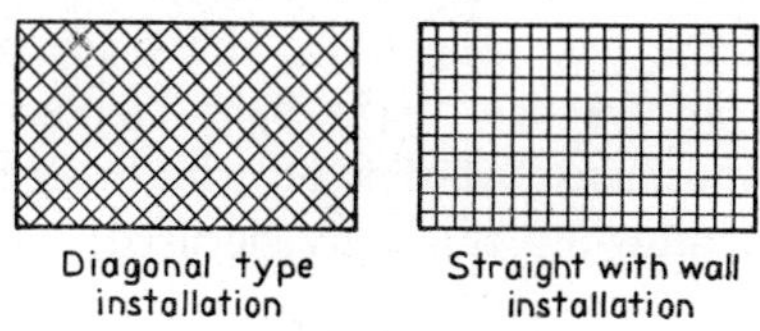

FIG. 3.—Diagonal and Straight Type Tile Arrangements.

Diagonal layouts such as that shown in Fig. 3 are often desirable from the standpoint of appearance and are suggested where floor boards are wider than 2″. In such layouts, approximately 4″ should be added to the length and width of the room when calculating the room area to allow for the additional trim needed.

WINDOW GLASS AND GLAZING

Common window glass is technically known as sheet glass or cylinder glass. It is usually set with putty and fastened with triangular pieces of zinc called glazier's points, driven into the wood over the glass and covered with putty.

Besides common window glass there are other kinds used in building construction, such as plate glass, wire glass, ornamental and colored glass, skylight glass, etc.

The best quality of window glass is specified as AA, the second as A, and the third as B. It is graded as double-thick or single-thick, and each thickness is further divided into three qualities, first, second, or third. This grading is based upon the color and brilliancy, and the presence or absence of flaws in the material. Single-thick window glass is approximately $1/16$ inch in thickness, double-thick being approximately $1/8$ inch.

Stock Sizes of Window Glass.—The regular stock sizes vary by inches from 6 inches to 16 inches in width. Above that they vary by even inches up to 60 inches in width and 70 inches in length for double thickness, and up to 30 inches by 50 inches for single thickness.

Cost Calculations.—Window glass is sold by the box containing about 50 square feet of glass. The price per square foot increases rapidly as the size of the pane increases.

To find the number of boxes of window glass of a given required size the following rule may be used:

Divide the product of 50×144 by the product obtained by multiplying the length and width of each pane.

ILLUSTRATION: Find the number of boxes of glass required to furnish glass for 15 windows consisting of 4 panes of glass, each 13 inches $\times$ 28 inches.

$$50 \times 144 = 7200$$

$$13 \times 28 = 364$$

$$7200 \div 364 = 20 \text{ (approximately)}$$

Therefore, 1 box of glass will contain 20 panes.

$$15 \times 4 = 60 \text{ panes required.}$$

$$60 \div 20 = 3$$

Thus, 3 boxes of glass are needed.

Glass Blocks.—Many modern architectural designs call for the use of glass blocks. This material provides a modern appearance

TABLE 6

Sizes and Number of Panes in a Box of Window Glass

Size in Inches	Panes in Box	Size in Inches	Panes in Box	Size in Inches	Panes in Box	Size in Inches	Panes in Box
6 × 8	150	12 × 19	32	16 × 20	23	24 × 44	7
7 × 9	115	12 × 20	30	16 × 22	20	24 × 50	6
8 × 10	90	12 × 21	29	16 × 24	19	24 × 56	5
8 × 11	82	12 × 22	27	16 × 30	15	26 × 36	8
8 × 12	75	12 × 23	26	16 × 36	12	26 × 40	7
9 × 10	80	12 × 24	25	16 × 40	11	26 × 48	6
9 × 11	72	13 × 14	40	18 × 20	20	26 × 54	5
9 × 12	67	13 × 15	37	18 × 22	18	28 × 34	8
9 × 13	62	13 × 16	35	18 × 24	17	28 × 40	6
9 × 14	57	13 × 17	33	18 × 26	15	28 × 46	6
9 × 15	53	13 × 18	31	18 × 34	12	28 × 50	5
9 × 16	50	13 × 19	29	18 × 36	11	30 × 40	6
10 × 10	72	13 × 20	28	18 × 40	10	30 × 44	5
10 × 12	60	13 × 21	26	18 × 44	9	30 × 48	4
10 × 13	55	13 × 22	25	20 × 22	16	30 × 54	4
10 × 14	52	13 × 24	23	20 × 24	15	32 × 42	5
10 × 15	48	14 × 15	34	20 × 25	14	32 × 44	5
10 × 16	45	14 × 16	32	20 × 26	14	32 × 46	5
10 × 17	42	14 × 18	29	20 × 28	13	32 × 48	5
10 × 18	40	14 × 19	27	20 × 30	12	32 × 50	4
11 × 11	59	14 × 20	26	20 × 34	11	32 × 54	4
11 × 12	55	14 × 22	23	20 × 36	10	32 × 56	4
11 × 13	50	14 × 24	22	20 × 40	9	34 × 60	4
11 × 14	47	14 × 28	18	20 × 44	8	34 × 40	5
11 × 15	44	14 × 32	16	20 × 50	7	34 × 44	5
11 × 16	41	14 × 36	14	22 × 24	14	34 × 46	5
11 × 17	39	14 × 40	13	22 × 26	13	34 × 50	4
11 × 18	36	15 × 16	30	22 × 28	12	34 × 52	4
12 × 12	50	15 × 18	27	22 × 36	9	34 × 56	4
12 × 13	46	15 × 20	24	22 × 40	8	36 × 44	5
12 × 14	43	15 × 22	22	22 × 50	7	36 × 50	4
12 × 15	40	15 × 24	20	24 × 28	11	36 × 56	4
12 × 16	38	15 × 30	16	24 × 30	10	36 × 60	3
12 × 17	35	15 × 32	15	24 × 32	9	36 × 64	3
12 × 18	33	16 × 18	25	24 × 36	8	40 × 60	3

and admits diffused daylight over a wide area, at the same time affording complete privacy.

Glass blocks are hollow "all-glass" units which have fused seals. The interior of the blocks is relatively free of water vapor, and the dry dead air space acts as an effective heat insulator. A single cavity glass block has an insulating value greater than that of an 8-inch brick wall, and more than twice that of ordinary windows. Double cavity blocks which have a fibrous glass screen insert that divides the dead air space into two pockets also are available. These provide even better thermal insulation.

A special resilient plastic coating on all mortar edges forms a permanent bond between glass and mortar, insuring a high degree of wind resistance and weather-tightness. The glass block edge construction forms a "key-lock" mortar joint which allows a full bed of mortar and a visible joint of only about ¼ inch.

Glass blocks are made in various patterns and in three sizes: 5¾″ × 5¾″, 7¾″ × 7¾″, and 11¾″ × 11¾″ (generally referred to as 6″, 8″, and 12″). All units are 3⅞″ thick. Special shapes are available for turning corners and for building curved panels. Typical estimating data for glass blocks are shown in Table 7.

TABLE 7

GLASS BLOCK ESTIMATING DATA

(For 100 sq. ft. of panel laid with ¼″ visible mortar joints)

Size of block	6″	8″	12″
Number of blocks	400	225	100
Weight of panel	2000 lbs.	1800 lbs.	1900 lbs.
Volume of mortar	4.3 cu. ft.	3.2 cu. ft.	2.2 cu. ft.

†**Plate Glass and Thermopane®.**—The trend toward the construction of buildings with larger portions of their external structures composed of glass, including picture windows with wide and high panes, has resulted in the use in construction of various types of plate glass, as well as double-glass with insulating air spaces between the two panes of glass. In this way one obtains the greater strength necessary to withstand the greater wind and thermal stresses on the greater areas of glass, and also the lower transmission of heat through the window, that is necessary to reduce heat losses in winter and heat gains in summer.

The calculation of the thickness of plate glass required to withstand wind loadings can be computed from tables which show the relation between the maximum area of glass in square feet and the wind loading in pounds per square foot. This loading is established by local building codes and ordnances, and must be found out for the local area from the building authorities. Assume that in a given locality the maximum permissible load is 30 lbs. per square foot, one would use the graphs of Fig. 4 to find the maximum area that could be covered with a given thickness of glass. Thus for a $9.7' \times 10'$ window, there would be an area of 97 sq. ft. Find on the bottom of Fig. 4 the vertical coordinate for 30 lbs. per sq. ft. loading, and run up that line until it intersects the horizontal coordinate for 97 sq. ft. This point is above the graph for ⅜″ plate, but below that for ½″ plate, so the latter is specified.

Reference to Fig. 4 also discloses the words "Design Factor 2.5." This means that the graphs in this figure have been constructed for the average pressure divided by 2.5, which is thus the factor of safety. When a design factor other than 2.5 is chosen, the graphs in Fig. 4 may be used providing the design load is adjusted as follows:

$$\left[\begin{array}{c}\text{design load for}\\ \text{use with graphs}\end{array}\right] = \left[\frac{\text{actual design load}}{2.5}\right] \times \left[\begin{array}{c}\text{chosen design}\\ \text{factor}\end{array}\right]$$

† Acknowledgement is made to the Libbey-Owens-Ford Glass Co. for the information on which this section is based.

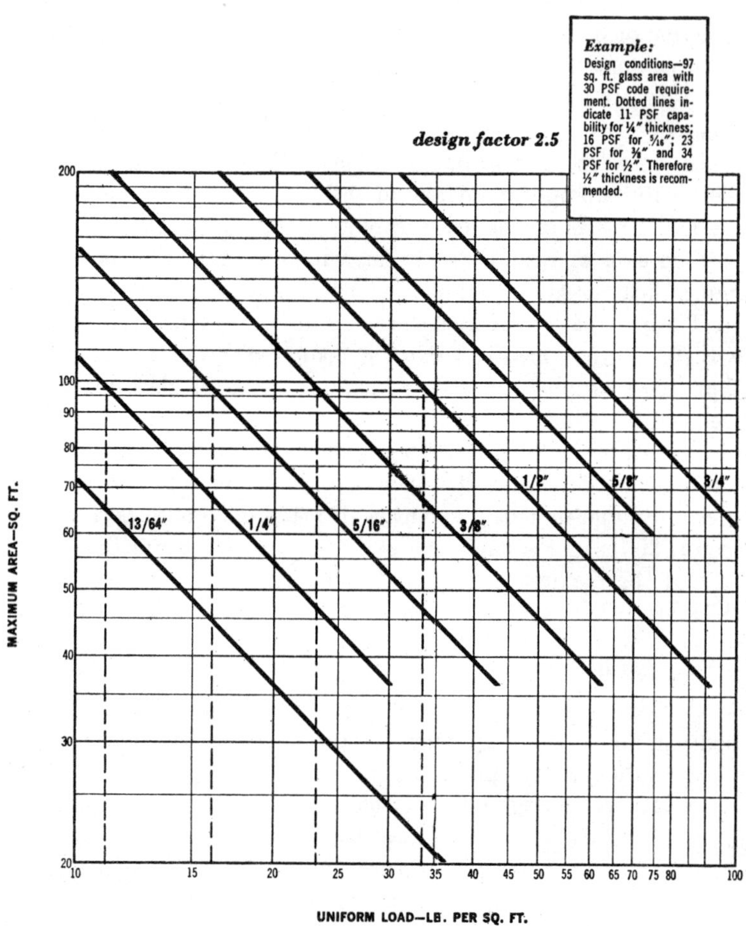

FIG. 4.—Thickness of Glass Required for Various Areas. (Graphs apply to square and rectangular lights of glass when the length is not more than 5 times the width.)

For example: Assume a building code requires the choice of glass areas and thicknesses to be based on a design load of 25 psf and the architect or engineer decides the appropriate design factor is 2.0. The adjusted design load for use with the graphs would be determined as follows:

$$\text{design load for use with graphs} = \left[\frac{25}{2.5}\right] \times [2.0]$$

$$= 20 \text{ lbs. per sq. ft.}$$

The types of glass for which these graphs were prepared are Libbey-Owens-Ford Polished Plate Glass for the $^{13}\!/_{64}''$ and $\frac{1}{4}''$ thicknesses, and L-O-F Heavy-Duty-Plate Glass® for the greater thicknesses. The data can be used for other types of glass by multiplying the strength found by suitable factors, as follows:

Type of Glass	Multiplying Factor
Tuf-flex®	4.0
Thermopane®	1.5
Vitrolux®	2.0
Wired	0.5
Laminated	0.6
Rough Plate	1.0
Sandblasted	0.4

The complete data on the varieties of L-O-F polished plate glass for residential glazing are given in Table 8.

Laminated safety glass is made in two lights of polished plate or sheet glass bonded by a plastic, which is a polyvinyl butyral resin selected for strength, durability and adhesion. It is suitable for use not only in automobiles, but also in bathroom partitions and doors because of the greater thermal stresses and mechanical shocks to which they may be subjected. Table 9 gives selection data on laminated safety glass.

Thermopane® is manufactured in two styles, differing in the method of sealing the two panels together. In the GlasSeal® edge the two panes are fused together, and therefore the sizes of the lights are more restricted than are those of the Bonder-

TABLE 8

Plate Glass	Thickness	Quality	Thickness Tolerance	Maximum Size: Standard	Maximum Size: Special	Approx. Weight Lbs. per Sq. Ft.	Luminous Illuminant C (Average Daylight) Transmitt.	Average Solar Radiation Transmittance: Ultraviolet Radiation	Average Solar Radiation Transmittance: Total Solar Radiation
	1/4″	Silvering	±1/32″	up to 25 sq. ft.					
Parallel-O-Plate	1/4″	Mir. Glazing	±1/32″	up to 75 sq. ft.		3.27	89.1	67.8	79.9
	1/4″	Glazing	±1/32″	124″ × 170″	124″ × 252″				
Regular Plate	3/8″	Glazing	±1/32″	72″ × 74″	74″ × 120″	1.64	90.6	75.9	86.1
Parallel-O-Grey	13/64″	Glazing	±1/32″	84″ × 120″		2.66	50.0	44.0	52.2
	1/4″			96″ × 138″	120″ × 192″	3.27	44.2	39.0	46.6
Parallel-O-Bronze	13/64″	Glazing	±1/32″	84″ × 120″		2.66	54.4	32.5	47.0
	1/4″		±1/32″	96″ × 138″	120″ × 192″	3.27	48.5	27.5	41.0
Heat Absorbing	1/4″	Glazing	±1/32″	84″ × 120″	96″ × 138″	3.27	74.7	44.9	46.3
Regular Plate	1/4″	Glazing	±1/32″	±1/16″	72″ × 120″	3.27	89.1	67.8	79.9
Grey	1/4″	Glazing	±1/32″	±1/16″	72″ × 120″	3.27	44.2	39.0	46.6
Bronze	1/4″	Glazing	±1/32″	±1/16″	72″ × 120″	3.27	48.5	27.5	41.0
Heat Absorbing	1/4″	Glazing	±1/32″	±1/16″	72″ × 120″	3.27	74.7	44.9	46.3

TABLE 9

Types	Nominal Thickness	Thickness Range	Max. Size	Net Wt. Lbs. per Sq. Ft.
Thin Safety Sheet	9/64″	.120–.160	7 sq. ft.	1.92
S. S. Safety Sheet	7/32″	.185–.250	15 sq. ft.	2.49
Combination Safety Sheet (S.S. & D.S.)	15/64″	.200–.260	15 sq. ft.	2.91
D. S. Safety Sheet	1/4″	.220–.280	15 sq. ft.	3.32
E-Z-Eye Safety Sheet	15/64″	.200–.260	15 sq. ft.	2.91
Safety Plate (Clear and E-Z-Eye)	1/4″	.220–.280	72″ × 138″	3.16
Heavy Safety Plate	3/8″	±1/32″	72″ × 138″	4.88
	1/2″	±1/16″	72″ × 138″	6.54
	5/8″	±1/16″	72″ × 138″	8.13
	3/4″	±1/16″	72″ × 138″	9.81
	7/8″	±1/16″	72″ × 138″	11.45
	1″	±1/16″	72″ × 138″	13.08

metic® style, which have a glass-to-metal edge. These sizes are shown in Table 10. (Note that SSA means "single-strength A-quality" and DSA means "double-strength A-quality.")

In computing the heat savings in winter, the factor used is represented by U_t which is the heat transfer coefficient for Thermopane®. Under winter conditions, it is 0.69 B.T.U. per hour per sq. ft. per °F. temperature difference for GlasSeal® Thermopane (with 3/16″ air space), 0.58 for Bondermetic® Thermopane (with 1/2″ air space), both as against a single glass pane value of 1.13 in the same units. Under summer conditions, U_t is 0.64 B.T.U. per hour per sq. ft. per °F. temperature difference for the GlasSeal® Thermopane, 0.56 for the Bondermetic® Thermopane, and 1.06 for a single glass pane.

Using these the applicable figures above, the reduction in heating load due to the use of Thermopane can be calculated by the following equation:

$$\text{Reduction in Btu per hr.} = (U_s - U_t) \times (t_i - t_0)$$

TABLE 10

Bondermetic

2 panes of $\frac{1}{4}''$ polished Parallel-O-Plate $\frac{1}{2}''$ air space. 6.5 lbs. per sq. ft.

Width	Height	Width	Height
$33''$	$\times\ 76\frac{3}{4}''$	$64\frac{1}{2}''$	$\times\ 50''$
$35\frac{1}{2}''$	$\times\ 36''$	$64\frac{1}{2}''$	$\times\ 58''$
$35\frac{1}{2}''$	$\times\ 48\frac{1}{8}''$	$64\frac{1}{2}''$	$\times\ 66''$
$35\frac{1}{2}''$	$\times\ 60\frac{3}{8}''$	$66\frac{5}{8}''$	$\times\ 47\frac{3}{4}''$
$42''$	$\times\ 66''$	$66\frac{5}{8}''$	$\times\ 60\frac{1}{8}''$
$42''$	$\times\ 72''$	$68\frac{3}{4}''$	$\times\ 36''$
$44\frac{1}{2}''$	$\times\ 36''$	$68\frac{3}{4}''$	$\times\ 48\frac{1}{8}''$
$44\frac{1}{2}''$	$\times\ 46''$	$68\frac{3}{4}''$	$\times\ 60\frac{3}{8}''$
$44\frac{1}{2}''$	$\times\ 48\frac{1}{8}''$	$68\frac{3}{4}''$	$\times\ 72\frac{3}{4}''$
$44\frac{1}{2}''$	$\times\ 60\frac{3}{8}''$	$70\frac{1}{8}''$	$\times\ 52\frac{1}{2}''$
$44\frac{1}{2}''$	$\times\ 72\frac{3}{4}''$	$70\frac{1}{8}''$	$\times\ 56\frac{1}{2}''$
$45''$	$\times\ 76\frac{3}{4}''$	$72''$	$\times\ 48''$
$45\frac{3}{8}''$	$\times\ 52''$	$72''$	$\times\ 60''$
$46\frac{1}{8}''$	$\times\ 52\frac{1}{2}''$	$72\frac{1}{2}''$	$\times\ 46''$
$47\frac{3}{4}''$	$\times\ 50\frac{3}{8}''$	$72\frac{1}{2}''$	$\times\ 50''$
$48''$	$\times\ 48''$	$72\frac{1}{2}''$	$\times\ 58''$
$48''$	$\times\ 60''$	$72\frac{1}{2}''$	$\times\ 66''$
$48\frac{1}{2}''$	$\times\ 42''$	$75''$	$\times\ 36''$
$48\frac{1}{2}''$	$\times\ 46''$	$75''$	$\times\ 48\frac{1}{8}''$
$48\frac{1}{2}''$	$\times\ 50''$	$75''$	$\times\ 60\frac{3}{8}''$
$48\frac{1}{2}''$	$\times\ 58''$	$80\frac{1}{2}''$	$\times\ 50''$
$50\frac{3}{8}''$	$\times\ 47\frac{3}{4}''$	$80\frac{1}{2}''$	$\times\ 58''$
$50\frac{3}{8}''$	$\times\ 60\frac{1}{8}''$	$84''$	$\times\ 66''$
$55\frac{1}{4}''$	$\times\ 36''$	$84''$	$\times\ 72''$
$55\frac{1}{4}''$	$\times\ 48\frac{1}{8}''$	$93''$	$\times\ 36''$
$55\frac{1}{4}''$	$\times\ 60\frac{3}{8}''$	$93''$	$\times\ 48\frac{1}{8}''$
$56\frac{1}{2}''$	$\times\ 42''$	$93''$	$\times\ 60\frac{3}{8}''$
$56\frac{1}{2}''$	$\times\ 46\frac{1}{8}''$	$93''$	$\times\ 72\frac{3}{4}''$
$56\frac{1}{2}''$	$\times\ 50''$	$96''$	$\times\ 66''$
$56\frac{1}{2}''$	$\times\ 58\frac{1}{8}''$	$96''$	$\times\ 72''$
$56\frac{1}{2}''$	$\times\ 66''$	$96\frac{1}{2}''$	$\times\ 50''$
$57''$	$\times\ 76\frac{3}{4}''$	$96\frac{1}{2}''$	$\times\ 58''$
$58\frac{1}{8}''$	$\times\ 52\frac{1}{2}''$	$116\frac{1}{2}''$	$\times\ 58''$
$64\frac{1}{2}''$	$\times\ 46''$		

2 panes of DSA window glass $\frac{1}{4}''$ air space. 3.25 per sq. ft.

Width	Height	Width	Height
$21\frac{3}{4}''$	$\times\ 62\frac{3}{4}''$	$42\frac{1}{2}''$	$\times\ 22\frac{1}{2}''$
$25\frac{3}{4}''$	$\times\ 62\frac{3}{4}''$	$45\frac{1}{2}''$	$\times\ 25\frac{1}{2}''$

2 panes of $\frac{3}{16}''$ "A" heavy sheet glass with $\frac{1}{2}''$ air space. 5 lbs. per sq. ft.

Width	Height	Width	Height
$35\frac{1}{2}''$	$\times\ 60\frac{3}{8}''$	$56\frac{1}{2}''$	$\times\ 46\frac{1}{8}''$
$48\frac{1}{2}''$	$\times\ 42''$	$56\frac{1}{2}''$	$\times\ 50''$
$48\frac{1}{2}''$	$\times\ 50''$	$56\frac{1}{2}''$	$\times\ 42''$

Glasseal

SSA window glass with $\frac{3}{16}''$ air space (thickness $.375'' \pm .050''$) 2.4 lbs. per sq. ft.

Width	Height	Width	Height
$16''$	$\times\ 20''$	$28''$	$\times\ 16''$
$16''$	$\times\ 22''$	$28''$	$\times\ 18''$
$16''$	$\times\ 24''$	$28''$	$\times\ 20''$
$16''$	$\times\ 32''$	$28''$	$\times\ 22''$
$16''$	$\times\ 36''$	$28''$	$\times\ 24''$
$16''$	$\times\ 48''$	$32''$	$\times\ 16''$
$16''$	$\times\ 60''$	$32''$	$\times\ 20''$
$16\frac{9}{16}''$	$\times\ 24\frac{5}{8}''$	$32''$	$\times\ 22''$
$16\frac{9}{16}''$	$\times\ 30\frac{13}{16}''$	$32''$	$\times\ 24''$
$16\frac{9}{16}''$	$\times\ 36\frac{13}{16}''$	$36''$	$\times\ 14''$
$16\frac{9}{16}''$	$\times\ 49''$	$36''$	$\times\ 16''$
$16\frac{9}{16}''$	$\times\ 61\frac{3}{16}''$	$36''$	$\times\ 18''$
$19''$	$\times\ 15''$	$36''$	$\times\ 20''$
$19\frac{1}{2}''$	$\times\ 53''$	$36''$	$\times\ 22''$
$20''$	$\times\ 16''$	$36''$	$\times\ 24''$
$20''$	$\times\ 20''$	$36''$	$\times\ 30''$
$20''$	$\times\ 22''$	$36\frac{5}{8}''$	$\times\ 14\frac{1}{4}''$
$20''$	$\times\ 24''$	$36\frac{5}{8}''$	$\times\ 18\frac{1}{4}''$
$20''$	$\times\ 32''$	$36\frac{5}{8}''$	$\times\ 22\frac{1}{4}''$
$20''$	$\times\ 36''$	$36\frac{5}{8}''$	$\times\ 30\frac{1}{4}''$
$20''$	$\times\ 48''$	$39''$	$\times\ 14''$
$20''$	$\times\ 60''$	$39''$	$\times\ 18$
$21\frac{1}{16}''$	$\times\ 24\frac{5}{8}''$	$39''$	$\times\ 22''$
$21\frac{1}{16}''$	$\times\ 30\frac{13}{16}''$	$39''$	$\times\ 30''$
$21\frac{1}{16}''$	$\times\ 36\frac{13}{16}''$	$39\frac{5}{8}''$	$\times\ 14\frac{1}{4}''$
$21\frac{1}{16}''$	$\times\ 49''$	$39\frac{5}{8}''$	$\times\ 18\frac{1}{4}''$
$21\frac{1}{16}''$	$\times\ 61\frac{3}{16}''$	$39\frac{5}{8}''$	$\times\ 22\frac{1}{4}''$
$22''$	$\times\ 18''$	$39\frac{5}{8}''$	$\times\ 30\frac{1}{4}''$
$22''$	$\times\ 55\frac{9}{16}''$	$40''$	$\times\ 16''$
$24''$	$\times\ 16''$	$40''$	$\times\ 20''$
$24''$	$\times\ 20''$	$40''$	$\times\ 24''$
$24''$	$\times\ 22''$	$42\frac{1}{2}''$	$\times\ 22\frac{1}{2}''$
$24''$	$\times\ 24''$	$44''$	$\times\ 14''$
$24''$	$\times\ 32''$	$44''$	$\times\ 16''$
$24''$	$\times\ 36''$	$44''$	$\times\ 18''$
$24''$	$\times\ 48''$	$44''$	$\times\ 22''$
$24''$	$\times\ 60''$	$44''$	$\times\ 30''$
$24\frac{1}{4}''$	$\times\ 15\frac{1}{4}''$	$44\frac{5}{8}''$	$\times\ 14\frac{1}{4}''$
$27\frac{1}{4}''$	$\times\ 14\frac{1}{4}''$	$44\frac{5}{8}''$	$\times\ 18\frac{1}{4}''$
$27\frac{1}{4}''$	$\times\ 18\frac{1}{4}''$	$44\frac{5}{8}''$	$\times\ 22\frac{1}{4}''$
$27\frac{1}{4}''$	$\times\ 22\frac{1}{4}''$	$44\frac{5}{8}''$	$\times\ 30\frac{1}{4}''$
$27\frac{1}{4}''$	$\times\ 30\frac{1}{4}''$	$45\frac{1}{2}''$	$\times\ 25\frac{1}{2}''$
$28''$	$\times\ 14''$		

where U_s = overall coefficient of heat transmission for single glass,

Btu per hr. per sq. ft. per deg. F.

U_t = overall coefficient of heat transmission for Thermopane®, Btu per hr.—sq. ft. per deg. F.

A = Area of glass, sq. ft.

t_i = indoor design temperature, °F.

t_0 = winter outdoor design temperature, °F.

Installation.—The use of plate glass, Thermopane® and other heavy types of glass requires special attention to its installation. While putty is still one of the important glazing compounds, it is essential that the wooden groove be clean and dry and then coated with a good priming paint before the glass is placed to prevent absorption into the wood of the oil from the putty. Window frames that are not to be painted should be primed with varnish. Other than putty, the principal glazing compounds, which are often superior for heavy glass installations, are the mastics. There are two groups: the elastic glazing compounds which are formulated to remain plastic over long periods of time, and the non-skinning compounds which have a resin base of polybutene. This type, however, is not to be used in combination glazing with curing type sealants. These sealants are available in three major types: (1) the chemical curing type, which is based on silicones or urethane resins and which cures by a chemical reaction to form a firm resilient seal; (2) the solvent release type, which dries by evaporation of its solvent, leaving behind the resin content, which may be of the acrylic, Neoprene® or vinyl type in a soft and pliable state; and (3) the two-part polymer base sealants, which cure more rapidly than the other types. These types are available both for use by pouring and in "sealing guns." For heavy work the preformed sealants which are made of natural or synthetic rubber and from various plastics are particularly useful. They are available in preformed tapes, beads, ribbons or mastics, and in both resilient and non-resilient types.

XV

Plumbing

Introduction.—Plumbing is defined as the art of installing in buildings the pipes, fixtures, and other apparatus for bringing in the water supply and removing liquid and water-carried wastes. It has developed within the span of a generation from a task which could be handled by a handy man or a lead wiper to a trade which requires a sound fundamental knowledge of hydraulics, mechanics, and building.

In new building construction the plumber locates the pipes and fixtures as shown on the plans, but he must be on the alert to insure that the installations do not violate the local plumbing code or the principles of good practice. When installing plumbing in an old building, even greater responsibility rests on him. Here it may be his lot to determine all of the pipe sizes and locations. In either case his work may even include bringing water from an independent source of supply and disposing of the wastes by an independent system.

As we have suggested, a complete plumbing installation consists of two mutually independent systems, one the water supply, the other the disposal of wastes, often called the *sanitary plumbing*. These two systems will be considered separately.

Figure 1 shows a typical water supply and drainage layout for a two-story house with a basement. This layout conforms to the Uniform Plumbing Code for Housing. The plumbing layout for a bathroom in a one-story house without a basement is shown in Fig. 2.

Water Systems.—A water system is made up of a source of supply (city main, well and pump, etc.), a distribution and drain-

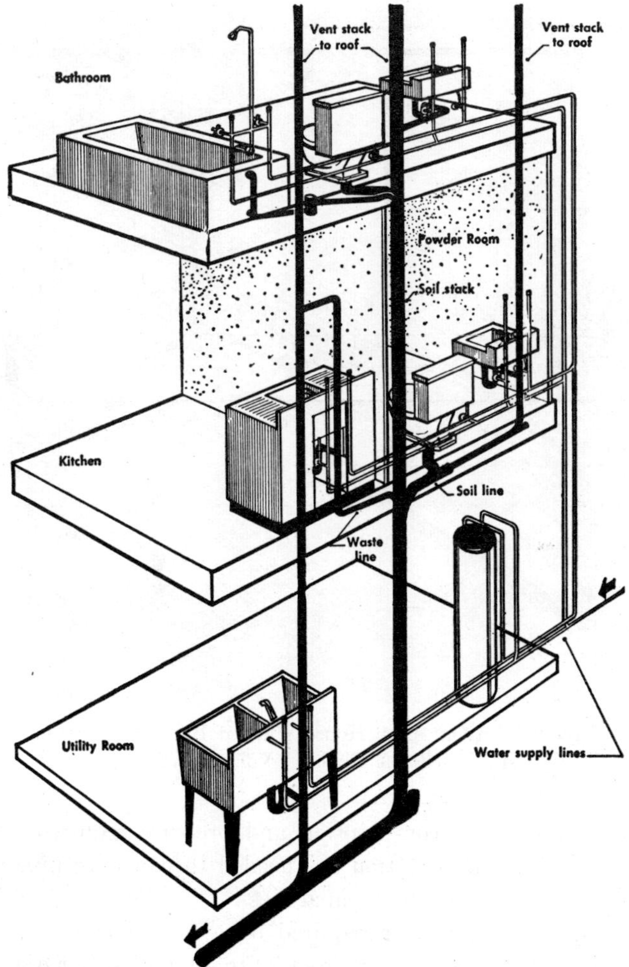

FIG. 1.—A Typical Water Supply and Drainage Layout for a Two-story House with a Basement. This layout conforms with the Uniform Plumbing Code for Housing. (From *University of Illinois Bulletin,* Vol. 48, No. 15.)

age system (Figs. 1 and 2), and various types of fixtures (closet, bathtub, shower, sink, etc.). The size of all these elements de-

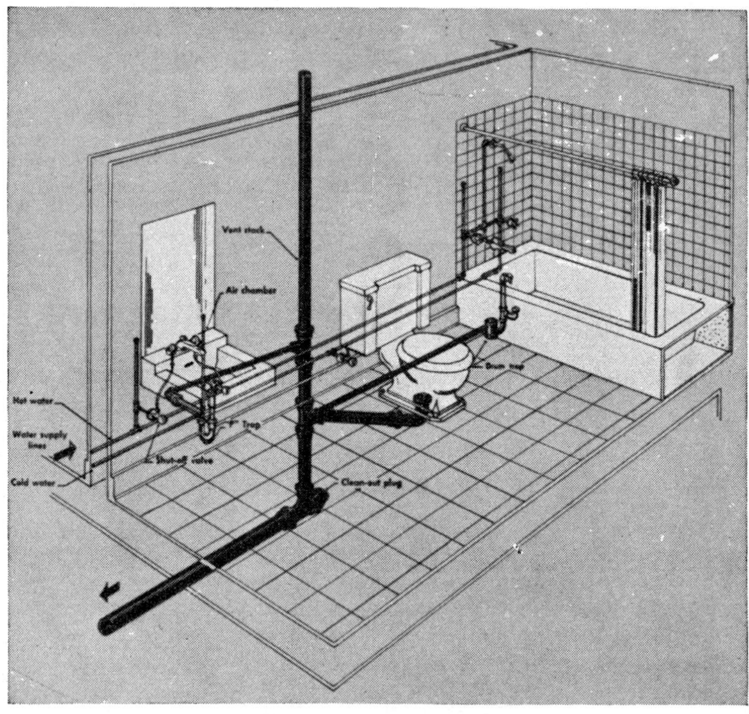

Fig. 2.—Layout for a One-story House Without Basement. (From *University of Illinois Bulletin,* Vol. 48, No. 15.)

pends on the water consumption and pressure required. Estimates on water consumption are needed to calculate pipe sizes, tanks, reservoirs, and pump capacities.

The quantity of water required by a family can readily be computed. Each person uses from 20 to 40 gallons per day; this includes requirements for cooking and laundry. The amounts required for some specific uses are shown in Table 1.

In some plumbing installations such as sprinkler systems for theaters, office buildings, and warehouses, it is required to have a storage unit capable of holding a large quantity of water.

TABLE 1

WATER CONSUMPTION (COMBINED HOT AND COLD WATER) IN U. S. GALLONS PER FIXTURE *

	Type	Size of Inlet	Pressure at Outlet in Psi				
			10	20	30	40	50
Water closets Tank types	All types		18				
Flush valve types	Wash down Siphon jet Blow-out		30	 30 40			
Urinals Tank types	All types		5				
Flush valve types	Stall Wash-down Siphon jet Blow-out	¾	5 7	 25 30			
Showers	2¼″ head 4 " 5 " 6 " 8 " 8″, tubular	½ " " " ¾ "		4 5½ 6½ 7¼ 9 16			
Lavatory	18 x 20 21 x 24		5 9				
Kitchen sink			7				
Bathtub			15				
Laundry tray			9				
Hoses, lawn	Solid stream with spray	¾ ¾	6 4				

* *Plumbing Practice and Design,* Vol. II, S. Plum, John Wiley & Sons, Inc., p. 110.

Cisterns and reservoirs are sometimes used for the storage of rainwater in rural areas. The water is collected from the roofs of buildings by a system of gutters or downspouts which lead the water to the cistern. Table 2 gives the capacities of plain cylindrical cisterns and tanks and may be used to calculate the size of such containers.

TABLE 2

CAPACITIES OF PLAIN CYLINDRICAL CISTERNS AND TANKS

Depth of Cistern or Tank in Feet	Diameter of Cistern or Tank in Feet								
	4	5	6	7	8	9	10	11	12
	Capacity of Cistern or Tank in Gallons								
4	376	588	846	1152	1504	1904	2350	2844	3384
5	470	735	1058	1439	1880	2380	2938	3555	4230
6	564	881	1269	1727	2256	2855	3525	4265	5076
7	658	1028	1481	2015	2632	3331	4113	4976	5922
8	752	1175	1692	2303	3008	3807	4700	5687	6768
9	846	1322	1904	2591	3384	4283	5288	6398	7614
10	940	1469	2115	2879	3760	4759	5875	7109	8460
11	1034	1616	2327	3167	4132	5235	6463	7820	9306
12	1128	1763	2537	3455	4512	5711	7050	8531	10152

ILLUSTRATION: In a certain sprinkler system it is required that 7,000 gallons of water be stored on the roof of the building. What size tank is required?

Referring to Table 2, the size of the tank with the next larger capacity is selected. This tank is 11 ft. in diameter and 10 ft. deep.

Hydropneumatic Water Systems.—In rural and suburban areas where there is no central source of water (reservoirs), hydropneumatic water systems are used to supply the required water.

The water is usually obtained from a deep or shallow well and is raised to the surface and distributed by a motor- or engine-driven pump. A tank, originally filled with air, is used with the pump (see Fig. 3). When pumped partly full with water, the air compresses and the pressure thus created is used to force the water through the pipes. The effective capacity of the hydropneumatic tank is increased if an initial air pressure is used.

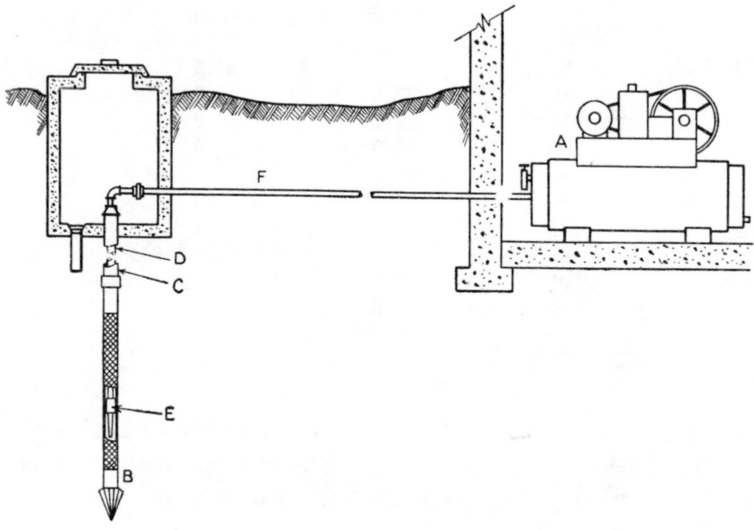

Fig. 3.—A Typical Hydropneumatic Water System. *A*—motordriven pump mounted directly on a pressure tank; *B*—well point (driven into ground); *C*—drive pipe or well casing; *D*—suction pipe inside well casing; *E*—foot valve which acts like a check valve to prevent loss of head in casing; *F*—horizontal suction pipe from well to pump. In this type of installation, the vertical distance from the pump to the water level must not exceed 22 ft.

Table 3 gives the capacity of a tank with and without initial air pressure. Then, if a pump is to be run every other day to supply the requirements of a family using 120 gallons per day with a working range of pressures from 10 pounds to 50 pounds gage pressure, the tank capacity may be figured as follows: At

TABLE 3

WATER CAPACITY OF A HYDROPNEUMATIC TANK WITH AND WITHOUT AN INITIAL AIR PRESSURE *

Gage Pressure, Pounds per Square Inch	Water in Tank When No Initial Air Pressure Is Provided, Percent	Water in Tank With 10 Lb Initial Air Pressure, Percent
50	76.9	61.5
45	75.0	58.3
40	72.7	54.5
35	70.0	50.0
30	66.7	44.4
25	62.5	37.5
20	57.1	28.6
15	50.0	16.7
10	40.0	....
5	25.0	....

* From Circular 303, University of Illinois, College of Agriculture.

50 pounds pressure the tank will hold 61.5 percent of its capacity and at 10 pounds it will be empty. The total quantity of water needed between pumpings is $120 \times 2 = 240$ gallons. Then the total capacity of tank required is $240 \div 61.5$ percent which is 390 gallons. From Table 4, the dimensions of the tank next larger in capacity are 30 inches in diameter by 12 feet long. When a pump on a hydropneumatic system is operated by an electric motor, provision is usually made for starting the motor automatically when the pressure drops to a certain point.

Some idea of the delivery capacity, suction lift, and size of motors required to drive this type of pumping unit can be gained from Table 5. This table lists the performance of Montgomery Ward's shallow well jet pump (a compact, centrifugal type pump specially designed for water systems).

TABLE 4

STANDARD SIZES OF TANKS FOR HYDROPNEUMATIC WATER SYSTEMS

Diameter, Inches	Length, Feet	Size, Gallons	Diameter, Inches	Length, Feet	Size, Gallons
24	6	140	42	8	575
24	8	190	42	10	720
24	10	235	42	12	865
30	6	220	42	14	1000
30	8	295	48	10	940
30	10	365	48	12	1128
30	12	440	48	14	1300
36	6	315	48	16	1500
36	8	420	48	18	1700
36	10	525	48	20	1880
36	12	630	48	24	2260
36	14	735			

TABLE 5

PERFORMANCE OF SHALLOW WELL JET PUMPS

Suction Lift	Discharge Pressure					
	20 lbs.		30 lbs.		40 lbs.	
	¼-HP	½-HP	¼-HP	½-HP	¼-HP	½-HP
	Delivery Capacity in Gallons per Hour					
5 ft.	500	930	460	880	260	530
10 ft.	430	820	400	785	230	440
15 ft.	360	700	340	690	180	360
20 ft.	300	590	280	585	150	290

Municipal Water Supply Connections.—City plumbing installations usually connect with a water main in the street as a source of supply. This connection is made in some cities by the water company or the water department, and in others, by the plumber.

There are three types of connections: taps, wet-connections and three-way branches. With the use of special tools, a hole may be bored in a water main and a tap inserted without interrupting the service. Taps are of brass and are made in the following sizes: ⅝ in., ¾ in., 1 in., 1½ in., and 2 in. The tap connects with the service pipe by a lead gooseneck as shown in Fig. 4 so bent that settlement of either pipe will not loosen the

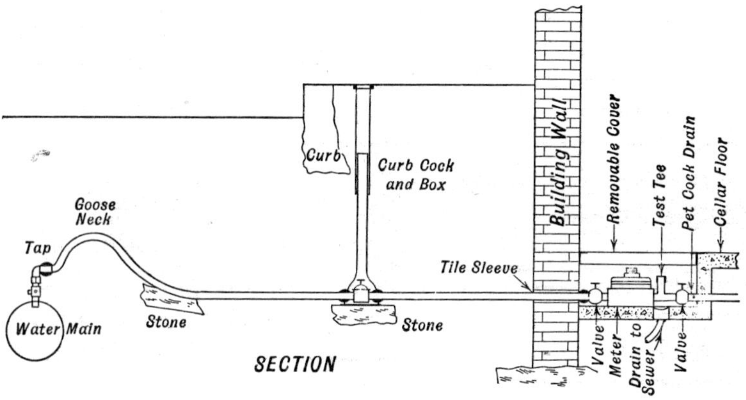

Fig. 4.—House Service Connection.

connection or break the pipe. Taps are commonly used for buildings requiring less than 200 gallons per minute. For larger connections such as for apartment buildings, factories or office buildings, several taps may be used leading to a common service pipe. Wet-connections and three-way branches are also only used for these larger demands.

Distributing Systems.—After delivery to the building, water is distributed to the fixtures by branch pipes and risers as shown in Fig. 5. In dwellings, small apartment buildings and low struc-

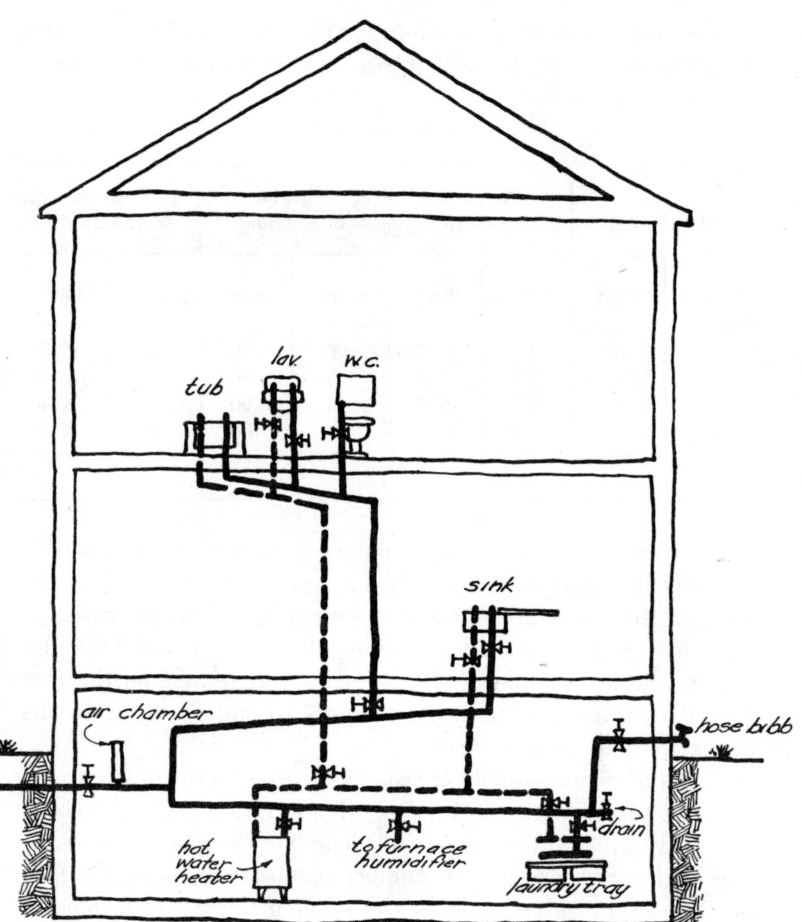

FIG. 5.—Sketch of Typical Hot and Cold Water Distribution System for a Private Dwelling. This house has a full bathroom *A* (tub, sink and water closet) on the second floor; a half bathroom *B* (sink and water closet only), and a kitchen with sink *C* on the first floor. The basement equipment includes a water heater *D*, laundry tub *E*, and two hose connections *F* and *G* (for lawn sprinkling, car washing, etc.). A jet pump with pressure tank *H* supplies water from an outside well. Where city water is available, this pump would not be needed.

tures of any kind, the pressure from the city mains is relied upon to deliver the water to the fixtures. Table 6 shows the pressure

TABLE 6

Height of Building		Pressures at Curb		Height of Building		Pressures at Curb	
Stories	Feet	Pounds	Feet	Stories	Feet	Pounds	Feet
2	20	15	34.5	7	70	40	92.0
3	30	20	46.0	8	80	45	103.5
4	40	25	57.5	9	90	50	115.0
5	50	30	69.0	10	100	55	126.5
6	60	35	80.5	11	110	60	138.0

at the curb necessary to supply buildings of various heights with properly designed plumbing. Where automatic flush valves are used on the upper floors, these pressures should be increased at least five pounds or roof tanks installed. Very tall buildings require the installation of pumps and storage tanks on the roof (see Fig. 6) or at intermediate points to supply water to the upper floors at proper pressures.

Sizes of Water-Supply Pipes.—The size of wrought-iron pipe required to deliver a certain flow of water to a fixture depends on the available water pressure, the length of the pipe, the smoothness of its interior and the number of obstructions to the flow in the form of valves, elbows, and other fittings. Only some of these factors can be determined with any degree of accuracy and the deficiency must be supplied by experience and the exercise of good judgment.

Information as to the pressure of water available at a building site can usually be obtained from the city water department's office. Water pressure is measured in pounds per square inch and

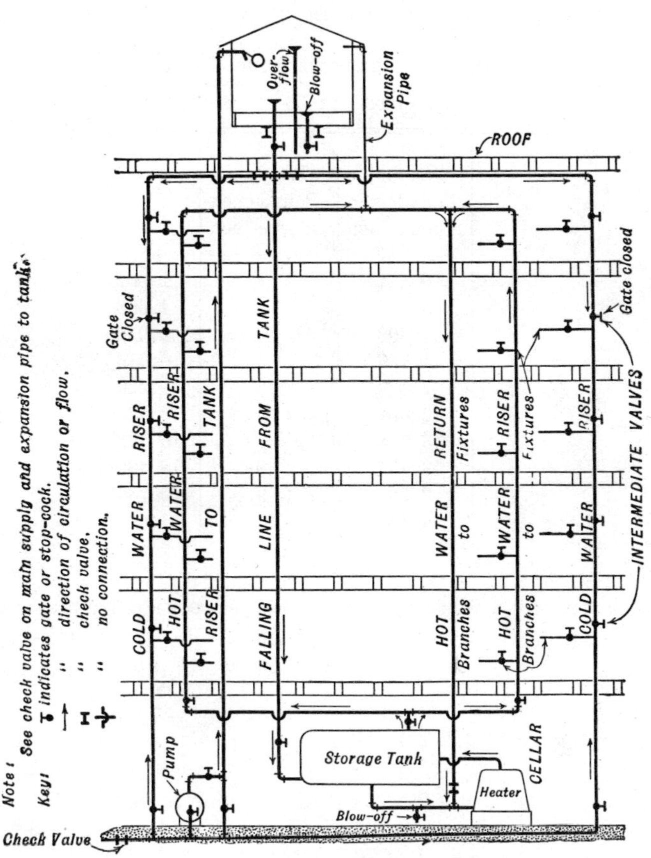

FIG. 6.

in feet of head. Table 7 may be used to convert values in one unit to those in the other.

TABLE 7

HEAD AND PRESSURE EQUIVALENTS

(Water Assumed at 62.5 Lb per Cu Ft)

Head Feet	Pressure, Pounds per Square Inch	Pressure, Pounds per Square Inch	Head Feet
1	0.434	1	2.304
2	0.868	2	4.608
3	1.302	3	6.912
4	1.736	4	9.216
5	2.170	5	11.520
6	2.604	6	13.824
7	3.038	7	16.128
8	3.472	8	18.432
9	3.906	9	20.736
10	4.340	10	23.040

ILLUSTRATION: What pressure in pounds per square inch corresponds to a head of 85.3 feet?

From table (left half)

$$
\begin{aligned}
80 = 10 \times 3.472 &= 34.72 \\
5 &= 2.170 \\
.3 = .1 \times 1.302 &= .1302 \\
\hline
\text{Head of 85.3 ft.} \quad & 37.02 \quad \text{lb. per sq. in.} \quad \text{(Ans.)}
\end{aligned}
$$

ILLUSTRATION: How many feet of head is the equivalent of a water pressure of 45 pounds per square inch?

From table (right half)

$$
\begin{aligned}
40 = 10 \times 9.216 &= 92.16 \\
5 &= 11.52 \\
\hline
\text{Pressure} &= 103.68 \text{ ft. of head} \quad \text{(Ans.)}
\end{aligned}
$$

Table 8 gives the recommended rates of supply to plumbing fixtures, Table 9 the recommended sizes of water-supply pipes to fixtures, and Table 10 the sizes of branch water-supply pipes to

TABLE 8

RECOMMENDED RATES OF SUPPLY TO PLUMBING FIXTURES

(Gallons per Minute)

Fixture	Recommended by						
	H. E. Babbitt *	A. Buenger †			W. S. Timmis ‡	Copper and Brass §	Plum **
		Fair	Good	Excel-lent			
Bath tub	10	3	4	6	15	10	15
Wash basin	2	2	3	4	4	5	7
Manicure table	....	1	1½	2	....	....	....
Slop sink	5	3	4	6	15	....	10
Pantry sink	1	2	4	6	....	....	....
Kitchen sink	5	....	....	....	15	10	9
Shower bath	6	4	6	8	8	5	9
Bidet	1	....	....	....	....	....	....
Drinking fountain	1	....	....	....	....	....	....
Laundry tray	5	4	6	8	....	10	9
Urinal	4	....	....	....	6	....	7
Hot-water heater	5	....	...	....	....	....	....
Water-closet	5	....	....	....	8	5	18
Water-closet flush valve	50	....	....	....	30	30	40
Garden hose	12	....	....	....	....	10	....

* *Plumbing*, McGraw-Hill, Second Edition.
† *Jour. Am. Soc. Heat. Vent. Engrs.*, Vol. 26, p. 701.
‡ *Ibid.*, Vol. 28, p. 397.
§ *Practical Brass Plumbing*, Copper and Brass Research Association.
** *Plumbing Practice and Design*, Vol. II, John Wiley & Sons, Inc., p. 110.

fixtures. There should be neither an increase nor a decrease in the size of a branch pipe between the fixture it serves and the riser or branch from which it obtains its water.

TABLE 9

Recommended Sizes of Water-Supply Pipes to Fixtures *

(Standard Wrought Pipe)

Sizes based on pressure drop of 30 lb per 100 ft.
Hot-water faucets to be disregarded when estimating sizes of risers and mains.

Fixture	Number of Fixtures								
	1	2	4	8	12	16	24	32	40
Water closet:									
Tank:									
Gpm	8	16	24	48	60	80	96	128	150
Pipe size, inches	½	¾	1	1¼	1½	1½	2	2	2
Flush valve:									
Gpm	30	50	80	120	140	160	200	250	300
Pipe size, inches	1	1¼	1½	2	2	2	2½	2½	2½
Urinal:									
Tank:									
Gpm	6	12	20	32	42	56	72	90	120
Pipe size, inches	½	¾	1	1¼	1¼	1¼	1½	2	2
Flush valve:									
Gpm	25	37	45	75	85	100	125	150	175
Pipe size, inches	1	1¼	1¼	1½	1½	2	2	2	2
Wash basin: †									
Gpm	4	8	12	24	30	40	48	64	75
Pipe size, inches	½	½	¾	1	1	1¼	1¼	1½	1½
Bath tub:									
Gpm	15	30	40	80	96	112	144	192	240
Pipe size, inches	¾	1	1¼	1½	2	2	2	2½	2½
Shower bath:									
Gpm	8	16	32	64	96	128	192	256	320
Pipe size, inches	½	¾	1¼	1½	2	2	2½	2½	3
Sinks,† slop, kitchen:									
Gpm	15	25	40	64	84	96	120	150	200
Pipe size, inches	¾	1	1¼	1½	1½	2	2	2	2½

* W. S. Timmis, *Jour. Am. Soc. Heat. Vent. Engrs.*, Vol. 28, p. 397.
† Each faucet.

The determination of the proper size of a riser or branch to serve a number of fixtures involves a consideration of the probability of simultaneous use of these fixtures. Thus, in the case of the one-family house shown in Fig. 5, it is conceivable that the bath tub, the water-closet, the sink, and the garden hose might be used simultaneously. What should then be the size of the riser beyond the water pump? Referring to Table 10, next to the last

TABLE 10

SIZES OF BRANCH WATER-SUPPLY PIPES TO FIXTURES *

Description of the fixture	U. S. Department of Commerce recommendation, minimum size	Recommendation by W. S. L. Cleverdon †		Recommendation for sizes for different rates of pressure loss		
		Pressure, 5–15 lb.	Pressure, over 15 lb.	$H =$ $0.5L$ ‡	$H =$ L ‡	$H =$ $5L$ ‡
	In.	In.	In.	In.	In.	In.
Water closet	⅜	½	⅜	½	⅜	⅜
Urinal	..	½	⅜	⅜	⅜	⅜
Bathtub 4 ft. long	½	¾	½ to ⅝	1	¾	½
Bathtub 7 ft. long	..	....	....	1¼	1	¾
Wash basin	⅜	½	⅜	⅜	⅜	⅜
Laundry tray	½	¾ to 1	½ to ¾	½	½	⅜
Kitchen sink, small	½	⅝ to ¾	½ to ⅝	¾	½	⅜
Kitchen sink, hotel	..	....	....	¾	½	⅜
Slop sink	..	¾	½ to ⅝	¾	½	⅜
Hot-water heater	½	¾ to 1	⅝ to ¾	¾	½	½
Bidet	..	⅝ to ¾	½	⅜	⅜	⅜
Shower	..	....	....	¾	½	½
Garden hose	..	....	....	¾	¾	½
Drinking fountain	..	....	....	⅜	⅜	⅜
Water-closet flushometer	..	1¼ to 1½	1	1½	1¼	1
Pantry sink	..	½	⅜	⅜	⅜	⅜
Urinal flush valve	..	⅝ to ¾	½			
Foot bath	..	⅝ to ¾	½	½	⅜	⅜

* *Plumb ng*, H. E. Babbitt, Second Edition, 1950, McGraw-Hill, p. 55.
† *P'umbers Trade Journal*, Vol. 72, p. 867.
‡ H = head loss per unit length; L = length.

column (except for the garden hose), we find that the individual pipe sizes for the fixtures named would be ¾ in., ½ in., ¾ in., and

½ in., respectively. The next problem is then to determine what size pipe will carry as much water as these four combined. From Table 10 we note that one ¾-inch pipe is equivalent in capacity to 2.8 ½-inch pipes. Using this table we can reduce each branch pipe to terms of equivalent ½-inch pipes. Thus,

$$
\begin{aligned}
1\text{—}\tfrac{3}{4}\text{-inch pipe} &= 2.8\text{—}\tfrac{1}{2}\text{-inch pipes}\\
1\text{—}\tfrac{1}{2}\text{-inch pipe} &= 1.0\text{—}\tfrac{1}{2}\text{-inch pipe}\\
1\text{—}\tfrac{3}{4}\text{-inch pipe} &= 2.8\text{—}\tfrac{1}{2}\text{-inch pipes}\\
1\text{—}\tfrac{1}{2}\text{-inch pipe} &= 1.0\text{—}\tfrac{1}{2}\text{-inch pipe}
\end{aligned}
$$

Sum equivalent to 7.6 ½-inch pipes

Referring this sum to Table 11 we find that the corresponding single pipe would be between 1 inch and 1¼ inches. Experience would probably dictate that the 1-inch size would be ample.

TABLE 11

EQUIVALENT PIPE SIZES

(The number of ½-in. pipes which will discharge as a single pipe of another size for the same pressure loss.)

Size of Pipe, Inches	½	⅝	¾	1	1¼	1½	2	2½	3	4	5	6
Number of ½-in. pipes with same capacity....	1	1.7	2.8	5.7	10.0	15.6	32.0	55.8	88.3	181	316	498

Let us consider another example, that of a riser leading to a theater washroom and serving 4 water-closets with flush valves, 8 urinals with flush valves, and 2 wash basins. During the intermission of a performance, these facilities would be so heavily taxed that simultaneous usage would be the safest assumption in computing the size of the riser. Referring this time to Table 9, we note that the sizes of pipes to serve these groups of fixtures are

1½ in., 1½ in., and ½ in., respectively. Referring next to Table 11, the equivalent number of ½-inch pipes are,

1½-in. = 15.6 ½-inch pipes
1½-in. = 15.6 ½-inch pipes
½-in. = 1.0 ½-inch pipe

Total 32.2 ½-inch pipes (or one 2-inch pipe)

Sizes of Copper and Brass Pipe.—Iron pipe carrying soft or corrosive water will rust even if galvanized. The rust forms a spongy mass which seriously impedes the flow of water. The calculations for the wrought-iron pipe sizes above, took a certain amount of this reduction into account. Therefore, if copper, brass, or lead pipe is used a smaller pipe size than that arrived at by these calculations may be used. Table 12 gives the equivalent non-ferrous pipe and tubing sizes which may be used.

TABLE 12

NON-FERROUS PIPES OF CAPACITIES EQUAL TO WROUGHT-IRON OR STEEL PIPES, INCHES

Copper Tubing, Any Water	Copper, Brass or Lead Pipe, Any Water	Wrought-Iron or Steel Pipe		
		Hard Water	Soft Water	Corrosive Water or Softened Water
....	⅜	⅜	½	⅝
⅜	½	½	⅝	¾
....	⅝	⅝	¾	1
½	¾	¾	1	1¼
¾	1	1	1¼	1½

Piping Installation.—Every plumbing installation should follow a building plan or a sketch showing the pipe sizes and the locations and general arrangement of fixtures. Figures 7 and 8

Plumbing symbols

Symbol	Plan	Initials	Item
		D	Drainage line
		VS	Vent line
			Tile pipe
		C.W.	Cold water line
		H.W.	Hot water line
		H.W.R.	Hot water return
		G	Gas pipe
		D.W.	Ice water supply
		D.R.	Ice water return
		F.L.	Fire line
		I.W.	Indirect waste
		I.S.	Industrial sewer
		AW	Acid waste
		A	Air line
		V	Vacuum line
		R	Refrigerator waste
			Gate valves
			Check valves
C.O. C.O.		C.O.	Cleanout
F.D.		F.D.	Floor drain
R.D.		R.D.	Roof drain
R.E.F.		R.E.F.	Refrigerator drain
		S.D.	Shower drain
G.T.		G.T.	Grease trap
S.C.		S.C.	Sill cock
G		G	Gas outlet
VAC.		VAC.	Vacuum outlet
		M	Meter
			Hydrant
H.R.		H.R.	Hose rack
H.R.		H.R.	Hose rack, built in
L		L	Leader
		H.W.T.	Hot water tank
		W.H.	Water heater
		W.M.	Washing machine
		R.B.	Range boiler

Freestanding Corner Built-in
Shower Rod
BATH TUBS
Pedestal Wall hung Corner Manicure Dental
LAVATORIES
Kitchen Double drain board Pantry Kitchen Single drain board
Combination sink and tub Vegetable Slop sink
SINKS
SHOWER LAUNDRY TUBS
Tank Flush valve Bidet
TOILETS
Stall types Wall hung
URINALS
Pedestal Wall Recessed
DRINKING FOUNTAINS

Fig. 7.—Standard Plumbing Symbols. (From *Plumbing,* H. E. Babbitt, Second Edition, 1950, McGraw-Hill, p. 13.)

illustrate the standard conventions by which piping and fixtures are shown on building drawings. In new buildings, the pipes that are to be concealed should go in after the framing is erected.

Piping	
Plumbing	
100 *Soil, Waste, or Leader (above Grade)*	
101 *Soil, Waste, or Leader (below Grade)*	
102 *Vent*	
103 *Cold Water*	
104 *Hot Water*	
105 *Hot Water Return*	
106 *Fire Line*	—F—F—
107 *Gas*	—G—G—
108 *Acid Waste*	Acid
109 *Drinking Water Flow*	
110 *Drinking Water Return*	
111 *Vacuum Cleaning*	—V—V—
112 *Compressed Air*	—A—
Sprinklers	
120 *Main Supplies*	—S—
121 *Branch and Head*	
122 *Drain*	—S---S—
Pneumatic Tubes	
123 *Tube Runs*	

Pipe Fittings and Valves	Screwed
130 *Joint*	
131 *Elbow—90 deg.*	
132 *Elbow—45 deg.*	
133 *Elbow—Turned Up*	
134 *Elbow—Turned Down*	
135 *Elbow—Long Radius*	LR
136 *Side Outlet Elbow—Outlet Down*	
137 *Side Outlet Elbow—Outlet Up*	
138 *Base Elbow*	
139 *Double Branch Elbow*	

Pipe Fittings and Valves (cont'd)	Screwed
140 *Single Sweep Tee*	
141 *Double Sweep Tee*	
142 *Reducing Elbow*	2, 4
143 *Tee*	
144 *Tee—Outlet Up*	
145 *Tee—Outlet Down*	
146 *Side Outlet Tee Outlet Up*	
147 *Side Outlet Tee Outlet Down*	
148 *Cross*	
149 *Reducer*	
150 *Eccentric Reducer*	
151 *Lateral*	
152 *Gate Valve*	
153 *Globe Valve*	
154 *Angle Globe Valve*	
155 *Angle Gate Valve*	
156 *Check Valve*	
157 *Angle Check Valve*	
158 *Stop Cock*	
159 *Safety Valve*	
160 *Quick Opening Valve*	
161 *Float Operating Valve*	
162 *Motor Operated Gate Valve*	M
163 *Expansion Joint Flange*	
164 *Reducing Flange*	
165 *Union*	
166 *Bushing*	

Fig. 8.—Graphical Symbols for Use on Drawings.

If the drawings do not show the " roughing in " dimensions of the fixtures, these should be obtained. Piping should be so located that there will be no danger of water freezing it in a building normally heated. " Horizontal " water pipes should be pitched

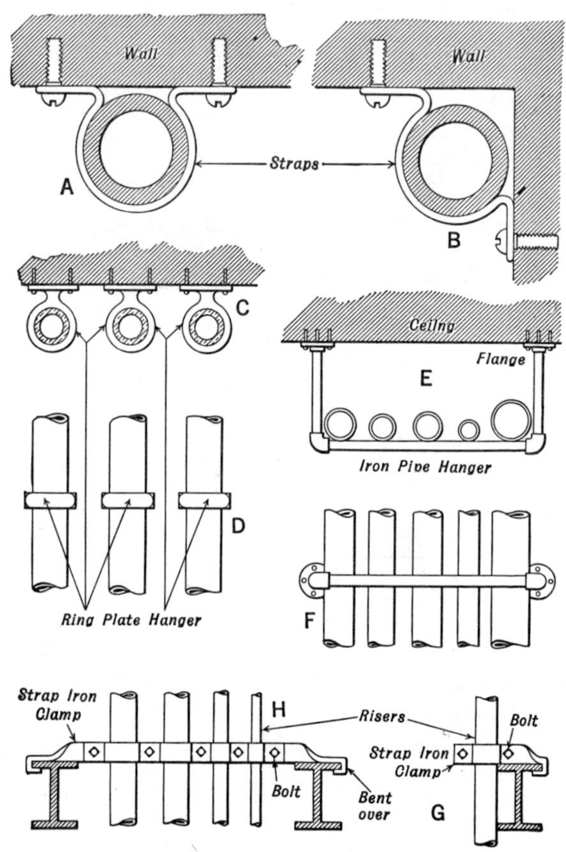

Copper and Brass Research Ass'n.

FIG. 9.—Types of Pipe Hangers.

$\frac{1}{10}$ inch per foot towards the supply pipe so that the entire system may be drained by a stop-and-waste valve just inside the cellar wall. Soil and waste pipes should be sloped at least ¼ inch per foot toward the sewer. No sags or pockets in which water will freeze should be permitted.

Pipe Supports.—A sure way of insuring proper slope on alignment of pipes is adequate support. For horizontal pipes ¾-inch

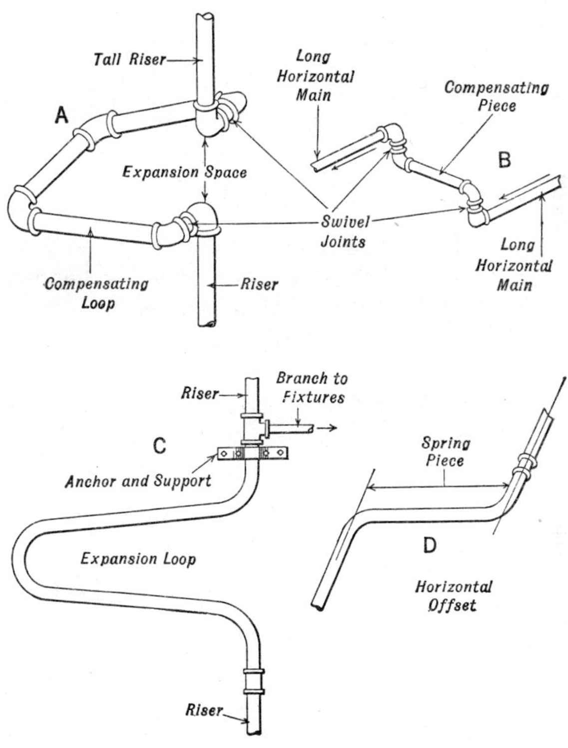

Copper and Brass Research Ass'n.

FIG. 10.—Loops for Expansion.

and larger, pipe hangers should be about 10 feet apart; for ½-inch and ⅜-inch pipe not more than 6 or 8 feet apart. Figure 9 shows several different types of pipe hangers.

Expansion of Pipe.—Pipes expand with an increase in temperature, and when a pipe is long or the change in temperature apt to be great, definite provision must be made to care for this expansion. Pipes passing through concrete or plastered walls should be given freedom of movement by passing them through a sleeve. Either expansion joints, loops, or swing joints (Fig. 10) must be used between fixed supports when the movement is apt to be great.

The change in length or the linear expansion in inches for a pipe 100 feet long can readily be computed from the formula

$$E = 100 \times 12 \times k \times (t_1 - t_2)$$

when E = expansion in inches per 100 feet of length;
k = the rate of increase per degree of temperature, called the coefficient of expansion (this varies with the material, see Table 13);
t_1 = the highest temperature the pipe will reach;
t_2 = the lowest temperature the pipe will reach.

TABLE 13

COEFFICIENTS OF LINEAR EXPANSION

Material	Coefficient of Expansion per degree Fahrenheit	Change in length per 100 ft per 100 degree change in temperature, in.
Wrought iron	.0000067	$\frac{13}{16}$
Copper	.0000093	$1\frac{1}{8}$
Brass	.0000104	$1\frac{1}{4}$
Cast iron	.0000059	$\frac{11}{16}$
Steel	.0000067	$\frac{13}{16}$
Lead	.0000159	$1\frac{15}{16}$

ILLUSTRATION: In Fig. 11 what will be the change in length of the riser between the first and the fifth floors if its temperature changes from 32° F to 212° F?

$E = 100 \times 12 \times k \times (t_1 - t_2)$
$E = 100 \times 12 \times 0.0000104 \times (212 - 32)$
$E = 2.25$ inches per 100 ft.

Then the expansion for 50 ft. is $\frac{2.25}{2} = 1\frac{1}{8}$ in. (Ans.)

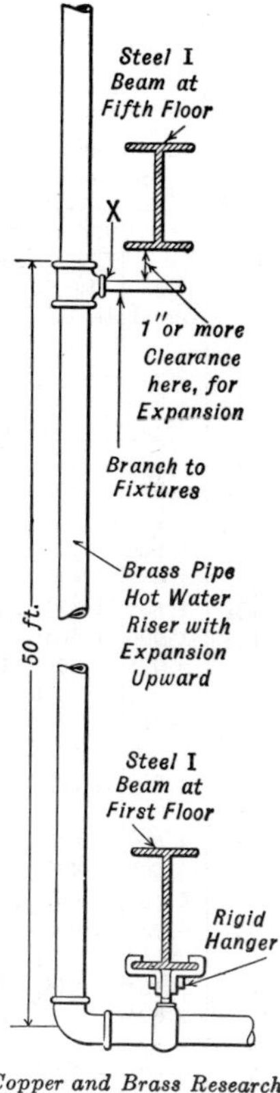

Copper and Brass Research Ass'n.

FIG. 11.—Clearance for Expansion.

ILLUSTRATION: What would be the change of length of a wrought-iron riser in place of the brass in the preceding illustration?

$E = 100 \times 12 \times k \times (t_1 - t_2)$
$E = 100 \times 12 \times 0.0000067 \times (212 - 32)$
$E = 1.447$ inches per 100 ft

Then the expansion per 50 ft is

$\frac{1.447}{2} = \frac{23}{32}$ in. (Ans.)

Wrought Pipe.—Pipes for water supply and waste disposal are made of wrought iron, wrought steel, cast iron, copper, brass, and lead. Wrought-iron pipe is, whether galvanized or black, most commonly used for the water supply plumbing of buildings. Wrought steel is less commonly used because it has a greater tendency to rust. However, where very high pressures are encountered its use may be the more desirable. Wrought pipe is specified by nominal inside diameters up to twelve inches; above twelve inches the pipe is known as O.D., or outside diameter pipe and is specified accordingly with the desired thickness of walls. "Standard" pipe is used for pressures up to 125 pounds per square inch. "Extra strong" and "double extra

TABLE 14

Standard Wrought Pipe

Size, In.	Diameters, Inches		Thickness, Inches	Weight per Foot, Pounds		Threads per Inch	Length of Thread, Inches (Distance E in Fig. 17, p. 441)		Taper per Foot, Inch	Hydrostatic Test, Pounds
	External	Internal		Plain ends	Threads and couplings					
1/8	.405	.265	.070	.244	.245	27	3/8	.375	3/4	750
1/4	.540	.360	.090	.424	.425	18	9/16	.569	3/4	750
3/8	.675	.489	.093	.567	.568	18	9/16	.574	3/4	750
1/2	.840	.618	.111	.850	.852	14	3/4	.748	3/4	750
3/4	1.050	.820	.115	1.130	1.134	14	3/4	.760	3/4	750
1	1.315	1.043	.136	1.678	1.684	11 1/2	15/16	.944	3/4	750
1 1/4	1.660	1.374	.143	2.272	2.281	11 1/2	1	.968	3/4	750
1 1/2	1.900	1.604	.148	2.717	2.731	11 1/2	1	.984	3/4	750
2	2.375	2.059	.158	3.652	3.678	11 1/2	1	1.017	3/4	1000
2 1/2	2.875	2.459	.208	5.793	5.819	8	1 1/2	1.512	3/4	1000
3	3.500	3.058	.221	7.575	7.616	8	1 9/16	1.575	3/4	1000
3 1/2	4.000	3.538	.231	9.109	9.202	8	1 5/8	1.625	3/4	1000
4	4.500	4.016	.242	10.790	10.889	8	1 11/16	1.675	3/4	1000
4 1/2	5.000	4.496	.252	12.538	12.642	8	1 3/4	1.725	3/4	1000
5	5.563	5.037	.263	14.617	14.810	8	1 3/4	1.781	3/4	1000
6	6.625	6.053	.286	18.974	19.185	8	1 7/8	1.887	3/4	1000
7	7.625	7.011	.307	23.544	23.769	8	2	1.987	3/4	1000
*8	8.625	8.059	.283	24.696	25.000	8	2 1/16	2.087	3/4	800
8	8.625	7.967	.329	28.554	28.809	8	2 1/16	2.087	3/4	1000
9	9.625	8.927	.349	33.907	34.188	8	2 3/16	2.187	3/4	900
*10	10.750	10.182	.284	31.201	32.000	8	2 5/16	2.300	3/4	600
*10	10.750	10.124	.313	34.240	35.000	8	2 5/16	2.300	3/4	800
10	10.750	10.006	.372	40.483	41.132	8	2 5/16	2.300	3/4	900
11	11.750	10.986	.382	45.557	46.247	8	2 3/8	2.400	3/4	800
*12	12.750	12.078	.336	43.773	45.000	8	2 1/2	2.500	3/4	600
12	12.750	11.986	.382	49.562	50.706	8	2 1/2	2.500	3/4	800
14	14.000	13.250	.375	53.510	55.712	8	2 5/8	2.625	3/4	700
15	15.000	14.250	.375	57.437	59.859	8	2 3/4	2.725	3/4	700
16	16.000	15.250	.375	61.364	63.927	8	2 13/16	2.825	3/4	600
17	17.000	16.250	.375	65.292	69.436	8	2 15/16	2.925	3/4	550
18	18.000	17.250	.375	69.219	73.681	8	3	3.025	3/4	550
20	20.000	19.250	.375	77.073	82.078	8	3 1/4	3.225	3/4	550

* Unless specified the lighter weight will not be furnished.

strong " pipes are used for higher pressures. The extra thickness is gained by making the bore smaller, the nominal diameter remaining the same. Figure 12 shows a comparison of the cross-sections of ¾-inch pipe of the three different weights. Table 14 gives the dimensions of standard wrought pipe. This pipe is sold

Fig. 12.—Full Size Sections of ¾-in. Pipe.

in random lengths averaging 20 feet, threaded unless otherwise ordered and with one coupling on each length. Extra strong and double extra strong are also sold in random lengths but generally with plain ends. Table 15 gives the dimensions of these sizes.

TABLE 15

Extra Strong and Double Extra Strong Pipe

Nominal Size, In.	External Diameter, Inches	Internal Diameter		Nominal Size, In.	External Diameter, Inches	Internal Diameter	
		Extra Strong	Double Extra Strong			Extra Strong	Double Extra Strong
⅛	0.405	0.215	....	1½	1.900	1.500	1.100
¼	0.540	0.302	....	2	2.375	1.939	1.503
⅜	0.675	0.423	....	2½	2.875	2.323	1.771
½	0.840	0.546	0.252	3	3.500	2.900	2.300
¾	1.050	0.742	0.434	3½	4.00	3.364	2.728
1	1.315	0.957	0.599	4	4.50	3.826	3.152
1¼	1.660	1.278	0.896				

Cast-Iron Pipe.—Pipe of cast iron is universally used in municipal water distribution systems. In the case of a large building or factory, the service pipe may be of this material. Cast-iron pipe of lighter weight is also generally used for the drainage plumbing of buildings. Sizes of the water pipe range from 3 inches to 84 inches nominal inside diameter and the standard length of bell-and-spigot sections is 12 feet. The dimensions of this pipe and fittings have been standardized. Some of these dimensions are shown in Tables 16 and 17.

PIPE OF OTHER MATERIALS

Copper and Brass Pipe.—Copper and brass pipe are excellent plumbing materials, but rather expensive. Therefore they are not generally used except where corrosive water conditions or local Code requirements make their use necessary. Copper pipe is made of 99.9% pure copper in standard and extra-heavy wrought pipe sizes as given in Tables 14 and 15. It is sometimes used for the water service (the pipe joining the city main or well with the house distribution system), because of its great durability. Its resistance to corrosion permits the use of a smaller pipe size than would be needed if wrought iron or steel pipe were used, for a given capacity as indicated in Table 12.

Brass is an alloy of copper and zinc. It has all the advantages of iron pipe plus the fact that it does not rust. Commonly available brasses are made in various grades from "yellow brass," containing about 67% copper, to " red brass " containing about 85% copper.

Brass pipe is replacing galvanized iron pipe to a large degree because of its greater durability. A system using brass pipe will generally last the life of the building in which it is installed. The smooth interior surface of brass pipe permits it to carry water with less friction. This means that a smaller brass pipe size can be used to carry a given volume of water than that which would be required if steel pipe were used.

TABLE 16

Standard Dimensions of Bells, Sockets, Spigot Beads, and Outside Diameters of Pit Cast Pipe

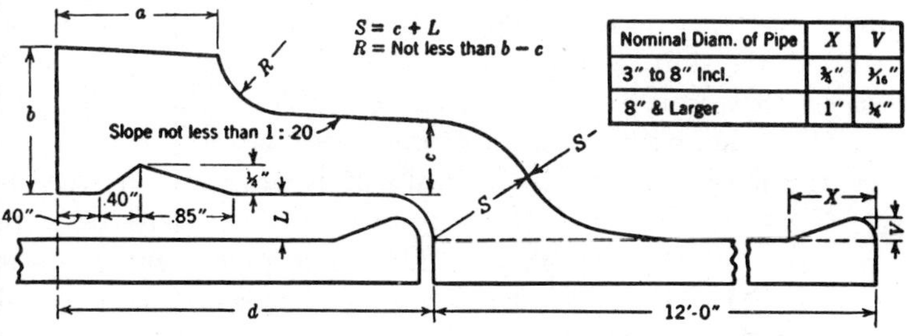

Nominal Diam	Thickness of Pipe		Outside Diam of Pipe	Dimensions of Bells					
	From	To		Diam of Socket	Thickness of Joint L	Depth of Socket d	a	b	c
3	0.37	0.45	3.80	4.60	0.40	3.50	1.25	1.30	0.65
	0.46	0.53	3.96	4.76	0.40	3.50	1.25	1.30	0.65
4	0.40	0.45	4.80	5.60	0.40	3.50	1.50	1.30	0.65
	0.46	0.55	5.00	5.80	0.40	3.50	1.50	1.30	0.65
6	0.43	0.50	6.90	7.70	0.40	3.50	1 50	1 40	0.70
	0.51	0.60	7.10	7.90	0.40	3.50	1.50	1 40	0.70
	0.61	0.66	7.22	8.02	0.40	4.00	1.50	1.75	0.75
	0.67	0.74	7.38	8.18	0.40	4.00	1.50	1.85	0.85
8	0.46	0.57	9.05	9.85	0.40	4.00	1.50	1.50	0.75
	0.58	0.70	9.30	10.10	0.40	4.00	1.50	1.50	0.75
	0.71	0 76	9.42	10.22	0.40	4.00	1.50	1.85	0.85
	0 77	0.85	9.60	10.40	0.40	4.00	1.50	1.95	0.95
10	0.50	0.60	11.10	11.90	0.40	4.00	1.50	1.50	0.75
	0 61	0 75	11.40	12.20	0.40	4.00	1.50	1.60	0.80
	0 76	0 85	11.60	12.40	0 40	4.50	1.75	1.95	0.95
	0.86	0.97	11.84	12.64	0.40	4.50	1.75	2.05	1.05
12	0.54	0.65	13.20	14.00	0.40	4.00	1.50	1.60	0.80
	0.66	0 80	13.50	14.30	0.40	4.00	1.50	1.70	0.85
	0 81	0 94	13.78	14.58	0.40	4.50	1.75	2.05	1.05
	0 95	1.09	14.08	14.88	0.40	4.50	1.75	2.20	1.20

All dimensions given in inches.

From American Standard A21.2. Complete tables are available from American Water Works Association or American Standards Association.

A comparison of pipe sizes required for different water conditions is shown in Table 12. Brass pipe is manufactured in 12-ft. lengths in standard pipe sizes.

Lead Pipe.—The use of lead pipe is decreasing. Factors which have contributed to its unpopularity include the high cost of the material, the skill required to install lead pipe, and the possible dangers from lead poisoning. Some of its desirable features are flexibility, durability, vibration resistance, and its great resistance to corrosion. The size of lead pipe is designated by the inside diameter. Service and supply pipes in three grades range in size from ½″ ID through 2″ ID. Waste pipes range in size from 1¼″ ID through 6″ ID. Lead pipe is smooth, lasting, and pliable, but requires skilled workmanship to achieve smooth soldered or wiped joints.

Vitrified-Clay Pipe.—This type of pipe is generally used for draining sewage, industrial wastes, and storm water. It is manufactured with nominal internal diameters ranging from 4″ to 36″ ID. The *Handbook of Vitrified Clay Sewer Pipe and Kindred Clay Products* of the Clay Sewer Pipe Association, Inc., contains detailed dimensions and weights of fittings for this type of pipe.

Plastic Pipe.—Plastic pipe offers a number of valuable features. They include freedom from corrosion; less likelihood of cracking when water freezes within the pipe; lower weight; and greater ease of handling and assembly. A number of kinds are available, including polyvinyl chloride (PVC), unplasticized polyvinyl chloride (UPVC), polyethylene (PE), polypropylene (PP), chlorinated polyether (trade name Penton®), acrylonitrile-butadiene-styrene (ABS), cellulose-acetate-butyrate (CAB), polycarbonate, polyacetal, polyvinyl chloride (trade name High-Temp Geon®) and polyvinylidene fluoride (trade name Kynar®).

These types of piping differ somewhat in their strength, corrosion resistance and other properties, so that for industrial applications, especially in the chemical industry, data on the resistance to specific conditions of temperature, pressure and chemical action of the various types of plastic pipe, as well as plastic-lined pipe, should be obtained from a general manufacturer, such as Tube Turns Plastics, Inc., 30th and Magazine Sts.,

Louisville, Ky., 40211. For residential use, all the kinds mentioned here have suitable properties, so that selection is largely a question of availability and cost. It is to be noted that for conditions in which the pipe is exposed to mechanical shock, the PVC and UPVC types are available in both normal impact and high impact types.

A number of methods are available for joining plastic pipe and fittings. Like metallic pipe, plastic pipe may be threaded and joined directly to fittings; in the larger industrial sizes flanged connections are used. The thermoplastic types, which include polyethylene, polypropylene and chlorinated polyether, can be joined (for most fittings, after screwing the threaded ends together by hand) by thermal bonding, a process in which they are fused together by a special heating unit called the THERMO-SEAL® tool, which may be obtained from Cabot Piping Systems, Louisville, Ky., 40201. The same company provides solvent cements for joining plastic piping and fittings by cementing them together, although these cements can also be obtained from other suppliers of plastic piping.

Plastic piping of the PE and PP types, which are among those which may be connected by thermal sealing, is available in the sizes shown in Table 16A, in the end dimensions shown in Table 16B, and in the fitting types shown in Fig. 13-14. The fittings for cemented plastic pipe are similar, but are not threaded.

TABLE 16A

POLYETHYLENE (PE) AND POLYPROPYLENE (PP) PIPE IN STANDARD 10′ AND 20′ LENGTHS

Pipe Size	Outside Diam.	Wall Thickness	Approx. Wt./Ft.
1½	1.900	0.145	.32
2	2.375	0.154	.43
3	3.500	0.216	.90
4	4.500	0.237	1.26
6	6.625	0.280	2.24

TABLE 16B

END DIMENSIONS OF THERMO-SEAL® FITTINGS

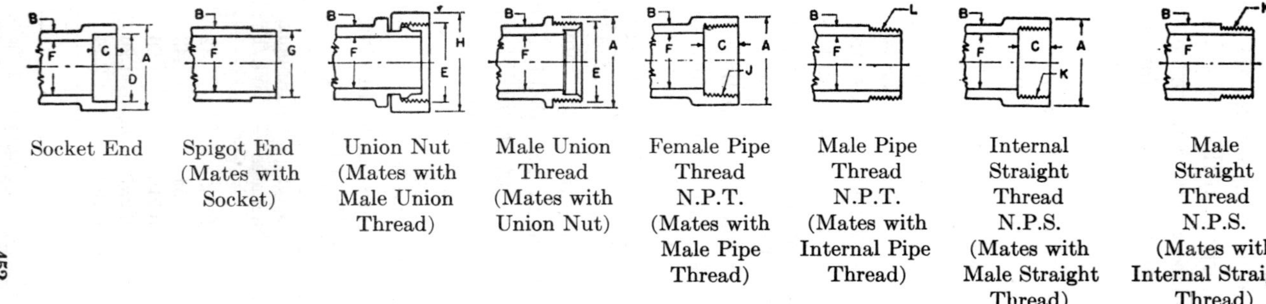

Socket End | Spigot End (Mates with Socket) | Union Nut (Mates with Male Union Thread) | Male Union Thread (Mates with Union Nut) | Female Pipe Thread N.P.T. (Mates with Male Pipe Thread) | Male Pipe Thread N.P.T. (Mates with Internal Pipe Thread) | Internal Straight Thread N.P.S. (Mates with Male Straight Thread) | Male Straight Thread N.P.S. (Mates with Internal Straight Thread)

Nom. Pipe Size	Outside Diam.			Inside Diam.		Wall Thickness	Depth of Socket	Threads			
	A	G	H	D	F	B	C	E	J	K	L
1½	2½	1.900	2¹³⁄₁₆	1.902	1.610	.218	¹¹⁄₁₆	2⅜–12N2	1½–11½NPT	1½–11½NPS	1½–11½NPT
2	3	2.375	3¼	2.380	2.067	.231	¾	2¾–12N2	2 –11½NPT	2 –11½NPS	2 –11½NPT
3	4¼	3.500	4¹¹⁄₁₆	3.495	3.068	.324	1.200	4 – 8N2	3 – 8 NPT	3 – 8 NPS	3 – 8 NPT
4	5²¹⁄₆₄	4.500	5⅞	4.495	4.093	.375	1.300	5⅛– 8N2	4 – 8 NPT	4 – 8 NPS	4 – 8 NPT

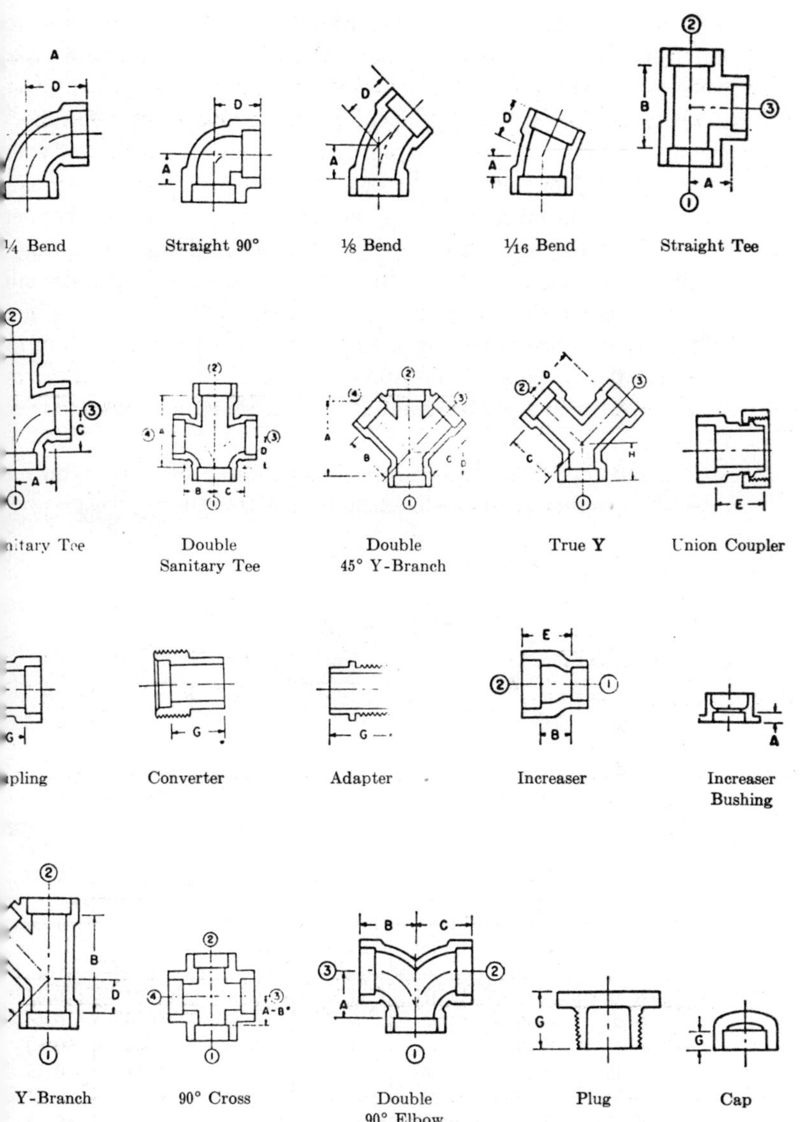

FIG. 13-14. Types of Plastic Pipe Thermo-Seal® Fittings.

Other Materials.—A number of other materials are used to manufacture special types of pipe. Cement pipe, concrete pipe, asbestos-cement pipe ("Transite"), and bituminized-fiber pipe, as well as a wide variety of specially coated pipes such as rubber- or glass-lined pipe are commercially available. All of these types have special properties which fit them for special applications.

Pipe Joints.—Wrought-iron, wrought-steel, brass, and copper pipes are joined to each other by threaded or flanged couplings, unions, or fittings, or by welding. These pipes are usually cut and threaded on the job to suit the requirements. Bell-and-spigot cast-iron pipes are joined by first ramming a strand of oakum or jute into the bottom of the joint space and then filling the remainder of the space with poured lead or calked lead wool. (See Table 17.)

The cutting, threading, calking, and joining of pipes requires a certain number of specialized plumbing tools.

TABLE 17

Standard Thicknesses of Cast-iron Pit Cast Pipe

Size Inches	Class 50	Class 100	Class 150	Class 200	Class 250	Class 300	Class 350
	50 Lb Pressure	100 Lb Pressure	150 Lb Pressure	200 Lb Pressure	250 Lb Pressure	300 Lb Pressure	350 Lb Pressure
	115 Ft Head	231 Ft Head	346 Ft Head	462 Ft Head	577 Ft Head	693 Ft Head	808 Ft Head
	Thickness Inches	Thickness Inches	Thickness Inches	Thickness Inches	Thickness Inches	Thickness Inches	Thickness Inches
3	0.37	0.37	0.37	0.37	0.37	0.37	0.37
4	0.40	0.40	0.40	0.40	0.40	0.40	0.40
6	0.43	0.43	0.43	0.43	0.43	0.46	0.50
8	0.46	0.46	0.46	0.46	0.50	0.54	0.58
10	0.50	0.50	0.54	0.58	0.63	0.68	0.73
12	0.54	0.54	0.58	0.63	0.68	0.73	0.79

Pipe Threads.—Pipes and fittings are threaded to the American Standard Pipe Thread shown in cross-section in Fig. 16.

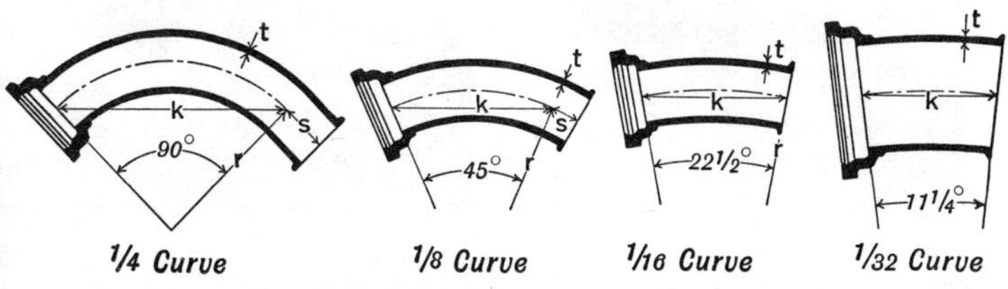

Fig. 15.—Standard Bell-and-spigot Curves.

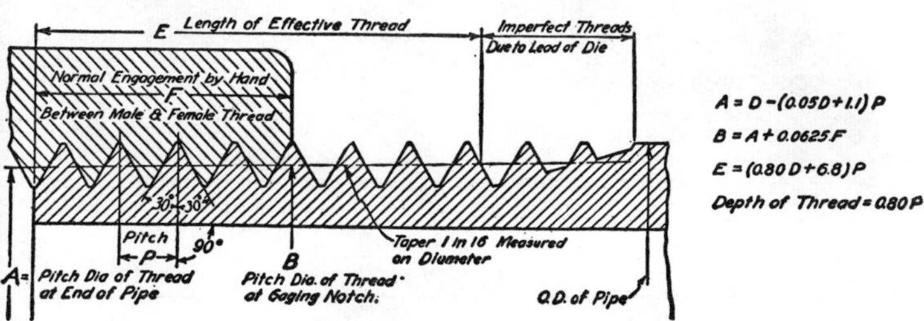

Fig. 16.—American Standard Pipe Thread (ASA B2.1).

Female threads are cut on the fittings by the manufacturer so that the plumber's only practical concern is the cutting of male threads on pipes. This he does by the use of a die selected for the proper pipe size and held in a stock. When the full length of the die is run onto the pipe so that the pipe end is flush with the face of the die, the correct length of thread is cut. The number of threads per inch and the effective length *E* (Fig. 16) for pipes of various sizes is given in columns 7, 8, and 9 in Table 14. The latter figures will be used in connection with piping measurements.

Hubless Cast-Iron Soil Pipe.—In addition to the bell-and-spigot types of cast-iron pipe joints which are joined by the older lead or combination calking methods, a method superior in many respects has been developed by the Cast Iron Pipe Institute. Known by their trade name of No-Hub® pipe, it provides

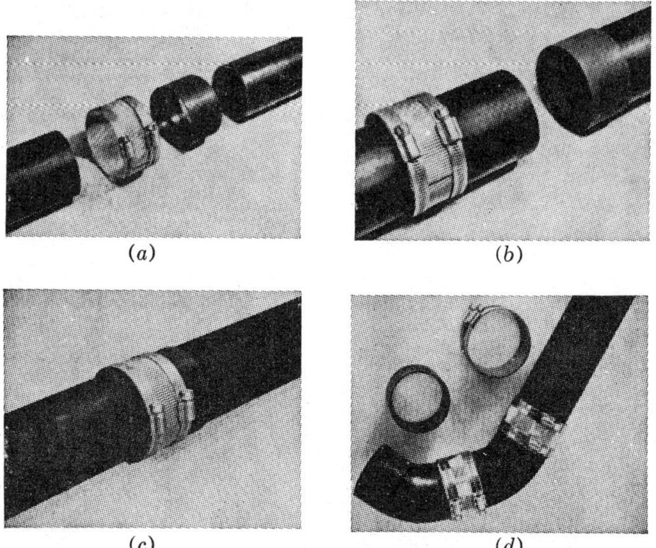

(*a*) (*b*)

(*c*) (*d*)

Fig. 17.

a mechanical joint composed of a Neoprene sleeve gasket and stainless steel shield-clamp rather than the lead and oakum just described. The method is shown in Fig. 17 in four illustrations. Fig. 17(*a*) shows the parts of the joint: the two pipe-ends, the shield-clamp and the Neoprene sleeve gasket. Fig. 17(*b*) shows the sleeve-gasket placed on the end of the right-hand pipe-end and the shield-clamp on the left-hand pipe. The joint is then completed by butting the pipe ends against an integrally-molded shoulder inside the sleeve-gasket, sliding the shield-clamp into position, and tightening it to make the completed joint of Fig. 17(*c*). Fittings are jointed and fastened in the same way (Fig. 17(*d*)).

Pipe Fittings.—As already shown for cast iron and plastic pipe, the couplings tees, elbows, crosses, etc., used for joining pipes, making branches, turns, etc., are called fittings. Small pipes are usually "made up" with screwed or threaded fittings. The sizes of fittings are identified by the nominal pipe size. In specifications for reducing tees and crosses, the size of the

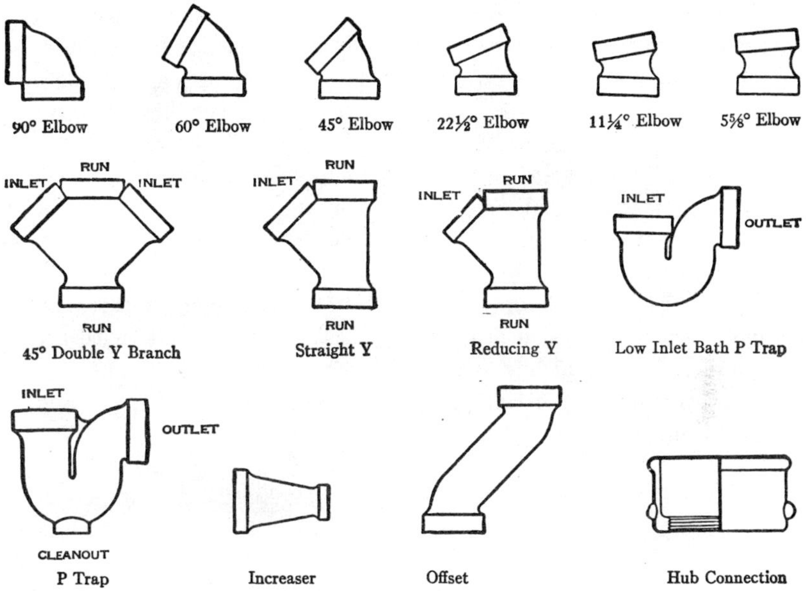

FIG. 18.—Cast-iron Drainage Fittings. Screwed for wrought pipe.

largest run opening is given first, followed by the size of the opening at the other end of the run. Where the fitting is a tee,

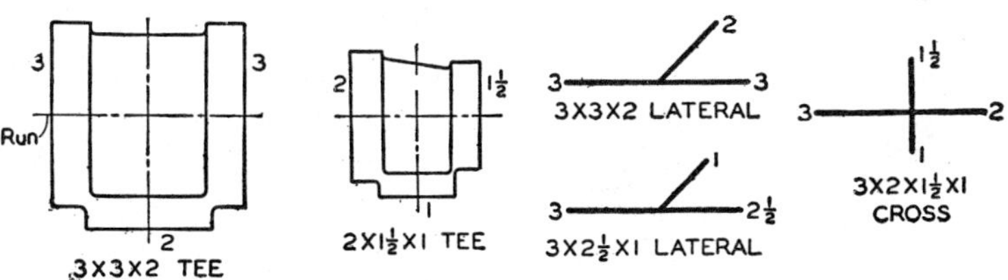

FIG. 19.—Specifications of Fittings.

the size of the outlet is given next. Where the fitting is a cross, the largest opening is the third dimension followed by the opposite opening. Fig. 19 illustrates these conventions.

The assembling of pipes larger than 4 inches with screwed fittings is often cumbersome. For the larger pipe sizes it is, therefore, customary to use flanged fittings. In this case the pipes are threaded and flanges screwed onto them, but the fittings have their flanges cast into place.

Two pieces of straight pipe may be joined either by a threaded coupling or, when making up the last joint or where it is desirable to have a joint which may readily be dissembled, by a union, either screwed or flanged.

A short piece of pipe (usually less than 12 inches) threaded at the ends is called a *nipple*. A *close nipple* is about twice the length E in Fig. 17 and is threaded all the way.

Many of these fittings are illustrated in Figs. 20 to 27 and their dimensions are given in Tables 18 to 20.

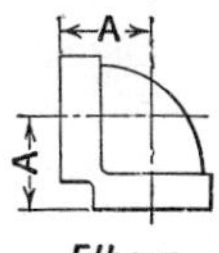

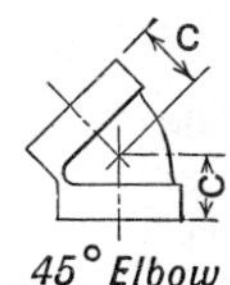

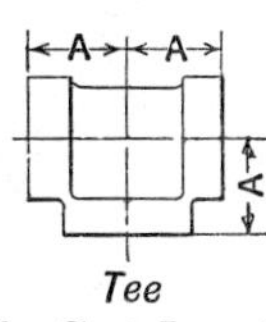

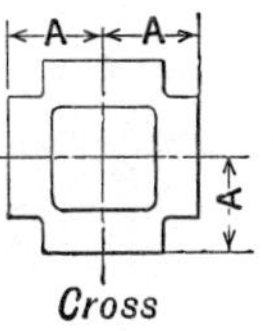

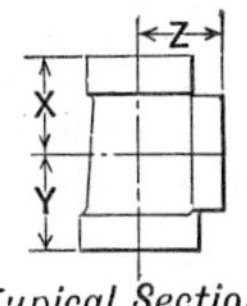

Fig. 20.—125-lb. Cast-Iron Screwed Fittings.

TABLE 18

Dimensions of Cast-Iron Screwed Fittings *

(See Fig. 20)

All dimensions given in inches.

Nominal pipe size, in.	Standard (125 lb)		Extra Heavy (250 lb)		Nominal pipe size, in.	Standard (125 lb)		Extra Heavy (250 lb)	
	A	C	A	C		A	C	A	C
1/4	13/16	3/4	15/16	13/16	3 1/2	3 7/16	2 3/8	3 3/4	2 5/8
3/8	15/16	13/16	1 1/16	7/8	4	3 13/16	2 5/8	4 1/8	2 13/16
1/2	1 1/8	7/8	1 1/4	1	5	4 1/2	3 1/16	4 7/8	3 3/16
3/4	1 5/16	1	1 7/16	1 1/8	6	5 1/8	3 7/16	5 5/8	3 1/2
1	1 1/2	1 1/8	1 5/8	1 5/16	8	6 9/16	4 1/4	7	4 5/16
1 1/4	1 3/4	1 5/16	1 15/16	1 1/2	10	8 1/16	5 3/16	8 5/8	5 3/16
1 1/2	1 15/16	1 7/16	2 1/8	1 11/16	12	9 1/2	6	10	6
2	2 1/4	1 11/16	2 1/2	2	14 O.D.	10 3/8	.....	11	
2 1/2	2 11/16	1 15/16	2 15/16	2 1/4	16 O.D.	11 13/16	.....	12 1/2	
3	3 1/16	2 3/16	3 3/8	2 1/2					

* U. S. Govt. master specifications.

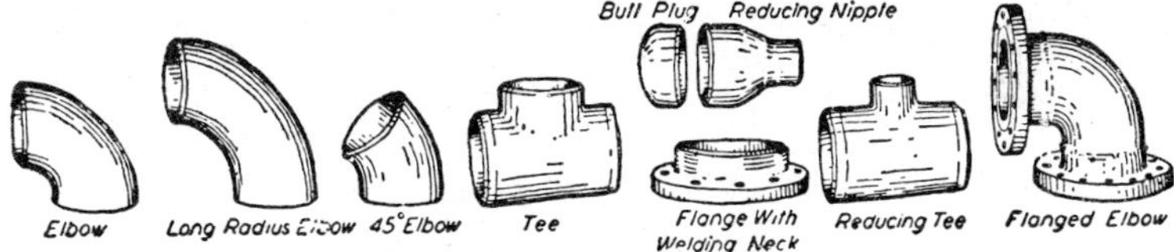

Fig. 21.—Welded Pipe Fittings.

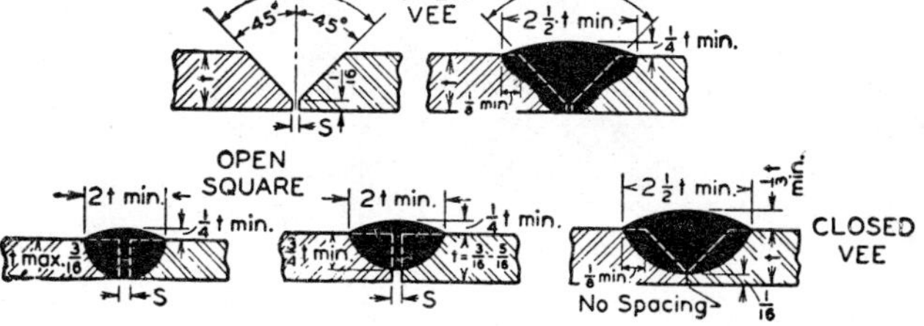

Fig. 22.—Pipe Welds.

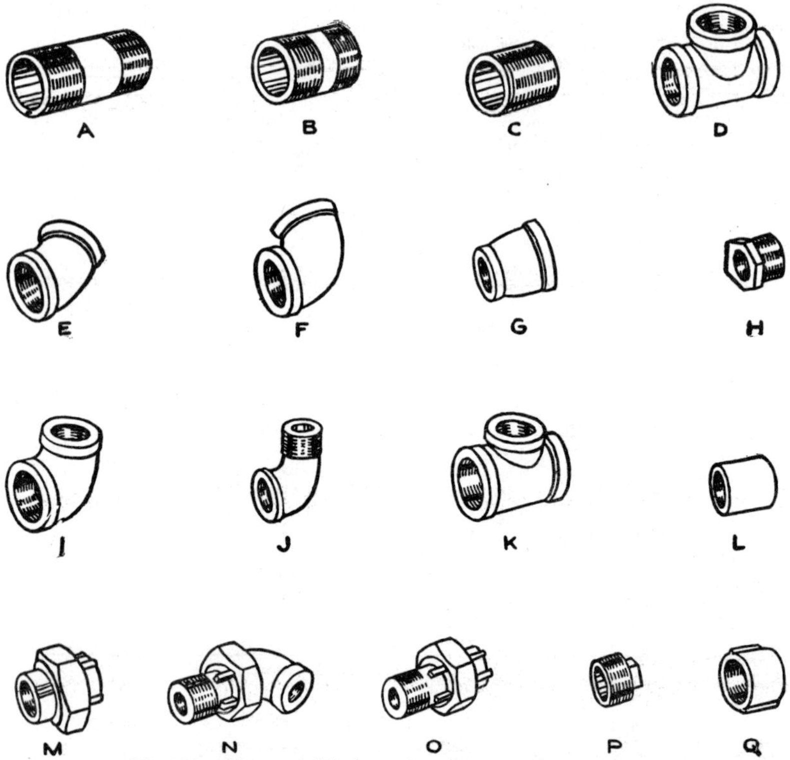

FIG. 23.—Screwed Fittings for Iron and Brass Pipe.

A. Long nipple, usually 3, 4, 5 or 6″ long.

B. Short nipple (length varies with diameter of pipe).

C. Close nipple (length varies with diameter of pipe).

D. Tee or "T" fitting.

E. Forty-five degree elbow or 45° "L."

F. Ninety-degree elbow or "L."

G. Reducing coupling or reducer.

H. Bushing (hex head bushing).

I. Reducing elbow.

J. Street elbow.

K. Reducing tee.

L. Coupling.

M. Union.

N. Elbow union.

O. Tank union.

P. Pipe plug.

Q. Pipe cap.

Reducing tees: Give size A, then B, and last C.
Either size of B or C may be reduced or both.
Example: 1 × ¾ × ¾ "T"

Hex head bushings: Give size A and B.
Example: 1¼ × ¾ bushing.

Reducing couplings: Give size A and B.
Example: 1 × ¾ coupling.

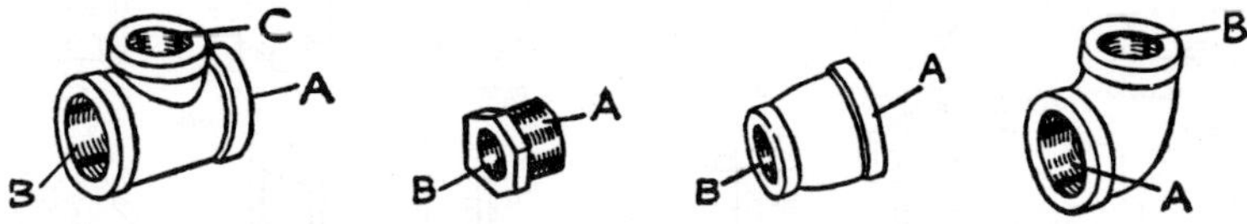

Fig. 24.—Designating the Size of Reducing Fittings.

Reducing fittings usually reduce only one size, from a given size to the next size smaller. However, bushings which reduce two or more sizes are quite common.

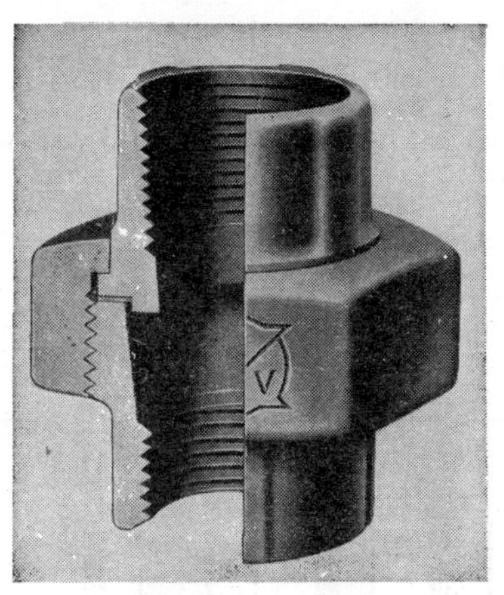

Fig. 25.—Common Gasket Type Union. *Courtesy The Kennedy Valve Mfg. Co.*

Fig. 26.—Union with Ground Brass Seat. *Courtesy E. M. Dart Mfg. Co.*

Unions are made in two general types: (a) the old style gasket type, which requires a fiber gasket at the joint of the two members and (b) the brass-seated union, which requires no gasket, as a watertight joint is obtained by the brass inserts at the seat of the fitting. The latter has replaced the former in most plumbing systems, and is used universally in systems made up of brass pipe and fittings.

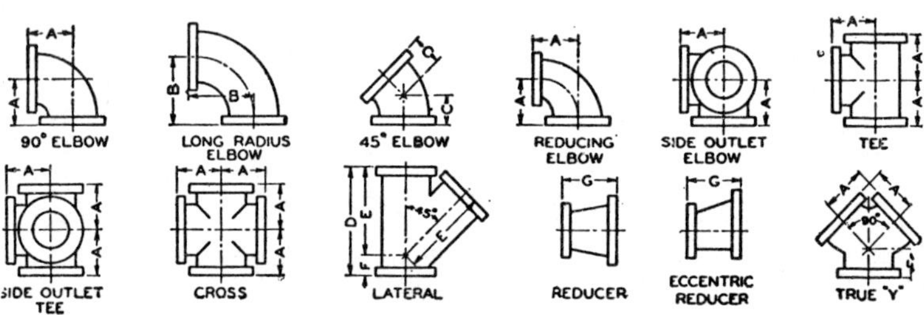

Fig. 27.—Flanged Fittings.

TABLE 19

Dimensions of American Standard 125 lb. Cast Iron Flanged Fittings

Nominal Pipe Size	*A*	*B*	*C*	*D*	*E*	*F*	*G*
1	$3\frac{1}{2}$	5	$1\frac{3}{4}$	$7\frac{1}{2}$	$5\frac{3}{4}$	$1\frac{3}{4}$	....
$1\frac{1}{4}$	$3\frac{3}{4}$	$5\frac{1}{2}$	2	8	$6\frac{1}{4}$	$1\frac{3}{4}$	....
$1\frac{1}{2}$	4	6	$2\frac{1}{4}$	9	7	2	....
2	$4\frac{1}{2}$	$6\frac{1}{2}$	$2\frac{1}{2}$	$10\frac{1}{2}$	8	$2\frac{1}{2}$	5
$2\frac{1}{2}$	5	7	3	12	$9\frac{1}{2}$	$2\frac{1}{2}$	$5\frac{1}{2}$
3	$5\frac{1}{2}$	$7\frac{3}{4}$	3	13	10	3	6
$3\frac{1}{2}$	6	$8\frac{1}{2}$	$3\frac{1}{2}$	$14\frac{1}{2}$	$11\frac{1}{2}$	3	$6\frac{1}{2}$
4	$6\frac{1}{2}$	9	4	15	12	3	7
5	$7\frac{1}{2}$	$10\frac{1}{4}$	$4\frac{1}{2}$	17	$13\frac{1}{2}$	$3\frac{1}{2}$	8
6	8	$11\frac{1}{2}$	5	18	$14\frac{1}{2}$	$3\frac{1}{2}$	9
8	9	14	$5\frac{1}{2}$	22	$17\frac{1}{2}$	$4\frac{1}{2}$	11
10	11	$16\frac{1}{2}$	$6\frac{1}{2}$	$25\frac{1}{2}$	$20\frac{1}{2}$	5	12
12	12	19	$7\frac{1}{2}$	30	$24\frac{1}{2}$	$5\frac{1}{2}$	14
14 O.D.	14	$21\frac{1}{2}$	$7\frac{1}{2}$	33	27	6	16
16 O.D.	15	24	8	$36\frac{1}{2}$	30	$6\frac{1}{2}$	18
18 O.D.	$16\frac{1}{2}$	$26\frac{1}{2}$	$8\frac{1}{2}$	39	32	7	19

TABLE 20

DIMENSIONS OF 125-LB CAST-IRON REDUCING TEES (See Fig. 20)

Nominal Pipe Sizes	Center to End			Nominal Pipe Sizes	Center to End		
	X	Y	Z		X	Y	Z
1/2 × 1/2 × 3/4	1 1/4	1 1/4	1 1/16	1 1/2 × 3/4 × 1 1/2	1 15/16	1 3/4	1 15/16
1/2 × 1/2 × 3/8	1 1/16	1 1/16	1	1 1/2 × 3/4 × 1 1/4	1 [illegible]	1 5/8	1 7/8
3/4 × 3/4 × 1	1 7/16	1 7/16	1 3/8	1 1/2 × 1/2 × 1 1/2	1 9/16	1 11/16	1 15/16
3/4 × 3/4 × 1/2	1 3/16	1 1/4	1 1/4	2 × 2 × 3	2 7/8	2 7/8	2 1/2
3/4 × 3/4 × 3/8	1 1/8	1 1/8	1 1/8	2 × 2 × 2 1/2	2 5/8	2 5/8	2 3/8
3/4 × 1/2 × 3/4	1 5/16	1 1/4	1 5/16	2 × 2 × 1 1/2	2	2	2 3/16
3/4 × 1/2 × 1/2	1 3/16	1 1/8	1 1/4	2 × 2 × 1 1/4	1 7/8	1 7/8	2 1/8
1 × 1 × 1 1/2	1 13/16	1 13/16	1 5/8	2 × 2 × 1	1 3/4	1 3/4	2
1 × 1 × 1 1/4	2 1/8	2 1/8	1 7/8	2 × 2 × 3/4	1 5/8	1 5/8	2
1 × 1 × 3/4	1 3/8	1 3/8	1 7/16	2 × 2 × 1/2	1 1/2	1 1/2	1 7/8
1 × 1 × 1/2	1 1/4	1 1/4	1 3/8	2 × 1 1/2 × 2 1/2	2 5/8	2 1/2	2 3/8
1 × 1 × 3/8	1 3/16	1 3/16	1 1/4	2 × 1 1/2 × 2	2 1/4	2 3/16	2 1/4
1 × 3/4 × 1	1 1/2	1 7/16	1 1/2	2 × 1 1/2 × 1 1/2	2	1 15/16	2 3/16
1 × 3/4 × 3/4	1 3/8	1 5/16	1 7/16	2 × 1 1/2 × 1 1/4	1 7/8	1 13/16	2 1/8
1 × 3/4 × 1/2	1 1/4	1 3/16	1 3/8	2 × 1 1/2 × 1	1 3/4	1 5/8	2
1 × 1/2 × 1	1 1/2	1 3/8	1 1/2	2 × 1 1/2 × 3/4	1 5/8	1 1/2	2
1 × 1/2 × 3/4	1 3/8	1 1/4	1 7/16	2 × 1 1/2 × 1/2	1 1/2	1 7/16	1 7/8
1 × 3/8 × 1	1 1/2	1 1/4	1 1/2	2 × 1 1/4 × 2	2 1/4	2 1/8	2 1/4
1 1/4 × 1 1/4 × 2	1 11/16	1 11/16	1 9/16	2 × 1 1/4 × 1 1/2	2	1 7/8	2 3/16
1 1/4 × 1 1/4 × 1 1/2	1 7/8	1 7/8	1 13/16	2 × 1 1/4 × 1 1/4	1 7/8	1 3/4	2 1/8
1 1/4 × 1 1/4 × 1	1 9/16	1 9/16	1 11/16	2 × 1 1/4 × 1	1 3/4	1 9/16	2
1 1/4 × 1 1/4 × 3/4	1 7/16	1 7/16	1 5/8	2 × 1 × 2	2 1/4	2	2 1/4
1 1/4 × 1 1/4 × 1/2	1 5/16	1 5/16	1 1/2	2 × 1 × 1 1/2	2	1 13/16	2 3/16
1 1/4 × 1 × 1 1/2	1 7/8	1 13/16	1 13/16	2 × 3/4 × 2	2 1/4	2	2 1/4
1 1/4 × 1 × 1 1/4	1 3/4	1 11/16	1 3/4	2 1/2 × 2 1/2 × 4	3 1/2	3 1/2	3 1/16
1 1/4 × 1 × 1	1 9/16	1 1/2	1 11/16	2 1/2 × 2 1/2 × 3	3	3	2 13/16
1 1/4 × 1 × 3/4	1 7/16	1 3/8	1 5/8	2 1/2 × 2 1/2 × 2	2 3/8	2 3/8	2 5/8
1 1/4 × 1 × 1/2	1 5/16	1 1/4	1 1/2	2 1/2 × 2 1/2 × 1 1/2	2 3/16	2 3/16	2 1/2
1 1/4 × 3/4 × 1 1/4	1 3/4	1 5/8	1 3/4	2 1/2 × 2 1/2 × 1 1/4	2 1/16	2 1/16	2 7/16
1 1/4 × 3/4 × 1	1 9/16	1 7/16	1 11/16	2 1/2 × 2 1/2 × 1	1 7/8	1 7/8	2 3/8
1 1/4 × 1/2 × 1 1/4	1 3/4	1 1/2	1 3/4	2 1/2 × 2 1/2 × 3/4	1 3/4	1 3/4	2 5/16
1 1/2 × 1 1/2 × 2	2 3/16	2 3/16	2	2 1/2 × 2 × 2 1/2	2 11/16	2 5/8	2 11/16
1 1/2 × 1 1/2 × 1 1/4	1 13/16	1 13/16	1 7/8	2 1/2 × 2 × 2	2 3/8	2 1/4	2 5/8
1 1/2 × 1 1/2 × 1	1 5/8	1 5/8	1 13/16	2 1/2 × 2 × 1 1/2	2 3/16	2	2 1/2
1 1/2 × 1 1/2 × 3/4	1 1/2	1 1/2	1 3/4	2 1/2 × 2 × 1 1/4	2 1/16	1 7/8	2 7/16
1 1/2 × 1 1/2 × 1/2	1 7/16	1 7/16	1 11/16	2 1/2 × 2 × 1	1 7/8	1 3/4	2 3/8
1 1/2 × 1 1/4 × 1 1/2	1 15/16	1 7/8	1 15/16	2 1/2 × 2 × 3/4	1 3/4	1 5/8	2 5/16
1 1/2 × 1 1/4 × 1 1/4	1 13/16	1 3/4	1 7/8	2 1/2 × 2 × 1/2	1 5/8	1 1/2	2 1/4
1 1/2 × 1 1/4 × 1	1 5/8	1 9/16	1 13/16	2 1/2 × 1 1/2 × 2 1/2	2 11/16	2 1/2	2 11/16
1 1/2 × 1 1/4 × 3/4	1 1/2	1 7/16	1 3/4	2 1/2 × 1 1/2 × 2	2 3/8	2 3/16	2 5/8
1 1/2 × 1 1/4 × 1/2	1 7/16	1 5/16	1 11/16	2 1/2 × 1 1/2 × 1 1/2	2 3/16	1 15/16	2 1/2
1 1/2 × 1 × 2	2 3/16	2	2	2 1/2 × 1 × 2 1/2	2 11/16	2 3/8	2 11/16
1 1/2 × 1 × 1 1/2	1 15/16	1 13/16	1 15/16	3 × 3 × 4	3 5/8	3 5/8	3 5/16
1 1/2 × 1 × 1 1/4	1 13/16	1 11/16	1 7/8	3 × 3 × 3 1/2	3 5/16	3 5/16	3 3/16
1 1/2 × 1 × 1	1 [illegible]	1 1/2	1 13/16	3 × 3 × 2 1/2	2 13/16	2 13/16	3

All dimensions given in inches.

TABLE 20—*Continued*

Nominal Pipe Sizes	Center to End			Nominal Pipe Sizes	Center to End		
	X	Y	Z		X	Y	Z
3 × 3 × 2	2 1/2	2 1/2	2 7/8	4 × 2 × 4	3 13/16	3 7/16	3 13/16
3 × 3 × 1 1/2	2 5/16	2 5/16	2 13/16	4 × 1 1/2 × 4	3 13/16	3 5/16	3 13/16
3 × 3 × 1 1/4	2 3/16	2 3/16	2 3/4	4 × 1 1/4 × 4	3 13/16	3 1/4	3 13/16
3 × 3 × 1	2	2	2 5/8	5 × 5 × 6	5	5	4 5/8
3 × 3 × 3/4	1 7/8	1 7/8	2 5/8	5 × 5 × 4	4	4	4 7/16
3 × 2 1/2 × 3	3 1/16	3	3 1/16	5 × 5 × 3 1/2	3 3/4	3 3/4	4 5/16
3 × 2 1/2 × 2 1/2	2 13/16	2 11/16	3	5 × 5 × 3	3 1/2	3 1/2	4 1/4
3 × 2 1/2 × 2	2 1/2	2 3/8	2 7/8	5 × 5 × 2 1/2	3 1/4	3 1/4	4 1/8
3 × 2 1/2 × 1 1/2	2 5/16	2 3/16	2 13/16	5 × 5 × 2	2 15/16	2 15/16	4
3 × 2 1/2 × 1 1/4	2 3/16	2 1/16	2 3/4	5 × 5 × 1 1/2	2 3/4	2 3/4	3 15/16
3 × 2 1/2 × 1	2	1 7/8	2 11/16	5 × 4 × 5	4 1/2	4 7/16	4 1/2
3 × 2 × 3	3 1/16	2 7/8	3 1/16	5 × 4 × 4	4	3 13/16	4 7/16
3 × 2 × 2 1/2	2 13/16	2 5/8	3	5 × 4 × 3	3 1/2	3 5/16	4 1/4
3 × 2 × 2	2 1/2	2 1/4	2 7/8	5 × 4 × 2	2 15/16	2 3/4	4
3 × 2 × 1 1/2	2 5/16	2	2 13/16	5 × 3 × 5	4 1/2	4 1/4	4 1/2
3 × 1 1/2 × 3	3 1/16	2 13/16	3 1/16	5 × 3 × 4	4	3 5/8	4 7/16
3 × 1 × 3	3 1/16	2 11/16	3 1/16	5 × 3 × 3	3 1/2	3 1/16	4 1/4
3 1/2 × 3 1/2 × 3	3 3/16	3 3/16	3 5/16	5 × 2 × 5	4 1/2	4	4 1/2
3 1/2 × 3 1/2 × 2 1/2	2 15/16	2 15/16	3 1/4	6 × 6 × 8	6 3/8	6 3/8	5 9/16
3 1/2 × 3 1/2 × 2	2 5/8	2 [illegible]/8	3 1/8	6 × 6 × 5	4 7/16	4 7/16	5
3 1/2 × 3 1/2 × 1 1/2	2 3/8	2 3/8	3 1/16	6 × 6 × 4	4 1/8	4 1/8	4 15/16
3 1/2 × 3 1/2 × 1 1/4	2 1/4	2 1/4	3	6 × 6 × 3	3 5/8	3 5/8	4 3/4
3 1/2 × 3 × 3	3 3/16	3 1/16	3 5/16	6 × 6 × 2 1/2	3 3/8	3 3/8	4 11/16
3 1/2 × 3 × 2 1/2	2 15/16	2 13/16	3 1/4	6 × 6 × 2	3 1/16	3 1/16	4 9/16
3 1/2 × 3 × 2	2 5/8	2 1/2	3 1/8	6 × 5 × 5	4 5/8	4 1/2	5
3 1/2 × 3 × 1 1/2	2 3/8	3 5/16	3 1/16	6 × 5 × 4	4 1/8	4	4 15/16
3 1/2 × 2 × 3 1/2	3 7/16	3 1/8	3 7/16	6 × 4 × 6	5 1/8	4 15/16	5 1/8
3 1/2 × 1 1/2 × 3 1/2	3 7/16	3 1/16	3 7/16	6 × 4 × 4	4 1/8	3 13/16	4 15/16
4 × 4 × 6	4 15/16	4 15/16	4 1/8	6 × 3 × 6	5 1/8	4 3/4	5 1/8
4 × 4 × 5	4 7/16	4 7/16	4	6 × 2 × 6	5 1/8	4 9/16	5 1/8
4 × 4 × 3 1/2	3 9/16	3 9/16	3 11/16	8 × 8 × 6	5 9/16	5 9/16	6 3/8
4 × 4 × 3	3 5/16	3 5/16	3 5/8	8 × 8 × 5	5	5	6 1/4
4 × 4 × 2 1/2	3 1/16	3 1/16	3 1/2	8 × 8 × 4	4 1/2	4 1/2	6 3/16
4 × 4 × 2	2 3/4	2 3/4	3 7/16	8 × 8 × 3	4	4	6 1/16
4 × 4 × 1 1/2	2 1/2	2 1/2	3 5/16	8 × 8 × 2 1/2	3 11/16	3 11/16	6
4 × 4 × 1 1/4	2 3/8	2 3/8	3 1/4	8 × 8 × 2	3 7/16	3 7/16	5 13/16
4 × 4 × 1	2 1/4	2 1/4	3 3/16	8 × 6 × 8	6 9/16	6 3/8	6 9/16
4 × 3 1/2 × 3	3 5/16	3 3/16	3 5/8	8 × 6 × 6	5 9/16	5 1/8	6 3/8
4 × 3 1/2 × 2 1/2	3 1/16	2 15/16	3 1/2	10 × 10 × 8	7	7	7 7/8
4 × 3 1/2 × 2	2 3/4	2 5/8	3 7/16	10 × 10 × 6	6	6	7 11/16
4 × 3 × 4	3 13/16	3 5/8	3 13/16	10 × 10 × 4	4 15/16	4 15/16	7 1/2
4 × 3 × 3	3 5/16	3 1/16	3 5/8	12 × 12 × 8	7 7/16	7 7/16	9 1/16
4 × 3 × 2	2 3/4	2 1/2	3 7/16	12 × 12 × 6	6 7/16	6 7/16	8 7/8
4 × 2 1/2 × 4	3 13/16	3 1/2	3 13/16				
4 × 2 1/2 × 2 1/2	3 1/16	2 11/16	3 1/2				

All dimensions given in inches.

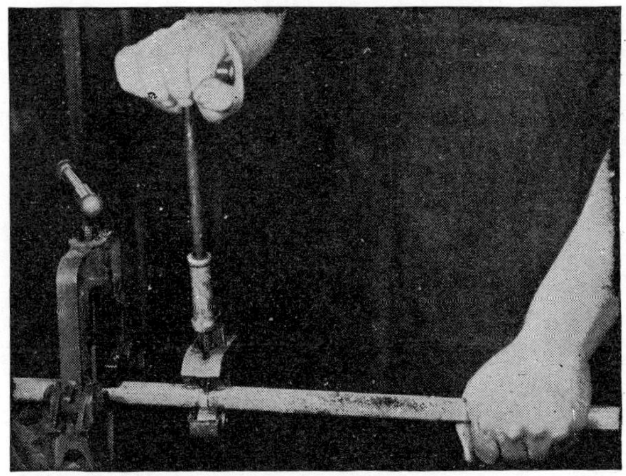

Cutting pipe with a pipe cutter.

TABLE 21

NORMAL ENGAGEMENT BETWEEN MALE AND FEMALE THREADS TO MAKE TIGHT JOINTS †

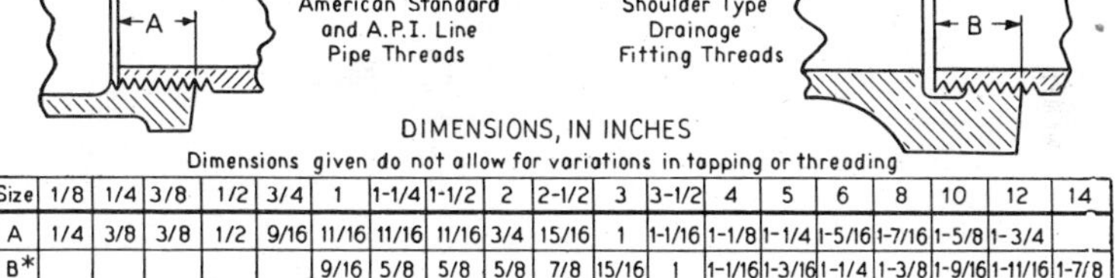

DIMENSIONS, IN INCHES

Dimensions given do not allow for variations in tapping or threading

Size	1/8	1/4	3/8	1/2	3/4	1	1-1/4	1-1/2	2	2-1/2	3	3-1/2	4	5	6	8	10	12	14
A	1/4	3/8	3/8	1/2	9/16	11/16	11/16	11/16	3/4	15/16	1	1-1/16	1-1/8	1-1/4	1-5/16	1-7/16	1-5/8	1-3/4	
B*						9/16	5/8	5/8	5/8	7/8	15/16	1	1-1/16	1-3/16	1-1/4	1-3/8	1-9/16	1-11/16	1-7/8

* Using American Standard Taper Male Thread with Crane Shoulder Type Drainage Fittings. The external thread, however, should not be threaded small to gage and not more than one turn large.

† Crane Co., Chicago, Illinois.

Measuring Pipes.—When making a piping installation it is important that the pipes be cut to the proper lengths to insure obtaining proper slopes, locations of the fittings, and to eliminate the possibility of undue strain on the fittings. Piping drawings, as shown in Fig. 28, give the sizes of pipes and fittings and the distances from center to center of fittings and pipes. Determining the lengths of pipes to be cut from these center-line dimensions is done by applying the dimensions of the pipe fittings and the

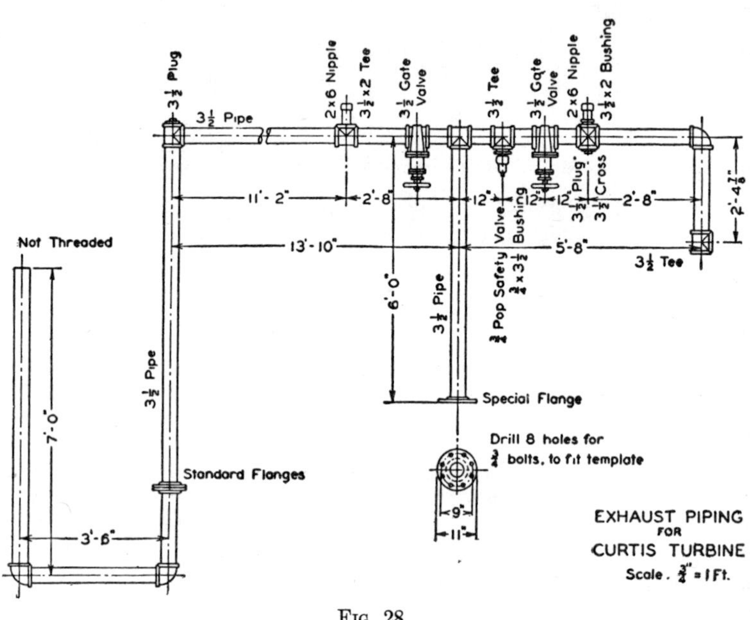

FIG. 28.

length of thread as given in the preceding tables. This is illustrated in Fig. 29 where D is the center-to-center distance between pipes, A_1 and A_2 the dimensions of screwed fittings as given in Tables 18 to 20, E the length of thread as given in Table 14, and L the desired length of pipe. Then,

$$L = D - (A_1 + A_2 - 2E)$$

ILLUSTRATION: If the pipe in Fig. 29 is $2\frac{1}{2}$-inch, the fittings

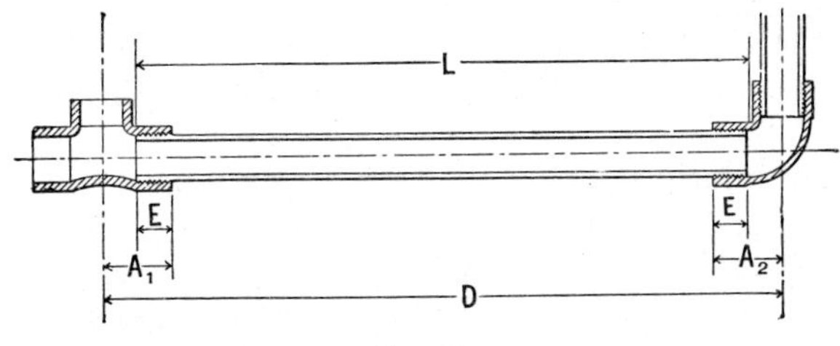

FIG. 29.

125-pound cast iron, and the distance D is 8 feet 6 inches, what length L should the pipe be cut?

$$D = 8 \times 12 + 6 = 102 \text{ inches}$$
$$A_1 = 2\tfrac{11}{16} \text{ inches. From Table 18}$$
$$A_2 = 2\tfrac{11}{16} \text{ inches. From Table 18}$$
$$E = 1\tfrac{1}{2} \text{ inches. From Table 14}$$

Then,

$$L = D - (A_1 + A_2 - 2E)$$
$$L = 102 - (2\tfrac{11}{16} + 2\tfrac{11}{16} - 2 \times 1\tfrac{1}{2})$$
$$L = 102 - (5\tfrac{3}{8} - 3)$$
$$L = 102 - 2\tfrac{3}{8} = 99\tfrac{5}{8} \text{ inches} = 8 \text{ feet } 3\tfrac{5}{8} \text{ inches (Ans.)}$$

ILLUSTRATION: What is the actual length of the pipe in Fig. 28 situated between the two tees whose center-to-center distance is 11 feet 2 inches? The pipe is standard wrought and the fittings are 125-pound cast iron.

In this problem

$$D = 11 \times 12 + 2 = 134 \text{ inches}$$
$$A_1 = 3\tfrac{7}{16} \text{ inches. From Table 18}$$
$$A_2 = 2\tfrac{5}{8} \text{ inches. From Table 18}$$
$$E = 1\tfrac{5}{8} \text{ inches. From Table 14}$$

Then,

$$L = D - (A_1 + A_2 - 2E)$$
$$L = 134 - (3\tfrac{7}{16} + 2\tfrac{5}{8} - 2 \times 1\tfrac{5}{8})$$
$$L = 134 - (3\tfrac{7}{16} + 2\tfrac{10}{16} - 3\tfrac{4}{16}) = 134 - (5\tfrac{17}{16} - 3\tfrac{4}{16})$$
$$L = 134 - 2\tfrac{13}{16} = 131\tfrac{3}{16} \text{ inches} = 10 \text{ feet } 11\tfrac{3}{16} \text{ inches (Ans.)}$$

Similar principles are used in measuring pipes when flanged couplings and flanged fittings are used. For example, if the distance D in Fig. 30 is fixed, then the distance L between the faces of the fittings is $D - (A_1 + A_2)$. The dimensions A_1 and A_2 may both be found from Table 27. Then, if the pipe B is cut $\frac{1}{4}$ inch shorter than the distance L, that quarter inch may be distributed as follows: $\frac{1}{16}$ inch clearance between each end of the pipe and the face of its screwed flange, and $\frac{1}{16}$ inch space for each of two gaskets. There is no substitute for experience and judgment in making the proper allowances for clearances.

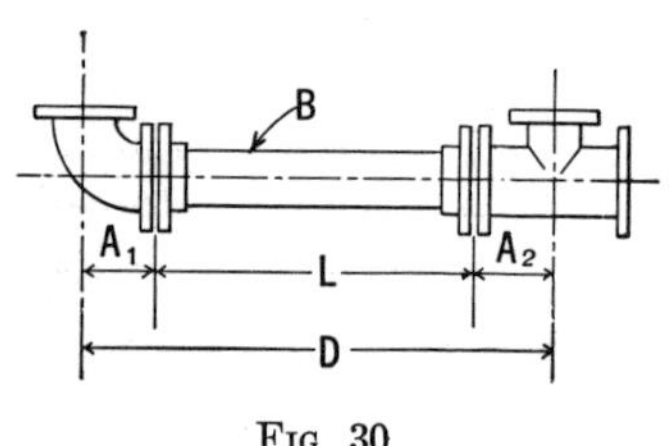

FIG. 30.

ILLUSTRATION: If the pipe shown in Fig. 30 has a nominal diameter of 6 inches and the fitting to the left is a long radius elbow, what is the length L if the center-to-center distance D is 7 feet 9 inches?

$$D = 7 \times 12 + 9 = 93 \text{ inches}$$
$$A_1 = 11\tfrac{1}{2} \text{ inches. ``B'' for 6-in. pipe in Table 27}$$
$$A_2 = 8 \text{ inches. ``A'' for 6-in. pipe in Table 27}$$

Then,

$$L = D - (A_1 + A_2)$$
$$L = 93 - (11\tfrac{1}{2} + 8)$$
$$L = 93 - 19\tfrac{1}{2} = 73\tfrac{1}{2} \text{ inches} = 6 \text{ feet } 1\tfrac{1}{2} \text{ inches (Ans.)}$$

The principles are again applied to the measurement of bell-and-spigot type cast-iron water pipe.

ILLUSTRATION: What is the length L of the pipe in Fig. 31 if the distance D is 9 feet 3 inches and the pipe and fittings are standard cast-iron water pipe of a nominal diameter of 6 inches?

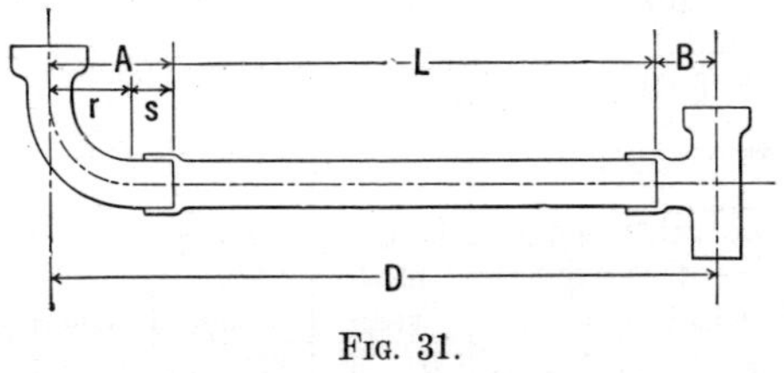

FIG. 31.

From the figure, $L = D - (A + B)$. A is made up of the two dimensions r and s. From standard references, these turn out to be 16 inches and 24 inches, respectively; and B is found to be 12 inches. Then we may write

$$L = D - (A + B)$$
$$L = 9 \times 12 - (16 + 24 + 12)$$
$$L = 108 - 52 = 56 \text{ inches} = 4 \text{ feet } 8 \text{ inches (Ans.)}$$

Measuring Diagonal Pipe.—If two offset pipes are to be connected by a diagonal pipe and the angle of the fittings and one of the dimensions A, B, or C (Fig. 32) are known, the other dimensions may readily be found. Without going into the principles of trigonometry back of it we offer Table 22 as a short-cut to these calculations. The table applies equally well to offsets from Y-connections. The numbers in the table are calculated from the trigonometry of the triangle A, B, C.

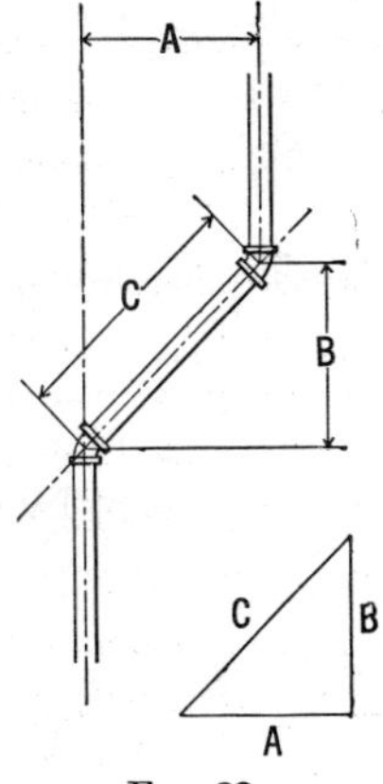

FIG. 32.

Knowing the angle of the fitting and the length of either A or B, the other dimensions can be found by multiplying the known length by the proper figure in the table.

Illustration: If the fittings in Fig. 32 are $22\frac{1}{2}$-degree elbows and the offset A is 2 feet 6 inches, what are the lengths of B and C? A is then $2 \times 12 + 6 = 30$ inches.

TABLE 22

Angle of Fittings (See Fig. 32)		Length of B when $A = 1$	Length of A when $B = 1$	Length of C when $A = 1$	Length of C when $B = 1$	Length of A when $C = 1$	Length of B when $C = 1$
1/64 curve	5⅝°	10.1531	0.0985	10.2033	1.0048	0.098	0.9952
1/32 curve	11¼°	5.0273	0.1989	5.1258	1.0196	0.1951	0.9809
1/16 curve	22½°	2.4142	0.4142	2.6131	1.0828	0.3826	0.9239
1/12 curve	30°	1.7320	0.5773	2.0000	1.1547	0.5000	0.866
⅛ curve	45°	1.0000	1.0000	1.4142	1.4142	0.7071	0.7071
⅙ curve	60°	0.5773	1.7320	1.1547	2.0000	0.866	0.5000
3/16 curve	67½°	0.4142	2.4142	1.0824	2.6131	0.9239	0.3826

In column 3 of Table 22 opposite $22\frac{1}{2}$ degrees we find the factor 2.4142. Then

$$B = 30 \times 2.4142 = 76.43 \text{ in.} = 6 \text{ ft. } 0\tfrac{7}{16} \text{ in.}$$

(Use Table 1, page 19, for conversion.) (Ans.)

Similarly, from column 5,

$$C = 30 \times 2.6131 = 78.393 \text{ in.} = 6 \text{ ft. } 6\tfrac{3}{8} \text{ in.} \quad \text{(Ans.)}$$

Having found the length of C, the actual length of the pipe may be found by the method of the preceding paragraphs.

Illustration: If the fittings in Fig. 32 are 60-degree elbows and B is 10 inches, what are the lengths A and C?

From column 4 of Table 22,

$$A = 10 \times 1.7320 = 17.320 \text{ in.} = 17\tfrac{5}{16} \text{ in.} \quad \text{(Ans.)}$$

Then, from column 6 of Table 22,

$$C = 10 \times 2 = 20 \text{ in.} \quad \text{(Ans.)}$$

Copper Tubing.—A plumbing material that has come into widespread use recently is copper tubing. Although copper tubes with flanged fittings have been used in automotive, gasoline, and oil lines for many years, and copper water tubing has been used successfully in Europe and other parts of the world, its use in plumbing systems in the United States has been comparatively recent.

One of the main reasons for the popularity of copper tubing is its ease of installation. It is soft enough to be bent easily around obstructions, and it can be run between studding and flooring beams without the use of fittings. Another advantage is that there is no weakening of the tubing at the joint as in threaded pipe, where the original thickness of the pipe is reduced. Copper tubes with sweated or flanged fittings can be much thinner than threaded pipe without being weak at the joints.

Copper tubing is also rigid enough to be strung up in long lengths without undue sagging, if hard temper tubing is employed. It can withstand the rough usage that pipes usually get when being installed. Under normal conditions, it is resistant to corrosion. Hard-drawn copper tubes, like iron pipe, can be damaged by freezing water. Soft or annealed copper tubes, however, resist the bursting pressure caused by freezing water.

Because of its flexibility and ductility, soft or annealed copper tubing is especially suitable for underground water service. In such installations, there is always a certain amount of settlement of the ground, and the flexible tubing readily accommodates itself to any ground movements. If used under normal conditions, and if installed with ordinary skill and care, copper tubes can last the life of the building in which they are installed.

There are two widely used types of copper tubing, and these are classified by wall thicknesses. Type L (light) is furnished in either hard or soft temper for general plumbing and heating applications. Type K (heavy) is available in either hard or soft temper and is used for underground as well as for general plumbing and heating purposes. The hard tubes are usually installed with solder fittings, while the soft tubes can be installed with solder or flared fittings.

Type K and Type L hard temper tubes are furnished in straight lengths of 20 ft. The soft temper tubes are available in 20-ft. lengths, and for sizes up to $1\frac{1}{4}$ in. inclusive, in coils of 60 ft. A thinner copper tube known as Type M also is available,

TABLE 23

SIZES AND WEIGHTS OF COPPER WATER TUBE

Standard Water Tube Size	Actual Outside Diameter	Nominal Wall Thickness			Theoretical Weight		
		Type K	Type L	Type M	Type K	Type L	Type M
Inches	Inches	Inches	Inches	Inches	Lb./Ft.	Lb./Ft.	Lb./Ft.
3/8	0.500	0.049	0.035	.....	0.269	0.198	
1/2	.625	.049	.040	.....	.344	.285	.....
5/8	.750	.049	.042	.....	.418	.362	.....
3/4	.875	.065	.045	.....	.641	.455	.....
1	1.125	.065	.050	.....	.839	.655	.....
1¼	1.375	.065	.055	.....	1.04	.884	.....
1½	1.625	.072	.060	.....	1.36	.114	.....
2	2.125	.083	.070	.....	2.06	1.75	.....
2½	2.625	095	.080	0.065	2.93	2.48	2.03
3	3.125	.109	.090	.072	4.00	3.33	2.68
3½	3.625	.120	.100	.083	5.12	4.29	3.58
4	4.125	.134	.110	.095	6.51	5.38	4.66
5	5.125	.160	.125	.109	9.67	7.61	6.66
6	6.125	.192.	.140	.122	13.9	10.2	8.92
8	8.125	.271	.200	.170	25.9	19.3	16.5
10	10.125	.338	.250	.212	40.3	30.1	25.6
12	12.125	.405	.280	.254	57.8	40.4	36.7

but only in sizes above 2 in. Table 23 shows the sizes and weights of copper water tubing for these three types.

TABLE 24

SIZES OF COPPER TUBE WATER SUPPLY BY SHORT BRANCHES TO PLUMBING FIXTURES *

Fixture	Pressures		
	High Over 60 lbs.	Medium 30 to 60 lbs	Low Under 30 lbs.
	Inch	Inch	Inch
To Baths	½	¾	¾
Lavatories	⅜	½	½
Tank Closets	⅜	⅜	½
Valve Closets	1	1	1¼
Pantry Sinks	½	½	½
Kitchen Sinks	½	½	¾
Slop Sinks	½	¾	¾
Showers	½	½	¾
Urinals (Flush Tank)	½	¾	¾
Urinals (Flush Valve)	¾	¾	¾
Drinking Fountains	⅜	⅜	½

* *Copper Tube,* T. N. Thomson, Copper & Brass Research Assn., 1949, p. 24.

TABLE 25

RELATIVE SIZES OF BRANCHES AND WATER MAINS *

Size of Main	Number & Size of Branches Main Will Supply—Running Full
½″	Two—⅜″
¾″	Two—½″
1″	Two—¾″
1¼″	Two—1″ *or* One—1″ and Two—¾″
1½″	Two—1¼″ *or* One—1¼″ and Two ¾″
2″	Two—1½″ *or* One—1½″ and Two—1¼″
2½″	Two—1½″ and Two—1¼″ *or* One—2″ and Two—1¼″
3″	One—2½″ and One—2″ *or* Two—2″ and Two—1½″
3½″	Two—2½″ *or* One—3″ and One—2″ *or* Four—2″
4″	One—3½″ and One—2½″ *or* Two—3″ *or* Three—2½″ and One—2″ *or* Six—2″

* *Copper Tube,* T. N. Thomson, Copper & Brass Research Assn., 1949. p. 24.

The sizes of tubes commonly used as branches to fixtures vary with the pressures at the fixtures and are shown in Table 24. Table 25 is used to determine the size of a main that will service a given number and size of branches.

Copper Tube Fittings.—A wide variety of fittings for copper tubing are available. They come in practically every form in which screwed fittings are supplied. There are also available

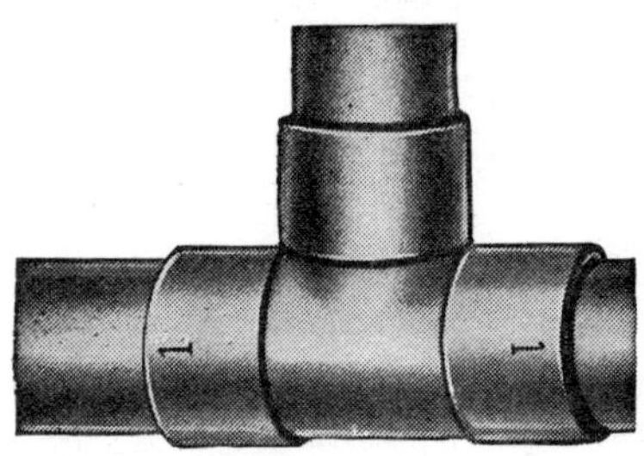

FIG. 33.—Soldered Type Fitting for Copper Tube.

special adapters for joining copper tube to threaded pipe of iron pipe size. A typical soldered type fitting is shown in Fig. 33 and a flanged type fitting appears in Fig. 34.

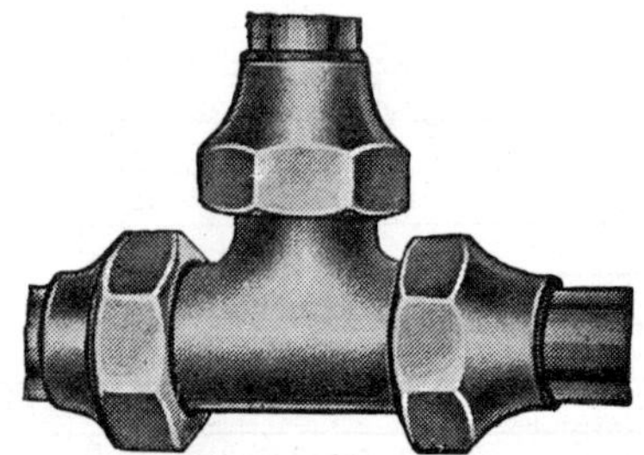

FIG. 34.—Flanged Type Fitting for Copper Tube.

Hot Water Supply.—There are two main ways of providing hot water for small dwellings or apartments. In one system, water is heated directly in a separate unit by gas, coal, oil, or electricity, and then is stored in a hot water storage tank. In

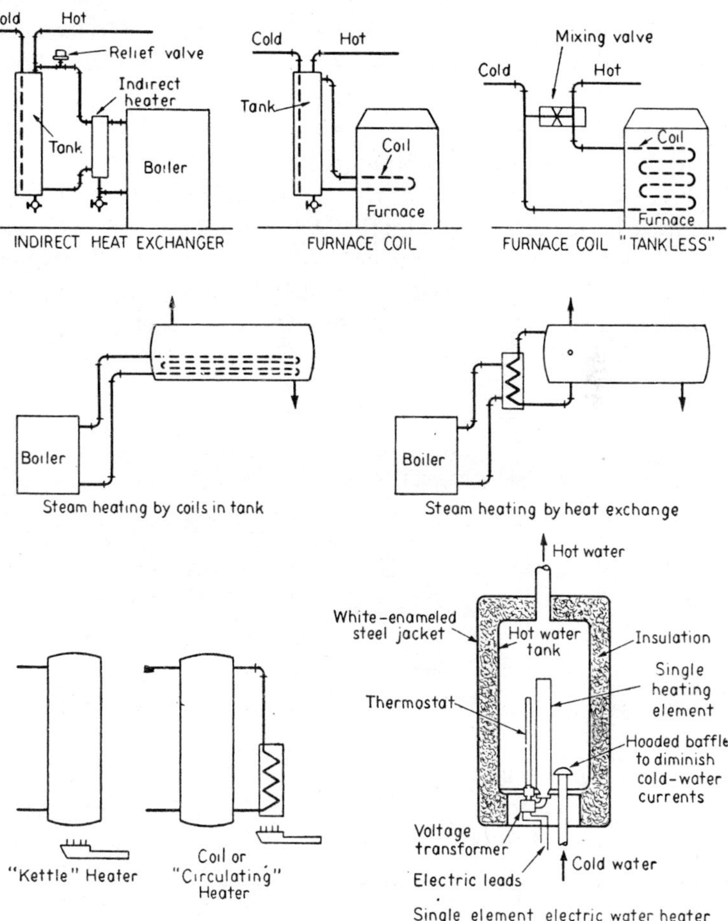

Fig. 35.—Various Methods of Obtaining Hot Water. At the top are shown three ways in which the boiler or furnace is used as the source of heat. In the center, steam is used as the source. At the bottom are shown three types of separate hot water heaters.

the second method, the heating system boiler is used to heat the water indirectly by contact heaters consisting of copper coils surrounded by the boiler water. This type of installation eliminates the need for a separate water heater, and in certain types of installations also eliminates the storage tank. Fig. 35 shows some typical methods for obtaining hot water.

Small tanks which stand vertically are called range boilers and have been standardized by the Division of Simplified Practice of the Department of Commerce to provide one shell tapping 6 inches from the top and one 6 inches up from the bottom (measurements to be made from the edge of the shell plate), and two tappings in the top and one in the bottom. All tappings are

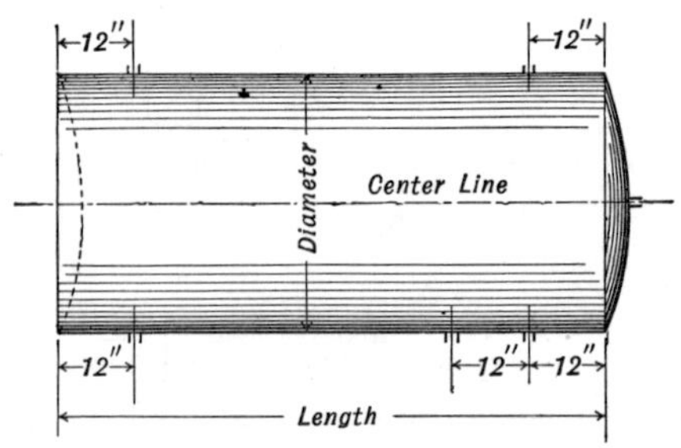

Fig. 36.—Standard Location of Openings for Hot Water Storage Tanks.

one inch. On special order, tanks with four side openings may be obtained. These are placed in line 6, 18, and 26 inches from the bottom and 6 inches from the top.

Range boilers are made in two classes, "standard" for 85 pounds working pressure, and "extra heavy" for 150 pounds working pressure. Dimensions are given in Table 26.

Storage tanks of larger capacity are mounted horizontally and have been standardized as to tappings as shown in Fig. 36. The standard dimensions of these tanks are also given in Table 26. They are made for a "standard" working pressure of 65 pounds and "extra heavy" of 100 pounds.

The hot-water-supply system may consist of one branch pipe leading to each fixture as shown in Fig. 5. In this case an interval of time elapses after opening a faucet before the hot water arrives and a certain amount of water is wasted. This can be obviated by providing a loop, as shown in Fig. 6, through which the hot water constantly circulates. However, in this system a great

TABLE 26

DIMENSIONS OF RANGE BOILERS AND HOT WATER STORAGE TANKS

Range Boilers			Storage tanks		
Diameter, inches	Length, inches	Capacity, gallons	Diameter, inches	Length, feet	Capacity, gallons
12	36	18	20	5	82
12	48	24	24	5	118
12	60	30	24	6	141
14	48	32	30	6	220
14	60	40	30	8	294
16	48	42	36	6	318
16	60	52	36	8	423
18	60	66	42	7	504
20	60	82	42	8	576
22	60	100	42	10	720
24	60	120	42	14	1008
24	72	144	48	10	940
24	96	192	48	16	1504
			48	20	1880

Diameters refer to inside measurements; lengths are mean lengths of sheets, not over-all length of tank.

amount of heat is lost and fuel wasted which overcomes the advantages.

The circulation of water in a loop or between the heater and the storage tank depends on the fact that water is slightly less in weight when hot than when cold. Therefore, the hot water tends to rise and the cold to sink, thus providing the circulation. However, a free circulation requires that the pipes be pitched properly, and humps or air pockets must be avoided.

In selecting a water heater it must be borne in mind that the capacity of the heater depends on the grate area and the efficiency with which the heat may be absorbed from the fuel by the water.

A certain type of heating unit known as an instantaneous or tankless heater is sometimes employed to furnish hot water. These units have large heat absorbing surfaces and large heaters which permit very rapid heating of water to meet peak demands without the need for a storage tank. They are always automatic in operation and depend upon either the flow of water or a thermostatic control to maintain desired water temperature.

A mixing valve may be placed in the hot water output line to maintain constant water temperature during the heating season when the boiler temperature may be higher than is required for domestic hot water. This valve mixes cold water with the hot water to give the desired temperature.

TABLE 27 *

COST COMPARISON FOR WATER HEATERS ON YEARLY BASIS

	COAL	OIL	GAS			ELECTRICITY
			Natural	Manufactured	L.P. (Bottled)	
Fuel Used Per Year	3600 pounds	270 gallons	32,450 cubic feet	61,800 cubic feet	1475 pounds	6658 kwh
Cost @	\$ 9/ton - \$16.20 \$12/ton - 21.60 \$15/ton - 27.00 \$18/ton - 32.40 \$21/ton - 37.80	10¢/gal. - \$27.00 11¢/gal. - 29.70 12¢/gal. - 32.40 13¢/gal. - 35.10 14¢/gal. - 37.80	5¢/100 cu. ft. - \$16.22 6¢/100 cu. ft. - 19.46 7¢/100 cu. ft. - 22.70 8¢/100 cu. ft. - 25.94 9¢/100 cu. ft. - 29.18	7¢/100 cu. ft. - \$43.26 8¢/100 cu. ft. - 49.44 9¢/100 cu. ft. - 55.62 10¢/100 cu. ft. - 61.80 11¢/100 cu. ft. - 67.98	5¢/lb. - \$ 73.75 6¢/lb. - 88.50 7¢/lb. - 103.25 8¢/lb. - 118.00 9¢/lb. - 132.75	1½¢/kwh - \$ 99.87 2¢ /kwh - 133.16 2½¢/kwh - 166.45 3¢ /kwh - 199.74 3½¢/kwh - 233.03
Efficiency Assumed	50%	60%	70%	70%	70%	100%
Fuel Value	12,500 Btu/lb.	140,000 Btu/gal.	1000Btu/cu. ft.†	525 Btu/cu. ft.	22,000 Btu/lb.	3412 Btu/kwh

* *University of Illinois Small Homes Council Circular* G5.0.

Computations based on 50 gal. per day, 100° F. temperature rise, and 50% stand-by and piping loss. Efficiency of burners indicated above by fuel.

† A therm is equal to 100,000 Btu.

The cost of obtaining hot water varies widely depending upon geographical location, fuel cost, and type of heating system employed. Table 27 shows the annual cost for operating water heaters using a variety of fuels (heaters not connected to the house system). It can be seen that the most economical type of heater depends upon the most economical fuel for a given part of the country. No one fuel is the "best" for the job.

Although electric water heaters have the greatest thermal efficiency, this advantage is offset by the greater cost of electricity. In certain areas this is compensated for by obtaining favorable electric rates for water heating. Another method of reducing the cost of operating electric water heaters is to have them controlled to receive current only during certain periods of the day or night when the demand for electric power from the public utility is normally at a minimum.

A large, well-insulated storage tank is needed with this arrangement to provide hot water between periods when electric power is not available.

Drainage Plumbing.—Drainage plumbing, using the term in a broad sense, consists of the waste pipes which carry the used water not containing human excrement from such fixtures as sinks, wash basins, etc., soil pipes which carry the wastes from water-closets and urinals, vent pipes which admit air to the system, and traps which prevent the foul air in the pipes from entering the house. A vertical drainage pipe is known as a "stack." Fig. 37 illustrates the elements of the drainage plumbing for a one-family house.

The physics of drainage plumbing is rather complicated, but recommendations based on experimental work done largely by the Bureau of Standards are easy to understand.

Fixture Units.—In order to compare the discharges of various fixtures for determining the sizes of traps and pipes the so-called *fixture unit* has been devised. This unit is equivalent to a discharge of about 7.5 gallons per minute. The rate of discharge for various fixtures in terms of fixture units is given in Table 28.

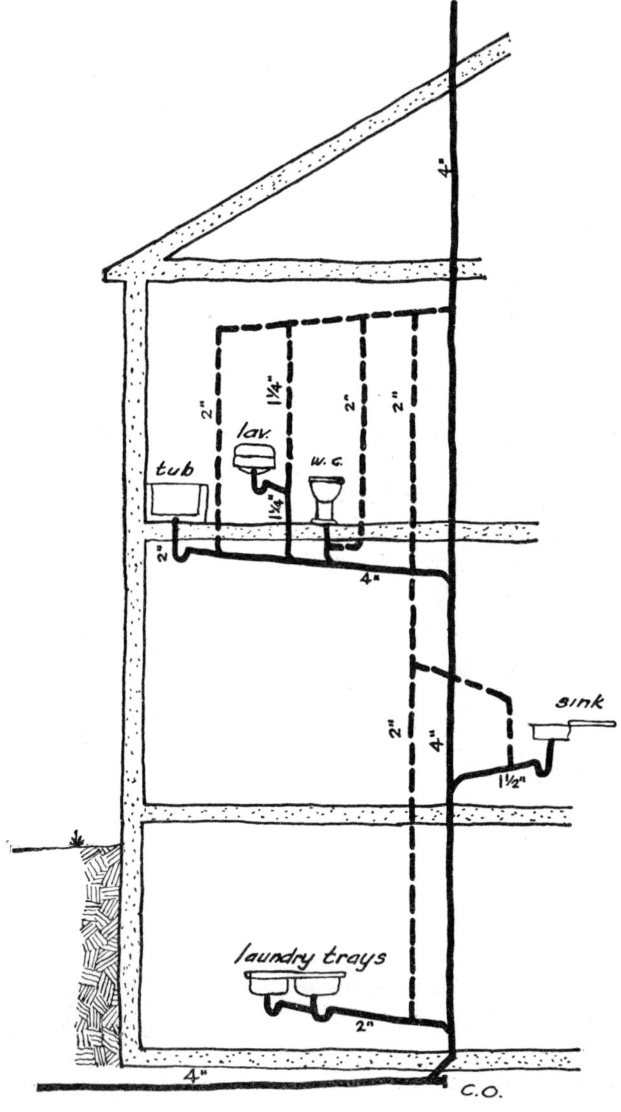

FIG. 37.—Sketch of Typical Drainage Plumbing for a One-family House.

TABLE 28

RATE OF DISCHARGE OF PLUMBING FIXTURES IN FIXTURE UNITS

Fixture	Units	Fixture	Units
One lavatory or wash basin. .	1	One urinal.	3
One kitchen sink.	1½	One floor drain.	3
One bathtub.	2	One shower bath.	3
One laundry tray.	3	One slop sink.	4
One combination fixture.	3	One water-closet.	6

One bathroom group consisting of one water-closet, one lavatory, and one bathtub and overhead shower; or one water-closet, one lavatory, and one shower compartment is regarded as having a combined discharge of eight fixture units. One hundred eighty square feet of roof or drained area in horizontal projection counts as one fixture unit.

Capacities of Vertical and Horizontal Drains.—The capacity of vertical soil stacks depending on the type of inlet fittings has been determined experimentally as given in Table 29. It is evi-

TABLE 29

CAPACITY OF SOIL STACKS IN FIXTURE UNITS

Diameter, inches	Single or double sanitary T fittings	Single or double Y, combination Y, and one-eighth bend fittings
2	6	12
3	13.5	27
4	24	48

dent from this table that a three-inch soil stack is adequate for any ordinary dwelling. It also emphasizes the effect of type of inlet fixture on the capacity of the stack. These figures presume, of course, that the outlet at the bottom is clear and of sufficient capacity to carry off the discharge without backing it up into the soil stack.

The capacity of horizontal drains depends on the slope as well as the size of pipes. Slopes flatter than one-quarter inch fall per foot are not recommended. Table 30 gives capacities of horizontal drains.

TABLE 30

CAPACITIES OF HORIZONTAL DRAINS IN FIXTURE UNITS

Diameter of drain, inches	Slope, $\frac{1}{8}$-in. fall per foot	Slope, $\frac{1}{4}$-in. fall per foot	Slope, $\frac{1}{2}$-in. fall per foot	Diameter of drain, inches	Slope, $\frac{1}{8}$-in. fall per foot	Slope, $\frac{1}{4}$-in. fall per foot	Slope, $\frac{1}{2}$-in. fall per foot
3	15	18	21	8	990	1,392	2,220
4	84	96	114	10	1,800	2,520	3,900
5	162	216	264	12	3,084	4,320	6,912
6	300	450	600				

Traps.—Good practice and most plumbing codes provide that each fixture must have an individual trap except that laundry trays may have a common trap. In general, these traps must provide a seal of at least one inch under all operating conditions. The minimum trap diameters and drain sizes for various fixtures are given in Table 31. Class 1 applies to private installations, residences, apartments, etc.; class 2 applies to semipublic installations, office buildings, factories, dormitories, etc.; and class 3 applies to public installations, schools, railroad stations, public comfort stations, etc.

Vent Pipes.—The main purpose of vents in plumbing systems is to release the suction which results when water flows through the drainage pipes and thus prevents the water seals in the traps from being siphoned out. Common practice is to make the vent a continuation of the soil stack as shown in Fig. 37. Most building codes require that any fixture more than five feet from the

TABLE 31

MINIMUM TRAP DIAMETERS, MINIMUM DRAIN SIZES, AND FIXTURE UNIT VALUES

Fixture and class of installation	Minimum nominal trap diameter, inches	Minimum nominal diameter, inches, individual drain	Fixture units
1 lavatory or washbasin, class 1	1¼	1¼	1
1 lavatory or washbasin, class 2 or 3	1¼	1¼	2
1 water-closet, class 1	3	3	3
1 water-closet, class 2	3	3	5
1 water-closet, class 3	3	3	6
1 bathtub, class 1	1½	1½	3
1 bathtub, class 2 or 3	2	2	4
1 shower stall, shower head only, class 1	1½	1½	2
1 shower stall, multiple spray, class 1	2	2	4
1 shower stall, head only, class 2 or 3	2	2	3
1 shower stall, multiple spray, class 2 or 3	3	3	6
1 urinal, lip, or each 2 feet of trough or gutter	1½	1½	2
1 urinal, stall or wall hung with tank or flush-valve supply	2	2	4
1 urinal, pedestal or blow out	3	3	5

soil stack must have a separate vent. Figure 38 illustrates good and bad practice in such venting. The line xy is the hydraulic gradient when the bowl is full and $x'y$ the gradient when the bowl is almost empty. The vent connection should come above the line xy to prevent back-flow into the vent pipe. Figure 39 illustrates types of plumbing installations including venting recommended by the U. S. Department of Commerce.

The size of vent required depends on its length and the load on the soil stack. Experiments have shown that for a three-inch soil stack carrying a capacity load of 40 fixture units a 2-inch vent 80 feet long or a 1½-inch vent 20 feet long is satisfactory.

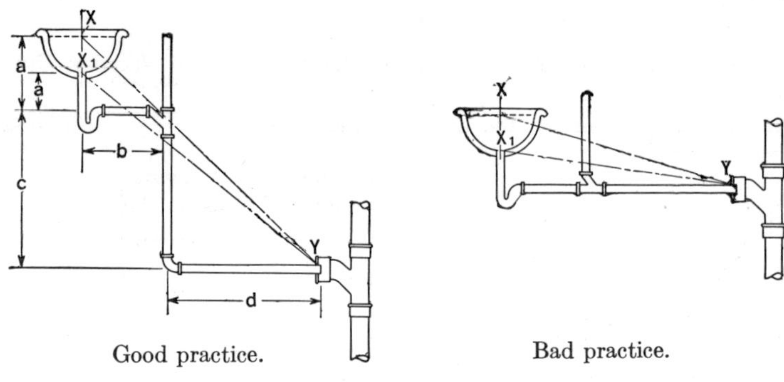

Good practice. Bad practice.

Fig. 38.

The use of a vent stack less than 1¼ inches in diameter is not recommended. Table 32 gives size and length of main vents for variously loaded soil stacks.

TABLE 32

Size and Length of Main Vents

Diameter of Soil or Waste Stack (Inches)	Number of Fixture Units on Soil or Waste Stack	Maximum Permissible Developed Length of Vent								
		1¼-in. vent	1½-in. vent	2-in. vent	2½-in. vent	3-in. vent	4-in. vent	5-in. vent	6-in. vent	8-in. vent
		Feet	Feet	Feet	Feet	Feet	Feet	Feet	Feet	Feet
1¼	2	75								
1½	8	70	150							
2	24	28	70	300						
3	40	..	20	80	260	650				
3	80	..	18	75	240	600				
4	310	..	..	30	95	240	1000			
4	620	..	..	22	70	180	750			
5	750	..	..	..	28	70	320	1000		
5	1500	..	..	..	20	50	240	750		
6	1440	..	..	..	..	20	95	240	1000	
6	2880	..	..	..	..	18	70	180	750	
8	3100	..	..	..	..	..	30	80	350	1100
8	6200	..	..	..	..	..	25	60	250	800

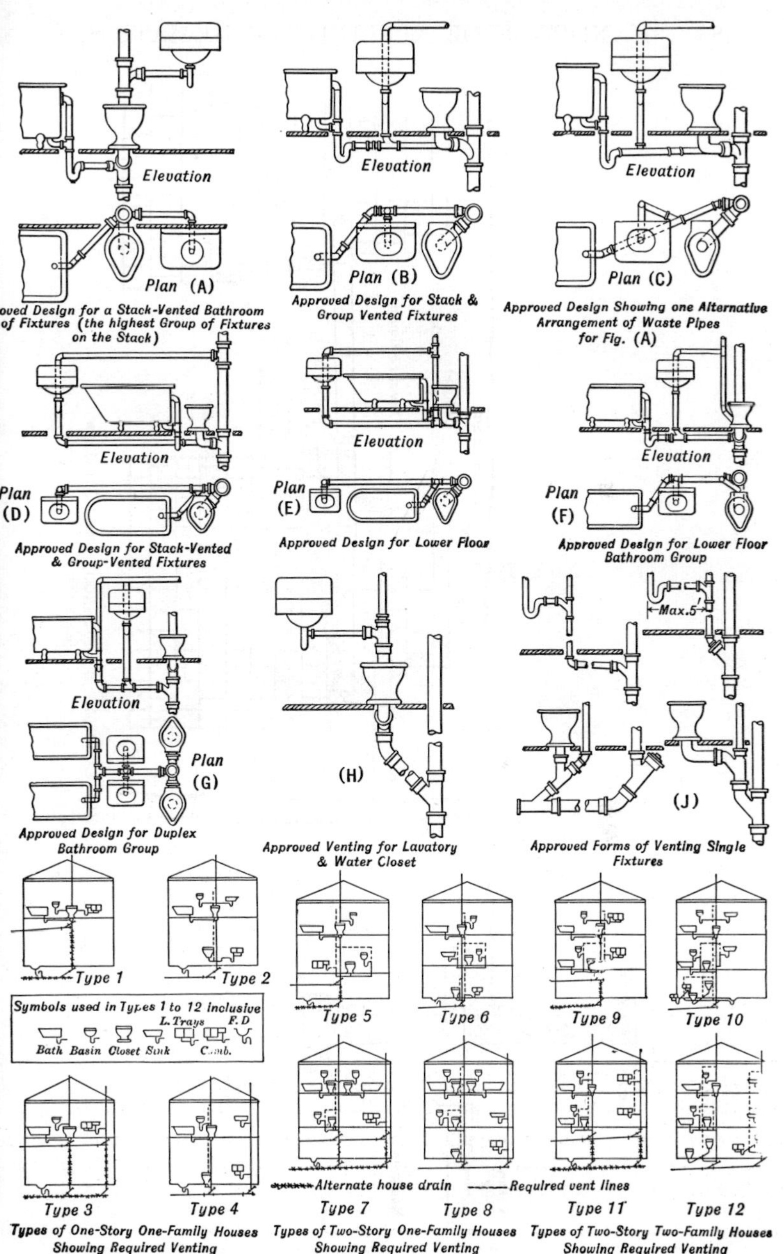
Elevation
Plan (A)
oved Design for a Stack-Vented Bathroom
of Fixtures (the highest Group of Fixtures
on the Stack)
Elevation
Plan (B)
Approved Design for Stack &
Group Vented Fixtures
Elevation
Plan (C)
Approved Design Showing one Alternative
Arrangement of Waste Pipes
for Fig. (A)
Elevation
Plan
(D)
Approved Design for Stack-Vented
& Group-Vented Fixtures
Elevation
Plan
(E)
Approved Design for Lower Floor
Elevation
Plan
(F)
Approved Design for Lower Floor
Bathroom Group
Elevation
Plan
(G)
Approved Design for Duplex
Bathroom Group
(H)
Approved Venting for Lavatory
& Water Closet
Max.5'
(J)
Approved Forms of Venting Single
Fixtures
Type 1
Type 2
Symbols used in Types 1 to 12 inclusive
L. Trays
F. D
Bath
Basin
Closet
Sink
Comb.
Type 3
Type 4
Types of One-Story One-Family Houses
Showing Required Venting
Type 5
Type 6
Type 7
Type 8
Alternate house drain
Required vent lines
Types of Two-Story One-Family Houses
Showing Required Venting
Type 9
Type 10
Type 11
Type 12
Types of Two-Story Two-Family Houses
Showing Required Venting

FIG. 39.—Plumbing Installations Recommended by "Hoover Report."

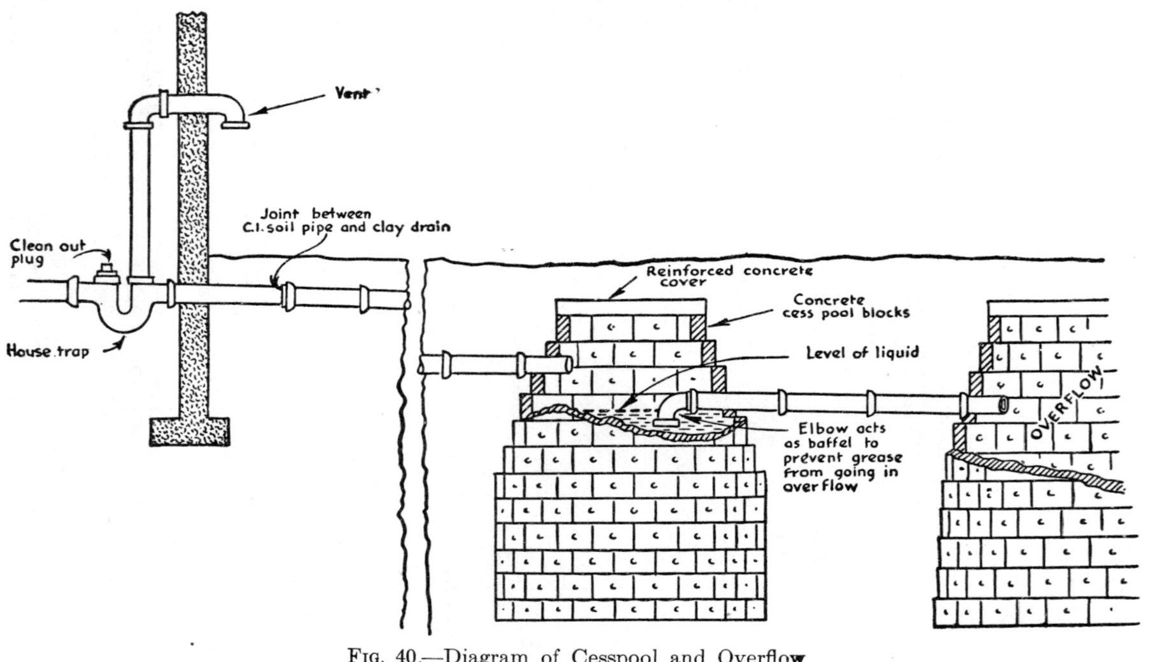

FIG. 40.—Diagram of Cesspool and Overflow.

Essential Parts

- **HOUSE SEWER — Pipe line which carries household wastes from the house to the septic tank.**
- **SEPTIC TANK — A water-tight container in which sewage is disintegrated by bacterial action. (The enlarged view of the single-chamber tank shows how sewage decomposes.)**
- **OUTLET SEWER LINE — Tile line which carries the liquid wastes from the septic tank to the disposal area.**
- **DISTRIBUTION BOX (Needed only for multiple disposal lines) — A small box, with outlets, which discharges an equal amount of the liquid wastes simultaneously into several disposal lines.**
- **DISPOSAL LINE — Drain-tile lines which allow the liquid wastes to seep into the soil through open joints of tile.**

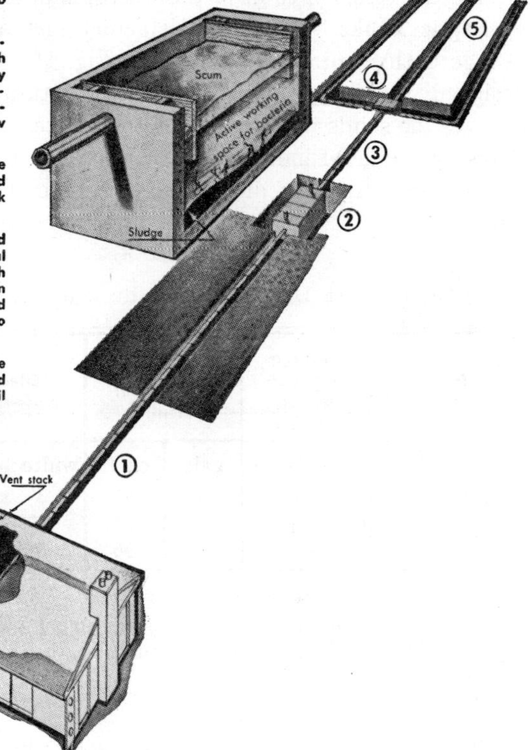

Fig. 41.—A Typical Septic Tank System. (*Univ. of Illinois Small Homes Council Circular* G5.5.)

Cesspools.—When access to public sewer systems is not available, some means must be provided to dispose of wastes. One common method is the use of a "leeching" cesspool, shown in Fig. 40. This is a pit in the ground, about 15 or 20 ft. from the house, made of stone, wood, brick, or concrete blocks, whose function is to drain liquid wastes through its walls into the surrounding ground. It slowly fills with solid matter that must eventually be removed.

Mortar is not used in the construction of the cesspool, since openings must be left for drainage purposes. A jacket of sand around the cesspool to filter the wastes is desirable.

Septic Tanks.—A more satisfactory method for disposing of sewage is the septic tank, shown in Fig. 41. This is a specially designed water-tight tank in which natural bacterial action changes the solids to liquid which then are easily drained. Septic tanks are made of metal, brick, concrete, or masonry. Table 33 lists the capacity and dimensions of single chamber rectangular septic tanks.

TABLE 33 ‡

CAPACITY AND DIMENSIONS FOR RECTANGULAR SEPTIC TANKS

No. of people	Sewage flow per person per day	Capacity* for 72-hour retention		Effective cross-section		Length*
	gals.	gals.	cu. ft.	width	depth†	
7 (or less)	26 −	540	72	3′	x 4′	6′ 0″
9	23 +	630	84	3′	x 4′	7′ 0″
12	20	720	96	3′	x 4′	8′ 0″

‡ *University of Illinois Small Homes Council Circular* G5.5.

* Capacity and length apply to single-chamber tank and first chamber of double-chamber tank. Capacity and length of second chamber are half that of the first.

† Depth is from the bottom of tank to the outlet tile.

Plumbing Fixtures.—A discussion of plumbing fixtures and other appurtenances is beyond the scope of this book. Such information, together with roughing-in dimensions, is available from fixture manufacturers, as well as in books on plumbing and pipe-fitting.

XVI

Heating

Heat and Temperature.—The problem of the man devising or constructing a heating system is to transfer the heat energy of a fuel into the air of the building as efficiently as possible. A knowledge of some of the basic properties of heat is extremely helpful in understanding the principles which govern the operation of house heating plants, and particularly in diagnosing the trouble when these fail to function properly. A physicist would define heat as "molecular energy" but we are not as much concerned with its definition as we are with the effects it produces.

Intensity of heat is measured in terms of degrees Fahrenheit, the freezing point of water being 32° F. and the boiling point 212° F.

Quantity of heat is measured by the British thermal unit (Btu) and one Btu is that quantity of heat which will raise the temperature of one pound of water one degree Fahrenheit.

Quantity of heat must not be confused with intensity. For example, a cupful of water at 150° F. will contain a *smaller* quantity of heat than a pailful of water at 70° F.

Effects of Heat on Fluids.—When air is heated it expands. When water above 39.2° F. is heated it also expands. Both of these substances are fluids. When fluids expand they become less dense, that is, they weigh less per cubic foot, and if they are free to move, the lighter fluid will rise to the top and the denser fluid will flow to the bottom to take its place. This principle is employed in hot air heating plants and hot water heating plants. In either case the lighter heated fluid rises and loses its heat in the rooms of a house and upon cooling becomes more dense and descends to the heating plant. In such a system a pound of water going into a radiator at a temperature of 180° F. and coming out of it at a temperature of 90° F. has given up 90 Btu of heat.

If the temperature of a pound of water in a steam boiler under atmospheric pressure is 150° F. and 62 Btu of heat are added to it, the water will increase in temperature to about 212° F. Adding a small additional amount of heat to this water will neither increase the temperature of the water nor convert the whole pound of it to steam. As much as 970 Btu must be added to this pound of water at 212° F. to change it to steam at 212° F. This additional heat is called the *latent heat of vaporization.* In heating houses by steam, most of the heat in the rooms is derived from this latent heat of vaporization which is given up by the steam in the radiators in changing back to water.

Heat Transfer.—In a heating plant, for example, a hot water system, the heat from the fuel is transferred to the casing of the boiler, then to the water in the boiler, then to the water in the radiator, to the casing of the radiator, to the room which is being heated, then finally through the walls and windows to the outdoors where it is dissipated. There are three ways by which heat is transferred, by radiation, by conduction, and by convection.

Radiation.—Heat travels in direct rays from a source much the same as light does. This is best illustrated by the heat which comes from the sun or from a fire in an open fireplace. In either case if the direct rays are cut off by an object, a heat shadow is formed and the same intensity of the heat is not felt in the shadow.

Conduction.—If the end of an iron rod is heated, the heat will be transferred from one iron particle to the next until the heat has traveled the whole length of the rod. This is called conduction. Some materials conduct heat more readily than others. Copper is a particularly good conductor. Materials which are poor conductors, such as asbestos, and mineral wool, are used for insulation.

Convection.—Heat transfer by convection depends on the circulation of a fluid, the warmed particles of the fluid mingling with the unwarmed particles. Thus the circulation of warm air from a hot air furnace is an example of heat transfer by convection. So is also the circulation of water in a hot water heating system.

Estimating Heating Requirements.—When a public utility company builds an electric power plant, a gas plant, or a water supply system it must first estimate the probable *demand* which the consumer will place upon the service. The design of a heating plant is approached from much the same angle. First the heat *demand* of the building must be determined and then the radiators, pipes, boilers, etc., must be selected to satisfy this demand completely yet economically. The heat demand of a building depends on the following factors: *

1. Outside temperature. 2. Rain or snow. 3. Sunshine or cloudiness. 4. Wind velocity.	*Outside Conditions (The Weather)*
5. Heat transmission of exposed parts of buildings. 6. Infiltration of air through cracks, crevices, and open doors and windows. 7. Heat capacity of materials. 8. Rate of absorption of solar radiation of exposed material.	*Building Construction*
9. Inside temperatures. 10. Stratification. 11. Type of heating system. 12. Ventilation requirements. 13. Period and nature of occupancy. 14. Temperature regulation.	*Inside Conditions*

It will be noticed that many of these factors are variable and this leads to a great many combinations of circumstances. Values for many of these factors have been established by the American Society of Heating and Ventilating Engineers, the Heating and Piping Contractors National Association and university research bureaus so that the heat required by a room or a house in terms of Btu per hour can be set up in practically a single equation.

* From the Guide of A.S.H. & V.E.

Needless to say this equation is long and its solution tedious. The Heating and Piping Contractors National Association has therefore compiled a Standard Radiation Estimating Table which shortens the work materially. By the use of this table the heat requirements of a room may be translated directly into square feet of steam radiation (see below) without going into the intermediate step of estimating the number of Btu's required. Before describing this method of estimating, we shall discuss some of the factors entering into the estimate and define what is meant by radiation.

Temperature, Wind and Exposure.—The amount of heat lost from a room depends partly on the difference between the inside and the outside temperatures. The average outside temperature during the heating season varies, of course, with the locality. Experience has shown that periods of intense cold are generally of short duration so that the factor which is used as the base temperature in the calculations is several degrees higher.

Desirable inside temperatures have been fairly well standardized. These are listed in Table 1.

TABLE 1

INSIDE TEMPERATURES

Type of room or building	Temperature, degrees F
Warm air baths	120
Steam baths	110
Hospital operating rooms	85
Bath rooms	80
Paint shops	80
Hospitals	72 to 75
Public buildings	68 to 72
Residences	70
Schools	70
Factories	65
Stores	65
Gymnasia	55 to 60
Machine shops	60 to 65
Foundries, boiler shops, etc.	50 to 60

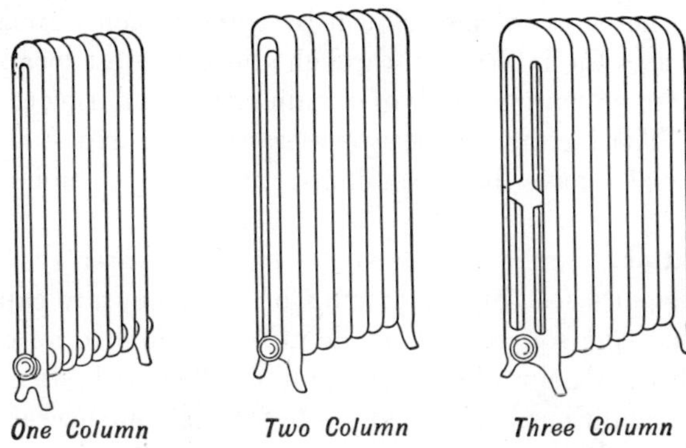

Fig. 1.—Typical Old-style Cast-iron Radiators No Longer Manufactured. (*Courtesy American Oil Burner Association.*)

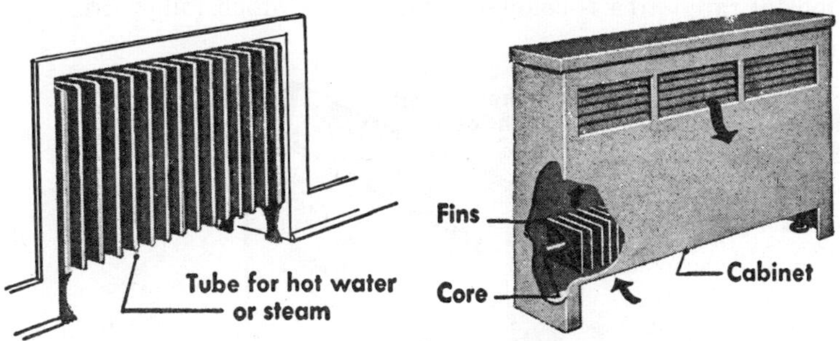

Fig. 2.—Recessed Radiator. (From *University of Illinois Small Homes Council Circular.*)

Fig. 3.—*Convector.* (From *University of Illinois Small Homes Council Circular.*)

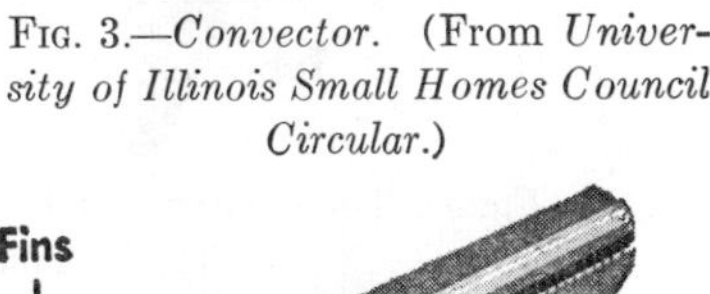

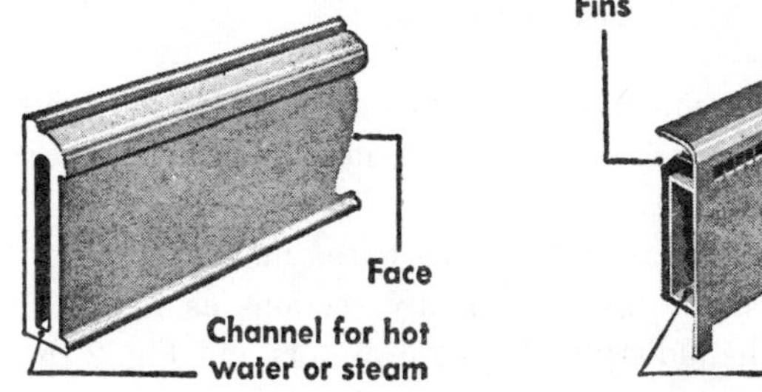

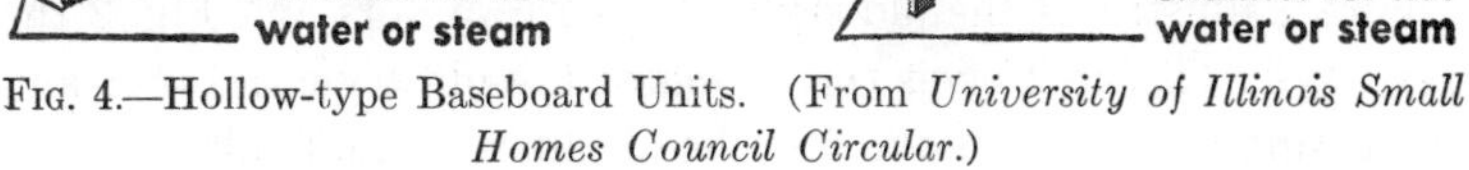

Fig. 4.—Hollow-type Baseboard Units. (From *University of Illinois Small Homes Council Circular.*)

Wind increases the loss of heat by transmission through walls and increases the infiltration through cracks. Most localities are subjected to prevailing winds of certain intensity during the winter months. The factors for base temperature and exposure to prevailing winds have been combined in a single tabulation in Table 16.

Radiation.—Radiators for steam and hot water systems are rated in square feet by the amount of heat they are capable of giving off. One square foot of radiation is equal to an emission of 240 Btu per hour when a radiator is filled with steam under standard conditions. However, radiators seldom operate under standard conditions and manufacturers' ratings sometimes vary, so that in actual practice an emission of only 225 Btu per square foot of rated area is counted. The tables which follow are made up on this basis.

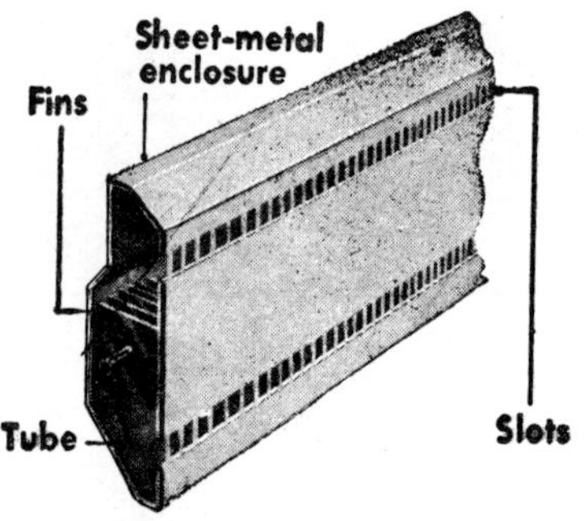

Fig. 5.—Finned-Tube Baseboard Unit. (From *University of Illinois Small Homes Council Circular.*)

A square foot of steam radiation gives off 150 Btu per hour when a radiator is used in a hot water heating system at a mean temperature of 170 degrees.

The most common radiators are of cast iron manufactured in several heights and made up of as many sections as required. Figure 1 illustrates the old style of such radiators and Fig. 2 the new style. Table 2 gives the rating of the old style radiator and the rating of the new style may be obtained from Table 3.

TABLE 2

Approximate Rating of Old-Style Radiators, Square Feet per Section

No. of Columns	Height in inches										
	13	15	16	18	20	22	23	26	32	38	45
1	.....	.....	.....	.....	1.5	.....	1.7	2.0	2.5	3.0	.....
2	.....	1.5	.....	.....	2.0	2.3	2.3	2.7	3.3	4.0	5.0
3	.....	.....	.....	2.3	.....	3.0	.....	3.8	4.5	5.0	6.0
4	.....	.....	.....	3.0	.....	4.0	.....	5.0	6.5	8.0	10.0
5	.....	.....	.....	4.7	.....	.....	.....	7.0	.....	10.0	.....
6	3.0	.....	3.8	4.5	5.0	.....	.....	.....	.....	.....	.....

Illustration: A computation shows that a certain room will require 45 square feet of radiation and it is desired to use 3-column old-style radiators 32 inches high. How many sections will be required?

From Table 2 the radiating surface of one section of 3-column, 32-inch old-style radiator is 4.5 square feet. Then,

$$\frac{45}{4.5} = 10 \text{ sections required (Ans.)}$$

Illustration: Another room requires 18 square feet of radiation and it is desired to use a small-tube cast-iron radiator 25 inches high. How many sections and tubes per section will be required?

The rating of a section depends upon the number of tubes per section used. From Table 3, the rating of a 25-inch high section is 1.6 square feet for 3 tubes per section, 2.0 square feet for 4 tubes per section, 2.4 square feet for 5 tubes per section, and 3.0 square

TABLE 3

SMALL-TUBE CAST-IRON RADIATORS *

Number of Tubes per Section	Catalog Rating per Section[a]		Section Dimensions: A Height[c]	Section Dimensions: B Width, Minimum	Section Dimensions: B Width, Maximum	Section Dimensions: C Spacing[b]	Section Dimensions: D Leg Height[c]
	Sq Ft	Btu/hr	In.	In.	In.	In.	In.
3[d]	1.6	384	25	3¼	3½	1¾	2½
4[d]	1.6 1.8 2.0	384 432 480	19 22 25	4 7/16 4 7/16 4 7/16	4 13/16 4 13/16 4 13/16	1¾ 1¾ 1¾	2½ 2½ 2½
5[d]	2.1 2.4	504 576	22 25	5⅝ 5⅝	6 5/16 6 5/16	1¾ 1¾	2½ 2½
6[d]	2.3 3.0 3.7	552 720 888	19 25 32	6 13/16 6 13/16 6 13/16	8 8 8	1¾ 1¾ 1¾	2½ 2½ 2½

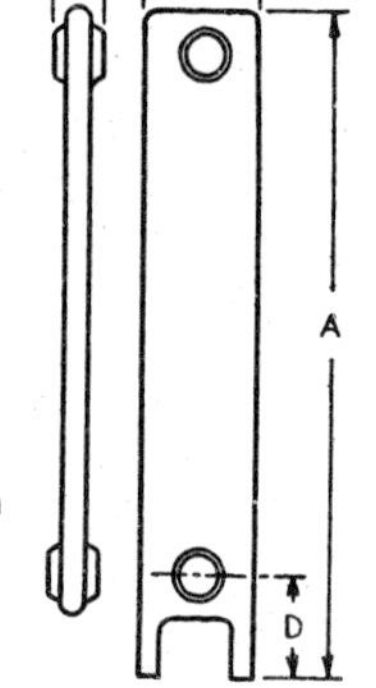

* *A.S.H.V.E.*

[a] These ratings are based on steam at 215° F. and air at 70° F. They apply only to installed radiators exposed in a normal manner; not to radiators installed behind enclosures, grilles, or under shelves.

[b] Length equals number of sections times 1¾ in.

[c] Overall height and leg height, as produced by some manufacturers, are one inch (1 in.) greater than shown in Columns A and D. Radiators may be furnished without legs. Where greater than standard leg heights are required this dimension shall be 4½ in.

[d] Or equal.

feet for 6 tubes per section. In this problem 18 square feet of radiation are required. There are several possibilities:

(1) $18/1.6 = 1\frac{1}{8}$ (or 2) sections of 3 tubes per section.
(2) $18/2 = 9$ sections of 4 tubes per section.
(3) $18/2.4 = 7\frac{1}{2}$ (or 8) sections of 5 tubes per section.
(4) $18/3 = 6$ sections of 6 tubes per section.

In these solutions, (1) or (3) gives a larger radiation than s necessary. The answers arrived at in (2) or (4)—9 sections of 4 tubes per section or 6 sections of 6 tubes per section—can be used. The final selection will depend upon the space available and the cost of the unit.

Another style of radiator is a so-called "wall radiator" which s hung on the walls or ceiling to conserve space. These are usually two-column affairs and both have several coils cast as one unit or are made up in units from separate sections.

Table 4 gives the radiation areas of such units.

TABLE 4

RATINGS OF CAST-IRON WALL RADIATOR UNITS, SQUARE FEET

Height, inches	Length or width, inches	Thickness, inches	Heating surface, square feet
13¼	16½	3	6½
13¼	22	3	8
13¼	29	3	11
22	13¼	3	8
29	13¼	3	11

Heating coils are sometimes also made up from standard pipe or a pipe riser may be used to heat a small room. Table 5 gives the heating surface of standard pipe.

ILLUSTRATION: A bathroom requires five square feet of radiation. If the headroom available is 8 feet, how large a pipe riser will be required to provide the necessary radiation?

Since 5 square feet of radiation are required from 8 feet of pipe, then $5/8 = 0.625$ square foot is required per foot of pipe. Referring this per-foot figure to Table 5 to obtain the diameter of pipe, we find that the 2-inch pipe fills the need very closely. (Ans.)

TABLE 5

HEATING SURFACE OF STANDARD PIPE, SQUARE FEET

Length of pipe, feet	Nominal diameter of pipe, inches									
	¾	1	1¼	1½	2	2½	3	4	5	6
1	.275	.346	.434	.494	.622	.753	.916	1.175	1.455	1.73[illegible]

Estimating Radiation.—Estimating radiation requirements i[illegible] simple with the aid of the tables. Let us take as an example th[illegible] room shown in Fig. 6. This represents the dining room of a hous[illegible]

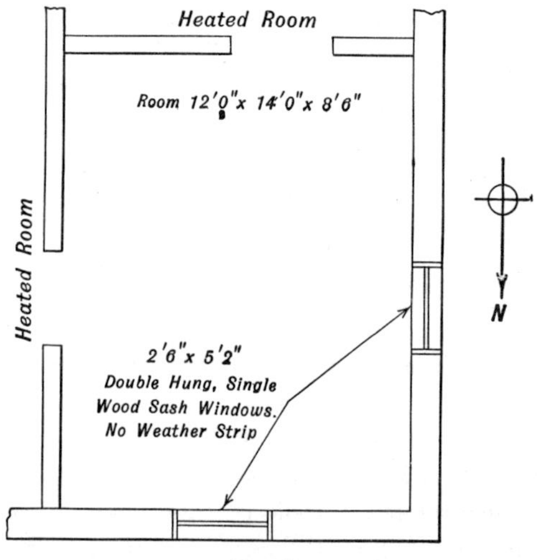

FIG. 6.

in Philadelphia. The outside walls of frame construction, 1-inch sheathing, and brick veneer. Inside walls are plastered on wood lath on studding. The floor above has heated rooms. The problem is now to find how many square feet of steam radiation will

TABLE 6

Summarizing the whole estimate we have:	Area or lin. ft.	Sq. ft. radiation
North wall, 12 × 8.5 = 102 sq. ft. − 12.9 sq. ft. (window)	89.1 sq. ft.	8 sq. ft.
1 window, 2 ft. 6 in. by 5 ft. 2 in.	12.9 sq. ft.	4 sq. ft.
Infiltration, cracks	17.8 ft.	5 sq. ft.
Total for north wall without exposure factor		17 sq. ft.
Exposure factor		0.94 sq. ft.
Total		15.98 sq. ft.
West wall, 14 × 8.5 = 119 sq. ft. − 12.9 sq. ft. (window)	106.1 sq. ft.	9 sq. ft.
1 window, 2 ft. 6 in. by 5 ft. 2 in.	12.9 sq. ft.	4 sq. ft.
Infiltration, cracks	17.8 lin. ft.	5 sq. ft.
Total for west wall without exposure factor		18 sq. ft.
Exposure factor		0.94 sq. ft.
Total		16.92 sq. ft.
The total for the room is		
North wall		15.98 sq. ft.
West wall		16.92 sq. ft.
Total		32.90 sq. ft.

be required to heat this room. This may then be translated into other terms as desired.

The area of the north wall is 12 × 8.5 = 102 square feet. The area of the window (2′6″ × 5′2″) is 12.9 square feet. The net wall area is 102 — 12.9 = 89.1 square feet.

Then we look through Tables 7 to 18 to find the one which has the figures for this type of construction. In Table 7, Wall No. 27, we find "Brick Veneer, 1-inch Wood Sheathing." Following this line across to Column C, which represents plaster on wood lath on studding, we find a figure 0.27 which is called a

TABLE 7

Coefficients of Transmission (*U*) of Frame Walls *

Exterior Finish	Interior Finish	Type of Sheathing: Gypsum (½ in. thick) A	Type of Sheathing: Plywood (5/16 in. thick) B	Type of Sheathing: Wood[f] (25/32 in. thick) Bldg. Paper C	Type of Sheathing: Insulating Board (25/32 in. thick) D	Wall Number
Wood Siding (Clapboard)	Metal Lath and Plaster[b]	0.33	0.32	0.26	0.20	1
	Gypsum Board (⅜ in.) Decorated	0.32	0.32	0.26	0.20	2
	Wood Lath and Plaster	0.31	0.31	0.25	0.19	3
	Gypsum Lath (⅜ in.) Plastered[c]	0.31	0.30	0.25	0.19	4
	Plywood (⅜ in.) Plain or Decorated	0.30	0.30	0.24	0.19	5
	Insulating Board (½ in.) Plain or Decorated	0.23	0.23	0.19	0.16	6
	Insulating Board Lath (½ in.) Plastered[c]	0.22	0.22	0.19	0.15	7
	Insulating Board Lath (1 in.) Plastered[c]	0.17	0.17	0.15	0.12	8
Wood[d] Shingles	Metal Lath and Plaster[b]	0.25	0.25	0.26	0.17	9
	Gypsum Board (⅜ in.) Decorated	0.25	0.25	0.26	0.17	10
	Wood Lath and Plaster	0.24	0.24	0.25	0.16	11
	Gypsum Lath (⅜ in.) Plastered[c]	0.24	0.24	0.25	0.16	12
	Plywood (⅜ in.) Plain or Decorated	0.24	0.24	0.24	0.16	13
	Insulating Board (½ in.) Plain or Decorated	0.19	0.19	0.19	0.14	14
	Insulating Board Lath (½ in.) Plastered[c]	0.19	0.18	0.19	0.13	15
	Insulating Board Lath (1 in.) Plastered[c]	0.14	0.14	0.15	0.11	16
Stucco	Metal Lath and Plaster[b]	0.43	0.42	0.32	0.23	17
	Gypsum Board (⅜ in.) Decorated	0.42	0.41	0.31	0.23	18
	Wood Lath and Plaster	0.40	0.39	0.30	0.22	19
	Gypsum Lath (⅜ in.) Plastered[c]	0.39	0.39	0.30	0.22	20
	Plywood (⅜ in.) Plain or Decorated	0.39	0.38	0.29	0.22	21
	Insulating Board (½ in.) Plain or Decorated	0.27	0.27	0.22	0.18	22
	Insulating Board Lath (½ in.) Plastered[c]	0.26	0.26	0.22	0.17	23
	Insulating Board Lath (1 in.) Plastered[c]	0.19	0.19	0.16	0.14	24
Brick Veneer[e]	Metal Lath and Plaster[b]	0.37	0.36	0.28	0.21	25
	Gypsum Board (⅜ in.) Decorated	0.36	0.36	0.28	0.21	26
	Wood Lath and Plaster	0.35	0.34	0.27	0.20	27
	Gypsum Lath (⅜ in.) Plastered[c]	0.34	0.34	0.27	0.20	28
	Plywood (⅜ in.) Plain or Decorated	0.34	0.33	0.27	0.20	29
	Insulating Board (½ in.) Plain or Decorated	0.25	0.25	0.21	0.17	30
	Insulating Board Lath (½ in.) Plastered[c]	0.24	0.24	0.20	0.16	31
	Insulating Board Lath (1 in.) Plastered[c]	0.18	0.18	0.15	0.13	32

* Table footnotes on p. 504.

TABLE 8

COEFFICIENTS OF TRANSMISSION (U) OF FRAME WALLS WITH INSULATION BETWEEN FRAMING *, a

COEFFICIENT WITH *NO* INSULATION BETWEEN FRAMING	COEFFICIENT WITH INSULATION BETWEEN FRAMING				NUMBER
	MINERAL WOOL OR VEGETABLE FIBERS IN BLANKET OR BAT FORM[b] (Thickness below)			3⅝ IN. MINERAL WOOL BETWEEN FRAMING[c]	
	1 IN.	2 IN.	3 IN.		
	A	B	C	D	
0.11	0.078	0.063	0.054	0.051	33
0.13	0.088	0.070	0.058	0.055	35
0.15	0.097	0.075	0.062	0.059	37
0.17	0.10	0.080	0.066	0.062	39
0.19	0.11	0.084	0.069	0.065	41
0.21	0.12	0.088	0.072	0.067	43
0.23	0.12	0.091	0.074	0.069	45
0.25	0.13	0.094	0.076	0.071	47
0.27	0.14	0.097	0.078	0.073	49
0.29	0.14	0.10	0.080	0.075	51
0.31	0.14	0.10	0.081	0.076	53
0.33	0.15	0.10	0.083	0.077	55
0.35	0.15	0.11	0.084	0.078	57
0.37	0.16	0.11	0.085	0.080	59
0.39	0.16	0.11	0.086	0.081	61
0.41	0.16	0.11	0.087	0.082	63
0.43	0.17	0.11	0.088	0.082	65

These coefficients are expressed in Btu per (hour)(square feet)(Fahrenheit degree difference in temperature between the air on the two sides), and are based on an outside wind velocity of 15 mph.

* *A.S.H.V.E.*

a This table may be used for determining the coefficients of transmisson of frame constructions with the types and thicknesses of insulation indicated in Columns A to D inclusive between framing. Columns A, B and C may be used for walls, ceilings or roofs with only one air space between framing but are not applicable to ceilings with no flooring above. (See Table 13.) Column D is applicable to walls only. Example: Find the coefficient of transmission of a frame wall consisting of wood siding, 25⁄32 in. insulating board sheathing studs, gypsum lath and plaster, with 2 in. blanket insulation between studs. According to Table 7, a wall of this construction with no insulation between studs has a coefficient of 0.19 (Wall No. 4D). Referring to Column B above it will be found that a wall of this value with 2 in. blanket insulation between the studs has a coefficient of 0.084.

Coefficients corrected for 2 × 4 framing, 16 in. on centers—15 percent of surface area.

b Based on one air space between framing.

c No air space.

TABLE 9

Coefficients of Transmission (U) of Masonry Walls *

Type of Masonry	Thickness of Masonry Inches	Plain Walls—No Interior Finish	Plaster (1/2 in.) on Walls	Metal Lath and Plaster[f] Furred[h]	Gypsum Board (3/8 in.) Decorated—Furred[h]	Gypsum Lath (3/8 in.) Plastered[g]—Furred[h]	Insulating Board (1/2 in.) Plain or Decorated—Furred[h]	Insulating Board Lath (1/2 in.) Plastered[g]—Furred[h]	Insulating Board Lath (1 in.) Plastered[g]—Furred[h]	Gypsum Lath[g] Plastered Plus 1 in. Blanket Insulation—Furred[h]	Wall Number
		Interior Finish (Plus Insulation Where Indicated)									
		A	B	C	D	E	F	G	H	I	
Solid[a] Brick	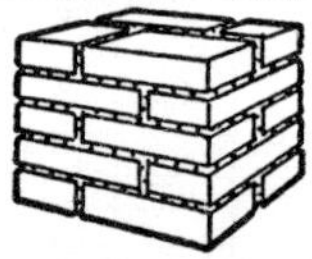 8	0.50	0.46	0.32	0.31	0.30	0.22	0.22	0.16	0.14	67
	12	0.36	0.34	0.25	0.25	0.24	0.19	0.19	0.14	0.13	68
	16	0.28	0.27	0.21	0.21	0.20	0.17	0.16	0.13	0.12	69
Hollow[b] Tile (Stucco Exterior Finish)	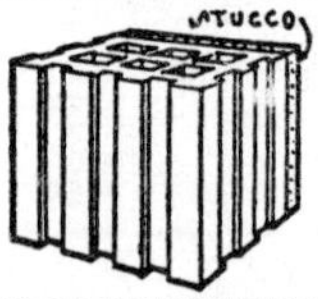 8	0.40	0.37	0.27	0.27	0.26	0.20	0.20	0.15	0.13	70
	10	0.39	0.37	0.27	0.27	0.26	0.20	0.19	0.15	0.13	71
	12	0.30	0.28	0.22	0.22	0.21	0.17	0.17	0.13	0.12	72
	16	0.24	0.24	0.19	0.19	0.18	0.15	0.15	0.12	0.11	73
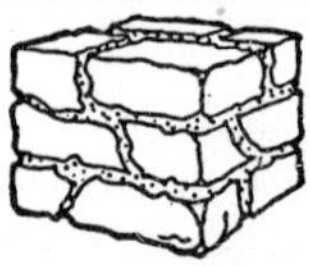 Stone[c]	8	0.70	0.64	0.39	0.38	0.36	0.26	0.25	0.18	0.16	74
	12	0.57	0.53	0.35	0.34	0.33	0.24	0.23	0.17	0.15	75
	16	0.49	0.45	0.31	0.31	0.29	0.22	0.22	0.16	0.14	76
	24	0.37	0.35	0.26	0.26	0.25	0.19	0.19	0.15	0.13	77
Poured Concrete[d]	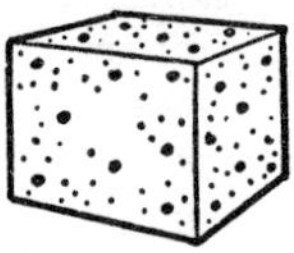 6	0.79	0.71	0.42	0.41	0.39	0.27	0.26	0.19	0.16	78
	8	0.70	0.64	0.39	0.38	0.36	0.26	0.25	0.18	0.16	79
	10	0.63	0.58	0.37	0.36	0.34	0.25	0.24	0.18	0.15	80
	12	0.57	0.53	0.35	0.34	0.33	0.24	0.23	0.17	0.15	81
Hollow Concrete Blocks		**Gravel Aggregate**									
	8	0.56	0.52	0.34	0.34	0.32	0.24	0.23	0.17	0.15	82
	12	0.49	0.46	0.32	0.31	0.30	0.22	0.22	0.16	0.14	83
		Cinder Aggregate									
	8	0.41	0.39	0.28	0.28	0.27	0.21	0.20	0.15	0.13	84
	12	0.38	0.36	0.26	0.26	0.25	0.20	0.19	0.15	0.13	85
		Light Weight Aggregate[e]									
	8	0.36	0.34	0.26	0.25	0.24	0.19	0.19	0.15	0.13	86
	12	0.34	0.33	0.25	0.24	0.24	0.19	0.18	0.14	0.13	87

* Table footnotes on p. 504.

TABLE 10

Coefficients of Transmission (U) of Brick and Stone Veneer Masonry Walls *

Typical Construction	Facing	Backing	Interior Finish (Plus Insulation Where Indicated)									Wall Number
			Plain Walls—no Interior Finish	Plaster (½ in.) on Walls	Metal Lath and Plaster—Furred[e]	Gypsum Board (⅜ in). Decorated —Furred[e]	Gypsum Lath (⅜ in.) Plastered[f]—Furred[e]	Insulating Board (½ in.) Plain or Decorated—Furred[e]	Insulating Board Lath (½ in.) Plastered[f]—Furred[e]	Insulating Board Lath (1 in.) Plastered[f]—Furred[e]	Gypsum Lath Plastered[f] Plus 1 in. Blanket Insulation—Furred[e]	
			A	B	C	D	E	F	G	H	I	
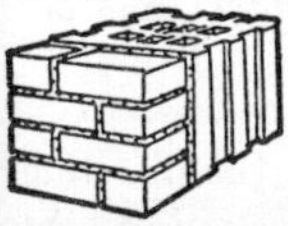	4 in. Brick Veneer[a]	6 in. Hollow Tile[b]	0.35	0.34	0.25	0.25	0.24	0.19	0.18	0.14	0.13	88
		8 in. Hollow Tile[b]	0.34	0.32	0.25	0.24	0.23	0.19	0.18	0.14	0.13	89
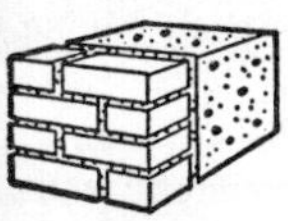		6 in. Concrete	0.59	0.54	0.35	0.35	0.33	0.24	0.23	0.17	0.15	90
		8 in. Concrete	0.54	0.50	0.33	0.33	0.31	0.23	0.23	0.17	0.15	91
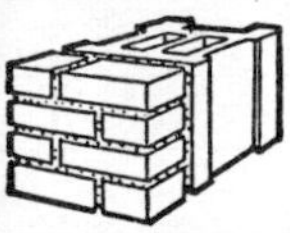		8 in. Concrete Blocks[c] (Gravel Aggregate)	0.44	0.41	0.29	0.29	0.28	0.21	0.21	0.16	0.14	92
		8 in. Concrete Blocks[c] (Cinder Aggregate)	0.34	0.33	0.25	0.24	0.24	0.19	0.18	0.14	0.13	93
		8 in. Concrete Blocks[c] (Light Weight Aggregate)[d]	0.31	0.29	0.23	0.23	0.22	0.18	0.17	0.14	0.12	94
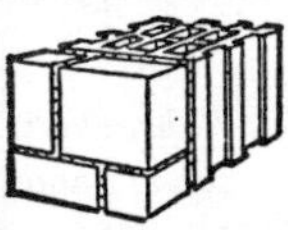	4 in. Cut Stone Veneer[a]	6 in. Hollow Tile[b]	0.37	0.35	0.26	0.26	0.25	0.19	0.19	0.15	0.13	95
		8 in. Hollow Tile[b]	0.36	0.34	0.25	0.25	0.24	0.19	0.19	0.14	0.13	96
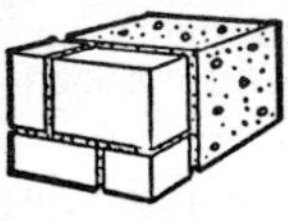		6 in. Concrete	0.63	0.58	0.37	0.36	0.34	0.25	0.24	0.18	0.15	97
		8 in. Concrete	0.57	0.53	0.35	0.34	0.33	0.24	0.23	0.17	0.15	98
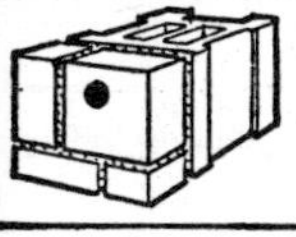		8 in. Concrete Blocks[c] (Gravel Aggregate)	0.47	0.44	0.30	0.30	0.29	0.22	0.21	0.16	0.14	99
		8 in. Concrete Blocks[c] (Cinder Aggregate)	0.36	0.34	0.25	0.25	0.24	0.19	0.19	0.15	0.13	100
		8 in. Concrete Blocks[c] (Light Weight Aggregate)[d]	0.32	0.30	0.23	0.23	0.22	0.18	0.17	0.14	0.12	101

* Table footnotes on p. 504.

Table footnotes for Table 7, p. 500

These coefficients are expressed in Btu per (hour)(square feet)(Fahrenheit degree difference in temperature between the air on the two sides), and are based on an outside wind velocity of 15 mph.

No insulation between studs[a] (see Table 8).

[a] Coefficients not weighted; effect of studding neglected.

[b] Plaster assumed ¾ in. thick.

[c] Plaster assumed ½ in. thick.

[d] Furring strips (1 in. nominal thickness) between wood shingles and all sheathings except wood.

[e] Small air space and mortar between building paper and brick veneer neglected.

[f] Nominal thickness, 1 in.

Table footnotes for Table 9, p. 502

Coefficients are expressed in Btu per (hour)(square foot)(Fahrenheit degree difference in temperature between the air on the two sides), and are based on an outside wind velocity of 15 mph.

[a] Based on 4 in. hard brick and remainder common brick.

[b] The 8 in. and 10 in. tile figures are based on two cells in the direction of heat flow. The 12 in. tile is based on three cells in the direction of heat flow. The 16 in. tile consists of one 10 in. and one 6 in. tile, each having two cells in the direction of heat flow.

[c] Limestone or sandstone.

[d] These figures may be used with sufficient accuracy for concrete walls with stucco exterior finish.

[e] Expanded slag, burned clay or pumice.

[f] Thickness of plaster assumed ¾ in.

[g] Thickness of plaster assumed ½ in.

[h] Based on 2 in. furring strips; one air space.

Table footnotes for Table 10, p. 503

Coefficients are expressed in Btu per (hour)(square foot)(Fahrenheit degree difference in temperature between the air on the two sides), and are based on an outside wind velocity of 15 mph.

[a] Calculation based on ½ in. cement mortar between backing and facing, except in the case of the concrete backing which is assumed to be poured in place.

[b] The hollow tile figures are based on two air cells in the direction of heat flow.

[c] Hollow concrete blocks.

[d] Expanded slag, burned clay or pumice.

[e] Thickness of plaster assumed ¾ in.

[f] Thickness of plaster assumed ½ in.

[g] Based on 2 in. furring strips; one air space.

TABLE 11

COEFFICIENTS OF TRANSMISSION (*U*) OF FRAME PARTITIONS OF INTERIOR WALLS *, a

INTERIOR FINISH	SINGLE PARTITION (Finish on one side only of studs)	DOUBLE PARTITION (Finish on both sides of studs)		PARTITION NUMBER
		NO INSULATION BETWEEN STUDS	1 IN. BLANKET[d] BETWEEN STUDS. ONE AIR SPACE.	
	A	B	C	
Metal Lath and Plaster[b]	0.69	0.39	0.16	1
Gypsum Board (⅜ in.) Decorated	0.67	0.37	0.16	2
Wood Lath and Plaster	0.62	0.34	0.15	3
Gypsum Lath (⅜ in.) Plastered[c]	0.61	0.34	0.15	4
Plywood (⅜ in.) Plain or Decorated	0.59	0.33	0.15	5
Insulating Board (½ in.) Plain or Decorated	0.36	0.19	0.11	6
Insulating Board Lath (½ in.) Plastered[c]	0.35	0.18	0.11	7
Insulating Board Lath (1 in.) Plastered[c]	0.23	0.12	0.082	8

Coefficients are expressed in Btu per (hour)(square foot)(Fahrenheit degree difference in temperature between the air on the two sides), and are based on still air (no wind) conditions on both sides.

* *A.S.H.V.E.*

[a] Coefficients not weighted; effect of studding neglected.

[b] Plaster assumed ¾ in. thick.

[c] Plaster assumed ½ in. thick.

[d] For partitions with other insulations between studs refer to Table 8, using values in Column B of above table, in left-hand column of Table 8. Example: What is the coefficient of transmission (*U*) of a partition consisting of gypsum lath and plaster on both sides of studs with 2 in. blanket between studs? Solution: According to above table, this partition with no insulation between studs (No. 4B) has a coefficient of 0.34. Referring to Table 8, it will be found that a wall having a coefficient of 0.34 with no insulation between studs, will have a coefficient of 0.10 with 2 in. of blanket insulation between studs (No. 56B).

TABLE 12

Coefficients of Transmission (U) of Masonry Partitions *

Type of Partition		Thickness of Masonry (Inches)	Type of Finish: No Finish (Plain walls) A	Plaster One Side B	Plaster Both Sides[a] C	Partition Number
Hollow Clay Tile		3	0.50	0.47	0.43	9
		4	0.45	0.42	0.40	10
Hollow Gypsum Tile		3	0.35	0.33	0.32	11
		4	0.29	0.28	0.27	12
Hollow Concrete Tile or Blocks	Cinder Aggregate	3	0.50	0.47	0.43	13
		4	0.45	0.42	0.40	14
	Light Weight Aggregate[b]	3	0.41	0.39	0.37	15
		4	0.35	0.34	0.32	16
Common Brick		4	0.50	0.46	0.43	17

Coefficients are expressed in Btu per (hour)(square foot)(Fahrenheit degree difference between the air on the two sides) and are based on still air (no wind) conditions on both sides.

* *A.S.H.V.E.*

[a] 2 in. solid plaster partition, U = 0.53.

[b] Expanded slag, burned clay or pumice.

TYPE OF CEILING	INSULATION BETWEEN, OR ON TOP OF, JOISTS (No Flooring Above)												WITH FLOORING[a] (On Top of Ceiling Joists)		Number
	None	Insulating Board on Top of Joists		Blanket or Bat Insulation[f] Between Joists[g]			Vermiculite Insulation Between Joists[g]			Mineral Wool Insulation Between Joists[g]			Single Wood Floor[b]	Double Wood Floor[c]	
		½ In.	1 In.	1 In.	2 In.	3 In.	2 In.	3 In.	4 In.	2 In.	3 In.	4 In.			
	A	B	C	D	E	F	G	H	I	J	K	L	M	N	
No Ceiling		0.37	0.24										0.45	0.34	1
Metal Lath and Plaster[d]	0.69	0.26	0.19	0.19	0.12	0.093	0.18	0.14	0.11	0.12	0.093	0.077	0.30	0.25	2
Gypsum Board (⅜ in.) Plain or Decorated	0.67	0.26	0.18	0.19	0.12	0.092	0.18	0.13	0.10	0.12	0.092	0.077	0.30	0.24	3
Wood Lath and Plaster	0.62	0.25	0.18	0.19	0.12	0.091	0.17	0.13	0.10	0.12	0.091	0.076	0.28	0.24	4
Gypsum Lath (⅜ in.) Plastered[e]	0.61	0.25	0.18	0.19	0.12	0.091	0.17	0.13	0.10	0.12	0.091	0.076	0.28	0.24	5
Plywood (⅜ in.) Plain or Decorated	0.59	0.24	0.18	0.19	0.12	0.091	0.17	0.13	0.10	0.12	0.091	0.076	0.28	0.23	6
Insulating Board (½ in.) Plain or Decorated	0.36	0.19	0.15	0.16	0.10	0.082	0.14	0.12	0.097	0.10	0.082	0.069	0.22[h]	0.19[h]	7
Insulating Board Lath (½ in.) Plastered[e]	0.35	0.19	0.15	0.15	0.10	0.081	0.14	0.11	0.096	0.10	0.081	0.068	0.21	0.18	8
Insulating Board Lath (1 in.) Plastered[e]	0.23	0.15	0.12	0.12	0.089	0.072	0.12	0.097	0.084	0.089	0.072	0.061	0.16	0.14	9

Coefficients are expressed in Btu per (hour)(square foot)(Fahrenheit degree difference between the air on the two sides) and are based on still air (no wind) conditions on both sides.

* *A.S.H.V.E.*

[a] Coefficients corrected for framing on basis of 15 percent area, 2 in. × 4 in. (nominal) framing, 16 in. on centers.

[b] 25/32 in. yellow pine or fir.

[c] 25/32 in. pine or fir sub-flooring plus 13/16 in. hardwood flooring.

[d] Plaster assumed ¾ in. thick.

[e] Plaster assumed ½ in. thick.

[f] Based on insulation in contact with ceiling, and consequently no air space between.

[g] For coefficients for constructions in Columns M and N (except No. 1) with insulation between joists, refer to Table 8. Example: The coefficient for No. 3-N of Table 12 is 0.24. With 2 in. blanket insulation between joists, the coefficient will be 0.093. (See Table 8.) (Column D of Table 8 applicable only for 3⅝ in. joists.)

[h] For 25/32 in. insulating board sheathing applied to the under side of the joists, the coefficient for single wood floor (Column M) is 0.18 and for double wood floor (Column N) is 0.16. For coefficients with insulation between joists, see Table 8.

TABLE 14

COEFFICIENTS OF TRANSMISSION (U) OF CONCRETE CONSTRUCTION FLOORS AND CEILINGS *

TYPE OF CEILING	THICKNESS OF CONCRETE (INCHES)	TYPE OF FLOORING					NUMBER
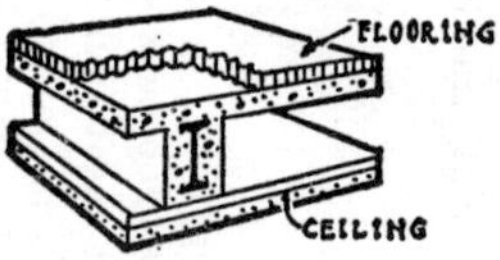		No Flooring (Concrete Bare)	Tile[a] or Terrazzo Flooring on Concrete	⅛ In. Asphalt Tile[b] Directly on Concrete	Parquett[c] Flooring in Mastic on Concrete	Double Wood Floor on Sleepers[d]	
		A	B	C	D	E	
No Ceiling	3	0.68	0.65	0.66	0.45	0.25	1
	6	0.59	0.56	0.58	0.41	0.23	2
	10	0.50	0.48	0.49	0.36	0.22	3
½ in. Plaster Applied to Underside of Concrete	3	0.62	0.59	0.60	0.43	0.24	4
	6	0.54	0.52	0.53	0.39	0.22	5
	10	0.46	0.44	0.45	0.34	0.21	6
Metal Lath and Plaster[e]—Suspended or Furred	3	0.38	0.37	0.37	0.30	0.19	7
	6	0.35	0.34	0.35	0.28	0.18	8
	10	0.32	0.31	0.32	0.26	0.17	9
Gypsum Board (⅜ in.) and Plaster[f]—Suspended or Furred	3	0.36	0.35	0.35	0.28	0.19	10
	6	0.33	0.32	0.33	0.27	0.18	11
	10	0.30	0.29	0.30	0.24	0.17	12
Insulating Board Lath (½ in.) and Plaster[f]—Suspended or Furred	3	0.25	0.24	0.25	0.21	0.15	13
	6	0.23	0.23	0.23	0.20	0.15	14
	10	0.22	0.21	0.22	0.19	0.14	15

Coefficients are expressed in Btu per (hour)(square foot)(Fahrenheit degree difference in temperature between the air on the two sides), and are based on still air (no wind) conditions on both sides.

* *A.S.H.V.E.*

[a] Thickness of tile assumed to be 1 in.

[b] Conductivity of asphalt tile assumed to be 3.1.

[c] Thickness of wood assumed to be 13/16 in.; thickness of mastic, ⅛ in. ($k = 4.5$). Column D may also be used for concrete covered with carpet.

[d] Based on 25/32 in. yellow pine or fir sub-flooring and 13/16 in. hardwood finish flooring with an air space between sub-floor and concrete.

[e] Thickness of plaster assumed to be ¾ in.

[f] Thickness of plaster assumed to be ½ in.

coefficient. Then turning to Table 19 and looking along the top line of the center section for a column headed by 0.27 we find 0.26 and then 0.28. Either column may be used with sufficient accuracy. For our purpose, let us look down the column headed

TABLE 15

COEFFICIENTS OF TRANSMISSION (U) OF CONCRETE BASEMENT FLOORS ON GROUND WITH VARIOUS TYPES OF FINISH FLOORING *

$U = 0.10^a$ Btu per (hr) (sq ft) (Fahrenheit degree temperature difference between the ground and the air over the floor).

---

* *A.S.H.V.E.*

[a] Since few data are available, a coefficient of 0.10 is frequently used for all types of basement concrete floors on the ground, with or without insulation. For basement wall below grade, use the same average coefficient (0.10). A lower ground temperature should, however, be used for walls than floors. For further data see A.S.H.V.E. RESEARCH REPORT No. 1213—Heat Loss Through Basement Walls and Floors, by F. C. Houghten, S. I. Taimuty, Carl Gutberlet and C. J. Brown (A.S.H.V.E. TRANSACTIONS, Vol. 48, 1942, p. 369).

by 0.28 to find the figure coming closest to 89.1, the net wall area. We find the figure 92.0 and following this line to the extreme left we arrive at the figure 8 which represents the square feet of radiation required for the heat lost by transmission through the wall.

Next considering the loss of heat through the window we look down the second column from the left in the same table for the figure closest to 12.9, the area of the window. This is 11.7, and following this line to the left we find that 4 square feet of radiation will be required to care for the heat lost by transmission through the window.

This double hung window has 17.8 lineal feet of cracks (the sum of the lengths of two vertical and three horizontal cracks). Referring to the small table on infiltration on page 481 we find, opposite "double hung wood sash" the figure 50 which represents cubic feet per hour per lineal foot of crack. Then referring to Table 19 under infiltration, finding the column headed by 50 and looking for the figure in this column closest to 17.8 we find 17.9. Following this to the left we find that 5 square feet of radiation are required for the infiltration.

TABLE 16

Coefficients of Transmission (U) of Flat Roofs Covered with Built-up Roofing. No Ceiling—Under Side of Roof Exposed *

(See Table 17 for Flat Roofs with Ceilings)

Type of Roof Deck	Thickness of Roof Deck (Inches)	No Insulation	Insulation on Top of Deck (Covered with Built-Up Roofing)							Number
			Insulating Board (Thickness Below)				Corkboard (Thickness Below)			
			½ In.	1 In.	1½ In.	2 In.	1 In.	1½ In.	2 In.	
		A	B	C	D	E	F	G	H	
Flat Metal Roof Deck[a]		0.94	0.39	0.24	0.18	0.14	0.23	0.17	0.13	
Precast Cement Tile	1⅝ in.	0.84	0.3	0.24	0.17	0.14	0.22	0.16	0.13	2
Concrete	2 in.	0.82	0.36	0.24	0.17	0.14	0.22	0.16	0.13	3
	4 in.	0.72	0.34	0.23	0.17	0.13	0.21	0.16	0.12	4
	6 in.	0.65	0.33	0.22	0.16	0.13	0.21	0.15	0.12	5
Gypsum Fiber Concrete[b] on ½ in. Gypsum Board	2½ in.	0.38	0.24	0.18	0.14	0.12	0.17	0.13	0.11	6
	3½ in.	0.31	0.21	0.16	0.13	0.11	0.15	0.12	0.10	7
Wood[c]	1 in.	0.49	0.28	0.20	0.15	0.12	0.19	0.14	0.12	8
	1½ in.	0.37	0.24	0.17	0.14	0.11	0.17	0.13	0.11	9
	2 in.	0.32	0.22	0.16	0.13	0.11	0.16	0.12	0.10	10
	3 in.	0.23	0.17	0.14	0.11	0.096	0.13	0.11	0.091	11

* Table footnotes on p. 512.

TABLE 17

Coefficients of Transmission (U) of Flat Roofs Covered with Built-up Roofing. With Lath and Plaster Ceilings [a] *

(See Table 16 for Flat Roofs with No Ceilings)

Type of Roof Deck	Thickness of Roof Deck (Inches)	No Insulation	Insulation on Top of Deck (Covered with Built-Up Roofing)							Number
			Insulating Board (Thickness Below)				Corkboard (Thickness Below)			
			½ In.	1 In.	1½ In.	2 In.	1 In.	1½ In.	2 In.	
		A	B	C	D	E	F	G	H	
Flat Metal Roof Deck		0.46	0.27	0.19	0.15	0.12	0.18	0.14	0.11	12
Precast Cement Tile	1⅝ in.	0.43	0.26	0.19	0.15	0.12	0.18	0.14	0.11	13
Concrete	2 in.	0.42	0.26	0.19	0.14	0.12	0.18	0.14	0.11	14
	4 in.	0.40	0.25	0.18	0.14	0.12	0.17	0.13	0.11	15
	6 in.	0.37	0.24	0.18	0.14	0.11	0.17	0.13	0.11	16
Gypsum Fiber Concrete[b] on ½ in. Gypsum Board	2½ in.	0.27	0.19	0.15	0.12	0.10	0.14	0.12	0.097	17
	3½ in.	0.23	0.17	0.14	0.11	0.097	0.13	0.11	0.091	18
Wood[c]	1 in.	0.31	0.21	0.16	0.13	0.11	0.15	0.12	0.10	19
	1½ in.	0.26	0.19	0.15	0.12	0.10	0.14	0.11	0.095	20
	2 in.	0.24	0.17	0.14	0.11	0.097	0.13	0.11	0.092	21
	3 in.	0.18	0.14	0.12	0.10	0.087	0.11	0.095	0.082	22

* Table footnotes on p. 512.

Table footnotes for Table 16, p. 510

Coefficients are expressed in Btu per (hour)(square foot)(Fahrenheit degree difference in temperature between the air on two sides), and are based on an outside wind velocity of 15 mph.

* *A.S.H.V.E.*

[a] Coefficient of transmission of bare corrugated iron (no roofing) is 1.50 Btu per (hr.)(sq. ft. of projected area)(F. deg. difference in temperature) based on an outside wind velocity of 15 mph.

[b] 87½ percent gypsum, 12½ percent wood fiber. Thickness indicated includes ½ in. gypsum board.

[c] Nominal thicknesses specified—actual thicknesses used in calculations.

Table footnotes for Table 17, p. 511

Coefficients are expressed in Btu per (hour)(square foot)(Fahrenheit degree difference in temperature between the air on two sides), and are based on an outside wind velocity of 15 mph.

* *A.S.H.V.E.*

[a] Calculations based on metal lath and plaster ceilings, but coefficients may be used with sufficient accuracy for gypsum lath or wood lath and plaster ceilings. It is assumed that there is an air space between the under side of the roof deck and the upper side of the ceiling.

[b] 87½ percent gypsum, 12½ percent wood fiber. Thickness indicated includes ½ in. gypsum board.

[c] Nominal thicknesses specified—actual thicknesses used in calculations.

We have now found three separate radiation figures 8, 4, and 5 which total 17 square feet. This must now be multiplied by a factor for exposure and temperature. This is found in Table 20 in the "N" column, since this is a north wall, and opposite Philadelphia. The factor is 0.94. Then $17 \times 0.94 = 15.98$ square feet.

The radiation for the west wall is estimated in a similar manner. It happens that in this particular case the temperature and exposure factor is the same as for the north wall.

The radiation required for each room in the house may be estimated in a similar manner.

Coefficients of Transmission (U) of Pitched Roofs *

Type of Ceiling (Applied Directly to Roof Rafters)	Wood Shingles (on 1 x 4 wood strips[c] spaced 2 in. apart)				Asphalt Shingles or Roll Roofing (on solid wood sheathing)[c]				Slate or Tile[b] (on solid wood sheathing)[c]				Number
	Insulation Between Rafters				Insulation Between Rafters				Insulation Between Rafters				
	None	Blanket or Bat (Thickness Below)			None	Blanket or Bat (Thickness Below)			None	Blanket or Bat (Thickness Below)			
		1 In.	2 In.	3 In.		1 In.	2 In.	3 In.		1 In.	2 In.	3 In.	
	A	B	C[a]	D[a]	E	F	G[a]	H[a]	I	J	K[a]	L[a]	
No Ceiling Applied to Rafters	0.48[f]	0.15	0.10	0.081	0.52[f]	0.15	0.11	0.084	0.55[f]	0.16	0.11	0.085	1
Metal Lath and Plaster[d]	0.31	0.14	0.10	0.081	0.33	0.15	0.10	0.083	0.34	0.15	0.10	0.083	2
Gypsum Board (⅜ in.) Decorated	0.30	0.14	0.10	0.080	0.32	0.15	0.10	0.082	0.33	0.15	0.10	0.083	3
Wood Lath and Plaster	0.29	0.14	0.10	0.080	0.31	0.14	0.10	0.081	0.32	0.15	0.10	0.082	4
Gypsum Lath (⅜ in.) Plastered[e]	0.29	0.14	0.10	0.079	0.31	0.14	0.10	0.081	0.32	0.15	0.10	0.082	5
Plywood (⅜ in.) Plain or Decorated	0.29	0.14	0.099	0.079	0.30	0.14	0.10	0.081	0.31	0.15	0.10	0.081	6
Insulating Board (½ in.) Plain or Decorated	0.22	0.12	0.090	0.072	0.23	0.12	0.091	0.074	0.24	0.13	0.092	0.074	7
Insulating Board Lath (½ in.) Plastered[e]	0.22	0.12	0.088	0.072	0.22	0.12	0.090	0.073	0.23	0.12	0.091	0.074	8
Insulating Board Lath (1 in.) Plastered[e]	0.16	0.10	0.078	0.064	0.17	0.10	0.079	0.065	0.17	0.10	0.080	0.066	9

Coefficients are expressed in Btu per (hour)(square foot)(Fahrenheit degree difference in temperature between the air on the two sides), and are based on an outside wind velocity of 15 mph.

* *A.S.H.V.E.*

[a] Coefficients corrected for framing on basis of 15 percent area, 2 in. × 4 in. (nominal), 16 in. on centers.

[b] Figures in Columns I, J, K and L may be used with sufficient accuracy for rigid asbestos shingles on wood sheathing. Layer of slater's felt neglected.

[c] Sheathing and wood strips assumed 25/32 in. thick.

[d] Plaster assumed ¾ in. thick.

[e] Plaster assumed ½ in. thick.

[f] No air space included in 1-A, 1-E or 1-I; all other coefficients based on one air space.

Special Cases.—If the east wall of the room in Fig. 6 ha been a solid partition of wood lath and plaster on each side o the studding and the room on the other side *unheated,* additiona radiation would be necessary for the loss of heat through thi wall. This radiation is estimated by the use of Table 11 which for the particular conditions of this problem, gives in column "B the coefficient 0.34. This coefficient is now referred to the *right hand* portion of Table 19. We find there a column headed by 0.35, which is sufficiently close. Looking down this column til we reach a number equal to the area of the wall ($14 \times 8.5 =$ 119 sq. ft.) we note that this lies midway between 110 and 12 and that the radiation (last column) is, therefore, 6½ square fee for the loss through this wall.

Then there is the case when the space below or the space abov a room is unheated. Let us take the case of the room shown in Fig. 6 if this room has an unheated attic above it and a ceilin of plaster on insulating board lath with no insulation betweer the joists and no flooring in the attic. This case is covered in

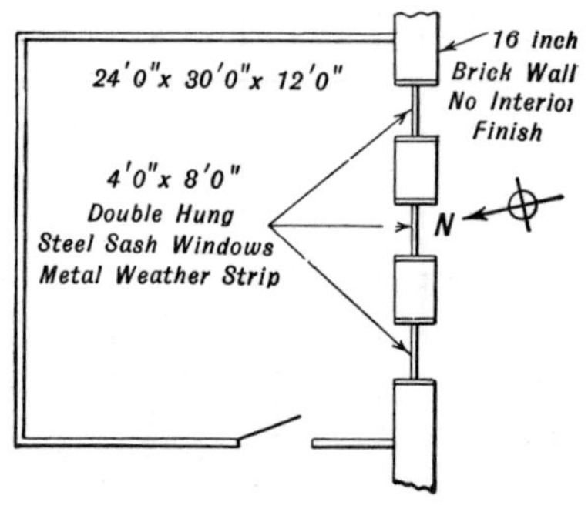

Fig. 7.

Table 13 and we find in line 8, column "A" that the coefficien is 0.35. Then again consulting the *right-hand* portion of Table 1 we find a column headed by this figure. Looking down it for a figure representing the area of the ceiling ($12 \times 14 = 168$) we

TABLE 19.—HEATING AND PIPING CONTRACTORS NATIONAL

Showing Radiation Required

Col. 38″ team Rad.	Glass		Infiltration																					
	Window or Door	Skylight	Rate per Lineal Foot																					
			25	50	100	200																		
	1.1	1.3	0.45	0.9	1.8	3.6	0.08	0.10	0.12	0.14	0.16	0.18	0.20	0.22	0.24	0.26	0.28	0.30	0.35	0.40	0.45	0.50	0.55
225	77.0	91.0	31.5	63.0	126	252	5.60	7.00	8.40	9.80	11.2	12.6	14.0	15.4	16.8	18.2	19.6	21.0	24.5	28.0	31.5	35.0	38.5
1	2.92	2.47	7.14	3.57	1.79	0.89	40.2	32.1	26.8	23.0	20.1	17.9	16.1	14.6	13.4	12.4	11.5	10.7	9.18	8.04	7.14	6.43	5.84
2	5.84	4.94	14.3	7.14	3.58	1.78	80.4	64.2	53.6	46.0	40.2	35.8	32.2	29.2	26.8	24.8	23.0	21.4	18.4	16.1	14.3	12.9	11.7
3	8.76	7.41	21.4	10.7	5.37	2.67	121	96.3	80.4	69.0	60.3	53.7	48.3	43.8	40.2	37.2	34.5	32.1	27.5	24.1	21.4	19.3	17.5
4	11.7	9.88	28.6	14.2	7.16	3.56	161	128	107	92.0	80.4	71.6	64.4	58.4	53.6	49.6	46.0	42.8	36.7	32.2	28.6	25.7	23.3
5	14.6	12.4	35.7	17.9	8.95	4.45	201	160	134	115	101	89.5	80.5	73.0	67.0	62.0	57.5	53.5	45.9	40.2	35.7	32.1	29.2
6	17.5	14.8	42.8	21.4	10.7	5.34	241	193	161	138	121	107	96.6	87.6	80.4	74.4	69.0	64.2	55.1	48.2	42.8	38.6	35.0
7	20.4	17.3	49.9	25.0	12.5	6.23	281	225	188	161	141	125	113	102	93.8	86.8	80.5	74.9	64.3	56.3	50.0	45.0	40.9
8	23.4	19.8	57.1	28.6	14.3	7.12	322	257	214	184	161	143	129	117	107	99.2	92.0	85.6	73.4	64.3	57.1	51.4	46.7
9	26.3	22.2	64.3	32.1	16.1	8.01	362	289	241	207	181	161	145	131	121	112	103	96.3	82.6	72.4	64.3	57.9	52.6
10	29.2	24.7	71.4	35.7	17.9	8.90	402	321	268	230	201	179	161	146	134	124	115	107	91.8	80.4	71.4	64.3	58.4
11	32.1	27.2	78.5	39.3	19.7	9.79	442	353	295	253	221	197	177	161	147	136	126	118	101	88.4	78.5	70.7	64.2
12	35.0	29.6	85.7	42.8	21.5	10.7	482	385	322	276	241	215	193	175	161	149	138	128	110	96.5	85.7	77.2	70.1
13	38.0	32.1	92.8	46.4	23.2	11.6	523	417	348	299	261	233	209	190	174	161	149	139	119	105	92.8	83.6	75.9
14	40.9	34.6	100	50.0	25.1	12.5	563	449	375	322	281	251	225	204	188	174	161	150	129	113	100	90.0	81.8
15	43.8	37.1	107	53.6	26.9	13.4	603	481	402	345	301	268	241	219	201	186	172	160	138	121	107	96.4	87.6
16	46.7	39.5	114	57.1	28.6	14.2	643	514	429	368	322	286	258	234	214	198	184	171	147	129	114	103	93.4
17	49.6	42.0	121	60.7	30.4	15.1	683	546	456	391	342	304	274	248	228	211	195	182	156	137	121	109	99.3
18	52.6	44.5	129	64.3	32.2	16.0	724	578	482	414	362	322	290	263	241	223	207	193	165	145	129	116	105
19	55.5	46.9	136	67.9	34.0	16.9	764	610	509	437	382	340	306	277	255	236	218	203	174	153	136	122	111
20	58.4	49.4	143	71.4	35.8	17.8	804	642	536	460	402	358	322	292	268	248	230	214	184	161	143	129	117
21	61.3	51.9	150	75.0	37.6	18.7	844	674	563	483	422	376	338	307	281	260	241	225	193	169	150	135	123
22	64.2	54.3	157	78.5	39.4	19.6	884	706	590	506	442	394	354	321	295	273	253	235	202	177	157	141	128
23	67.2	56.8	164	82.1	41.2	20.5	925	738	616	529	462	412	370	336	308	285	264	246	211	185	164	148	134
24	70.8	59.3	171	85.7	43.0	21.4	965	770	643	552	482	430	386	350	322	298	276	257	220	193	171	154	140
25	73.0	61.8	179	89.3	44.8	22.3	1005	802	670	575	502	447	402	365	335	310	287	267	229	201	178	161	146

Association Standard Radiation Estimating Table

for Quantities Indicated

0.60	0.65	0.70	0.75	0.80	0.85	0.90	1.00	0.10	0.15	0.20	0.25	0.30	0.35	0.40	0.50	0.60	0.70	0.80	0.90	U
42.0	45.5	49.0	52.5	56.0	59.5	63.0	70.0	3.50	5.25	7.00	8.75	10.5	12.3	14.0	17.5	21.0	24.5	28.0	31.5	$U\ (T_i$
5.36	4.95	4.59	4.29	4.02	3.78	3.57	3.21	64.3	42.9	32.1	25.7	21.4	18.3	16.1	12.9	10.7	9.18	8.04	7.14	[illegible]
10.7	9.90	9.18	8.58	8.04	7.56	7.14	6.42	129	85.8	64.2	51.4	42.8	36.6	32.2	25.8	21.4	18.4	16.1	14.3	[illegible]
16.1	14.8	13.8	12.9	12.1	11.3	10.7	9.63	193	129	96.3	77.1	64.2	54.9	48.3	38.7	32.1	27.5	24.1	21.4	[illegible]
21.4	19.8	18.4	17.2	16.1	15.1	14.3	12.8	257	172	128	103	85.6	73.2	64.4	51.6	42.8	36.7	32.2	28.6	[illegible]
26.8	24.7	22.9	21.4	20.1	18.9	17.8	16.0	321	214	160	128	107	91.5	80.5	64.5	53.5	45.9	40.2	35.7	[illegible]
32.2	29.7	27.5	25.7	24.1	22.7	21.4	19.3	386	257	193	154	128	110	96.6	77.4	64.2	55.1	48.2	42.8	[illegible]
37.5	34.6	32.1	30.0	28.1	26.5	25.0	22.5	450	300	225	180	150	128	113	90.3	74.9	64.3	56.3	50.0	[illegible]
42.9	39.6	36.7	34.3	32.2	30.2	28.6	25.7	514	343	257	205	171	146	129	103	85.6	73.4	64.3	57.1	[illegible]
48.2	44.5	41.3	38.6	36.2	34.0	32.1	28.9	579	386	289	231	193	165	145	116	96.3	82.6	72.4	64.3	[illegible]
53.6	49.5	45.9	42.9	40.2	37.8	35.7	32.1	643	429	321	257	214	183	161	129	107	91.8	80.4	71.4	[illegible]
59.0	54.4	50.5	47.2	44.2	41.6	39.3	35.3	707	472	353	282	235	201	177	142	118	101	88.4	78.5	[illegible]
64.3	59.4	55.1	51.5	48.2	45.4	42.8	38.5	772	515	385	308	257	220	193	155	128	110	96.5	85.7	[illegible]
69.7	64.3	59.7	55.8	52.3	49.1	46.4	41.7	836	558	417	334	278	238	209	168	139	119	105	92.8	[illegible]
75.0	69.3	64.3	60.1	56.3	52.9	50.0	44.9	900	601	449	359	300	256	225	181	150	129	113	100	[illegible]
80.4	74.2	68.8	64.3	60.3	56.7	53.6	48.1	964	643	481	385	321	274	241	193	160	138	121	107	[illegible]
85.8	79.2	73.4	68.6	64.3	60.5	57.1	51.4	1029	686	514	411	342	293	258	206	171	147	129	114	[illegible]
91.1	84.1	78.0	72.9	68.3	64.3	60.7	54.6	1093	729	546	436	364	311	274	219	182	156	137	121	[illegible]
96.5	89.1	82.6	77.2	72.4	68.0	64.3	57.8	1157	772	578	462	385	329	290	232	193	165	145	129	[illegible]
102	94.0	87.2	81.5	76.4	71.8	67.8	61.0	1222	815	610	488	407	348	306	245	203	174	153	136	[illegible]
107	99.0	91.8	85.8	80.4	75.6	71.4	64.2	1286	858	642	514	428	366	322	258	214	184	161	143	[illegible]
113	104	96.4	90.1	84.4	79.4	75.0	67.4	1350	901	674	539	449	384	338	271	225	193	169	150	[illegible]
118	109	101	94.4	88.4	83.2	78.5	70.6	1415	944	706	565	471	403	354	283	235	202	177	157	[illegible]
123	114	106	98.7	92.5	86.9	82.1	73.8	1479	987	738	591	492	421	370	296	246	211	185	164	[illegible]
129	119	110	103	96.5	90.7	85.7	77.0	1543	1030	770	616	514	439	386	309	257	220	193	171	[illegible]
134	124	115	107	100	94.5	89.2	80.2	1607	1072	802	642	535	457	402	322	267	229	201	178	[illegible]

← These Items Figured on Basis of $U\left(\frac{T_i - T_o}{2}\right)$ →

TABLE 20

Combined Temperature and Exposure Factors

City	Base Temp.	Temp. Factor	POINTS OF COMPASS							
			N	NE	E	SE	S	SW	W	NW
lbany, N. Y.	+ 5°	0.93	1.02	1.02	0.97	0.93	0.93	0.93	1.02	1.02
altimore, Md.	+30°	0.57	0.80	0.80	0.74	0.57	0.74	0.74	0.80	0.80
irmingham, Ala.	+30°	0.57	0.66	0.66	0.57	0.57	0.57	0.60	0.66	0.66
oston, Mass.	+15°	0.79	1.02	0.86	0.79	0.79	0.79	1.02	1.02	1.02
uffalo, N. Y.	0°	1.00	1.00	1.00	1.00	1.00	1.25	1.40	1.40	1.40
hicago, Ill.	+10°	0.86	1.07	0.86	0.86	0.86	0.99	1.16	1.16	1.16
incinnati, Ohio	+15°	0.79	0.86	0.79	0.79	0.79	1.06	1.06	1.06	0.94
leveland, Ohio	+ 5°	0.93	1.07	1.00	1.00	0.93	1.00	1.07	1.07	1.07
olumbus, Ohio	+15°	0.79	0.94	0.90	0.79	0.94	1.07	1.07	1.07	0.94
enver, Colo.*	+20°	0.80	1.04	1.04	0.96	1.00	1.00	1.00	0.80	1.04
etroit, Mich.	0°	1.00	1.10	1.00	1.00	1.00	1.10	1.10	1.10	1.10
astport, Me.	+10°	0.86	1.24	1.03	1.03	0.86	0.86	1.24	1.24	1.24
rand Rapids, Mich.	+15°	0.79	0.87	0.79	0.79	0.79	0.84	0.87	0.87	0.87
reen Bay, Wis.	− 5°	1.07	1.07	1.07	1.07	1.07	1.12	1.18	1.18	1.18
reensboro, N. C.	+35°	0.50	0.60	0.60	0.60	0.50	0.50	0.60	0.60	0.60
ouston, Texas	+40°	0.43	0.81	0.56	0.51	0.43	0.43	0.43	0.81	0.81
ndianapolis, Ind.	+15°	0.79	1.03	0.84	0.84	0.79	0.90	0.99	1.03	1.03
thaca, N. Y.	+15°	0.79	0.87	0.79	0.84	0.84	0.84	0.84	0.87	0.87
ansas City, Mo.	+15°	0.79	1.14	1.06	0.79	0.79	0.86	0.86	1.14	1.14
os Angeles, Cal.	+50°	0.29	0.43	0.43	0.43	0.29	0.29	0.29	0.43	0.43
ouisville, Ky.	+20°	0.71	0.93	0.93	0.71	0.75	0.75	1.04	1.04	1.04
adison, Wis.	+ 5°	0.93	1.16	1.07	1.02	0.93	1.02	1.16	1.16	1.16
emphis, Tenn.	+30°	0.57	0.80	0.68	0.63	0.57	0.74	0.74	0.80	0.80
ilwaukee, Wis.	+10°	0.86	1.07	0.86	0.86	0.86	0.99	1.16	1.16	1.16
ew Orleans, La.	+45°	0.36	0.54	0.50	0.45	0.36	0.36	0.36	0.54	0.54
ew York, N. Y.	+10°	0.86	1.29	1.07	0.86	0.86	0.86	1.14	1.29	1.29
orfolk, Va.	+30°	0.57	0.86	0.74	0.68	0.57	0.57	0.68	0.86	0.86
hiladelphia, Pa.	+15°	0.79	0.94	0.86	0.86	0.79	0.79	0.79	0.94	0.94
ittsburgh, Pa.	+15°	0.79	1.02	0.79	0.79	0.79	1.02	1.06	1.06	1.06
ortland, Ore.	+25°	0.64	0.64	0.64	0.64	0.64	0.64	0.64	0.64	0.64
rovidence, R. I.	+15°	0.79	1.18	0.98	0.79	0.79	0.86	0.98	1.18	1.18
ichmond, Va.	+30°	0.57	0.77	0.71	0.71	0.57	0.74	0.74	0.77	0.77
ochester, N. Y.	+10°	0.86	0.90	0.86	0.86	0.86	1.07	1.11	1.11	1.11
t. Louis, Mo.	+20°	0.71	0.93	0.86	0.71	0.86	0.86	0.86	0.93	0.93
t. Paul, Minn.	− 5°	1.07	1.28	1.07	1.07	1.07	1.07	1.18	1.28	1.28
acramento, Cal.	+45°	0.35	0.45	0.42	0.42	0.42	0.42	0.35	0.45	0.45
alt Lake City, Utah	+25°	0.64	0.71	0.64	0.71	0.71	0.71	0.64	0.71	0.71
an Antonio, Texas	+45°	0.36	0.61	0.61	0.50	0.36	0.36	0.36	0.61	0.61
an Diego, Cal.	+55°	0.20	0.20	0.23	0.23	0.23	0.23	0.23	0.27	0.27
an Francisco, Cal.	+45°	0.36	0.43	0.43	0.43	0.36	0.36	0.36	0.36	0.41
eattle, Wash.	+25°	0.64	0.64	0.64	0.64	0.80	0.80	0.80	0.64	0.64
yracuse, N. Y.	0°	1.00	1.10	1.00	1.00	1.00	1.05	1.10	1.10	1.10
Vashington, D. C.	+20°	0.71	0.86	0.71	0.71	0.71	0.71	0.71	0.86	0.86
Vichita, Kans.	+10°	0.86	1.03	1.03	0.94	0.86	0.86	0.86	0.86	1.03

* Denver base temperature and exposure factors based on actual Weather Bureau ecords, but due to rapid changes and high altitude both temperature factors and combined emperature and exposure factors have been corrected to care for these conditions.

find 165 and to the right in this line is the corresponding radiatio of 9 sq. ft. This must be added to the total radiation alread computed.

The tables which have been used are based on an insid temperature of 70 degrees. If a different inside temperature i desired, the radiation is computed by the tables in the regula manner and then multiplied by the proper factor from Table 21 Let us illustrate by an example. Figure 7 represents a room i Grand Rapids, Michigan, used as a gymnasium in which a tem perature of 60 degrees is desired. The problem is to find th radiation required.

Solution: (proceed as in the previous problem).

	Area or lin. ft.	Sq. ft. radiations
South wall 30 × 12 = 360 sq. ft. − 96 sq. ft. (3 windows at 32 sq. ft.)	264 sq. ft.	23 sq. ft.
3 windows 4 ft. × 8 ft.	96 sq. ft.	33 sq. ft.
Infiltration	84 lin. ft.	24 sq. ft.
Total without either exposure or temperature factors		80 sq. ft.
Exposure and temperature factor		0.84
Total		67.2 sq. ft.

INFILTRATION TABLE

Stationary Wood Sash	25	Rolled Section Steel Windows	100*
Double Hung Wood Sash	50	French Doors	100
Double Hung Steel Sash	100	Outside Doors, Residences	100
Casement Windows, Wood	100	Same with Storm Doors	50
Casement Windows, Steel	50	Same with Inner Vestibule Doors	50
		Outside Doors, Store, etc.	200

Metal Weather Strip Deducts 50 per cent

* Per foot of crack of Ventilating Sash.

Up to this point the procedure has been the same as in th previous estimate, that is, the radiation has been estimated for room to be kept at a temperature of 70 degrees. This radiatio is now multiplied by a factor found in Table 21. We look firs

TABLE 21

Conversion Factors

Base Temperature	Room Temperature								
	80	75	70	65	60	55	50	45	40
− 5°	1.219	1.104	1	0.093	0.811	0.725	0.646	0.572	0.498
0°	1.228	1.111	1	0.896	0.801	0.712	0.623	0.549	0.472
+ 5°	1.239	1.119	1	0.892	0.791	0.698	0.608	0.525	0.447
+10°	1.253	1.123	1	0.886	0.780	0.680	0.586	0.498	0.415
+15°	1.269	1.13	1	0.878	0.765	0.659	0.569	0.465	0.375
+20°	1.289	1.14	1	0.870	0.748	0.634	0.528	0.427	0.332
+25°	1.312	1.151	1	0.859	0.728	0.604	0.489	0.380	0.277
+30°	1.343	1.166	1	0.845	0.702	0.566	0.44	0.312	0.207
+35°	1.380	1.183	1	0.829	0.669	0.519	.....	.....	.....
+40°	1.433	1.21	1	0.806	0.627	0.453	.....	.....	.....
+45°	1.504	1.243	1	0.773	0.561	0.363	.....	.....	.....

FORMULA

$$\text{Factor} = \frac{Tr - Tb}{70 - Tb} \times \frac{Ts - 70}{Ts - Tr}$$

Tr = Room Temperature
Tb = Base Temperature
Ts = 215 deg.

To calculate amount of radiation required for other room temperatures than 70 deg., compute the amount for 70 deg. and multiply by the factor shown corresponding to room temperature desired and *proper base temperature*.

for the proper column (60 degrees) and then follow down till we reach the line of the proper base temperature (+15°, see Table 20). The factor is 0.765.

Then, $67.2 \times 0.765 = 51.41$ sq. ft. of radiation required. (Ans.)

Similarly, if the amount of radiation is wanted when the room is to be heated by wall coils, by indirect steam radiation, by vapor radiation, by hot water radiation, etc., the factors for conversion given in Table 22 are used.

TABLE 22

RADIATOR TRANSMISSION FACTORS

For Room Temperature of	70 Deg. F.
And Steam Pressure of	1 Lb. Gage
Direct Steam Radiation (Standard 3 Col. 38 in. High)	225 Btu per square foot.

Multiply by the following factors for the equivalent of 3 Col. 38 in. radiation of the following types.

Wall Coil	0.75
Double Wall Coil	0.90
Ceiling Coil	1.00
Wall Radiator	0.82
Double Wall Radiators	1.00
Wall Radiator (Ceiling)	1.00

	Increase Surface
Indirect Steam Radiation	50 percent
Direct Indirect Steam Radiation	25 percent

Vapor Radiation: Open return line vapor systems, on which thermostatic traps are not used, require 10 percent to 20 percent additional surface in each radiator to act as a condenser and prevent the flow of steam into the return main.

Hot Water Radiation: In figuring hot water radiators, assume mean temperature of the water in the radiators to be 170 deg. Under this condition the amount of hot water radiating surface may be determined by adding 50 percent to the amount of steam radiating surface figured.

Approximate Method of Estimating Radiation.—A method of estimating radiation known as the "2–20–200 Method," formerly widely used, is not accurate but is presented here because it may be used for quick rough estimates. It calls for one square foot of radiating surface for each 2 square feet of glass surface, one for each 20 square feet of net outside walls and one for each 200 cubic feet of room contents. We may express it in this fashion.

$$\text{Steam radiation, sq. ft.} = \frac{G}{2} + \frac{W}{20} + \frac{C}{200}$$

where G = glass area in sq. ft.

W = net exposed wall area in sq. ft.

C = cubical contents of room in cu. ft.

Applying this to the problem of Fig. 6 solved on page 483 we have

$$\text{Steam radiation} = \frac{26}{2} + \frac{195}{20} + \frac{1428}{200} = 30 \text{ sq. ft.} \quad \text{(Ans.)}$$

In designing a heating system the equivalent steam radiation for each room of a house must, of course, be estimated. These figures are used not only to determine the sizes of the radiators required but also the sizes of pipes and boilers.

Selecting Size of Boiler.—The boiler of a heating system must provide capacity for:

1. The radiators which heat the rooms.
2. The heat lost in the pipes.
3. The heat consumed in water heaters and other appliances.
4. Reserve capacity needed for starting up a cold system, for intermittent firing, and careless operation.

Since the capacities of commercial heating boilers are rated in terms of square feet of equivalent direct radiation it is convenient to reduce all of the factors to these terms. The radiation for a house is, of course, the sum of the radiation required by each room.

The loss of heat from the pipes varies with the installation and the degree and kind of pipe covering, if any. However, a flat allowance of 25 percent for steam systems and 35 percent for hot water systems, of the total radiation for the house is considered good practice for general installations.

The allowance of equivalent radiation for water-heating appliances is made on the following basis: * for water boilers with coil

* From *The Ideal Fitter,* American Radiator Co.

in firebox, $2\frac{1}{2}$ sq. ft. equivalent direct water radiation per gallon of storage tank capacity; for externally attached water heaters below water-line of steam boilers, $1\frac{1}{2}$ sq. ft. of equivalent direct steam radiation per gallon of tank capacity; and for externally attached water heaters below the water line of steam boilers *without* storage tank, 4 square feet for each gallon of water heated per hour.

The reserve capacity needed for small boilers is from 50 to 65 percent of the total capacity needed for other purposes.

ILLUSTRATION: A building has an estimated direct radiation requirement for steam heating of 440 square feet and an externally attached water heater connected to a 120-gallon storage tank.

What rated boiler capacity will it require if a reserve capacity of 60 percent is deemed ample?

440 sq. ft. steam radiation............	=	440 sq. ft. edr.*
Piping loss (440 × 0.25).............	=	110 sq. ft. edr.
Water heater (120 × $1\frac{1}{2}$)............	=	180 sq. ft. edr.
Total..........................	=	730 sq. ft. edr.
Capacity for warming up (730 × 0.60).	=	438 sq. ft. edr.
Required capacity of boiler.......	=	1168 sq. ft. edr. (Ans.)

Reputable boiler manufacturers have accurate performance records for all of their boilers and are prepared to guarantee the rated capacities.

Warm Air Heating Systems.—The design of warm air heating systems has been more or less standardized and published in the form of a Code by the National Warm Air Heating Association. It is possible here to illustrate only the essential steps. These involve the determination of the following items: †

* Equivalent direct radiation.

† From the *Guide,* American Society of Heating and Ventilating Engineers.

1. The heat loss in Btu per hour from each room in the building.

2. Area and diameter in inches of warm-air pipes in basement (known as leaders).

3. Area and dimensions in inches of vertical pipes (known as wall stacks).

4. Free and gross area and dimensions in inches of warm-air registers.

5. Area and dimensions of recirculating or outside air ducts in inches.

6. Free and gross area and dimensions in inches of recirculating registers.

7. Size of furnace necessary to supply the warm air required to overcome the heat loss from the building. This *size* should include square inches of leader pipe area which furnace must supply. It is also desirable to call for a minimum bottom fire-pot diameter in inches, which is the nominal grate diameter.

8. Area and dimensions in inches of chimney and smoke pipe.

Heat Loss in Btu per Hour.—The heat loss in Btu per hour can be arrived at conveniently and with sufficient accuracy by using the tables for estimating steam radiation and multiplying by 225, the equivalent Btu emission per hour on which the tables are based. Thus, in the problem of estimating the heat requirements for the room in Fig. 6, we found that the direct steam radiation required was 32.9 square feet. Multiplying this by 225 we find the heat loss to be about 7400 Btu per hour.

Size of Leader Pipes.—When a warm air system is designed to give an air temperature of 175 degrees Fahrenheit at the registers, and H represents the heat in Btu per hour to be supplied to a room, then the approximate area of the leaders should be:

For the first floor, $0.009H$ sq. in.
For the second floor, $0.006H$ sq. in.
For the third floor, $0.005H$ sq. in.

ILLUSTRATION: A first-floor room requires 7400 Btu per hour. What size of leader pipe will it require?

$$0.009 \times 7400 = 66.6 \text{ sq. in.} = \text{area of leader}$$

$$2\sqrt{\frac{66.6}{\pi}} = 9\tfrac{1}{4} \text{ inches (approx.)} = \text{diameter of leader}$$

A 9-inch leader would be used in this case. They are installed only to the nearest inch and no leaders smaller than 8 inches in diameter should be used.

Stacks and Registers.—The sizes of wall stacks and registers do not lend themselves to mathematical determination. However, accepted practice is to make the area of stacks greater than 70 percent of the area of the leaders to which they are connected. Registers should have a net area not less than the area of the leader which connects with it.

FIG. 8.—Different Types of Registers and Intakes Used in Warm Air Systems. (From *University of Illinois Small Homes Council Circular* G3.1.)

Gravity Warm Air.—In a gravity warm air heating system the warm air rises upward through the house from a centrally located furnace. The cooler air in the rooms flows downward through the return air intakes back to the furnace. This type of system is economical to install and lends itself to low-cost homes. It is simple to operate, has no electric motors or controls, and responds rapidly to changes in outdoor temperatures. A house with a compact floor plan will heat well with gravity warm air since the leader pipes and return air ducts will be very short.

This type of system is not suitable for heating basementless homes or for heating basement rooms since the furnace must be below the level of the rooms to be heated in order to function

Univ. of Illinois Circular

Fig. 9.—Gravity Warm-air Heating System.

properly. By adding a blower it can be converted to a forced-air system. When making such a conversion, filters are usually added and some minor changes made in the duct system.

Forced Warm-Air System.—Similar in operation to the gravity warm-air system, the difference lies in that the air is circulated by a blower. The cooler air in the rooms is sucked down through the return air intakes and ducts to the furnace where it is heated and then recirculated to the house under slight pressure. The system responds quickly to outside temperature changes and can be used to heat basementless houses, large structures, and basement rooms since air circulation is maintained by the blower.

This type of installation costs more than the gravity system, but requires less space for the ducts and furnace. Also the furnace

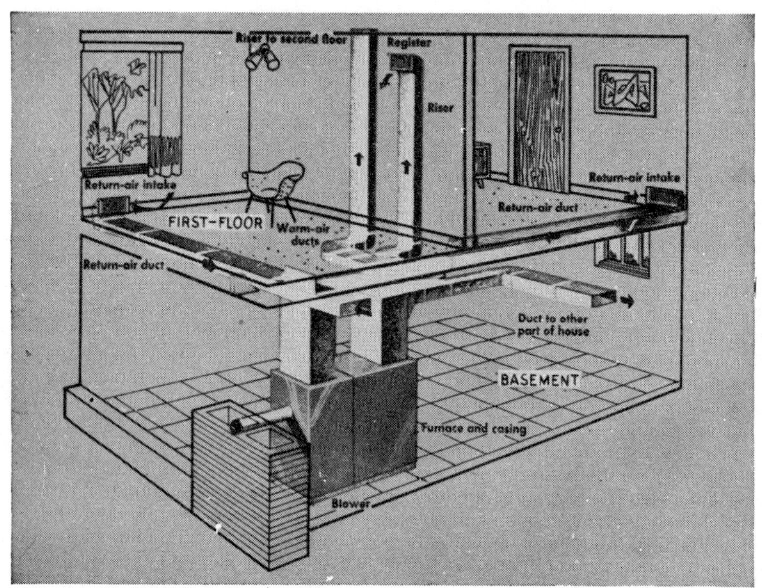

Univ. of Illinois Circular

FIG. 10.—A Forced Warm-air Heating System.

does not have to be centrally located to be assured of proper heat distribution. A humidifier and filters can be easily used, whereas in a gravity system filters would restrict the flow of air.

Perimeter Heating System.—Perimeter heating lends itself best to basementless houses built on a concrete slab. The duct system, embedded in the concrete, encircles the slab. Warm air in the ducts enters the rooms through floor registers while the cool air returns to the furnace through intakes at high locations on the inside walls. As with the forced warm-air system a blower is required to maintain proper circulation of the air.

A perimeter heating system is economical to install and needs very little floor space if a furnace specifically designed for it is used. It eliminates cold floors, can be used with a humidifier and filters, and lends itself to standard thermostatic controls. A well-constructed concrete slab laid on proper fill is essential

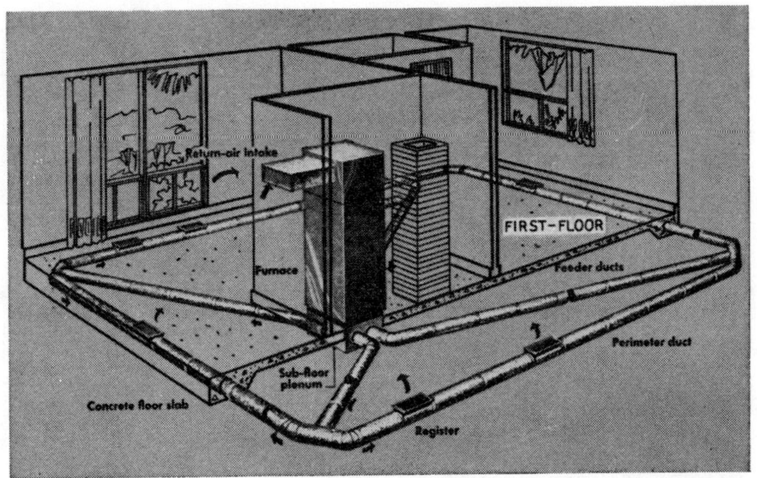

Univ. of Illinois Circular

Fig. 11.—Warm-air Perimeter Installation.

for proper operation. Several duct arrangements are possible. The ducts can be one of the standard precast forms such as concrete pipe or vitrified tile, or sheet metal.

Warm-Air Panel System.—A hollow sheet metal panel is placed in the ceiling, floor, or wall, through which warm air is circulated from the furnace by a blower and duct system. Heat is transmitted from the panel to the room by radiation and convection currents. No registers are required, but they are sometimes used.

Since a panel is part of the house, it can only be installed in a new building. The system can be used with or without a basement, but has definite temperature limitations. Floors cannot exceed 85° F., while walls and ceilings should not go higher than approximately 120° F. Proper insulation in back of the panel is essential for economical operation of this type of system. While well suited for standard thermostatic control, panels are subject to overheating or underheating where rapid temperature changes may occur.

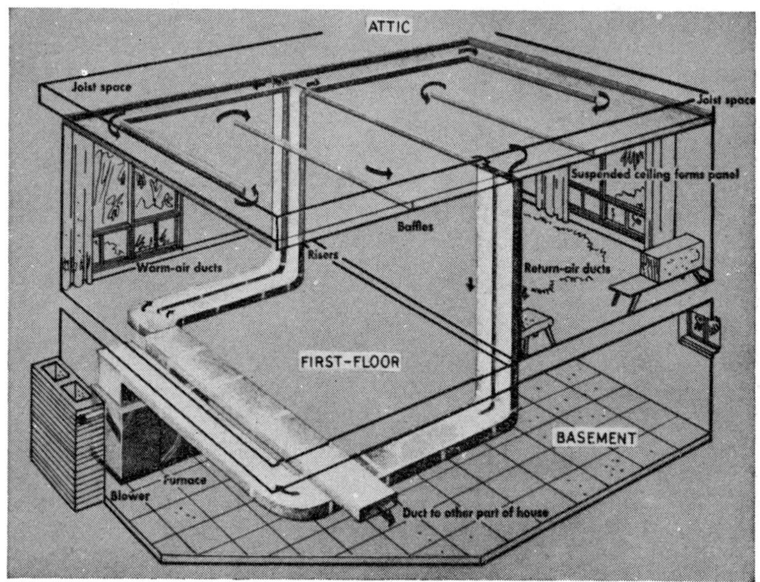

Univ. of Illinois Circular

FIG. 12.—Ceiling Type Warm-air Panel Installation.

One-Pipe Steam System.—Steam generated in the boiler rises to the room heating units where it condenses, forms water, and returns to the boiler. When it condenses it gives up its "heat of vaporization" to the radiator, which in turn heats the room. In a one-pipe system the same pipe which carries the steam to the room also returns the water (condensed steam) to the boiler. The pipes must therefore be larger than those used in other types of systems.

This type of system is simple and economical to install. No auxiliary motors, pumps, or blowers are required other than those needed for operation of an automatic fuel burner (if one is used). Domestic hot water can be supplied year-round if an automatic burner is used. Radiator temperatures cannot be controlled and they must be completely on, or off, for proper operation. Steam

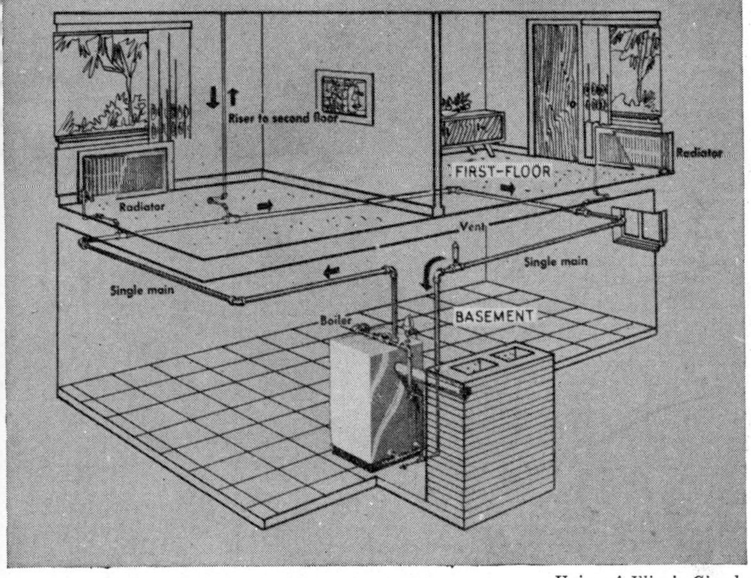

Univ. of Illinois Circular

FIG. 13.—A One-pipe Steam Heating System.

heat is not recommended for heating basementless homes, or basement rooms, because the boiler must be below the level of the radiators.

Gravity Hot-Water System.—The entire system is full of water. When the water in the boiler is heated, it rises to the room heating units. The cooler water in the rooms flows downward back to the boiler. Circulation is maintained by *gravity* or natural convection.

In an *open* system the expansion tank is placed above the highest radiator and the water is exposed to the air, or "open." In a *closed* system, such as that shown in Fig. 14, the expansion tank is located near the boiler. The heated water compresses the air in the tank, raising the pressure which in turn raises the boiling point of the water. This permits a higher temperature system and smaller heating units.

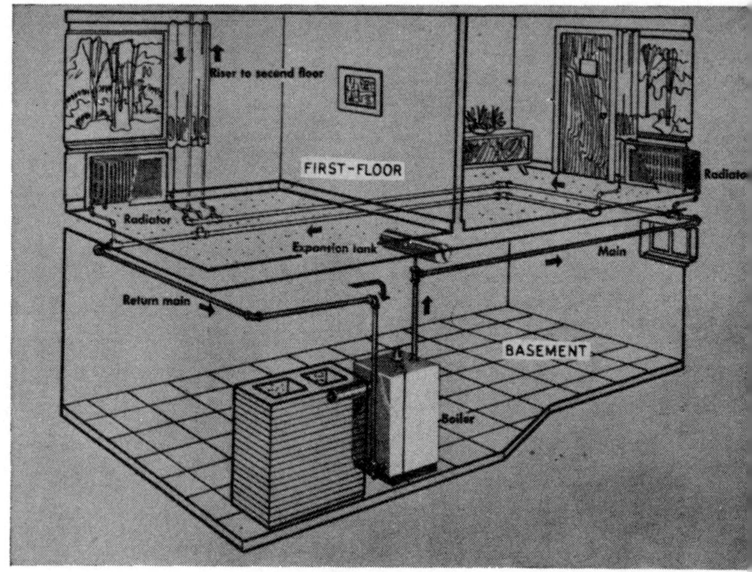

Univ. of Illinois Circula

Fig. 14.—A Gravity Hot-water Heating System (Closed Type).

It is economical to install, requiring no auxiliary motors other than those used for an automatic burner. Large supply and return mains are required for good circulation. It has a slow response to temperature changes and is not recommended for basementless houses or heating of basement rooms.

Forced Hot-Water Heating System.—As with the gravity system all pipes and heating units are full of water. The hot water from the boiler is forced through the system by a circulating pump. The one-pipe system shown in Fig. 15 has one main which supplies hot water to the heating units, and also returns the cold water to the boiler. In a two-pipe system one main supplies hot water, while the other returns the cold water back to the boiler.

This type of system responds rapidly to temperature changes and can be used for domestic hot water year-round if an automatic fuel burner is used. Since circulation is maintained by a pump,

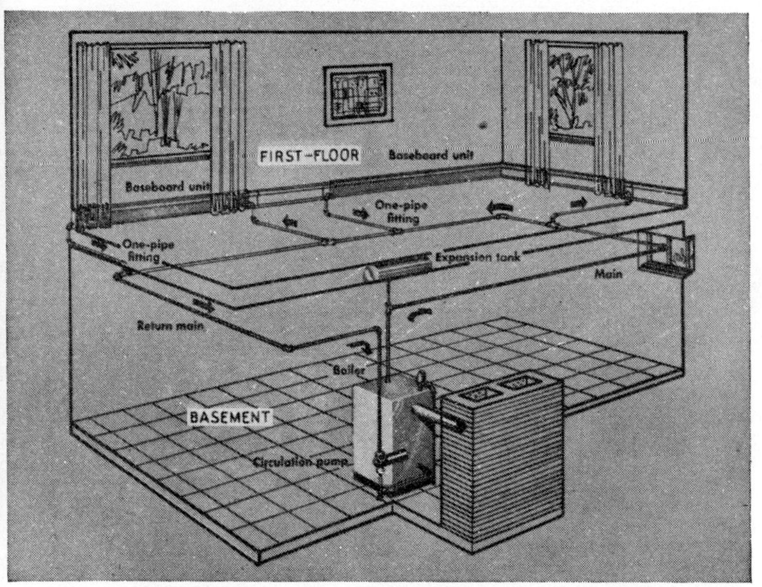

Univ. of Illinois Circular

FIG. 15.—A Forced Hot-water Heating System (One-pipe Type).

it lends itself to heating basementless houses and basement rooms. Smaller pipes can also be used because the pump is capable of maintaining circulation against relatively high friction heads. It is more expensive to install than the gravity system because of the pump and need for special fittings with the one-pipe system. The smaller pipes will compensate for this, however.

Hot-Water Panel System.—With this type of system hot water from the boiler is circulated by a pump through pipe coils (usually tubing) which are buried in the floor, ceiling, or wall. No radiators, convectors, or baseboard units are required because the tubing acts as the heating unit, transmitting heat to the room by direct radiation and indirectly by convection.

Hot-water panels can be used in houses with or without basements, but they must be built into the house. As with warm air panels temperatures cannot exceed 85° F. on floors and 120° F.

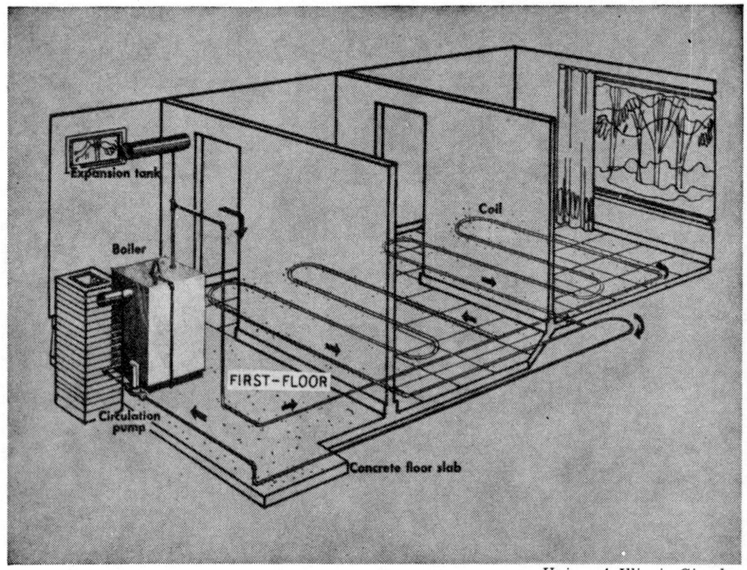

Univ. of Illinois Circular

FIG. 16.—Floor Installation of a Hot-water Panel System.

for walls or ceilings. For economical operation proper insulation in back of the panel is essential to keep heat losses down. While well suited to standard thermostatic control, the panels may overheat or underheat. This is particularly true where rapid temperature changes may occur, such as in houses with large glass areas or in certain regions of the country.

Insulation.—While several references are made in this Chapter to the effect of insulation upon heat transfer coefficients of walls, roofs, etc., and thus upon heating requirements, the subject of insulation itself has been deferred to this section. Here it can be discussed as a unit.

There are a number of materials from which insulation is made, metallurgical products like rock wool, which is made from slag; glass fiber products, such as Fiberglas®; vegetable products such as kapok; mineral products such as asbestos; and a number of blown plastic products such as polystyrene. All these materials possess the common property of having small air spaces which have low heat transmission rates. The differences between them in heat transmission are usually not the decisive consideration in choosing a type of insulation. Preference is given to other considerations, for example, vegetable materials may be eaten by rodents or vermin, and are objectionable from this point of view. Some materials absorb moisture more readily than others—from this point of view the glass and slag products are superior. In general, however, the decisive consideration in choosing an insulation is its physical form, which may be that of (1) particles (that is, granules or flocs); (2) blankets of various widths which are made by placing the particles or loosely woven fibers of insulation within containing materials, such as paper or metal foil; (3) batts, which are similar to blankets but prepared in standard lengths and widths, so they are sealed on all edges; and (4) slab insulation, which is prepared from the various insulating materials mentioned by combining them with a binder to form the slab and yet to retain their insulating properties.

The choice of one of these four forms of insulation depends somewhat upon the type of installation. The loose materials, i.e., fibrous or granular, have the advantage that they can be blown by use of a blower and hose into spaces within walls. In new house construction, however, one has a considerable choice of the type of insulation to be used. In the following tables of Fiberglas® building insulation, the various types are classified by their R values. The symbol "R" is a measure of the resistance to heat transmission, and therefore the higher the R value the better the performance. These values are specified for various types of insulation for various Fiberglas® products as shown in Table 4.

The types of insulation cited in the table are described in Table 5.

Vapor Barriers.—To enhance the effectiveness of insulation the use of vapor barriers is strongly recommended. A vapor barrier near the inside of all exposed walls, floors and ceilings can keep moisture vapor from penetrating to a point where it can condense. Two factors are vital in making vapor barriers effective:

(1) Vapor permeance must be less than 1 perm (a measure of the ability of materials to stop moisture).
(2) Vapor barrier must be kept warm enough with insulation on cold side to be sure vapor cannot condense on it.

There are a number of types of vapor barriers, including the following:

Insulation Vapor Barriers. Insulation with a kraft-asphalt vapor barrier provides good condensation protection if carefully applied.

Polyethylene Film. Use of unfaced Friction-Fit insulation with large sheets of film applied inside offers installation economies and a separate vapor barrier.

Foil-Backed Gypsum Board. Aluminum foil adhered to gypsum board in the factory provides a continuous noncombustible vapor barrier on the warm side of the wall construction and is installed in the standard manner.

Paint and Other Films. Vapor resistant paints on interior walls and ceilings offer some condensation protection as do vinyl wall coverings and other vapor-impervious interiors.

TABLE 4

	Air Conditioned Electrically Heated Quality Home Standard	Heated Only Moderate Comfort and Economy Standard	Minimum Comfort Standard
Recommended for	All electrically heated dwellings (including resistance heating, and heat pump systems) and all air conditioned dwellings. Also in Alaska (and Canadian provinces) and wherever fuel costs economically justify it. Also for homes in a price and quality bracket where the finest of products and standards prevail.	All homes heated but not air conditioned in moderate climate and rate areas. These standards help assure good comfort levels in the North and Central areas in winter and in the South in the summer. There is marked improvement in comfort and substantial reductions in heating costs compared to minimum standard at right.	Minimum housing in mild climates where low cost fuels and minimum income buyers make improved insulation unprofitable or unsalable to builder. Since low-income buyers need the fuel savings better insulation can bring, and more buyers can qualify when housing expense is cut, this standard is rarely desirable to the builder.
Roof—Ceilings	Fiberglas 6 inch Batts (R = 19), or when air conditioned but conventionally heated, Fiberglas Thick Double Foil Batts (R = 19). R = 24 also available for severe climate and high rate areas.	Fiberglas Thick Standard Batts or Roll Blankets (R = 13), or Fiberglas Medium Thick Double Foil (R = 16).	Fiberglas Medium Standard Batts or Roll Blankets (R = 9).

TABLE 4 (Cont.)

	Air Conditioned Electrically Heated Quality Home Standard	Heated Only Moderate Comfort and Economy Standard	Minimum Comfort Standard
WALLS frame	Fiberglas Thick Standard Batts or Roll Blankets (R = 11), or Fiberglas Medium Foil Faced Batts or Roll Blankets (R = 11s). R = 13 also available.	Fiberglas Medium Standard Batts or Roll Blankets (R = 8), or Fiberglas Economy Foil Faced Batts or Roll Blankets (R = 8s).	Fiberglas Economy Batts or Roll Blankets (R = 7).
masonry	Fiberglas Masonry Wall Batts (R = 3) between nominal 1″ furring strips. In cavity walls, Fiberglas Prescored Perimeter Insulation; $1\frac{5}{8}$″ thick (R = 7) in 2″ or wider cavity.	Fiberglas Masonry Wall Batts (R = 3) between nominal 1″ furring strips. In cavity walls, Fiberglas Prescored Perimeter Insulation; 1″ thick (R = 4) in 2″ or wider cavity.	Fiberglas Masonry Wall Batts (R = 3), between nominal 1″ furring strips.
sill detail	Fiberglas Sill Sealer under wood sills on foundation.	Fiberglas Sill Sealer under wood sills on foundation.	Fiberglas Sill Sealer (To save caulking expense only) under wood sills on foundation.

TABLE 4 (Cont.)

	Air Conditioned Electrically Heated Quality Home Standard	Heated Only Moderate Comfort and Economy Standard	Minimum Comfort Standard
FLOORS over basements	With heated basement, no insulation. Over unheated basement areas, Fiberglas Reverse Flange Batts, Thick (R = 13), installed between floor joists.	With heated basement, no insulation. Over unheated basement areas, Fiberglas Reverse Flange Batts, Medium (R = 9), installed between floor joists.	With heated basement, no insulation. Over unheated basement areas Fiberglas Economy Batts or Roll Blankets (R = 7), installed between floor joists.
vented crawl spaces	If used as heating plenum, Fiberglas Prescored Perimeter Insulation, 1⅝″ thick (R = 7), installed around inside of crawl space wall to 12″ below outside grade.	If used as heating plenum, Fiberglas Prescored Perimeter Insulation 1″ thick (R = 4), installed around inside of crawl space to 12 inches below outside grade.	If used as heating plenum, Fiberglas Perimeter Insulation, ¾″ thick (R = 3), installed around inside of crawl space walls to 12 inches below outside grade.
	If vented, Fiberglas Reverse Flange Batts, Thick (R = 13) installed between floor joists.	If vented, Fiberglas Reverse Flange Batts, Medium (R = 9) installed between floor joists.	If vented, Fiberglas Economy Batts or Roll Blankets (R = 7) installed between floor joists.
slab-on-grade	Fiberglas Prescored Perimeter Insulation, 1⅝″ Thick (R = 7), installed between slab edge and foundation and under slab for a total width of 24 inches.	Fiberglas Prescored Perimeter Insulation, 1″ thick (R = 4), installed between slab edge and foundation and under slab for a total width of 24 inches.	None in the South, unless to separate exposed sun-lit slab from house slab. In Central and North States, Fiberglas Perimeter Insulation ¾″ thick (R = 3) installed between slab edge and foundation and under slab for a total width of 24 inches.

TABLE 5

<table>
<tr><th>Application</th><th>Product</th><th>Description</th><th colspan="5">Performance Data</th></tr>
<tr><td rowspan="5">Ceilings
Floors
Walls</td><td rowspan="5">Paper Blanket Insulation</td><td rowspan="5">Kraft faced product for nailing to studs or joists or unrolling in open joist spaces from above. Strong asphalted kraft paper provides a vapor barrier and folded nailing flanges for attachment. Packaged: rolls, batts in rolls and batts in tubes.</td><td colspan="5">Installed Thermal Resistance</td></tr>
<tr><td>Product</td><td>↕</td><td colspan="2">↔</td><td>Widths</td></tr>
<tr><td>Thick</td><td>R = 13</td><td colspan="2">R = 11</td><td>15″, 19″, 23″</td></tr>
<tr><td>Medium</td><td>R = 9</td><td colspan="2">R = 8</td><td>15″, 19″, 23″</td></tr>
<tr><td>Economy</td><td>R = 7</td><td colspan="2">R = 7</td><td>15″, 19″, 23″</td></tr>
<tr><td rowspan="4">Floors</td><td rowspan="4">Paper Blanket Insulation with nailing flanges</td><td rowspan="4">Nailing flange on breather-paper face of insulation opposite vapor barrier for installation from the cold side between floor joists in crawl spaces or over garages or other unheated spaces. Can be used in wall constructions but requires a separate vapor barrier on warm side.</td><td colspan="5">Installed Thermal Resistance</td></tr>
<tr><td>Product</td><td colspan="3">↕</td><td>Widths</td></tr>
<tr><td>Thick</td><td colspan="3">R = 13</td><td>15″, 23″</td></tr>
<tr><td>Medium</td><td colspan="3">R = 9</td><td>15″, 23″</td></tr>
<tr><td rowspan="7">Ceilings
Floors
Walls</td><td rowspan="7">Foil-Faced Insulation</td><td rowspan="7">An insulation product with a genuine aluminum foil vapor barrier asphalted to one side. Full insulating value is obtained when installed with an air space of at least ¾″ facing the foil.
Medium and Standard for walls, and thicknesses for floors and ceiling. Packaged in rolls. Super Thick packaged batts in tubes only.</td><td colspan="5">Installed Thermal Resistance</td></tr>
<tr><td>Product</td><td>↑</td><td>↓</td><td>↔</td><td>Widths</td></tr>
<tr><td>Super Thick</td><td>R = 23s</td><td>R = 25s</td><td>. . .</td><td>15″, 23″</td></tr>
<tr><td>Extra</td><td>R = 14s</td><td>R = 16s</td><td>R = 11</td><td>15″, 23″</td></tr>
<tr><td>Full</td><td>R = 13s</td><td>R = 15s</td><td>R = 10</td><td>15″, 23″</td></tr>
<tr><td>Medium</td><td>R = 10s</td><td>R = 13s</td><td>R = 11s</td><td>15″, 23″</td></tr>
<tr><td>Standard</td><td>R = 9s</td><td>R = 11s</td><td>R = 9s</td><td>15″, 23″</td></tr>
</table>

TABLE 5 (Cont.)

Application	Product	Description	Performance Data			
			Installed Thermal Resistance			
			Product	↑	↓	Widths
Ceilings	Double-Foil Insulation	Utilizing the heat reflective value of genuine aluminum foil, this product has a solid foil sheet asphalted to one side, providing a vapor barrier, and a vapor porous foil breather paper on the other side. Specifically designed to give maximum thermal performance in minimum space. Packaged: rolls and batts in tubes.	Thick	R = 15s	R = 19s	15″, 19″, 23″
			Medium	R = 11s	R = 16s	15″, 23″
			Installed Thermal Resistance			
			Product	↕		Widths
Ceilings Floors	Six-inch Batts	For ceilings or floors between joists especially for air-conditioned homes or other applications where high thermal efficiency is desired. Has an asphalted kraft paper vapor barrier with nailing flanges on one side only. Special heavier density product called R = 24 provides top thermal effectiveness. Packaged: batts in tubes.	6″	R = 19		15″, 23″
			R = 24	R = 24		15″, 23″

XVII

Ventilating and Air Conditioning

Ventilation is defined * in part as the process of supplying air to, or removing air from, any space by natural or mechanical means. The word itself implies quantity, but air must be of the proper quality also. The American Society of Heating and Ventilating Engineers Code of Minimum Requirements for Comfort Air Conditioning defines air conditioning ". . . as the process by which simultaneously the temperature, moisture content, movement and quality of the air in enclosed spaces intended for human occupancy may be maintained within required limits."

Fans.—A fan is the most economical method of mechanically removing air from, or supplying air to, a room or building. Fans are also used to "move air around," thus providing a desired local cooling effect. Fans are available in a wide variety of sizes and shapes, depending on where they are to be installed and how they are to be used. They are rated by the amount of air they can move in a given time: "cfm" or cubic feet per minute. Table 1 gives ventilation standards which may be used as the basis of minimum fan selection. Night air cooling by an attic fan is a popular way of cooling a house at relatively little cost. The size of the fan to be used depends on the floor area of the house and how many air changes are desired. Table 2 gives the approximate capacities of attic fans for houses of different floor areas.

* *A.S.H.V.E. Guide.*

TABLE 1

FAN CAPACITIES RECOMMENDED *

Floor Area of House, sq. ft.	Amount of Air Fan Should Deliver † For Regions of Cool Nights, cu. ft./min.		For Regions of Warm Nights, cu. ft./min.
800	3000	to	6500
1000	4000	to	8000
1200	5000	to	9500
1400	5500	to	11000
1600	6500	to	13000
1800	7000	to	14500

* *University of Illinois Small Homes Council Circular G6.0.*

† Capacities are for so-called "free air delivery." If obstructions or resistances to flow of air exist (such as louvers, screens, ducts), then the actual deliveries will be less than the "free air delivery." In general, low-speed fans provide quieter operation than small, high-speed units.

Air Conditioning.—Air conditioning systems cool and dehumidify warm, humid air by passing it over a cooling coil. All air contains moisture (water vapor). When the air is warm it can hold more moisture than when it is cold. In an air conditioning unit warm air, from the building or outside, is cooled by passing it over a cold evaporator coil through which refrigerant constantly circulates (the same as in an ordinary refrigerator). Since cool air cannot hold as much moisture as warm air the excess water vapor condenses out on the evaporator coil. The cooled and dehumidified air is then sent into the room or building by a fan.

The size unit needed for a particular installation will depend upon several factors such as the prevailing temperatures in the region, the area of the room or building to be cooled, the number of windows exposed to the sun or in the shade, the height of the rooms, thickness of the walls, number of people who will use the rooms, how the rooms will be used (apartment, office, shop, etc.), whether or not the rooms are directly under the roof, how the

TABLE 2

VENTILATION STANDARDS [a]

Application	Smoking	Cfm per Person [b]		Cfm per Sq. Ft. of Floor [b]
		Recommended	Minimum [c]	Minimum [c]
Apartment, Average	Some	20	10	....
Deluxe	Some	30	25	0.33
Banking Space	Occasional	10	7½	....
Barber Shops	Considerable	15	10	....
Beauty Parlors	Occasional	10	7½	....
Brokers' Board Rooms	Very Heavy	50	20	....
Cocktail Bars	Heavy	40	25	....
Corridors (Supply or Exhaust)		..	..	0.25
Department Stores	None	7½	5	0.05
Directors' Rooms	Extreme	50	30	....
Drug Stores [e]	Considerable	10	7½	....
Factories [d,f]	None	10	7½	0.10
Five and Ten Cent Stores	None	7½	5	....
Funeral Parlors	None	10	7½	....
Garages [d]		..	..	1.0
Hospitals, Operating Rooms [f,g]	None	..	..	2.0
Private Rooms	None	30	25	0.33
Wards	None	20	10	....
Hotel Rooms	Heavy	30	25	0.33
Kitchens, Restaurant		..	..	4.0
Residence		..	..	2.0
Laboratories [e]	Some	20	15	....
Meeting Rooms	Very Heavy	50	30	1.25
Offices, General	Some	15	10	....
Private	None	25	15	0.25
Private	Considerable	30	25	0.25
Restaurant, Cafeteria [e]	Considerable	12	10	....
Dining Room [e]	Considerable	15	12	....
School Rooms [d]	None	..	..	....
Shop, Retail	None	10	7½	....
Theater [d]	None	7½	5	....
Theater	Some	15	10	....
Toilets [d] (Exhaust)		..	..	2.0

[a] *A.S.H.V.E. Guide.* Taken from present-day practice or large air conditioning companies. [b] This is contaminant-free air and may be either outdoor air or recirculated air which has been appropriately purified. [c] When minimum is used, take the larger of the two. [d] See local codes which may govern. [e] May be governed by exhaust. [f] May be governed by special sources of contamination or local codes. [g] All outside air recommended to overcome explosion hazard of anesthetics. See *National Board of Fire Underwriters'* pamphlet No. 56.

building is insulated, etc. All these factors contribute to what is known as the *cooling load* which the air conditioner must be able to handle. Cooling load is expressed in *tons* (of refrigeration) or in Btu per hour. A ton of refrigeration is the amount of cooling equal to the melting of a ton of ice per day. It is also equal to 200 Btu per minute, or 12,000 Btu per hour.

Calculation of Cooling Load.—The air-conditioning unit must be large enough to absorb all the heat and moisture generated in a room or building. To determine the proper size unit it is therefore necessary to calculate the maximum amount of heat it is anticipated must be removed. A detailed procedure for doing this can be found in the *A.S.H.V.E. Guide*.

A simpler method, particularly suitable for room air-conditioners, is suggested by the Air-Conditioning and Refrigeration Institute. Table 3 shows a cooling load estimate form which, by proper use of the factors, will make it possible to calculate the cooling load of a room. For example, it is desired to calculate the cooling load and select a room air-conditioner for the bedroom shown in Fig. 1. It will be occupied by three people.

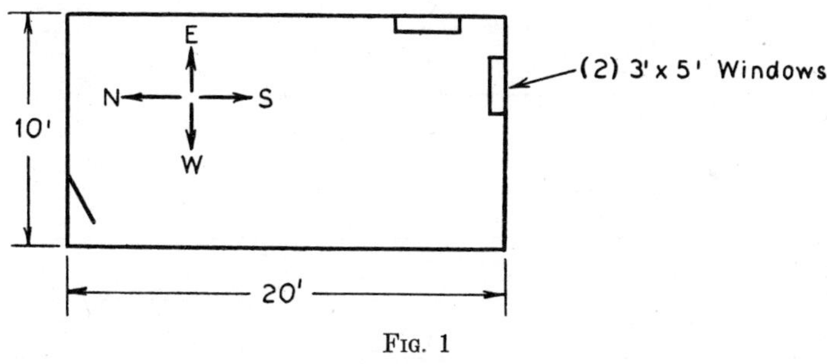

Fig. 1

The appropriate places in Table 3 are filled out as shown in the table, in the quantity column. They are then multiplied by the indicated factors and cooling units result in the last column. These values are totaled for the cooling load of the room in Btu per hour. For this bedroom the cooling load is 8,375 Btu per hour. To determine what size unit is required divide this by 12,000 Btu per hour to get tons of refrigeration. Thus, 8,375/12,000 = 0.698. The closest standard unit is 0.75 or ¾ ton.

TABLE 3

COOLING LOAD ESTIMATE FORM FOR ROOM AIR-CONDITIONERS

Recommended Practice for Members of Room Air-Conditioner Section of Air-Conditioning and Refrigeration Institute. This estimate is suitable for comfort air-conditioning jobs not requiring specific conditions of temperature and humidity.

Customer.................................... Buyer....................................
Address.................................... Space to be used for....................................
Estimate by.................... Date.................... Approval.................... Date....................

ITEM	Quantity	Factor		Cooling Units
1. WINDOWS EXPOSED TO SUN, FACING*		Inside Shades	Outside Awnings	
a. East, Southeast, or South	15 sq ft	45	25	
b. Southwest	15 sq ft	65	40	975
c. West	.... sq ft	100	60	
d. Northwest	.... sq ft	35	25	
*Use only the exposure with the largest load.				
2. WINDOWS FACING NORTH OR IN SHADE (Include all windows not included in Item 1.)	0 sq ft	14		
3. WALLS (based on lineal feet of wall)				
a. Light construction, exposed to sun*	10 ft	90		900
b. Heavy construction, exposed to sun*	.... ft	50		
c. Shaded walls or partitions (Include all walls not included in 3a or 3b.)	30 ft	30		900
*Use for only that exposure used in Item 1.				
4. ROOF OR CEILING (Use one only.)				
a. Roof, uninsulated	.... sq ft	16		
b. Roof, with one inch or more insulation	200 sq ft	7		1400
c. Ceiling, with occupied space above	.... sq ft	3		
d. Ceiling, with attic space above	.... sq ft	10		
5. FLOOR (Neglect floor directly on ground or over unheated basement.)	200 sq ft	3		600
6. PEOPLE AND VENTILATION—Number of people	3	900		2700
7. LIGHTS AND ELECTRICAL EQUIPMENT IN USE	300 watts	3		900
8. DOORS AND ARCHES CONTINUOUSLY OPEN TO UNCONDITIONED SPACE (lineal feet of width)	0 ft	300		
9. TOTAL LOAD in cooling units to be used for selection of room air conditioner (s)	xxxx	xxxx		8375

Air Conditioning Systems.—Several types of systems are available, depending on the cooling load and the type of installation desired. Central air conditioning is available in many different sizes and shapes for summer use only or for year-round use. They usually require water-cooled condensers and the use of a water-conserving device known as a cooling tower. Fig. 2 shows how a central air conditioner operates to keep the entire house cool.

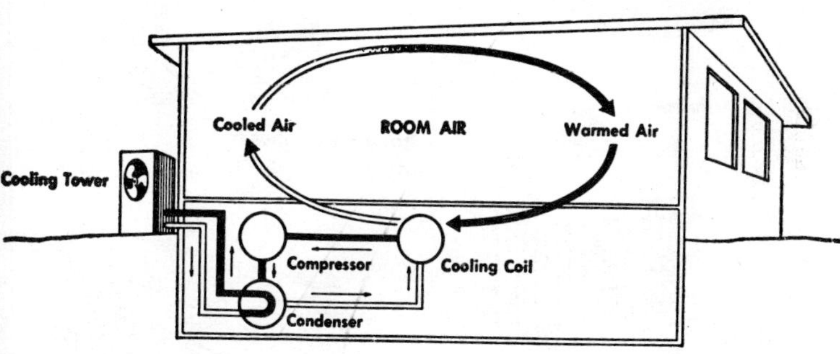

FIG. 2.—Passage of Heat and Cooled Air in a Central Air-conditioning System. In an air-conditioned house, the warm air from the rooms is passed over a cooling coil and is then returned to the rooms as cooled air. The heat removed from the air is eventually discharged from the house by means of a condenser. With water-cooled condensers, a cooling tower makes re-use of water possible. (From *Univ. of Illinois Circular.*)

If a warm air system, with ducts, exists it may be possible to install air conditioner into this system. It will be necessary to check to be certain that the ducts have ample capacity for cooling. If the ducts do not have sufficient capacity it will be necessary to add additional ducts and supply outlets. The most popular types of outlets are diffusers, and registers with fixed or adjustable vanes. Fig. 3 shows different types of outlets and how they are installed.

Room air conditioners of the window and cabinet type are used where it is desired to cool one room, or a small section of a house. The window units fit into a window opening and are available in sizes from ⅓ to 1 ton of cooling capacity. The room air is circulated through the unit where it is cooled, dehumidified, and filtered. Since the condenser is air cooled no water pipes are necessary. Similarly since the moisture condensed from the room air is evaporated directly to the outdoors no drainage system is required. Cabinet units stand on the floor and require a wall

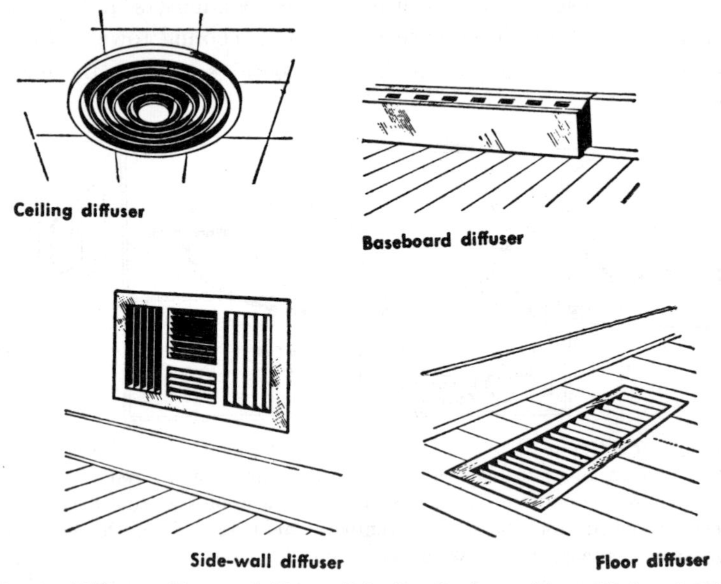

FIG. 3.—Different Types of Air-conditioning Outlets. (From *Univ. of Illinois Circular.*)

opening through which the air to cool the condenser is brought in and discharged. Units of these types can usually be plugged into existing house wiring if it is of adequate capacity. If the wiring is inadequate it will be necessary to run a special branch circuit for the air conditioner.

Heat pumps are a new type of year-round air conditioning system which remove heat from the house during the summer and transfer it to the outside ground, well, air, river, or lake. During the winter heat is withdrawn from the ground (or other source) and pumped into the house. Since the heat pump is electrically operated it is economical at the present time only in areas where the cost of electricity is low.

XVIII

Machine Shop Work

Measuring Instruments.—A knowledge of measuring instruments is one of the first things needed in machine shop work because the foundation of present-day machinery manufacture is based upon measuring instruments. All types of calipers, outside, inside, hermaphrodite, thread, vernier, micrometer, etc., and all types of gages, caliper, collar and plug, limit, internal and external threads, etc., are required to make possible the modern production system of interchangeable parts.

Methods of Measuring

Measuring is an art which must be mastered in order to produce good machine work. It can be mastered only by patient and careful practice. Proficiency in measurement will save many a job from being spoiled or rejected.

Calipers.—A steel scale and outside calipers are commonly used to measure diameters on lathe work. First, the calipers are accurately set to the required measurement as shown in Fig. **1**. Then they are tried on the work, care being taken that they are held in a plane perpendicular to the longitudinal axis of the work. When the work is of the desired size the calipers should slide snugly across the cylinder without forcing. If a number of pieces of work of the same diameter are to be measured, it is often desirable to set the calipers on a standard test cylinder. The "feel" of the calipers on the test cylinder should be carefully gaged and when the work is measured the same "feel" should be obtained when the work is of the required size.

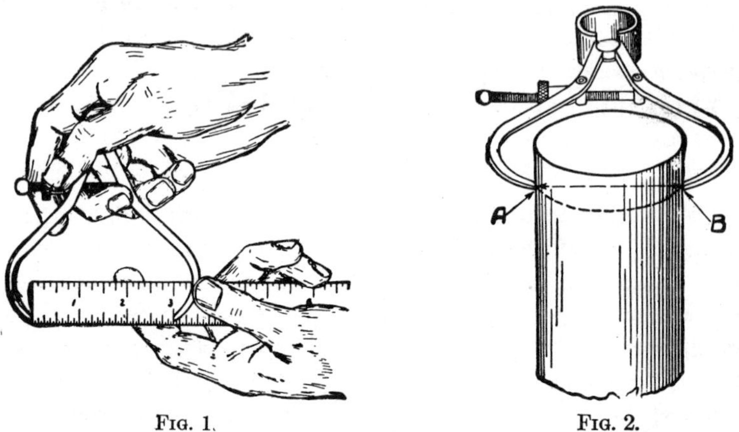

Fig. 1. Fig. 2.

Inside calipers (Fig. 3) are used for measuring the diameters of holes. They, too, are set on steel scales and require a sensitive feel for accuracy. When a shaft is being turned to fit a certain

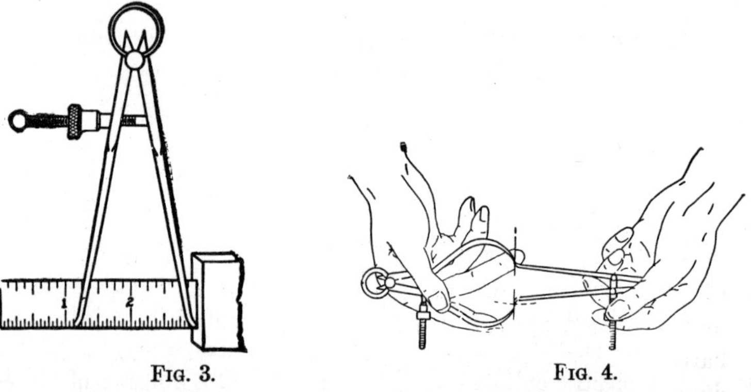

Fig. 3. Fig. 4.

hole, the measurement of the hole may be transferred to a pair of outside calipers as shown in Fig. 4. Micrometers may also be set from the inside calipers in a similar manner.

Micrometer Calipers.—When greater accuracy is required than can be obtained by using a caliper and scale, a micrometer caliper such as shown in Fig. 5 is used. This is a precision instrument which requires the most careful treatment. The micrometer screw has 40 threads to the inch. If the sleeve or thimble is turned one complete revolution, the spindle will advance $\frac{1}{40}$ inch, which is equivalent to 0.025 inch. The sleeve has 25 graduations so that if it is turned one graduation or

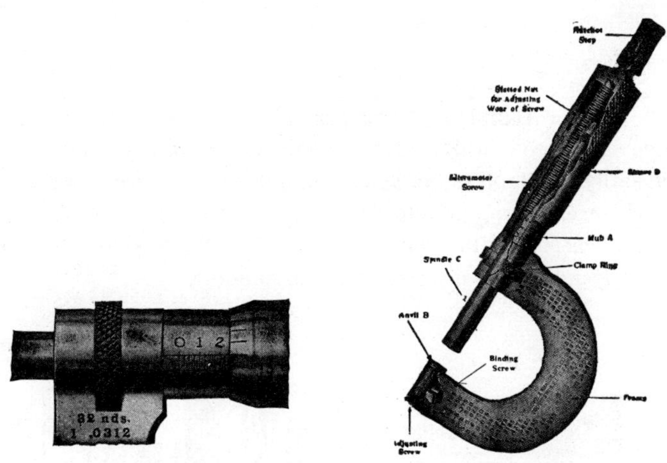

Fig. 5.—Micrometer.

$\frac{1}{25}$ of a revolution the spindle will move $\frac{1}{25}$th of $\frac{1}{40}$th of an inch or 0.001 inch. Four complete turns of the sleeve will move the spindle 4×0.025 inch or 0.100 inch. The hub or barrel of the micrometer has a number at each 0.100 inch division and four unnumbered spaces between, each representing 0.025 inch.

To read a micrometer, add the readings on the hub and the sleeve.

ILLUSTRATION: What is the reading of the micrometer shown in Fig. 6?

Numbered graduations........	0.200 inch
Unnumbered graduations......	0.050 inch
Sleeve graduations............	0.008 inch
Total......................	0.258 inch (Ans.)

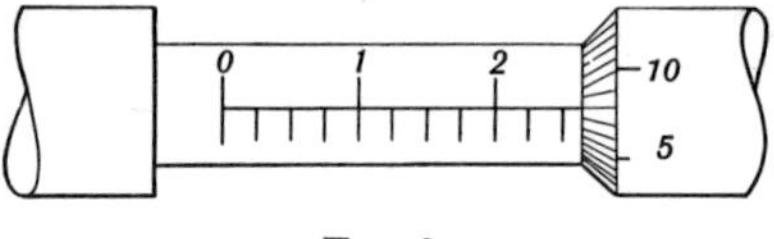

Fig. 6.

Vernier Caliper.—A caliper based on the principle that the eye can more readily judge the coincidence of two lines than it can visually interpolate between graduations on a scale is known as a vernier caliper. A simple form is shown in Fig. 7. A more complicated adaptation is used to measure gear teeth and is usually known as a gear-tooth micrometer.

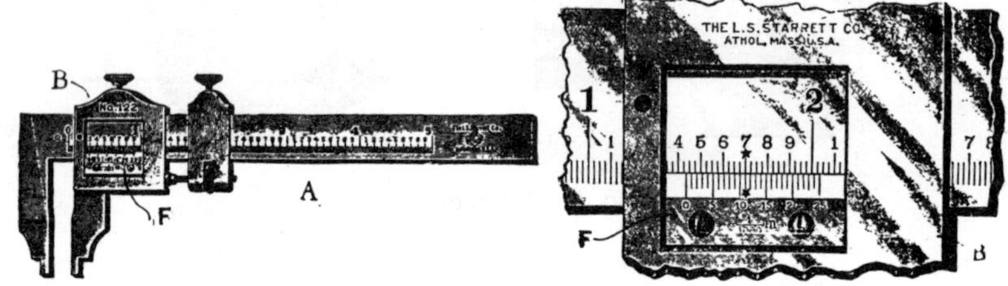

Fig. 7.—Vernier Caliper.

A vernier has a fixed scale and a sliding scale so graduated that the sum total length of its divisions is exactly equal to the sum of the lengths of one fewer divisions on the fixed scale. We will illustrate by a specific example.

On the fixed or (A) scalc of Fig. 8 let the major numbered divisions represent inches. Then the minor numbered divisions are tenths of inches and the unnumbered divisions are quarters of tenths, $\frac{1}{40}$th or 0.025 inch each point. It will be noted that the movable scale has twenty-five equal divisions which aggregate a

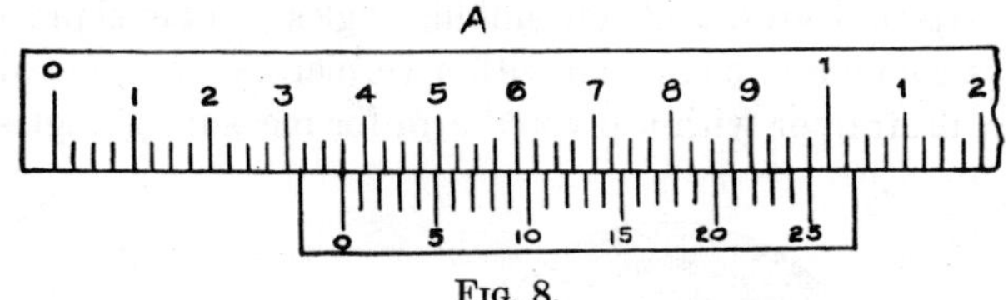

FIG. 8.

length exactly equal to twenty-four divisions on the fixed scale. The difference in length between one division on the fixed scale and one division on the sliding scale is $\frac{1}{25}$ of $\frac{1}{40}$ or 0.001 inch. Then if the sliding scale in Fig. 8 moves slightly to the right so that its second line will coincide with (4) it will have moved 0.001 inch.

To read a vernier, note the number of divisions and calibrated parts of divisions up to the zero or index of the sliding scale.

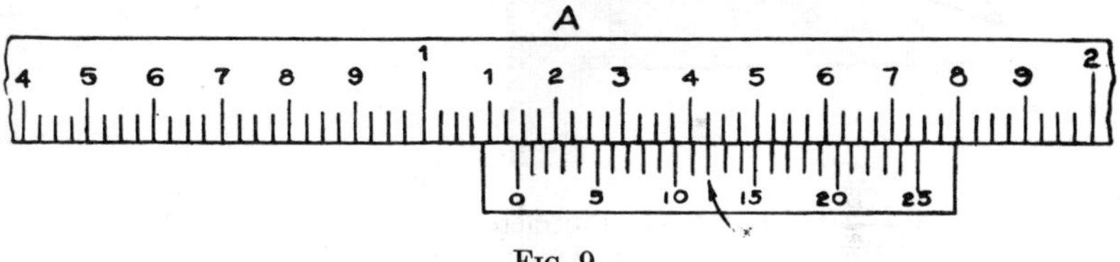

FIG. 9.

The fraction of the space on which the zero rests may be read directly on the sliding scale at the point of coincidence of any two lines.

ILLUSTRATION: What is the reading of the vernier shown in Fig. 9?

Whole inches	1.000
Tenths	0.100
Thousandths	0.025
Vernier	0.012
Total	1.137 in. (Ans.)

Vernier Bevel Protractor

A protractor is used for dividing circles into any number of equal parts or degrees and determining angles. A bevel protractor, which is commonly combined with a vernier, is a graduated semicircular protractor with a pivoted arm for measuring angles.

Fig. 10—Universal Bevel Protractors with Vernier and Acute Angle Attachment.
(Courtesy of The L. S. Starrett Company).

The disc of a vernier bevel protractor is graduated in degrees from 0 to 90° each way. The vernier plate is graduated so that 12 divisions on the vernier occupy the same space as 23 degrees on the disc. The difference between the width of one of the 12 spaces on the vernier and two of the 23 spaces on the disc is therefore $\frac{1}{12}$ of a degree.

Each space on the vernier is $\frac{1}{12}$ of a degree, or five minutes shorter than two spaces on the disc. If a line on the vernier coincides with a line on the disc and the protractor is rotated until the next line on the vernier coincides with the next line but one on the disc, the vernier has been moved through an arc of $\frac{1}{12}$ of a degree, or 5 minutes.

To read the protractor, note on the disc the number of whole degrees between 0 on the disc and 0 on the vernier. Then count in the same direction the number of spaces from 0 on the vernier to a line that coincides with a line on the disc. Multiply this number by 5 and the product will be the number of minutes to be added to the number of whole degrees.

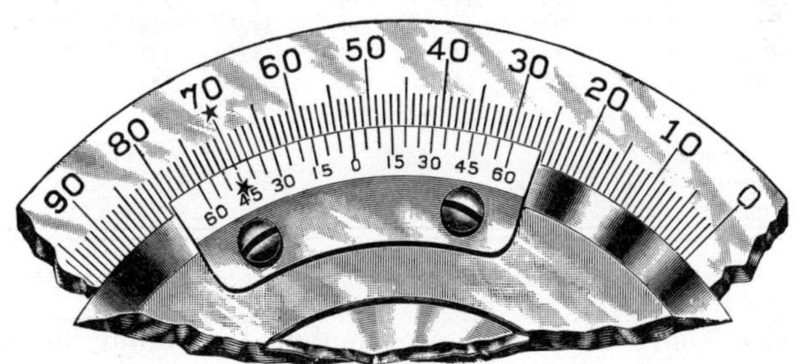

Fig. 11.—How to Read Universal Bevel Protractor with Vernier. (Courtesy of The L. S. Starrett Company)

Illustration: What is the reading of the vernier bevel protractor in Fig. 11.

Whole degrees....................	52°
Minutes, 9 × 5..................	45′
	52° 45′

The starred 45 line, the 9th from zero, is the one that coincides with a line on the disc.

Gage Blocks.—Johansson Gage Blocks are rectangular pieces of tool steel, hardened, ground, stabilized, and finished to an accuracy of a few millionths of an inch. They are used to check micrometers, snap gages, sine bar settings, and any other place where extremely precise measurements are required.

A full set of eighty-one blocks, which have surfaces flat and parallel within .000008 of an inch, is made up of four series:

First Series

.1001″	.1002″	.1003″	.1004″	.1005″	.1006″	.1007″	.1008″	.1009″

Second Series

.101″	.102″	.103″	.104″	.105″	.106″	.107″	.108″	.109″	.110″
.111″	.112″	.113″	.114″	.115″	.116″	.117″	.118″	.119″	.120″
.121″	.122″	.123″	.124″	.125″	.126″	.127″	.128″	.129″	.130″
.131″	.132″	.133″	.134″	.135″	.136″	.137″	.138″	.139″	.140″
.141″	.142″	.143″	.144″	.145″	.146″	.147″	.148″	.149″	

Third Series

				.050″				
.100″	.200″	.300″	.400″	.500″	.600″	.700″	.800″	.900″
.150″	.250″	.350″	.450″	.550″	.650″	.750″	.850″	.950″

Fourth Series

1.000″	2.000″	3.000″	4.000″

Blocks from these series are combined to build up to the dimension which is being checked. It is always desirable to build the combination with the fewest number of blocks. To do this begin with the right-hand figure of the specified size and continue working from right to left. For example, it is desired to build a combination of blocks to check a 1.2721-inch dimension. The following block combinations will do the job:

		.1008	.1006	.1007
	.1009	.1003	.1005	.1004
.1001	.1002	.139	.138	.141
.149	.147	.132	.133	.130
.123	.124	.100	.500	.600
.900	.800	.700	.300	.200
1.2721	1.2721	1.2721	1.2721	1.2721

These blocks are made so accurately that it is necessary to wipe each of the blocks together when assembling them.

Metal Cutting

Cutting Tools.—Machine work practice has developed a number of different tools for cutting metal, all of which have, however, several points of similarity. (See Fig. 12.) They consist in general of a shank by which they are held in the cutting machine

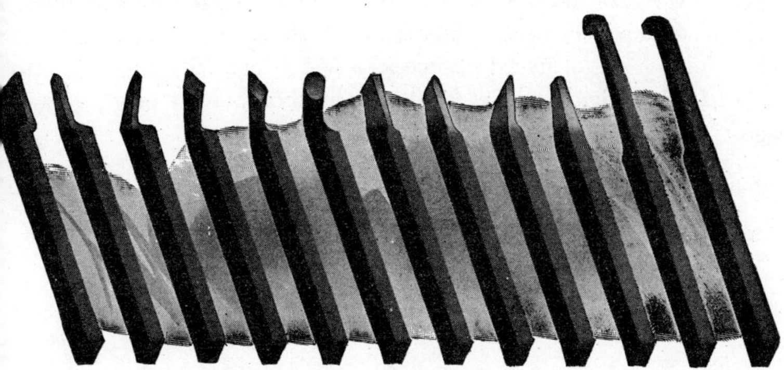

FIG. 12.

and a cutting edge which engages the metal being cut, and shears off the shaving. Figure 13 shows the shape of cutting edge of a standard forged lathe tool. The shape to which a tool is ground depends on the machine in which it is to be used, the type of cut

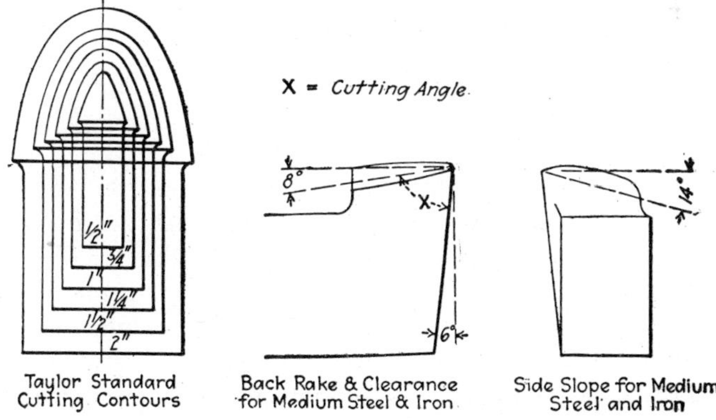

FIG. 13.—Standard Lathe Tool.

it is to make and the hardness of the metal which is to be cut. The tool illustrated in Fig. 13 is a round-nose roughing tool. The angle marked 8° is called the *back rake* or *front top rake*; the angle marked 6° is known as the *clearance* or *front rake*; and the angle marked 14° the *side slope* or top *side rake.* Forged lathe tools such as we have discussed are used mainly for large work involving heavy cutting. For more delicate work the cutting edge is ground on a small piece of metal known as a *tool bit* which is inserted into a *tool holder* (Fig. 14) which replaces the shank of the larger forged tool.

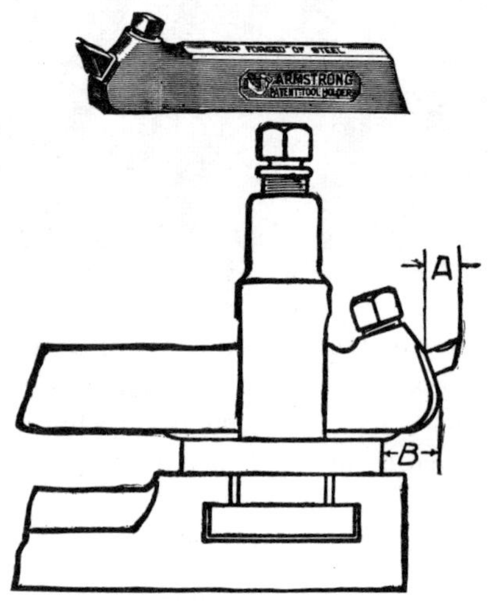

Fig. 14.

Tool steels or high carbon steels contain from 0.60 percent to 1.50 percent carbon, the hardness increasing with the amount of carbon.

High speed steels contain several other ingredients such as tungsten, chromium, manganese, silicon, molybdenum, vanadium and nickel. Tungsten and chromium in particular give the steel the property of retaining its cutting ability under very high speeds or heat.

Cutting Speed.—Cutting speed is the velocity with which a cutting tool engages the work and is always given in feet per minute (f.p.m.). The term feet per minute has somewhat different meanings for different machines. In turning work on a lathe it means the number of linear feet, measured on the surface of the work, which passes the edge of a cutting tool in one minute.

On a shaper it means the rate in f.p.m. at which the tool passes the work, while on a planer it means the rate in f.p.m. at which the work passes the tool.

On a milling machine it means the surface speed of the cutter, i.e., the speed of a point on the rim of the cutter.

The following formula may be used to calculate cutting speeds of lathes and milling machines.

$$C = \frac{\pi R D}{12}$$

where C = cutting speed
R = revolutions per minute
D = diameter of work, or diameter of cutter in inches
$\pi = 3\frac{1}{7}$ or 3.1416

Illustration: What is the cutting speed if a piece of work $\frac{1}{2}$ inch in diameter is turning at 458 revolutions per minute?

$$C = \frac{\pi D R}{12}$$

$$= \frac{22 \times 1 \times 458}{7 \times 2 \times 12} = 60 \text{ feet per minute} \quad \text{(Ans.)}$$

If a certain cutting speed is wanted, the proper revolutions per minute may be found by the following transposition of the preceding formula:

$$R = \frac{12C}{\pi D}$$

Illustration: A cutting speed of 80 feet per minute is desired on a piece of work whose average diameter is 2 inches. What speed of the machine will be required?

TABLE 1

Table of Speeds

Diam. In.	Cutting Speeds in Feet per Minute								
	20	30	40	50	60	70	80	90	100
	Revolutions per Minute								
¼	306	458	611	764	916	1070	1222	1376	1528
⅜	204	306	407	509	612	712	814	916	1019
½	153	229	306	382	458	534	612	688	764
⅝	122	183	244	306	366	428	488	550	611
¾	102	153	204	255	306	356	408	458	509
⅞	87	131	175	218	262	306	350	392	437
1	76	115	153	191	230	268	306	344	382
1⅛	68	102	136	170	204	238	272	306	340
1¼	61	92	122	153	184	214	244	274	306
1⅜	56	83	111	139	167	194	222	250	278
1½	51	76	102	127	152	178	204	228	255
1⅝	47	71	94	118	141	165	188	212	235
1¾	44	65	87	109	130	152	174	196	218
1⅞	41	61	82	102	122	143	163	183	204
2	38	57	76	95	114	134	152	172	191
2⅛	36	54	72	90	108	126	144	162	180
2¼	34	51	68	85	102	119	136	153	170
2⅜	32	48	64	80	97	112	129	145	161
2½	31	46	61	76	92	106	122	134	153
2⅝	29	44	58	73	88	102	117	130	146
2¾	28	42	56	70	83	97	111	125	139
2⅞	27	40	53	67	80	93	106	119	133
3	25	38	51	64	76	90	102	114	127

Diam. In.	Cutting Speeds in Feet per Minute							
	110	120	130	140	150	160	170	180
	Revolutions per Minute							
¼	1681	1833	1986	2139	2292	2462	2615	2780
⅜	1120	1222	1324	1426	1528	1632	1735	1836
½	840	917	993	1070	1146	1221	1298	1374
⅝	672	733	794	856	917	976	1036	1098
¾	560	611	662	713	764	816	867	918
⅞	480	524	568	611	655	699	742	786
1	420	458	497	535	573	611	649	687
1⅛	373	407	441	475	509	542	576	610
1¼	336	367	397	428	458	489	520	551
1⅜	306	333	361	389	417	444	472	500
1½	280	306	331	357	382	407	433	458
1⅝	259	282	306	329	353	377	400	423
1¾	240	262	284	306	327	349	371	393
1⅞	224	244	265	285	306	326	346	366
2	210	229	248	267	287	306	324	344
2⅛	198	216	234	252	270	288	306	323
2¼	187	204	221	238	255	272	289	306
2⅜	177	193	210	225	241	257	273	290
2½	168	183	199	214	229	244	260	275
2⅝	160	175	189	204	218	233	248	262
2¾	153	167	181	194	208	222	236	250
2⅞	146	159	173	186	199	213	226	239
3	140	153	166	178	191	204	216	229

$$R = \frac{12C}{\pi D}$$

$$= \frac{12 \times 80 \times 7}{2 \times 22} = \frac{1680}{11} = 153 \text{ r.p.m.} \quad \text{(Ans.)}$$

Illustration: Find the cutting speed of a side facing milling cutter 6 inches in diameter running at 30 revolutions per minute.

$$C = \frac{\pi RD}{12}$$

$$= \frac{22 \times 30 \times 6}{7 \times 12} = 47 \text{ feet per minute} \quad \text{(Ans.)}$$

Cutting speeds may conveniently be found directly by reference to Table 1.

Proper Cutting Speed.—It can readily be seen that if the cutting speed in machine work is too slow, the parts produced per day will be fewer and the costs will mount. In competitive manufacturing it is, then, necessary to run the cutting operations at the *maximum safe cutting speed.* What this speed is cannot be definitely stated. In general the maximum safe cutting speed may be defined as a speed slightly lower than that at which the tool or the work may be injured by excessive heat and the cutting edge dulled too rapidly.

Cutting speeds depend on the following conditions:

1. Kind of steel used, whether tool steel or high-speed steel.
2. Shape of tool, whether narrow or broadnosed.
3. Lip angle of tool or inclined angle of nose.
4. Position of tool in the tool post.
5. Sharpness of tool.
6. Depth of cut and amount of feed.
7. Material to be cut, whether soft, medium or hard, or whether brass, cast iron, or steel.
8. Cooling medium, whether used or not, the amount of cooling and lubricating effect produced.
9. Heat treatment of steel.

10. Elasticity of work or tool, which causes chattering.
11. Rigidity with which work is held.
12. Condition of machine to be used.

The proper cutting speed of a lathe with modern high speed tools, can be found by using the following empirical formula:

$$V = \frac{H \times S}{(\sqrt[3]{D} + Y)(\sqrt[2]{F} - Z)}$$

when V = cutting speed in feet per minute
D = depth of cut, taking $\frac{1}{64}$ inch as a unit
F = feed, taking $\frac{1}{64}$ inch per revolution as a unit
H = constant for hardness of material to be cut:

Hard cast iron or steel, 0.6
Medium cast iron or steel, 1.0
Soft cast iron or steel, 2.0

S = constant for size of tool:

232 for $\frac{3}{4}$ in. sq. tool on cast iron
215 for $\frac{1}{2}$ in. sq. tool on cast iron
325 for $\frac{3}{4}$ in. sq. tool on steel
288 for $\frac{1}{2}$ in. sq. tool on steel

Y = constant:

3 for $\frac{3}{4}$ in. sq. tool on cast iron
8 for $\frac{1}{2}$ in. sq. tool on cast iron
−2 for $\frac{3}{4}$ in. sq. tool on steel
0 for $\frac{1}{2}$ in. sq. tool on steel

Z = constant:

0 for $\frac{3}{4}$ in. sq. tool on cast iron
0.3 for $\frac{1}{2}$ in. sq. tool on cast iron
0.3 for $\frac{3}{4}$ in. sq. tool on steel
0.5 for $\frac{1}{2}$ in. sq. tool on steel

With carbon tool steel, the cutting speed is one-half of the above amount.

TABLE 2

Chart Showing Approximate Cutting Speeds in Feet per Minute for Various Machines and Materials

Material	Machine	High Speed Steel Tools	Tool Steel Tools
		Speed in Feet per Minute	Speed in Feet per Minute
Tool steel	Drill press	50–60	20–30
	Lathe	50–70	25–35
	Miller	50–60	20–30
	Shaper	40–50	20–25
	Gear cutter	....	....
	Planer	40–50	20–50
	Screw machine	60–70	25–35
Cast iron	Drill press	100–170	40–80
	Lathe	75–175	40–80
	Miller	100–150	60–80
	Shaper	80–100	50–60
	Gear cutter	60–80	30–50
	Planer	70–90	40–50
	Screw machine	100–150	50–70
Machine steel	Drill press	100–120	50–60
	Lathe	100–150	50–70
	Miller	100–125	50–70
	Shaper	60–80	50–60
	Gear cutter	60–80	30–40
	Planer	50–70	40–50
	Screw machine	100–150	50–70
Brass, bronze	Drill press	200–300	100–150
	Lathe	150–300	70–150
	Miller	150–250	80–125
	Shaper	100–120	60–80
	Gear cutter	100–125	50–60
	Planer	90–100	60–70
	Screw machine	200–300	100–150
Aluminum	Drill press	200–300	100–150
	Lathe	200–300	100–150
	Miller	200–350	100–175
	Shaper	125–200	80–125
	Gear cutter	150–200	70–100
	Planer	150–200	75–100
	Screw machine	200–300	100–150

The above speeds should be increased or decreased according to the nature of the work, tool, lubricant, machine, etc.

ILLUSTRATION: What is the proper cutting speed of a $\frac{1}{2}$ in. square high speed tool in a lathe when the depth of cut is $\frac{1}{32}$ inch and the feed per revolution is $\frac{1}{64}$ inch upon a piece of medium steel?

$$H = 1.0 \qquad D = 2 \qquad F = 1$$

$$S = 288 \qquad Y = 0 \qquad Z = 0.5$$

$$V = \frac{H \times S}{(\sqrt[3]{D + Y})\,(\sqrt[2]{F - Z})} = \frac{1 \times 288}{(\sqrt[3]{2 + 0})\,(\sqrt[2]{1 - 0.5})}$$

$$V = \frac{288}{(\sqrt[3]{2})\,(\sqrt[2]{0.5})} = \frac{288}{1.26 \times 0.71}$$

$$V = \frac{288}{0.895} = 322 \text{ ft. per minute} \quad \textbf{(Ans.)}$$

Table 2 shows approximate cutting speed for various machines and materials.

Estimating Time of Making Cut.—To find the time in minutes required to take one complete cut over a part to be turned, the following formula may be used.

$$T = \frac{L}{R \times F}$$

when
T = time in minutes
L = total length of cut in inches
R = revolutions per minute
F = feed per revolution of the machine

Feed may be expressed in terms of the distance which the cutting tool advances along the work for each revolution or stroke of the machine; for example, a feed of 0.020 inch.

This is the form to be used in the above formula. Feed may also be expressed in terms of number of revolutions per inch of side motion of the cutting tool; for instance, a feed of 100 means that the cutting tool moves one inch for each 100 revolutions of the machine or the motion per revolution is 0.010 inch

Illustration: What will be the time required to make a cut 8 inches long if the speed of the machine is 60 r.p.m. and the feed is 0.008 inch?

$$T = \frac{L}{R \times F}$$

$$= \frac{8}{60 \times 0.008} = 17 \text{ minutes} \quad \text{(Ans.)}$$

Power Required for Cutting.—The power required to remove a given amount of metal depends on the shape and sharpness of the cutting tool, hardness of the work, depth and feed of cut, lubrication of cutting point, and also upon the kind and condition of machine.

The average horsepower required to drive the machine can be determined by the product of the amount of chips (W) multiplied by two constants (Y, Z). The quantity (Y) varies with the kind of material to be cut and (Z) with the kind of machine to be used.

$$\text{Horsepower required} = YZW$$

When W = weight of metal removed in pounds per hour.

Y = constant, 1.0 for cast iron
1.3 for mild steel
2.0 for tool steel
0.7 for bronze

Z = constant, 0.035 for lathe
0.030 for shaper
0.025 for miller
0.030 for drill

Illustration: What power will be required to run a lathe at 80 r.p.m. to turn a piece of cast iron 6 inches in diameter with a $\frac{1}{32}$ inch feed and $\frac{3}{32}$ inch depth of cut?

The first problem will be to find the amount of metal removed per hour. This is represented by a ribbon $\frac{1}{32}$ inch wide, $\frac{3}{32}$ inch thick and a length represented by the cutting speed in inches per hour.

Cutting speed = $\pi \times D \times$ r.p.m. $\times 60 = \pi \times 6 \times 80 \times 60 = 28{,}800\pi$ inches per hour. This represents the length of the ribbon. The volume of the ribbon is then,

$$28{,}800\pi \times \tfrac{1}{32} \times \tfrac{3}{32} = \tfrac{3}{1024} \times 28{,}800\pi = 84.37\pi \text{ cubic inches.}$$

The weight of one cubic foot of cast iron is 450 pounds. The weight of one cubic inch is $\frac{450}{1728} = 0.26$ pound.

Weight (W) removed per hour is $84.37 \times 0.26 \times \pi = 68.91$ pounds per hour. $Y = 1.0$. $Z = 0.035$.
Then,

$$\text{horsepower} = YZW = 1 \times 0.035 \times 68.91 = 2.41 \text{ hp. (Ans.)}$$

TABLE 3

APPROXIMATE HORSEPOWER ELECTRIC MOTOR REQUIRED TO DRIVE VARIOUS TYPES OF MACHINES

Drill Presses		Shapers	
Sensitive drill up to ½ in.	¼ to ¾ hp	10 in. to 14 in. stroke....	1 to 2 hp
12 in. to 20 in..........	1 hp	16 in. to 18 in. stroke....	2 to 3 hp
24 in. to 28 in..........	2 hp	20 in. to 24 in. stroke....	3 to 5 hp
30 in. to 32 in..........	3 hp	30 in. stroke............	5 to 7½ hp

Lathes		Planers	
6 in. to 10 in..........	1 hp	22 in..................	3 hp
12 in. to 14 in..........	1 to 2 hp	24 in. to 27 in..........	3 to 5 hp
16 in. to 20 in..........	2 to 3 hp	30 in..................	5 to 7½ hp
22 in. to 27 in..........	3 to 5 hp	36 in	10 to 15 hp
30 in. to 36 in..........	7½ to 10 hp	42 in..................	15 to 20 hp

Universal Milling Machines		Gear Cutters	
No. 1.................	1 to 2 hp	36 in. × 9 in..........	2 to 3 hp
" 1½................	2 to 3 hp	48 in. × 10 in..........	3 to 5 hp
" 2.................	3 to 5 hp	60 in. × 12 in..........	5 to 7½ hp
" 3.................	5 to 7½ hp	72 in. × 14 in..........	7½ to 10 hp
" 4.................	7½ to 10 hp		

Grinders	
8 in. to 10 in. wheel..........	5 hp
12 in. to 14 in. wheel..........	7½ hp
16 in. to 20 in. wheel..........	10 hp

Taper Calculations

A piece of work is said to taper when there is a gradual and uniform increase or decrease in its diameter or thickness. Examples are, a wedge which has two plane surfaces separating at a uniform rate from the edge to the base, and a cone or lathe center (Fig. 15) whose diameter increases at a uniform rate from the apex to the base.

FIG. 15.—Wedge and Lathe Center.

Wedge-shaped pieces are used in machine design for keys to attach wheels to shafts and as tapered gibs for adjusting sliding bearings.

Conical tapers, in addition to their use on lathe centers, find a wide use on shanks of twist drills, reamers, etc. (Fig. 16.)

FIG. 16.

Amount of Taper.—The amount of taper is expressed as a certain number of inches or parts of an inch per foot and indicates a variation in diameter or thickness of that amount in twelve inches of length. For example, if a truncated cone twelve inches long is 4 inches in diameter at the small end and 5 inches in diameter at the large end, the taper is $\frac{5 - 4}{1} = 1$ inch. If another cone has end diameters of 4 inches and 5 inches, respectively, but is only six inches long, the taper is $\frac{5 - 4}{\frac{1}{2}} = 2$ inches. Tapers are also expressed in terms of degrees of the angle which one side makes with the center line axis of the work.

Standard Tapers.—Lathe centers, drilling machine spindles, tapered-shank milling cutters, and many other machine shop tools have tapers. In order to provide a degree of interchangeability of parts, machine and tool manufacturers have standardized on a few tapers which we will define.

TABLE 4

MORSE TAPERS

ANGLE OF KEY 8° 19′ = TAPER $1\frac{1}{4}$ IN 12

DETAIL DIMENSIONS

NUMBER OF TAPER			0	1	2	3	4	5	6	7
	DIAM. OF PLUG AT SMALL END	D	.252	.369	.572	.778	1.020	1.475	2.116	2.750
	DIAM. AT END OF SOCKET	A	.3561	.475	.700	.938	1.231	1.748	2.494	3.270
SHANK	WHOLE LENGTH OF SHANK	B	$2\frac{11}{32}$	$2\frac{9}{16}$	$3\frac{1}{8}$	$3\frac{7}{8}$	$4\frac{7}{8}$	$6\frac{1}{8}$	$8\frac{9}{16}$	$11\frac{5}{8}$
SHANK	SHANK DEPTH	S	$2\frac{7}{32}$	$2\frac{7}{16}$	$2\frac{15}{16}$	$3\frac{11}{16}$	$4\frac{5}{8}$	$5\frac{7}{8}$	$8\frac{1}{4}$	$11\frac{1}{4}$
	DEPTH OF HOLE	H	$2\frac{1}{32}$	$2\frac{3}{16}$	$2\frac{5}{8}$	$3\frac{1}{4}$	$4\frac{1}{8}$	$5\frac{1}{4}$	$7\frac{3}{8}$	$10\frac{1}{8}$
	STANDARD PLUG DEPTH	P	2	$2\frac{1}{8}$	$2\frac{9}{16}$	$3\frac{3}{16}$	$4\frac{1}{16}$	$5\frac{3}{16}$	$7\frac{1}{4}$	10
TONGUE	THICKNESS OF TONGUE	t	$\frac{5}{32}$	$\frac{13}{64}$	$\frac{1}{4}$	$\frac{5}{16}$	$\frac{15}{32}$	$\frac{5}{8}$	$\frac{3}{4}$	$1\frac{1}{2}$
TONGUE	LENGTH OF TONGUE	T	$\frac{1}{4}$	$\frac{3}{8}$	$\frac{7}{16}$	$\frac{9}{16}$	$\frac{5}{8}$	$\frac{3}{4}$	$1\frac{1}{8}$	$1\frac{3}{8}$
TONGUE	DIAMETER OF TONGUE	d	.235	.343	$\frac{17}{32}$	$\frac{23}{32}$	$\frac{31}{32}$	$1\frac{13}{32}$	2	$2\frac{5}{8}$
KEYWAY	WIDTH OF KEYWAY	W	.160	.213	.260	.322	.478	.635	.760	1.135
KEYWAY	LENGTH OF KEYWAY	L	$\frac{9}{16}$	$\frac{3}{4}$	$\frac{7}{8}$	$1\frac{3}{16}$	$1\frac{1}{4}$	$1\frac{1}{2}$	$1\frac{3}{4}$	$2\frac{5}{8}$
	END OF SOCKET TO KEYWAY	K	$1\frac{15}{16}$	$2\frac{1}{16}$	$2\frac{1}{2}$	$3\frac{1}{16}$	$3\frac{7}{8}$	$4\frac{15}{16}$	7	$9\frac{1}{2}$
TAPER PER FOOT			.625	.600	.602	.602	.623	.630	.626	.625
TAPER PER INCH			.05208	.05	.05016	.05016	.05191	.0525	.05216	.05208
NUMBER OF KEY			0	1	2	3	4	5	6	7

SOUTH BEND LATHE WORKS

Fig. 17.—Morse Tapers.

The *Morse* standard has a taper of approximately $\frac{5}{8}$ inch per foot. This taper is further defined as No. 1, No. 2, etc., depending on the diameter at the small end. Figure 17 and Table 4 give the chief characteristics of this taper.

Brown & Sharpe is another standard, with a taper of $\frac{1}{2}$ inch per foot. This is also specified by numbers as follows:

No. of taper...........	4	5	7	9
Diameter at small end..	0.35 in.	0.45 in.	0.60 in.	0.90 in.

Three other tapers are: *Jarno*, 0.6 inch per foot; *Sellers and Pipe* taper, $\frac{3}{4}$ inch per foot; and *Pratt & Whitney* pins, $\frac{1}{4}$ inch per foot.

Formulas for Calculating Tapers

$$\text{T.P.I.} = \frac{\text{T.P.F.}}{12} \qquad \text{T.P.F.} = \frac{12(D - d)}{l}$$

$$\text{T.P.L.} = \frac{l \times \text{T.P.F.}}{12} \qquad l = \frac{12(D - d)}{\text{T.P.F.}}$$

$$D = d + \frac{(l \times \text{T.P.F.})}{12} \qquad d = D - \frac{(l \times \text{T.P.F.})}{12}$$

when T.P.I. = taper per inch; D = larger diameter
T.P.F. = taper per foot; d = smaller diameter
T.P.L. = taper in any length; l = length of taper

ILLUSTRATION: In a taper bushing, $D = 2$ inches, $d = 1\frac{1}{2}$ inches, and $1 = 3$ inches. Find the taper per foot.

$$\text{T.P.F.} = \frac{12(D - d)}{l}$$

$$= \frac{12(2 - 1\frac{1}{2})}{3}$$

$$= \frac{12 \times 1}{2 \times 3} = 2$$

Therefore, the taper per foot is 2 inches.

ILLUSTRATION: If the taper of the shank of an end mill is 0.625 inch per foot and $D = \frac{3}{4}$ inch and $d = \frac{1}{2}$ inch, find the length of the taper.

$$l = \frac{12(D - d)}{\text{T.P.F.}}$$

$$= \frac{12(\frac{3}{4} - \frac{1}{2})}{\text{T.P.F.}}$$

$$= \frac{12 \times 1}{4 \times .625} = 4.8.$$

Therefore, the length of the taper is 4.8 inches.

Taper Turning in Lathe.—There are three ways of turning tapers on a lathe, (1) by offsetting the tailstock, (2) by using a taper attachment, and (3) by using the compound rest.

Offsetting the Tailstock.—The tailstock or dead center is moved out of alignment with the line center by means of screws on the base of the tailstock.

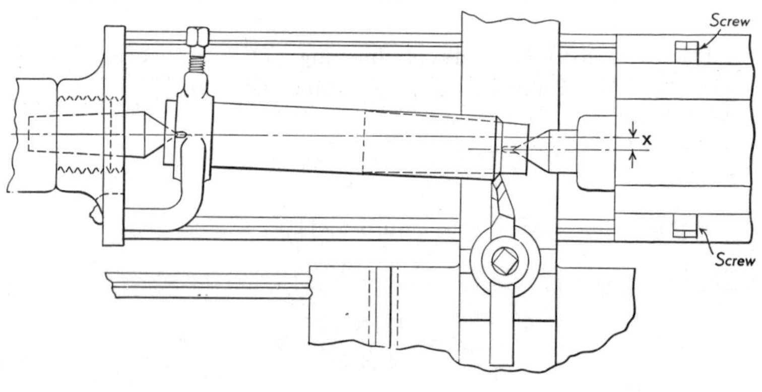

FIG. 18.

Formulas for Calculating the Amount of Offset

(*a*) When the taper runs the entire length of the bar.

$$O = \frac{D - d}{2}$$

where O = offset

ILLUSTRATION: Find the offset if a bar is to be turned taper to diameters of $1\frac{1}{2}$ inches and $\frac{7}{8}$ inch respectively.

$$O = \frac{D - d}{2}$$

$$= \frac{1\frac{1}{2} - \frac{7}{8}}{2}$$

$$= \frac{5}{8 \times 2} = \frac{5}{16}$$

Therefore the offset is $\frac{5}{16}$ inches

(*b*) When the taper runs only part of the length of a bar.

$$O = \frac{(D - d)L}{2l}$$

where L is the total length of the bar in inches or the total distance between the centers of the lathe.

ILLUSTRATION: Find the offset if a taper 3 inches long with diameter of 2 inches and $1\frac{1}{2}$ inches respectively is to be turned on a bar 12 inches long.

$$O = \frac{(D - d)L}{2l}$$

$$= \frac{(2 - 1\frac{1}{2})12}{2 \times 3}$$

$$= \frac{1 \times 12}{2 \times 2 \times 3} = 1$$

Therefore the tailstock offset is 1 inch.

(*c*) When a bar is tapered to a given taper per foot.

$$O = \frac{\text{T.P.F.} \times L}{24}$$

ILLUSTRATION: Find the offset if a T.P.F. of $\frac{1}{2}$ inch is to be turned on a bar 6 inches long.

$$O = \frac{\text{T.P.F.} \times L}{24}$$

$$= \frac{1 \times 6}{2 \times 24} = \tfrac{1}{8}$$

Therefore, the tailstock offset is $\frac{1}{8}$ inch.

NOTE: The above formulas are only exact between the ends of the centers. As Fig. 18 shows, the centers penetrate a short distance into the stock, thus the formulas give only a close approximation.

(*d*) By using a taper attachment. This device permits the tool to feed transversely at the same time that it feeds longitudinally, thus turning a taper. The guide bar is swiveled on a central pin an amount proportional to the taper, without considering the length of the stock to be turned. There are graduations at either end of the plate upon which the guide swivels indicate the amount of taper. Thus, in setting a taper attachment, only the taper per foot must be obtained.

(*e*) By using the compound rest. This part of a lathe permits the cutting tool to be set at any desired angle, thus making possible the turning of very steep tapers. The slide of the compound rest is set at the complementary angle to the angle which the taper makes with the center line of the lathe.

Taper Angle.—A steep taper is usually referred to as an angle. Angles up to 10° are commonly designated as tapers, while a larger angle is stated either as the included angle or as the angle with the center line.

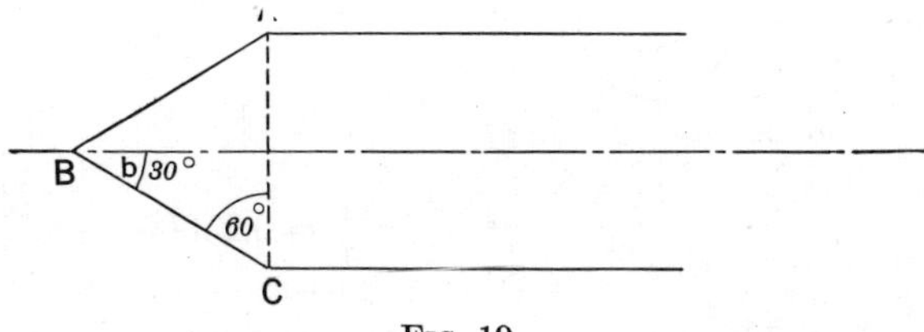

FIG. 19.

In the above sketch ABC is the included angle and the angle b is the angle with the center line.

The following formulas may be used to calculate b, the angle with the center line.

when the T.P.F. is known

$$\tan b = \frac{\text{T.P.F.}}{24}$$

when the diameters and length of the taper are known

$$\tan b = \frac{D - d}{2l}$$

ILLUSTRATION: If the taper per foot is $\frac{1}{2}$ inch, find the angle with the center line and the included angle.

$$\tan b = \frac{\text{T.P.F.}}{24}$$

$$= \frac{\frac{1}{2}}{24}$$

$$= \frac{1}{48} = 0.02083$$

From the table of tangents, 0.02083 = 1° 11′ 35″.

Therefore, b, the angle with the center line is 1° 11′ 35″ and the included angle is 2 × b or 2° 23′ 10″.

ILLUSTRATION: If $D = 1\frac{1}{8}$ inch, $d = \frac{1}{2}$ inch, and $1 = 1\frac{1}{4}$ inch, find the angle with the center line and the included angle.

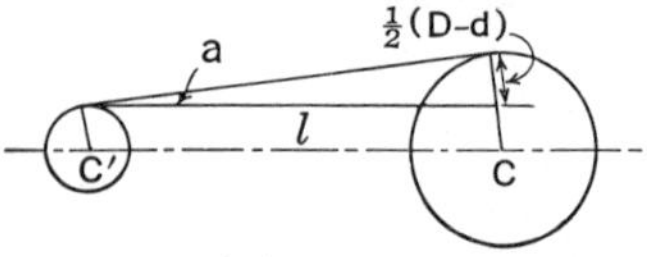

FIG. 20

$$\tan b = \frac{D - d}{2l}$$

$$= \frac{1\frac{1}{8} - \frac{1}{2}}{2 \times 1\frac{1}{4}}$$

$$= \frac{5 \times 1 \times 4}{8 \times 2 \times 5} = \frac{1}{4} = 0.25000$$

From the table of tangents 0.2500 = 14° 2′.

Therefore, b the angle with the center line is 14° 2′ and the included angle is $2 \times b$ or 28° 4′.

The table on page 533 shows tapers per foot and corresponding angles.

Measuring Tapers with a Sine Bar.—An instrument known as a *sine bar* is often used to measure the angle of a taper.

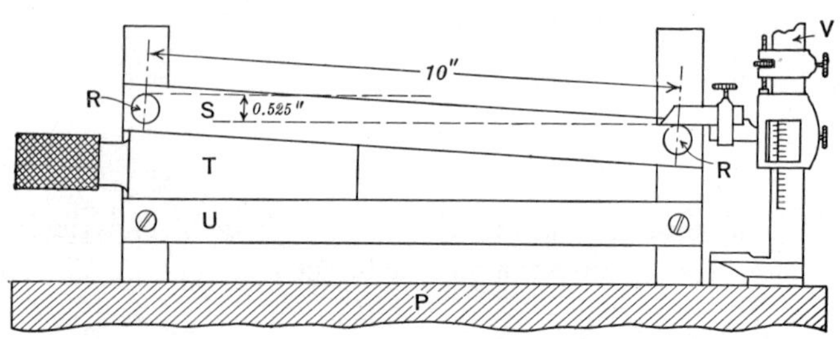

FIG. 21

P, scraped surface plate; *R*, *R*, plugs; *S*, hardened-steel sine bar; *T*, taper plug gage; *U*, straight edge; *V*, vernier height gage

The taper to be measured is placed on the straight edge U, which is parallel to the surface plate P, and the sine bar S, which has two plugs R, R set 10″ apart, is clamped along the taper. Then r, the difference in height in inches between the plugs, is found by means of the height gage V. Letting A be the included angle, we have the following formulas:

$$\sin A = \frac{r}{10} \qquad r = 10 \sin A$$

For example, in the above figure $r = 0.525''$, and we have

$$\sin A = \frac{r}{10} = \frac{0.525}{10} = 0.0525; \text{ whence } A = 3° \ 1'.$$

Therefore the included angle of the taper plug gage is 3° 1′.

TABLE 5

TAPERS AND ANGLES

Taper per Foot	Included			With Center Line			Taper	Taper per Inch from Center Line
	Deg.	Min.	Sec.	Deg.	Min.	Sec.		
1/8	0	35	48	0	17	54	0.010416	0.005203
3/16	0	53	44	0	26	52	.015625	.007812
1/4	1	11	36	0	35	48	.020833	.010416
5/16	1	29	30	0	44	45	.026042	.013021
3/8	1	47	24	0	53	42	.031250	.015625
7/16	2	5	18	1	2	39	.036458	.018229
1/2	2	23	10	1	11	35	.041667	.020833
9/16	2	41	4	1	20	32	.046875	.023438
5/8	2	59	42	1	29	51	.052084	.026042
11/16	3	16	54	1	38	27	.057292	.028646
3/4	3	34	44	1	47	22	.062500	.031250
13/16	3	52	38	1	56	19	.067708	.033854
7/8	4	10	32	2	5	16	.072917	.036456
15/16	4	28	24	2	14	12	.078125	.039063
1	4	46	18	2	23	9	.083330	.041667
1 1/4	5	57	48	2	58	54	.104666	.052084
1 1/2	7	9	10	3	34	35	.125000	.062500
1 3/4	8	20	26	4	10	13	.145833	.072917
2	9	31	36	4	45	48	.666666	.083332
2 1/2	11	53	36	5	56	48	.208333	.104166
3	14	15	0	7	7	30	.250000	.125000
3 1/2	16	35	40	8	17	50	.291666	.145833
4	18	55	28	9	27	44	.333333	.166666
4 1/2	21	14	2	10	37	1	.375000	.187500
5	23	32	12	11	46	6	.416666	.208333
6	28	4	2	14	2	1	.500000	.250000

Testing Tapers.—To test a taper for a given angle, the difference in height r of the plugs is found from the second formula, and bar S is set to this distance by means of the height gage. The taper is then tested between bars S and U.

For example, what should be the difference in height of the plugs for testing a taper which is to have an included angle of 26° 30′?

We have

$$r = 10 \sin A = 10 \times 0.4462 = 4.462.$$

Hence the difference in height of the plugs should be 4.462″. This result can be found in the table on page 576 under the column headed constant 26 degrees and opposite 30 in the column headed minutes.

Measuring Tapers with Discs. The angle of a taper may also be measured by means of two discs of unequal diameters.

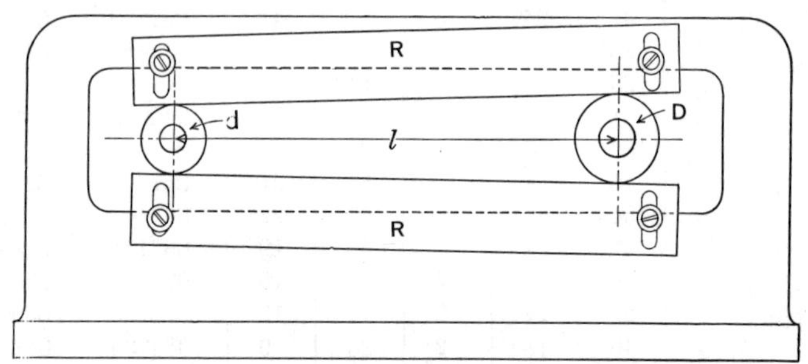

FIG. 22.—Measuring Tapers with Discs.
R, *R*, hardened-steel edges; *D*, *d*, discs of different diameters; *l*, distance between centers of discs.

The discs are placed as shown above, and the straight edges R, R, which are made of hardened steel and carefully ground, are adjusted so that the tangent lines form the taper.

Taking a as the angle with the center axis, D as the larger liameter, d as the smaller diameter, and l as the distance between he centers, as shown in the figure below, we have

$$\sin a = \frac{\frac{1}{2}(D - d)}{l},$$

whence $$\sin a = \frac{D - d}{2l}.$$

Angle a can then be found from a table of sines, and from it we can find $2a$, the included angle of the taper.

Furthermore, from the formula for $\sin a$ we have

$$l = \frac{D - d}{2 \sin a},$$

so that, given D, d, and the angle with the axis, we can find l

Screw Threads

Screw threads are familiar to every mechanic. They are used on bolts to hold pieces of machinery together, on testing machines to transmit power and in the micrometer caliper for measuring purposes. The threads on a bolt are known as " outside threads " and those in a nut as " inside threads."

The principal parts of a thread, established by the National Screw Thread Commission are shown in Fig. 25. The pro-

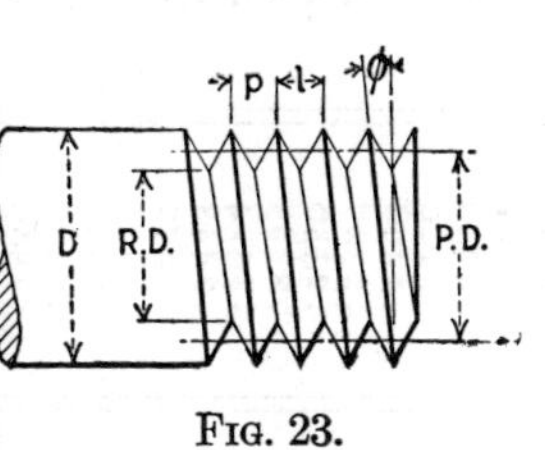

FIG. 23.

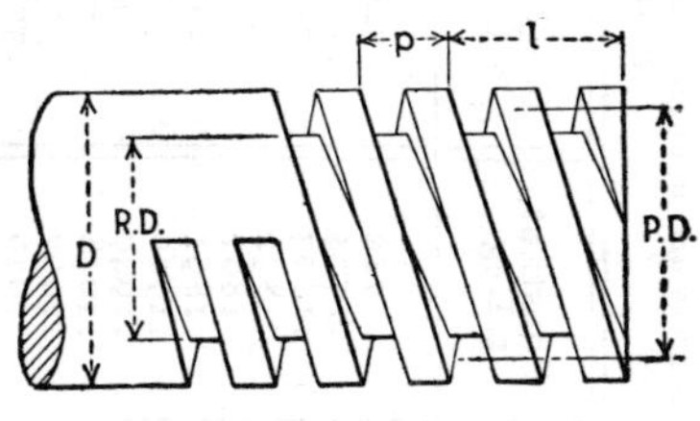

FIG. 24.—Double Square Thread.

truding edge is known as the *crest*. The base of the groove is called the *root*. The *depth of thread*, i.e., the perpendicular dis-

Table 6.—Sine Bar Table

Table 6 is calculated for degrees and minutes with sines based on a radius of 10. In the preceding problem, instead of looking up the sine of A and multiplying by 10, the table gives the result without computation.

Min.	Constant, 0 Deg.	Constant, 1 Deg.	Constant, 2 Deg.	Constant, 3 Deg.	Constant, 4 Deg.	Constant, 5 Deg.	Constant, 6 Deg.	Constant, 7 Deg.	Constant, 8 Deg.
0	.0000	.1745	.3490	.5234	.6976	.8716	1.0453	1.2187	1.3917
1	.0029	.1774	.3519	.5263	.7005	.8745	1.0482	1.2216	1.3946
2	.0058	.1803	.3548	.5292	.7034	.8774	1.0511	1.2245	1.3975
3	.0087	.1832	.3577	.5321	.7063	.8803	1.0540	1.2274	1.4004
4	.0116	.1862	.3606	.5350	.7092	.8831	1.0569	1.2302	1.4033
5	.0145	.1891	.3635	.5379	.7121	.8860	1.0597	1.2331	1.4061
6	.0175	.1920	.3664	.5408	.7150	.8889	1.0626	1.2360	1.4090
7	.0204	.1949	.3693	.5437	.7179	.8918	1.0655	1.2389	1.4119
8	.0233	.1978	.3723	.5466	.7208	.8947	1.0684	1.2418	1.4148
9	.0262	.2007	.3752	.5495	.7237	.8976	1.0713	1.2447	1.4177
10	.0291	.2036	.3781	.5524	.7266	.9005	1.0742	1.2476	1.4205
11	.0320	.2065	.3810	.5553	.7295	.9034	1.0771	1.2504	1.4234
12	.0349	.2094	.3839	.5582	.7324	.9063	1.0800	1.2533	1.4263
13	.0378	.2123	.3868	.5611	.7353	.9092	1.0829	1.2562	1.4292
14	.0407	.2132	.3897	.5640	.7382	.9121	1.0858	1.2591	1.4320
15	.0436	.2181	.3926	.5669	.7411	.9150	1.0887	1.2620	1.4349
16	.0465	.2211	.3955	.5698	.7440	.9179	1.0916	1.2649	1.4378
17	.0495	.2240	.3984	.5727	.7469	.9208	1.0945	1.2678	1.4407
18	.0524	.2269	.4013	.5756	.7498	.9237	1.0973	1.2706	1.4436
19	.0553	.2298	.4042	.5785	.7527	.9266	1.1002	1.2735	1.4464
20	.0582	.2327	.4071	.5814	.7556	.9295	1.1031	1.2764	1.4493
21	.0611	.2356	.4100	.5844	.7585	.9324	1.1060	1.2793	1 4493
22	.0640	.2385	.4129	.5873	.7614	.9353	1.1089	1.2822	1.4522
23	.0669	.2414	.4159	.5902	.7643	.9382	1.1118	1.2851	1.4551
24	.0698	.2443	.4188	.5931	.7672	.9411	1.1147	1.2880	1.4580
25	.0727	.2472	.4217	.5960	.7701	.9440	1.1176	1.2908	1.4608
26	.0756	.2501	.4246	.5980	.7730	.9469	1.1205	1.2937	1.4637
27	.0785	.2530	.4275	.6018	.7739	.9498	1.1234	1.2966	1.4666
28	.0814	.2560	.4304	.6047	.7788	.9527	1.1269	1.2995	1.4695
29	.0844	.2589	.4333	.6076	.7817	.9556	1.1291	1.3024	1.4723
30	[illegible]	[illegible]	[illegible]	[illegible]	[illegible]	[illegible]	[illegible]	[illegible]	1.4752

31	.0902	.2647	.4391	.6134	.7875	.9614	1.1349	1.3081	1.4810
32	.0931	.2676	.4420	.6163	.7904	.9642	1.1378	1.3110	1.4838
33	.0960	.2705	.4449	.6192	.7933	.9671	1.1407	1.3139	1.4867
34	.0989	.2734	.4478	.6221	.7962	.9700	1.1436	1.3168	1.4896
35	.1018	.2763	.4507	.6250	.7993	.9729	1.1465	1.3197	1.4925
36	.1047	.2792	.4536	.6279	.8020	.9758	1.1494	1.3226	1.4954
37	.1076	.2821	.4565	.6308	.8049	.9787	1.1523	1.3254	1.4982
38	.1105	.2850	.4594	.6337	.8078	.9816	1.1552	1.3283	1.5011
39	.1134	.2879	.4623	.6360	.8107	.9845	1.1580	1.3312	1.5040
40	.1164	.2908	.4653	.6395	.8136	.9874	1.1609	1.3341	1.5069
41	.1193	.2938	.4682	.6424	.8165	.9903	1.1638	1.3370	1.5097
42	.1222	.2967	.4711	.6453	.8194	.9932	1.1667	1.3399	1.5126
43	.1251	.2996	.4740	.6482	.8223	.9961	1.1696	1.3427	1.5155
44	.1280	.3025	.4769	.6511	.8252	.9990	1.1725	1.3456	1.5184
45	.1309	.3054	.4798	.6540	.8281	1.0019	1.1754	1.3485	1.5212
46	.1338	.3083	.4827	.6569	.8310	1.0048	1.1783	1.3514	1.5241
47	.1367	.3112	.4856	.6598	.8339	1.0077	1.1812	1.3543	1.5270
48	.1396	.3141	.4885	.6627	.8368	1.0106	1.1840	1.3572	1.5299
49	.1425	.3170	.4914	.6656	.8397	1.0135	1.1869	1 3600	1.5327
50	.1454	.3199	.4943	.6685	.8426	1.0164	1.1898	1.3629	1.5356
51	.1483	.3228	.4972	.6714	.8455	1.0192	1.1927	1.3658	1.5385
52	.1513	.3257	.5001	.6743	.8484	1.0221	1.1956	1.3687	1.5414
53	.1542	.3286	.5030	.6773	.8513	1.0250	1.1985	1.3716	1.5442
54	.1571	.3316	.5059	.6802	.8542	1.0279	1.2014	1.3744	1.5471
55	.1600	.3345	.5088	.6831	.8571	1.0308	1.2043	1.3773	1.5500
56	1629	.3374	.5117	.6860	.8600	1.0337	1.2071	1.3802	1.5529
57	.1658	.3403	.5146	.6889	.8629	1.0366	1.2100	1.3831	1.5557
58	.1687	.3432	.5175	.6918	.8658	1.0395	1.2129	1.3860	1.5586
59	.1716	.3461	.5205	.6947	.8687	1.0424	1.2158	1.3889	1.5615
60	.1745	.3490	.5234	.6976	.8716	1.0453	1.2187	1.3917	1.5643

TABLE 6.—SINE BAR TABLE—*Continued*

Min.	Constant, 9 Deg.	Constant, 10 Deg.	Constant, 11 Deg.	Constant, 12 Deg.	Constant, 13 Deg.	Constant, 14 Deg.	Constant, 15 Deg.	Constant, 16 Deg.	Constant, 17 Deg.
0	1.5643	1.7365	1.9081	2.0791	2.2495	2.4192	2.5882	2.7564	2.9237
1	1.5672	1.7393	1.9109	2.0820	2.2523	2.4220	2.5910	2.7592	2.9265
2	1.5701	1.7422	1.9138	2.0848	2.2552	2.4249	2.5938	2.7620	2.9293
3	1.5730	1.7451	1.9167	2.0877	2.2580	2.4277	2.5966	2.7648	2.9321
4	1.5758	1.7479	1.9195	2.0905	2.2608	2.4305	2.5994	2.7676	2.9348
5	1.5787	1.7508	1.9224	2.0933	2.2637	2.4333	2.6022	2.7704	2.9376
6	1.5816	1.7537	1.9252	2.0962	2.2665	2.4362	2.6050	2.7731	2.9404
7	1.5845	1.7565	1.9281	2.0990	2.2693	2.4390	2.6079	2.7759	2.9432
8	1.5873	1.7594	1.9309	2.1019	2.2722	2.4418	2.6107	2.7787	2.9460
9	1.5902	1.7623	1.9338	2.1047	2.2750	2.4446	2.6135	2.7815	2.9487
10	1.5931	1.7651	1.9366	2.1076	2.2778	2.4474	2.6163	2.7843	2.9515
11	1.5959	1.7680	1.9395	2.1104	2.2807	2.4503	2.6191	2.7871	2.9543
12	1.5988	1.7708	1.9423	2.1132	2.2835	2.4531	2.6219	2.7899	2.9571
13	1.6017	1.7737	1.9452	2.1161	2.2863	2.4559	2.6247	2.7927	2.9599
14	1.6046	1.7766	1.9481	2.1189	2.2892	2.4587	2.6275	2.7955	2.9626
15	1.6074	1.7794	1.9509	2.1218	2.2920	2.4615	2.6303	2.7983	2.9654
16	1.6100	1.7823	1.9538	2.1246	2.2948	2.4644	2.6331	2.8011	2.9682
17	1.6132	1.7852	1.9566	2.1275	2.2977	2.4672	2.6359	2.8039	2.9710
18	1.6160	1.7880	1.9595	2.1303	2.3005	2.4700	2.6387	2.8067	2.9737
19	1.6189	1.7909	1.9623	2.1331	2.3033	2.4728	2.6415	2.8095	2.9765
20	1.6218	1.7937	1.9652	2.1360	2.3062	2.4756	2.6443	2.8123	2.9793
21	1.6246	1.7966	1.9680	2.1388	2.3090	2.4784	2.6471	2.8150	2.9821
22	1.6275	1.7995	1.9709	2.1417	2.3118	2.4813	2.6500	2.8178	2.9849
23	1.6304	1.8023	1.9737	2.1445	2.3146	2.4841	2.6528	2.8206	2.9876
24	1.6333	1.8052	1.9766	2.1474	2.3175	2.4869	2.6556	2.8234	2.9904
25	1.6361	1.8081	1.9794	2.1502	2.3203	2.4897	2.6584	2.8262	2.9932
26	1.6390	1.8109	1.9823	2.1530	2.3231	2.4925	2.6612	2.8290	2.9960
27	1.6419	1.8138	1.9831	2.1559	2.3260	2.4954	2.6640	2.8318	2.9987
28	1.6447	1.8166	1.9880	2.1587	2.3288	2.4982	2.6668	2.8346	3.0015
29	1.6476	1.8195	1.9908	2.1616	2.3316	2.5010	2.6696	2.8374	3.0043
30	1.6505	1.8224	1.9937	2.1644	2.3345	2.5038	2.6724	2.8402	3.0071

31	1.6533	1.8252	1.9965	2.1672	3.3373	2.5066	2.6752	2.8429	3.0098
32	1.6562	1.8281	1.9994	2.1701	2.3401	2.5094	2.6780	2.8457	3.0126
33	1.6591	1.8309	2.0022	2.1729	2.3429	2.5122	2.6808	2.8485	3.0154
34	1.6620	1.8338	2.0031	2.1758	2.3458	2.5151	2.6836	2.8513	3.0182
35	1.6648	1.8367	2.0079	2.1786	2.3486	2.5179	2.6864	2.8541	3.0209
36	1.6677	1.8395	2.0108	2.1814	2.3514	2.5207	2.6892	2.8569	3.0237
37	1.6706	1.8424	2.0136	2.1843	2.3542	2.5235	2.6920	2.8597	3.0265
38	1.6734	1.8452	2.0165	2.1871	2.3571	2.5263	2.6948	2.8625	3.0292
39	1.6763	1.8481	2.0193	2.1899	2.3599	2.5291	2.6976	2.8652	3.0320
40	1.6792	1.8509	2.0222	2.1928	2.3627	2.5320	2.7004	2.8680	3.0348
41	1.6820	1.8538	2.0250	2.1956	2.3656	2.5348	2.7032	2.8708	3.0376
42	1.6849	1.8567	2.0279	2.1985	2.3684	2.5376	2.7060	2.8736	3.0403
43	1.6878	1.8595	2.0307	2.2013	2.3712	2.5404	2.7088	2.8764	3.0431
44	1.6906	1.8624	2.0336	2.2041	2.3740	2.5432	2.7116	2.8792	3.0459
45	1.6935	1.8652	2.0364	2.2070	2.3769	2.5460	2.7144	2.8820	3.0486
46	1.6964	1.8681	2.0393	2.2098	2.3797	2.5488	2.7172	2.8847	3.0514
47	1.6992	1.8710	2.0421	2.2126	2.3825	2.5516	2.7200	2.8875	3.0542
48	1.7021	1.8738	2.0450	2.2155	2.3853	2.5545	2.7228	2.8903	3.0570
49	1.7050	1.8767	2.0478	2.2183	2.3882	2.5573	2.7256	2.8931	3.0597
50	1.7078	1.8795	2.0507	2.2212	2.3910	2.5601	2.7284	2.8959	3.0625
51	1.7107	1.8824	2.0535	2.2240	2.3938	2.5629	2.7312	2.8987	3.0653
52	1.7136	1.8852	2.0563	2.2268	2.3966	2.5657	2.7340	2.9015	3.0680
53	1.7164	1.8881	2.0592	2.2297	2.3995	2.5685	2.7368	2.9042	3.0708
54	1.7193	1.8910	2.0620	2.2325	2.4023	2.5713	2.7396	2.9070	3.0736
55	1.7222	1.8938	2.0649	2.2353	2.4051	2.5741	2.7424	2.9098	3.0763
56	1.7250	1.8967	2.0677	2.2382	2.4073	2.5769	2.7452	2.9126	3.0791
57	1.7279	1.8995	2.0706	2.2410	2.4128	2.5798	2.7480	2.9154	3.0819
58	1.7308	1.9024	2.0734	2.2438	2.4136	2.5826	2.7508	2.9182	3.0846
59	1.7336	1.9052	2.0763	2.2467	2.4164	2.5854	2.7536	2.9209	3.0874
60	1.7365	1.9081	2.0791	2.2495	2.4192	2.5882	2.7564	2.9237	3.0902

Table 6.—Sine Bar Table—*Continued*

Min.	Constant, 18 Deg.	Constant, 19 Deg.	Constant, 20 Deg.	Constant, 21 Deg.	Constant, 22 Deg.	Constant, 23 Deg.	Constant, 24 Deg.	Constant, 25 Deg.	Constant, 26 Deg.
0	3.0902	3.2557	3.4202	3.5837	3.7461	3.9073	4.0674	4.2262	4.3837
1	3.0929	3.2584	3.4229	3.5864	3.7488	3.9100	4.0700	4.2288	4.3863
2	3.0957	3.2612	3.4257	3.5891	3.7515	3.9127	4.0727	4.2315	4.3889
3	3.0986	3.2639	3.4284	3.5918	3.7542	3.9153	4.0753	4.2341	4.3916
4	3.1012	3.2667	3.4311	3.5945	3.7569	3.9180	4.0780	4.2367	4.3942
5	3.1040	3.2694	3.4339	3.5973	3.7595	3.9207	4.0806	4.2394	4.3968
6	3.1068	3.2722	3.4366	3.6000	3.7622	3.9234	4.0833	4.2420	4.3994
7	3.1095	3.2749	3.4393	3.6027	3.7649	3.9260	4.0860	4.2446	4.4020
8	3.1123	3.2777	3.4421	3.6054	3.7676	3.9287	4.0886	4.2473	4.4046
9	3.1151	3.2804	3.4448	3.6081	3.7703	3.9314	4.0913	4.2499	4.4072
10	3.1178	3.2832	3.4475	3.6108	3.7730	3.9341	4.0939	4.2525	4.4098
11	3.1206	3.2859	3.4503	3.6135	3.7757	3.9367	4.0966	4.2552	4.4124
12	3.1233	3.2887	3.4530	3.6162	3.7784	3.9394	4.0992	4.2578	4.4151
13	3.1261	3.2914	3.4557	3.6190	3.7811	3.9421	4.1019	4.2604	4.4177
14	3.1289	3.2942	3.4584	3.6217	3.7838	3.9448	4.1045	4.2631	4.4203
15	3.1316	3.2969	3.4612	3.6244	3.7865	3.9474	4.1072	4.2657	4.4229
16	3.1344	3.2997	3.4639	3.6271	3.7892	3.9501	4.1098	4.2683	4.4255
17	3.1372	3.3024	3.4666	3.6298	3.7919	3.9528	4.1125	4.2709	4.4281
18	3.1399	3.3051	3.4694	3.6325	3.7946	3.9555	4.1151	4.2736	4.4307
19	3.1427	3.3079	3.4721	3.6352	3.7973	3.9581	4.1178	4.2762	4.4333
20	3.1454	3.3106	3.4748	3.6379	3.7999	3.9608	4.1204	4.2788	4.4359
21	3.1482	3.3134	3.4775	3.6406	3.8026	3.9635	4.1231	4.2815	4.4385
22	3.1510	3.3161	3.4803	3.6434	3.8053	3.9661	4.1257	4.2841	4.4411
23	3.1537	3.3189	3.4830	3.6461	3.8080	3.9688	4.1284	4.2867	4.4437
24	3.1565	3.3216	3.4857	3.6488	3.8107	3.9715	4.1310	4.2894	4.4464
25	3.1593	3.3244	3.4884	3.6515	3.8134	3.9741	4.1337	4.2920	4.4490
26	3.1620	3.3271	3.4912	3.6542	3.8161	3.9768	4.1363	4.2946	4.4516
27	3.1648	3.3298	3.4939	3.6569	3.8188	3.9795	4.1390	4.2972	4.4542
28	3.1675	3.3326	3.4966	3.6586	3.8215	3.9822	4.1416	4.2999	4.4568
29	3.1703	3.3353	3.4993	3.6623	3.8241	3.9848	4.1443	4.3025	4.4594
30	3.1730	3.3381	3.5021	3.6650	3.8268	3.9875	4.1469	4.3051	4.4620

31	3.1758	3.3408	3.5048	3.6677	3.8295	3.9902	4.1496	4.3077	4.4646
32	3.1786	3.3436	3.5075	3.6704	3.8322	3.9928	4.1522	4.3104	4.4672
33	3.1813	3.3463	3.5102	3.6731	3.8349	3.9955	4.1549	4.3130	4.4698
34	3.1841	3.3490	3.5130	3.6758	3.8376	3.9982	4.1575	4.3156	4.4724
35	3.1868	3.3518	3.5157	3.6785	3.8403	4.0008	4.1602	4.3182	4.4750
36	3.1896	3.3545	3.5184	3.6812	3.8430	4.0035	4.1628	4.3209	4.4776
37	3.1923	3.3573	3.5211	3.6839	3.8456	4.0062	4.1655	4.3235	4.4802
38	3.1951	3.3600	3.5239	3.6867	3.8483	4.0088	4.1681	4.3261	4.4828
39	3.1979	3.3627	3.5266	3.6894	3.8510	4.0115	4.1707	4.3287	4.4854
40	3.2006	3.3655	3.5293	3.6921	3.8537	4.0141	4.1734	4.3313	4.4880
41	3.2034	3.3682	3.5320	3.6948	3.8564	4.0168	4.1760	4.3340	4.4906
42	3.2061	3.3710	3.5347	3.6975	3.8591	4.0195	4.1787	4.3366	4.4932
43	3.2089	3.3737	3.5375	3.7002	3.8617	4.0221	4.1813	4.3392	4.4958
44	3.2116	3.3764	3.5402	3.7029	3.8644	4.0248	4.1840	4.3418	4.4984
45	3.2144	3.3792	3.5429	3.7056	3.8671	4.0275	4.1866	4.3445	4.5010
46	3.2171	3.3819	3.5456	3.7083	3.8698	4.0301	4.1892	4.3471	4.5036
47	3.2199	3.3846	3.5484	3.7110	3.8725	4.0328	4.1919	4.3497	4.5062
48	3.2227	3.3874	3.5511	3.7137	3.8752	4.0355	4.1945	4.3523	4.5088
49	3.2254	3.3901	3.5538	3.7164	3.8778	4.0381	4.1972	4.3549	4.5114
50	3.2282	3.3929	3.5565	3.7191	3.8805	4.0408	4.1998	4.3575	4.5140
51	3.2309	3.3956	3.5592	3.7218	3.8832	4.0434	4.2024	4.3602	4.5166
52	3.2337	3.3983	3.5619	3.7245	3.8859	4.0461	4.2051	4.3628	4.5192
53	3.2364	3.4011	3.5647	3.7272	3.8886	4.0488	4.2077	4.3654	4.5218
54	3.2392	3.4038	3.5674	3.7299	3.8912	4.0514	4.2104	4.3680	4.5243
55	3.2419	3.4065	3.5701	3.7326	3.8939	4.0541	4.2130	4.3706	4.5269
56	3.2447	3.4093	3.5728	3.7353	3.8966	4.0567	4.2156	4.3733	4.5295
57	3.2474	3.4120	3.5755	3.7380	3.8993	4.0594	4.2183	4.3759	4.5321
58	3.2502	3.4147	3.5782	3.7407	3.9020	4.0621	4.2209	4.3785	4.5347
59	3.2529	3.4175	3.5810	3.7434	3.9046	4.0647	4.2235	4.3811	4.5373
60	3.2557	3.4202	3.5837	3.7461	3.9073	4.0674	4.2262	4.3837	4.5399

TABLE 6.—SINE BAR TABLE—*Continued*

Min.	Constant, 27 Deg.	Constant, 28 Deg.	Constant, 29 Deg.	Constant, 30 Deg.	Constant, 31 Deg.	Constant, 32 Deg.	Constant, 33 Deg.	Constant, 34 Deg.	Constant, 35 Deg.
0	4.5399	4.6947	4.8481	5.0000	5.1504	5.2992	5.4464	5.5919	5.7358
1	4.5425	4.6973	4.8506	5.0025	5.1529	5.3017	5.4488	5.5943	5.7381
2	4.5451	4.6999	4.8532	5.0050	5.1554	5.3041	5.4513	5.5968	5.7405
3	4.5477	4.7024	4.8557	5.0076	5.1579	5.3066	5.4537	5.5992	5.7429
4	4.5503	4.7050	4.8583	5.0101	5.1604	5.3091	5.4561	5.6016	5.7453
5	4.5529	4.7076	4.8608	5.0126	5.1628	5.3115	5.4586	5.6040	5.7477
6	4.5554	4.7101	4.8634	5.0151	5.1653	5.3140	5.4610	5.6064	5.7501
7	4.5580	4.7127	4.8659	5.0176	5.1678	5.3164	5.4635	5.6088	5.7524
8	4.5606	4.7153	4.8684	5.0201	5.1703	5.3189	5.4659	5.6112	5.7548
9	4.5632	4.7178	4.8710	5.0227	5.1728	5.3214	5.4683	5.6136	5.7572
10	4.5658	4.7204	4.8730	5.0252	5.1753	5.3238	5.4708	5.6160	5.7596
11	4.5684	4.7229	4.8761	5.0277	5.1778	5.3263	5.4732	5.6184	5.7619
12	4.5710	4.7255	4.8786	5.0302	5.1803	5.3288	5.4756	5.6208	5.7643
13	4.5736	4.7281	4.8811	5.0327	5.1828	5.3312	5.4781	5.6232	5.7667
14	4.5762	4.7306	4.8837	5.0352	5.1852	5.3337	5.4805	5.6256	5.7691
15	4.5787	4.7332	4.8852	5.0377	5.1877	5.3361	5.4829	5.6280	5.8715
16	4.5813	4.7358	4.8888	5.0403	5 1902	5.3386	5.4854	5.6305	5.7738
17	4.5839	4.7383	4.8913	5.0428	5.1927	5.3411	5.4878	5.6329	5.7762
18	4.5865	4.7409	4.8938	5.0453	5.1952	5.3435	5.4902	5.6353	5.7786
19	4.5891	4.7434	4.8964	5.0478	5.1977	5.3460	5.4927	5.6377	5.7810
20	4.5917	4.7460	4.8989	5.0503	5.2002	5.3484	5.4951	5.6401	5.7833
21	4.5942	4.7486	4.9014	5.0528	5.2096	5.3509	5.4975	5.6425	5.7857
22	4.5968	4.7511	4.9040	5.0553	5.2051	5.3534	5.4999	5.6449	5.7881
23	4.5994	4.7537	4.9065	5.0578	5.2076	5.3558	5.5024	5.6473	5.7904
24	4.6020	4.7562	4.9090	5.0603	5.2101	5.3583	5.5048	5.6497	5.7928
25	4 6046	4.7588	4.9116	5.0628	5.2126	5.3607	5.5072	5.6521	5.7952
26	4.6072	4.7614	4.9141	5.0654	5.2151	5.3632	5.5097	5.6545	5.7976
27	4.6097	4.7639	4.9166	5.0679	5.2175	5.3656	5.5121	5.6569	5.7999
28	4.6123	4.7665	4.9192	5.0704	5.2200	5.3681	5.5145	5.6593	5.8023
29	4.6149	4.7690	4.9217	5.0729	5.2225	5.3705	5.5169	5.6617	5.8047
30	4.6175	4.7716	4.9242	5.0754	5.2250	5.3730	5.5194	5.6641	5.8070

31	4.6201	4.7741	4.9268	5.0779	5.2275	5.3754	5.5218	5.6665	5.8094
32	4.6226	4.7767	4.9293	5.0804	5.2299	5.3779	5.5242	5.6689	5.8118
33	4.6252	4.7793	4.9318	5.0829	5.2324	5.3804	5.5266	5.6713	5.8141
34	4.6278	4.7818	4.9344	5.0854	5.2349	5.3828	5.5291	5.6736	5.8165
35	4.6304	4.7844	4.9369	5.0879	5.2374	5.3853	5.5315	5.6760	5.8189
36	4.6330	4.7869	4.9394	5.0904	5.2399	5.3877	5.5339	5.6784	5.8212
37	4.6355	4.7895	4.9419	5.0929	5.2423	5.3902	5.5363	5.6808	5.8236
38	4.6381	4.7920	4.9445	5.0954	5.2448	5.3926	5.5388	5.6832	5.8260
39	4.6407	4.7946	4.9470	5.0979	5.2473	5.3951	5.5412	5.6856	5.8283
40	4.6433	4.7971	4.9495	5.1004	5.2498	5.3975	5.5436	5.6880	5.8307
41	4.6458	4.7997	4.9521	5.1029	5.2522	5.4000	5.5460	5.6904	5.8330
42	4.6484	4.8022	4.9546	5.1054	5.2547	5.4024	5.5484	5.6928	5.8354
43	4.6510	4.8048	4.9571	5.1079	5.2572	5.4049	5.5509	5.6952	5.8378
44	4.6536	4.8073	4.9596	5.1104	5.2597	5.4073	5.5533	5.6976	5.8401
45	4.6561	4.8099	4.9622	5.1129	5.2621	5.4097	5.5557	5.7000	5.8426
46	4.6587	4.8124	4.9647	5.1154	5.2646	5.4122	5.5581	5.7024	5.8449
47	4.6613	4.8150	4.9672	5.1179	5.2671	5.4146	5.5605	5.7047	5.8472
48	4.6639	4.8175	4.9697	5.1204	5.2696	5.4171	5.5630	5.7071	5.8396
49	4.6664	4.8201	4.9723	5.1229	5.2720	5.4195	5.5654	5.7095	5.8519
50	4.6690	4.8226	4.9748	5.1254	5.2745	5.4220	5.5678	5.7119	5.8543
51	4.6716	4.8252	4.9773	5.1279	5.2770	5.4244	5.5702	5.7143	5.8567
52	4.6742	4.8277	4.9798	5.1304	5.2794	5.4269	5.5726	5.7167	5.8590
53	4.6767	4.8303	4.9824	5.1329	5.2819	5.4293	5.5750	5.7101	5.8614
54	4.6793	4.8328	4.9849	5.1354	5.2844	5.4317	5.5775	5.7215	5.8637
55	4.6819	4.8354	4.9874	5.1379	5.2869	5.4342	5.5799	5.7238	5.8661
56	4.6844	4.8379	4.9899	5.1404	5.2893	5.4366	5.5823	5.7262	5.8684
57	4.6870	4.8405	4.9924	5.1429	5.2918	5.4391	5.5847	5.7286	5.8708
58	4.6896	4.8430	4.9950	5.1454	5.2943	5.4415	5.5871	5.7310	5.8731
59	4.6921	4.8456	4.9975	5.1479	5.2967	5.4440	5.5895	5.7334	5.8755
60	4.6947	4.8481	5.0000	5.1504	5.2992	5.4464	5.5919	5.7358	5.8779

TABLE 6.—SINE BAR TABLE—*Continued*

Min.	Constant, 36 Deg.	Constant, 37 Deg.	Constant, 38 Deg.	Constant, 39 Deg.	Constant, 40 Deg.	Constant, 41 Deg.	Constant, 42 Deg.	Constant, 43 Deg.	Constant, 44 Deg.
0	5.8779	6.0182	6.1566	6.2932	6.4279	6.5606	6.6913	6.8200	6.9466
1	5.8802	6.0205	6.1589	6.2955	6.4301	6.5628	6.6935	6.8221	6.9487
2	5.8826	6.0228	6.1612	6.2977	6.4323	6.5650	6.6956	6.8242	6.9508
3	5.8849	6.0251	6.1635	6.3000	6.4346	6.5672	6.6978	6.8264	6.9529
4	5.8873	6.0274	6.1658	6.3022	6.4368	6.5694	6.6999	6.8285	6.9549
5	5.8896	6.0298	6.1681	6.3045	6.4390	6.5716	6.7021	6.8306	6.9570
6	5.8920	6.0321	6.1704	6.3068	6.4412	6.5738	6.7043	6.8327	6.9591
7	5.8943	6.0344	6.1726	6.3090	6.4435	6.5759	6.7064	6.8349	6.9612
8	5.8967	6.0367	6.1749	6.2113	6.4457	6.5781	6.7086	6.8370	6.9633
9	5.8990	6.0390	6.1772	6.3136	6.4479	6.5803	6.7107	6.8391	6.9654
10	5.9014	6.0414	6.1795	6.3158	6.4501	6.5825	6.7129	6.8412	6.9675
11	5.9037	6.0437	6.1818	6.3180	6.4524	6.5847	6.7151	6.8434	6.9690
12	5.9061	6.0460	6.1841	6.3203	6.4546	6.5869	6.7172	6.8455	6.9717
13	5.9084	6.0483	6.1864	6.3225	6.4568	6.5891	6.7194	6.8476	6.9737
14	5.9108	6.0506	6.1887	6.3248	6.4590	6.5913	6.7215	6.8497	6.9758
15	5.9131	6.0529	6.1909	6.3271	6.4612	6.5935	6.7237	6.8518	6.9779
16	5.9154	6.0553	6.1932	6.3293	6.4635	6.5956	6.7258	6.8539	6.9800
17	5.9178	6.0576	6.1965	6.3316	6.4657	6.5978	6.7280	6.8561	6.9821
18	5.9201	6.0599	6.1978	6.3338	6.4679	6.6000	6.7301	6.8582	6.9842
19	5.9225	6.0622	6.2001	6.3361	6.4701	6.6022	6.7323	6.8603	6.9862
20	5.9248	6.0645	6.2024	6.3383	6.4723	6.6044	6.7344	6.8624	6.9885
21	5.9272	6.0668	6.2046	6.3400	6.4746	6.6066	6.7366	6.8645	6.9904
22	5.9295	6.0691	6.2069	6.3428	6.4768	6.6088	6.7387	6.8666	6.9925
23	5.9318	6.0714	6.2092	6.3451	6.4790	6.6109	6 7409	6.8688	6.9946
24	5.9342	6.0738	6.2115	6.3473	6.4812	6.6131	6.7430	6.8709	6.9966
25	5.9365	6.0761	6.2138	6.3496	6.4834	6.6153	6.7452	6.8730	6.9987
26	5.9389	6.0784	6.2160	6.3518	6.4856	6.6175	6.7473	6.8751	7.0008
27	5.9412	6.0807	6.2183	6.3540	6.4878	6.6197	6.7495	6.8772	7.0029
28	5.9436	6.0830	6.2206	6.3563	6.4901	6.6218	6.7516	6.8793	7 0049
29	5.9459	6.0853	6.2229	6.3585	6.4923	6.6240	6.7538	6.8814	7 0070
30	5.9482	6.0876	6.2251	6.3608	6.4945	6 6262	6.7559	6 8835	7.0091

31	5.9506	6.0899	6.2274	6.3630	6.4967	6.6284	6.7580	6.8857	7.0112
32	5.9529	6.0922	6.2297	6.3653	6.4989	6.6306	6.7602	6.8878	7.0132
33	5.9552	6.0945	6.2320	6.3675	6.5011	6.6327	6.7623	6.8899	7.0153
34	5.9576	6.0968	6.2342	6.3698	6.5033	6.6349	6.7645	6.8920	7.0174
35	5.9599	6.0991	6.2365	6.3720	6.5055	6.6371	6.7666	6.8941	7.0195
36	5.9622	6.1015	6.2388	6.3742	6.5077	6.6393	6.7688	6.8962	7.0215
37	5.9646	6.1038	6.2411	6.3765	6.5100	6.6414	6.7709	6.8983	7.0236
38	5.9669	6.1061	6.2433	6.3787	6.5122	6.6436	6.7730	6.9004	7.0257
39	5.9693	6.1084	6.2456	6.3810	6.5144	6.6458	6.7752	6.9025	7.0277
40	5.9716	6.1107	6.2479	6.3832	6.5166	6.6480	6.7773	6.9046	7.0298
41	5.9739	6.1130	6.2502	6.3854	6.5188	6.6501	6.7795	6.9067	7.0319
42	5.9763	6.1153	6.2524	6.3877	6.5210	6.6523	6.7816	6.9088	7.0339
43	5.9786	6.1176	6.2547	6.3899	6.5232	6.6545	6.7837	6.9109	7.0360
44	5.9809	6.1199	6.2570	6.3922	6.5254	6.6566	6.7859	6.9130	7.0381
45	5.9832	6.1222	6.2592	6.3944	6.5276	6.6588	6.7880	6.9151	7.0401
46	5.9856	6.1245	6.2615	6.3966	6.5298	6.6610	6.7901	6.9172	7.0422
47	5.9879	6.1268	6.2638	6.2989	6.5320	6.6632	6.7923	6.9193	7.0443
48	5.9902	6.1291	6.2660	6.4011	6.5342	6.6653	6.7944	6.9214	7.0463
49	5.9926	6.1314	6.2683	6.4033	6.5364	6.6675	6.7965	6.9235	7.0484
50	5.9949	6.1337	6.2706	6.4056	6.5386	6.6697	6.7987	6.9256	7.0505
51	5.9972	6.1360	6.2728	6.4078	6.5408	6.6718	6.8008	6.9277	7.0525
52	5.9995	6.1383	6.2751	6.4100	6.5430	6.6740	6.8029	6.9298	7.0546
53	6.0019	6.1406	6.2774	6.4123	6.5452	6.6762	6.8051	6.9319	7.0567
54	6.0042	6.1429	6.2796	6.4145	6.5474	6.6783	6.8072	6.9340	7.0587
55	6.0065	6.1451	6.2819	6.4167	6.5496	6.6805	6.8093	6.9361	7.0608
56	6.0089	6.1474	6.2842	6.4190	6.5518	6.6827	6.8115	6.9382	7.0628
57	6.0112	6.1497	6.2864	6.4212	6.5540	6.6848	6.8136	6.9403	7.0649
58	6.0135	6.1520	6.2887	6.4234	6.5562	6.6870	6.8157	6.9424	7.0670
59	6.0158	6.1543	6.2909	6.4256	6.5584	6.6891	6.8179	6.9445	7.0690
60	6.0182	6.1566	6.2932	6.4279	6.5606	6.6913	6.8200	6.9466	7.0711

tance from the crest to the bottom of the groove, is represented by H in Fig. 25. Twice the depth is called the double depth. The diameter measured over the crests is the *outside diameter* or *major diameter* (indicated by D in Fig. 23). The diameter measured at the root is the *root diameter* or *minor diameter*. The *pitch diameter* (indicated by PD in Fig. 23) is the diameter measured between the mid-points between the crest and the root of

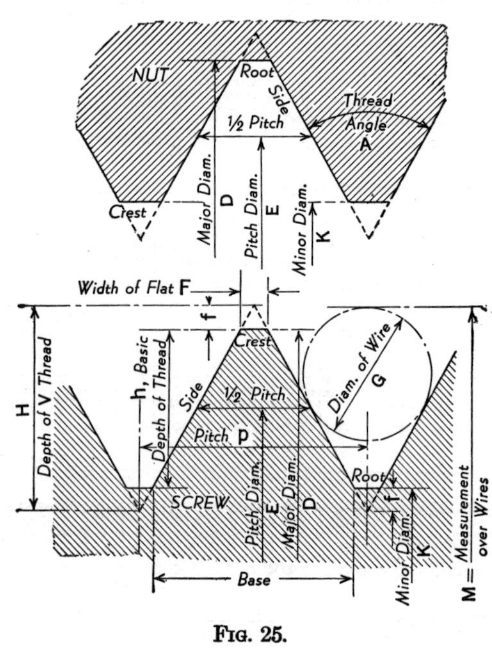

FIG. 25.

the thread. It is equal to $D-d$. The *pitch*, P, is the longitudinal distance between any point on one thread and the corresponding point on the adjacent thread. *Lead*, l, is the distance which a screw advances when turned one complete revolution. In a single-thread screw it is equal to the pitch; in a double-thread screw it is twice the pitch, etc. (See Figs. 23 and 24.)

Symbols

For use in formulas for expressing relations of screw threads, and for use on drawings and for similar purposes, the following symbols should be used:

Major diameter	D
Corresponding radius	d
Pitch diameter	E
Corresponding radius	e
Minor diameter	K
Corresponding radius	k
Angle of thread	A
One half angle of thread	a
Number of turns per inch	N
Number of threads per inch	n
Lead	$L = \frac{1}{N}$
Pitch or thread interval	$p = \frac{1}{n}$
Helix angle	s
Tangent of helix angle	$S = \frac{L}{3.14159 \times E}$
Width of basic flat at top, crest, or root	F
Depth of basic truncation	f
Depth of sharp V thread	H
Depth of American National form of thread	h
Length of engagement	Q
Included angle of taper	Y
One half included angle of taper	y

There are different forms of screw threads—Sharp V, American National, Whitworth, Square, Acme, American National Pipe, etc. The methods of calculating the elements of these threads are shown in the following pages.

Sharp V Thread.—This thread is shown in Figs. 23 and 26. The sides of the thread form an angle of 60 degrees with each other and are theoretically sharp at the top and bottom.

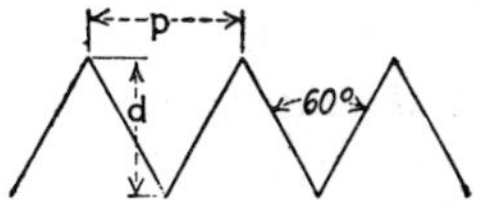

p
d
60°

Fig. 26.

The pitch and depth of the thread are found by the following formulas:

$$\text{Pitch} = \frac{1}{\text{No. of threads per inch}}$$

$$\text{Depth} = \frac{0.866}{\text{No. of threads per inch}} \text{ or } 0.8660 \times \text{pitch}$$

ILLUSTRATION: What is the depth of a V-thread of $\frac{1}{8}$-inch pitch?

$$\text{Depth} = 0.866 \text{ pitch} = 0.866 \times \tfrac{1}{8} = 0.108.$$

Therefore the depth is 0.108 inch.

ILLUSTRATION: What is the root diameter of a $\frac{3}{4}$ inch × 10-V thread? ($\frac{3}{4}$ inch × 10 means 1 inch in diameter and 10 threads to the inch). From the figures 17 and 18 it is evident that the root diameter is equal to the outside or major diameter minus the double depth of the thread.

$$\text{depth} = 0.866 \times \tfrac{1}{10} = 0.0866,\ 0.0866 \times 2 = 0.173$$

$$\text{R.D.} = 0.750 - 0.173 = 0.577 \text{ in. (Ans.)}$$

Unified and American Thread.—After World War II the need was felt for a screw thread system which would permit interchangeability of parts with other nations. Representatives of the United Kingdom, Canada, and the United States set up what is known as a Unified Thread Series to obtain this interchangeability.

The thread system agreed upon permits some minor variations in British and American practice but still permits the desired interchangeability. The Unified Thread Series can be identified in Tables 7 and 8 by the bold face type. The sides of the thread

TABLE 7a

Unified and American Standard Coarse-thread Series—Basic Dimensions

Sizes	Basic Major Diameter,	Thds. per Inch,	Basic Pitch Diameter,	Minor Diameter Ext. Thds.	Minor Diameter Int. Thds.	Lead Angle at Basic Pitch Diameter,		Section at Minor Diameter	Tensile Stress Area
	Inches		Inches	Inches	Inches	Deg	Min	Sq In.	Sq In.
1 *(.073)*	0.0730	64	0.0629	0.0538	0.0561	4	31	0.0022	0.0026
2 *(.086)*	0.0860	56	0.0744	0.0641	0.0667	4	22	0.0031	0.0036
3 *(.099)*	0.0990	48	0.0855	0.0734	0.0764	4	26	0.0041	0.0048
4 *(.112)*	0.1120	40	0.0958	0.0813	0.0849	4	45	0.0050	0.0060
5 *(.125)*	0.1250	40	0.1088	0.0943	0.0979	4	11	0.0067	0.0079
6 *(.138)*	0.1380	32	0.1177	0.0997	0.1042	4	50	0.0075	0.0090
8 *(.164)*	0.1640	32	0.1437	0.1257	0.1302	3	58	0.0120	0.0139
10 *(.190)*	0.1900	24	0.1629	0.1389	0.1449	4	39	0.0145	0.0174
12 *(.216)*	0.2160	24	0.1889	0.1649	0.1709	4	1	0.0206	0.0240
1/4	**0.2500**	**20**	**0.2175**	**0.1887**	**0.1959**	**4**	**11**	**0.0269**	**0.0317**
5/16	**0.3125**	**18**	**0.2764**	**0.2443**	**0.2524**	**3**	**40**	**0.0454**	**0.0522**
3/8	**0.3750**	**16**	**0.3344**	**0.2983**	**0.3073**	**3**	**24**	**0.0678**	**0.0773**
7/16	**0.4375**	**14**	**0.3911**	**0.3499**	**0.3602**	**3**	**20**	**0.0933**	**0.1060**
1/2	**0.5000**	**13**	**0.4500**	**0.4056**	**0.4167**	**3**	**7**	**0.1257**	**0.1416**
9/16	**0.5625**	**12**	**0.5084**	**0.4603**	**0.4723**	**2**	**59**	**0.1620**	**0.1816**
5/8	**0.6250**	**11**	**0.5660**	**0.5135**	**0.5266**	**2**	**56**	**0.2018**	**0.2256**
3/4	**0.7500**	**10**	**0.6850**	**0.6273**	**0.6417**	**2**	**40**	**0.3020**	**0.3340**
7/8	**0.8750**	**9**	**0.8028**	**0.7387**	**0.7547**	**2**	**31**	**0.4193**	**0.4612**
1	**1.0000**	**8**	**0.9188**	**0.8466**	**0.8647**	**2**	**29**	**0.5510**	**0.6051**
1 1/8	**1.1250**	**7**	**1.0322**	**0.9497**	**0.9704**	**2**	**31**	**0.6931**	**0.7627**
1 1/4	**1.2500**	**7**	**1.1572**	**1.0747**	**1.0954**	**2**	**15**	**0.8898**	**0.9684**
1 3/8	**1.3750**	**6**	**1.2667**	**1.1705**	**1.1946**	**2**	**24**	**1.0541**	**1.1538**
1 1/2	**1.5000**	**6**	**1.3917**	**1.2955**	**1.3196**	**2**	**11**	**1.2938**	**1.4041**
1 3/4	**1.7500**	**5**	**1.6201**	**1.5046**	**1.5335**	**2**	**15**	**1.7441**	**1.8983**
2	**2.0000**	**4 ½**	**1.8557**	**1.7274**	**1.7594**	**2**	**11**	**2.3001**	**2.4971**
2 1/4	**2.2500**	**4 ½**	**2.1057**	**1.9774**	**2.0094**	**1**	**55**	**3.0212**	**3.2464**
2 1/2	**2.5000**	**4**	**2.3376**	**2.1933**	**2.2294**	**1**	**57**	**3.7161**	**3.9976**
2 3/4	**2.7500**	**4**	**2.5876**	**2.4433**	**2.4794**	**1**	**46**	**4.6194**	**4.9326**
3	**3.0000**	**4**	**2.8376**	**2.6933**	**2.7294**	**1**	**36**	**5.6209**	**5.9659**
3 1/4	**3.2500**	**4**	**3.0876**	**2.9433**	**2.9794**	**1**	**29**	**6.7205**	**7.0992**
3 1/2	**3.5000**	**4**	**3.3376**	**3.1933**	**3.2294**	**1**	**22**	**7.9183**	**8.3268**
3 3/4	**3.7500**	**4**	**3.5876**	**3.4433**	**3.4794**	**1**	**16**	**9.2143**	**9.6546**
4	**4.0000**	**4**	**3.8376**	**3.6933**	**3.7294**	**1**	**11**	**10.6084**	**11.0805**

Courtesy American Society of Mechanical Engineers.

In British practice, the term "Effective Diameter" is used instead of "Pitch Diameter."

The area designated as "Stress Area" is based upon a diameter that is a mean of the pitch and minor diameters to allow for the strengthening effect of the threads.

Figures in bold type indicate Unified threads.

TABLE 7b

UNIFIED AND AMERICAN STANDARD FINE-THREAD SERIES—BASIC DIMENSIONS

Sizes	Basic Major Diameter,	Thds. per Inch,	Basic Pitch Diameter,	Minor Diameter Ext. Thds.	Minor Diameter Int. Thds.	Lead Angle at Basic Pitch Diameter,		Section at Minor Diameter	Tensile Stress Area
	Inches		Inches	Inches	Inches	Deg	Min	Sq In.	Sq In.
0 (.060)	0.0600	80	0.0519	0.0447	0.0465	4	23	0.0015	0.0018
1 (.073)	0.0730	72	0.0640	0.0560	0.0580	3	57	0.0024	0.0027
2 (.086)	0.0860	64	0.0759	0.0668	0.0691	3	45	0.0034	0.0039
3 (.099)	0.0990	56	0.0874	0.0771	0.0797	3	43	0.0045	0.0052
4 (.112)	0.1120	48	0.0985	0.0864	0.0894	3	51	0.0057	0.0065
5 (.125)	0.1250	44	0.1102	0.0971	0.1004	3	45	0.0072	0.0082
6 (.138)	0.1380	40	0.1218	0.1073	0.1109	3	44	0.0087	0.0101
8 (.164)	0.1640	36	0.1460	0.1299	0.1339	3	28	0.0128	0.0146
10 (.190)	0.1900	32	0.1697	0.1517	0.1562	3	21	0.0175	0.0199
12 (.216)	0.2160	28	0.1928	0.1722	0.1773	3	22	0.0226	0.0257
1/4	0.2500	28	0.2268	0.2062	0.2113	2	52	0.0326	0.0362
5/16	0.3125	24	0.2854	0.2614	0.2674	2	40	0.0524	0.0579
3/8	0.3750	24	0.3479	0.3239	0.3299	2	11	0.0809	0.0876
7/16	0.4375	20	0.4050	0.3762	0.3834	2	15	0.1090	0.1185
1/2	0.5000	20	0.4675	0.4387	0.4459	1	57	0.1486	0.1597
9/16	0.5625	18	0.5264	0.4943	0.5024	1	55	0.1888	0.2026
5/8	0.6250	18	0.5889	0.5568	0.5649	1	43	0.2400	0.2555
3/4	0.7500	16	0.7094	0.6733	0.6823	1	36	0.3513	0.3724
7/8	0.8750	14	0.8286	0.7874	0.7977	1	34	0.4805	0.5088
1	1.0000	14	0.9536	0.9124	0.9227	1	22	0.6464	0.6791
1	1.0000	12	0.9459	0.8978	0.9098	1	36	0.6245	0.6624
1 1/8	1.1250	12	1.0709	1.0228	1.0348	1	25	0.8118	0.8549
1 1/4	1.2500	12	1.1959	1.1478	1.1598	1	16	1.0237	1.0721
1 3/8	1.3750	12	1.3209	1.2728	1.2848	1	9	1.2602	1.3137
1 1/2	1.5000	12	1.4459	1.3978	1.4098	1	3	1.5212	1.5799

TABLE 8

Selected Combinations of Special Diameter and Pitch—Class 2B *

(UN, N, NEF, and NS Threads)

Designation			L_e	Internal Thread Limits of Size						
				Minor Diameter			Pitch Diameter			Major Diameter
Size	Threads per Inch	Thread Symbol	9 X Pitch	Limits		Tolerance	Limits		Tolerance	
				Min	Max		Min	Max		Min
(.190)	28	NS-2B	0.3214	0.1513	0.1604	0.0091	0.1668	0.1711	0.0043	0.1900
(.190)	36	NS-2B	0.2500	0.1599	0.1669	0.0070	0.1720	0.1759	0.0039	0.1900
(.190)	40	NS-2B	0.2250	0.1629	0.1691	0.0062	0.1738	0.1775	0.0037	0.1900
(.190)	48	NS-2B	0.1875	0.1674	0.1725	0.0051	0.1765	0.1799	0.0034	0.1900
(.190)	56	NS-2B	0.1607	0.1707	0.1749	0.0042	0.1784	0.1816	0.0032	0.1900
2 (.216)	32	NEF-2B	0.2812	0.1822	0.1895	0.0073	0.1957	0.1998	0.0041	0.2160
2 (.216)	36	NS-2B	0.2500	0.1859	0.1923	0.0064	0.1980	0.2019	0.0039	0.2160
2 (.216)	40	NS-2B	0.2250	0.1889	0.1946	0.0057	0.1998	0.2035	0.0037	0.2160
2 (.216)	48	NS-2B	0.1875	0.1934	0.1981	0.0047	0.2025	0.2059	0.0034	0.2160
2 (.216)	56	NS-2B	0.1607	0.1967	0.2006	0.0039	0.2044	0.2076	0.0032	0.2160
1/4	24	NS-2B	0.3750	0.2049	0.2139	0.0090	0.2229	0.2277	0.0048	0.2500
1/4	32	NEF-2B	0.2812	0.2162	0.2229	0.0067	0.2297	0.2339	0.0042	0.2500
1/4	**36**	**UN-2B**	**0.2500**	**0.2199**	**0.2258**	**0.0059**	**0.2320**	**0.2360**	**0.0040**	**0.2500**
1/4	40	NS-2B	0.2250	0.2229	0.2282	0.0053	0.2338	0.2376	0.0038	0.2500
1/4	48	NS-2B	0.1875	0.2274	0.2317	0.0043	0.2365	0.2401	0.0036	0.2500
1/4	56	NS-2B	0.1607	0.2307	0.2343	0.0036	0.2384	0.2417	0.0033	0.2500
5/16	20	NS-2B	0.4500	0.2584	0.2680	0.0096	0.2800	0.2852	0.0052	0.3125
5/16	28	NS-2B	0.3214	0.2738	0.2807	0.0069	0.2893	0.2937	0.0044	0.3125
5/16	32	NEF-2B	0.2812	0.2787	0.2847	0.0060	0.2922	0.2964	0.0042	0.3125
5/16	**36**	**UN-2B**	**0.2500**	**0.2824**	**0.2877**	**0.0053**	**0.2945**	**0.2985**	**0.0040**	**0.3125**
5/16	40	NS-2B	0.2250	0.2854	0.2902	0.0048	0.2963	0.3001	0.0038	0.3125
5/16	48	NS-2B	0.1875	0.2899	0.2940	0.0041	0.2990	0.3026	0.0036	0.3125
3/8	18	NS-2B	0.5000	0.3149	0.3246	0.0097	0.3389	0.3445	0.0056	0.3750
3/8	20	NS-2B	0.4500	0.3209	0.3297	0.0088	0.3425	0.3479	0.0054	0.3750
3/8	28	NS-2B	0.3214	0.3363	0.3426	0.0063	0.3518	0.3564	0.0046	0.3750
3/8	32	NEF-2B	0.2812	0.3412	0.3469	0.0057	0.3547	0.3591	0.0044	0.3750
3/8	**36**	**UN-2B**	**0.2500**	**0.3449**	**0.3501**	**0.0052**	**0.3570**	**0.3612**	**0.0042**	**0.3750**
3/8	40	NS-2B	0.2250	0.3479	0.3527	0.0048	0.3588	0.3628	0.0040	0.3750
7/16	16	NS-2B	0.5625	0.3698	0.3800	0.0102	0.3969	0.4028	0.0059	0.4375
7/16	18	NS-2B	0.5000	0.3774	0.3865	0.0091	0.4014	0.4070	0.0056	0.4375
7/16	24	NS-2B	0.3750	0.3924	0.3994	0.0070	0.4104	0.4153	0.0049	0.4375
7/16	**28**	**UNEF-2B**	**0.3214**	**0.3988**	**0.4051**	**0.0063**	**0.4143**	**0.4189**	**0.0046**	**0.4375**
7/16	32	NS-2B	0.2812	0.4037	0.4094	0.0057	0.4172	0.4216	0.0044	0.4375
1/2	12	N-2B	0.7500	0.4098	0.4223	0.0125	0.4459	0.4529	0.0070	0.5000
1/2	14	NS-2B	0.6429	0.4227	0.4336	0.0109	0.4536	0.4601	0.0065	0.5000
1/2	16	NS-2B	0.5625	0.4323	0.4419	0.0096	0.4594	0.4655	0.0061	0.5000
1/2	18	NS-2B	0.5000	0.4399	0.4485	0.0086	0.4639	0.4697	0.0058	0.5000
1/2	24	NS-2B	0.3750	0.4549	0.4619	0.0070	0.4729	0.4780	0.0051	0.5000
1/2	**28**	**UNEF-2B**	**0.3214**	**0.4613**	**0.4676**	**0.0063**	**0.4768**	**0.4816**	**0.0048**	**0.5000**
1/2	32	NS-2B	0.2812	0.4662	0.4719	0.0057	0.4797	0.4842	0.0045	0.5000
9/16	14	NS-2B	0.6429	0.4852	0.4956	0.0104	0.5161	0.5226	0.0065	0.5625
9/16	16	NS-2B	0.5625	0.4948	0.5040	0.0092	0.5219	0.5280	0.0061	0.5625
9/16	20	NS-2B	0.4500	0.5084	0.5162	0.0078	0.5300	0.5355	0.0055	0.5625
9/16	24	NEF-2B	0.3750	0.5174	0.5244	0.0070	0.5354	0.5405	0.0051	0.5625
9/16	**28**	**UN-2B**	**0.3214**	**0.5238**	**0.5301**	**0.0063**	**0.5393**	**0.5441**	**0.0048**	**0.5625**
9/16	32	NS-2B	0.2812	0.5287	0.5344	0.0057	0.5422	0.5467	0.0045	0.5625
5/8	12	N-2B	0.7500	0.5348	0.5463	0.0115	0.5709	0.5780	0.0071	0.6250
5/8	14	NS-2B	0.6429	0.5477	0.5577	0.0100	0.5786	0.5852	0.0066	0.6250
5/8	16	NS-2B	0.5625	0.5573	0.5662	0.0089	0.5844	0.5906	0.0062	0.6250
5/8	20	NS-2B	0.4500	0.5709	0.5787	0.0078	0.5925	0.5981	0.0056	0.6250
5/8	24	NEF-2B	0.3750	0.5799	0.5869	0.0070	0.5979	0.6031	0.0052	0.6250
5/8	**28**	**UN-2B**	**0.3214**	**0.5863**	**0.5926**	**0.0063**	**0.6018**	**0.6067**	**0.0049**	**0.6250**
5/8	32	NS-2B	0.2812	0.5912	0.5969	0.0057	0.6047	0.6093	0.0046	0.6250
11/16	12	N-2B	0.7500	0.5973	0.6085	0.0112	0.6334	0.6405	0.0071	0.6875
11/16	24	NEF-2B	0.3750	0.6424	0.6494	0.0070	0.6604	0.6656	0.0052	0.6875

* American Society of Mechanical Engineers. Internal thread limits of size applicable to lengths of engagement of from 5 to 15 times the pitch, inclusive. For 8-, 12- and 16-thread series sizes not shown, use dimensions

TABLE 8

Selected Combinations of Special Diameter and Pitch—Class 2B *—*Continued*

(UN, N, NEF, and NS Threads)

Designation			L_e	Internal Thread Limits of Size						
				Minor Diameter			Pitch Diameter			Major Diameter
Size	Threads per Inch	Thread Symbol	9 X Pitch	Limits		Tolerance	Limits		Tolerance	
				Min	Max		Min	Max		Min
3/4	12	N-2B	0.7500	0.6598	0.6707	0.0109	0.6959	0.7031	0.0072	0.7500
3/4	14	NS-2B	0.6429	0.6727	0.6822	0.0095	0.7036	0.7103	0.0067	0.7500
3/4	18	NS-2B	0.5000	0.6899	0.6980	0.0081	0.7139	0.7199	0.0060	0.7500
3/4	20	UNEF-2B	0.4500	0.6959	0.7037	0.0078	0.7175	0.7232	0.0057	0.7500
3/4	24	NS-2B	0.3750	0.7049	0.7119	0.0070	0.7229	0.7282	0.0053	0.7500
3/4	28	UN-2B	0.3214	0.7113	0.7176	0.0063	0.7268	0.7318	0.0050	0.7500
3/4	32	NS-2B	0.2812	0.7162	0.7219	0.0057	0.7297	0.7344	0.0047	0.7500
13/16	12	N-2B	0.7500	0.7222	0.7329	0.0106	0.7584	0.7656	0.0072	0.8125
13/16	16	UN-2B	0.5625	0.7448	0.7533	0.0085	0.7719	0.7782	0.0063	0.8125
13/16	20	UNEF-2B	0.4500	0.7584	0.7662	0.0078	0.7800	0.7857	0.0057	0.8125
7/8	10	NS-2B	0.9000	0.7667	0.7789	0.0122	0.8100	0.8178	0.0078	0.8750
7/8	12	N-2B	0.7500	0.7848	0.7952	0.0104	0.8209	0.8281	0.0072	0.8750
7/8	16	UN-2B	0.5625	0.8073	0.8158	0.0085	0.8344	0.8407	0.0063	0.8750
7/8	18	NS-2B	0.5000	0.8149	0.8230	0.0081	0.8389	0.8449	0.0060	0.8750
7/8	20	UNEF-2B	0.4500	0.8209	0.8287	0.0078	0.8425	0.8482	0.0057	0.8750
7/8	24	NS-2B	0.3750	0.8299	0.8369	0.0070	0.8479	0.8532	0.0053	0.8750
7/8	28	UN-2B	0.3214	0.8363	0.8426	0.0063	0.8518	0.8568	0.0050	0.8750
7/8	32	NS-2B	0.2812	0.8412	0.8469	0.0057	0.8547	0.8594	0.0047	0.8750
15/16	12	UN-2B	0.7500	0.8473	0.8575	0.0102	0.8834	0.8908	0.0074	0.9375
15/16	16	UN-2B	0.5625	0.8698	0.8783	0.0085	0.8969	0.9034	0.0065	0.9375
15/16	20	UNEF-2B	0.4500	0.8834	0.8912	0.0078	0.9050	0.9109	0.0059	0.9375
1	10	NS-2B	0.9000	0.8917	0.9037	0.0120	0.9350	0.9430	0.0080	1.0000
1	16	UN-2B	0.5625	0.9323	0.9408	0.0085	0.9594	0.9659	0.0065	1.0000
1	18	NS-2B	0.5000	0.9399	0.9480	0.0081	0.9639	0.9701	0.0062	1.0000
1	20	UNEF-2B	0.4500	0.9459	0.9537	0.0078	0.9675	0.9734	0.0059	1.0000
1	24	NS-2B	0.3750	0.9549	0.9619	0.0070	0.9729	0.9784	0.0055	1.0000
1	28	UN-2B	0.3214	0.9613	0.9676	0.0063	0.9768	0.9820	0.0052	1.0000
1	32	NS-2B	0.2812	0.9662	0.9719	0.0057	0.9797	0.9846	0.0049	1.0000
1 1/16	12	UN-2B	0.7500	0.9723	0.9823	0.0100	1.0084	1.0158	0.0074	1.0625
1 1/16	16	UN-2B	0.5625	0.9948	1.0033	0.0085	1.0219	1.0284	0.0065	1.0625
1 1/16	18	NEF-2B	0.5000	1.0024	1.0105	0.0081	1.0264	1.0326	0.0062	1.0625
1 1/8	10	NS-2B	0.9000	1.0167	1.0287	0.0120	1.0600	1.0680	0.0080	1.1250
1 1/8	14	NS-2B	0.6429	1.0477	1.0565	0.0088	1.0786	1.0855	0.0069	1.1250
1 1/8	16	UN-2B	0.5625	1.0573	1.0658	0.0085	1.0844	1.0910	0.0066	1.1250
1 1/8	18	NEF-2B	0.5000	1.0649	1.0730	0.0081	1.0889	1.0951	0.0062	1.1250
1 1/8	20	UN-2B	0.4500	1.0709	1.0787	0.0078	1.0925	1.0984	0.0059	1.1250
1 1/8	24	NS-2B	0.3750	1.0799	1.0869	0.0070	1.0979	1.1034	0.0055	1.1250
1 1/8	28	UN-2B	0.3214	1.0863	1.0926	0.0063	1.1018	1.1070	0.0052	1.1250
1 3/16	12	UN-2B	0.7500	1.0973	1.1073	0.0100	1.1334	1.1409	0.0075	1.1875
1 3/16	16	UN-2B	0.5625	1.1198	1.1283	0.0085	1.1469	1.1535	0.0066	1.1875
1 3/16	18	NEF-2B	0.5000	1.1274	1.1355	0.0081	1.1514	1.1577	0.0063	1.1875
1 1/4	10	NS-2B	0.9000	1.1417	1.1537	0.0120	1.1850	1.1932	0.0082	1.2500
1 1/4	14	NS-2B	0.6429	1.1727	1.1815	0.0088	1.2036	1.2106	0.0070	1.2500
1 1/4	16	UN-2B	0.5625	1.1823	1.1908	0.0085	1.2094	1.2160	0.0066	1.2500
1 1/4	18	NEF-2B	0.5000	1.1899	1.1980	0.0081	1.2139	1.2202	0.0063	1.2500
1 1/4	20	UN-2B	0.4500	1.1959	1.2037	0.0078	1.2175	1.2236	0.0061	1.2500
1 1/4	24	NS-2B	0.3750	1.2049	1.2119	0.0070	1.2229	1.2285	0.0056	1.2500
1 5/16	12	UN-2B	0.7500	1.2223	1.2323	0.0100	1.2584	1.2659	0.0075	1.3125
1 5/16	16	UN-2B	0.5625	1.2448	1.2533	0.0085	1.2719	1.2785	0.0066	1.3125
1 5/16	18	NEF-2B	0.5000	1.2524	1.2605	0.0081	1.2764	1.2827	0.0063	1.3125

and symbols shown in tables for UNC, NC, UNF, and NF thread series. Bold type indicates Unified threads-UN.

In British practice, the term "Effective Diameter" is used instead of "Pitch Diameter."

TABLE 8

Selected Combinations of Special Diameter and Pitch—Class 2B—*Continued*

(UN, N, NEF, and NS Threads)

Designation			L_e	Internal Thread Limits of Size						
				Minor Diameter			Pitch Diameter			Major Diameter
Size	Threads per Inch	Thread Symbol	9 X Pitch	Limits		Tolerance	Limits		Tolerance	
				Min	Max		Min	Max		Min
1 3/8	10	NS-2B	0.9000	1.2667	1.2787	0.0120	1.3100	1.3182	0.0082	1.3750
1 3/8	14	NS-2B	0.6429	1.2977	1.3065	0.0088	1.3286	1.3356	0.0070	1.3750
1 3/8	16	UN-2B	0.5625	1.3073	1.3158	0.0085	1.3344	1.3410	0.0066	1.3750
1 3/8	18	NEF-2B	0.5000	1.3149	1.3230	0.0081	1.3389	1.3452	0.0063	1.3750
1 7/16	12	UN-2B	0.7500	1.3473	1.3573	0.0100	1.3834	1.3910	0.0076	1.4375
1 7/16	16	UN-2B	0.5625	1.3698	1.3783	0.0085	1.3969	1.4037	0.0068	1.4375
1 7/16	18	NEF-2B	0.5000	1.3774	1.3855	0.0081	1.4014	1.4079	0.0065	1.4375
1 1/2	10	NS-2B	0.9000	1.3917	1.4037	0.0120	1.4350	1.4433	0.0083	1.5000
1 1/2	14	NS-2B	0.6429	1.4227	1.4315	0.0088	1.4536	1.4608	0.0072	1.5000
1 1/2	16	UN-2B	0.5625	1.4323	1.4408	0.0085	1.4594	1.4662	0.0068	1.5000
1 1/2	18	NEF-2B	0.5000	1.4399	1.4480	0.0081	1.4639	1.4704	0.0065	1.5000
1 1/2	20	UN-2B	0.4500	1.4459	1.4537	0.0078	1.4675	1.4737	0.0062	1.5000
1 1/2	24	NS-2B	0.3750	1.4549	1.4619	0.0070	1.4729	1.4787	0.0058	1.5000
1 9/16	16	N-2B	0.5625	1.4948	1.5033	0.0085	1.5219	1.5287	0.0068	1.5625
1 9/16	18	NEF-2B	0.5000	1.5024	1.5105	0.0081	1.5264	1.5329	0.0065	1.5625
1 5/8	6	NS-2B	1.5000	1.4446	1.4646	0.0200	1.5167	1.5272	0.0105	1.6250
1 5/8	7	NS-2B	1.2857	1.4704	1.4875	0.0171	1.5322	1.5420	0.0098	1.6250
1 5/8	10	NS-2B	0.9000	1.5167	1.5287	0.0120	1.5600	1.5683	0.0083	1.6250
1 5/8	12	N-2B	0.7500	1.5348	1.5448	0.0100	1.5709	1.5785	0.0076	1.6250
1 5/8	14	NS-2B	0.6429	1.5477	1.5565	0.0088	1.5786	1.5858	0.0072	1.6250
1 5/8	16	N-2B	0.5625	1.5573	1.5658	0.0085	1.5844	1.5912	0.0068	1.6250
1 5/8	18	NEF-2B	0.5000	1.5649	1.5730	0.0081	1.5889	1.5954	0.0065	1.6250
1 5/8	20	NS-2B	0.4500	1.5709	1.5787	0.0078	1.5925	1.5987	0.0062	1.6250
1 5/8	24	NS-2B	0.3750	1.5799	1.5869	0.0070	1.5979	1.6037	0.0058	1.6250
1 11/16	16	N-2B	0.5625	1.6198	1.6283	0.0085	1.6469	1.6538	0.0069	1.6875
1 11/16	18	NEF-2B	0.5000	1.6274	1.6355	0.0081	1.6514	1.6580	0.0066	1.6875
1 3/4	6	NS-2B	1.5000	1.5696	1.5896	0.0200	1.6417	1.6523	0.0106	1.7500
1 3/4	7	NS-2B	1.2857	1.5954	1.6125	0.0171	1.6572	1.6671	0.0099	1.7500
1 3/4	8	UN-2B	1.1250	1.6147	1.6297	0.0150	1.6688	1.6781	0.0093	1.7500
1 3/4	10	NS-2B	0.9000	1.6417	1.6537	0.0120	1.6850	1.6934	0.0084	1.7500
1 3/4	12	UN-2B	0.7500	1.6598	1.6698	0.0100	1.6959	1.7037	0.0078	1.7500
1 3/4	14	NS-2B	0.6429	1.6727	1.6815	0.0088	1.7036	1.7109	0.0073	1.7500
1 3/4	16	UNEF-2B	0.5625	1.6823	1.6908	0.0085	1.7094	1.7163	0.0069	1.7500
1 3/4	18	NS-2B	0.5000	1.6899	1.6980	0.0081	1.7139	1.7205	0.0066	1.7500
1 3/4	20	UN-2B	0.4500	1.6959	1.7037	0.0078	1.7175	1.7238	0.0063	1.7500
1 13/16	16	N-2B	0.5625	1.7448	1.7533	0.0085	1.7719	1.7788	0.0069	1.8125
1 7/8	6	NS-2B	1.5000	1.6946	1.7146	0.0200	1.7667	1.7773	0.0106	1.8750
1 7/8	7	NS-2B	1.2857	1.7204	1.7375	0.0171	1.7822	1.7921	0.0099	1.8750
1 7/8	8	NS-2B	1.1250	1.7397	1.7547	0.0150	1.7938	1.8031	0.0093	1.8750
1 7/8	10	NS-2B	0.9000	1.7667	1.7787	0.0120	1.8100	1.8184	0.0084	1.8750
1 7/8	12	N-2B	0.7500	1.7848	1.7948	0.0100	1.8209	1.8287	0.0078	1.8750
1 7/8	14	NS-2B	0.6429	1.7977	1.8065	0.0088	1.8286	1.8359	0.0073	1.8750
1 7/8	16	N-2B	0.5625	1.8073	1.8158	0.0085	1.8344	1.8413	0.0069	1.8750
1 7/8	18	NS-2B	0.5000	1.8149	1.8230	0.0081	1.8389	1.8455	0.0066	1.8750
1 7/8	20	NS-2B	0.4500	1.8209	1.8287	0.0078	1.8425	1.8488	0.0063	1.8750
1 15/16	16	N-2B	0.5625	1.8698	1.8783	0.0085	1.8969	1.9039	0.0070	1.9375

TABLE 8

Selected Combinations of Special Diameter and Pitch—Class 2B—*Continued*

(UN, N, NEF, and NS Threads)

Designation			L_e	Internal Thread Limits of Size						
				Minor Diameter			Pitch Diameter			Major Diameter
Size	Threads per Inch	Thread Symbol	9 X Pitch	Limits Min	Limits Max	Tolerance	Limits Min	Limits Max	Tolerance	Min
2	6	NS-2B	1.5000	1.8196	1.8396	0.0200	1.8917	1.9025	0.0108	2.0000
2	7	NS-2B	1.2857	1.8454	1.8625	0.0171	1.9072	1.9172	0.0100	2.0000
2	8	UN-2B	1.1250	1.8647	1.8797	0.0150	1.9188	1.9282	0.0094	2.0000
2	10	NS-2B	0.9000	1.8917	1.9037	0.0120	1.9350	1.9435	0.0085	2.0000
2	12	UN-2B	0.7500	1.9098	1.9198	0.0100	1.9459	1.9538	0.0079	2.0000
2	14	NS-2B	0.6429	1.9227	1.9315	0.0088	1.9536	1.9610	0.0074	2.0000
2	16	UNEF-2B	0.5625	1.9323	1.9408	0.0085	1.9594	1.9664	0.0070	2.0000
2	18	NS-2B	0.5000	1.9399	1.9480	0.0081	1.9639	1.9706	0.0067	2.0000
2	20	UN-2B	0.4500	1.9459	1.9537	0.0078	1.9675	1.9739	0.0064	2.0000
2 1/16	16	N-2B	0.5625	1.9948	2.0033	0.0085	2.0219	2.0289	0.0070	2.0625
2 1/8	12	N-2B	0.7500	2.0348	2.0448	0.0100	2.0709	2.0788	0.0079	2.1250
2 1/8	16	N-2B	0.5625	2.0573	2.0658	0.0085	2.0844	2.0914	0.0070	2.1250
2 3/16	16	N-2B	0.5625	2.1198	2.1283	0.0085	2.1469	2.1539	0.0070	2.1875
2 1/4	6	NS-2B	1.5000	2.0696	2.0896	0.0200	2.1417	2.1525	0.0108	2.2500
2 1/4	7	NS-2B	1.2857	2.0954	2.1125	0.0171	2.1572	2.1672	0.0100	2.2500
2 1/4	8	UN-2B	1.1250	2.1147	2.1297	0.0150	2.1688	2.1782	0.0094	2.2500
2 1/4	10	NS-2B	0.9000	2.1417	2.1537	0.0120	2.1850	2.1935	0.0085	2.2500
2 1/4	12	UN-2B	0.7500	2.1598	2.1698	0.0100	2.1959	2.2038	0.0079	2.2500
2 1/4	14	NS-2B	0.6429	2.1727	2.1815	0.0088	2.2036	2.2110	0.0074	2.2500
2 1/4	16	UN-2B	0.5625	2.1823	2.1908	0.0085	2.2094	2.2164	0.0070	2.2500
2 1/4	18	NS-2B	0.5000	2.1899	2.1980	0.0081	2.2139	2.2206	0.0067	2.2500
2 1/4	20	UN-2B	0.4500	2.1959	2.2037	0.0078	2.2175	2.2239	0.0064	2.2500
2 5/16	16	N-2B	0.5625	2.2448	2.2533	0.0085	2.2719	2.2791	0.0072	2.3125
2 3/8	12	N-2B	0.7500	2.2848	2.2948	0.0100	2.3209	2.3290	0.0081	2.3750
2 3/8	16	N-2B	0.5625	2.3073	2.3158	0.0085	2.3344	2.3416	0.0072	2.3750
2 7/16	16	N-2B	0.5625	2.3698	2.3783	0.0085	2.3969	2.4041	0.0072	2.4375
2 1/2	6	NS-2B	1.5000	2.3196	2.3396	0.0200	2.3917	2.4026	0.0109	2.5000
2 1/2	7	NS-2B	1.2857	2.3454	2.3625	0.0171	2.4072	2.4174	0.0102	2.5000
2 1/2	8	UN-2B	1.1250	2.3647	2.3797	0.0150	2.4188	2.4284	0.0096	2.5000
2 1/2	10	NS-2B	0.9000	2.3917	2.4037	0.0120	2.4350	2.4437	0.0087	2.5000
2 1/2	12	UN-2B	0.7500	2.4098	2.4198	0.0100	2.4459	2.4540	0.0081	2.5000
2 1/2	14	NS-2B	0.6429	2.4227	2.4315	0.0088	2.4536	2.4612	0.0076	2.5000
2 1/2	16	UN-2B	0.5625	2.4323	2.4408	0.0085	2.4594	2.4666	0.0072	2.5000
2 1/2	18	NS-2B	0.5000	2.4399	2.4480	0.0081	2.4639	2.4708	0.0069	2.5000
2 1/2	20	UN-2B	0.4500	2.4459	2.4537	0.0078	2.4675	2.4741	0.0066	2.5000
2 5/8	12	N-2B	0.7500	2.5348	2.5448	0.0100	2.5709	2.5790	0.0081	2.6250
2 5/8	16	N-2B	0.5625	2.5573	2.5658	0.0085	2.5844	2.5916	0.0072	2.6250
2 3/4	6	NS-2B	1.5000	2.5696	2.5896	0.0200	2.6417	2.6526	0.0109	2.7500
2 3/4	7	NS-2B	1.2857	2.5954	2.6125	0.0171	2.6572	2.6674	0.0102	2.7500
2 3/4	8	UN-2B	1.1250	2.6147	2.6297	0.0150	2.6688	2.6784	0.0096	2.7500
2 3/4	10	NS-2B	0.9000	2.6417	2.6537	0.0120	2.6850	2.6937	0.0087	2.7500
2 3/4	12	UN-2B	0.7500	2.6598	2.6698	0.0100	2.6959	2.7040	0.0081	2.7500
2 3/4	14	NS-2B	0.6429	2.6727	2.6815	0.0088	2.7036	2.7112	0.0076	2.7500
2 3/4	16	UN-2B	0.5625	2.6823	2.6908	0.0085	2.7094	2.7166	0.0072	2.7500
2 3/4	18	NS-2B	0.5000	2.6899	2.6980	0.0081	2.7139	2.7208	0.0069	2.7500
2 7/8	12	N-2B	0.7500	2.7848	2.7948	0.0100	2.8209	2.8291	0.0082	2.8750
2 7/8	16	N-2B	0.5625	2.8073	2.8158	0.0085	2.8344	2.8417	0.0073	2.8750

TABLE 8

Selected Combinations of Special Diameter and Pitch—Class 2B—*Continued*

(UN, N, NEF, and NS Threads)

Designation			L_e	Internal Thread Limits of Size						
				Minor Diameter			Pitch Diameter			Major Diameter
Size	Threads per Inch	Thread Symbol	9 X Pitch	Limits Min	Limits Max	Tolerance	Limits Min	Limits Max	Tolerance	Min
3	6	NS-2B	1.5000	2.8196	2.8396	0.0200	2.8917	2.9028	0.0111	3.0000
3	7	NS-2B	1.2857	2.8454	2.8625	0.0171	2.9072	2.9176	0.0104	3.0000
3	8	UN-2B	1.1250	2.8647	2.8797	0.0150	2.9188	2.9286	0.0098	3.0000
3	10	NS-2B	0.9000	2.8917	2.9037	0.0120	2.9350	2.9439	0.0089	3.0000
3	12	UN-2B	0.7500	2.9098	2.9198	0.0100	2.9459	2.9541	0.0082	3.0000
3	14	NS-2B	0.6429	2.9227	2.9315	0.0088	2.9536	2.9613	0.0077	3.0000
3	16	UN-2B	0.5625	2.9323	2.9408	0.0085	2.9594	2.9667	0.0073	3.0000
3	18	NS-2B	0.5000	2.9399	2.9480	0.0081	2.9639	2.9709	0.0070	3.0000
3 1/8	12	N-2B	0.7500	3.0348	3.0448	0.0100	3.0709	3.0791	0.0082	3.1250
3 1/8	16	N-2B	0.5625	3.0573	3.0658	0.0085	3.0844	3.0917	0.0073	3.1250
3 1/4	6	NS-2B	1.5000	3.0696	3.0896	0.0200	3.1417	3.1528	0.0111	3.2500
3 1/4	7	NS-2B	1.2857	3.0954	3.1125	0.0171	3.1572	3.1676	0.0104	3.2500
3 1/4	8	UN-2B	1.1250	3.1147	3.1297	0.0150	3.1688	3.1786	0.0098	3.2500
3 1/4	10	NS-2B	0.9000	3.1417	3.1537	0.0120	3.1850	3.1939	0.0089	3.2500
3 1/4	12	UN-2B	0.7500	3.1598	3.1698	0.0100	3.1959	3.2041	0.0082	3.2500
3 1/4	14	NS-2B	0.6429	3.1727	3.1815	0.0088	3.2036	3.2113	0.0077	3.2500
3 1/4	16	UN-2B	0.5625	3.1823	3.1908	0.0085	3.2094	3.2167	0.0073	3.2500
3 1/4	18	NS-2B	0.5000	3.1899	3.1980	0.0081	3.2139	3.2209	0.0070	3.2500
3 3/8	12	N-2B	0.7500	3.2848	3.2948	0.0100	3.3209	3.3293	0.0084	3.3750
3 3/8	16	N-2B	0.5625	3.3073	3.3158	0.0085	3.3344	3.3419	0.0075	3.3750
3 1/2	6	NS-2B	1.5000	3.3196	3.3396	0.0200	3.3917	3.4030	0.0113	3.5000
3 1/2	7	NS-2B	1.2857	3.3454	3.3625	0.0171	3.4072	3.4177	0.0105	3.5000
3 1/2	8	UN-2B	1.1250	3.3647	3.3797	0.0150	3.4188	3.4287	0.0099	3.5000
3 1/2	10	NS-2B	0.9000	3.3917	3.4037	0.0120	3.4350	3.4440	0.0090	3.5000
3 1/2	12	UN-2B	0.7500	3.4098	3.4198	0.0100	3.4459	3.4543	0.0084	3.5000
3 1/2	14	NS-2B	0.6429	3.4227	3.4315	0.0088	3.4536	3.4615	0.0079	3.5000
3 1/2	16	UN-2B	0.5625	3.4323	3.4408	0.0085	3.4594	3.4669	0.0075	3.5000
3 1/2	18	NS-2B	0.5000	3.4399	3.4480	0.0081	3.4639	3.4711	0.0072	3.5000
3 5/8	12	N-2B	0.7500	3.5348	3.5448	0.0100	3.5709	3.5793	0.0084	3.6250
3 5/8	16	N-2B	0.5625	3.5573	3.5658	0.0085	3.5844	3.5919	0.0075	3.6250
3 3/4	6	NS-2B	1.5000	3.5696	3.5896	0.0200	3.6417	3.6530	0.0113	3.7500
3 3/4	7	NS-2B	1.2857	3.5954	3.6125	0.0171	3.6572	3.6677	0.0105	3.7500
3 3/4	8	UN-2B	1.1250	3.6147	3.6297	0.0150	3.6688	3.6787	0.0099	3.7500
3 3/4	10	NS-2B	0.9000	3.6417	3.6537	0.0120	3.6850	3.6940	0.0090	3.7500
3 3/4	12	UN-2B	0.7500	3.6598	3.6698	0.0100	3.6959	3.7043	0.0084	3.7500
3 3/4	14	NS-2B	0.6429	3.6727	3.6815	0.0088	3.7036	3.7115	0.0079	3.7500
3 3/4	16	UN-2B	0.5625	3.6823	3.6908	0.0085	3.7094	3.7169	0.0075	3.7500
3 3/4	18	NS-2B	0.5000	3.6899	3.6980	0.0081	3.7139	3.7211	0.0072	3.7500
3 7/8	12	N-2B	0.7500	3.7848	3.7948	0.0100	3.8209	3.8294	0.0085	3.8750
3 7/8	16	N-2B	0.5625	3.8073	3.8158	0.0085	3.8344	3.8420	0.0076	3.8750
4	6	NS-2B	1.5000	3.8196	3.8396	0.0200	3.8917	3.9031	0.0114	4.0000
4	7	NS-2B	1.2857	3.8454	3.8625	0.0171	3.9072	3.9178	0.0106	4.0000
4	8	UN-2B	1.1250	3.8647	3.8797	0.0150	3.9188	3.9288	0.0100	4.0000
4	10	NS-2B	0.9000	3.8917	3.9037	0.0120	3.9350	3.9441	0.0091	4.0000
4	12	UN-2B	0.7500	3.9098	3.9198	0.0100	3.9459	3.9544	0.0085	4.0000
4	14	NS-2B	0.6429	3.9227	3.9315	0.0088	3.9536	3.9616	0.0080	4.0000
4	16	UN-2B	0.5625	3.9323	3.9408	0.0085	3.9594	3.9670	0.0076	4.0000
4 1/4	4	UN-2B	2.2500	3.9794	4.0094	0.0300	4.0876	4.1014	0.0138	4.2500
4 1/4	6	NS-2B	1.5000	4.0696	4.0896	0.0200	4.1417	4.1531	0.0114	4.2500
4 1/4	7	NS-2B	1.2857	4.0954	4.1125	0.0171	4.1572	4.1678	0.0106	4.2500
4 1/4	8	UN-2B	1.1250	4.1147	4.1297	0.0150	4.1688	4.1788	0.0100	4.2500
4 1/4	10	NS-2B	0.9000	4.1417	4.1537	0.0120	4.1850	4.1941	0.0091	4.2500
4 1/4	12	UN-2B	0.7500	4.1598	4.1698	0.0100	4.1959	4.2044	0.0085	4.2500
4 1/4	14	NS-2B	0.6429	4.1727	4.1815	0.0088	4.2036	4.2116	0.0080	4.2500
4 1/4	16	UN-2B	0.5625	4.1823	4.1908	0.0085	4.2094	4.2170	0.0076	4.2500

TABLE 8

Selected Combinations of Special Diameter and Pitch—Class 2B—*Continued*

(UN, N, NEF, and NS Threads)

Designation			L_e	Internal Thread Limits of Size						
				Minor Diameter			Pitch Diameter			Major Diameter
				Limits			Limits			
Size	Threads per Inch	Thread Symbol	9 X Pitch	Min	Max	Tolerance	Min	Max	Tolerance	Min
4 1/2	4	UN-2B	2.2500	4.2294	4.2594	0.0300	4.3376	4.3514	0.0138	4.5000
4 1/2	6	NS-2B	1.5000	4.3196	4.3396	0.0200	4.3917	4.4031	0.0114	4.5000
4 1/2	7	NS-2B	1.2857	4.3454	4.3625	0.0171	4.4072	4.4178	0.0106	4.5000
4 1/2	8	UN-2B	1.1250	4.3647	4.3797	0.0150	4.4188	4.4288	0.0100	4.5000
4 1/2	10	NS-2B	0.9000	4.3917	4.4037	0.0120	4.4350	4.4441	0.0091	4.5000
4 1/2	12	UN-2B	0.7500	4.4098	4.4198	0.0100	4.4459	4.4544	0.0085	4.5000
4 1/2	14	NS-2B	0.6429	4.4227	4.4315	0.0088	4.4536	4.4616	0.0080	4.5000
4 1/2	16	UN-2B	0.5625	4.4323	4.4408	0.0085	4.4594	4.4670	0.0076	4.5000
4 3/4	4	UN-2B	2.2500	4.4794	4.5094	0.0300	4.5876	4.6016	0.0140	4.7500
4 3/4	6	NS-2B	1.5000	4.5696	4.5896	0.0200	4.6417	4.6533	0.0116	4.7500
4 3/4	7	NS-2B	1.2857	4.5954	4.6125	0.0171	4.6572	4.6681	0.0109	4.7500
4 3/4	8	UN-2B	1.1250	4.6147	4.6297	0.0150	4.6688	4.6791	0.0103	4.7500
4 3/4	10	NS-2B	0.9000	4.6417	4.6537	0.0120	4.6850	4.6944	0.0094	4.7500
4 3/4	12	UN-2B	0.7500	4.6598	4.6698	0.0100	4.6959	4.7046	0.0087	4.7500
4 3/4	14	NS-2B	0.6429	4.6727	4.6815	0.0088	4.7036	4.7119	0.0083	4.7500
4 3/4	16	UN-2B	0.5625	4.6823	4.6908	0.0085	4.7094	4.7173	0.0079	4.7500
5	4	UN-2B	2.2500	4.7294	4.7594	0.0300	4.8376	4.8516	0.0140	5.0000
5	6	NS-2B	1.5000	4.8196	4.8396	0.0200	4.8917	4.9033	0.0116	5.0000
5	7	NS-2B	1.2857	4.8454	4.8625	0.0171	4.9072	4.9181	0.0109	5.0000
5	8	UN-2B	1.1250	4.8647	4.8797	0.0150	4.9188	4.9291	0.0103	5.0000
5	10	NS-2B	0.9000	4.8917	4.9037	0.0120	4.9350	4.9444	0.0094	5.0000
5	12	UN-2B	0.7500	4.9098	4.9198	0.0100	4.9459	4.9546	0.0087	5.0000
5	14	NS-2B	0.6429	4.9227	4.9315	0.0088	4.9536	4.9619	0.0083	5.0000
5	16	UN-2B	0.5625	4.9323	4.9408	0.0085	4.9594	4.9673	0.0079	5.0000
5 1/4	4	UN-2B	2.2500	4.9794	5.0094	0.0300	5.0876	5.1016	0.0140	5.2500
5 1/4	6	NS-2B	1.5000	5.0696	5.0896	0.0200	5.1417	5.1533	0.0116	5.2500
5 1/4	7	NS-2B	1.2857	5.0954	5.1125	0.0171	5.1572	5.1681	0.0109	5.2500
5 1/4	8	UN-2B	1.1250	5.1147	5.1297	0.0150	5.1688	5.1791	0.0103	5.2500
5 1/4	10	NS-2B	0.9000	5.1417	5.1537	0.0120	5.1850	5.1944	0.0094	5.2500
5 1/4	12	UN-2B	0.7500	5.1598	5.1698	0.0100	5.1959	5.2046	0.0087	5.2500
5 1/4	14	NS-2B	0.6429	5.1727	5.1815	0.0088	5.2036	5.2119	0.0083	5.2500
5 1/4	16	UN-2B	0.5625	5.1823	5.1908	0.0085	5.2094	5.2173	0.0079	5.2500
5 1/2	4	UN-2B	2.2500	5.2294	5.2594	0.0300	5.3376	5.3516	0.0140	5.5000
5 1/2	6	NS-2B	1.5000	5.3196	5.3396	0.0200	5.3917	5.4033	0.0116	5.5000
5 1/2	7	NS-2B	1.2857	5.3454	5.3625	0.0171	5.4072	5.4181	0.0109	5.5000
5 1/2	8	UN-2B	1.1250	5.3647	5.3797	0.0150	5.4188	5.4291	0.0103	5.5000
5 1/2	10	NS-2B	0.9000	5.3917	5.4037	0.0120	5.4350	5.4444	0.0094	5.5000
5 1/2	12	UN-2B	0.7500	5.4098	5.4198	0.0100	5.4459	5.4546	0.0087	5.5000
5 1/2	14	NS-2B	0.6429	5.4227	5.4315	0.0088	5.4536	5.4619	0.0083	5.5000
5 1/2	16	UN-2B	0.5625	5.4323	5.4408	0.0085	5.4594	5.4673	0.0079	5.5000
5 3/4	4	UN-2B	2.2500	5.4794	5.5094	0.0300	5.5876	5.6018	0.0142	5.7500
5 3/4	6	NS-2B	1.5000	5.5696	5.5896	0.0200	5.6417	5.6535	0.0118	5.7500
5 3/4	7	NS-2B	1.2857	5.5954	5.6125	0.0171	5.6572	5.6683	0.0111	5.7500
5 3/4	8	UN-2B	1.1250	5.6147	5.6297	0.0150	5.6688	5.6793	0.0105	5.7500
5 3/4	10	NS-2B	0.9000	5.6417	5.6537	0.0120	5.6850	5.6946	0.0096	5.7500
5 3/4	12	UN-2B	0.7500	5.6598	5.6698	0.0100	5.6959	5.7049	0.0090	5.7500
5 3/4	14	NS-2B	0.6429	5.6727	5.6815	0.0088	5.7036	5.7121	0.0085	5.7500
5 3/4	16	UN-2B	0.5625	5.6823	5.6908	0.0085	5.7094	5.7175	0.0081	5.7500
6	4	UN-2B	2.2500	5.7294	5.7594	0.0300	5.8376	5.8518	0.0142	6.0000
6	6	NS-2B	1.5000	5.8196	5.8396	0.0200	5.8917	5.9035	0.0118	6.0000
6	7	NS-2B	1.2857	5.8454	5.8625	0.0171	5.9072	5.9183	0.0111	6.0000
6	8	UN-2B	1.1250	5.8647	5.8797	0.0150	5.9188	5.9293	0.0105	6.0000
6	10	NS-2B	0.9000	5.8917	5.9037	0.0120	5.9350	5.9446	0.0096	6.0000
6	12	UN-2B	0.7500	5.9098	5.9198	0.0100	5.9459	5.9549	0.0090	6.0000
6	14	NS-2B	0.6429	5.9227	5.9315	0.0088	5.9536	5.9621	0.0085	6.0000
6	16	UN-2B	0.5625	5.9323	5.9408	0.0085	5.9594	5.9675	0.0081	6.0000

form an angle of 60 degrees with each other as shown in Figs. 27 and 28. The root contour may be either flat or rounded, the exact

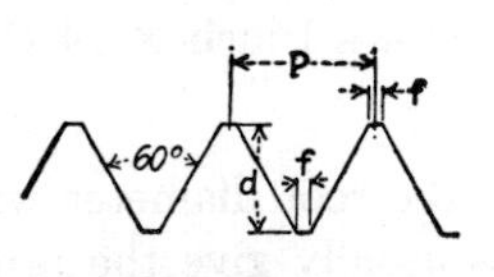

FIG. 27.

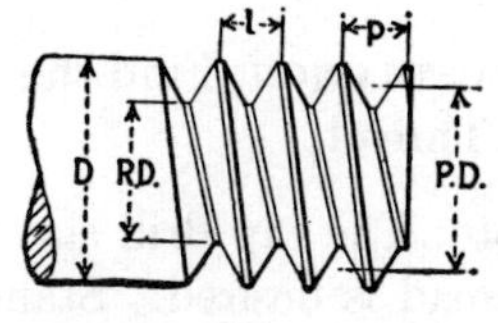

FIG. 28.

shape of the rounding not being specified since in practice it will vary with tool wear.

General formulas for the basic dimensions of the Unified and American Standard Screw Threads are:

$$\text{Pitch} = \frac{1}{\text{No. of threads per inch}}$$

$$\text{Depth external thread} = 0.61343 \times \text{pitch}$$

$$\text{Depth internal thread} = 0.54127 \times \text{pitch}$$

$$\text{Flat at crest, external thread} = 0.125 \times \text{pitch}$$

$$\text{Flat at crest, internal thread} = 0.25 \times \text{pitch}$$

$$\text{Flat at root, internal thread} = 0.125 \times \text{pitch}$$

ILLUSTRATION: Find the depth, pitch, and the flat of a $\frac{1}{4}$ inch $\times$ 20 Unified Screw Thread.

$$\text{Pitch} = \frac{1}{\text{No. of threads per inch}} = \frac{1}{20}$$

$$\text{Depth external thread} = 0.61343 \times \text{pitch}$$

$$= 0.61343 \times \tfrac{1}{20}$$

$$= 0.03077$$

$$\text{Depth internal thread} = 0.54127 \times \tfrac{1}{20}$$

$$= 0.02706$$

$$\text{Flat at crest, external thread} = 0.125 \times \tfrac{1}{20}$$

$$= 0.00625$$

$$\text{Flat at crest, internal thread} = 0.25 \times \tfrac{1}{20} = 0.0125$$

$$\text{Flat at root, internal thread} = 0.125 \times \tfrac{1}{20} = 0.00625$$

ILLUSTRATION: Find the tap drill size for a 1 inch × 14 Unified Screw Thread.

Note: The tap drill size is equal to the root diameter when a full thread is desired. Standard tables usually give the tap drill size as 75% of a full thread.

$$\text{Depth external thread} = 0.61343 \times \text{pitch} = 0.61343 \times \tfrac{1}{14}$$

$$= 0.0438$$

$$\text{Double depth} = 0.0438 \times 2 = 0.0876$$

$$\text{Root diameter} = 1.000 - 0.0876 = 0.9124.$$

Therefore the depth is 0.0438 inch and tap drill size, root diameter, is 0.9124 inch.

Whitworth Standard Thread.—This is the British Standard thread. As shown in Fig. 29, the roots and the crests are rounded and the sides form an angle of 55° with each other. If the thread were carried to a sharp point top and bottom, the rounded part would take $\frac{1}{6}$ at the top and $\frac{1}{6}$ at the bottom. Thus, $\frac{2}{3}$ is left for the depth of the thread. In such a thread the pitch, the depth and the radius are found by the following formulas.

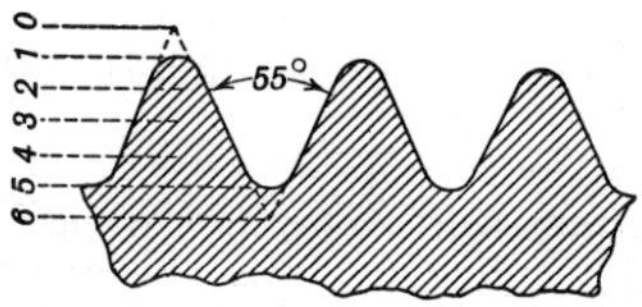

FIG. 29.

$$\text{Pitch} = \frac{1}{\text{No. threads per inch}}$$

$$\text{Depth} = \frac{0.6403}{\text{No. threads per inch}} \quad \text{or} \quad 0.6403 \times \text{pitch}$$

$$\text{Radius} = \frac{0.1373}{\text{No. threads per inch}} \quad \text{or} \quad 0.1373 \times \text{pitch}$$

ILLUSTRATION: Find the pitch, depth, and radius for a $\frac{11}{16}$ inch $\times$ 11 Whitworth Standard Screw Thread.

$$\text{Pitch} = \frac{1}{\text{No. threads per inch}} = \frac{1}{11}$$

$$\text{Depth} = 0.6403 \times \text{pitch} = 0.6403 \times \tfrac{1}{11} = 0.0582 \text{ in.} \quad \text{(Ans.)}$$

$$\text{Radius} = 0.1373 \times \text{pitch} = 0.1373 \times \tfrac{1}{11} = 0.0125 \text{ in.} \quad \text{(Ans.)}$$

Square Thread.—The sides of the square thread are parallel and the depth of the thread is equal to the width of the space between the teeth. (See Figs. 30, 31.) This space is theoretically equal to one-half of the pitch. It is necessary in practice to make the space in the nut a trifle wider than the thread so as to have a running fit between the screw and the nut.

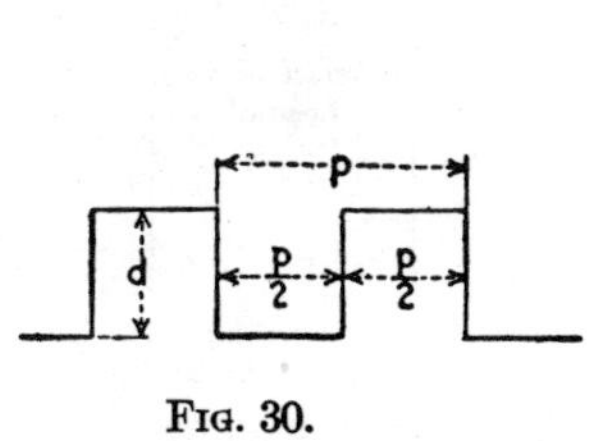

FIG. 30.

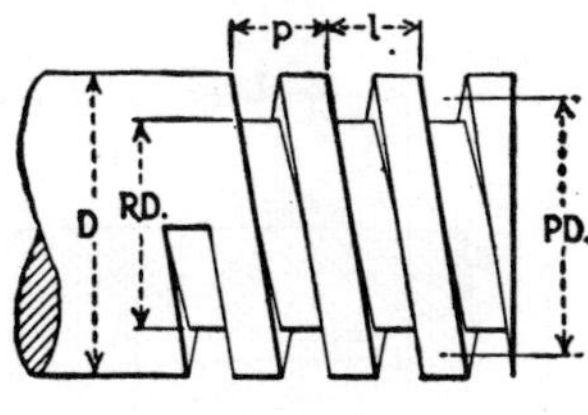

FIG. 31.—Single Square Thread.

Acme Thread.—The Acme Thread (see Fig. 32) has to a large extent replaced the square thread because of greater ease in cutting and of the greater strength secured. The angle between the sides is 29°, 14½° on each side of the vertical.

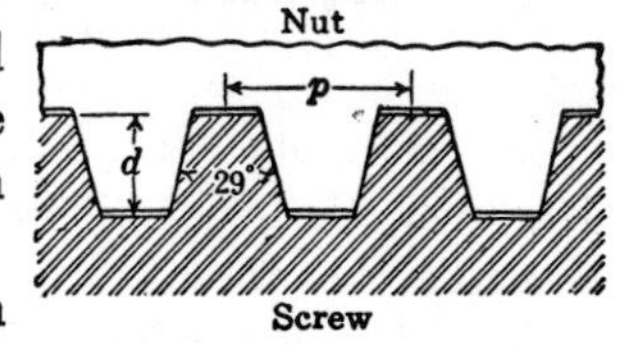

FIG. 32.—Acme Thread.

The following formulas are used in calculating measurements of Acme Screw Threads and tap threads.

For Screws:

$d = \frac{1}{2}p + 0.0100$
$f = 0.3707p$
$c = 0.3707p - 0.0052$ in.

For Taps:

$d = \frac{1}{2}p + 0.0200$ in.
$f = 0.3707p - 0.0052$ in.
$c = 0.3707p - 0.0052$ in.

when d = depth of thread, f = width of flat at top of thread, and c = width of flat at root of thread.

Diameter of tap = diameter of screw + 0.200 inch

Diameter at root of thread (tap and screw) = diameter of screw − (p + 0.020 inch)

American Standard Taper Pipe Thread.—This was formerly known as the Briggs standard pipe thread. These threads are similar to the American National Thread, the sides making an angle of 60 degrees, but the root and crest are slightly rounded.

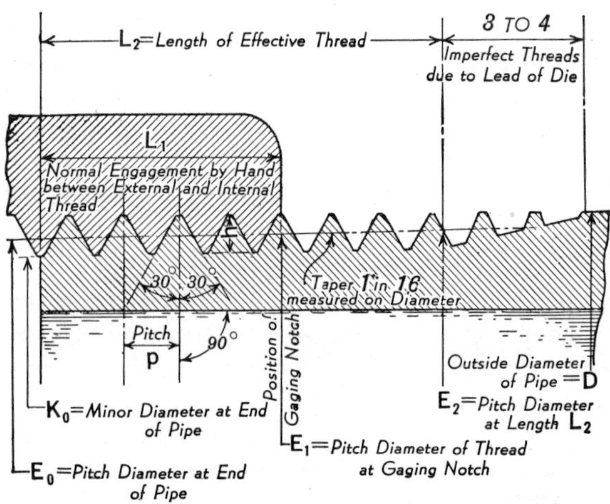

FIG. 33.

However, the chief difference between pipe thread and ordinary thread is that there is a taper on the diameter equal to $\frac{1}{16}$ inch per inch or $\frac{3}{4}$ inch per foot. The thread depth equals 0.8 × pitch of thread. The number of threads per inch for various pipe sizes is given in Table 9 with the other elements.

Pipe threads are employed on standard iron, steel, and brass pipe where the joint is subject to internal pressure from the liquid or gas which the pipe is carrying. These joints are usually made up with a special joint compound and "plumber's cotton."

TABLE 9.

size of ...ches	Number of threads per inch, n	Pitch, p	Depth of thread, h	Outside diameter of pipe, D	Length of normal engagement by hand, L_1	Length of effective thread, L_2	Increase in diameter per thread, $\frac{0.0625}{n}$	Pitch diameters: At end of pipe, or at length L_1 from end of coupling, $E_0=D-\frac{0.05D+1.1}{n}$	Pitch diameters: At length L_1 on pipe, or at end of coupling, $E_1=E_0+\frac{L_1}{16}$			Basic minor diameter at small end of pipe,[1] K_0
								Basic	Maximum	Basic	Minimum	
1	2	3	4	5	6	7	8	9	10	11	12	13
		Inch	*Inch*	*Inches*	*Inches*	*Inches*	*Inch*	*Inches*	*Inches*	*Inches*	*Inches*	*Inches*
	27	0.03704	0.02963	0.405	0.180	0.26385	0.00231	0.36351	0.37823	0.37476	0.37129	0.33388
	18	.05556	.04444	540	200	40178	.00347	47739	49510	48989	48468	.43294
	18	.05556	.04444	.675	240	.40778	.00347	.61201	.63222	62701	.62181	.56757
	14	.07143	.05714	840	.320	53371	.00446	75843	.78513	77843	.77173	70129
	14	.07143	.05714	1.050	.339	54571	.00446	96768	.99556	98887	98217	91054
	11½	.08696	.06957	1.315	400	.68278	.00543	1.21363	1.24678	1.23863	1.23048	1.14407
	11½	.08696	.06957	1.660	.420	70678	.00543	1.55713	1.59153	1.58338	1.57523	1.48757
	11½	.08696	.06957	1.900	.420	.72348	.00543	1.79609	1.83049	1.82234	1.81418	1.72652
	11½	.08696	.06957	2.375	.436	.75652	.00543	2.26902	2.30442	2.29627	2.28812	2.19946
	8	12500	.10000	2.875	.682	1.13750	.00781	2.71953	2.77388	2.76216	2.75044	2.61953
	8	12500	10000	3.500	.766	1.20000	.00781	3.34062	3.40022	3.38850	3.37678	3.24063
	8	.12500	10000	4.000	821	1.25000	.00781	3.83750	3.90053	3.88881	3.87709	3.73750
	8	.12500	10000	4.500	.844	1.30000	.00781	4.33438	4.39884	4.38712	4.37541	4.23438
	8	.12500	.10000	5.000	.875	1.35000	.00781	4.83125	4.89766	4.88594	4.87422	4.73125
	8	12500	10000	5.563	.937	1.40630	.00781	5.39073	5.46101	5.44929	5.43757	5.29073
	8	.12500	.10000	6.625	.958	1.51250	.00781	6.44609	6.51769	6.50597	6.49425	6.34609
	8	.12500	.10000	7.625	1.000	1.61250	.00781	7.43984	7.51406	7.50234	7.49062	7.33984
	8	.12500	.10000	8.625	1.063	1.71250	.00781	8.43359	8.51175	8.50003	8.48831	8.33359
	8	.12500	.10000	9.625	1.130	1.81250	.00781	9.42734	9.50969	9.49797	9.48625	9.32734
		.12500	.10000	10.750	1.210	1.92500	.00781	10.54531	10.63266	10.62094	10.60922	10.44531
	8	0.12500	0.10000	11.750	1.285	2.02500	0.00781	11.53906	11.63109	11.61938	11.60766	11.43906
	8	.12500	.10000	12.750	1.360	2.12500	.00781	12.53281	12.62953	12.61781	12.60609	12.43281
	8	.12500	.10000	14.000	1.562	2.25000	.00781	13.77500	13.88434	13.87262	13.86091	13.67500
	8	.12500	.10000	15.000	1.687	2.35000	.00781	14.76875	14.88591	14.87419	14.86247	14.66875
	8	.12500	.10000	16.000	1.812	2.45000	.00781	15.76250	15.88747	15.87575	15.86403	15.66250
	8	.12500	.10000	17.000	1.900	2.55000	.00781	16.75625	16.88672	16.87500	16.86328	16.65625
	8	.12500	.10000	18.000	2.000	2.65000	.00781	17.75000	17.88672	17.87500	17.86328	17.65000
	8	.12500	.10000	20.000	2.125	2.85000	.00781	19.73750	19.88203	19.87031	19.85859	19.63750
	8	.12500	.10000	22.000	2.250	3.05000	.00781	21.72500	21.87734	21.86562	21.85391	21.62500
	8	.12500	.10000	24.000	2.375	3.25000	.00781	23.71250	23.87266	23.86094	23.84922	23.61250
	8	.12500	.10000	26.000	2.500	3.45000	.00781	25.70000	25.86797	25.85625	25.84453	25.60000
	8	.12500	.10000	28.000	2.625	3.65000	.00781	27.68750	27.86328	27.85156	27.83984	27.58750
	8	.12500	.10000	30.000	2.750	3.85000	.00781	29.67500	29.85859	29.84688	29.83516	29.57500

[1] Given as information for use in selecting tap drills.

Measuring Screw Threads.—There are several methods of measuring screw threads, depending on what instruments are available. The number of threads per inch may be determined by means of a steel scale as shown in Fig. 34 or by a pitch gage as shown in Fig. 35.

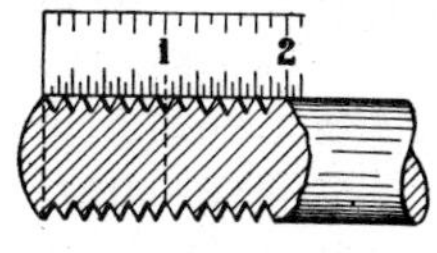

FIG. 34.

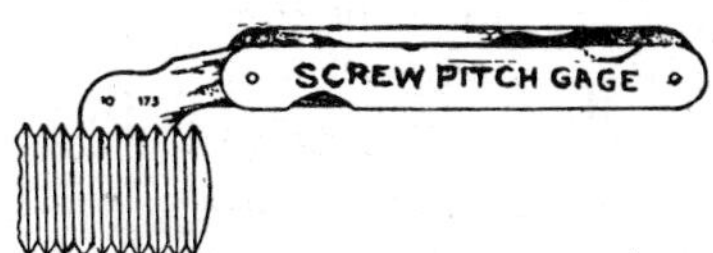

FIG. 35.

Pitch diameter is one of the most important measurements o[f] a screw. This may be read directly from a special thread microm[-]eter caliper. However, if such an instrument is not available, a[n] accurate measurement may be obtained with an ordinary microm[-]eter by the three-wire method. Three wires of equal diamete[r]

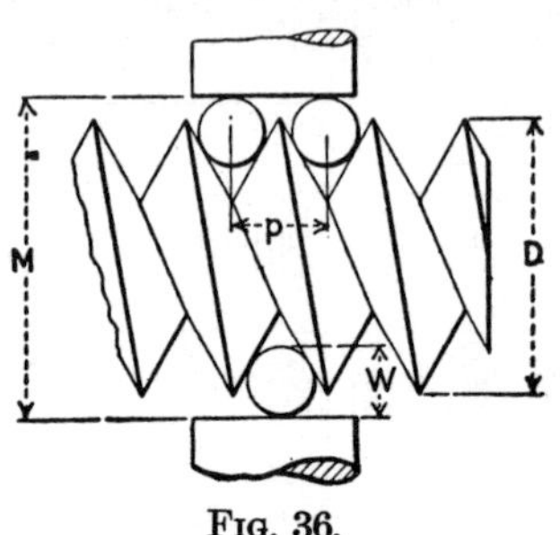

FIG. 36.

are arranged as shown in Fig. 36, one wire being placed in the angl[e] of thread on one side of the screw and the other two on the oppo[-]site side, then measuring over the whole with a micrometer.

When W = diameter of wire,

M = micrometer reading,

pitch diameter of the American National thread is:

$$\text{P.D.} = M - 3W + \frac{0.8660}{N}$$

Other equations derived from substitution of relations pertaining to this thread are:

$$D = M - 3W + 1.5155p$$
$$M = D - 1.5155p + 3W$$

ILLUSTRATION: What will be the correct micrometer reading of a $\frac{1}{2}$ in. $\times$ 12 (American National) thread if the three-wire system is used and the diameter of the wires is 0.070 in.?

$W = 0.070;\ p = \frac{1}{12};\ D = \frac{1}{2}$

$$M = D - 1.5155p + 3W = \frac{1}{2} - \frac{1.5155}{12} + 3 \times 0.070$$

$$M = 0.5 - 0.1263 + 0.210 = 0.5837 \text{ in.} \quad \text{(Ans.)}$$

Illustration: What is the pitch diameter of the threads in the above illustration?

$$\text{P.D.} = M - 3W + \frac{0.8660}{N} = 0.5837 - 3 \times 0.070 + \frac{0.8660}{12}$$

$$\text{P.D.} = 0.5837 - 0.210 + 0.0723 = 0.4460 \text{ in. (Ans.)}$$

Similar equations have been developed for use when measuring the Sharp V thread by the three-wire system. They are:

$$\text{P.D.} = M - 3W + \frac{0.8660}{N} \quad \text{(as before)}$$

But, $D = M - 3W + 1.7320p,$

and $M = D - 1.7320p + 3W$

In the three-wire system of measurement, any wire which will project above the crest of the thread and which has a diameter less than the pitch may be used. However, the best results will be obtained when the wire is of such size that it is tangent to the sides of the thread at the mid-points between the root and the crests. A wire which meets this qualification is known as the *best wire.* It can readily be demonstrated that the best-wire diameter is equal to two-thirds of the depth of a V thread. Since the depth of the V thread equals $\frac{0.866}{N}$, the best-thread diameter W is $\frac{2}{3} \times \frac{0.866}{N} = \frac{0.57735}{N}$. This formula also holds true for the proper size of wire for measuring American National threads.

Illustration: What is the best-wire size for measuring a $\frac{1}{2}$ in. $\times$ 13 American National thread bolt by the three-wire method?

$$W = \frac{0.57735}{N} = \frac{0.57735}{13} = 0.04441 \text{ inch diameter (Ans.)}$$

Screw Thread Angle.—The *angle of the helix* is designated by ϕ in Fig. 42. This angle varies with the pitch diameter and the lead of the screw.

$$\text{Tangent of helix angle} = \frac{\text{lead}}{\text{P.D.} \times \pi}$$

ILLUSTRATION: What is the helix angle of a $\frac{5}{8}$ in. $\times$ 11 American National thread?

$$\text{lead} = l = \tfrac{1}{11}$$

$$\text{P.D.} = D - \frac{0.6495}{N} = \frac{5}{8} - \frac{0.6495}{11}$$

$$\tan \phi = \frac{\frac{1}{11}}{\pi\left(\frac{5}{8} - \frac{0.6495}{11}\right)} = \frac{1}{\pi \times 11 \times (0.625 - 0.059)}$$

$$\tan \phi = \frac{7 \times 1}{22 \times 11 \times 0.566} = 0.051125$$

$$\phi = 2° \, 56' \text{ (Ans.)}$$

Taps and Tap Drills.—Internal threads less than three-quarter inch in diameter may be cut by the use of taps, shown in Fig. 37,

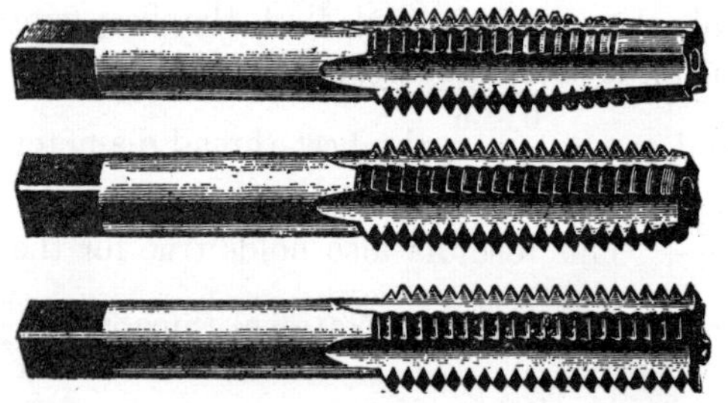

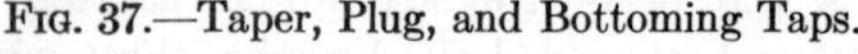

FIG. 37.—Taper, Plug, and Bottoming Taps.

FIG. 38.

and the corresponding external threads may be cut with a die. (Fig. 38.)

When drilling a hole preparatory to tapping, the theoretical size of the drill is the root diameter of the screw which is to fit the

TABLE 10

Sizes of Twist Drills with Decimal Equivalents

Size	Decimal Equivalent	Size	Decimal Equivalent	Size	Decimal Equivalent	Size	Decimal Equivalent
1/2"	0.5000"	1/4"	0.2500"	#26	0.1470"	#56	0.0465"
31/64"	.4844	E	.2500	#27	.1440	#57	.0430
15/32"	.4688	D	.2460	9/64"	.1406	#58	.0420
29/64"	.4531	C	.2420	#28	.1405	#59	.0410
7/16"	.4375	B	.2380	#29	.1360	#60	.0400
27/64"	.4219	15/64"	.2344	#30	.1285	#60½	.0390
Z	.4130	A	.2340	1/8"	.1250	#61	.0380
13/32"	.4063	#1	.2280	#31	.1200	#62	.0370
Y	.4040	#2	.2210	#32	.1160	#63	.0360
X	.3970	7/32"	.2188	#33	.1130	#64	.0350
25/64"	.3906	#3	.2130	#34	.1110	#65	.0330
W	.3860	#4	.2090	#35	.1100	#66	.0320
V	.3770	#5	.2055	7/64"	.1094	1/32"	.0313
3/8"	.3750	#6	.2040	#36	.1065	#67	.0310
U	.3680	13/64"	.2031	#37	.1040	#68	.0300
23/64"	.3594	#7	.2010	#38	.1015	#68½	.0295
T	.3580	#8	.1990	#39	.0995	#69	.0290
S	.3480	#9	.1960	#40	.0980	#69½	.0280
11/32"	.3438	#10	.1935	#41	.0960	#70	.0270
R	.3390	#11	.1910	3/32"	.0938	#71	.0260
Q	.3320	#12	.1890	#42	.0935	#71½	.0250
21/64"	.3281	3/16"	.1875	#43	.0890	#72	.0240
P	.3230	#13	.1850	#44	.0860	#73	.0230
O	.3160	#14	.1820	#45	.0820	#73½	.0225
5/16"	.3125	#15	.1800	#46	.0810	#74	.0220
N	.3020	#16	.1770	#47	.0785	#74½	.0210
19/64"	.2969	#17	.1730	5/64"	.0781	#75	.0200
M	.2950	11/64"	.1719	#48	.0760	#76	.0180
L	.2900	#18	.1695	#49	.0730	#77	.0160
9/32"	.2813	#19	.1660	#50	.0700	1/64"	.0156
K	.2810	#20	.1610	#51	.0670	#78	.0150
J	.2770	#21	.1590	#52	.0635	#78½	.0145
I	.2720	#22	.1570	1/16"	.0625	#79	.0140
H	.2660	5/32"	.1563	#53	.0595	#79½	.0135
17/64"	.2656	#23	.1540	#54	.0550	#80	.0130
G	.2610	#24	.1520	#55	.0520	...	....
F	.2570	#25	.1495	3/64"	.0469	...	....

tapped hole. In actual practice the drill is a little larger t prevent excessive strain on the tap and facilitate productio Table 10 gives sizes of twist drills with decimal equivalents Table 11 the proper tap drill sizes of American National Thread

TABLE 11

TAP DRILL SIZES FOR THREADS OF AMERICAN STANDARD FORM *

Thread Diameter	Threads to 1″	Diameter of Commercial Drill	Thread Diameter	Threads to 1″	Diameter of Commercial Drill
¼″	20	0.2010″	1″	8	0.8750″
5/16	18	.2570	1⅛	7	0.9844
⅜	16	.3125	1¼	7	1.1094
7/16	14	.3680	1⅜	6	1.2187
½	13	.4210	1½	6	1.3437
9/16	12	.4844	1⅝	5½	1.4531
⅝	11	.5312	1¾	5	1.5625
¾	10	.6562	1⅞	5	1.6875
⅞	9	.7656	2	4½	1.7812

* These tap drill diameters permit approximately 75% of a full thread.

Cutting Threads on a Lathe.—If a thread-cutting tool is brought up to a piece of work previously turned and the feed is thrown in, the threads cut will correspond to the threads on the lead screw. If the lead screw has six threads per inch and makes six revolutions, the carriage will travel one inch. The threading tool will travel the same distance along the piece to be threaded. If the spindle and the lead screw are geared one to one, the spindle will make the same number of revolutions as the lead screw. If the gear on the stud is one-half that on the lead screw, the spindle makes twice as many revolutions as the feed screw, the spindle revolving twelve times while the tool moves one inch. Therefore twelve threads will be cut.

The rate of the feed may be changed by inserting different

gears to transmit the motion from the stud to the lead screw. These gears, of which a number are provided for each machine are called "change gears" and are arranged as shown in Fig. 39. When a single idler gear connects the gears, the spindle and the lead screw, the arrangement is called *simple gearing.* To find the gear ratio between the stud and lead screw in simple gearing, the following formula is used:

$$\frac{\text{threads per in. of lead screw}}{\text{threads per in. to be cut}} = \frac{\text{teeth in gear on stud}}{\text{teeth in gear on lead screw}}$$

Illustration: What gear ratio will be required to cut 16 threads per inch on a lathe which has a lead screw with 6 threads per inch?

$$\frac{\text{threads per in. of lead screw}}{\text{threads per in. to be cut}} = \frac{6}{16} = \frac{3}{8} \text{ (Ans.)}$$

Having obtained the ratio of the gears, it is necessary to multiply the numerator and denominator by some number so that the result will represent gears in stock. From the above illustration,

$$\frac{3}{8} \times \frac{8}{8} = \frac{24}{64} \begin{array}{l} = \text{teeth in gear on stud} \\ = \text{teeth in gear on lead screw} \end{array}$$

If gears with 24 and 64 teeth are not available, some other number must be tried. Gears with 30 and 80 teeth, respectively, would serve equally well as seen below.

$$\frac{3}{8} \times \frac{10}{10} = \frac{30}{80} \begin{array}{l} = \text{teeth in gear on stud} \\ = \text{teeth in gear on lead screw} \end{array}$$

Sometimes it is not possible to obtain the correct ratio with two gears, particularly when a very small or a very large number of threads per inch are to be cut. Then it is necessary to insert two additional gears keyed to the same shaft, either replacing the idler as shown in Fig. 39 or in addition to the idler. This is called *compound gearing.*

For compound gearing, the same formula as given for simple gearing may be used except that both the numerator and the denominator are divided into two factors.

ILLUSTRATION: What change gears are required to cut a screw thread with 30 threads per inch on a lathe with a lead screw of 5 threads per inch?

$$\frac{5}{30} = \frac{1}{6} = \frac{1 \times 1}{3 \times 2}$$

These factors are then multiplied separately by numbers which will give suitable gear teeth numbers as follows:

$$\frac{1}{3} \times \frac{30}{30} = \frac{30}{90} \quad \text{and} \quad \frac{1}{2} \times \frac{25}{25} = \frac{25}{50}$$

The numbers 30 and 25 represent the teeth on driving gears; 90 and 50 the teeth on driven gears.

The number of threads per inch on the lead screw varies with the make of machine, hence the necessity of having different sets of change gears. The following are the standard gears supplied with a Reed lathe having a 5-pitch screw:

25–30–35–40–40–45–50–55–60–65–69–70–75–80–90

The following are the standard gears supplied with the Pratt & Whitney lathe having a 6-pitch screw:

30–40–50–60–65–70–75–80–90–95–100–105–110–115–120

ILLUSTRATION: What change gears can be used to cut 13 threads to the inch with a lathe that has a lead screw with four threads to the inch, using a stud gear of 20 teeth? From the proportion on page 560.

$$x : 20 = 13 : 4$$

$$x = \frac{20 \times 13}{4} = 65$$

Therefore, a 65 T gear is used on lead screw.

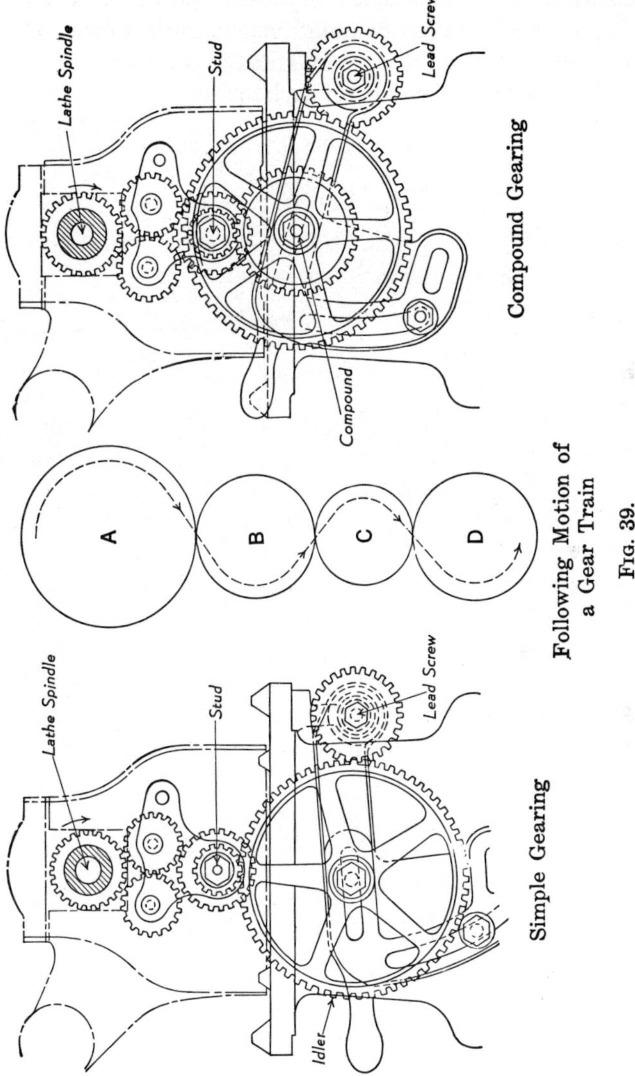

Compound Gearing

Following Motion of a Gear Train

Simple Gearing

FIG. 39.

ILLUSTRATION: Using a 110-tooth gear on the lead screw and a 75 on the stud, with compound driven and driver gears of 50 and 80 teeth, respectively, how many threads per inch will be cut if the lead screw has 6 threads per inch?

$$x : 6 = (110 \times 50) : (75 \times 80)$$

$$x = \frac{6 \times 110 \times 50}{75 \times 80} = 5.5$$

Therefore 5.5 threads per inch will be cut.

Milling

Simple Indexing.—In machine shop milling it is often necessary to machine a piece of work on several faces with considerable accuracy. This is usually accomplished by attaching the work to a dividing or index head so that it may be rotated into any position. (See Fig. 40.) On all standard dividing heads it requires 40 turns of the index crank to revolve the dividing head spindle once.

If a piece of work is to be cut at any number of points equidistant apart on its periphery, then to find the number of turns of the index crank for these divisions, divide the number of turns required for one revolution of the dividing head (40) by the number of divisions wanted.

$$R = \frac{40}{N}$$

when N = number of divisions required;
R = number of turns of the crank for given division.

ILLUSTRATION: A 57-toothed gear is to be turned on a milling machine. How many turns of the crank will be required to turn the work from one tooth to the next?

$$R = \frac{40}{N} = \frac{40}{57} \text{ revolution (Ans.)}$$

FIG. 40.—Above, Simple Indexing; Below, Differential Indexing.

The last illustration brings up the question of how the crank is to be stopped accurately at fractional revolutions, in this case $\frac{40}{57}$ of a revolution. This is accomplished by the perforated index plate shown in Fig. 40. This plate has small holes evenly spaced along concentric circles. There are generally three interchangeable plates with each dividing head. The following list gives the number of holes per circle on the three plates used on a standard machine.

Plate	Number of Holes in the Various Circles
1.........	15 — 16 — 17 -- 18 — 19 — 20
2.........	21 — 23 — 27 — 29 — 31 — 33
3.........	37 — 39 — 41 — 43 — 47 — 49

Some dividing heads have only one plate. In this case the plate has holes on each side as follows:

one side 24–25–28–30–34–37–38–39–41–42–43

and on the other side

46–47–49–51–53–54–57–58–59–62–66

The crank is provided with an index pin which engages the desired hole and holds the crank stationary.

ILLUSTRATION: What is the simple indexing for 330 divisions?

$R = \dfrac{40}{N} = \dfrac{40}{330} = \dfrac{4}{33}$ revolution, or 4 spaces on the circle with 33 holes or 8 spaces on the circle with 66 holes. (Ans.)

In order to obtain a number of divisions that cannot be obtained with ordinary index plates a process of differential indexing is used. By this process the index plate is revolved by suitable gears which connect it to the dividing head spindle, the stop pin holding the index plate being disengaged altogether. See Fig. 40.)

The rotary or differential motion of the index plate takes place when the crank is turned, which turns the plate either forward or backward as may be required. The result is that the actual movement of the crank, in indexing, is either more or less than the movement in relation to the index plate.

The differential method cannot be used in connection with spiral milling, because the dividing head spindle is geared to the lead screw of the milling machine.

The amount of rotation of the index plate may be regulated by the difference in velocity ratios of the change gears.

ILLUSTRATION: Find the indexing required for 81 divisions. By simple indexing the index crank would be rotated through $\frac{40}{81}$ of a turn for each division, but as there is no plate with 81 divisions, the spacing is impossible: therefore, another fraction is selected whose value is near $\frac{40}{81}$, for example, $\frac{40}{81}$ or $\frac{10}{21}$, then a 21-hole circle can be used.

In simple indexing for 80 divisions the movement of the index crank is $\frac{40}{80}$ or $\frac{1}{2}$ of a turn for each cut.

If the crank is given $\frac{1}{2}$ of a turn eighty-one times, it makes $40\frac{1}{2}$ turns or $\frac{1}{2}$ of a turn more than the 40 turns required for one revolution of the work. Hence the index plate must move backward $\frac{1}{2}$ of a revolution while the work revolves once

$$\frac{40}{80} = \frac{1}{2}, \quad 81 \times \tfrac{1}{2} = 40\tfrac{1}{2}$$

$$41\tfrac{1}{2} - 40 = 1\tfrac{1}{2}$$

Hence the ratio of the gears is 1 : 2.

$$\frac{1}{2} \times \frac{24}{24} = \frac{24}{48}$$

A 24 T gear (driving) is placed on the special differential indexing center in the spindle of the dividing head; and the 48 T gear (driven) is placed on the worm shaft which turns the index plate (See gear on spindle and gear on worm in Fig. 40.)

TABLE 12

Leads, Change Gears and Angles for Cutting Spirals

Lead of Spiral, Inches	Gear on Worm	1st Intermediate Gear	2d Intermediate Gear	Gear on Screw	Diameter of Work, Inches									
					⅛	¼	⅜	½	⅝	¾	⅞	1	1¼	1½
0.67	24	86	24	100	30¼	...	...	...	Approximate Angles for Milling Machine Table					
0.78	24	86	28	100	26	44½	...	...						
0.89	24	86	32	100	23½	41	...	...						
1.12	24	86	40	100	19	34½	...	...						
1.34	24	86	48	100	16	30¼	41½	...	...	...	...	...	...	...
1.46	24	64	28	72	14¾	28	38½	...	...	...	...	...	...	...
1.56	24	86	56	100	13¾	26½	37	...	...	...	...	...	...	...
1.67	24	64	32	72	12¾	25	34¾	43¼	...	...	...	...	...	...
1.94	32	64	28	72	11¼	21¾	31	39	45	...	...	...	...	...
2.08	24	64	40	72	10¼	20½	29½	37	43¼	...	...	...	...	...
2.22	32	56	28	72	9¾	19¼	27½	35	41¼	...	...	...	...	...
2.50	24	64	48	72	8¾	17	25	32	38	43¼	...	...	...	...
2.78	40	56	28	72	8	15½	23	29½	35¼	40½	44¾	...	...	...
2.92	24	64	56	72	7½	15	21¾	28¼	34	39	43¼	...	...	...
3.24	40	48	28	72	6¾	13¼	19¾	25¾	31¼	36	40½	44¼	...	...
3.70	40	48	32	72	6	11¾	17½	23	28	32½	36½	40½	...	...
3.89	56	48	24	72	5½	11¼	16¾	22	26¾	31¼	35¼	39	...	...
4.17	40	72	48	64	5¼	10½	15¾	20½	25¼	29½	33½	37	43¼	...
4.46	48	40	32	86	4¾	9¾	14¾	19¼	23¾	27¾	31½	35	41½	...
4.86	40	64	56	72	4½	9	13½	17¾	22	25¾	29½	33	39	44¼
5.33	48	40	32	72	4	8¼	12¼	16½	20¼	23¾	27¼	30½	36½	41½
5.44	56	40	28	72	4	8	12	16	20	23½	26¾	30	36	41
6.12	56	40	28	64	3½	7¼	11	14½	17¾	21	24¼	27	33	37¾
6.22	56	40	32	72	3½	7	10¾	14¼	17½	20¾	23¾	26¾	32½	37¼
6.48	56	48	40	72	3¼	6¾	10¼	13½	16¾	20	23	25¾	31½	36¼
6.67	64	48	28	56	3¼	6½	10	13¼	16½	19½	22½	25¼	30¾	35¼
7.29	56	48	40	64	3	6¼	9¼	12¼	15	18	20½	23½	28½	33
7.41	64	48	40	72	3	6	9	12	14¾	17¾	20¼	22¾	28¼	32½
7.62	64	48	32	56	2¾	5¾	8¾	11½	14½	17¼	19¾	22¼	27½	32
8.33	48	32	40	72	2½	5¼	8	10½	13¼	15¾	18¼	20½	25½	29½
8.95	86	48	28	56	2½	5	7½	10	12½	14¾	17	19¼	24	28
9.33	56	40	48	72	2¼	4¾	7¼	9½	11¾	14	16¼	18½	23	27
9.52	64	48	40	56	2¼	4½	7	9¼	11½	13¾	16	18¼	22½	26½
10.29	72	40	32	56	2	4¼	6½	8¾	10¾	12¾	15	17¼	21	24¾
10.37	64	48	56	72	2	4¼	6½	8½	10½	12¾	14¾	17	20¾	24½
10.50	48	40	56	64	2	4¼	6¼	8½	10½	12½	14½	16¾	20½	24¼
10.67	64	40	48	72	2	4	6¼	8¼	10¼	12¼	14¼	16½	20¼	24
10.94	56	32	40	64	2	4	6	8¼	10¼	12	14	16¼	20	23½
11.11	64	32	40	72	2	4	6	8	10	11¾	13¾	16	19¾	23
11.66	56	32	48	72	1¾	3¾	5¾	7½	9½	11¼	13¼	15¼	18¾	22
12.00	72	40	32	48	1¾	3¾	5½	7¼	9¼	11	12¾	15	18¼	21½
13.12	56	32	48	64	1½	3½	5¼	6¾	8½	10¼	11¾	13½	16¾	20
13.33	56	28	48	72	1½	3¼	5	6½	8¼	10	11½	13¼	16½	19½
13.71	64	40	48	56	1½	3¼	4¾	6½	8	9¾	11¼	13	16	19
15.24	64	28	48	72	1½	3	4½	5¾	7¼	8¾	10¼	11¾	14½	17¼
15.56	64	32	56	72	1¼	2¾	4¼	5¾	7¼	8¾	10	11½	14¼	17
15.75	56	64	72	40	1¼	2¾	4¼	5½	7	8½	9¾	11¼	14	16¾
16.87	72	32	48	64	1¼	2½	4	5¼	6¾	7¾	9¼	10½	13¼	15¾
17.14	64	32	48	56	1¼	2½	4	5¼	6½	7¾	9	10¼	13	15½
18.75	72	32	40	48	1	2¼	3½	4¾	6	7¼	8¼	9½	12	14¼
19.29	72	32	48	56	1	2¼	3½	4½	5¾	7	8	9¼	11½	13¾
19.59	64	28	48	56	1	2¼	3¼	4½	5¾	6¾	8	9¼	11½	13½
19.69	72	32	56	64	1	2¼	3¼	4½	5¾	6¾	8	9	11½	13½
21.43	72	24	40	56	1	2	3¼	4¼	5¼	6¼	7½	8½	10½	12½
22.50	72	28	56	64	1	2	3	4	5	6	7	8	10	12
23.33	64	32	56	48	1	2	3	4	5	5¾	6¾	7¾	9¾	11½
26.25	72	24	56	64	1	1¾	2¾	3½	4¼	5	6	7	8½	10¼
26.67	64	28	56	48	¾	1¾	2¾	3½	4¼	5	6	6¾	8½	10
28.00	64	32	56	40	¾	1¾	2½	3¼	4	4¾	5¾	6½	8	9½
30.86	72	28	48	40	¾	1½	2¼	3	3¾	4½	5	5¾	7¼	8¾

As the motion of the index plate must be in the direction opposite to the movement of the index crank, idler gears must be used. These do not affect the ratio.

The following gears are generally available for differential indexing: 24–24–28–32–40–44–48–56–64–72–86–100.

Angular Indexing.—Sometimes a milling job calls for making cuts at intervals of a certain number of degrees around the periphery of a piece of work. With a standard index head, where 40 turns of the index crank are required for one revolution of the work one turn of the crank equals $\frac{1}{40}$ of 360 degrees or 9 degrees.

Thus, if one complete turn of the crank equals 9 degrees, 2 holes in the 18 circle or 3 holes in the 27 circle must equal 1 degree, or 1 hole in the 18 circle will equal $\frac{1}{2}$ degree or 30 minutes, and 1 hole in the 27 circle will equal $\frac{1}{3}$ of a degree or 20 minutes.

ILLUSTRATION: What is the angular indexing for 19 degrees?

If 1 turn equals 9 degrees, 2 turns equal 18 degrees. Add 2 holes on 18 circle or 3 holes on 27 circle.

Indexing for 19 degrees is then, 2 turns + 2 holes on 18 circle or 2 turns + 3 holes on 27 circle. (Ans.)

ILLUSTRATION: What is the angular indexing for 7 degrees 40 minutes?

$$40 \text{ minutes} = \tfrac{2}{3} \text{ degree.}$$

then

$$7\tfrac{2}{3} \div 9 = \tfrac{23}{27}.$$

Therefore, the indexing for 7 deg. 40 min. is 23 holes on 27 circle. (Ans.)

Table 13 gives the plain and differential indexing of the numbers up to 370.

Spiral Milling.—Cutting a helical milling cutter as shown in Fig. 41, or a twist drill, is called *spiral milling* and can be attained by the use of an index head so geared to the longitudinal feed screw of the milling machine as to impart a rotary motion to the work as it is fed along under the cutter by the action of a train of gears.

The *lead of a helix* or *spiral* is the distance, measured along the axis of the work, which the spiral makes in one full turn around the work.

TABLE 13

Number of Divisions	Index Circle	No. of Turns of Index	Gear on Worm	No. 1 Hole		Gear on Spindle	Idlers	
				First Gear on Stud	Second Gear on Stud		No. 1 Hole	No. 2 Hole
2	Any	20						
3	39	$13\frac{13}{39}$						
4	Any	10						
5	Any	8						
6	39	$6\frac{26}{39}$						
7	49	$5\frac{35}{49}$						
8	Any	5						
9	27	$4\frac{12}{27}$						
10	Any	4						
11	33	$3\frac{21}{33}$						
12	39	$3\frac{13}{39}$						
13	39	$3\frac{3}{39}$						
14	49	$2\frac{42}{49}$						
15	39	$2\frac{26}{39}$						
16	20	$2\frac{10}{20}$						
17	17	$2\frac{6}{17}$						
18	27	$2\frac{6}{27}$						
19	19	$2\frac{2}{19}$						
20	Any	2						
21	21	$1\frac{19}{21}$						
22	33	$1\frac{27}{33}$						
23	23	$1\frac{17}{23}$						
24	39	$1\frac{26}{39}$						
25	20	$1\frac{12}{20}$						
26	39	$1\frac{21}{39}$						
27	27	$1\frac{13}{27}$						
28	49	$1\frac{21}{49}$						
29	29	$1\frac{11}{29}$						
30	39	$1\frac{13}{39}$						
31	31	$1\frac{9}{31}$						
32	20	$1\frac{5}{20}$						
33	33	$1\frac{7}{33}$						
34	17	$1\frac{3}{17}$						
35	49	$1\frac{7}{49}$						
36	27	$1\frac{3}{27}$						
37	37	$1\frac{3}{37}$						
38	19	$1\frac{1}{19}$						
39	39	$1\frac{1}{39}$						
40	Any	1						
41	41	$\frac{40}{41}$						
42	21	$\frac{20}{21}$						

Table 13.—*Continued*

Number of Divisions	Index Circle	No. of Turns of Index	Gear on Worm	No. 1 Hole		Gear on Spindle	Idlers	
				First Gear on Stud	Second Gear on Stud		No. 1 Hole	No. 2 Hole
43	43	$\frac{40}{43}$						
44	33	$\frac{30}{33}$						
45	27	$\frac{24}{27}$						
46	23	$\frac{20}{23}$						
47	47	$\frac{40}{47}$						
48	18	$\frac{15}{18}$						
49	49	$\frac{40}{49}$						
50	20	$\frac{16}{20}$						
51	17	$\frac{14}{17}$	24			48	24	44
52	39	$\frac{30}{39}$						
53	49	$\frac{35}{49}$	56	40	24	72		
54	27	$\frac{20}{27}$						
55	33	$\frac{24}{33}$						
56	49	$\frac{35}{49}$						
57	21	$\frac{15}{21}$	56			40	24	44
58	29	$\frac{20}{29}$						
59	39	$\frac{26}{39}$	48			32	44	
60	39	$\frac{26}{39}$						
61	39	$\frac{26}{39}$	48			32	24	44
62	31	$\frac{20}{31}$						
63	39	$\frac{26}{39}$	24			48	24	44
64	16	$\frac{10}{16}$						
65	39	$\frac{24}{39}$						
66	33	$\frac{20}{33}$						
67	21	$\frac{12}{21}$	28			48	44	
68	17	$\frac{10}{17}$						
69	20	$\frac{12}{20}$	40			56	24	44
70	49	$\frac{28}{49}$						
71	18	$\frac{10}{18}$	72			40	24	
72	27	$\frac{15}{27}$						
73	21	$\frac{12}{21}$	28			48	24	44
74	37	$\frac{20}{37}$						
75	15	$\frac{8}{15}$						
76	19	$\frac{10}{19}$						
77	20	$\frac{10}{20}$	32			48	44	
78	39	$\frac{20}{39}$						
79	20	$\frac{10}{20}$	48			24	44	
80	20	$\frac{10}{20}$						
81	20	$\frac{10}{20}$	48			24	24	44
82	41	$\frac{20}{41}$						
83	20	$\frac{10}{20}$	32			48	24	44

TABLE 13.—*Continued*

Number of Divisions	Index Circle	No. of Turns of Index	Gear on Worm	No. 1 Hole: First Gear on Stud	No. 1 Hole: Second Gear on Stud	Gear on Spindle	Idlers: No. 1 Hole	Idlers: No. 2 Hole
84	21	10/21						
85	17	8/17						
86	43	20/43						
87	15	7/15	40			24	24	44
88	33	15/33						
89	18	8/18	72			32	44	
90	27	12/27						
91	39	18/39	24			48	24	44
92	23	10/23						
93	18	8/18	24			32	24	44
94	47	20/47						
95	19	8/19						
96	21	9/21	28			32	24	44
97	20	8/20	40			48	44	
98	49	20/49						
99	20	8/20	56	28	40	32		
100	20	8/20						
101	20	8/20	72	24	40	48		24
102	20	8/20	40			32	24	44
103	20	8/20	40			48	24	44
104	39	15/39						
105	21	8/21						
106	43	16/43	86	24	24	48		
107	20	8/20	40	56	32	64		24
108	27	10/27						
109	16	6/16	32			28	24	44
110	33	12/33						
111	39	13/39	24			72	32	
112	39	13/39	24			64	44	
113	39	13/39	24			56	44	
114	39	13/39	24			48	44	
115	23	8/23						
116	29	10/29						
117	39	13/39	24			24	56	
118	39	13/39	48			32	44	
119	39	13/39	72			24	44	
120	39	13/39						
121	39	13/39	72			24	24	44
122	39	13/39	48			32	24	44
123	39	13/39	24			24	24	44
124	31	10/31						

Table 13.—*Continued*

Number of Divisions	Index Circle	No. of Turns of Index	Gear on Worm	No. 1 Hole		Gear on Spindle	Idlers	
				First Gear on Stud	Second Gear on Stud		No. 1 Hole	No. 2 Hole
125	39	$\frac{13}{39}$	24			40	24	44
126	39	$\frac{13}{39}$	24			48	24	44
127	39	$\frac{13}{39}$	24			56	24	44
128	16	$\frac{5}{16}$						
129	39	$\frac{13}{39}$	24			72	24	44
130	39	$\frac{12}{39}$						
131	20	$\frac{6}{20}$	40			28	44	
132	33	$\frac{10}{33}$						
133	21	$\frac{6}{21}$	24			48	44	
134	21	$\frac{6}{21}$	28			48	44	
135	27	$\frac{8}{27}$						
136	17	$\frac{5}{17}$						
137	21	$\frac{6}{21}$	28			24	56	
138	21	$\frac{6}{21}$	56			32	44	
139	21	$\frac{6}{21}$	56	32	48	24		
140	49	$\frac{14}{49}$						
141	18	$\frac{5}{18}$	48			40	44	
142	21	$\frac{6}{21}$	56			32	24	44
143	21	$\frac{6}{21}$	28			24	24	44
144	18	$\frac{5}{18}$						
145	29	$\frac{8}{29}$						
146	21	$\frac{6}{21}$	28			48	24	44
147	21	$\frac{6}{21}$	24			48	24	44
148	37	$\frac{10}{37}$						
149	21	$\frac{6}{21}$	28			72	24	44
150	15	$\frac{4}{15}$						
151	20	$\frac{5}{20}$	32			72	44	
152	19	$\frac{5}{19}$						
153	20	$\frac{5}{20}$	32			56	44	
154	20	$\frac{5}{20}$	32			48	44	
155	31	$\frac{8}{31}$						
156	39	$\frac{10}{39}$						
157	20	$\frac{5}{20}$	32			24	56	
158	20	$\frac{5}{20}$	48			24	44	
159	20	$\frac{5}{20}$	64	32	56	28		
160	20	$\frac{5}{20}$						
161	20	$\frac{5}{20}$	64	32	56	28		24
162	20	$\frac{5}{20}$	48			24	24	44
163	20	$\frac{5}{20}$	32			24	24	44
164	41	$\frac{10}{41}$						
165	33	$\frac{8}{33}$						

TABLE 13.—*Continued*

Number of Divisions	Index Circle	No. of Turns of Index	Gear on Worm	No. 1 Hole		Gear on Spindle	Idlers	
				First Gear on Stud	Second Gear on Stud		No. 1 Hole	No. 2 Hole
166	20	$\frac{5}{20}$	32			48	24	44
167	20	$\frac{5}{20}$	32			56	24	44
168	21	$\frac{5}{21}$						
169	20	$\frac{5}{20}$	32			72	24	44
170	17	$\frac{4}{17}$						
171	21	$\frac{5}{21}$	56			40	24	44
172	43	$\frac{10}{43}$						
173	18	$\frac{4}{18}$	72	56	32	64		
174	18	$\frac{4}{18}$	24			32	56	
175	18	$\frac{4}{18}$	72	40	32	64		
176	18	$\frac{4}{18}$	72	24	24	64		
177	18	$\frac{4}{18}$	72			48	24	
178	18	$\frac{4}{18}$	72			32	44	
179	18	$\frac{4}{18}$	72	24	48	32		
180	18	$\frac{4}{18}$						
181	18	$\frac{4}{18}$	72	24	48	32		24
182	18	$\frac{4}{18}$	72			32	24	44
183	18	$\frac{4}{18}$	48			32	24	44
184	23	$\frac{5}{23}$						
185	37	$\frac{8}{37}$						
186	18	$\frac{4}{18}$	48			64	24	44
187	18	$\frac{4}{18}$	72	48	24	56		24
188	47	$\frac{10}{47}$						
189	18	$\frac{4}{18}$	32			64	24	44
190	19	$\frac{4}{19}$						
191	20	$\frac{4}{20}$	40			72	24	
192	20	$\frac{4}{20}$	40			64	44	
193	20	$\frac{4}{20}$	40			56	44	
194	20	$\frac{4}{20}$	40			48	44	
195	39	$\frac{8}{39}$						
196	49	$\frac{10}{49}$						
197	20	$\frac{4}{20}$	40			24	56	
198	20	$\frac{4}{20}$	56	28	40	32		
199	20	$\frac{4}{20}$	100	40	64	32		
200	20	$\frac{4}{20}$						
201	20	$\frac{4}{20}$	72	24	40	24		24
202	20	$\frac{4}{20}$	72	24	40	48		24
203	20	$\frac{4}{20}$	40			24	24	44
204	20	$\frac{4}{20}$	40			32	24	44
205	41	$\frac{8}{41}$						
206	20	$\frac{4}{20}$	40			48	24	44

TABLE 13.—*Continued*

Number of Divisions	Index Circle	No. of Turns of Index	Gear on Worm	No. 1 Hole		Gear on Spindle	Idlers	
				First Gear on Stud	Second Gear on Stud		No. 1 Hole	No. 2 Hole
207	20	$\frac{4}{20}$	40			56	24	44
208	20	$\frac{4}{20}$	40			64	24	44
209	20	$\frac{4}{20}$	40			72	24	44
210	21	$\frac{4}{21}$						
211	16	$\frac{8}{16}$	64			28	44	
212	43	$\frac{8}{43}$	86	24	24	48		
213	27	$\frac{5}{27}$	72			40	44	
214	20	$\frac{4}{20}$	40	56	32	64		24
215	43	$\frac{8}{43}$						
216	27	$\frac{5}{27}$						
217	21	$\frac{4}{21}$	48			64	24	44
218	16	$\frac{8}{16}$	64			56	24	44
219	21	$\frac{4}{21}$	28			48	24	44
220	33	$\frac{6}{33}$						
221	17	$\frac{3}{17}$	24			24	56	
222	18	$\frac{3}{18}$	24			72	44	
223	43	$\frac{8}{43}$	86	48	24	64		24
224	18	$\frac{3}{18}$	24			64	44	
225	27	$\frac{5}{27}$	24			40	24	44
226	18	$\frac{3}{18}$	24			56	44	
227	49	$\frac{8}{49}$	56	64	28	72		
228	18	$\frac{3}{18}$	24			48	44	
229	18	$\frac{3}{18}$	24			44	48	
230	23	$\frac{4}{23}$						
231	18	$\frac{3}{18}$	32			48	44	
232	29	$\frac{5}{29}$						
233	18	$\frac{3}{18}$	48			56	44	
234	18	$\frac{3}{18}$	24			24	56	
235	47	$\frac{8}{47}$						
236	18	$\frac{3}{18}$	48			32	44	
237	18	$\frac{3}{18}$	48			24	44	
238	18	$\frac{3}{18}$	72			24	44	
239	18	$\frac{3}{18}$	72	24	64	32		
240	18	$\frac{3}{18}$						
241	18	$\frac{3}{18}$	72	24	64	32		24
242	18	$\frac{3}{18}$	72			24	24	44
243	18	$\frac{3}{18}$	64			32	24	44
244	18	$\frac{3}{18}$	48			32	24	44
245	49	$\frac{8}{49}$						
246	18	$\frac{3}{18}$	24			24	24	44
247	18	$\frac{3}{18}$	48			56	24	44

Table 13.—*Continued*

Number of Divisions	Index Circle	No. of Turns of Index	Gear on Worm	No. 1 Hole: First Gear on Stud	No. 1 Hole: Second Gear on Stud	Gear on Spindle	Idlers: No. 1 Hole	Idlers: No. 2 Hole
248	31	$\frac{5}{31}$						
249	18	$\frac{3}{18}$	32			48	24	44
250	18	$\frac{3}{18}$	24			40	24	44
251	18	$\frac{8}{18}$	48	44	32	64		24
252	18	$\frac{3}{18}$	24			48	24	44
253	33	$\frac{5}{33}$	24			40	56	
254	18	$\frac{3}{18}$	24			56	24	44
255	18	$\frac{8}{18}$	48	40	24	72		24
256	18	$\frac{3}{18}$	24			64	24	44
257	49	$\frac{8}{49}$	56	48	28	64		24
258	43	$\frac{7}{43}$	32			64	24	44
259	21	$\frac{3}{21}$	24			72	44	
260	39	$\frac{6}{39}$						
261	29	$\frac{4}{29}$	48	64	24	72		
262	20	$\frac{3}{20}$	40			28	44	
263	49	$\frac{3}{49}$	56	64	28	72		24
264	33	$\frac{5}{33}$						
265	21	$\frac{8}{21}$	56	40	24	72		
266	21	$\frac{3}{21}$	32			64	44	
267	27	$\frac{4}{27}$	72			32	44	
268	21	$\frac{3}{21}$	28			48	44	
269	20	$\frac{3}{20}$	64	32	40	28		24
270	27	$\frac{4}{27}$						
271	21	$\frac{8}{21}$	56			72	24	
272	21	$\frac{3}{21}$	56			64	24	
273	21	$\frac{3}{21}$	24			24	56	
274	21	$\frac{8}{21}$	56			48	44	
275	21	$\frac{8}{21}$	56			40	44	
276	21	$\frac{8}{21}$	56			32	44	
277	21	$\frac{3}{21}$	56			24	44	
278	21	$\frac{3}{21}$	56	32	48	24		
279	27	$\frac{4}{27}$	24			32	24	44
280	49	$\frac{7}{49}$						
281	21	$\frac{3}{21}$	72	24	56	24		24
282	43	$\frac{6}{43}$	86	24	24	56		
283	21	$\frac{3}{21}$	56			24	24	44
284	21	$\frac{8}{21}$	56			32	24	44
285	21	$\frac{8}{21}$	56			40	24	44
286	21	$\frac{3}{21}$	56			48	24	44
287	21	$\frac{3}{21}$	24			24	24	44
288	21	$\frac{3}{21}$	28			32	24	44

TABLE 13.—*Continued*

Number of Divisions	Index Circle	No. of Turns of Index	Gear on Worm	No. 1 Hole		Gear on Spindle	Idlers	
				First Gear on Stud	Second Gear on Stud		No. 1 Hole	No. 2 Hole
289	21	$\frac{3}{21}$	56			72	24	44
290	29	$\frac{4}{29}$						
291	15	$\frac{2}{15}$	40			48	44	
292	21	$\frac{3}{21}$	28			48	24	44
293	15	$\frac{2}{15}$	48	32	40	56		
294	21	$\frac{3}{21}$	24			48	24	44
295	15	$\frac{2}{15}$	48			32	44	
296	37	$\frac{5}{37}$						
297	33	$\frac{4}{33}$	28	48	24	56		
298	21	$\frac{3}{21}$	28			72	24	44
299	23	$\frac{3}{23}$	24			24	56	
300	15	$\frac{2}{15}$						
301	43	$\frac{6}{43}$	24			48	24	44
302	16	$\frac{2}{16}$	32			72	24	
303	15	$\frac{2}{15}$	72	24	40	48		24
304	16	$\frac{2}{16}$	24			48	44	
305	15	$\frac{2}{15}$	48			32	24	44
306	15	$\frac{2}{15}$	40			32	24	44
307	15	$\frac{2}{15}$	72	48	40	56		24
308	16	$\frac{2}{16}$	32			48	44	
309	15	$\frac{2}{15}$	40			48	24	44
310	31	$\frac{4}{31}$						
311	16	$\frac{2}{16}$	64	24	24	72		
312	39	$\frac{5}{39}$						
313	16	$\frac{2}{16}$	32			28	56	
314	16	$\frac{2}{16}$	32			24	56	
315	16	$\frac{2}{16}$	64			40	24	
316	16	$\frac{2}{16}$	64			32	44	
317	16	$\frac{2}{16}$	64			24	44	
318	16	$\frac{2}{16}$	56	28	48	24		
319	29	$\frac{4}{29}$	48	64	24	72		24
320	16	$\frac{2}{16}$						
321	16	$\frac{2}{16}$	72	24	64	24		24
322	23	$\frac{3}{23}$	32			64	24	44
323	16	$\frac{2}{16}$	64			24	24	44
324	16	$\frac{2}{16}$	64			32	24	44
325	16	$\frac{2}{16}$	64			40	24	44
326	16	$\frac{2}{16}$	32			24	24	44
327	16	$\frac{2}{16}$	32			28	24	44
328	41	$\frac{5}{41}$						
329	16	$\frac{2}{16}$	64	24	24	72		24

TABLE 13.—*Continued*

Number of Divisions	Index Circle	No. of Turns of Index	Gear on Worm	No. 1 Hole		Gear on Spindle	Idlers	
				First Gear on Stud	Second Gear on Stud		No. 1 Hole	No. 2 Hole
330	33	$\frac{4}{33}$						
331	16	$\frac{2}{16}$	64	44	24	48		24
332	16	$\frac{2}{16}$	32			48	24	44
333	18	$\frac{2}{18}$	24			72	44	
334	16	$\frac{2}{16}$	32			56	24	44
335	33	$\frac{4}{33}$	72	48	44	40		24
336	16	$\frac{2}{16}$	32			64	24	44
337	43	$\frac{5}{43}$	86	40	32	56		
338	16	$\frac{2}{16}$	32			72	24	44
339	18	$\frac{2}{18}$	24			56	44	
340	17	$\frac{2}{17}$						
341	43	$\frac{5}{43}$	86	24	32	40		
342	18	$\frac{2}{18}$	32			64	44	
343	15	$\frac{2}{15}$	40	64	24	86		24
344	43	$\frac{5}{43}$						
345	18	$\frac{2}{18}$	24			40	56	
346	18	$\frac{2}{18}$	72	56	32	64		
347	43	$\frac{5}{43}$	86	24	32	40		24
348	18	$\frac{2}{18}$	24			32	56	
349	18	$\frac{2}{18}$	72	44	24	48		
350	18	$\frac{2}{18}$	72	40	32	64		
351	18	$\frac{2}{18}$	24			24	56	
352	18	$\frac{2}{18}$	72	24	24	64		
353	18	$\frac{2}{18}$	72			56	24	
354	18	$\frac{2}{18}$	72			48	24	
355	18	$\frac{2}{18}$	72			40	24	
356	18	$\frac{2}{18}$	72			32	24	
357	18	$\frac{2}{18}$	72			24	44	
358	18	$\frac{2}{18}$	72	32	48	24		
359	43	$\frac{5}{43}$	86	48	32	100		24
360	18	$\frac{2}{18}$						
361	19	$\frac{2}{19}$	32			64	44	
362	18	$\frac{2}{18}$	72	28	56	32		24
363	18	$\frac{2}{18}$	72			24	24	44
364	18	$\frac{2}{18}$	72			32	24	44
365	20	$\frac{2}{20}$	32	48	24	56		
366	18	$\frac{2}{18}$	48			32	24	44
367	18	$\frac{2}{18}$	72			56	24	24
368	18	$\frac{2}{18}$	72	24	24	64		24
369	41	$\frac{4}{41}$	32	56	28	64		
370	37	$\frac{4}{37}$						

TABLE 13.—*Continued*

Number of Divisions	Index Circle	No. of Turns of Index	Gear on Worm	No. 1 Hole		Gear on Spindle	Idlers	
				First Gear on Stud	Second Gear on Stud		No. 1 Hole	No. 2 Hole
371	21	$\frac{2}{21}$	32	56	24	64		
372	18	$\frac{2}{18}$	48			64	24	44
373	20	$\frac{2}{20}$	40	48	32	72		
374	18	$\frac{2}{18}$	72	64	32	56		24
375	18	$\frac{2}{18}$	24			40	24	44
376	47	$\frac{5}{47}$						
377	29	$\frac{8}{29}$	24			24	56	
378	18	$\frac{2}{18}$	32			64	24	44
379	20	$\frac{2}{20}$	48	56	40	72		
380	19	$\frac{2}{19}$						
381	18	$\frac{2}{18}$	24			56	24	44
382	20	$\frac{2}{20}$	40			72	24	
383	20	$\frac{2}{20}$	40			68 [1]	44	
384	20	$\frac{2}{20}$	40			64	44	
385	20	$\frac{2}{20}$	32			48	44	
386	20	$\frac{2}{20}$	40			56	44	
387	43	$\frac{4}{43}$	32	56	28	64		
388	20	$\frac{2}{20}$	40			48	44	
389	20	$\frac{2}{20}$	40			44	56	
390	39	$\frac{4}{39}$						
391	20	$\frac{2}{20}$	48	24	40	72		
392	49	$\frac{5}{49}$						
393	20	$\frac{2}{20}$	40			28	44	
394	20	$\frac{2}{20}$	40			24	56	
395	20	$\frac{2}{20}$	64			32	44	
396	20	$\frac{2}{20}$	56	28	40	32		
397	20	$\frac{2}{20}$	64	24	40	32		
398	20	$\frac{2}{20}$	100	40	64	32		
399	21	$\frac{2}{21}$	32			64	44	
400	20	$\frac{2}{20}$						
401	21	$\frac{2}{21}$	56	32	24	76 [1]		
402	21	$\frac{2}{21}$	28			48	44	
403	20	$\frac{2}{20}$	64	24	40	32		24
404	20	$\frac{2}{20}$	72	24	40	48		24
405	20	$\frac{2}{20}$	64			32	24	44
406	20	$\frac{2}{20}$	40			24	24	44
407	20	$\frac{2}{20}$	40			28	24	44
408	20	$\frac{2}{20}$	40			32	24	44
409	20	$\frac{2}{20}$	40	24	32	48		24
410	41	$\frac{4}{41}$						

NOTE. Special gears in this and following tables are 46, 47, 52, 58, 68, 70, 76, 84.

[1] Special gear.

TABLE 13.—*Continued*

Number of Divisions	Index Circle	No. of Turns of Index	Gear on Worm	No. 1 Hole: First Gear on Stud	No. 1 Hole: Second Gear on Stud	Gear on Spindle	Idlers: No. 1 Hole	Idlers: No. 2 Hole
411	21	$\frac{2}{21}$	28			24	56	
412	20	$\frac{2}{20}$	40			48	24	44
413	21	$\frac{2}{21}$	48			32	44	
414	21	$\frac{2}{21}$	56			32	44	
415	20	$\frac{2}{20}$	32			48	24	44
416	20	$\frac{2}{20}$	40			64	24	44
417	21	$\frac{2}{21}$	56	32	48	24		
418	20	$\frac{2}{20}$	40			72	24	44
419	33	$\frac{3}{33}$	44	28	24	72		
420	21	$\frac{2}{21}$						
421	20	$\frac{2}{20}$	48	56	40	72		24
422	20	$\frac{2}{20}$	40	44	32	64		24
423	21	$\frac{2}{21}$	72	24	56	48		24
424	43	$\frac{4}{43}$	86	24	24	48		
425	21	$\frac{2}{21}$	72	48	56	40		24
426	21	$\frac{2}{21}$	56			32	24	44
427	20	$\frac{2}{20}$	40	48	32	72		24
428	20	$\frac{2}{20}$	40	56	32	64		24
429	21	$\frac{2}{21}$	28			24	24	44
430	43	$\frac{4}{43}$						
431	21	$\frac{2}{21}$	72	44	28	48		24
432	20	$\frac{2}{20}$	40	56	28	64		24
433	20	$\frac{2}{20}$	40	44	24	72		24
434	21	$\frac{2}{21}$	48			64	24	44
435	21	$\frac{2}{21}$	28			40	24	44
436	20	$\frac{2}{20}$	40	48	24	72		24
437	23	$\frac{2}{23}$	32			64	44	
438	21	$\frac{2}{21}$	28			48	24	44
439	43	$\frac{4}{43}$	86	24	24	72		24
440	33	$\frac{3}{33}$						
441	21	$\frac{2}{21}$	32			64	24	44
442	20	$\frac{2}{20}$	40	56	24	72		24
443	20	$\frac{2}{20}$	40	48	24	86		24
444	21	$\frac{2}{21}$	56	48	24	64		24
445	33	$\frac{3}{33}$	64	32	44	40		24
446	33	$\frac{3}{33}$	44			24	24	48
447	21	$\frac{2}{21}$	28			72	24	44
448	20	$\frac{2}{20}$	40	64	24	72		24
449	33	$\frac{3}{33}$	64	32	44	72		24
450	33	$\frac{3}{33}$	44			40	24	32

TABLE 13.—*Continued*

Number of Divisions	Index Circle	No. of Turns of Index	Gear on Worm	No. 1 Hole: First Gear on Stud	No. 1 Hole: Second Gear on Stud	Gear on Spindle	Idlers: No. 1 Hole	Idlers: No. 2 Hole
451	33	$\frac{3}{33}$	24			24	24	44
452	33	$\frac{3}{33}$	44			48	24	40
453	33	$\frac{3}{33}$	44			52 [1]	24	40
454	49	$\frac{4}{49}$	56	64	28	72		
455	49	$\frac{4}{49}$	28	40	32	64		
456	21	$\frac{2}{21}$	56	64	24	72		24
457	33	$\frac{3}{33}$	44			68 [1]	24	40
458	33	$\frac{3}{33}$	44			72	24	24
459	27	$\frac{2}{27}$	24	48	24	72		
460	23	$\frac{2}{23}$						
461	33	$\frac{3}{33}$	44	28	24	72		24
462	33	$\frac{3}{33}$	32			64	24	44
463	21	$\frac{2}{21}$	56	64	24	86		24
464	33	$\frac{3}{33}$	44	48	28	56		24
465	33	$\frac{3}{33}$	44	24	24	100		24
466	49	$\frac{4}{49}$	56	48	28	64		
467	33	$\frac{3}{33}$	44	48	32	72		24
468	39	$\frac{3}{39}$	28	48	24	56		
469	49	$\frac{4}{49}$	28			48	44	
470	47	$\frac{4}{47}$						
471	49	$\frac{4}{49}$	56	32	28	76 [1]		
472	49	$\frac{4}{49}$	56	32	28	72		
473	33	$\frac{3}{33}$	48	64	32	72		24
474	49	$\frac{4}{49}$	56	32	28	64		
475	49	$\frac{4}{49}$	56	40	28	48		
476	49	$\frac{4}{49}$	56			64	24	
477	27	$\frac{2}{27}$	24	48	24	56		
478	49	$\frac{4}{49}$	56	24	28	64		
479	49	$\frac{4}{49}$	56	32	28	44		
480	49	$\frac{4}{49}$	56	32	28	40		
481	37	$\frac{3}{37}$	24			24	56	
482	33	$\frac{3}{33}$	44	56	24	72		24
483	49	$\frac{4}{49}$	56			32	44	
484	49	$\frac{4}{49}$	56	24	28	32		
485	23	$\frac{2}{23}$	46 [1]	24	24	100		24
486	27	$\frac{2}{27}$	32	56	28	64		
487	39	$\frac{3}{39}$	24	72	52 [1]	44		
488	33	$\frac{3}{33}$	44	64	24	72		24
489	23	$\frac{2}{23}$	46 [1]	58 [1]	32	64		24
490	49	$\frac{4}{49}$						

[1] Special gear.

TABLE 13.—*Continued*

Number of Divisions	Index Circle	No. of Turns of Index	Gear on Worm	No. 1 Hole		Gear on Spindle	Idlers	
				First Gear on Stud	Second Gear on Stud		No. 1 Hole	No. 2 Hole
491	33	$\frac{3}{33}$	44	68[1]	24	72		24
492	41	$\frac{3}{41}$	28	48	24	56		
493	29	$\frac{2}{29}$	32	64	24	72		
494	39	$\frac{3}{39}$	32			64	44	
495	27	$\frac{2}{27}$	32	40	24	64		
496	49	$\frac{4}{49}$	56	24	28	32		24
497	49	$\frac{4}{49}$	56			32	24	44
498	27	$\frac{2}{27}$	48	56	24	64		
499	49	$\frac{4}{49}$	56	24	28	48		24
500	49	$\frac{4}{49}$	56	32	28	40		24
501	49	$\frac{4}{49}$	56	32	28	44		24
502	49	$\frac{4}{49}$	56	32	28	48		24
503	23	$\frac{2}{23}$	46[1]	64	32	86		24
504	49	$\frac{4}{49}$	56			64	24	24
505	49	$\frac{4}{49}$	56	40	28	48		24
506	49	$\frac{4}{49}$	56	32	28	64		24
507	39	$\frac{3}{39}$	24			24	56	
508	49	$\frac{4}{49}$	56	32	28	72		24
509	49	$\frac{4}{49}$	56	32	28	76[1]		24
510	49	$\frac{4}{49}$	56	40	28	64		24
511	49	$\frac{4}{49}$	28			48	24	44
512	49	$\frac{4}{49}$	56	44	28	64		24
513	27	$\frac{2}{27}$	32			64	44	
514	49	$\frac{4}{49}$	56	48	28	64		24
515	27	$\frac{2}{27}$	72	32	24	100		
516	43	$\frac{8}{43}$	32	56	28	64		
517	49	$\frac{4}{49}$	56	48	28	72		24
518	49	$\frac{4}{49}$	28			64	24	44
519	27	$\frac{2}{27}$	72	56	32	64		
520	39	$\frac{3}{39}$						
521	27	$\frac{2}{27}$	72	76[1]	48	64		
522	29	$\frac{2}{29}$	48	64	24	72		
523	27	$\frac{2}{27}$	72	68[1]	48	64		
524	27	$\frac{2}{27}$	72	32	24	64		
525	27	$\frac{2}{27}$	72	40	32	64		
526	49	$\frac{4}{49}$	56	64	28	72		24
527	31	$\frac{2}{31}$	32	64	24	72		
528	27	$\frac{2}{27}$	72	24	24	64		
529	27	$\frac{2}{27}$	72	44	48	64		
530	15	$\frac{1}{15}$	24	56	32	64		

[1] Special gear.

TABLE 13.—*Continued*

Number of Divisions	Index Circle	No. of Turns of Index	Gear on Worm	No. 1 Hole		Gear on Spindle	Idlers	
				First Gear on Stud	Second Gear on Stud		No. 1 Hole	No. 2 Hole
531	27	$\frac{2}{27}$	72			48	24	
532	27	$\frac{2}{27}$	72	32	48	64		
533	27	$\frac{2}{27}$	72	32	48	56		
534	27	$\frac{2}{27}$	72			32	44	
535	27	$\frac{2}{27}$	72	32	48	40		
536	39	$\frac{3}{39}$	52 [1]			64	24	44
537	27	$\frac{2}{27}$	72	28	56	32		
538	29	$\frac{2}{29}$	58 [1]	56	24	72		
539	49	$\frac{4}{49}$	28	48	24	56		24
540	27	$\frac{2}{27}$						
541	39	$\frac{3}{39}$	52 [1]	56	32	48		24
542	39	$\frac{3}{39}$	52 [1]	44	32	64		24
543	27	$\frac{2}{27}$	72	24	48	32		24
544	15	$\frac{1}{15}$	40	56	24	64		
545	15	$\frac{1}{15}$	32	44	24	64		
546	39	$\frac{3}{39}$	32			64	24	44
547	27	$\frac{2}{27}$	72	32	48	56		24
548	27	$\frac{2}{27}$	72	32	48	64		24
549	27	$\frac{2}{27}$	72			48	24	24
550	15	$\frac{1}{15}$	32	40	24	64		
551	29	$\frac{2}{29}$	32			64	44	
552	27	$\frac{2}{27}$	72	24	24	64		24
553	49	$\frac{4}{49}$	28	48	24	72		24
554	27	$\frac{2}{27}$	72	56	48	64		24
555	15	$\frac{1}{15}$	24			72	44	
556	15	$\frac{1}{15}$	24	44	40	64		
557	15	$\frac{1}{15}$	40	32	24	86		
558	27	$\frac{2}{27}$	48			64	24	44
559	39	$\frac{3}{39}$	24			72	24	44
560	43	$\frac{3}{43}$	86	40	32	64		
561	27	$\frac{2}{27}$	72	56	32	64		24
562	27	$\frac{2}{27}$	72	44	24	64		24
563	29	$\frac{2}{29}$	58 [1]			68 [1]	44	
564	43	$\frac{3}{43}$	86	24	24	56		
565	15	$\frac{1}{15}$	24			56	44	
566	43	$\frac{3}{43}$	86	24	24	44		
567	15	$\frac{1}{15}$	32	44	40	64		
568	15	$\frac{1}{15}$	40	32	24	64		
569	29	$\frac{2}{29}$	58 [1]			44	24	
570	15	$\frac{1}{15}$	32			64	44	

[1] Special gear.

TABLE 13.—*Continued*

Number of Divisions	Index Circle	No. of Turns of Index	Gear on Worm	No. 1 Hole: First Gear on Stud	No. 1 Hole: Second Gear Stud	Gear on Spindle	Idlers: No. 1 Hole	Idlers: No. 2 Hole
571	43	$\frac{3}{43}$	86	28	64	32		
572	15	$\frac{1}{15}$	40	28	24	64		
573	15	$\frac{1}{15}$	40			72	24	
574	41	$\frac{3}{41}$	32			64	24	44
575	15	$\frac{1}{15}$	24			40	44	
576	15	$\frac{1}{15}$	40			64	24	
577	43	$\frac{3}{43}$	86	32	64	44		24
578	15	$\frac{1}{15}$	48	44	40	64		
579	15	$\frac{2}{29}$	40			56	44	
580	29	$\frac{1}{15}$						
581	15	$\frac{1}{15}$	48	32	40	76 [1]		
582	15	$\frac{2}{27}$	40			48	44	
583	27	$\frac{1}{15}$	72	64	24	86		24
584	15	$\frac{1}{15}$	48	32	40	64		
585	15	$\frac{1}{15}$	24			24	56	
586	15	$\frac{2}{29}$	72	48	40	56		
587	29	$\frac{1}{15}$	58 [1]			28	24	44
588	15	$\frac{1}{15}$	40			32	44	
589	15	$\frac{1}{15}$	72	44	40	48		
590	15	$\frac{1}{15}$	48			32	44	
591	15	$\frac{1}{15}$	40			24	44	
592	16	$\frac{1}{16}$	24			72	44	
593	15	$\frac{1}{15}$	72	28	40	48		
594	33	$\frac{2}{33}$	32	56	28	64		
595	15	$\frac{1}{15}$	72			24	44	
596	15	$\frac{1}{15}$	72	24	40	32		
597	33	$\frac{2}{33}$	44	56	24	72		
598	16	$\frac{1}{16}$	64	56	24	72		
599	43	$\frac{3}{43}$	86	44	24	84		24
600	15	$\frac{1}{15}$						
601	29	$\frac{2}{29}$	58 [1]	56	48	72		24
602	43	$\frac{3}{43}$	32			64	24	44
603	15	$\frac{1}{15}$	72	24	40	24		24
604	16	$\frac{1}{16}$	32			72	24	
605	15	$\frac{1}{15}$	72			24	24	44
606	15	$\frac{1}{15}$	72	24	40	48		24
607	15	$\frac{1}{15}$	72	28	40	48		24
608	16	$\frac{1}{16}$	32			64	44	
609	15	$\frac{1}{15}$	40			24	24	44
610	15	$\frac{1}{15}$	48			32	24	44

[1] Special gear.

Number of Divisions	Index Circle	No. of Turns of Index	Gear on Worm	No. 1 Hole		Gear on Spindle	No. 2 Hole	
				First Gear on Stud	Second Gear on Stud		No. 1 Hole	No. 2 Hole
611	15	1/15	72	44	40	48		24
612	15	1/15	40			32	24	44
613	16	1/16	64	48	32	72		
614	15	1/15	72	48	40	56		24
615	15	1/15	24			24	24	44
616	16	1/16	32			48	44	
617	33	2/33	44	32	24	86		
618	15	1/15	40			48	24	44
619	16	1/16	48	28	32	72		
620	31	2/31						
621	15	1/15	40			56	24	44
622	16	1/16	64	24	24	72		
623	16	1/16	64	24	24	68[1]		
624	16	1/16	24			24	56	
625	15	1/15	24			40	24	44
626	16	1/16	32			28	56	
627	15	1/15	40			72	24	44
628	16	1/16	32			24	56	
629	16	1/16	64			44	24	
630	16	1/16	64			40	24	
631	16	1/16	64	28	56	72		
632	16	1/16	64			32	44	
633	16	1/16	64			28	44	
634	16	1/16	64			24	44	
635	15	1/15	24			56	24	44
636	16	1/16	56	28	48	24		
637	49	8/49	24			24	56	
638	29	2/29	48	64	24	72		24
639	33	2/33	44	28	32	64		
640	16	1/16						
641	33	2/33	44	32	48	76[1]		
642	16	1/16	72	24	64	24		24
643	16	1/16	64	28	56	24		24
644	49	8/49	56			32	44	
645	15	1/15	24			72	24	44
646	16	1/16	64			24	24	44
647	16	1/16	64			28	24	44
648	16	1/16	64			32	24	44
649	33	2/33	72			48	24	
650	16	1/16	64			40	24	44

[1] Special gear.

TABLE 13.—*Continued*

Number of Divisions	Index Circle	No. of Turns of Index	Gear on Worm	No. 1 Hole		Gear on Spindle	No. 2 Hole	
				First Gear on Stud	Second Gear on Stud		No. 1 Hole	No. 2 Hole
651	16	$\frac{1}{16}$	64			44	24	24
652	16	$\frac{1}{16}$	32			24	24	44
653	33	$\frac{2}{33}$	72	28	44	48		
654	16	$\frac{1}{16}$	64			56	24	44
655	16	$\frac{1}{16}$	64	40	32	48		24
656	16	$\frac{1}{16}$	24			24	24	44
657	18	$\frac{1}{18}$	32	48	24	56		
658	16	$\frac{1}{16}$	64	24	24	72		24
659	16	$\frac{1}{16}$	64	24	24	76 [1]		24
660	33	$\frac{2}{33}$						
661	16	$\frac{1}{16}$	64	56	48	72		24
662	16	$\frac{1}{16}$	64	44	24	48		24
663	17	$\frac{1}{17}$	24			24	56	
664	16	$\frac{1}{16}$	32			48	24	44
665	49	$\frac{3}{49}$	56			40	24	44
666	18	$\frac{1}{18}$	24			72	44	
667	16	$\frac{1}{16}$	64	48	32	72		24
668	16	$\frac{1}{16}$	32			56	24	44
669	33	$\frac{2}{33}$	44			24	24	24
670	33	$\frac{2}{33}$	72	48	44	40		24
671	33	$\frac{2}{33}$	72			48	24	24
672	18	$\frac{1}{18}$	24			64	44	
673	16	$\frac{1}{16}$	48	44	32	72		24
674	33	$\frac{2}{33}$	72	56	44	48		24
675	33	$\frac{2}{33}$	44			40	24	24
676	16	$\frac{1}{16}$	32			72	24	44
677	18	$\frac{1}{18}$	48	32	24	86		
678	18	$\frac{1}{18}$	24			56	44	
679	49	$\frac{3}{49}$	28			44	24	40
680	17	$\frac{1}{17}$						
681	33	$\frac{2}{33}$	44			56	24	24
682	33	$\frac{2}{33}$	48			64	24	24
683	16	$\frac{1}{16}$	32			86	24	44
684	18	$\frac{1}{18}$	32			64	44	
685	18	$\frac{1}{18}$	24	56	48	40		
686	15	$\frac{1}{15}$	40	64	24	86		24
687	18	$\frac{1}{18}$	24			44	48	
688	16	$\frac{1}{16}$	24			72	24	44
689	39	$\frac{2}{39}$	24	48	24	56		
690	18	$\frac{1}{18}$	24			40	56	

[1] Special gear.

TABLE 13.—*Continued*

Number of Divisions	Index Circle	No. of Turns of Index	Gear on Worm	No. 1 Hole		Gear on Spindle	Idlers	
				First Gear on Stud	Second Gear on Stud		No. 1 Hole	No. 2 Hole
691	18	$\frac{1}{18}$	48	32	24	58[1]		
692	18	$\frac{1}{18}$	72	56	32	64		
693	18	$\frac{1}{18}$	32			48	44	
694	17	$\frac{1}{17}$	68[1]			56	24	44
695	18	$\frac{1}{18}$	72	24	24	100		
696	18	$\frac{1}{18}$	24			32	56	
697	17	$\frac{1}{17}$	24			24	24	44
698	18	$\frac{1}{18}$	72	44	24	48		
699	18	$\frac{1}{18}$	48			56	44	
700	18	$\frac{1}{18}$	72	40	32	64		
701	17	$\frac{1}{17}$	68[1]	48	32	56		24
702	18	$\frac{1}{18}$	24			24	56	
703	19	$\frac{1}{19}$	24			72	44	
704	18	$\frac{1}{18}$	72	24	24	64		
705	18	$\frac{1}{18}$	48			40	44	
706	18	$\frac{1}{18}$	72			56	24	
707	18	$\frac{1}{18}$	72			52[1]	24	
708	18	$\frac{1}{18}$	72			48	24	
709	18	$\frac{1}{18}$	72			44	24	
710	18	$\frac{1}{18}$	72			40	24	
711	18	$\frac{1}{18}$	64			32	44	
712	18	$\frac{1}{18}$	72			32	24	
713	18	$\frac{1}{18}$	72			28	44	
714	18	$\frac{1}{18}$	72			24	44	
715	18	$\frac{1}{18}$	72	32	64	40		
716	18	$\frac{1}{18}$	72	28	56	32		
717	18	$\frac{1}{18}$	72	24	64	32		
718	33	$\frac{2}{33}$	44	58[1]	24	64		24
719	17	$\frac{1}{17}$	68[1]	52[1]	24	72		24
720	18	$\frac{1}{18}$						
721	21	$\frac{1}{21}$	24	64	32	68[1]		
722	19	$\frac{1}{19}$	32			64	44	
723	18	$\frac{1}{18}$	72	24	64	32		24
724	18	$\frac{1}{18}$	72	28	56	32		24
725	18	$\frac{1}{18}$	72	24	48	40		24
726	18	$\frac{1}{18}$	72			24	24	44
727	18	$\frac{1}{18}$	72			28	24	44
728	18	$\frac{1}{18}$	72			32	24	44
729	18	$\frac{1}{18}$	64			32	24	44
730	20	$\frac{1}{20}$	32	48	24	56		

[1] Special gear.

By the *lead of the milling machine* is meant the distance the table will travel while the index head spindle makes one complete revolution when the gear ratio between the feed screw and the worm gear stud is 1 to 1.

Fig. 41.—Spiral Milling. A—Gear on worm (driven). B—First gear on stud (driver). C—Second gear on stud (driven). D—Gear on screw (driver).

The lead of the milling machine equals the revolutions of the feed screw required for one revolution of the index head spindle with equal gears, times the lead of the feed screw.

$$\frac{\text{Lead of spiral}}{\text{Lead of machine}} = \frac{\text{product of driven gears}}{\text{product of driving gears}}$$

In finding the change gears to be used in a compound train, place the lead to be cut in the numerator, and the lead of milling machine in the denominator, then resolve the fraction into its factors and multiply each pair of factors by the same number until suitable numbers of teeth in change gears are obtained.

The following change gears are available on most milling machines: 24–24–28–32–40–44–48–56–64–72–86–100.

Illustration: What change gears are required for a spiral index head to cut a 36-inch lead with a 10-inch lead milling machine?

$$\frac{36}{10} = \frac{4 \times 9}{2 \times 5} = \frac{4}{2} \times \frac{16}{16} = \frac{64}{32}$$

$$\frac{9}{5} \times \frac{8}{8} = \frac{72}{40}$$

The 64 and 72 are driven gears and 32 and 40 are driving gears. Then place the 72 T gear on worm, the 40 T gear on screw, the 32 T first gear on stud and 64 T second gear on stud. (See Fig. 41.)

Illustration: What lead or spiral can be cut with the following gears if the lead on the machine is 10 inches; gear on worm, 40; first gear on stud, 24; second gear on stud, 24: gear on screw, 32?

$$\text{Driven gears} = 40 \times 24 = 960$$

$$\text{Driving gears} = 24 \times 32 = 768$$

Then,

$$\frac{\text{Lead of spiral}}{10} = \frac{960}{768}$$

$$\text{Lead of spiral} = \frac{10 \times 960}{768} = \frac{10 \times 5}{4} = 12.5 \text{ in. (Ans.)}$$

The Angle of Helix.—This is the angle which the spiral makes with the axis of the work. The swiveled milling machine table must be set to this angle when cutting a helix. This angle (π) may be found by the following formula:

$$\text{tangent of helix angle} = \frac{\pi \times \text{diameter of work}}{\text{lead of helix}}$$

ILLUSTRATION: A helix with a 24-inch lead is to be cut on a piece of work 3 inches in diameter. What is the angle of helix?

$$\tan \phi = \frac{\pi \times \text{diameter of work}}{\text{lead of helix}} = \frac{3.1416 \times 3}{24}$$

$$\tan \phi = \frac{3.1416}{8} = 0.3927$$

$$\phi = 21^\circ\ 26' \text{ (Ans.)}$$

NOTE: Because the scale by which angle ϕ is set is usually graduated only to fourths of a degree, the table would be set $21\frac{1}{2}^\circ$.

The angle of the helix may be found graphically as follows: Draw a base line equivalent to the lead and a vertical line equal to the circumference. If the two lines are then connected by a hypotenuse the helix angle (ϕ) which the hypotenuse makes with the base may be measured with a protractor.

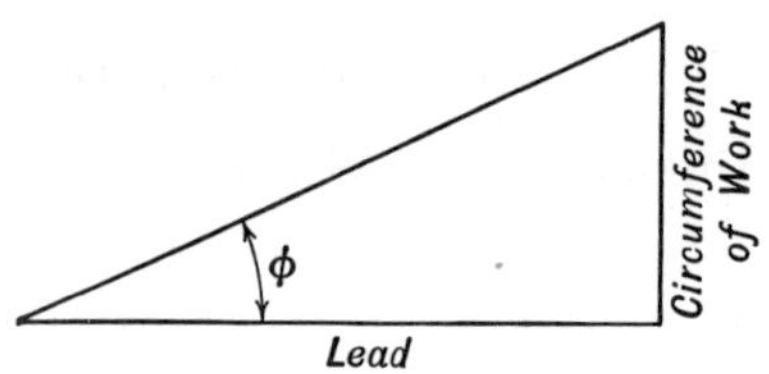

If the drawing is made on paper, the triangle may be cut out and wrapped around the work with the side representing the circumference encircling the work. The hypotenuse will trace out the helix on the work.

TABLE 14

Pitches and Approximate Angles for Cutting Spirals on the Universal Milling Machine

To find the angle for cutters of a larger diameter than given in the table, make a drawing as shown in the diagram; the angle b being a right angle. Let $b\,c$ equal the circumference. Let $a\,b$ equal the pitch. Connect $c\,a$ by a line, and measure the angle a with a protractor; or divide the circumference by the lead and the quotient will be the tangent of the angle. Find the angle in a table of tangents.

Diameter of Mill, Cutter, or Drill to be Cut

Inches

Values Given Under Diameters are Angles in Degrees

Gear on Worm	First Gear on Stud	Second Gear on Stud	Gear on Screw	Pitch in Inches to one Turn	1/8	1/4	3/8	1/2	5/8	3/4	7/8	1	1¼	1½	1¾	2	2¼	2½	2¾	3	3¼	3½	3¾	4
24	64	24	72	1.25	17½	32¼																		
24	64	28	72	1.46	14¾	28	38½																	
24	64	30	72	1.56	14¼	26¾	37																	
24	64	32	72	1.67	12¾	25	34¾	43¼																

$$\text{The lead in inches in one turn} = \frac{10 \times \text{Gear on Worm} \times \text{2nd Gear on Stud}}{\text{Gear on Screw} \times \text{1st Gear on Stud}}$$

32	64	28	72	1.94	11¼	21¾	31	39	45														
24	64	40	72	2.08	10¼	20½	29½	37	43¼														
32	56	28	72	2.22	9¾	19¼	27½	35	41¼														
24	64	48	72	2.50	8¾	17	25	32	38	43¼													
40	56	28	72	2.78	8	15½	23	29½	35¼	40½	44¾												
24	64	56	72	2.92	7½	15	21¾	28¼	34	39	43¼												
40	48	28	72	3.24	6¾	13½	19¾	25¾	31¼	36	40½	44¼											
40	48	32	72	3.70	6	11¾	17½	23	28	32½	36½	40½											
56	48	24	72	3.89	5½	11¼	16¾	22	26¾	31¼	35¼	39											
40	72	48	64	4.17	5¼	10½	15¾	20½	25¼	29½	33½	37	43¼										
56	48	28	72	4.53	5	9¾	14¾	19¼	23¾	27¾	31½	35	41½										
40	64	56	72	4.86	4½	9	13½	17¾	22	25¾	29½	33	39	44¼									
48	40	32	72	5.33	4	8¼	12¼	16¼	20¼	23¾	27¼	30½	36½	41½									
56	40	28	72	5.44	4	8	12	16	20	23½	26¾	30	36	41									
56	40	28	64	6.12	3½	7¼	11	14½	17¾	21	24¼	27	33	37¾	42								
56	40	32	72	6.22	3½	7	10¾	14¼	17½	20¾	23¾	26¾	32½	37¼	41½								
56	48	40	72	6.48	3¼	6¾	10¼	13½	16¾	20	23	25¾	31½	36¼	40¼	44¼							
64	48	28	56	6.67	3¼	6½	10	13¼	16½	19½	22½	25¼	30¾	35¼	39½	43½							
56	48	40	64	7.29	3	6¼	9¼	12¼	15	18	20½	23½	28½	33	37	41	44¼						
64	48	40	72	7.41	3	6	9	12	14¾	17¾	20¼	22¾	28¼	32½	36½	40¼	43¾						
64	48	32	56	7.62	2¾	5¾	8¾	11½	14½	17¼	19¾	22¼	27½	32	36	39½	43						
48	32	40	72	8.33	2½	5¼	8	10½	13¼	15¾	18¼	20½	25½	29½	33½	37	40½	43½					
72	30	24	64	9.00	2½	5	7½	10	12½	14¾	17	19¼	24	28	31¾	35¼	38½	41¼	44				
56	40	48	72	9.33	2¼	4¾	7¼	9½	11¾	14	16¼	18½	23	27	30½	34	37¼	40¼	43				
64	48	40	56	9.52	2¼	4½	7	9¼	11½	13¾	16	18¼	22½	26½	30	33½	36½	39½	42¼	45			
72	40	32	56	10.29	2	4¼	6½	8¾	10¾	12¾	15	17¼	21	24¾	28¼	31½	34½	37½	40	42½	45		
64	48	56	72	10.37	2	4¼	6½	8½	10½	12¾	14¾	17	20¾	24½	28	31¼	34¼	37¼	39¾	42¼	44¾		
48	40	56	64	10.50	2	4¼	6¼	8½	10½	12½	14½	16¾	20½	24¼	27¾	31	34	36¾	39½	42	44¼		

TABLE 14

Pitches and Approximate Angles for Cutting Spirals on the Universal Milling Machine—*Concluded*

To find the angle for cutters of a larger diameter than given in the table, make a drawing as shown in the diagram; the angle *b* being a right angle. Let *b c* equal the circumference. Let *a b* equal the pitch. Connect *c a* by a line, and measure the angle *a* with a protractor; or divide the circumference by the lead and the quotient will be the tangent of the angle. Find the angle in a table of tangents.

Gear on Worm	First Gear on Stud	Second Gear on Stud	Gear on Screw	Pitch in Inches to one Turn	Diameter of Mill, Cutter, or Drill to be Cut — Inches — Values Given Under Diameters are Angles in Degrees																			
					1/8	1/4	3/8	1/2	5/8	3/4	7/8	1	1 1/4	1 1/2	1 3/4	2	2 1/4	2 1/2	2 3/4	3	3 1/4	3 1/2	3 3/4	4
64	40	48	72	10.67	2	4	6 1/4	8 1/4	10 1/4	12 1/4	14 1/4	16 1/2	20 1/4	24	27 1/4	30 1/2	33 1/2	36 1/2	39	41 1/2	43 3/4			
56	32	40	64	10.94	2	4	6	8 1/4	10 1/4	12	14	16 1/4	20	23 1/2	26 3/4	30	33	35 3/4	38 1/4	40 3/4	43			
64	32	40	72	11.11	2	4	6	8	10	11 3/4	13 3/4	16	19 3/4	23	26 1/2	29 1/2	32 1/2	35 1/4	38	40 1/4	42 1/2	44 3/4		
56	32	48	72	11.66	1 3/4	3 3/4	5 3/4	7 1/2	9 1/2	11 1/4	13 1/4	15 1/4	18 3/4	22	25 1/4	28 1/2	31 1/4	34	36 1/2	39	41 1/4	43 1/2		

72	40	32	48	12.00	1¾	3¾	5½	7¼	9¼	11	12¾	15	18¼	21½	24¾	27¾	30½	33¼	35¾	38	40¼	42½	44¾	
56	32	48	64	13.12	1½	3½	5¼	6¾	8½	10¼	11¾	13½	16¾	20	22¾	25¾	28¼	31	33¼	35¾	37¼	40	42	43¾
56	28	48	72	13.33	1½	3¼	5	6½	8¼	10	11½	13¼	16½	19½	22½	25½	28	30½	33	35¼	37½	39½	41½	43¼
64	40	48	56	13.71	1½	3¼	4¾	6½	8	9¾	11¼	13	16	19	22	24¾	27¼	30	32¼	34½	36½	38¾	40¾	42½
64	28	48	72	15.24	1½	3	4½	5¾	7¼	8¾	10¼	11¾	14½	17¼	20	22½	25	27¼	29½	31¾	34	35¾	37¾	39½
64	32	56	72	15.56	1¼	2¾	4¼	5¾	7¼	8¾	10	11½	14¼	17	19½	22	24½	27	29	31¼	33¼	35¼	37	39
56	64	72	40	15.75	1¼	2¾	4¼	5½	7	8½	9¾	11¼	14	16¾	19¼	21¾	24¼	26½	28¾	31	33	35	36¾	38½
72	32	48	64	16.87	1¼	2½	4	5¼	6¾	7¾	9¼	10½	13¼	15¾	18¼	20½	22¾	25	27	29¼	31¼	33¼	35	36½
64	32	48	56	17.14	1¼	2½	4	5¼	6½	7¾	9	10¼	13	15½	17¾	20¼	22¼	24¼	26¾	29	30¾	32¾	34½	36
72	32	40	48	18.75	1	2¼	3½	4¾	6	7¼	8¼	9½	12	14¼	16¼	18½	20¾	22¾	25	26¾	28½	30¼	32	33¾
72	32	48	56	19.29	1	2¼	3½	4½	5¾	7	8	9¼	11½	13¾	16	18¼	20¼	22¼	24	26	28	29¾	31½	33
64	28	48	56	19.59	1	2¼	3¼	4½	5¾	6¾	8	9¼	11½	13½	15¾	18	20	22	23¾	25¾	27½	29¼	31	32¾
72	32	56	64	19.69	1	2¼	3¼	4½	5¾	6¾	8	9	11½	13½	15¾	17¾	20	21¾	23¾	25½	27½	29¼	31	32½
72	24	40	56	21.43	1	2	3¼	4¼	5¼	6¼	7½	8½	10½	12½	14½	16½	18½	20¼	22	23¾	25½	27¼	29	30¼
72	28	56	64	22.50	1	2	3	4	5	6	7	8	10	12	13¾	15¾	17½	19¼	21	22¾	24½	26	27¾	29¼
64	32	56	48	23.33	1	2	3	4	5	5¾	6¾	7¾	9¾	11½	13¼	15¼	17	18¾	20¼	22	23½	25¼	27	28¼
72	24	56	64	26.25	1	1¾	2¾	3½	4¼	5	6	7	8½	10¼	12	13½	15	16¾	18¼	19¾	21¼	22¾	24¼	25½
64	28	56	48	26.67	¾	1¾	2¾	3½	4¼	5	6	6¾	8½	10	11¾	13¼	14¾	16½	18	19½	21	22¼	23¾	25¼
64	32	56	40	28.00	¾	1¾	2½	3¼	4	4¾	5¾	6½	8	9½	11¼	12¾	14¼	15¾	17¼	18¾	20	21½	22¾	24
72	28	48	40	30.86	¾	1½	2¼	3	3¾	4½	5	5¾	7¼	8¾	10	11½	13	14¼	15½	17	18½	19½	21	22
72	32	56	40	31.50	¾	1½	2¼	3	3½	4¼	5	5¾	7¼	8½	10	11¼	12¾	14	15¼	16½	18	19¼	20½	21¾
72	32	64	40	36.00	½	1¼	2	2½	3¼	3¾	4½	5	6¼	7½	8¾	10	11	12¼	13½	14¾	16	17	18¼	19¼
72	28	64	40	41.14	½	1	1¾	2¼	2¾	3¼	4	4½	5½	6½	7¾	8¾	9¾	10¾	11¾	13	14	15	16	17
72	28	56	32	45.00	½	1	1½	2	2½	3	3½	4	5	6	7	8	9	10	11	11¾	12¾	13¾	14¾	15½
72	24	64	40	48.00	½	1	1½	2	2½	3	3¼	3¾	4¾	5½	6½	7½	8½	9¼	10¼	11¼	12	13	13¾	14½
72	28	64	32	51.43	½	¾	1½	1¾	2¼	2¾	3	3½	4½	5¼	6	7	7¾	8¾	9½	10½	11¼	12	12¾	13¾
72	24	64	32	60.00	½	¾	1¼	1½	2	2¼	2½	3	3¾	4½	5¼	6	6¾	7½	8¼	9	9½	10¼	11	11¾
72	24	64	28	68.57	¼	¾	1	1¼	1¾	2	2¼	2½	3¼	4	4¼	5¼	5¾	6½	7¼	8	8½	9	9¾	10¼

Gears

Types of Gears.—There is a great variety of gears with regard to shapes, sizes, and uses. They may, however, be classified under four general groups: spur gears, bevel gears, worm gearing, and spiral or helical gears.

Spur Gears are the most commonly used gears and are used to transmit positive rotary motion between parallel shafts. They are cylindrical in shape and the teeth are cut parallel with the axis.

FIG. 42. Spur Gear.

FIG. 43.—Bevel Gears.

Bevel Gears are used to transmit positive rotary motion to shafts at an angle to each other, and in the same plane.

The teeth of a bevel gear are made on a frustum of a cone whose apex is the same point as the intersection of the axes of the shafts.

Bevel gears usually connect shafts running at right angles. When the angle of the shafts is 90 degrees and the velocity ratio is 1 to 1, then both gears are of the same size and are called *miter gears*. If the velocity ratio between two gears is other than 1 to 1, the smaller gear is called the *pinion*.

Worm Gearing is used to transmit power between two shafts at 90 degrees to each other, but not in the same plane, and is generally used when it is desired to obtain smoothness of action and great speed reduction from one shaft to another.

Fig. 44.

The greatest objection to worm-gear drive is the excessive friction between the teeth, making them very inefficient and subject to heating.

A *worm* is a screw so cut as to mesh properly with the teeth of a worm wheel, the included angle of the sides being 29 degrees.

The *worm wheel* is similar to a spiral spur gear. It usually has a concave face and the tooth spaces are concave and at an angle other than 90 degrees to the side of the gear.

Spiral or Helical Gears are used to drive shafts parallel to each other and in the same plane, or shafts at angles to each other but not in the same plane.

Herringbone Gears conform to two spiral gears fastened to each other, one right hand and the other left hand, thus equalizing the side thrust. They are used to transmit power between two parallel shafts. Herringbone gears are very quiet in action because some parts of their teeth are always in full action.

Efficiencies of Gears.—In relative efficiency, the different styles of gearing rank as follows, from the most efficient to the least efficient: spur, herringbone, bevel, spiral or helical, and worm.

Gearing Definitions.—The *center distance* of a pair of gears is the shortest distance between the centers of the shafts on which they are mounted.

The *pitch circles* of a pair of gears have the same diameters as a pair of friction rolls which would fill the same center distance and revolve at the same velocity ratio.

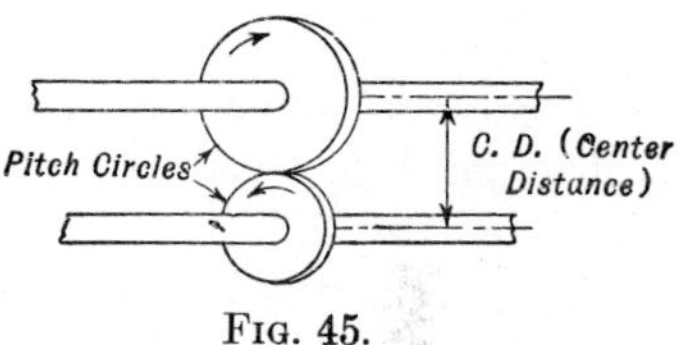

Fig. 45.

The *pitch diameter* of a gear is the diameter of its pitch circle.

The *diametral pitch* is the number of teeth a gear has per inch of pitch diameter. To find the diametral pitch, divide the number of teeth by the pitch diameter. The pitch diameter may in turn be found by dividing the number of teeth by the diametral pitch.

The *circular pitch* is the distance from the center of one tooth to the center of the next, measured along the pitch line. To find the circular pitch, divide the pitch circle by the number of teeth, or divide π by the diametral pitch.

The *addendum* is the height of the tooth above the pitch line.

The *dedendum* is the depth below the pitch line to which the tooth of the mating gear extends.

The *size of gear tooth* is designated by its pitch; thus, a 10-pitch tooth has an addendum of $\frac{1}{10}$ inch and a dedendum of $\frac{1}{10}$ inch.

Note: The term "pitch" when used alone always refers to the diametral pitch.

The *tooth thickness* is measured along the pitch line and is one-half the circular pitch.

The *working depth* is the depth in the tooth space to which the tooth of the mating gear extends, and is equal to the addendum plus the dedendum.

The *clearance* is the distance from the point of the tooth to the bottom of the space in the mating gear.

The *whole depth* is the distance from the top of the tooth to the bottom of the same tooth and consists of the addendum, dedendum, and clearance.

The *outside diameter* is found by adding twice the addendum to the pitch diameter.

The *root diameter* is the diameter at the bottom of the tooth space.

The *face* of the gear tooth is that part of the gear tooth outline which extends above the pitch line.

The *flank* is that part of the gear tooth outline below the pitch line.

The *fillet* is the rounded corner where the flank of the tooth runs to the bottom of the tooth space.

The *base circle* is the circle from which the involute curve is generated. It is drawn tangent to the pressure line. Its position will vary according to the pressure angle used. The two common pressure angles are $14\frac{1}{2}$ degrees and 20 degrees. The former is the more common, while the latter is used on the so-called "stub-tooth." For a $14\frac{1}{2}$-degree pressure angle tooth gear, the base circle will lie inside the pitch circle a distance equal to $\frac{1}{60}$ of the pitch diameter.

Tooth Curves.—The shape of gear teeth is usually either involute or cycloidal.

The *involute curve* is the more desirable because it will allow a certain amount of variation in the center distance, and is for this reason used almost universally.

The way actually to draw this curve on paper with drawing instruments is explained on page 137.

Cycloid Gear Teeth will not be described in detail at this point since this principle is used mostly in large cast gears of one-inch circular pitch or more. These gears must always meet on the pitch line in both gears and racks. This means that there can be no variation in the pitch diameter.

Cycloidal teeth are constructed by making the outline of the face a part of an epicycloid and the flanks a part of a hypocycloid.

With these definitions in mind, we may proceed to a study of the characteristics of individual gear types.

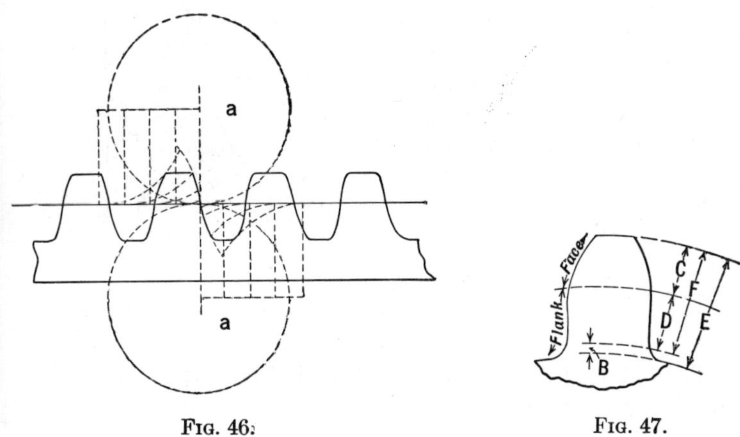

FIG. 46.　　　　FIG. 47.

Characteristics of the Spur Gear.—The preceding definitions as applied to the spur gear and as illustrated in Figs. 47 and 48 are:

A = Circular pitch or distance from center of one tooth to the next, measured on the pitch line.

B = Clearance.

C = Addendum—the height of a tooth above the pitch circle.

D = Dedendum—bottom of tooth between pitch diameter and clearance.

E = Whole depth—addendum, dedendum, and clearance.

F = Working depth—addendum and dedendum.

G = Thickness of tooth—width of tooth from outside to outside on pitch line.

H = Outside diameter.

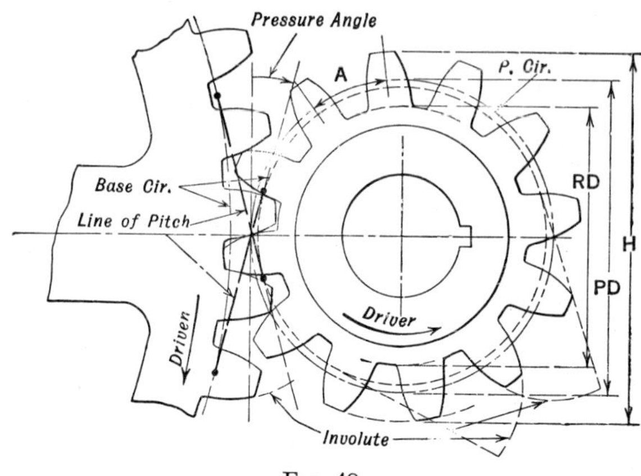

FIG. 48.

The following is a list of symbols and abbreviations used in the formulas of spur gear relationships:

P = Diametral pitch, or pitch.
O.D. = Outside diameter.
N = Number of teeth.
Np = Number of teeth in pinion.
Ng = Number of teeth in gear.
N.R. = Number of teeth in rack.
L = Length of rack.
P.D. = Pitch diameter.
C.D. = Center distance.
C.P. = Circular pitch.
Wh.D. = Whole depth.
Wg.D. = Working depth.
Add. = Addendum.
Ded. = Dedendum.
C = Clearance.
Th. = Thickness of tooth.
R.D. = Root diameter.

The following are formulas for dimensions of spur gears.

$P = \pi \div$ C.P. or N $\div$ P.D.

O.D. = (N + 2) $\div P$ or (N + 2) $\times$ C.P. $\div \pi$ or P.D. + (2 $\times$ Add.)

C.P. = $\pi \div P$ or P.D. $\times \pi \div N$

P.D. = $N \div P$ or $N \times$ C.P. $\div \pi$ or O.D. − (2 $\times$ Add.)

C.D. = $(Ng + Np) \div 2P$ or $(Ng + Np) \times$ C.P. $\div$ 6.2832.

Clear. = 0.157 $\div P$ or C.P. $\div$ 20

Add. = 1 $\div P$ or C.P. $\div \pi$ or C.P. $\times$ 0.318

Ded. = 1 $\div P$ or C.P. $\div \pi$ or C.P. $\times$ 0.318

Wh.D. = 2.157 $\div P$ or 0.6866 $\times$ C.P.

Th. = 1.5708 $\div$ P or C.P. $\div$ 2

$N = P \times$ P.D. or $\pi \times$ P.D. $\div$ C.P.

$L = \pi \times$ N.R. $\div P$ or N. $\times$ C.P.

R.D. = O.D. − 2 Wh.D. or P.D. − 2(Ded. + C)

Illustration: How many teeth are there in a gear of 4 pitch 8-in. pitch diameter?

P = 4 = no. teeth per in. of pitch diameter.

P.D. = 8 in.

Then, $N = P \times$ P.D. $= 4 \times 8 = 32$ teeth (Ans.)

Illustration: What are the addendum, dedendum and clearance of a 4-pitch gear?

$$\text{Addendum} = \frac{1}{P} = \frac{1}{4} = 0.25 \text{ in. (Ans.)}$$

$$\text{Dedendum} = \frac{1}{P} = \frac{1}{4} = 0.25 \text{ in. (Ans.)}$$

$$\text{Clearance} = \frac{0.157}{P} = \frac{0.157}{4} = 0.0392 \text{ in. (Ans.)}$$

ILLUSTRATION: What is the approximate outside diameter of a gear whose circular pitch is 0.500 in. and which has 60 teeth

$$\text{C.P.} = 0.500,$$

and
$$P = \frac{\pi}{0.500} = 6.2832$$

Then $\text{O.D.} = \dfrac{N+2}{p} = \dfrac{62}{6.2832} = 10$ inches (approx.) (Ans.)

ILLUSTRATION: What is the center distance of two gears of 40 and 60 teeth, 10 pitch?

$$\text{Center distance} = \frac{Np + Ng}{2P} = \frac{40 + 60}{2 \times 10} = 5 \text{ in. (Ans.)}$$

ILLUSTRATION: Given approximate center distance of two gears of $5\frac{1}{8}$ in., ratio 15 to 26, 8 pitch; find pitch diameter, outside diameter and number of teeth in each gear.

NOTE: The subscripts g for "gear" and p for "pinion" are added to indicate the symbol applies to the gear or to the pinion. Thus, P.D._g is the pitch diameter of the gear.

$$\begin{aligned} \text{P.D.}_g &= 2V_p \times \frac{\text{C.D.}}{V_g + V_p} \\ &= 2 \times 26 \times \frac{5.125}{15 + 26} \\ &= 52 \times 0.125 = 6.5 \text{ in.} \\ N &= 8 \times 6.5 = 52 \text{ teeth} \\ \text{O.D.} &= \frac{52 + 2}{8} = \frac{54}{8} = 6.75 \text{ in.} \end{aligned}$$

$$\begin{aligned} \text{P.D.}_p &= 2V_g \times \frac{\text{C.D.}}{V_g + V_p} \\ &= 2 \times 15 \times \frac{5.125}{15 + 26} \\ &= 30 \times 0.125 = 3.75 \text{ in.} \\ N &= 8 \times 3.75 = 30 \text{ teeth} \\ \text{O.D.} &= \frac{30 + 2}{8} = \frac{32}{8} = 4 \text{ in.} \end{aligned}$$

Cutting Spur Teeth.—Smooth-running involute gear teeth may be cut on a milling machine by the use of standard gear cutters. A separate set is required for each pitch and there are eight cutters to each set. These cutters are adapted to cut gears ranging from 12-tooth to a rack. The following table can be used to select the proper number of cutter when the number of teeth to be cut is known:

No. of cutter	No of teeth	No. of cutter	No. of teeth
1	135 to rack	5	21 to 25
2	55 to 134	6	17 to 20
3	35 to 54	7	14 to 16
4	26 to 34	8	12 to 13

ILLUSTRATION: What number of cutter should be used to cut (*a*) an 18-tooth gear; (*b*) a 48-tooth gear?

(*a*) No. 6. (Ans.)

(*b*) No. 3. (Ans.)

The depth to which the slot between the teeth is cut depends upon the diametral pitch. All gears of one pitch have the same depth of slot. Table 15 gives the depths to which the spaces should be cut in gears of various pitch.

TABLE 15

DEPTH OF SPACES IN GEARS

Diametral pitch	Depth to be cut in gear, inches	Thickness of tooth on pitch line, in.	Diametral pitch	Depth to be cut in gear, inches	Thickness of tooth on pitch line, in.
2	1.078	0.785	12	0.180	0.131
2½	0.863	0.628	14	0.154	0.112
3	0.719	0.523	16	0.135	0.098
3½	0.616	0.448	18	0.120	0.087
4	0.539	0.393	20	0.108	0.079
5	0.431	0.314	22	0.098	0.071
6	0.359	0.262	24	0.090	0.065
7	0.307	0.224	26	0.083	0.060
8	0.270	0.196	28	0.077	0.056
9	0.240	0.175	30	0.072	0.052
10	0.216	0.157	32	0.067	0.049
11	0.196	0.143			

Characteristics of Bevel Gears.—When the pitch of two bevel gears is the same, they will mesh properly regardless of the number of teeth, providing they have twelve or more teeth. A gear with less than twelve teeth must be specially cut to avoid interference of teeth while rolling.

The pitch, outside diameter, and pitch diameter of a bevel gear are always calculated on the large end of the tooth.

Figure 49 shows a cross section of a bevel gear and pinion.

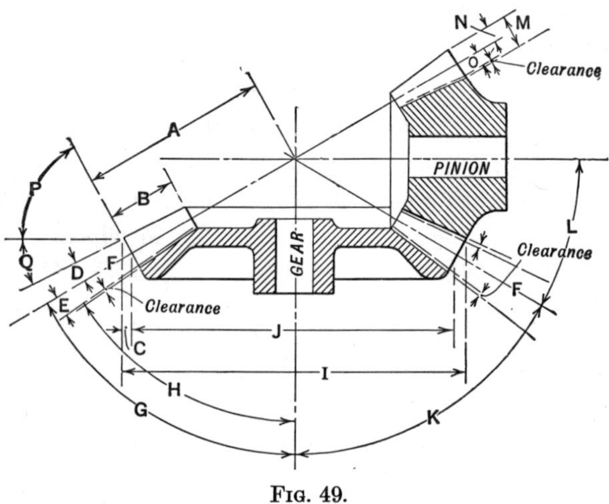

FIG. 49.

The following is a list of symbols and abbreviations of the bevel gear parts and a key to these parts in the figure:

P.C.R.	= Pitch cone radius........	= A
W. of F.	= Width of face...........	= B
Ang. add.	= Angular addendum......	= C
Add. ang.	= Addendum angle........	= D
Ded. ang.	= Dedendum angle........	= E
P. line	= Pitch line...............	= F
P.C. ang.	= Pitch cone angle.........	= G
Cut. ang.	= Cutting angle...........	= H
O.D.	= Outside diameter........	= I
P.D.	= Pitch diameter..........	= J

P.C. ang. G. = Pitch cone angle of gear . . = K
P.C. ang. P. = Pitch cone angle of pinion = L
Wh.D. = Whole depth. . . . . = M
Add. = Addendum. . . . . = N
Ded. = Dedendum. . . . . = O
E. ang. = Edge angle. . . . . = P
F. ang. = Face angle. . . . . = Q
Ng = Number of teeth in gear
Np = Number of teeth in pinion
N = Number of teeth
P = Diametral pitch, or pitch
T = Thickness of tooth
N' = Number of teeth for which to select cutter

The principal bevel gear formulas are:

Tangent of P.C. ang. of pinion = $Np \div Ng$
Tangent of P.C. ang. of gear = $Ng \div Np$
Pitch diameter. . . . . = $N \div P$
Addendum. . . . . = $1 \div P$ or C.P. $\times$ 0.318 or C.P. $\div \pi$
Dedendum. . . . . = $1 \div P$ or C.P. $\times$ 0.318 or C.P. $\div \pi$
Whole depth of tooth. . . . . = $2.157 \div P$ or C.P. $\times$ 0.687
Pitch cone radius. . . . . = P.D. $\div$ (2 $\times$ sin P.C. ang.)
Thickness of tooth. . . . . = $1.571 \div P$ or C.P. 2
Small addendum. . . . . = (P.C.R $- B$) $\div$ P.C.R. $\times$ Add.
Small thickness of tooth. . . . . . = (P.C.R. $- B$) $\div$ P.C.R. $\times$ thickness
Tangent ang. of addendum. . . = Add. $\div$ P.C.R.
Tangent ang. of dedendum. . . = Ded. $\div$ P.C.R.
Face angle. . . . . = 90 deg. $-$ (P.C. ang. $+$ Add. ang.)
Cutting angle. . . . . = P.C. ang. $-$ Ded. ang.
Angular addendum. . . . . = Cos. of P.C. ang. $\times$ Add.
Outside diameter. . . . . = Ang. add. $\times$ 2 $+$ P.D.
No. of teeth for which to select cutter. . . . . = $\dfrac{N}{\text{Cos. of P.C. Ang.}}$

ILLUSTRATION: What is the pitch cone radius, addendum angle, and outside diameter of a bevel gear whose pitch diameter is 4 inches, pitch cone angle 60 degrees, and which is 10 pitch?

Summarizing the known factors:

$$\text{P.D.} = 4 \text{ in.}$$
$$\text{P.C. ang.} = 60^\circ$$
$$P = 10$$

Then, pitch cone radius (P.C.R.) = P.D. ÷ (2 × sin P.C. ang.)

$$\text{P.C.R.} = \frac{4}{2 \times \sin 60^\circ} = \frac{2}{0.866} = 2.309 \text{ in. (Ans.)}$$

$$\text{Addendum} = \frac{1}{P} = \frac{1}{10} = 0.10 \text{ in.}$$

$$\text{Tangent addendum angle} = \text{Add.} \div \text{P.C.R.} = \frac{0.100}{2.309} = 0.04331$$

$$\text{Addendum angle} = 2^\circ\ 29' \text{ (Ans.)}$$
$$\text{Angular addendum} = \cos \text{P.C. ang.} \times \text{Add.}$$
$$\text{Ang. add.} = \cos 60^\circ \times 0.10 = 0.50 \times 0.10 = 0.050 \text{ in.}$$
$$\text{Outside diameter} = \text{Ang. add.} \times 2 + \text{P.D.}$$
$$\text{O.D.} = 0.05 \times 2 + 4 = 4.10 \text{ in. (Ans.)}$$

ILLUSTRATION: What is the whole depth of tooth at the small end of a bevel gear with 30 teeth, 6 pitch and a pitch cone angle of 54 degrees and a width of face of 1 inch?

Since all of the dimensions of a gear tooth (except width of face) gradually decrease until they are zero at the intersection of the centerline axes, we can best solve this problem by finding the whole depth at the large end of the tooth and multiplying this by $\frac{\text{P.C.R.} - B}{\text{P.C.R.}}$. (See Fig. 49.)

$$\text{Whole depth at large end} = \frac{2.157}{P} = \frac{2.157}{6} = 0.3595 \text{ in.}$$

$$\text{Pitch diameter} = \frac{N}{P} = \frac{30}{6} = 5 \text{ in.}$$

$$\text{Pitch cone radius} = \frac{\text{P.D.}}{2 \times \sin \text{P.C. ang.}} = \frac{5}{2 \times \sin 54°}$$

$$\text{P.C.R.} = \frac{5}{2 \times 0.809} = 3.09$$

$$\text{P.C.R.} - B = 3.09 - 1 = 2.09$$

$$\text{Then, whole depth at small end} = 0.3595 \times \frac{2.09}{3.09} = 0.2432 \text{ in.}$$

(Ans.)

ILLUSTRATION: A pair of 2-pitch bevel gears with shafts at 90 degrees have a velocity ratio of $2\frac{1}{2}$ to 1 and the pinion has 24 teeth. What is the face angle, pitch cone angle and cutting angle of the larger gear?

$$Np = 24 \text{ teeth}$$

then,

$$Ng = 2\tfrac{1}{2} \times 24 = 60 \text{ teeth}$$

$$\text{Tangent pitch cone angle of gear} = \frac{Ng}{Np} = \frac{60}{24} = 2.50$$

$$\text{Pitch cone angle} = 68° \, 12' \quad \text{(Ans.)}$$

$$\text{Pitch diameter} = \frac{N}{P} = \frac{60}{2} = 30 \text{ in.}$$

$$\text{Pitch cone radius} = \frac{\text{P.D.}}{2 \times \sin \text{P.C. ang.}} = \frac{30}{2 \times \sin 68° \, 12'}$$

$$\text{P.C.R.} = \frac{30}{2 \times 0.9285} = 16.1551$$

$$\text{Addendum} = \frac{1}{P} = \frac{1}{2} = 0.50$$

$$\text{Tan Angle of Addendum} = \frac{\text{Add.}}{\text{P.C.R.}} = \frac{0.50}{16.1551} = 0.03095$$

Angle of Addendum $= 1° 46'$

Face angle $= 90° - (\text{P.C. ang.} + \text{Add. ang.}) = 90° - (68° 12' + 1° 46')$

Face angle $= 90° - 69° 58' = 20° 2'$ (Ans.)

Add. angle = Ded. angle

Then, Cutting angle = P.C. ang. − Add. ang. $= 68° 12' - 1° 46' = 66° 26'$ (Ans.)

Characteristics of Worms and Worm Wheels.—The worm wheel or gear is similar to a spiral spur gear. It usually has a concave face and the tooth spaces are concave and at an angle other than 90 degrees to the side of the gear.

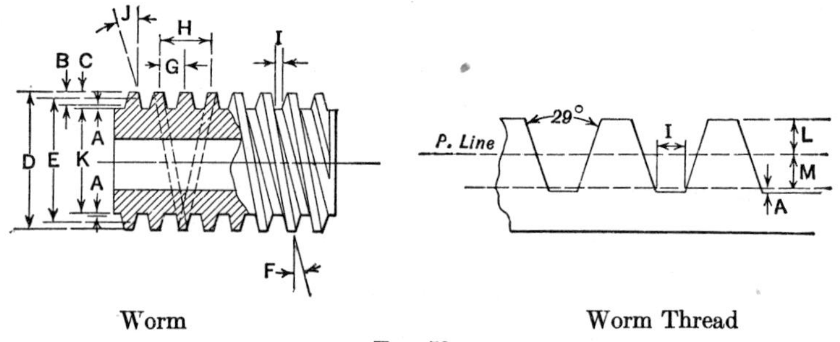

Worm

Worm Thread

Fig. 50.

The *linear pitch* is the distance from the center of one tooth to the center of the next, measured on the pitch circle. The ratio between the linear pitch and the diameter of the worm is arbitrary. It may be four times the circular pitch of a worm gear for single thread; five times the circular pitch of the worm gear for double thread; six times the circular pitch of the worm gear for triple thread.

The *lead* sometimes differs from the pitch and it is the distance a tooth on the worm would advance in one revolution, or the distance the worm wheel advances in one complete turn of a worm.

Parts of the worm with reference to Fig. 50 are:

A = Clearance
B = Working depth of tooth
C = Whole depth of tooth
D = Outside diameter of worm
E = Pitch diameter of worm
F = Angle of helix
G = Linear pitch
H = Lead
I = Thickness of end of tool at bottom of space
J = Half angle of tooth
K = Root diameter of worm
L = Addendum
M = Dedendum

Worm relations are:

Lead = linear pitch × no. of separate threads on the worm
Linear pitch = lead ÷ no. of separate threads on the worm
Addendum = linear pitch × 0.3183
Whole depth of thread = linear pitch × 0.6866

Width of threading tool at end or width of bottom of space = linear pitch × 0.31

O.D. = P.D. + (2 × Add.)
P.D. = O.D. − (2 × Add.)
P.D. = 2 × center distance − P.D. of gear
Root diameter = O.D. − 2 × whole depth of tooth

Cotangent of angle of worm tooth or gashing angle of wheel = P.D. × π ÷ lead.

Illustration: What is the root diameter of a worm whose outside diameter is $1\frac{1}{4}$ inches and whose linear pitch is 0.25 inch?

Whole depth of tooth $= P \times 0.6866 = 0.25 \times 0.6866 = 0.17165$ in.

Root diameter $=$ O.D. $- 2 \times$ whole depth

Root diameter $= 1.25 - 2 \times 0.17165 = 0.9067$ in. (Ans.)

ILLUSTRATION: What is the width of a thread tool at its cutting edge for a worm whose linear pitch is 0.215 inch?

Width of thread tool $= P \times 0.31 = 0.215 \times 0.31 = 0.06665$ in. (Ans.)

ILLUSTRATION: What is the angle of worm tooth or gashing angle of wheel, if the outside diameter of the worm is $1\frac{3}{4}$ inches, the linear pitch is 0.60 inch and the screw is double thread?

Addendum $= P \times 0.3183 = 0.60 \times 0.3183 = 0.1910$ in.

P.D. $=$ O.D. $- (2 \times$ Add.$) = 1.75 - 2 \times 0.1910 = 1.368$ in.

Lead $=$ pitch $\times$ no. of separate threads

Lead $= 0.60 \times 2 = 1.20$ inches

$$\text{Cotangent of angle of worm} = \frac{\text{P.D.} \times \pi}{\text{lead}} = \frac{1.368 \times \pi}{1.20}$$

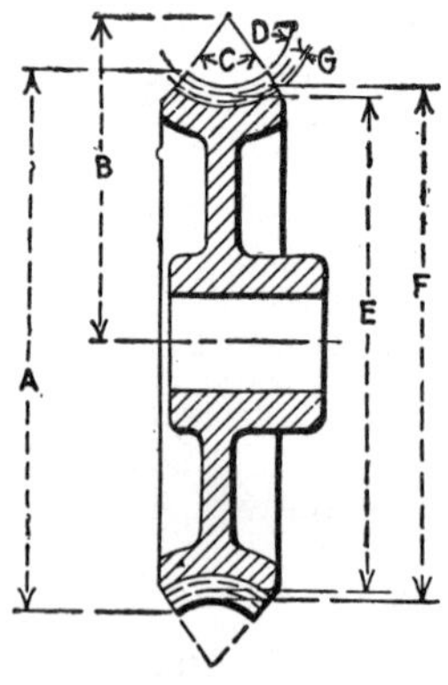

FIG. 51.—Worm Wheel.

Cotangent $= 3.5814$

Angle of worm $= 15° 36'$ (Ans.)

The following list indicates the meaning of the symbols used as dimensions in Fig. 51.

A = O.D. of worm wheel
B = Center distance of worm and worm wheel
C = Angle of face
D = Throat radius
E = Pitch diameter
F = Throat diameter
G = Clearance

Worm wheel formulas are:

$$\text{P.D.} = (\text{no. of teeth in wheel} \times \text{linear pitch of worm}) \div \pi$$
$$\text{Throat diameter} = \text{P.D. of worm wheel} + 2 \times \text{Add.}$$
$$\text{Radius of throat} = \tfrac{1}{2} \text{ of O.D. of worm} - (2 \times \text{Add. of worm})$$
$$\text{Center distance} = (\text{P.D. of worm} + \text{P.D. of wheel}) \div 2$$
$$\text{O.D.} = (\text{throat radius} - \text{throat radius} \times \text{cosine } \tfrac{1}{2} \text{ face angle}) \times 2 + \text{throat diameter of wheel.}$$

Addendum of worm wheel = addendum of worm.

ILLUSTRATION: What is the pitch diameter of worm wheel with 48 teeth and a linear pitch of 0.350 inch?

$$\text{P.D.} = (\text{no. teeth} \times \text{linear pitch}) \div \pi$$
$$\text{P.D.} = \frac{48 \times 0.350}{3.1416} = 5.3475 \text{ in.} \quad \text{(Ans.)}$$

ILLUSTRATION: What is the radius of curvature of worm wheel throat if the pitch of the worm is 0.150 inch and the outside diameter is 1 inch?

$$\text{Addendum of worm} = \text{linear pitch} \times 0.3183$$
$$\text{Addendum} = 0.150 \times 0.3183 = 0.04775 \text{ in.}$$

Then, radius of throat $= \frac{1}{2}$ of O.D. of worm $- (2 \times$ Add. of worm)

$$\text{radius of throat} = \tfrac{1}{2} \times 1 - 2 \times 0.04775$$
$$\text{radius of throat} = 0.5000 - 0.0955 = 0.4045 \text{ in.} \quad \text{(Ans.)}$$

ILLUSTRATION: What is the outside diameter of a worm wheel whose face angle is 70 degrees, throat radius 0.500 inch, number of teeth 32 and linear pitch of worm 0.200 inch?

$$\text{Addendum} = \text{linear pitch} \times 0.3183 = 0.200 \times 0.3183 = 0.07366 \text{ in.}$$

Pitch diameter of gear = (no. of teeth in wheel × linear pitch of worm) ÷ π

$$\text{Pitch diameter} = \frac{32 \times 0.200}{\pi} = \frac{6.4}{3.1416} = 2.0372$$

Throat diameter = P.D. of worm wheel + 2 × Add.

Throat diameter = 2.0372 + 2 × 0.07366

Throat diameter = 2.0372 + 0.1473 = 2.1845 in.

Outside diameter, diameter to sharp corners, } = (throat radius − throat radius × cosine of $\frac{1}{2}$ face angle) × 2 + throat diameter of wheel.

Outside Diameter = (0.5 − 0.5 × cos 35°) × 2 + 2.1845

Outside Diameter = (0.5 − 0.5 × 0.8192) × 2 + 2.1845

Outside Diameter = (0.5000 − 0.4096) × 2 + 2.1845

Outside Diameter = 0.0904 × 2 + 2.1845 = 2.3653 in. (Ans.)

Planing and Shaping

Dovetail.—One of the problems of planing and shaping which lends itself to mathematical solution is the measurement of dovetail slides. The dimensions of these are usually given as shown in Fig. 52, but it is difficult to make these measurements on the work with any great accuracy because the edges are not uniformly

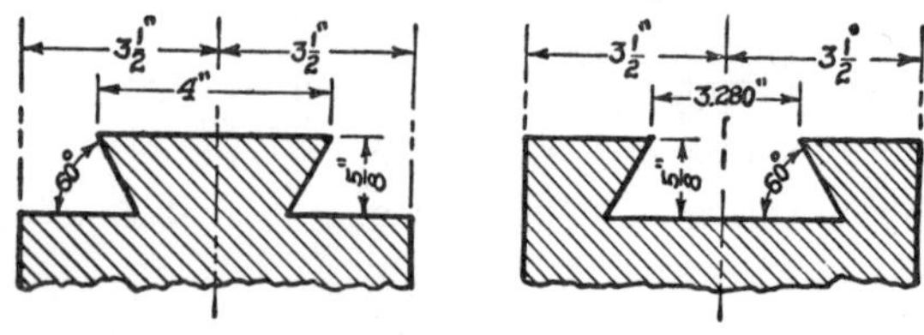

Fig. 52.

sharp. The method used is to measure between rods of equal diameter in the case of the slot, as shown in Fig. 53, and over rods on its counterpart.

To obtain X and Y (Fig. 53) which are used in the practical measuring of dovetail slides, the following formulas may be used.

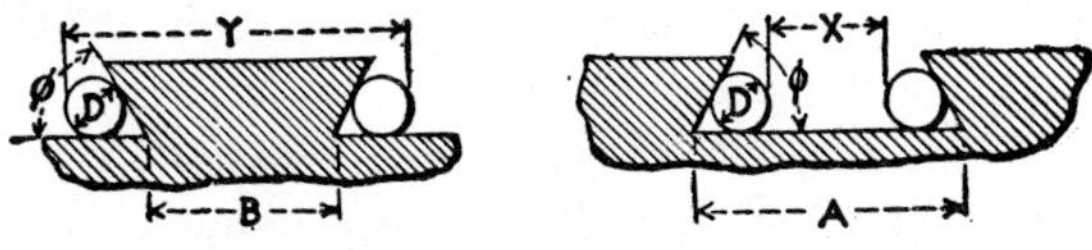

Fig. 53.

$$X = A - [D(1 + \cot \tfrac{1}{2}\phi)]$$
$$Y = D(1 + \cot \tfrac{1}{2}\phi) + B$$

The best size of plug or rod to use is one whose diameter is two-thirds the depth of the slot.

Illustration: What is the overall length in measuring a male dovetail, if the following data are given on the blue print: angle 66°, width at bottom 2.956 inches, if plugs $\frac{3}{4}$ inch in diameter are used?

$$Y = D(1 + \cot \tfrac{1}{2}\varphi) + B = 0.75(1 + \cot 33°) + 2.956$$
$$Y = 0.75(1 + 1.5399) + 2.956 = 4.861 \text{ in.} \quad \text{(Ans.)}$$

Illustration: What is the distance between $\frac{5}{8}$ inch plugs placed in a female dovetail which is cut to a 2.125 inch width at the bottom and has an included angle of 50 degrees?

$$X = A - [D(1 + \cot \tfrac{1}{2}\phi)] = 2.125 - [0.625(1 + \cot 25°)]$$
$$X = 2.125 - [0.625(1 + 2.1445)] = 0.160 \text{ inch} \quad \text{(Ans.)}$$

Grinding

Finishing by Grinding.—Machine work is often turned or planed oversize by an amount of from 0.002 to 0.010 inch and the excess removed by grinding. In the grinding operation, cuts of 0.001 or less can easily be made and the result is a finish of greater smoothness and accuracy than can readily be obtained with a cutting tool. Wheels of emery or silicon carbide are most commonly used in finishing metal work.

Speed of Grinding Wheel.—Grinding wheels do good work at surface speeds of 5000 to 6000 feet per minute. The surface speed depends on the speed of revolution and the diameter of the wheel.

The following formulas may be used to find the surface speed in feet per minute of a wheel.

$$S = \frac{\pi RD}{12}$$

S = Surface speed

R = Revolutions per minute

D = Diameter in inches

$\pi = 3\frac{1}{7}$ or 3.14

ILLUSTRATION: What is the surface speed of a 9-inch grinding wheel revolving at a speed of 2500 revolutions per minute?

$$S = \frac{\pi RD}{12}$$

$$= \frac{22 \times 2500 \times 9}{7 \times 12} = 5893 \text{ feet per minute.}$$

If a certain surface speed of a given wheel is desired, to find the number of revolutions of the wheel spindle,

$$R = \frac{12S}{\pi D}$$

ILLUSTRATION: A surface speed of 5500 feet per minute is desired from an 18 inch grinding wheel. How many revolutions per minute should it turn?

$$R = \frac{12S}{\pi D}$$

$$= \frac{5500 \times 12}{3.14 \times 18} = 1168 \text{ revolutions per minute} \quad \text{(Ans.)}$$

Table 16 gives the necessary revolutions per minute for obtaining certain surface speeds from wheels of various diameters.

TABLE 16

GRINDING WHEEL SPEEDS

Diameter of wheel, inches	Revolutions per minute for surface speed of 4000 feet	Revolutions per minute for surface speed of 5000 feet	Revolutions per minute for surface speed of 6000 feet	Revolutions per minute for surface speed of 9000 feet
1	15,279	19,099	22,918	34,377
2	7,639	9,549	11,459	17,188
3	5,093	6,366	7,639	11,459
4	3,820	4,775	5,730	8,595
5	3,056	3,820	4,584	6,876
6	2,546	3,183	3,820	5,729
7	2,183	2,728	3,274	4,911
8	1,910	2,387	2,865	4,297
10	1,528	1,910	2,292	3,438
12	1,273	1,592	1,910	2,864
14	1,091	1,364	1,637	2,455
16	955	1,194	1,432	2,149
18	849	1,061	1,273	1,910
20	764	955	1,146	1,719
22	694	868	1,042	1,562
24	637	796	955	1,433
30	509	637	764	1,146
36	424	531	637	954

The revolutions per minute at which wheels are run is dependent on conditions and style of machine and the work to be ground.

Fits

Types of Fits.—In the mating of two parts of a machine, the perfection of the mating is called the *fit*. Sometimes the pieces are assembled so that there may be motion between them, as, for instance, a shaft in a bearing or an engine crosshead in its frame. In other cases two parts may be assembled so that they can act only in unison, as a flywheel on a shaft or a tire on a locomotive wheel.

Fits may be classified broadly as *running fits*, *wringing fits*, *pressed fits* and *shrinking fits*.

Running Fit.—To make a running fit, like a bearing, an allowance may be made of about two thousandths of an inch for a shaft one inch in diameter, and one thousandth more for each inch the shaft is increased in diameter.

If D = diameter of the hole in inches and d = diameter of the shaft, then, $d = D - [(D - 1) \times 0.001 + 0.002]$

ILLUSTRATION: A shaft is to run in a self-aligning and self-oiling bearing 6 inches in diameter. What should be the diameter of the shaft?

$$d = D - [(D - 1) \times 0.001 + 0.002] = 6 - (5 \times 0.001 + 0.002)$$

$$d = 5.9975 \text{ inches} \quad \text{(Ans.)}$$

Wringing Fit.—In a fit of this type, the shaft is made the same size as the hole into which it is to fit.

Pressed Fit.—The force required to press a shaft into a hole made for a press fit will depend not only on the allowance made on the fit, but also on the kind of material, the length of the fit, the finish, etc. Press fits are frequently made so that a pressure of from 5 to 10 tons per inch diameter is required to force the shaft into its hole.

When the length of the fit is from two to three times its diameter, and the finish is good and smooth, an allowance of three-quarters to one and one-quarter of a thousandth of an inch may do well for pressing a one-inch shaft of machinery steel into a hole in cast iron or machinery steel, and as the shaft increases in size, the allowance may be increased about half of one-thousandth for each inch the shaft is increased in diameter. There is no hard and fast rule for making these allowances; judgment and experience alone will dictate what modifications to make.

Setting up the above rule in equation form with average values when D = diameter of hole in inches and d = diameter of shaft, we get

$$d = D + [(D - 1) \times 0.0005 + 0.001]$$

ILLUSTRATION: A shaft is to be turned for a press fit into a 3-inch hole. To what diameter should it be turned?

$$d = D + [(D - 1) \times 0.0005 + 0.001] = 3 + [2 \times 0.0005 + 0.001]$$

$$d = 3 + 0.002 = 3.002 \text{ in.} \quad \text{(Ans.)}$$

Shrinking Fit.—The allowance to be made for a shrinking fit will vary more or less according to the nature of work and the judgment of the designer.

When shrinking a collar on a shaft or similar work, an allowance of 0.002 inch to 0.003 inch will do for a shaft of one inch diameter, and as the shaft increases in diameter add 0.0005 inch to the allowance for each inch the diameter is increased.

$$d = D + [(D - 1) \times 0.0005 + 0.0025]$$

ILLUSTRATION: A shaft is to have a collar shrunk onto it with a 7-inch hole. What should be the diameter of the shaft?

$$d = D + [(D - 1) \times 0.0005 + 0.0025]$$

$$= 7 + [(7 - 1) \times 0.0005 + 0.0025]$$

$$d = 7 + 0.0030 + 0.0025 = 7.0055 \text{ in.} \quad \text{(Ans.)}$$

References.—ENGINEERING TOOLS AND PROCESSES by H. C. Hesse; BLUEPRINT READING: UNDERSTANDING SHOP PRACTICES by F. Nicholson and F. Jones; THE MACHINISTS' AND DRAFTSMEN'S HANDBOOK and MACHINE SHOP: THEORY AND PRACTICE, both by A. Wagener and H. R. Arthur; all published by D. Van Nostrand Company, contain much valuable information on the subject of Machine Shop Work.

XIX

Sheet Metal Work

Sheet metal work makes abundant use of geometry in that flat sheets must be made into the common geometrical shapes of cones, cylinders, etc. The plans or drawings usually give the dimensions of the finished shape, and the problem is one of laying out a design on the flat metal so that when it is cut and bent it will result in the desired shape.

When the surface of a solid is thus opened or flattened out it is said to be developed. The following figures will indicate the meaning of the term development as applied to the surfaces of different solids. Moreover, a knowledge of volume is necessary in sheet metal work, for example; a tinsmith is required to make some cylinder shaped cans to hold one gallon each and to be 6 inches high. What radius should be used in laying out the base? Practically all formulas of surface and cubic measure apply to problems in sheet metal work. Some are given on pages 115–125.

A cube is shown in Fig. 1 together with its development.

The following formulas may be used:

Volume, $V = S^3$

Side, $S = \sqrt[3]{V}$

Lateral Surface, L = area of two ends + areas of all side faces.

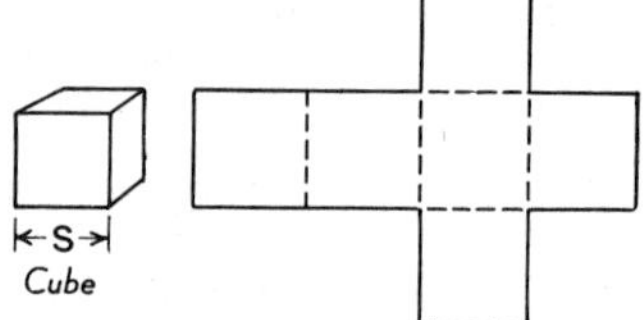

FIG. 1.

ILLUSTRATIONS: 1. Find the volume of a cube whose side is 9.5 inches.

$$V = S^3 = 9.5^3 = 9.5 \times 9.5 \times 9.5 = 857.38 \text{ cu. in.}$$

2. If the volume of a cube is 231 cubic inches, find the length of the side.

$$S = \sqrt[3]{V} = \sqrt[3]{231}$$

from the table on page 29 find 6.136 in.

3. If in Fig. 1 each side is 1 foot, find the lateral surface.

$$L = \text{area of two ends} + \text{areas of all side faces.}$$
$$= 2 \text{ sq. ft.} + 4 \text{ sq. ft.} = 6 \text{ sq. ft.}$$

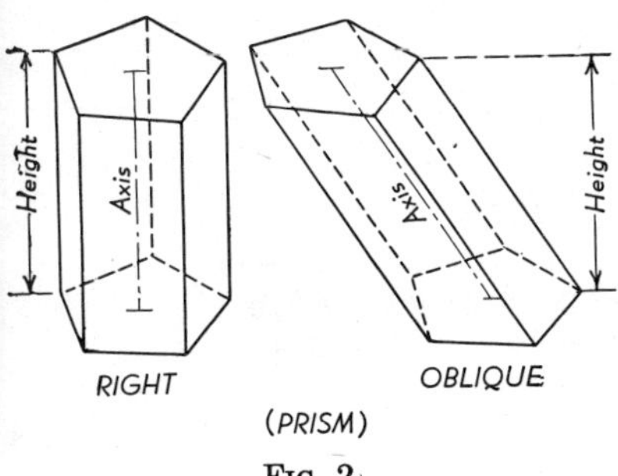

(PRISM)

FIG. 2.

A pentagonal prism is shown on the left, Fig. 2, and its development in Fig. 3. Prisms are named triangular, square, pentagonal, etc., in accordance with the shape of the base.

The following formulas may be used which apply to all prisms:

Volume, $V = A$, area of end surface $\times$ h, height

Lateral Surface, L = area of two ends = areas of all side faces

The area of a pentagon can be found by multiplying the length of the side by 1.7204. See table, page 651, for constants to determine the area of polygons.

ILLUSTRATIONS: 1. If a pentagonal prism measures 1.5 feet on a side 6 feet in height, find the volume.

Area of end surface = 1.7204 $\times$ 1.5 = 2.58 sq. ft.

$$V = Ah$$
$$= 2.58 \times 6 = 15.48 \text{ cu. ft.}$$

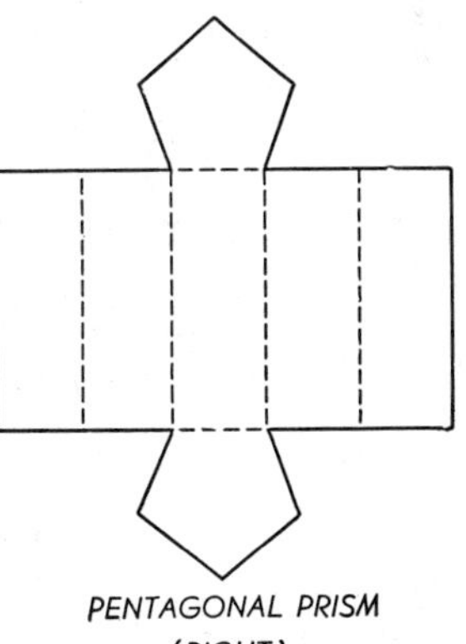
PENTAGONAL PRISM

(RIGHT)

FIG. 3.

2. Find the lateral surface in the above illustration.

$$L = \text{area of two ends} + \text{area of all side faces}$$

from preceding problem, area of one end surface = 2.58 square feet, therefore,

$$2 \times 2.58 = 5.16 \text{ square feet} = \text{area of two ends}$$

In a right prism the area of the side faces = perimeter of base × height.

$$\text{Area of one side face} = 1.5 \times 6 = 9.0 \text{ sq. ft.}$$

$$\text{Area of 5 faces} = 9 \times 5 = 45 \text{ sq. ft.}$$

or, perimeter of base, $(1.5 \times 5) = 7.5 \times \text{height } (6) = 45$ sq. ft.

$$L = 5.16 + 45 = 50.16 \text{ sq. ft.}$$

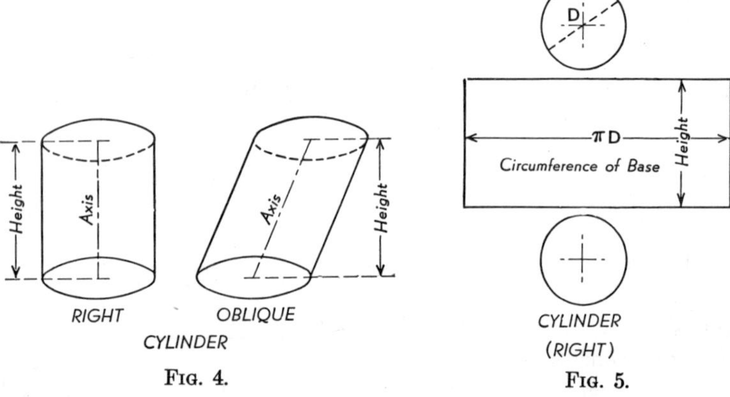

FIG. 4. FIG. 5.

A cylinder is shown in Fig. 4 and its development in Fig. 5. The following formulas may be used:

Volume, $V = 3.1416r^2h = 0.7854d^2h$.

Lateral or cylindrical surface = perimeter of base × height $= 3.1416dh$.

Total area, A, lateral or cylindrical surface and end surfaces $= 6.2832r(r + h)$.

ILLUSTRATIONS: 1. Find the volume of a cylinder whose diameter is $2\frac{1}{2}$ inches and height is 20 inches.

$$V = 0.7854d^2h = 0.7854\ (2.5)^2 \times 20$$
$$= 0.7854 \times 6.25 \times 20 = 98.18 \text{ cu. in.}$$

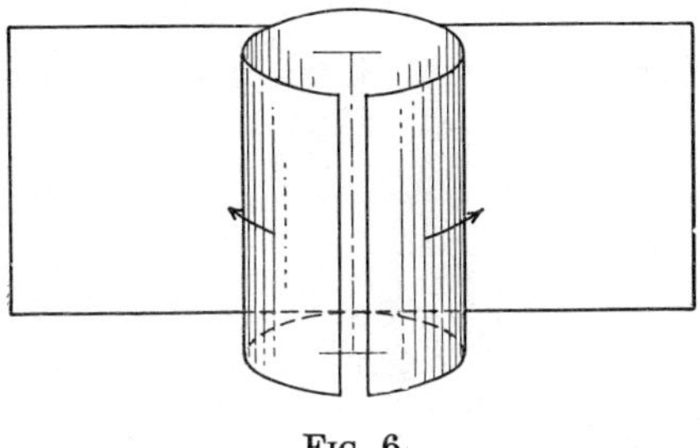

FIG. 6.

2. Find the lateral surface in the above illustration.

$$L = 3.1416dh = 3.1416 \times 2.5 \times 20 = 157.08 \text{ sq. in.}$$

3. In illustration 1 find the total lateral area.

$$A = 6.2832r\ (r + h) = 6.2832 \times 1.25\ (1.25 + 20) = 166.9 \text{ sq. in.}$$

When the volume of a prism and the area of the base is known the height may be found by the following formula:

$$\text{Height} = \frac{\text{volume}}{\text{area of base}}$$

ILLUSTRATION: Find the height of a cylinder 2 feet 2 inches in diameter to contain 6500 cubic inches.

$$\text{Height} = \frac{6500}{531} = 12.24 \text{ in.}$$

Aid. Area of base $= \pi r^2 = 3.14 \times 13 \times 13 = 531$ sq. in.

A square pyramid is shown in Fig. 7 and its development in Fig. 8. Pyramids are named triangular, square, pentagonal, etc. in accordance with the shape of the base. The following formulas may be used.

Volume, $V = \frac{1}{3}h \times$ area of base

Lateral surface, $L =$ area of the base + areas of all the triangular faces, or, $\frac{1}{2} \times$ perimeter of base $\times$ slant height.

NOTE. In a right pyramid all triangular faces are isosceles.

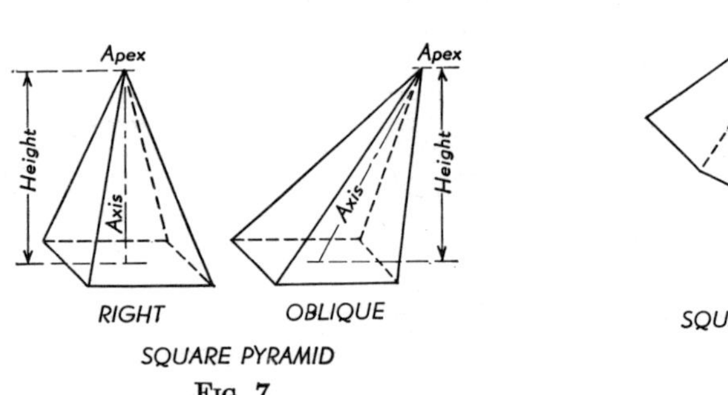

FIG. 7. FIG. 8.

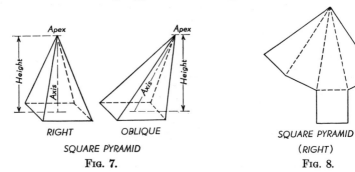

ILLUSTRATIONS: 1. A pyramid whose base is 2 feet square has a height of 6 feet. Find the volume.

Area of base $= 2 \times 2 = 4$ sq. ft.

$V = \frac{1}{3}h \times$ area of base $= \frac{1}{3} \times 6 \times 4 = 8$ cu. ft.

2. In illustration 1 find the lateral surface.

Area of base, from above, $=4$ square feet

Fig. 9 indicates that the lateral surface is made of four isosceles triangles similar to ADE. The base of each triangle is a side of the base of the pyramid. The lateral surface of the pyramid is obtained by multiplying the area of one of the triangles by 4.

Area of a triangle $= \frac{1}{2}$ base $\times$ height

If the pyramid is 6 feet high and the base is 2 feet square, then the base of the triangle ADE = 2 feet but the height line of the triangle is the line AB, called the slant height.

Figure 10 shows the pyramid with one quarter removed so that the actual height AC and the slant height AB can be seen. From this figure it is evident that triangle ABC is a right triangle with the slant height AB as the hypotenuse. The height of altitude, AC, of this right triangle is 6 feet, the height of the pyramid. The base, BC, of the triangle is half the distance across the square or $\frac{1}{2} \times 2 = 1$. The hypotenuse

$$AB = \sqrt{AC^2 + BC^2} = \sqrt{6^2 + 1^2} = \sqrt{37} = 6.08 \text{ feet}$$

the slant height, which is the height or altitude of the triangle ADE.

$$\text{Area of triangle } ADE = \tfrac{1}{2} \times \text{base} \times \text{height}$$
$$= \tfrac{1}{2} \times 2 \times 6.08 = 6.08 \text{ sq. ft.}$$

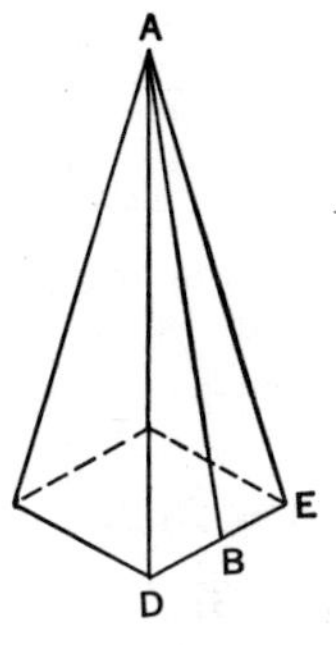

FIG. 9.

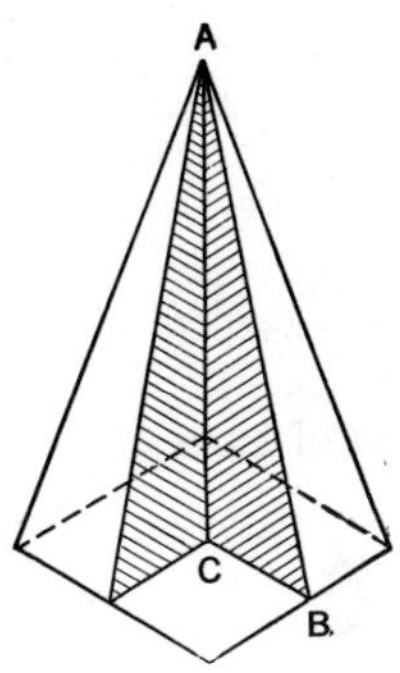

FIG. 10.

The lateral surface of the square pyramid is four times the area of one of the sides, therefore $4 \times 6.08 = 24.32$ square feet.

By the other formula,

$$\text{Lateral surface} = \tfrac{1}{2} \times \text{perimeter of base} \times \text{slant height}$$
$$= \tfrac{1}{2} \times (4 \times 2) \times 6.08 = 24.32 \text{ sq. ft.}$$

A cone is shown in Fig. 11 and its development in Fig. 12. The volume of a cone. like that of a pyramid. is one-third the volume of a cylinder of the same size, thus the formulas are similar to those used in pyramids.

$$\text{Volume, } V = \tfrac{1}{3}h \times \text{area of base}$$

$$\text{Lateral surface} = \tfrac{1}{2} \times \text{perimeter of base} \times \text{slant height}$$

ILLUSTRATIONS: 1. Find the volume and the lateral surface of a cone the base of which is a circle 6 feet in diameter and whose height is 4 feet.

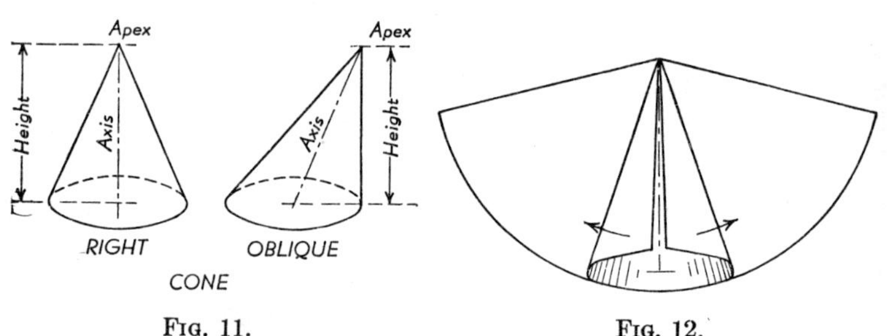

FIG. 11. FIG. 12.

$$\begin{aligned}
\text{Area of base} &= 0.7854d^2 \\
&= 0.7854 \times 6 \times 6 = 28.27 \\
V &= \tfrac{1}{3} \times 4 \times 28.27 = 37.7 \text{ cu. ft.} \\
\text{Perimeter of base} &= 3.1416 \times 6 = 18.8496 \\
\text{Lateral surface} &= \text{perimeter of base} \times \text{slant height} \\
&= \tfrac{1}{2} \times 18.8496 \times \sqrt{3^2 + 4^2} \\
&= 9.4248 \times \sqrt{25} \\
&= 9.4248 \times 5 \\
&= 47.124 \text{ sq. ft.}
\end{aligned}$$

2. Making a conical ventilator top which will be 24 inches in diameter and 6 inches high. What shape of metal should be cut, allowing $1\frac{1}{2}$ inches for a lap joint?

First it is necessary to determine two other dimensions of the cone, the slant height and the circumference of the base.

The slant height $= \sqrt{6^2 + 12^2} = \sqrt{36 + 144}$

$= \sqrt{180} = 13.4164 = 13\frac{7}{16}$ in.

The circumference $= \pi \times 24 = 3.1416 \times 24 = 75.3984 = 75\frac{3}{8}$ in.

Then, using the slant height as a radius, draw a circle on the metal to be cut. The length $75\frac{3}{8}$ inches plus the $1\frac{1}{2}$ inches for lap may be measured off on the circumference of this circle and the metal cut.

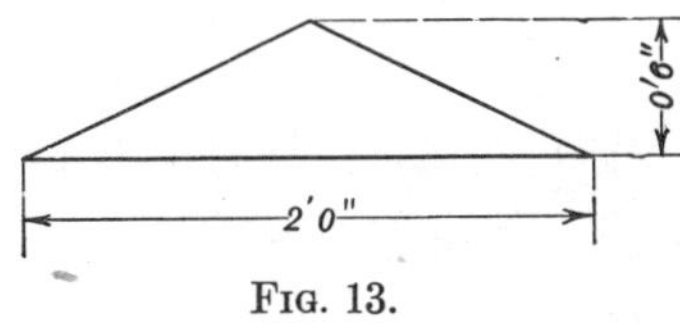

Fig. 13.

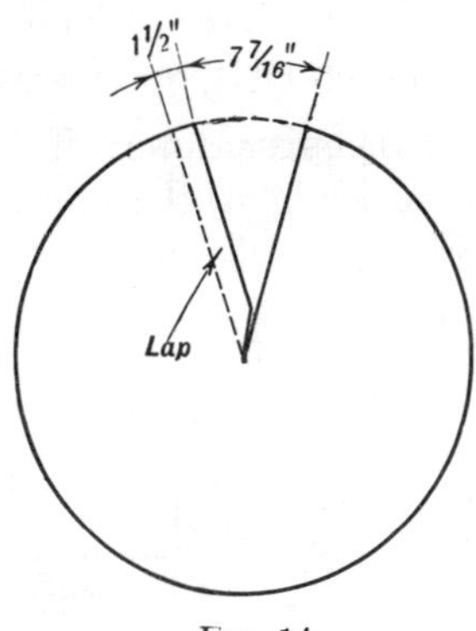

Fig. 14.

However, if the length of the circumference of the circle just drawn is computed, and the $75\frac{3}{8}$ plus $1\frac{1}{2}$ inches subtracted from this length, the difference provides a shorter measurement along the circumference. Thus

Circumference of flat circle

$$= 13.4164 \times 2 \times \pi = 84.294 = 84\frac{5}{16} \text{ in.}$$

Then, $84\frac{5}{16} - (75\frac{3}{8} + 1\frac{1}{2}) = 7\frac{7}{16}$ inches as the distance to be measured along the circumference.

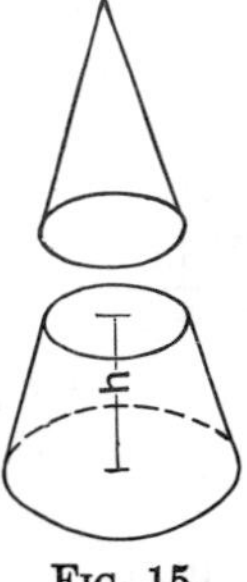

Fig. 15.

The part of a regular pyramid or of a cone which is left after its top has been cut off by a plane parallel to its base is called the frustum of the pyramid or cone. In practical work the frustum of a pyramid or cone has more applications than the pyramid or cone. The height is the shortest distance between the bases which are the base of the pyramid or cone and

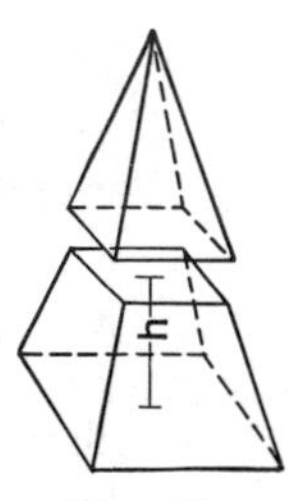

Fig. 16.

the section made by the cutting plane. The lateral faces of a frustum of a regular pyramid are trapezoids.

The following formulas may be used:

$$\text{Volume, } V = \frac{h}{3} \quad (B + b + \sqrt{B \times b})$$

where B = area of large base and b = area of small base.

Lateral surface = average perimeter of bases $\times$ slant height

ILLUSTRATIONS: 1. Find the volume of the frustum of a cone 5 inches high, the upper base being 4 inches and the lower base, 8 inches in diameter.

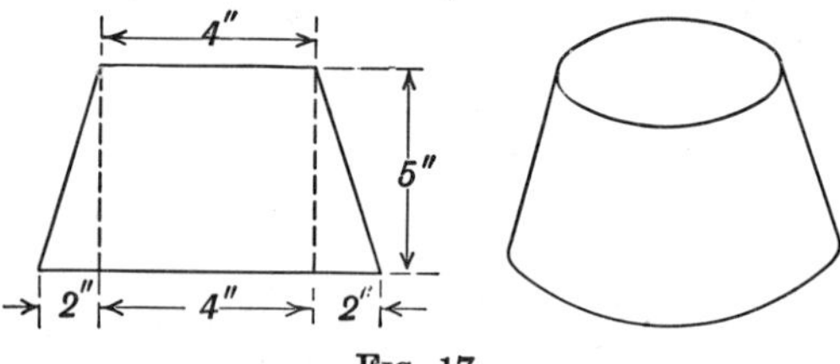

FIG. 17.

$$\text{Area of upper base, } b = 3.14 \times 2 \times 2 = 12.56$$
$$\text{Area of lower base, } B = 3.14 \times 4 \times 4 = 50.24$$

$$V = \frac{h}{3} (B + b + \sqrt{B \times b})$$

$$= \frac{5}{3} (50.24 + 12.56 + \sqrt{50.24 \times 12.55})$$

$$= \frac{5}{3} \times 87.92$$

$$= 146.6 \text{ cu. in.}$$

In illustration 1, find the lateral surface.

$$\text{Perimeter of upper base, } b = 3.14 \times 8 = 25.12 \text{ inches}$$
$$\text{Perimeter of lower base, } B = 3.14 \times 4 = \underline{12.56 \text{ inches}}$$
$$37.68 \text{ inches}$$

$$\text{Average perimeter} = \frac{37.68}{2} = 18.84 \text{ in.}$$

$$\text{Slant height} = \sqrt{5^2 + 2^2} = \sqrt{29} = 5.38 \text{ in.}$$
$$\text{Lateral surface} = 18.84 \times 5.38 = 101.36 \text{ sq. in.}$$

Frequently a sheet metal pattern maker is required to design a container of a certain capacity and is required to calculate the height.

The volume formula is transposed to read:

$$h = \frac{3 \times \text{volume}}{B + b + \sqrt{B \times b}}$$

ILLUSTRATION: 2. A container shaped like the frustum of a cone is to contain 1 cubic foot. If the upper base is 12 inches in diameter and the lower base 16 inches in diameter find the height.

$$\text{Area of } B = 3.14 \times 6 \times 6 = 113.04 \text{ sq. in.}$$

$$\text{Area of } b = 3.14 \times 8 \times 8 = 200.96 \text{ sq. in.}$$

then,

$$h = \frac{3 \times \text{volume (in cubic inches)}}{B + b + \sqrt{B \times b}}$$

$$= \frac{3 \times 17.28}{200.96 + 113.10 + \sqrt{200.96 \times 113.04}}$$

$$= \frac{5184}{314.00 + \sqrt{22{,}739.886}} = \frac{5184}{464.07} = 11.19 \text{ in.}$$

Figure 18 is the frustum of a hexagonal pyramid and its development is shown in Fig. 19. The formulas used are the same as those used for the frustum of a cone, i.e.;

$$\text{Volume, } V = \frac{h}{3}(B + b + \sqrt{B \times b})$$

Lateral surface = Average perimeter of bases × slant height.

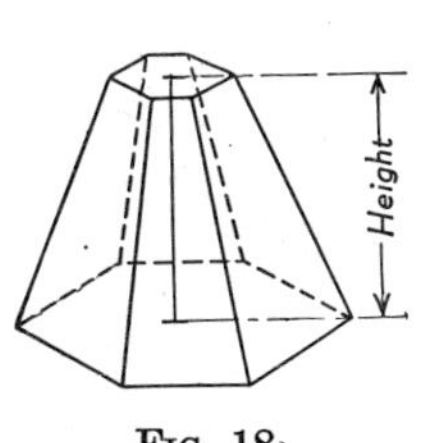

FIG. 18.

ILLUSTRATION: Find the volume of the frustum of a hexagonal pyramid 9 inches high, the side of the upper base being 2 inches and the side of the lower base 4 inches.

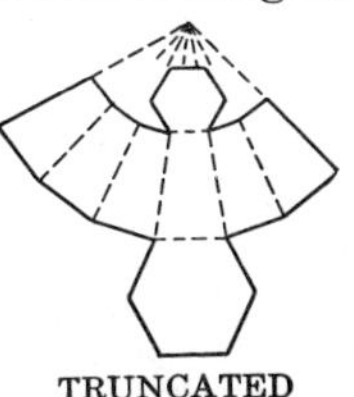

TRUNCATED HEXAGONAL PYRAMID

FIG. 19.

$$\text{Area of upper base, } b = 2.5980 \times 2 \times 2 = 10.392$$

$$\text{Area of lower base, } B = 2.5980 \times 4 \times 4 = 41.568$$

$$V = \frac{9}{3}(41.568 + 10.392 + \sqrt{10.392 \times 41.568})$$

$$= \frac{9}{3}(51.96 + \sqrt{431.9747})$$

$$= \frac{9}{3}(51.96 + 20.78)$$

$$= 218.22 \text{ cu. in.}$$

The following table may be used to lay out regular polygons and to calculate their area. Notice that 2.5980 used for finding the area of the bases in preceding problem is a constant taken from this table.

TABLE 1

ELEMENTS OF REGULAR POLYGONS

Number of sides	Name of figure	Diameter of circle that will just enclose when side is 1	Diameter of circle that will just go inside when side is 1	Length of side where diameter of enclosure circle equals 1	Length of side where inside circle equals 1	Angle formed by lines drawn from center to corners	Angle formed by outer sides of figures	To find area of figure multiply side by itself and by the number in this column
3	Triangle....	1.1546	0.5774	0.8660	1.7320	120°	60°	0.4330
4	Square.....	1.4142	1.000	0.7071	1.0000	90	90	1.0000
5	Pentagon...	1.7012	1.3764	0.5878	0.7265	72	108	1.7204
6	Hexagon...	2.0000	1.7320	0.5000	0.5774	60	120	2.5980
7	Heptagon...	2.3048	2.0766	0.4338	0.4815	51°–26′	128°–34′	3.6339
8	Octagon...	2.6132	2.4142	0.3827	0.4142	45°	135°	4.8284
9	Nonagon..	2.9238	2.7474	0.3420	0.3639	40	140	6.1818
10	Decagon..	3.2360	3.0776	0.3090	0.3247	36	144	7.6942
11	Undecagon..	3.5494	3.4056	0.2817	0.2936	32°–43′	147°–17′	9.3656
12	Dodecagon..	3.8638	3.7320	0.2858	0.2679	30°	150°	11.1961

A sphere is a solid in which all points on the surface are at the same distance from an internal point called the center. The volume and lateral surface may be found by the following formulas:

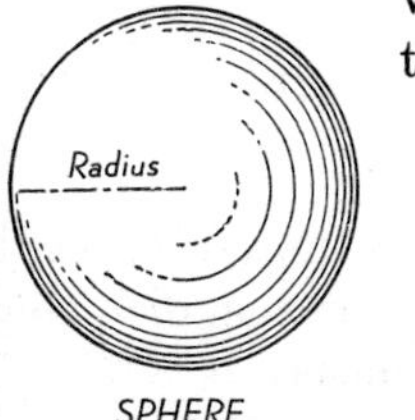

SPHERE

FIG. 20.

$$\text{Volume, } V = \frac{4\pi r^3}{3} = 4.1888r^3 \quad \text{or,}$$

$$\frac{\pi d^3}{6} = 0.5236d^3$$

$$\text{Lateral surface, } L = 4\pi r^2 = 3.1416d^2$$

ILLUSTRATION: Find the volume and lateral surface of a sphere $6\frac{1}{2}$ inches in diameter.

$$V = 0.5236d^3 = 0.5236 \times 6.5 \times 6.5 \times 6.5 = 143.79 \text{ cu. in.}$$
$$L = 3.1416d^2 = 3.1416 \times 6.5 \times 6.5 = 132.73 \text{ sq. in.}$$

There are problems in sheet metal work that occur in daily practice in which the rules of mensuration must be used before the pattern draftsman can make the development. Among these problems are transition pieces for heating, ventilating, blower and exhaust work together with the sizes and areas of outlets.

ILLUSTRATIONS:

1. Find the radius a tinsmith should use in laying out a circular hole for a pipe, the cross-section of which is 166 square inches. The formula $r = \sqrt{0.32A}$, which is derived from $A = \pi r^2$, can be used.

$$\text{radius} = \sqrt{0.32 \times \text{area}}$$
$$= \sqrt{0.32 \times 166}$$
$$= \sqrt{53.12} = 7\tfrac{7}{32} \text{ in.}$$

2. A tinsmith is required to make some cylindric cans to hold 1 gallon (231 cu. in.) each and to be 8 inches high. What radius should be used in laying out the base?

Allowance for seams are neglected.

$$\text{radius} = \sqrt{\frac{231}{8} \times 0.32}$$

$$= \sqrt{9.24} = 3\tfrac{1}{32} \text{ in.}$$

3. Find the height of a flaring measure required to hold 3 gallons and whose top diameter is 7 inches, the bottom diameter $11\frac{1}{2}$ inches and the diameter in the center is $9\frac{1}{4}$ inches.

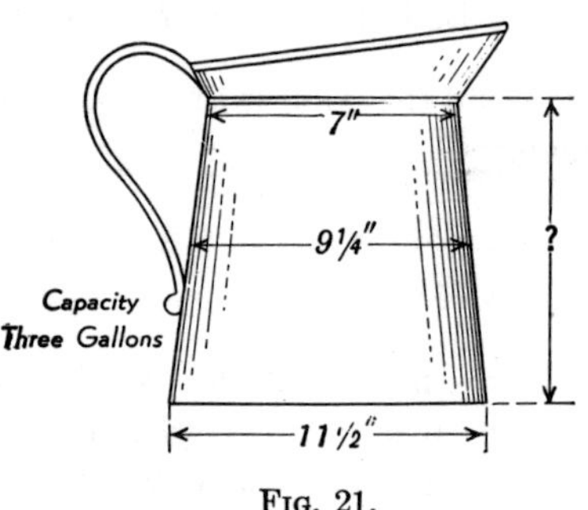

Fig. 21.

Rule.—Divide the capacity in cubic inches by the sum of the areas of the top and bottom diameters plus 4 times the area of the center section. Then multiply the quotient by 6.

Capacity $= 3 \times 231 = 693$ cu. in.

Area of top $= 7 \times 7 \times 0.7854 = 38.485$ sq. in.

Area of bottom $= 11.5 \times 11.5 \times 0.7854 = 103.87$ sq. in.

Area of middle section $= \dfrac{7 + 11.5}{2} = 9.25,\ 9.25 \times 9.25 \times 0.7854$

$= 67.20$ sq. in.

$$67.20 \times 4 = 268.80$$

$$38.485 + 103.87 + 268.80 = 411.155$$

$693 \div 411.55 = 1.684,\ 1.684 \times 6 = 10.11$ in. or $10\frac{1}{8}$ inches, the required height of the measure.

The table of areas and circumferences of circles page 683 is convenient for finding the area of circular-shaped vessels without computation. In the preceding case to find the area of the top, look under the column headed diameter and after 7 to the right read 38.4846 in the area column.

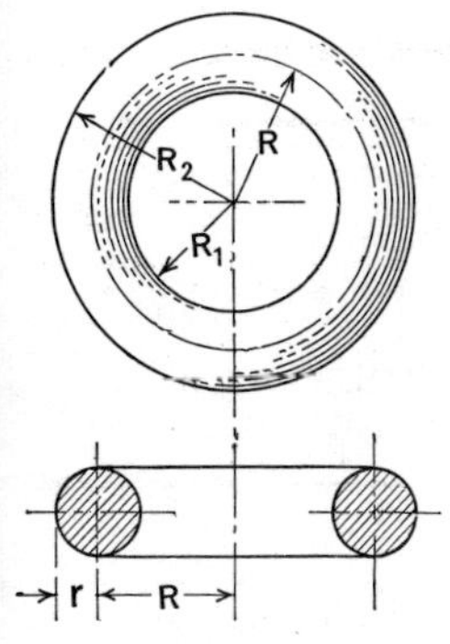

Fig. 22.

4. Find the volume and lateral surface of a cylindric ring whose outside diameter is 12 inches and whose inside diameter is 8 inches.

$$\text{Volume} = \pi r^2 \times (2\pi R)$$

where,
$$R = \frac{R_1 + R_2}{2} = \frac{4 + 6}{2} = 5$$

$$r = \frac{R_2 - R_1}{2} = \frac{6 - 4}{2} = 1$$

Then,
$$V = (3.14 \times 1 \times 1) \times (2 \times 3.14 \times 5)$$
$$= 3.14 \times 31.4 = 98.6 \text{ cu. in.}$$

$$\text{Lateral Surface, } L = (2\pi r) \times (2\pi R)$$
$$= (2 \times 3.14 \times 1) \times (2 \times 3.14 \times 5)$$
$$= 6.28 \times 31.4 = 197.2 \text{ sq. in.}$$

5. In the offset boot shown in Fig. 23, find the length of the rectangular pipe in order that its dimension will equal the area of the 10-inch round pipe if the width of the rectangular pipe is 4 inches.

Area of round pipe $= \pi r^2 = 3.14 \times 5 \times 5 = 78.5$ sq. in.

Then, $78.5 \div 4 = 19.625$ or $19\frac{5}{8}$ in.

Therefore, the size of the rectangular riser of equal area to the 10-inch round pipe is 4 in. $\times$ $19\frac{5}{8}$ in.

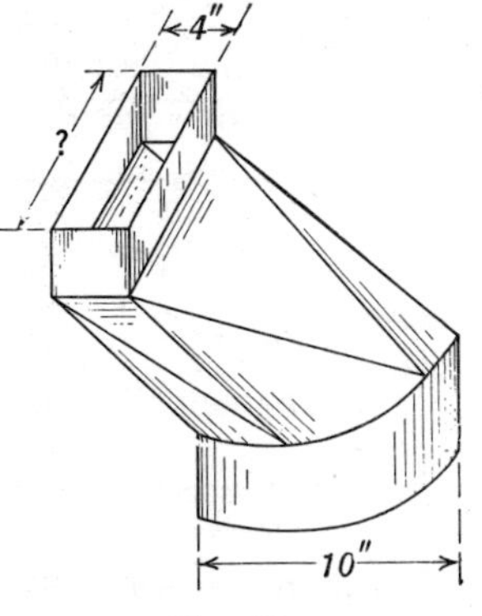

Fig. 23.

6. Find the volume and lateral surface of a ring with a 6-inch diameter and a square cross-section of 1 inch.

$$\begin{aligned}\text{Volume} &= \pi H(R_2^2 - R_1^2)\\ &= 3.14 \times 1(3 \times 3 - 2 \times 2)\\ &= 3.14 \times 5 = 15.7 \text{ cu. in.}\\ \text{Lateral surface} &= 2(B + H)(2\pi R)\\ &= 2(1 + 1)(2 \times 3.14 \times 2.5)\\ &= 4 \times 15.7 = 62.8 \text{ sq. in.}\end{aligned}$$

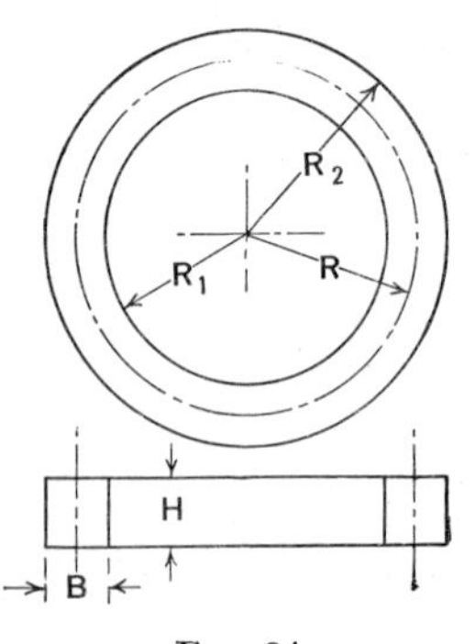

FIG. 24.

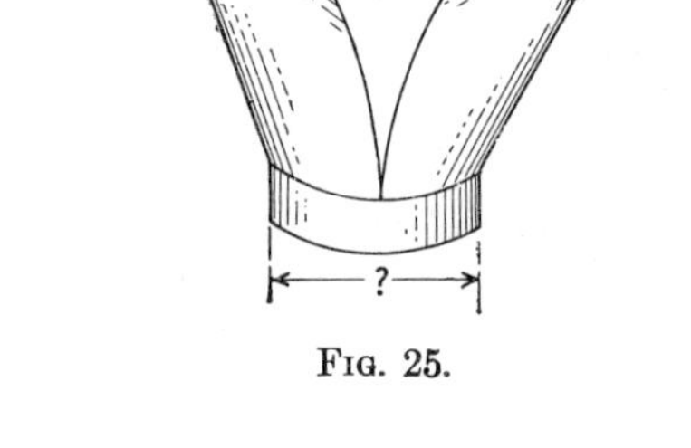

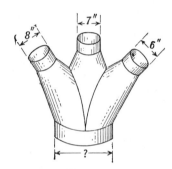

FIG. 25.

7. Find the diameter of a main pipe whose capacity will equal the combined capacity of three branches whose diameters are 6 inches, 7 inches, and 8 inches respectively.

Areas of circles vary as the squares of their diameters, therefore,

$$\begin{aligned}\text{Diameter of main pipe} &= \sqrt{8^2 + 7^2 + 6^2}\\ &= \sqrt{149} = 12.2 \text{ or } 12\tfrac{1}{4} \text{ in.}\end{aligned}$$

The square root can be found in table of squares, square roots, etc. on page 29.

Another method which makes use of the tables of areas of circles on page 683 is:

Area of 6″ pipe =	28.2744	square inches
Area of 7″ pipe =	38.4846	square inches
Area of 8″ pipe =	50.2656	square inches
Combined areas =	117.0246	square inches

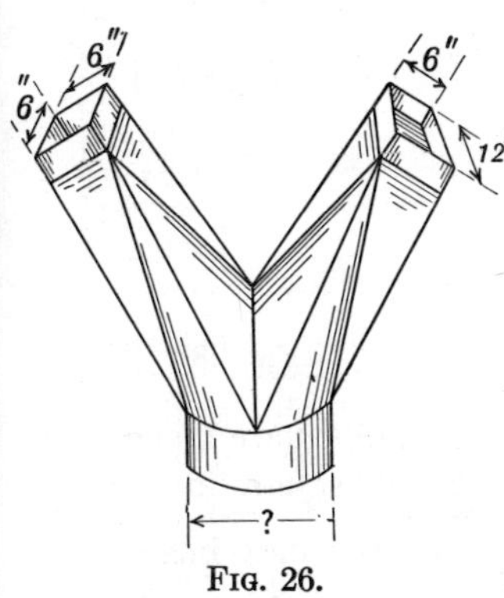

Fig. 26.

Taking the nearest number in the table to 117.0246, i.e., 117.859, which is the diameter of a pipe $12\frac{1}{4}$ inches.

8. Find the diameter of a round main to equal the area of one square and one rectangular branch of a two-branched prong, Fig. 26, when one branch is 6 inches square and the other measures 6 inches × 12 inches.

Area of square branch	= 6 × 6 =	36	square inches
Area of rectangular branch	= 6 × 12 =	72	square inches
	Combined area =	108	square inches

The nearest calculation from the table on page 683 is 108.43, the diameter of a $11\frac{3}{4}$-inch circle.

9. Find the missing dimension of a rectangular pipe of area equal to that of two round branches, 8 inches and 12 inches in diameter when one side of the rectangular pipe measures 10 inches. See Fig. 27.

From the table,

The area of the 8-inch pipe =	50.265	square inches
The area of the 12-inch pipe =	113.098	square inches
The combined area =	163.363	square inches

The missing dimension will be, 163.363 ÷ 10 or 16.33 = $16\frac{5}{16}$ in.

10. Find the size of a square pipe having an area equal to that of two round branches whose diameters are $13\frac{5}{8}$ inches and 16 inches respectively. See Fig. 28.

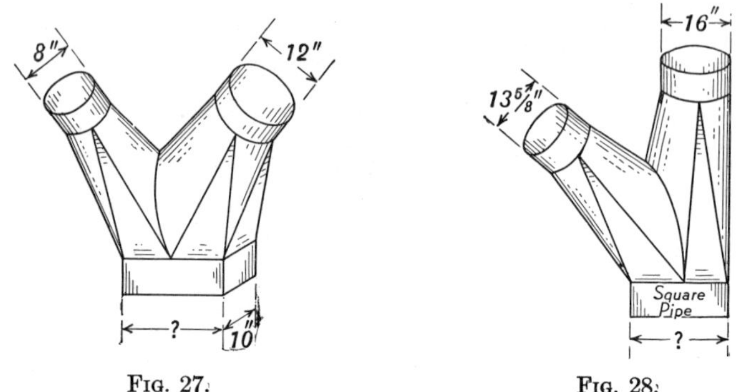

FIG. 27.　　FIG. 28.

The area of the $13\frac{5}{8}$-inch pipe = 145.80 square inches
The area of the 16-inch pipe = 201.06 square inches

The combined area = 346.86 square inches

The $\sqrt{346.86} = 18.6$ or $18\frac{5}{8}$ inches, thus making the required size of the square main pipe, $18\frac{5}{8}$ in. $\times$ $18\frac{5}{8}$ in.

11. Find the increased sizes of the ducts. ***A*, *B*, and *C*,** shown in the ventilating system, in order to take care of the 8-inch, 10-inch and 15-inch branches respectively.
From the tables of areas of circles, page 683.

Area of 6-inch pipe = 28.2743 square inches
Area of 8-inch pipe = 50.2655 square inches

Combined area = 78.5398 square inches

this is the area of a 10-inch circle. Thus pipe *A* should have a diameter of 10 inches.

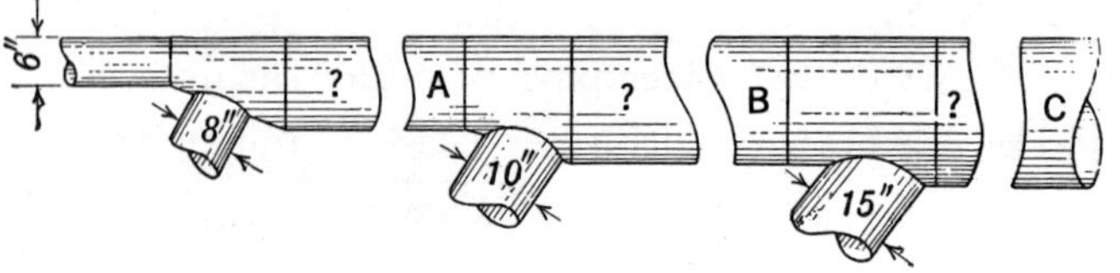

FIG. 29.

Area of B = area of pipe A + area of 10-inch branch. Because both are 10 inches in diameter, $78.539 \times 2 = 157.078$ square inches, the combined areas. From the table, page 683, the nearest number to 157.078 is 159.48, the area of a $14\frac{1}{4}$-inch circle. Thus, pipe B should have a diameter of $14\frac{1}{4}$ inches.

$$
\begin{aligned}
\text{Area of } C &= \text{area of pipe } B + \text{area of 15-inch branch} \\
&= 159.48 + 176.71 \\
&= 336.19 \text{ sq. in.}
\end{aligned}
$$

From the table, page 683, find 336.19 equal to $20\frac{11}{16}$. Thus pipe C should have a diameter of $20\frac{11}{16}$ inches.

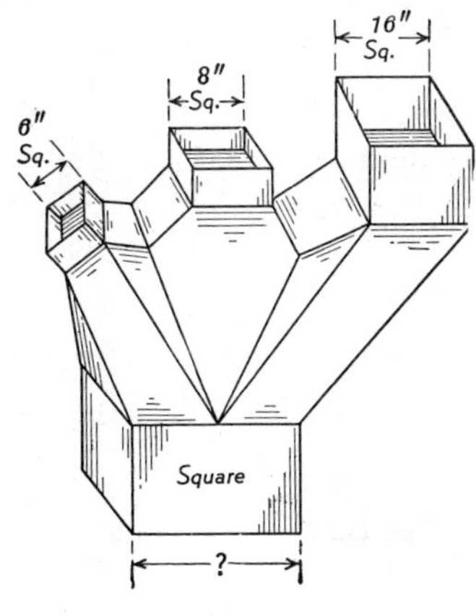

Fig. 30.

Find the size of a square main in a three-branched fitting whose outlets are, 16 inches × 16 inches, 8 inches × 8 inches and 6 inches × 6 inches respectively. From the tables, page 29,

Area of 6-inch × 6-inch outlet	=	36 square inches
Area of 8-inch × 8-inch outlet	=	64 square inches
Area of 16-inch × 16-inch outlet	=	256 square inches
Combined area	=	356 square inches

Side of square main = $\sqrt{356} = 18.86$.

In practical work the size of the square would be taken as $18\frac{7}{8}$ inches $\times$ $18\frac{7}{8}$ inches.

13. Find the dimensions of a rectangular vertical flue having an equal area to the combined areas of five horizontal vent ducts shown in Fig. 31.

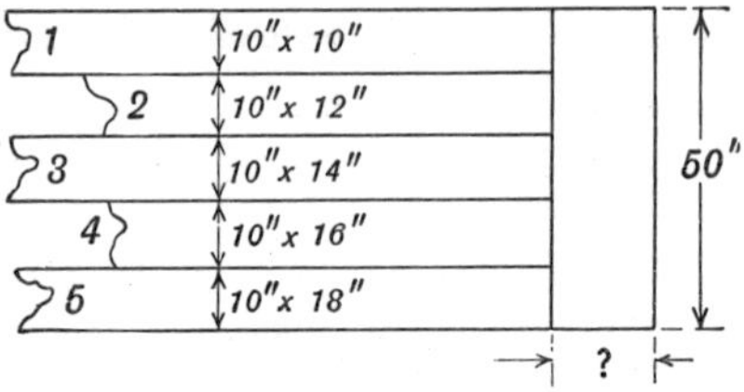

Fig. 31.

Area of first duct $= 10 \times 10 = 100$ square inches
Area of second duct $= 10 \times 12 = 120$ square inches
Area of third duct $= 10 \times 14 = 140$ square inches
Area of fourth duct $= 10 \times 16 = 160$ square inches
Area of fifth duct $= 10 \times 18 = 180$ square inches

Combined area $= 700$ square inches

As all ducts are set in 10-inch way, the space taken up is $5 \times 10 = 50$ inches.

Therefore, $700 \div 50 = 14$. Thus the flue with the required area will be 14 inches $\times$ 50 inches.

14. Find the amount of tin required for the funnel shown in the sketch if the slant height of the upper piece is 4.5 inches, and the slant height of the lower piece is 3.5 inches. Allow $\frac{1}{2}$ inch on the length and width of each piece for locks.

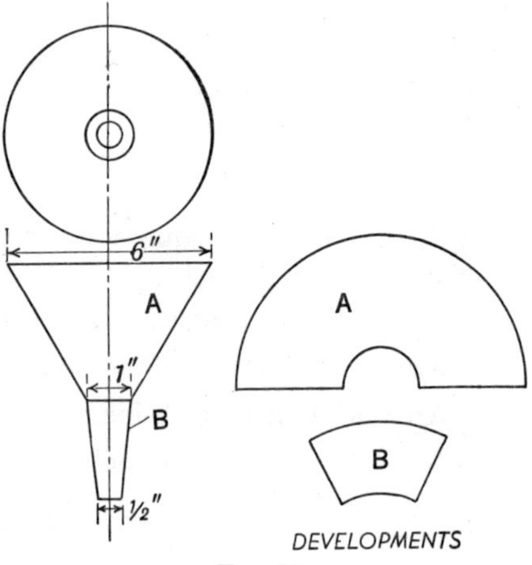

Fig. 32.

The formula from page 651.

Lateral surface = average perimeter of bases × slant height

Perimeter of upper large base = 3.14 × 6 = 18.84 inches

Perimeter of upper small base = 3.14 × 1 = 3.14 inches

21.98 inches

$$\text{Average perimeter} = \frac{21.98}{2} = 10.99 \text{ inches}$$

Perimeter of lower large base = 3.14 × 1 = 3.14 inches

Perimeter of lower small base = 3.14 × 0.5 = 1.57 inches

$$\text{Average perimeter} = \frac{4.71}{2} = 2.35 \text{ inches}$$

Upper slant height $4\frac{1}{2}$ inches + $\frac{1}{2}$ inch = 5 inches

Lower slant height $3\frac{1}{2}$ inches + $\frac{1}{2}$ inch = 4 inches

10.99 + 0.5 = 11.49 × 5 = 57.45 square inches

2.35 + 0.5 = 2.85 × 4 = 11.40 square inches

68.85 square inches

Therefore 68.85 square inches of tin are required.

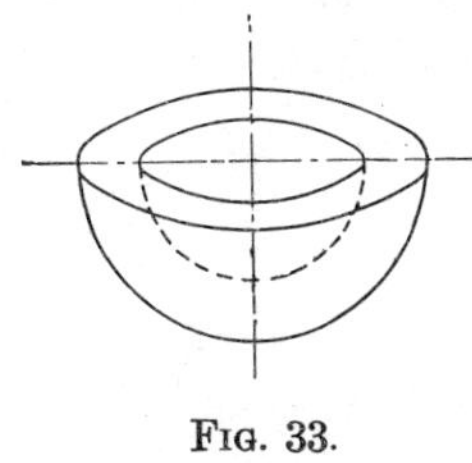

Fig. 33.

15. Find the volume of the hemispherical bowl shown in Fig. 33, when the outside diameter is 12 inches and the inside diameter is 8 inches.

Aid. Treat the solid as the differences of two hemispheres.

From page 654. Volume of a sphere = $0.5236d^3$

whence $V = 0.5236 \times 12 \times 12 \times 12 = 904.78$ cubic inches

$V = 0.5236 \times 8 \times 8 \times 8 = 268.08$ cubic inches

904.72 ÷ 2 = 452.36 cubic inches in outside hemisphere

268.08 ÷ 2 = 134.04 cubic inches in inside hemisphere

318.32 cubic inches, the volume of the bowl.

TABLE 2.—CAPACITY OF TANKS IN UNITED STATES GALLONS

Decimal Equivalents of Fractional Parts of a Gallon

0.03125 of a gallon	= 1	gill	0.53125 of a gallon	= 17	gills
0.06250 " " "	= ½	pint	0.56250 " " "	= 4½	pints
0.09375 " " "	= 3	gills	0.62500 " " "	= 5	pints
0.12500 " " "	= 1	pint	0.59375 " " "	= 19	gills
0.15625 " " "	= 5	gills	0.65625 " " "	= 21	gills
0.18750 " " "	= 1½	pints	0.68750 " " "	= 5½	pints
0.21875 " " "	= 7	gills	0.71875 " " "	= 23	gills
0.2500 " " "	= 1	quart	0.75000 " " "	= 3	quarts
0.28125 " " "	= 9	gills	0.78125 " " "	= 25	gills
0.31250 " " "	= 2½	pints	0.81250 " " "	= 6½	pints
0.34375 " " "	= 11	gills	0.84375 " " "	= 27	gills
0.37500 " " "	= 3	pints	0.87500 " " "	= 7	pints
0.40625 " " "	= 13	gills	0.90625 " " "	= 29	gills
0.43750 " " "	= 3½	pints	0.93750 " " "	= 7½	pints
0.46875 " " "	= 15	gills	0.968750 " " "	= 31	gills
0.50000 " " "	= ½	gallon	1.00000 " " "	= 1	gallon

Tin Roofing.—Pure block tin is not used for common building purposes; but thin plates of sheet iron covered with it on both sides constitute the *tinned plates*, or, as they are called, the *tin*, used for covering roofs, rain pipes and many domestic utensils. For roofs it is laid on boards.

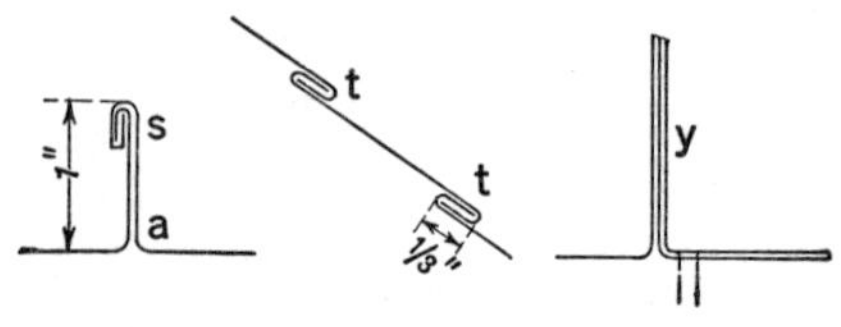

FIG. 34.

The sheets of tin are united as shown in Fig. 34. First, several sheets are joined together in the shop, end for end, as at *tt*, by being first bent over, then hammered flat, and then soldered. These are then formed into a roll to be carried to the roof, a roll being long enough to reach from the peak to the eaves. Different rolls being spread up and down the roof are then united along

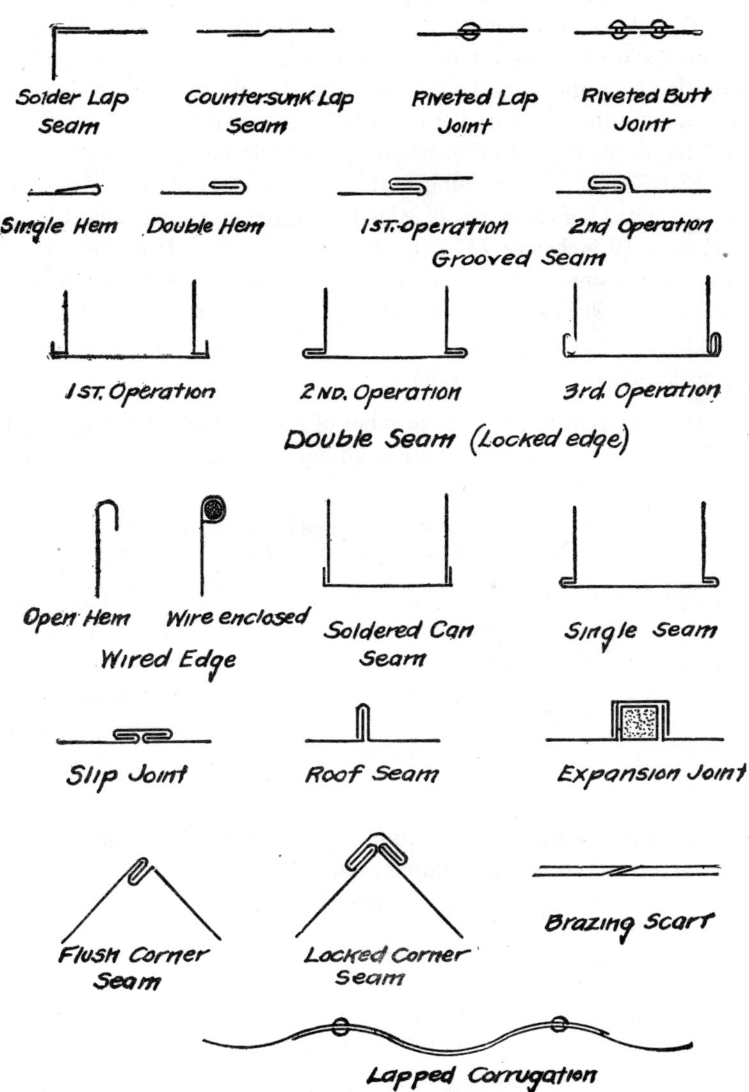

FIG. 35.—Types of Joints and Edges for Sheet Metal.

their sides by simply being bent as at *a* and *s*, by a tool for that purpose. The roofers call the bending at *s* a *double groove*, or *double lock*; and the more simple ones at *t*, a *single groove*, or lock.

To hold the tin securely to the sheeting boards, pieces of the tin 3 or 4 inches long by 2 inches wide, called cleats, are nailed to the boards at about every 18 inches along the joints of the rolls that are to be united, and are bent over with the double groove *s*. This will be understood from *y*, where the middle piece is the cleat.

Flat-Seam Tin Roofing.—When a sheet of tin 14×20 inches with $\frac{1}{2}$-inch edges is edged or folded, it will measure 13 inches $\times$ 19 inches or 247 square inches of area. However, when this sheet is joined to other sheets on the roof, its covering capacity is only $12\frac{1}{2}$ inches $\times$ $18\frac{1}{2}$ inches or 231.25 square inches. A box of 112 sheets, 14 inches $\times$ 20 inches, laid this way will cover, approximately, 180 square feet.

Illustration: Find the number of sheets of tin 14 inches $\times$ 20 inches required for one square (100 square feet) using flat seams with $\frac{1}{2}$-inch edge.

$$100 \times 144 \text{ (the number of square inches in a square foot)} = 14{,}400$$
$$14{,}400 \div 231.25 = 63$$

Standing-Seam Tin Roofing.—When standing-seams edged $1\frac{1}{4}$ inches and $1\frac{1}{2}$ inches are used $2\frac{3}{4}$ inches is taken off the width; and the flat cross-seams edged $\frac{3}{8}$ inches take $1\frac{1}{8}$ inches off the length of the sheet. The covering capacity of each 14×20-inch sheet is, therefore, $1\frac{1}{4}$ inches $\times$ $18\frac{7}{8}$ inches or 212.34 square inches. A box of 112 sheets 14 inches $\times$ 20 inches laid in this will cover 165 square feet.

Illustration: Find the number of sheets of tin required for one square when using standing seams.

$$14{,}400 \div 212.34 = 68$$

Note: The weight of sheet metal is calculated on page 272 in the chapter on weights and measures.

Corrugated sheets of iron and steel are used not only for roofing but also for siding of sheds, mills and other structures. These sheets are carried in stock in 4-foot, 5-foot, 6-foot, 8-foot, 9-foot and 10-foot lengths, the 8-foot length being the most commonly used. The usual width of sheets is 24 inches between the centers of the outer corrugations, so that the covering width is 24 inches when one corrugation is used for the side lap. Ordinary corrugated sheets should have a lap 1½ or 2 corrugations side lap for roofing in order to secure water-tight side seams. For covering roofs, either 3-inch, 2½-inch or 2-inch corrugations should be used, the 2-inch corrugation being the most common size. No. 28 gage corrugated iron is generally used for applying to wooden buildings. When laid on a roof, corrugated sheets should have a lap on the lower end from 3 to 6 inches, according to the pitch of the roof.

TABLE 3

NUMBER OF SQUARE FEET OF CORRUGATED SHEETS TO COVER 100 SQUARE FEET OF ROOF

End Laps................	1 inch	2 inches	3 inches	4 inches	5 inches
	Sq. Ft.	Sq. Ft.	Sq. Ft.	Sq. Ft.	Sq. Ft.
Side lap, 1 corrugation...	110	111	112	113	114
Side lap, 1½ corrugations...	116	117	117	119	120
Side lap, 2 corrugations...	123	124	125	126	127

CORRUGATED IRON ROOFING

B. W. Gauge	Weight per square (100 square feet). Plain	Galvanized
Number	Pounds	
28	97	Weights from 5 to 15 per cent heavier than plain, according to the number B. W. G.
26	105	
24	128	
22	150	
20	185	
18	270	
16	340	

Allow one-third the net width for lapping and for corrugations. From 2½ to 3½ pounds for rivets will be required per square.

The best plates, both for tinning and for ternes, are made of charcoal iron, which, being tough, bears bending better. Coke is used for making cheaper plates, but is inferior as regards bending.

Much use is made of what is called leaded tin, or ternes, for roofing. It is simply sheet iron coated with an alloy of lead and tin, lead being less expensive. In one standard brand the alloy is 32% tin, 68% lead.

TABLE 4

GALVANIZED SHEET IRON

Am. Galv. Iron Assn. B. W. G.

No.	Ounces avoir. per square foot	Square feet per 2240 pounds	No.	Ounces avoir. per square foot	Square feet per 2240 pounds	No.	Ounces avoir. per square foot	Square feet per 2240 pounds
29	12	2987	24	17	2108	19	33	1084
28	13	2757	23	19	1886	18	38	943
27	14	2560	22	21	1706	17	43	833
26	15	2389	21	24	1493	16	48	746
25	16	2240	20	28	1280	14	60	597

TABLE 5.—TIN REQUIRED FOR FLAT SEAMS

No. of square feet..	**100**	**110**	**120**	**130**	**140**	**150**	**160**	**170**	**180**	**190**	**200**
Sheets required.....	**63**	**69**	**75**	**81**	**88**	**94**	**100**	**106**	**112**	**119**	**125**
No. of square feet..	**210**	**220**	**230**	**240**	**250**	**260**	**270**	**280**	**290**	**300**	**310**
Sheets required.....	**131**	**137**	**144**	**150**	**156**	**162**	**169**	**175**	**181**	**187**	**193**
No. of square feet..	**320**	**330**	**340**	**350**	**360**	**370**	**380**	**390**	**400**	**410**	**420**
Sheets required.....	**200**	**206**	**212**	**218**	**224**	**231**	**237**	**243**	**249**	**256**	**262**
No. of square feet..	**430**	**440**	**450**	**460**	**470**	**480**	**490**	**500**	**510**	**520**	**530**
Sheets required.....	**268**	**274**	**281**	**287**	**293**	**299**	**305**	**312**	**318**	**324**	**330**
No. of square feet..	**540**	**550**	**560**	**570**	**580**	**590**	**600**	**610**	**620**	**630**	**640**
Sheets required.....	**337**	**343**	**349**	**355**	**362**	**368**	**374**	**380**	**386**	**393**	**396**
No. of square feet..	**650**	**660**	**670**	**680**	**690**	**700**	**710**	**720**	**730**	**740**	**750**
Sheets required.....	**405**	**411**	**418**	**424**	**430**	**436**	**442**	**448**	**455**	**461**	**467**
No. of square feet..	**760**	**770**	**780**	**790**	**800**	**810**	**820**	**830**	**840**	**850**	**860**
Sheets required.....	**474**	**480**	**486**	**492**	**499**	**505**	**511**	**517**	**523**	**530**	**536**
No. of square feet..	**870**	**880**	**890**	**900**	**910**	**920**	**930**	**940**	**950**	**960**	**970**
Sheets required.....	**542**	**548**	**554**	**561**	**567**	**573**	**579**	**586**	**592**	**598**	**604**
No. of square feet..	**980**	**990**	**1000**	....	....	....	....	....	....	....	....
Sheets required.....	**610**	**617**	**625**	....	....	....	....	....	....	....	....

A box of 112 sheets 14 by 20 in laid in this way will cover 180 sq ft.

TABLE 5.—TIN REQUIRED FOR FLAT SEAMS—*Continued*

No. of square feet..	100	110	120	130	140	150	160	170	180	190	200
Sheets required.....	30	33	36	39	42	45	47	50	53	56	59
No. of square feet..	210	220	230	240	250	260	270	280	290	300	310
Sheets required.....	62	65	68	71	74	77	80	83	86	89	92
No. of square feet..	320	330	340	350	360	370	380	390	400	410	420
Sheets required.....	94	97	100	103	106	109	112	115	118	121	124
No. of square feet..	430	440	450	460	470	480	490	500	510	520	530
Sheets required.....	127	130	133	136	139	141	144	147	150	153	156
No. of square feet..	540	550	560	570	580	590	600	610	620	630	640
Sheets required.....	159	162	165	168	171	174	177	180	183	186	188
No. of square feet..	650	660	670	680	690	700	710	720	730	740	750
Sheets required.....	191	194	197	200	203	206	209	212	215	218	221
No. of square feet..	760	770	780	790	800	810	820	830	840	850	860
Sheets required.....	224	227	230	233	235	238	241	244	247	250	253
No. of square feet..	870	880	890	900	910	920	930	940	950	960	970
Sheets required...	256	259	262	265	268	271	274	277	280	282	285
No. of square feet..	980	990	1000	....	....	....	....	....	....	....	....
Sheets required.....	288	291	294	....	....	....	....	....	....	....	....

A box of 112 sheets 28 by 20 in laid in this way will cover 381 sq ft.

TABLE 6.—TIN REQUIRED FOR STANDING SEAMS

No. of square feet..	100	110	120	130	140	150	160	170	180	190	200
Sheets required.....	68	75	82	89	95	102	109	116	123	129	136
No. of square feet..	210	220	230	240	250	260	270	280	290	300	310
Sheets required.....	143	150	156	163	170	177	184	190	197	204	211
No. of square feet..	320	330	340	350	360	370	380	390	400	410	420
Sheets required.....	218	224	231	238	245	251	258	265	271	279	285
No. of square feet..	430	440	450	460	470	480	490	500	510	520	530
Sheets required.....	292	299	306	312	319	326	333	340	346	353	360
No. of square feet..	540	550	560	570	580	590	600	610	620	630	640
Sheets required.....	367	374	379	387	393	401	407	414	421	428	435
No. of square feet..	650	660	670	680	690	700	710	720	730	740	750
Sheets required.....	441	447	455	462	468	475	482	489	495	501	509
No. of square feet..	760	770	780	790	800	810	820	830	840	850	860
Sheets required.....	515	523	529	536	543	550	557	563	570	577	584
No. of square feet..	870	880	890	900	910	920	930	940	950	960	970
Sheets required.....	590	597	604	611	618	623	630	637	644	651	658
No. of square feet..	980	990	1000	....	....	....	....	....	....	....	....
Sheets required.....	665	672	679	....	....	....	....	....	....	....	....

A box of 112 sheets 14 by 20 in laid in this way will cover 165 sq ft.

TABLE 6.—TIN REQUIRED FOR STANDING SEAMS—*Continued*

No. of square feet..	100	110	120	130	140	150	160	170	180	190	200
Sheets required.....	32	35	38	41	44	47	50	53	56	59	62
No. of square feet..	210	220	230	240	250	260	270	280	290	300	310
Sheets required.....	65	68	71	74	77	80	84	87	90	94	97
No. of square feet..	320	330	340	350	360	370	380	390	400	410	420
Sheets required.....	100	103	106	109	112	115	118	121	125	128	131
No. of square feet..	430	440	450	460	470	480	490	500	510	520	530
Sheets required.....	134	137	141	144	147	150	153	156	159	162	165
No. of square feet...	540	550	560	570	580	590	600	610	620	630	640
Sheets required.....	168	171	174	177	180	184	187	190	193	196	19[illegible]
No. of square feet..	650	660	670	680	690	700	710	720	730	740	75
Sheets required.....	202	205	208	211	214	218	221	224	227	230	23[illegible]
No. of square feet..	760	770	780	790	800	810	820	830	840	850	860
Sheets required.....	236	239	242	245	249	252	255	258	261	265	268
No. of square feet..	870	880	890	900	910	920	930	940	950	960	970
Sheets required.....	271	274	277	280	283	286	289	292	296	299	302
No. of square feet...	980	990	....	....	....	....	....	....	....	....	....
Sheets required.....	305	308	....	....	....	....	....	....	....	....	....

A box of 112 sheets 28 by 20 in laid in this way will cover 360 sq ft.

In giving orders, it is important to specify whether charcoal plates or coke ones are required; also whether *tinned* plates, or *ternes*.

Tinned and leaded sheets of Bessemer and other cheap steel are now much used. They are sold at about the price of charcoal tin and terne plates.

If the tin is laid with a flat-seam or flat lock, the roof should have an incline of $\frac{1}{2}$ inch or more to a foot. If laid with a standing seam, there should be an incline of not less than 2 inches to a foot.

This is put up in rolls 14, 20, and 28 inches wide for the convenience of roofers. Each roll contains 108 square feet. The following table shows the number of sheets required per lineal foot for 20- and 28-inch widths.

Roof Flashings.—Flashings are pieces of tin, lead or copper, let into the joints of a wall so as to lap over gutters and in places where leaks are likely to occur such as around chimneys, dormers, skylights and in valleys. In shingle work, the valley flashings are usually 14 inches wide, while the length depends upon the length

TABLE 7.—Sizes and Weight of Sheet Tin

Mark	Number of sheets in box	Dimension		Weight of box, pounds
		Length, inches	Breadth, inches	
1C	225	13¾	10	112
11C	225	13¼	9¾	105
111C	225	12¾	9½	98
1X	225	13¾	10	140
1XX	225	13¾	10	161
1XXX	225	13¾	10	182
1XXXX	225	13¾	10	203
DC	100	16¾	12½	105
DX	100	16¾	12½	126
DXX	100	16¾	12½	147
DXXX	100	16¾	12½	168
DXXXX	100	16¾	12½	189
5DC	200	15	11	168
5DX	200	15	11	189
5DXX	200	15	11	210
5DXXX	200	15	11	231
1CW	225	13¾	10	112

A box containing 225 sheets, 13¾ by 10, contains 214.84 square feet; but allowing for seams it will cover only 150 square feet of roof.

of the valley. The sides of dormers, chimneys and all intersections are flashed with tin cut so as to turn up 3½ inches on the vertical and 3 inches on the roof. Flashings are measured by the number of square feet.

There are also in use for roofing, certain compound metals which resist tarnish better than either lead, tin, or zinc but which are so fusible as to be liable to be melted by large burning cinders falling on the roof from a neighboring conflagration.

A roof covered with tin or other metal should, if possible, slope not much *less* than five degrees, or about an inch to a foot;

TABLE 8.—TIN IN ROLLS OR GUTTER STRIPS

Feet	Widths		Feet	Widths		Feet	Widths		Hundred feet	Widths	
	20	28		20	28		20	28		20	28
1	1	1	35	16	23	69	31	44	2	89	128
2	1	2	36	16	23	70	32	45	3	134	192
3	2	2	37	17	24	71	32	45	4	178	256
4	2	3	38	17	24	72	32	46	5	223	320
5	3	4	39	18	25	73	33	47	6	267	384
6	3	4	40	18	26	74	33	47	7	312	444
7	4	5	41	19	27	75	34	48	8	356	512
8	4	5	42	19	27	76	34	48	9	401	576
9	4	6	43	20	28	77	35	49	10	445	640
10	5	7	44	20	28	78	35	50	11	495	704
11	5	7	45	20	29	79	36	50	12	540	768
12	6	8	46	21	29	80	36	51	13	585	832
13	6	9	47	21	30	81	36	52	14	630	896
14	7	9	48	22	31	82	37	52	15	675	960
15	7	10	49	22	31	83	37	53	16	720	1 024
16	8	11	50	23	32	84	38	54	17	765	1 088
17	8	11	51	23	33	85	38	54	18	810	1 152
18	8	12	52	24	33	86	39	55	19	855	1 216
19	9	12	53	24	34	87	39	55	20	900	1 280
20	9	13	54	24	34	88	40	56	21	945	1 344
21	10	14	55	25	35	89	40	57	22	990	1 408
22	10	14	56	25	36	90	40	57	23	1 035	1 472
23	11	15	57	26	36	91	41	58	24	1 080	1 536
24	11	16	58	26	37	92	41	59	25	1 135	1 600
25	12	16	59	27	38	93	42	59	26	1 170	1 664
26	12	17	60	27	38	94	42	60	27	1 215	1,738
27	12	18	61	28	39	95	43	61	28	1 260	1 792
28	13	18	62	28	40	96	43	62	29	1 305	1 856
29	13	19	63	28	40	97	44	62	30	1 350	1 920
30	14	19	64	29	41	98	44	63	31	1 395	1 984
31	14	20	65	29	41	99	44	64	32	1 440	2 048
32	15	21	66	30	42	100	45	64	33	1 485	2 112
33	15	21	67	30	43	.....	.....	....	34	1 530	2 176
34	16	22	68	31	43	.....	.....	....	35	1 575	2 240

and at the eaves there should be a sudden fall into the rain-gutter, to prevent rain from backing up so as to overtop the double-groove joint *s*, and thus cause leaks. When coal is used for fuel, tin roofs should receive two coats of paint when first put up, and a coat every 2 or 3 years after. Where wood only is used, this is not necessary; and a tin roof with a good pitch will last 20 or 30 years.

Two good workmen can put on, and paint outside, from 250 to 300 square feet of tin roof, per day of 8 hours.

Tinned iron plates are sold by the box. These boxes, unlike glass, have *not* equal areas of contents. They may be designated or ordered either by their names or sizes. Many makers, however, have their private brands in addition; and some of these have a much higher reputation than others.

TABLE 9.—Weights of Sheet Steel and Iron

United States Standard Gage

Number of Gage	App. Thickness	Weight per Sq. Foot		No. of Gage	App. Thickness	Weight per Sq. Foot	
		Steel	Iron			Steel	Iron
0000000	.5	20.320	20.00	17	.05625	2.286	2.25
000000	.46875	19.050	18.75	18	.05	2.032	2.
00000	.4375	17.780	17.50	19	.04375	1.778	1.75
0000	.40625	16.510	16.25	20	.0375	1.524	1.50
000	.375	15.240	15.00	21	.03437	1.397	1.375
00	.34375	13.970	13.75	22	.03125	1.270	1.25
0	.3125	12.700	12.50	23	.02812	1.143	1.125
1	.28125	11.430	11.25	24	.025	1.016	1.
2	.26562	10.795	10.625	25	.02187	.903	.875
3	.25	10.160	10.00	26	.01875	.762	.75
4	.23437	9.525	9.375	27	.01718	.698	.687
5	.21875	8.890	8.75	28	.01562	.635	.623
6	.20312	8.255	8.125	29	.01406	.571	.562
7	.1875	7.620	7.5	30	.0125	.508	.5
8	.17187	6.985	6.875	31	.01093	.440	.437
9	.15625	6.350	6.25	32	.01015	.413	.406
10	.14062	5.715	5.625	33	.00937	.381	.375
11	.125	5.080	5.00	34	.00859	.349	.343
12	.10937	4.445	4.375	35	.00781	.317	.312
13	.09375	3.810	3.75	36	.00703	.285	.281
14	.07812	3.175	3.125	37	.00664	.271	.265
15	.0703	2.857	2.812	38	.00625	.254	.25
16	.0625	2.540	2.50				

Weight of 1 cubic foot is assumed to be 487.7 lbs. for steel plates and 480 lbs. for iron plates.

Aluminum Roofing.—Aluminum roofing and siding have become very popular, especially for farm and industrial buildings. As a roofing and siding material, aluminum offers a number of important advantages. It resists normal outdoor exposure, it does not need painting, it does not stain or streak painted buildings, and for decorative purposes it can be painted after about three months' exposure to weather. The lighter weight of aluminum roofing also makes possible economies in the supporting framework of a building.

In the summer, aluminum reflects the sun's rays and helps to keep the building interior cooler. In the winter it helps to retain heat inside buildings. Another useful feature of aluminum is that it will not adversely affect the purity of rain water for cistern collection. In installing aluminum roofing or siding, aluminum nails with neoprene washers are generally used.

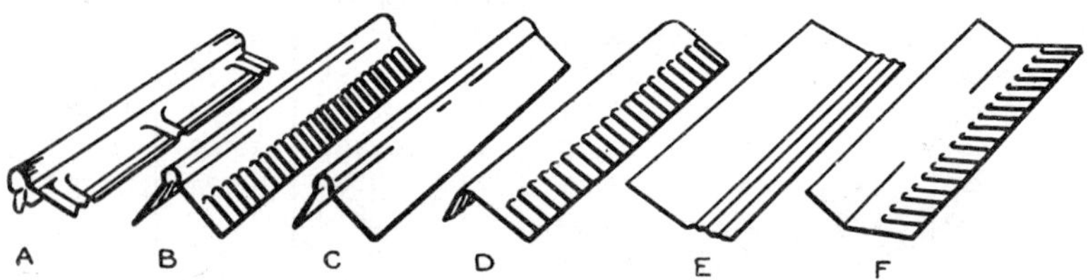

Fig. 36.—Typical Accessories for Finishing Roofing Jobs.

For roofing and siding applications, aluminum sheet is corrugated or crimped to add rigidity and strength. It comes in sheets of various sizes and thicknesses. Typical types available are $1\frac{1}{4}''$ corrugated, 0.019″ thick (equivalent to 26 gage steel); $2\frac{1}{2}''$ corrugated, 0.024″ thick (equivalent to 24 gage steel); and a V-crimp style, 0.019″ thick (equivalent to 26 gage steel). All of these types are available in sheets 26 inches wide by 6, 7, 8, 10, and 12 feet long. A wide variety of accessories such as ridge rolls to cover the top of a ridge on finished roofing jobs, side wall and end wall flashings, and narrow rolls of aluminum for valleys and flashings are available to simplify roofing finishing. Some of these are shown in Fig. 36.

TABLE 10

Gauge No.	American or Brown & Sharpe's	Birmingham or Stubs	Wash.&Moen	Imperial S. W. G.	London or Old English	United States Standard	Gauge No.
0000000			.490	.500		.500	0000000
000000	.5800		.460	.464		.46875	000000
00000	.5165		.430	.432		.4375	00000
0000	.4600	.454	.3938	.400	.454	.40625	0000
000	.4096	.425	.3625	.372	.425	.375	000
00	.3648	.380	.3310	.348	.38	.34375	00
0	.3249	.340	.3065	.324	.34	.3125	0
1	.2893	.300	.2830	.300	.3	.28125	1
2	.2576	.284	.2625	.276	.284	.265625	2
3	.2294	.259	.2437	.252	.259	.25	3
4	.2043	.238	.2253	.232	.238	.234375	4
5	.1819	.220	.2070	.212	.22	.21875	5
6	.1620	.203	.1920	.192	.203	.203125	6
7	.1443	.180	.1770	.176	.18	.1875	7
8	.1285	.165	.1620	.160	.165	.171875	8
9	.1144	.148	.1483	.144	.148	.15625	9
10	.1019	.134	.1350	.128	.134	.140625	10
11	.09074	.120	.1205	.116	.12	.125	11
12	.08081	.109	.1055	.104	.109	.109375	12
13	.07196	.095	.0915	.092	.095	.09375	13
14	.06408	.083	.0800	.080	.083	.078125	14
15	.05707	.072	.0720	.072	.072	.0703125	15
16	.05082	.065	.0625	.064	.065	.0625	16
17	.04526	.058	.0540	.056	.058	.05625	17
18	.04030	.049	.0475	.048	.049	.05	18
19	.03589	.042	.0410	.040	.040	.04375	19
20	.03196	.035	.0348	.036	.035	.0375	20
21	.02846	.032	.03175	.032	.0315	.034375	21
22	.02535	.028	.0286	.028	.0295	.03125	22
23	.02257	.025	.0258	.024	.027	.028125	23
24	.02010	.022	.0230	.022	.025	.025	24
25	.01790	.020	.0204	.020	.023	.021875	25
26	.01594	.018	.0181	.018	.0205	.01875	26
27	.01420	.016	.0173	.0164	.0187	.0171875	27
28	.01264	.014	.0162	.0148	.0165	.015625	28
29	.01126	.013	.0150	.0136	.0155	.0140625	29
30	.01003	.012	.0140	.0124	.01372	.0125	30
31	.008928	.010	.0132	.0116	.0122	.0109375	31
32	.007950	.009	.0128	.0108	.0112	.01015625	32
33	.007080	.008	.0118	.0100	.0102	.009375	33
34	.006305	.007	.0104	.0092	.0095	.00859375	34
35	.005615	.005	.0095	.0084	.009	.0078125	35
36	.005000	.004	.0090	.0076	.0075	.00703125	36
37	.004453		.0085	.0068	.0065	.006640625	37
38	.003965		.008	.0060	.0057	.00625	38
39	.003531		.0075	.0052	.005		39
40	.003145		.007	.0048	.0045		40
41	.002800			.0044			41
42	.002494			.004			42
43	.002221			.0036			43
44	.001978			.0032			44
45	.001761			.0028			45
46	.001568			.0024			46
47	.001397			002			47
48	.001244			.0016			48
49	.001018			.0012			49
50	.0009863			.001			50

TABLE 11

Weights of Steel, Wrought Iron, Brass and Copper Plates

BIRMINGHAM OR STUBS' GAGE

No. of Gage	Thickness in Inches	Weight in Lbs. per Square Foot			
		Steel	Iron	Brass	Copper
0000	.454	18.52	18.16	19.431	20.556
000	.425	17.34	17.00	18.190	19.253
00	.380	15.30	15.20	16.264	17.214
0	.340	13.87	13.60	14.552	15.402
1	.300	12.24	12.00	12.840	13.590
2	.284	11.59	11.36	12.155	12.865
3	.259	10.57	10.36	11.085	11.733
4	.238	9.71	9.52	10.186	10.781
5	.220	8.98	8.80	9.416	9.966
6	.203	8.28	8.12	8.689	9.196
7	.180	7.34	7.20	7.704	8.154
8	.165	6.73	6.60	7.062	7.475
9	.148	6.04	5.92	6.334	6.704
10	.134	5.47	5.36	5.735	6.070
11	.120	4.90	4.80	5.137	5.436
12	.109	4.45	4.36	4.667	4.938
13	.095	3.88	3.80	4.066	4.303
14	.083	3.39	3.32	3.552	3.769
15	.072	2.94	2.88	3.081	3.262
16	.065	2.65	2.60	2.782	2.945
17	.058	2.37	2.32	2.482	2.627
18	.049	2.00	1.96	2.097	2.220
19	.042	1.71	1.68	1.797	1.902
20	.035	1.43	1.40	1.498	1.585
21	.032	1.31	1.28	1.369	1.450
22	.028	1.14	1.12	1.198	1.270
23	.025	1.02	1.00	1.070	1.132
24	.022	.898	.88	.941	.997
25	.020	.816	.80	.856	.906
26	.018	.734	.72	.770	.815
27	.016	.653	.64	.685	.725
28	.014	.571	.56	.599	.634
29	.013	.530	.52	.556	.589
30	.012	.490	.48	.514	.544
31	.010	.408	.40	.428	.453
32	.009	.367	.36	.385	.408
33	.008	.326	.32	.342	.362
34	.007	.286	.28	.2996	.317
35	.005	.204	.20	.214	.227
36	.004	.163	.16	.171	.181

TABLE 12.

WEIGHTS OF STEEL, WROUGHT IRON, BRASS AND COPPER PLATES

AMERICAN OR BROWN & SHARPE GAGE

No. of Gage	Thickness in Inches	Weight in Lbs. per Square Foot			
		Steel	Iron	Brass	Copper
0000	.46	18.77	18.40	19.688	20.838
000	.4096	16.71	16.38	17.533	18.557
00	.3648	14.88	14.59	15.613	16.525
0	.3249	13.26	13.00	13.904	14.716
1	.2893	11.80	11.57	12.382	13.105
2	.2576	10.51	10.30	11.027	11.670
3	.2294	9.39	9.18	9.819	10.392
4	.2043	8.34	8.17	8.745	9.255
5	.1819	7.42	7.28	7.788	8.242
6	.1620	6.61	6.48	6.935	7.340
7	.1443	5.89	5.77	6.175	6.536
8	.1285	5.24	5.14	5.499	5.821
9	.1144	4.67	4.58	4.898	5.183
10	.1019	4.16	4.08	4.361	4.616
11	.0908	3.70	3.63	3.884	4.110
12	.0808	3.30	3.23	3.458	3.660
13	.0720	2.94	2.88	3.080	3.260
14	.0641	2.62	2.56	2.743	2.903
15	.0571	2.33	2.28	2.442	2.585
16	.0508	2.07	2.03	2.175	2.302
17	.0453	1.85	1.81	1.937	2.050
18	.0403	1.64	1.61	1.725	1.825
19	.0359	1.46	1.44	1.536	1.626
20	.0320	1.31	1.28	1.367	1.448
21	.0285	1.16	1.14	1.218	1.289
22	.0253	1.03	1.01	1.085	1.148
23	.0226	.922	.904	.966	1.023
24	.0201	.820	.804	.860	.910
25	.0179	.730	.716	.766	.811
26	.0159	.649	.636	.682	.722
27	.0142	.579	.568	.608	.643
28	.0126	.514	.504	.541	.573
29	.0113	.461	.452	.482	.510
30	.0100	.408	.400	.429	.454
31	.0089	.363	.356	.382	.404
32	.0080	.326	.320	.340	.360
33	.0071	.290	.284	.303	.321
34	.0063	.257	.252	.269	.286
35	.0056	.228	.224	.240	.254
36	.0050	.190	.188	.214	.226
37	.0045	.169	.167	.191	.202
38	.0040	.151	.149	.170	.180
39	.0035	.134	.132	.151	.160
40	.0031	.119	.118	.135	.142

TABLE 13.

Diam. of Circle, D	Side of Square, S	Area of Circle or Square	Diam. of Circle, D	Side of Square, S	Area of Circle or Square	Diam. of Circle, D	Side of Square, S	Area of Circle or Square
½	0.44	0.196	20½	18.17	330.06	40½	35.89	1288.25
1	0.89	0.785	21	18.61	346.36	41	36.34	1320.25
1½	1.33	1.767	21½	19.05	363.05	41½	36.78	1352.65
2	1.77	3.142	22	19.50	380.13	42	37.22	1385.44
2½	2.22	4.909	22½	19.94	397.61	42½	37.66	1418.63
3	2.66	7.069	23	20.38	415.48	43	38.11	1452.20
3½	3.10	9.621	23½	20.83	433.74	43½	38.55	1486.17
4	3.54	12.566	24	21.27	452.39	44	38.99	1520.53
4½	3.99	15.904	24½	21.71	471.44	44½	39.44	1555.28
5	4.43	19.635	25	22.16	490.87	45	39.88	1590.43
5½	4.87	23.758	25½	22.60	510.71	45½	40.32	1625.97
6	5.32	28.274	26	23.04	530.93	46	40.77	1661.90
6½	5.76	33.183	26½	23.49	551.55	46½	41.21	1698.23
7	6.20	38.485	27	23.93	572.56	47	41.65	1734.94
7½	6.65	44.179	27½	24.37	593.96	47½	42.10	1772.05
8	7.09	50.265	28	24.81	615.75	48	42.54	1809.56
8½	7.53	56.745	28½	25.26	637.94	48½	42.98	1847.45
9	7.98	63.617	29	25.70	660.52	49	43.43	1885.74
9½	8.42	70.882	29½	26.14	683.49	49½	43.87	1924.42
10	8.86	78.540	30	26.59	706.86	50	44.31	1963.50
10½	9.31	86.590	30½	27.03	730.62	50½	44.75	2002.96
11	9.75	95.033	31	27.47	754.77	51	45.20	2042.82
11½	10.19	103.87	31½	27.92	779.31	51½	45.64	2083.07
12	10.64	113.10	32	28.36	804.25	52	46.08	2123.72
12½	11.08	122.72	32½	28.80	829.58	52½	46.53	2164.75
13	11.52	132.73	33	29.25	855.30	53	46.97	2206.18
13½	11.96	143.14	33½	29.69	881.41	53½	47.41	2248.01
14	12.41	153.94	34	30.13	907.92	54	47.86	2290.22
14½	12.85	165.13	34½	30.57	934.82	54½	48.30	2332.83
15	13.29	176.71	35	31.02	962.11	55	48.74	2375.83
15½	13.74	188.69	35½	31.46	989.80	55½	49.19	2419.22
16	14.18	201.06	36	31.90	1017.88	56	49.63	2463.01
16½	14.62	213.82	36½	32.35	1046.35	56½	50.07	2507.19
17	15.07	226.98	37	32.79	1075.21	57	50.51	2551.76
17½	15.51	240.53	37½	33.23	1104.47	57½	50.96	2596.72
18	15.95	254.47	38	33.68	1134.11	58	51.40	2642.08
18½	16.40	268.80	38½	34.12	1164.16	58½	51.84	2687.83
19	16.84	283.53	39	34.56	1194.59	59	52.29	2733.97
19½	17.28	298.65	39½	35.01	1225.42	59½	52.73	2780.51
20	17.72	314.16	40	35.45	1256.64	60	53.17	2827.43

TABLE 14.

Gauge Numbers and Millimeter Equivalents

Gauge No.	American or Brown & Sharpe's		Birmingham or Stubs	
	Inches	Millimeters	Inches	Millimeters
000000	.5800	14.732		
00000	.5165	13.119		
0000	.4600	11.684	.454	11.532
000	.4096	10.404	.425	10.795
00	.3648	9.266	.380	9.652
0	.3249	8.252	.340	8.636
1	.2893	7.348	.300	7.620
2	.2576	6.543	.284	7.214
3	.2294	5.827	.259	6.579
4	.2043	5.189	.238	6.045
5	.1819	4.620	.220	5.588
6	.1620	4.115	.203	5.156
7	.1443	3.665	.180	4.572
8	.1285	3.264	.165	4.191
9	.1144	2.906	.148	3.759
10	.1019	2.588	.134	3.404
11	.09074	2.305	.120	3.048
12	.08081	2.053	.109	2.769
13	.07196	1.828	.095	2.413
14	.06408	1.628	.083	2.108
15	.05707	1.450	.072	1.829
16	.05082	1.291	.065	1.651
17	.04526	1.150	.058	1.473
18	.04030	1.024	.049	1.245
19	.03589	.912	.042	1.067
20	.03196	.812	.035	.889
21	.02846	.723	.032	.813
22	.02535	.644	.028	.711
23	.02257	.573	.025	.635
24	.02010	.511	.022	.559
25	.01790	.455	.020	.508
26	.01594	.405	.018	.457
27	.01420	.361	.016	.406
28	.01264	.321	.014	.356
29	.01126	.286	.013	.330
30	.01003	.255	.012	.305
31	.008928	.227	.010	.254
32	.007950	.202	.009	.229
33	.007080	.180	.008	.203
34	.006305	.160	.007	.178
35	.005615	.143	.005	.127
36	.005000	.127	.004	.102
37	.004453	.113		
38	.003965	.101		
39	.003531	.090		
40	.003145	.080		
41	.002800	.071		
42	.002494	.063		
43	.002221	.056		
44	.001978	.050		

TABLE 15.

1 gallon = 231 cu in. 1 cu ft = 7.4805 gal

Diameter in inches *	For 1 ft in length		Diameter in inches	For 1 ft in length		Diameter in inches	For 1 ft in length	
	Cu ft, also area in sq ft	U. S. gal 231 cu in		Cu ft, also area in sq ft	U. S. gal, 231 cu in		Cu ft, also area in sq ft	U. S. gal, 231 cu in
¼	0.0003	0.0025	6¾	0.2485	1.859	19	1.969	14.73
5/16	0.0005	0.0040	7	0.2673	1.999	19½	2.074	15.51
⅜	0.0008	0.0057	7¼	0.2867	2.145	20	2.182	16.32
7/16	0.0010	0.0078	7½	0.3068	2.295	20½	2.292	17.15
½	0.0014	0.0102	7¾	0.3276	2.450	21	2.405	17.99
9/16	0.0017	0.0129	8	0.3491	2.611	21½	2.521	18.86
⅝	0.0021	0.0159	8¼	0.3712	2.777	22	2.640	19.75
11/16	0.0026	0.0193	8½	0.3941	2.948	22½	2.761	20.66
¾	0.0031	0.0230	8¾	0.4176	3.125	23	2.885	21.58
13/16	0.0036	0.0269	9	0.4418	3.305	23½	3.012	22.53
⅞	0.0042	0.0312	9¼	0.4667	3.491	24	3.142	23.50
15/16	0.0048	0.0359	9½	0.4022	3.682	25	3.409	25.50
1	0.0055	0.0408	9¾	0.5185	3.879	26	3.687	27.58
1¼	0.0085	0.0638	10	0.5454	4.080	27	3.976	29.74
1½	0.0123	0.0918	10¼	0.5730	4.286	28	4.276	31.99
1¾	0.0167	0.1249	10½	0.6013	4.498	29	4.587	34.31
2	0.0218	0.1632	10¾	0.6303	4.715	30	4.909	36.72
2¼	0.0276	0.2066	11	0.6600	4.937	31	5.241	39.21
2½	0.0341	0.2550	11¼	0.6903	5.164	32	5.585	41.78
2¾	0.0412	0.3085	11½	0.7213	5.396	33	5.940	44.43
3	0.0491	0.3672	11¾	0.7530	5.633	34	6.305	47.16
3¼	0.0576	0.4309	12	0.7854	5.875	35	6.681	49.98
3½	0.0668	0.4998	12½	0.8522	6.375	36	7.069	52.88
3¾	0.0767	0.5738	13	0.9218	6.895	37	7.467	55.86
4	0.0873	0.6528	13½	0.9940	7.436	38	7.876	58.92
4¼	0.0985	0.7369	14	1.0690	7.997	39	8.296	62.06
4½	0.1134	0.8263	14½	1.1470	8.578	40	8.727	65.28
4¾	0.1231	0.9206	15	1.2270	9.180	41	9.168	68.58
5	0.1364	1.0200	15½	1.3100	9.801	42	9.621	71.97
5¼	0.1503	1.1250	16	1.3960	10.440	43	10.085	75.44
5½	0.1650	1.2340	16½	1.4850	11.110	44	10.559	78.99
5¾	0.1803	1.3490	17	1.5760	11.790	45	11.045	82.62
6	0.1963	1.4690	17½	1.6700	12.490	46	11.541	86.33
6¼	0.2131	1.5940	18	1.7680	13.220	47	12.048	90.13
6½	0.2304	1.7240	18½	1.8670	13.960	48	12.566	94.00

* Actual.

TABLE 16.

Number of U. S. Gallons in Rectangular Tanks

For One Foot in Depth

1 cu ft = 7.4805 gal

Width, ft	Length of tank, ft										
	2	2.5	3	3.5	4	4.5	5	5.5	6	6.5	7
2	29.92	37.40	44.88	52.36	59.84	67.32	74.81	82.29	89.77	97.25	104.73
2.5	....	46.75	56.10	65.45	74.80	84.16	93.51	102.80	112.21	121.56	130.91
3	....	....	67.32	78.54	89.77	100.99	112.21	123.43	134.65	145.87	157.09
3.5	....	....	....	91.64	104.73	117.82	130.91	144.00	157.09	170.18	183.27
4	....	....	....	....	119.69	134.65	149.61	164.57	179.53	194.49	209.45
4.5	....	....	....	....	....	151.48	168.31	185.14	201.97	218.80	235.63
5	....	....	....	....	....	....	187.01	205.71	224.41	243.11	261.82
5.5	....	....	....	....	....	....	....	226.28	246.86	267.43	288.00
6	....	....	....	....	....	....	....	....	269.30	291.74	314.18
6.5	....	....	....	....	....	....	....	....	....	316.05	340.36
7	....	....	....	....	....	....	....	....	....	....	366.54

Width, ft	Length of tank, ft									
	7.5	8	8.5	9	9.5	10	10.5	11	11.5	12
2	112.21	119.69	127.17	134.65	142.13	149.61	157.09	164.57	172.05	179.53
2.5	140.26	149.61	158.96	168.31	177.66	187.01	196.36	205.71	215.06	224.41
3	168.31	179.53	190.75	202.97	213.19	224.41	235.63	246.86	258.07	269.30
3.5	196.36	209.45	222.54	235.63	248.73	261.82	274.90	288.00	301.09	314.18
4	224.41	239.37	254.34	269.30	284.26	299.22	314.18	329.14	344.10	359.06
4.5	252.47	269.30	286.13	302.96	319.79	336.62	353.45	370.28	387.11	403.94
5	280.52	299.22	317.92	336.62	355.32	374.03	392.72	411.43	430.13	448.83
5.5	308.57	329.14	349.71	370.28	390.85	411.43	432.00	452.57	473.14	493.71
6	336.62	359.06	381.50	403.94	426.39	448.83	471.27	493.71	516.15	538.59
6.5	364.67	388.98	413.30	437.60	461.92	486.23	510.54	534.85	550.16	583.47
7	392.72	418.91	445.09	471.27	497.45	523.64	549.81	575.00	602.18	628.36
7.5	420.78	448.83	476.88	504.93	532.98	561.04	589.08	617.14	645.19	673.24
8	....	478.75	508.67	538.59	568.51	598.44	628.36	658.28	688.20	718.12
8.5	....	....	540.46	572.25	604.05	635.84	667.63	690.42	731.21	703.00
9	....	....	....	605.92	639.58	673.25	706.90	740.56	774.23	807.89
9.5	....	....	....	.....	675.11	710.65	746.17	781.71	817.24	852.77
10	....	....	....	.....	.....	748.05	785.45	822.86	860.26	897.66
10.5	....	....	....	.....	.....	.....	824.73	864.00	903.26	942.56
11	....	....	....	.....	.....	.....	.....	905.14	946.27	987.43
11.5	....	....	....	.....	.....	.....	.....	....	989.29	1032.3
12	....	....	....	.....	.....	.....	.....	....	.....	1077.2

To find weight of water in pounds at 62° F., multiply the number of gallons by 8⅓.

References.—GENERAL METAL WORK by Alfred B. Grayshon, published by the D. Van Nostrand Company, contains additional material on sheet metal work.

TABLE 17

Circumference and Areas of Circles

Diameter.	Circumference.	Area.	Diameter.	Circumference.	Area.
$\frac{1}{64}$	0.0491	0.00019	$\frac{43}{64}$	2.1108	0.35454
$\frac{1}{32}$	0.0982	0.00077	$\frac{11}{16}$	2.1598	0.37122
$\frac{3}{64}$	0.1473	0.00173	$\frac{45}{64}$	2.2089	0.38829
$\frac{1}{16}$	0.1964	0.00307	$\frac{23}{32}$	2.2580	0.40574
$\frac{5}{64}$	0.2454	0.00479	$\frac{47}{64}$	2.3071	0.42357
$\frac{3}{32}$	0.2945	0.00690	$\frac{3}{4}$	2.3562	0.44179
$\frac{7}{64}$	0.3436	0.00940	$\frac{49}{64}$	2.4053	0.46039
$\frac{1}{8}$	0.3927	0.01227	$\frac{25}{32}$	2.4544	0.47937
$\frac{9}{64}$	0.4418	0.01553	$\frac{51}{64}$	2.5035	0.49874
$\frac{5}{32}$	0.4909	0.01918	$\frac{13}{16}$	2.5525	0.51849
$\frac{11}{64}$	0.5400	0.02320	$\frac{53}{64}$	2.6016	0.53862
$\frac{3}{16}$	0.5890	0.02761	$\frac{27}{32}$	2.6507	0.55914
$\frac{13}{64}$	0.6381	0.03241	$\frac{55}{64}$	2.6998	0.58004
$\frac{7}{32}$	0.6872	0.03758	$\frac{7}{8}$	2.7489	0.60132
$\frac{15}{64}$	0.7363	0.04314	$\frac{57}{64}$	2.7980	0.62299
$\frac{1}{4}$	0.7854	0.04909	$\frac{29}{32}$	2.8471	0.64504
$\frac{17}{64}$	0.8345	0.05542	$\frac{59}{64}$	2.8962	0.66747
$\frac{9}{32}$	0.8836	0.06213	$\frac{15}{16}$	2.9452	0.69029
$\frac{19}{64}$	0.9327	0.06922	$\frac{61}{64}$	2.9943	0.71349
$\frac{5}{16}$	0.9818	0.07670	$\frac{31}{32}$	3.0434	0.73708
$\frac{21}{64}$	1.0308	0.08456	$\frac{63}{64}$	3.0925	0.76105
$\frac{11}{32}$	1.0799	0.09281	1	3.1416	0.78540
$\frac{23}{64}$	1.1290	0.10144	$1\frac{1}{64}$	3.1907	0.81013
$\frac{3}{8}$	1.1781	0.11045	$1\frac{1}{32}$	3.2398	0.83525
$\frac{25}{64}$	1.2272	0.11984	$1\frac{3}{64}$	3.2889	0.86075
$\frac{13}{32}$	1.2763	0.12962	$1\frac{1}{16}$	3.3379	0.88664
$\frac{27}{64}$	1.3254	0.13979	$1\frac{5}{64}$	3.3870	0.91291
$\frac{7}{16}$	1.3744	0.15033	$1\frac{3}{32}$	3.4361	0.93956
$\frac{29}{64}$	1.4235	0.16126	$1\frac{7}{64}$	3.4852	0.96660
$\frac{15}{32}$	1.4726	0.17258	$1\frac{1}{8}$	3.5343	0.99402
$\frac{31}{64}$	1.5217	0.18427	$1\frac{9}{64}$	3.5834	1.02182
$\frac{1}{2}$	1.5708	0.19635	$1\frac{5}{32}$	3.6325	1.05001
$\frac{33}{64}$	1.6199	0.20881	$1\frac{11}{64}$	3.6816	1.07858
$\frac{17}{32}$	1.6690	0.22166	$1\frac{3}{16}$	3.7306	1.10753
$\frac{35}{64}$	1.7181	0.23489	$1\frac{13}{64}$	3.7797	1.13687
$\frac{9}{16}$	1.7671	0.24850	$1\frac{7}{32}$	3.8288	1.16659
$\frac{37}{64}$	1.8162	0.26250	$1\frac{15}{64}$	3.8779	1.19670
$\frac{19}{32}$	1.8653	0.27688	$1\frac{1}{4}$	3.9270	1.22718
$\frac{39}{64}$	1.9144	0.29165	$1\frac{17}{64}$	3.9761	1.25806
$\frac{5}{8}$	1.9635	0.30680	$1\frac{9}{32}$	4.0252	1.28931
$\frac{41}{64}$	2.0126	0.32233	$1\frac{19}{64}$	4.0743	1.32095
$\frac{21}{32}$	2.0617	0.33824	$1\frac{5}{16}$	4.1233	1.35297

TABLE 17—*Continued*

Diameter.	Circumference.	Area.	Diameter.	Circumference.	Area.
$1\frac{21}{64}$	4.1724	1.38538	$2\frac{1}{8}$	6.6759	3.5466
$1\frac{11}{32}$	4.2215	1.41817	$2\frac{3}{16}$	6.8722	3.7584
$1\frac{23}{64}$	4.2706	1.45134	$2\frac{1}{4}$	7.0686	3.9761
$1\frac{3}{8}$	4.3197	1.48489	$2\frac{5}{16}$	7.2649	4.2
$1\frac{25}{64}$	4.3688	1.51883	$2\frac{3}{8}$	7.4613	4.4301
$1\frac{13}{32}$	4.4179	1.55316	$2\frac{7}{16}$	7.6576	4.6664
$1\frac{27}{64}$	4.4670	1.58786	$2\frac{1}{2}$	7.8540	4.9087
$1\frac{7}{16}$	4.5160	1.62295	$2\frac{9}{16}$	8.0503	5.1573
$1\frac{29}{64}$	4.5651	1.65843	$2\frac{5}{8}$	8.2467	5.4119
$1\frac{15}{32}$	4.6142	1.69428	$2\frac{11}{16}$	8.4430	5.6727
$1\frac{31}{64}$	4.6633	1.73052	$2\frac{3}{4}$	8.6394	5.9396
$1\frac{1}{2}$	4.7124	1.76715	$2\frac{13}{16}$	8.8357	6.2126
$1\frac{33}{64}$	4.7615	1.80415	$2\frac{7}{8}$	9.0321	6.4918
$1\frac{17}{32}$	4.8106	1.84154	$2\frac{15}{16}$	9.2284	6.7772
$1\frac{35}{64}$	4.8597	1.87932	3	9.4248	7.0686
$1\frac{9}{16}$	4.9087	1.91748	$3\frac{1}{16}$	9.6211	7.3662
$1\frac{37}{64}$	4.9578	1.95602	$3\frac{1}{8}$	9.8175	7.6699
$1\frac{19}{32}$	5.0069	1.99494	$3\frac{3}{16}$	10.0138	7.9798
$1\frac{39}{64}$	5.0560	2.03425	$3\frac{1}{4}$	10.2102	8.2958
$1\frac{5}{8}$	5.1051	2.07394	$3\frac{5}{16}$	10.4066	8.6179
$1\frac{41}{64}$	5.1542	2.11402	$3\frac{3}{8}$	10.6029	8.9462
$1\frac{21}{32}$	5.2033	2.15448	$3\frac{7}{16}$	10.7992	9.2807
$1\frac{43}{64}$	5.2524	2.19532	$3\frac{1}{2}$	10.9956	9.6211
$1\frac{11}{16}$	5.3014	2.23654	$3\frac{9}{16}$	11.1919	9.9678
$1\frac{45}{64}$	5.3505	2.27815	$3\frac{5}{8}$	11.3883	10.3206
$1\frac{23}{32}$	5.3996	2.32015	$3\frac{11}{16}$	11.5846	10.6796
$1\frac{47}{64}$	5.4487	2.36252	$3\frac{3}{4}$	11.7810	11.0447
$1\frac{3}{4}$	5.4978	2.40528	$3\frac{13}{16}$	11.9773	11.4160
$1\frac{49}{64}$	5.5469	2.44843	$3\frac{7}{8}$	12.1737	11.7933
$1\frac{25}{32}$	5.5960	2.49195	$3\frac{15}{16}$	12.3701	12.1768
$1\frac{51}{64}$	5.6450	2.53586	4	12.5664	12.5664
$1\frac{13}{16}$	5.6941	2.58016	$4\frac{1}{16}$	12.7628	12.9622
$1\frac{53}{64}$	5.7432	2.62483	$4\frac{1}{8}$	12.9591	13.3641
$1\frac{27}{32}$	5.7923	2.66989	$4\frac{3}{16}$	13.1554	13.7721
$1\frac{55}{64}$	5.8414	2.71534	$4\frac{1}{4}$	13.3518	14.1863
$1\frac{7}{8}$	5.8905	2.76117	$4\frac{5}{16}$	13.5481	14.6066
$1\frac{57}{64}$	5.9396	2.80738	$4\frac{3}{8}$	13.7445	15.0330
$1\frac{29}{32}$	5.9887	2.85397	$4\frac{7}{16}$	13.9408	15.4656
$1\frac{59}{64}$	6.0377	2.90095	$4\frac{1}{2}$	14.1372	15.9043
$1\frac{15}{16}$	6.0868	2.94831	$4\frac{9}{16}$	14.3335	16.3492
$1\frac{61}{64}$	6.1359	2.99606	$4\frac{5}{8}$	14.5299	16.8002
$1\frac{31}{32}$	6.1850	3.04418	$4\frac{11}{16}$	14.7262	17.2573
$1\frac{63}{64}$	6.2341	3.0927	$4\frac{3}{4}$	14.9226	17.7206
2	6.2832	3.1416	$4\frac{13}{16}$	15.1189	18.19
$2\frac{1}{16}$	6.4795	3.3410	$4\frac{7}{8}$	15.3153	18.6655

TABLE 17—*Continued*

Diameter.	Circumference.	Area.	Diameter.	Circumference.	Area.
4 15/16	15.5116	19.1472	10½	32.9868	86.5903
5	15.7080	19.6350	10⅝	33.3795	88.6643
5⅛	16.1007	20.6290	10¾	33.7722	90.7625
5¼	16.4934	21.6476	10⅞	34.1649	92.8858
5⅜	16.8861	22.6907	11	34.5576	95.0334
5½	17.2788	23.7583	11⅛	34.9503	97.2055
5⅝	17.6715	24.8505	11¼	35.343	99.4019
5¾	18.0642	25.9673	11⅜	35.7357	101.6234
5⅞	18.4569	27.1084	11½	36.1284	103.8691
6	18.8496	28.2744	11⅝	36.5211	106.1394
6⅛	19.2423	29.4648	11¾	36.9138	108.4338
6¼	19.635	30.6797	11⅞	37.3065	110.7537
6⅜	20.0277	31.9191	12	37.6992	113.098
6½	20.4204	33.1831	12¼	38.4846	117.859
6⅝	20.8131	34.4717	12½	39.2700	122.719
6¾	21.2058	35.7848	12¾	40.0554	127.677
6⅞	21.5985	37.1224	13	40.8408	132.733
7	21.9912	38.4846	13¼	41.6262	137.887
7⅛	22.3839	39.8713	13½	42.4116	143.139
7¼	22.7766	41.2826	13¾	43.1970	148.490
7⅜	23.1693	42.7184	14	43.9824	153.938
7½	23.5620	44.1787	14¼	44.7678	159.485
7⅝	23.9547	45.6636	14½	45.5532	165.130
7¾	24.3474	47.1731	14¾	46.3386	170.874
7⅞	24.7401	48.7071	15	47.1240	176.715
8	25.1328	50.2656	15¼	47 9094	182.655
8⅛	25.5255	51.8487	15½	48.6948	188.692
8¼	25.9182	53.4561	15¾	49.4802	194.828
8⅜	26.3109	55.0884	16	50.2656	201.062
8½	26.7036	56.7451	16¼	51.051	207.395
8⅝	27.0963	58.4264	16½	51.8364	213.825
8¾	27.489	60.1319	16¾	52.6218	220.354
8⅞	27.8817	61.8625	17	53.4072	226.981
9	28.2744	63.6174	17¼	54.1926	233.706
9⅛	28.6671	65.3968	17½	54.9780	240.529
9¼	29.0598	67.2008	17¾	55.7634	247.450
9⅜	29.4525	69.0293	18	56.5488	254.470
9½	29.8452	70.8823	18¼	57.3342	261.587
9⅝	30.2379	72.7599	18½	58.1196	268.803
9¾	30.6306	74.6619	18¾	58.905	276.117
9⅞	31.0233	76.5888	19	59.6904	283.529
10	31.4160	78.5400	19¼	60.4758	291.040
10⅛	31.8087	80.5158	19½	61.2612	298.648
10¼	32.2014	82.5158	19¾	62.0466	306.355
10⅜	32.5941	84.5409	20	62.8320	314.16

TABLE 17—*Continued*

Diameter.	Circumference.	Area.	Diameter.	Circumference.	Area.
21	65.9736	346.361	66	207.34	3421.19
22	69.1152	380.134	67	210.49	3525.65
23	72.2568	415.477	68	213.63	3631.68
24	75.3984	452.39	69	216.77	3739.28
25	78.540	490.87	70	219.91	3848.45
26	81,681	530.93	71	223.05	3959.19
27	84.823	572.56	72	226.19	4071.50
28	87.965	615.75	73	229.34	4185.39
29	91.106	660.52	74	232.48	4300.84
30	94.248	706.86	75	235.62	4417.86
31	97.389	754.77	76	238.76	4536.46
32	100.53	804.25	77	241.90	4656.63
33	103.67	855.30	78	245.04	4778.36
34	106.81	907.92	79	248.19	4901.67
35	109.96	962.11	80	251.33	5026.55
36	113.10	1017.88	81	254.47	5153.00
37	116.24	1075.21	82	257.61	5281.02
38	119.38	1134.11	83	260.75	5410.61
39	122.52	1194.59	84	263.89	5541.77
40	125.66	1256.64	85	267.04	5674.50
41	128.81	1320.25	86	270.18	5808.80
42	131.95	1385.44	87	273.32	5944.68
43	135.09	1452.20	88	276.46	6082.12
44	138.23	1520.53	89	279.60	6221.14
45	141.37	1590.43	90	282.74	6361.73
46	144.51	1661.90	91	285.88	6503.88
47	147.65	1734.94	92	289.03	6647.61
48	150.80	1809.56	93	292.17	6792.91
49	153.94	1885.74	94	295.31	6939.78
50	157.08	1963.50	95	298.45	7088.22
51	160.22	2042.82	96	301.59	7238.23
52	163.36	2123.72	97	304.73	7389.81
53	166.50	2206.18	98	307.88	7542.96
54	169.65	2290.22	99	311.02	7697.69
55	172.79	2375.83	100	314.16	7853.98
56	175.93	2463.01	101	317.30	8011.85
57	179.07	2551.76	102	320.44	8171.28
58	182.21	2642.08	103	323.58	8332.29
59	185.35	2733.97	104	326.73	8494.87
60	188.50	2827.43	105	329.87	8659.01
61	191.64	2922.47	106	333.01	8824.73
62	194.78	3019.07	107	336.15	8992.02
63	197.92	3117.25	108	339.29	9160.88
64	201.06	3216.99	109	342.43	9331.32
65	204.20	3318.31	110	345.58	9503.32

XX

Electricity

Electricity has more useful and universal application than any other natural phenomenon, and the end of the range of its applications is not yet in sight. Its increasing importance need not be emphasized here, but it is significant to note that many even recent developments have been the result of new study of the fundamental principles of the subject. It is also significant that the applications of electricity which enter into the daily life of the average person range from the simple heating elements and the dry cell which operates the door bell, to the more intricate motors and vacuum tubes.

Hence, it is important for the practical man to understand the fundamental principles of the subject in order to appreciate the rules which have been laid down for the applications. This section devotes a substantial amount of space to these fundamentals and with each step shows how they are applied and how the calculations pertaining to them are made.

The Nature of Electricity.—Electricity is, as we have suggested, a phenomenon of nature. It exists all about us like the air we breathe, but why it exists or what it actually is, we are unable to say. We do know, however, something of what it can do and how it acts under certain conditions, and that is the more important concern in adapting it to the uses of mankind. Since electricity already exists, it is obvious that we cannot create it. We can, however, create a *flow* of electricity as we create a flow of water through a pipe by means of a pump. This flow or current of electricity is created by mechanical, thermal, or chemical means. The energy of these agents is transformed into electrical energy capable of doing work. The work of the man dealing with electricity may be epitomized as the proper control of electrical energy while performing useful service.

Units.—To cause a current of electricity to flow, there must be a pressure. This is known as electromotive force and is measured in *volts*; it is therefore often referred to as *voltage*. The current flow is measured in *amperes* and the resistance to such a flow is measured in *ohms*. These are the fundamental units, and they are defined as follows:

A *volt* is a unit of electrical pressure or potential difference (pd) or the electromotive force (emf) required to cause a current of one ampere to flow through a resistance of one ohm.

An *ampere* is a unit of current strength, or the quantity of flow, or the quantity of current which will flow through a resistance of one ohm under an electromotive force of one volt.

An *ohm* is a unit of resistance, or the resistance of a conductor through which a current of one ampere will pass under an electromotive force of one volt.

Thus the three units depend on one another. One of the three must therefore be stated independently, and this one is the *ohm*. The ohm is usually defined as the resistance of a certain conductor of a particular material, size and form.

Ohm's Law.—This is a statement of the relation between volts, amperes and ohms. It may be expressed as

$$I = \frac{E}{R}, \quad \text{or} \quad E = IR, \quad \text{or} \quad R = \frac{E}{I}$$

where I = current in amperes

E = electromotive force in volts

R = resistance in ohms

These are the standard algebraic letter symbols and are not to be confused with the abbreviations for these quantities.

ILLUSTRATION: How many amperes of current are flowing in a circuit with a resistance of 25 ohms when the pressure is 110 volts?

$$I = \frac{E}{R}, \quad I = \frac{110}{25} = 4.4 \text{ amperes} \quad \text{(Ans.)}$$

ILLUSTRATION: A circuit has a resistance of 200 ohms. What is the applied voltage when 0.03 ampere of current is flowing?

$$E = IR,\ E = 200 \times 0.03 = 6 \text{ volts} \quad \text{(Ans.)}$$

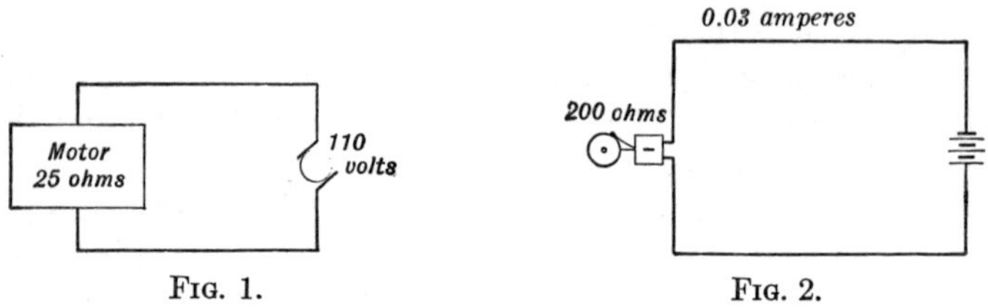

FIG. 1. FIG. 2.

Electric Circuits

Series Connections.—Two or more pieces of electrical apparatus in a circuit one after the other are said to be in series.

The resistance of a series combination in a circuit is the sum of the resistances of the separate parts.

Current through a series combination is the same as the current through each part.

Voltage across a series combination is the sum of the voltages across separate parts.

ILLUSTRATION: It is desired to use 110-volt lights in a street car which operates on a current of 550 volts. How many lights must be placed in series?

By the last rule above, the voltage across a series combination is the sum of the voltages across separate parts.

Number of lights (separate parts) $= \frac{550}{110} = 5$ lights (Ans.)

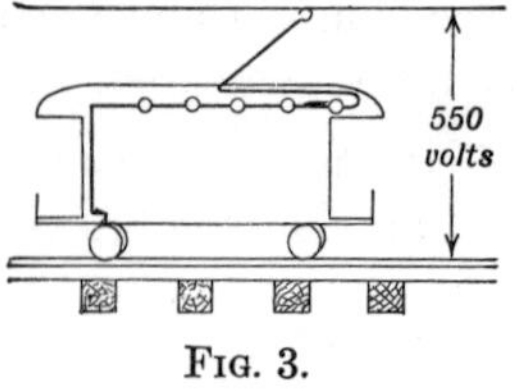

FIG. 3.

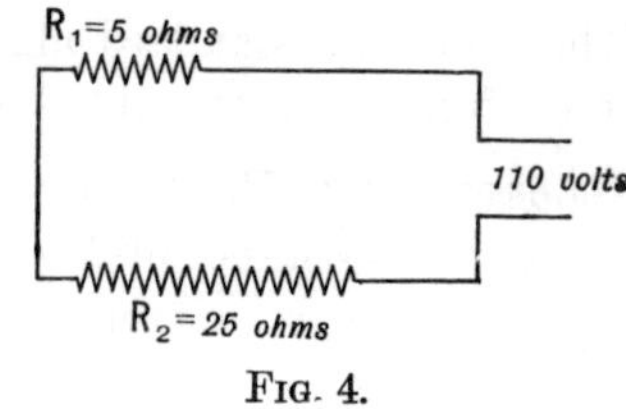

FIG. 4.

ILLUSTRATION: A 110-volt circuit has a resistance R_1 of 5 ohms and a resistance R_2 of 25 ohms. What current flows through it?

$$\text{Total resistance} = R_1 + R_2$$

$$I = \frac{110}{R_1 + R_2}, \quad I = \frac{110}{5 + 25}, \quad I = \frac{110}{30} = 3.67 \text{ amperes} \quad \text{(Ans.)}$$

ILLUSTRATION: A current of 4 amperes flows through the circuit in Fig. 5. What is the voltage at the terminals?

Volts across 8-ohm resistance = 4 × 8 = 32 volts
Volts across 30-ohm resistance = 4 × 30 = 120 volts
Volts across 15-ohm resistance = 4 × 15 = 60 volts

Volts across circuit at terminals = 212 volts (Ans.)

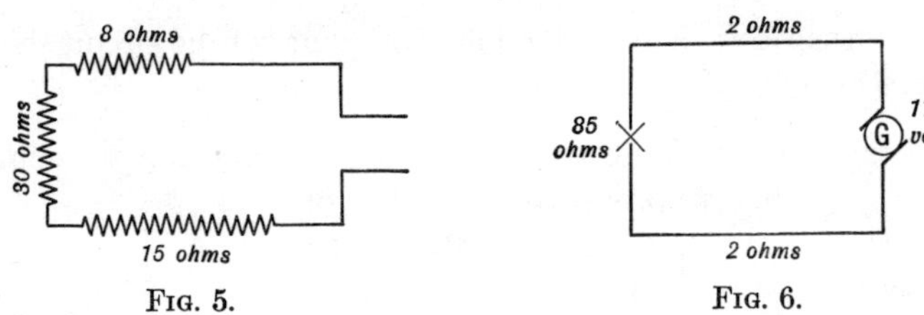

FIG. 5. FIG. 6.

ILLUSTRATION: In Fig. 6 a generator is producing an electric current with a pressure of 188 volts. An arc light on the circuit has a resistance of 85 ohms and the resistance of each wire leading to it is 2 ohms. What is the voltage at the lamp terminals?

Total current

$$= I = \frac{E}{R_1 + R_2 + R_3} = \frac{118}{2 + 85 + 2} = \frac{118}{89} = 1.3 \text{ amperes}$$

Volts lost through R_1 = 1.3 × 2 = 2.6 volts
Volts lost through R_3 = 1.3 × 2 = 2.6 volts

Total potential drop through wires = 5.2 volts

Voltage at lamp terminals then = 118 − 5.2 = 112.8 volts (Ans.)

Parallel Connections.—Two or more pieces of electrical apparatus in a circuit so connected that the current is divided between them are said to be in parallel.

The resistance of a parallel circuit is the reciprocal of the sum of the reciprocals for the various resistances. This joint resistance is less than any branch resistance.

Because mathematical difficulties arise when finding joint resistance by using reciprocals, another method is to find the product divided by the sum of two resistances at a time. This is explained later.

The current through a parallel circuit is the sum of the currents through the separate branches.

Voltage across a parallel circuit is the same as the voltage across each branch.

ILLUSTRATION: What is the joint resistance of the circuit shown in Fig. 7?

By the rule, the joint resistance is the reciprocal of the sum of the reciprocals of the parts, or

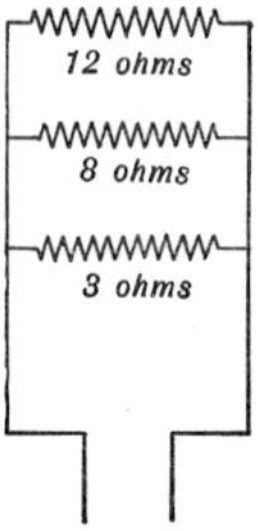

$$\frac{1}{R} = \frac{1}{R_1} + \frac{1}{R_2} + \frac{1}{R_3}$$

Then, $\frac{1}{R} = \frac{1}{12} + \frac{1}{8} + \frac{1}{3}$, and, reducing to common denominator,

$$\frac{1}{R} = \frac{2}{24} + \frac{3}{24} + \frac{8}{24} = \frac{13}{24}$$

12 ohms

8 ohms

3 ohms

FIG. 7.

R is then the reciprocal of $\frac{13}{24}$ or $\frac{24}{13}$ which reduces to 1.85 ohms (Ans.)

The problem can be set up in one equation as follows:

$$\text{Joint resistance} = \frac{1}{\frac{2+3+8}{24}} = \frac{1}{\frac{13}{24}} = \frac{24}{13} = 1.85 \text{ ohms} \quad \text{(Ans.)}$$

Joint resistance of the same circuit found by the product over he sum method:

$$\text{Joint resistance of 12 ohms and 8 ohms} = \frac{12 \times 8}{12 + 8} = \frac{96}{20} = 4.8 \text{ ohms}$$

Joint resistance of 4.8 ohms and 3 ohms

$$= \frac{4.8 \times 3}{4.8 + 3} = \frac{14.4}{7.8} = 1.85 \text{ ohms (Ans.)}$$

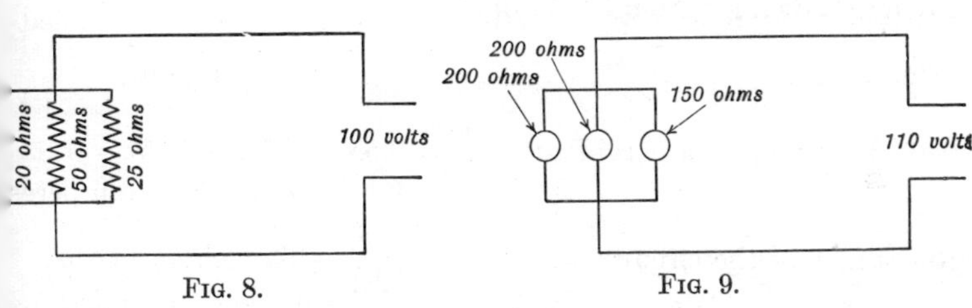

FIG. 8. FIG. 9.

ILLUSTRATION: A current is flowing in the circuit shown in Fig. 8 under a pressure of 100 volts. How great is this current?

Current flowing through 20-ohm resistance.................... $= \frac{100}{20} = 5$ amperes

Current flowing through 50-ohm resistance.................... $= \frac{100}{50} = 2$ amperes

Current flowing through 25-ohm resistance.................... $= \frac{100}{25} = 4$ amperes

Current through the combination.. = 11 amperes (Ans.)

ILLUSTRATION: What is the total or joint resistance of the above circuit?

According to the rule, $\frac{1}{R} = \frac{1}{R_1} + \frac{1}{R_2} + \frac{1}{R_3}$. Then,

$$\frac{1}{R} = \frac{1}{20} + \frac{1}{50} + \frac{1}{25} = \frac{5 + 2 + 4}{100} = \frac{11}{100}$$

and, joint resistance $R = \frac{100}{11} = 9.1$ ohms (Ans.)

ILLUSTRATION: The circuit in Fig. 9 has two lamps with resistances of 200 ohms each and one lamp with a resistance of 150 ohms. What is the joint resistance of the combination and total current if the pressure is 110 volts?

Current flowing through each 200-ohm lamp. $= \frac{110}{200} = 0.55$ ampere

Current flowing through 150-ohm lamp. . . . . $= \frac{110}{150} = 0.73$ ampere

Current through the combination. . $= 0.55 + 0.55 + 0.73 = 1.83$ amperes

Resistance of the circuit. . . . . $= \frac{110}{1.83} = 60.1$ ohms (Ans.)

Series-Parallel Connections.—In many actual installations electrical apparatus instead of being in a simple parallel or series connection, is in a combination of these.

In a series-parallel circuit each part must be considered separately when computing the current, voltage and resistance of the entire circuit.

ILLUSTRATION: In the circuit shown in Fig. 10 a string of 110-volt lamps is connected to a 220-volt circuit. Two lamps are placed in series and each set of two is in parallel in the circuit. What is the total current if the resistance of each lamp is 200 ohms?

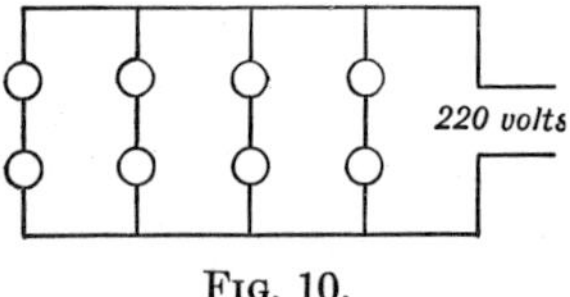

FIG. 10.

The resistance of each set of two lamps in series $= 200 + 200 = 400$ ohms.

The current flowing through each set of two-series lamps is,

$$\tfrac{220}{400} = 0.55 \text{ ampere}$$

Since there are four identical sets in parallel the total current of the circuit is,

$$4 \times 0.55 = 2.2 \text{ amperes (Ans.)}$$

ILLUSTRATION: In the circuit shown in Fig. 11, what is the total resistance and the total current?

Resistance of the parallel combination =

$$\frac{1}{\frac{1}{16}+\frac{1}{24}} = \frac{1}{\frac{3}{48}+\frac{2}{48}} = \frac{1}{\frac{5}{48}} = \frac{48}{5} = 9.6 \text{ ohms}$$

Total resistance of the circuit.. $= 9.6 + 30 + 10 = 49.6$ ohms (Ans.)

Total current................ $= \frac{42}{49.2} = 0.86$ ampere (Ans.)

Line Drop.—Wire used in electirc circuits offers a certain resistance to the passage of the electric current and results in loss of pressure or voltage which must often be taken into account. The amount of the resistance varies with the material and the temperature of the conductor. This will be treated more specifically in a later section. Now we are concerned only with the computation of typical "line drops."

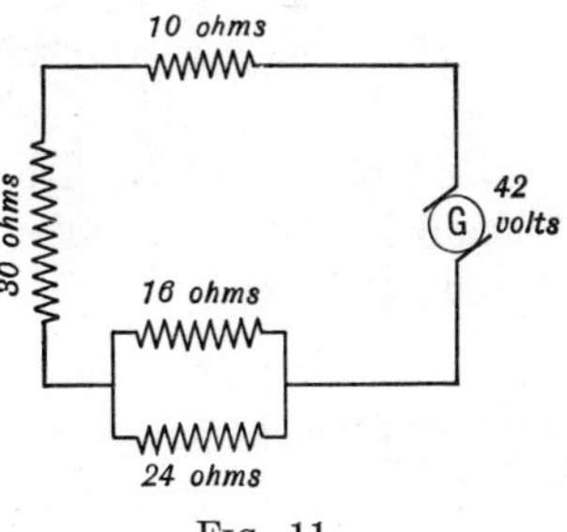

FIG. 11.

The loss of voltage is equal to the product of the current flowing in a conductor and the resistance of the conductor between any two points.

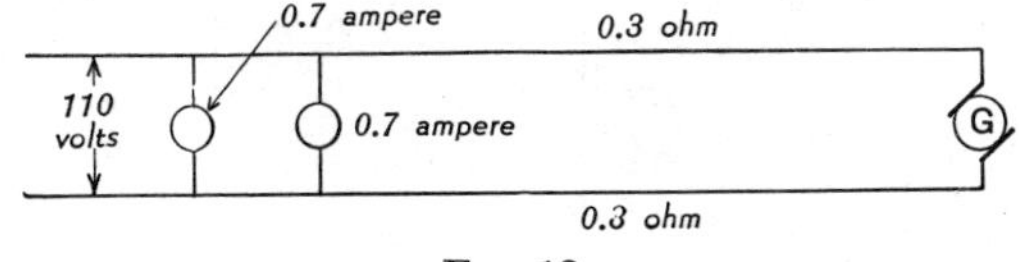

FIG. 12.

ILLUSTRATION: Two lamps shown in Fig. 12 each require 0.7 ampere of current. It is desired that they operate at 110 volts. What voltage must the generator produce if the resistance of each conductor is 0.8 ohm?

Amperes in the circuit	$= 2 \times 0.7$	$=$ 1.4 amperes
Volts lost in upper wire	$= 1.4 \times 0.8$.....	$=$ 1.12 volts
Volts lost in lower wire	$= 1.4 \times 0.8$.....	$=$ 1.12 volts
Volts required across lamps.............		$=$ 110.00 volts
Voltage which must be produced at the generator..........................		112.24 volts (Ans.)

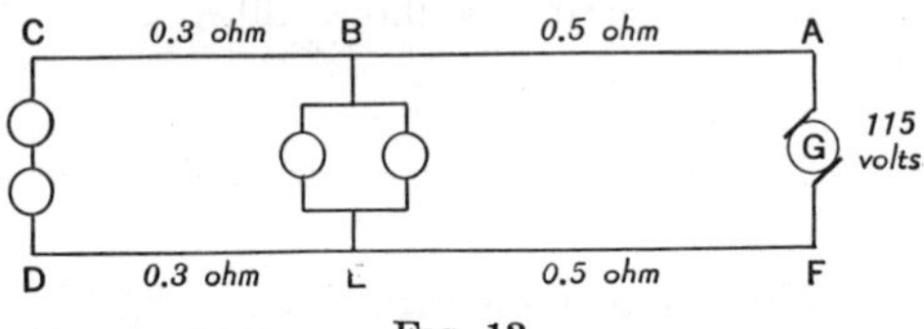

FIG. 13.

ILLUSTRATION: In the circuit shown in Fig. 13, each lamp takes 0.5 ampere of current. What is the voltage drop between A and B, B and C, D and E, and E and F?

$$\text{Current through } BE = 2 \times 0.5 = 1.0 \text{ ampere}$$
$$\text{Current through } CD = 0.5 \text{ ampere}$$
$$\text{Current through } AB \text{ and } EF = 1.5 \text{ amperes}$$

$$\text{Line drop } A \text{ to } B = 0.5 \times 1.5 = 0.75 \text{ volt (Ans.)}$$
$$\text{Line drop } E \text{ to } F = 0.5 \times 1.5 = 0.75 \text{ volt (Ans.)}$$
$$\text{Line drop } B \text{ to } C = 0.3 \times 0.5 = 0.15 \text{ volt (Ans.)}$$
$$\text{Line drop } D \text{ to } E = 0.3 \times 0.5 = 0.15 \text{ volt (Ans.)}$$

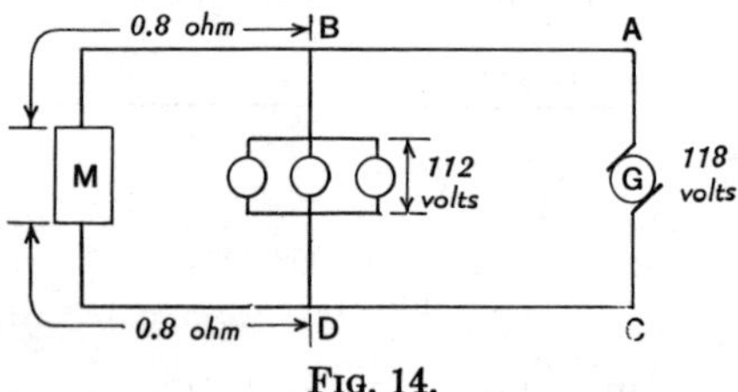

FIG. 14.

ILLUSTRATION: In the circuit shown in Fig. 14 find the resistance of the line between the generator and the lamps if each lamp takes 0.6 ampere of current and the motor 3.2 amperes.

Loss of voltage between generator and lamps

$$= 118 - 112 = 6 \text{ volts}$$

Loss of voltage in each wire $= \frac{6}{2} = 3$ volts

Total amperes in line $AB = 1.8 + 3.2 = 5$ amperes

$$\text{Resistance of line } AB = \frac{E}{I} = \frac{3}{5} = 0.60 \text{ ohm}$$

Resistance of both lines AB and $DC = 0.60 \times 2 = 1.2$ ohm

Electric Insulators and Conductors.—Resistance has been defined as opposition to the flow of electricity. Some materials or substances do not permit the passage of electricity through them at all, and are called *insulators*. Such materials as glass, porcelain, dry wood, paper, wax, rubber, most gases, many liquids and minerals, plastics, etc., are insulators. These materials cover, separate, or support parts of electrical apparatus and circuits to prevent the escape or undesired flow of electricity.

Substances which allow the passage of electricity are called *conductors*. Most metals and some liquids, gases and minerals are conductors. Conductors which require high electric pressure (voltage) to send an electric current through them are said to be poor conductors, or to have high resistance. Otherwise, they are said to be good conductors, or to have low resistance.

Most metals are good conductors, the best (those of lowest resistance) among the common metals being silver, copper, gold, aluminum, tungsten, zinc, and brass, in the order named. Of these copper is the most plentiful and consequently the most used.

The resistance of a conductor varies with its temperature, increasing as the conductor is heated and decreasing as it is cooled. The resistance of a particular conductor of any one material depends also upon the size and shape of the conductor.

For most purposes conductors in the form of wires or rods are used. The calculations of resistances and sizes of wires form an important subject in themselves and are dealt with in the following section.

Electrical Measuring Instruments

Current and Voltage.—A *voltmeter* is an instrument which shows the electromotive force impressed upon its terminals.

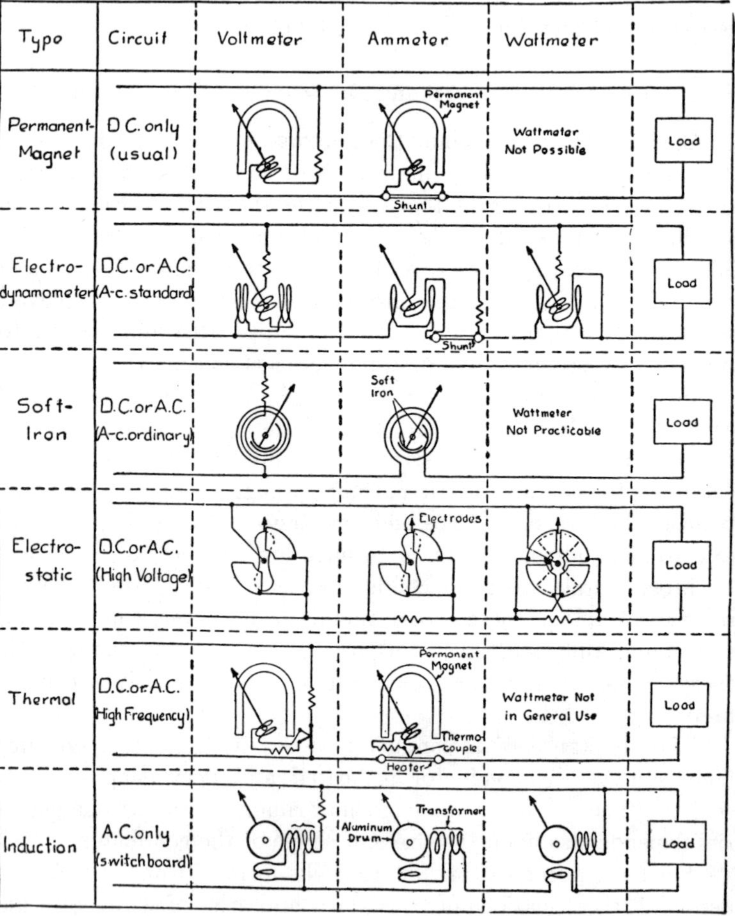

Type
Circuit
Voltmeter
Ammeter
Wattmeter
Permanent-Magnet
D.C. only (usual)
Permanent Magnet
Shunt
Wattmeter Not Possible
Load
Electro-dynamometer
D.C. or A.C. (A-c. standard)
Shunt
Load
Soft-Iron
D.C. or A.C. (A-c. ordinary)
Soft Iron
Wattmeter Not Practicable
Load
Electro-static
D.C. or A.C. (High Voltage)
Electrodes
Load
Thermal
D.C. or A.C. (High Frequency)
Permanent Magnet
Thermo-couple
Heater
Wattmeter Not in General Use
Load
Induction
A.C. only (switchboard)
Aluminum Drum
Transformer
Load

From "Electrical Engineering," by L. A. Hazeltine. Reproduced by special permission of The Macmillan Company, publishers.

FIG. 15.—Types of Voltmeter, Ammeter and Wattmeter.

An *ammeter* is an instrument which shows the strength of the current flowing through it.

The different types of voltmeters and ammeters are illustrated in Fig. 15. The similarity in construction and principle of operation of the two meters will be noted in the case of one type which is based on the D'Arsonval galvanometer. It consists fundamentally of a coil mounted on a pivot placed between the poles of a horseshoe magnet. When a current flows through the coil, the coil has a magnetic field and attempts

Fig. 16.—A Direct Current Milliammeter.

to turn so that its lines of force are in line with those of the horseshoe magnet. A pointer attached to the coil moves along a graduated scale.

In an *ammeter* the angle through which the coil turns is proportional to the strength of the current. Doubling the strength of the current makes the coil turn through twice as great an angle. An ammeter may be constructed and calibrated to measure thousandths of an ampere. Such an instrument is known as a *milliammeter*. The prefix " milli " always means one-thousandth.

To convert amperes into milliamperes, multiply by 1000.

Illustration: How many milliamperes are 0.055 ampere?

$$0.055 \times 1000 = 55 \text{ milliamperes} \quad \text{(Ans.)}$$

An ammeter must always be connected on one side of a circuit in series with one conductor, so that all of the current passes through it. *Never connect an ammeter across the line.*

A voltmeter differs from an ammeter chiefly in that it has a high resistance in series with the coil and that its scale is, of course, calibrated for volts. A voltmeter measures the voltage or fall of potential between the two points to which it is connected. It should therefore be connected *across the line,* not in the line. A voltmeter may be calibrated to measure millivolts and it is then known as a millivoltmeter.

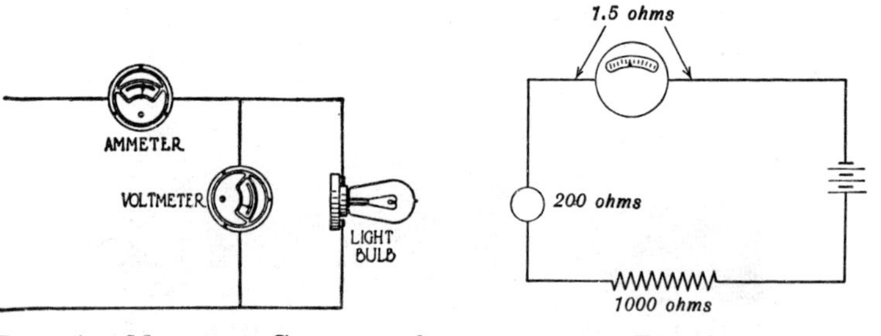

Fig. 17.—Measuring Current and Voltage of an Electric Light Bulb.

Fig. 18.

To convert volts into millivolts, multiply by 1000.

Illustration: How many millivolts are 0.28 volt?

$$0.28 \times 1000 = 280 \text{ millivolts} \quad \text{(Ans.)}$$

Illustration: In the circuit shown in Fig. 18 the battery generates current at a pressure of 4.224 volts. What is the current flowing through the circuit and what is the potential drop through the meter in millivolts?

Total resistance of circuit $= 1000 + 200 + 1.5 = 1201.5$ ohms.

$$I = \frac{E}{R} \qquad I = \frac{4.224}{1201.5} = 0.0035 \text{ ampere} \quad \text{(Ans.)}$$

Potential drop through meter $= IR = 0.0035 \times 1.5 = 0.00525$

$$0.00525 \times 1000 = 5.25 \text{ millivolts} \quad \text{(Ans.)}$$

Resistance.—The simplest method of measuring resistance is by the use of an ammeter and a voltmeter. It is necessary only to obtain the voltage and the current and then apply Ohm's Law, $R = \frac{E}{I}$.

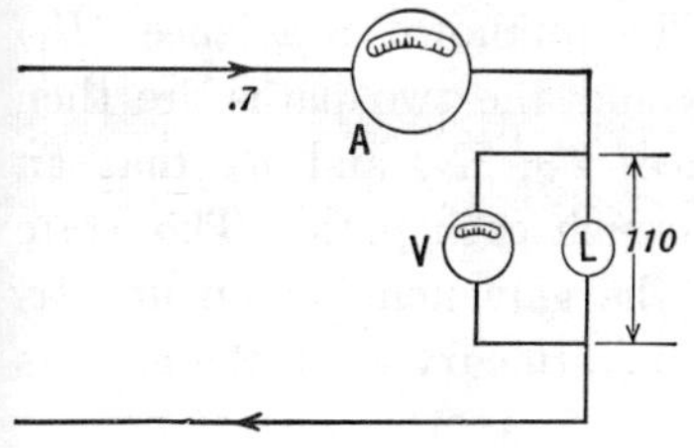

Fig. 19.

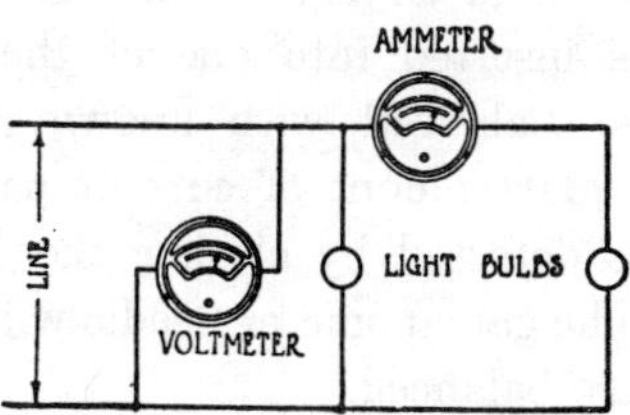

Fig. 20.—Two Light Bulbs in Parallel. Measuring Resistance of One.

Illustration: In the partial circuit shown in Fig. 19 the potential measures 110 volts and the current 0.7 ampere. What is the resistance of the light?

$$R = \frac{E}{I} \qquad R = \frac{110}{0.7} = 157 \text{ ohms} \quad \text{(Ans.)}$$

Illustration: If the potential measures 110 volts and the current is 0.5 ampere, what are the resistances of the two lights in Fig. 20?

Inspection will reveal that the ammeter measures only the current used by the right-hand bulb. The resistance of this bulb is, therefore,

$$R = \frac{E}{I} = \frac{110}{0.5} = 220 \text{ ohms} \quad \text{(Ans.)}$$

The resistance of the left-hand bulb if the same size as the other, will also take 0.5 ampere and would have the same resistance.

The *Wheatstone bridge* is the most accurate instrument for measuring resistances. It consists essentially of a device for providing two parallel paths for an electric current to pass through as shown in Fig. 21. The unknown resistance (R_1) is inserted into one of the paths and the two paths are then so balanced with known resistances R_2, R_3, and R_4 that an equal amount of current passes through each path. This state is detected by closing the key to the galvanometer on line *bc*. The galvanometer needle will remain stationary when the circuits are balanced.

However, R_2, R_3, and R_4 varied until the galvanometer needle is at zero. Then the voltage drop from *a* to *b* is the same as from *a* to *c*, and from *b* to *d* is the same as from *c* to *d*.

Expressed mathematically:

Let I = amperes in the line from the battery

I_x = amperes in R_1 and R_2 branch

I_y = amperes in R_3 and R_4 branch

Then for balance, or when the needle is at zero

$$R_1 I_x = R_3 I_y \quad (1)$$

and

$$R_2 I_x = R_4 I_y \quad (2)$$

Dividing equation 1 by equation 2

$$\frac{R_1 I_x}{R_2 I_x} = \frac{R_3 I_y}{R_4 I_y}$$

whence

$$\frac{R_1}{R_2} = \frac{R_3}{R_4}$$

and

$$R_1 = \frac{R_2 R_3}{R_4}$$

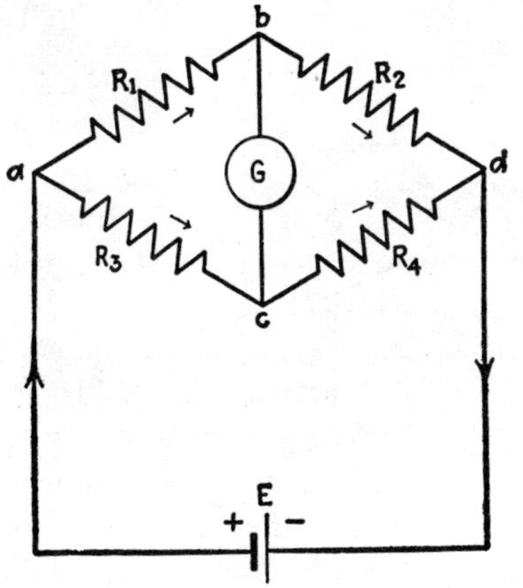

Fig. 21.—Wheatstone Bridge.

Illustration: What is the value of R_1 if $R_2 = 1000$ ohms, $R_3 = 783$ ohms, and $R_4 = 100$ ohms?

$$R_1 = \frac{R_2 R_3}{R_4}$$

$$R_1 = \frac{1000 \times 783}{100} = 7830 \text{ ohms} \quad \text{(Ans.)}$$

The Wheatstone bridge is made up in a variety of forms for commercial testing, one of which is shown in Fig. 22.

Fig. 22.—Dial Type Wheatstone Bridge or Testing Set.

Fig. 23.—Weston Ohmmeter.

An *ohmmeter* is an instrument for the direct measurement of electrical resistances. It is based on the principles of the Wheatstone bridge.

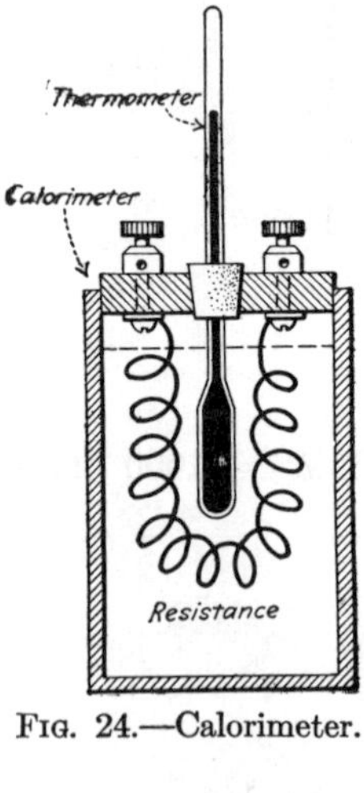

Thermometer
Calorimeter
Resistance

Fig. 24.—Calorimeter.

Electrical Heat.—Heat is generated by the passage of electricity through a conductor, and the amount of heat so generated can be measured by a calorimeter such as that shown in Fig. 24. In this device a coil of wire of known resistance is immersed in a known weight of water, and the resulting rise in temperature of the water is measured. It has been determined experimentally from such an apparatus that one ampere flowing through one ohm for one second always develops 0.24 *calorie* of heat—a calorie being the amount of heat required to raise the temperature of one gram of water one degree centigrade.

Electrical Work and Power

Force.—Force may be defined as that which changes or tends to change the state of motion of a body. We immediately recognize a number of applications of the common conceptions of force which fall within this definition. A pitcher exerts a force on a ball and causes it to fly through the air. The catcher exerts a force which arrests the motion of the ball. The batter transmits a force through the bat which causes a change in direction in the flight of the ball. The batter may lean on his bat and exert a considerable force on it without producing any motion.

Force is manifested in several different forms. There is the *force of gravity* resulting from the attraction between the earth and other bodies; *muscular force,* such as exerted by a horse pulling a wagon; and *mechanical force,* such as that exerted by a locomotive pulling a train. When dynamite breaks up rock in a quarry the action is due to *chemical force;* a force which tends to produce a flow of electricity is *electromotive force;* and the attraction between a magnet and a piece of iron is *magneto-mechanical force. The ordinary unit of force is the pound.*

Mass and Weight.—The *mass* of a body is the quantity of matter in it. The *weight* of a body is the force exerted by gravity on the mass of the body. A ball of iron may weigh five pounds at sea level, but if it is taken to an altitude of seven miles in a balloon, the gravitational force on it will be much decreased and hence it will weigh less. However, the mass will in both cases be the same. *The unit of weight is the pound.*

Work.—Work is done when force overcomes a resistance and moves a body on which it acts. *Work is force acting through space.* The amount of work done is measured by the product of the magnitude of the force and the distance through which it acts:

$$\text{work} = \text{force} \times \text{distance}$$

$$\text{work} = \text{pounds} \times \text{feet} = \text{foot-pounds.}$$

ILLUSTRATION: It takes a force of 125 pounds to pull a cable through a conduit. What work is done if a section between two manholes 200 feet apart is pulled through?

$$\text{Work} = F \times D = 125 \times 200 = 25{,}000 \text{ foot-pounds} \quad \text{(Ans.)}$$

ILLUSTRATION: A man weighing 150 pounds climbs to the top of a 22-foot telegraph pole. What work is done?

$$\text{Work} = F \times D = 150 \times 22 = 3{,}300 \text{ foot-pounds.} \quad \text{(Ans.)}$$

It must be noted that work is not always done when a force acts. For instance, if a man exerts all of his force to lift a 500-pound weight, no work is done if the resistance is not overcome and no motion results.

We have seen that force manifests itself in several different forms. There is a corresponding variety of work. A steam engine does work when it operates a derrick; dynamite does work when it throws stone through a distance; a gasoline engine does work when it propels an automobile; chemical action in a battery sets up a force which causes an electrical current to flow. However, whether the work is done mechanically, chemically, thermally, or electrically, it can be expressed in foot-pounds. Work done takes no account of time. A man may lift a 25-pound weight two feet in a minute or in an hour. The work done in each case is 50 foot-pounds.

Power.—*Power is the rate of doing work.* It is not to be confused with the total work done.

$$\text{Power} = \frac{\text{work}}{\text{time}}$$

or

$$\frac{\text{foot-pounds}}{\text{time}} = \text{foot-pounds per unit of time.}$$

Any convenient unit of time may be used. In mechanical work it may be foot-pounds per minute or foot-pounds per second. An arbitrarily selected unit of power is the *horsepower*, (H. P.).

$$\text{One horsepower} = 33{,}000 \text{ foot-pounds per minute}$$

or

$$\frac{33{,}000}{60} = 550 \text{ foot-pounds per second.}$$

Illustration: A mine hoist raises a cage weighing 800 pounds at the rate of 300 feet per minute. What horsepower is it expending?

$$\text{H.P.} = \frac{800 \times 300}{33,000} = \frac{240,000}{33,000} = 7.3 \quad \text{(Ans.)}$$

Illustration: A tractor exerts a pull of 500 pounds in pulling a motor on skids at a rate of five miles per hour. What horsepower is it exerting?

Since there are 5280 feet in a mile and 60 minutes in an hour,

$$5 \text{ mph} = \frac{5 \times 5280}{60} = 440 \text{ feet per minute.}$$

$$\text{Power} = 500 \times 440 = 220,000 \text{ foot-pounds per minute.}$$

$$\text{Horsepower} = \frac{220,000}{33,000} = 6.7 \quad \text{(Ans.)}$$

Difference between Energy, Force, Work, and Power.—It is important to have a clear understanding of the differences of the meanings of these terms. *Energy* is the capacity to do work. *Force* is one of the factors of work and has to be exerted through a distance to do work. *Work* is done when energy is expended and is reckoned as the product of the magnitude of a force and the distance through which it acts in overcoming a resistance. *Power* is the rate of doing work.

Electrical Work.—Work is, as we have seen, force acting through space, or energy expended in overcoming resistance. However, force may exist without work being done, as for example, when a man pushes against a table but does not move it.

An electrical force exists between the two terminals of a battery tending to send an electrical current through the air. However, the resistance of the air is too great and no current flows, hence no work is done. The same is true when a generator is running on an open circuit. When a wire is connected across the terminals, the force is able to overcome the resistance of the wire, a current flows, and electrical work is done. In this case, the work takes the form of generation of heat. If an electric lamp is connected in the circuit, the work manifests itself as heat and light. If a motor is connected in the circuit, the work is that done by the motor in turning its shaft and pulley or gear under load.

The unit of electrical work is the amount of work performed by a current of one ampere flowing for one second under a pressure of one volt, and is called a joule.

Electrical work = volts × amperes × seconds

One joule has been found by experiment to be equivalent to 0.7375 foot-pound of mechanical work.

Electrical work is a subordinate factor in applied electricity to electrical *power,* which we shall now consider.

Electrical Power.—Power is, as we have seen, the rate at which work is done, and is independent of the total amount of work accomplished. A few paragraphs back we saw that

$$\text{Power} = \frac{\text{work}}{\text{time}}$$

Then, as we may expect,

$$\text{Electrical power} = \frac{\text{electrical work}}{\text{time}}$$

The unit of electrical power is the *watt.* It is equivalent to one joule of electrical work per second. Then,

$$\text{Watts} = \frac{\text{joules}}{\text{seconds}} = \frac{\text{volts} \times \text{amperes} \times \text{seconds}}{\text{seconds}}$$

The " seconds " cancel out and the equation becomes,

$$\text{Watts} = \text{volts} \times \text{amperes}.$$

One watt, therefore, equals one volt multiplied by one ampere.

To find the rate in watts at which energy is expended in a circuit:

Multiply the current in amperes by the pressure in volts causing it to flow.

In general, electrical power = voltage × current.

When P = watts expended

I = current in amperes

E = pressure in volts

Then $P = E \times I$.

ILLUSTRATION: A 110-volt circuit has fifty incandescent lamps connected in parallel, each with a resistance of 220 ohms and two electric toasters each with resistances of 18.33 ohms. How many watts of power are consumed by the circuit?

By Ohm's Law $I = \frac{E}{R}$. Then $I = \frac{110}{220} = \frac{1}{2}$ ampere current for each lamp.

$50 \times \frac{1}{2} = 25$ amperes of current for fifty lamps.

$$I = \frac{110}{18.33} = 6 \text{ amperes for each toaster.}$$

Then

$$P = E \times I = 110 \times (25 + 6 + 6) = 4070 \text{ watts} \quad \text{(Ans.)}$$

ILLUSTRATION: If it requires 12 amperes of current to operate the heaters on a street car from a 550-volt circuit, what power will be required?

$$P = E \times I. \quad P = 550 \times 12 = 6{,}600 \text{ watts.} \quad \text{(Ans.)}$$

To find the current when the power and the voltage are known: *Divide the watts expended by the voltage causing the current to flow.*

$$\text{Current} = \frac{\text{watts}}{\text{volts}} \qquad I = \frac{P}{E}$$

ILLUSTRATION: What current does a 75-watt lamp require when operating on a 110-volt circuit?

$$I = \frac{P}{E} = \frac{75}{110} = 0.68 \text{ ampere} \quad \text{(Ans.)}$$

Illustration: What current does a 550-watt electric flatiron require when connected to a 115-volt circuit?

$$I = \frac{P}{E} = \tfrac{550}{115} = 4.78 \text{ amperes} \quad \text{(Ans.)}$$

To find the voltage when the power and current are known: *Divide the watts expended by the current flowing.*

$$\text{Volts} = \frac{\text{watts}}{\text{amperes}}. \qquad E = \frac{P}{I}$$

Illustration: A 1200-watt motor requires a current of 10 amperes. What voltage is necessary to operate it?

$$E = \frac{P}{I} = \tfrac{1200}{10} = 120 \text{ volts} \quad \text{(Ans.)}$$

Electrical Horsepower.—The relationship between mechanical work and the expenditure of electrical energy has been determined by calorimeter experiments. From the results thus obtained, the following relationship has been established.

$$1 \text{ watt} = 0.7375 \text{ foot-pound per second}$$

or

$$1 \text{ foot-pound per second} = \frac{1}{0.7375} = 1.356 \text{ watts}$$

Since 550 foot-pounds per second are equivalent to 1 mechanical horsepower, an equivalent rate of electrical power would be:

$$\frac{550}{0.7375} = 746 \text{ watts} = 1 \text{ electrical horsepower.}$$

The electrical horsepower is a convenient unit since the watt is very small.

To find the electrical horsepower maintained in any circuit or part of a circuit:

Multiply the volts causing the current to flow by the current expressed in amperes and divide this product by 746.

$$\text{H.P.} = \frac{\text{watts}}{746} = \frac{\text{volts} \times \text{amperes}}{746} = \frac{E \times I}{746}$$

ILLUSTRATION: A motor on a 220-volt circuit requires 28 amperes of current. What horsepower is it using?

$$\text{H.P.} = \frac{E \times I}{746} = \frac{220 \times 28}{746} = \frac{6160}{746} = 8.26 \quad \text{(Ans.)}$$

ILLUSTRATION: A generator maintains a pressure of 110 volts across an electric light circuit and the ammeter indicates 75 amperes. What horsepower is being generated by the generator?

$$\text{H.P.} = \frac{E \times I}{746} = \frac{110 \times 75}{746} = \frac{8250}{746} = 11.06 \quad \text{(Ans.)}$$

The Kilowatt.—The *kilowatt* is a still larger unit of power. *One kilowatt equals 1000 watts.* The following relations are immediately obvious:

$$\text{Kilowatts (kw.)} = \frac{\text{watts}}{1000} = \frac{E \times I}{1000}$$

$$\text{Watts} = \text{kw.} \times 1000$$

$$1 \text{ h.p.} = 0.746 \text{ kw.}$$

$$1 \text{ kw.} = \frac{1}{0.746} = 1.34 \text{ h.p.}$$

ILLUSTRATION: What is the capacity in kilowatts of a generator carrying a load of 400 amperes at a pressure of 220 volts?

$$\text{Kw.} = \frac{E \times I}{1000} = \frac{220 \times 400}{1000} = 88 \quad \text{(Ans.)}$$

ILLUSTRATION: At full load how many amperes can be delivered by a 60-kilowatt generator at a pressure of 110 volts?

$$\text{Watts} = \text{kw.} \times 1000 = 60 \times 1000 = 60{,}000 \text{ watts} = P$$

$$I = \frac{P}{E} = \frac{60{,}000}{110} = 545 \text{ amperes} \quad \text{(Ans.)}$$

The Watt-hour and Kilowatt-hour.—Electrical energy or power when sold to a consumer is usually measured in terms of *watt-hours* or *kilowatt-hours* since the joule is too small a unit for practical use in this connection. A watt-hour is one watt expended for one hour. It is equivalent to 3600 watt-seconds (or joules) and also to 60 watt-minutes.

$$\text{Watt-hours} = \text{watts} \times \text{hours}$$

A *kilowatt-hour* is a larger unit of electrical work and equal to 1000 watts maintained for one hour, or 500 watts maintained for two-hours, etc.

$$\text{Kilowatt-hours} = \text{kw.} \times \text{hours}$$

An electrical *horsepower-hour* is one electrical horsepower maintained for one hour or 746 watts maintained for one hour.

$$\text{Horsepower-hours} = \text{h.p.} \times \text{hour}$$

The dials of a consumer's meter, by which the electrical energy used for light and power is measured, generally record kilowatt-hours. In Fig. 25 the dial face of a watt-hour meter has four

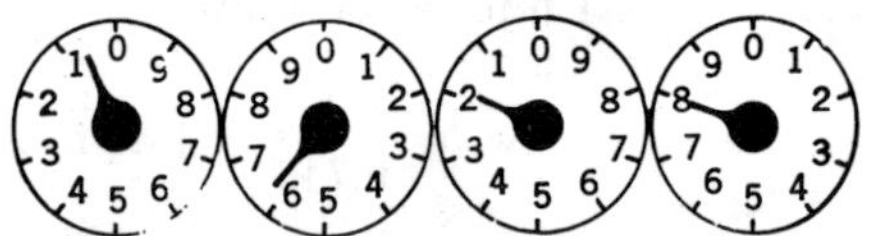

FIG. 25.—Dial of Watt-hour Meter.

circles. In the preceding figure each denotes the amount of energy in kilowatt-hours measured by the movement of the pointer over one division of the corresponding scale. One complete revolution of the pointer of any scale moves the pointer of the next scale immediately to its left over one division. It will be noted that some of the pointers turn in a clockwise direction and that the others turn counter-clockwise. In reading a pointer on a circle, it is necessary to look at the pointer immediately to its right to determine whether or not it has reached the point on which it appears to rest. For example, in Fig. 25 it is almost impossible to tell by looking at the second pointer from the right whether it has passed the "2" or failed to reach it. However, by looking at the circle to its right, it is apparent that it has not yet reached the "2."

Illustration: How many kilowatt-hours does the meter in Fig. 25 read? 0618 kilowatt-hours. (Ans.)

Electrical Power Calculations.—A number of formulas are here presented which have a great variety of practical applications in the calculation of electrical power. These rules and formulas have been derived either by transposition of the formulas presented on the preceding pages or by combining them with the formulas expressing Ohm's Law. They are applicable equally well to the whole or a part of a circuit. *Caution must be exercised in the use of these formulas to use the volts lost in only the particular part of the circuit considered, and also the resistance of, and the current through, this part only.*

Given current and pressure, to find the watts expended:

The watts lost or expended in any circuit equal the product of the current and the voltage causing it to flow.

$$P = E \times I$$

when P = watts expended

I = current in amperes

E = pressure in volts

The use of this formula is illustrated on page 707.

Given current and resistance, to find the energy expended in watts:

The watts lost or expended in any circuit are equal to the current squared multiplied by the resistance. This is often called the "I-square R loss."

$$P = I^2 \times R$$

ILLUSTRATION: The resistance of the field magnets of a dynamo is 430 ohms and the magnetizing current is 3 amperes. What power is used in magnetizing the field?

$$P = I^2R = 3 \times 3 \times 430 = 3870 \text{ watts} \quad \text{(Ans.)}$$

ILLUSTRATION: An electric light has a resistance of 121 ohms and uses 0.909 ampere. How many watts does it use?

$$P = I^2R = 0.909 \times 0.909 \times 121 = 100 \text{ watts} \quad \text{(Ans.)}$$

Given resistance and pressure, to find the watts expended:

The watts lost or expended in any circuit are equal to the square of the voltage divided by the resistance.

$$P = \frac{E^2}{R}$$

ILLUSTRATION: An electromagnet with a resistance of 40 ohms is operated on the current from a 6-volt storage battery. What power is expended and what current does the magnet require?

$$P = \frac{E^2}{R} = \frac{6 \times 6}{40} = 0.90 \text{ watt} \quad \text{(Ans.)}$$

$$I = \frac{E}{R} = \frac{6}{40} = 0.15 \text{ ampere} \quad \text{(Ans.)}$$

ILLUSTRATION: The resistance of a solenoid is 60 ohms. What power will be expended in it if a current passes through it under a pressure of 110 volts?

$$P = \frac{E^2}{R} = \frac{110 \times 110}{60} = 201.67 \text{ watts} \quad \text{(Ans.)}$$

Given watts expended and current, to find the resistance:

The resistance is equal to watts expended divided by the square of the current.

$$R = \frac{P}{I^2}$$

ILLUSTRATION: An electric flatiron uses 660 watts of power and draws 6 amperes current. What is its resistance?

$$R = \frac{P}{I^2} = \frac{660}{6 \times 6} = 18.33 \text{ ohms} \quad \text{(Ans.)}$$

ILLUSTRATION: A 75-watt incandescent lamp requires 0.682 ampere. What is its resistance?

$$R = \frac{P}{I^2} = \frac{75}{0.682 \times 0.682} = \frac{75}{0.465} = 161.3 \text{ ohms} \quad \text{(Ans.)}$$

Given watts expended and resistance, to find the current:

The current equals the square root of the quotient of the watts divided by the resistance.

$$I = \sqrt{\frac{P}{R}}$$

ILLUSTRATION: The resistance of a 55-watt lamp is 220 ohms. What current will it require?

$$I = \sqrt{\frac{P}{R}} = \sqrt{\frac{55}{220}} = \sqrt{\frac{1}{4}} = \tfrac{1}{2} \text{ ampere} \quad \text{(Ans.)}$$

ILLUSTRATION: A printshop glue pot has a resistance of 88 ohms and draws 8,800 watts of power. What current does it require?

$$I = \sqrt{\frac{P}{R}} = \sqrt{\frac{8{,}800}{88}} = \sqrt{100} = 10 \text{ amperes} \quad \text{(Ans.)}$$

Given watts expended and pressure, to find the resistance:

The resistance equals the square of the voltage divided by the watts expended.

$$R = \frac{E^2}{P}$$

ILLUSTRATION: What is the resistance of a 55-watt, 110-volt incandescent lamp?

$$R = \frac{E^2}{P} = \frac{110 \times 110}{55} = 220 \text{ ohms} \quad \text{(Ans.)}$$

ILLUSTRATION: An electric furnace uses 7.2 kilowatts of power when operating at a pressure of 80 volts. What is its resistance?

$$R = \frac{E^2}{P} = \frac{80 \times 80}{7{,}200} = 0.889 \text{ ohm} \quad \text{(Ans.)}$$

All of the above formulas are applicable to problems where the power is given or wanted in electrical horsepower, by remembering that 1 horsepower = 746 watts = 0.746 kw, and 1 kw = 1.34 hp.

ILLUSTRATION: A calcium carbide electric furnace uses 3,500 amperes of current at 110 volts. How much horsepower is expended?

$$P = 3{,}500 \times 110 = 385{,}000 \text{ watts}$$

Since

$$1 \text{ hp} = 746 \text{ watts,}$$

$$\text{hp} = \frac{385{,}000}{746} = 516 \quad \text{(Ans.)}$$

ELECTRIC CELLS AND BATTERIES

Dry Cells.—The simplest method of producing electric current is by chemical means. A simple primary cell may be made by placing two dissimilar metals into an acid or alkaline solution. If the two metals are then connected by a piece of wire, a chemical reaction will result between the metals and the liquid solution and an electric current will flow through the wire.

Primary Cells—Carbon-Zinc.—The primary cells on the market, however, are all various types of dry cells. The standard carbon-zinc Leclanché type dry cell is now, and is likely to remain, the most widely used system because of low cost and reliable performance. Its electrochemical system uses a zinc anode, a manganese-dioxide cathode, and an electrolyte of ammonium chloride and zinc chloride dissolved in water. Powdered carbon is used in the depolarizing mix, usually in the form of acetylene black, to improve conductivity of the mix and to retain moisture.

The standard carbon-zinc dry battery is considered a primary type, i.e., not efficiently reversible. The basic cell is made in many shapes and sizes but two general categories exist:

1. Round Cells—available as unit cells or in assembled batteries.
2. Flat Cells—available in multi-cell batteries only.

The difference between the round and flat cells is mostly physical. The chemical ingredients are the same in both cases—carbon, depolarizing mix separator, electrolyte and zinc. The round-type cell is shown in Fig. 26, and the flat-type in Fig. 26A.

In flat cells, carbon is coated on a zinc plate to form a duplex electrode—a combination of the zinc of one cell and the carbon of the adjacent one. The "Mini-Max" cell (Figure 26A) contains no expansion chambers or carbon rod as does the round cell. This increases the amount of depolarizing mix available per unit cell volume and therefore the energy content. In addition the flat cell, because of its rectangular form, reduces waste space in assembled batteries. The energy to volume ratio of a battery utilizing round cells is inherently poor because of the voids occurring between cells. These two factors account for an energy to volume improvement of nearly 100% for "Mini-Max" cells compared to round cell assemblies.

The nominal voltage of a carbon-zinc cell is 1.5 v. Present battery types are available in voltages ranging from 1.5 to 510 v. Cells and batteries may, of course, be connected in series to obtain higher voltages, in parallel to achieve greater service capacity, or in series-parallel to obtain higher voltage and greater service capacity.

Primary Cells—Alkaline Type.—The electrochemical system of alkaline cells is comprised of a zinc anode of large surface area, a manganese-dioxide cathode of high density, and a potassium-hydroxide electrolyte. This alkaline primary differs from

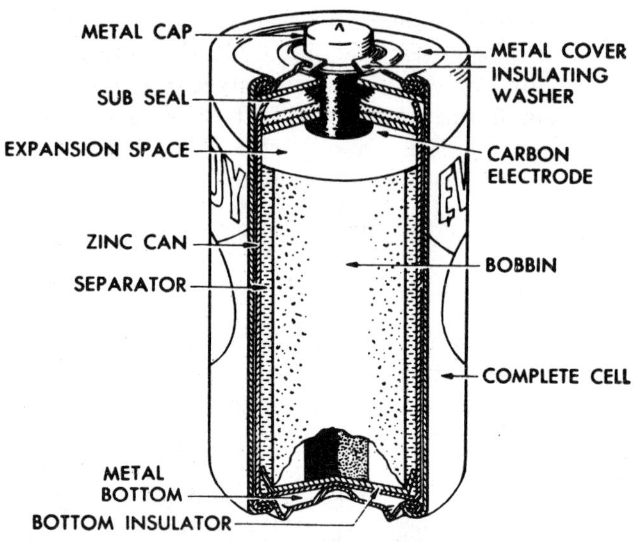

Fig. 26.—Standard Round Carbon-Zinc Cell.

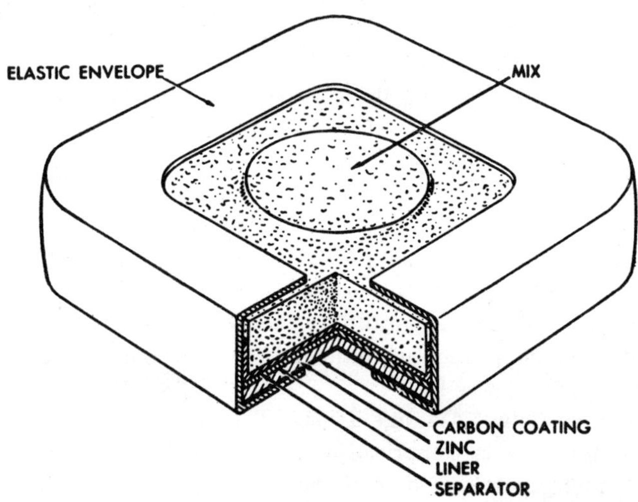

Fig. 26A.—"Mini-Max"® Flat Carbon-Zinc Cell.

the conventional Leclanché cell primarily in the highly alkaline electrolyte that is used. The cell is a high rate source of electrical energy. Its outstanding advantages are derived from the unique assembly of components and construction methods. (See Figure 27.)

Two principal features are a manganese-dioxide cathode of high density in conjunction with a steel can which serves as a cathode current collector, and a zinc anode of extra high surface area in contact with the electrolyte. These features, coupled with the use of a potassium-hydroxide electrolyte of high conductivity give these cells their very low internal resistance and impedance and high service capacity. Nominal voltage of an alkaline-manganese dioxide primary cell is 1.5 v in standard N, AA, C, ½D, D, and G cell sizes. Batteries are available with voltages up to 6 v and in a number of different service capacities. The alkaline-manganese primary cell is for applications requiring more power or longer life than can be obtained from carbon-zinc batteries. Alkaline cells contain 50 to 100 per cent more total energy than a carbon-zinc cell of the same size.

Primary Cells—Mercury Type.—The mercury battery consists essentially of a depolarizing mercuric-oxide cathode, an anode of pure amalgamated zinc, and a concentrated aqueous electrolyte of potassium hydroxide saturated with zincate. The nominal voltage is 1.35 v for a mercury cell with a depolarizer of 100 percent mercuric oxide and 1.4 v for a cell with a mixture of mercuric oxide and manganese dioxide.

The fundamental components of the mercury cell are a pressed mercuric-oxide cathode (in sleeve or pellet form) and pressed cylinders, or pellets, of powdered zinc with steel enclosures. These provide precise mechanical assemblies having maximum dimensional stability and marked improvements in performance over dry batteries of Leclanché (carbon-zinc) type. Cells are currently produced in two different designs using either flat or cylindrical types of pressed powder electrodes. Electrochemically both cells are the same, differing only in case design and internal electrode arrangements.

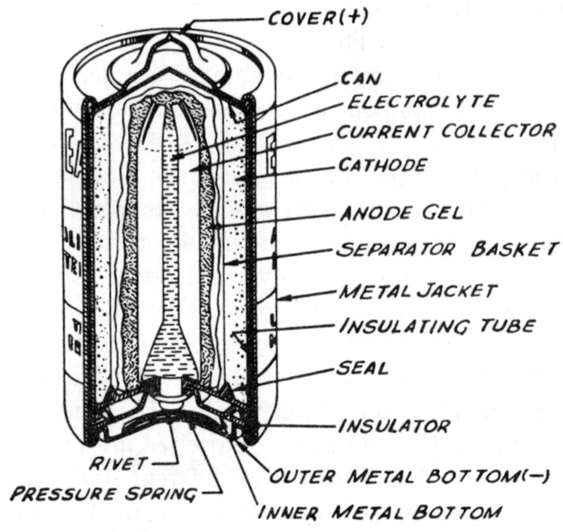

Fig. 27.—Primary Alkaline Cell.

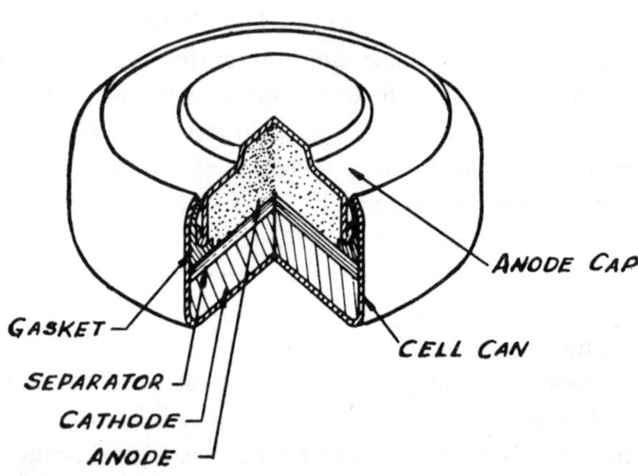

Fig. 27A.—Silver Oxide Cell.

1. Depolarizing cathodes of mercuric oxide, to which a small percentage of graphite is added, are shaped and either consolidated to the cell case (for flat electrode types), or pressed into the cases of the cylindrical types.

2. Anodes are formed of amalgamated zinc powder of high purity, pressed into either flat or cylindrical shapes.

3. A permeable barrier of specially selected material prevents migration of any solid particles in the cell, thereby contributing effectively to long shelf and service life.

4. Insulating and sealing gaskets are molded of polyethylene or Neoprene, depending on the application for which the cell or battery will be used.

5. Inner cell tops are plated with materials which provide an internal surface to which zinc will form a zinc amalgam bond.

6. Cell cases and outer tops of nickel-plated steel are used to resist corrosion, and to provide greatest passivity to internal cell materials.

7. An outer, nickel-plated, steel jacket is generally used for single cell "A" battery types. This outer jacket is a necessary component for the "self-venting construction" which provides a means of releasing excessive gas in the cell. Venting occurs if operating abnormalities, such as reverse currents or short circuits, produce excessive gas in the cell.

Primary Cells—Silver Type.—The silver oxide battery consists of a depolarizing silver oxide cathode, a zinc anode of high surface area and a highly alkaline electrolyte. The electrolyte is potassium hydroxide in hearing aid batteries. This is used to obtain maximum power density at hearing aid current drains. Sodium hydroxide is used in watch batteries for long term reliability. Mixtures of silver oxide and manganese dioxide may be tailored to provide a flat discharge curve or increased service hours.

Silver oxide batteries are well suited, for example, for use in hearing aids, instruments, photo electric exposure devices, electric watches and as reference voltage sources.

A cutaway of a silver oxide cell is shown in Figure 27A.

The milliampere-hour capacity is the same or a little greater than that of mercury batteries, but the silver oxide battery operates at 1.5 v while mercury batteries operate at about 1.3 v. Thus, silver oxide batteries offer 15 to 20 percent more power than mercury batteries of the same size. Present types are available in 1.5 v cells in capacities of 60 to 165 na-hr. A 9 volt battery rated at 165 milliampere-hours is also obtainable. Maximum current output ranges up to 10 ma.

Connecting Cells in Groups.—Primary cells are often connected in groups to produce greater voltages or currents. There are three methods of connection—series, parallel and series parallel.

In series connection the positive terminal of one cell is connected to the negative of the next succeeding cell and the line is connected to the remaining terminals. When the cells are so connected, the voltage of the battery is the sum of the voltages of all the cells. The current, or amperage, of the battery is equal to the amperage of one cell.

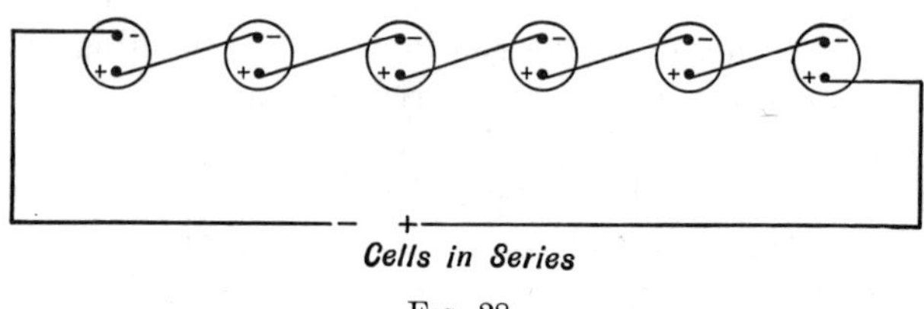

FIG. 28.

ILLUSTRATION: What current and what voltage may be obtained from the battery shown in Fig. 28 if the voltage of each cell is $1\frac{1}{2}$ volts and the maximum amperage 25 amperes?

$$\text{Voltage} = 1\tfrac{1}{2} + 1\tfrac{1}{2} + 1\tfrac{1}{2} + 1\tfrac{1}{2} + 1\tfrac{1}{2} + 1\tfrac{1}{2}$$

or

$$6 \times 1\tfrac{1}{2} = 9 \text{ volts} \quad \text{(Ans.)}$$

$$\text{Amperage} = 25 \text{ amperes} \quad \text{(Ans.)}$$

In *parallel connection* the positive terminals of all cells are connected to one line and the negative terminals to the other line. When cells are connected in parallel, every cell should be of the same voltage. The voltage of the battery is the same as the voltage of one cell. The amperage is the sum of the amperages of each cell.

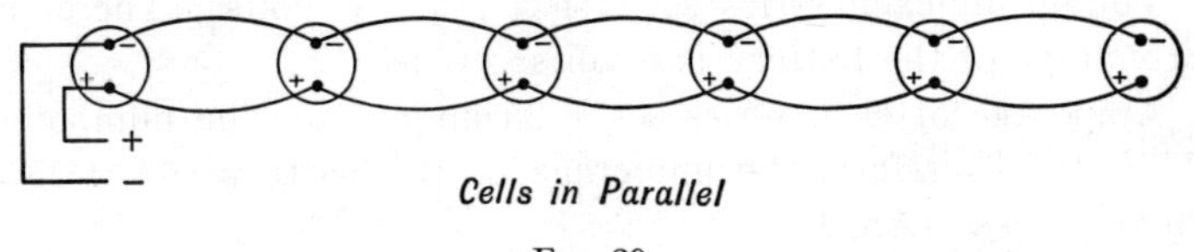

Cells in Parallel

Fig. 29.

Illustration: What is the voltage and amperage of the battery shown in Fig. 29 if the voltage of each cell is $1\frac{1}{2}$ volts and the amperage of each cell 25 amperes?

Voltage of battery $= 1\frac{1}{2}$ volts (Ans.)

Amperage of battery $= 6 \times 25 = 150$ amperes (Ans.)

A series-parallel connection is made up of sets of cells connected in series with each set connected in parallel in the circuit as shown in Fig. 30. Each set must have the same voltage or an equal number of similar cells.

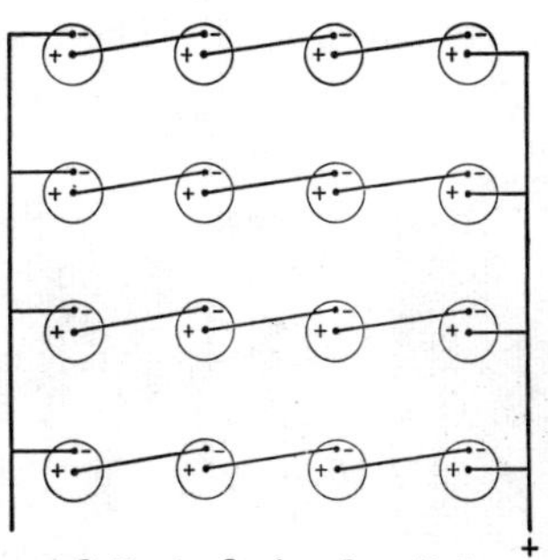

Cells in Series Parallel

Fig. 30.

The voltage of a series-parallel battery is equal to the voltage of each set connected in series, and the amperage is equal to the sum of the amperes delivered by each set.

ILLUSTRATION: In the battery shown in Fig. 30 the voltage of each cell is $1\frac{1}{2}$ volts and the amperage is 25 amperes. What is the voltage and amperage of the battery?

Voltage of each series set $= 4 \times 1\frac{1}{2} = 6$ volts. Therefore, the voltage of the battery is 6 volts. (Ans.)

Amperage of each series set $= 25$ amperes. The number of sets $= 4$. Therefore, the amperage of the battery $= 4 \times 25 = 100$ amperes. (Ans.)

Secondary Cells—Lead Type.—Secondary cells are distinguished from primary cells in that the latter are designed to be recharged, while the former are not. The rechargeable feature is a great advantage in cells operating as parts of prime movers and other moving machines, because it is often feasible to install a generator to be driven by the machinery, as in the automobile, and thus to keep the battery charged (if the load on it is not too great). The type of storage battery used on automobiles consists usually of three lead-type secondary cells connected to each other in a case, as shown in Fig. 31. The lead secondary cell consists of two plates in an electrolyte of diluted sulfuric acid.

When a lead storage cell has a full charge, the plate which forms the positive terminal is lead peroxide and the plate which forms the negative terminal is lead. The lead peroxide is made

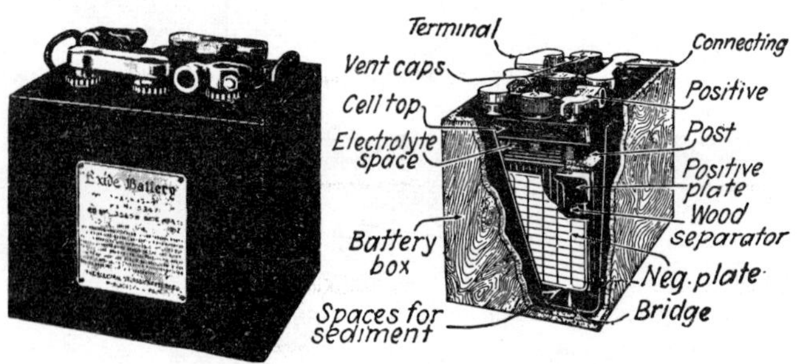

FIG. 31.

into a paste by mixing it with sulphuric acid and the lead in the negative plate is spongy to facilitate the chemical action. Neither of these substances is hard enough to be made into plates so they are pressed into the gridwork of cast lead plates as shown in Fig. 32.

When current is being drawn from a storage cell, the sulphuric acid (H_2SO_4) breaks up into its component parts of ions of $H^{+}{}_2$

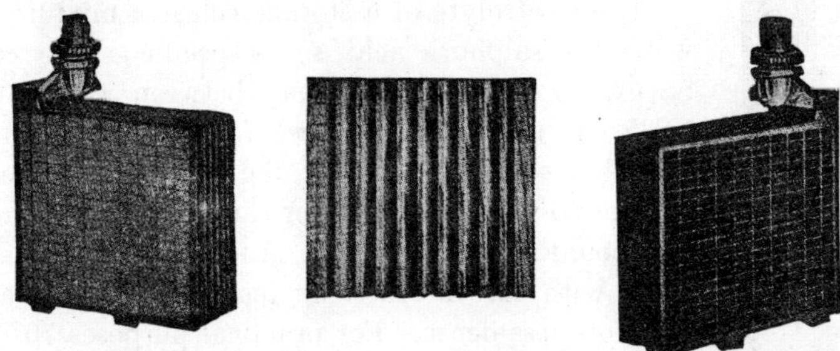

Fig. 32.—Plates of a Storage Battery. A Separator Is Shown in the Center.

and $SO^{-}{}_4$. The SO_4 goes to the lead plate giving it a negative charge and combining with the lead to form lead sulphate. The hydrogen goes to the lead peroxide plate giving it a positive charge and combining with the oxygen of the hydrogen peroxide to form water (H_2O).

It is obvious that as the discharging process continues, the electrolyte increases its proportion of water and decreases its proportion of sulphuric acid. It is possible to take advantage of this to measure the strength of a cell because sulphuric acid is heavier than water.

Specific Gravity.—Sulphuric acid is 1.835 times as heavy as water. This ratio is called the specific gravity of sulphuric acid The *specific gravity* of any substance is the ratio of its weight to the weight of an equal volume of water at 39.1 degrees Fahrenheit.

Any floating object displaces its own weight of the fluid in which it floats. This principle is employed in the instrument known as the *hydrometer* (Fig. 33). It consists of a sealed glass tube weighted at one end and having a graduated scale in the other end. When this tube is floated in water it will sink to a certain depth. When it is floated in a liquid heavier than water, such as sulphuric acid, it will displace its own weight without sinking so deeply into the liquid. The scale, which is read at the level of the liquid on the tube, can be calibrated to give the specific gravity directly.

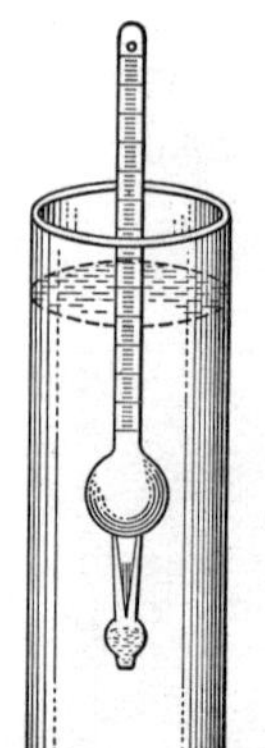

FIG. 33.—Hydrometer.

The electrolyte of a storage cell is a mixture of water and sulphuric acid; so its specific gravity can be expected to be something between 1.000 and 1.835. In a fully charged cell for automobile ignition the specific gravity of the electrolyte should not exceed 1.300; in a cell for radio work, the reading should not exceed 1.280. Liquids generally expand with an increase in temperature and are therefore less dense. For practical purposes 70° F. has been set as the standard temperature for the comparison of specific gravities of storage battery electrolytes.

When the temperature of an electrolyte is greater than 70° F add one point to the fourth figure of the measured specific gravity for each 3° above 70° to obtain the actual specific gravity.

ILLUSTRATION: The temperature of an electrolyte is 94° F. and the hydrometer reading is 1.280. What is the correct specific gravity?

$$94^\circ - 70^\circ = 24^\circ. \quad 24/3 = 8$$

Therefore, the actual specific gravity is

$$1.280 + 0.008 = 1.288 \quad \text{(Ans.)}$$

Similarly, when the electrolyte is *colder* than 70°, *subtract* one point from the fourth place of the measured specific gravity for each 3° below 70° to obtain the actual specific gravity.

ILLUSTRATION: The temperature of an electrolyte is 40° and the hydrometer reading is 1.270. What is the actual specific gravity?

$$70^\circ - 40^\circ - 30^\circ. \quad 30/3 = 10$$

The actual specific gravity is

$$1.270 - 0.010 = 1.260 \quad \text{(Ans.)}$$

The readings of the hydrometer show the condition of the battery in accordance with the following table:

Reading	Condition
1.280–1.300	Full charge
1.250	¼ Discharged
1.215	½ Discharged
1.180	¾ Discharged
1.150	Discharged

Other Types of Secondary Cells—The Edison Cell.—The storage cell devised by Thomas A. Edison employs nickel oxide for the positive electrode and finely divided iron for the negative electrode. These materials are packed into pockets carried by steel grids. The electrolyte is a solution of potassium hydroxide. As the cell discharges, the nickel oxide becomes reduced and the iron becomes oxidized, but the electrolyte remains unchanged. The chemical reactions are complex and not fully understood.

This cell is lighter than the lead storage cell, less subject to mechanical derangement, and it not injured by freezing. Moreover, cells of this type can be discharged and left in that condition without injury. They are therefore particularly suited for application in vehicles such as delivery trucks, fork lift trucks, subways, etc. The voltage output is lower than that of the lead storage cell and drops somewhat during discharge, having an average value of about 1.2 volts.

The Alkaline Secondary Cell.—While the Edison Cell employs an alkaline electrolyte the term *alkaline cell* is now applied to the newer Eveready® alkaline-manganese dioxide cell. The alkaline-manganese dioxide batteries, already discussed for their primary cell types, are also designed as secondary (storage) batteries, and are maintenance free, hermetically sealed, and will operate in any position. These batteries have been specifically designed for electronic and electrical applications where low initial cost and a low operating cost are of paramount importance.

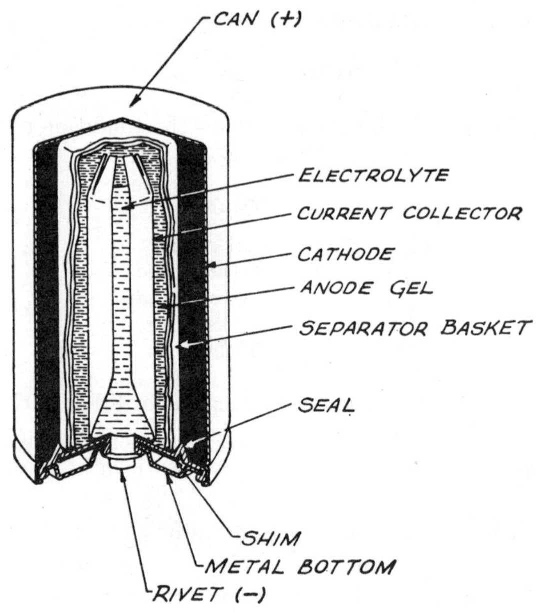

FIG. 33A.—Alkaline Secondary Cell.

Like the primary alkaline cells, these secondary (storage) cells also have zinc and manganese dioxide electrodes, and a potassium hydroxide electrolyte, as shown in Fig. 33A.

Nickel-Cadmium Secondary Cells.—In the uncharged condition the positive electrode of a nickel-cadmium cell is nickelous hydroxide, the negative cadmium hydroxide. In the charged condition the positive electrode is nickelic hydroxide, the negative metallic cadmium. The electrolyte is potassium hydroxide. The average operating voltage of the cell under normal discharge conditions is about 1.2 volts. The over-all chemical reaction of the nickel-cadmium system can be considered as:

$$\underset{\text{Charged}}{Cd + 2NiOOH + 2KOH} \rightleftarrows \underset{\text{Discharged}}{Cd(OH)_2 + 2NiO + 2KOH}$$

During the latter part of a recommended charge cycle and during overcharge, nickel-cadmium batteries generate gas. Oxygen is generated at the positive (nickel) electrode after it becomes fully charged and hydrogen is formed at the negative (cadmium) electrode when it reaches full charge.

A conventional vented type nickel-cadmium battery will liberate oxygen and hydrogen plus entrained electrolyte fumes through a valve. In order to hermetically seal a nickel-cadmium cell it is necessary to develop means of using up this gas inside the cell. This is acomplished as follows:

1. The battery is constructed with excess ampere-hour capacity in the cadmium electrode.

2. Starting with both electrodes in the fully discharged state, charging the battery causes the positive (nickel) electrode to reach full charge first and it starts oxygen generation. Since the negative (cadmium) electrode has not yet reached full charge it cannot cause hydrogen to be generated.

3. The cell is designed so that the oxygen formed can reach the surface of the metallic cadmium electrode where it reacts, forming electrochemical equivalents of cadmium oxide.

4. Thus in overcharge the cadmium electrode is oxidized at a rate just sufficient to offset input energy, keeping the cell in equilibrium at full charge.

This process can continue for long periods. The level of oxygen pressure thus established in the cell is determined by the charge rate used.

"Eveready"® nickel-cadmium cells are available in three basic configurations—button, cyclindrical and rectangular. The range of capacities for each type is as follows:

Button: (See Figs. 33B and 33C.) 20–300 milliampere-hours
Cylindrical: 450–8000 milliampere-hours
Rectangular: 1.6–23 ampere-hours

When required, two to ten button cells may be assembled into a higher voltage series stack by special factor welding techniques. (Under special conditions more than ten cells may be stacked.)

Rating of Storage Batteries.—Storage batteries are rated in ampere-hours. A current of one ampere flowing for one hour is an ampere-hour. Batteries are rated on the basis of the current which they can deliver continuously for a period of 8 hours. In other words, a 120-ampere-hour battery will deliver 15 amperes of current for eight hours. It will not, however, deliver 120 amperes for one hour or 60 amperes for two hours. The ampere-

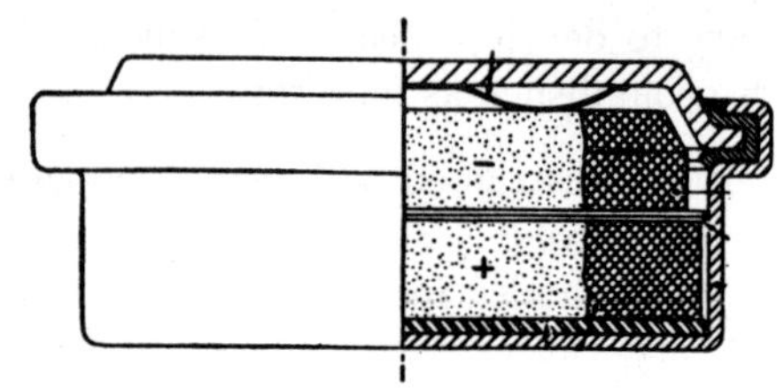

FIG. 33B.—Standard Nickel-Cadmium Button Cell.

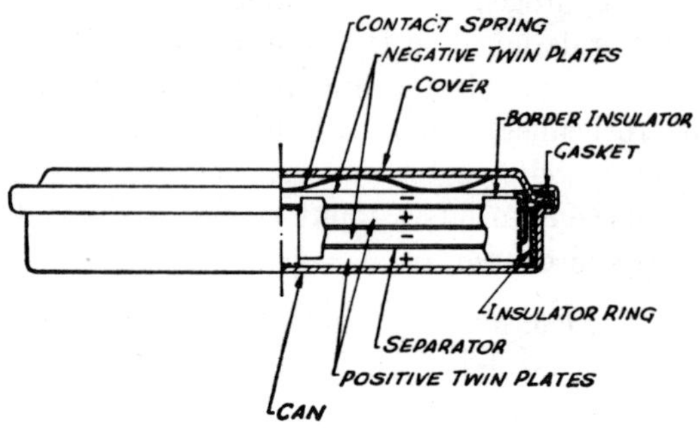

FIG. 33C.—High-Rate Nickel-Cadmium Button Cell.

hour life of a battery is governed by the rate at which it is discharged. If it is permitted to discharge at a very low rate its total ampere-hours of life will probably exceed its rated capacity. If, however, a heavy demand for current is placed upon it, such as when operating an automobile-engine starting motor, its life will be very short if the period of the demand is for any considerable length of time.

Starting batteries for automobiles have rated capacities from 80 to 160 ampere-hours.

Illustration: A battery delivers 5 amperes of current for 22 hours. What is its capacity?

$$\text{Capacity} = 5 \times 22 = 110 \text{ ampere-hours} \quad \text{(Ans.)}$$

Storage Battery Voltage.—The voltage of a lead storage cell when fully charged is about 2.2 volts. During discharge it will give current at a nearly constant pressure of 2 volts. The difference is lost in internal resistance.

Illustration: What voltage will a cell give when discharging 15 amperes of current if its internal resistance is 0.013 ohm and its open circuit electromotive force 2.195 volts?

Volts required to send 15 amperes through a resistance of 0.013 ohm $= 15 \times 0.013 = 0.195$.

$$\text{Terminal voltage} = 2.195 - 0.195 = 2.000 \text{ volts.}$$

Storage cells are connected in series to give batteries which will give higher voltage than a single cell. Three separate cells are commonly made with only one jar, which is provided with partitions that divide it into three separate compartments. The three-cell six-volt battery is the most common in automotive use.

Charging Storage Batteries.—A storage battery may be charged by connecting the positive wire of a direct-current 110-volt circuit to the positive terminal of the battery and the negative wire to the negative terminal, provided that suitable resistances are placed in the circuit to reduce the voltage and control the charging rate. A 6-ampere rate is satisfactory for small batteries and a 10-ampere rate for 100-ampere-hour batteries.

Illustration: A 3-cell battery is to be charged at a 10-ampere rate with a charging voltage of 2.5 volts per cell. What resistance will be required if the battery is being charged from a 110-volt line?

Total charging voltage of battery $= 3 \times 2.5 = 7.5$ volts.

Voltage through external resistances $= 110 - 7.5 = 102.5$ volts. Then, by Ohm's law

$$R = \frac{102.5}{10} = 10.25 \text{ ohms resistance required} \quad \text{(Ans.)}$$

Current from Dry Cells.—*The current which a cell will deliver to a circuit is equal to the voltage divided by the sum of the internal and external resistances.*

ILLUSTRATION: In the circuit shown in Fig. 34a the internal resistance of the cell is 0.1 ohm and the resistance of the bell is 25 ohms. What current flows in the circuit?

$$I = \frac{1.5}{0.1 + 25} = \frac{1.5}{25.1} = 0.060 \text{ ampere} \quad \text{(Ans.)}$$

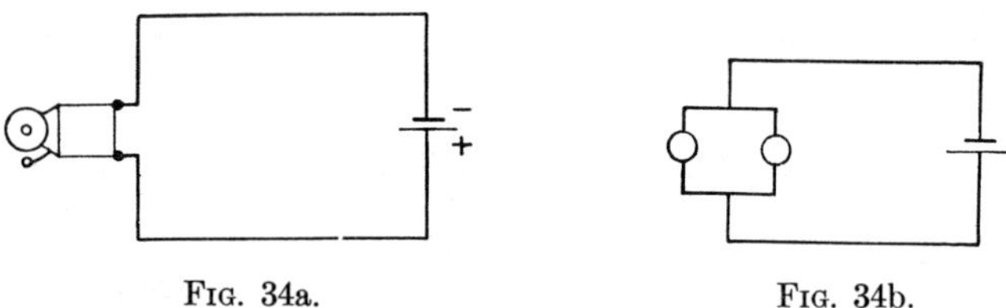

FIG. 34a. FIG. 34b.

ILLUSTRATION: In the circuit in Fig. 34b the internal resistance of a 1.6-volt cell is 0.085 ohm. What current flows in the circuit if the resistance of each light is 5 ohms?

The joint resistance of the lamps by the law of the reciprocal of the sum of the reciprocals of the individual resistances is,

Reciprocal of 5 ohms $= \frac{1}{5}$

Sum of reciprocals $= \frac{1}{5} + \frac{1}{5} = \frac{2}{5}$

Joint resistance is reciprocal of sum $= \frac{5}{2} = 2.5$ ohms

Current in circuit

$$= \frac{1.6}{2.5 + 0.085} = \frac{1.6}{2.585} = 0.619 \text{ ampere} \quad \text{(Ans.)}$$

Voltage from Primary Cells.—The electrical pressure or voltage produced by a primary cell must overcome both the internal resistance and the external resistance. The voltage *rating*, or that usually referred to, is known as the "open circuit voltage" and is that obtained across the terminals when the cell is not delivering current. When delivering current the terminal voltage, that available for external resistance, is equal to the open circuit voltage minus the voltage across the internal resistance.

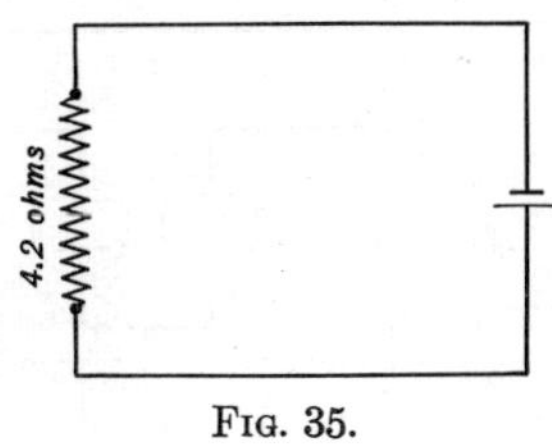

FIG. 35.

ILLUSTRATION: The cell shown in the circuit in Fig. 35 has an open circuit voltage of 1.52 volts and an internal resistance of 0.09 ohm. How many volts are used up by the internal resistance and what is the voltage across the resistance?

By Ohm's Law, total current in the circuit is

$$I = \frac{E}{r + R}. \qquad I = \frac{1.52}{0.09 + 4.2} = \frac{1.52}{4.29} = 0.354 \text{ ampere}$$

Voltage required to force 0.354 ampere, the internal resistance

$$E = Ir. \qquad E = 0.354 \times 0.09 = 0.032 \text{ volt}$$

Therefore, voltage across external resistance =

$$1.52 - 0.032 = 1.498 \text{ volts} \quad \text{(Ans.)}$$

Cells Arranged in Series.—When a number of cells are connected in series and to an external circuit, the current flowing through the external circuit will also pass through each cell. Since each cell has a certain resistance, a portion of the total electromotive force will be used up in overcoming the internal resistance.

To find the total internal resistance of a number of similar cells connected in series:

Multiply the resistance of one cell by the number of cells in the series.

The total internal resistance $= r \times ns$

When r = internal resistance
ns = number of cells in series

ILLUSTRATION: Eight dry cells (Fig. 36) each with an internal resistance of 0.095 ohm are connected in series. What is the total internal resistance?

$$\text{Total resistance} = r \times ns = 0.095 \times 8 = 0.76 \text{ ohm} \quad \text{(Ans.)}$$

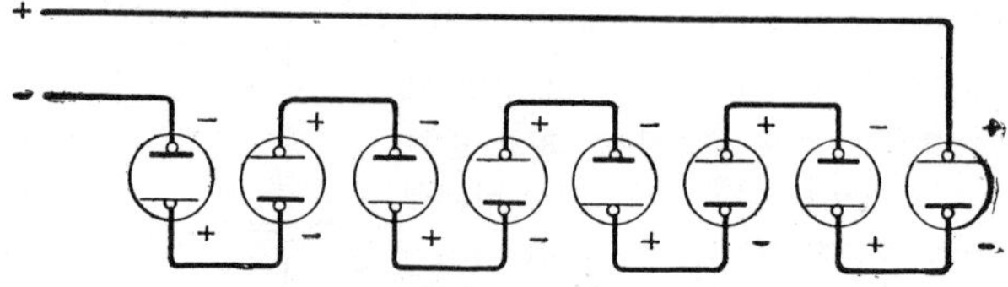

FIG. 36.

Current from Cells in Series.—To find the current that will be maintained in an external circuit by a number of cells in series:

Multiply the electromotive force of one cell by the number connected in series. Find the total internal resistance as above. Then by Ohm's Law the current is equal to the total electromotive force divided by the total resistance.

Expressed as an equation, this rule is,

$$I = \frac{E \times ns}{(r \times ns) + R}$$

When E = electromotive force of one cell

r = internal resistance of one cell

ns = number of cells in series

R = external resistance.

ILLUSTRATION: If the cells shown in Fig. 36 each have an electromotive force of 1.45 volts and an internal resistance of 0.095 ohm, what is the total current if a 15-ohm resistance is connected in the circuit?

$$I = \frac{E \times ns}{(r \times ns) + R} = \frac{1.45 \times 8}{(0\ 095 \times 8) + 15} = \frac{11.60}{15.76} = 0.74 \text{ ampere}$$

(Ans.)

Cells in Parallel.—When a number of similar cells are connected in parallel, as in Fig. 37, and to an external circuit, the total current does not have to overcome the total internal resistance of all the cells as is the case when they are connected in series, but is divided evenly among the cells in parallel. With a path for the current through each cell, the total resistance is much less than that for one cell.

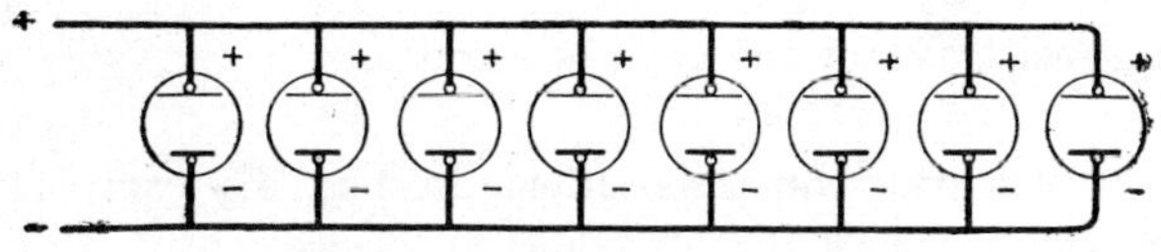

FIG. 37.

To find the internal resistance of a number of cells in parallel: *Divide the resistance of one cell by the number connected in parallel.*

$$\text{Total resistance} = \frac{r}{nq}$$

When r = internal resistance of one cell

nq = number of cells in parallel

ILLUSTRATION: What is the total internal resistance of the cells shown in Fig. 37 if the resistance of each cell is 0.2 ohm?

$$\text{Total resistance} = \frac{r}{nq} = \frac{0.2}{8} = 0.025 \text{ ohm} \quad \text{(Ans.)}$$

Current from Cells in Parallel.—To find the current that will be maintained in an external circuit by a number of cells connected in parallel:

Divide the electromotive force of one cell by the sum of the external and internal resistances of a circuit.

This rule set up as an equation becomes,

$$I = \frac{E}{\dfrac{r}{nq} + R}$$

When E = total electromotive force of one cell

r = internal resistance of one cell

R = total external resistance

nq = number of cells in parallel

The quantity $\frac{r}{nq}$ in this equation will be recognized as the total internal resistance of the cells in parallel.

ILLUSTRATION: The cells shown in Fig. 37, each having an internal resistance of 0.2 ohm and an electromotive force of 1.5 volts, are connected to an external circuit with a total resistance of 12 ohms. What is the total current in the external circuit?

By the above rule,

$$I = \frac{E}{\frac{r}{nq} + R} = \frac{1.5}{\frac{0.2}{8} + 12} = \frac{1.5}{12.025} = 0.125 \text{ ampere} \quad \text{(Ans.)}$$

Advantage of Cells in Parallel Connection.—Cells are connected in parallel when it is desired to obtain the maximum current through an external circuit of low resistance. When cells are connected in parallel their zinc or negative plates are all connected to each other and their carbon or positive elements are also connected to each other. The result is that the group of cells is the equivalent of one large cell, the positive and negative plates of which are equal in area to the sum of the areas of the respective plates in the separate cells. This grouping is, therefore, capable of giving a large quantity of electrical current. When the external resistance is small the strength of the current will be great; when the resistance is large, it will be small.

ILLUSTRATION: If a dry cell with an internal resistance of 0.3 ohm and an electromotive force of 1.5 volts is connected to an external circuit with a total resistance of 0.1 ohm, what will be the resultant flow of current?

$$\text{Current} = \frac{E}{r + R} = \frac{1.5}{0.3 + 0.1} = 3.75 \text{ amperes} \quad \text{(Ans.)}$$

ILLUSTRATION: If eight dry cells with similar characteristics are substituted for the single cell in the above illustration, what is then the current in the external circuit?

$$\text{Current} = \frac{E}{\dfrac{r}{nq} + R} = \frac{1.5}{\dfrac{0.3}{8} + 0.1} = \frac{1.5}{0.1375} = 10.91 \text{ amperes} \quad \text{(Ans.)}$$

Advantage of Cells in Series Connection.—A series grouping of cells is employed when the external resistance is the principal one to be overcome and a maximum current strength in the circuit is desired. The advantage of this type of connection is shown by the following examples.

ILLUSTRATION: A dry cell with an electromotive force of 1.5 volts and an internal resistance of 0.3 ohm is connected to an external circuit with a total resistance of 100 ohms. What current will flow through the circuit?

$$I = \frac{E}{r + R} = \frac{1.5}{0.3 + 100} = \frac{1.5}{100.3} = 0.014955 \text{ ampere} \quad \text{(Ans.)}$$

ILLUSTRATION: What current will flow in the external circuit if ten cells with similar characteristics are connected in parallel in the circuit of the above illustration instead of the single cell?

$$I = \frac{E}{\dfrac{r}{nq} + R} = \frac{1.5}{\dfrac{0.3}{10} + 100} = \frac{1.5}{100.03} = 0.014994 \text{ ampere} \quad \text{(Ans.)}$$

It will be seen, therefore, that there is little to be gained in the amount of current produced by substituting ten cells in parallel for the single cell. However, let us substitute ten cells in series in the next example.

ILLUSTRATION: Substitute ten cells of like characteristics in series connection for the cells in the above example. What will then be the current in the external circuit?

$$I = \frac{E \times ns}{(r \times ns) + R} = \frac{1.5 \times 10}{(0.3 \times 10) + 100} = \frac{15}{103} = 0.14563 \text{ ampere}$$

(Ans.)

It is evident from these illustrations that nearly ten times the current from cells in parallel connection passes through the circuit when the same cells are connected in series.

Cells Grouped in Parallel-Series.—It is sometimes desirable to group cells in a combination of series and parallel to give either the

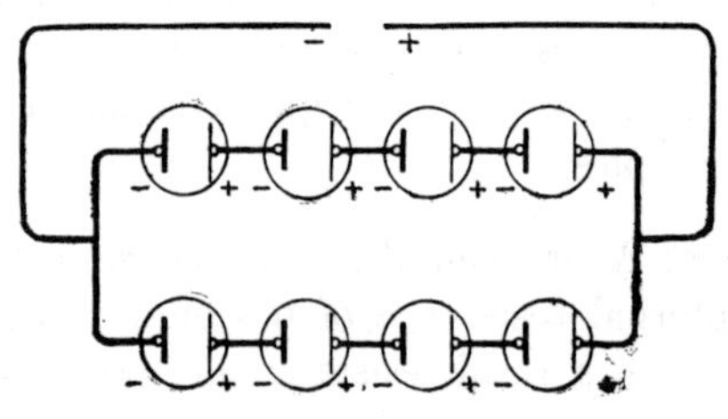

FIG. 38.

maximum current through an external resistance or to increase the capacity of the cells for maintaining a current in a circuit for a long period of time. Figure 38 shows such a connection consisting of two parallel sets of four cells in series. This is sometimes called a *multiple-series* combination.

If 6 volts are required to light a small lamp, four dry cells of 1.5 volts each connected in series will produce (neglecting internal resistance) the required 6 volts and will operate the lamp for a period of possibly 4 hours. If, however, eight cells are connected in parallel series as in Fig. 38 the total electromotive force will still be 6 volts, but the lamp will now be illuminated for a period of 8 hours.

To find the internal resistance of any multiple-series combination of cells:

Multiply the resistance of one cell by the number of cells in one group and divide the product by the number of groups in parallel or multiple.

$$\text{The total resistance} = \frac{r \times ns}{nq}$$

when, r = resistance of one cell

ns = number of cells in series in one group

nq = number of groups in parallel

ILLUSTRATION: What is the internal resistance of the combination of eight cells shown in Fig. 38 if the resistance of each cell is 0.2 ohm?

$$\text{Total resistance} = \frac{r \times ns}{nq} = \frac{0.2 \times 4}{2} = 0.4 \text{ ohm} \quad \text{(Ans.)}$$

Current Strength from Cells in Parallel-Series Combinations.— To find the current that will be maintained in an external circuit by any parallel-series combination of cells:

Divide the total electromotive force of one series group by the sum of the combined internal and external resistances.

Expressed as an equation, this rule becomes

$$I = \frac{E \times ns}{\dfrac{r \times ns}{nq} + R}$$

when, I = current in the external circuit

E = electromotive force of one cell

ns = number of cells in series in one group

nq = number of groups in parallel

r = internal resistance of one cell

R = external resistance

Illustration: Fifteen cells are so connected that five cells are in series and three sets of five are in parallel. What current will flow through an external circuit connected to these cells if the total external resistance is 8 ohms and the electromotive force of each cell is 1.5 volts and the internal resistance 0.1 ohm?

$$I = \frac{E \times ns}{\frac{r \times ns}{nq} + R} = \frac{1.5 \times 5}{\frac{0.1 \times 5}{3} + 8} = \frac{7.5}{0.167 + 8} = 0.918 \text{ ampere}$$

(Ans.)

Groups of cells in parallel may also be connected in series as shown in Fig. 39. This is called a series-parallel or a series-multiple connection.

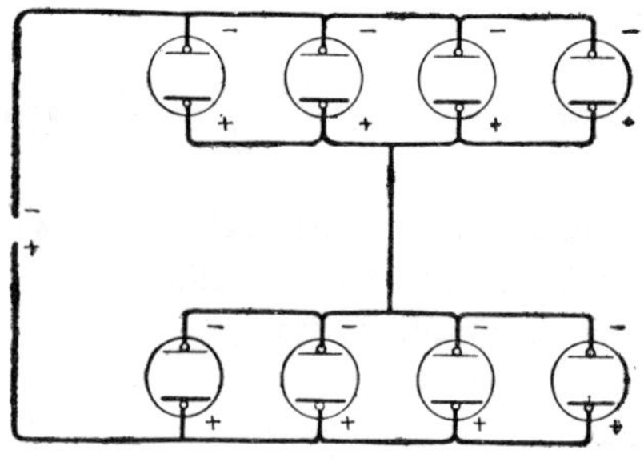

Fig. 39.

To find the current that will be maintained in an external circuit from any series-parallel combination of cells, several progressive steps are necessary as follows:

Find the internal resistance of one parallel group and consider the result as data for one "equivalent" cell (group). Calculate the total electromotive force and resistance for the parallel groups in series and determine the current by Ohm's Law.

Illustration: The series-parallel combination shown in Fig. 39 is connected to a circuit with an external resistance of 2 ohms. The cells are four in parallel, two groups in series. Each has an electromotive force of 1.5 volts and an internal resistance of 0.1 ohms. What current will flow through the circuit?

The electromotive force of 1 group = 1.5 volts

The electromotive force of 2 groups in series = 1.5 × 2 = 3.0 volts

$$\text{The internal resistance of 1 group} = \frac{r}{nq} = \frac{0.1}{4} = 0.025 \text{ ohm}$$

The internal resistance of 2 groups in series

$$= r \times ns = 0.025 \times 2 = 0.05 \text{ ohm.}$$

By Ohm's Law

$$I = \frac{E}{r + R} = \frac{3}{0.05 + 2} = \frac{3}{2.05} = 1.463 \text{ amperes} \quad \text{(Ans.)}$$

Secondary Cells.—Primary cells (dry cells) produce electric currents as a result of chemical action. Secondary cells do not in themselves produce current but have the property of "storing" electric current with which they may be charged and will later give up the current which has been accumulated. Such cells are called *storage cells* or *accumulators.* When these cells are connected in groups of two or more, the group is called a *storage battery.*

Storage batteries are widely used for stand-by emergency service in power substations and in telephone and telegraph work. However, to a great majority of people the storage battery connotes an ignition unit of the automobile or internal combustion engine. For this reason a full treatment of this subject will be found on page 708 of this book.

ELECTROMAGNETS

Magnetization of Iron and Steel by an Electric Current.—When a number of turns of insulated wire are wound around a soft iron bar and a current is sent through the wire, the bar will attract iron filings. This property is called *magnetism.* The wire wound around the iron core is called an *electromagnet.* An electromagnet differs from a permanent iron magnet in that it has magnetic properties only when a current flows through the wire. If a piece of cardboard is fitted around the longitudinal axis of an electromagnet and iron filings sprinkled around generously (Fig. 40), the filings will not only be attracted by both ends of the magnet, called the *poles*, but will also arrange themselves in a regular order at some distance from the coil. The lines which these filings form represent the *lines of force* of the *magnetic field* about the magnet. If a piece of iron is laid in the magnetic field the lines of force will converge to it at both ends. It is apparent, therefore, that the lines of force find it easier to pass through the piece of iron than through the air or through the filings. *The capability of any substance for conducting magnetic lines of force is termed its permeability.*

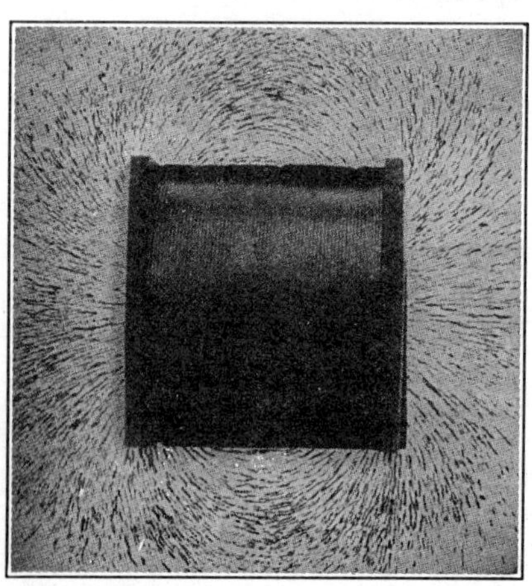

FIG. 40.—Lines of Force Around a Coil.

Solenoids.—A coil of wire wound on an insulating spool is called a *solenoid.* The winding is always in the same direction, layer upon layer, similar to the thread on a spool. If a solenoid is suspended by a thread from the midpoint of its longitudinal axis it will swing into a position with one end pointing north and the other end south. The pole to the north, or the north-seeking-pole is called the N-pole and the opposite end the S-pole. It is a phenomenon of magnetism that unlike poles of magnets attract each other and like poles repel each other.

A solenoid with an iron core is, as we have seen, an *electromagnet.* If the core is fitted loosely so that it may be pulled out it will be subjected to a strong "sucking" action when a current is passed through the coil (Fig. 42). This principle is extensively employed to operate the feeding mechanism of electric arc lights, to close switches at a distance for remote control purposes, and in automatic circuit breakers.

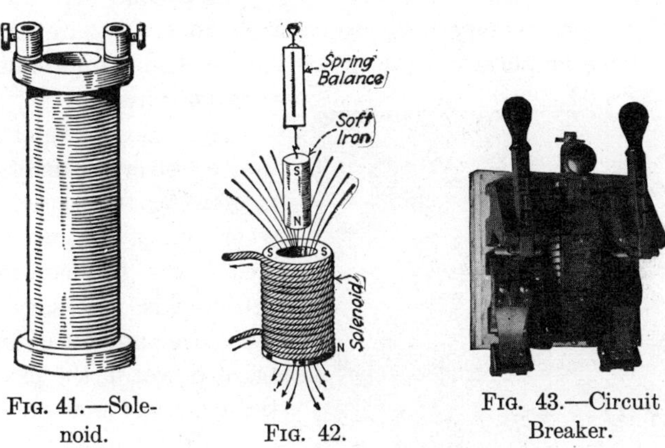

Fig. 41.—Solenoid.

Fig. 42.

Fig. 43.—Circuit Breaker.

A circuit breaker is used to protect electrical circuits against abnormal conditions arising therein. The most common form is the overload type which opens the circuit when the current becomes excessive. Circuit breakers are also used for opening the circuit if the voltage falls below a certain value or if the polarity of a direct-current circuit is reversed.

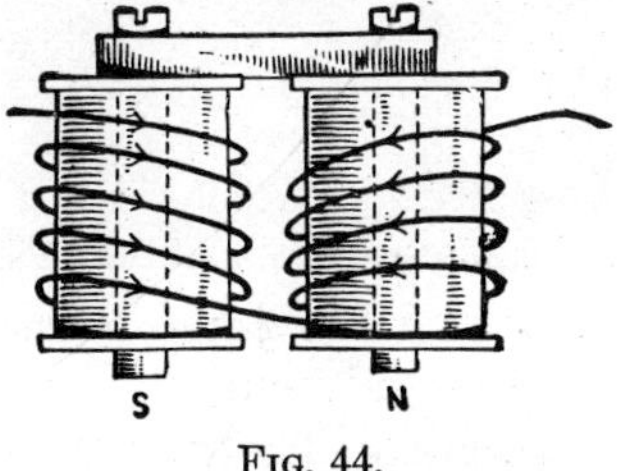

Fig. 44.

Applications of the Electromagnet.—If, instead of winding a coil around a straight bar, a bar in the form of a horseshoe is used, bringing the N-pole and S-pole close together, a much stronger magnet will result. In actual practice, the wire is not wound onto the bar but is wound onto spools which are slipped over the bar. The bar need not be in one piece and is commonly made up of three pieces as shown in Fig. 44. These are called the pole pieces and the yoke. The electromagnet finds many applications in this form in electric bells, buzzers, telegraph sounders and relays, etc. Electromagnets of very powerful attractions are built for industrial use in handling iron and steel. These consist of a steel casting having a groove turned to receive the exciting coil. The lifting power of an electromagnet is proportional to the square of the product of the amperes flowing in the magnet and the number of turns of wire.

FIG. 45.—Lifting Magnet.

The most important use of electromagnets is in generators and motors, where they are used to create the intense magnetic fields necessary for the development of electrical power in the case of the generator and the rotation of the armature in the case of a motor.

Magnetomotive Force.—Magnetism or *magnetic flux* (total number of lines of force) depends upon the number of turns of wire in the coil of an electromagnet as well as upon the current strength; the current and number of turns being jointly responsible for the force that drives the magnetic flux around the magnetic circuit, just as an electromotive force drives an electric current around an electric circuit. The magnetizing force set up by a current flowing through a solenoid or any coil of wire is called the *magnetomotive force* (abbreviated mmf). It is directly proportional to the current and to the number of turns on a solenoid. The magnetomotive force is, therefore, proportional to the product of the number of turns and the current strength. That is, one ampere flowing through ten coils or turns will produce the same magnetomotive force as ten amperes flowing through one turn. The magnetomotive force may be expressed in a unit called the *ampere-turn.* The relationship may be expressed by the formula,

$$\text{mmf in ampere-turns} = I \times T$$

when I = current in amperes
T = number of turns on the coil.

ILLUSTRATION: What is the magnetomotive force of a coil with 50 turns through which a current of 3 amperes is passing?

$$\text{mmf} = I \times T = 3 \times 50 = 150 \text{ ampere-turns (Ans.)}$$

It is evident from this relationship that a magnet with a certain magnetomotive force can be made with heavy wire of low resistance and few turns or with smaller wire of high resistance and many turns. Electric bell, telephone, and telegraph instruments are usually made of fine wire since they are usually located some distance from the battery so that the current may be very small. When it is desired to operate a small magnet on a 110-volt circuit it is wound with fine wire so that its resistance will be high and the current consumed small.

Field Intensity.—In magnetic calculations, the magnetomotive force per unit length of the magnetic circuit is called the *intensity of the magnetic field.* This field intensity is the magnetomotive force divided by the length (l) of the magnetic path and is represented by the letter $\mathscr{H}$. It has been determined experimentally that one ampere-turn will produce 1.257 lines of force through an air-path one centimeter in length and one square centimeter in cross-sectional area.

Therefore, the field intensity is,

$$\mathscr{H} = \frac{\text{mmf}}{l} = \frac{1.257 \times I \times T}{l}$$

where l is the length of the path in centimeters and T the number of turns.

If the length (l) of the magnetic path of a solenoid is known, the mmf necessary to produce a desired field intensity ($\mathscr{H}$), is obtained by multiplying $\mathscr{H} \times l$.

Illustration: The coil shown in Fig. 46 has a core which forms a complete ring so that there are no free poles. Each line of force has then a complete path inside the core so that the

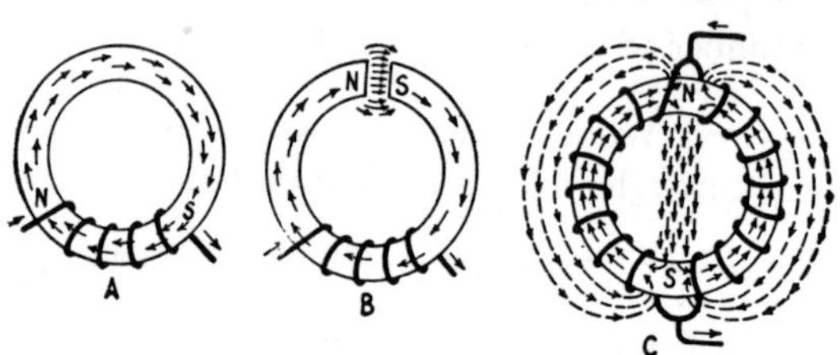

Fig. 46.—Magnetic Polarity of an Iron Ring.

length of the magnetic circuit can easily be measured. If the coil has 30 turns and the current is 15 amperes then

$$\text{mmf} = I \times T = 15 \times 30 = 450 \text{ ampere-turns.}$$

If the mean length of the magnetic circuit is 18 centimeters, then the magnetomotive force per centimeter length is

$$\mathscr{H} = \frac{1.257 \times I \times T}{l} = \frac{1.257 \times 15 \times 30}{18} = 31.4$$

This means that a uniform magnetic field is produced in the solenoid of 31.4 lines per square centimeter.

The difference between the two formulas which have been given should be kept distinctly in mind. The quantity $\mathscr{H}$ represents the force magnetizing a unit length of the core of a solenoid, or the strength of field in lines of force per square centimeter within a coil with an air coil. The quantity mmf represents the force (magnetic pressure) that tends to drive the lines of force throughout the entire path of any kind of material.

If l is given in inches, then $\mathscr{H}$ becomes

$$\mathscr{H} = \frac{.495 \times I \times T}{l}, \text{ in which } l, \text{ is in inches.}$$

Law of the Magnetic Circuit.—Just as electric pressure (emf) is the force that moves electricity through an electric circuit, so magnetic pressure (mmf) drives lines of force through a magnetic circuit. All magnetic substances offer more or less resistance to the passage through them of magnetic lines of force. This magnetic "resistance" is called *reluctance* and its symbol is $\mathscr{R}$. The *total number* of lines of force set up in a magnetic substance is termed the *magnetic flux*. Magnetic flux, or total number of lines of force, is treated as a *magnetic* current flowing in a magnetic circuit.

The calculation of the magnetic flux, which will be represented by N, is similar to the calculation of current in an electric circuit by Ohm's Law. In the latter case, the strength of the electric current equals the electromotive force divided by the resistance, or, $I = \frac{E}{R}$. Similarly, in a magnetic circuit

$$\text{magnetic flux} = \frac{\text{magnetomotive force}}{\text{reluctance}}$$

or,

$$N = \frac{\text{mmf}}{\mathscr{R}}$$

Magnetic Density, Permeability and Reluctance.—It is sometimes necessary to specify the flux density in any part of a magnetic circuit, that is, the number of lines passing through a unit area measured at right angles to their direction, whether that part of the circuit is air or some other substance. This number is termed the *magnetic density* or *magnetic induction* of a substance and is denoted by the letter $\mathscr{B}$. If the total flux N is known, and the area A through which it is uniformly distributed, is also known, then the flux density is

$$\mathscr{B} = \frac{N}{A}$$

If A is expressed in square inches, then the flux density will be in number of lines per square inch.

The magnetic density produced in *air* by a solenoid depends upon the magnetic field alone. The magnetic density or induction $\mathscr{B}$ produced in a *magnetic* substance when placed in a solenoid depends also upon the *permeability* of the substance.

The permeability of a magnetic substance is the ratio of the magnetic density $\mathscr{B}$ in the substance to the intensity of the magnetic field $\mathscr{H}$ acting upon the substance; that is a ratio of the number of lines of force per unit area, set up in the material, to the number that would be set up in air under the same conditions. The symbol for permeability is the Greek letter μ (pronounced mu), and its value for any magnetic substance is expressed in the equation

$$\mu = \frac{\mathscr{B}}{\mathscr{H}}$$

If the value of μ and $\mathscr{H}$ are known, the magnetic density is

$$\mathscr{B} = \mu \times \mathscr{H}$$

The permeability of air or nonmagnetic substances is unity or 1; since through air the flux density $\mathscr{B} = \mathscr{H}$, or $\dfrac{\mathscr{B}}{\mathscr{H}} = 1$.

Soft iron under a field intensity of $\mathscr{H} = 10$ (this corresponds to 20.3 ampere-turns per inch) has a flux density $\mathscr{B} = 14{,}000$ lines per square centimeter. Consequently, the permeability is

$$\mu = \frac{\mathscr{B}}{\mathscr{H}} = \frac{14{,}000}{10} = 1400$$

In magnetic materials, the value of the permeability does not remain the same for all flux densities. It varies as shown in Table 1 below:

TABLE 1

Flux Density and Permeability

Flux Density		Permeability		
Lines per square inch	Lines per square centimeter	Annealed sheet steel	Cast steel	Cast iron
20,000	3,100	2600	1400	280
30,000	4,650	2900	1500	230
40,000	6,200	3100	1400	160
50,000	7,750	3200	1350	110
60,000	9,300	3100	1250	80
70,000	10,850	2400	1100	65
80,000	12,400	1800	750	50
90,000	14,000	1400	500	
100,000	15,500	750	280	
110,000	17,400	320	145	
120,000	18,600	160	70	
130,000	20,150	75		

The reluctance of a magnetic circuit depends upon three quantities: the *length* of the circuit, the cross-sectional *area* of the circuit, and the *permeability* of the material of the circuit. The reluctance *increases* as the length of the magnetic circuit increases, and *decreases* as the cross-sectional area is increased and the permeability increases. That is, the reluctance is directly proportional to the length of the magnetic circuit, is inversely proportional to the cross-sectional area and varies as the material of the circuit. This may be expressed by the following formula:

$$\mathscr{R} = \frac{l}{A \times \mu}$$

when $\mathscr{R}$ represents the reluctance, l the length of the magnetic circuit in inches, A the sectional area of the circuit in square inches, and μ the permeability of the material constituting the circuit.

Attractive Force of an Electromagnet.—The magnetism of an electromagnet increases as the current through it is increased, up to a saturation point, but is not directly proportional to the current; that is, if when one ampere is passed through a certain magnet, a force of 56 pounds is required to detach its keeper, then when two amperes are passed through it, not twice the force, or 112 pounds is required, but usually much less.

The lifting or adhesive power of an electromagnet is called its *tractive force.* The tractive force is proportional to the square of the density of lines of force per square inch, and the area of surface contact. To determine the tractive force or " pull " in pounds of an electromagnet, let

$\mathscr{B}$ = flux density or lines of force per square inch.
A = area of contact in square inches.

Then, the pull in pounds is

$$P = \frac{\mathscr{B}^2 \times A}{72{,}134{,}000}$$

ILLUSTRATION: What is the tractive force of a magnet if the density of the lines of force per square inch is 96,750 and the area of contact is one square inch?

$$P = \frac{\mathscr{B}^2 \times A}{72{,}134{,}000} = \frac{(96{,}750)^2}{72{,}134{,}000} \times A$$

$$P = \frac{9{,}360{,}562{,}500}{72{,}134{,}000} = 129.7 \text{ lb.} \quad \text{(Ans.)}$$

Table 2 gives the traction of electromagnets for various degrees of magnetizations.

GENERATORS AND MOTORS

Dynamo.—A dynamo is a machine which converts either mechanical energy into electrical energy or electrical energy into mechanical energy. A dynamo which converts mechanical energy into electrical energy is called a *generator.* A dynamo which converts electrical energy into mechanical energy in the form of rotation is called a *motor.*

TABLE 2

MAGNETIZATION AND TRACTION OF ELECTROMAGNETS

$\mathscr{B}$ Lines per Sq. Cm.	$\mathscr{B}''$ Lines per Sq. Inch	Dynes per Sq. Cm.	Grammes per Sq. Cm.	Kilogs per Sq. Cm.	Pounds per Sq. Inch.
1,000	6,450	39,790	40.56	.04056	.577
2,000	12,900	159,200	162.3	.1623	2.308
3,000	19,350	358,100	365.1	.3651	5.190
4,000	25,800	636,600	648.9	.6489	9.228
5,000	32,250	994,700	1,014	1.014	14.39
6,000	38,700	1,432,000	1,460	1.460	20.75
7,000	45,150	1,950,000	1,987	1.987	28.26
8,000	51,600	2,547,000	2,596	2.596	36.95
9,000	58,050	3,223,000	3,286	3.286	46.72
10,000	64,500	3,979,000	4,056	4.056	57.68
11,000	70,950	4,815,000	4,907	4.907	69.77
12,000	77,400	5,730,000	5,841	5.841	83.07
13,000	83,850	6,725,000	6,855	6.855	97.47
14,000	90,300	7,800,000	7,550	7.550	113.1
15,000	96,750	8,953,000	9,124	9.124	129.7
16,000	103,200	10,170,000	10,390	10.390	147.7
17,000	109,650	11,500,000	11,720	11.720	166.6
18,000	116,100	12,890,000	13,140	13.140	186.8
19,000	122,550	14,360,000	14,630	14.630	208.1
20,000	129,000	15,920,000	16,230	16.230	230.8

The electric generator operates on the principle of *electromagnetic induction.* This principle is illustrated in Fig. 47. A loop of wire revolving between the two poles of a magnet cuts the magnetic lines of force. This sets up an electromotive force in the loop and causes a current to flow around it. In the position *ABCD* (Fig. 47) the loop cuts no lines of force and therefore no current is induced. However, during the first quarter of the revo-

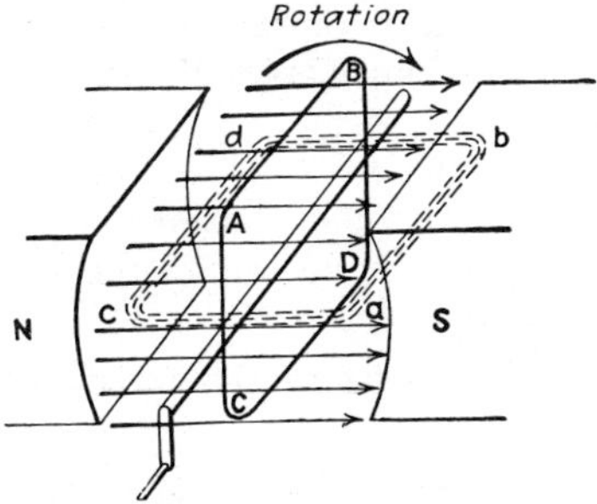

FIG. 47.—Direction and Magnitude of the Induced emf in a Generator.

lution from this point the lines of force are cut at a gradually increasing rate till the loop is in the position *abcd* when the rate of change, and also the electromotive force is a maximum. During the next quarter revolution the cutting of the lines of force gradually decreases until at the end of a half revolution the electromotive force is again zero. During the course of this half revolution the current flows in only one direction, from *a* to *c*, to *d*, to *b*, but the strength has constantly changed from zero to a maximum and back to zero again. During the second half revolution, the same variations in electromotive force occur but the induced current is in the opposite direction. The current is, therefore, reversed twice in every revolution, or an *alternating current flows around the loop.*

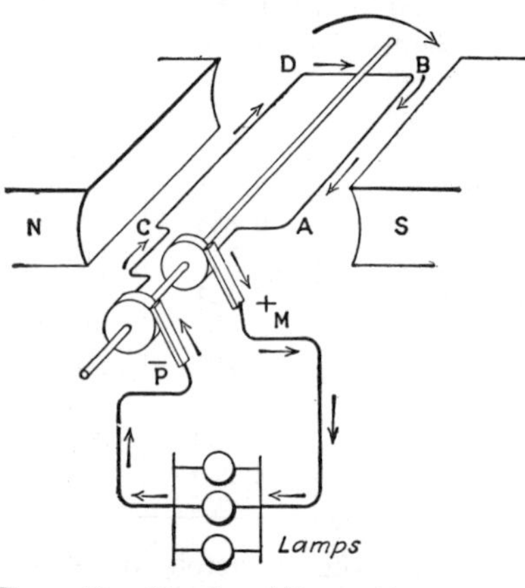

Fig. 48.—Simple Alternating-current Generator. At the instant depicted in the revolution, brush *M* is positive.

Simple Alternating Current Generators.—In order to use in an external circuit the current generated in the revolving loop it is necessary to employ a connecting device consisting of two *collector rings* and brushes insulated from the shaft and from each other. Figures 48 and 49 show the elements of an alternating current generator and the two positions of the loop illustrate the reversal of current in the circuit.

The magnets between which the loop revolves are called the *field magnets* or simply the *field.* The revolving loop is called the *armature.*

The electromotive force produced by a generator depends upon:

1. *The number of lines of force cut by the armature wires.*
2. *The number and length of the cutting wires,*

3. *The speed at which the armature revolves and the lines of force are cut.*

It is apparent, therefore, that if instead of the single loop shown in Fig. 48 an iron core with many turns of wire is substituted, the lines of force between the field magnets will be in-

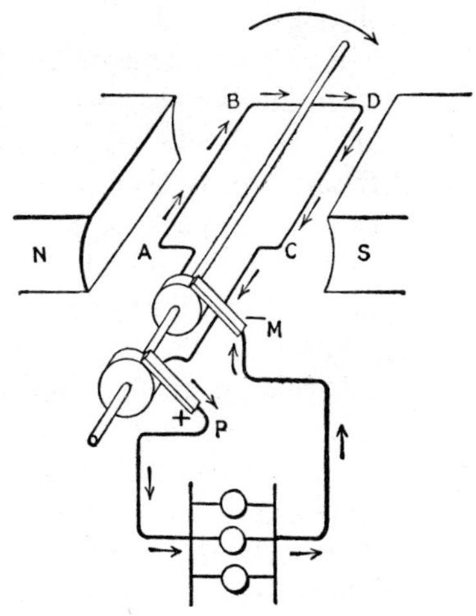

FIG. 49.—Simple Alternating-current Generator. Direction of current in coil at one-half revolution from the positive in Fig. 48; brush *M* is now negative.

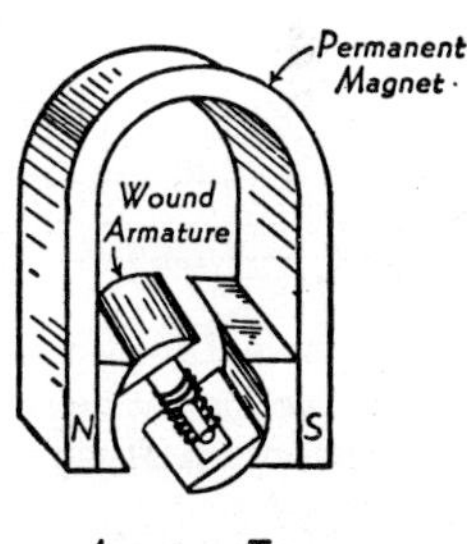

FIG. 50.—Principle of Armature-type Magneto.

creased and the number of wires cutting these lines will be increased. The result is that the electromotive force is greatly increased. The *magneto generator* (Fig. 50) is constructed on this principle. It consists of a coil of wire revolving between permanent magnets.

Simple Direct Current Generators.—In order to obtain current flowing in only one direction from a generator, it is necessary to intercept the current from the revolving loop in such a manner that the electromotive force generated by each half revolution is transmitted to separate branches of the external circuit. This is accomplished by substituting one split ring for the two collector rings as shown in Fig. 51. This split ring is called the *commutator*. Brushes rest on the ring at diametrically opposite points, one having a *positive* polarity and the other *negative*.

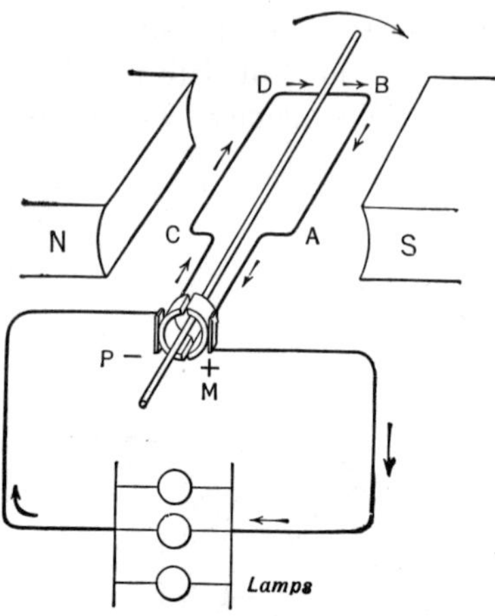

Fig. 51.—Simple Direct-current Generator. At the instant depicted in the revolution, brush M is positive.

Principle of the Motor.—If an electric current is passed through a coil or a loop it will create a magnetic field with an N-pole on one side and an S-pole on the other. If this loop is then placed between the poles of a magnet as in Fig. 52, it will tend to turn until its lines of force are in line with the lines of force of the field magnets. When it reaches this point the rotation stops. In order to obtain continuous rotation it is necessary to reverse the current in the loop at the instant that the turning effect ceases. These reversals are automatically performed by the commutator when the brushes are correctly set and adjusted.

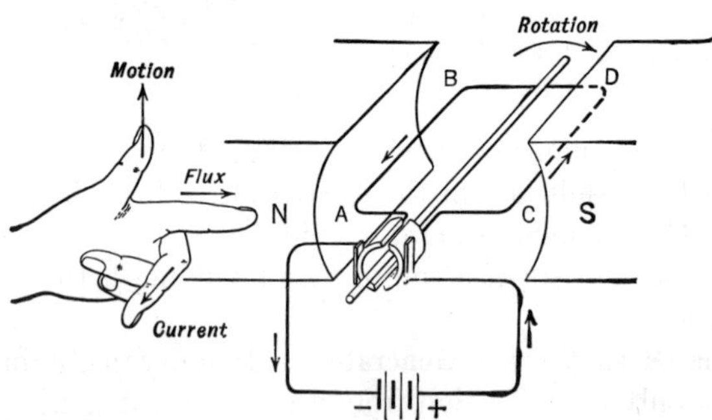

Fig. 52.—Single Loop Armature Driven as a Motor.

The direction of rotation of a motor can be found by the left-hand rule as illustrated in Fig. 52. When the polarity of the field magnets and the direction of the current through the armature have been determined, place the left hand so that the fingers correspond with the polarity and direction of current in the single armature coil motor, and it is found that the loop will rotate in the direction of the hands of a clock. The direction of rotation of a motor can be changed by reversing the current either through the armature or through the fields, but not through both.

Classification of Dynamos According to Their Field Excitation.—Practical dynamos are different in several respects from the elemental forms which have been discussed in the preceding paragraphs. Instead of the revolving loop, the armature consists of a number of coils; instead of a split ring, the commutator consists of a number of segments or sections; and instead of permanent magnets, the field consists of electromagnets. The field magnets may be magnetized by current from a separate generator or by the machine itself and the generator would be styled a *separately-excited* or a *self-excited* generator, respectively. Generators may be classified according to methods used to excite the field magnets as follows:

(*a*) *Magneto Machines* (Fig. 50).—The field magnets are permanent magnets of horseshoe form and the armature is designed for either direct or alternating current. Such machines supply limited power and are used chiefly in gasoline engine work, telephone signalling, testing of circuits, and firing electric blasting detonators.

(*b*) *Series Machines* (Fig. 53) (*Constant Current*).—The field magnets are connected in series with the armature and wound with a few turns of heavy wire having a low resistance, so as to present little opposition to the main current flowing through them. Series generators are used only for series arc street-lighting circuits and in the Thury system of high-voltage direct-current power transmission.

In a *constant-current* circuit supplied by a series generator, the current is maintained constant through the external circuit while the electromotive force varies with each change in the resistance of the circuit. The series constant-current generator is now little used.

(c) *Shunt Machines* (Fig. 54) (*Constant Potential.*)—The field magnets are connected in parallel or shunt with the armature and are wound with many turns of small wire; they have a high resistance, compared with the armature, since only a small portion of the current need flow through them.

(d) *Separately-excited Machines* (Figs. 55 and 56) (*Constant*

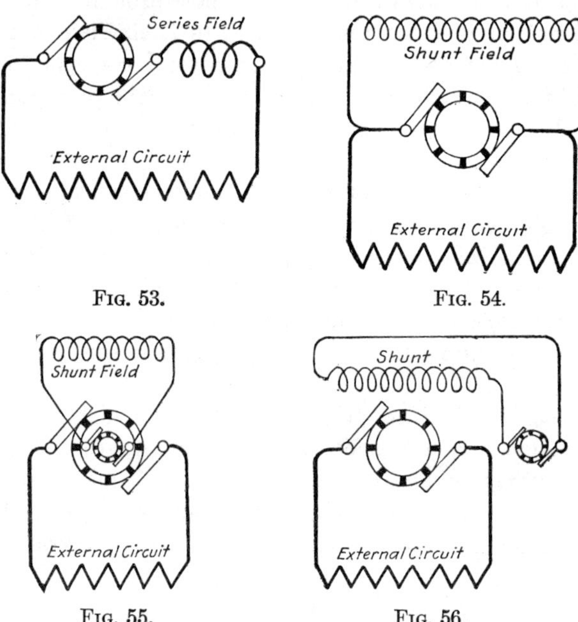

Fig. 53.

Fig. 54.

Fig. 55.

Fig. 56.

Figs. 53–56.—Classification of Generators according to the Method of Exciting the Field Magnets.

Potential).—Current for the field magnets is supplied from a separate generator. In Fig. 55 this generator forms a part of the main machine by having a separate armature on the same shaft, while in Fig. 56 the field is supplied by a distinct machine called an *exciter*.

(e) *Compound Short-shunt Machines* (Fig. 57) (*Constant Potential*).—The field cores contain two independent spools. One is und with a few turns of heavy wire, forming the *series coil,* d connected in series with the main circuit; the other with a eat many turns of smaller wire, forming the *shunt coil,* and nnected in shunt with the armature.

(*f*) *Compound Long-shunt Machines* (Fig. 58) (*Constant Potenl*).—The same as (*e*) except that the shunt field bridges not

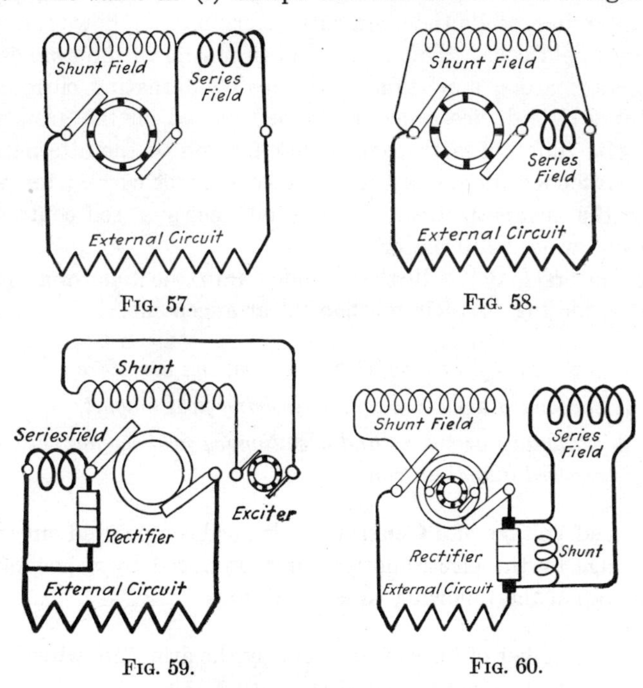

FIG. 57.

FIG. 58.

FIG. 59.

FIG. 60.

FIGS. 57–60.—Classification of Generators according to the Method of Exciting the Field Magnets.

nly the armature but also the series field; hence it is called a *ong shunt.*

(*g*) *Separately-excited Alternating-current Generators* (Figs. 55 nd 56).—The field magnets are excited by direct current from separate exciter. Alternating current generators, or alternaors, always require an exciter, since the alternating current cannot be employed to excite the fields. The exciter may be eithe a separate generator or an independent direct-current windin upon the alternator shaft.

(*h*) *Compound Separately-excited Alternating-current Genera tors* (Fig. 59).—Two independent field windings correspond t the series and shunt coils of Fig. 57. The shunt coil is supplie from an exciter, while the main current, commuted, flows throug the series field coils. This method is employed in the *composite wound* alternators, a portion of the main alternating current i commuted by a device called a *rectifier*, located on the armatur shaft. Its function is to change that portion of the alternatin current intended for the series coils into a direct current for pro ducing the magnetization. Figure 60 shows a self-containe composite wound alternator.

Generators may be further divided into the following thre classes according to their mechanical arrangement:

1. *A stationary field magnet and a revolving armature,*
2. *A stationary armature and a revolving field magnet,*
3. *A stationary armature and a stationary field magnet, betwee which is revolved a toothed iron core.*

Induced Voltage of a Generator.—It has been pointed out tha the voltage or the electromotive force produced by a generato depends upon the following three conditions:

1. The number of lines of force cut by the armature wires,
2. The number and length of the cutting wires,
3. The speed at which the armature revolves.

It has been determined experimentally that an electromotive force of one volt is generated when one turn of wire cuts 100,000,000 (usually written 10^8) lines of force in one second. The induced voltage is then the product of the number of lines of force, or flux, and the number of times these are cut by the wire in one second, divided by 10^8.

ILLUSTRATION: How many volts are generated in a wire which cuts 4,000,000 lines of force 1,200 times a minute?

The rate of cutting is $\frac{1{,}200}{60} = 20$ times a second.

Then,

$$\text{Induced voltage} = \frac{4{,}000{,}000 \times 20}{100{,}000{,}000} = 0.8 \text{ volt} \quad \text{(Ans.)}$$

From these relationships it is possible to develop the following formula for the volts developed in the armature of a generator when the number of poles is the same as the number of paths through the armature:

$$E = \frac{CNR}{10^8}$$

when

E = generated electromotive force in volts
C = the number of active armature conductors
N = the flux per pole
R = the speed of the armature in revolutions per second.

ILLUSTRATION: What voltage is generated by a dynamo having 175 active conductors on its armature if the flux per pole is 4,000,000 lines and the speed of rotation 1500 rpm?

In this case, $C = 175$, N = 4,000.000, and $R = \frac{1500}{60}$

Then,

$$E = \frac{CNR}{10^8}$$

$$E = \frac{175 \times 4{,}000{,}000 \times 1500}{100{,}000{,}000 \times 60}$$

$$E = 175 \text{ volts} \quad \text{(Ans.)}$$

Illustration: An armature generates 220 volts of electromotive force when rotating at a speed of 1200 rpm. What is the flux per pole if there are 250 active armature conductors?

In this case, $E = 220$, $C = 250$, $R = \dfrac{1200}{60} = 20$

The formula $E = \dfrac{CNR}{10^8}$ may be transposed to

$$N = \frac{E \times 10^8}{CR}$$

Substituting known values,

$$N = \frac{220 \times 100{,}000{,}000}{250 \times 20}$$

$$N = 220 \times 20{,}000$$

$$N = 4{,}400{,}000 \quad \text{(Ans.)}$$

Action of a Shunt Generator.—Since a part of the current generated by a shunt generator is used to energize the field magnets, the voltage in the external circuit is something less than the induced electromagnetic force. If the potential of the external circuit measures 112 volts, the induced electromotive force will be $112 + I \times r$, where I equals the current through the fields and r equals the armature resistance.

A field rheostat is used to adjust the voltage in the external circuit. If this is set with the main circuit open so that the voltage will be, for example, 112 volts and the switch is closed so that more current flows from the armature, the voltmeter will at once indicate a lower potential of about 108 volts. If the speed is the same as before, the loss is due to two causes: first, there is an increased drop in the armature due to the additional current flowing through it, which lowers the potential difference at the brushes; second, the potential difference at the brushes being lowered, less current flows around the field so that there are not quite as many lines of force as before.

A statement of the voltage of a generator at no load and when carrying full load is spoken of as its *voltage regulation.* The *percentage regulation* is the ratio of the change in voltage between no-load and full-load to the voltage at full-load.

$$\% \text{ voltage regulation} = \frac{(\text{no-load voltage}) - (\text{full-load voltage}) \times 100}{\text{full-load voltage}}$$

ILLUSTRATION: The voltage of a shunt generator when operating at no load is 112 and when operating at full load is 108. What is its voltage regulation?

$$\text{Percent regulation} = \frac{112 - 108}{108} = \frac{4}{108} = 0.037 = 3.7 \text{ percent} \quad \text{(Ans.)}$$

Shunt generators are adapted only to installations where the load is fairly constant, when they require very little attention after the proper adjustment of the field rheostat has been made.

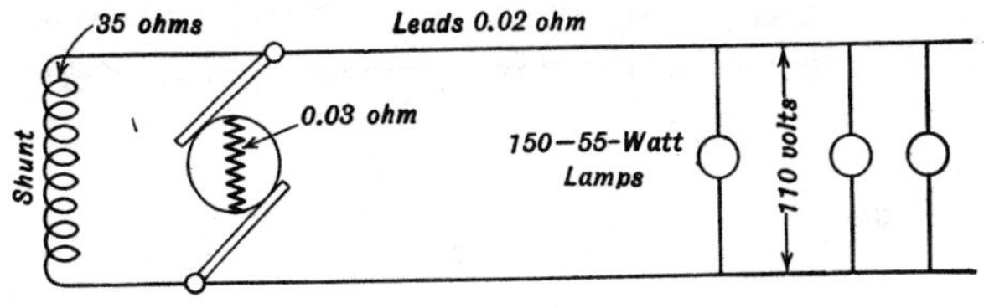

FIG. 61.

SHUNT GENERATOR PROBLEM: A shunt generator, Fig. 61, maintains 110 volts across 150 incandescent lamps joined in parallel, requiring 55 watts and 110 volts each. The lamps are located a distance from the generator and the resistance of the leads is 0.02 ohm. Resistance of the armature is 0.03 ohm and of the field coils is 35 ohms.

1. What is the potential difference at the brushes?

$$I = \frac{P}{E} = \frac{55}{110} \times 150 = 75 \text{ amperes for lamps}$$

$$E = I \times R = 75 \times 0.02 = 1.5 \text{ volt drop in leads}$$

$$110 + 1.5 = 111.5 \text{ volts potential difference at brushes (Ans.)}$$

2. What is the total electromotive force generated?

$$I = \frac{E}{R} = \frac{111.5}{35} = 3.19 \text{ amperes through the fields}$$

$$75 + 3.19 = 78.19 \text{ amperes through armature}$$

$$E = I \times R = 78.19 \times 0.03 = 2.35 \text{ volts drop in armature}$$

$$111.5 + 2.35 = 113.85 \text{ volts total emf (Ans.)}$$

3. What are the watts lost in the armature?

$$P = E \times I = 2.35 \times 78.19 = 183.7 \text{ watts lost in armature (Ans.)}$$

4. What are the watts lost in the field?

$$P = E \times I = 111.5 \times 3.19 = 355.7 \text{ watts lost in the field (Ans.)}$$

5. What watts are lost in the leads?

$$P = E \times I = 1.5 \times 75 = 112.5 \text{ watts lost in leads (Ans.)}$$

6. What power is supplied to lamps?

$$P = E \times I = 110 \times 75 = 8{,}250 \text{ watts supplied to lamps (Ans.)}$$

Compound Machines.—The compound-wound generator possesses the characteristics of both the series and the shunt dynamos. It is designed to give automatically a better regulation of voltage on constant-potential circuits than is possible with a shunt machine. The shunt field is the same as in the shunt generator and independent series field spools are added, through which the main current flows. When current flows in the external circuit, the voltage at the brushes is not lowered, as in the shunt generator, since the series winding strengthens the field by the current flowing through it and thus raises the voltage in proportion to the increased current. By a proper selection of the number of turns in the series coils, the voltage is thus kept automatically constant for wide fluctuations in load. If a greater number of turns is used in the series coil than required for constant terminal voltage at all loads, the voltage will rise as the load is increased and thus make up for the loss on the transmission lines, so that a constant voltage will be maintained at some point distant from the generator. The machine is then said to be *over-compounded.*

Compound-wound direct-current generators are used extensively in electric lighting and power stations and in electric railway power stations where the load is very fluctuating.

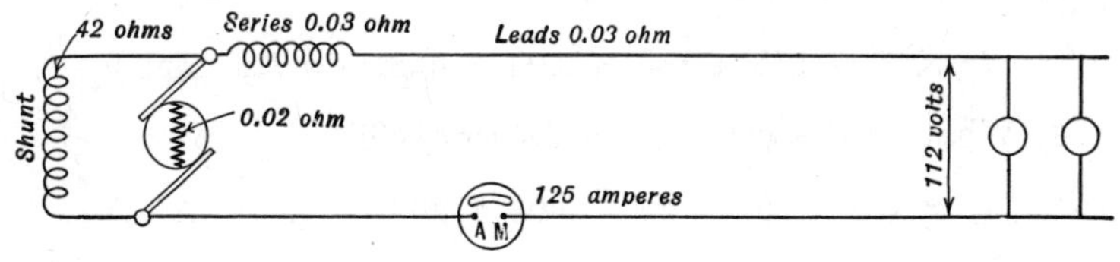

FIG. 62.

COMPOUND GENERATOR PROBLEM: A compound generator, Fig. 62, supplies 125 amperes at 112 volts to a group of lamps located a distance from the generator. The resistances are: Leads, 0.03 ohm; armature, 0.02 ohm; series coil, 0.03 ohm; and shunt coil, 42 ohms.

1. What is the potential difference at the brushes?

$E = I \times R = 125 \times 0.03 = 3.75$ volts drop in leads.

$112 + 3.75 = 115.75$ volts potential difference at terminals

$E = I \times R = 125 \times 0.03 = 3.75$ volts drop in series field.

$115.75 + 3.75 = 119.50$ volts pd at brushes (Ans.)

2. What is the total electromotive force generated?

$$I = \frac{E}{R} = \frac{119.50}{42} = 2.8 \text{ amperes through shunt field}$$

$125 + 2.8 = 127.8$ amperes total current through armature

$E = I \times R = 127.8 \times 0.02 = 2.556$ volts drop in armature

Total emf $= 112$ volts (lamps) $+ 3.75$ volts (leads) $+ 3.75$ volts (series coil) $+ 2.556$ volts (armature) $=$

122.06 volts (Ans.)

3. What are the watts lost in the leads?

$P = I^2 \times R = 125 \times 125 \times 0.03 = 468.75$ watts (Ans.)

4. What are the watts lost in the series coil?

$P = I^2 \times R = 125 \times 125 \times 0.03 = 468.75$ watts (Ans.)

5. What are the watts lost in the shunt coil?

$P = I^2 \times R = 2.8 \times 2.8 \times 42 = 329.28$ watts (Ans.)

6. What are the watts lost in the armature?

$P = I^2 \times R = 127.8 \times 127.8 \times 0.02 = 326.66$ watts (Ans.)

7. What is the power supplied to the external circuit?

$P = E \times I = 115.75 \times 125 = 14{,}468.75$ watts (Ans.)

Losses in a Dynamo.—The losses of power in a dynamo fall into two general classes:

(1) *Mechanical Losses,*
(2) *Electrical Losses.*

(1) The mechanical losses include the friction between the armature shaft and its bearings, windage, and the friction of the brushes on the commutator. These friction losses are practically constant for all speeds.

(2) The electrical losses include the I^2R losses in the armature and fields and at the brush contacts, the losses due to eddy currents and hysteresis. The losses in the field rheostat when it is in series with the field magnets of a generator should be included, even in separately-excited machines.

All the losses may then be summed up as due to:

(1) Mechanical friction
(2) Electrical friction (resistance)
(3) Magnetic friction (hysteresis)

Efficiency of a Generator.—The efficiency of a generator is the ratio of the power output to the power input. When specific load conditions are not referred to it is always understood that the efficiency is expressed as of full or rated load. Instead of attempting to determine the mechanical power input of a generator, it is sometimes more convenient to obtain an equivalent figure indirectly by adding the value of the losses to the output. We may then state,

$$\text{efficiency} = \frac{\text{output}}{\text{input}} = \frac{\text{output}}{\text{output \& losses}}$$

$$\text{efficiency} = \frac{P}{P + p}$$

when P = output of generator in watts

p = total losses of generator in watts

ILLUSTRATION: If it requires 57 kw to drive a 50-kw generator, what is its efficiency?

Here $P + p = 57$

Then, $\text{eff} = \frac{P}{P + p} = \frac{50}{57} = 0.88 = 88 \text{ percent}$ (Ans.)

Two efficiencies are recognized with electrical machinery, *conventional efficiency* and *directly-measured efficiency*. Unless otherwise specified, conventional efficiency is the one employed. Conventional efficiency of machinery is the ratio of the output to the sum of the output and the losses; or of the input minus the losses, to the input. In either case conventional values are assigned to one or more of the losses. This is necessary because it is practically impossible to measure some of the losses in electrical machinery.

The efficiency of a generator varies with the size of the machine and the load it is supplying. For example, a 5-kw dynamo may have as low an efficiency as 80 percent; a well-designed 40-kw machine, 90 percent, and a 500-kw generator, 94 percent. Again, a certain 200-kw generator may have an efficiency of 93 percent at full load, 92 percent at three-quarter load, 90 percent at half load, and 84 percent at one-quarter load.

Direct Current Motors.—The principle of the operation of a motor is described on page 744 and much of the descriptive matter in the preceding paragraphs on generators applies equally well to motors. Motors may be classified as (a) *series* wound, (b) *shunt wound*, and (c) *compound wound.*

Counter Electromotive Force of a Motor.—The wires of a motor armature, rotating in its own magnetic field, cut the lines of force just as if the armature were being driven as in a generator. Hence, there is an induced electromotive force in the wires. This induced pressure is in a direction opposite to that of the current applied to the armature. It is called the *counter electromotive force* and is always in such direction as to oppose the current applied at the terminals. A motor with no load will run at such a speed that the counter electromotive force is nearly equal to the applied pressure.

The counter electromotive force of a motor running at any speed will be the same as when it is run as a generator at this speed, provided the field strength is the same in both cases. Hence, to find the counter emf of a motor at any speed, run it as a generator at this speed and measure the induced emf by a voltmeter.

The counter emf in a motor can never equal the applied emf.

but is less by an amount equal to the drop in the motor armature. To find the current flowing through the armature of a motor:

Subtract the counter emf from the applied emf and divide this result by the armature resistance. This, Ohm's Law applied to a motor, may be expressed:

$$I = \frac{E - \mathcal{E}}{r} = \frac{\text{voltage drop in armature}}{r}$$

when E = emf applied at motor brushes
$\mathcal{E}$ = counter emf developed by motor
I = current through motor armature
r = internal resistance of motor armature

ILLUSTRATION: A motor is connected to a 110-volt circuit. Its counter emf is 105 volts at a particular speed. What current is being supplied to the motor if the resistance of the armature is 1 ohm?

$$I = \frac{E - \mathcal{E}}{r} = \frac{110 - 105}{1} = 5 \text{ amperes (Ans.)}$$

The speed which any motor attains is such that the counter emf developed and the drop in the armature are exactly equal to the applied emf. This may be expressed by a transposition of the preceding formula.

Counter emf + $(I \times r)$ = applied emf,

or $\mathcal{E} + (I \times r) = E.$

The voltage drop in the armature of a motor is a small percentage of the applied pressure, perhaps 2 percent of the terminal pressure in a 500-kw motor and about 5 percent in a 1-kw motor, so that the counter emf is not much different from the applied emf. Since the power driving a motor equals the applied pressure times the current, most of which is usefully expended in mechanical output, the counter emf is an essential and valuable feature of a motor rather than a detriment.

To find the counter electromotive force of a motor:

Multiply the resistance of the armature by the current flowing through it and subtract this product from the emf applied to the motor brushes. This may be expressed as follows by again transposing the preceding formulas:

$$\mathscr{E} = E - (I \times r)$$

ILLUSTRATION: The armature resistance of a shunt-wound motor is 0.7 ohm; and at a certain load 10 amperes flow through it; the voltage at the motor brushes is 112 volts. What is the counter emf?

$$\mathscr{E} = E - (I \times r) = 112 - (10 \times 0.7) = 105 \text{ volts} \quad \text{(Ans.)}$$

When a motor is just starting, it is obvious that it has no counter emf. Then, if it were directly connected to the supply

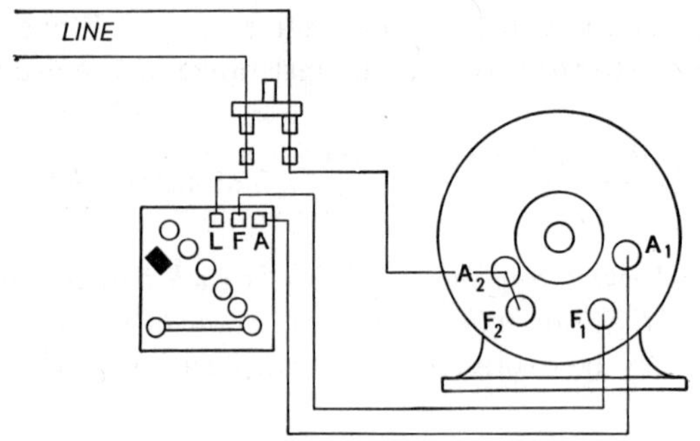

A_1 *Armature Terminal* A_2–F_3–F_1– *Field Terminal*
L–*Line Terminal,* F–*Field Terminal* A–*Armature Terminal*

FIG. 62-A.

mains, a tremendous amount of current would flow through the armature since its resistance is very low. This might result in considerable damage to the windings before a sufficient counter emf has been built up to check the flow. The problem is solved by using a rheostat called a *starting box* to limit the current or lower the voltage until the motor attains its proper running speed. Such starting boxes are always used in the armature circuits of large shunt motors.

Mechanical Power of a Motor.—To find the mechanical power developed by a motor:

Multiply the counter emf by the current through the armature.

$$P = \mathscr{E} \times I$$

The mechanical power developed includes that dissipated as mechanical friction losses and the power which is expended in eddy currents and hysteresis.

Illustration: A small 110-volt motor whose armature resistance is 0.5 ohm runs at a speed to develop a counter emf of 105 volts.

1. What power is developed by this motor?

$$I = \frac{E - \mathscr{E}}{r} = \frac{110 - 105}{0.5} = 10 \text{ amperes}$$

then $P = \mathscr{E} \times I = 105 \times 10 = 1050$ watts (Ans.)

2. What power is supplied to this motor?

$$P = E \times I = 110 \times 10 = 1100 \text{ watts} \quad \text{(Ans.)}$$

Large motors are tested for output by coupling them to generators and measuring the power which is developed by the latter.

Output and Efficiency of Motors.—The capacity of motors to perform useful work is rated according to the amount of power they will maintain at full load at their pulleys, within the limit of permissible heating. The efficiency of a motor, as in the case of the generator, is the ratio of output to input. The energy furnished to a motor is readily measured and from this must be subtracted the losses in the motor to obtain the available energy. These losses are, (1) the I^2R losses in the armature and fields, and the stray power loss, which includes friction, eddy currents and hysteresis.

$$\text{Efficiency} = \frac{\text{output}}{\text{input}} = \frac{\text{input} - \text{losses}}{\text{input}}$$

ILLUSTRATION: A 6-H.P. 110-volt shunt-wound motor has an armature resistance of 0.2 ohm and a field resistance of 40 ohms. The counter emf for a certain speed under load is 100 volts and the stray power loss is 300 watts.

(1) What is the efficiency?

$$\text{Armature current} = I = \frac{E - \mathscr{E}}{r} = \frac{110 - 100}{0.2} = 50 \text{ amperes}$$

$$\text{Field current} = I = \frac{E}{R} = \frac{110}{40} = 2.75 \text{ amperes}$$

Voltage drop in armature $= 110 - 100 = 10$ volts

Power loss in armature $= P = E \times I = 10 \times 50 = 500$ watts

Power loss in field $= P = E \times I = 110 \times 2.75 = 302.5$ watts

Stray power loss $= 300$ watts

Total loss $= 500 + 302.5 + 300 = 1102.5$ watts

Power input in armature $= P = E \times I = 110 \times 50 =$ 5500 watts

Power input in field $= 302.5$ watts

Total power input $= 5500 + 302.5 = 5802.5$ watts

$$\text{Efficiency} = \frac{\text{input} - \text{losses}}{\text{input}} = \frac{5802.5 - 1102.5}{5802.5}$$

$$= \frac{4700}{5802.5} = 0.81 = 81\% \quad \text{(Ans.)}$$

(2) What is the power output?

$$\text{Motor output} = \frac{4700}{1000} = 4.7 \text{ kw. or } \frac{4700}{746} = 6.3 \text{ H.P.} \quad \text{(Ans.)}$$

Current Required by Motor.—When the output, efficiency and voltage are known, the current required by the motor can be determined by the following rule:

If the output of the motor is expressed in kilowatts (kw.), *mul-*

tiply the kw. rating by 1000 *and divide by the voltage of a motor and its efficiency.* Expressing this as a formula,

$$I = \frac{\text{kw.} \times 1000}{E \times \%M}$$

when E = voltage required by the motor,

kw. = kilowatt rating of the motor,

$\%M$ = efficiency of the motor expressed as a decimal.

Illustration: What current is required by a 30-kw., 220-volt motor whose efficiency is 85%?

$$I = \frac{\text{kw.} \times 1000}{E \times \%M} = \frac{30 \times 1000}{220 \times 0.85} = 160 \text{ amperes} \quad \text{(Ans.)}$$

When the rating is given in horsepower (H.P.), *multiply the H.P. by* 746 *and divide this product by the voltage of the motor and by its efficiency.* This becomes,

$$I = \frac{\text{H.P.} \times 746}{E \times \%M},$$

when H.P. = horsepower of the motor and the other factors are as above.

Illustration: What current will be required by a 2-H.P. 110-volt motor whose efficiency is 90%?

$$I = \frac{\text{H.P.} \times 746}{E \times \%M} = \frac{2 \times 746}{110 \times 0.90} = 15 \text{ amperes} \quad \text{(Ans.)}$$

ALTERNATING CURRENTS

Advantages of Alternating Current.—An *alternating current* of electricity is a current which changes its direction of flow at regular intervals of time, usually much shorter than one second.

Alternating current has several advantages over direct current principally in transmission and distribution and for this reason, nearly all of the current generated today is alternating current.

The following problem illustrates the economy which can be effected in the transmission of power by the use of high voltages obtainable only with alternating current.

ILLUSTRATION: 50,000 watts (50 kw.) of power are to be transmitted with a line drop of 2 percent. If the weight of copper required when the energy is delivered at 100 volts is assumed to be 1000 pounds, then the amounts of copper necessary for other voltages are as follows:

Line Voltage, E	Line Current, I amperes	Line Drop, e, volts	Power Loss, Ie, watts	Line Resistance, $R = \frac{e}{I}$, ohms	Copper, Pounds
100	500	2	1000	0.004	1000
200	250	4	1000	0.016	250
500	100	10	1000	0.100	40
1000	50	20	1000	0.400	10

These figures show that the *weight of copper wire required for conducting a certain amount of energy with the same percentage loss on the line is inversely proportional to the square of the transmitting voltage.*

It can also be observed that for the transmission of the same amount of power, the increase in line voltage, E, is accompanied by a proportionate decrease in line current. For the same power loss of 1000 watts the reduction of the current from 500 amperes to 250 amperes effects a saving in wire size. This is shown in the resistance column. For 500 amperes a line of 0.004 ohm resistance is used and for 250 amperes a line four times this resistance or 0.016 ohm is used. This indicates that wire of only one-quarter the weight is used to transmit 250 amperes as compared with 500 amperes. The figures in the last column show this fact.

It is not feasible to build direct-current generators to deliver current at higher than 5000 volts, the limitation being in insulation and commutation. Therefore, in order to obtain the economies of high-voltage power transmission, it is necessary to use alternating current. Alternators can be designed for as much as 20,000 volts because the stationary armature can be more readily insulated. Another factor in this consideration is that the voltage of direct current can be changed only by the coupling of two machines in a motor-generator set. On the other hand, transformers can be used to change alternating current efficiently over a wide range.

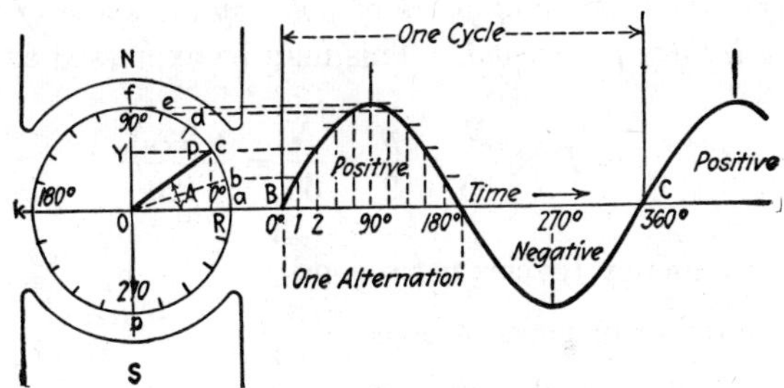

FIG. 63.—Plotting a Sine Curve.

Cycles and Frequency of Alternating Current.—In the discussion of elemental generators it was seen that the electromotive force produced in each coil of an armature rises from zero to a maximum, then declines gradually to zero again, reverses in direction, gradually attaining a maximum in the reversed direction, and then returning to zero. If the value of the electromotive force of one revolution is plotted as the ordinate and time as abscissa, the resulting curve will be as shown in Fig. 63. This is called a *sinusoid* or *sine curve*.

When the alternating current or emf has passed from zero to its maximum value in one direction, to zero. then to its maximum value in the other direction, and back to zero, the complete set of values passed through in that time is called a *cycle*. This cycle of changes takes place in a certain length of time called a *period*. The number of complete cycles in one second is called the *frequency* of the voltage or current. Frequency is, then, *cycles per second* and is sometimes spoken of merely as *cycles*. That is, if an alternator performs the cycle of events depicted in Fig. 63 from B to C sixty times a second, it is said to have a *frequency* of 60 *cycles*. This would mean 120 changes in direction or *alternations* per second. Frequencies of 25 and 60 cycles are standard in the United States.

To find the frequency in cycles of the voltage or current from any alternating current generator:

Multiply the number of pairs of poles by the speed of the armature in revolutions per second. This may be expressed as

$$f = P \times \frac{N}{60} = \frac{p}{2} \times \frac{N}{60} = \frac{p \times N}{120}$$

when f = frequency (cycles per second)

P = number of *pairs* of poles

N = speed in revolutions per *minute*

p = number of *poles*.

ILLUSTRATION: What is the frequency of the current furnished by an alternator having 24 poles and running at a speed of 300 revolutions per minute?

$$f = P \times \frac{N}{60} = \frac{24}{2} \times \frac{300}{60} = 12 \times 5 = 60 \text{ cycles. (Ans.)}$$

With both the current and emf of alternating current constantly fluctuating, instantaneous values of these qualities are not of great practical concern. Meters used to measure alternating current voltage, amperage and wattage, measure only the average or *effective* values. Alternating currents are expressed in terms of the value of the direct current which would produce the same heating effect and this is called the *effective* value.

Phase and Polyphase.—When the current and the voltage of an alternating current both reach a maximum at the same time they are said to be *in phase.* (Fig. 64*a*.) If they do not reach a maximum at the same time they are said to be *out of phase.* Figure 64*b*, *c*, *d*, shows three cases of the current being out of phase; in *b* it is said to *lag* behind, in *c* it is said to *lead* the voltage, and in *d* the curves are in *opposite* phase. This lag or lead may be

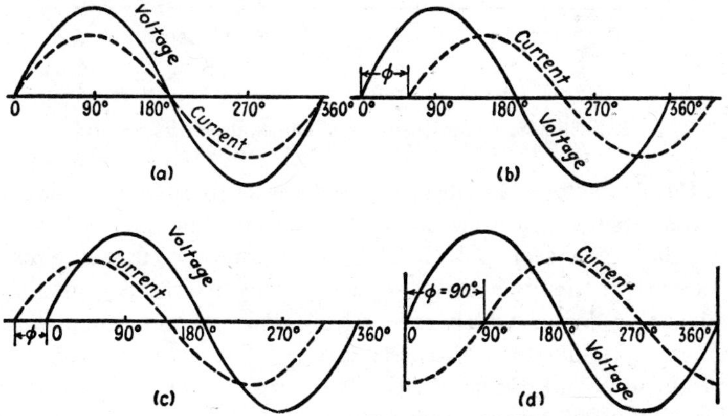

Fig. 64.—Current and Voltage Relations in Alternating-current Circuits. (*a*) Current in phase with voltage, (*b*) current lags behind impressed voltage, (*c*) current leads the voltage, (*d*) current lags 90 degrees.

expressed as an angle and is usually represented by ϕ, and is called *angular displacement* or *difference in phase.* The angle ϕ is then called the *phase angle.*

"Phase" is also used to express the displacement of two or more different emf's or currents of equal frequency but lacking coincidence in time of rise and fall. An alternator which generates a single voltage is called a *single phase* alternator; a machine which generates two or more separate emf's is called a *polyphase* generator.

Three-phase generators are very widely used. In this case three single-phase currents 120 degrees apart, as shown in Fig. 65, are generated. Theoretically three sets of two wires are required for the conduction of the current, but since the algebraic sum of the currents in the three circuits (if balanced) is at every instant equal to zero, the three return wires, one on each circuit may be dispensed with, leaving but three wires.

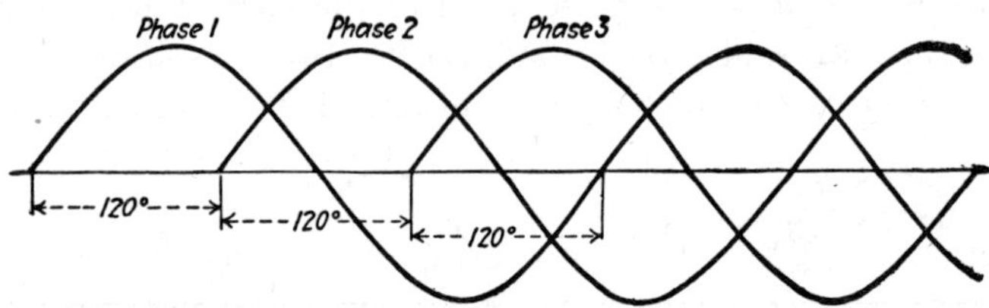

Fig. 65.—Sine emf Curves of a Three-phase Alternator.

Power Factor.—In the study of direct current we saw that the power expended in a circuit was the product of the applied emf and the current or $E \times I$. In the alternating current circuits met with in practice there exists not only resistance, but other influencing forces which are called *inductance* and *capacitance.* (These are defined and discussed later.) The latter two cause the current to be out of phase with the impressed emf. As a result of this, the actual power is reduced.

If we let P = power, E = effective voltage, and I = effective current, then $E \times I$ is called the *apparent power* and is expressed in volt-amperes or kilovolt-amperes (kva). However, if the current I has a lag of ϕ degrees behind the emf, the *actual power* expended in the circuit is

$$P = E \times I \times \cos \phi$$

The factor $\cos \phi$ is called the power factor of the circuit and is usually expressed in percent. Transposing the above equation,

$$\text{Power factor} = \cos \phi = \frac{P}{E \times I},$$

from which we may define power factor as the ratio of the actual power to the apparent power.

ILLUSTRATION: What current will a 220-volt alternator produce in a circuit which has a power factor of 85 percent and takes 1 kilowatt?

In this problem $P = 1000$, $E = 220$, and $\cos\phi = 0.85$. Then,

$$P = E \times I \times \cos\phi$$

$$1000 = 220 \times I \times 0.85$$

$$I = \frac{1000}{220 \times 0.85} = 5.35 \text{ amperes} \quad \text{(Ans.)}$$

Inductance.—We have already referred to the fact that the flow of alternating current depends not only on the resistance but also on *inductance.*

When a current flows through a wire it sets up a magnetic field about the wire. If the current is broken the change in the magnetic field is capable of inducing an emf in a nearby wire. This property is called *inductance*: its symbol is L and the unit is the *henry.* In a wire carrying an alternating current the current is broken many times a second. Not only does this tend to induce current in nearby wires, but in the current-carrying wire itself. This is called self induction and, moreover, the induced emf is opposite in direction to the current emf. The resulting opposition to the flow of the current may be considered as an apparent additional resistance and is called *inductive reactance* to distinguish it from the resistance of the conductor.

The value of inductive reactance is expressed in ohms and it depends on the factors given in the following formula:

$$X_L = 2\pi \times f \times L$$

Where X_L = inductive reactance in ohms

f = frequency (cycles per second)

L = inductance (henrys)

$\pi = 3.1416$

This formula is also useful in the transposed form

$$L = \frac{X_L}{2\pi \times f}$$

ILLUSTRATION: What would be the inductive reactance of a coil of wire having an inductance of 0.03 henry when connected to an emf of 60 cycles?

$$X_L = 2\pi \times f \times L = 2\pi \times 60 \times 0.03 = 11.32 \text{ ohms} \quad \text{(Ans.)}$$

ILLUSTRATION: What is the inductance of a coil which has an inductive reactance of 2.5 ohms when connected to an emf of 25 cycles?

$$L = \frac{X_L}{2\pi \times f} = \frac{2.5}{2\pi \times 25} = \frac{0.1}{2\pi} = 0.0159 \text{ henry} \quad \text{(Ans.)}$$

Resistance.—Resistance in an alternating-current circuit has exactly the same effect as it has in a direct current circuit. This property of an electric circuit always occasions a loss which appears as heat. If an alternating current of I amperes (effective value) flows through a resistance of R ohms, the loss will be I^2R watts.

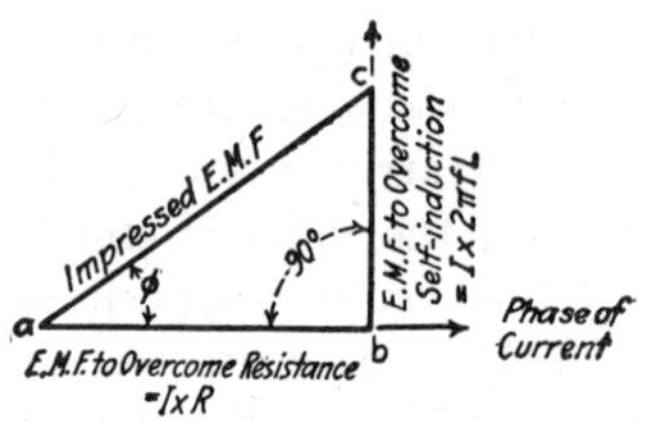

FIG. 66.—Components of emf Impressed on Inductive Circuit.

Components of Impressed emf.—The emf of a circuit must be sufficiently large to overcome the resistance and to overcome the inductive reactance. It may be regarded as having two components, one devoted to each of these functions, as shown in Fig. 66. The relationship between these components is given in the following definitions:

Resistance is that quantity which, when multiplied by the current, gives that component of the impressed emf which is in phase with the current.

Reactance is that quantity which, when multiplied by the current, gives that component of the impressed emf which is at right angles to the current. Then, when

$$ab = E_r = RI = \text{resistance drop}$$

$$bc = E_L = 2\pi fLI = \text{reactance drop}$$

$$ac = E = \text{impressed emf}$$

According to the "hypotenuse square" rule of right triangles, therefore

$$E = \sqrt{E_r^2 + E_L^2}$$

or

$$E = \sqrt{(I \times R)^2 + (I \times 2\pi fL)^2}$$

In the following circuit the various elements are represented:

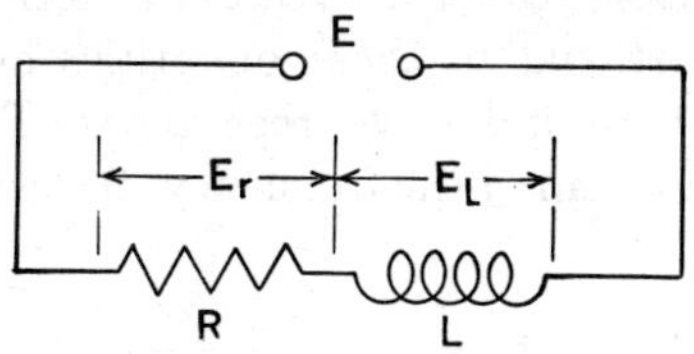

Fig. 67-A.

These equations show that the voltage drop due to resistance and that due to reactance cannot be added arithmetically, but must be added geometrically at right angles to each other to obtain the total voltage on the circuit.

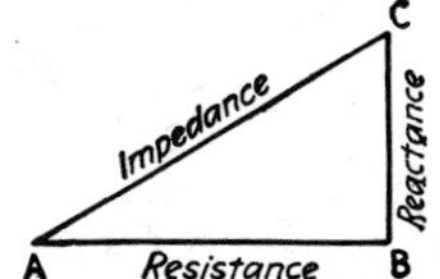

Fig. 67.—Graphical Representation of Impedance

Impedance.—The combined effect of resistance and reactance is called impedance to distinguish it from its two components which may also be represented graphically at right angles as in Fig. 67. Impedance has the symbol Z and is expressed in ohms.

Then,

$$Z = \sqrt{R^2 + X_L^2}$$

Other variations of the above formula are also useful. Thus:

Given the impedance and resistance, to find the reactance, use:

$$X_L = \sqrt{Z^2 - R^2}$$

Given the impedance and reactance, to find resistance, use:

$$R = \sqrt{Z^2 - X_L{}^2}$$

Capacitance.—Most circuits have the faculty of storing an electrical charge and a momentary flow of current takes place after the circuit is opened. This property is called *capacitance* and is utilized in condensers. It has been found that the current increases directly with the increase in capacitance and also with the increase of frequency. Therefore the apparent resistance due to the condenser, called *capacitive reactance*, decreases with, that is, is inversely proportional to these quantities and hence directly opposite in effect to inductive reactance. Then, if C is the capacitance in farads, and f the frequency, the capacitive reactance will be

$$X = \frac{1}{2\pi \times f \times C}$$

ILLUSTRATION: What is the capacitive reactance of a 40-microfarad condenser to an alternating current of 60 cycles? (1 microfarad = one-millionth part of a farad)

$$40 \text{ microfarads} = 0.000040 \text{ farad.}$$

$$X_c = \frac{1}{2\pi \times f \times c} = \frac{1}{2 \times 3.1416 \times 60 \times 0.000040} = \frac{1}{0.0151}$$

$$= 66.3 \text{ ohms} \quad \text{(Ans.)}$$

Circuits Having Inductance, Capacitance and Resistance.—When a circuit contains both inductance and capacitance, the net reactance, X, is equal to the arithmetical difference between the inductive reactance, X_L, and the capacitive reactance, X_c, or $X = X_L - X_c$.

Therefore the impedance of a circuit containing inductance, capacitance and resistance is equal to the square root of the quantity

[resistance² + (inductive reactance − capacitive reactance)²],

or

$$Z = \sqrt{R^2 + X^2} = \sqrt{R^2 + (X_L - X_c)^2}$$

ILLUSTRATION: What would be the combined impedance of a circuit, having a coil of 3 ohms resistance and of 0.01 henry inductance in series with a condenser of 60-microfarad capacity to an alternating current of 60 cycles?

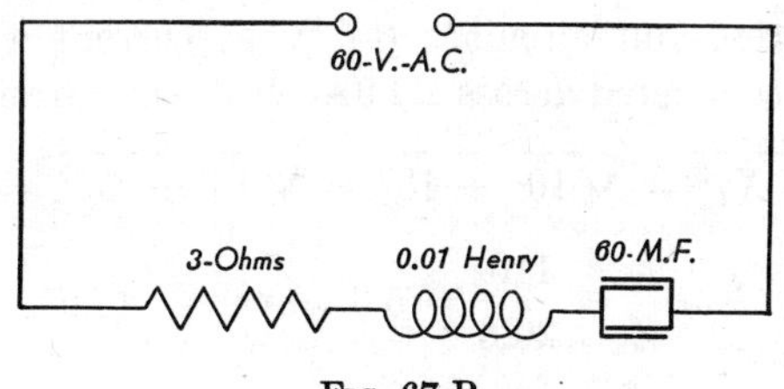

FIG. 67-B

$$X_L = 2\pi \times f \times L = 2 \times 3.1416 \times 60 \times 0.01 = 3.77 \text{ ohms}$$

$$X_c = \frac{1}{2\pi \times f \times C} = \frac{1}{2 \times 3.1416 \times 60 \times 0.000060} = 44.2 \text{ ohms}$$

$$Z = \sqrt{R^2 + (X_L - X_c)^2} = \sqrt{3^2 + (3.77 - 44.2)^2}$$

$$= \sqrt{3^2 + (-40.43)^2}$$

$$Z = \sqrt{9 + 1635.36} = 40.55 \text{ ohms} \quad \text{(Ans.)}$$

Ohm's Law for Alternating-Current Circuits.—In the early pages of this section Ohm's Law applying to direct currents was stated in the three forms,

$$I = \frac{E}{R}, \quad E = I \times R, \quad \text{and} \quad R = \frac{E}{I}$$

We have seen that instead of simple resistance we have in the case of alternating current a number of influences which when grouped together are called impedance and designated by the letter Z. Then Ohm's Law for alternating currents may be expressed:

$$I = \frac{E}{Z}, \quad E = I \times Z, \quad \text{and} \quad Z = \frac{E}{I}$$

When E = emf or the pressure applied to any circuit

Z = impedance of the circuit expressed in ohms

I = current strength in that circuit

ILLUSTRATION: (*a*) What current will flow through a coil with a resistance of 10 ohms and a reactance of 18 ohms when connected to a 60-cycle 110-volt circuit? (*b*) What current would flow if this coil were connected across a 110-volt direct-current circuit?

$$Z = \sqrt{R^2 + X_L{}^2} = \sqrt{10^2 + 18^2} = \sqrt{100 + 324} = 20.6 \text{ ohms}$$

$$(a)\ I = \frac{E}{Z} = \frac{110}{20.6} = 5.3 \text{ amperes} \quad \text{(Ans.)}$$

$$(b)\ I = \frac{E}{R} = \frac{110}{10} = 11.0 \text{ amperes} \quad \text{(Ans.)}$$

Impedance may be measured by the volt ammeter method in the same way as resistance is measured in a direct-current circuit, using, of course, an alternating-current voltmeter and ammeter, the impedance being calculated from $Z = E \div I$.

Transformers.—It has already been pointed out that one of the advantages of alternating current is that its voltage may be transformed at will to higher or to lower potentials. This is accomplished by a device called a *transformer* which consists of two windings insulated from each other, but so situated that the magnetic flux developed by one of the windings threads through the other. By running an alternating current through the first winding, there is a constant change in the magnetic flux which induces a current in the second. The two windings are called the *primary* and the *secondary*, the primary being the winding which receives the energy from the supply circuit and the secondary that which receives the energy by induction from the primary.

Figure 68, illustrating three types of power transformers, shows the relation of the two windings to each other and to the core built up from annealed punchings of thin sheet steel. Small transformers such as are placed on poles in power distribution circuits are contained in a cast iron or sheet steel case which is then filled with oil. The oil serves the double purpose of adding further insulation to the windings and of transmitting the heat to the case, where it is dissipated by radiation and air circulation. Such transformers are called self-cooled. Larger transformers such as are used in substations may have the oil cooled by circulating water or air or may be cooled by a blast of air circulated through the windings.

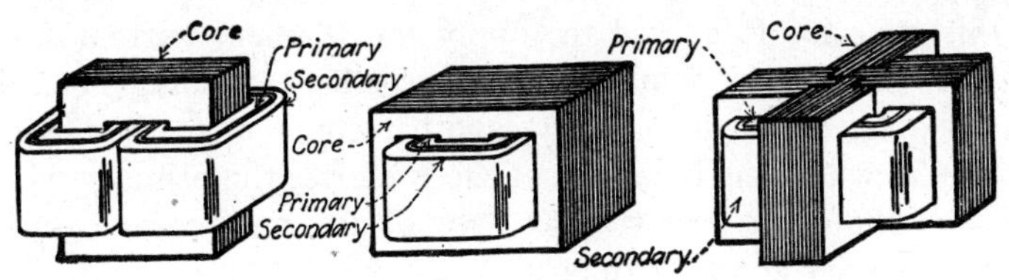

Fig. 68.—Types of Transformers. Left—core type; center—shell type; right—combined core and shell type.

The signal transformers, which are used in radio, television and other communication circuits are available in a considerably larger number of types than the power transformers. These types range from simple air-wound coils, which may consist of only a hollow tube of paperboard, plastic or ceramic on which one coil is wound upon or beside the other, with paper or other insulation between the coils. On the other hand, many types of signal transformers require cores of high permeability material, such as a stack of steel laminations, a mass iron powder or, less commonly in this type of transformer, a solid mass of iron.

The reason for these many different designs is because of the many purposes for which transformers are used in communication circuits. Thus transformers operating at radio frequencies (several hundred kilocycles per second or greater) do not require cores, since the inductive reactance is proportional to the frequency. For the same reason transformers operating at audio frequencies (100 to 15,000 cycles per second) do require cores to transmit the magnetic flux from one winding (the primary) to the other (the secondary). Among the other special requirements are those of the audio output transformer, which links the output stage of a receiver to the input of a speaker, and must therefore operate at a high impedance ratio (roughly 5000 ohms on the receiver side to 5 ohms in the speaker circuit.) Still another type is the "transformer coupler" which links the output of one amplifier stage to the input of the next. As a final example there is the transformer which supplies two or more secondary windings from a single primary winding. A transformer of this type is often used to supply the filament current for a number of tubes. While several secondary windings can be placed about a single primary, another common method is the E type, in which the inner bar of the E carries the primary winding, and the outer bars the two secondaries, with cores between each secondary and the primary.

The transformation of the current from one voltage to another is accomplished by having more turns on one winding than the other. Thus, if the primary winding has 250 turns and the secondary has 1000 turns, then the voltage available at the secondary terminals will be $1000 \div 250 = 4$ times as great as the voltage impressed upon the primary. If we let n_2 represent the number of turns on the high-voltage winding and n_1 the number of turns on the low-voltage winding, then the ratio $n_2 \div n_1 = r$ is called the *ratio of transformation,* and

$$r = \frac{n_2}{n_1} = \frac{E_2}{E_1}$$

when E_2 and E_1 are the respective voltages of the two windings. When a transformer is used to deliver a current at a voltage higher than that it receives, it is called a *step-up* transformer, and when it delivers a current at a lower potential it is called a *step-down* transformer.

Transformers are very efficient in their operation, often rating over 98 percent, so that for many practical calculations the losses may be ignored and the power output regarded as equal to the power input. Then, since power equals volts times amperes we may write

$$P = E_1 \times I_1 = E_2 \times I_2$$

where I_1 and I_2 are the currents in the low and high voltage windings, respectively. From this we may derive the following ratios:

$$\frac{E_1}{E_2} = \frac{I_2}{I_1}$$

which states in effect that the ratio of the voltage is the inverse ratio of the currents in the two windings.

ILLUSTRATION: The primary voltage of a 15-kw. transformer used to supply electricity to a 220-volt circuit is 2200 volts. What is the ratio of this transformer and what are the full-load currents in the two windings, neglecting losses?

This is, of course, a step-down transformer and $E_1 = 220$ volts, $E_2 = 2200$ volts, $P = 15{,}000$ watts. Then

$$r = \frac{E_2}{E_1} = \frac{2200}{220} = 10 \quad \text{(Ans.)}$$

$$P = E_1 \times I_1$$

then $$I_1 = \frac{P}{E_1} = \frac{15{,}000}{220} = 68 \text{ amperes} \quad \text{(Ans.)}$$

and $$I_2 = \frac{P}{E_2} = \frac{15{,}000}{2200} = 6.8 \text{ amperes} \quad \text{(Ans.)}$$

ILLUSTRATION: What are the full-load currents in the two windings of a 30-kw. transformer used to supply electricity to a 110-volt circuit if the primary voltage is 3300 volts?

$$I_1 = \frac{P}{E_1} = \frac{30,000}{110} = 273 \text{ amperes} \quad \text{(Ans.)}$$

$$I_2 = \frac{P}{E_2} = \frac{30,000}{3300} = 9.1 \text{ amperes} \quad \text{(Ans.)}$$

Transformer Design Calculations.—While the calculations in the precise design of a transformer are quite lengthy, involving consideration of temperature effects and several other variables, the major steps are outlined below for those who may wish to wind a small transformer.

1. Compute the voltage and current requirements of the secondary circuit from the components in it.

2. Choose a core size which is suitable for the volt-ampere rating just calculated. This choice requires comparison with other small transformers, since transformers for radio and television receivers are made as small as possible without overheating (200° F. is maximum for paper and phenolic resin insulation).

3. Determine the flux density in the core. This depends upon the material of the core (70,000 lines per sq. in. for silicon steel).

4. Calculate the number of turns required in the primary to produce this flux.

5. From the volt-ampere rating of the secondary as calculated in (1), calculate the volt-ampere rating of the primary by multiplying by an efficiency factor (85–95%) for iron-core small transformers. From this figure and the primary circuit, compute the primary current when full load is on the secondary.

6. Calculate the no-load current on the primary to avoid overloading.

7. Choose a wire size from Table 4 to take the full-load current.

8. By the equations given in this section, calculate the number of turns and wire sizes for the secondary.

Alternators.—The principles of the alternating current generator have already been discussed and the three principal types classified and described. Revolving field alternators are used practically to the exclusion of all other types in power generating stations. Their field magnets wound in slots revolve inside a stationary armature similarly wound. This results in a well-balanced machine of low resistance which can be successfully operated in connection with high-speed turbines. The revolving field magnets are energized by direct current which reaches them through slip rings. This current is often of a much lower potential than that received from the stationary armature.

Alternating-current generators are usually rated in kilovolt-amperes (kva) instead of kilowatts, since it is impossible for the manufacturer to know in advance the amount of inductance and capacitance of the circuits to which the alternator is required to furnish power.

Engine and hydraulic turbine-driven alternators are in the slow-speed class, and are characterized by large diameter, short length, and many poles. The steam turbine-driven alternator is a high-speed machine having a length larger than its diameter. Standard speeds of turbine-driven alternators range from 1200 to 3600 rpm, with 1800 rpm very common practice. To supply the d-c for the field, a source of d-c at 110–250 volts is necessary.

Conversion.—While practically all electric power is generated as alternating current, some functions are best served by direct current and it is convenient to have some means of changing the alternating current to direct current. This is called *conversion.*

Urban subways usually operate on direct current and the power supplied is most frequently converted to direct current by a machine called a *rotary converter.* This is essentially an alternator and a direct-current generator combined in one machine. Its revolving armature receives alternating current through slip rings and by tapping the armature coils at proper points and connecting them with a segmented commutator, direct current may be taken off by means of brushes.

When only a small amount of direct current is required from an alternating-current source, a device known as a rectifier may be used. This permits the current to pass in only one direction. The four common types in use are, the mercury-arc rectifier, the vibrating rectifier, the tungar rectifier and an electrolytic rectifier. These find use in electroplating, storage-battery charging and radio work.

Alternating Current Motors.—A detailed description of alternating current motors is beyond the scope of this work because the mathematical problems connected with these machines are the concern chiefly of the designer and engineer. However, for the sake of completeness we will list the important types.

1. The *polyphase induction motor* of the squirrel-cage armature type is the most widely used alternating current motor in industrial service. It consists of a wound stationary part called the *stator*, which corresponds to the field magnets of a direct-current motor, and a rotating member called the *rotor*, which corresponds to the armature. Polyphase alternating currents flowing through the stator set up a rotating magnetic field which induces a current in copper bars parallel to the axis of the rotor and the reaction of the magnetic flux of these rotor conductors against the rotating field produces rotation of the rotor. Some motors of this type have a wound rotor to inject resistance into the rotor winding and obtain a higher starting torque.

2. The *single-phase induction motor* differs from the polyphase motor chiefly in that provision must be made for starting the motor and bringing it up to a speed corresponding to the frequency in the stator windings. This is done by one of three methods. (1) the split-phase methods in which an auxiliary stator winding is provided for starting purposes only, (2) an auxiliary winding may be connected to the single-phase line through an external inductance to split the phase, and (3) by providing a wound rotor and a commutator for starting as a repulsion motor.

3. *Single-phase commutator motors* may be divided into three sub-types: plain repulsion, single-phase series, and repulsion induction motors. Of these the second is the simplest form and in general design is practically the same as the direct-current series

motor. It may be operated on either direct current or alternating current and for this reason it is widely used for operating household appliances and small tools.

4. The *synchronous motor* is constructed in practically the same manner as a corresponding alternator, and any alternator may be run as a synchronous motor. However, some auxiliary means must be provided for bringing this type of motor up to synchronous speed before it is connected to the alternating current. This is usually accomplished by attaching to the rotor an auxiliary cage winding similar to the rotor winding of a squirrel-cage induction motor.

WIRE CALCULATIONS

Mil-foot.—In calculating the resistance of wire, the standard unit used is a wire $\frac{1}{1000}$ inch in diameter and one foot long. Such a piece of wire is called a *mil-foot.* The word "mil," however used, means one-thousandth. The cross-sectional area of a wire whose diameter is one mil is one *circular mil.* Since areas of like-shaped surfaces vary as the squares of their dimensions, it follows that the cross-sectional area of a circle whose diameter is 2 mils, is 4 circular mils (See Fig. 69); one whose diameter is 3 mils has an area of 9 circular mils, etc. From this we may devise the rule that:

When d represents the diameter of a wire in mils, d^2 is its cross-sectional area in circular mils.

The resistance of one mil-foot of copper wire is 10.79 ohms at 75° Fahrenheit. The resistance of ten feet will be 107.9 ohms. One foot of copper wire $\frac{2}{1000}$ inch in diameter will have one-fourth the resistance, or 10.79 divided by 4 = 2.70 ohms. *Resistance of a conductor varies directly as the length, inversely as the cross-sectional area, with the material of the conductor, and with its temperature.*

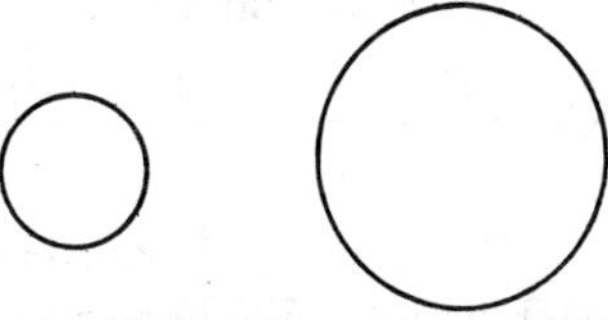

FIG. 69.—The Diameter of the Larger Circle is Twice as Great as that of the Smaller, but the Area is Four Times as Great.

Calculating Resistance of Wires.—Given the length and are of any wire, to find its resistance:

The resistance of any wire at a given temperature is equal to i length in feet multiplied by the resistance of a mil-foot (K) and th product divided by its area in circular mils.

$$R = \frac{K \times L}{d^2}$$

When
R = resistance in ohms
K = resistance of one mil-foot in ohms
L = length of wire in feet
d = diameter in mils
d^2 = area in circular mils

ILLUSTRATION: What is the resistance of 500 feet of coppe wire having a cross-sectional area of 4107 circular mils?

$$K \text{ for copper} = 10.79$$
$$d^2 = 4107$$

Then $R = \frac{K \times L}{d^2} = \frac{10.79 \times 500}{4107} = 1.31$ ohms (Ans.)

ILLUSTRATION: Find the resistance of a copper wire 10.03 mil in diameter and 85 feet long.

$K = 10.79$
$d = 10.03$
$d^2 = 100.5$

Then $R = \frac{K \times L}{d^2} = \frac{10.79 \times 85}{100.5} = 9.13$ ohms (Ans.)

The value K is constant for the same wire, but different for each metal. We have seen that it is 10.79 ohms for copper at 75° Fahrenheit. The value of K for other metals is given in Table 3. The variation of resistance with temperature is roughly proportional to the absolute temperature. The following table is based on a temperature of 68° Fahrenheit.

TABLE 3

RESISTANCE OF A MIL-FOOT OF METALS (VALUES OF K)

Silver, 9.84	Zinc, 36.69	German Silver, 128.29
Copper, 10.79	Platinum, 59.02	Platinoid, 188.93
Aluminum, 17.21	Iron, 63.35	Mercury, 586.24

ILLUSTRATION: Substitute iron wire for copper wire in the preceding illustration. It then calls for the resistance of an iron wire 10.03 mils in diameter and 85 feet long.

$$K = 63.35$$
$$d = 10.03$$
$$d^2 = 100.5$$

$$\text{Then} \quad R = \frac{K \times L}{d^2} = \frac{63.35 \times 85}{100.5} = 53.57 \text{ ohms} \quad \text{(Ans.)}$$

From this it is seen that the resistance of iron is about six times that of copper.

Wire Gage.—This is the term used in describing the size of wire. There are a number of wire gages which have been developed by different manufacturers. The American standard for electrical purposes is the B. & S. gage (Brown & Sharpe Manufacturing Company).

Wires larger than No. 0000 B. & S. are seldom made solid but are built up of a number of small wires. The group of wires is called a " strand "; the term " wire " being reserved for the individual wires of the strand. Strands are usually built up of wires of such a size that the cross-section of the metal in the strand is the same as the cross-section of a solid wire having the same gage number. The sizes of wire larger than No. 0000 are given only in circular mils.

Wire Calculations.—Given the resistance and area of a wire, to find the length.

The length of any wire is equal to its resistance multiplied by its circular mil area, and this product divided by the resistance of a mil-foot (K).

$$L = \frac{R \times d^2}{K}$$

TABLE 4

Wire Table, Standard Annealed Copper at a Temperature of 25° Centigrade (77° Fahrenheit)

American Wire Gage (Brown & Sharpe)

Gage No.	Diam. in Mils d	Area	Weight		Length		Resistance	
		Cir. Mils d^2	Lbs. per 1000 ft.	Lbs. per ohm	Feet per lb.	Feet per ohm	Ohms per 1000 ft.	Ohms per lb.
0000	460.0	211660.	640.5	12810.	1.561	20010.	0.04998	0.00007805
000	409.6	167800.	507.9	8057.	1.968	15870.	.06303	.0001217
00	364.8	133100.	402.8	5067.	2.482	12580.	.07947	.0001935
0	324.9	105500.	319.5	3187.	3.130	9979.	.1002	.0003138
1	289.3	83690.	253.3	2004.	3.947	7913.	.1264	.0004990
2	257.6	66370.	200.9	1260.	4.977	6276.	.1594	.0007934
3	229.4	52640.	159.3	792.7	6.276	4977.	.2009	.001262
4	204.3	41740.	126.4	498.6	7.914	3947.	.2534	.002006
5	181.9	33100.	100.2	313.5	9.980	3130.	.3195	.003189
6	162.0	26250.	79.46	197.2	12.58	2482.	.4029	.005071
7	144.3	20820.	63.02	124.0	15.87	1968.	.5080	.008064
8	128.5	16510.	49.98	77.99	20.01	1561.	.6406	.01282
9	114.4	13090.	39.63	49.05	25.23	1238.	.8078	.02039
10	101.9	10380.	31.43	30.85	31.82	981.8	1.019	.03242
11	90.74	8234.	24.92	19.40	40.12	778.5	1.284	.05155
12	80.81	6530.	19.77	12.20	50.59	617.4	1.620	.08196
13	71.96	5178.	15.68	7.673	63.80	489.6	2.042	.1303
14	64.08	4107.	12.43	4.826	80.44	388.3	2.576	.2072
15	57.07	3257.	9.858	3.035	101.4	307.9	3.248	.3295
16	50.82	2583.	7.818	1.909	127.9	244.2	4.095	.5239
17	45.26	2048.	6.200	1.200	161.3	193.7	5.164	.8330
18	40.30	1624.	4.917	0.7549	203.4	153.6	6.512	1.325
19	35.89	1288.	3.899	.4748	256.5	121.8	8.210	2.106
20	31.96	1022.	3.092	.2986	323.4	96.59	10.35	3.349
21	28.46	810.1	2.452	.1878	407.8	76.60	13.06	5.325
22	25.35	642.4	1.945	.1181	514.2	60.74	16.46	8.467
23	22.57	509.5	1.542	.07427	648.4	48.17	20.76	13.46
24	20.10	404.0	1.223	.04671	817.7	38.20	26.18	21.41
25	17.90	320.4	0.9699	.02938	1031.	30.30	33.01	34.04
26	15.94	254.1	.7692	.01847	1300.	24.02	41.62	54.13
27	14.20	201.5	.6100	.01162	1639.	19.05	52.48	86.07
28	12.64	159.8	.4837	.007307	2067.	15.11	66.18	136.8
29	11.26	126.7	.3836	.004595	2607.	11.98	83.46	217.6
30	10.03	100.5	.3042	.002890	3287.	9.503	105.2	346.0
31	8.928	79.70	.2413	.001818	4145.	7.536	132.7	550.2
32	7.950	63.21	.1913	.001143	5227.	5.976	167.3	874.8
33	7.080	50.13	.1517	.0007189	6591.	4.739	211.0	1391.
34	6.305	39.75	.1203	.0004521	8310.	3.759	266.1	2212.
35	5.615	31.52	.09542	.0002843	10480.	2.981	335.5	3517.
36	5.000	25.00	.07568	.0001788	13210.	2.364	423.0	5592.
37	4.453	19.83	.06001	.0001125	16660.	1.874	533.5	8892.
38	3.965	15.72	.04759	.00007074	21010.	1.487	672.7	14140.
39	3.531	12.47	.03774	.00004448	26500.	1.179	848.2	22480.
40	3.145	9.888	.02993	.00002798	33410.	0.9349	1070.	35740.

Illustration: What is the length of a German silver wire wound on a spool if its resistance is 30 ohms and the size of the wire is No. 20 B. & S.?

$$K = 128.29 \text{ for German silver (See Table 3)}$$

$$\text{No. 20 B. \& S.} = 1022 \text{ circular mils (See Table 4)}$$

$$\text{Then} \quad L = \frac{R \times d^2}{K} = \frac{30 \times 1022}{128.29} = 239 \text{ feet} \quad \text{(Ans.)}$$

Given the length and resistance of a wire, to find the area:

The area in circular mils of any wire is equal to its length multiplied by the resistance of a mil-foot (K) and this product divided by its resistance.

$$d^2 = \frac{L \times K}{R}$$

Illustration: A reel of 800 feet of copper wire has a resistance of 5 ohms at 75° F. What is its circular mil area?

$$K = 10.79 \text{ for copper at } 75° \text{ F.}$$

$$\text{Then } d^2 = \frac{L \times K}{R} = \frac{800 \times 10.79}{5} = 1{,}726 \text{ circular mils} \quad \text{(Ans.)}$$

Illustration: A mile of aluminum wire on a power line has a resistance of 1.086 ohms. What is its circular mil area?

$$1 \text{ mile} = 5280 \text{ feet}$$

$$K = 17.21 \text{ for aluminum (See Table 3)}$$

$$\text{Then } d^2 = \frac{L \times K}{R} = \frac{5280 \times 17.21}{1.086} = 83{,}673 \text{ circular mils} \quad \text{(Ans.)}$$

This is evidently a No. 1 wire whose area is 83,690 circular mils. When the area in circular mils is known, the square root of this number is the diameter in mils, or thousandths of an inch.

Given the area of a wire, to find its weight:

The weight per mile (5280 feet) *of any bare copper wire in pounds is equal to the area in circular mils divided by the constant 62.5.*

$$\text{Pounds per mile} = \frac{d^2}{62.5}$$

ILLUSTRATION: Copper telegraph wire 14-gage B. & S. is furnished in coils containing 1.20 miles. What is the weight of such a coil?

d^2 = 4107 circular mils for 14-gage wire (See Table 4)

$$\text{Then, weight of 1 mile} = \frac{d^2}{62.5} = \frac{4107}{62.5} = 66 \text{ pounds}$$

$$\text{weight of coil} = 66 \times 1.2 = 79.2 \text{ pounds} \quad \text{(Ans.)}$$

Copper weighs about 555 pounds per cubic foot and iron about 480 pounds. Therefore, the weight of a length of iron wire would be $\frac{480}{555} = \frac{32}{37}$ times that of a corresponding length of copper wire.

ILLUSTRATION: If the wire in the preceding illustration were iron instead of copper, what would be its weight?

$$79.2 \times \tfrac{32}{37} = 68.5 \text{ pounds} \quad \text{(Ans.)}$$

Finding Size of Wire Required.—The formula given on page 781 for the determination of the area of a wire needed, $d^2 = \frac{L \times K}{R}$, may be transformed for more practical application by expressing the resistance (R) in terms of current and voltage drop. From Ohm's Law we have $R = \frac{E}{I}$

$$\text{whence,} \qquad d^2 = \frac{L \times K}{\frac{E}{I}} = \frac{L \times K \times I}{E}$$

ILLUSTRATION: A power line 800 feet long is run to a motor requiring 25 amperes. The voltage drop in the line must not exceed 20 volts. What size wire will be required?

$$d^2 = \frac{L \times K \times I}{E} = \frac{800 \times 2 \times 10.79 \times 25}{20} = 21{,}580 \text{ circular mils}$$

Referring to Table 4, the wire size next larger than this area is gage 6, and this should, therefore, be the wire used.

Wire Calculations Using Tables.—Where many wire size calculations are required, use of the foregoing formulas requires much tedious work. Tables 5, 6, and 7 have been devised to simplify and reduce this work by giving the proper copper wire size and circuit length for a given load.

These tables are based on a maximum voltage drop of 1%, and practical conductor operating temperatures of 50° C. and 60° C. This means that the wire size and wire length arrived at by using these tables will not cause a voltage drop greater than 1% at the load. The operating temperature of the circuit also will not exceed 60° C. on lighting circuits and 50° C. on other circuits.

Table 5 covers lighting circuits on 3-wire and 4-wire, balanced 115-volt lighting loads. Branch circuit sizes are shown. Table 6 covers single-phase, a-c, 3-wire, 115/230-volt, 60-cycle, 100% power factor systems. Feeder sizes generally used for lighting loads are listed. Table 7 covers 3-phase, a-c, 230-volt, 60-cycle, delta, 85% power factor systems, usually power loads such as motors.

It is necessary to know the load requirement in amperes to use the tables. The ampere load is located in the first column of the proper system voltage table. On this line going horizontally to the right, the required circuit length is found, and at the top of the latter column the proper wire size is found.

Illustration: A 115-volt, 3-wire lighting circuit, 35 feet long, has to carry a balanced lighting load of 15 amperes. What wire size should be used so that the voltage drop will not exceed 1% (1.15 volts)?

Using Table 5, 15 amperes is found in the first column. Reading across horizontally, it can be seen that 35 feet does not appear in the table. The next higher value, 40 feet, is therefore selected. The top of this column indicates that No. 12 wire is the proper size to use.

Installation of Interior Wiring.—All interior wiring must be installed in such a manner that it will be protected from mechanical injury and be safe as regards fire hazard or danger to life.

TABLE 5

Table of average circuit lengths for 3 and 4 wire balanced 115 v. lighting loads with 1%. (1.15 volts) drop from supply cabinet to first outlet supplying permanently connected current consuming appliance or lighting fixture. All conductors grouped in same conduit.

Copper resistance "R" — 13 ohms per CM ft. at 60°C (140°F)

AMPERES (A), WATTS (W), WITH CONDUIT CONDUCTOR (C), FILLS (F)

MAXIMUM OVERCURRENT CIRCUIT PROTECTION	INTERMITTENT LOADS			CONTINUOUS LOADS		
	100% F 2-3 C	80% F 4-6 C	70% F 7-9 C	100% F 2-3 C	80% F 4-6 C	70% F 7-9 C
15 A	15 A 1725 W	12 A 1380 W	10.5 A 1207 W	12 A 1380 W	9.6 A 1104 W	8.4 Amps. 966 Watts
20 A	20 A 2300 W	16 A 1840 W	14 A 1610 W	16 A 1840 W	12.8 A 1472 W	11.2 A 1288 W

LOADS AND LENGTHS IN FEET FOR 1% DROP ON 3 AND 4 WIRE 115 V. CIRCUITS

AMPERE LOAD	#10 WIRE	#12 WIRE	#14 WIRE
1	946	596	374
2	474	298	188
3	316	198	124
4	236	148	94
5	190	120	76
6	158	100	62
7	136	86	54
8	118	74	46
9	106	66	42
10	94	60	38
11	86	54	34
12	78	50	32
13	72	46	28
14	68	42	26
15	64	40	24
16	60	38	
17	56	36	
18	52	34	
19	50	32	
20	48	30	
21	46		
22	44		
23	42		
24	40		
25	38		
26	36		
27	36		
28	34		
29	32		
30	32		

On 4-wire, 3-phase "Y" power and light service with 2-ph, 3-wire circuits use ⅔ table circuit lengths for 2-ph circuits when tapped off 4-wire system.

For 2-wire, 115-volt circuits, use one-half table circuit lengths.

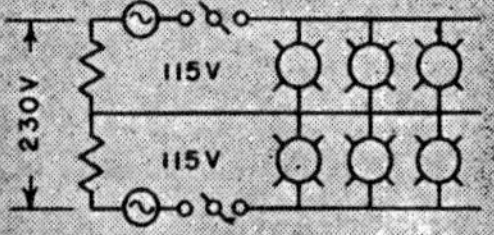

230V

115V

115V

Example: 2-Wire, 115-volt, No. 14 wire circuit

$$\text{Length} = \frac{1.15\ (\text{Volt drop}) \times 4107\ (\text{CM of \#14W})}{15\ (\text{Amps.}) \times 2 \times 13\ (\text{R of wire @ }60^\circ\text{C})}$$

$$= 12.1 \text{ feet}$$

TABLE 6

Table of average circuit lengths for single-phase, A.C., 3-wire (230-115-230 v.), 60 cycle, 100% power factor system. Feeders with 1% (1.15 volts) drop for balanced 3-wire loads. For 2-wire, 230-volt single-phase loads, voltage drop is 2.3 volts. Circuit lengths in feet.

Copper resistance "R"—12.5 ohms per "CM" ft. at 50°C. (122°F.)
Reactance and impedance losses calculated for each wire.
Conductors closely grouped in metallic conduit.

AMPERE LOAD	WIRE SIZE—CIRCULAR MILS					WIRE SIZE—B & S or A.W.G.									
	500	400	350	300	250	4/0	3/0	2/0	1/0	1	2	3	4	6	8
40	1106	898	788	669	558	475	378	299	239	188	150	119	94	59	38
50	885	719	630	535	447	380	303	240	191	150	120	91	75	47	30
60	737	599	525	446	372	317	252	200	159	125	100	79	62	39	
70	632	513	450	382	319	271	216	171	136	107	86	68	53	34	
80	553	449	394	334	279	238	189	150	119	94	75	59	47		
90	491	399	350	297	248	211	168	133	106	83	67	53	42		
100	442	359	315	267	223	190	151	120	95	75	60	47			
110	402	327	286	243	203	173	138	109	87	68	55				
120	369	299	263	223	186	158	126	100	79	63					
130	340	276	242	206	172	146	116	92	73	58					
140	316	257	225	191	159	136	108	86	68						
150	295	240	210	178	149	127	101	80	64						
160	276	225	197	167	140	119	95	75	60						
170	260	211	185	157	131	112	89	70							
180	246	200	175	148	124	106	84	66							
190	233	189	166	140	117	100	80								
200	221	180	157	134	112	95	76								
210	211	171	150	127	106	90									
220	201	163	143	122	101	86									
230	192	156	137	116	97	83									
240	184	150	131	111	93										
250	177	144	126	107	89										
260	170	138	121	103	86										
270	164	133	117	99											
280	158	128	112	96											
290	152	124	109	92											
300	147	120	105												
310	143	116	102												
320	138	112													
330	134	109													
340	130	106													
350	126														
360	123														
370	119														
380	116														

AMPERES PER KW. at 100% to 70% P.F.

POWER FACTOR	1-PHASE	3-PHASE
1.00	4.35	2.51
.95	4.58	2.65
.90	4.83	2.80
.85	5.12	2.90
.80	5.44	3.14
.75	5.80	3.35
.70	6.21	3.59

APPROX. COPPER "R"—OHMS PER "CM" FT.

Temp. °C	Temp. °F	"R"—per CM FT.
20	68	10.4 ohms
25	77	10.8 ohms
30	86	11.0 ohms
40	104	11.5 ohms
50	122	12.5 ohms
60	140	13.0 ohms
75	167	13.6 ohms

FORMULAE FOR AMP. PER KW. AT 100% P.F.

$$\text{1-PHASE } A = \frac{1\ (KW) \times 1000\ (W)}{230\ (V) \times 1.00\ (P.F.)}$$

$$\text{3-PHASE } A = \frac{1\ (KW) \times 1000\ (W)}{230\ (V) \times 1.00\ (P.F. \times 1.732}$$

Conductivity of Copper

Soft Annealed—all sizes	98.3%
Medium Hard—.32 to .04 Dia.	96.7%
Hard Drawn—.32 to .04 Dia.	96.3%

FORMULAE FOR CALCULATING DISTANCE:

Copper "R"—12.5 ohms/CM ft. at 50°C.
A.C. Distance Formulae give only approximate lengths.

$$\text{D.C. Distance} = \frac{\text{Volts drop} \times \text{wire size (circular mils)}}{\text{Ampere load} \times 2 \times R\ (\text{Resistance per CM ft.} = 12.5)}$$

$$\text{1-PHASE Distance} \doteq \frac{\text{Volts drop} \times \text{wire size (circular mils)}}{\text{Amps. (at given PF)} \times 2 \times R\ (12.5 + \text{Reactance} + \text{Impedance})}$$

For 3-Phase distance on 3-Phase Delta Circuits — use 1-Phase amperes x .866 $\left(\frac{\sqrt{3}}{2}\right)$

3-Phase Distance = 1.15 x 1-Phase distance, approximately

TABLE 7

Table of average circuit lengths for a three-phase, a.c., 230-volt, 60 cycle, Delta, 85% power factor system. Feeders with 1% (2.3 volts) drop with balanced loads. Circuit lengths in feet.

Copper resistance "R"—12.5 ohms per "CM" ft. at 50°C. (122°F.)
Reactance and impedance losses calculated for each wire.
Conductors closely grouped in metallic conduit.

AMPERE LOAD	WIRE SIZE—CIRCULAR MILS					WIRE SIZE—B & S or A.W.G.									
	500	400	350	300	250	4/0	3/0	2/0	1/0	1	2	3	4	6	8
40	710	625	584	530	475	429	364	303	253	208	173	139	113	75	49
50	568	500	467	424	380	343	291	242	203	167	139	111	90	60	39
60	473	417	389	353	317	286	243	202	169	139	115	93	75	50	
70	406	357	333	303	271	245	208	173	145	119	99	79	64	43	
80	355	312	292	265	238	214	182	151	127	104	87	69	56		
90	316	278	259	235	211	191	162	134	113	93	77	62	45		
100	284	250	233	212	190	172	146	121	101	83	69	55			
110	258	227	212	193	173	156	132	110	92	76	63				
120	237	208	195	177	158	143	121	101	84	69	58				
130	218	192	180	163	146	132	112	93	78	64					
140	203	179	167	151	136	123	104	86	72						
150	189	168	156	141	127	114	97	81	67						
160	177	156	146	132	119	107	91	76							
170	167	147	137	125	112	101	86	71							
180	158	139	130	118	106	95	81	67							
190	149	132	123	112	100	90	77								
200	142	125	117	106	95	86	73								
210	135	119	111	101	90	82									
220	129	114	106	96	86	78									
230	123	109	101	92	83	75									
240	118	104	97	88	79										
250	114	100	93	85	76										
260	109	96	90	81	73										
270	105	93	86	78											
280	101	89	83	76											
290	98	86	80	73											
300	95	83	78												
310	92	81	75												
320	89	78													
330	86	76													
340	83	73													
350	81														
360	79														
370	77														
380	75														

Approx. Resistance—Ohms per "CM" Ft. at Copper Temperatures of 50°C to 75°C.

WIRE SIZE	R @ 50°C	R @ 60°C	R @ 75°C
No. 14 to 6	11.6	12.1	12.7
No. 4 to 1/0	12.0	12.5	13.1
2/0 to 4/0	12.4	12.9	13.5
300 MCM	12.7	13.2	13.8
400 MCM	13.0	13.5	14.2
500 MCM	13.4	13.9	14.6
AVERAGE "R"	12.5	13.0	13.6

NOTES:

For 208-volt, 4-wire, "Y" feeders — multiply table circuit lengths by 90%

For 230 volt, 1-phase feeders—multiply table circuit lengths by 85%

For 460-volt, 3- or 4-wire feeders — multiply table circuit lengths by 2

For 2 percent voltage dop — double 1% table circuit lengths

For 3 percent voltage drop—triple 1% table circuit lengths

For aluminum wire which is same size as copper, use 70% of table circuit lengths; or use circuit length of copper wire which is two sizes smaller than aluminum wire under consideration. (Aluminum wire—ohms per CM ft. @ 60°C is 17)

Example:

A 4/0 aluminum wire with 40-ampere load has a circuit length of 303 ft. same as a 2/0 copper wire in table with a voltage drop of one percent.

Only approved materials may be used and the work must conform to the local building codes or fire ordinances and to the rules of the National Board of Fire Underwriters as set forth in its "National Electrical Code." This code is in effect throughout the United States and Canada, and gives definite rules for the installation of all kinds of wiring. It also specifies carefully the kind of material, such as wire, conduit, fuses, etc., that may be installed. Copies of the code may be obtained by applying to the National Fire Protection Association, 60 Batterymarsh St., Boston, Massachusetts, 02110.

Installation of wiring for light or power service, at voltages not exceeding 600 volts, may be done by any of the following plans, all of which are approved by the code, but the use of some of them is restricted in special places.

Open or Exposed Wiring.—Wires are supported on porcelain knobs or cleats; the knobs or cleats should separate the wires about 2½ inches and should be ½ inch from the surface along which they run.

Molding Work.—Wires are run in a wood or metal molding. The metal molding consists of a sheet steel trough or backing and a steel cover which is snapped on the backing after the wires are in place. Wood molding consists of a backing with grooves for the wires and a capping which is nailed to the backing after the wires are in place; this molding is made for two and three wires. Molding work is particularly adapted to the wiring of buildings after their completion and has the advantage of cheapness, simplicity and accessibility.

Rigid Conduit.—Wires are run in unlined conduits which are free from scale on the inside and are coated with enamel on the inside and outside; the outside is sometimes galvanized when used where the pipe is exposed to the weather. Conduits must be continuous from outlet to outlet, at which places metal junction boxes made for the purpose are located; the conduit must properly enter and be secured to all fittings, and the system must be mechanically strong. Conduit affords the best protection to the wires from mechanical injury and may be used for all classes of service. It is chiefly used in buildings of fireproof construction where wires are concealed; it is also frequently used for circuits run exposed in power houses and industrial establishments. Conduit systems must be grounded, that is, connected to the earth, by connecting the conduit to a water pipe (on the street side of the meter); grounding is necessary so that in the case of a breakdown of the wire insulation, the conduit will not be charged to a dangerous potential. Table 8 applying to complete conduit systems shows the size of conduit required for several wires.

Flexible Conduit.—Wires are installed in a flexible conduit that is made of steel strips wound spirally to form a tube; the edges of the strip interlock in such a manner that the tube can be concealed work where rigid conduit could not be used. It is not water-tight and therefore is not as suitable as the rigid conduit where exposed to moisture.

Armored Cable.—A flexible armor similar to the above flexible conduit is placed directly upon the wire. The wire is rubber insulated and covered with a braid the same as the wire used in metal conduit systems. This armored cable is made with either single, double, or triple conductors and is used for the same classes of service as the flexible conduit; in fact, it is used more frequently than the flexible conduit since it is cheaper and easier to install.

Plastic Insulation.—A number of types of wiring insulated with thermoplastic materials are available, some of which are classed as UF (Underground Feeder) in the 1965 National Electric Code. For example, there is a small diameter building wire insulated with polyvinylchloride; designated by the National Electric Code letters TW, made by the Triangle Conduit and Cable Co., which is approved for both wet and dry locations for circuits not exceeding 600 volts and temperatures not exceeding 75° F. The Code permits the use of this type of wire for underground applications in concrete slabs or other masonry in direct contact with earth, in wet locations, and where condensation and moisture accumulation within the raceway may occur. This wire

TABLE 8*

Maximum Number of Conductors in Trade Sizes of Conduit or Tubing

New Work or Rewiring—Types RF-2, RFH-2, R, RH, RW, RHH, RHW, RH-RW†
New Work—FEP, FEPB, RUH, RUW, T, TF, THHN, THW, THWN, TW

Size AWG or MCM	Maximum Number of Conductors in Conduit or Tubing (Based upon % conductor fill, Table 3, Chap. 9, for new work)											
	½ Inch	¾ Inch	1 Inch	1¼ Inch	1½ Inch	2 Inch	2½ Inch	3 Inch	3½ Inch	4 Inch	5 Inch	6 Inch
18	7	12	20	35	49	80	115	176				
16	6	10	17	30	41	68	98	150				
14	4	6	10	18	25	41	58	90	121	155		
12	3	5	8	15	21	34	50	76	103	132	208	
10	1	4	7	13	17	29	41	64	86	110	173	
8	1	3	4	7	10	17	25	38	52	67	105	152
6	1	1	3	4	6	10	15	23	32	41	64	93
4	1	1	1	3‡	5	8	12	18	24	31	49	72
3		1	1	3	4	7	10	16	21	28	44	63
2		1	1	3	3	6	9	14	19	24	38	55
1		1	1	1	3	4	7	10	14	18	29	42
0			1	1	2	4	6	9	12	16	25	37
00			1	1	1	3	5	8	11	14	22	32
000			1	1	1	3	4	7	9	12	19	27
0000				1	1	2	3	6	8	10	16	23
250				1	1	1	3	5	6	8	13	19
300				1	1	1	3	4	5	7	11	16
350				1	1	1	1	3	5	6	10	15
400					1	1	1	3	4	6	9	13
500					1	1	1	3	4	5	8	11
600						1	1	1	3	4	6	9
700						1	1	1	3	3	6	8
750						1	1	1	3	3	5	8
800						1	1	1	2	3	5	7
900						1	1	1	1	3	4	7
1000						1	1	1	1	3	4	6
1250							1	1	1	1	3	5
1500								1	1	1	3	4
1750								1	1	1	2	4
2000								1	1	1	1	3

*1965 National Electrical Code.

†For symbols, see Table 10.

‡Where an existing service run of conduit or electrical metallic tubing does not exceed 50 ft. in length and does not contain more than the equivalent of two quarter bends from end to end, two No. 4 insulated and one No. 4 bare conductors may be installed in 1-inch conduit or tubing.

TABLE 9

Allowable Ampacities of Insulated Copper Conductors†
Not More than Three Conductors in Raceway or Cable or Direct Burial (Based on Room Temperature of 30° C. 86° F.

Size AWG MCM	Temperature Rating of Conductor 60° C (140° F)	75° C (167° F)	85°–90° C (185° F)	110° C (230° F)	125° C (257° F)	200° C (392° F)
14	15	15	25†	30	30	30
12	20	20	30†	35	40	40
10	30	30	40†	45	50	55
8	40	45	50	60	65	70
6	55	65	70	80	85	95
4	70	85	90	105	115	120
3	80	100	105	120	130	145
2	95	115	120	135	145	165
1	110	130	140	160	170	190
0	125	150	155	190	200	225
00	145	175	185	215	230	250
000	165	200	210	245	265	285
0000	195	230	235	275	310	340
250	215	255	270	315	335	...
300	240	285	300	345	380	...
350	260	310	325	390	420	...
400	280	335	360	420	450	...
500	320	380	405	470	500	...
600	355	420	455	525	545	...
700	385	460	490	560	600	...
750	400	475	500	580	620	...
800	410	490	515	600	640	...
900	435	520	555	...	...	...
1000	455	545	585	680	730	...
1250	495	590	645	...	...	...
1500	520	625	700	785	...	...
1750	545	650	735	...	...	...
2000	560	665	775	840	...	...
Correction Factors, Room Temps. Over 30° C. 86° F.						
C. F.						
40 104	.82	.88	.90	.94	.95	...
45 113	.71	.82	.85	.90	.92	...
50 122	.58	.75	.80	.87	.89	...
55 131	.41	.67	.74	.83	.86	...
60 140	...	.58	.67	.79	.83	.91
70 158	...	.35	.52	.71	.76	.87
75 167	...	...	.43	.66	.72	.86
80 176	...	...	.30	.61	.69	.84
90 194	...	...	...	.50	.61	.80
100 212	...	...	...	...	.51	.77
120 248	...	...	...	...	...	.69
140 284	...	...	...	...	...	.59

†National Electric Code, 1965. Note that "ampacity" means capacity in amperes. Note also that the types of cable to which the various temperature ratings apply are given in Table 10.

TABLE 9A

CONDUCTOR APPLICATION*

Trade Name	Type Letter	Max. Operating Temp.	Application Provisions
Rubber-Covered Fixture Wire	†RF-1	60°C 140°F	Fixture wiring. Limited to 300 V.
Solid or 7-Strand	†RF-2	60°C 140°F	Fixture wiring, and as permitted in Section 310-8.
Rubber-Covered Fixture Wire	†FF-1	60°C 140°F	Fixture wiring. Limited to 300 V.
Flexible Stranding	†FF-2	60°C 140°F	Fixture wiring, and as permitted in Section 310-8.
Heat-Resistant Rubber-Covered Fixture Wire	†RFH-1	75°C 167°F	Fixture wiring. Limited to 300 V.
Solid or 7-Strand	†RFH-2	75°C 167°F	Fixture wiring, and as permitted in Section 310-8.
Heat-Resistant Rubber-Covered Fixture Wire	†FFH-1	75°C 167°F	Fixture wiring. Limited to 300 V.
Flexible Stranding	†FFH-2	75°C 167°F	Fixture wiring, and as permitted in Section 310-8.
Thermoplastic-Covered Fixture Wire—Solid or Stranded	†TF	60°C 140°F	Fixture wiring, and as permitted in Section 310-8, and for circuits as permitted in Article 725.
Thermoplastic-Covered Fixture Wire—Flexible Stranding	†TFF	60°C 140°F	Fixture wiring, and as permitted in Section 310-8, and for circuits as permitted in Article 725.
Cotton-Covered, Heat-Resistant, Fixture Wire	*CF	90°C 194°F	Fixture wiring. Limited to 300 V.
Asbestos-Covered Heat-Resistant, Fixture Wire	*AF	150°C 302°F	Fixture wiring. Limited to 300 V. and Indoor Dry Location.

TABLE 9A (Cont'd.)

CONDUCTOR APPLICATION*

Trade Name	Type Letter	Max. Operating Temp.	Application Provisions
Silicone Rubber Insulated Fixture Wire	*SF-1	200°C 392°F	Fixture wiring. Limited to 300 V.
Solid or 7 Strand	*SF-2	200°C 392°F	Fixture wiring and as permitted in Section 310-8.
Silicone Rubber Insulated Fixture Wire	*SFF-1	150°C 302°F	Fixture wiring. Limited to 300 V.
Flexible Stranding	*SFF-2	150°C 302°F	Fixture wiring and as permitted in Section 310-8.
Code Rubber	R	60°C 140°F	Dry locations.
Heat-Resistant Rubber	RH	75°C 167°F	Dry locations.
Heat Resistant Rubber	RHH	90°C 194°F	Dry locations.
Moisture-Resistant Rubber	RW	60°C 140°F	Dry and wet locations. For over 2000 volts, insulation shall be ozone-resistant.
Moisture and Heat Resistant Rubber	RH-RW	60°C 140°F	Dry and wet locations. For over 2000 volts, insulation shall be ozone-resistant.
		75°C 167°F	Dry locations. For over 2000 volts, insulation shall be ozone-resistant.
Moisture and Heat Resistant Rubber	RHW	75°C 167°F	Dry and wet locations. For over 2000 volts, insulation shall be ozone-resistant.
Latex Rubber	RU	60°C 140°F	Dry locations.
Heat Resistant Latex Rubber	RUH	75°C	Dry locations.

TABLE 9A (Cont'd.)

CONDUCTOR APPLICATION*

Trade Name	Type Letter	Max. Operating Temp.	Application Provisions
Moisture Resistant Latex Rubber	RUW	60°C 140°F	Dry and wet locations.
Thermoplastic	T	60°C 140°F	Dry locations.
Moisture-Resistant Thermoplastic	TW	60°C 140°F	Dry and wet locations.
Heat-Resistant Thermoplastic	THHN	90°C 194°F	Dry locations.
Moisture and Heat-Resistant Thermoplastic	THW	75°C 167°F	Dry and wet locations.
Moisture and Heat-Resistant Thermoplastic	THWN	75°C 167°F	Dry and wet locations.
Thermoplastic and Asbestos	TA	90°C 194°F	Switchboard wiring only.
Thermoplastic and Fibrous Outer Braid	TBS	90°C 194°F	Switchboard wiring only.
Synthetic Heat-Resistant	SIS	90°C 194°F	Switchboard wiring only.
Mineral Insulation (Metal Sheathed)	MI	85°C 185°F 250°C 482°F	Dry and wet locations with Type O termination fittings. For special application.
Silicone-Asbestos	SA	90°C 194°F 125°C 257°F	Dry locations. For special application.

TABLE 9A (Cont'd.)

CONDUCTOR APPLICATION*

Trade Name	Type Letter	Max. Operating Temp.	Application Provisions
Fluorinated Ethylene Propylene	FEP or FEPB	90°C 194°F	Dry locations.
		200°C 392°F	Dry locations—special applications.
Varnished Cambric	V	85°C 185°F	Dry locations only. Smaller than No. 6 by special permission.
Asbestos and Varnished Cambric	AVA	110°C 230°F	Dry locations only.
Asbestos and Varnished Cambric	AVL	110°C 230°F	Dry and wet locations.
Asbestos and Varnished Cambric	AVB	90°C 194°F	Dry locations only.
Asbestos	A	200°C 392°F	Dry locations only. In raceways, only for leads to or within apparatus. Limited to 300 V.
Asbestos	AA	200°C 392°F	Dry locations only. Open wiring. In raceways, only for leads to or within apparatus. Limited to 300 V.
Asbestos	AI	125°C 257°F	Dry locations only. In raceways, only for leads to or within apparatus. Limited to 300 V.
Asbestos	AIA	125°C 257°F	Dry locations only. Open wiring. In raceways, only for leads to or within apparatus.
Paper		85°C 185°F	For underground service conductors, or by special permission.

*1965 National Electric Code.

†Fixture wires are not intended for installation as branch circuit conductors nor for the connection of portable or stationary appliances.

is available in American Wire Gauge sizes ranging from 8 to 14 (single strand), as well as in seven-strand, 19-strand, 37-strand and 61-strand types as needed for the high current ratings required in industry or feeder lines.

Pertinent Code provisions applying to this type of wire are:

(a) Underground feeder and branch circuit cable may be used underground, including direct burial in the earth, as feeder or branch circuit cable when provided with overcurrent protection of the rated ampacity.

(b) Where single conductor cables are installed, all cables of the feeder circuit, sub-feeder circuit, or branch circuit, including the neutral conductor, if any, shall be run together in the same trench or raceway.

(c) A minimum depth of 18 inches shall be maintained for conductors and cables buried directly in the earth, when supplementary protection from physical injury such as a covering board, concrete pad, raceway, etc., is not provided.

Demand Calculations for Feeder or Service Wires.—Sizes of feeder wires to supply both light and power loads are determined on a basis of the type of building they are to serve and the floor areas. For example, the minimum watts per unit area and demand factors for single-family dwellings are:

Three watts per square foot, plus 1500 watts for appliances.

For area of 3000 or less square feet, demand 100 percent; for the next 117,000 square feet, 35 percent.

No demand shall be applied in connection with appliance loads.

Calculations for Single Family Dwelling.—The requirements of the National Electrical Code (which is distributed by the National Fire Protection Association, 60 Batterymarsh St., Boston, Mass., 02110) govern the entire operation of planning and installation of wiring systems. They give specifications for wire sizes and circuit loading in terms of the power requirements of the lights and appliances to be installed. As a first step in computing the power requirements of appliances, the following list of average requirements of representative appliances shown in Table 9B should be used:

TABLE 9B

Appliance	Average Wattage	Appliance	Average Wattage
Blanket	150	Immersion heater	300
Bread mixer	200	Iron, household	1000
Clocks	3	Iron, travelers'	330
Cigar lighter	100	Ironer	1320
Coffee maker	550	Kitchen mixer with grinder	200
Coffee percolator	450	Mechanical exerciser	500
Curling iron	20	Phonograph	40
Chafing dish	600	Piano player	125
Cream whipper	75	Range	8000
Dish washer	100	Refrigerator	300
Egg boiler	250	Radio	100
Fan, 8-inch	30	Roaster	1320
Fan, 10-inch	35	Sewing machine	75
Fan, 12-inch	50	Soldering iron	200
Frying pan	600	Sun lamp	450
Griddle	450	Teakettle	400
Grill	600	Teapot	400
Hair drier	50	Toaster	450
Heater (radiant)	1000	Vacuum cleaner	160
Heating pad	50	Vibrator	50
Hot plate	660	Washing Machine	175
Humidifier	500	Water heater	2000
Ice-cream freezer	300	Waffle iron	660

To illustrate the method of calculation, it will be applied to a small family house, consisting of two floors and a basement, of which the plan views are shown in Figs. 70, 71 and 72.

Beginning with the problem of general lighting, we can apply the following Code regulation:

General Illumination. . . . not less than one branch circuit be installed for each 500 square feet of floor area. . . .

Since the floor space referred to here means floor space intended for occupancy, the first step is to determine the number of square feet contained in the first and second floors of the house. The first floor, without the dinette, measures $31' \times 15'\ 6''$, outside dimensions. Thus the first floor, without the dinette,

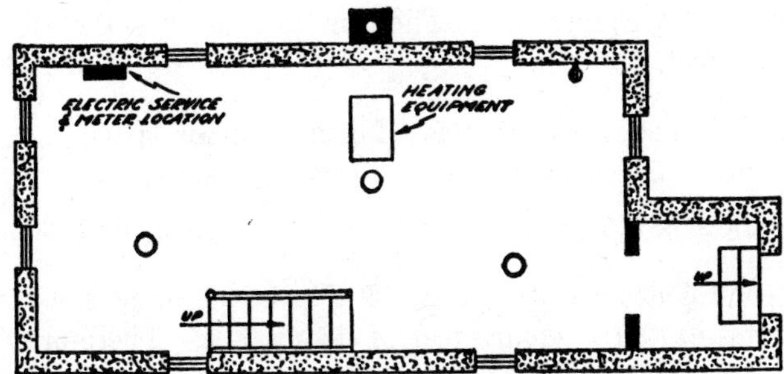

FIG. 70.—Floor Plan—Basement.

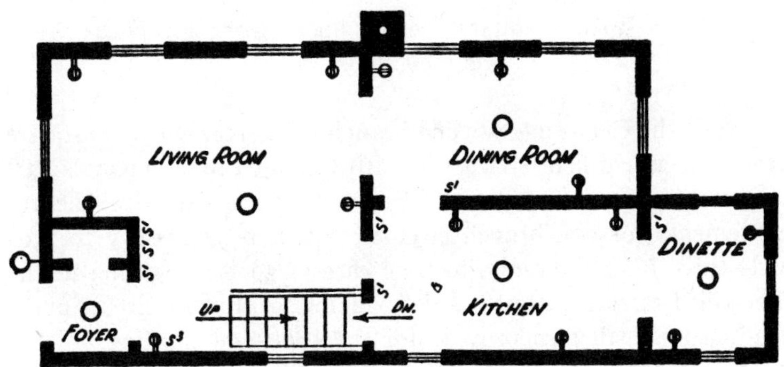

FIG. 71.—Floor Plan—First Floor.

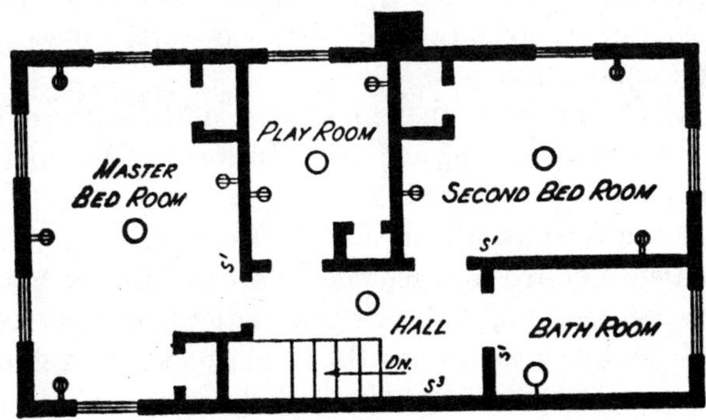

FIG. 72.—Floor Plan—Second Floor.

contains 480½ square feet. The dinette measures 6′ 6″ × 8′, or 52 square feet.

	480½	square feet of general floor space
plus	52	square feet of dinette floor space
makes	532½	square feet, total floor space, first floor

The second floor measures 31′ × 15′ 6″, outside dimensions, making a total of 480½ square feet of floor space. Therefore:

	480½	square feet of floor space, second floor
plus	532½	square feet of floor space, first floor
makes	1013	square feet of floor space, for both occupied floors

Now, the Code suggests one branch circuit for each 500 square feet of occupied floor space. So: 1013 divided by 500 equals 2.02 circuits for general lighting. Since we have exceeded the requirements for two branch circuits, it will be necessary to provide three branch circuits to take care of general lighting needs. The third circuit is thus a safety factor to prevent circuit overloading and will also provide for the basement lighting.

So far, then, our calculations call for three branch circuits for general lighting. In addition, however, we must make provision for a small-appliance circuit, as required by the Code. Our calculated small appliance load is 12,000 watts, so we plan a 20 ampere appliance circuit.

Our total provision, therefore, must include three branch circuits for general lighting and one branch circuit for appliances, or a total of four branch circuits.

We can now go ahead and list the location, kind, and number of outlets required for each floor, starting with the basement.

Basement (see Fig. 70). 1. One outlet (ceiling) near the foot of the stairway connecting the basement with the first floor. This outlet is to be controlled by a switch located in the kitchen, near the entrance to the basement stairway.

2. One outlet (ceiling) over the heating equipment, so placed that it will illuminate the gauges. This outlet will have key-socket control.

3. One outlet (ceiling) in the open space near the rear entrance to the basement; this will be key-socket controlled.

4. One receptacle outlet (side-wall) for an electric washing machine, located in the corner of the basement. This outlet is to be of the 3-pole type.

First Floor (see Fig. 71). 1. One outside outlet (side-wall) at the main entrance door to the house, to be controlled by a switch located in the living room, near the entrance door.

2. One outlet (ceiling) in the foyer, to be controlled by a switch located in the living room, near the entrance door.

3. One outlet (ceiling) in the living room, to be controlled by a switch located near the living-room main entrance door.

4. One switch located at the foot of the stairs leading to the second floor to control the landing hall outlet on the second floor.

5. One outlet (ceiling) in the dining room, to be controlled by a switch located near the door between the dining room and the kitchen, on the dining-room side.

6. One switch to be located in the kitchen, near the entrance door leading to the basement, to control the basement light at the foot of the basement stairs.

7. One outlet (ceiling) in the kitchen, to be controlled by a switch located near the door between the kitchen and the living room, on the kitchen side.

8. One outlet (ceiling) in the dinette, to be controlled by a switch located near the opening between the kitchen and the dinette, on the dinette side.

Receptacles for First Floor. Living room: 16′ × 14′ 6″; total (gross) distance around room is 61′. Thus, 61′ divided by 12 equals 5.0, or 5 receptacles.

Dining room: 13′ 6″ × 7′; total (gross) distance around room is 41′. Thus, 41′ divided by 12 equals 3.4, or 3 receptacles.

Kitchen: 13′ 6″ × 7′; total (gross) distance around room is 41′. Thus, 41′ divided by 12 equals 3.4, or 3 receptacles.

Dinette: 6′ × 7′; total (gross) distance around room is 26′. Thus, 26′ divided by 12 equals 2.1, or 2 receptacles.

Second Floor (see Fig. 72). 1. One outlet (ceiling) in the master bedroom, to be controlled by a switch located in the bedroom, near the entrance door.

2. One outlet (ceiling) in the playroom, to be pull-chain controlled at the fixture.

3. One outlet (ceiling) in the second bedroom, to be controlled by a switch located on the bedroom side of the entrance door.

4. One outlet (wall-bracket) in the bathroom, to be located over the medicine chest and to be controlled by a switch located on the bathroom side of the entrance door.

5. One outlet (ceiling) in the hall, to be controlled by a switch located near the stair landing, and the same fixture to be also controlled by the switch on the first floor, located near the foot of the stairs.

Receptacles for Second Floor—Master bedroom: 9′ 6″ × 14′ 6″; total (gross) distance around room is 48′. Thus, 48′ divided by 12 equals 4.0, or 4 receptacles.

Playroom: 6′ 6″ × 8′ 6″; total (gross) distance around room is 30′. Thus, 30′ divided by 12 equals 2.5, or 2 receptacles.

Second bedroom: 13′ × 8′ 6″; total (gross) distance around room is 43′. Thus, 43′ divided by 12 equals 3.5, or 3 receptacles.

The total of the various kinds of electrical outlets needed for the installation is:

Ceiling outlets and wall-brackets	14
Receptacle outlets	23

We are providing four branch circuits for fourteen ceiling and wall-bracket outlets and twenty-three receptacle outlets. Remember, the Code requires that receptacle outlets in the dining room, kitchen, and dinette, and also the washing machine outlet in the basement, be placed on a special circuit. Thus nine of the twenty-three receptacle outlets must be placed on an independent circuit. That leaves fourteen ceiling and wall-bracket outlets and fourteen receptacle outlets to be divided among three branch circuits.

Proportioning the Load. The next step, then, is to proportion the load—to divide the fourteen lighting outlets and the fourteen convenience receptacle outlets among three branch circuits so that no single branch circuit will be overloaded. There are more ways than one of proportioning the load. However, it is necessary to bear in mind one special point: It is advisable, where possible, to place a ceiling outlet on one circuit, and the receptacle outlets in the same room on another circuit, so that a fuse blow-out in either circuit will not cut off the room entirely from electric service. This method may involve the use of a little more armored cable, but it will prove to be worth the extra cost.

A wiring diagram showing every outlet, floor by floor, will be helpful, because by means of such wiring diagrams we can plan the distribution of the load, circuit by circuit. Let us then examine Figure 73.

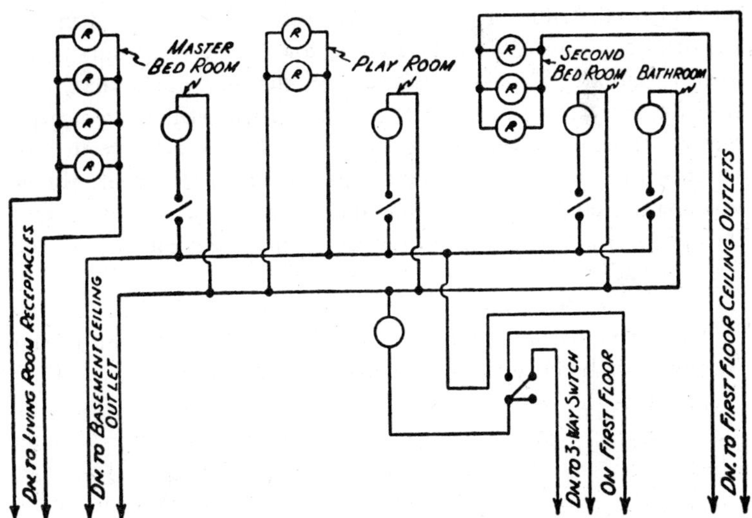

FIG. 73.—Wiring Diagram for Second Floor.

According to the specification, the following outlets are to be installed on the second floor:

Master bedroom	ceiling outlet
	switch to control ceiling outlet
	4 base-receptacles
Playroom	ceiling outlet, pull-chain control
	2 base-receptacles
Second bedroom	ceiling outlet
	switch to control ceiling outlet
	3 base-receptacles
Bathroom	wall-bracket
	switch to control wall-bracket
Hall	ceiling outlet
	3-way switch to control ceiling outlet

Note that the four master bedroom receptacles are connected to the circuit which picks up the living room receptacles (Figure 73), the playroom receptacles are connected to the second floor ceiling outlet circuit, and the second bedroom receptacle outlets are connected to the first floor ceiling outlet circuit. By this method of distribution—placing the convenience outlets in the various rooms on different circuits, and the ceiling outlets on another circuit (if possible)—we gain the advantage that, if a fuse for one circuit blows out, each room will still have electric service, unless, of course, the main fuse blows out. Note also (Figure 73) that a 3-wire cable must be run between the three-way switch controlling the hall outlet on the second floor and the three-way switch controlling the same outlet located on the first floor near the stairway.

Figure 74 shows the circuit distribution for the first floor as expressed in a wiring diagram. According to the specification, the following outlets are to be installed on the first floor:

Outside main entrance door ...	wall-bracket
	switch to control wall-bracket
Foyer	ceiling outlet
	switch to control ceiling outlet

Living room . . . . . . ceiling outlet
switch to control ceiling outlet
5 base-receptacles
3-way switch for second-floor hall outlet

Dining room . . . . . ceiling outlet
switch to control ceiling outlet
3 base-receptacles (appliance circuit)

Kitchen . . . . ceiling outlet
switch to control ceiling outlet
3 receptacles (appliance circuit)
switch to control basement outlet

Dinette . . . . ceiling outlet
switch to control ceiling outlet
2 base-receptacles (appliance circuit)

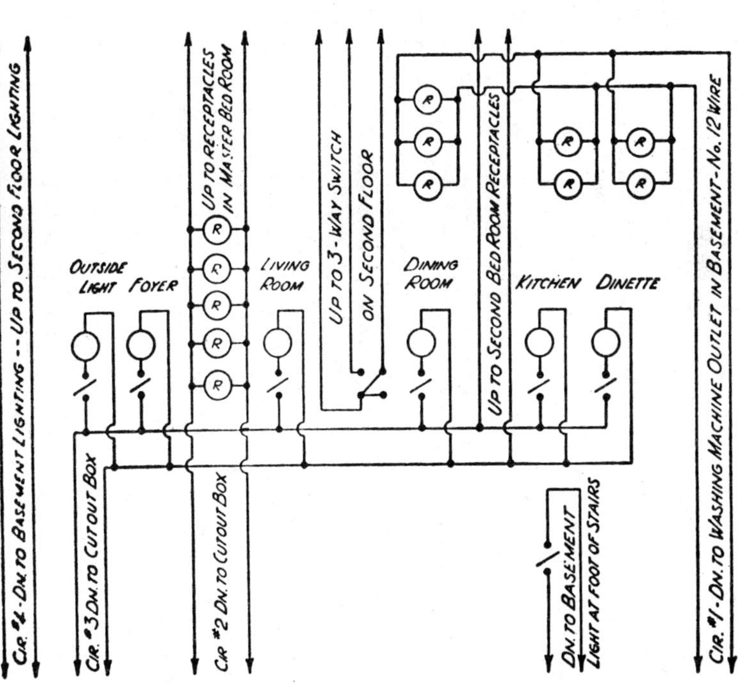

Fig. 74.—Wiring Diagram for First Floor.

Note, in Figure 74, that each circuit has been numbered to assist in clarifying the wiring diagram. Circuit 1 has been chosen as the appliance circuit, and the receptacles in dining room, kitchen, and dinette have therefore been placed on Circuit, which is continued down to the washing machine outlet in the basement (see Fig. 75) and then to the branch circuit cutout box at the service installation. This circuit, which services receptacles in dining room, kitchen, dinette, and laundry only, must be installed with not less than No. 12 wire, to comply with the National Electrical Code.

Circuit 2 runs directly from the branch cutout box in the basement service installation to the living-room base-receptacles, where it is tide-in with the four base-receptacles in the master bedroom on the second floor. Thus Circuit 2 includes nine convenience receptacles in the living room and the master bedroom.

Circuit 3 includes all the ceiling lighting on the first floor and

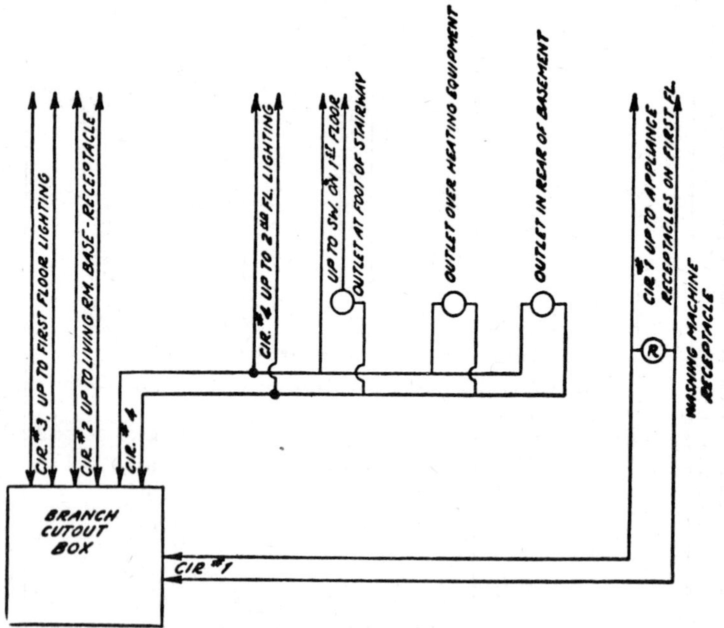

FIG. 75.—Wiring Diagram for Basement.

the three convenience receptacles in the second-floor second bedroom. This includes a total of nine outlets: the outside wall-bracket, foyer ceiling outlet, living room ceiling outlet, dining room ceiling outlet, kitchen ceiling outlet, dinette ceiling outlet, and the three convenience receptacles in the second bedroom on the second floor. It must be remembered and understood, of course, that, since switches consume no current, they are naturally not included in these calculations for circuit loading.

In distributing the general lighting load on the first and second floors among three branch circuits, it is necessary to keep in mind the three basement outlets, which must also be cared for in one or another of these three branch circuits. Living-room, dining-room, and kitchen ceiling outlets generally use higher-wattage lamps than do those in bedrooms and bathrooms. We therefore consider it advisable to make one circuit of all the ceiling outlets on the first floor (that is, Circuit 3), plus the three convenience receptacles in the second bedroom, and place all the basement outlets and the ceiling outlets on the second floor, plus the two playroom convenience receptacles, on Circuit 4.

To make the wiring diagram complete for the first floor, we have included the wiring for the switch in the kitchen which controls the basement outlet at the foot of the basement stairs and have also included that for the three-way switch in the living room near the foot of the stairway leading to the second floor, that is, the switch that controls the hall outlet on the second floor. This completes, then, the picture of the total wiring needs for the first floor.

Figure 75 shows the circuit distribution for the basement. According to the specification, a total of three ceiling outlets and one appliance receptacle outlet is to be installed, as follows:

Basement .... ceiling outlet near the foot of the stairs leading from the first floor, controlled by a switch on the first floor
ceiling outlet over heating equipment, key-socket controlled
ceiling outlet in open cellar space, key-socket controlled
appliance receptacle outlet service equipment

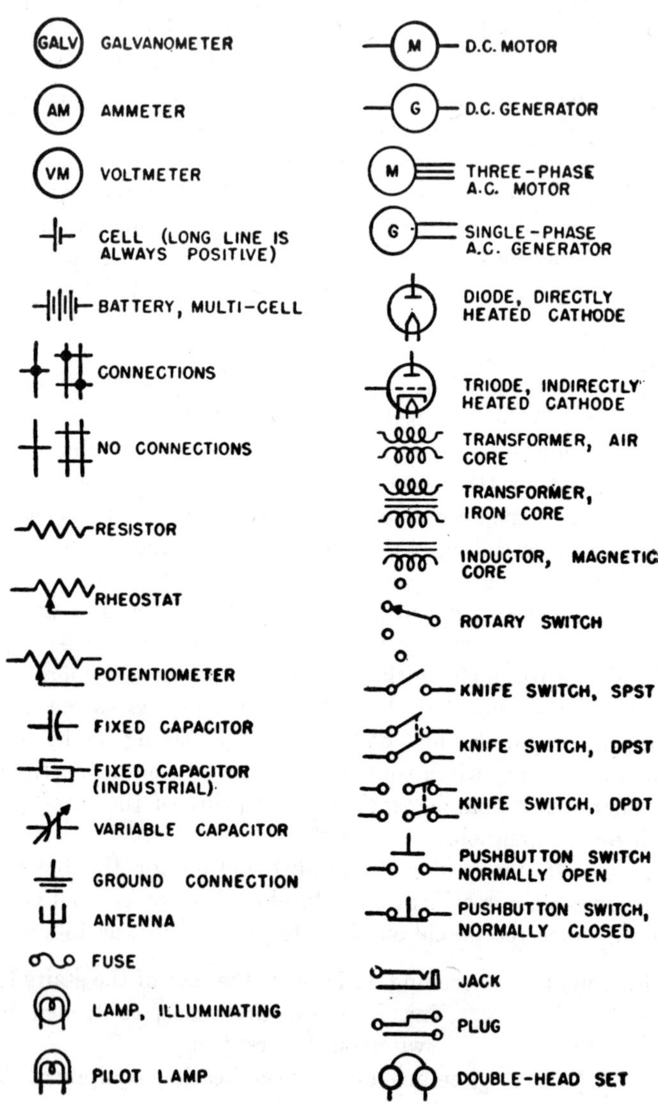

FIG. 70.—Electrical Symbols.

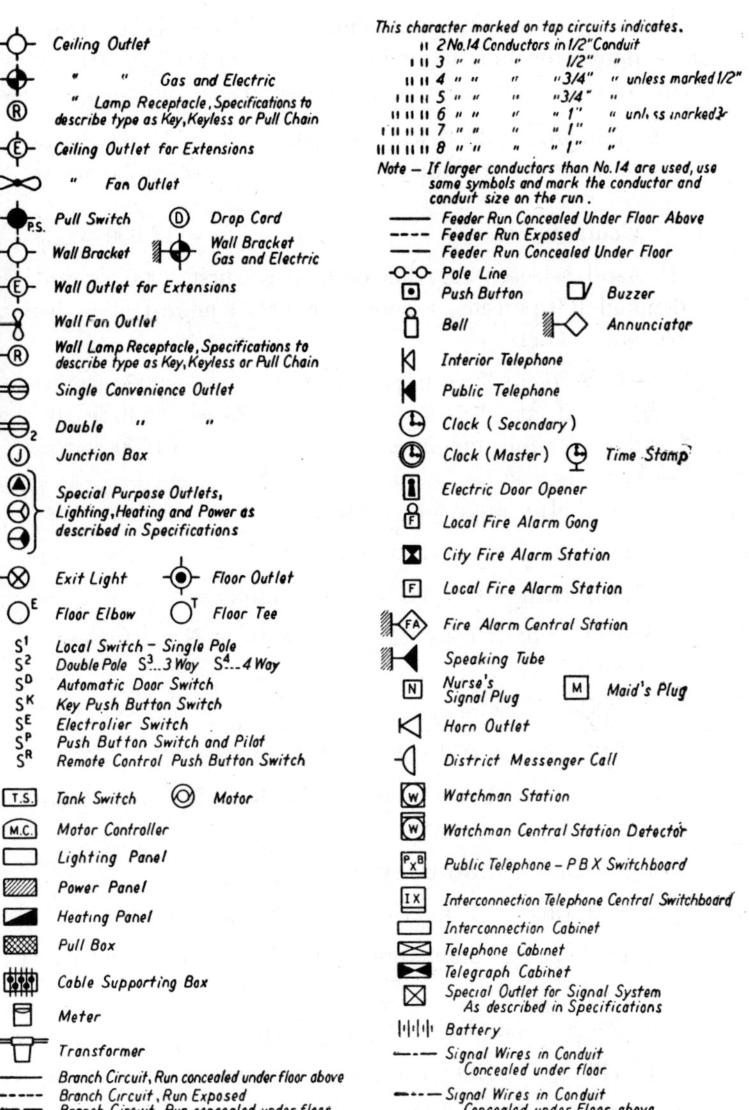
Ceiling Outlet
" " Gas and Electric
" Lamp Receptacle, Specifications to describe type as Key, Keyless or Pull Chain
Ceiling Outlet for Extensions
" Fan Outlet
Pull Switch
P.S.
Drop Cord
Wall Bracket
Wall Bracket Gas and Electric
Wall Outlet for Extensions
Wall Fan Outlet
Wall Lamp Receptacle, Specifications to describe type as Key, Keyless or Pull Chain
Single Convenience Outlet
Double " "
Junction Box
Special Purpose Outlets, Lighting, Heating and Power as described in Specifications
Exit Light
Floor Outlet
Floor Elbow
Floor Tee
S1 Local Switch – Single Pole
S2 Double Pole S3...3 Way S4...4 Way
SD Automatic Door Switch
SK Key Push Button Switch
SE Electrolier Switch
SP Push Button Switch and Pilot
SR Remote Control Push Button Switch
T.S. Tank Switch
Motor
M.C. Motor Controller
Lighting Panel
Power Panel
Heating Panel
Pull Box
Cable Supporting Box
Meter
Transformer
Branch Circuit, Run concealed under floor above
Branch Circuit, Run Exposed
Branch Circuit, Run concealed under floor
This character marked on tap circuits indicates.
2 No. 14 Conductors in 1/2" Conduit
3 " " " 1/2" "
4 " " " "3/4" " unless marked 1/2"
5 " " " "3/4" "
6 " " " " 1" " unless marked 3/4
7 " " " " 1" "
8 " " " " 1" "
Note – If larger conductors than No. 14 are used, use same symbols and mark the conductor and conduit size on the run.
Feeder Run Concealed Under Floor Above
Feeder Run Exposed
Feeder Run Concealed Under Floor
Pole Line
Push Button
Buzzer
Bell
Annunciator
Interior Telephone
Public Telephone
Clock (Secondary)
Clock (Master)
Time Stamp
Electric Door Opener
Local Fire Alarm Gong
City Fire Alarm Station
Local Fire Alarm Station
Fire Alarm Central Station
Speaking Tube
Nurse's Signal Plug
Maid's Plug
Horn Outlet
District Messenger Call
Watchman Station
Watchman Central Station Detector
Public Telephone – P B X Switchboard
Interconnection Telephone Central Switchboard
Interconnection Cabinet
Telephone Cabinet
Telegraph Cabinet
Special Outlet for Signal System As described in Specifications
Battery
Signal Wires in Conduit Concealed under floor
Signal Wires in Conduit Concealed under Floor above

FIG. 71.—Electrical Symbols for Architectural Plans.

ILLUSTRATION: What minimum size of feeder is required for a single-family dwelling having a floor area of 3800 square feet exclusive of unoccupied cellars, unfinished attics, and open porches?

Area in sq. ft., 3800 $\times$ 3 watt per sq. ft.	=	11,400 watts
Allowance for appliances	=	1500 watts
Computed load	=	12,900 watts

Demand selected for this occupancy, first 3000 square feet = demand 100 percent; excess over 3000 square feet = demand 35 percent. Then

3000 sq. ft. at 3 watts per sq. ft. $\times$ 1	=	9000 watts
800 sq. ft. at 3 watts per sq. ft. $\times$ 0.35	=	840 watts
Allowance for appliances	=	1500 watts
Load after applying demand	=	11,340 watts

For 115-volt, 2-wire system:

11,340 watts $\div$ 115 volts = 98.8 amperes

Size of Type R conductors (Table 9) = No. 1 for each wire (Ans.)

For 230-volt, 2-wire system:

11,340 watts $\div$ 230 volts = 49.4 amperes

Size of Type R conductors (Table 9) = No. 6 for each wire (Ans.)

For 115–230-volt, 3-wire system:

11,340 watts $\div$ 2 $\times$ 115 volts = 49.4 amperes

Size of Type R conductors (Table 9) = No. 6 for each wire (Ans.)

For 120–208 volt, 4-wire, 3-phase system:

11,340 $\div$ 3 $\times$ 120 volts = 31.5 amperes

Size of Type R conductors (Table 9) = No. 8 for each wire (Ans.)

The specifications for buildings other than single-family dwellings are contained in the "National Electrical Code" and the computations are carried out in the same manner as above.

Grounding.—In installing electric wiring and appliances, the matter of grounding is so important that the pertinent provisions of the 1965 National Electrical Code should always be followed. Therefore selected provisions of the Code are reproduced below.

250-32. Service Conductor Enclosures. Service raceways, service cable sheaths or armoring, when of metal, shall be grounded.

250-33. Other Conductor Enclosures. Metal enclosures for conductors shall be grounded, except they need not be grounded in runs of less than 25 feet which are free from probable contact with ground, grounded metal, metal lath or conductive thermal insulation and which, where within reach from grounded surfaces, are guarded against contact by persons.

250-42. Fixed Equipment-General. Under any of the following conditions, exposed, noncurrent-carrying metal parts of fixed equipment, which are liable to become energized, shall be grounded:

(a) Where equipment is supplied by means of metal-clad wiring;

(b) Where equipment is located in a wet location and is not isolated;

(c) Where equipment is located within reach of a person who can make contact with any grounded surface or object;

(d) Where equipment is located within reach of a person standing on the ground;

(e) Where equipment is in a hazardous location;

(f) Where equipment is in electrical contact with metal or metal lath;

(g) Where equipment operates with any terminal at more than 150 volts to ground, except as follows:

(1) Enclosures for switches or circuit breakers where accessible to qualified persons only;

(2) Metal frames of electrically heated devices, exempted by special permission, in which case the frames shall be permanently and effectively insulated from ground;

(3) Transformers mounted on wooden poles at a height of more than 8 feet from the ground.

250-43. Fixed Equipment—Specific. Exposed, noncurrent-carrying metal parts of the following kinds of equipment, regardless of voltage, shall be grounded:

(a) Frames of motors;

(b) Controller cases for motors, except lined covers of snap switches;

(c) Electric equipment of elevators and cranes;

(d) Electric equipment in garages, theatres and motion picture studios, except pendant lampholders on circuits of not more than 150 volts to ground;

(e) Motion-picture projection equipment;

(f) Electric signs and associated equipment, unless these are inaccessible to unauthorized persons and are also insulated from ground and from other conductive objects;

250-45. Portable Equipment. Under any of the following conditions, exposed noncurrent-carrying metal parts of portable equipment, which are liable to become energized, shall be grounded:

(a) In hazardous locations;

(b) When operated at more than 150 volts to ground, except:

(1) Motors, where guarded;

(2) Metal frames of electrically heated appliances;

(c) In residential occupancies, (1) clothes-washing, clothes-drying, and dish-washing machines, sump pumps, and (2) portable, hand held, motor operated tools and appliances of the following types: drills, hedge clippers, lawn mowers, wet scrubbers, sanders and saws.

Exception: Such tools and appliances protected by an approved system of double insulation, or its equivalent, need not be grounded. Where such an approved system is employed the equipment shall be distinctively marked.

Portable tools or appliances not provided with special insulating or grounding protection are not intended to be used in damp, wet or conductive locations.

(d) In other than residential occupancies, (1) portable appliances used in damp or wet locations, or by persons standing on the ground or on metal floors or working inside of metal tanks or boilers, and (2) portable tools which are likely to be used in wet and conductive locations shall be grounded except that they need not be grounded where supplied through an insulating transformer with ungrounded secondary of not over 50 volts.

It is recommended that the frames of all portable motors which operate at more than 50 volts to ground be grounded.

Methods of Grounding. 250-51. Effective Grounding. The path to ground from circuits, equipment, and conductor enclosures shall (1) be permanent and continuous and (2) shall have ample carrying capacity to conduct safely any currents liable to be imposed on it, and (3) shall have impedance sufficiently low to limit the potential above ground and to facilitate the operation of the overcurrent devices in the circuit.

250-52. Location of System Ground Connection. The grounding conductor may be connected to the grounded conductor of the wiring system at any convenient point on the premises on the supply side of the service disconnecting means.

It is recommended that high capacity services have the grounding conductor connected to the grounded conductor of the system within the service entrance equipment enclosure.

250-53. Common Use of Grounding Conductor. The grounding conductor of a wiring system shall also be used for grounding equipment, conduit and other metal raceways or enclosures for conductors, including service conduit or cable sheath and service equipment.

250-56. Short Sections of Raceway. Isolated sections of metal raceway or cable armor, where required to be grounded, shall preferably be grounded by connecting to other grounded raceway or armor, but may be grounded in accordance with Section 250-57.

250-57. Fixed Equipment.

(a) Metal boxes, cabinets and fittings, or noncurrent-carrying metal parts of other fixed equipment, where metallically connected to grounded cable armor or metal raceway, are considered to be grounded by such connection.

(b) Where not so connected they may be grounded in one of the following ways:

 (1) By a grounding conductor run with circuit conductors; this conductor may be uninsulated, but where it is provided with an individual covering, the covering shall be finished a continuous green color or a continuous green color with a yellow stripe.

 (2) By a separate grounding conductor installed the same as a grounding conductor for conduit and the like;

250-58. Equipment on Structural Metal.

(a) Electric equipment secured to and in contact with the grounded structural metal frame of a building, shall be deemed to be grounded.

(b) Metal car frames supported by metal hoisting cables attached to or running over sheaves or drums of elevator machines shall be deemed to be grounded where the machine is grounded in accordance with this Code.

250-59. Portable Equipment. Noncurrent-carrying metal parts of portable equipment may be grounded in any one of the following ways:

(a) By means of the metal enclosure of the conductors feeding such equipment, provided an approved grounding-type attachment plug is used, one fixed contacting member being for the purpose of grounding the metal enclosure, and provided, further, that the metal enclosure of the conductors is attached to the attachment plug and to the equipment by connectors approved for the purpose;

Exception: The grounding contacting member of grounding type attachment plugs on the power supply cord of hand-held tools or hand-held appliances may be of the movable self-restoring type.

Attachment plug caps are not intended to be used as terminations for metal-clad cable or flexible metal conduit.

(b) By means of a grounding conductor run with the power supply conductors in a cable assembly or flexible cord that is properly terminated in an approved grounding-type attachment plug having a fixed grounding contacting member. The grounding conductor in a cable assembly may be uninsulated; but where an individual covering is provided for such conductors it shall be finished a continuous green color or a continuous green color with a yellow stripe.

Exception: The grounding contacting member of grounding type attachment plugs on the power supply cord of hand-held tools or hand-held appliances may be of the movable self-restoring type.

(c) A separate flexible wire or strap, insulated or bare, protected as well as practicable against physical damage may be used only by special permission except where a part of an approved portable equipment.

250-60. Frames of Electric Ranges and Electric Clothes Dryers. Frames of electric ranges and electric clothes dryers shall be grounded by any of the means provided for in Sections 250-57 and 250-59 or where served by 120-240 volt, three-wire branch circuits, they may be grounded by connection to the grounded circuit conductors, provided the grounded circuit conductors are not smaller than 10 AWG. The frames of wall-mounted ovens and counter-mounted cooking units shall be grounded and may be grounded in the same manner as electric ranges.

Wire Sizes for Branch Circuits.—That portion of the supply conductors which extends from the street or duct or transformers to the service switch of the building supplied is called the *service circuit.* That portion of the wiring system which extends beyond the final automatic overload protective device (fuse box) is called the *branch circuit.*

The sizes of wire required for lighting circuits or combination lighting and power circuits for dwellings and apartments connected to separate meters may be computed as above. However, most local codes and good practice require a minimum size of No. 14 wire for these circuits while No. 18 flexible wire is permitted in fixtures and drop cords.

These minimum sizes are usually the governing factors for ordinary requirements. Where, however, special heating or power units are to be used, the sizes of wire must be computed or obtained from a table. If a circuit is to be run for a motor, and the voltage and the current which the motor will use are known, the size of the wire required may be found in Table 10.

ILLUSTRATION: What minimum size of Type R insulated copper wire would be required for a motor with a full-load current rating of 40 amperes?

Running down column 1 of Table 10 until 40 is reached, size of Type R wire is found in column 2 to be No. 6. (Ans.)

If the wiring is being done for a motor whose power requirements are not known, but whose horsepower is known, the current required may be found from Table 11, 12, or 13.

ILLUSTRATION: What size of Type RH wire would be required for a 50-horsepower, 220-volt, 3-phase, induction-type, alternating-current motor?

Current required (from Table 13) = 125 amperes

Size of wire (from Table 10) = No. 00 (Ans.)

TABLE 10

Conductor Sizes and Overcurrent Protection for Motors *

Col. No. 1	2	3	5	6	7	8	9	10
					Maximum Allowable Rating of Branch Circuit Fuses			
Full load current rating of motor amperes	Minimum size conductor in raceways For conductors in air or for other insulations see tables 1 and 2 AWG and MCM		For Running Protection of Motors****		With Code Letters Single-phase and squirrel cage and synchronous. Full voltage, resistor and reactor starting, Code letters F to V inc. Without Code Letters Same as above.	With Code Letters Single-phase, squirrel cage and synchronous. Full voltage, resistor or reactor start, Code letters B–E inc. Autotransformer start, Code letters F to V inc. Without Code Letters Squirrel cage and synchronous, autotransformer start, high reactance squirrel cage.*** Both not more than 30 amperes	With Code Letters Squirrel cage and synchronous Autotransformer start, Code letters B–E inc. Without Code Letters Squirrel cage and synchronous autotransformer start, high reactance squirrel cage.*** Both more than 30 amperes	With Code Letters All motors. Code letter A.* Without Code Letters DC and wound-rotor motors.
			Maximum rating of non-adjustable protective devices	Maximum setting of adjustable protective device				
	Type R Type T	Type RH	Amperes	Amperes				
1**	14	14	2*	1.25*	15	15	15	15
2**	14	14	3*	2.50*	15	15	15	15
3**	14	14	4*	3.75*	15	15	15	15
4**	14	14	6*	5.0 *	15	15	15	15
5**	14	14	8*	6.25*	15	15	15	15
6**	14	14	8*	7.50*	20	15	15	15
7	14	14	10*	8.75*	25	20	15	15
8	14	14	10*	10.0 *	25	20	20	15
9	14	14	12*	11.25*	30	25	20	15
10	14	14	15*	12.50*	30	25	20	15
11	14	14	15*	13.75*	35	30	25	20
12	14	14	15	15.00	40	30	25	20
13	12	12	20	16.25	40	35	30	20
14	12	12	20	17.50	45	35	30	25
15	12	12	20	18.75	45	40	30	25
16	12	12	20	20.00	50	40	35	25
17	10	10	25	21.25	60	45	35	30
18	10	10	25	22.50	60	45	40	30
19	10	10	25	23.75	60	50	40	30
20	10	10	25	25.0	60	50	40	30
22	10	10	30	27.50	70	60	45	35
24	10	10	30	30.00	80	60	50	40
26	8	10	35	32.50	80	70	60	40
28	8	10		35.00	90	70	60	45

* 1953 *National Electrical Code.* For further information on the use of this table, see the latest edition of the Code.

TABLE 10

Conductor Sizes and Overcurrent Protection for Motors *—*Continued*

Col. No. 1	2	3	5	6	7	8	9	10
					Maximum Allowable Rating of Branch Circuit Fuses			
Full load current rating of motor amperes	Minimum size conductor in raceways For conductors in air or for other insulations see tables 1 and 2 AWG and MCM		For Running Protection of Motors****		With Code Letters Single-phase and squirrel cage and synchronous. Full voltage, resistor and reactor starting, Code letters F to V inc. Without Code Letters Same as above.	With Code Letters Single-phase, squirrel cage and synchronous. Full voltage, resistor or reactor start, Code letters B-E inc. Autotransformer start, Code letters F to V inc. Without Code Letters Squirrel cage and synchronous, autotransformer start, high reactance squirrel cage.*** Both not more than 30 amperes	With Code Letters Squirrel cage and synchronous Autotransformer start, Code letters B-E inc. Without Code Letters Squirrel cage and synchronous autotransformer start, high reactance squirrel cage.*** Both more than 30 amperes.	With Code Letters All motors. Code letter A. Without Code Letters DC and wound-rotor motors.
	Type R Type T	Type RH	Maximum rating of non-adjustable protective devices Amperes	Maximum setting of adjustable protective device Amperes				
30	8	8	40	37.50	90	70	60	45
32	8	8	40	40.00	100	80	70	50
34	6	8	45	42.50	110	90	70	60
36	6	8	45	45.00	110	90	80	60
38	6	6	50	47.50	125	100	80	60
40	6	6	50	50.00	125	100	80	60
42	6	6	50	52.50	125	110	90	70
44	6	6	60	55.0	125	110	90	70
46	4	6	60	57.50	150	125	100	70
48	4	6	60	60.0	150	125	100	80
50	4	6	60	62.50	150	125	100	80
52	4	6	70	65.0	175	15	110	80
54	4	4	70	67.50	175	150	110	90
56	4	4	70	70.00	175	150	120	90
58	3	4	70	72.50	175	150	120	90
60	3	4	80	75.00	200	150	120	90
62	3	4	80	77.50	200	175	125	100
64	3	4	80	80.00	200	175	150	100
66	2	4	80	82.50	200	175	150	100
68	2	4	90	85.00	225	175	150	110
70	2	3	90	87.50	225	175	150	110
72	2	3	90	90.00	225	200	150	110
74	2	3	90	92.50	225	200	150	125
76	2	3	100	95.00	250	200	175	125
78	1	3	100	97.50	250	200	175	125
80	1	3	100	100.00	250	200	175	125
82	1	2	110	102.50	250	225	175	125
84	1	2	110	105 00	250	225	175	150
86	1	2	110	107 50	300	225	175	150
88	1	2	110	110.00	300	225	200	150
90	0	2	110	112.50	300	225	200	150
92	0	2	125	115.00	300	250	200	150
94	0	1	125	117.50	300	250	200	150
96	0	1	125	120.00	300	250	200	150
98	0	1	125	122.50	300	250	200	150
100	0	1	125	125.00	300	250	200	150
105	00	1	150	131.5	350	300	225	175

* 1953 *National Electrical Code*. For further information on the use of this table, see the latest edition of the Code.

TABLE 10

Conductor Sizes and Overcurrent Protection for Motors—*Continued*

Col. No. 1	2	3	5	6	7	8	9	10
					Maximum Allowable Rating of Branch Circuit Fuses			
Full load current rating of motor amperes	Minimum size conductor in raceways For conductors in air or for other insulations see tables 1 and 2 AWG and MCM		For Running Protection of Motors****		With Code Letters Single-phase and squirrel cage and synchronous. Full voltage, resistor and reactor starting, Code letters F to V inc. Without Code Letters Same as above.	With Code Letters Single-phase, squirrel cage and synchronous. Full voltage, resistor or reactor start, Code letters B–E inc. Autotransformer start, Code letters F to V inc. Without Code Letters Squirrel cage and synchronous, autotransformer start, high reactance squirrel cage.*** Both not more than 30 amperes	With Code Letters Squirrel cage and synchronous Autotransformer start, Code letters B–E inc. Without Code Letters Squirrel cage and synchronous autotransformer start, high reactance squirrel cage.*** Both more than 30 amperes	With Code Letters All motors. Code letter A. Without Code Letters DC and wound-rotor motors.
	Type R Type T	Type RH	Maximum rating of non-adjustable protective devices Amperes	Maximum setting of adjustable protective device Amperes				
110	00	0	150	137.5	350	300	225	175
115	00	0	150	144.0	350	300	250	175
120	000	0	150	150.0	400	300	250	200
125	000	00	175	156.5	400	350	250	200
130	000	00	175	162.5	400	350	300	200
135	0000	00	175	169.0	450	350	300	225
140	0000	00	175	175.0	450	350	300	225
145	0000	000	200	181.5	450	400	300	225
150	0000	000	200	187.5	450	400	300	225
155	0000	000	200	194.0	500	400	350	250
160	250	000	200	200.0	500	400	350	250
165	250	0000	225	206.	500	450	350	250
170	250	0000	225	213.	500	450	350	300
175	300	0000	225	219.	600	450	350	300
180	300	0000	225	225.	600	450	400	300
185	300	0000	250	231.	600	500	400	300
190	300	250	250	238.	600	500	400	300
195	350	250	250	244.	600	500	400	300
200	350	250	250	250.	600	500	400	300
210	400	300	250	263.	...	600	450	350
220	400	300	300	275.	...	600	450	350
230	500	300	300	288.	...	600	500	350
240	500	350	300	300.	...	600	500	400
250	500	350	300	313.	...	...	500	400
260	600	400	350	325.	...	...	600	400
270	600	400	350	338.	...	...	600	450
280	600	500	350	350.	...	...	600	450
290	700	500	350	363.	...	...	600	450
300	700	500	400	375.	...	...	600	450
320	750	600	400	400.	...	...	...	500
340	900	600	450	425.	...	...	...	600
360	1000	700	450	450.	...	...	...	600
380	1250	750	500	475.	...	...	...	600
400	1500	900	500	500.	...	...	...	600
420	1750	1000	600	525.	...	...	...	...
440	2000	1250	600	550.	...	...	...	...
460	....	1250	600	575.	...	...	...	...
480	....	1500	600	600.	...	...	...	...
500	....	1500	...	625.	...	...	...	...

TABLE 11

FULL-LOAD CURRENT * DIRECT-CURRENT MOTORS †

HP	115V	230V	550V
½	4.6	2.3	
¾	6.6	3.3	1.4
1	8.6	4.3	1.8
1½	12.6	6.3	2.6
2	16.4	8.2	3.4
3	24.	12.	5.0
5	40.	20.	8.3
7½	58.	29.	12.0
10	76.	38.	16.0
15	112.	56.	23.0
20	148.	74.	31.
25	184.	92.	38.
30	220.	110.	46.
40	292.	146.	61.
50	360.	180.	75.
60	430.	215.	90.
75	536.	268.	111.
100		355.	148.
125		443.	184.
150		534.	220.
200		712.	295.

* These values for full-load current are average for all speeds.
† 1953 *National Electrical Code.*

TABLE 12

FULL-LOAD CURRENT * SINGLE-PHASE A.C. MOTORS †

HP	115V	230V	440V
1/6	3.2	1.6	
¼	4.6	2.3	
½	7.4	3.7	
¾	10.2	5.1	
1	13.	6.5	
1½	18.4	9.2	
2	24.	12.	
3	34.	17.	
5	56.	28.	
7½	80.	40.	21.
10	100.	50.	26.

* These values of full-load current are for motors running at speeds usual for belted motors and motors with normal torque characteristics. Motors built for especially low speeds or high torques may require more running current, in which case the nameplate current rating should be used.

† 1953 *National Electrical Code.*

For full-load currents of 208- and 200-volt motors, increase corresponding 230-volt motor full-load current by 10 and 15 percent, respectively.

TABLE 13 *

FULL-LOAD CURRENT ‡

	Induction Type Squirrel Cage and Wound Rotor Amperes					Synchronous Type †Unity Power Factor Amperes			
HP	110V	220V	440V	550V	2300V	220V	440V	550V	2300V
½	4	2	1	.8					
¾	5.6	2.8	1.4	1.1					
1	7	3.5	1.8	1.4					
1½	10	5	2.5	2.0					
2	13	6.5	3.3	2.6					
3		9	4.5	4					
5		15	7.5	6					
7½		22	11	9					
10		27	14	11					
15		40	20	16					
20		52	26	21					
25		64	32	26	7	54	27	22	5.4
30		78	39	31	8.5	65	33	26	6.5
40		104	52	41	10.5	86	43	35	8
50		125	63	50	13	108	54	44	10
60		150	75	60	16	128	64	51	12
75		185	93	74	19	161	81	65	15
100		246	123	98	25	211	106	85	20
125		310	155	124	31	264	132	106	25
150		360	180	144	37		158	127	30
200		480	240	192	48		210	168	40

* 1953 *National Electrical Code.*

For full-load currents of 208- and 200-volt motors, increase the corresponding 200-volt motor full-load current by 6 and 10 percent, respectively.

‡ These values of full-load current are for motors running at speeds usual for belted motors and motors with normal torque characteristics. Motors built for especially low speeds or high torques may require more running current, in which case the nameplate current rating should be used.

† For 90 and 80 percent P.F. the above figures should be multiplied by 1.1 and 1.25 respectively.

TABLE 14

Summary of Basic Electrical Calculations

DC CIRCUIT CHARACTERISTICS

Ohm's Law:

$$E = IR \qquad I = \frac{E}{R} \qquad R = \frac{E}{I}$$

E = voltage impressed on circuit (volts)
I = current flowing in circuit (amperes)
R = circuit resistance (ohms)

Resistances in series:

$$R_t = R_1 + R_2 + R_3 + \cdots \text{ etc.}$$

R_t = total resistance (ohms)

R_1, R_2, etc. = individual resistances (ohms)

Resistances in parallel:

$$R_t = \frac{1}{\dfrac{1}{R_1} + \dfrac{1}{R_2} + \dfrac{1}{R_3} + \cdots \text{ etc.}}$$

Formulas for the conversion of electrical and mechanical power:

$$\text{HP} = \frac{\text{watts}}{746} = \text{watts} \times .00134$$

$$= \frac{\text{kilowatts}}{.746} = \text{kilowatts} \times 1.34$$

Kilowatts = watts ÷ 1000 = HP × .746
Watts = HP × 746

In direct current circuits, electrical power is equal to the product of the voltage and current:

$$P = EI = I^2R = \frac{E^2}{R}$$

P = power (watts)
E = voltage (volts)
I = current (amperes)
R = resistance (ohms)

TABLE 14.—SUMMARY OF BASIC ELECTRICAL CALCULATIONS—*Continued*

Solving the basic formula for I, E, and R gives

$$I = \frac{P}{E} = \sqrt{\frac{P}{R}} \qquad E = \frac{P}{I} = \sqrt{RP} \qquad R = \frac{E^2}{P} = \frac{P}{I^2}$$

ENERGY

Energy is the capacity for doing work. Electrical energy is expressed in kilowatt-hours (kwhr), one kilowatt-hour representing the energy expended by a power source of 1 kw over a period of 1 hour.

EFFICIENCY

Efficiency of a machine, motor or other device is the ratio of the energy output (useful energy delivered by the machine) to the energy input (energy delivered to the machine), usually expressed as a percentage:

$$\text{Efficiency} = \frac{\text{output}}{\text{input}} \times 100\%$$

or

$$\text{Output} = \frac{\text{input} \times \text{efficiency}}{100\%}$$

TORQUE

Torque may be described as the measure of the tendency of a body to rotate. It is expressed in pound-feet or pounds of force acting at a certain radius:

Torque (pound-feet) = force tending to produce rotation (pounds)
× distance from center of rotation to point at which force is applied (feet)

Relations between torque and horsepower:

$$\text{Torque} = \frac{33{,}000 \times \text{HP}}{6.28 \times \text{rpm}}$$

$$\text{HP} = \frac{6.28 \times \text{rpm} \times \text{torque}}{33{,}000}$$

rpm = speed of rotating part (rev. per minute)

TABLE 14.—SUMMARY OF BASIC ELECTRICAL CALCULATIONS—*Continued*

AC CIRCUIT CHARACTERISTICS

The instantaneous values of an alternating current or voltage vary from zero to a maximum value each half cycle. In the practical formulae which follow, the "effective value" of current and voltage is used, defined as follows:

$$\text{Effective value} = 0.707 \times \text{maximum instantaneous value}$$

Impedance:

Impedance is the total opposition to the flow of alternating current. It is a function of resistance, capacitive reactance and inductive reactance. The following formulae relate these circuit properties:

$$X_L = 2\pi fL \qquad X_C = \frac{1}{2\pi fC} \qquad Z = \sqrt{R^2 + (X_L - X_C)^2}$$

X_L = inductive reactance (ohms)
X_C = capacitive reactance (ohms)
Z = impedance (ohms)
f = frequency (cycles per second)
C = capacitance (farads)
L = inductance (henrys)
R = resistance (ohms)
π = 3.14

In circuits where one or more of the properties L, C or R is absent, the impedance formula is simplified as follows:

Resistance only:	Inductance only:	Capacitance only:
$Z = R$	$Z = X_L$	$Z = X_C$
Resistance and Inductance only:	Resistance and Capacitance only:	Inductance and Capacitance only:
$Z = \sqrt{R^2 + X_L{}^2}$	$Z = \sqrt{R^2 + X_C{}^2}$	$Z = \sqrt{(X_L - X_C)^2}$

Ohm's Law for AC circuits:

$$E = IZ \qquad I = \frac{E}{Z} \qquad Z = \frac{E}{I}$$

Capacitances in parallel:

$$C_t = C_1 + C_2 + C_3 + \cdots \text{ etc.}$$

C_t = total capacitance (farads)
C_1, C_2, etc. = individual capacitances (farads)

TABLE 14.—SUMMARY OF BASIC ELECTRICAL CALCULATIONS—*Continued*

Capacitances in series:

$$C_t = \frac{1}{\frac{1}{C_1} + \frac{1}{C_2} + \frac{1}{C_3} + \cdots \text{ etc.}}$$

Inductances in series and parallel:

The resulting circuit inductance of several inductances in series or parallel is determined exactly as the sum of resistances in series and parallel as described above.

PHASE ANGLE

An alternating current through an inductance lags the voltage across the inductance in time by an angle computed as follows:

$$\text{Tangent of angle of lag} = \frac{X_L}{R}$$

An alternating current through a capacitance leads the voltage across the capacitance in time by an angle computed as follows:

$$\text{Tangent of angle of lead} = \frac{X_C}{R}$$

The resultant angle by which a current leads or lags the voltage in an entire circuit is called the phase angle and is computed as follows:

$$\text{Cosine of phase angle} = \frac{R}{Z}$$

POWER FACTOR

Power factor of a circuit or system is the ratio of actual power (watts) to apparent power (volt-amperes), and is equal to the cosine of the phase angle of the circuit:

$$\text{PF} = \frac{\text{actual power}}{\text{apparent power}} = \frac{\text{watts}}{\text{volts} \times \text{amperes}} = \frac{\text{KW}}{\text{KVA}} = \frac{R}{Z}$$

KW = kilowatts

KVA = kilovolt-amperes = volt-amperes × 1000

PF = power factor (expressed as decimal)

TABLE 14.—SUMMARY OF BASIC ELECTRICAL CALCULATIONS—*Continued*

SINGLE PHASE CIRCUITS

$$KVA = \frac{EI}{1000} = \frac{KW}{PF} \qquad KW = KVA \times PF$$

$$I = \frac{P}{E \times PF} \qquad E = \frac{P}{I \times PF} \qquad PF = \frac{P}{E \times I}$$

$P = E \times I \times PF$
P = power (watts)

TWO-PHASE CIRCUITS

$$KVA = \frac{2EI}{1000} = \frac{KW}{PF} \qquad KW = KVA \times PF$$

$$I = \frac{P}{2E \times PF} \qquad E = \frac{P}{2I \times PF} \qquad PF = \frac{P}{E \times I}$$

$P = 2E \times I \times PF$
E = phase voltage (volts)

THREE-PHASE CIRCUITS, BALANCED STAR OR WYE

$$I_N = O \qquad I = I_p \qquad E = \sqrt{3}\,E_p = 1.73E_p$$

$$E_p = \frac{E}{\sqrt{3}} = \frac{E}{1.73} = 0.577E$$

I_N = current in neutral (amperes)
I = line current per phase (amperes)
I_p = current in each phase winding (amperes)
E = voltage, phase to phase (volts)
E_p = voltage, phase to neutral (volts)

THREE-PHASE CIRCUITS, BALANCED DELTA

$$I = \sqrt{3}\,I_p = 1.73I_p \qquad I_p = \frac{I}{\sqrt{3}} = 0.577I \qquad E = E_p$$

TABLE 14.—SUMMARY OF BASIC ELECTRICAL CALCULATIONS—*Continued*

POWER: BALANCED 3-WIRE, 3-PHASE CIRCUIT, DELTA OR WYE

For unity power factor (PF = 1.0):

$$P = \sqrt{3}\, EI = 1.73EI$$

$$I = \frac{P}{\sqrt{3}\, E} = \frac{0.577P}{E} \qquad E = \frac{P}{\sqrt{3}\, I} = \frac{0.577P}{I}$$

P = total power (watts)

For any load:

$$P = 1.73EI \times \text{PF} \qquad VA = 1.73EI$$

$$E = \frac{P}{\text{PF} \times 1.73 \times I} = \frac{0.577 \times P}{\text{PF} \times I}$$

$$I = \frac{P}{\text{PF} \times 1.73 \times E} = \frac{0.577 \times P}{\text{PF} \times E}$$

$$\text{PF} = \frac{P}{1.73 \times I \times E} = \frac{0.577 \times P}{I \times E}$$

VA = apparent power (volt-amperes)
P = actual power (watts)
E = line voltage (volts)
I = line current (amperes)

POWER LOSS: ANY AC OR DC CIRCUIT

$$P = I^2R \qquad I = \sqrt{\frac{P}{R}} \qquad R = \frac{P}{I^2}$$

P = power heat loss in circuit (watts)
I = effective current in conductor (amperes)
R = conductor resistance (ohms)

XXI

Electronics

Many kinds of mathematics find application in the field of electronics. They range from simple arithmetic and algebra to highly complex calculations which are beyond the scope of this book. In this chapter we will cover some of the simpler appli-

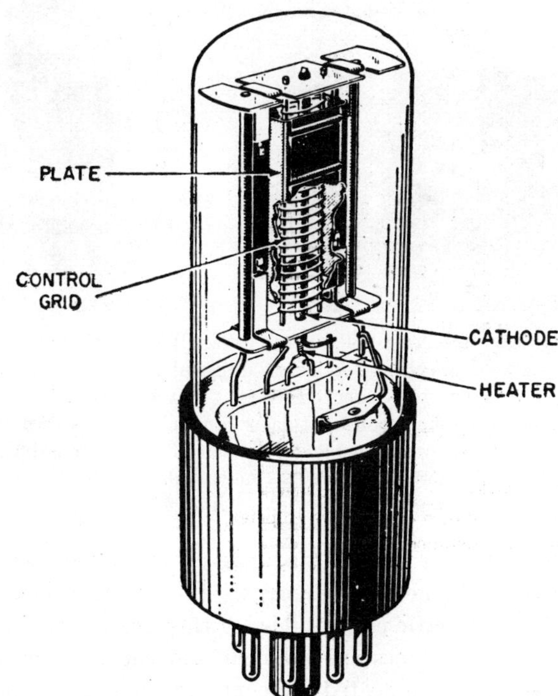

Fig. 1.—Drawing Showing the Construction of a Typical Electron Tube.

cations of mathematics in electronics. One of these is a special technique which is very useful for rapid solution of many electronic tube performance calculations. This technique is known as graphical calculation. Before discussing this technique, we need to know a few basic facts about electron tubes in general.

Electron Tubes.—An electron tube is a device in which electrons flow from some electron-emitting surface (*cathode*) to a collector surface (*anode* or *plate*), through a vacuum or gas in a closed glass or metal container. (See Figs. 1 and 2.)

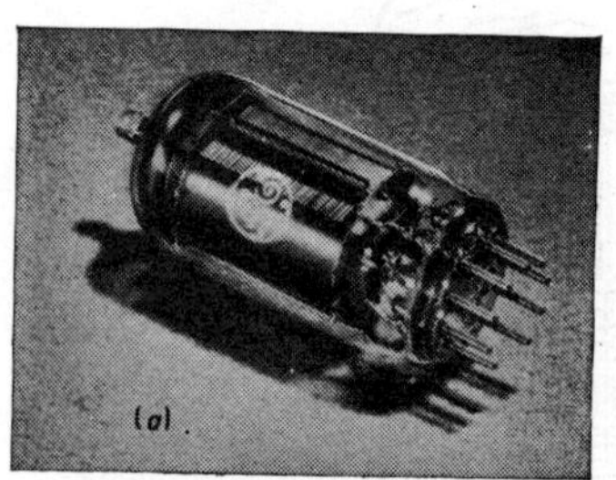

FIG. 2.—Typical Receiving Electron Tubes. (*a*) is a miniature type (*Courtesy General Electric Company*) designed for high frequency applications such as in television receivers. (*b*) is a typical low power receiving tube type whose construction is similar to that shown in Fig. 1 except that these are two plates and two control grids. It is known as a "twin triode" and is actually two tubes in one glass envelope. (*Courtesy Radio Corporation of America*)

Certain materials when heated give off or emit electrons. In electron tubes, the heat which causes this electron emission is produced by an electric current flowing through a wire called the *filament*. In some tubes the filament and the cathode are the same element, and the heated filament is therefore the source of electrons. In other tubes the filament heats a separate cathode which is made of a much better electron-emitting material. This latter type of tube has what is known as an *indirectly heated cathode* and is probably the most common type employed in electronic circuits.

The electron flow can be controlled by varying the voltage applied to the plate. Making the plate more positive with respect to the cathode causes more of the negative electrons being emitted from the cathode to be attracted to the plate, resulting in an increased current flow. When the plate is made negative with respect to the cathode, the negative electrons are repelled and current flow ceases.

Greatly increased control of the electron flow can be achieved by means of a third element known as the *grid* which is placed between the cathode and the plate. The grid is a mesh or coil of fine wires through which electrons can easily pass. By applying different voltages to the grid, the flow of electrons can be increased, decreased, or stopped. Making the grid positive with respect to the cathode increases the flow of electrons to the plate. When the grid is negative with respect to the cathode, the flow of electrons to the plate is reduced. A point can be reached when the grid is sufficiently negative to stop the flow of electrons to the plate completely. Thus the grid acts as a "valve" to control the flow of electrons or current between the cathode and the plate.

The simplest type of electron tube consists of two elements, a cathode and a plate, and is called a *diode*. Its ability to permit current flow only when the plate is positive makes it useful as a *rectifier* to change alternating current into pulsating direct current.

When a grid is added, the tube is known as a *triode*. This type of tube has a very wide variety of uses, the most important of which is amplification of small signals.

Characteristic Curves.—The effect of the grid voltage on the plate current flow for a given tube can be shown by a set of *characteristic curves* as in Fig. 4. In order to obtain these curves, the triode is connected as shown in Fig. 3. One battery is used to furnish plate voltage for the tube, and the other for the grid.

Filament connections are not shown in order to simplify the circuit.

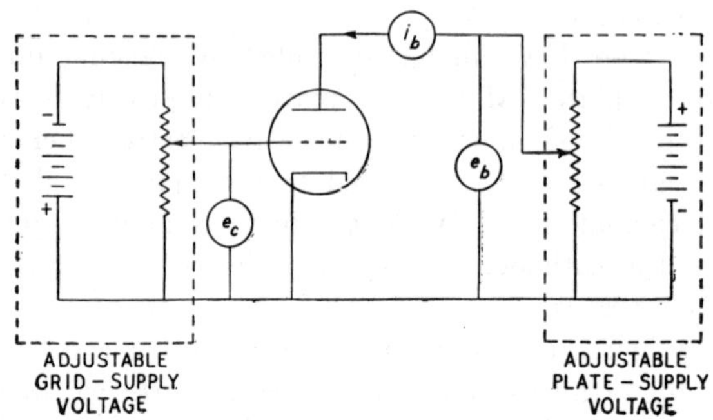

FIG. 3.—Circuit Used to Obtain Triode Characteristic Curves. i_b = plate current; e_c = grid voltage; e_b = plate voltage.

The procedure is as follows: First the plate voltage is set at a certain value, say, 250 volts. Then the grid voltage is varied and the corresponding plate current readings are plotted on a graph. The plate voltage is then set at another value, say, 200 volts. The grid voltage again is varied, and the resulting plate current curves are plotted. Repeating this process for several values of plate voltage results in a group or *family* of characteristic curves. Note that, for any given plate voltage, there is a value of negative grid voltage that will stop the plate current flow. This is referred to as the *cutoff voltage.*

The grid then acts like a valve, controlling the flow of plate current. Varying the grid voltage has a much greater effect upon plate current flow than is obtained when the plate voltage is varied. This means that plate current changes caused by relatively large changes in plate voltage can be produced by relatively small changes in grid voltage.

Because a small voltage on the grid has the same effect as a large voltage on the plate, the triode can be used as an *amplifier* which magnifies or amplifies a small signal on the grid to a large signal on the plate. The amplified output does not come from the tube itself, but from the power source (in this case a battery), connected between the cathode and the plate of the tube. By action of the grid, the tube controls the power from this source and converts it into the form required to do a particular task.

A *load resistor* or *load impedance* must be connected in the plate or output circuit of the tube in order to make use of the controlled power. With this load impedance in the circuit the action is something like this: as the grid voltage is varied by a small amount, the plate current varies, and the resulting varying plate current flowing through the plate load impedance produces large varying voltage drops. These varying voltage drops are magnified or amplified versions of the varying grid voltages. In a typical case a change of one volt in the grid might cause a variation in the voltage across the load impedance of perhaps 50 volts. This would be the same as saying that the input voltage (change in grid voltage) was magnified or *amplified* 50 times in the output circuit.

The performance of tubes depends upon their design, specifically upon the geometry of the different elements. This includes separation of the elements, the shape and size of the elements, the number of turns and the spacing of the grid wires, and several other physical factors. All these factors govern the maximum voltages that can be applied, the cutoff grid voltage, the maximum plate current, etc. They are expressed by a group of numbers known as the *tube constants.*

Tube Constants.—Tube constants differ from tube characteristics. The latter are graphical representations of tube behavior under a given set of conditions. The former are individual numerical ratings resulting from the geometry of the tube. Tubes with similar constants demonstrate similar relationships, but specific values of grid voltage, plate voltage, and plate current necessary to make the tube perform properly may be different for the various tubes. The three most important tube constants are *amplification factor, a-c plate resistance,* and *transconductance.* These constants can be measured directly, or the technique of graphical calculation can be used to compute them from the characteristic curves for a given tube. Characteristic curves for tubes are readily available in tube manuals, and by graphical calculation it is a simple matter to find the constants for a given set of operating conditions.

Amplification Factor.—The *amplification factor* or *amplification constant* of a tube is the ratio between a small change in plate voltage and a small change in grid voltage which results in the same change in plate current. It is a measure of the effectiveness of the control grid voltage in controlling the plate current as compared with the plate voltage. In formula form,

$$\text{Amplification Factor} = \frac{\text{small change in plate voltage}}{\text{small change in grid voltage}}$$

The amplification factor is usually designated by the Greek letter μ (mu), pronounced "mew." If a tube has a μ of 100, then the grid voltage change required to produce a certain change in plate current is 100 times less than the plate voltage change required to cause the same change in plate current. For example, suppose that it takes a plate voltage change of 10 volts to produce a 1-milliampere change in plate current, and it takes a grid voltage change of 0.1 volt to produce the same 1-milliampere change in plate current. Then

$$\mu = \frac{\text{change in plate voltage}}{\text{change in grid voltage}} = \frac{10}{0.1} = 100$$

It is important to remember that the changes measured have to be small and that it is the changes in the grid voltage and in plate voltage that are considered in these calculations and not the individual values of plate and grid voltage.

The amplification factor of triodes ranges from about 3 to 100. Low-μ triodes have amplification factors of less than 7 or 8; medium-μ triodes have amplification factors of about 8 to 30; and the amplification factor for high-μ triodes is 30 or more.

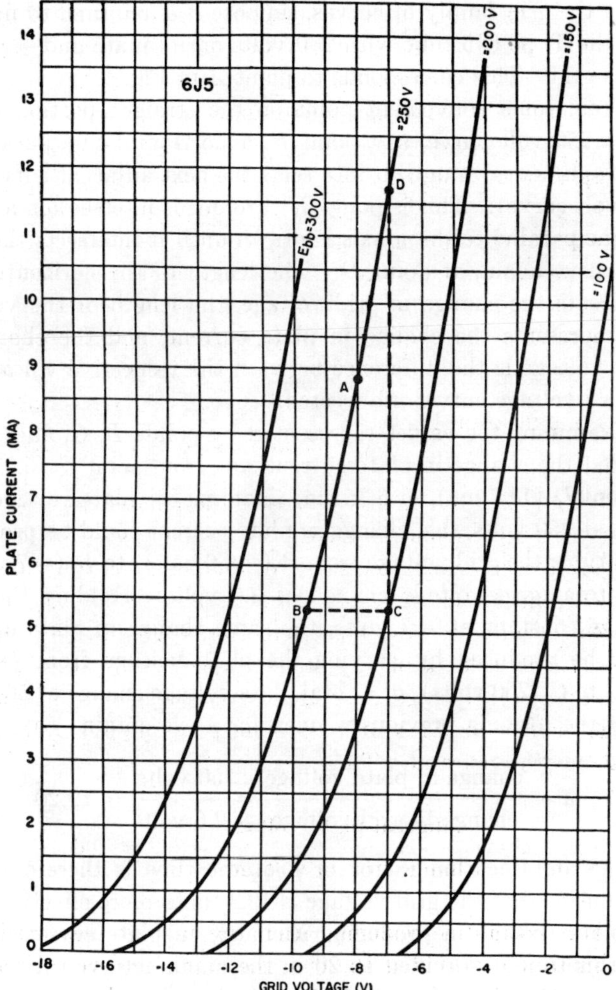

FIG. 4.—Grid Characteristics or Grid Family of Curves of a type 6J5 Triode. These curves can be used to determine the amplification factor of the table by graphical calculations.

Graphical Calculation of μ.—The amplification factor of a tube can be found directly from the characteristic curves for the tube.

Using the grid family of curves, suppose it is required to find the μ of the Type 6J5 tube with 250 volts on its plate and —8 volts on its grid. This corresponds to point A in Fig. 4.

From some convenient point on the straight portion of the E_{bb} = 250-volt curve, say point B, a horizontal line parallel to the x axis is drawn until it intersects the next adjacent curve (the 200-volt curve). This is point C. From this intersection a vertical line parallel to the y axis is drawn until it intersects the 250-volt curve again, at point D. The length of the horizontal line represents the change in grid voltage, the length of the vertical line represents the change in plate current, and the change in plate voltage is the difference between the respective voltages of the two voltage curves intersected.

Examining the values represented by points B, C, and D, we see that the change in plate current was from point C (5.35 ma) to point D (11.7 ma), or 6.35 ma. Holding the plate voltage constant at 250 volts, this change in plate current could be produced by varying the grid voltage from B (—9.6 volts) to D (—7 volts), or a total grid voltage change of 2.6 volts. Holding the grid voltage constant at —7 volts, the same change in plate current could be produced by varying the plate voltage from D (250 volts) to C (200 volts), or a total plate voltage change of 50 volts. Using the formula previously given for μ, we obtain

$$\mu = \frac{\text{change in plate voltage}}{\text{change in grid voltage}} = \frac{50 \text{ volts}}{2.6 \text{ volts}} = 19.20$$

The amplification factor or μ of the tube is therefore 19.2. This means that the grid voltage is 19.2 times more effective than the plate voltage in producing a change in plate current. The μ of this tube is specified as 20 in the manufacturer's literature, which means that the graphical calculation produced a result about 4% from the listed value. This is close enough for practical purposes, especially since the μ will vary slightly on the nonlinear portions of the curves anyway.

Plate Resistance.—The *plate resistance* (r_p), another important vacuum tube constant, is a measure of the internal resistance of the tube to the flow of electrons from the cathode to the plate. This constant is expressed in two ways. One of these is the *d-c resistance,* which is the opposition to current flow when steady values of voltage are applied to the tube elements. It is determined by applying Ohm's Law:

$$r_p(\text{ohms}) = \frac{\text{steady-state plate voltage in volts}}{\text{steady-state plate current in amperes}}$$

at any point on the plate current characteristic curve. For example, on the plate characteristic curves or the plate family of curves in Fig. 5, when the grid voltage is —8 volts, we note that

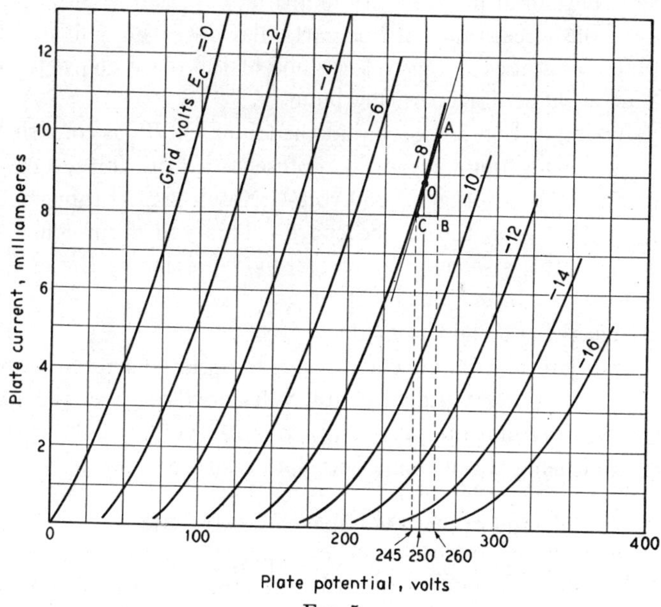

FIG. 5.

for a plate voltage of 250 volts the plate current is 8.9 ma. Then, using Ohm's Law, we obtain

$$r_p = \frac{250 \text{ volts}}{0.0089 \text{ amp.}} = 28{,}100 \text{ ohms}$$

The *a-c plate resistance* is the ratio of a small change in plate current at a given grid voltage. In formula form it is:

$$r_p(\text{ohms}) = \frac{\text{small change in plate voltage (volts)}}{\text{small change in plate current (amperes)}}$$

Graphical Calculation of r_p.—Like the amplification factor, the plate resistance can also be found from the tube's characteristic curves. In this instance the plate characteristics are used.

The operating point is selected, a tangent to the curve is drawn at this point, and a small right triangle is constructed using the tangent line as the hypotenuse. The ratio of the base of this triangle, measured in volts, to the altitude, measured in amperes, gives the plate resistance at this particular operating point. Thus the plate resistance is really the slope of the plate characteristic curve at a particular operating point.

Referring to Fig. 5, we see that the triangle *ABC* is constructed at the operating point *O* which represents a grid voltage of —8 volts and a plate voltage of 250 volts. Note that the tangent line runs practically on top of the curve. This will occur when the operating point is selected on the straight portion of the characteristic curve. In such cases the curve itself can be used as the hypotenuse of the right triangle.

Point *A* on the triangle corresponds to a plate current of about 10 ma and a plate voltage of 260 volts. At point *C* the plate current is 8 ma and the plate voltage is 245 volts. Taking these values and using the formula for plate resistance, we obtain

$$r_p = \frac{\text{change in plate voltage}}{\text{change in plate current}} = \frac{260 - 245}{0.010 - 0.008}$$

$$= \frac{15}{0.002} = 7{,}500 \text{ ohms}$$

The a-c plate resistance will vary, depending upon the operating point selected. A few sample calculations at different operating points will show that the higher the plate voltage, the lower will be the plate resistance; and the more positive the grid voltage (operating at a smaller negative grid voltage), the lower will be the plate resistance.

Transconductance.—The third important tube constant is called *transconductance* or the *mutual conductance*. It expresses the change in plate current caused by a change in grid voltage with the plate voltage held constant. In formula form it is:

$$\text{Transconductance (mhos)} = \frac{\text{change in plate current (amperes)}}{\text{change in grid voltage (volts)}}$$

The symbol commonly used for transconductance is g_m. The unit of transconductance is the *mho* (ohm spelled backward), which represents a condition where a grid voltage change of 1 volt produces a plate current change of 1 amp. This value is too large for use with vacuum tubes and therefore the *micromho,* or the millionth part of a mho, is usually employed. It corresponds to a 1-microampere change in plate current for a 1-volt change in grid voltage. Thus, an electron tube which operates so that a 1-volt change in grid voltage produces a 1-ma change in plate current (1000 microamperes) has a rating of 1000 μmhos, or we say that it has a transconductance of 1000 (micromhos being understood).

Graphical Calculation of g_m.—The transconductance also can be found by graphical means. The plate family of curves is used, and the procedure is quite simple.

On the grid voltage curve, a point is selected in such a way that a line projected upward or downward vertically from this point will intersect an adjacent grid voltage curve. This is the initial point *A* in Fig. 6, which represents a plate voltage of 235 volts, a grid voltage of —8 volts, and a plate current of 7.1 ma. Going up vertically from this point the next curve is intersected at point *B*, which represents the same plate voltage, a grid voltage of —6 volts, and a plate current of 12 ma. Using the formula for g_m, we obtain

$$g_m = \frac{\text{change in plate current}}{\text{change in grid voltage}} = \frac{0.012 - 0.0071}{8-6} = \frac{0.0049}{2}$$

$$= 0.00245 \text{ mho, or } 2450\ \mu\text{mho}$$

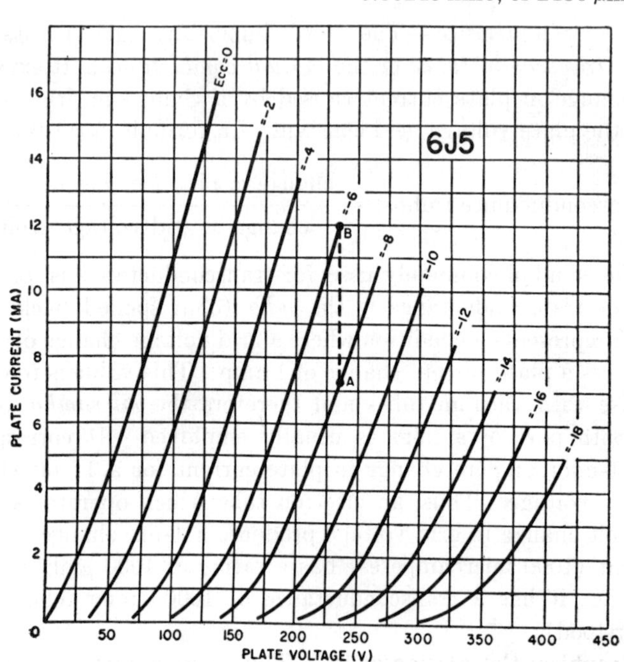

Fig. 6.—Graphically Calculating Transconductance Using the Plate Characteristics.

In terms of plate current change per 1-volt change on the grid, this figure means that a change of 2.45 ma will be obtained for a change of 1 volt on the grid.

Relation of μ, r_p, and g_m.—At a given operating point, the three tube constants: μ, r_p, and g_m are related to one another by the following formulas:

$$\mu = g_m/r_p, \quad r_p = \mu/g_m, \quad g_m = \mu/r_p$$

Therefore if two of the constants are known, the third can be readily found.

The foregoing examples are only a few ways in which graphical techniques are used to simplify and speed up calculations in designing electronic devices. Some of the other applications require a more comprehensive knowledge of electron tubes and their associated circuits and therefore are beyond the scope of this book. The reader is referred to the standard textbooks for additional reading on this subject.

Transistors.—The transistor is an electronic device for amplification and/or control which consists of a semiconducting material to which contact is made by two or more electrodes which are usually metal points or metal surfaces soldered to the semiconductor. The transistor is capable of performing many of the functions filled by the electron tube. Although its operating characteristics depend on temperature, the transistor has many advantages relative to the electron tube since it requires no heater (or filament) current, is small and light weight, can be made mechanically rigid and long lasting, and operates at low voltages with comparatively high efficiencies.

Semiconductors are materials which have electrical conduction properties much poorer than good conductors but, on the other hand, much better than materials classed as insulators. The most common semiconductors in use at present for transistors are germanium and silicon. In a perfect crystal state an atom of germanium (or silicon) has four electrons bounded to electrons of neighboring atoms in which the two electrons in each bond are shared equally with the two atoms at its ends. This bond structure is three dimensional, but the action is indicated in a two dimensional form in Fig. 6. Absorption of energy either as a result of light or heat will cause rupture of a covalent bond releasing an electron and leaving a defect, of "hole," in the otherwise periodic structure. For germanium an absorbed energy of 0.72 electron volts will release an electron from the bond while 1.12 electron volts is required for the same effect in silicon. Once released, the electron can move about the crystal with considerable freedom. Furthermore, it is also possible for an electron in bond adjacent to the bond which has just lost an electron to jump

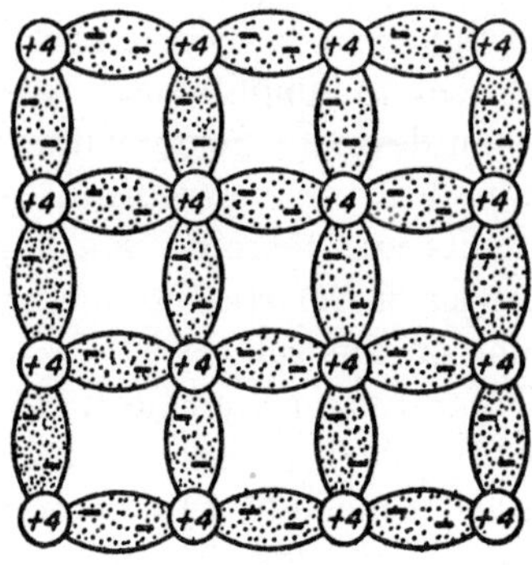

FIG. 6.

into the hole, leaving an electron deficit in the bond which was vacated. By this mechanism it is possible for a hole to move through the crystal at the same time free electrons move.

The hole concept represents a means of describing an incomplete assemblage of electrons and with this fact understood, holes may be considered to have a positive charge equal in magnitude to that possessed by an electron as well as an appropriate mass. These two properties enable calculation of the hole movement to be expected in the presence of electric and magnetic fields. The conductivity (σ mho-cm.$^{-1}$) of a semiconductor can be expressed in terms of the density of holes (p per cubic centimeter) and of electrons (n per cubic centimeter) and the electron and hole mobilities μ_n and μ_p-cm./sec. per volt/cm.), the latter being the drift velocity in a unit electric field. The result is

$$\sigma = q(n\mu_n + p\mu_p)$$

where q is the electron charge. It is clear that an increase in either hole or electron concentration will change the conductivity of the semiconductor although the two densities have different numerical effects due to the difference between hole and electron mobilities ($\mu_p = 1700$, $\mu_n = 3600$ cm./sec. per volts/cm. for germanium; $\mu_p = 400$, $\mu_n = 1200$ cm./sec. per volts/cm. for silicon).

In a pure sample of a semiconductor the holes and electrons are produced in pairs so that the two densities are equal (n_i). The conductivity that results for such an intrinsic semiconductor is called the intrinsic conductivity (σ_i). It is expressible as

$$\sigma_i = qn_i(\mu_n + \mu_p)$$

Transistors depend for their operation on either an excess of free electrons or a deficit of electrons (excess of holes) in atomic bonds. A semiconductor in which the first condition obtains is spoken of as an *n* type material (excess of negative carriers) whereas one in which the second situation exists is referred to as a *p* type semiconductor (excess of positive carriers). An excess of electrons or holes is produced by adding minute amounts of impurities to an intrinsic semiconductor. Some atoms (such as arsenic and antimony) have five bonding electrons, and if small amounts of these atoms are added to an otherwise pure sample of germanium or silicon (which, as stated above, have four bonding electrons), four of the five electrons of the impurity atom are shared with neighboring germanium atoms. The fifth valence electron is free to move throughout the crystal. An excess of electrons is thus created by the addition of the impurities; *n* type germanium now exists.

There are also atoms, such as aluminum and gallium which have only three bonding electrons. When such atoms are added in small amount to pure germanium (or silicon), only three germanium atoms can be bonded to the impurity atom by sharing of electrons. Since the impurity atom occupies a position in the crystal structure normally filled by a germanium atom (which has four bonds), the addition of the impurity will result in the deficiency of one electron in a germanium bond in the proximity of the impurity. A hole has thus been created and *p* type germanium has resulted. It should be noted that if both five-electron and three-electron atoms are present, the material will exhibit the properties of *n* or *p* type material depending on whether the electron density exceeds the hole density or vice versa.

A rod of single crystal germanium with alternating n and p type regions can be obtained by adding suitable impurities to the molten germanium as the rod is withdrawn from the crucible used for melting in the process of forming a crystal. If the two types of impurities are added in succession, it is possible to obtain n-p-n and p-n-p configurations along the rod. By this method, there will be a region created where n type conductivity predominates followed by one where p type prevails in the same germanium crystal. A p-n junction will be produced where the two regions of different conductivity come together. In contrast with these so-called grown junctions, it is also possible to obtain a pair of p-n junctions for transistor action by starting with a thin wafer of germanium of a prescribed conductivity type and fusing impurity regions of the opposite type in localized areas on opposite sides of the wafer, leaving a thin section of the original material in the center of the wafer cross section. Junctions prepared in this manner are called fused alloy junctions.

Before considering the transistor itself which consists of a combination of two p-n junctions (arranged as p-n-p or n-p-n), it is of interest to examine conditions which exist in a single p-n junction. (The three-bonding electron atoms like aluminum are called acceptor atoms, since they are "electron-deficient," while the five-bonding-electron atoms like arsenic are called donor atoms.) Fig. 7 shows a p-n junction including the acceptor and

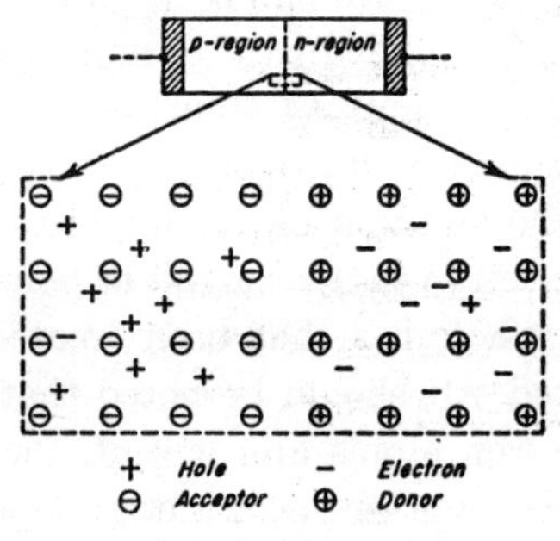

FIG. 7.

donor atoms in the n and p regions, respectively, as well as the holes and electrons which are free to move. It is noted that even in the n type region there are some holes, although a relatively small number, and a corresponding situation exists in the p type region. These carriers, which are of opposite sign to the intended or majority carriers, come into existence due to the rupture of atomic bonds as the result of the thermal energy of the atoms and are called minority carriers. If no external potential difference is applied across the electrodes shown in the figure, an equilibrium condition will be established wherein no electrons or holes exist in the immediate vicinity of the junction. Because of the arrangement of the fixed, oppositely charged donor and acceptor atoms at the boundary of the n and p type materials, an electric dipole is formed yielding an appropriate electric field and corresponding difference of potential across the boundary between the two regions.

If a potential difference of suitable magnitude is applied to the junction such that the electrode connected to the p region is positive with respect to the other electrode (producing a forward biased junction), the majority carriers in each region will be forced to flow across the junction, the net current flow being the sum of the current carried by the holes plus that carried by the electrons. The respective densities of holes and electrons (p and n) are determined by the impurity concentrations in the p and n type regions. By suitable control of the relative amounts of impurities used, the current that flows across the junction in the forward biased condition can be arranged to be composed principally of holes, principally of electrons, or some suitable combination of both. In any event, the flow of conventional current is into the p region and out of the n region. If the polarity of the external potential difference is reversed, the only current that will flow into the terminals of the device will be that contributed by the few minority carriers that are free in the material, which will be a small one indeed compared to the current obtained in the forward biased condition. This current rapidly reaches a limiting value called the saturation current as the reverse bias voltage across the device is increased. Because the current that flows is practically independent of the applied voltage beyond a certain minimum value, the impedance of the reversed bias junction is extremely high. By contrast, the impedance in the forward biased direction is very low.

A transistor of the *n-p-n* type is shown diagrammatically in Fig. 8a. A small *p* type region is contained between *n* type regions and connections are made to all three regions. The terminals are labelled emitter, base, and collector for reasons that will become apparent shortly. In normal operation as an amplifier, the emitter to base junction is forward biased and the collector to base junction is reverse bias. The emitter to base impedance is low whereas the collector to base impedance is high for reasons considered above. Fig. 8b indicates the current flow existing across the junctions. A small saturation current composed of holes and electrons flows across the collector junction; this current is not controlled by the emitter. On the other hand, the electrons coming into the base from the emitter due to the forward biased emitter-base junction represent minority carriers for the collector-base junction and for a thin enough base section these carriers pass through the base layer, cross the collector

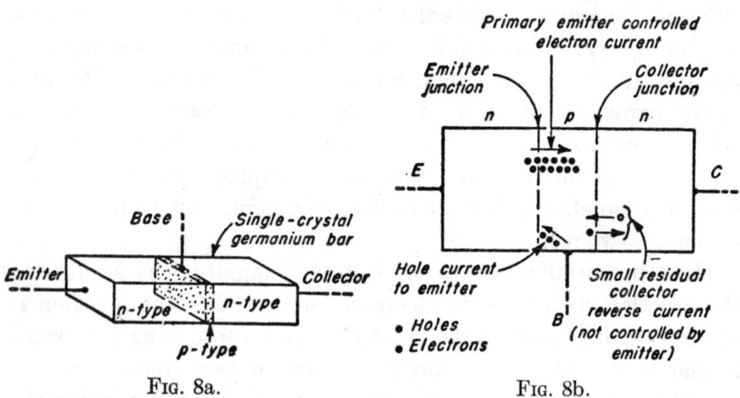

FIG. 8a.

FIG. 8b.

junction, and contribute to the collector current. A hole current from the base region also flows across the emitter junction to add to the electron current in producing the total emitter current. The essential action is the emitting of carriers from the emitter region and the collection of practically all of these carriers by the collector.

By proper design of the impurity concentrations and base layer width, the ratio of the emitter electron current to the emitter-base hole current can be made very large (values in the order of 100 are attainable). If the input of the amplifier is taken to be between base and emitter with the output taken between collector and emitter, then the input current is the small base-emitter hole current and the output current is the emitter-collector electron current. A significant current gain is thus achieved. If, on the other hand, the input current is applied between emitter and base and the output circuit is connected between collector and base, then no current gain results since the current entering the emitter terminal (sum of hole and electron currents) is only slightly less than the current leaving the collector (electron current from emitter plus collector saturation current). This latter amplifier connection does permit a voltage gain, however, because the high output impedance permits use of a large load resistor through which a current essentially equal to the input current can flow. In contrast with vacuum tubes, transistors control output currents as a result of changes in input current rather than input voltage.

Another type of transistor is shown in Fig. 9. This device is known as a point contact transistor. Rectifying action occurs at both emitter and collector contacts with the germanium surface. A *p-n* barrier exists at both contacts as shown by the dotted lines as a result of contact between the bulk *n* type germanium and *p* type inserts. The base connection is made through a large area contact at the bottom of the material. If the emitter is forward biased and the collector reverse biased with respect to the base, holes are injected from the *p* region at the emitter into the *n* type region and are swept to the collector electrode under the influence of the electric field between collector and base. In addition to the high impedance of the collector-circuit which permits voltage gain with the device, there is also a current gain effected. As a result of the current gain, the input resistance of a point contact transistor amplifier can become negative permitting relaxation oscillations under certain conditions.

Characteristic curves are plotted for transistors as they are for electron tubes, with the added property that transistor de-

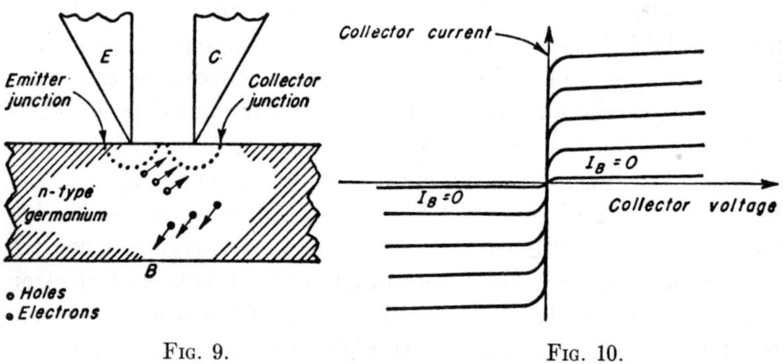

FIG. 9.

FIG. 10.

vices are available in opposite conductivity types, for example, *n-p-n* and *p-n-p* transistors. If two of these devices are made to have identical characteristics except for their conductivity types, it is found that they are symmetrical counterparts of one another in that one has positive currents and voltages whereas the other has identical negative characteristics. One set of collector voltage-current characteristics appears in the first quadrant while the other has odd function symmetry being located in the third quadrant. Fig. 10. shows typical collector characteristics. This property is called complementary symmetry. It is of advantage in arranging for biasing currents and in providing certain circuit functions.

Diodes.—Any two-electrode device, having an anode and a cathode, which has marked unidirectional characteristics is a diode. Many types of diodes are known. Some are electron tubes, so connected as to function as two-electrode devices. There are also many other types. For example:

A *crystal diode* is a diode consisting of a semiconducting material such as a germanium or silicon, as one electrode, and a fine wire "whisker" resting on the semiconductor as the other electrode. Because of its low capacitance, the device finds considerable application as a rectifier or detector of microwave frequencies.

A *semiconductor diode* is a two-electrode semiconductor device having an asymmetrical voltage-current characteristic. A double-base diode is a semiconductor diode in which a potential gradient is produced across the base region by the application of a voltage between two electrodes at either end of the base. The correct polarity and magnitude of this voltage causes the diode to exhibit a controllable, negative resistance between one of the base electrodes and the anode.

A *junction diode* is a semiconductor diode whose nonsymmetrical volt-ampere characteristics are manifested as the result of the junction found between *n*-type and *p*-type semiconductor materials. This junction may be either diffused, grown or alloyed.

The Zener diode. A Zener diode is a special type of junction diode in which the reverse breakdown occurs more sharply than in an ordinary diode. In the breakdown region the voltage across the diode is almost independent of current over a wide range of current; a typical value for the incremental resistance, from the slope of the characteristic, is a few ohms. In small low-power diodes the minimum current I_{min} is about 1 milliampere, and the maximum current I_{max} is about 1 ampere. When biased within this range, the Zener diode acts essentially as a battery of voltage V_Z with a low internal resistance.

A *tunnel diode* has a *p-n* junction with very high densities of both acceptor and donor impurities. A consequence of the high impurity density is to produce a very narrow transition region. This arises because the ionized impurity atoms are close together, and so produce a high space charge density; hence the equilibrium difference in potential is built up in a very short distance.

The resulting current characteristic has a sharp declining portion, which enables the tunnel diode to act as a high speed switch.

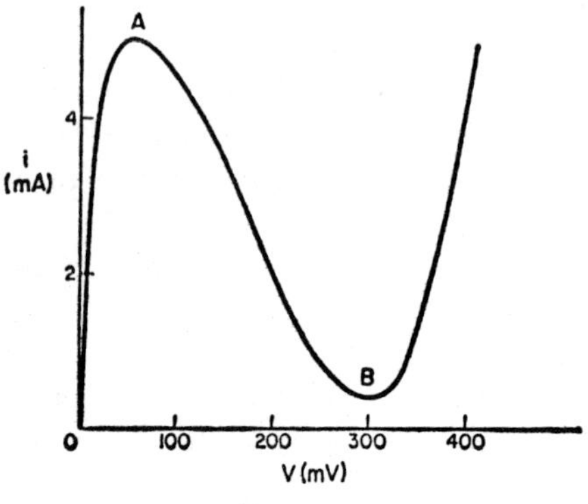

FIG. 11.

Properties of Circuits.—Before treating the circuits used in radio and television equipment, some topics pertaining to electric circuits generally require somewhat more extended treatment than they were given in the preceding chapter. This discussion does not repeat the treatment given there of such fundamentals as resistance, capacitance, inductance, Ohm's Law and similar elementary topics.

Circuit Elements in Multiple.—*Multiple Resistances.*

1. When connected in series, as shown in Figs. 12 and 13, the total resistance of two or more resistors is higher than any of the units.

Expressed as a formula:

$$R_{\text{total}} = R_1 + R_2 + R_3$$

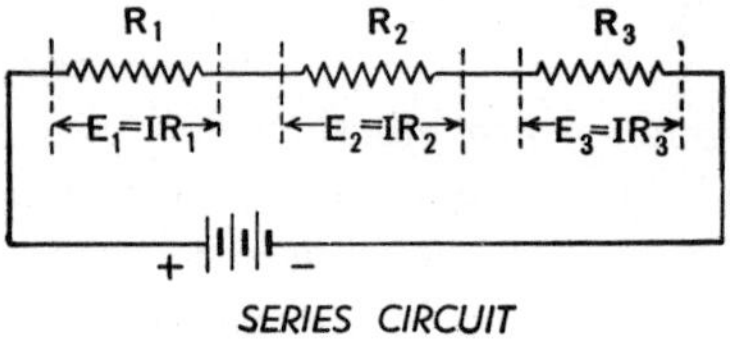

R_1 R_2 R_3

$E_1 = IR_1$ $E_2 = IR_2$ $E_3 = IR_3$

\+ −

SERIES CIRCUIT

FIG. 12.

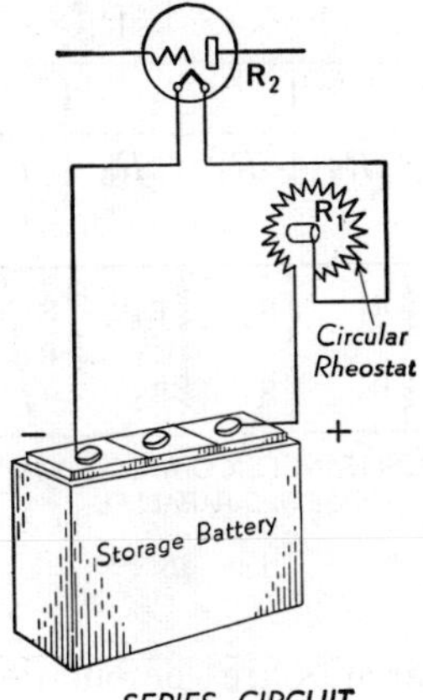

SERIES CIRCUIT

FIG. 13.

2. When connected in parallel as in Figs. 14 and 15, the total resistance of two or more resistors is decreased. Expressed as a formula:

$$R_{\text{total}} = \frac{1}{\dfrac{1}{R_1} + \dfrac{1}{R_2} + \dfrac{1}{R_3}}$$

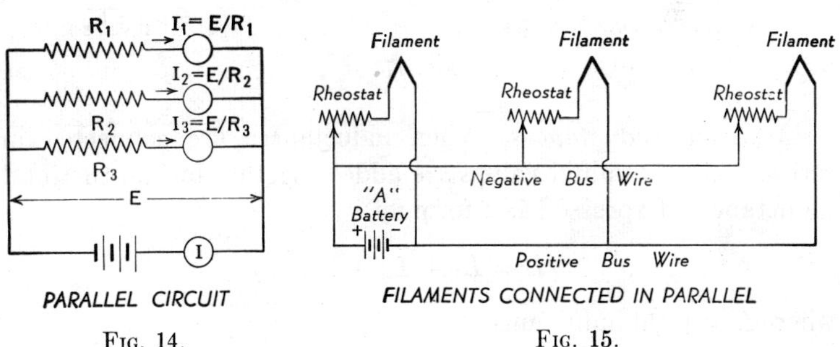

PARALLEL CIRCUIT

FIG. 14.

FILAMENTS CONNECTED IN PARALLEL

FIG. 15.

3. When connected in series parallel as shown in Fig. 16, the total resistance is shown by the following formula:

$$R_{\text{total}} = \cfrac{1}{\cfrac{1}{R_1 + R_2} + \cfrac{1}{R_3 + R_4} + \cfrac{1}{R_5 + R_6} + \cfrac{1}{R_7 + R_8 + R_9}}$$

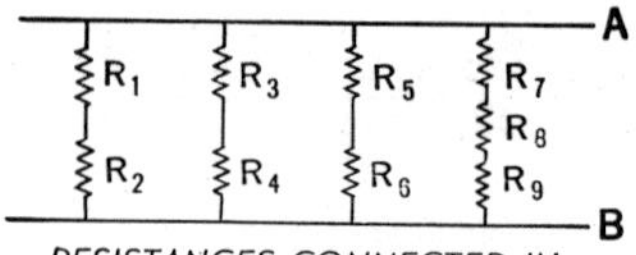

RESISTANCES CONNECTED IN SERIES-PARALLEL

Fig. 16.

Nearly all radio circuits are combinations of series and parallel circuits. The problems that arise in these complicated circuits can be solved by the use of Ohm's Law. Figure 17 shows a combined circuit.

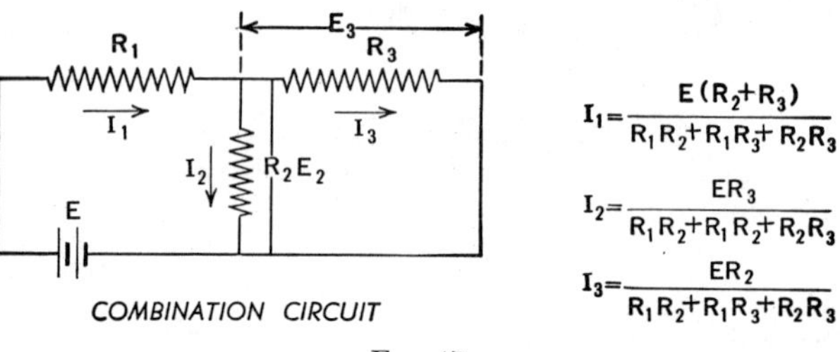

COMBINATION CIRCUIT

Fig. 17.

Multiple Inductances. When inductances are connected in series, their individual values are added together to find the total inductance. Expressed as a formula:

$$L = L_1 + L_2 + 2M$$

where L = total inductance

M = mutual inductance (usually measured on a Wheatstone bridge)

When inductances are connected in parallel the total inductance may be calculated from the following formula:

$$L = \frac{L_1 \times L_2 - M^2}{L_1 + L_2 - 2M}$$

To find an inductance to match a known capacity the following formula may be used:

$$L = \frac{1}{(2\pi)^2 \times f^2 \times C \times 10^8}$$

where $(2\pi)^2 = 39.47$
f = frequency
C = capacity in microfarads
$10^8 = 100{,}000{,}000$

The lumped inductance of coils for transmitting and receiving may be calculated from the following formula:

$$L = \frac{0.2A^2N^2}{3A + 9B + 10C}$$

where L = inductance in microhenries
A = mean diameter of coil in inches
B = length of winding in inches
C = radial depth of winding in inches
N = number of turns of wire

The quantity, C, may be neglected if the coil is a single-layer solenoid.

ILLUSTRATION: Find the inductance of a coil with 35 turns of No. 30 D.S.C. wire on a receiving coil form that has a diameter of 1 5 inches. (From the copper wire table it can be seen that 35 turns of No. 30 D.S.C. wire will give a length of one-half inch.)

$$L = \frac{0.2A^2N^2}{3A + 9B}$$

$$= \frac{0.2 \times (1.5)^2 \times (35)^2}{(3 \times 1.5) \times (9 \times 0.5)}$$

$$= \frac{5512.5}{9} = 61.25 \text{ microhenries.} \quad \text{(Ans.)}$$

TABLE 1

COPPER WIRE TABLE

Gauge No. B. & S.	Diam. in Mils [1]	Circular Mil Area	Turns per Linear Inch [2]				Turns per Square Inch [2]			Feet per Lb.		Ohms per 1000 ft. 250 C.	Current-Carrying Capacity at 1500 C.M. per Amp. [3]	Diam. in mm.	Nearest British S.W.G. No.
			Enamel	S.S.C.	D.S.C. or S.C.C.	D.C.C.	S.C.C.	Enamel S.C.C.	D.C.C.	Bare	D.C.C.				
1	289.3	82690	.....	.....	.....	.....	.....	.....	.....	3.947	.....	.1264	55.7	7.348	1
2	257.6	66370	.....	.....	.....	.....	.....	.....	.....	4.977	.....	.1593	44.1	6.544	3
3	229.4	52640	.....	.....	.....	.....	.....	.....	.....	6.276	.....	.2009	35.0	5.827	4
4	204.3	41740	.....	.....	.....	.....	.....	.....	.....	7.914	.....	.2533	27.7	5.189	5
5	181.9	33100	.....	.....	.....	.....	.....	.....	.....	9.980	.....	.3195	22.0	4.621	7
6	162.0	26250	.....	.....	.....	.....	.....	.....	.....	12.58	.....	.4028	17.5	4.115	8
7	144.3	20820	.....	.....	.....	.....	.....	.....	.....	15.87	.....	.5080	13.8	3.665	9
8	128.5	16510	7.6	.....	7.4	7.1	.....	.....	.....	20.01	19.6	.6405	11.0	3.264	10
9	114.4	13090	8.6	.....	8.2	7.8	.....	.....	.....	25.23	24.6	.8077	8.7	2.906	11
10	101.9	10380	9.6	.....	9.3	8.9	87.5	84.8	80.0	31.82	30.9	1.018	6.9	2.588	12
11	90.74	8234	10.7	.....	10.3	9.8	110	105	97.5	40.12	38.8	1.284	5.5	2.305	13
12	80.81	6530	12.0	.....	11.5	10.9	136	131	121	50.59	48.9	1.619	4.4	2.053	14
13	71.96	5178	13.5	.....	12.8	12.0	170	162	150	63.80	61.5	2.042	3.5	1.828	15
14	64.08	4107	15.0	.....	14.2	13.3	211	198	183	80.44	77.3	2.575	2.7	1.628	16
15	57.07	3257	16.8	.....	15.8	14.7	262	250	223	101.4	97.3	3.247	2.2	1.450	17
16	50.82	2583	18.9	18.9	17.9	16.4	321	306	271	127.9	119	4.094	1.7	1.291	18
17	45.26	2048	21.2	21.2	19.9	18.1	397	372	329	161.3	150	5.163	1.3	1.150	18
18	40.30	1624	23.6	23.6	22.0	19.8	493	454	399	203.4	188	6.510	1.1	1.024	19
19	35.89	1288	26.4	26.4	24.4	21.8	592	553	479	256.5	237	8.210	.86	.9116	20
20	31.96	1022	29.4	29.4	27.0	23.8	775	725	625	323.4	298	10.35	.68	.8118	21
21	28.46	810.1	33.1	32.7	29.8	26.0	940	895	754	407.8	370	13.05	.54	.7230	22
22	25.35	642.4	37.0	36.5	34.1	30.0	1150	1070	910	514.2	461	16.46	.43	.6438	23
23	22.57	509.5	41.3	40.6	37.6	31.6	1400	1300	1080	648.4	584	20.76	.34	.5733	24
24	20.10	404.0	46.3	45.3	41.5	35.6	1700	1570	1260	817.7	745	26.17	.27	.5106	25
25	17.90	320.4	51.7	50.4	45.6	38.6	2060	1910	1510	1031	903	33.00	.21	.4547	26
26	15.94	254.1	58.0	55.6	50.2	41.8	2500	2300	1750	1300	1118	41.62	.17	.4049	27
27	14.20	201.5	64.9	61.5	55.0	45.0	3030	2780	2020	1639	1422	52.48	.13	.3606	29
28	12.64	159.8	72.7	63.6	60.2	48.5	3670	3350	2310	2067	1759	66.17	.11	.3211	30
29	11.26	126.7	81.6	74.8	65.4	51.8	4300	3900	2700	2607	2207	83.44	.084	.2859	31
30	10.03	100.5	90.5	83.3	71.5	55.5	5040	4660	3020	3287	2534	105.2	.067	.2546	33
31	8.928	79.70	101.	92.0	77.5	59.2	5920	5280	.....	4145	2768	132.7	.053	.2268	34
32	7.950	63.21	113.	101.	83.6	62.6	7060	6250	.....	5227	3137	167.3	.042	.2019	36
33	7.080	50.13	127.	110.	90.3	66.3	8120	7360	.....	6591	4697	211.0	.033	.1798	37
34	6.305	39.75	143.	120.	97.0	70.0	9600	8310	.....	8310	6168	266.0	.026	.1601	38
35	5.615	31.52	158.	132.	104.	73.5	10900	8700	.....	10480	6737	335.0	.021	.1426	38–39
36	5.000	25.00	175.	143.	111.	77.0	12200	10700	.....	13210	7877	423.0	.017	.1270	39–40
37	4.453	19.83	198.	154.	118.	80.3	.....	.....	.....	16660	9309	533.4	.013	.1131	41
38	3.965	15.72	224.	166.	126.	83.6	.....	.....	.....	21010	10666	672.6	.010	.1007	42
39	3.531	12.47	248.	181.	133.	86.6	.....	.....	.....	26500	11907	848.1	.008	.0897	43
40	3.145	9.88	282.	194.	140.	89.7	.....	.....	.....	33410	14222	1069	.006	.0799	44

[1] A mil is 1/1000 (one thousandth) of an inch.
[2] The figures given are approximate only, since the thickness of the insulation varies with different manufacturers.
[3] The current-carrying capacity at 1000 C.M. per ampere is equal to the circular-mil area (Column 3) divided by 1000.

Inductive Reactance. The combined effect of frequency and inductance in coils is termed inductive reactance. To find inductive reactance the following formula may be used:

where $X_L = 2\pi fL$

X_L = the inductive reactance in ohms

$\pi = 3.1416$

f = frequency in cycles per second

L = inductance in henries

ILLUSTRATION: A coil has an inductance of 1000 microhenries. If the frequency of the current is 500 kilocycles per second find the inductive reactance.

$$
\begin{aligned}
X_L &= 2\pi fL \\
&= 2 \times 3.1416 \times (500 \times 1000) \times \frac{1000}{1{,}000{,}000} \\
&= 3141.6 \text{ ohms} \quad \text{(Ans.)}
\end{aligned}
$$

Multiple Condensers.—Capacitances may be connected in series or in parallel like resistances or inductances. When they are connected in parallel the resultant capacity is the sum of the individual capacities. Expressed as a formula:

$$C_{\text{parallel}} = C_1 + C_2 + C_3$$

The sketch, Fig. 18, illustrates the method of connecting condensers in parallel.

The sketches indicate condensers connected in series (Fig. 19 and 20).

When condensers are connected in series the capacity may be found by the following formula:

$$C_{\text{series}} = \frac{1}{\dfrac{1}{C_1} + \dfrac{1}{C_2} + \dfrac{1}{C_3}}$$

If but two condensers are considered the formula becomes

$$C_{\text{series}} = \frac{C_1 \times C_2}{C_1 + C_2}$$

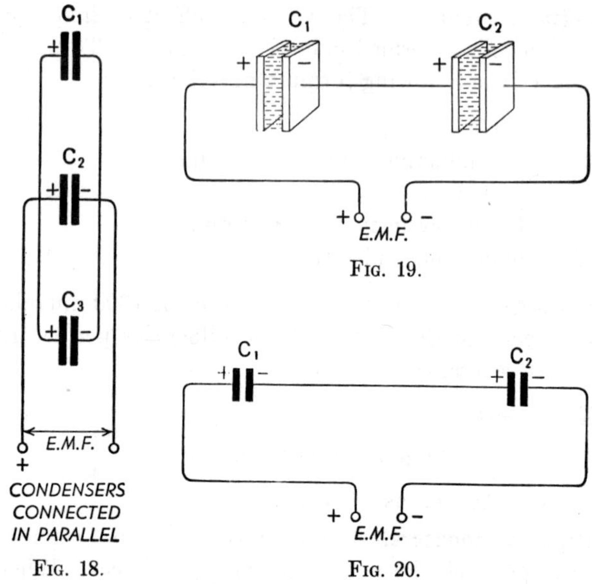

FIG. 18.

FIG. 19.

FIG. 20.

ILLUSTRATION: Find the resultant capacity when condensers of 0.002 microfarad and 0.0015 microfarad capacity are connected in parallel.

$$C = 0.002 + 0.0015 = 0.0035 \ \mu\text{fd.} \quad \text{(Ans.)}$$

ILLUSTRATION: If the condensers in the preceding problem are connected in series find the total capacity.

$$C = \frac{0.002 \times 0.0015}{0.002 + 0.0015} = \frac{0.000003}{0.0035} = 0.00086 \ \mu\text{fd.} \quad \text{(Ans.)}$$

ILLUSTRATION: A radio circuit requires a 0.00035 microfarad condenser but only a 0.0005 microfarad variable condenser is available. What fixed condenser may be used to reduce the maximum capacity in the circuit to the required value? How shall the condenser be connected?

Because the total capacity is to be reduced the fixed condenser must be connected in series with the variable condenser. Then,

$$C = \frac{1}{\frac{1}{C_1} + \frac{1}{C_2}}$$

Removing the reciprocal from the right side and placing it at the left side;

$$\frac{1}{C} = \frac{1}{C_1} + \frac{1}{C_2} \quad \text{and} \quad \frac{1}{C_2} = \frac{1}{C} - \frac{1}{C_1}$$

$$\frac{1}{C_2} = \frac{1}{0.00035} - \frac{1}{0.0005} = \frac{1 - 0.7}{0.00035} = \frac{0.3}{0.00035}$$

$$C = \frac{0.00035}{0.3} = 0.001166 \quad \text{or} \quad 1166\ \mu\mu\text{fd.} \quad \text{(Ans.)}$$

Capacitive Reactance. Condensers have a reactance that is inversely proportional to the condenser size and to the frequency of the applied voltage.

$$X_c = \frac{1}{2\pi fC}$$

where X_c = capacitive reactance in ohms
π = 3.1416
f = frequency in cycles per second
C = condenser capacitance in farads

However, the capacitance, in most practical cases, is given in microfarads (μfd.). Then the formula becomes:

$$X_c = \frac{1{,}000{,}000}{2\pi fC_{\mu\text{fd.}}}$$

ILLUSTRATION: A 3-plate fixed air condenser with a capacity of 0.0001 microfarad is used in an antenna circuit that is operated on a frequency of 3750 kilocycles. Find the capacitive reactance.

$$X_c = \frac{1{,}000{,}000}{2\pi fC_{\mu\text{fd.}}}$$

$$X_c = \frac{1{,}000{,}000}{2 \times 3.1416 \times (3{,}750 \times 1{,}000) \times 0.0001}$$

$$= \frac{100}{0.1356} = 725 \text{ ohms} \quad \text{(Ans.)}$$

Impedance.—The parts of a radio circuit, such as coils and condensers, are never pure reactances. Some resistance, however small, is always present. The reactance and resistance combined together are called impedance, which is influenced almost entirely by the frequency of the alternating voltages impressed upon the circuit.

To find impedance the following formula may be used:

$$Z = \sqrt{X^2 + R^2}$$

where Z = impedance in ohms
X = total reactance
R = resistance

ILLUSTRATION: What is the impedance in a circuit of 4 ohms resistance and 3 ohms reactance?

$$\begin{aligned} Z &= \sqrt{x^2 + R^2} \\ &= \sqrt{3^2 + 4^2} \\ &= \sqrt{9 + 16} = \sqrt{25} = 5 \text{ ohms} \quad \text{(Ans.)} \end{aligned}$$

Ohm's Law for Alternating Currents.—If inductances did not have any resistance it could be assumed that the current would be equal to the voltage divided by the reactance. However, this is not the case and Ohm's Law for alternating current becomes:

$$I = \frac{E}{Z}, \quad Z = \frac{E}{I}, \quad E = IZ$$

ILLUSTRATION: A 60 cycle alternating current of 5 amperes flows through a coil whose inductance in 4000 microhenries and whose resistance is 2 ohms. Find the voltage across the coil.

$$I = \frac{E}{Z}, \quad Z = \sqrt{R^2 + x^2}$$

$$\begin{aligned} X &= 2\pi fL \\ &= 6.28 \times 60 \times 0.004 = 1.507 \text{ ohms} \end{aligned}$$

$$\begin{aligned} Z &= \sqrt{4^2 + 1.507^2} = \sqrt{18.25} = 4.27 \text{ ohms} \\ E &= IZ = 5 \times 4.27 = 21.35 \text{ volts} \quad \text{(Ans.)} \end{aligned}$$

Resonance.—A condition of resonance is obtained when a capacity reactance and an inductive reactance of equal magnitude are connected either in series or parallel. The most important circuits in radio are those in which either series or parallel resonance occurs.

The resonant frequency may be determined by the following formula which is frequently called the fundamental equation in radio:

$$f = \frac{1}{2\pi \sqrt{LC}}$$

where f = frequency in cycles per second
$2\pi = 6.2832$
L = inductance in henries
C = capacitance in farads

Because it is more convenient to use smaller units such as microhenries and microfarads, the formula becomes:

$$f = \frac{1{,}000{,}000}{2\pi \sqrt{LC}}$$

where L = microhenries and C = microfarads.

Illustration: To what frequency will a circuit tune that has an inductance of 9 microhenries and a capacity of 0.0002 microfarad?

$$f = \frac{1{,}000{,}000}{2\pi \sqrt{LC}}$$

$$f = \frac{1{,}000{,}000}{6.28 \sqrt{0.0002 \times 9}}$$

$$= \frac{1{,}000{,}000}{0.2663} = 3{,}755{,}000 \text{ cycles} \quad \text{(Ans.)}$$

$$\text{kilocycles} = \frac{3{,}755{,}000}{1000} = 3{,}755 \quad \text{(Ans.)}$$

Wavelength.—The wavemeter which depends on the principles of resonance is an important instrument in radio measurements. The wavelength is equal to the speed at which electric waves travel (186,000 miles a second, approx.) divided by the frequency in cycles. Expressed as a formula:

$$\lambda = \frac{V}{f}$$

where λ = (pronounced lambda) = wavelength in meters
f = frequency in cycles
V = the velocity of propagation of electro-magnetic waves —approx. 300,000,000 meters per second

ILLUSTRATION: What frequency in cycles and in kilocycles corresponds to a wavelength of 200 meters?

$$f = \frac{V}{\lambda}$$

$$= \frac{300{,}000{,}000}{200} = 1{,}500{,}000 \text{ cycles} \quad \text{(Ans.)}$$

$$\text{kilocycles} = \frac{1{,}500{,}000}{1000} = 1500 \quad \text{(Ans.)}$$

ILLUSTRATION: What wavelength corresponds to 1500 kilocycles?

$$\lambda = \frac{V}{f}$$

$$= \frac{300{,}000{,}000}{1{,}500{,}000} = 200 \text{ meters} \quad \text{(Ans.)}$$

The resonance of a tuned circuit is expressed in terms of wavelength as follows:

$$\lambda = 1885\sqrt{LC}$$

where λ = wavelength in meters
L = inductance in microhenries
C = capacitance in microfarads

ILLUSTRATION: A radio circuit has an inductance of 900 microhenries and a capacity of 0.001 microfarad. Find the wavelength for which this circuit will be resonant.

$$\begin{aligned}\lambda &= 1885 \sqrt{LC} \\ &= 1885 \sqrt{900 \times 0.001} \\ &= 1885 \times 0.95 = 1791 \text{ meters.} \quad \text{(Ans.)}\end{aligned}$$

Electron Tube Circuits.—Electron tube circuits that is connections of resistors, inductors, capacitors, sources of power, and electron tubes assume a wide variety of forms. One convenient method of classifying electron tube circuits employs the amplitude of input signal as a basis. Reference to certain fundamental properties of electron tubes is desirable before describing the circuit classifications. For definiteness, the triode vacuum tube will be used as an example. If the grid to cathode voltage of such a tube is not allowed to become negative, the quantities associated with the device which are of principal interest in describing its behavior are the plate to cathode voltage drop, the plate current, and the grid to cathode voltage drop. The relations among these three variables are usually shown by a graph of two of the three quantities with the third assigned convenient (constant) values. The intersections of the plate characteristics and the load line (the locus of all combinations of the plate voltage and current permitted by the constraints imposed by the elements external to the tube) establish a relation between grid voltage and plate current known as the transfer characteristic. A typical transfer characteristic is shown in Figure 21.

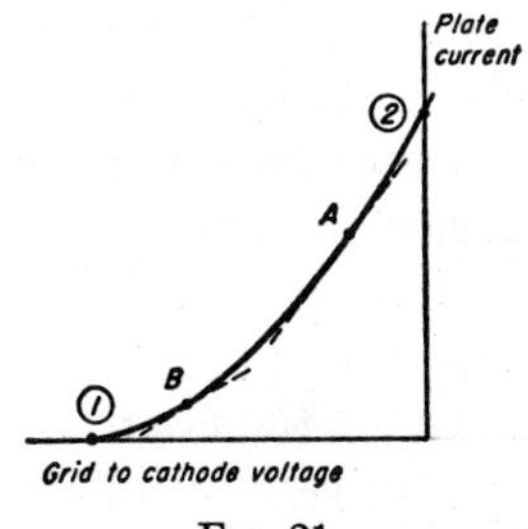

FIG. 21.

Point (1) is commonly referred to as "cut off" for the tube since the plate current is reduced to zero. Point (2) is identified as the "zero bias" point. The difference in grid to cathode voltage between zero bias and cut off is often referred to as the "grid base" of the tube.

This characteristic tells at a glance the plate current to be expected for a given grid to cathode voltage. In practice a grid bias is chosen so that in conjunction with the assigned plate to cathode voltage an appropriate operating point is established about which changes in grid voltage and plate current will take place when an output signal is applied between grid and cathode. The points *A* and *B* in Figure 21 are typical operating points. The choice of operating point is an important consideration in the operation of a tube as an amplifier. It is seen from Figure 21 that although the transfer characteristic is not linear, there are certain regions where the slope of the curve does not change appreciably over a fairly large increment of grid voltage. Depending on the location of the operating point, it is possible to choose an amplitude of input signal so that the change in plate current for this amplitude and all smaller values can be considered to be linearly related to the input signal within a prescribed small amount of error. Operation with such a combination of bias and input signal amplitude is called linear operation of the tube. It is clear from the figure that the range of input signal amplitude permitted to achieve a prescribed degree of linearity between grid voltage and plate current depends on the choice of operating point. Point *A* permits a larger value of input signal for a prescribed degree of Linearity, for example, than does point *B*.

Vacuum tube circuits can be classified as to whether the tubes operate in a linear fashion corresponding to small signal inputs or whether the input signals are large enough to cause excursions over such a large portion of the transfer characteristic that the tangent to curve at the operating point is not a good approximation to the tube behavior over the entire amplitude range of the input signal. There are thus two principal categories of vacuum tube operation, namely, linear or small signal behavior and nonlinear or large signal action. Large signal operation is usually analyzed by the use of graphical constructions utilizing plate characteristics, whereas small signal operation is treated by equivalent circuits by means of which changes in plate current and plate voltage from the quiescent values may be computed.

A trigger or flip-flop circuit is a typical vacuum tube circuit. It has two conditions of stable equilibrium and which can be switched from one stable state to the other by some external influence such as the application of an input pulse. Figure 22

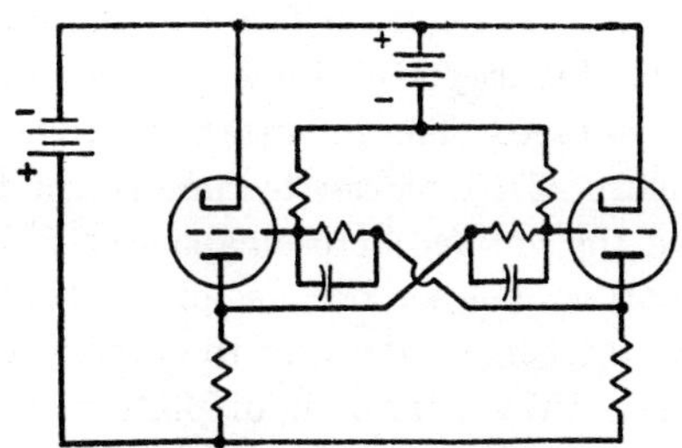

FIG. 22.—Eccles-Jordan circuit.

shows one of the most common flip-flop circuits, the Eccles-Jordan circuit. Two triodes are used in a symmetrical circuit so arranged that one tube is at zero bias and the other is cut off. A positive trigger pulse of sufficient amplitude applied to the grid of the tube cut off, or a negative pulse applied between grid and cathode of the conducting tube, causes a positive feedback action which results in a rapid change of state in each tube. The conducting tube is forced to the cut off state, while at the same time the tube previously cut off is forced into conduction near zero bias. This change of states is effected by the connection from the plate of one tube to the grid of the other. If a tube conducts plate current, the potential at the plate is reduced below the battery voltage by the voltage drop in the resistor connected from the plate to the battery. The grid voltage of the other tube is thus forced by voltage divider action to be below the potential previously attained when plate current was cut off in the first tube. By proper design of the voltage divider and the negative battery supply, the grid to cathode voltage can be made to swing from a slightly positive value (approximately zero bias) to a negative value sufficient to cut the tube off as a result of the variation of the plate voltage of the other tube as the latter changes from the cut off value to that at the full conduction condition.

Another pulse type circuit, which utilizes the transient voltage produced in a resistor-capacitor circuit, is the sweep circuit employed for deflecting the electron beam in the cathode ray tube of a television set. Figure 23(a) shows a simple circuit for generating a sweep voltage. The important voltage waveforms are illustrated in Figure 23(b). Prior to time t_1, the triode conducts at zero bias, and the voltage across capacitor C is only a small fraction of the battery voltage because of the voltage drop in the large resistor R. At time t_1 the grid to cathode voltage is made negative enough to cut off the flow of plate current. This condition is maintained until time t_2. During the interval of time

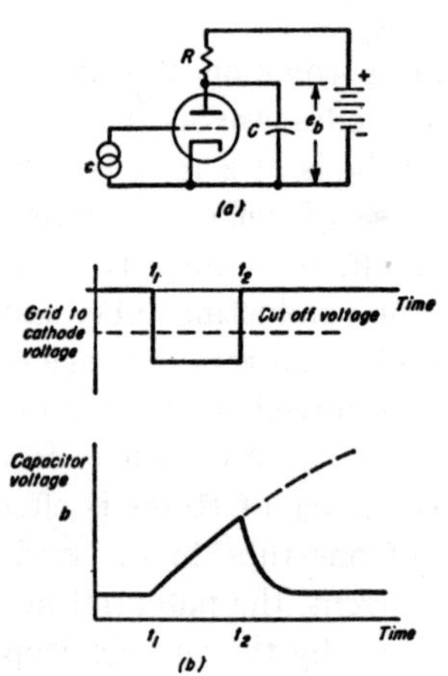

FIG. 23.—(a) Typical sweep generator circuit; (b) Input and output voltage waveforms for sweep circuits.

that the tube is cut off, the capacitor C is charged from the battery through the resistor R. The voltage across the capacitor increases along the exponential curve associated with the transient in a resistor-capacitor circuit. Ultimately e_b would become equal to the battery voltage. The exponential rise is not permitted to continue after time t_2, however, for when the tube is returned to zero bias at that time, the capacitor discharges through the tube and its voltage rapidly assumes the value which it had at the start of the charging process. Many refinements, including feedback, may be added to the circuit of Figure 23(a) to increase the degree of linearity of the voltage change between t_1 and t_2.

Transistor Circuits.—Transistors may be combined with sources of power and passive elements (resistors, inductors, and capacitors) to form transistor circuits of many forms which are used for the generation, amplification, shaping, and control of electrical signals. Many transistor circuits are similar in form to vacuum tube arrangements designed to perform corresponding functions, but, on the other hand, the unusual properties of transistors also lead to circuit arrangements which have no vacuum tube counterparts. Significant differences between vacuum tubes and transistors make the latter far more attractive in applications where the principal objective is the amplification of low level signals where comparatively little power is associated with the signal itself, the combining and processing of signals as required in electrical computation, and the switching of electrical signals to various paths in accordance with appropriate command signals. The fact that extremely low power is required for a transistor to effect these functions combined with its property of much smaller size and weight than a vacuum tube make possible the fabrication of electronic circuits to perform very complicated operations in a small space with low power consumption. The availability of *n-p-n* and *p-n-p* transistors which have the principal charge carriers of opposite sign leads to advantages in cascading transistor amplifiers or data processing circuits on a d-c basis and to simplifications in attaining push-pull operation with its concomitant benefits.

One fundamental difference between vacuum tubes and transistors is the manner in which control of the output current is effected. Vacuum tubes are voltage operated devices, i.e., the plate current is controlled by the voltage impressed between the grid and cathode. The collector current, the usual output current in a transistor, on the other hand, is controlled by the flow of current between base and emitter. The potential difference across the emitter-base junction is a non-linear function of the junction current. As a consequence, to avoid introducing amplitude distortion, the input signal current must be furnished from a high impedance source. This factor alone often results in differences between vacuum tube circuits and the corresponding transistor versions.

As with vacuum tube circuits, when a transistor is employed in conjunction with other circuit elements, a suitable combination of operating parameters must be chosen to establish a quiescent operating point about which the input signal will cause variations in the various electrode currents. The operating point is established by providing appropriate sources of potential in series with resistors in order to obtain the desired steady emitter, base, and collector currents referred to as bias currents.

Operation of transistors may take place in a linear or nonlinear fashion depending on the amplitude of the input current. When the variations in the input current are small (defined in the same sense as "small signal" operation with electron tube circuits), equivalent circuits may be used for the calculation of the changes in currents about the operating point. Fig. 24(a) illustrates common symbols for the p-n-p and n-p-n transistors. Emitter, base, and collector terminals are indicated by e, b, and c, respectively. Fig. 24 (b) shows a p-n-p unit connected in a common base amplifier circuit, and Figs. 24(c) and 24(d) indicate (in the dashed outline) one form of equivalent circuit that may be used to compute incremental changes in transistor currents. The representation is spoken of as the T equivalent circuit, and the parameters are defined in terms of the steady emitter-base and collector-base potential differences and currents (V_e, V_c, I_e, and I_c) shown on the following page.

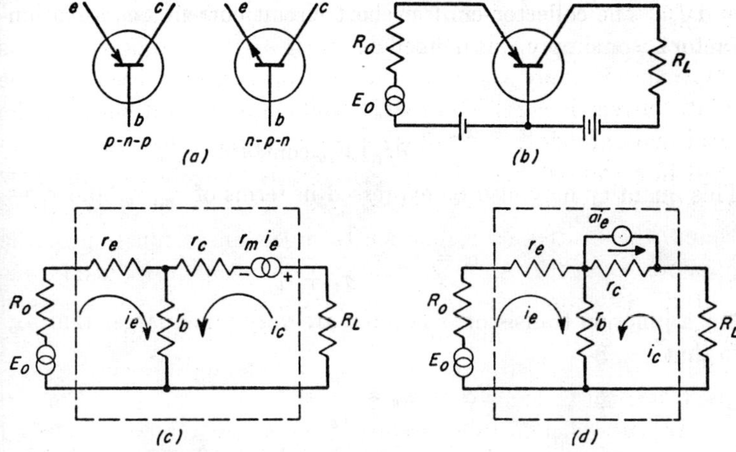

FIG. 24.—(*a*) Symbols for transistor; (*b*) Common base amplifier circuit; (*c*) and (*d*) Two transistor T equivalent circuit.

Base resistance, $r_b = \left.\dfrac{\partial V_e}{\partial I_c}\right| I_e \text{ constant}$

Emitter resistance, $r_e = \left.\dfrac{\partial V_e}{\partial I_e}\right| I_e \text{ constant} - \left.\dfrac{\partial V_e}{\partial I_e}\right| I_c \text{ constant}$

Collector resistance, $r_c = \left.\dfrac{\partial V_c}{\partial I_c}\right| I_c \text{ constant} - \left.\dfrac{\partial V_e}{\partial I_c}\right| I_c \text{ constant}$

Mutual resistance, $r_m = \left.\dfrac{\partial V_c}{\partial I_e}\right| I_c \text{ constant} - \left.\dfrac{\partial V_e}{\partial I_e}\right| I_e \text{ constant}$

$$a = \frac{r_m}{r_e}$$

The partial derivatives * appearing in the above expressions may be determined from suitable graphs of the parameters V_e, V_c, I_e,

* The partial derivative is obtained when the process of differentiation (Chapter VI) is applied to expressions in more than one variable, by treating the excess variable or variables (written after the vertical rule in the expression) as constants.

and I_c. The collector-emitter short circuit current amplification factor α_{ce} or simply α, is defined as

$$\alpha = \alpha_{ce} = \left.\frac{\partial I_c}{\partial I_e}\right|_{V_{eb} \text{ constant}}$$

This quantity may also be expressed in terms of r_m, r_b, and r_c as

$$\alpha = \alpha_{ce} = \frac{r_m + r_b}{r_c + r_b}$$

For a junction transistor, r_m and r_c are very much larger than r_b, so that

$$\alpha = \alpha_{ce} \approx \frac{r_m}{r_c} = a$$

For this reason a and α are often used interchangeably in the equivalent circuits of Figs. 1(c) and (d). Another quantity of interest is the collector-base short circuit current amplification factor α_{cb} defined as

$$\alpha_{cb} = \left.\frac{\partial I_c}{\partial I_b}\right|_{V_c \text{ constant}}$$

For a junction transistor the following approximate relation is valid

$$\alpha_{cb} \approx \frac{a}{1-a} \approx \frac{\alpha}{1-\alpha}$$

The quantity α_{cb} appears in the quivalent T circuit for the common emitter connection. Fig. 25 indicates the equivalence for

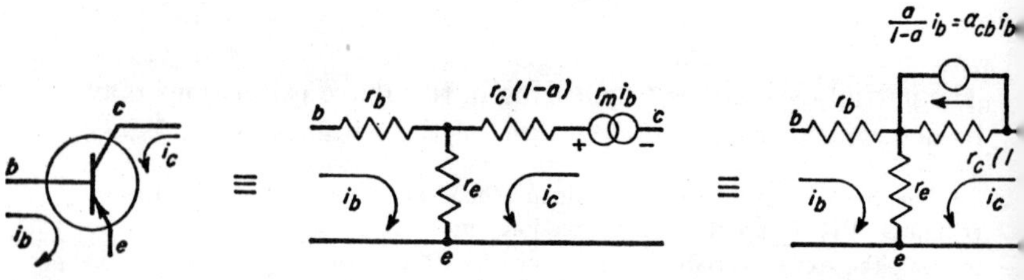

FIG. 25.

this arrangement. It is to be noted that the values of the various parameters in the equivalent circuits vary with the electrode currents as well as with frequency. The circuits presented are valid only at low frequencies. There are effects which occur at high frequencies which require assigning a frequency dependence to α as well as the introduction of capacitors across various elements. Typical low frequency values of r_c and r_m are in megohms; or r_b, several hundred to a thousand ohms; or r_e, less than one hundred ohms; and of α and α_{cb} 0.98 to 0.99 and 50 to 100, respectively.

There are several forms of equivalent circuits that are used to represent transistor small signal operation in addition to the T network form just described. Another representation that finds frequent use is one employing the so-called hybrid parameters. The parameters are "hybrid" in the sense that they do not all have the same dimensions. Fig. 26 indicates the equivalent

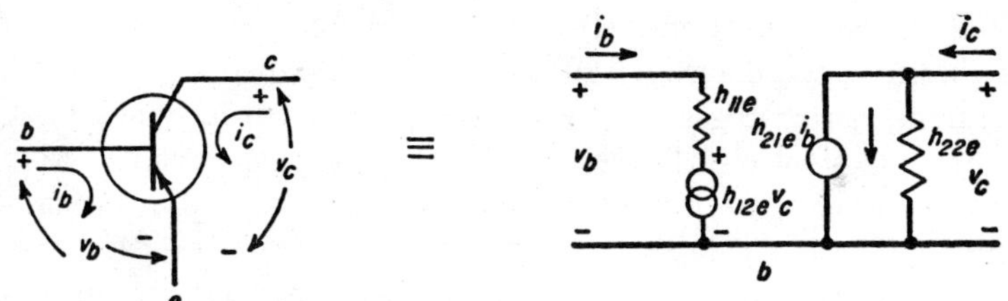

FIG. 26.

circuit for the common emitter connected transistor using these parameters. The base-emitter circuit is represented by an impedance h_{11e} in series with a voltage generator. The collector-emitter circuit is represented by an admittance h_{12e} in parallel with a current generator. The parameters are defined in terms of static currents and voltages as follows:

h_{11e} = input impedance (short circuit across output terminals)

$$= \left.\frac{\partial V_{be}}{\partial I_b}\right| V_{ce} \text{ constant}$$

h_{22e} = output admittance (no connection across input terminals)

$$= \left.\frac{\partial I_e}{\partial V_{ce}}\right|_{I_b \text{ constant}}$$

h_{12e} = reverse open circuit voltage amplification factor

$$= \left.\frac{\partial V_{be}}{\partial V_{ce}}\right|_{I_b \text{ constant}}$$

h_{21e} = forward short circuit current amplification factor

$$= \left.\frac{\partial I_c}{\partial I_e}\right|_{V_{ce} \text{ constant}} = \alpha_{cb}$$

These parameters may be defined for any form of transistor connection. The subscript e denotes common emitter connection; a corresponding labelling is used to distinguish the parameters for other connections.

The A-M Radio Receiver.—A type of radio receiver that continues to be used extensively is the superheterodyne receiver. It is shown in the diagram of Fig. 27 in a simplified circuit which contains fewer stages than would be provided in an actual set, in order to show clearly the major functional part-circuits. Also, transistors have been used in place of electron tubes, in accordance with present-day design.

This circuit differs from others in that it converts all incoming radio-frequency signals to a common carrier frequency. This is accomplished in the first detector, mixer or converter as it is variously called. The signal from the antenna is fed by a tuned coupled circuit to the mixer unit (or in more elaborate sets a tuned radio-frequency stage may be inserted between the antenna and the mixer). In the mixer stage the incoming signal is heterodyned with a locally generated signal so a beat frequency signal, called the intermediate frequency, is produced. This new frequency signal is radio frequency, ranging from around 450 kc. to several megacycles depending upon the purpose for which the receiver is designed. The intermediate frequency has exactly the same modulation as the original signal. In many

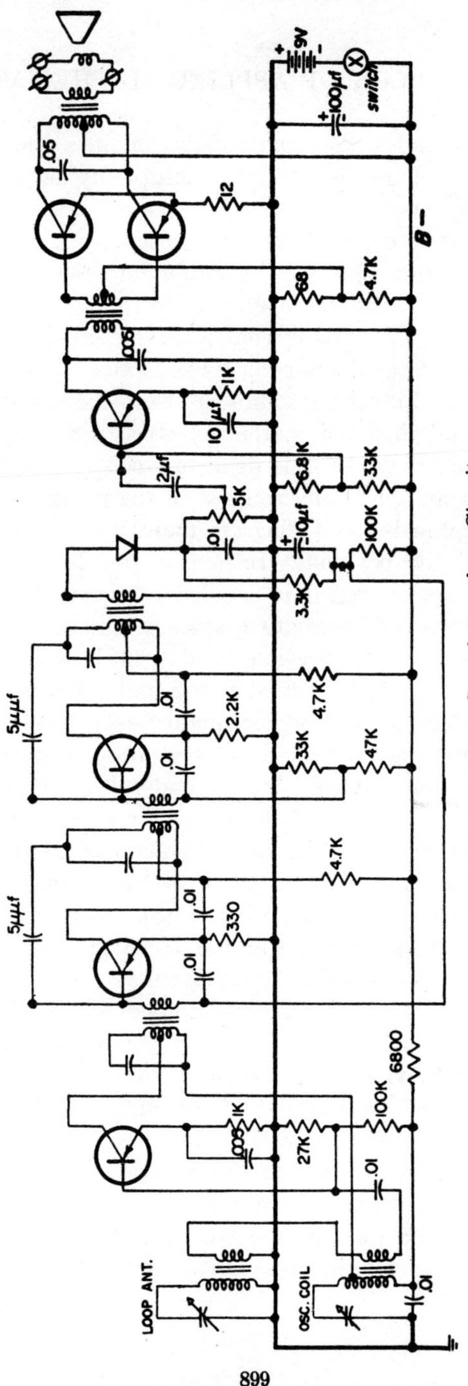

Fig. 27.—Transistor Superheterodyne Circuit.

broadcast receivers the mixer unit combines the functions of mixer and oscillator by using a multiplicity of transistors (in vacuum tube sets, a pentagrid tube is used). However, at higher frequencies it is desirable or even necessary to use a separate unit for oscillator and feed its output into the mixer. Regardless of how the oscillator operates, its frequency is always adjusted by the main tuning control of the receiver so the beat frequency output of the mixer is a fixed value. This intermediate frequency signal is then amplified by fixed-tuned radio-frequency amplifiers and then fed to the detector (commonly called the second detector) where it is demodulated. The audio signal is then further amplified and coupled to the speaker.

The simplified-circuit diagram indicates the various circuits and their relative positions. In some of the cheaper superheterodynes the antenna signal is coupled to the first transistor detector, then the intermediate frequency output of this coupled without further amplification to a grid bias or regenerative detector and hence to the final power unit. Various refinements are often added to the higher quality sets. Among these are automatic volume control, automatic frequency control, noise suppression, tone control, fidelity and selectivity controls, etc. The superheterodyne gives much greater selectivity by its system of frequency changing and also permits the use of circuits having a more uniform response to the sidebands.

Television.—Television is the transmission of scenes, either still or motion, by electrical means, commonly by radio, for instantaneous viewing without permanent recording. For a practical system certain fundamental components or functions are necessary:

1. Camera device to pick up the scene.
2. Tranducer to convert the light impulses of the scene to a corresponding electrical signal.
3. Transmitter to convert the electrical signals into proper form to be transmitted to the receiver.
4. Receiver to pick up the transmitted signals and convert them to the proper form to apply to a transducer.

5. Transducer to convert the electrical signals back into light in a reproduction of the original scene.

The first three of these topics are outside the scope of this book, which deals with signal receivers. Therefore this discussion deals with the equipment required for reception of television signals, and their transformation into sound and pictures on the screen of a cathode-ray tube.

The received scene must be reconstructed from the electrical pulses reaching the receiver. If we assume that the original scanner broke the picture down into ten lines and each line had ten elements side by side (this is determined by the width of the hole, being the width of the picture divided by the hole width), then we have 10 times 10 or 100 elements in the scene and our reproduced picture must be built of 100 blocks. It can readily be seen that this would give a very coarse mosaic effect to the picture since each element is fixed in intensity of light. More lines would give more elements and finer detail in the receiver scene. A close comparison may be drawn between this and the printed pictures of newspapers and magazines. The relatively coarse newspaper pictures are lacking in detail while the usual magazine half-tones give very good detail. The difference between these pictures is the number of dots or elements (clearly visible if the printed picture is viewed through a magnifying glass) of which they are composed.

Besides the number of lines per scene, the rate at which the scenes are repeated is also important. The television pictures must be repeated at a rate high enough to give the illusion of smooth motion. While the motion picture rate of 24 frames per sec. would be satisfactory for this rate, the value should be harmonically related to the power line frequency in order to minimize certain interference effects. The standard power supply frequency of 60 cycles dictates the use of 30 frames per sec. for television. Furthermore, to improve the quality of the reproduced picture, the scanning is not done for adjacent lines in order, but the picture is scanned over alternate lines first and then scanned again over those missed the first time. This double scanning, known as interlaced scanning, is done in the thirtieth-of-a-second period of one frame.

A block diagram of a complete television system is shown in Fig. 28. The original scene is focused on the camera tube by a light lens system. The camera tube converts this light picture into the sequence of electrical elements necessary for transmission.

The very minute electrical signals coming from the camera tube are amplified by wide band video amplifiers, the wide band being necessitated by the great range of frequencies produced by the modern multi-line systems. This wide band amplifier feeds a monitor circuit which reproduces the televised scene on a picture tube so the operator can check the camera circuit operation continuously. It also furnished the input for the modulator which modulates the picture signals on the radio-frequency carrier in a manner very similar to that of the audio modulation of conventional broadcasting. The modulated radio frequency is then supplied to the antenna and radiated into space. At the same time the microphone picks up the sound associated with the scene. This signal is amplified and used to frequency modulate the sound carrier of the television transmitter. The sound-modulated carrier is radiated simultaneously with the picture carrier. Both are then picked up by the receiving antenna, amplified and fed through the first detector and intermediate frequency amplifiers of a superheterodyne receiver. The two types of signal are then separated and each is fed to its proper detector or demodulator. The sound signals, now at audio frequency, are further amplified and drive the loudspeaker. The picture signal circuits are much more complex. In order to reproduce the scene at the receiver, it is necessary for the receiver transducer, whatever its nature, to follow exactly the operation of the camera tube. Thus when the electron beam in the camera scans the scene at a given rate, each scanning line starting at a definite time after the preceding one, the scanning beam in the picture tube in the receiver must retrace the scene in exactly the same order, each line starting at the same time interval after the

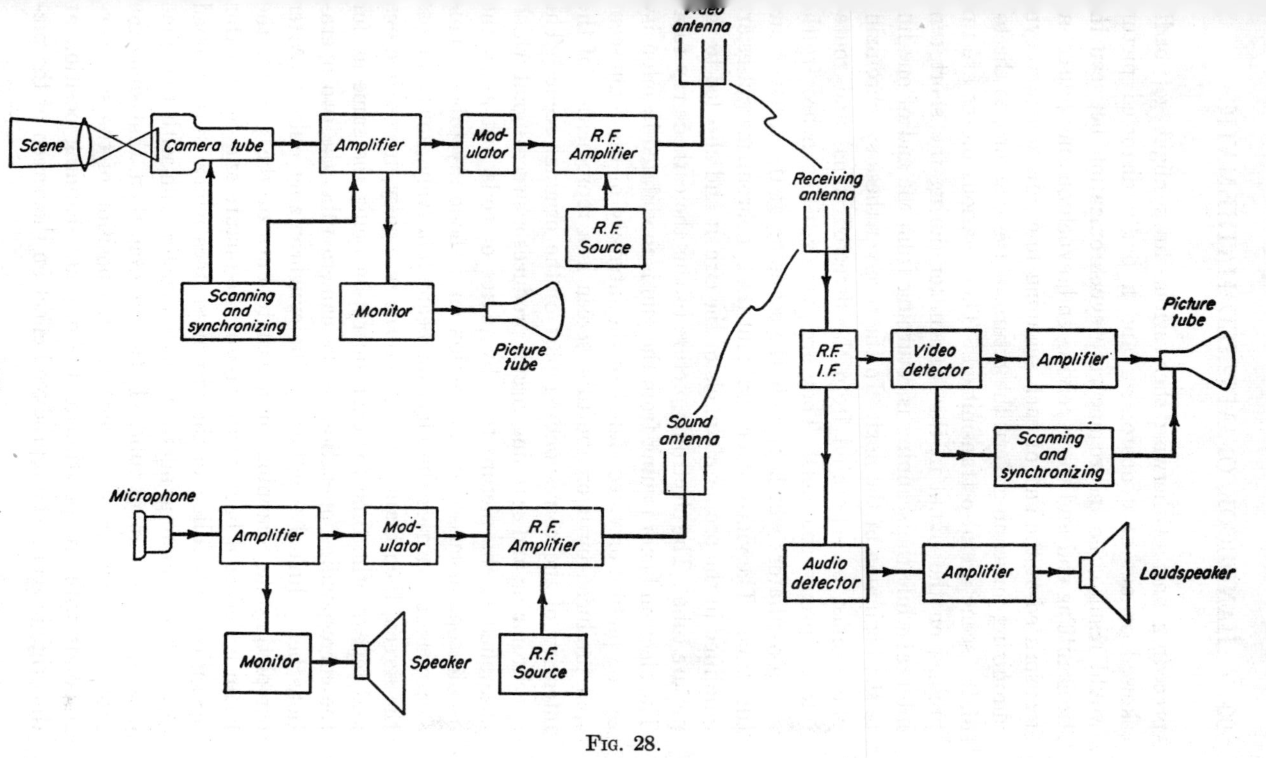
Scene
Camera tube
Amplifier
Mod-ulator
R.F. Amplifier
antenna
R.F. Source
Scanning and synchronizing
Monitor
Picture tube
Receiving antenna
R.F. I.F.
Video detector
Amplifier
Picture tube
Scanning and synchronizing
Sound antenna
Microphone
Amplifier
Mod-ulator
R.F. Amplifier
Monitor
Speaker
R.F. Source
Audio detector
Amplifier
Loudspeaker

FIG. 28.

preceding one. Otherwise the various lines might get badl skewed or out of synchronism and a badly distorted pictur would result. To insure accurate synchronization between th transmitting end and the receiving end, synchronizing pulses ar transmitted at the end of each scanning line. In addition, syn chronizing pulses to govern the return of the scanning to the to of the scene are also transmitted. These various pulses are im pressed on the signal in the transmitter during the short tim interval while the scanning is returning from the end of one lin to the beginning of the next. In the receiver these synchronizin pulses must be separated from the detected signal and route into the proper channels. They are then used to synchronize th sweep oscillators which provide the scanning at the receiver pic ture tube. The video signals without the synchronizing pulses ar amplified in the proper channels of the circuit and also fed to th picture tube. The electronic picture tube is the cathode ray tube The electron beam issuing from the gun is modulated in intensity by the picture or video signal so the intensity of the fluorescen spot produced by it on the tube screen is a reproduction of the intensity of the corresponding part of the original scene. The synchronizing and scanning circuit produces a sweep signal which is applied to the picture tube by plates or coils just as in the oscilloscope discussed in the section on cathode ray tubes. This sweep action carries the electron beam relatively slowly across the screen, then blanks it and returns it rapidly, moves it down and repeats, the time for each operation being the same as for the corresponding operation in the camera tube, the two operations being linked together by the synchronizing pulses. After completing the scanning of alternate lines of the picture, the beam is deflected back to the top of the picture and repeats this operation, now filling in the alternate lines which were skipped on the first scanning, again in exact synchronism with the same process in the camera tube. It can be seen, then, that since the intensity of the original scene and the position of the spot corresponds with the position of the original scanning position at the original scene, the reproduced effect on the screen of the picture tube is the scene which was picked up by the camera.

As has been discussed in detail above, television provides for the reproduction of pictures upon the face of a cathode ray tube by control of the brightness of the white light emanating from the tube face. To reproduce scenes in color, that is, to provide color television, not only is transmitting and receiving apparatus of additional complexity needed, but a device capable of providing a colored light output at the scanning location is also required. To provide the basis for the subsequent explanation of a color television system, some attention must be given first to color representation.

Human vision can be expressed in three separate color sensations, each of which may be stimulated in various degrees. Although these sensations always appear to act together, it is believed that if they could be stimulated separately they would be found to be the sensations produced by red, blue, and green. The respective amount of these three primaries needed to match a color of particular wavelength are called the tristimulus coefficients. Since the total of the three quantities is 100%, if any two are known, the other is known also. Thus the three quantities used to specify the amounts of the primaries can be reduced to two, allowing a two-dimensional plot to represent any color. Letting X, Y, and Z be the amount of the primaries needed to match a given color, the trichromatic coefficients x, y, and z are defined as

$$x = \frac{X}{X + Y + Z}$$

$$y = \frac{Y}{X + Y + Z}$$

$$z = \frac{Z}{X + Y + Z}$$

but z is equal to $100\% - (x + y)$. The two-dimensional plot formed by the use of two trichromatic coefficients to represent colors is known as a chromaticity diagram.

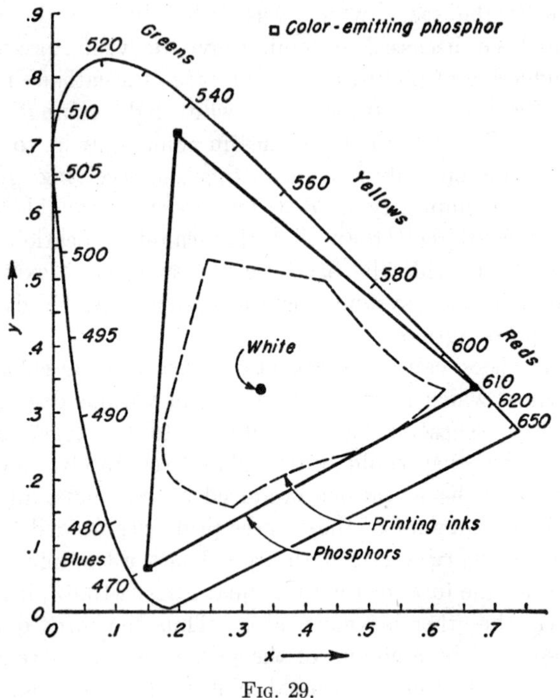

FIG. 29.

The diagram using x and y coordinates is shown in Fig. 29. The horseshoe shaped curve shown contains pure spectral colors; the numbers on the diagram are spectral wavelengths in millimicrons. All colors observable with the human eye are contained within the boundary formed by the horseshoe shaped figure. A color known as equal energy white is one containing equal intensities of the three CIE primaries. (CIE is the French abbreviation for the International Commission on Illumination.) It is shown with the coordinates (0.333, 0.333). The principle of using three arbitrary primary colors to represent a given color can be interpreted in terms of the chromaticity diagram. If the three physical colors are represented by points on the diagram, then the triangle formed by connecting the points contains all possible colors that can be formed from the primaries. In Fig. 4, points are shown for red, blue, and green cathode ray tube phosphors taken as primary color sources. The area of the triangle formed by the use of these points as vertices, compared with the region included by the boundary of the pure spectral colors, indicates the proportion of visible colors that can be produced by the light from three cathode ray tube phospors. Any color within the triangle indicated in the figure can be reproduced by the phosphors. The possible range of color reproduction provided by modern printing inks is shown in the diagram for comparison.

It follows from the foregoing discussion that televising a colored scene required furnishing to the receiving location not only information concerning the color or chromaticity of each picture element in the scene but also data concerning the brightness or luminance of each televised element. One method of generating electrical signals which may be used to reproduce a color picture is shown in Fig. 30. The scene to be televised is viewed through a lens system by three image orthicon tubes, each of which is preceded by a color filter, to separate the required color. Each camera tube is supplied with the same scanning and synchronizing information as is depicted for the camera tube in the black and white television system shown in Fig. 28. The outputs from the three camera tubes provide at each instant

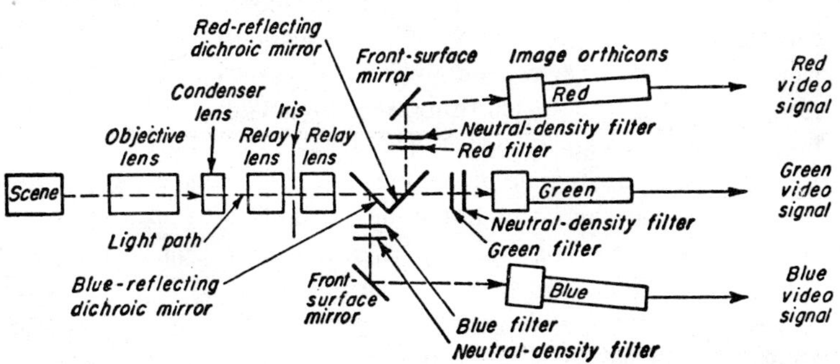

Fig. 30.

of time, just as in the black and white system, signals which are proportional to the light intensity reaching the camera tube from the scene being televised. Because of the color filters that are inserted in front of the camera tubes, the signal from the red channel is proportional to the red content of the various picture elements in the scene being scanned, a corresponding situation exists for the blue and green channels. The three color signals thus produced contain the information from which the scene being televised can be synthesized from sources of red, blue, and green light which are capable of being varied in intensity in accordance with the individual camera signals.

FM Stereo.—The principle of all radio transmission of speech, music, and television pictures is based upon the production of a uniform electromagnetic wave (the carrier) by the sending station and the imposition upon that wave of another wave representing the information to be transmitted, a process called modulation. There are two common methods by which this is done, as shown in Fig. 31. Consider that (*a*) in that figure represents the carrier wave. Then the signal wave can be added to it by changing the amplitude of its oscillations, as shown in part (*b*) of the figure. This is Amplitude Modulation (AM). The other method is to vary the frequency of its oscillations, as shown in part (*c*). At one time all home radio receivers operated on AM.

The number of stations in the frequencies used for AM receivers became so great, however, that their allotted bandwidths were as little as 5000 or 7000 cycles. This meant that the audio frequencies they could transmit were curtailed, especially in the higher frequencies. However, the frequency bands allotted to FM were in the region of megacycles (millions of cycles per second) rather than the hundreds or thousands of cycles per second allotted to AM. Therefore each FM station was allotted a band wide enough to transmit the full range of audible sound. Moreover, FM provides higher signal-to-noise ratios, another characteristic which is advantageous for high fidelity transmission. In stereophonic systems, in which a binaural effect is pro-

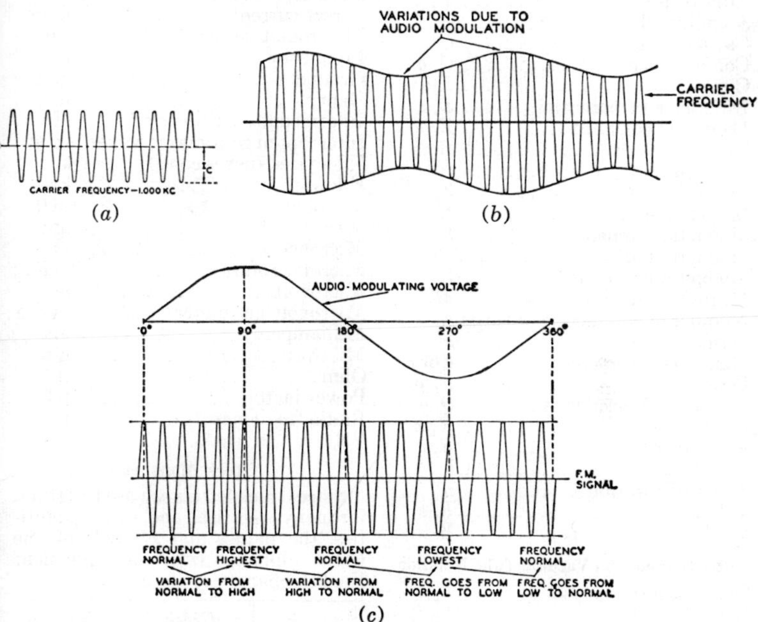

FIG. 31.

duced by transmitting signals from two microphones so spaced so that the home receiver can reproduce from two speakers effects that give the sound depth sensation of two human ears, fidelity of the sound received to that in the studio is especially important. (It is to be noted that because of the much higher frequencies that are broadcast, FM signals are not appreciably reflected by the atmosphere under normal conditions, so that, unlike AM, effective reception is normally restricted to points that are essentially in line-of-sight of the sending station.)

Because of the much higher signal-to-noise ratio desired, FM receivers (or separate tuners and amplifiers) have many more stages than AM receivers. Moreover, FM stereo equipment requires an entire additional section, called the multiplex section, to separate the inputs to the two speakers.

Standard Letter Symbols for Electrical Quantities

Admittance................ Y, y
Angular velocity ($2\pi f$)....... ω
Capacitance................ C
Conductance............... G, g
Current.................. I, i
Difference of potential....... E, e
Dielectric constant.......... K or ϵ
Energy.................. W
Frequency................ f
Impedance................ Z, z
Inductance................ L
Magnetic intensity.......... H
Magnetic flux.............. Φ
Magnetic flux density........ B
Mutual inductance.......... M
Number of conductors or turns N
Permeability............... μ
Phase displacement.......... θ or ϕ
Power.................... P, p
Quantity of electricity....... Q, q
Reactance................ X, x
Resistance................ R, r
Susceptance............... b
Speed of rotation........... n
Voltage.................. E, e
Work.................... W

Letter Symbols for Vacuum Tube Notation

Grid potential............. E_g, e_g
Grid current............... I_g, i_g
Grid conductance........... g_g
Grid resistance............. r_g
Grid bias voltage........... E_c
Plate potential............. E_p, e_p
Plate current.............. I_p, i_p
Plate conductance........... g_p
Plate resistance............ r_p
Plate supply voltage......... E_b
Emission current............ I_s
Mutual conductance.......... g_m
Amplification factor......... 1
Filament terminal voltage.... E_f
Filament current............ I_f
Filament supply voltage..... E_a
Grid-plate capacity.......... C_{gp}
Grid-filament capacity....... C_{gf}
Plate-filament capacity...... C_{pf}
Grid capacity ($C_{gp} + C_{gf}$).... C_g
Plate capacity ($C_{gp} + C_{pf}$)... C_p
Filament capacity ($C_{gf} + C_{pf}$). C_f

NOTE.—Small letters refer to instantaneous values.

Abbreviations Commonly Used in Radio

Alternating current.......... a.c.
Antenna.................. ant.
Audio frequency............ a.f.
Continuous waves........... c.w.
Cycles per second........... ~
Decibel.................. db.
Direct current............. d.c.
Electromotive force......... e.m.f.
Frequency................ f.
Ground.................. gnd.
Henry.................. h.
Intermediate frequency...... i.f.
Interrupted continuous waves. i.c.w.
Kilocycles (per second)...... kc.
Kilowatt................. kw.
Megohm.................. MΩ
Microfarad................ μfd.
Microhenry................ μh.
Micromicrofarad............ μμfd.
Microvolt................. μv.
Microvolt per meter......... μv/m
Milliampere............... ma.
Milliwatt................. mw.
Ohm.................... Ω
Power factor............... p.f.
Radio frequency............ r.f.
Volt..................... v.

Greek Alphabet

Since Greek letters are used to stand for many electrical and radio quantities, the names and symbols of the Greek alphabet with the equivalent English characters are given.

Greek Letter	*Greek Name*	*English Equivalent*
Α α	Alpha	a
Β β	Beta	b
Γ γ	Gamma	g
Δ δ	Delta	d
Ε ϵ	Epsilon	e
Ζ ζ	Zeta	z
Η η	Eta	ē
Θ ϑ	Theta	th
Ι ι	Iota	i
Κ κ	Kappa	k
Λ λ	Lambda	l
Μ μ	Mu	m
Ν ν	Nu	n
Ξ ξ	Xi	x
Ο ο	Omicron	ŏ
Π π	Pi	p
Ρ ρ	Rho	r
Σ σ	Sigma	s
Τ τ	Tau	t
Υ υ	Upsilon	u
Φ φ	Phi	ph
Χ χ	Chi	ch
Ψ ψ	Psi	ps
Ω ω	Omega	ō

XXII

Print Shop

The printer's mathematical problems consist mainly of estimating the amount of composition which will be required to set up certain copy, the printing space which this will consume, and estimating quantities of printing and paper. The computations he makes are entirely conventional, but the units he uses are peculiar to his trade.

Type Measure.—Previous to the year 1886 type sizes were indicated by names, *brevier*, *bourgeois*, *pica*, etc. Some of these names have continued in use although the *point* system is now used exclusively. The following is a list of the standard sizes of type with the old name designations:

TABLE 1

Old Name	Point Size	Old Name	Point Size
Excelsior	3	2-line Brevier or Columbian	16
Brilliant	3½	3-line Nonpareil or Great Primer	18
Semi-Brevier	4	2-line Long Primer or Paragon	20
Diamond	4½	2-line Small Pica	22
Pearl	5	2-line Pica	24
Agate	5½	2-line English	28
Nonpareil	6	5-line Nonpareil	30
Minion	7	3-line Pica or Double Great Primer	36
Brevier	8	7-line Nonpareil	42
Bourgeois	9	4-line Pica or Canon	48
Long Primer	10	9-line Nonpareil	54
Small Pica	11	5-line Pica	60
Pica	12	6-line Pica	72
English	14		

A few of these types sizes are illustrated in Fig. 1.

A *point*, as used in printing, is 0.0138 inch or approximately $\frac{1}{72}$ inch. Thus, an 8-point type has a body (Fig. 2) $\frac{8}{72}$ inch and nine lines of this type measure one inch vertically on a page; a 12-point type has a body $\frac{12}{72}$ inch and equals six lines to the inch.

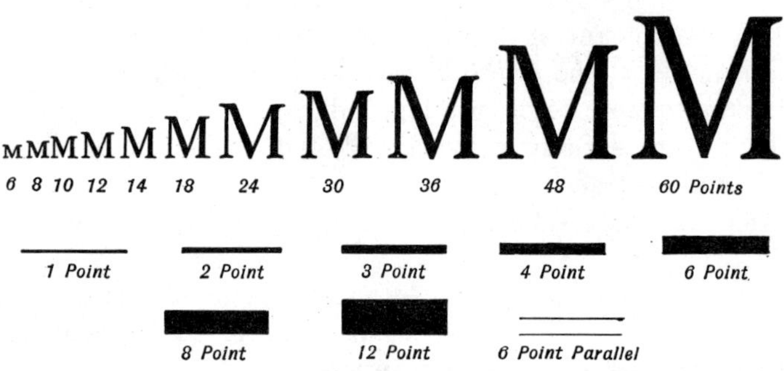

FIG. 1.

The number of lines and fractions of lines (in points) to the inch for the more common sizes of type are given below:

TABLE 2

	Lines	Points	
6 point equals	12		to the inch
7 point equals	10	2	to the inch
8 point equals	9		to the inch
9 point equals	8		to the inch
10 point equals	7	2	to the inch
11 point equals	6	6	to the inch
12 point equals	6		to the inch

The size of type as we have defined it is really the size of the block to which the letters are attached. The sizes of the letters themselves depend on the *face* or the cut and shape of the type *family* in question. One 12-point face may have a small body with long descenders while another may have a full heavy body.

Thus:

This line is twelve-point Cloister.

This line is twelve-point Bodoni.

It will be noted that while these two faces are of the same point size, one requires more lateral space for a word than the other. This brings us up to the consideration of the *pica* which is a unit used to measure the length of a line of type and also the size of a type page. A *pica* is 12 points or $\frac{1}{6}$ inch. Thus, a line of type four inches long is said to be 24 picas long. Similarly, a type page 7 inches by $4\frac{1}{2}$ inches is 42 by 27 picas.

ILLUSTRATION: A type page is 54 picas long. How many lines of 12-point type will it accommodate?

54×12 (no. of points in a pica) $\div$ 12 (no. points of type) =

54 lines (Ans.)

ILLUSTRATION: How many lines of 10-point type will go on a page 42 picas long?

$$42 \times 12 \div 10 = 504 \div 10 = 50\tfrac{4}{10} \text{ or } 50 \text{ lines} \quad \text{(Ans.)}$$

ILLUSTRATION: A type page is to be 6 inches high. How many lines of 8-point type will it hold?

From Table 2, 8-point type gives 9 lines per inch.

Then, $9 \times 6 = 54$ lines (Ans.)

Another solution is to change the 6 inches to the equivalent number of points and divide by 8. This may be done in one operation as follows:

$$\frac{6 \times \overset{9}{\cancel{72}}}{\cancel{8}} = 6 \times 9 = 54 \quad \text{(Ans.)}$$

Space Measure.—A low type without printing surface is set between words, following the period of a sentence, and between letters of a word when justifying a short line. A *quad* (or *em quad*) is a space type which is just as wide as it is thick. Thus in a 10-point font an em quad is 10 points or four-fifths pica wide. In a 12-point font an em is 12 points or one pica wide. A space half an em in width is called an *en*. Types 5 *to em*, 4 *to em*, and 3 *to em* are also used in spacing.

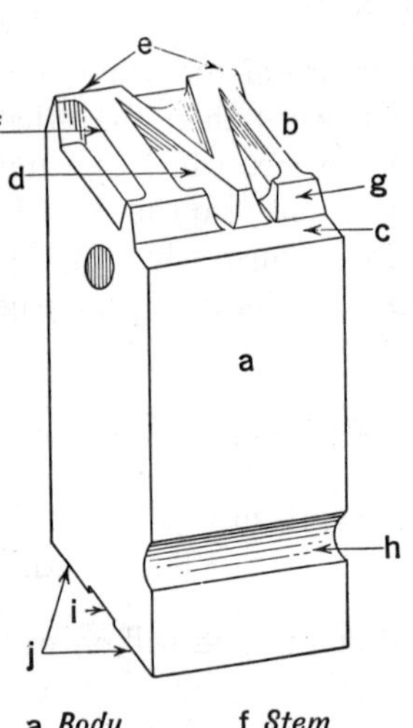

a *Body* f *Stem*
b *Face* g *Beard*
c *Shoulder* h *Nick*
d *Counter* i *Groove*
e *Serifs* j *Feet*

Fig. 2. Metal Type

FIG. 2.

The space between lines of type is often increased for the sake of legibility by the insertion of *lead* rules of any desired thickness. They are furnished in 1-point, 2-point, 3-point, etc., thicknesses as shown in Fig. 1. Type so spaced is said to be leaded. When type is set on a Linotype it may be leaded by having a shoulder cast directly on the slug. Thus, a 10-point type may be cast on a 12-point slug, the effect being the same as inserting a 2-point lead into 10-point type set "solid". Type so set is referred to as "10 on 12", "8 on 10", or whatever the combination of type size to slug size may be. If type is designated as "leaded", without further qualification, a 2-point lead is meant. When computing the number of lines of leaded type to a page, the leading must, of course, be taken into account.

ILLUSTRATION: How many lines of type 9-point on 11 will go onto a page 36 picas long?

$$36 \text{ picas} \times 12 = 432 \text{ points}$$

$$\tfrac{432}{11} = 39\tfrac{3}{11} \text{ or } 39 \text{ lines.} \quad \text{(Ans.)}$$

Estimating Type Space—*Line Method.*—We noted in the discussion of type sizes that considerable difference exists between the lateral space required by different type faces of the same point size. If any considerable amount of type is to be set, this difference may aggregate many pages. The most accurate method of estimating is based on a character count of the copy and the use of a space factor for the type face selected.

Standard office typewriters made in the U. S. (but not usually foreign machines, or electric machines with proportional spacing) produce either 10 or 12 characters to the inch (pica or elite type). These machines write six lines to the inch "single spaced" or three lines to the inch "double spaced." Each punctuation mark takes the same space as a letter. The number of characters on a page of typed manuscript may then be found very readily by measuring the length of an average line in inches and multiplying by ten or twelve, as the case may be, then measuring the length of the type page in inches and multiplying by six if it is single spaced, three if it is double spaced. The product of these products is the number of characters on the page.

Illustration: A manuscript page typed single spaced on a "pica" typewriter has an average line length of six inches and a typed page length of eight and one-half inches. How many characters are there to the line and how many characters per page?

$6 \times 10 = 60$ characters per line. (Ans.)
$8\frac{1}{2} \times 6 = 51$ lines per page
$60 \times 51 = 3060$ characters per page (Ans.)

Illustration: A manuscript page is typed double spaced on an "elite" typewriter to a depth of nine inches with an average line length of six and one-half inches. How many characters are there to the line and how many characters to the page?

$6\frac{1}{2} \times 12 = 78$ characters per line (Ans.)
$9 \times 3 = 27$ lines per page
$78 \times 27 = 2106$ characters per page (Ans.)

TABLE 3

AVERAGE NUMBER OF CHARACTERS TO ONE PICA

(*Each letter, space, and punctuation point is counted as a character*)

Type Face	Type Size, Points						
	6	7	8	9	10	11	12
Antique No. 1	3.35	.....	2.75	.....	2.4	.....	2.1
Benedictine and Benedictine Book	3.9	3.5	3.1	2.8	2.5	2.35	2.2
Benedictine Bold	.....	.....	2.85	.....	2.35	.....	2.0
Bodoni	3.9	3.4	3.0	.....	2.55	.....	2.4
Bodoni Book	3.95	3.6	3.2	2.9	2.75	.....	2.5
Bodoni Bold	3.6	.....	2.8	.....	2.4	.....	2.2
Caslon	.....	3.5	3.2	2.9	2.75	2.4	2.2
Caslon Old Face	4.05	.....	3.45	3.1	3.0	2.75	2.4
Caslon No. 3	3.7	.....	3.1	.....	2.45	.....	2.2
Century Expanded	3.45	3.1	2.8	2.6	2.4	2.3	2.1
Century Bold	.....	.....	2.9	.....	2.35	.....	2.1
Cheltenham	.....	.....	3.45	3.15	2.9	2.7	2.55
Cheltenham Wide	3.1	.....	2.9	.....	2.5	.....	2.2
Cheltenham Condensed	.....	.....	3.5	.....	2.85	.....	2.6
Cheltenham Bold	3.25	.....	2.8	.....	2.3	2.2	2.1
Cloister	4.0	.....	3.45	.....	3.1	2.95	2.85
Cloister Wide and Cloister Bold	3.6	.....	3.0	.....	2.7	.....	2.5
De Vinne	3.5	.....	3.0	2.85	2.6	2.3	2.1
Elzevir No. 3	3.75	.....	3.0	2.8	2.65	2.4	2.25
Franklin	3.7	3.55	3.2	2.9	2.75	2.6	2.35
Garamond	4.2	.....	3.45	3.1	2.9	2.6	2.4
Garamond Bold	3.7	.....	3.1	.....	2.55	.....	2.1
Granjon	.....	.....	3.45	.....	2.9	2.75	2.5
Ionic No. 5	3.16	2.9	2.63	2.45			
Narciss	.....	.....	.....	.....	2.4	.....	2.0
Number 16	3.15	2.85	2.6	2.4	2.3	.....	2.0
Old Style No. 1	3.55	3.25	3.0	2.8	2.65	2.55	2.3
Old Style No. 7	3.8	3.45	3.2	3.0	2.75	2.55	2.35
Original Old Style	3.7	.....	3.05	.....	2.75	.....	2.45
Scotch	3.35	.....	3.0	.....	2.7	2.55	2.25

In these computations, the short lines at the ends of the paragraphs are regarded as full lines, for if the type is set 20 to 30 picas wide approximately the same number of similar short lines will occur in the type.

Table 3 shows the number of characters in a line for a width of one pica of a number of common Linotype faces. Each letter, space, and punctuation point is counted as a character. Having selected the type face and the width of line, the amount of space which this will take can be computed as shown in the following illustration.

ILLUSTRATION: A manuscript of 75 typewritten pages averaging 2500 characters per page is to be set in Caslon 11-point on 13, 26 picas wide and 40 picas depth of page. How many type pages will this make?

From Table 3 we note that Caslon 11-point type averages 2.4 characters per pica. Then the number of characters per line is,

$$2.4 \times 26 = 63.4 \text{ characters per line of type}$$

The number of lines of type on a 13-point slug which a 40-pica page will hold is,

$$\frac{12 \times 40}{13} = 37 \text{ lines.}$$

Then the number of characters per page of type is the product of the number of characters per line and the number of lines,

$$63.4 \times 37 = 2346 \text{ characters of type per page.}$$

The manuscript of 75 pages averaging 2500 characters per page consists of,

$$75 \times 2500 = 187{,}500 \text{ characters}$$

The number of type pages is then simply,

$$\frac{187{,}500}{2346} = 80 \text{ pages} \quad \text{(Ans.)}$$

Estimating Type Space—*Area Method.*—A less accurate method of estimating type space consists of estimating the number of words in a manuscript and using a figure for the number of words per square inch of a certain type size. It takes no account of the variations of the different types of one size and we present it here only because it is still used in a number of shops.

The number of words of a manuscript may be estimated by counting a few lines on a representative page. To obtain the average length of a line in words, count ten lines and move the decimal point one place to the left. Thus, if ten lines of a manuscript contain 115 words, the average line has 11.5 words. The number of lines per page is determined as above, and the number of words per page is then the product of the average number of words per line and the number of lines.

ILLUSTRATION: Ten lines of a manuscript consists of 92 words. How many words are there on the page if there are thirty lines?

$$92 \div 10 = 9.2 \text{ average words per line}$$

$$9.2 \times 30 = 276 \text{ words per page} \quad \text{(Ans.)}$$

Knowing the number of words of a piece of copy, Table 4 may then be used to determine the space which this will cover.

ILLUSTRATION: A manuscript of 40,000 words is to be set in 10-point on 12 type. How much space will this cover in square inches and how many pages will it cover if the type page is 6 inches by 4 inches?

From Table 4, 10-point leaded type averages 16 words per square inch. Then, the copy will cover

$$\frac{40{,}000}{16} = 2{,}500 \text{ square inches} \quad \text{(Ans.)}$$

A page 6 inches by 4 inches is 24 square inches. Then the number of type pages will be

$$\frac{2{,}500}{24} = 104\tfrac{1}{6} \text{ pages} \quad \text{(Ans.)}$$

TABLE 4

WORDS AND EMS TO THE SQUARE INCH

(*Approximate number of words of average length for type of average width*)

Size	Old Name	Leaded (2-point)	Solid	No. of Ems
4½ point	Diamond	74	98	256
5 "	Pearl	46	66	208
5½ "	Agate	40	60	172
6 "	Nonpareil	32	44	144
7 "	Minion	26	34	106
8 "	Brevier	22	30	81
9 "	Bourgeois	19	24	64
10 "	Long Primer	16	20	52
11 "	Small Pica	14	17	43
12 "	Pica	11	14	36
14 "	English	9	11	26
18 "	Great Primer	7	8	16
22 "	Double Small Pica	4	5	11

Book Paper.—The paper generally used for books, magazines, circulars, catalogues, etc., is designated as *book paper*. This is a broad classification which includes a variety of finishes, colors, and weights, both coated and uncoated. The *substance weight* of paper is the weight in pounds of one *ream* (500 sheets) 25 in. by 38 in. in size. Thus, if a 60-pound paper is specified for a job, it means paper of such weight that 500 sheets of it 25 in. by 38 in. weigh 60 pounds.

The price of book paper used to be quoted by the pound in 1000 sheet lots. This has gone into discard by paper dealers, most of whom now quote prices per pound in ream lots. A higher price is demanded for quantities less than 500 sheets.

The weight of paper required for a job may, of course, be computed by arithmetic if the substance weight and sheet size are known. However, this is needless work since most jobs involve the use of standard sizes, and weights may be obtained from Table 5 for 1000 sheets and from Table 7 for the ream.

TABLE 5

WEIGHT IN POUNDS OF 1000 SHEETS OF BOOK PAPER

Sheet size, inches	Substance weight, pounds										
	30	35	40	45	50	60	70	80	90*	100	120*
22×32	44	52	59	67	74	89	104	119	133	148	178
24×36	54	64	72	82	90	110	128	146	164	182	218
25×38	60	70	80	90	100	120	140	160	180	200	240
26×29	48	56	64	72	80	96	112	126	142	158	190
26×40	66	76	88	98	110	132	154	176	198	218	262
28×42	74	86	100	112	124	148	174	198	222	248	298
28×44	78	90	104	118	130	156	182	208	234	260	312
29×52	96	112	128	144	160	192	224	252	284	316	380
30½×41	78	92	106	118	132	158	184	210	236	264	316
32×44	88	104	118	134	148	178	208	238	266	296	356
33×46	96	112	128	144	160	192	224	256	288	320	384
34×44	94	110	126	142	158	188	220	252	284	314	378
35×45	100	116	132	150	166	198	232	266	298	332	398
36×48	108	128	144	164	180	220	256	292	328	364	436
38×50	120	140	160	180	200	240	280	320	360	400	480
41×61	156	184	212	236	264	316	368	420	472	528	632
42×56	148	172	200	224	248	296	348	396	444	496	596
44×56	156	180	208	232	260	312	364	416	468	520	624
44×64	176	208	236	268	296	356	416	476	532	582	712

* Applies only to coated papers.

Sometimes a paper is required which is not one of the standard sizes included in Tables 5 or 7. In that case the weight may be found by determining the weight per square inch per 1000 sheets of paper of the same substance weight and multiplying this by the *area* of the sheet size in question. The product will be the weight of 1000 sheets of that paper. For example, the area of a standard sized sheet is 25 in. × 38 in. = 950 sq. in. The weight of 1000 sheets of 60-pound paper is, from Table 5, 120 pounds. Then the weight of 1000 sheets per square inch of this paper is 120 ÷ 950 = 0.12632 pound. Table 6 is a list of the unit weights per 1000 sheets of several substance weights.

TABLE 6

WEIGHT PER SQUARE INCH OF 1000 SHEETS OF BOOK PAPER, POUNDS

Substance weight, pounds	Unit weight, pound
50	0.10526
60	0.12632
70	0.14737
80	0.16842
100	0.21053

ILLUSTRATION: 4000 sheets of 70-pound substance weight paper 32 inches by 48 inches are needed for a job. What is the weight of this paper?

This sheet size does not appear in either Tables 5 or 7. Then Table 6 may be used.

Area of sheet = 32 × 48 = 1536 sq. in.
Weight of 1000 sheets = 1536 × .14737 = 226.4 pounds
Weight of 4000 sheets = 4 × 226.4 = 906 whole pounds (Ans.)

ILLUSTRATION: Five reams of 70-pound paper 28 inches by 44 inches are needed for a job. What is the actual weight of this paper?

From Table 7, one ream of 70-pound paper, 28 inches by 44 inches weighs 91 pounds. Five reams then weigh.

5 × 91 = 455 pounds (Ans.)

Cover Papers.—Cover papers are designated by substance weights which refer to the weight of one ream (500 sheets) of a

TABLE 7

Weight in Pounds of One Ream of Book Paper

Size, inches	Substance weight, pounds												
	25	28	30	35	40	45	50	60	70	80	90	100	120
22×32	....	....	22	26	30	34	37	45	52	60	67	74	89
24×36	23	26	27	32	36	41	45	55	64	73	82	91	109
25×38	25	28	30	35	40	45	50	60	70	80	90	100	120
26×29	20	22	24	28	32	36	40	48	56	63	71	79	95
26×40	27	31	33	38	44	49	55	66	77	88	99	109	131
28×42	31	35	37	43	50	56	62	74	87	99	111	124	149
28×44	33	36	39	45	52	58	65	78	91	104	117	130	156
29×52	40	45	48	56	64	72	80	96	112	126	142	158	190
30½×41	33	37	39	46	53	59	66	79	92	105	118	132	158
32×44	37	42	44	52	59	67	74	89	104	119	133	148	178
33×46	40	45	48	56	64	72	80	96	112	128	144	160	192
34×44	39	44	47	55	63	71	79	94	110	126	142	157	189
35×45	42	47	50	58	66	75	83	99	116	133	149	166	199
36×48	46	51	54	64	72	82	90	110	128	146	164	182	218
38×50	50	56	60	70	80	90	100	120	140	160	180	200	240
41×61	66	74	78	92	106	118	132	158	184	210	236	264	316
42×56	62	70	74	86	100	112	124	148	174	198	222	248	298
44×56	66	73	78	90	104	116	130	156	182	208	234	260	312
44×64	74	83	88	104	118	134	148	178	208	238	266	296	356

TABLE 8

Weight of 1000 Sheets of Cover Paper, Pounds

Sheet size, inches	Substance weight, pounds						
	25	35	40	50	65	80	130
20×26	50	70	80	100	130	160	260
23×35	78	109	124	155	201	248	402
26×40	100	140	160	200	260	320	520

sheet size 20 in. by 26 in. They, too, are now usually quoted at so much per pound per 1000 sheets. Table 8 gives the weights per 1000 sheets for three standard sizes and Table 9 the corresponding weights per ream.

TABLE 9

WEIGHT OF ONE REAM OF COVER PAPER, POUNDS

Size, inches	Substance weight, pounds						
	25	35	40	50	65	80	90
20×26	25	35	40	50	65	80	90
23×35	39	54½	62	77½	100½	124	139½
26×40	50	70	80	100	130	160	180

Writing Papers.—Bond, writing and ledger papers are referred to a substance weight per ream of a sheet size 17 in. by 22 in. Table 10 gives the weight per 1000 sheets of the more common sheet sizes of writing papers. Many more sizes than those listed in this table are, of course, manufactured and the weights of odd sizes may be determined by obtaining the unit weights per square inch per 1000 sheets as was illustrated with the book papers.

TABLE 10

WEIGHT OF 1000 SHEETS OF WRITING PAPER, POUNDS

Sheet size, inches	Substance weight, pounds						
	13	16	20	24	28	32	36
17×22	26	32	40	48	56	64	72
17×28	33	41	51	61	71	81	92
19×24	32	39	49	59	68	78	88
22×34	52	64	80	96	112	128	144
24×38	64	78	98	118	137	156	176
28×34	66	82	102	122	143	162	184
34×44	104	128	160	192	224	256	288

Table 11 gives the unit weights per 1000 sheets for the more common substance weights.

TABLE 11

WEIGHT PER SQUARE INCH OF 1000 SHEETS OF WRITING PAPER, POUNDS

Substance weight, pounds	Unit weight, pound
16	0.08556
20	0.10695
24	0.12834
28	0.14973
32	0.17112

TABLE 12

SIZES OF PAPER AND COVER PAPER ACCOMMODATING DIFFERENT PAGE SIZES WITH MINIMUM WASTE

Page size, inches	Sheet size, inches	Number of pages	Cover paper size, inches	Number of covers
$3\frac{1}{4}\times 5\frac{1}{8}$	28×44	8, 16, 32	23×35	18
$3\frac{3}{4}\times 5\frac{1}{8}$	32×44	8, 16, 32	23×35	16
3×6	25×38	24	20×26	12
$3\frac{7}{8}\times 5\frac{3}{8}$	33×46	8, 16, 32	23×35	16
$3\frac{3}{4}\times 7$	32×44	24	23×35	12
$4\frac{1}{2}\times 5\frac{7}{8}$	25×38	8, 16, 32	20×26	8
$3\frac{7}{8}\times 7\frac{1}{4}$	33×46	24	23×35	12
$4\times 9\frac{1}{8}$	25×38	24	20×26	6
$5\times 6\frac{5}{8}$	28×42	8, 16, 32	23×35	9
$4\frac{7}{8}\times 7\frac{1}{4}$	$30\frac{1}{2}\times 41$	8, 16, 32	20×26	6
$5\frac{1}{4}\times 6\frac{5}{8}$	28×44	8, 16, 32	23×35	9
$4\frac{1}{2}\times 8$	25×38	24	20×26	6
$5\frac{1}{4}\times 7\frac{1}{2}$	32×44	8, 16, 32	23×35	8
$5\frac{1}{2}\times 7\frac{7}{8}$	33×46	8, 16, 32	23×35	8
$4\frac{3}{4}\times 8\frac{7}{8}$	$30\frac{1}{2}\times 41$	24	20×26	4
$5\frac{1}{4}\times 10\frac{1}{4}$	32×44	24	23×35	9
$6\times 9\frac{1}{8}$	25×38	8, 16, 32	20×26	4
$6\frac{1}{4}\times 9\frac{1}{4}$	26×29	24	20×26	4
$7\frac{1}{2}\times 10\frac{5}{8}$	32×44	8, 16, 32	23×35	4
$8\times 11\frac{1}{8}$	33×46	8, 16, 32	23×35	4
$9\frac{1}{4}\times 12\frac{1}{8}$	25×38	8, 16	20×26	2

Selecting Paper.—When a job involves printing more than one page at a time, and particularly if a considerable number of impressions are to be made, the selection of paper of suitable size is of utmost importance. This is not only a question of reducing waste of paper, but also of reducing presswork. Thus, an 8-page booklet may be run through a press printing all eight pages at one time. Then the sheet is reversed and run through again with the pages so arranged that when it is cut in half and folded, two complete booklets result. If the page size can be so fitted to the sheet size that no waste beyond the necessary trim occurs, the greatest economy is effected.

Table 12 has been compiled to aid the printer in selecting a size of paper and cover stock which will accommodate 8, 16, 24, and 32 pages of various sizes with a sufficient allowance for folding and trim. The method of determining the number of pieces or pages which may be obtained from a sheet is to find what multiples or near-multiples the dimensions of the piece are of the dimensions of the sheet. The product of these multiples is the number of pieces which may be obtained. Thus, if we have a sheet 32 in. by 44 in. and wish to find how many pieces $7\frac{1}{2}$ in. by $10\frac{1}{2}$ in. we may obtain from it, we write the dimensions as follows:

$$\begin{array}{c} \cancel{32} \times \cancel{44} \\ \cancel{7\frac{1}{2}} \times \cancel{10\frac{1}{2}} \\ \hline 4 \times 4 = 16 \text{ pieces.} \end{array}$$

Cancelling out the dimensions of the smaller into those of the larger to find the number of *whole* times they are contained therein, we obtained 4 as the multiple in each case in this example. The product of these, 16, is the number of pieces which may be obtained.

ILLUSTRATION: How many pieces 5 inches by $6\frac{1}{2}$ inches may be obtained from a sheet 28 inches by 42 inches in size?

$$\begin{array}{c} \cancel{28} \times \cancel{42} \\ \cancel{6\frac{1}{2}} \times \cancel{5} \\ \hline 4 \times 8 = 32 \text{ pieces.} \quad \text{(Ans.)} \end{array}$$

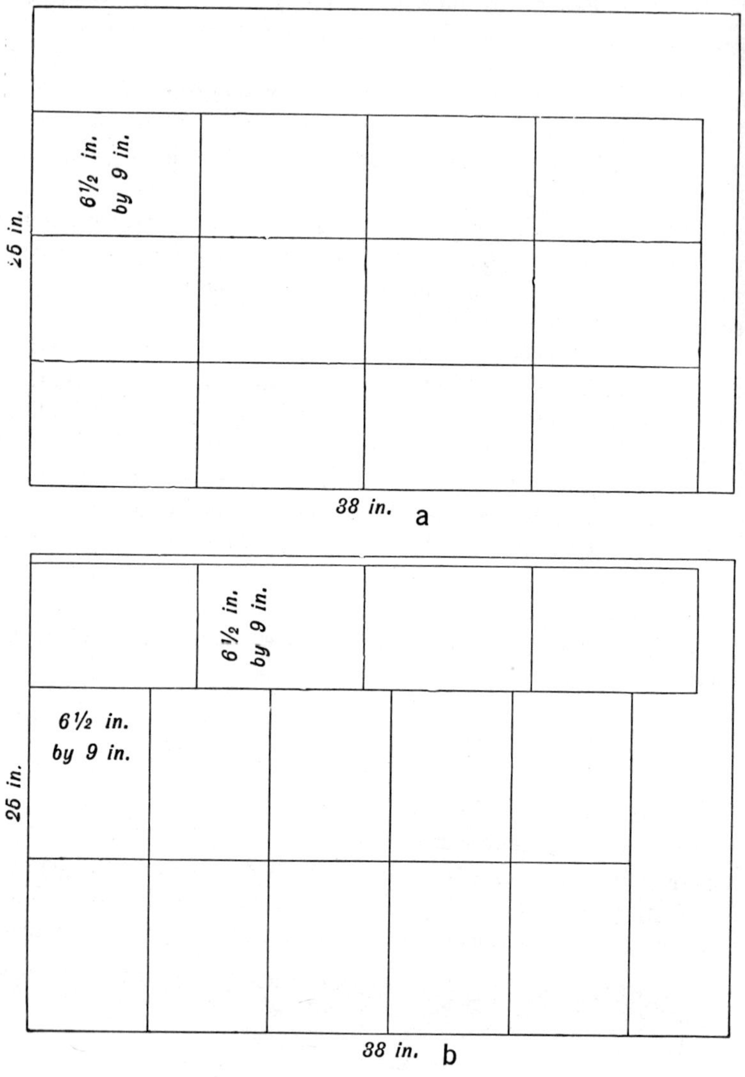

FIG. 3.

The above discussion has concerned itself only with lay-outs which permit straight cuts across the paper in either direction. Sometimes, however, it is necessary to use up a quantity of paper on hand for a certain job which, if it were cut straight across, would entail considerable waste. For example, it is desired to get as many pieces $6\frac{1}{2}$ inches by 9 inches as possible from a sheet 25 inches by 38 inches. By the ordinary computation,

$$\begin{array}{c}25 \times 38 \\ 6\frac{1}{2} \times 9 \\ \hline 3 \times 4 = 12 \text{ pieces,}\end{array}$$

the yield is found to be only 12 pieces as shown in Fig. 3a. However, if we transpose the dimensions of the piece and again cancel, we have,

$$\begin{array}{c}25 \times 38 \\ 9 \times 6\frac{1}{2} \\ \hline 2 \times 5 = 10 \text{ pieces.}\end{array}$$

There then remains a waste piece 7 inches by 38 inches which is large enough for use. This yields,

$$\begin{array}{c}7 \times 38 \\ C \times 9 \\ \hline 1 \times 4 = 4 \text{ pieces.}\end{array}$$

Thus, by cutting the sheet as shown in Fig. 3b, 14 pieces may be obtained.

Paper Allowance for Spoilage.—In each printing and binding operation a certain amount of paper is spoiled for further use. This must be taken into account when ordering stock. As the number of impressions increases the percentage of spoilage decreases. The following are safe values to use in estimating:

Number of copies	Percent spoilage		
	One color	Each add. color	Binding
100 to 250	10	5	5
250 to 500	6	4	4
500 to 1,000	5	2½	2½
1,000 to 5,000	4½	2½	2
5,000 to 10,000	3½	2½	2
Over 10,000	2	2	2

Estimating Quantity of Paper.—ILLUSTRATION: A job calls for 12,000 copies of a 64-page magazine trimmed flush to 6 inches by 9 inches; body stock to be 60-pound machine finished paper; cover stock 80-pound; and one color throughout. How much paper will be required for the job?

Referring to Table 12 we note that a sheet size of 25 in. by 38 in. will accommodate a 6 in. by 9 in. page size economically and conveniently. As we have seen by previous computations, it will take 16 pages on each side or a total of 32 pages. Two such sheets will then be needed for each copy of the magazine. With 12,000 magazines wanted, the sheets needed will be $12{,}000 \times 2 = 24{,}000$. This does not allow for waste. From the foregoing table we note that $3\frac{1}{2}$ percent for printing and 2 percent for binding must be added for waste. Then the total sheets required is,

$$24{,}000 + 24{,}000 \times (0.035 + 0.02) = 25{,}320 \text{ sheets}$$

Referring now to Table 5 we find that this paper weighs 120 pounds per 1000 sheets. Then,

$$120 \times 25.32 = 3038 \text{ pounds paper required.} \quad \text{(Ans.)}$$

From Table 12 we also note that 20 in. by 26 in. cover stock will make 4 covers for a trim size of 6 in. by 9 in. Then, for 12,000 copies, $12{,}000 \div 4 = 3000$ sheets will be needed. Again adding a total of $5\frac{1}{2}$ percent for waste for printing and binding we find that it will be prudent to provide

$$3000 + 3000 \times 0.055 = 3165 \text{ sheets}$$

Referring to Table 8 we see that this cover stock weighs 160 pounds per 1000 sheets. Then the weight required will be,

$$160 \times 3.165 = 507 \text{ pounds.} \quad \text{(Ans.)}$$

Illustrations.—When pictures are to be reproduced in letterpress or type printing, a plate called a *halftone* is made by a photochemical process which maintains the highlights and shadows of the picture by the use of variable size dots known as a *screen.* These screens, which are placed in the camera between the plate and the object to be photographed, vary in the number of lines to the inch. The standard screens for general commercial work range from 110 lines to the inch to 150 lines to the inch.

A reproduction which shows no highlights or shadows is known as a *line plate* or *line cut* and prints only black solid lines. The purpose of a line cut is to obtain a perfect reproduction of a drawing where lines only are used to suggest the tones that actually exist. The purpose of a halftone is to present a reproduction or drawing in which there are a quantity of tones.

The photoengraver prints the photograph of the original drawing on a smooth metal surface which has been sensitized in a manner similar to photographic paper. The portions which are not to print are eaten away by chemicals, leaving the printing surface composed of a large number of black lines, some large and some small. Tones are produced by the patterns of white and black formed by these lines and the associated white space.

Since photoengraving is a photographic reproduction process, it requires either a drawing, painting, or another photograph as the original material. Reductions in photoengraving are preferable because they reduce any imperfections in the same proportion.

Enlargements are undesirable because they increase imperfections of copy. Drawings usually are drawn twice the size of the reproduction, both dimensions being 100 percent larger. Thus a line cut 3″ × 3″ would require a drawing 6″ × 6″. When the job is made, a one-half reduction in each dimension would be ordered. The 50 percent reduction includes both dimensions.

Offset Printing.—The foregoing discussion of photoengravings and electrotypes refers, of course, to letterpress printing. There are now widely available a number of printing processes in which the plates for printing are made by photographic processes, so that the artwork can be photographed directly, without the necessity for making electrotypes. These processes are known collectively as offset methods, and four well-known varieties of them are Albumin Offset, Deep Etch Offset, Web Offset and Sheet Fed Gravure.

Albumin Offset, so named because albumin plates are used, is best suited to small quantity runs. The reason is that while these plates give good reproduction they do not wear well enough for long runs, and they cannot be stored. In this case, the printer stores the negatives, so that new prints can be made for them for future runs.

Deep Etch Offset uses deep etch plates which have a longer life expectancy than albumin plates, and which may be stored. This method is generally used for runs of intermediate size, approximately 15,000 to 40,000.

Web Offset operates on a web press and is used for runs of large size, most of which are on the order of 100,000 or greater, although the method can be extended to quantities as small as 40,000. It is especially useful where printing is done in more than one color. Unlike the other offset methods, the paper for Web Offset is used in rolls.

Sheet Fed Gravure is used for top quality reproduction of books in limited editions. It is the most costly process and is only used for printing books which are to command a premium price.

Photoengraving.—The dimensions of a desired photoengraving may be found, when both dimensions of the drawing or photograph and one dimension of the plate are known, by using a direct proportion as follows:

$$P_h : P_w = D_h : D_w \quad \text{or} \quad P_h/P_w = D_h/D_w$$

where P_h = height of the plate; P_w = width of the plate; D_h = height of the drawing or photograph; D_w = width of the drawing or photograph.

ILLUSTRATION: A photograph 12″ wide × 6″ high is to be used for a halftone 6″ wide. Find the height of the plate.

$$P_h/P_w = D_h/D_w$$

$$P_h/6 = \frac{6}{12}$$

$$P_h = \frac{6 \times 6}{12} = 3$$

Therefore the halftone will be 6″ wide × 3″ high.

When the process is used for line drawings (that is, drawings without gradation of color) the result is called zinc etching. Where such gradation exists, as in photographs, the work is usually done on copper by re-photographing the original photograph or other artwork through a wire screen, which breaks up the light rays so that the metal plate is sensitized in a pattern of dots which vary in size with the darkness of the area in the photograph. Screens vary from 65 to 150 wires ("lines") per inch.

Halftone and line plate photoengravings are priced on the printing face measurement or area of the plate or proof, exclusive of the bevel or margin of the plate as in the case of electrotypes.

The Standard Scale for Photo-Engravers (adopted February 1, 1940) evaluates plate areas from 1 × 1 to 17 × 22 with basic unit values for each ¼″ area. These basic unit values for halftone plates extend from 100 to 1348 and for line plates or zinc from 73 to 809. Some basic unit values are shown in the table on the next page.

Inches	Halftones	Line plates, zinc	Inches	Halftones	Line plates, zinc
3 × 4	128	97	10 × 11	456	270
5 × 6	189	132	11 × 12	528	313
6 × 7	229	151	12 × 13	607	359
7 × 8	275	173	13 × 14	691	409
9 × 10	388	231	14 × 15	784	463

Rule.—To find the selling value of halftones and line plates multiply the basic unit values by the individual shop selling rate.

ILLUSTRATION: A line plate proof measures 12″ × 13″. Find its selling price at $0.040 per unit.

From the table a 12″ × 13″ area is 359 units
359 × $0.040 = $14.36
Therefore the selling price is $14.36

Electrotypes. When a job is to be printed in multiples of the same plate a number of duplicates are made, then locked up in the form and printed on the press. The electrotype is one kind of such duplication. Whenever necessary, a type-set job may be set up several times and duplicated in this manner. Forms set in linotype and monotype have been known to print as many as 25,000 impressions of satisfactory quality. When runs are greater than the accepted standard of impressions from type, electrotypes may be used for duplicates.

Electrotypes are made in several styles, i.e., wax mold, lead mold, nickel types, wax plates, etc. They are sold by the square inch, the area being measured from the printing surface of the plate. When the electrotype surface area is measured from a proof, $\frac{1}{4}$″ is added to provide for a bevel on patent base plates or for a space to drive nails through the electrotype to fasten it to the wooden block.

The Standard Electrotype Scale (adopted 1941) gives complete information regarding the kinds and methods of pricing electrotypes. This scale is built on a ratio unit value of production which is provided for each square inch area of the various kinds of electrotype plates.

Unit values of production extend from 29 for one square inch to 940 for 216 square inches for lead mold steel face electrotypes. Some basic unit values of production are as follows:

Square inches	Line copper	Halftone copper	Halftone steel	Lead mold
1	29	36	44	58
4	36	45	54	72
12	52	65	78	104
24	76	95	114	152
30	89	111	134	178
40	109	136	164	218
44	117	146	176	234
49	128	160	192	256
56	142	178	213	284
64	158	198	237	316

ILLUSTRATION: A copper face electrotype of a type form measures $6\frac{3}{4}'' \times 7\frac{3}{4}''$. If the electrotype has a basic unit value of 142 production units, find its cost at $0.029 per unit.

$$6\tfrac{3}{4}'' + \tfrac{1}{4}'' = 7''$$

$$7\tfrac{3}{4}'' + \tfrac{1}{4}'' = 8''$$

$$7 \times 8 = 56$$

From the table, 56 sq. in. is 142 units

$142 \times \$0.029 = \$4.12$

Therefore the cost is $4.12.

BUSINESS MATHEMATICS

BY

PETER L. AGNEW, A.M., Ed.M.
New York University School of Education
New York, N. Y.

XXIII

Business Mathematics

Business has been defined as "the commercial activity of a community." Naturally, mathematics plays a very important role in the diverse transactions that are executed in this commercial activity.

Invoice.—Perhaps the most common of all business transactions is the buying and selling of commodities, which transaction is generally represented by an invoice which is an itemized list of goods sold by one party to another. The invoice ordinarily carries the following information:

1. Date
2. Name and address of person or firm selling the goods
3. Name and address of person or firm buying the goods
4. Order numbers of both the buyer and the seller
5. Terms and manner of shipment
6. Terms of payment
7. Items, or list of the goods sold, including (*a*) quantity, (*b*) name or brief description of goods sold, (*c*) unit price, (*d*) extension representing the total cost of each article, (*e*) total.

(1) New York, N. Y., Jan. 5, 1955

(2) THE AMERICAN CANDY COMPANY
125 Broadway
NEW YORK CITY

(3) Sold to Fred R. Sterlings
100 Main Street
Stamford, Connecticut

(4) Your order No. 6792
Our order No. D873

(5) Delivery: Our Truck

(6) Terms 2/10, n/30

	a	*b*	*c*	*d*
(7)	10	1# Boxes Peppermints	.55	$5.50
	25	2# " Cherries	.84	21.00
				$26.50 *e*

Calculations.—The mathematical phase of the invoice primarily has to do with lower part, the quantity, the unit price, the extension, and the total. In preparing the invoice, the various types of goods covered by the invoice are listed separately with the quantity and unit price of each. The quantity is multiplied by the unit price in order to get what is known as the extension. The extensions are then added to get the total.

ILLUSTRATION: A shoe manufacturer sold a customer 36 pairs of women's pumps at $2.95 per pair, and 36 pairs of women's oxfords at $2.75 per pair. What is the amount of each extension and what is the total?

36 pr.	Women's Pumps	@ $2.95 per pair = 36 × 2.95 =	$106.20	
36 pr.	Women's Oxfords	@ $2.75 per pair = 36 × 2.75 =	99.00	
		106.20 plus 99.00	$205.20	(Ans.)

Unit Price.—It should be noted that unit prices are sometimes quoted in terms of price per dozen or price per cwt. (hundredweight) or in some other common quantity, while the goods are listed in terms of so many units or so many pounds. Calculations are then slightly more complicated. When price is quoted as so much per special quantity, the price is usually multiplied by the total number of units or pounds and the resulting answer is divided by the number of units in the special quantity, i.e., divided by 100 if the price is quoted per cwt., divided by 2,000 if the price is quoted per ton, etc.

ILLUSTRATION: An invoice lists 6789 lbs. of goods at $6.75 per cwt. What is the amount of the extension?

$$\$6.75 \times 6789 = \$45{,}825.75$$
$$\$45{,}825.75 \div 100 = \$458.26 \quad \text{(Ans.)}$$

When the unit price is by the dozen, the quantity is frequently

GEORGE M. SPINNEY, INC.
WHOLESALE GROCERIES
475 FOURTH AVENUE
NEW YORK

TELEPHONE
MURRAY HILL 4-6930

TERMS 2/10, n/30.

SOLD TO Good Purchaser
123 Cash Street,
South Orange, N. Y.

SHIPPED BY ________ YOUR ORDER NO. ________
DATE ________ REQUISITION NO. ________

		PRICE	DISC.	EXTENSION	TOTAL
2	Bbls. Potatoes	4.—	1/4	6.00	
5	Bu. Beans	2.00	1/5	8.00	
3	Bags Flour	3.00	1/3	6.00	
					20.00

Fig. 1.—Invoice

expressed in dozens and fractions thereof. In such cases, the extension is made in the usual way.

ILLUSTRATION: What is the cost of 12½ doz. black nylon hose @ $8.65 per doz.?

$$\$8.65 \times 12.5 = \$108.13 \quad \text{(Ans.)}$$

Aliquot Parts.—Any number that is contained in another number an equal number of times is called an aliquot part of that number. Aliquot parts are used extensively in making business calculations, particularly in connection with extensions on invoices as well as discounts and interest calculations. The aliquot parts of a number are the fractional parts of that number. The commonly used aliquot parts of the dollar are 50¢ (1/2), 25¢ (1/4), 20¢ (1/5), 16 2/3¢ (1/6), 12½¢ (1/8), 10¢ (1/10), 8 1/3¢ (1/12), 6¼¢ (1/16), 5¢ (1/20), 2¢ (1/50).

ILLUSTRATION: How would the following items of an invoice be calculated using aliquot parts?

428 lbs.	@ 25¢ =	$107.00
192 lbs.	@ 37½¢ =	72.00
280 lbs.	@ 70¢ =	196.00
	Total	$375.00

These extensions should be calculated as follows:

$25 = \frac{1}{4}$	$\frac{1}{4}$ of 428 = $107.00
$37\frac{1}{2} = \frac{3}{8}$	$\frac{3}{8}$ of 192 = 72.00
$70 = \frac{7}{10}$	$\frac{7}{10}$ of 280 = 196.00

Invoice and Bill.—These terms are used more or less synonymously; however, the term bill is more frequently applied to a bill for services such as a telephone bill or a lawyer's bill, while the term invoice is almost invariably applied to an itemized listing of goods sold.

Discounts.—Closely associated with invoices are discounts. There are two kinds of commercial discounts: cash discount and trade discount. (There is another type of discount known as "Bank Discount" which is really a form of interest and, therefore, is discussed under the general heading of interest.)

NEW YORK TELEPHONE COMPANY

Addresses: See back of Stub

Telephone: Dial 811 or Call "Business Office"

BELL SYSTEM — NEW YORK TELEPHONE COMPANY — AMERICAN TELEPHONE & TELEGRAPH CO.

SEPT. 1, 19

BU 7 4375 — C J WOJTAK
1233 FLATBUSH AVE
BROOKLYN 26 N Y

LOCAL SERVICE for One Month in Advance	Message Units Included	75	6 73 ★
ADDITIONAL MESSAGE UNITS to Date of Bill	See back of Bill		★
TOLL CALLS AND TELEGRAMS. List enclosed			
OTHER CHARGES OR CREDITS. Explanation enclosed			
BALANCE FROM LAST BILL. Please disregard this amount if paid			
		TOTAL	6 73

★ Includes 10% U. S. tax and 3% N. Y. City tax.
See back of bill to determine amount of tax.

FIG. 2.

Cash Discount is a percent of a bill that may be deducted if the bill is paid within a certain specified time. The rate of this discount is stated in the terms of the invoice which includes the rate and the number of days within which the discount may be deducted. Some rather common terms are: 2/10, n/30 (meaning that two percent of the total of the bill may be deducted if it is paid within 10 days. If not paid within 10 days, no discount will be allowed and the full amount of the bill must be paid within 30 days), 5/30, n/60 (meaning that five percent of the total of this bill may be deducted if it is paid within thirty days. If not paid within 30 days, no discount will be allow and the full amount of the bill must be paid within 60 days.)

ILLUSTRATION: The total of an invoice is \$897.50 and the terms of payment are 5/30, n/60. How much discount may be deducted and how much must be paid if the invoice is paid within 30 days?

Total of invoice.	\$897.50
Less 5% discount.	44.88
Net amount to be paid.	\$852.62 (Ans.)

Applying the principle of aliquot parts mentioned previously, this discount should be calculated as follows:

$$5\% = \tfrac{1}{20} \qquad \tfrac{1}{20} \text{ of } \$897.50 = \$44.88$$

Trade Discount is a discount granted to a purchaser and is deducted at the time the bill is made out. It is used largely in connection with catalogue and list prices in order that these prices may be brought in line with true market values. Trade discount is also used at times in connection with purchases in large quantities being offered as a special inducement to attract large orders.

These discounts are sometimes in the form of a single rate of discount and sometimes in the form of a series of discounts, each one of the series being deductable from the net amount remaining after the preceding discount has been deducted.

ILLUSTRATION: If an order were placed for 100 hats, the quotation on which was \$1.75 less 10%, how would the invoice read?

100 Hats	\$1.75	\$175.00
	Less 10% discount	17.50
	Net Amount	\$157.50 (Ans.)

Chain Discounts.—Frequently the discount quotation is in the form of a series in which case the quotation might be \$1.75 less 25, 10, and 5%. The basic principle involved in chain discounts is that each succeeding discount is based on what is left after the preceding discount is deducted.

ILLUSTRATION: If the quotation on 100 hats was $1.75 less 25, 10, and 5%, what would be the net amount of the invoice? This item could be calculated as follows:

100 Hats	$1.75	$175.00
	Less 25%	43.75
		131.25
	Less 10%	13.13
		118.12
	Less 5%	5.90
		$112.22 (Ans.)

Chain Discount Tables.—Rather than use this long arithmetic process, most business organizations use decimal equivalents for chain discounts. A table of the most common equivalents is shown in Table I. By consulting this table, you will find the decimal equivalent of almost any combination.

ILLUSTRATION: How would the invoice for 100 hats at $1.75 less 25, 10, and 5% be calculated when a table of decimal equivalents is used? By consulting the table of decimal equivalents, one will find that the decimal equivalent of the series 25, 10, and 5%, as listed on the table is 0.64125.

100 less 0.64125 = 0.35875

100 Hats @ $1.75	= $175.00
Less 0.35875 (175 × 0.35875)	= 62.78
Net Amount	$112.22

Calculating Decimal Equivalents.—When a table of decimal equivalents is not available or when the decimal equivalent of a particular series of chain discounts does not appear on an available table, it may be necessary to calculate the equivalent.

TABLE I.

Table of Net Decimal Equivalents of Chain Discounts

Multiplying the gross amount by the net decimal equivalent for a chain discount gives the net amount of the invoice. To obtain the discount only, subtract the decimal equivalent given below from 100 and multiply the gross amount by the remainder.

The net equivalent of a chain discount is the same regardless of the sequence of the separate discounts. Example: 60–10–5% is the same as 10–5–60%.

Rate %	**5**	**7½**	**10**	**12½**	**15**	**16⅔**	**20**	**25**	**30**	**33⅓**	**35**	**37½**
	.95	.925	.90	.875	.85	.83333	.80	.75	.70	.66667	.65	.625
2½	.92625	.90188	.8775	.85313	.82875	.8125	.78	.73125	.6825	.65	.63375	.60938
5	.9025	.87875	.855	.83125	.8075	.79166	.76	.7125	.665	.63333	.6175	.59375
5 2½	.87994	.85678	.83363	.81047	.78731	.77187	.741	.69469	.64838	.6175	.60206	.57891
5 5	.85738	.83481	.81225	.78969	.76713	.75208	.722	.67688	.63175	.60167	.58663	.56406
5 5 2½	.83594	.81394	.79194	.76925	74795	73328	.70395	.65995	.61596	.58663	.57196	.54996
7½	.87875	.85563	.8325	.80938	.78625	.77083	.74	.69375	.6475	.61667	.60125	.57813
7½ 2½	.85678	.83423	.81169	.78914	.76659	.75156	.7215	.67641	.63131	.60125	.58622	.56367
7½ 5	.83481	.81284	.79088	.76891	.74694	.73229	.703	.65906	.61513	.58583	.57119	.54922
10	.855	.8325	.81	.7875	.765	.75	.72	.675	.63	.6	.585	.5625
10 2½	.83363	.81169	.78975	.76781	.74588	.73125	.702	.65813	.61425	.585	.57038	.54844
10 5	.81225	.79088	.7695	.74813	.72675	.7125	.684	.64125	.5985	.57	.55575	.53438
10 5 2½	.79194	.7711	.75026	.72942	.70858	.69469	.6669	.62522	.58354	.55575	.54186	.52102
10 7½	.79088	.77006	.74925	.72844	.70763	.69375	.666	.62438	.58275	.555	.54113	.52031
10 10	.7695	.74925	.729	.70875	.6885	.675	.648	.6075	.567	.54	.5265	.50625
10 10 5	.73103	.71179	.69255	.67331	.65408	.64125	.6156.	.57713	.53865	.513	.50018	.48094
10 10 5 2½	.71275	.69399	.67524	.65648	.63772	.62522	.60021	.5627	.52518	.50018	.48767	.46891

Rate %	**40**	**50**	**60**	**62½**	**65**	**66⅔**	**70**	**75**	**80**	**85**	**87½**	**90**
	.60	.50	.40	.375	.35	.33333	.30	.25	.20	.15	.125	.10
2½	.585	.4875	.39	.36563	.34125	.325	.2925	.24375	.195	.14625	.12188	.0975
5	.57	.475	.38	.35625	.3325	.31667	.285	.2375	.19	.1425	.11875	.095
5 2½	.55575	.46313	.3705	.34734	.32419	.30875	.27788	.23156	.18525	.13894	.11578	.09263
5 5	.5415	.45125	.361	.33844	.31588	.30083	.27075	.22563	.1805	.13538	.11281	.09025
5 5 2½	.52796	.43997	.35198	.32998	.30798	.29331	.26398	.21998	.17599	.13199	.10999	.08799
7½	.555	.4625	.37	.34688	.32375	.30833	.2775	.23125	.185	.13875	.11563	.0925
7½ 2½	.54113	.45094	.36075	.3382	.31566	.30063	.27056	.22547	.18038	.13528	.11273	.09019
7½ 5	.52725	.43938	.3515	.32953	.30756	.29292	.26363	.21969	.17575	.13181	.10984	.08788
10	.54	.45	.36	.3375	.315	.3	.27	.225	.18	.135	.1125	.09
10 2½	.5265	.43875	.351	.32906	.30713	.2925	.26325	.21938	.1755	.13163	.10969	.08775
10 5	.513	.4275	.342	.32063	.29925	.285	.2565	.21375	.171	.12825	.10688	.0855
10 5 2½	.50018	.41681	.33345	.31261	.29177	.27788	.25009	.20841	.16673	.12504	.1042	.08336
10 7½	.4995	.41625	333	.31219	.29138	.2775	.24975	.20813	.1665	.12488	.10406	.08325
10 10	.486	.405	.324	.30375	.2835	.27	.243	.2025	.162	.1215	.10125	.081
10 10 5	.4617	.38475	.3078	.28856	.26933	.2565	.23085	.19238	.1539	.11543	.09619	.07695
10 10 5 2½	.45016	.37513	.30011	.28135	.26259	.25009	.22508	.18757	.15005	.11254	.09378	.07503

From: Instruction Manual, "Burroughs Typewriter Billing Machine," published by Burroughs Adding Machine Company, Detroit, Michigan.

The decimal equivalent of any combination may be calculated by using 100% as the original base, and basing each successive discount on the percent left after the preceding discount has been deducted and finally deducting the final rate from the original 100%.

ILLUSTRATION: What is the decimal equivalent of discount series 25, 10, 5%, and 1% calculated by the above described method?

	100%
Less	25
	75
Less 10%	7.5
	67.5
Less 5%	3.375
	64.125
Less 1%	.64125
	63.48375

100% less 63.48375 = 36.51625% (Ans.)

INTEREST

Interest is money paid for the use of money. The sum upon which the interest is charged, the base amount owed, is called the *principal*. The amount of interest per dollar of principal charged to a borrower or paid to a lender, depends upon (1) the percentage rate per interest period, (2) the length of the period, (3) the time the interest is charged, and (4) the time or times during the loan period when the principal and interest are repaid, in a single payment or in installments.

ILLUSTRATION: A man borrows $1000 for one year, interest to be charged at the rate of 6%. How much interest will be due at the end of the year? What will be the total amount to be paid?

Principal.	$1000.00
Interest @ 6% (1000 × 0.06).	60.00
Total Amount.	$1060.00

Bankers' Time.—Most interest calculations are not quite that simple because funds are not usually used for a year; rather are they usually used for a period of days or months, and the interest must be calculated for that length of time. In order to simplify somewhat this calculation, most business organizations, including banks, have adopted the policy of treating the year as if it included 360 days, 12 months of 30 days each. This is usually called bankers' time.

Using bankers' time, one may calculate the interest by multiplying the principal by the number of days that the money was used over 360; by the rate of interest expressed in the form of a fraction. Because of the possibilities for cancelling, this is known as the *cancellation method* of calculating interest.

ILLUSTRATION: $2000 is borrowed for 10 days with interest at the rate of 6% per annum. How much interest must be paid? What amount (principal plus interest) must be paid at the end of 10 days:

$$\$2000 \times \tfrac{10}{360} \times \tfrac{6}{100} = \tfrac{10}{3} = \$3.33 \text{ Interest}$$

$$\$2000 \text{ plus } \$3.33 = \$2003.33 \quad \text{(Amount)}$$

60-Day Method.—As suggested previously, most loans are made for a relatively short time. Because of this, business has evolved a simple technique centered around 60 days for calculating interest for short terms. $1000 at interest for one year at the rate of 6% per annum would yield $60.00. For 60 days, (one-sixth of a year $\frac{60}{360} = \frac{1}{6}$) the yield would be $10.00, $\frac{1}{6}$ of $60.00. $10.00 is 1% of $1000 and the same figure could have been determined by merely moving the decimal point two places to the left, $10.00.

Thus we evolve the rule that: To find interest at six per cent for sixty days, move the decimal point two places to the left.

ILLUSTRATION: How much interest must be paid on $1768.47 for 60 days with interest at the rate of 6%?

Interest on $1768.47 for 60 days at 6% = $17.6848 or $17.68.

This was determined merely by moving the decimal point in $1768.47 two places to the left, the result being $17.68.47

Interest for Other Terms.—Interest for terms other than 60 days may be calculated by applying the principle of aliquot parts. The common aliquot parts of 60 are: 30 (1/2), 20 (1/3), 15 (1/4), 12 (1/5), 10 (1/6), 6 (1/10), 5 (1/12), 4 (1/15). Interest is first determined for 60 days and then the proper fractional part or combination of fractional parts is determined.

ILLUSTRATION: $875.00 is borrowed for 30 days with interest at the rate of 6% per annum. What is the amount of the interest?

Interest on $875.00 for 60 days at 6% = $8.75
30 days equals $\frac{1}{2}$ of 60 days.
Interest on $875.00 for 30 days ($\frac{1}{2}$ of $8.75) = $4.38 (Ans.)

Interest at Other Rates.—Quite frequently the rate of interest is not 6% but some other rate agreed upon by the parties involved. One method of calculating this interest is by applying the principle of aliquot parts. The aliquot parts of six are 3 (1/2), 2 (1/3), $1\frac{1}{2}$ (1/4), 1 (1/6), $\frac{1}{2}$ (1/12). In calculating interest at a rate other than 6%, the interest is first calculated at 6% by the 60-day method and then the proper fractional part is determined from that.

ILLUSTRATION: $1000 was borrowed for 30 days at 8%. What is the amount of the interest?

Interest on $1000 @ 6% for 60 days =	$10.00
Interest on $1000 @ 6% for 30 days =	5.00
Interest on $1000 @ 2% (1/3 of 6%) =	1.67
Interest on $1000 @ 8% =	$ 6.67 (Ans.)

Interest Tables.—If much of a firm's business involves interest, precomputed tables are used to avoid the necessity of calculating the interest for every transaction. Table II shows simple interest on amounts from $1.00 to $9.00 for various periods of time and at various rates. In using this table to find the interest on a given principal at a given rate, one should

a. Run down the side of the table until he comes to the given rate.

b. If the principal in question is divisible to one figure by 10 or a multiple of 10, use the resulting quotient as a basic principal, that is, for $900 use 9, for $60 use 6, for $8000 use 8. If the principal is an odd number use one.

c. After selecting the basic principal in the correct interest rate group, follow along the line to the left until you reach the column headed by the number of days or months for which you are computing the interest.

d. Multiply the figure thus found by the true principal if you are using one for a base principal or move the decimal point to right the correct number of times if you are using a one-figure quotient determined by dividing by 10, or a multiple of 10.

ILLUSTRATION: $500 is borrowed for 20 days with interest at 5%. What amount of interest will have to be paid?

Using the interest table:

a. Run down the side of the table to the 5% section.

b. As $500 divided by 100 equals 5, use $5 as a basic principal.

c. Following along the $5 line to the 20-day column, it will be noted that the interest on $5 at 5% for 20 days equals 0.01388.

d. Moving the decimal point two places to the right to multiply by 100, it will be found that interest on $500 at 5% for 20 days equals $1.38. (Ans.)

ILLUSTRATION: $463.75 was borrowed for 3 months with interest at 7%. What amount of interest will have to be paid?

TABLE 2

Simple Interest

Rate	Principal	Time.									
		1 Year	6 Mo.	5 Mo.	4 Mo.	3 Mo.	2 Mo.	1 Mo.	20 d.	10 d.	1 d.
	$1	.040	.0200	.01666˅6	.013˅3	.01000	.0066˅6	.00333˅3	.0022˅2	.00111˅1	.000111˅1
	2	.080	.0400	.03333˅3	.026˅6	.02000	.0133˅3	.00666˅6	.0044˅4	.00222˅2	.000222˅2
	3	.120	.0600	.05000	.040	.03000	.0200	.01000	.0066˅6	.00333˅3	.000333˅3
	4	.160	.0800	.06666˅6	.053˅3	.04000	.0266˅6	.01333˅3	.0088˅8	.00444˅4	.000444˅4
4%	5	.200	.1000	.08333˅3	.066˅6	.05000	.0333˅3	.01666˅6	.0111˅1	.00555˅5	.000555˅5
	6	.240	.1200	.10000	.080	.06000	.0400	.02000	.0133˅3	.00666˅6	.000666˅6
	7	.280	.1400	.11666˅6	.093˅3	.07000	.0466˅6	.02333˅3	.0155˅5	.00777˅7	.000777˅7
	8	.320	.1600	.13333˅3	.106˅6	.08000	.0533˅3	.02666˅6	.0177˅7	.00888˅8	.000888˅8
	9	.360	.1800	.15000	.120	.09000	.0600	.03000	.0200	.01000	.001000
	$1	.045	.0225	.01875	015	.01125	.0075	.00375	.0025	.00125	.000125
	2	.090	.0450	.03750	.030	.02250	.0150	.00750	.0050	.00250	.000250
	3	.135	.0675	.05625	.045	.03375	.0225	.01125	.0075	.00375	.000375
	4	.180	.0900	.07500	.060	.04500	.0300	.01500	.0100	.00500	.000500
4½%	5	.225	.1125	.09375	.075	.05625	.0375	.01875	.0125	.00625	.000625
	6	.270	.1350	.11250	.090	.06750	.0450	.02250	.0150	.00750	.000750
	7	.315	.1575	.13125	.105	.07875	.0525	.02625	.0175	.00875	.000875
	8	.360	.1800	.15000	.120	.09000	.0600	.03000	.0200	.01000	.001000
	9	.405	2025	.16875	.135	.10125	0675	.03375	.0225	·01125	.001125
	$1	.050	.0250	.02083˅3	.016˅6	.01250	.0083˅3	.00416˅6	.0027˅7	.00138˅8	.000138˅8
	2	.100	.0500	.04166˅6	.033˅3	.02500	.0166˅6	.00833˅3	.0055˅5	.00277˅7	.000277˅7
	3	.150	.0750	.06250	.050	.03750	.0250	.01250	.0083˅3	.00416˅6	.000416˅6
	4	.200	.1000	.08333˅3	.066˅6	.05000	.0333˅3	.01666˅6	.0111˅1	.00555˅5	.000555˅5
5%	5	.250	.1250	.10416˅6	.083˅3	.06250	.0416˅6	.02083˅3	.0138˅8	.00694˅4	.000694˅4
	6	.300	.1500	.12500	.100	.07500	.0500	.02500	.0166˅6	.00833˅3	.000833˅3
	7	.350	.1750	.14583˅3	.116˅6	.08750	.0583˅3	.02916˅6	.0194˅4	.00972˅2	.000972˅2
	8	.400	.2000	.16666˅6	.133˅3	.10000	.0666˅6	.03333˅3	0222˅2	.01111˅1	.001111˅1
	9	.450	.2250	.18750	.150	.11250	.0750	.03750	.0250	.01250	.001250
	$1	.060	.0300	.02500	.020	.01500	.0100	.00500	.0033˅3	.00166˅6	.000166˅6
	2	.120	.0600	.05000	.040	.03000	.0200	.01000	.0066˅6	.00333˅3	.000333˅3
	3	.180	.0900	.07500	.060	.04500	.0300	.01500	.0100	.00500	.000500
	4	.240	.1200	.10000	.080	.06000	.0400	.02000	.0133˅3	.00666˅6	.000666˅6
6%	5	.300	.1500	.12500	.100	.07500	.0500	.02500	.0166˅6	.00833˅3	.000833˅3
	6	.360	.1800	.15000	.120	.09000	.0600	.03000	.0200	.01000	.001000
	7	.420	.2100	.17500	.140	.10500	.0700	.03500	.0233˅3	.01166˅6	.001166˅6
	8	.480	.2400	.20000	.160	.12000	.0800	.04000	.0266˅6	.01333˅3	.001333˅3
	9	.540	.2700	.22500	.180	.13500	.0900	.04500	.0300	.01500	.001500
	$1	.070	.0350	.02916˅6	.023˅3	.01750	.0116˅6	.00583˅3	.0038˅8	.00194˅4	.000194˅4
	2	.140	.0700	.05833˅3	.046˅6	.03750	.0233˅3	.01166˅6	.0077˅7	.00388˅8	.000388˅8
	3	.210	.1050	.08750	.070	.05250	.0350	.01750	.0116˅6	.00583˅3	.000583˅3
	4	.280	.1400	.11666˅6	.093˅3	.07000	.0466˅6	.02333˅3	.0155˅5	.00777˅7	.000777˅7
7%	5	.350	.1750	.14583˅3	.116˅6	.08750	.0583˅3	.02916˅6	.0194˅4	.00972˅2	.000972˅2
	6	.420	.2100	.17500	.140	.10500	.0700	.03500	.0233˅3	.01166˅6	.001166˅6
	7	.490	.2450	.20416˅6	.163˅3	.12250	.0816˅6	.04083˅3	.0272˅2	.01361˅1	.001361˅1
	8	.560	.2800	.23333˅3	.186˅6	.14000	.0933˅3	.04666˅6	.0311˅1	.01555˅5	.001555˅5
	9	.630	.3150	.26250	.210	.15750	.1050	.05250	0350	.01750	.001750

★ Note that *all* repeating decimals may be extended indefinitely. Thus, the interest on $1.00 at 4% for 4 months is given as .013˅3 or 1⅓ cents, because the decimal .013˅3 = .01333333. . .; hence the interest on $1,000,000, at the same rate and for the same time, is $13,333.33⅓. Decimals which are not repeating decimals are *exact*.

Using the interest table:

a. Run down the side of the table to the 7% section.

b. As this principal may not be reduced to a single figure by dividing by 10 or a multiple of 10, use $1.00 as a basic principle.

c. Following along the $1.00 line to the 3 months column, it will be noted that interest on $1.00 at 7% for 3 months equals 0.01750.

d. Multiplying this amount by $463.75, the true principal, it will be found that interest on $463.75 at 7% for 3 months equals $8.115625 or $8.12. (Ans.)

If the number of days in a given problem does not appear in the table, the amount of interest for various numbers of days may be combined; thus, interest for 70 days may be determined by adding together the interest for 60 days (2 months) and the interest for 10 days.

ILLUSTRATION: $5000 was borrowed for 80 days with interest at 6%. What amount of interest must be paid when the obligation is due?

Using the interest table:

a. Run down the side of the table to the 6% section.

b. Eliminate the zeros by pointing off 3 places and thus adopt $5 as the basic principal.

c Follow along the $5 line to the 2 months (60-day) column and note that

Interest on $5 at 6% = 0.0500

also that in the 20-day column

Interest on $5 at 6% = 0.016666

Therefore Interest on $5 @ 6% for 80 days = 0.066666

d. Move the decimal point 3 places to the right to multiply by 1000, and thus we find that

Interest on $5000 @ 6% for 80 days = $66.6666, or $66.67 (Ans.)

If the number of days does not readily lend itself to such combinations, it is frequently more simple to find the interest for one day and then multiply by the number of days.

ILLUSTRATION: $750 was borrowed for 17 days with interest @ 7%. What amount of interest must be paid when the obligation is due?

Using the table:

a. Run down the column to the 7% section.

b. As the principal cannot be reduced to one figure, use $1 as the basic principal.

c. Follow along the $1 line to the 1 day column and note that

Interest on $1 at 7% for 1 day = 0.0001944

Therefore

Interest on $1 @ 7% for 15 days
(0.0001944 × 15) = 0.0029160

d. The interest on $750 at 7% for 15 days equals $750 × 0.0029160 = $2.187 or $2.19 (Ans.)

It may be noted in the foregoing illustrations in which the interest table was used, the calculations in some instances were rather awkward. This is due to the fact that the particular table being used is not necessarily the best for all interest computations. Firms making use of precomputed interest tables will usually have those that best fit their particular needs.

Legal and Lawful Rates of Interest.—The legal rate of interest is the rate that may legally be charged in the absence of any definite agreement between the parties. This is particularly true of judgments and overdue accounts where interest is to be charged but it may also apply in other situations where interest is applicable but where no specific rate has been agreed upon.

The lawful rate (sometime called the contract rate) is the maximum rate that can be charged when a definite agreement has been made. In some states the legal and the lawful rates are the same. In other states they vary widely, while in still other states certain conditions are attached to the contract rate. The charging of a rate of interest above the lawful or contract rate is known as "usury," which in some states is a crime, in others a misdemeanor. In either case, it is punishable by a variety of penalties. New York and Maryland do not permit corporations to plead "usury" as a defense.

Automobile Loans.—As in the case of smaller consumer loans, the borrower of money toward the purchase of an automobile is not charged an interest rate computed for each payment period on the unpaid balance at the beginning of that period. Instead, the amount he is to pay over the period is taken from tables, such as Table 2A, and he makes a uniform payment at each period as is shown in the second column of this table. These tables are computed for various rates of interest. This particular table is based on the very low return to the bank of 5½ percent. However, by reference to the figures in the table, such as the finance charge of $55.24 per year for $1000 to be paid during the year, it can be computed that the actual interest rate is between 10 and 11 percent. This does not mean, however, that the lending institution earns that percentage on the money loaned. The overhead costs of these loans include the costs of preparing and sending periodic bills, the cost of bookkeeping for periodical payments and part of the costs of repossession and resale of automobiles on which the payments are defaulted. Thus, of the $55.24 annual charge the interest earned by the lending institution is only 5.5 percent, even on the unpaid balances, a sum which can be calculated to amount to $28.13. The difference between the $55.24 and this figure, which is $27.11, represents these costs of making and collecting the automobile loans of the institution. Note that in this table one of these charges, the life insurance premium on the owner of the car, which must be carried by the finance institution, is shown separately, but all the other costs, including the insurance policy, which the bank must carry on the automobile itself, are contained in the charge of $27.11.

Personal Finance.—Of course, smaller loans, especially those made on articles which deteriorate more rapidly than automobiles, carry higher overhead charges. Another reason for this is the fact that the smaller the loan the higher the proportion

TABLE 2A

Sample Automobile Loan Schedule

Unpaid Bal.	12 Months				18 Months			
	Amt. Per Mo.	Amt. of Note	Life Prem.	Finance Charge	Amt. Per Mo.	Amt. of Note	Life Prem.	Finance Charge
1	.08	.96	.00	-.04	.06	1.08	.01	.07
2	.17	2.04	.01	.03	.12	2.16	.02	.14
3	.26	3.12	.02	.10	.18	3.24	.03	.21
4	.35	4.20	.03	.17	.24	4.32	.04	.28
5	.44	5.28	.03	.25	.30	5.40	.04	.36
10	.88	10.56	.06	.50	.60	10.80	.08	.72
15	1.32	15.84	.08	.76	.90	16.20	.12	1.08
20	1.76	21.12	.11	1.01	1.21	21.78	.16	1.62
25	2.20	26.40	.13	1.27	1.51	27.18	.20	1.98
30	2.65	31.80	.16	1.64	1.81	32.58	.24	2.34
35	3.09	37.08	.19	1.89	2.12	38.16	.28	2.88
40	3.53	42.36	.21	2.15	2.42	43.56	.31	3.25
45	3.97	47.64	.24	2.40	2.72	48.96	.35	3.61
50	4.41	52.92	.26	2.66	3.03	54.54	.39	4.15
55	4.86	58.32	.29	3.03	3.33	59.94	.43	4.51
60	5.30	63.60	.32	3.28	3.63	65.34	.47	4.87
65	5.74	68.88	.34	3.54	3.93	70.74	.51	5.23
70	6.18	74.16	.37	3.79	4.24	76.32	.55	5.77
75	6.62	79.44	.39	4.05	4.54	81.72	.59	6.13
80	7.06	84.72	.42	4.30	4.84	87.12	.62	6.50
85	7.51	90.12	.45	4.67	5.15	92.70	.66	7.04
90	7.95	95.40	.47	4.93	5.45	98.10	.70	7.40
95	8.39	100.68	.50	5.18	5.75	103.50	.74	7.76
1,000	88.37	1,060.44	5.20	55.24	60.60	1,090.80	7.75	83.05
1,100	97.21	1,166.52	5.72	60.80	66.66	1,199.88	8.52	91.36
1,200	106.04	1,272.48	6.24	66.24	72.72	1,308.96	9.30	99.66
1,300	114.88	1,378.56	6.76	71.80	78.78	1,418.04	10.07	107.97
1,400	123.72	1,484.64	7.28	77.36	84.84	1,527.12	10.85	116.27
1,500	132.56	1,590.72	7.80	82.92	90.90	1,636.20	11.62	124.58
1,600	141.39	1,696.68	8.32	88.36	96.96	1,745.28	12.40	132.88
1,700	150.23	1,802.76	8.84	93.92	103.02	1,854.36	13.17	141.19
1,800	159.07	1,908.84	9.36	99.48	109.08	1,963.44	13.95	149.49
1,900	167.90	2,014.80	9.88	104.92	115.14	2,072.52	14.72	157.80

TABLE 2A (Cont.)

Unpaid Bal.	12 Months				18 Months			
	Amt. Per Mo.	Amt. of Note	Life Prem.	Finance Charge	Amt. Per Mo.	Amt. of Note	Life Prem.	Finance Charge
2,000	176.74	2,120.88	10.40	110.48	121.20	2,181.60	15.49	166.11
2,100	185.58	2,226.96	10.92	116.04	127.26	2,290.68	16.27	174.41
2,200	194.42	2,333.04	11.44	121.60	133.33	2,399.94	17.04	182.90
2,300	203.25	2,439.00	11.96	127.04	139.39	2,509.02	17.82	191.20
2,400	212.09	2,545.08	12.48	132.60	145.45	2,618.10	18.59	199.51
2,500	220.93	2,651.16	13.00	138.16	151.51	2,727.18	19.37	207.81
2,600	229.77	2,757.24	13.52	143.72	157.57	2,836.26	20.14	216.12
2,700	238.60	2,863.20	14.03	149.17	163.63	2,945.34	20.92	224.42
2,800	247.44	2,969.28	14.55	154.73	169.69	3,054.42	21.69	232.73
2,900	256.28	3,075.36	15.07	160.29	175.75	3,163.50	22.47	241.03
3,000	265.12	3,181.44	15.59	165.85	181.81	3,272.58	23.24	249.34
3,100	273.95	3,287.40	16.11	171.29	187.87	3,381.66	24.01	257.65
3,200	282.79	3,393.48	16.63	176.85	193.93	3,490.74	24.79	265.95
3,300	291.63	3,499.56	17.15	182.41	199.99	3,599.82	25.56	274.26
3,400	300.46	3,605.52	17.67	187.85	206.05	3,708.90	26.34	282.56
3,500	309.30	3,711.60	18.19	193.41	212.11	3,817.98	27.11	290.87
3,600	318.14	3,817.68	18.71	198.97	218.17	3,927.06	27.89	299.17
3,700	326.98	3,923.76	19.23	204.53	224.23	4,036.14	28.66	307.48
3,800	335.81	4,029.72	19.75	209.97	230.29	4,145.22	29.44	315.78
3,900	344.65	4,135.80	20.27	215.53	236.35	4,254.30	30.21	324.09
4,000	353.49	4,241.88	20.79	221.09	242.41	4,363.38	30.98	332.40

formed by these costs. This situation can be shown clearly by an illustration.

ILLUSTRATION: A purchase is made of furniture totalling \$580.50. The agreement is that one-third of the total is to be paid at the time of purchase and the remainder is to be paid off in monthly installments. The rate on the unpaid balance is 8% and it is to be handled by a finance company. One-third of \$580.50 (\$193.50) was paid when the contract was executed, leaving a balance of \$387 to be paid in monthly installments. How much must be paid monthly?

The true rate of interest (i.e., the rate without other charges) can be computed approximately by assuming payments of $\frac{1}{12}$ of \$387 (\$32.25) plus accrued interest each month for the year. This would work out as follows:

	Principal		*Interest*		
1st	\$32.25	+	\$0.22	=	\$32.47
2nd	32.25	+	0.43	=	32.68
3rd	32.25	+	0.65	=	32.90
4th	32.25	+	0.86	=	33.11
5th	32.25	+	1.08	=	33.33
6th	32.25	+	1.29	=	33.54
7th	32.25	+	1.51	=	33.76
8th	32.25	+	1.72	=	33.97
9th	32.25	+	1.94	=	34.19
10th	32.25	+	2.15	=	34.40
11th	32.25	+	2.37	=	34.62
12th	32.25	+	2.58	=	34.83
	\$387.00		\$16.80		\$403.80

In the standard method of computing finance charges, however, a flat 8% is charged on the entire balance of \$387. This amounts to \$30.96 in interest so that the customer pays this sum instead of \$16.80, a difference of \$14.16 which represents the expenses and profit to the lending institution. Compare this figure of \$14.16 excess on a \$387 advance for one year against the \$27.11 charge by the bank for nearly three times as much money for the same period, and it will be seen how much greater the charges are on the smaller loans and the more perishable security. Such credit loans, by the way, are not usually handled by banks.

Small Loans.—While in the two examples cited the excess finance charge over normal interest, which represents the expenses of the lending institution, were not excessive, it is obvious and is often true that small loan rates have been excessive in the past and are often so at present. For that reason, a number of states in the United States and provinces of Canada have established regulations governing the amount which may be charged on various types of small loans. These regulations have been summarized for a few states in Table 2B.

TABLE 2B

Small Loan Regulations Summarized for Four States

MASSACHUSETTS—6%/no limit	
	Tender Act: A right to prepay with int. at 18% p.a., plus $5.00. Ceiling $1,000
Loan & Investment Cos. . . .	Morris or similar plan companies. Less than $500, 12% discount. More than $500, 9% discount.
Retail Sales Finance	Disclosure. Refund: Rule of 78ths, after deducting acq. chg. Motor Vehicles (other than those subject to motor vehicle sales finance statute)—$12.50 Other goods—$5.00
Motor Vehicle Sales Finance	Retail Instalment Sales of Motor Vehicles N. & U. C., model year of year of sale or prior $8/$100 N. & U. C., not in above and not more than 2 yrs. old $10/$100 O. U. C. $12/$100 Del. chg: $5 or 5% Refund: Rule of 78ths, after deducting acq. chg. of $12.50
NEW YORK—6%/6%	
	Bank Instalment Loans $6/$100 discount. Ceiling $5,000. Max. 25 mos. if $1,200 or less. 37 mos. max. over $1,200, not in excess of $5,000. 37 mos. max. for R. P. improvement, min. $10. Del. chg: 5¢/$1, max. $5 or an aggregate of 2% of loan, but not in excess of $25. Refund: Rule of 78ths, subject to $10 min. chg.
Loan & Investment Cos. . . .	Industrial Banks $6/$100 discount. Ceiling $5,000. 25 mos. max., if less than $1,200. 37 mos. if more than $1,200 or if R. P. Min. chg: $10. Fee: $1/$50 up to $250. Max: $5. More than $250: $5 plus 1% of excess above $250. Max: $20. Del. chg: 5¢/$1. Max: $5 per instalment. Total Max: 2% or $25. Refund: Pro-rata as of following date, subject to $10 min. chg.

TABLE 2B (Cont.)

Retail Sales Finance (All Goods Act)	Retail Instalment Sales Act Unpaid principal balance $500 $10/$100 p.a. Unpaid principal balance over $500 8/$100 p.a. Revolving credit: on that amount not in excess of $500—1½% per mo. Del. chg: 5% or $5. Refund: Rule of 78ths Merchandise certificate refund: pro-rata.
Motor Vehicle Sales Finance	N. C., current model $7/$100 N. & U. C., not more than 2 yrs. old $10/$100 O. U. C. $13/$100 Insurance premium $7/$100 Del. chg: 5% or $5. Ext. or Refinance chg: service fee not to exceed $5, plus total additional chg. not to exceed 1% per mo. Refund: Rule of 78ths, after deducting acq. chg. $15. Refinance chg: 1% per mo. or refund and charge as per contrat on unpaid balance.

NOTE: In this table, p.a. means per annum, N.C. means new car, and U.C. means used car.

Commercial Partial Payments.—A partial payment is an amount that is not sufficient to liquidate an indebtedness. The finance plan for financing installment sales and the small loan payments already discussed are merely partial payment plans that apply in personal financing. Where business organizations borrow and make partial payments, other methods are applied. When such partial payments are made on interest-bearing items, a problem arises as to the amount due at the time of final settlement. There are two rules that are commonly followed: (1) the *Merchants' Rule*, and (2) the *United States Rule*.

Merchants' Rule.—Under this rule interest is charged on the principal for the full time and is credited on the payments from the date of each payment to the date of final payment. The interest on the principal less the interest credited on the periodic payments equals the interest charged.

ILLUSTRATION: On May 1, a man borrowed $5000 to be paid back at the rate of $1000 each month. The interest rate is 6%.

He pays $1000 on the first of June, July, August, and September. Applying the Merchants' Rule, how much must be paid on October 1 to settle the account?

Interest on $5000 for 5 months, May 1 to Oct. 1, @ 6% =		$125.00
Interest credited as follows:		
On $1000 for 4 months (from June 1 to Oct. 1) @ 6%,	$20.00	
On $1000 for 3 months (from July 1 to Oct. 1) @ 6%,	15.00	
On $1000 for 2 months (from Aug. 1 to Oct. 1) @ 6%,	10.00	
On $1000 for 1 month (from Sept. 1 to Oct. 1) @ 6%,	5.00	
Total interest credit......................		50.00
Interest due October 1....................		75.00
Unpaid principal October 1...............		1000.00
		$1075.00

United States Rule.—Under this rule, all interest accrued on the unpaid balance is deducted from the payment before the remainder is deducted from the principal or that part of the principal that is still unpaid at the time payment was made.

ILLUSTRATION: Applying the United States Rule to the problem cited in the Illustration under Merchants' Rule, how much would the man have to pay on October 1 to settle his account:

Original Principal..............................		$5000.00
Payment on June 1............................	$1000.00	
Less int. on $5000 @ 6% for 1 mo............	25.00	975.00
		4025.00
Payment on July 1............................	$1000.00	
Less int. on $4025 @ 6% for 1 mo............	20.13	979.87
		3045.13
Payment on August 1..........................	$1000.00	
Less int. on $3045.13 @ 6% for 1 mo..........	15.23	984.77
		2060.36
Payment on September 1......................	$1000.00	
Less int. on $2060.36 @ 6% for 1 mo..........	10.30	989.70
		1070.66
Interest on $1070.66 @ 6% for 1 mo............		5.35
Amount due October 1..............		$1076.01

TABLE 3

This table shows nine methods of borrowing $1000, showing the sources, arrangements made, amount at borrower's disposal, **total finance charge, which includes interest, apparent and real interest rate paid, and disadvantages of each main method of borrowing.**

Sources	Terms	Total Finance Charge You Pay	Interest Rate		Advantages	Disadvantages
			Real	Apparent		
1. Personal Loan Company	Loan must be repaid in monthly installments. Usually, borrower must be employed, or have collateral—such as a car or jewelry—worth the value of the loan. $500 is top amount you can borrow in many states.	$127.00	26.2%	12.7%	Informal atmosphere. Usually easy to get new loans once your credit has been established by promptly paying back one loan.	High interest rate; large amount of interest. By repaying in monthly installments, you steadily cut down the amount of loan money you can use.
2. Life Insurance Policy (Excluding term insurance)	Amount you can borrow grows larger each year you keep policy in force. This "Loan Value" is specified in policy—as is interest rate. You can repay whenever you wish—and in part or whole amount.	60.00	6.0%	6.0%	Low rate and mount of interest; easy availability; no limit on length of loan; defaulted interest payments merely added to amount of loan.	It takes 10 years before the average ordinary life policy taken out at age 25 will have a loan value of $1000.
3. Automobile Installment Finance Company	Available to help finance the purchase of new or used autos—or to pay for repairs or accessories.	92.31	18.7%	9.2%	Readily available, at time of car purchase. Gives dealer an added reason to give you good service.	Moderately high interest rate and amount of interest. By repaying in monthly installments you steadily cut down the loan money you can use.

TABLE 3 (Cont.)

Sources	Terms	Total Finance Charge You Pay	Interest Rate		Advantages	Disadvantages
			Real	Apparent		
4. Bank Personal Loan Dept. (Monthly repayment—collateral)	Borrower turns over good collateral—such as stocks, bonds, or chattel mortgage on car. Loan must be repaid in monthly installments. Typical face value of loan is $43.72, but the $43.72 is deducted in advance.	43.72	6.0%	8.9%	Moderate rate and total interest for this type of loan. Can be used to finance purchase of new or used car.	Ties up your stocks or bonds in bank's hands. By repaying in monthly installments, you steadily cut down amount of loan money you can use.
5. Bank Personal Loan Dept. (Lump sum repayment—collateral	Borrower turns over good collateral—such as stocks or bonds. Lump sum repayment at end of year permitted.	76.50	7.65%	7.65%	Low interest rate; low total amount of interest. Whole amount of loan available to you for whole year.	Your collateral tied up in bank and may have to be sold at an unfavorable time—if loan comes due and you have no other way to repay it.
6. Mortgage on House	Your house is security for loan, which must be repaid in monthly installments.	36.40	3.6%	7.5%	Low interest rate and total amount of interest.	By repaying in monthly installments you steadily cut down the amount of loan money you can use.
7. Credit Union	Membership in the credit union is necessary. Loan must be repaid in monthly installments. Typical face value of loan is $1,000, but $40 is deducted in advance.	50.00	4.0%	8.1%	Easy availability to members of the group.	Moderately high cost. By repaying in monthly in-installments you steadily cut down the amount of loan money you can use.

Mortgages.—A mortgage is a loan on property where conditional title to the property is given as a pledge of repayment. The payments on the mortgage and the interest charges are fixed according to the interest rate and the period of payment agreed upon with the lender. Most modern mortgages are based on a constant monthly payment plan (see Table 5, Plan III).

In addition to the mortgage payments it is usually possible to make other payments such as taxes, assessments, insurance, and upkeep, part of the same monthly payment plan. The bank or savings and loan association holds this money in escrow (the money is accumulated in a special fund). When the taxes or insurance come due they are paid by the bank for the homeowner.

Table 4 shows the major expenditures in financing a home after down payments and preliminary costs are paid. Expenses are assumed to be $18 for taxes and assessments, $3 for insurance, and $20 for upkeep for every $1000 loaned on the house.

TABLE 4

ANNUAL HOME OWNERSHIP COST FOR EACH $1000 BORROWED *

Payment Period	4% interest			5% interest			6% interest		
	10 years	15 years	20 years	10 years	15 years	20 years	10 years	15 years	20 years
Interest and payment on each $1000 of loan per year † (based on a systematic loan reduction plan)	$122	$ 89	$ 73	$127	$ 95	$ 79	$133	$101	$ 86
Taxes and assessments	18	18	18	18	18	18	18	18	18
Insurance	3	3	3	3	3	3	3	3	3
Upkeep	20	20	20	20	20	20	20	20	20
Total annual outlay on each $1000 borrowed	$163	$130	$114	$168	$136	$120	$174	$142	$127

* University of Ill. Small Homes Council Bulletin A1.3.

† To find the total payment on interest and principal to maturity for each $1000 of loan, multiply the annual payment by the number of years in the payment period. Subtract the $1000 principal from the total payment to find the total interest to maturity for each $1000 of loan.

TABLE 5

FOUR PLANS FOR REPAYMENT OF $10,000 IN 10 YEARS WITH INTEREST AT 6% *

	End of Year	Interest Due (6% of money owed at start of year)	Total Money Owed Before Year-End Payment	Year-End Payment	Money Owed After Year-End Payment
Plan I	0				$10,000
	1	$600	$10,600	$ 600	10,000
	2	600	10,600	600	10,000
	3	600	10,600	600	10,000
	4	600	10,600	600	10,000
	5	600	10,600	600	10,000
	6	600	10,600	600	10,000
	7	600	10,600	600	10,000
	8	600	10,600	600	10,000
	9	600	10,600	600	10,000
	10	600	10,600	10,600	0
Plan II	0				$10,000
	1	$600	$10,600	$1,600	9,000
	2	540	9,540	1,540	8,000
	3	480	8,480	1,480	7,000
	4	420	7,420	1,420	6,000
	5	360	6,360	1,360	5,000
	6	300	5,300	1,300	4,000
	7	240	4,240	1,240	3,000
	8	180	3,180	1,180	2,000
	9	120	2,120	1,120	1,000
	10	60	1,060	1,060	0
Plan III	0				$10,000.00
	1	$600.00	$10,600.00	$1,358.68	9,241.32
	2	554.48	9,795.80	1,358.68	8,437.12
	3	506.23	8,943.35	1,358.68	7,584.67
	4	455.08	8,039.75	1,358.68	6,681.07
	5	400.86	7,081.93	1,358.68	5,723.25
	6	343.40	6,066.65	1,358.68	4,707.98
	7	282.48	4,990.45	1,358.68	3,631.77
	8	217.91	3,849.68	1,358.68	2,491.00
	9	149.46	2,640.46	1,358.68	1,281.78
	10	76.90	1,358.68	1,358.68	0.00
Plan IV	0				$10,000.00
	1	$ 600.00	$10,600.00	$ 0.00	10,600.00
	2	636.00	11,236.00	0.00	11,236.00
	3	674.16	11,910.16	0.00	11,910.16
	4	714.61	12,624.77	0.00	12,624.77
	5	757.49	13,382.26	0.00	13,382.26
	6	802.94	14,185.20	0.00	14,185.20
	7	851.11	15,036.31	0.00	15,036.31
	8	902.18	15,938.49	0.00	15,938.49
	9	956.31	16,894.80	0.00	16,894.80
	10	1,013.69	17,908.49	17,908.49	0.00

* Grant, E. L., *Principles of Engineering Economy,* Revised Edition, The Ronald Press.

New Note Method.—Some banks in handling this problem avoid some of the involved calculation by having the debtor pay his thousand dollars each month plus accrued interest and give a new note for the balance. This greatly simplifies the problem for both the bank and the borrower.

Illustration: Using the same problem, find the amount to be paid and the amount of the new note to be given at the end of each month.

		Principal	*Interest*	*Amount Paid*	*New Note*
May 1.............	Borrowed	$5000			
June 1.............	Paid	1000 +	$25.00 =	$1025	$4000
July 1.............	Paid	1000 +	20.00 =	1020	3000
August 1...........	Paid	1000 +	15.00 =	1015	2000
September 1........	Paid	1000 +	10.00 =	1010	1000
October 1..........	Paid	1000 +	5.00 =	1005	0
Total Interest.....................			$75.00		

Series of Notes.—Still another method of handling this matter is by having the borrower make out five $1000 notes bearing interest at 6%, one due each month. This procedure is even more simple than the new note plan.

Illustration: Still using the same problem, assume that the borrower of the $5000 was asked to make out a series of five $1000 notes each bearing interest at 6%. How much must be paid when each note is due?

			Interest	*Amount*
May 1.............	Borrowed	$5000		
June 1.............	Paid	1000	$5.00	$1005
July 1.............	Paid	1000	10.00	1010
August 1...........	Paid	1000	15.00	1015
September 1.......	Paid	1000	20.00	1020
October 1..........	Paid	1000	25.00	1025
			$75.00	$1075

Relative Merits.—There is relatively little difference among the methods treated. The United States Method gives a slightly larger interest return to the lender than does the Merchants' Method; in the problem used to illustrate these various methods, this difference amounted to $1.01. Because of this, the United States Method is usually used where large sums are involved. In considering the relative merits of the *new note* and the *series of notes* plans, it should be noted that while the interest paid under the two methods is $75.00 in each case, the same as under the Merchants' Rule, if one considers the present worth of the interest in relation to the final due date of the obligation, one perceives that the "series of notes" plan tends in the direction of the Merchants' Rule, while the New Note Method plan approximately equals the United States Rule.

Negotiable Instruments.—Because interest is so closely associated with certain negotiable instruments, it seems advisable to give them some brief attention at this point. A negotiable instrument is usually defined as being an instrument the legal title of which may be passed from one party to another by endorsement and delivery or merely by delivery. According to the New York Negotiable Instruments Law which is standard, basic factors that make a business paper negotiable are: (1) It must be in writing signed by the one who is to pay, (2) It must contain an unconditional promise or order to pay a certain sum in money, (3) It must be payable on demand or at a fixed or determinable future time, (4) It must be payable to order or to bearer, (5) Where the instrument is addressed to a drawee, he must be named or otherwise indicated therein with reasonable certainty.

Negotiable Instruments differ from other contracts in two rather vital respects: (1) as to quality of the title, and (2) as to consideration. When a person receives title to a negotiable instrument in the absence of any knowledge of any infirmity in the title of the person delivering that title to him, he receives a good valid title. In ordinary contracts, the title passes by assignment and the assignee becomes subject to all the defenses that may exist between the original parties.

All contracts must have consideration, but in the case of negotiable instruments this quality is conclusively presumed between all others than the original parties.

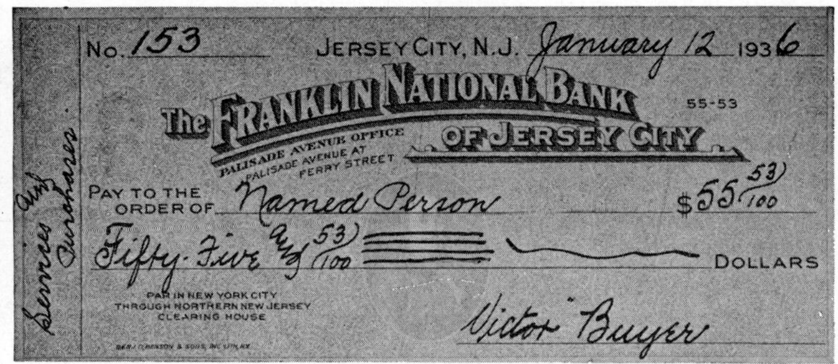

FIG. 3.—Check

Instruments of Exchange.—Broadly speaking, negotiable instruments fall into two classifications: (1) Instruments of Exchange, and (2) Instruments of Credit. An instrument of exchange

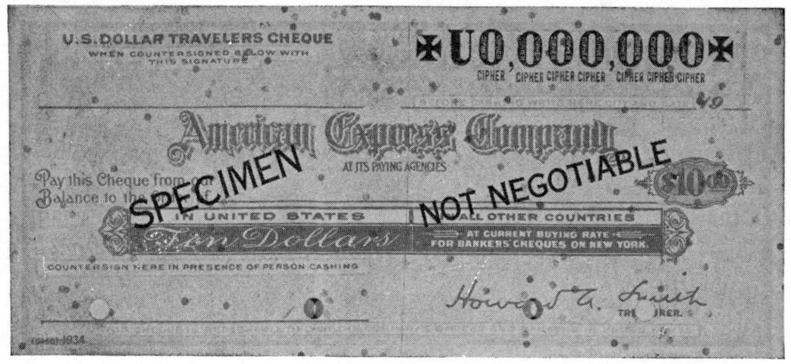

FIG. 4.—Travelers' Check

is an unconditional order in writing to pay to the order of a specified person or to bearer a certain sum of money. An instrument of exchange is used to transfer money without actually exchanging the cash, and bears no interest. The most commonly used negotiable instruments that fall into this category are: (1) check, (2) cashier's check, (3) certified check, (4) bank draft, (5) Post Office money order, (6) Express money order, and (7) travelers' check.

Instruments of Credit.—This type of instrument may be defined as being an agreement to pay at a later date a fixed sum of money to the order of a specified person or to bearer. It must be in writing and must be signed by the person who is to pay. This type of instrument is used in connection with various types of deferred payments and frequently, although not always, bears

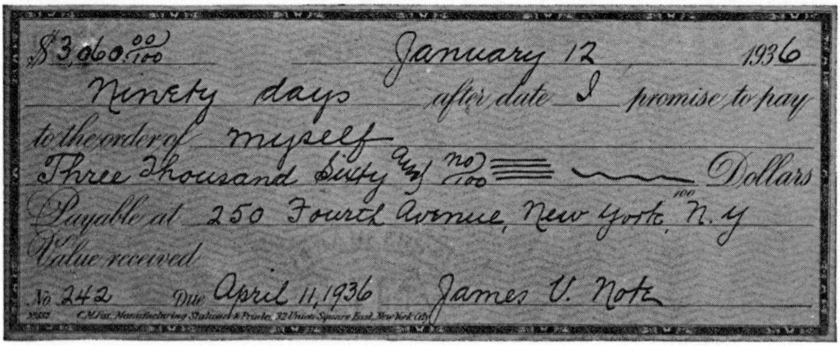

$3060.00/100 January 12 1936
Ninety days after date I promise to pay
to the order of myself
Three Thousand Sixty and no/100 Dollars
Payable at 250 Fourth Avenue, New York, N.Y.
Value received
No. 242 Due April 11, 1936 James V. Note

Fig. 5.—Promissory Note

interest. The most commonly used negotiable instruments that fall in this category are: (1) promissory notes, (2) commercial drafts, (3) trade acceptances, and (4) bonds. Interest on notes and other forms of negotiable instruments that bear interest is calculated the same as any ordinary interest, usually by using the 60-day method or by using an interest table.

Illustration: On August 15, $2500 is borrowed on a 90-day note bearing interest at 6%. What amount must be paid when the note is due:

Principal (Face of the Note)		$2500.00
Interest on $2500 @ 6% for 60 days	$25.00	
Interest on $2500 @ 6% for 30 days	12.50	
90		37.50
Total Amount to be Paid when Note is Due		$2537.50 (Ans.)

Bank Discount.—Promissory notes and other forms of instruments of exchange are used in connection with credit operations (1) between merchandising and industrial organizations as well as between persons, (2) between such individuals and business organizations and banks. If a person or firm receives a note, draft, or trade acceptance from another, he may hold it until it is due and then collect the face plus the interest if it happened to be an interest-bearing draft. If, however, he would like to have the money before it is due he may take it to the bank and receive an amount equal to its present value. This is known as discounting the paper at the bank.

Discounting a Non-Interest-Bearing Note.—The process of bank discount involves five steps (1) determining the value of the paper at maturity (when it is due), (2) determining the date of maturity, (3) counting the exact number of days between the day that the paper is taken over by the bank (called the day of discount) and the date of maturity. This is known as the term of discount. (4) Calculating the discount (really interest) for the term of discount based on the value at maturity, (5) determining the Net Proceeds by deducting the discount from the value at maturity.

ILLUSTRATION: On May 2, Harold Jones receives a 60-day non-interest-bearing note for $750 from one of his customers. He holds it until May 17 and then takes it to the bank and discounts it. The rate of discount at the bank is 6%. What is the net proceeds?

The five steps are as follows:

(1) Value at Maturity. In the case of a non-interest-bearing note, only the face of the note is due at maturity. In this case, the value at maturity is $750.

(2) Date of Maturity is the due date of the note. This note is due 60 days after May 2. There are 29 more days in May. Twenty-nine plus 30, in June, makes 59. Fifty-nine plus one in July makes 60. Therefore, the date of maturity is July 1. It might be noted here that when a note reads days, days are counted,

if it reads months, months are counted. If this had read "two months" the due date would be two months after May 2, or July 2. As it read 60 days, the due date is July 1.

(3) Term of Discount is unexpired time, the exact number of days between the date of discount and the date of maturity. As this note was discounted on May 17, there are 14 more days in May, 30 in June, and one in July, a total of 45 days. This could have been readily ascertained by consulting Table 6, a table for finding the number of days between dates.

TABLE 6

FOR FINDING NUMBER OF DAYS BETWEEN ANY TWO DATES IN TWO CONSECUTIVE YEARS.*

	First Year.													Second Year.											
Day Mo.	Jan.	Feb.	March.	April.	May.	June.	July.	Aug.	Sept.	Oct.	Nov.	Dec.	Day Mo.	Jan.	Feb.	March.	April.	May.	June.	July.	Aug.	Sept.	Oct.	Nov.	**Dec.**
1	1	32	60	91	121	152	182	213	244	274	305	335	**1**	366	397	425	456	486	517	547	578	609	639	670	**700**
2	2	33	61	92	122	153	183	214	245	275	306	336	**2**	367	398	426	457	487	518	548	579	610	640	671	**701**
3	3	34	62	93	123	154	184	215	246	276	307	337	**3**	368	399	427	458	488	519	549	580	611	641	672	**702**
4	4	35	63	94	124	155	185	216	247	277	308	338	**4**	369	400	428	459	489	520	550	581	612	642	673	**703**
5	5	36	64	95	125	156	186	217	248	278	309	339	**5**	370	401	429	460	490	521	551	582	613	643	674	**704**
6	6	37	65	96	126	157	187	218	249	279	310	340	**6**	371	402	430	461	491	522	552	583	614	644	675	**705**
7	7	38	66	97	127	158	188	219	250	280	311	341	**7**	372	403	431	462	492	523	553	584	615	645	676	**706**
8	8	39	67	98	128	159	189	220	251	281	312	342	**8**	373	404	432	463	493	524	554	585	616	646	677	**707**
9	9	40	68	99	129	160	190	221	252	282	313	343	**9**	374	405	433	464	494	525	555	586	617	647	678	**708**
10	10	41	69	100	130	161	191	222	253	283	314	344	**10**	375	406	434	465	495	526	556	587	618	648	679	**709**
11	11	42	70	101	131	162	192	223	254	284	315	345	**11**	376	407	435	466	496	527	557	588	619	649	680	**710**
12	12	43	71	102	132	163	193	224	255	285	316	346	**12**	377	408	436	467	497	528	558	589	620	650	681	**711**
13	13	44	72	103	133	164	194	225	256	286	317	347	**13**	378	409	437	468	498	529	559	590	621	651	682	**712**
14	14	45	73	104	134	165	195	226	257	287	318	348	**14**	379	410	438	469	499	530	560	591	622	652	683	**713**
15	15	46	74	105	135	166	196	227	258	288	319	349	**15**	380	411	439	470	500	531	561	592	623	653	684	**714**
16	16	47	75	106	136	167	197	228	259	289	320	350	**16**	381	412	440	471	501	532	562	593	624	654	685	**715**
17	17	48	76	107	137	168	198	229	260	290	321	351	**17**	382	413	441	472	502	533	563	594	625	655	686	**716**
18	18	49	77	108	138	169	199	230	261	291	322	352	**18**	383	414	442	473	503	534	564	595	626	656	687	**717**
19	19	50	78	109	139	170	200	231	262	292	323	353	**19**	384	415	443	474	504	535	565	596	627	657	688	**718**
20	20	51	79	110	140	171	201	232	263	293	324	354	**20**	385	416	444	475	505	536	566	597	628	658	689	**719**
21	21	52	80	111	141	172	202	233	264	294	325	355	**21**	386	417	445	476	506	537	567	598	629	659	690	**720**
22	22	53	81	112	142	173	203	234	265	295	326	356	**22**	387	418	446	477	507	538	568	599	630	660	691	**721**
23	23	54	82	113	143	174	204	235	266	296	327	357	**23**	388	419	447	478	508	539	569	600	631	661	692	**722**
24	24	55	83	114	144	175	205	236	267	297	328	358	**24**	389	420	448	479	509	540	570	601	632	662	693	**723**
25	25	56	84	115	145	176	206	237	268	298	329	359	**25**	390	421	449	480	510	541	571	602	633	663	694	**724**
26	26	57	85	116	146	177	207	238	269	299	330	360	**26**	391	422	450	481	511	542	572	603	634	664	695	**725**
27	27	58	86	117	147	178	208	239	270	300	331	361	**27**	392	423	451	482	512	543	573	604	635	665	696	**726**
28	28	59	87	118	148	179	209	240	271	301	332	362	**28**	393	424	452	483	513	544	574	605	636	666	697	**727**
29	29	...	88	119	149	180	210	241	272	302	333	363	**29**	394	...	453	484	514	545	575	606	637	667	698	**728**
30	30	...	89	120	150	181	211	242	273	303	334	364	**30**	395	...	454	485	515	546	576	607	638	668	699	**729**
31	31	...	90	...	151	...	212	243	...	304	...	365	**31**	396	...	455	...	516	...	577	608	...	669	...	**730**

* Subtract the number opposite the first date from the **number opposite the last. If the 29th of February is included, add one day.**

(4) Discount is really interest based on the value at maturity for the terms of discount. Bank discount is calculated precisely the same as is interest:

Interest on \$750 @ 6% for 60 days............	\$7.50
Interest on \$750 @ 6% for 15 days............	1.875
Interest on \$750 @ 6% for 45 days............	\$5.625 or \$5.63

(5) Net Proceeds is the amount due after the discount has been deducted from the value at maturity. In this case, \$750 less \$5.63, or \$744.37 (Ans.)

Discounting an Interest-Bearing Note.—The only difference between discounting an interest-bearing note and a non-interest bearing note is in the value at maturity. In a non-interest bearing note, the value at maturity is face only, in an interest bearing note the value at maturity is the face plus interest for the full life of the note.

ILLUSTRATION: On April 15, the Jones Manufacturing Company received a 90-day note from one of its customers. The note was for \$1200 with interest at 6%. On May 1, the Jones Company discounted it at the bank. What is the Net Proceeds?

The five steps are as follows:

1. Value at Maturity: Interest on \$1200 for 90 days is \$18.00. The value at maturity is \$1200 plus \$18.00, \$1218.00.

2. Date of maturity:

April	15 more days
May	31 more days
June 30	30 more days
	76
July	14 due date
	90

3. Terms of discount:

By actual count

May 1–31	30 days
June	30
July	14
	74 days

By using Table 6

July 14	195
May 1	121
	74 days

4. Discount:

Interest on $1218 @ 6% for 60 days....	$12.18
Interest on $1218 @ 6% for 12 days....	2.436
Interest on $1218 @ 6% for 2 days....	.406
	$15.022 = $15.02

5. Proceeds:

Value at maturity.......	$1218.00	
Less Discount..........	15.02	
Net Proceeds.....	$1202.98	(Ans.)

Exact Interest.—Various financial organizations when dealing with each other and governments as a general rule use the exact or "accurate" method of calculating interest. In this method the 365 day year (in leap year 366) is used as the time basis rather than the 360-day year so-called bankers' time. When large financial transactions are involved, the slight five- or six-day inaccuracy of bankers' time makes a decided difference. The amount of interest is determined by finding the exact number of days that the obligation remained unpaid and then multiplying the principal by the exact number of days over 365 by the rate of interest expressed in fractional form. Cancellation may be applied if possible.

Illustration: The state and county taxes of the City of Willbum amounting to \$347,689 were due and payable on June 30. The city was unable to meet this obligation until October 1, at which time payment was made in full, plus accrued interest at the rate of 6%. Find: (*a*) the amount of exact interest on the obligation; (*b*) the amount of interest if it were calculated on the basis of bankers' time (360-day year); (*c*) which is greater, and by how much:

(*a*) The obligation was due June 30 and paid October 1. Using Table 6 it may be noted that the exact time between these two dates is

October 1	274
June 30	181
	93 days

The exact interest equals

$$\$347{,}689 \times \frac{93}{365} \times \frac{6}{100} = \$5315.36 \quad \text{(Ans.)}$$

(*b*) If bankers' time had been used, this would have been calculated as follows:

Interest @ 6% for 60 days.......	\$3476.89
Interest @ 6% for 30 days.......	1738.445
Interest @ 6% for 3 days.......	173.8445
	\$5389.1795 = \$5389.18 (Ans.)

(*c*) The difference in the interest figured by the two methods equals:

Interest calculated on Bankers' Time...........	\$5389.18
Interest calculated or Exact Time.............	5315.36
Interest calculated on the basis of Bankers' Time greater than interest calculated on basis of Exact Time by.......................	\$73.82 (Ans.)

Compound Interest.—Interest that is earned on other interest earned in previous periods and added to the principal is called compound interest. Interest may be compounded annually semi-annually, quarterly, or at even more frequent intervals.

ILLUSTRATION: A man deposits $500 on January 2, 1954, in a savings bank which pays interest at the rate of 4% per annum compounded quarterly. Assume that the quarters correspond with the calendar year and that interest is credited to accounts as of March 31, June 30, September 30, and December 31. If the account was allowed to stand for two years, how much would be on deposit at the end of that time? This would work out as follows:

	Principal	*Interest*	*Amount*	
January 1, 1954, Deposit..	$500.00	.....	....	
March 31, 1954...........	500.00	$5.00	$505.00	
June 30, 1954............	505.00	5.05	510.05	
September 30, 1954.......	510.05	5.10	515.15	
December 31, 1954........	515.15	5.15	520.30	
March 31, 1955...........	520.30	5.20	525.50	
June 30, 1955............	525.50	5.25	530.75	
September 30, 1955.......	530.75	5.30	536.05	
December 31, 1955.......	536.05	5.36	541.41	(Ans.)

Compound interest earned over period of two years equals $41.41.

Pre-computing Compound Interest.—At times, an individual is interested for one reason or another in knowing how much a given sum of money might build up to if left at interest for a period of years. This may be calculated by

(1) Adding the interest rate per interest period to $1.00 and multiplying this by itself as many times as there are interest periods in the whole term of years.

(2) Multiplying this product by the amount to be deposited in the first place, the original principal, to ascertain the new amount. Because such problems usually involve a large number

of interest periods, compound interest tables are generally used. Such tables give the amount that $1.00 will amount to at compound interest for any given number of periods at various periodic rates. Table 7 is a compound interest table. To use it, determine the number of interest periods, (*a*) follow down the left column until that figure is reached, (*b*) follow the line across to the column headed by the periodic rate, (*c*) multiply the number thus determined by the principal.

ILLUSTRATION: A man deposits $1200 in a bank which pays interest at the rate of 4% per annum compounded semi-annually. If the deposit is allowed to remain in the bank, how much will have accumulated at the end of 15 years?

(*a*) If interest is paid semi-annually at the rate of 4% per annum, the semi-annual rate or periodic rate is 2%.

(*b*) If interest is paid semi-annually, there are two interest periods per year. In fifteen years, there are thirty interest periods.

(*c*) Turning to Table 7, it will be noted that interest on $1.00 compounded for 30 periods at 2% = 1.81134.

(*d*) Principal $1200.00
× 1.81134

$2173.60800 = $2173.61 (Ans.)

Calculated by Logarithms.—Compound interest may also be computed by using logarithms. This method is frequently used when compound interest tables are not available or when the periodic interest rate is now shown in tables that are available.

The formula followed when using logarithms is:

Sum = Amount Deposited × (1 + periodic interest rate)
number of periods

That is: $S = x(1 + i)^n$

TABLE 7

Compound Interest Table

Amount of $1 at compound interest for periods 1 to 50 at various *periodic rates.

Periods. n.	*Periodic Rate 2%	3%	3½%	4%	4½%	5%	6%	7%
1	1.02000	1.03000	1.03500	1.04000	1.04500	1.05000	1.06000	1.07000
2	1.04040	1.06090	1.07123	1.08160	1.09203	1.10250	1.12360	1.14490
3	1.06121	1.09273	1.10872	1.12486	1.14117	1.15763	1.19102	1.22504
4	1.08243	1.12551	1.14752	1.16986	1.19252	1.21551	1.26248	1.31080
5	1.10408	1.15927	1.18769	1.21665	1.24618	1.27628	1.33823	1.40255
6	1.12616	1.19405	1.22926	1.26532	1.30226	1.34010	1.41852	1.50073
7	1.14869	1.22987	1.27228	1.31593	1.36086	1.40710	1.50363	1.60578
8	1.17166	1.26677	1.31681	1.36857	1.42210	1.47746	1.59385	1.71819
9	1.19509	1.30477	1.36290	1.42331	1.48610	1.55133	1.68948	1.83846
10	1.21899	1.34392	1.41060	1.48024	1.55297	1.62889	1.79085	1.96715
11	1.24337	1.38423	1.45997	1.53945	1.62285	1.71034	1.89830	2.10485
12	1.26824	1.42576	1.51107	1.60103	1.69588	1.79586	2.01220	2.25219
13	1.29361	1.46853	1.56396	1.66507	1.77220	1.88565	2.13293	2.40985
14	1.31948	1.51259	1.61870	1.73168	1.85194	1.97993	2.26090	2.57853
15	1.34587	1.55797	1.67535	1.80094	1.93528	2.07893	2.39656	2.75903
16	1.37279	1.60471	1.73399	1.87298	2.02237	2.18287	2.54035	2.95216
17	1.40024	1.65285	1.79468	1.94790	2.11338	2.29202	2.69277	3.15882
18	1.42825	1.70243	1.85749	2.02582	2.20848	2.40662	2.85434	3.37993
19	1.45681	1.75351	1.92250	2.10685	2.30786	2.52695	3.02560	3.61653
20	1.48595	1.80611	1.98979	2.19112	2.41171	2.65330	3.20714	3.86968
21	1.51567	1.86029	2.05943	2.27876	2.52024	2.78596	3.39957	4.14057
22	1.54598	1.91610	2.13151	2.36991	2.63365	2.92523	3.60354	4.43041
23	1.57690	1.97358	2.20611	2.46471	2.75217	3.07152	3.81976	4.74054
24	1.60844	2.03279	2.28332	2.56330	2.87602	3.22510	4.04894	5.07237
25	1.64061	2.09378	2.36324	2.66583	3.00544	3.38635	4.29188	5.42744
26	1.67342	2.15659	2.44595	2.77246	3.14068	3.55567	4.54939	5.80736
27	1.70689	2.22129	2.53156	2.88336	3.28201	3.73346	4.82224	6.21388
28	1.74103	2.28792	2.62016	2.99870	3.42970	3.92013	5.11170	6.64885
29	1.77585	2.35656	2.71187	3.11864	3.58406	4.11614	5.41840	7.11427
30	1.81134	2.42726	2.80672	3.24339	3.74532	4.32194	5.74351	7.61227
31	1.84759	2.50008	2.90501	3.37312	3.91386	4.53804	6.08812	8.14513
32	1.88454	2.57508	3.00670	3.50805	4.08998	4.76494	6.45340	8.71529
33	1.92224	2.65233	3.11193	3.64837	4.27403	5.00319	6.84061	9.32536
34	1.96068	2.73190	3.22085	3.79430	4.46637	5.25335	7.25115	9.97813
35	1.99989	2.81386	3.33358	3.94608	4.66735	5.51600	7.68611	10.6766
36	2.03989	2.89827	3.45025	4.10392	4.87738	5.79182	8.14728	11.4240
37	2.08069	2.98518	3.57101	4.26806	5.09686	6.08141	8.63611	12.2236
38	2.12230	3.07478	3.69599	4.43880	5.32618	6.38548	9.15428	13.0793
39	2.16475	3.16702	3.82535	4.61635	5.56590	6.70475	9.70354	13.9948
40	2.20801	3.26203	3.95924	4.80100	5.81637	7.03999	10.2855	14.9745
41	2.25221	3.35989	4.09781	4.99306	6.07811	7.39199	10.9029	16.0227
42	2.29725	3.46069	4.24124	5.19276	6.35162	7.76159	11.5571	17.1443
43	2.34320	3.56451	4.38968	5.40047	6.63744	8.14967	12.2505	18.3444
44	2.39006	3.67144	4.54332	5.61649	6.93613	8.55715	12.9855	19.6285
45	2.43786	3.78159	4.70233	5.84115	7.24826	8.98504	13.7647	21.0025
46	2.48662	3.89503	4.86692	6.07480	7.57443	9.43426	14.5906	22.4727
47	2.53635	4.01188	5.03726	6.31779	7.91528	9.90597	15.4660	24.0458
48	2.58708	4.13224	5.21356	6.57050	8.27146	10.4013	16.3939	25.7290
49	2.63882	4.25621	5.39604	6.83330	8.64368	10.9213	17.3776	27.5300
50	2.69160	4.38389	5.58491	7.10665	9.03265	11.4674	18.4202	29.4571

* Periods may be annual, semi-annual or quarterly, etc. Periodic rates are proportioned to the length of the period. Thus, 4% annual = 2% semi-annual rate.

ILLUSTRATION: $642.80 was to be left on deposit for 12 years at a bank paying interest at the rate of 3½% compounded semi-annually. What amount will be on deposit at the end of 12 years?

Interest for 12 years at 3½% compounded semi-annually means that there will be 24 interest periods at the rate of 1¾% per period; therefore, the amount at maturity (S) will be

$$S = 642.80 \times (1.0175)^{24} = \log 642.80 + 24 \times \log 1.0175$$

$$\begin{aligned} \log 642.80 &= 3.808076 \\ 12 \log \quad 1.0175 &= 0.180840 \\ \hline \log S &= 3.988916 \end{aligned} \qquad \begin{aligned} \log 1.0175 = \quad & 0.007535 \\ & \times \quad 24 \\ \hline & 0.030140 \\ & 0.015070 \\ \hline & 0.180840 \end{aligned}$$

$$S = \$974.80 \quad \text{(Ans.)}$$

Interest on Bank Deposits.—It should be noted that there probably would be a slight discrepancy between the amount as worked out in the preceding solution and the amount as built up by the bank over the years. This would be due to the fact that banks usually ignore cents in the principal in calculating the interest at the end of each period.

Some other factors pertaining to interest in bank deposits that might be noted here are:

(1) While most interest is earned in savings accounts only, some banks pay interest on checking accounts. This practice varies widely, it usually being paid only when a reasonably good daily balance is maintained varying in different banks from $500 to $5000. The rate is usually 4% per annum.

(2) Savings banks usually have rules whereby money deposited on or before a specified day in the month, as the 5th or 10th, shall draw interest from the first of the month. Deposits made after that date will draw interest from the first of the following month.

(3) Money usually has to be on deposit for a minimum of three months before any interest is credited. If withdrawals are

made during an interest period, the withdrawal is usually deducted from money on deposit at the beginning of the period, and no interest is paid on such funds.

ILLUSTRATION: A man withdraws $1000 from a savings account 15 days before the end of the interest period. How much interest does he lose?

He loses all interest accrued on this sum for 2½ months—about $8.33 if the rate is 4% per annum compounded quarterly.

This and various other restrictive rules tend to reduce the actual rate of interest paid, especially if one makes deposits and withdrawals with any degree of frequency.

Some circumvent the above loss of interest by borrowing from the bank, using the savings account for security for the time that must elapse between the day the money is needed and the day the interest is due to be credited.

ILLUSTRATION: If the man mentioned in the previous illustration had followed this practice, how much of his interest would he have saved?

Interest on $1000 @ 4% for 3 months.....	$10.00	
Interest on $1000 @ 6% for 15 days......	2.50	
Net interest saved..............	$ 7.50	(Ans.)

Service Charges.—Many banks now make a charge for servicing checking accounts when an adequate balance is not maintained by the depositor. Here again practice varies in different banks, the balance to be maintained varying from $50 to $500 and the service charge ranging from 50¢ to $2.00. Some banks charge so much a check. Others permit the depositor to draw a minimum number of checks without making a charge, while still others use combinations of these various conditions.

Profit and Loss

Almost all business is organized for the purpose of making a profit. The profit (or the loss) for a fiscal period is usually shown in a statement prepared by the bookkeeper or accountant which is known as a Profit and Loss Statement. While the form of this statement will vary somewhat in terms of the specific business for which it is drawn up, it will fundamentally include sections which will set forth some analysis of (1) operating income, (2) operating costs, (3) non-operating income, and (4) non-operating cost. The net result of the statement will be the net profit for the period in question. It might be well to point out that the terms "income," "profit," "revenue," and "earnings" are used more or less synonymously by accountants in the preparation of profit and loss statements and that the term "fiscal period" means a financial period of any length of time. A few firms prefer to calculate their profit every week. Many calculate it once a month. Some use an arbitrarily adopted financial period of 4 or 5 weeks. Others use a fiscal period of 2 months, 3 months, 6 months, or a year.

Frequency in calculating profits or losses is a great aid to proper management. As a basic rule, profits or losses should be calculated as frequently as is commensurate with the value of such calculations to the management with due consideration given to the cost involved. In addition to having profits and losses calculated at frequent intervals, most firms have a definite summary of their financial affairs prepared at the end of their fiscal year and on the basis of this report they pay income taxes, divide profits, and make plans for the future. The fiscal year is a twelve-month period and may or may not coincide with the calendar year. Because of income tax and other reports that must be made, many firms have their fiscal year coincide with the calender year, but many others prefer to have the fiscal year end at a dull season when final inventory and other work necessary at the close of a fiscal year may be performed with the least possible disturbance to the business. The following is a profit and loss statement of a retail grocery store for the month ending January 31, 19—.

EDWIN S. HELLER

Profit and Loss Statement for Period Extending from January 1 to January 31, 19—

Income from Sales—				
Sales			$24,276.50	
Less Returns & Allowances			341.25	$23,935.25
Cost of Goods Sold—				
Mdse Inventory Jan. 1		$6,842.67		
Purchases	$18,482.20			
Less Ret. & All.	331.61			
Frgt. & Cartage In		18,150.59		
Less Inventory Jan. 31		141.17	$25,144.43	
			6,497.60	
Net Cost of Goods Sold				18,646.83
Gross Profit				$5,288.42
Operating Expenses—				
Selling Expenses—				
Salaries of Sales Force		$1,575.00		
Advertising		360.00		
Store Supplies		175.65		
Rent of Store		400.00		
Delivery Expenses		640.75		
Insurance on Stock		45.15		
Taxes		15.65		
Light, Heat & Power		75.20		
Repairs to Store Equipment		41.20		
Depr. on Store Equipment		27.49		
Depr. on Delivery Equipment		18.20		
Total Selling Expenses			$3,374.29	
General Administrative Expenses—				
Management & Off. Salaries		525.00		
Office Supplies & Postage		162.50		
Rent of Office		100.00		
Depr. on Office Equipment		15.20		
Total Adm. Exp.			802.70	
Total Operating Expenses				4,176.99
				$1,111.43
Add: Other Income:				
Discount on Purchases			$201.76	
Interest on Notes Receivable			22.16	
Interest on Bank Deposits			14.20	
Total Extraneous Income				238.12
Total Income				$1,349.55
Deduct: Other Costs:				
Discount on Sales			$321.60	
Interest on Notes Payable			41.16	
Total Extraneous Cost				362.76
Net Profit				$986.79

Percentage of Profit.—When talking about the percentage of profit, one must be sure to know what is being used as a base. If a man buys an article for \$100 and sells it for \$150, it is obvious that he made a profit of \$50, but what was the percentage of profit? There is much controversy as to what should be used as the base, the cost or the selling price. If we use the cost, \$100, we would immediately determine that the rate of profit was 50% ($\frac{50}{100}$). If we use the selling price as a base, we then would find that the rate of profit is $33\frac{1}{3}$ ($\frac{50}{150}$). Technically, the use of the selling price as basis for calculating profits is not correct because the selling price includes profit which will cause the base to vary. On the other hand, however, the selling price as above affords the business man an opportunity to calculate not only gross and net profits, but also many other relationships on the same base.

ILLUSTRATION: The Profit and Loss Statement of the business of Edwin S. Heller is shown above. (*a*) What percent of the sales represents Net Profit? (*b*) Cost of Goods Sold? (*c*) Gross Profit? (*d*) Operating expense? (*e*) Operating Profit? (*f*) Non Operating Income? (*g*) Non-Operating Cost?

Each of these percentages will be determined by using the net sales as a base (letting it equal 100%) and dividing it into the item in question. Thus the percent of (*a*) net profit based on the sales equals

$$\$986.79 \div \$23,935.25 = 4.12\% \quad \text{(Ans.)}$$

All the other percentages in question are determined in the same way. Thus we find that

(*b*) Cost of Goods Sold equals	77.91% of sales	(\$18,646.83 ÷ \$23,935.25)
(*c*) Gross Profit equals	22.09% of sales	(5,288.42 ÷ 23,935.25)
(*d*) Operating Expense equals	17.45% of sales	(4,176.99 ÷ 23,935.25)
(*e*) Operating Profit equals	4.64% of sales	(1,111.43 ÷ 23,935.25)
(*f*) Non-operating Income equals	0.99% of sales	(238.12 ÷ 23,935.25)
(*g*) Non-operating Cost equals	1.52% of sales	(362.76 ÷ 23,935.25)

Price Fixing.—In determining the price at which a commodity may be sold, the business man must keep in mind the cost to produce or procure that commodity, the cost of doing business, and a fair margin of profit.

Experience will usually show a man approximately what these percentages are and he may guide himself accordingly. If he finds that, for example, 28¢ of every dollar of sales must be used to pay the running expenses of the business and that 2¢ of every dollar of sales must be used to give him a fair return on his investment, he knows that 30¢ of every sales dollar must represent gross profit. He, therefore, in setting his selling price will let the cost of the article represent 70% of his selling price.

ILLUSTRATION: A shoe retailer can buy shoes at $2.45 per pair from the manufacturer and he must make a gross profit of 30% on the selling price. At what price should he sell the shoes?

The Cost $2.45 = 70\%$ or $\frac{7}{10}$ of the selling price
$.35 = \frac{1}{10}$ of the selling price
$3.50 = \frac{10}{10}$ or 100% of the selling price (Ans.)

In some lines of business it is possible to follow this rule and apply it to all commodities sold. However, a number of factors will frequently require the business man to vary this procedure. Competition in some lines may require him to cut his margin of gross profit, while the very nature of other lines may permit him to charge more.

If several lines of commodities are carried, as in a department store, the cost of operating each department should be calculated and price ratios adjusted accordingly. Fast moving commodities in departments which do not cost much to operate may be sold at a relatively low margin of gross profit, while slower moving commodities in more expensive departments will have to be sold at a higher margin of gross profit. Thus groceries may conceivably be sold at a mark-up of 15 to 25%, while furniture may require a mark-up of 30 to 40%.

Leaders.—Many business organizations, particularly retail stores, sell certain articles at cost or even below cost in order to attract customers with the hope that once these customers buy that particular article, they will also purchase some other regularly priced commodities. Such articles are called "leaders" and their prices are fixed in terms of their cost, price asked at other places, and the probability of a given price attracting profitable customers.

Need for Records.—Records of sales and cost of sales should be carefully kept in order that a business man may know how the business is progressing. Too frequently, the inclination is to watch the volume of sales and not pay enough attention to the cost. Carefully kept records will frequently assist in the adjustment of costs and selling prices so that business may be done most profitably and at the same time competition will be adequately met.

Price Marking.—Most stores find it advisable to mark the selling price of each article on the article itself or on a tag or label attached to the article. This reduces the number of errors in quoting prices to customers, it means that sales people do not have to depend so much upon their memories, and it makes it possible to shift sales people from one counter to another without fear that they will sell goods at incorrect prices.

Very often the tag or label contains not only the selling price but also the cost price, the latter usually in code. Such a procedure facilitates the work at inventory time and at the same time keeps the cost a secret from both the customer and the sales person. It also makes it possible for the manager to adjust intelligently prices downward on a commodity that is moving slowly.

The code used for marking the cost price usually consists of a word or group of words which contain ten different letters, each representing the figures from zero to nine. So that the secrecy of the code may be more completely preserved, extra letters such as x or y are usually used to represent digits that are repeated one or more times in the price. "Brown Chest," or "White Cloud," are words that may be used as codes.

ILLUSTRATION: What are two word groups that may be used as codes?

B R O W N	C H E S T	or	W H I T E	C L O U D
1 2 3 4 5	6 7 8 9 0		1 2 3 4 5	6 7 8 9 0

They may also be used in reverse.

ILLUSTRATION: A retailer bought shoes at \$3.30 per pair and had an established mark-up of 25% based on the selling price. How would the price tag read if "Brown Chest" with x as a repeater were used as a code for the cost?

\$3.30 equals 75% or $\frac{3}{4}$ of the selling price. Then the selling price will be \$4.40 per pair. The price tag would read as follows:

Cost	*oxt*	
Selling Price	4.40	(Ans.)

The words cost and selling price do not usually appear on the tag, the code for the cost price usually appearing above the line and the selling price listed below the line. The selling price may also be coded, but this is usually not done because there is no particular need for secrecy and there are fewer chances for error if the price is plainly marked.

Selling Price Based on Cost.—Some firms still base their percentage of mark-up on cost rather than selling price. When this is done, the percentage of mark-up is determined by noting the percentage the gross profit bears or must bear to the cost of goods sold. If this established percentage must be $33\frac{1}{3}\%$, then that percent of the cost is calculated and added to it to determine the selling price.

ILLUSTRATION: A hat costs \$1.65 and the mark-up is $33\frac{1}{3}\%$ based on the cost. What is the selling price:

$$
\begin{aligned}
\$1.65 &= 100\% \text{ cost} \\
.55 &= 33\tfrac{1}{3}\% \text{ of cost (gross profit)} \\
\hline
\$2.20 &= 133\tfrac{1}{3}\% \text{ of cost (selling price)} \quad \text{(Ans.)}
\end{aligned}
$$

Odd Figures.—Many stores prefer not to quote prices at odd figures. To take care of this problem, they frequently make a rule that articles will be priced at the next figure divisible by five or ten above the one actually determined by calculations.

ILLUSTRATION: A store established a rule that prices should be fixed at the next figure divisible by 5 or 10 above the one actually determined by calculation. At what price will the following goods be marked?

Unit Cost	*Mark-Up Based on Selling Price*
0.47	25%
2.25	35%
6.48	33⅓%

Unit Cost	*Mark-Up Based on Selling Price*	*Mark-Up*	*Actually Calculated Selling Price*	*Fixed Price 5 and 10 Rule*
0.47	25%	0.16	0.63	0.65
2.25	30%	0.66	2.91	2.95
6.48	33⅓%	3.24	9.72	9.75

Instead of using figures divisible by five or ten, business organitions sometimes use figures that are supposed to have a good psychological effect on the buying public such as 39¢, 49¢, 69¢, 98¢, etc. The calculations are made the same but the special price scale is applied.

Manufacturing Cost.—Manufacturing Costs are usually divided into three major items, (1) raw materials, (2) direct labor, and (3) expenses applied to production called overhead burden, or indirect costs. This last item would include expenses of supervision, light, heat, power, depreciation, factory supplies, taxes, rentals, etc.

In preparing a statement showing the cost of goods manufactured, the problem is relatively simple. One simply lists from the bookkeeping records the cost of all materials used in production, the cost of all labor directly applied to production, and the indirect costs such as those listed. This information, along with the proper adding in of old inventories of goods in process, and deducting new inventories, will give one the cost of goods manufactured for a given period.

ILLUSTRATION: Make up a statement showing the cost of goods manufactured by the Warren Shoe Company during the month of June, 19—.

WARREN SHOE COMPANY

Cost of Goods Manufactured June 1–June 30, 19—

Materials—		
Upper leather used	$4561.75	
Sole leather used	1321.73	
Lining material used	298.21	
Findings material used	327.62	
Cost of raw materials used		$6,509.31
Direct Labor		5,981.27
		$12,490.58
Manufacturing Expenses—		
Salaries and wages	$1327.61	
Rent	350.00	
Rentals and Royalties	157.62	
Depreciation on Lasts, Dies & Patterns	275.62	
Light, heat, & power	76.21	
Taxes	27.25	
Depreciation on Machine Equipment	42.57	
Total Manufacturing Expenses		$2,256.88
Total Cost of Manufacturing		$14,747.46
Add: Goods in Process, Inv. June 1		2,321.65
		$17,069.11
Deduct: Goods in Process, Inv. June 30		2,576.21
Total Cost of Goods Manufactured		$14,492.90

Estimating Cost.—The real problem in dealing with manufacturing cost is not that of looking back over records to find what goods did cost, but rather looking ahead and estimating what they are going to cost. Every manufacturer has to quote prices, frequently in advance of actually making the goods, and the price he quotes must be low enough to help him to compete favorably with other manufacturers and at the same time be high enough to cover the cost of the goods along with giving him a fair margin of profit.

Estimating the cost of materials and the cost of direct labor is relatively simple. A manufacturer can usually tell about how much the material going into a product will cost, and about how much the labor directly applied to the product will cost. The allowance for overhead, however, is quite a different problem because the volume of production causes the cost of producing any particular unit to vary. Overhead costs (rent, superintendence, depreciation, etc.) are about the same whether the factory is almost idle or running at capacity production, and will jump up perceptibly only when it is necessary to enlarge quarters, add to equipment, etc.

There are various ways of estimating the overhead to be added in as part of the estimated cost of a unit. One very popular method is that of determining by experience that ratio that has existed in the past between the prime cost (raw materials plus direct labor) and the factory expenses. By referring to the statement of the cost of goods manufactured by Warren Shoe Company shown previously, you will notice that this ratio is about one to six; in other words, the manufacturing expenses amount to a figure that is about one-sixth of the prime cost or about one-seventh of the total cost of manufacturing. If experience has shown that approximately this ratio has existed each month, it may be used as the standard and may be applied when estimating the cost of goods to be produced.

ILLUSTRATION: A manufacturer desires to fix a selling price on shoes he is planning to make. The raw materials going into the shoes (upper leather, sole leather, trimmings, linings, findings, etc.) are estimated to cost \$1.40 per pair. The direct labor required on the shoe (cutting, stitching, stock fitting, lasting, etc.) is estimated to cost \$1.25 per pair. $16\frac{2}{3}\%$ of the prime cost has been established as the standard factory overhead charge. In addition, a standard mark-up of 20% based on the selling price is applied to cover the cost of selling, office administration and other general overhead costs. What is the cost to manufacture the shoes, and at what may they be sold?

Raw materials	$1.40
Direct labor	1.25
Prime Cost	$2.65
Factory overhead ($16\frac{2}{3}\%$ of $2.65)	0.44
Cost to Manufacture	$3.09
Mark-up to cover general overhead (20% of Selling Price)	0.77
Calculated Selling Price	$3.86

NOTE: This price of $3.86 would probably be rounded off to $3.90 or $3.95 or if competition was particularly keen, it might be fixed at $3.85.

There is real danger in too much dependence on overhead standard rates that have been established solely on the basis of experience. Instead of accepting the figures as such, one should look behind the figures to determine why such a ratio exists, if it can be justified, and what improvements can be made to lower the relative cost of overhead. Are the factory costs too high? Can efficiency methods be adopted that will tend to reduce these costs or speed up production without necessitating expansion of the plant? These and many similar questions should be carefully thought of before one adheres too closely to overhead ratios and percentages established solely on the experiences of the past.

COMMISSIONS AND BROKERAGE

Agents are frequently used by growers and manufacturers who for some reason or other do not choose to undertake to market some or all of their goods themselves; or such agents are used when people desiring to procure certain merchandise find it inconvenient for them to do the buying themselves. These agents or factors are usually called *commission merchants,* their commission usually being a certain percent of the selling or buying price, or sometimes a flat rate per unit (bu, bbl, bale, ton, etc.) bought or sold.

When using the services of a commission merchant to market is goods, the grower or manufacturer simply consigns the goods) the merchant who receives them, pays any unpaid freight harges, has them hauled to his place of business, frequently isures them and pays other expenses incidental to handling them, nd sells them at the best price he can get. Sometimes the selling of the direct sale type where the agent contacts his customer r vice-versa, and a sale is consummated if the price and terms are greeable to both; in other lines, the goods are sold at auction o the highest bidder.

When the goods are finally sold, the commission merchant enders an "Account Sales" upon which he lists the number f units sold at given prices and these are extended and totalled, he total thus determined is called the gross proceeds.

Also on the Account Sales are listed the various incidentally ncurred expenses along with the commission which is usually 8 o 10 percent of the gross proceeds. The total of these charges is leducted from the gross proceeds to determine the net proceeds. The amount of the net proceeds is usually remitted with the ccount sales.

Illustration: A commission merchant receives a shipment of 50 cases of eggs, each case containing 30 dozen. He sold 40 cases (1200 dozen) at 18¢ per dozen and the remaining 10 cases (300 dozen) at 17¢ per dozen. He paid freight and cartage $15.27 and insurance $2.32. Commission was charged at the rate of 10% on the gross sales. How would the Account Sales appear?

The Account Sales would appear as follows:

40 cases Eggs, 1200 dozen @ 18¢ per dozen		$216.00
10 cases Eggs, 300 dozen @ 17¢ per dozen		51.00
		$267.00
Charges—		
Freight	$15.27	
Insurance	2.32	
Commission, 10%	26.70	44.29
Net Proceeds		$222.71

Southern Specialty Fruit
Produce

JAMES WILLIAMS
COMMISSION MERCHANT
2 WASHINGTON STREET

Shipping No. 39

Sold for Account of

NEW YORK, Dec. 21, 19

William Adams
Morgantown
West Virginia

References:
CHASE FRANKLIN NATIONAL BANK
HANOVER CENTRAL BANK & TRUST CO., of N. Y.

4 20		50 Yams	3	1.40	4 20	
			3	1.35	4 05	
			38	1.25	47 50	
			6 Lost Repacking			
			50			55 75
		Icing				
		Loading				
		Assorting				
		Express				
		Freight				
		Cartage				
		Commission		5 58		5 58
E. & O. E.		Net Proceeds				50 17
		TO AVOID ERRORS AND DELAYS, ALWAYS MAIL US INVOICE OF WHAT YOU SHIP				

Fig. 6.—Account Sales.

Account Purchase.—When a commission merchant is commissioned to buy merchandise for a client, the procedure is just the reverse. He buys it, sometimes at auction, sometimes through private purchase, and pays whatever expenses are necessary, such as insurance, freight, etc., in transferring the goods to the principal.

The report of the purchase is called an "Account Purchase" and lists the number or articles or units bought with the unit price paid, the extension and total, known as the "Prime Cost." To the prime cost are added the various costs involved in making the purchase including the commission. This final amount is called the "gross cost."

ILLUSTRATION: A commission merchant buys 1000 lbs. of raw silk at $1.39 per pound for a client, pays freight and cartage $27.62, and charges a 5% commission. How would the account purchases appear?

The account purchases would appear as follows:

July 22—1000# Raw Silk, $1.39		$1390.00
Prime Cost		$1390.00
Charges—		
Freight and Drayage	$27.62	
Commission	69.50	
Total Charges		97.12
		$1487.12

Salesmen's Commissions and Bonus.—There are a variety of systems used in paying salesmen, perhaps the most common of which are: (*a*) straight salary, (*b*) salary plus commission on all sales, (*c*) salary plus commission on certain items or groups of items, (*d*) salary plus commission on sales above a certain predetermined quota, (*e*) straight commission.

Straight Salary.—Many firms pay their salesmen on a straight salary basis feeling that their salesmen will work well without special commission or bonus incentives. This plan almost entirely eliminates a certain ruthless or high pressure type of salesmenship that so frequently destroys good will. The salary is almost always reasonably substantial. Sales work is looked upon as being the life-blood of any industry and successful salesmen are usually well paid.

Salary Plus Commission on All Sales.—Some firms prefer to have the salesmen have an opportunity to earn more if they can sell more, but at the same time like to give them the security of a regular salary regardless of business conditions. The salary as such is usually relatively small, set on what might be called a subsistence level and the rate of commission is set so that with normal effort a man should be able to earn a fairly substantial income.

ILLUSTRATION: A man receives a salary of $100 and a 5% commission on all sales. His sales for the month of October were $5000. What are his earnings for the month:

Salary....................	$100.00	
Commission, 5% of $5000...	250.00	
	$350.00	(Ans.)

Salary Plus Commission on Certain Items.—Some firms have a fundamental policy of paying a straight salary but use the commission as a special incentive to have salesmen sell certain items or groups of items. This may be and frequently is only a temporary arrangement and is used to move a special lot of slow-moving merchandise, to introduce a new item or line, to make salesmen "selling conscious" of articles that they have been neglecting or for other reasons.

ILLUSTRATION: A paint company noticed that its line of lacquers was not selling well and offered its salesmen a special commission of 5% on all sales in that line. One salesman whose salary was $275 per month sold $425.00 worth of lacquers during the month of May. What was his earning during the month?

Salary........................	$275.00
Commission, 5% of $425.00......	21.25
	$296.25

Salary Plus Commission.—A salesman's quota may be set in various ways, but is usually determined by experience in the past. One favorite method for establishing a quota is to average the three or four best months that the salesman is paid a commission (sometimes called bonus in such cases) on all sales above the established figure.

ILLUSTRATION: A company pays a commission (or bonus) of 1% on all sales above the salesmen's quota. The quota is established by averaging the total sales made by each salesman in the best four months that each had in the preceding year. A salesman whose best months in the preceding year were January $18,750.00, March $17,925.00, September $19,256.00, and December $16,225.00, who was paid a monthly salary of $325.00, made sales totalling $19,475.00 in a given month. What was his total earnings?

Quota equals $18,750.00
17,925.00
19,256.00
16,225.00

$72,156.00 ÷ 4 = $18,039.00, or, in round numbers, $18,000.00, established monthly quota for new year.

Earnings for the Month:

Salary....................................	$325.00
Commission of 1% on ($19,475 − $18,000).....	14.75
	$339.75 (Ans.)

Straight Commission.—Under this plan, the salesman receives no salary as such but is entirely dependent upon his commission. A straight commission usually means that a flat rate of commission is paid on all articles sold. In some cases, however, a difference of commission is paid on different lines of goods sold by the firm.

ILLUSTRATION: A firm handles office equipment and supplies. Salesmen are paid no salaries but receive a commission of 10% on all equipment sold and 15% on all supplies sold. In one month, a salesman sold equipment totalling \$2257.65 and supplies totalling \$926.18. What were his monthly earnings?

Commission on equipment	= 10% of \$2257.65	\$225.77
Commission on supplies	= 15% of 926.18	138.93
		\$364.70 (Ans.)

PAYROLLS

A list of employees and the amount to be paid to each for a specific time is called a *payroll.* When pay is calculated on a time basis, the number of hours worked and the rate per hour is usually included. When the employee is paid on a piece-work basis, the number of pieces completed and the rate per piece is frequently included. The total of the payroll is the amount to be paid to all employees for the time specified, which is usually a week but may be a longer period. When the total is determined, a check is made out payable to the order of payroll. If the company pays each employee by check, the payroll check which covers the whole payroll is usually deposited in a special payroll account maintained at the bank for the purpose and special individual checks for each employee are drawn against this particular account and are distributed to the employees.

If the company pays each employee in cash, it is necessary to prepare a currency memorandum in order to know just how many bills and various coins will be needed and from it a Payroll Currency slip which is taken to the bank with the payroll check so that the bank will know how many bills and coins of various denominations to give to the paymaster or his representative when cashing the check. This currency and change is then distributed among the various pay employees and these in turn are passed out at a given time to the employees. Most companies usually require a receipt from the employee when he is given his pay envelope.

Time Basis.—Many firms pay on a time basis which is usually fixed at so much per hour for so many hours per day and so many days per week. A very common time schedule is 7 hours per day and 5 days per week, making a total of 35 hours per week. While

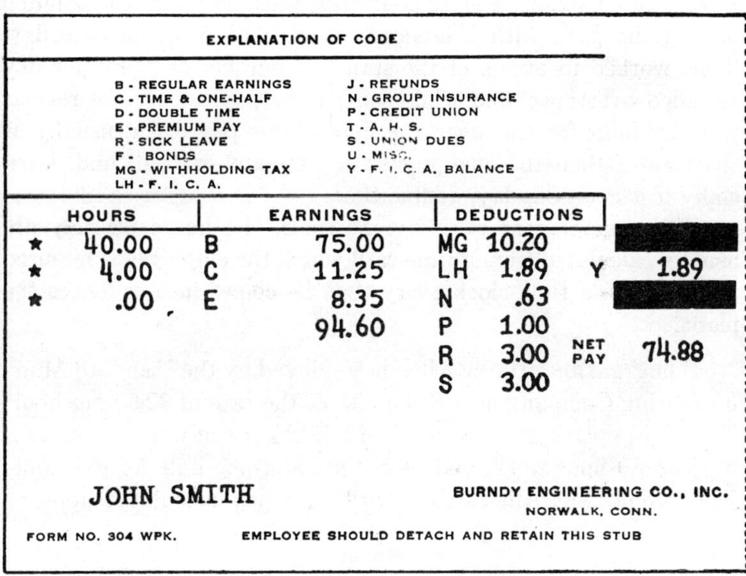

EXPLANATION OF CODE

B - REGULAR EARNINGS
C - TIME & ONE-HALF
D - DOUBLE TIME
E - PREMIUM PAY
R - SICK LEAVE
F - BONDS
MG - WITHHOLDING TAX
LH - F. I. C. A.
J - REFUNDS
N - GROUP INSURANCE
P - CREDIT UNION
T - A. H. S.
S - UNION DUES
U - MISC.
Y - F. I. C. A. BALANCE

HOURS		EARNINGS	DEDUCTIONS			
★ 40.00	B	75.00	MG	10.20		
★ 4.00	C	11.25	LH	1.89	Y	1.89
★ .00	E	8.35	N	.63		
		94.60	P	1.00		
			R	3.00	NET PAY	74.88
			S	3.00		

JOHN SMITH

BURNDY ENGINEERING CO., INC.
NORWALK, CONN.

FORM NO. 304 WPK. EMPLOYEE SHOULD DETACH AND RETAIN THIS STUB

FIG. 7.—A Typical Check or Payroll Envelope Stub Showing a Detailed Statement of Income and Deductions for an Employee in a Medium-sized Plant. Most of the code letters are self-explanatory. "Premium pay" (E) is a bonus paid the employee for exceeding the production quota established for his job. The amount varies, depending upon how much the quota is exceeded. "F.I.C.A." (LH) is Federal Insurance Contributions Act or Federal Old-Age and Survivors Insurance or the deduction for "Social Security." "AHS" (T) stands for Associated Hospital Service ("Blue Cross," "Blue Shield," etc.).

some firms have a standard week of more than 35 hours, many others, particularly during recent years, have tended to reduce the number of hours per week so that now we find plants operating on a basis of 30, 32, 36, and 40 hours per week, variously distributed among the days with a tendency toward no work on Saturday. Time worked in excess of the standard number of hours per day is called overtime; and while some firms pay merely the regular rate per hour for this overtime, most firms pay extra, usually at the rate of time-and-one-quarter or time-and-one-half and, especially if it is on Sunday, double time.

Where employees work by the hour, the hours per day are usually calculated from a time card which the employee is required to punch on a time clock every time he comes into or leaves the plant.

ILLUSTRATION: Harry Allen is employed by the Standard Manufacturing Company as a machinist at the rate of $2.00 per hour. The plant operates on the basis of 8 hours per day for five days a week, a 40-hour week, and pays time-and-one-half for overtime. How would Mr. Allen's time card with pay calculated appear?

No. 91

HARRY ALLEN

Employed as Machinist at 2.00 per hour, Week Ending Tuesday, May 21, 19—

Day	A.M.		P.M.		Overtime	Total Hours
	In	Out	In	Out		
Wednesday...	7:55	12:05	1:00	5:04	......	8
Thursday.....	7:56	12:01	12:55	5:02	......	8
Friday........	7:59	12:10	12:57	5:10	5:30–8:30	11
Saturday.....						
Sunday......						
Monday......	7:51	12:05	12:58	5:03	......	8
Tuesday......	8:00	12:02	12:55	5:07	......	8

Payroll credit:

Regular 40 hours at \$2.00 per hour	\$80.00
Overtime 3 hours at \$3.00 per hour	9.00
Total pay	\$89.00

The regular payroll for a company operating on a time basis is simply a summary of these individual time cards.

ILLUSTRATION: The Standard Manufacturing Company had working for it during the week ending May 21, 19— the employees listed below along with the hours per day each worked and his hourly rate. The company pays time-and-one-half for overtime. What is the total payroll for the week?

STANDARD MANUFACTURING COMPANY

Payroll for Week Ending May 21, 19—

Employee									Rate per hour	Reg. time	Over-time	Wages		
No.	Name	W	Th	F	S	S	M	T				Reg.	Over.	Total
91	Allen, Harry....	8	8	11	0	0	8	8	\$2.00	40	3	\$ 80.00	\$ 9.00	\$ 89.00
92	Moulton, James	8	8	8	0	0	7	8	\$2.10	39	0	81.90	......	81.90
93	Brown, Edward.	8	10	11	0	0	10	8	\$1.75	40	7	70.00	18.38	88.38
94	Paul, Samuel....	8	7	7	0	0	10	8	\$2.10	38	2	79.80	6.30	86.10
95	Garvis, John....	8	8	8	0	0	8	8	\$1.85	40	0	76.00	......	76.00
96	Young, James...	8	10	8	0	0	9	8	\$2.00	40	3	80.00	9.00	89.00
										Totals,........		467.70	42.68	510.38

(Ans.)

It should be noted that in calculating overtime hours, they are calculated in terms of day rather than week, and it is perfectly possible as in the case of Samuel Paul for a man not to have a full week of regular time to his credit and yet have overtime credit. It should further be noted referring back to the time card of Harry Allen that odd minutes are not counted unless they are more than fifteen. Also that most firms do not count time before the regular starting time in the morning or during the noon hour unless the employee has been specifically asked to work at either or both of these times.

Change Memorandum.—This memorandum is really an analysis of payroll made so that the paymaster will know how many bills and coins of various denominations will be needed for the pay envelopes.

ILLUSTRATION: Prepare a change memorandum for the payroll of the Standard Manufacturing Company for the week ending May 21, 19—.

CHANGE MEMORANDUM

For Payroll of Week Ending May 21, 19—

No. of Employee	Total Wages	Currency					Coins				
		20	10	5	2	1	50	25	10	5	1
91	$89.00	4	0	1	0	4	0	0	0	0	0
92	81.90	4	0	0	0	1	1	1	1	1	0
93	88.38	4	0	1	0	3	0	0	3	1	3
94	86.10	4	0	1	0	1	0	0	1	0	0
95	76.00	3	1	1	0	1	0	0	0	0	0
96	89.00	4	0	1	0	4	0	0	0	0	0
		23	1	5	0	14	1	1	5	2	3
Totals..	$510.38	460	10	25	0	14	.50	.25	.50	.10	.03

Payroll Currency Slip.—This slip is prepared so that it may accompany the payroll check to the bank. It is a summary of the change memorandum and lets the paying teller of the bank know how much money is wanted in various denominations of bills and coins.

THE MERCHANT'S BANK
Orange, N. J.
Payroll Currency Slip
Depositor: Standard Mfg. Co.
Date: May 24, 19—

Bills	Dollars	Cents
20	460	0
10	10	0
5	25	0
2	0	0
1	14	0
Coins		
Halves..........	0	50
Quarters........	0	25
Dimes..........	0	50
Nickels.........	0	10
Cents.........	0	3
Total........	509	1.38

It should be noted that there is a discrepancy between the date on the payroll and the date on the currency slip. This exists because most firms have their week end sometime during the week, as on Monday, Tuesday, or Wednesday, and they pay on the following Friday or Saturday. This is done in order to give the clerical staff in the office adequate time in which to make up the time cards and properly prepare the payroll records.

Piece Work.—Instead of paying by the hour, day, or week, many firms pay many of their employees so much for each unit of work they complete. The piece rate is usually established by determining how many pieces the average man can do in a given length of time. This means that a fast worker will usually be rewarded by being able to earn more than his slower co-workers.

The piece-work production is usually reported at the office on a slip. In some plants the worker keeps a record of the number of units he has completed during the day and has the foreman of the room counter-sign his slip; in other plants the worker holds the work he completes until a checker counts it and gives the worker credit, releasing the units so that they may go on to the next operation. In still other plants, the work goes through the plant in numbered cases or job lots and the worker simply records in his book the job number of each lot on which he performs his operation and reports this number together with the price at the end of each day. A checker then takes these slips and enters them in a book especially prepared for the purpose, checks the completion of the operation against the case or lot number. This record is usually made by placing the date and number or initial of the employee in the proper column in his checker's book. It thus prevents two people from being paid for doing the same work or one person from being paid twice for the job. It also helps to keep employees from claiming pay for work that they have not actually completed.

In preparing the payroll when the plant is on a piece-work basis, some organizations record on the payroll sheet the number of pieces or units completed each day by the worker, add the total for the week, multiply this total by the price paid per unit, and thus get the total amount to be paid to the worker.

ILLUSTRATION: Mr. L. Cook, employee number 61, completed 23 units of work on Wednesday, 24 Thursday, 20 Friday, 11 Saturday, 21 Monday, and 20 Tuesday. He is paid $37\frac{1}{2}$¢ per unit. What is the amount of his earnings for the week?

		W	Th	F	S	M	T	Total Units	Price per Unit	Total Pay
61	Cook, L.	23	24	20	11	21	20	119	$.37\frac{1}{2}$	44.63 (Ans.)

In other plants such a system is not possible. Any one worker may work on different operations and thus complicate the problem or may work on much the same type of operation but may get a different price because he is working types of materials. Such plants will usually have some form upon which are calculated the daily earnings of the worker and these daily earnings are entered on the payroll sheet. At the end of the week, the daily earnings are totalled to determine the workers' pay for the week.

ILLUSTRATION: The work slips of Mr. H. Sailer, employee number 48, indicate that he earned \$6.45 Wednesday, \$7.78 Thursday, \$5.01 Friday, \$3.78 Saturday, \$6.51 Monday, and \$7.02 Tuesday. What is the amount of his earnings for the week?

		W	Th	F	S	M	Tu	Total Pay
48	Sailer, H.	6.45	7.78	5.01	3.78	6.51	7.02	36.55 (Ans.)

Other Pay Bases.—While time and piece-work are by far the most widely used bases for paying workers, because of the many obvious possibilities for injustices to both the employer and the employee, several special systems have come into vogue in recent years. These systems are fundamentally time or piece-work systems or a combination of the two, but have special features that set them apart and which make them, at least in some places, much more satisfactory than the basic systems. They are generally known as incentive wage plans. One particularly interested in these special wage plans should make himself acquainted with the Halsey Premium Plan, the Taylor Differential Piece-Work Plan, the Gantt Task and Bonus Plan, along with various others.

INSURANCE

The basis of all insurance is risk. Insurance is an agreement between a professional risk taker (insurance company) and an individual or firm, whereby the insurance company agrees under certain specified conditions to indemnify an individual or his heirs in case of some certain type of loss. The principal risks that a man faces have to do with his life, his property, and his liability for losses caused to other persons through some legal fault of his.

Life Insurance.—The average man or woman carries life insurance for two reasons: first, to take care of the payment of expenses incurred in connection with his last illness and burial, and secondly, to leave at least some funds to his dependents so that they may not suffer too much of an economic strain in the event of his death. The contract between the insurance company and the insured is called a *policy*, and the fee paid to the insurance company is known as a *premium*. The person who receives the face of the policy upon the death of the insured is called the *beneficiary*.

Premium.—The amount of the premium of life insurance depends upon the type of insurance and the age of the person being insured. The actual payment made to the company by the insured is technically called the gross or office premium. The cost of the policy as such is based on the death rate given in the mortality table and at a given interest rate. This basic cost of the policy is called the net premium. The gross premium is determined by adding an amount to cover expenses of all kinds and profit to the net premium. This is called loading. Sometimes the premium of a policy is paid in one sum, this being called the net single premium. Usually it is calculated on an annual basis and is called the net annual premium. While many people pay their insurance premiums annually, many pay a slight extra charge for the privilege of paying semi-annually, quarterly or monthly.

Insurance Tables.—The actual mathematics of determining the various premiums is somewhat involved and as a result tables have been prepared that are used to calculate these premiums. Insurance companies usually issue a table which simply lists their rates by age per thousand dollars of face value of policy. Naturally, a separate table or a separate column in a composite table is devoted to each type of insurance. Table 8 is a composite table showing the rates per $1000 on various types of insurance.

TABLE 8

A Typical Table Showing Rates Per Thousand for Different Types of Insurance Without Accidental Death Benefit *

Age	ORDINARY LIFE: Life Pd. up at 85	ORDINARY LIFE: Life Pd. up at 65	20 Pay Life	PREFERRED RISK "Mod. Life 3": 1st 3 Yrs.	PREFERRED RISK "Mod. Life 3": There-after	ENDOWMENT: End't at 65	ENDOWMENT: End't at 60	ENDOWMENT: 20 Yr. End't	ENDOWMENT: 10 Yr. End't	TERM: Term to 65	TERM: 10 Yr. Term	TERM: 5 Yr. Term	Age
15	17.08	18.07	28.67	13.46	15.84	19.79	21.57	51.05	105.68	....	....	....	**15**
16	17.45	18.50	29.17	13.79	16.22	20.29	22.16	51.15	105.82	....	....	....	**16**
17	17.84	18.97	29.69	14.14	16.64	20.80	22.79	51.24	105.94	12.21	7.28	6.96	**17**
18	18.23	19.43	30.19	14.49	17.05	21.35	23.43	51.33	106.04	12.47	7.48	7.22	**18**
19	18.62	19.90	30.68	14.85	17.47	21.90	24.09	51.40	106.12	12.73	7.66	7.45	**19**
20	19.02	20.39	31.19	15.21	17.89	22.47	24.79	51.44	106.17	12.99	7.77	7.60	**20**
21	19.55	20.99	31.79	15.58	18.33	23.15	25.60	51.50	106.23	13.26	7.89	7.75	**21**
22	20.07	21.62	32.39	15.97	18.79	23.87	26.46	51.54	106.27	13.54	8.00	7.88	**22**
23	20.62	22.28	33.01	16.37	19.26	24.62	27.36	51.59	106.30	13.82	8.10	7.99	**23**
24	21.19	22.97	33.63	16.80	19.76	25.40	28.32	51.65	106.33	14.10	8.19	8.08	**24**
25	21.78	23.68	34.26	17.24	20.28	26.22	29.31	51.70	106.36	14.40	8.29	8.18	**25**
26	22.38	24.43	34.92	17.70	20.82	27.08	30.36	51.76	106.38	14.72	8.38	8.26	**26**
27	23.02	25.21	35.57	18.17	21.38	27.98	31.47	51.85	106.40	15.05	8.52	8.35	**27**
28	23.69	26.03	36.27	18.68	21.98	28.92	32.65	51.92	106.44	15.38	8.63	8.41	**28**
29	24.37	26.90	36.98	19.22	22.61	29.94	33.91	52.03	106.48	15.76	8.80	8.52	**29**
30	25.09	27.82	37.71	19.78	23.27	30.99	35.25	52.15	106.52	16.13	8.97	8.61	**30**
31	25.74	28.70	38.31	20.38	23.98	32.01	36.58	52.29	106.58	16.54	9.20	8.74	**31**
32	26.43	29.61	38.95	21.02	24.73	33.11	38.02	52.45	106.65	16.96	9.46	8.91	**32**
33	27.17	30.62	39.62	21.67	25.49	34.27	39.58	52.65	106.74	17.42	9.76	9.11	**33**
34	27.94	31.69	40.31	22.38	26.33	35.52	41.24	52.87	106.86	17.90	10.09	9.34	**34**
35	28.75	32.82	41.04	23.13	27.21	36.86	43.06	53.11	106.98	18.40	10.49	9.62	**35**
36	29.61	34.04	41.81	23.91	28.13	38.30	45.03	53.40	107.15	18.93	10.94	9.94	**36**
37	30.51	35.36	42.60	24.74	29.11	39.84	47.16	53.71	107.32	19.50	11.45	10.31	**37**
38	31.49	36.78	43.44	25.61	30.13	41.51	49.52	54.07	107.53	20.09	11.99	10.72	**38**
39	32.49	38.28	44.32	26.54	31.22	43.31	52.08	54.46	107.77	20.71	12.63	11.18	**39**
40	33.57	39.92	45.25	27.51	32.36	45.24	54.91	54.91	108.03	21.38	13.32	11.71	**40**
41	34.68	41.70	46 19	28.55	33.59	47.34	58.03	55.38	108.35	22.08	14.10	12.32	**41**
42	35.87	43.62	47.20	29.63	34.86	49.62	61.48	55.92	108.69	22.80	14.96	12.98	**42**
43	37.12	45.72	48.25	30.78	36.21	52.10	65.33	56.49	109.06	23.59	15.92	13.72	**43**
44	38.44	47.98	49.33	32.00	37.65	54.80	69.66	57.10	109.49	24.38	16.96	14.53	**44**
45	39.84	50.47	50.47	33.28	39.15	57.77	74.55	57.77	109.96	25.24	18.12	15.45	**45**
46	41.31	53.21	51.66	34.63	40.74	61.05	80.11	58.51	110.49	....	19.38	16.46	**46**
47	42.86	56.21	52.91	36.07	42.44	64.64	86.52	59.30	111.05	....	20.79	17.56	**47**
48	44.51	59.54	54.22	37.59	44.22	68.66	93.94	60.17	111.68	....	22.32	18.78	**48**
49	46.24	63.24	55.59	39.18	46.09	73.15	102.69	61.11	112.38	....	24.00	20.12	**49**
50	48.08	67.42	57.02	40.88	48.09	78.19	113.15	62.13	113.15	....	25.82	21.59	**50**
51	50.01	72.12	58.56	42.65	50.18	83.92	....	63.24	113.93	....	27.56	23.19	**51**
52	52.06	77.47	60.15	44.53	52.39	90.45	....	64.46	114.74	....	29.44	24.97	**52**
53	54.20	83.64	61.83	46.52	54.73	98.01	....	65.77	115.57	....	31.49	26.92	**53**
54	56.48	90.83	63.63	48.62	57.20	106.86	....	67.18	116.43	....	33.69	29.02	**54**
55	58.89	99.30	65.50	50.83	59.80	117.33	....	68.71	117.33	....	36.08	31.35	**55**
56	61.42	....	67.49	53.15	62.53	....	....	70.37	118.25	....	....	33.25	**56**
57	64.09	....	69.58	55.60	65.41	....	....	72.14	119.23	....	....	35.33	**57**
58	66.89	....	71.78	58.17	68.44	....	....	74.06	120.24	....	....	37.52	**58**
59	69.84	....	74.11	60.88	71.62	....	....	76.10	121.27	....	....	39.90	**59**
60	72.93	....	76.54	63.70	74.94	....	....	78.29	122.34	....	...	42.39	**60**
61	76.50	...	79.47	66.94	78.75	....	....	80.97	124.07				**61**
62	80.32	...	82.60	70.39	82.81	....	....	83.90	125.98	Minimum $5000			**62**
63	84.42	....	85.99	74.06	87.13	....	....	87.08	128.06				**63**
64	88.80	....	89.60	77.98	91.74	....	....	90.53	130.33	Convertible			**64**
65	93.49	...	93.49	82.16	96.66	....	....	94.27	132.81	Prior to 60	During Term Period		**65**
66	98.56	.	97.69	86.61	101.89	.	....	98.32	135.52				**66**
67	104.02	.	....	....	...	.	....	..	..				**67**
68	109.92		....	....	....	.	....	....	..				**68**
69	116.31	...	...	....	....	..	....	....	....				**69**
70	123.29	....		...	...	....	....	....	..				**70**

* *Courtesy The Prudential Life Insurance Co. of America.*

Types of Life Insurance.—Insurance companies issue several types of life insurance policies, the most commonly known of which are: (1) ordinary life, (2) endowment, (3) paid-up, and (4) term.

Ordinary life insurance is that type of policy which remains in force during the life of the insured, and upon death the face of the policy is paid to the beneficiary. Usually premiums are paid annually from the time the policy is taken out until death. (It should be noted that many companies do not require premiums to be paid on ordinary life policies after the age of 85.) To find the cost of such insurance, one should refer to the table of rates, run down the age column to the age of the person buying the insurance, and multiply the rate given in the ordinary life column by the the number of thousand dollars worth of insurance the purchaser desires to take. The amount thus determined is the rate to be paid annually from that date until the death of the individual.

ILLUSTRATION: A man, age 35, desires to take out a $5000 ordinary life insurance policy. What annual premium must he pay?

The rate for ordinary life insurance (paid up at 85) at the age of 35 (according to Table 8) is $28.75 per thousand. The rate for $5000 will be

$$\$28.75 \times 5 = \$143.75 \quad \text{(Ans.)}$$

Endowment policies call for the payment of a regular premium for a given period of years—usually twenty but may be fewer or more—at the end of which time the face of the policy becomes due and payable to the insured. In the event of the death of the insured at any time during which the policy is in force, the face of the policy is paid to the beneficiary. The procedure for calculating the premium is similar to that used in the preceding illustration, the only exception being that the rate is selected from the proper endowment column.

ILLUSTRATION: A man, age 25, desires to take out a policy that will pay him $2500 at age 45, providing he lives, or in the event of his death before that time, will pay a like sum to his beneficiary. How much must he pay annually?

The rate for 20-year endowment insurance at age 25 is $51.70 per $1000. The rate for $2500 will be

$$51.70 \times \tfrac{2500}{1000} = \$129.25 \quad \text{(Ans.)}$$

Paid-up policies are a compromise between the endowment policy and whole life policies. Under the paid-up policy the insured pays a regular premium for a given period of years (usually 10, 15, or 20 years) and pays no more, but the face of the policy is not payable to him as in the case of endowment policies but rather is payable to the beneficiary upon the death of the insured regardless of whether that death occurs before or after the premiums have all been paid. The value of this type of insurance is that a man may pay up all the premium charges during his healthiest and most productive years and still remain insured during his entire life. The procedure for calculating the premium is similar to that used in the preceding illustration.

ILLUSTRATION: A man, age 30, desires to take out a $10,000 policy, the premiums of which will be paid up when he is 50. The rate for insurance requiring 20 annual premiums at age 30 is $37.71 per $1000. The annual rate for $10,000 will be

$$\$37.71 \times 10 = \$377.10$$

Term insurance is purchased for only a given length of time (usually one, five, or ten years) but may be kept in force indefinitely by the regular payment of the premium. If the insured dies while the policy is in force, the face of the policy is paid to the beneficiary, but if the insured survives that given length of time and fails to renew his policy, he is no longer protected. This type of insurance is frequently carried during that period of a man's life when his family is most dependent upon him.

ILLUSTRATION: A man, age 34, has three small children and feels that he should carry additional protection, realizing that should he die within the next few years his family might be in unusually straitened circumstances. He decides to take out a ten year term policy for $5,000 to supplement other insurance which he is carrying. What will it cost him per year?

The rate for ten year term insurance at age 34 is $10.09. The cost of a $5000 policy will be

$$\$10.09 \times 5 = \$50.45$$

Cash Surrender.—In all other forms of insurance except term insurance (whole life, endowment, and paid-up) the insured is to some extent protected even though he fails to continue to pay his premiums and thus lets his policy lapse. If this takes place, he may do one of three things: he may ask the insurance company to pay him an amount in money that is equal to the present value of the policy. This is called the cash surrender value and is really that portion of the premium that has been set aside by the company as a reserve out of which to pay the policy, plus accumulated interest.

Loans.—If the insured chooses to do so, he may borrow at a fixed interest rate, the cash surrender value of his policy rather than accept the cash settlement and surrender his policy. The value of the loan plan is that the insured may get his money and at the same time keep his policy in full force. Should he die before the loan has been repaid, the beneficiary will receive the face of the policy less the unpaid loan and interest accrued thereon.

Extended Insurance.—If, in the case of a lapsed policy, the insured does not choose to apply for and receive the cash surrender value of his policy, he may simply receive extended insurance. This means that the cash surrender value will be used to keep his policy in force on a term basis until such value is exhausted. Should the insured during this period die, the beneficiary will receive the face value of the policy. Should he survive this period, however, the insurance company is no longer liable.

Paid-up Insurance.—The holder of a lapsed policy may, however, choose to accept paid-up insurance rather than extended insurance. Under such a plan, the cash surrender value will be used to buy insurance that will remain in force during life without his paying any more premiums. Naturally, the face of the new policy will be less than that of the older policy, the face being the amount of paid-up insurance that could ordinarily be bought for the cash surrender value.

Special Benefits.—Many life insurance policies carry clauses which cover special risks such as general accident, travel, accident, disability, or some combination of these. An additional fee is charged for these special coverages.

Other Personal Policies.—Practically all other policies covering risks to the person such as accident insurance, disability insurance, health insurance, are also based on more or less standard experience tables and the annual fees are calculated accordingly. As a general rule, such policies are on a term basis (usually one year) and expire at the time unless they are renewed. Naturally, they have no cash surrender value except a refund that may be claimed if the policy is cancelled during the year.

In quoting rates for this type of insurance, many companies classify people in terms of their occupation and give more favorable rates to those in the fields that are least hazardous.

Home Insurance

Types of Policies.—The earliest peril against which homes were insured was fire. In the course of time, however, the extended coverage fire policy developed, providing coverage against damage by such other agencies, natural and otherwise, as for example, lightning (since it can cause structural damage as well as fire damage), smoke, wind, hail, explosion, falling aircraft, civil commotion, vandalism and malicious mischief. In the past an additional policy was often written to cover losses incurred of the contents of the home (personal prcperty) by these perils, and still another to cover personal property losses by theft. Then a fourth type of policy would be written to insure the homeowner against the cost of defending, and settling, personal liability lawsuits by outsiders for injuries sustained on the property.

In the more recent past, there has been a major trend toward the writing of policies that combine these various types of coverage into single policies, which are usually called Homeowners Policies. There are three varieties in common use which are in order of increasing coverage and cost, the Standard Plan (also called the A policy), the Broad-Form Plan (also called the B policy) and the Combination Plan (also called the C policy). These policies differ, not only in the amount of coverage they provide for other risks than that on the house itself, but also in the number and kind of the perils and risks covered. Therefore, they are discussed here in order in the following paragraphs, in which the coverage of each is discussed in some detail. In order to arrive at comparative figures the rates of each variety of policy will be given for a homeowner living in a Class A section of Nassau County, New York, one who has a masonry house and the other having a frame house. It is necessary to assume specific localities because the insurance rate books divide homes into 18 premium groups, of which Group 1 pays the lowest rate and Group 18 pays the highest. The assignment of a home to these groups depends upon (1) the zone in which the home is located; thus Nassau County is zone 2 in New York State; (2) the fire protection facilities available in the neighborhood in which the home is located; all of Nassau County is listed as Class A in this matter of protection. It is then found from the group chart that a masonry home so located belongs in premium group 7, while a frame home similarly located belongs in premium group 9. It is also assumed that the insurable value of the home is $20,000. One should note that insurable value means cost of replacement of the home less a factor which depends upon its present state of depreciation. Thus a newly constructed house would have an insurable value of its full cost of replacement, while that figure would be decreased for an older home according to its condition. In other words, the insurable value has no direct relation to selling price, which includes the cost of land and is subject to many other elements than construction costs.

Coverage and Costs.—All three of these homeowners policies provide insurance for specified perils to the (1) dwelling and (2) out buildings, plus specified insurance coverage of (3) household goods on the premises, (4) household goods away from the premises and (5) additional living expenses incurred while damages to the home are being repaired. All three of the homeowners policies protect against lawsuits resulting from personal injuries of outsiders, the Standard Plan and the Broad Form Plan up to $10,000 and the Combination Plan up to $25,000. Moreover, these injuries include not only such property connected accidents as tripping on the stairs or slipping on the sidewalk, but also the acts of children or pets.

It is, however, on the types of perils for which buildings are insured that the risk covered by the three policies varies so widely. The Standard Plan insures against damages due to fire, lightning, wind storms, hail, explosion, smoke, aircraft (or vehicles), civil commotion, vandalism, malicious mischief, glass breakage, and theft. The Broad-Form Plan adds to these perils such others as damage due to collapse of building, destruction by weight of ice, snow and sleet, falling trees and other falling objects besides aircraft and rupture of steam or hot water heating systems.

The Combination Plan includes certain other damages, especially those which result from obvious carelessness on the part of the occupants of the home, such as rain or snow damage due to windows left open and damage due to objects dropped within the house, which would even include spillage of paint and similar material.

On a home with an insurable value of $20,000, and located in Nassau County the Standard Plan (Plan A) would provide coverage of $20,000 for the dwelling, $2,000 for the out buildings, $8,000 for household goods on the premises, $1,000 for household goods off the premises, and $2,000 for additional living expenses incurred during the time needed to repair the home damaged.

The premium for this coverage, on a three year prepaid basis would be $138 per year for the masonry home, and for the frame home it would be $147 per year.

The Broad-Form Plan (Plan B) would provide coverage for these homes of $20,000 for the dwelling, $2,000 for the out buildings, $8,000 for household goods on the premises, $1,000 for household goods off the premises, and $4,000 for additional living expenses. The cost of this coverage would be $177 for the masonry home and $192 for the frame home.

The Combination Plan (Plan C) would provide $20,000 for the dwelling, $2,000 for the out buildings, $10,000 for personal property, and $4,000 for additional living expenses. The rate for this coverage on the masonry home is $348 and on the frame home, $369, on a $100 deductible basis for all perils except fire and lightning. The reason for the major difference in rates between the Broad-Form Plan and the Combination Plan is due to the additional perils and articles insured. It is to be noted that this policy covers personal property instead of household goods. Personal property includes personal articles on the premises as well as household goods. Even this policy, however, does not include all the coverage available to the homeowner. Many other items can be added by the payment of increased premiums, so that it has become more necessary than ever to base the policy as finally written with specific reference to the home that is insured.

Automobile Insurance

There are three major types of automobile insurance coverage: (1) that covering the liability of the owner of the vehicle for bodily injury or property damage caused by his vehicle; (2) that covering the cost of repairing damages which his own vehicle may sustain due to accidents; and (3) damages which his vehicle may incur or losses which he may sustain due to other hazards, such as fire, theft and glass breakage. Of these, the first is by far the most possible.

The variables which govern automobile liability (casualty) insurance rates depend upon a number of considerations. The first of these is the territory in which the automobile is domiciled, and individual rate tables are prepared for each territory. The United States is divided into many territories, some of which may be quite small in area. Thus New York City is divided into eight territories: Bronx County North; Bronx County South; Kings County; Manhattan; Queens County; Queens County Suburban; Staten Island; and New York City Suburban (the only area within New York City limits that belongs in this territory is Governor's Island). Then each territorial table has columns for each amount of insurance, the amount being stated at the top of the column as two figures, one for the amount applying to injuries to one person, and the other for more than one person. The usual standard amounts so provided are: $10/20; $20/20; $20/40; $25/50; $50/100; and $100/300.) These figures mean, of course, the number of thousands of dollars of insurance. There are two columns for property damage, one up to $5,000 and another, on which the premiums are slightly greater, up to $10,000. The important part of the table, however, is the first column which lists 28 classes of risks, and the importance of these classes can be judged from the fact that the lowest premium in Manhattan, which is the annual rate for Class 1A-O EX for $50/100 coverage is $145.20, while the highest rate shown for this amount of coverage in Manhattan is for Class 2C-2X and amounts to $555.39. The method of determining these classifications is critically important in determining the premium which the owner of the private passenger automobile must pay. These classes are as follows:

CLASS 1—Owned by an individual or jointly by two or more relatives who are residents in the same household. None-Business Use—No Operator under 25.

Class 1A—Not used * to and from work.

Class 1B—Used * to or from work less than 10 road miles one way.

Class 1C—Used * to or from work 10 or more road miles one way.

* Used—Means customarily used in driving to and from work, including car pools, share the ride arrangements, and driving to and from railroad or bus station, whether or not parked all day—Class 1B or 1C used to and from work.

CLASS 2—Owned by an individual or jointly by two or more relatives who are residents in the same household. Business and None-Business Use—Operators under 25.

Class 2A—Male operator under 25 not owner or principal operator

OR

Male owner or operator *under* 25—married.

CLASS 2B—Male operator under 25 not owner or principal operator and a resident student at a school, college or educational institution over 100 road-miles from the place of principal garaging.

CLASS 3A—Customarily used in business—individually owned. No male operator or owner under 25.

Reference to the table, however, discloses that in addition to the entries of these primary classes of uses and drivers there are two other items in the class symbol. One of these is the use, following the classes 1A, 2B, etc., of a number which may be 0, 1 or 2. These numbers are the driving records subclassification and are determined from driving record (demerit) points assigned to the automobile. One point is assigned for each automobile accident involving the applicant, or any operator of the automobile resident in his household for every automobile accident within four years resulting in property damage above $50 or in bodily injury or death. Points are also assigned, of course, for such matters as driving while intoxicated, leaving the scene of an accident, and speeding which results in accidents or revocation of license.

The third item in the class designation consists of the letters EX or IN, the former denoting an experienced driver licensed for three years or more and the latter the inexperienced driver licensed for less than three years.

In connection with automobile liability rates it should also be noted that there has been made by many companies an extensive revision of this classification system. The classes mentioned have been replaced by some 260 "primary classifications" based on 52 separate "driver" categories—i.e., categories based on the age, sex and marital status of operators of the insured automobile—combined with five categories of car use: cars used for pleasure only, driven less than 10 miles to work, one-way; cars driven 10 miles or more to work, one-way; cars used for business; and cars used by farmers. The method of applying this method can be illustrated by means of the tables. Table 8A gives base rates for private passenger automobiles in various territories. Note that these rates cover not only bodily injury, but also property damage and medical payments. These rates are to be multiplied by two factors, the first of which depends upon the kind of driver and the second upon whether there is one or more cars in the family and whether the car is a compact or not. The factors for the drivers are given in Table 8B. While the factors for the cars are given in Table 8C, one determines the basic rate from Table 9, multiplies it by the driver factor from Table 8B, and then adds or deducts from it according to the factor in Table 8C. Then the final result is increased by a percentage according to the accident record of the driver on the point system described earlier in this section.

Risks in which the owner or principal operator of the automobile has been licensed less than three years are currently charged the base rate if no points have been assigned for accidents or convictions during the experience period. Under the proposed revisions, one driving record point will be charged to such risks. This one point will result in a surcharge of 30% of the "base premium." The "base" premium is the premium charged for subclass 0—that is, for a risk without driving record points. The surcharges for two, three or four or more points are 70%, 120% and 180% respectively, of the base premium.

TABLE 8A

SAMPLE BASE PREMIUM PAGE FOR THE
AUTOMOBILE CASUALTY MANUAL

NAME OF STATE
(State Code XX)

BASE PREMIUMS FOR
PRIVATE PASSENGER AUTOMOBILES

Bodily Injury		Territory Schedule							
Limits	Code	01	02	03	04	05	06	07	08
10/20	2	$46	$25	$33	$37	$31	$19	$29	$47
25/50	5	55	30	39	44	37	23	35	56
50/100	6	60	33	43	48	40	25	38	61
100/300	8	65	35	47	52	44	27	41	66

Property Damage		Territory Schedule							
Limits	Code	01	02	03	04	05	06	07	08
5,000	1	$22	$20	$19	$23	$20	$20	$20	$24
10,000	2	23	21	20	24	21	21	21	25
25,000	4	24	22	21	25	22	22	22	26

Medical Payments		Territory Schedule							
Limits	Code	01	02	03	04	05	06	07	08
500	1	$ 5	$ 4	$ 5	$ 5	$ 5	$ 4	$ 4	$ 5
750	2	6	5	6	6	6	5	5	6
1,000	3	7	6	7	7	7	6	6	7
2,000	4	9	8	9	9	9	8	8	9
5,000	6	12	11	12	12	12	11	11	12

Refer to company for the Base Premiums for limits other than those shown above.

TABLE 8B

Age, Sex and Marital Status	Pleasure Use	Work Less Than 10 Miles	Work 10 or More Miles	Business Use	Farm Use
		No Youthful Operator			
Only Operator in Household is a Female Age 30-64	.90	1.00	1.30	1.40	.65
One or More Operators Age 65 or Over	1.00	1.10	1.40	1.50	.75
All Other	1.00	1.10	1.40	1.50	.75
		Youthful Unmarried Female Operator —Without Driver Training (No Other Youthful Operator)			
Age of Youngest Female Operator is					
17 or under	1.55	1.65	1.95	2.05	1.30
18	1.40	1.50	1.80	1.90	1.15
19	1.25	1.35	1.65	1.75	1.00
20	1.10	1.20	1.50	1.60	.85
		Youthful Unmarried Female Operator —With Driver Training (No Other Youthful Operator)			
Age of Youngest Female Operator is					
17 or under	1.40	1.50	1.80	1.90	1.15
18	1.25	1.35	1.65	1.75	1.00
19	1.15	1.25	1.55	1.65	.90
20	1.05	1.15	1.45	1.55	.80
		Youthful Married Male Operator Without Driver Training			
Age of Youngest Male Operator is					
17 or under	1.80	1.90	2.20	2.30	1.55
18	1.70	1.80	2.10	2.20	1.45
19	1.60	1.70	2.00	2.10	1.35
20	1.50	1.60	1.90	2.00	1.25
		With Driver Training			
Age of Youngest Male Operator is					
17 or under	1.60	1.70	2.00	2.10	1.35
18	1.55	1.65	1.95	2.05	1.30
19	1.50	1.60	1.90	2.00	1.25
20	1.45	1.55	1.85	1.95	1.20
		With or Without Driver Training			
21	1.40	1.50	1.80	1.90	1.15
22	1.30	1.40	1.70	1.80	1.05
23	1.20	1.30	1.60	1.70	.95
24	1.10	1.20	1.50	1.60	.85
25					
26					
27					
28					
29					

TABLE 8B (Cont'd.)

Age, Sex and Marital Status	Pleasure Use	Work Less Than 10 Miles	Work 10 or More Miles	Business Use	Farm Use
	YOUTHFUL UNMARRIED MALE OPERATOR (Not Owner or Principal Operator) Without Driver Training				
Age of Youngest Male Operator is					
17 or under	2.30	2.40	2.70	2.80	2.05
18	2.10	2.20	2.50	2.60	1.85
19	1.90	2.00	2.30	2.40	1.65
20	1.70	1.80	2.10	2.20	1.45
	With Driver Training				
Age of Youngest Male Operator is					
17 or under	2.05	2.15	2.45	2.55	1.80
18	1.90	2.00	2.30	2.40	1.65
19	1.75	1.85	2.15	2.25	1.50
20	1.60	1.70	2.00	2.10	1.35
	With or Without Driver Training				
21	1.55	1.65	1.95	2.05	1.30
22	1.40	1.50	1.80	1.90	1.15
23	1.25	1.35	1.65	1.75	1.00
24	1.10	1.20	1.50	1.60	.85
25					
26					
27					
28					
29					
	YOUTHFUL UNMARRIED MALE OWNER OR PRINCIPAL OPERATOR Without Driver Training				
Age of Youngest Male Operator is					
17 or under	3.30	3.40	3.70	3.80	3.05
18	3.10	3.20	3.50	3.60	2.85
19	2.90	3.00	3.30	3.40	2.65
20	2.70	2.80	3.10	3.20	2.45
	With Driver Training				
Age of Youngest Male Operator is					
17 or under	2.70	2.80	3.10	3.20	2.45
18	2.65	2.75	3.05	3.15	2.40
19	2.60	2.70	3.00	3.10	2.35
20	2.55	2.65	2.95	3.05	2.30
	With or Without Driver Training				
21	2.50	2.60	2.90	3.00	2.25
22	2.30	2.40	2.70	2.80	2.05
23	2.10	2.20	2.50	2.60	1.85
24	1.90	2.00	2.30	2.40	1.65
25	1.70	1.80	2.10	2.20	1.45
26	1.50	1.60	1.90	2.00	1.25
27	1.35	1.45	1.75	1.85	1.10
28	1.20	1.30	1.60	1.70	.95
29	1.10	1.20	1.50	1.60	.85

TABLE 8C

		Sub-Class				
		0	1	2	3	4
		Single Car				
Non-Compact	Factor	+0.00	+0.30	+0.70	+1.20	+1.80
Compact	Factor	−0.10	+0.20	+0.60	+1.10	+1.70
		Multi-Car				
Non-Compact	Factor	−0.15	+0.00	+0.20	+0.45	+0.75
Compact	Factor	−0.25	−0.10	+0.10	+0.35	+0.65

Example. Assume that the named insured is a married male, 42 years old, with a daughter, age 17, who has successfully completed a driver training course, and a son, age 19, who has not had driver training. Neither child is married. The insured has one automobile which he uses to drive to and from work, six miles each way. The insured has incurred two driving points under the Safe Driver plan.

For this insured the primary rating classification factor is 2.00—the factor for a youthful unmarried male operator, not the owner or principal operator, age 19, without driver training, works less than 10 miles. To this primary factor is added the secondary classification factor of 0.70. (This is the factor for a single automobile, non-compact, under Safe Driver subclass 2.)

Thus, the final rating factor for this automobile is 2.70. The base premium for all coverages desired would be multiplied by this factor to determine the final premium.

Automobile Collision and Comprehensive Insurance.—Just as liability insurance is computed, as described above, by tables and factors, similar methods are used for computing collision and comprehensive insurance. The latter depends, of course, upon the car itself as well as territorial factors and are necessary for the cars of all manufacturers, all models and ages. For example, a collision policy on a two year old Chevrolet Impala in Manhattan would have the following premiums for a $50 deductible policy: $119 if the operator being over 25 (male) and if the operator had a four year accident free record, provided he used the car for pleasure only. If he had had one accident the rate would increase to $145; if he used the car for business, but had had no accidents the rate would increase to $165 (if there was a male operator under 25 in the family the accident free rate would increase to $224. In this case it might be desirable to carry the $100 deductible rate, which would only be $146. The comprehensive policy on insurance of the vehicle covering fire, theft and glass breakage, etc., for this two year old Chevrolet Impala would be $17 on a $50 deductible basis of $51 for full coverage.

Stock

Stock is the term applied to shares which represent ownership in a corporation. A corporation is an intangible person created by the state upon the request of individuals interested in organizing it. It operates under a charter granted by the state which, among other things, specifies the amount of stock that the corporation is authorized to issue, and the par value, if any.

Par Value.—The par value of stock is the face value, the value at which it must be originally issued. Most states now permit stock to be issued without par value. Such stock is known as no-par-value stock, each share merely representing a fractional part of the total ownership in the business.

Capital Stock.—The capital stock of a corporation represents the amount authorized and paid into the corporation as capital for conducting the business. The capital stock is divided into a certain number of shares which may have a par value.

Dividends.—When profit is made by the corporation, the board of directors may decide to retain some or all of it in the business for working capital or pay some or all of it to the stockholders as their share of the profit. This share of profit is known as a dividend. They are usually declared and paid quarterly, and are stated and paid in terms of a certain percent of the par value of the stock, as a 2% quarterly dividend, or they may be quoted as so much a share, as $2.00 a share quarterly dividend.

Common Stock.—The regular stock of a corporation is known as common stock. It carries with it no special preferences with regard to the distribution of profits or of assets. Whatever profits are left after dividends on preferred stocks are paid may be distributed among the common stockholders.

Preferred Stock.—Some stock, in order to attract the more conservative investor, carries a guarantee to pay dividends before any profits are distributed to the common stockholders. It also may carry preference with regard to the distribution of assets in case the business is dissolved. Some preferred stock carries the provision that, in case the dividends are not paid when they should be, the unpaid dividends will be allowed to accumulate and will be paid before any dividend is paid on common stock. This is known as cumulative preferred stock.

Market Value.—For those interested in buying or selling stocks, the market value is more important than is the par value. The market value is the price at which it may be bought or sold and depends upon a number of factors including whether or not it pays a good dividend. A stock exchange is a place where stocks are bought and sold, the members of the exchange acting as brokers for those who are the actual buyers or sellers.

Round Lots and Odd Lots.—Most trading is done in units of 100 shares, such lots being known as Round Lots. When fewer than 100 shares is traded, or if a total number of shares traded in a transaction is not divisible by 100, the fractional part of 100 shares is known as an odd lot. Federal and New York State taxes are charged to sellers in all transactions and the buyers of odd lots only.

Brokerage Charges.—The members of the stock exchange, acting as brokers for those trading stock, charge a fee known as "brokerage." The fees for stocks bought or sold on the New York Stock Exchange and the American Stock Exchange are:

On stocks selling at $1.00 per share and above, commission

rates are based upon the amount of money involved in a single transaction and are not less than the following:

On Round Lots (each single transaction not exceeding 100 shares, in a unit of trading; a combination of units of trading; or a combination of a unit or units, plus an odd lot):

Money Involved	Commission
Under $100	As mutually agreed
$100 to $399	2% plus $3
$400 to $2,399	1% plus $7
$2,400 to $4,999	½% plus $19
$5,000 and over	⅒% plus $39

A table of rates based on the above formulas for selected key round-dollar stock prices is given below.

For Odd Lot Transactions (sales of less than a round lot) the rates are the same as above, less $2.00. On odd lots, you also pay an extra differential for the odd lot dealer who handles the sale for your broker. On the vast majority of stocks, for which a round lot is 100 shares, the odd lot differential is ⅛ of a point (or 12½¢) per share for stock selling for less than $40 in round lots; and ¼ of a point (or 25¢) for stocks selling at $40 or more in round lots.

Minimum—When the amount involved is less than $100, commission is as mutually agreed; $100 or more, not more than $1.50 per share or $75 per single transaction, but in no case less than $6.

On stocks, rights, and warrants selling below $1 a share, commissions are on a per share basis and range from 10 cents per 100 shares for stocks selling at 1/256 to $5.25 per 100 shares for stocks selling for ⅞ or more but less than $1.

Minimum—Less than $100, as mutually agreed, an amount of $100 or more, not less than $6 or the rate per share, whichever is greater.

Commodities.—The unit trading of grains is 5,000 bushels on which the brokerage varies somewhat, being twenty-two dollars for wheat. On the other hand, for sugar the brokerage is twenty-five dollars for 112,000 pounds.

COMMISSION RATE TABLE
(Selected Round-Dollar Stock Prices)

Price	Rate per 100 Shares*	Price	Rate per 100 Shares*
$ 1	$ 6.00	19	$26.00
2	7.00	20	27.00
3	9.00	21	28.00
4	11.00	22	29.00
5	12.00	23	30.00
6	13.00	24	31.00
7	14.00	25	31.50
8	15.00	30	34.00
9	16.00	35	36.50
10	17.00	40	39.00
11	18.00	45	41.50
12	19.00	50	44.00
13	20.00	60	45.00
14	21.00	70	46.00
15	22.00	80	47.00
16	23.00	90	48.00
17	24.00	100	49.00
18	25.00	125	51.50

*Round lots only.

Stock Transfer Tax.—In addition to paying the brokerage charges, the seller, and sometimes the buyer, of stocks must pay a Federal tax and a state tax.

FEDERAL TRANSFER TAXES

For stocks, warrants, and rights to subscribe to stocks: $0.04 on each $100 (or major fraction thereof) of the actual sales price of the securities, with a maximum of $0.08 per share for stocks selling at more than $200 per share.

For bonds and debentures: $0.05 for each $100 (or fraction thereof) of par or face value, regardless of the actual selling price (equivalent to $0.50 per bond of $1,000 par value). There is no Federal transfer tax on the sale or transfer of rights to subscribe to bonds or debentures.

State Transfer Taxes

New York—$0.01 per share under $5; $0.02 per share betwee $5.00 and $10; $0.03 per share between $10 and $20; $0.04 pe share of $20 or more. New York State does not tax the sale o transfer of bonds, debentures, rights, or warrants.

Florida—$0.15 per $100 par value or per share on no pa stock, regardless of selling price.

South Car.—$0.04 per $100 par value or per share on no pa stock, regardless of selling price.

Texas—$0.033 per $100 par value or per share of no pa stock, regardless of selling price.

Computing the Cost of Stocks.—There are four steps in computing the cost of a lot of stock. They are: Base Cost, Brokerage Taxes (if any), and Total Cost.

Illustration: A man bought 225 shares of a stock, the market price of which was 135. What was the total cost?

The cost would be calculated as follows:

1. 225 shares are 2 round lots and one odd lot of 25.
2. Base cost = $135 × 100 (one round lot) = $13,500.

Cost of 2 round lots = $13,500 × 2 =	$27,000.00	
$135 × 25 (odd lot) =	3,375.00	
Total Base Cost		$30,375.00

3. Brokerage:

Round lot brokerage = ($\frac{1}{10}$% + $39) × No. of round lots		
($\frac{1}{10}$% × $13,500 + $39) × 2 =		
($13.50 + $39) × 2	$105.00	
Odd lot brokerage = ($\frac{1}{10}$% + $39) less $2.		
($\frac{1}{10}$% × $3375 + $39) − $2 =		
($3.38 + $39) − $2 =	$ 40.38	
Total Brokerage	$145.38	

4. Tax:

0.04% of $30,375 = $12.15		
0.04% × 225 = $9.00		
Total Tax	$ 21.15	

5. Total Charges .... $ 166.53
6. Total Cost $30,541.53

It should be noted that, if this transaction had been for a ound lot, the number of shares divisible by 100, no tax would ıave had to be paid.

Computing the Proceeds of a Sale of Stock.—There are also our steps in computing the proceeds from a sale of stock. They ıre:

1. Total proceeds. Multiply the selling price per share by the number of shares.
2. Brokerage. Multiply the number of shares by the brokerage rate as shown in the illustration below.
3. Taxes. Calculate stock transfer taxes on all shares sold.
4. Net Proceeds. Add the brokerage to the taxes to determine the total charges and deduct the total charges from the gross proceeds.

ILLUSTRATION: If a man sold 500 shares of a stock, the market value of which was $42\frac{3}{8}$ per share, how much did he receive?

The proceeds would be calculated as follows:

1. Total proceeds:
   42.375 × 100 × 5 = 4237.50 × 5 = ... $21,187.50
2. Round lot brokerage = (½% + $19)
   × No. round lots (½% × 4237.50
   + 19) × 5 = (21.19 + 19) × 5 =
   $40.19 × 5 = ... $ 200.95
3. Taxes:
   0.04% of $21,187.50 = 84.75
   500 × 0.04 = N.Y.S. Tax
   Total Tax .... $ 104.75
   Total Charges .... −305.70

   $20,881.80 (Ans.)

Listed and Unlisted Stocks.—Leading stock exchanges do not permit the trading of all stocks on the floor of the exchange, but rather restrict the trading to those that are listed by the exchange after an investigation into the corporation issuing the stock. These stocks are known as listed stocks. Other stock may be bought and sold with or without the services of a brokerage, but may not be traded on the floor of an exchange not listing them. These are known as unlisted stocks.

Short Selling.—The practice of selling stocks before the seller actually owns such stock is known as "short selling." The short seller hopes to make delivery of stock sold short by buying the stocks at a price lower than that at which he sold such stock. Because he is interested in having the price of the stock go down he is known as a "bear." The trader who buys first is said to be "long" on stock and hopes to sell at a price higher than he paid. As he is interested in the market going up, he is known as a "bull."

Bonds.—Governments and private corporations, when they borrow money for long periods, issue certificates of indebtedness known as bonds. When issued by private corporations, these bonds are usually secured by a mortgage on real estate, movable property, or by all the assets of the corporation. Bonds differ from stocks in that stocks represent a share of ownership in the corporation, while bonds represent indebtedness of the corporation, a liability.

There are many classes of bonds and various bases of classification. Most bonds carry a pledge of payment at a specified date and of a specified amount, the redemption date and redemption price, respectively. However, some bonds are subject to call at the demand of the bondholder, usually at specified dates before maturity. Serial bonds provide for redemption at a number of dates, instead of a single one. Annuity bonds provide for the repayment of both principal and interest, in equal installments at uniform periods.

Bonds are usually issued with a face value of $1000, although they sometimes are issued in smaller units such as $500.00, $100.00, or even $50.00. They bear interest which is usually payable semi-annually; also they have a due date. Some bonds have attached to them interest coupons, one for every six-month interest period during the life of the bond. These may be clipped every six months and cashed at any bank. This type of bond is known as a *coupon bond*.

Brokerage.—Bonds are bought and sold through brokers much the same as are stocks. Commission rates for the New York Stock Exchange are:

BOND COMMISSION RATES

(Except Government, Short Term or Called Bonds)

Price per $1000 (of Principal) Bond	Minimum Commission Rate per $1000 (of Principal) Bond: On orders of 1 or 2 bonds	On orders of 3 bonds	On orders of 4 bonds	On orders of 5 bonds or more
Selling at less than $10 (1%)................	$1.50	$1.20	$0.90	$0.75
Selling at $10 (1%) and above but under $100 (10%)	2.50	2.00	1.50	1.25
Selling at $100 (10%) and above............	5.00	4.00	3.00	2.50

A Federal Bond Transfer tax of 50¢ per $1000 of par value is charged the seller of all bonds. United States Government, state, municipal, and foreign government bonds are exempt from any tax.

Accrued Interest.—Bonds may be sold on the date that interest is due and paid or may be and frequently are sold at other times. When sold between interest dates, the seller of the bond is entitled to the interest that has been earned since the last interest date, but which is not yet paid. This is known as accrued interest.

Calculating the Proceeds of a Sale.—There are four or five steps in calculating the proceeds of a sale of bonds. They are:

1. Market Value. Multiply the quoted price by the par value of the bonds being sold.
2. Calculate the accrued interest and add to the market value.
3. Calculate the brokerage charges.
4. Calculate the Federal Tax and add it to the brokerage to determine the total charges.
5. Deduct the total charges from the market value plus accrued interest. If the bonds are sold on an interest date, the second step is eliminated as there would be no accrued interest.

ILLUSTRATION: Ten \$1000 Railroad 4½% bonds due in 1984 with interest, payable on April 1 and October 1, were sold on September 1, 1954, at 107¾. What were the proceeds?

1. Market Value		
10,000 × 1.0775		\$10,775.00
2. Accrued Interest		
Interest for 5 months @ 4½ on \$10,000 par value $10{,}000 \times \frac{4.5}{100} \times \frac{150}{360}$		187.50
Present Value		\$10,962.50
3. Brokerage		
\$1.25 per \$1000 of par value equals \$1.25 × 10	\$12.50	
4. Transfer Tax		
50¢ per 1000 of par value equals 50¢ × 10	\$5.00	
Total Charges		−17.50
Net Proceeds		\$10,945.00

Calculating the Cost of Bonds.—There are three or four steps (depending upon whether or not the bonds are bought on interest dates) that must be followed in calculating the cost of bonds.

1. Calculate the market value.
2. Calculate the accrued interest (if bond is bought between interest dates) and add to market value.
3. Calculate brokerage charges.
4. Add brokerage fees to present value to determine total cost.

ILLUSTRATION: One \$1000 5% bond due on June 1, 1959, was bought on September 1, 1954, at the rate of 105, the cost would be calculated as follows:

1. Market Value	
\$1000 × 1.05	\$1050.00
2. Accrued Interest, 5% on \$1000 for 3 months	
$\$1000 \times \frac{5}{100} \times \frac{90}{360}$	12.50
Present Value	\$1062.50
3. Brokerage, \$2.50 per \$1000 par value	2.50
4. Total Cost	\$1065.00

Annuities

Broadly speaking, an *annuity* is a number of equal payments made at equal periods of time. While technically the term annuity means "annual," practically the term annuity is applied whether the payments are made annually or at other intervals such as semi-annually, quarterly, or monthly.

TABLE 9A

Monthly Life Income from Retirement Annuity per $100 annual payment taken out by man Aged 40

	Monthly Income		
Retirement Age	Whole Life Annuity[1]	Installment-Refund Annuity[2]	Life Annuity Income Guaranteed for 10 years[3]
50	$ 5.25	$ 4.65	$ 5.16
55	8.76	7.58	8.41
60	13.92	11.27	12.56
65	19.94	15.72	17.65
70	27.48	20.50	24.17

[1] Terminates with death of annuitant. No further payment made by company.

[2] If annuitant dies before payments exceed purchase price, the difference is paid to beneficiary. If annuitant dies after payments exceed purchase price, beneficiary receives nothing.

[3] Payments guaranteed for life. Death of annuitant before 10 years results in payment to beneficiary until end of 10 year period.

Contingent Annuities: Basically there are two kinds of annuities: contingent annuities, and certain annuities. Contingent annuities are annuities in which the date of the last payment or the first payment or both cannot be foretold. Old age pension plans are contingent annuities.

Annuities Certain: Annuities certain are annuities in which the dates involved, beginning and ending, may be definitely established. There are many types of annuities that are certain, one of the most common being the type in which an individual is interested in investing or depositing a fixed sum annually (or at other intervals) and desires to know how much the money will be worth at the end of a given number of years.

ILLUSTRATION: Beginning on July 1, 1954, a man deposits $500 a year at a place where interest is paid annually at the rate of 5%. How much of a fund will he have built up after he has made his fifth deposit? The deposit date and the interest date are the same.

Calculated by Simple Arithmetic: There are several methods of determining the amount that will be on deposit when the last payment is made, the most cumbersome being the procedure whereby one calculates the interest at the end of the first period, adds it to the principal, adds the new deposit, calculates the interest on the total at the end of the second period, adds it to the balance, adds the new deposit, etc. Follow this procedure for each year.

This would work out as follows:

July 1, 1954	First Deposit	$ 500.00
July 1, 1955	Interest at 5% on $500	25.00
	Second Deposit	500.00
		$1025.00
July 1, 1956	Interest at 5% on $1025	51.25
	Third Deposit	500.00
		$1576.25
July 1, 1957	Interest at 5% on $1576.25	78.81
	Fourth Deposit	500.00
		$2155.06
July 1, 1958	Interest at 5% on $2155.06	107.75
	Fifth Deposit	500.00
	Total	$2762.81

Calculated by Logarithms: Naturally this method takes too long for practical purposes, especially if the number of deposits is high. If this type of problem is encountered frequently, it is best to be provided with annuity tables. These are pre-computed tables indicating how much $1.00 accumulates to if deposited at regular periods of time at regular interest rates. The amount given in the table is then multiplied by the amount of the regular deposit. If annuity tables are not available, this problem may be calculated by the use of logarithms. The formula is:

$$\frac{(1+i)^n - 1}{i} \times \text{regular payment} = \text{Amount of Annuity}$$

which, in this particular problem, would be:

$$\frac{(1+.05)^5 - 1}{0.05} \times 500 = \text{Amount of Deposit at end of 5th year.}$$

The solution by logarithms is as follows:

$$\begin{aligned} \text{Log} = 1.05 &= 0.021189 \\ &\quad \times 5 \\ \hline &0.105945 = 1.2762 \end{aligned}$$

$$\therefore \quad (1.05)^5 = 1.2762, \quad (1.05)^5 - 1 = 0.2762$$

$$\text{Amount} = \frac{500 \times 0\ 2762}{0.05}$$

$$\begin{aligned} &\text{Log} \quad 500 = 2.698970 \\ &\text{Log} \quad 0.2763 = 9.441381 - 10 \\ &\text{Co-log } 0.05 = 1\ 301030 \\ \hline &\qquad 3.441381 = \text{Total \$2763} \quad \text{(Ans.)} \end{aligned}$$

Note.—This total is 17¢ greater than the one determined by arithmetic. This is due to the slight inaccuracies resulting from the use of six-place logarithms.

Amount of Annual Payment: Somewhat the reverse of the problem just presented is that in which an individual desires to know how much of an annual deposit he must make at a given rate of interest in order to build up a given sum. This may be determined by calculating the amount of annuity of $1.00 for the given number of periods and dividing this into the amount desired. If annuity tables are available, this amount of an annuity of $1.00

at a given interest rate for a given number of periods may be readily determined. In the absence of such tables, the problem may be worked out by logarithms on the following formula:

$$\text{Total amount} \div \frac{(1+i)^n - 1}{i}$$

ILLUSTRATION: A man chooses to build up a sum of \$5000 over a period of 10 years by making an annual deposit. The interest rate is 4% and it may be assumed that the annual deposit will be made on the same date that interest is credited. How much must be paid annually? The formula applied to this problem will read:

$$\$5000 \text{ divided by } \frac{(1+.04)^{10} - 1}{0.04}$$

$$\begin{aligned} \text{Log } 1.04 &= .017033 \\ &\ \times 10 \\ \hline &\ .170330 = 1.4802 \end{aligned}$$

$$\therefore \quad (1.04)^{10} = 1.4802; \quad (1.04)^{10} - 1 = 0.4802$$

$$\begin{array}{ll} \text{Log } 0.4802 = & 9.681422 - 10 \\ \text{Co-log } 0.04 & 1.397940 \\ \hline & 1.079362 = 12.005 \end{array}$$

(Amount of annuity of \$1.00 will equal at end of 10-year period)

$$\begin{array}{rl} \text{Log } \$5{,}000 = & 3.698970 \\ \text{Less} & 1.079369 \\ \hline & 2.619608 = \$416.48 \end{array}$$

Amount of Annual Payment (Ans.)

Sinking Fund: A sinking fund is a fund established for the purpose of paying off a debt or of making some other necessary payment. Many industrial bond issues, and some others, are paid off on their due dates from a sinking fund; in fact, the terms of the bond quite frequently require this procedure. The sinking fund is established by placing in the fund each year an amount in cash that if invested immediately at a given rate of interest, will accumulate to a sum equal to the total indebtedness on the date that the obligation must be paid. The amount of the annual payment is determined in exactly the same manner as it was in the preceding problem.

ILLUSTRATION: A corporation issued $500,000 worth of bonds due in 20 years and desired or was required to set up a sinking fund, the amount of the annual payment, assuming interest at $4\frac{1}{2}\%$, would be determined by using the following formula:

$$\$500{,}000 \text{ divided by } \frac{(1.045)^{20} - 1}{0.045}$$

Present Value of an Annuity: The term annuity means annual payment. Another basic problem in dealing with annuities is that of determining the sum which, placed at a given rate of interest, will make it possible to pay out a given amount each year for a given number of years. This is known as the Present Value of an annuity.

ILLUSTRATION: A man chooses to place at interest a sum that will permit him to pay out $1200 per year for a period of four years, the first withdrawal to be made one year after the fund is established. The interest rate is 5%. How much must he deposit?

This, in reality, is a problem of calculating the compound discount on the sums to be paid. If annuity tables are available it will be found that an annuity of $1.00 for 4 periods at 5% is 3.545950. If an annuity of $1200 is to be available, 3.545950 × $1200 = $4255.14, the amount that must be deposited to establish the fund.

If the annuity tables are not available, this problem may be worked out with the use of logarithms by following the following formula:

$$\frac{1 - \dfrac{1}{(1 + i)^n}}{i} \times 1200$$

In terms of this problem, the formula would read:

$$1 - \frac{1}{(1 + .05)^n} \times \$1200$$

$$\begin{aligned} \text{Log } 1.05 &= 0.021189 \\ &\quad \times 4 \\ \hline &\quad 0.084756 \\ \text{Subtracted from log } 1 &= 1.0 \\ \hline &\quad 9.915244 - 10 = 0.82270 \\ &\quad 1 - 0.82270 = 0.17730 \end{aligned}$$

$$\begin{aligned} \text{Log } 0.17730 &= 9.248709 - 10 \\ \text{Co-log } 0.05 &= 1.301030 \\ \hline &\quad 0.549739 = 3.546, \text{ Amount necessary for annuity of } \$1 \\ \text{Log } \$1200 &= 3.079181 \\ \hline &\quad 3.628920 = \$4255.20, \text{ the amount of the fund to be established.} \end{aligned}$$

Deferred Annuities: This is an annuity, the payments on which do not begin for some time after the fund is established. Such an annuity might be established when a child is very young with a view toward paying his expenses through college.

ILLUSTRATION: Let us assume that in the previous problem, the fund was to be established 12 years earlier; or in other words, that payments were to be begun 13 years after the fund was originally set up. How much must be deposited if interest at the rate of 5% per annum will be earned?

This problem simply involves the calculation of present worth of the fund to be set up as of 12 years in advance. If the fund normally to be established was \$4255.20, a somewhat smaller sum will suffice if the deposit is to be made 12 years in advance and allowed to accumulate at compound interest (we will assume 5%) for all that time. This problem may be solved by determining the present value of $1.00 and multiplying at the present worth of the ordinary annuity which in this case is $4255.20.

$\dfrac{1}{(1+i)^n}$ = present worth of $1.00, therefore the formula as applied to this problem is:

$$\frac{1}{(1+.05)^{12}} \times \$4255.20$$

$$\begin{array}{rl}
\text{Log } 1.05 = & 0.021189 \\
 & \times 12 \\ \hline
 & 0.254268 \\
\text{Subtracted from log } 1 = & 1.0 \\ \hline
 & 9.745732 - 10 = 0.55684 \\
\text{Log } \$4255.20 = & 3.628920 \\ \hline
 & 3.374652 = \$2369.50
\end{array}$$

3.374652 = $2369.50, the amount that must be deposited if the annuity is to be established 12 years in advance.

Amortization: Strictly speaking, to amortize means to extinguish or liquidate a debt. Actually, however, there are two methods of disposing of debts: One, by the sinking fund method, which has already been described, and the other by the method known as amortization which, in common practice, means to liquidate the debt by making a series of equal periodic payments which include a part payment on the principal as well as interest on the principal outstanding.

Actually, this problem is another annuity problem in which the amount of the periodic payment is to be determined. It may be solved simply by dividing the present worth of an annuity of $1.00 at the given rate of interest for the given number of years into the full amount of the debt to be amortized. If annuity tables are not available, this problem which is called that of determining the amount of an annuity, may be solved by using logarithms, applying the following formula:

$$\frac{i}{1 - \dfrac{1}{(1+i)^n}} \times \text{debt}$$

ILLUSTRATION: A debt of \$12,000 is to be amortized over a period of 15 years, the interest rate being 5%. How much must be paid each year?

In terms of this problem, the formula would read:

$$\frac{0.05}{1 - \frac{1}{(1.05)^{15}}} \times \$12{,}000$$

$$\begin{aligned}
\text{Log of } 1.05 &= 0.021189 \\
&\quad\times 15 \\
&\overline{0.317835} \\
\text{Log of } 1 &= 1.0 \\
\text{Less} &\quad .317835 \\
&\overline{0.682165} = 0.48102 \\
1 - 0.48102 &= 0.51898 \\
\text{Log } 0.05 &= 8.698970 - 10 \\
\text{Co-log } 0.51898 &= .284848 \\
\text{Log } 12{,}000 &= 4.079181 \\
&\overline{3.062999} = \$1156.10,
\end{aligned}$$

the amount that must be paid annually (Ans.)

Depreciation.—There are a number of methods of computing depreciation charges. The method used in a particular instance is often a matter of the practice of the particular business, or for the particular type of asset. The choice of a method is not necessarily a unilateral decision of the individual business or accountant; the policy of governmental taxing agencies must frequently be considered, in view of the close relationship between depreciation charges and earnings. It is the policy of the U. S. Government to permit the annual depreciation deduction to be computed in any manner that is consistent with recognized trade practices. Three methods were specifically listed, and therefore these methods are discussed in this chapter. They are the straight-line method, the declining-balance method, and the sum of the years-digits method. There are, however, certain other general methods of computing depreciation charges, which are permitted if they are so used as to give results consistent with those obtained by the listed methods. Two of these methods—the unit of production method, the sinking fund method and a combination method are also included in this section.

Before discussing any of these methods, however, it is to be understood that their use in U. S. practice is subject to the general limitations upon permissible depreciation which are set forth in Publication 311 of the U. S. Treasury Department of Internal Revenue, which is part of the Code of Federal Regulations. That publication sets forth general regulations on depreciation which stipulate the general character of property which may be depreciated, and the general rule that experience with similar equipment in the same industry be the basis for the assumed life of equipment. There is also, however, Internal Revenue Service Publication #173 of the U. S. Treasury Department which is called Bulletin F, and which gives information and tables of the average useful life of depreciable equipment. The data in this bulletin are arranged by industries, and vary greatly in the detail with which individual industries are treated.

The Straight Line Method.—The straight line method is one of the most simple methods of computing depreciation charges. It consists of dividing the total depreciation expense (which is initial cost of the asset minus its scrap value) by the number of periods of useful life. The quotient is depreciation expense per period, which by this method is the same for each period. For example, an asset having an initial cost of \$1100.00; a scrap value of \$100.00; and a useful life of 10 years would have a depreciation expense of $\frac{\$1100 - \$100}{10} = \$100.00$ per year. Since by this method this figure is the same for each year of the life, the book value (which is initial cost − depreciation fund) decreases at the same rate from year to year, and its graph is a straight line, which gives the name to the method.

As another example, find the annual depreciation charge of a machine costing \$865.00, having a scrap value of \$61.25 at the end of a probable life of 5 years. Also find the book value at the end of 3 years.

$$\text{Depreciation expense} = \$865 - \$61.25 = \$803.75$$

$$\text{Depreciation charge per year} = \frac{\$803.75}{5} = \$160.75$$

$$\text{Depreciation fund after 3 years} = 3 \times \$160.75 = \$482.25$$

$$\text{Book value after 3 years} = \$865.00 - \$482.25 = \$372.75$$

It is to be noted that the straight-line method does not require the existance of a scrap value at the end of the period of depreciation, if the nature of the asset is such that its terminal value is then zero. In that case, the annual depreciation charge is the initial cost divided by the probable life in years.

However, by U. S. tax regulations if an asset is sold at the end of a depreciation period for more than the assumed scrap value, or if it is sold during the period for more than its current book value, the difference must be taken as a capital gain. On the other hand, failure to deduct the full amount of depreciation allowable each year does not prevent such amount from reducing the adjusted basis, nor does it entitle the taxpayer to a greater deduction in a subsequent year.

The Declining-Balance Method.—The declining-balance method is also designated by the more descriptive name of constant-percentage method. It is also referred to as a liberalized method, since it may be so applied as to give depreciation charges during the early years that exceed those calculated by the straight-line method. However, the extent of that excess is limited by U. S. Government regulations (generally to twice the depreciation charges for the first year calculated by the straight-line method). Also, the type of property to which the liberalized declining method may be applied is restricted; it cannot be applied to leases, patents and other intangible assets.

The basis of the declining-balance method is to depreciate the asset each year by a constant percentage of its book value.

This percentage may be assumed arbitrarily, or computed by the formula,

$$\text{Percentage Depreciation } d = 1 - \sqrt[n]{\frac{S.V.}{C.}} \qquad (1)$$

Where n is the period of depreciation in years, $S.V.$ is the Scrap Value and C is the original cost.

For example, an asset costing \$4,000.00 on January 1, 1961 estimated to have a scrap value of \$800.00 in ten years, would have an annual percentage depreciation:

$$d = 1 - \sqrt[10]{\frac{\$800}{\$4000}} = 1 - \sqrt[10]{0.2}$$

The figure $\sqrt[10]{0.2}$ is evaluated by logarithms to be .85.
So that $d = 1 - .85 = .15 = 15\%$

Then applying this annual depreciation rate, the asset would have the following depreciation charges and book value for the first 5 years:

Depreciation Rate 15%

(Values of Asset costing \$4,000.00 at Jan. 1 in various years)

Year	1961	1962	1963	1964	1965
Book Value	\$4,000.00	\$3,400.00	\$2,890.00	\$2,456.50	\$2,088.02
Depreciation Charge		600.00	510.00	433.50	368.48
Depreciation Fund		600.00	1,110.00	1,543.50	1,911.98

Since the asset was purchased on January 1, 1961, it has undergone one year's depreciation by January 1, 1962. The charge for this depreciation is 15% of its book value at the beginning of 1961, which is 15% of \$4,000.00, the cost. Therefore, the new book value as of January 1, 1962 is book value at beginning of year minus the depreciation charge, which is \$4,000.00 – \$600.00 = \$4,400.00, the \$600.00 going into the depreciation fund.

On January 1, 1963, the depreciation charge for 1962 is 15% of \$3,400.00 (book value at beginning of 1962) = \$510.00. Then the new book value as of January 1, 1963 is that at beginning of year minus the depreciation charge, which is \$3,400.00 minus \$510.00 = \$2,890.00, the \$510.00 going into the depreciation fund.

Thus, for each year, the depreciation charge is found by multiplying the book value at the beginning of the year by the depreciation rate, while the book value at the end of the year is the book value at beginning of the year less the depreciation charge. The depreciation fund is the total of the depreciation charges.

As explained there, machinery and other depreciable assets are often carried on the books and reported on the balance sheet at their initial cost, while the depreciation charges that have been made against these assets over the years are reported as a total "fund" or "reserve" which is subtracted from the initial cost to show the present value of the particular class of assets. It follows that a depreciation fund is merely a total of charges that have been made. It is thus not a "fund" at all in the ordinary everyday use of that term to denote a sum of money. The latter type of fund is called a sinking fund, and is treated in the last section of this chapter.

To derive general formulas, write these terms in symbols, using C for cost, BV for book value, D for depreciation charge, d for depreciation rate (decimal value), and $\sum D$ (the Greek letter $\sum$, sigma, means "sum of") for depreciation fund.

Then at the end of one year:

Depreciation charge for the year $= D_1 = Cd$

Book value $= (BV)_1 = C - Cd = C(1 - d)$

Depreciation fund $= \sum^{1} D_i = D_1 = Cd$

Then at the end of the second year:

Depreciation charge for the second year $= D_2 = (BV)_1 d = C(1 - d)d$

Book value at end of second year $= (BV)_2 = (BV)_1 - D_2 = C(1 - d) - C(1 - d)d = C(1 - d)^2$

Depreciation fund at end of second year $= \sum^{2} D_i = C - (BV)_2$

At the end of the third year;

Depreciation charge for the third year $= D_3 = (BV)_2 d = C(1 - d)^2 d$

Book value at end of third year $= (BV)_3 = (BV)_2 - D_3 = C(1 - d)^2 - C(1 - d)^2 d = C(1 - d)^3$

Depreciation fund at end of third year $= \sum^{3} D_i = C - (BV)_3$

Then at the end of the nth year

Depreciation charge for the nth year $= D_n = C(1 - d)^{n-1} d$ (2)

Book Value at end of nth year $= (BV)_n = C(1 - d)^n$ (3)

Depreciation fund at end of nth year $= \sum^{n} D_i = C - (BV)_n$ (4)

To apply these general formulas, solve the problem: An asset is purchased for \$8,000.00 on January 1, 1955. If it is depreciated by the declining-balance method at 14% annually, what is its book value on January 1, 1965, what is its depreciation fund on that date, and what is the depreciation charge from January 1, 1964 to January 1, 1965 (tenth year)?

Then by the Formula (3)

$$\begin{aligned}(BV)_{10} = C(1 - d)^{10} &= (\$8{,}000)(1 - .14)^{10} \\ &= (\$8{,}000)(.86)^{10}\end{aligned}$$

This computation is readily made by logarithms to give \$1,770.39, which is the book value of the asset after depreciation at 14% for 10 years.

Then by Formula (4), depreciation fund = cost − book value = \$8,000.00 − \$1,770.39 = \$6,229.61, and by Formula (2) the depreciation charge during the tenth year $= C(1 - d)^{n-1}d =$ $(\$8{,}000.00)(1 - .14)^9(.14)$. This computation is readily made by logarithms, as was that of the book value above, to give \$288.20, which is the depreciation during the tenth year.

Note that by the declining-balance method an asset depreciated at an annual rate of 14% had a book value after 10 years of \$1,770.39 or about 22% of its initial cost, whereas if the straight-line method had been used on the basis of a 10-year life and no scrap value, its first year depreciation would have been only 10% of the cost, but the book value would have been reduced to 0 in 10 years. Thus, the declining balance method gives more rapid initial depreciation, but slower depreciation later and never depreciates completely.

To compensate partly for this effect, the U. S. Government allows higher initial depreciation rates where the declining-balance method is used, extending up to twice the straight line rate, that is, an asset with an estimated life of ten years may be depreciated as much as 20% annually, under specified restrictions.

The advantage of using the accelerated declining-balance method, or other accelerated methods, is felt particularly in instances where depreciation is hastened by obsolescence. For example, the useful life of a machine may be terminated, not by wear to a point where maintenance becomes excessive, but by the introduction of new machines so much more efficient that the old machine can no longer operate competitively. In industries undergoing rapid development, therefore, methods of accelerated depreciation are highly advantageous.

It should be added here that the U. S. Tax Code states that the annual depreciation deduction could be increased by an allowance for extraordinary obsolescence, starting from the year in which it was reasonably certain that the property was affected by revolutionary inventions, abnormal growth, or radical economic changes, which cause its abandonment prior to the end of its normal useful life.

The Sum of Years—Digits Method.—This method is allowable, under U. S. Tax Regulations, for property similar to that for which the liberalized declining-balance method may be used.

In this method the depreciation in any year is found by multiplying the cost of the asset less its scrap value by a fraction, whose numerator is the number of the year in question (counting from date on which the asset is to be scrapped) and whose denominator is the values of all the years added together.

For example, an asset costing $6,000.00 purchased January

, 1960, with a useful life of 6 years, and a scrap value of $460.00, would have:

The number 6 assigned to the first year of its life.
The number 5 assigned to the second year of its life.
The number 4 assigned to the third year of its life.
The number 3 assigned to the fourth year of its life.
The number 2 assigned to the fifth year of its life.
The number 1 assigned to the sixth and last year of its life.

The number 21 as the total of its year numbers.

Then the depreciation charge during the first year

$$= \frac{6}{21}\text{(Cost} - \text{Scrap Value)} = \frac{6}{21}(\$6{,}000.00 - 460.)$$
$$= \frac{2}{7}(\$5540.00)$$
$$= \$1{,}582.86.$$

The depreciation charge during the second year

$$= \frac{5}{21}(\$5{,}540.00) = \$1{,}319.05$$

The depreciation charge during the third year

$$= \frac{4}{21}(\$5{,}540.00) = \$1{,}055.23$$

The depreciation charge during the fourth year

$$= \frac{3}{21}(\$5{,}540.00) = \$791.43$$

The depreciation charge during the fifth year

$$= \frac{2}{21}(\$5{,}540.00) = \$527.62$$

The depreciation charge during the sixth and last year

$$= \frac{1}{21}(\$5{,}540.00) = \$263.81$$

Note that the depreciation fund at the end of any period is found by adding the depreciation charges, and that the book value at the end of any period is the difference between initial cost and depreciation fund.

For example, the depreciation fund at the end of six years is

$1582.86, D_1
1319.05, D_2
1055.23, D_3
791.43, D_4
527.62, D_5
263.81, D_6

Depreciation or fund at end of 6 years $= \sum^{6} D_i = \$5540.00$

which checks the difference between cost and scrap value, as it should, since the calculation above was made on the basis of a useful life of 6 years.

Unit of Production Methods.—The basis of these unit of production methods, which apply most directly to machinery, is production. If a machine fabricates one article, or a limited number of kinds of articles, its output is most conveniently measured in terms of the number of articles produced. On the other hand, if the machine is used for working on a relatively small number of large articles, especially if several machines of the same type are so used, then the output is probably better measured in machine hours.

Whichever of these methods are used, they require the maintenance of records showing the number of hours for which the machine is operated daily, or the number of units which it produces daily, so that its annual output in machine hours or production units can be computed. There are also required estimates of the useful life of the machine in the same terms, that is, in terms of machine hours or units produced. In addition, of course, the initial cost and estimated salvage value of the machine are also required.

For example, a machine costing $6,000.00 has an estimated salvage value of $400.00, and an estimated production during its useful life of 70,000 units. What is its depreciation charge, depreciation fund and book value during its first four years of operation if its production is: First year, 10,000 units, second year, 6,000 units, third year, 8,000 units, fourth year, 7,000 units?

Then, first year; Depreciation charge $= D_1 = \left(\frac{10{,}000}{70{,}000}\right)$ ($6000.00 − $400.00) = $800.00; Depreciation fund $= \sum^{1} D_i =$ $800.00; Book value = $6000.00 − $800.00 = $5200.00.

Second year. Depreciation charge $= D_2 = \left(\frac{6{,}000}{70{,}000}\right)$ ($6000.00 − $400.00) = $480.00; Depreciation fund $= \sum^{2} D_i =$ $D_1 + D_2 =$ $1,280.00. Book value = $6000.00 − $1,280.00 = $4720.00.

Third year. Depreciation charge $= D_3 = \left(\frac{8{,}000}{70{,}000}\right)$ (\$6000.00 $-$ \$400.00) $=$ \$640.00; Depreciation fund $= \sum^{3} D_i = D_1 + D_2 + D_3 =$ \$1920.00. Book value $=$ \$6000.00 $-$ \$1920.00 $=$ \$4080.00.

Fourth year. Depreciation charge $= D_4 = \left(\frac{7{,}000}{70{,}000}\right)$ (\$6000.00 $-$ \$400.00) $=$ \$560.00; Depreciation fund $= \sum^{4} D_i = D_1 + D_2 + D_3 + D_4 =$ \$2480.00; Book value $=$ \$6000.00 $-$ \$2480.00 $=$ \$3520.00.

Note that the general formula for depreciation charge, as is apparent from the foregoing examples, can be written as

$$D_i = \left(\frac{N_i}{N_T}\right)(C - S.V.) \tag{5}$$

where D_i is the depreciation charge in any one year; N_i is the number of units produced in that year; N_T is the estimated total number of units produced during the useful life of the machine; and C and $S.V.$ are the initial cost of the machine and its estimated scrap value at the end of its useful life, respectively.

Depreciation calculations based upon machine hours are made in essentially the same way as those just described for the unit production basis. The annual depreciation charge can be computed by the relation $(C - S.V.)$ where N_i is the number of hours the machine is operated in a given year, N_T is the estimated total number of hours which the machine can operate in its useful life, C is initial cost and $S.V.$ is estimated scrap value at the end of useful life. Also, the unit cost method can be used for machine-hour calculations of depreciation in the form, $N_i = \left(\frac{C - S.V.}{N_T}\right)$ where N_i is again the number of hours the machine is operated in a given year, and the expression in parentheses is the depreciation charge per hour operated.

Sinking Fund Method.—The depreciation calculations made up to this point have been based upon the general usage of the term depreciation. In other words, the depreciation charges have been treated from the point of view of deductions from earnings to offset the gradual loss of assets by depreciation, which indeed they are. However, if sums of money are segregated to constitute a sinking fund to replace the depreciated assets, a somewhat infrequent practice, then the depreciation charges can be reduced by the interest which they will earn from the date of deposit to the date of retirement of the depreciated asset. The actual amount of the sinking fund can be computed by the method for finding the value of the rent of an ordinary annuity, as explained in Chapter 14.

As an example of the application of this method, find the depreciation charge for each year, and the value of the sinking fund at the end of each year for a machine costing \$865.00, having a scrap value of \$61.25 at the end of a probable life of 5 years, if interest is credited at 4%.

The total depreciation cost for the 5 years, which is the sum the sinking fund must then total, is Cost − Scrap Value = \$865.00 − \$61.25 = \$803.75.

The annual payment into the sinking fund is found by the formula for the periodic reserve payment into an annuity, which is

$$R = P \frac{i}{1 - \dfrac{1}{(1 - i)^n}} \tag{6}$$

where P is the principle, i is the decimal interest rate per period, and n is the number of periods. Substituting

$$R = (\$803.75) \frac{.04}{1 - \dfrac{1}{(1 - .04)^5}}$$

Solving by logarithms, we find that R is \$148.39 annually.

Then at the end of the first year, the depreciation sinking fund is \$148.39.

At the end of the second year, the amount of the depreciation sinking fund is (\$148.39) (1.04) + \$148.39 = \$154.32 + \$148.39 = \$302.71.

At the end of the third year, the amount of the depreciation sinking fund is (\$302.71) (1.04) + \$148.39 = \$314.81 + \$148.39 = \$463.30.

At the end of the fourth year, the amount of the depreciation sinking fund is (\$463.30) (1.04) + \$148.39 = \$481.82 + \$148.39 = \$630.21.

At the end of the fifth year, the amount of the depreciation sinking fund is (\$630.21) (1.04) + \$148.39 = \$655.42 + \$148.39 = \$803.81, which is six cents greater (due to the fractional cent rounding-off) than the required amount.

Note that the annual depreciation charges computed by this method resemble those by the straight-line method in that they do not change throughout the depreciation period. They differ, however, in that they are not only charges, but sums of money, and are therefore credited with interest, which reduces their amount. To gain an idea of the effect of this reduction, compare the annual charges on a \$865.00 asset, with a scrap value of \$61.25 and a life of 5 years. When this computation was made by the sinking fund method above, the annual depreciation charge and deposit was found to be \$148.39, against an annual charge of \$160.75 for the same asset when computed earlier in this chapter by the straight line method. The figure of \$148.39 was based upon 4% interest rate on the sinking fund; higher rates would increase the difference, and lower rates would decrease it.

Combination Methods.—In recent years the United States Internal Revenue Service has been willing to accept in many instances a combination of the straight line and the double declining balance methods. This combination can best be illustrated by an example. Consider an asset with an accepted life of 10 years, costing \$100. By the accepted method one would depreciate it by the straight line amount of \$10 for the first year. In later years, he could choose whichever of these two methods gave the highest depreciation. Thus the second year one would use the $9.00 figure of the declining balance method multiplied by 2, which is $18, leaving a value of $72. Likewise, on the third year he would use twice the declining balance amount of $7.20, which is $14.40, leaving a balance of $57.60. Simply the fourth year one would use a figure of $11.52, leaving a balance of $46.08. Then thereafter one could revert to the $10 annual basis of the straight line method, which gives greater deductions than the double declining balance method.

Foreign Exchange

In its conventional sense, the term Foreign Exchange means the commercial paper and instruments used in foreign trade and the problems attending the settling of them. The most commonly used items are *Bills of Exchange*.

Bill of Exchange.—A draft drawn by one party ordering a second party to pay to the first party or to a third party a given sum of money is a bill of exchange. When such drafts are used for domestic exchange, they are usually known as checks, or drafts. When they are used in foreign exchange, they are called foreign drafts or, more frequently, foreign bills of exchange. Bills of exchange may be payable on demand or they may be payable at a later date, the most common times being 30, 60, or 90 days after date.

Cables.—Orders for the transfer of money abroad that are transmitted by wire are known as telegraphic transfers or, more frequently in the United States, as cables. They are not bills of exchange in the strict sense of the term because they are not written bills and not negotiable instruments, but they are used extensively to take the place of bills of exchange because of the speed with which a transaction may be completed. If a regular demand draft is drawn and mailed, several days will elapse between the time it is mailed in this country and the time it is delivered in a European country or almost any other foreign country. The same transaction may be completed in a few minutes by using a "cable" instead of a regular draft.

Buying and Selling Foreign Exchange.—Buying and selling foreign exchange is handled through banks who sell exchange at whatever it will bring at a given time and who buy it for whatever it is worth in terms of current values. In many countries, there are two or more rates, only one of which applies to ordinary commercial transactions. It is therefore important when using listed exchange figures to verify with the bank that the rate used applies to the transaction in question. Because of the widespread use of cables, the basic rate is usually quoted in terms of the rate for cables with other forms of exchange (demand 30-, 60-, or 90-day drafts) being usually worth a little less. Newspapers such as the *New York Times* and the *Journal of Commerce* carry tables of quotations listing current foreign exchange values.

United States Money.—The dollar was established as the unit of United States money by an Act of Congress on August 8, 1786, and the subdivisions and multiples of this unit as then established are:

10 mills make 1 cent
10 cents make 1 dime
10 dimes make 1 dollar

English Money.—The pound sterling is the unit of English money. While the smaller denominations have always been computed on an arbitrary system as shown by the table below, the British Parliament has considered, in recent years, a proposal to change their currency to a decimal system based on the pound. The present system is as follows:

English Money

4 farthings	=	1 pence (d)
12 pence	=	1 shilling (s)
20 shillings	=	1 pound Sterling (£)

To find out how much English money will be exchanged for a given quantity of American money, it is necessary to divide the number of dollars by the value of the pound. If there is a decimal in the result equate it into shillings and pence.

ILLUSTRATION: A man wants to exchange \$500 for as much English money as he can get for it. Assume that the English pound is quoted at \$2.80. How many pounds, shillings, and pence will he receive?

$$500 \div 280 = 178.57 = £178, 11s, 5d. \quad \text{(Ans.)}$$

Other Foreign Money.—Most other foreign countries use a decimal system as a basis for their monetary systems. The following is a list of them:

Country	Principal Unit	Its One-Hundredth Part
France	franc	centime
Belgium	franc	centime
Switzerland	franc	centime
Italy	lira	centesimo
Greece	drachma	lepta
Spain	peseta	centimo
Finland	mark	penni
Germany	mark	pfennig
Brazil	cruzeiro	reis
Yugoslavia	dinar	para
Venezuela	bolivar	centimo
Argentina	peso	centavo

To find out how much foreign money will be exchanged for a given quantity of American money, it is necessary to divide the number of dollars by the value of the foreign unit. When dealing with the above-listed monies, the decimal equals the number of coins of smaller denomination.

ILLUSTRATION: A man wants to exchange \$50 for as much French money as he can get. Assume that the franc is quoted at 0.2025. How many francs and centimes will he receive?

$$50 \div 0.2025 = 246.85$$

.246 francs, 85 centimes (Ans.)

Method of Quoting.—Exchange may be quoted (1) on a premium and discount basis, (2) on a direct price basis, or (3) on an indirect price basis.

Domestic exchange is always quoted on a premium and discount basis, and some foreign monies of the same denominations, such as the Canadian dollar, are usually quoted that way.

Illustration: If Canadian dollars are quoted at 7.5% discount, how much would be paid for $1,000 exchange to Canada?

$$\begin{aligned} &\$1000 - 7.5\% \text{ per dollar} \\ &= \$1000 - 0.075(\$1000) \\ &= \$1000 - \$75 = \$925.00 \end{aligned}$$

Direct Price Basis.—This method means that exchange is quoted in terms of how many cents or dollars must be paid per unit or per 100 units of a foreign money. English money has always been quoted this way in the United States and at present practically all foreign monies are quoted this way. If the quotation for the English pound is 2.80, it means that two dollars and eighty cents must be paid for every English pound sterling one wants to buy or that that amount will be paid for any which a customer has to sell.

Illustration: A man owes a bill of 627 pounds in England and must buy sterling with which to pay the bill. How much must he pay if the pound is quoted at $2.80?

He will multiply the current price ($2.80) by 627 in order to determine how much he will have to pay. In this case, it would be $\$2.80 \times 627 = \$1753.60$ (Ans.)

Illustration: Another man sold goods of value $7827.62 to an English firm. How many pounds did he receive?

The English firm would divide the $7827.62 by 2.80 to determine the number of pounds sterling it would have to pay to settle. In this case, the answer would be:

$$\$7827.62 \div 2.80 = 2795.07 \text{ pounds} = £2795, 1s, 5d.$$

Indirect Price.—Many foreign monies are, or used to be, quoted indirectly, especially when the unit is less in value than the dollar. When the quotation states the number of foreign units that can be bought for a dollar, it is called the indirect price. England quotes all of its exchange on this indirect basis.

Methods of Payment.—Most foreign exchange transactions arise from transactions involving the sale of goods by an individual or firm in one country to an individual or firm in another country. There are three methods for arranging for the payment of such goods: (1) payment with the order, (2) establishing credit to be drawn on when goods are ready to be shipped, and (3) draft drawn at time of shipment. A fourth type, open account, could be mentioned, but it is used rarely and then by firms who have been doing business for extended periods.

Payment with Order.—Under this plan of payment, the purchaser sends exchange with his order. It really amounts to payment in advance and is seriously objected to by most purchasers. It is not generally used, but at times it seems justifiable. If an American firm were buying goods in Germany on such terms, it would be necessary for the American firm to buy a foreign draft (bill of exchange) payable on demand and attach it to the order for the goods. The draft would be for the total purchase price of the goods calculated in the foreign money.

ILLUSTRATION: An American firm orders goods from Germany. The goods were of such nature that the German firm demanded payment with the order. The order would amount to 325 marks. How much must be sent if the marks cost 25.00 cents?

$$325 \times 25 \text{ cents} =$$
$$325 \times 0.25 = \$81.25 \quad \text{(Ans.)}$$

Establishing Credit.—Establishing credit to be drawn against is required when a firm selling goods to a foreign customer desires to be certain that the goods will be paid for when they are ready to ship. This is used when the purchasing firm is not well known, and particularly when the goods must be made to order.

ILLUSTRATION: If a French firm ordered goods to cost approximately $1200 under such terms, it would be necessary for the French firm to have its bank establish the necessary credit in a New York bank against which the American firm may draw when the goods are ready for shipment. When the goods are ready for shipment, the selling firm draws a draft against the established credit. The draft the seller draws must be accompanied by the shipping documents.

How much will the necessary credit cost if the franc is worth $0.2025?

$$\$1200 \div \$0.2025 = 5925 \text{ francs, } 92 \text{ centimes.} \quad \text{(Ans.)}$$

Draft with Shipment.—The draft drawn at the time of shipment is the procedure in most general use. The draft is drawn on the foreign customer when the goods are ready to be shipped. The draft usually has all shipping documents attached. These documents are (1) Commercial Invoice, (2) Consular Invoice (not required in some countries), (3) Export Declaration, (4) Ocean Bill of Lading, (5) Marine Insurance Certificate or Policy. These drafts may be payable at sight or 30 days, 60 days, or 90 days after sight (or less frequently, at other periods of time). Sight drafts (or on demand) are usually marked D/P, meaning documents are endorsed and delivered at the time the buyer accepts the draft. The purchaser cannot claim the goods until these documents have been endorsed and delivered to him.

ILLUSTRATION: A firm in Italy sold an American firm an invoice of goods amounting to 172,600 lira. Upon shipment of the goods, a 60-day draft was drawn. How much did the American firm have to pay if the rate of exchange was $0.0016?

$$172{,}600 \times 0.0016 = 276.16 \quad \text{(Ans.)}$$

Posted or Nominal Rates.—Foreign exchange quotations show what are known as "Actual" rates and are used by bankers, regular traders in foreign exchange, and dealers in large transactions. Small letters of credit and small checks are sold at a slightly higher rate than the "actual." This is called the Posted or Nominal Rate.

Travelers' Money.—In order to safeguard his funds, the average traveler carries travelers' checks rather than regular cash. These checks are issued by the American Express, by some banks, and by some tourist companies. They are sold to prospective travelers in denominations from $10.00 up and are charged for at the rate of 75¢ per hundred dollars plus the face. Identification is established by having the traveler sign all checks in the presence of the person cashing them. These checks are issued in dollar denominations and may be cashed anywhere in the world at whatever may be the current rate at the time of cashing.

ILLUSTRATION: A traveler cashed a $75.00 travelers' check in England. The rate was $2.80. How much did he receive?

$75 ÷ 2.80 or 26.79 pounds, which equals £26, 15s, 10d. (Ans.)

FREIGHT SHIPMENTS

Freight is one of the least expensive methods of shipping goods and is used in the shipping of all sorts of commodities, the speed of delivery of which is not vitally important.

The principal document used in freight shipments is the bill of lading. It is issued in triplicate and contains (1) the name of the consignor (the one shipping the goods), (2) the consignee (the one to whom the goods are shipped), (3) weight, description, and other essential factors about the goods being shipped, and (4) directions pertaining to the route over which the goods should travel. It is the agreement between the shipper and the carrier whereby the carrier agrees to deliver the goods to the consignee.

The three copies of the bill of lading are called (1) original, (2) shipping order, and (3) memorandum copy. After being duly signed by both the shipper and the carrier, the original is forwarded to the consignee, the shipping order is retained by the carrier, and the memorandum copy is retained by the shipper for his files.

There are two kinds of bills of lading: (1) straight and (2) order. The straight bill of lading is the most commonly used and is not negotiable. It is used when the terms of payment have been satisfactorily adjusted between the consignor and consignee, and where delivery of goods is not contingent upon payment for them.

The order bill of lading is negotiable and is used when the consignee is to pay for the goods before they are delivered to him. When it is used, the original (instead of being sent direct to the consignee) is forwarded (usually with sight draft attached) through banks to his bank where he is asked to pay the draft. When this is done, the bill of lading is endorsed to him and with it he may claim his goods at the freight depot. This form of bill of lading is also used in connection with reconsignment and divergence procedures.

Freight Rates.—The charges made by the carrier are based on a hundred pound minimum, by less than carload shipments (l.c.l.) and per car rates for carload shipments. In boat freight the

(Uniform Domestic Straight Bill of Lading adopted by Carriers in Official, Southern and Western Classification territories, March 15, 1922, as amended August 1, 1930.)

THIS SHIPPING ORDER must be legibly filled in, in Ink, in Indelible Pencil, or in Carbon, and retained by the Agent.

Shipper's No. ______

Agent's No. ______

LACKAWANNA CENTRAL RAILROAD **Company**

RECEIVED, Subject to the classifications and tariffs in effect on the date of the issue of this Shipping Order,

AT ______, 19

FROM ______

the property described below, in apparent good order, except as noted (contents and condition of contents of packages unknown), marked, consigned, and destined as indicated below, which said company (the word company being understood throughout this contract as meaning any person or corporation in possession of the property under the contract) agrees to carry to its usual place of delivery at said destination, if on its own road or its own water line, otherwise to deliver to another carrier on the route to said destination. It is mutually agreed, as to each carrier of all or any of said property over all or any portion of said route to destination, and as to each party at any time interested in all or any of said property, that every service to be performed hereunder shall be subject to all the conditions not prohibited by law, whether printed or written, herein contained, including the conditions on back hereof, which are hereby agreed to by the shipper and accepted for himself and his assigns.

(Mail or street address of consignee—For purposes of notification only.)

Consigned to Ewill Receive

Destination Central Point State of Illinois County of Iowa

Route ______

______ Car Initial ______ Car No. ______

(Delivering Carrier)

NO. PACKAGES	DESCRIPTION OF ARTICLES, SPECIAL MARKS, AND EXCEPTIONS	*WEIGHT (SUBJECT TO CORRECTION)	CLASS OR RATE	CHECK COL.
6	cases specialized books	860	1st	
	25-30-17-28-44-19			
5	bdls sundry sheet stock	265	1st	

If this shipment is to be delivered to the consignee without recourse on the consignor, the consignor shall sign the following statement:

The carrier shall not make delivery of this shipment without payment of freight and all other lawful charges. (*See Section 7 of conditions.*)

______ (Signature of Consignor.)

If charges are to be prepaid, write or stamp here, "To be Prepaid."

Received $______ to apply in prepayment of the charges on the property described hereon.

______ *Agent or Cashier.*

Per ______ (The signature here acknowledges only the amount prepaid.)

Charges advanced:

$______

*If the shipment moves between two ports by a carrier by water, the law requires that the bill of lading shall state whether it is "carrier's or shipper's weight."

NOTE.—Where the rate is dependent on value, shippers are required to state specifically in writing the agreed or declared value of the property.

The agreed or declared value of the property is hereby specifically stated by the shipper to be not exceeding

Van Mac Pub. Co. per V.P.P. $1250.00

Will Van Mac Pub Co Shipper. Willard B. Jones Agent.

Per V. J. P. Per Paul V. Cronk

Permanent Post-office address of Shipper New York City

The fibre boxes used for this shipment conform to the specifications set forth in the box-maker's certificate thereon, and all other requirements of Rule 41 Consolidated Freight Classifications.

FIG. 10.—Bill of Lading.

charges are based on weight or cubic space occupied, depending on which is the most advantageous. As it is impossible for a railroad to know just how much it costs to transport goods, freight charges are established by endeavoring to determine how much it is worth to the customer to have the goods transported. Hence it costs much more to transport a carload of silk than it does to transport a carload of sand. There are two kinds of rates, commodity rates and class rates. Commodity rates are charged when goods are shipped in large quantities. Class rates are charged on all types of articles which are classified into several different groups.

Freight Tariff Book.—This gives the classification of all articles, the names of all railroad depots, and the rate to be charged between them.

To use this book, one must look up the classification of his article, then look up the cost of transporting that article from the shipping point to the destination.

ILLUSTRATION: A manufacturer of shoes finds that the rate to ship shoes from his point to the point of destination is \$2.75 per cwt. How much will it cost him to ship 10 cases weighing as follows: 141, 142, 141, 143, 145, 135, 135, 137, 138, 132 lbs.?

This will be calculated as follows:

10 cases weigh	141
	142
	141
	143
	145
	135
	135
	137
	138
	132
Total weight	1389

$$1389 \times 2.75 \div 100 = \$38.19 \text{ Total Charges}$$

Express Shipments

The cost of express shipments is calculated in about the same manner. The American Railway Express classifies goods as first, second, or third class, and has about 300 different scales of rates, each designated by a block number. The scale of rates applies to any particular commodity and depends upon distance, weight, size, value, and whether or not it is fragile or perishable. In truck shipment, rates are not so well standardized and depend on many factors, frequently competition having much to do with them. They are usually quoted in terms of weight, however, and are calculated just as are the other types of shipments.

Typical express rates are listed below. In this instance the minimum charge for any express shipment is $3.60. When figuring charges add 3% Federal tax for each shipment. For instance, if your order weighs 30 pounds and you live about 200 miles from shipping point, the charge would be $4.00. Add 3% ($0.12) Federal Tax, making the total charge $4.12.

Weight	100 Miles	200 Miles	300 Miles	400 Miles	500 Miles
5 Pounds	$3.60	$3.60	$3.60	$3.60	$3.60
10 Pounds	3.60	3.60	3.70	3.80	3.90
15 Pounds	3.70	3.80	3.95	4.15	4.35
20 Pounds	3.80	3.95	4.20	4.50	4.80
30 Pounds	3.85	4.00	4.30	4.65	4.95
40 Pounds	3.90	4.10	4.45	4.80	5.20
50 Pounds	3.95	4.25	4.60	5.00	5.60
75 Pounds	4.20	4.70	5.30	5.95	6.95
100 Pounds	4.40	5.20	6.05	7.15	8.50

Taxes

The amount of tax to be collected is determined by preparing a budget in the preparation of which each department or division estimates the amount it will expend during the next fiscal year. The various departmental estimates are then grouped in order to determine the budget for that governmental unit.

ILLUSTRATION: The officials of a small New England town estimated that the budgetary needs for the next year would be those listed below. They also estimated the income that would probably be received from various sources other than from real estate and personal property. How much must be raised from these sources?

Estimated Expenditures:

State Tax	$ 3,133.00	
County Tax	6,102.03	
Town Charges	3,700.00	
Town Maintenance	9,000.00	
State Aid Construction	1,758.00	
Public Health Nurse	1,000.00	
Interest	400.00	
Libraries	515.00	
Street Lighting	2,750.00	
Memorial Day	50.00	
Schools	12,068.07	
Abatements	100.00	
Police Department	300.00	
Elections and Registration	200.00	
		$41,076.10
Less:		
Bank Stock	92.75	
		92.75
		$40,983.35
Less Estimated Income from Other Sources:		
Auto Tax	$1,300.00	
Interest and Dividends	700.00	
R. R. Tax	2,000.00	
Savings Bank Tax	2,000.00	
		6,000.00
		$34,983.35
Plus Overlay		1,319.90
		$36,303.25
Less Local Exemption		72.22
Total to Raise		$36,231.03

Direct Taxes.—There are two forms of taxes, direct and indirect. Direct taxes are those paid by the individual as taxes. Most of these are paid by the individual directly to the governmental unit, such as poll taxes, property taxes, income taxes, and various license fees; while some direct taxes, such as taxes on gasoline, theatre tickets, and most sales taxes, are paid as tax by the purchaser but are collected by the seller and then turned over to the government levying the tax.

Indirect Taxes.—Indirect taxes are those that are paid at the source by importers of various merchandise as a "duty" or by manufacturers of various merchandise such as cigars, cigarettes, and alcoholic beverages as an "excise tax." While these taxes are ultimately passed along to the purchaser in the form of increased price, he is never as conscious of paying them as he is of paying direct taxes.

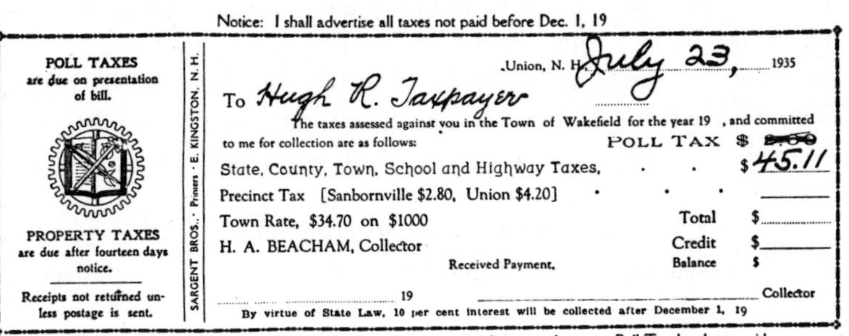
Notice: I shall advertise all taxes not paid before Dec. 1, 19

POLL TAXES are due on presentation of bill.

PROPERTY TAXES are due after fourteen days notice.

Receipts not returned unless postage is sent.

SARGENT BROS., Printers - E. KINGSTON, N. H.

Union, N. H. July 23, 1935

To Hugh R. Taxpayer

The taxes assessed against you in the Town of Wakefield for the year 19 , and committed to me for collection are as follows: POLL TAX $ 2.00

State, County, Town, School and Highway Taxes, . . $ 45.11

Precinct Tax [Sanbornville $2.80, Union $4.20] . . . ______

Town Rate, $34.70 on $1000 Total $______

H. A. BEACHAM, Collector Credit $______

Received Payment, Balance $

______ 19 ______ Collector

By virtue of State Law, 10 per cent interest will be collected after December 1, 19

It is illegal to register an automobile or secure an operating license unless your Poll Tax has been paid.

FIG. 11.—Tax Bill.

Property Taxes.—Taxes on real estate or personal property, called property taxes, are assessed by cities and towns, and paid by people owning such items. Some cities and towns pay relatively little attention to personal property, others endeavor to assess it as carefully as they assess real estate.

The property tax is the principal source of income for cities and towns, and the amount to be collected is determined by the amount required by the budget after deducting the estimated income from other sources such as poll taxes and license fees of one sort or another.

Tax Rate.—The rate at which property taxes are to be collected is usually quoted as so many mills on a dollar valuation, or so many dollars on a hundred or thousand dollars of property valuation. This is called the Tax Rate. Property values in a town or city is determined by officials known as assessors whose duty it is to place annually a value on all real estate and on taxable personal property. When the total value of taxable property in the community is determined, that total is divided into the total estimated budget (after the estimated income from other sources has been deducted) in order to determine the tax rate.

ILLUSTRATION: In a small town in New Hampshire the total assessed value of real estate and personal property was $1,163,636. The total budget, including taxes to be paid to the county and state by the town, called for expenditures of $41,076.00. After deducting estimated income from other sources such as auto tax, interest on investments, the amount to be raised was $33,601.06. What is the tax rate?

The tax rate is determined by dividing $33,601.06 by $1,163,636 (assessed value of real and personal property) and the result is $28.88 per thousand.

Tax Payment.—After the tax rate has been determined, the rate is applied to all assessed valuation and bills are sent to property owners. The amount is calculated by multiplying the assessed valuation by the rate. If there is a poll tax in force, this is included in the same bill.

ILLUSTRATION: A man owns real estate valued at $1400, and taxable personal property valued at $850. The property tax rate is 29.73 and there is a two-dollar poll tax in the state. What is his total tax?

The tax bill should be calculated as follows:

$$\$1400 \times \frac{29.73}{1000} = \$41.62$$

$$850 \times \frac{29.73}{1000} = 25.27$$

Poll Tax	2.00
Total Tax	$68.89

Income Taxes.—A tax on incomes is levied by the Federal Government and also by several state governments. All persons living in the United States whose income is in excess of certain minimums, and all corporations making a profit are required to pay income taxes. Income Tax Returns, which is a report of income for the year, must be filed with the Collector of Internal Revenue for the United States Government on or before April 15, and with tax laws at whatever date is specified by that law.

In preparing his return, the taxpayer should realize that in some matters he is permitted a choice of more than one method of procedure, so that it is clearly to his advantage to select the one that requires payment of the least tax. An obvious example is the question of whether or not to itemize deductions. If this is not done, the taxpayer is permitted to take a lump sum deduction (if the income reported exceeds $5000.) up to a maximum of $1000. of either 10% of his total income or (if married) $200. plus $100. for each exemption.

Another permissible choice is that permitted to married couples who are living together, of filing joint or separate returns. In general a joint return is the most advantageous. Under the following conditions it might pay to file separate returns:

1. If both husband and wife have capital losses totalling more than $1000. On the joint return only $1000. is deductible, which on the separate returns, both may deduct $1000. Note, however, that joint losses may be carried forward as deductions in future years.

2. If both husband and wife had income, but one of them had high medical expenses. Since 3% of the adjusted gross income must be deducted from the medical expense to find the allowable deduction, the deduction would be 3% of only the income of the spouse having the high medical expenses.

Of course, these two advantages of single returns may be outweighed by the advantages of joint returns, which include (1) lower rate schedule on joint returns than separate ones; (2) child care expenses usually not deductible from separate returns; (3) income of one spouse so low that his or her dependency deductions or other deductions exceed his or her income so that the full deductions can only be taken on a joint return.

The only way to prove which type of return is most advantageous in a given case is to compute the taxes by both methods.

Withholding Tax.—The United States Government and some state governments require employers to withhold a portion of the individual's income tax from his wages and pay it periodically to the government in question. Self-employed persons prepay their estimated taxes to the United States Government quarterly through the year, making a final adjustment payment

TABLE 10

WITHHOLDING RATES

Weekly Pay	Single Person	Married Person	
		No Dependents	Two Children
$75	$9.30	$6.40	$2.40
100	14.10	10.60	6.50
125	19.10	14.60	10.30
150	24.60	19.30	14.70
200	35.80	28.10	23.20
250	49.80	38.10	32.70
300	64.80	48.10	42.70
350	79.80	58.10	52.70
400	93.30	70.00	63.20
500	123.30	97.70	89.60
600	153.30	126.20	118.10

by April 15 of the following year. This practice is also followed, in addition to the withholding method, by employed people whose incomes exceed a specified amount. The United States withholding tax for employees is based upon a graduated scale as shown in Table 10.

Social Security.—Under the Federal Insurance Compensation Act 4.2% of the first $600 earned per year by an employee is paid by the employer to the Federal Government, along with an equal amount contributed by the employer. This money provides Social Security insurance.

Capital Stock Tax.—The corporation, in addition to the regular income tax, is required to make an additional report and pay an additional tax on the declared value of their stock.

Inter-corporate Dividends.—To assist in controlling holding companies, the Federal government has made 10% of dividends paid from one corporation to another taxable at the corporation income tax rate.

Estate Taxes.—The Federal government and several states levy taxes on estates of deceased persons, providing those estates are in excess of certain amounts. The government (largely to protect the estate tax) also levies taxes on sizable gifts. In the tax law, levies begin at two percent on that part of the estate that is in excess of $40,000 and range up to 70% of that part in excess of $50,000,000. These levies apply to the entire estate left by an individual, regardless of how many persons or institutions may inherit parts of it. New gift taxes are approximately three-fourths as high as the estate levies.

Sales Taxes.—Several states now levy a sales tax of one sort or another, the most common being a tax on retail sales. The tax is most manageable when there are few exempt items. The taxes are collected by the seller on fixed scales and paid to the government levying the tax as a certain percent of gross sales.

Duties on Imports.—Taxes are levied by the Federal government upon commodities imported. These taxes are called Duties, or Customs. There are two kinds, *ad valorem* and *specific*. Ad valorem duties are levied as a certain percent of the value. The value may be fixed by appraisal by U. S. Customs officials or may be determined by the invoice price in the country from which the article is imported.

Specific duty is a certain amount levied on certain articles. It may be per ton or per pound, bushel, yard, gallon, quart, or other unit measure. A duty may be either ad valorem or specific, but in some cases both types are levied on the same article. There are two forms of entry for imported goods: consumptive entry and warehouse entry. Under consumptive entry, the duty is paid on the goods at the time they come in. Under warehouse entry, the goods are placed in bonded warehouses and the duty must be paid when the goods are removed therefrom unless they are subjected to other regulations. Usually they must be removed within three years.

ILLUSTRATION: Assume that the duty on printing paper is 10% ad valorem, and ¼¢ per pound specific. A man bought 10,000 lbs. in England and paid the equivalent of $350. How much tax must be paid?

The duty would be as follows:

Ad valorem, 10% of $350.00	$35.00	
Specific ¼¢ × 10,000	25.00	
Total Duty	$60.00	(Ans.)

XXIV

Accounting

Purpose of Accounting.—Accounting is the maintenance of a systematic record of business transactions, and its use in preparing control reports or summaries. This information is very important, and often obligatory, in many of the external relations of any business. Every application for a bank-loan, and most efforts to secure additional capital or credit must be accompanied by accounting figures. Any claim for property loss made upon an insurance company must be based upon an evaluation of the assets lost or damaged. Whenever a business is sold, the accounting records play a significant and often a critical part in determining the total price. Last, but by no means least, these records are required by various tax agencies, including the U. S. Internal Revenue Service.

While accounting figures are so necessary in many of the outside relations of a business, they are, if anything, even more valuable in its internal operation. For their control summaries furnish the most practical means for (1) planning and keeping its operations profitable, and for (2) sustaining a favorable ratio of current assets to current liabilities. Both of these conditions must be maintained if the business is to continue in existence.

Accounting records are kept in "books," which may be bound or loose-leaf, or not books at all, but filing cabinets containing separate sheets. There are two kinds of books, those of first entry which are called journals and contain descriptions of the transactions; and those in which the journal entries are copied (posted) to record the effect upon the assets and liabilities of the business of the particular transaction; these books of later entry are called ledgers.

How to Start and Keep Books.—In a small business the bookkeeping system can be started with a single journal, which would then be called a day-book, and a single ledger, which would have separate pages for each account. The various transactions with their descriptions are recorded in the journal. However, transactions of the same kind are often, for practical reasons, grouped into a single entry—thus the cash sales of a retail store are commonly entered as a total figure for each day. As the business grows, some types of transactions become so numerous that time can be saved by having for them a separate journal, or a different place in the same journal or the filing cabinet. Among such frequently occurring transactions are sales, purchases, cash transactions, etc.

There is, however, one type of transaction for which a separate journal should be started at once. It is wages or salaries. As soon as the business grows to the point where employees are hired, a Payroll Record should be separated from the journal, or day-book. In this Payroll Record should be entered under the name of each employee his gross wages or salary, the amount of all deductions, including Social Security deductions and Withholding Tax deductions, and the net payments, which are the wages actually paid to the employee.

There are five kinds of ledger accounts: the two kinds of operating accounts are income entries, representing what is taken in, and expense entries, or what is paid out. The other three kinds of accounts are asset accounts (what is owned), liability accounts (what is owed), and capital accounts (amounting to the difference of what is owned and what is owed). Thus Sales is an income account; Purchases, an expense account; Accounts Receivable, an asset account; Accounts Payable, a liability account; and the Owner's Investment Account, a capital account.

The basic idea underlying accounting is expressed by the word balance, as used in the sense of balancing the books. Every transaction is regarded as a transfer from one account to another, in other words, an "addition" to one account is balanced by a "subtraction" from another, and since the two operations are theoretically done at the same time, the books are always in balance if no error has been made. However, instead of the words "addition" and "subtraction," which apply here only in the algebraic sense, the accountant uses the words debit and credit, and in the ledger, which has two columns for the purpose, debits are entered on the left and credits on the right. The meaning of these terms depends upon the nature of the account—for example, cash sales are debits to the cash account and credits to the sales account, payment of rent is a credit to the cash account and a debit to the expense account, etc. While the cash account entries are comparatively easy to reason out (receipts being debits and expenditures credits), this process is not nearly so easy for some of the other accounts. The best way to gain this knowledge is by following through a few steps in the accounting of a small business.

To start such a system, buy two books or types of forms—one for the single journal, or day-book, which would appear as shown in Figure 1; and the other for the ledger, which would appear as shown in Figure 2. Then take a group of representative transactions and record them in these books, beginning, of course, with the journal. The following list gives a description of a series of simple transactions:

October 1. Robert Jones invested in his new business $4,800.00 in cash and $2,000.00 in merchandise.

October 2. He paid $100.00 for rent, and $65.80 for stationery, account books and other office supplies.

October 3. He bought merchandise from K. Smith and Co. for $1945.52 on account, and he sold merchandise for $441.92 in cash.

October 4. He bought a safe for $160.00.

October 5. He sold merchandise to Walter Mitchell for $620.10, receiving $300.00 in cash.

October 6. He sold $460.48 merchandise to John Peters, receiving $100.00 cash, and Peters' 60-day note for the balance.

October 8. He bought merchandise in the amount of $958.20 from United Metals Co., paying $500.00 in cash.

October 9. Walter Mitchell returned $20.50 merchandise bought on October 5.

October 10. He paid postage $10.00; he bought merchandise at auction for cash $56.75; and he drew from business for personal use $50.00.

These transactions would be recorded in a single journal as shown in Figures 1a and 1b.

Note how these entries are made. The capital investment of $4800.00 is an acquisition of cash by the business; therefore, it is entered in the journal as a debit to Cash (right-hand column) and a credit (left-hand column) to the owner's Investment Account. Similarly, his merchandise investment of $2000.00 is a debit to Merchandise Inventory and a credit to his Investment Account. Rent paid is an expenditure of cash, and is therefore credited to Cash and debited to a General Expense account. Office supplies purchased are entered in the same way, except that a special account, separate from General Expense, called Expense Supplies, is commonly used for them. Purchases on account are obviously a credit to the supplier, entered to his account, and a debit to the Purchases of the business. Conversely, sales are a credit to the Sales account of the business, and if made for cash, as was the entry of $441.92, they are a debit to the Cash account.

Continuing with these transactions, the acquisition of a safe for $160.00 in cash is a credit to the Cash account and a debit to Furniture and Fixtures. The sale to Walter Mitchell for part cash is a credit to Sales, balanced by a debit to Cash and by another debit to Mitchell's account for the unpaid balance. Similarly, the sale to John Peters for part cash and a note for the balance is a credit to Sales, balanced by a debit to Cash and another debit to Notes Payable. The name of the maker of the note, its date and maturity are noted in the journal, as indeed is any fact of major importance about an entry. The purchase for part cash from the United Metals Co., is a debit to Purchases, balanced by a credit to Cash and another credit to the account of the United Metals Co. The merchandise returned by Walter

Date		Account	L.F.	Debit	Credit
Oct.	1	Cash	2	4800 00	
		Robert Jones, Investment	1		4800 00
Oct.	1.	Merchandise Inventory	3	2000 00	
		Robert Jones, Investment	1		2000 00
Oct.	2	General Expense	4	100 00	
		Cash	2		100 00
		Rent for October			
Oct.	2	Expense Supplies	5	65 80	
		Cash	2		65 80
Oct.	3	Purchases	6	1945 52	
		K. Smith + Co.	7		1945 52
Oct.	3	Cash	2	441 92	
		Sales	8		441 92
Oct.	4	Furniture + Fixtures	9	160 00	
		Cash	2		160 00
		Cost of safe			
Oct.	5	Cash	2	300 00	
		Walter Mitchell	10	320 10	
		Sales			620 10
Oct.	6	Cash	2	100 00	
		Notes Receivable	11	360 48	
		Sales	8		460 48
		Note of John Peters			
		for 60 days dated 10/6			
Oct.	8	Purchases	6	958 20	
		Cash	2		500 00
		United Metals Co.	12		458 20

FIG. 1a.

Month	Day	Account / Explanation	L.F.	Debit	Credit
Oct.	9	Sales Returns	13	20 50	
		Walter Mitchell	10		20 50
		Returned mdse.			
		claimed defective			
Oct.	10	Expense Supplies	5	10 00	
		Cash	2		10 00
		Postage			
Oct.	10	Purchases	6	56 75	
		Cash	2		56 75
Oct.	10	Robert Jones, Personal	14	50 00	
		Cash	2		50 00

Fig. 1b.

Mitchell is obviously entered as a credit to his account, and a debit to Sales Returns. (The debit could be made to sales, but this is contrary to accounting practice, and less informative.) Stamps bought for $10.00 on October 10 are a credit to Cash, and a debit to the Expense Supplies account. Similarly, merchandise bought for cash, $56.75, is a credit to Cash and a debit to the Purchases account. Finally, the withdrawal of $50.00 on October 10 by the proprietor of the business is a credit to the Cash account, and a debit to his Personal Account. Note that two accounts are kept for the proprietor: his Investment Account (the first of the above entries) and his Personal Account. This also is proper accounting practice, important in the preparation of statements and tax returns.

By following through the foregoing simple transactions, item by item, and comparing them with the journal entries shown in Figures 1a and 1b, one can easily gain insight into the handling of accounting transactions. Its value becomes clear at once from a consideration of the next step, which is the posting of these journal entries into the ledger.

The first step in starting a ledger is to begin accounts, each on a separate page, for every account that appears in the journal. These will include every customer who owes money to the business, every creditor to whom the business owes money, and every internal account of the business. Posting the foregoing transactions from the journal will yield the ledger accounts and entries which are arranged in Figures 2 to 15, in the order they would appear by posting them just as they occur in the journal.

By comparing these ledger accounts, one by one, with the journal entries (Figures 1a and 1b) one can see how the posting is done. The order of the ledger accounts here, which results from taking them, for purposes of demonstration, in the journal order, is not one that an accountant would follow. However, this matter of order is not important, because the page number of the ledger account is shown on the journal entry. It appears in a column between the title of the entry and its amount, as can be seen by referring to Figures 1a and 1b.

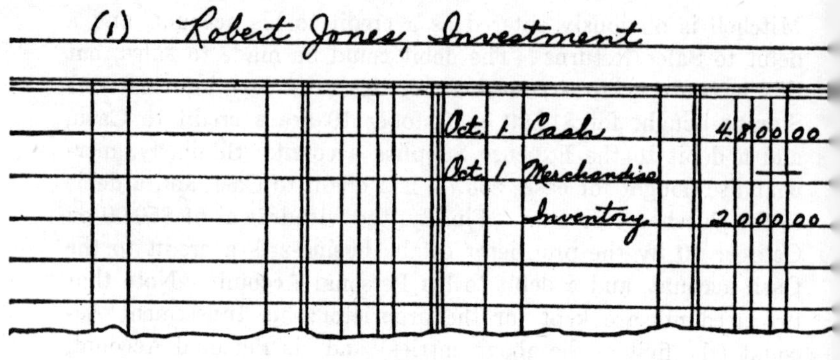

(1) Robert Jones, Investment
Oct. 1 Cash 4800 00
Oct. 1 Merchandise Inventory 2000 00

Fig. 2.

(2) Cash

Date		Item	Amount	Date		Item	Amount
Oct.	1	Robert Jones, Investment	4800 00	Oct.	2	Gen. Expense	100 00
Oct.	3	Sales	441 92	Oct.	2	Expense Supplies	65 80
Oct.	5	Sales	300 00	Oct.	4	Furn. + Fixt.	160 00
Oct.	6	Sales	100 00	Oct.	8	Purchases	500 00
				Oct.	10	Expense Supplies	10 00
				Oct.	10	Purchases	56 75
				Oct.	10	Robert Jones, Personal	50 00

Fig. 3.

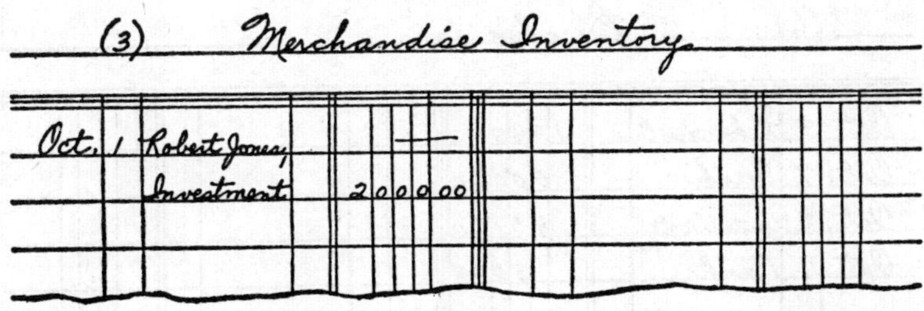

(3) Merchandise Inventory
Oct. 1 Robert Jones,
Investment 2000 00

FIG. 4.

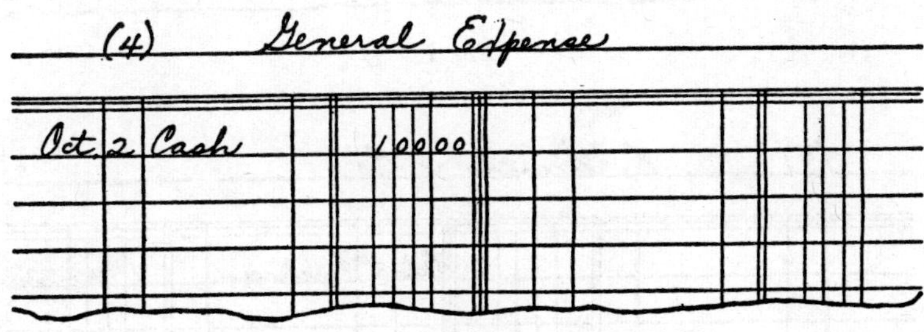

FIG. 5.

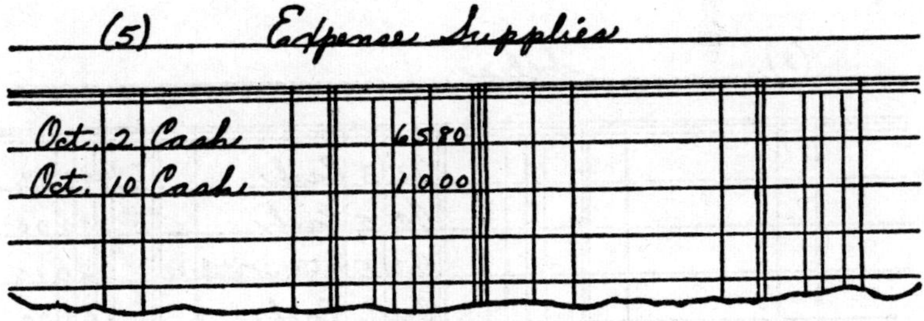

(5) Expense Supplies
Oct. 2 Cash 65 80
Oct. 10 Cash 10 00

FIG. 6.

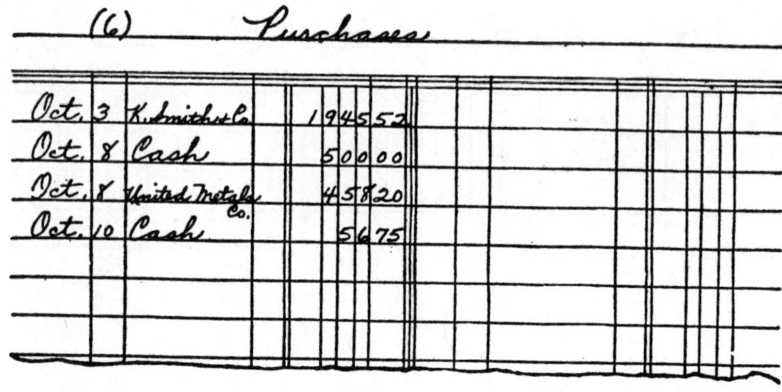

(6) Purchases

Date		Item	Amount	Date		Item	Amount
Oct.	3	K. Smith & Co.	1945 52				
Oct.	8	Cash	500 00				
Oct.	8	United Metals Co.	458 20				
Oct.	10	Cash	56 75				

Fig. 7.

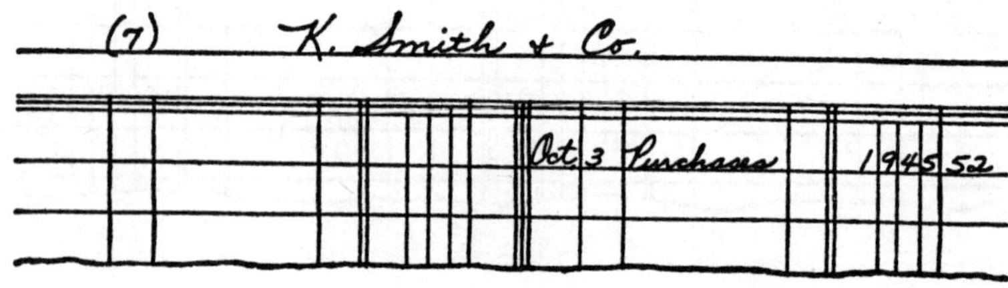

(7) K. Smith & Co.

Date		Item	Amount	Date		Item	Amount
				Oct.	3	Purchases	1945 52

Fig. 8.

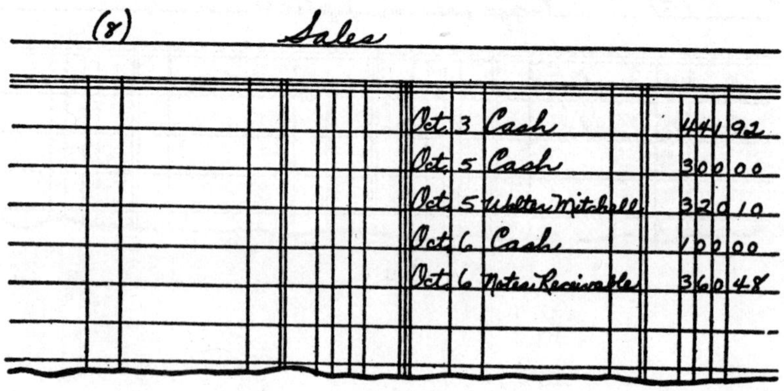

(8) Sales

Date		Item	Amount	Date		Item	Amount
				Oct.	3	Cash	441 92
				Oct.	5	Cash	300 00
				Oct.	5	Walter Mitchell	320 10
				Oct.	6	Cash	100 00
				Oct.	6	Notes Receivable	360 48

Fig. 9.

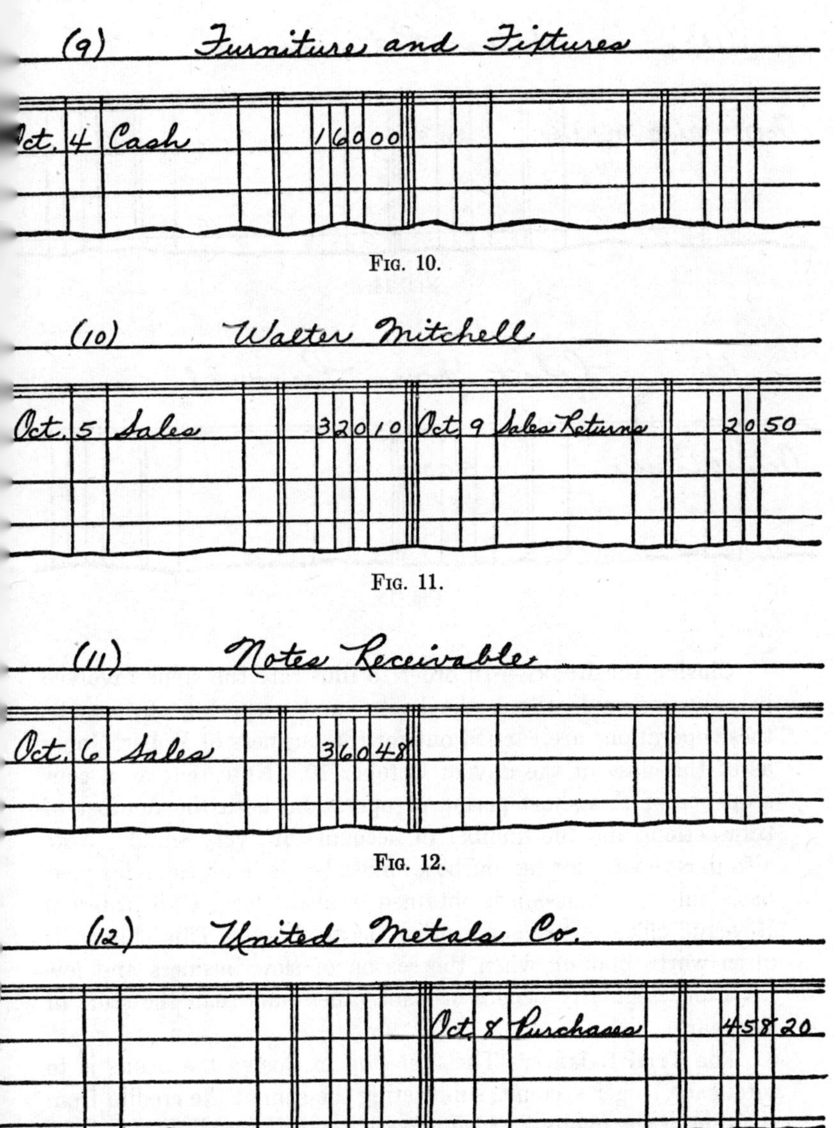

(9) Furniture and Fixtures

Date		Item	Fo.	Amount	Date		Item	Fo.	Amount
Oct.	4	Cash		160 00					

Fig. 10.

(10) Walter Mitchell

Date		Item	Fo.	Amount	Date		Item	Fo.	Amount
Oct.	5	Sales		320 10	Oct.	9	Sales Returns		20 50

Fig. 11.

(11) Notes Receivable

Date		Item	Fo.	Amount	Date		Item	Fo.	Amount
Oct.	6	Sales		360 48					

Fig. 12.

(12) United Metals Co.

Date		Item	Fo.	Amount	Date		Item	Fo.	Amount
					Oct.	8	Purchases		458 20

Fig. 13.

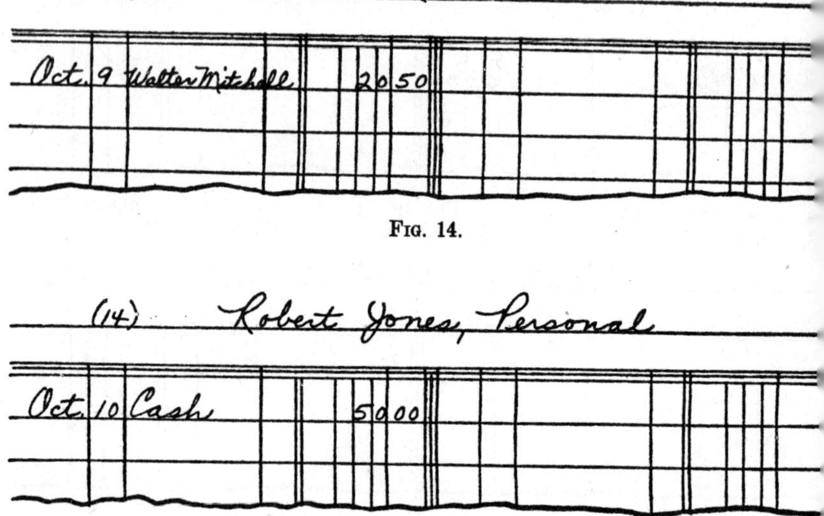

(13) Sales Returns
Oct. 9 Walter Mitchell 20 50

Fig. 14.

Fig. 15.

Closing the Books.—In order to illustrate the steps involved in adjusting and closing the books, and preparing statements, these operations are carried out for the business of Robert Jones as of the close of the day of October 10. Note that as a consequence of this short period of operation, both the number of transactions and the number of accounts are very small. Note also that reports for tax purposes must be made on calendar year basis, unless permission is obtained from the local U. S. Internal Revenue office to use some other fiscal period. This change is often worth making when the season of slow business and low inventory regularly occurs at some other time than the start of the year.

The Trial Balance.—The first step in closing the books is to total each ledger account, subtracting the sum of the credits from the sum of the debits, or vice versa if the total credits exceed the total debits, ruling off the accounts, and entering the net balance of each account in the debit or credit column, as the case may be, of the ledger. If one were to perform these operations for the fourteen accounts that were prepared for the business of Robert Jones through October 10, the following Trial Balance would be obtained:

TRIAL BALANCE, October 10

(1) Robert Jones, Investment		$6,800.00
(2) Cash	$4,699.37	
(3) Merchandise Inventory (Start of Period)	2,000.00	
(4) General Expense	100.00	
(5) Expense Supplies	75.80	
(6) Purchases	2,960.47	
(7) K. Smith & Co.		1,945.52
(8) Sales		1,522.50
(9) Furniture and Fixtures	160.00	
(10) Walter Mitchell	299.60	
(11) Notes Receivable	360.48	
(12) United Metals Co.		458.20
(13) Sales Returns	20.50	
(14) Robert Jones, Personal	$10,726.22	$10,726.22

The trial balance "balances" as indeed it must in a double-entry system unless there has been an error in posting or in arithmetic.

Adjustment Entries.—The next step in closing the books is to make the adjustment entries. They are required in accounting for various purposes, for example, to take care of expenses incurred but not payable, such as accrued wages, or of the unused period of prepaid expenses, such as unexpired insurance. The only adjustment information required in the business situation under discussion is the change in merchandise inventory. This is determined by taking a new inventory as of the close of business on October 10. This inventory is found to have a value of $3,974.60, which is posted in the Merchandise Inventory Account as a credit. Then the purchases are also posted to this account as a credit, and the difference between the sum of the beginning inventory and purchases, less the final inventory, is the cost of goods sold. Figure 16 shows how these entries appear on

(3) Merchandise Inventory

Oct.	1	Robert Jones, Investment	1	2000 00	Oct.	10	Inventory		3974 60
Oct.	10	Purchases	L6	2960 47	Oct.	10	Cost of Goods Sold (To Profit + Loss)		985 87
				4960 47					4960 47
Oct.	10	Inventory		3974 60					

Fig. 16.

the Merchandise Inventory Account, and it also shows the method of ruling-off an account and entering its new balance, which you should do at the close of an accounting period for all asset, liability and capital accounts. Another operation in closing these books is to post sales returns to the Sales Account, and to enter the difference as net sales.

(15) Profit and Loss

Oct.	10	Cost of Goods Sold	L3	985 87	Oct	10	Net Sales	L8	1502 00
		(From Mdse. Inventory)							
Oct.	10	General Expense	L4	100 00					
Oct.	10	Expense Supplies	L5	75 80					
Oct.	10	Robert Jones,							
		Investment	L1	340 33					
				1502 00					1502 00

Fig. 17.

(1) Robert Jones, Investment

Date		Item	F.	Amount	Date		Item	F.	Amount
Oct	10	Robert Jones Personal	14	50 00	Oct	1	Cash	2	4800 00
					Oct	1	Merchandise Inventory	3	2000 00
Oct	10	Capital + Profit		7090 33					
				7140 33	Oct	10	From Profit + Loss	15	340 33
									7140 33
					Oct	10	Capital + Profit		7090 33

Fig. 18.

Profit and Loss Account.—The next step in closing the books is to set up a new account in the ledger, the Profit and Loss Account. To it is transferred the cost-of-goods-sold adjustment entry from the Merchandise Inventory Account, the net sales entry from the Sales Account, and the various expense accounts. The difference found in this account, that is, profit or loss, is balanced by a transfer to the proprietor's Investment Account. This particular Profit and Loss Account is shown in Figure 17, and the proprietor's Investment Account as of October 10 (including the transfers from his Personal Account and from Profit and Loss) is shown in Figure 18.

Profit and Loss Statement.—The Profit and Loss Account which has just been posted is useful, not only in closing the books at the end of a period, but also in preparing the Profit and Loss Statement for that period. This statement has great value, as stated at the beginning of the chapter, not only in the external relations of the business, as with banks and taxing agencies, but also in the internal management. For that reason, it is presented in sufficient detail to show figures important to efficient control.

A Profit and Loss Statement for the Robert Jones business would be prepared as shown on following page.

Financial Statement.—The other important statement is the Financial Statement, which is also called the Balance Sheet. It shows the financial condition of the business. The information may be arranged in various ways, one of the most common shown on next page, in which the liabilities (and proprietorship entries) are grouped below the assets, or on a facing page.

Simplified Methods for Small Business.—To save time and money in keeping books, there are a number of short-cuts which accountants have approved for small business. Some are limited to particular circumstances or kinds of business, and those must be developed with an accountant. There is one short-cut method that is often found useful. A simple file is used for accounts payable, placing in it all the invoices grouped according to your creditors' names. This file saves the work of entering each invoice because one is, in effect, using the file as an accounts-payable journal.

Related to this method, there is another policy that should always be followed as a matter of sound accounting practice. It is this: to make the bank account of the business an accurate income-record by depositing in full each day's income. If wages, invoices or other expenses are paid in cash, checks are drawn for them to cash, the purpose of the payment being noted in the checkbook, and they are then endorsed and deposited in the bank with the remainder of that day's income. If this procedure is followed the cancelled checks will give all in one place, "receipts" or vouchers for every expenditure; while the deposits will constitute a complete record of income. These records will be of the greatest value, not only in any possible discussion with tax examiners or collectors, but also for future uses in obtaining bank loans or even proving volume of sales to a prospective purchaser of your business.

Accounting for Representative Types of Business.—The next step in the explanation of accounting methods is to review actual practices followed in specific types of enterprise. Therefore, to cover the needs of the small business as broadly as possible, ac-

ROBERT JONES

PROFIT AND LOSS STATEMENT, October 10

Sales		**$1,522.50**	
Less Sales Returns		**20.50**	
Net Sales			**$1,502.00**
Merchandise Inventory, Oct. 1	**$2,000.00**		
Purchases	**2,960.47**		
		4,960.47	
Merchandise Inventory		**3,974.60**	
Cost of Goods Sold			**985.87**
Gross Profit			**$ 516.13**
Expenses:			
General Expense		**100.00**	
Expense Supplies		**75.80**	
			175.80
Net Profit for the Period			**$ 340.33**

ROBERT JONES

FINANCIAL STATEMENT, October 10

		Assets	
Cash			$4,699.37
Notes Receivable			360.48
Accounts Receivable			299.60
Merchandise Inventory			3,974.60
Furniture and Fixtures			160.00
Total Assets			$9,494.05
		Liabilities	
Accounts Payable		$2,403.72	
Total Liabilities			$2,403.72
Net Worth			$7,090.33
		Proprietorship	
Robert Jones, Investment, Oct. 1		6,800.00	
Profit for the Period	$340.33		
Less Withdrawals	50.00		
		290.33	
New Worth			$7,090.33

counting systems will be discussed for (1) the mail-order business; (2) the retail business; (3) the small manufacturing business; and (4) the service business.

The Mail-Order Business.—In considering the accounting of a mail-order business, one will discover at once that two types of records are necessary. The first, which is the overall accounting of the business, is designed to produce the same statements and records already discussed. The second is the accounting of the individual sales operations; that is, each mailing, each newspaper or magazine advertisement, and as far as possible each radio or television broadcast or each short-term series of broadcasts. Such records are indispensable to the profitable operation of a mail-order business. Therefore, examples are now given of an accounting record for each of these three types of operation.. See Examples 1, 2 and 3.

These three examples cited are given to show how to keep the accounting records of individual mail-order advertisements. The process can be shortened somewhat with accumulated experience with some of these operations. For example, the shipping cost entries in all three cases are repetitive and can be combined, because the chief purpose of these figures is to determine whether these advertisements are profitable. For that reason also it was assumed that the mailing and shipping operations were done by outside agencies, for otherwise the labor charges for them would have to be found by analysis of the operating expenses of the mail-order company.

The critical question is that of whether the remaining operating expenses and administrative expenses are covered by the percentages of profit shown. These expenses include Clerical Salaries, Administrative Salaries, Rent, Insurance, Taxes on Payrolls, Other Taxes, Transportation on Goods Received. Moreover, if the product is manufactured on the premises and not purchased ready to ship, its unit cost, which was taken at $1.84, must include its share of all general and administrative expenses, as well as those directly chargeable. The answer to the question of whether the profit from the operation covers the overhead de-

EXAMPLE 1

Mailing to Customer List A
Mailing Period: March 12-15

Nature of List:	Purchasers of Cake Assortment A
Nature of Offer:	Autumn-Leaf Maple Candy Assortment ($4.95) with premium offer of maple syrup to be sent C.O.D. (or cash with order prepaid)
Nature of Mailing:	Two color circular, 17″ × 22″ in size; multigraphed letter on lithographed letterhead; business reply postcard; all in a #10 envelope, marked for "Bulk Mailing" under Section 34.66 Postal Laws & Regulations. (Such mailings pay 1½¢ postage up to the weight limit. They require a permit from the local postoffice, and must be sorted as directed.)
Quantity of Mailing:	16,430

I. *Cost of Mailing:		
Circulars and Letters:		
Delivered cost of circulars, including paper and printing	$191.30	
Cost of lithographing and paper for letterheads	53.35	
Cost of multigraphing letters	44.62	
Cost of paper-stock and printing of postcards	32.90	
Total cost of circulars and letters	$322.17	
Mailing Expense.		
†Addressing Envelopes from List	65.72	
†Folding, inserting, sealing and stamping	49.29	
†Sorting Mail	16.43	
Postage	246.50	
Postage due on orders received	13.65	
Total mailing expense	$391.30	
Total Cost of Mailing		$713.47

* These costs are shown more in detail than is consistent with the other accounting schedules in this chapter. This variation has been made to show you some of the specialized operations in direct-mail work.

† The labor operations are charged as if they were done by an outside mailing service. If the work is performed in the mail-order business, then these charges are not made directly; but there is a wage cost to pay the workers. Note also that this addressing charge is for labor only; if an outside list is used it will be larger to include list rental.

(Continued)

EXAMPLE 1 (Cont'd)

II. Income from Mailing			
Orders Received	525		
Shipments Not Accepted	66		
Sales	459		
Gross Income from Sales			$2,272.05
III. Shipping Expenses:			
C.O.D. charges on prepaid orders		$ 35.17	
Express charges on shipments not accepted		26.52	
C.O.D. collection charges		150.30	
Shipping cartons		60.65	
Shipping labor (see ** above)		50.14	
Total shipping expenses			323.05
IV. Cost of Goods Sold:			
459 assortments @ $1.84		844.56	
12 assortments returned in unsaleable condition		22.08	
Total Cost of Goods Sold		866.64	
Recapitulation:			
Gross Income from Sales			2,272.05
Total cost of mailing		713.47	
Total shipping expenses		323.05	
Total cost of goods sold		866.64	
Total costs			1,903.16
Profit before overhead			$ 368.89

EXAMPLE 2

MAGAZINE ADVERTISEMENT

Medium: Cosmopolitan
Date of Issue: March, 19——
Space Used: 53 lines in outside column on Page 11.
Nature of Offer: Autumn-Leaf Maple Candy Assortment ($4.95) with premium offer of maple syrup to be sent C.O.D. (or cash with order prepaid)

I. Cost of Advertisement	
Space	$355.00
Art-work	24.00
Halftones	33.60
Total Cost of Advertisement	412.60

(Continued)

EXAMPLE 2 (Cont'd)

II. Income from Advertisement			
Orders Received	322		
Shipments not accepted	53		
Sales	269		
Gross Income From Sales			$1,331.55
III. Shipping Expenses			
C.O.D. charges on prepaid orders		17.22	
Expenses charged on shipments not accepted		21.20	
C.O.D. collection charges		80.70	
Shipping cartons		38.10	
*Shipping labor		30.90	
(*See note in report of MAILING TO CUSTOMER LIST A)			
Total Shipping Expenses			188.12
IV. Cost of Goods Sold			
269 Assortments $1.84		$495.96	
10 Assortments Returned in Unsaleable Condition		18.40	
Total Cost of Goods Sold			514.36
Recapitulation			
Gross Income From Sales			1,331.55
Total Cost of Advertisement		426.00	
Total Shipping Expenses		188.12	
Total Cost of Goods Sold		514.36	
Total Costs			1,128.48
Profit before Overhead			$ 203.07

EXAMPLE 3

Radio Advertisement

Station: WINS in New York City

Dates of Broadcast: Five consecutive days, beginning March 3 and ending March 7.

Time Used: One-minute spot announcements each day at 3:30 P.M.

Nature of Offer: Autumn-Leaf Maple Candy Assortment ($4.95) with premium offer of maple syrup to be sent C.O.D. (or cash with order prepaid)

I. Cost of Broadcasting: Five announcements at $55.00 each, giving a total of $275.00

(Continued)

EXAMPLE 3 (Cont'd)

II. Income from Broadcasting			
Orders Received	214		
Shipments Not Accepted	39		
Sales	175		
Gross Income From Sales			$866.25
III. Shipping Expenses			
C.O.D. Charges on Prepaid Orders		$10.54	
Express Charges on Shipments Not Accepted		15.60	
C.O.D. Collection Charges		52.50	
Shipping Cartons		25.30	
*Shipping Labor		20.40	
(* See note in report of MAILING TO CUSTOMER LIST)			124.[illegible]
IV. Cost of Goods Sold			
175 Assortments @ $1.84		322.00	
8 Assortments Returned in Unsaleable Condition		14.72	
Total Cost of Goods Sold			336.72
Recapitulation			
Gross Income From Sales			$866.25
Total Cost of Broadcasting		$275.00	
Total Shipping Expenses		124.34	
Total Cost of Goods Sold		336.72	
Total Costs			736.06
Profit Before Overhead			$130.19

pends upon its amount, and shows why the overhead of a small-scale mail-order business must be held to an absolute minimum. As the business expands, the savings made possible by larger printings of letters and circulars, and the increased proportion of orders from the use of larger units of advertising space, can be hoped to take care of increased overhead, and even provide increased net profit.

It is to be emphasized that the three examples shown were purposely chosen as hypothetical cases, with correspondingly hypothetical figures. They are included to show accounting methods—not relative returns from different types of media.

Retail Store Accounting.—For the purposes of this treatment, a representative retail store has been taken as a general bookstore, that is, a bookstore carrying current fiction and nonfiction, without specializing in any particular kind of books, and with certain other merchandise, such as stationery, greeting cards, pens and pencils, etc. A special feature of bookstore accounting arises from the practice of carrying books on consignment, which should be shown in a separate inventory (not included in the regular inventory as part of the assets) and as these books are sold they should be credited to the account of the company to whom they belong, in a special Consignment Account Payable.

The balance shcet accounts to be kept in a retail bookstore would then be classified as follows:

Assets

Current Assets

1. Cash in Bank
2. Cash on Hand
3. Accounts Receivable (one for each account)
4. Notes Receivable
5. Bad Debt Reserve
6. Merchandise Inventory
7. Rental Library Stock
8. Other current assets

Fixed and Other Assets

9. Real Property
10. Furniture and Fixtures
11. Depreciation Reserve
12. Prepaid Expenses

Liability and Capital Accounts

Current Liabilities

20. Accounts Payable (one for each account)
21. Notes Payable
22. Library Deposits
23. Expenses Accrued but Not Due

Deferred Liabilities

24. Long Term Debts

Capital Accounts

30. Invested Capital
31. Surplus
32. Profit and Loss

The balance sheet of a bookstore is prepared by summarizing the foregoing accounts.

The operating accounts of a bookstore would consist of the following:

Operating Accounts

Income Accounts

40. Book Sales
41. Returned Books
42. Other Merchandise Sales
43. Returns of Other Merchandise
44. Income from Library Rentals
45. Income from Library Sales

Expense Accounts

50. Book Purchases
51. Other Merchandise Purchases
52. Library Purchases
53. Salaries
54. Rent
55. Telephone
56. Light and Heat
57. Taxes
58. Office Supplies and Office Postage
59. Shipping and Packaging Supplies and Shipping Charges
60. Charges on Incoming Shipments
61. Insurance
62. Bad Debts

The operating statement is prepared in the usual way, as explained in the early part of this chapter, by adding the income accounts (less returns), and subtracting the cost of goods sold to determine gross profit, and then subtracting total expenses to determine net profit. The cost of goods sold is determined for each of the three categories of income by adding purchases and charges on incoming shipments to inventory at beginning of the period, and subtracting inventory at end of period.

For the control interpretation of these figures in bookstore operation, the reader is referred to HANDBOOK OF ACCOUNTING METHODS, Second Edition, by J. K. Lasser. There will be seen the application of various ratios in the control of a business, particularly in detecting certain unsound conditions before they result in a net loss, rather than a net profit. For the retail bookstore business, for example, the suggested control figures are: (1) the ratio of total payroll (including owners' drawings) to sales, which should not exceed 13–14%; the ratio of total payroll to gross profit, which should not exceed 40–50%; (2) the ratio of rent to sales, which should not exceed 6–8%; and (3) the ratio of advertising to sales, which should rarely exceed 2%.

Of course, the figures given above may vary somewhat according to the special circumstances of a particular bookstore. They should, however, be watched closely because these three items, payroll, rent and advertising are usually the largest expenses of the bookstore. In exercising accounting control over other types of business, the same principle—that of watching ratios of the largest expense items to sales, or to gross profits—should be followed. Moreover, one other accounting figure that is so important to the bookstore is also very significant in many other types of retail business. That figure is inventory. In arriving at its true value, one needs to recognize differences in rate of depreciation or obsolescence of different classes of merchandise. For example, in a bookstore, the current fiction titles in general lose saleability more rapidly than the non-fiction titles, and therefore as books grow older the inventory-value of the former should be "written-down" more rapidly. The same principle should be applied in the accounting of other retailing businesses handling timely or seasonal products.

Accounting for the Small Manufacturing Business.—The business of food packing includes a wide range of possible enterprises for small-scale operation. Such enterprises are not, of course, necessarily small, but may vary in size from the home-kitchen production of jams and jellies to the giant cannery. The accounting methods suggested are, therefore, sufficiently flexible to meet the needs of the somewhat larger business, but not, of course, the truly large enterprise.

There is, however, one peculiarity of all enterprises of this nature, large and small. That is the limited seasonal availability of the raw material, which requires a short but intensive period of manufacturing activity which, if not planned carefully in advance, may leave in its wake an inventory of high-cost product that cannot be marketed profitably.

Holding this major consideration in mind, one can then set up the balance sheet accounts to be kept in a food packing business as follows:

Assets

Current Assets

1. Cash in Bank
2. Cash on Hand
3. Accounts Receivable
4. Notes Receivable
5. Marketable Investments
6. Bad Debt Reserve
7. Inventories of Raw Materials and Finished Products

This account includes a considerable number of items, such as delivered cost of raw materials, less discounts and allowances, and all the other costs, including manufacturing labor, which enter into the total cost of production of the finished product. This inventory also includes various supplies which are used in the manufacturing process, but are not considered to be raw materials.

8. Other Current Assets

Fixed Assets

9. Real Property
10. Machinery
11. Furniture and Fixtures
12. Depreciation Reserve
13. Prepaid Expenses

Liability and Capital Accounts

Current Liabilities

20. Accounts Payable
21. Notes Payable
22. Expenses Accrued but Not Due

Deferred Liabilities

23. Long Term Debt

Capital Accounts

30. Invested Capital
31. Profit and Loss

The balance sheet of a food packing business is prepaid by summarizing the foregoing accounts.

The operating accounts of a food packing business would consist of the following:

OPERATING ACCOUNTS

Income Accounts
- 40. Sales
- 41. Sales Allowances
- 42. Trade Discounts
- 43. Returned Goods

Expense Accounts

I. Direct Manufacturing Costs
- 50. Raw Materials

  In addition to showing the cost of raw materials, plus delivery cost, less discounts and allowances, a further breakdown if often made, for better cost control, into cost of raw product, cost of additives (sugar, seasonings, flavors, etc.), cost of containers, etc. This type of breakdown, which is essentially for cost accounting purposes, is important in the control of costs.
- 51. Manufacturing Labor, Including Labor Taxes and Insurance
- 52. Rent on Machinery Not Owned
- 53. Fuel
- 54. Manufacturing Supplies
- 55. Packing and Shipping Labor

II. Indirect Manufacturing Costs
- 60. Bookkeeping and Other Manufacturing Office Salaries
- 61. Indirect Labor (loading trucks, etc.)
- 62. Light, Heat and Power
- 63. Water
- 64. Repairs
- 65. Insurance
- 66. Property Taxes
- 67. Depreciation
- 68. Miscellaneous

III. Sales Costs
- 70. Salesmen's Salaries
- 71. Salesmen's Expense Accounts
- 72. Commissions and Brokerage
- 73. Advertising

IV. General and Administrative Expenses
- 80. Executive Salaries

81. Office Salaries (other than charged in account #60 above)
82. Expense Accounts (other than charged in account #71 above)
83. Office Expense
84. Insurance
85. Rent
86. Taxes (other than charged in accounts #51 and 61 above)
87. Depreciation (other than charged to Manufacturing in account #67 above)

A special record of great usefulness in a manufacturing business is the sales-production statement. It helps to avoid overproduction by showing the following groups of figures: (1) orders booked during period, orders shipped, balance unshipped orders; and (2) the inventory at beginning of period, the production during the period, and the orders shipped. From these two sets of figures, two figures are obtained for the balance of unshipped orders, which should check if this important figure has been computed correctly.

Accounting for the Small Service Business.—From the accounting standpoint, a garage is an interesting type of service business, because however small it may be, it requires departmentalized accounting methods. Only by computing separately the results of the various services and types of products sold can you know which of them are most profitable, or what is more serious, whether some of them are actually losing money.

The method by which this separate accounting is accomplished is by summarizing first the overhead expenses, which are then allocated among departments, each of which has its own operating statement showing income, expenses and net profit. One can see how this work is done by reviewing actual lists of accounts for the garage as a whole, and then for its individual departments.

OVERHEAD EXPENSES FOR GARAGE

121. Advertising
122. Association Dues and Assessments
123. Auto Expense
124. Bad Debts and Allowances

125. Collection Expense
126. Depreciation
    (a) Building*
    (b) Operating Equipment
    (c) Office and General
127. Donations
128. Insurance
    (a) Building*
    (b) Operating Equipment
    (c) Employer's Liability—Operating
    (d) Office and General
129. Legal and Accounting
130. Light and Heat
131. Maintenance and Repairs
    (a) Building*
    (b) Office and General
132. Miscellaneous
133. Office Expense
134. Postage
135. Rent
136. Salaries—Office and General
137. Stationery and Supplies
138. Taxes
    (a) Land and Building
    (b) Operating Equipment
    (c) General
    (d) Social Security
139. Telephone and Telegraph
140. Travel and Entertainment
150. Distribution (Cr.)

Nonoperating Accounts

161. Purchase Discounts
162. Interest Income
163. Miscellaneous Income
171. Sales Discounts
172. Interest Expense
173. Miscellaneous Charges
181. Employees' Bonuses—Profit Sharing
182. Income Taxes

* If all property is rented, this account may not be required.

The method of allocation of these overhead expenses among departments is as follows:

1. Building Expense is allocated by using the floor space occupied as the basis.

2. Fixed Charges are based on book value of fixed assets, including depreciation computed upon the basis of a property ledger or analysis of fixed assets.

3. Payroll Expense is computed upon compensation paid at applicable rates.

4. General Expense includes all other overhead expenses which cannot be applied on a more direct basis, and which are therefore charged to operating departments upon the basis of sales or gross income.

Also needed for the preparation of separate accounts for the different departments is a departmental breakdown of the income and current expense accounts, which, together with the distributed overhead, yields the various accounts to be kept.

Index of Tables

Index